North American Agroforestry

3rd Edition

EDITORS

Harold E. "Gene" Garrett, Shibu Jose, and Michael A. Gold

CONTRIBUTORS

Janaki R. R. Alavalapati, School of Forestry and Wildlife Sciences, Auburn University; **James A. Allen**, School of Forestry, Northern Arizona University; **Stephen H. Anderson**, School of Natural Resources, University of Missouri; **J. Arbuckle**, Department of Sociology, Iowa State University; **Carla Barbieri**, College of Natural Resources, North Carolina State University; **Gary Bentrup**, USDA National Agroforestry Center, Lincoln, Nebraska; **Thomas W. Bonnot**, School of Natural Resources, University of Missouri; **James R. Brandle**, School of Natural Resources, University of Nebraska; **Dave Brauer**, Conservation and Production Research Laboratory, USDA Agricultural Research Service; **Louise E. Buck**, College of Agriculture and Life Sciences, Cornell University; **Catherine J. Bukowski**, Virginia Polytechnic Institute and State University; **Dirk Burhans**, U.S. Forest Service, University of Missouri; **Zhen Cai**, School of Natural Resources, University of Missouri; **Michaela M. "Ina" Cernusca**, North Dakota State University; **J. L. Chamberlain**, USDA Forest Service; **Terry R. Clason**, Agricultural Center, Louisiana State University; **Brent R. W. Coleman**, School of Environmental Sciences, University of Guelph; **Dean Current**, Center for Integrated Natural Resource and Agricultural Management, University of Minnesota; **John Davis**, Applied Ecology, North Carolina State University; **Daniel C. Dey**, U.S. Forest Service, University of Missouri; **Stewart A. W. Diemont**, College Environmental Science and Forestry, State University of New York; **J. H. Fike**, School of Plant Environmental Science, Virginia Polytechnic Institute and State University; **Cornelia B. Flora**, Department of Sociology, Iowa State University; **Jie Gao**, San Jose State University; **Harold E. "Gene" Garrett**, School of Natural Resources, University of Missouri; **Larry D. Godsey**, Division of Business, Missouri Valley College; **Michael A. Gold**, School of Natural Resources, University of Missouri; **Andrew M. Gordon**, School of Environmental Sciences, University of Guelph; **Stephen C. Grado**, College of Forest Resources, Mississippi State University; **Robert K. Grala**, College of Forest Resources, Mississippi State University; **Hannah L. Hemmelgarn**, School of Natural Resources, University of Missouri; **Eric J. Holzmueller**, College of Agricultural Sciences, Southern Illinois University; **Thomas M. Isenhart**, College of Agriculture and Life Sciences, Iowa State University; **Guillermo Jimenez-Ferrer**, El Colegio de LA Frontera Sur; **Shibu Jose**, College of Agriculture, Food and Natural Resources, University of Missouri; **Robert J. Kremer**, School of Natural Resources, University of Missouri; **James P. Lassoie**, College of Agriculture and Life Sciences, Cornell University; **Teng Teeh Lim**, College of Agriculture, Food and Natural Resources, University of Missouri; **Chung-Ho Lin**, School of Natural Resources, University of Missouri; **Sarah T. Lovell**, School of Natural Resources, University of Missouri; **Robert L. McGraw**, College of Agriculture, Food and Natural Resources, University of Missouri; **D. Evan Mercer**, Southern Research Station, USDA Forest Service; **Joshua J. Millspaugh**, W.A. Franke College of Forestry & Conservation, University of Montana; **John F. Munsell**, College of Natural Resources and Environment, Virginia Polytechnic Institute and State University; **Joseph N. Orefice**, Forest & Agricultural Operations, Yale University School of Forestry & Environmental Studies; **Gabriel J. Pent**, College of Agriculture and Life Sciences, Virginia Polytechnic Institute and State University; **P. K. Ramachandran Nair**, School of Forest Resources and Conservation, University of Florida; **Richard C. Schultz**, College of Agriculture and Life Sciences, Iowa State University; **Peter L. Schultz**, Target, Inc., Headquarters; **John H. Schulz**, School of Natural Resources, University of Missouri; **Steven H. Sharrow**, Oregon State University; **William W. Simpkins**, Department of Geological and Atmospheric Sciences, Iowa State University; **Lorena Soto-Pinto**, El Colegio de LA Frontera Sur; **Erik Stanek**, Balzac Brothers & company, Charlston; **Eugene Takle**, Department of Agronomy, Iowa State University; **Naresh V. Thevathasan**, School of Environmental Sciences, University of Guelph; **Ranjith P. Udawatta**, School of Natural Resources, University of Missouri; **Corinne B. Valdivia**, College of Agriculture, Food and Natural Resources, University of Missouri; **W. D. "Dusty" Walter**, College of Agriculture, Food and Natural Resources, University of Missouri; **Eric E. Weber**, School of Natural Resources, University of Missouri; **Kevin J. Wolz**, Savannah Institute; **Mario Yanez**, Overtown Foodworks Office, Inhabit Earth; **Lisa Zabek**, Interior of British Columbia, Ministry of Agriculture; **Xinhua Zhou**, Campbell Scientific, Logan, Utah

EDITORIAL CORRESPONDENCE

American Society of Agronomy
Crop Science Society of America
Soil Science Society of America
5585 Guilford Road, Madison, WI 53711-5801, USA

North American Agroforestry

3rd Edition

Edited by

Harold E. "Gene" Garrett, Shibu Jose,
and Michael A. Gold

WILEY

Copublication by American Society of Agronomy and John Wiley & Sons, Inc.

Edition History
American Society of Agronomy, Inc. (1e, 2000 and 2e, 2009)

Editorial Correspondence:
American Society of Agronomy
5585 Guilford Road, Madison, WI 53711-58011, USA
agronomy.org

Registered Offices:
John Wiley & Sons, Inc., 111 River Street, Hoboken, NJ 07030, USA

For details of our global editorial offices, customer services, and more information about Wiley products, visit us at www.wiley.com.

Wiley also publishes its books in a variety of electronic formats and by print-on-demand. Some content that appears in standard print versions of this book may not be available in other formats.

Library of Congress Cataloging-in-Publication Data

Names: Garrett, H. E., editor. | Jose, Shibu, editor. | Gold, Michael Alan, editor. | John Wiley & Sons, publisher.
Title: North american agroforestry / edited by Harold E. "Gene" Garrett, Shibu Jose, and Michael A. Gold.
Description: 3rd edition. | Hoboken, NJ : Wiley [2022] | Includes bibliographical references and index.
Identifiers: LCCN 2021052140 (print) | LCCN 2021052141 (ebook) | ISBN 9780891183778 (hardback) | ISBN 9780891183846 (adobe pdf) | ISBN 9780891183839 (epub)
Subjects: LCSH: Agroforestry–United States. | Forest management–United States.
Classification: LCC S494.5.A45 N69 2022 (print) | LCC S494.5.A45 (ebook) | DDC 634.9/9–dc23/eng/20211028
LC record available at https://lccn.loc.gov/2021052140
LC ebook record available at https://lccn.loc.gov/2021052141

Cover Design: Wiley
Cover Image: Horticulture and Agroforestry Research Farm at the University of Missouri College of Agriculture, Food and Natural Resources in New Franklin, Missouri; © University of Missouri

Set in 9/11pt Palatino by Straive, Pondicherry, India

10 9 8 7 6 5 4 3 2 1

Contents

Perhaps more so than at any time in the history of mankind, we are faced with problems that threaten our very existence. Climate change, regardless of its cause, is resulting in unprecedented changes in events ranging from dramatic shifts in world weather patterns and the consequences thereof, to the melting of the earth's glaciers and its ultimate effects on world populations. Humanity has become a victim of its own success. We have conquered the wilderness, and in our attempt to meet the needs of the world's ever-growing population, we have endangered many of the ecosystem services upon which our livelihood depends. Our streams and rivers are contaminated with sediment, nutrients and pesticides, mostly products of our success in producing more food at cheaper prices, but also from the privileges success brings such as subdivisions with luxurious lawns requiring large amounts of agrochemicals. Our oceans become the final destination for these and other contaminants, and in combination with the warming of their waters, our oceans too are in a transition towards endangered.

With today's understanding of the consequences of current land-use systems, it is time for a new approach—alternatives must be found. One alternative that was quickly adopted in tropical regions in the 1970's and 80's but has been slower in gaining support in the temperate regions of the world, is agroforestry. Agroforestry exploits the positive interactions between trees and crops (including livestock) when they are carefully designed and integrated, they bridge the gap between production agriculture and natural resource management. Supported by four decades of research and demonstration agroforestry practices have been found to provide environmentally and economically sound alternatives to many of our unsustainable forestry and agricultural systems. While if offers opportunities for small farms to regain their relevance and viability it also provides humanity the opportunity to heal our planet by constructively addressing climate change, improving the quality of our air, our waters, and protecting and enhancing soil health.

The 3rd edition of North American Agroforestry--An Integrated Science and Practice (now shortened to North American Agroforestry) comes at a critical time as the nations of the world debate the pros and cons of making forestry and agriculture production system decisions based more on economics than on the future health of our planet. In addition to updating the topics found in the 2nd edition, this edition adds to the science with a new 6th practice of "Urban Food Forests" and chapters on: Agroforestry for Air Quality Benefits; Agroforestry for Soil Health; Agroforestry at the Landscape Level; An Overview of Agroforestry and its Relevance in the Mexican Context and Agroforestry Training and Education.

The chapter authors are all recognized authorities. Their writings, when taken collectively, are meant to provide a state-of-the-art understanding of agroforestry. Agroforestry is rapidly becoming recognized as a science that has great merit in helping address many of our nations' environmental problems while serving to the economic benefit of our nations' small farms.

Harold E. "Gene" Garrett
School of Natural Resources
The Center for Agroforestry
University of Missouri
Columbia, Missouri

Shibu Jose
College of Agriculture, Food and Natural Resources
University of Missouri
Columbia, Missouri

Michael A. Gold
School of Natural Resources
The Center for Agroforestry
University of Missouri
Columbia, Missouri

Acknowledgments

The 3rd edition of North American Agroforestry would not have been possible were it not for the assistance of two supporting co-editors in the preparation of the 1st edition. The current editors express their thanks to W. J. "Bill" Rietveld and R. F. "Dick" Fisher for making this 3rd edition possible. Many hours are spent preparing chapters of the nature found in this text, as is also the case for reviewers who give freely of their time. The editors wish to extend a very special thanks to the authors and dedicated reviewers for their contributions. We also are grateful to Caroline Todd at the University of Missouri Center for Agroforestry for the logistical support she provided during this project.

The senior editor would be remiss if he did not acknowledge his wife, Joyce, who provided guidance and computer skills in revising and formatting chapters. Without her very capable assistance and encouragement, his job as a co-editor would have been many times more difficult. And last but not least, to the many contributors to the temperate zone, agroforestry literature, the editors express acknowledgement and appreciation for a job well done!

This work is partially supported by the University of Missouri Center for Agroforestry and USDA ARS Dale Bumpers Small Farm Research Center, Agreement number 58-6020-6-001 from the USDA ARS.

Section I

Agroforestry Fundamentals

North American Agroforestry, Third Edition. Edited by Harold E. "Gene" Garrett, Shibu Jose, and Michael A. Gold.

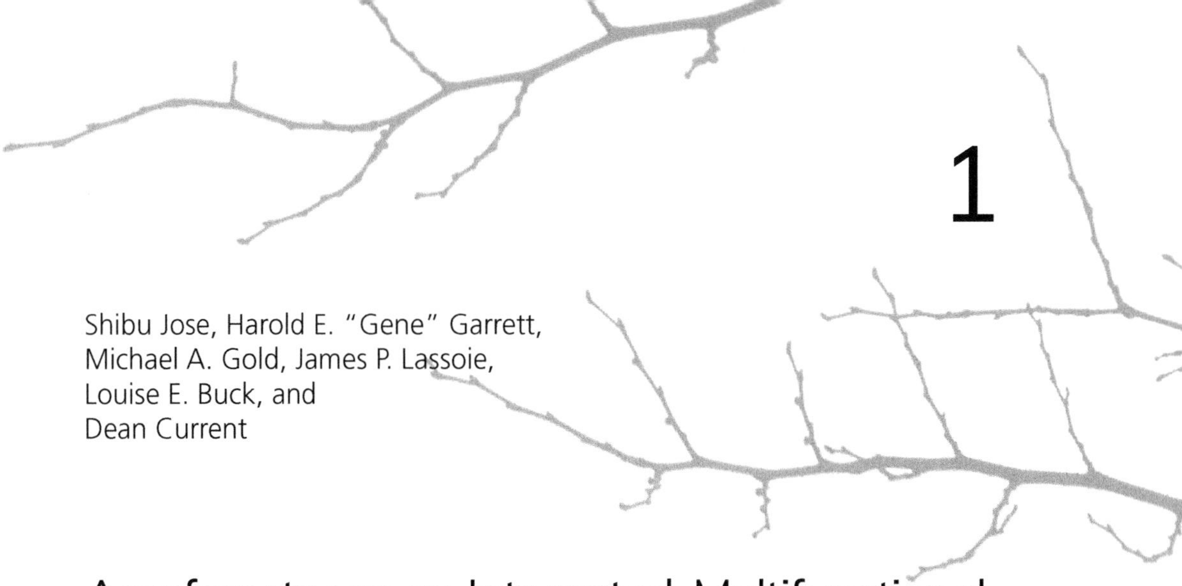

Shibu Jose, Harold E. "Gene" Garrett,
Michael A. Gold, James P. Lassoie,
Louise E. Buck, and
Dean Current

Agroforestry as an Integrated, Multifunctional Land Use Management Strategy

Agriculture is in the midst of a 21st century technological revolution, and we are well into the digital age of farming. The development of agriculture over 10,000 yr, including the technological advancements of the 20th century, has helped push the world population to 7.5 billion, with projections of 9.8 billion by 2050 (Searchinger et al., 2018). While the Green Revolution has helped to feed billions of people, the global environmental footprint of modern agriculture threatens the very existence of the socio-ecological system in which we live (Funabashi, 2018). The natural resource base, including soil and water, that supports agriculture is experiencing immense pressure. The world is looking for sustainable solutions not only for food security but also for environmental security for the burgeoning population (Searchinger et al., 2018).

The United States led the agricultural revolution with a massive commitment to enhancing food and fiber production capabilities. The overall strategy was to become self-sufficient with respect to agricultural crops and timber while improving the health and welfare of rural Americans. Obviously, this was successful within well-defined limits—today, food remains plentiful and relatively inexpensive, the timber famine was averted, and forest and farm lands abound. Such gains, however, did not come without some high environmental costs, and by the 1970s the public was demanding more environmentally benign land use practices (Laurence, 1987).

As a consequence of the environmental transgressions committed during the construction of our industrialized nation, new criteria for defining successful land use management strategies were identified during the late 1980s (Turner, 1988). Sustainability, stability, and equability have now joined increased production efficiency as objectives for agriculture (Conway, 1987), and forestry is developing new management strategies that optimize the yield of many products and multiple uses rather than merely maximizing the production of one—timber (Coufal & Webster, 1996; Gillis, 1990; Maser, 1994). But what about the hybrid between agriculture and forestry that is practiced worldwide—integrative management systems far more common than the developed world's often myopic approach to the production of a limited number of monocultures? Many professionals believe that agroforestry is a strategy for sustainable land use management that might be useful throughout North America (Garrett et al., 1994; Gold & Hanover, 1987; Kremen & Merenlender, 2018; Wiersum, 1990).

As we have moved into the 21st century, concerns have been raised about our dependence on foreign sources of fossil fuels, and the CO_2 emissions from our past and continued

North American Agroforestry, Third Edition. Edited by Harold E. "Gene" Garrett, Shibu Jose, and Michael A. Gold.

use of fossil fuels have increasingly been linked to global warming issues. The production of biofuels and energy from herbaceous and woody biomass has become a major interest with increasing amounts of research funding and private investment. Agroforestry practices combining herbaceous and woody species could play an important role in both the production of biomass for biofuels and energy as well as systems that enhance the ability of agricultural cropping to sequester and store carbon without the ecological problems of currently utilized agricultural systems (Downing, Volk, & Schmidt, 2005; Feliciano, Ledo, Hiller, & Nayak, 2018; Gruenewald et al., 2007; Holzmueller & Jose, 2012; Peichl, Thevathasan, Huss, & Gordon, 2006; Schoeneberger, 2005; Volk et al., 2006; White et al., 2007).

Because of its diversity, defining agroforestry could easily occupy an entire article—in fact on a number of occasions, it has (see Atangana, Khasa, Chang, & Degrande, 2013, pp. 35–47; Elevitch, Mazaroli, & Ragon, 2018; Lundgren, 1982; Nair, Viswanath, & Lubina, 2017). Presently, the concepts and practices of agroforestry in the United States are reasonably well understood within most professional circles to include ". . . intensive land management that optimizes the benefits (physical, biological, ecological, economic, social) arising from biophysical interactions created when trees and/or shrubs are deliberately combined with crops and/or livestock" (revised from Garrett et al., 1994). In identifying a niche for domestic agroforestry, emphasis must be directed toward a practice meeting the requirements of the four I's–that is, it must be intentional, intensive, integrative, and interactive. As discussed below, the options available under this definition are many (also see Chapter 2; Gold & Hanover, 1987; Campbell, Lottes, & Dawson, 1991; Schultz, Colletti, & Faltonson, 1995). Agroforestry practices in the North America involve more than the production of single products (e.g., monoculture field crops, livestock feedlots, forest plantations, biomass plantings, etc.), the extensive collection of special forest products (e.g., floral greens, mushrooms, wild game, etc.), or the extensive grazing of livestock in woodlots or on open ranges. This is not to minimize the importance of such land uses, but each one is already well supported by an established knowledge base and a well-educated group of practicing management professionals. Combining such practices into agroforestry arrangements that are ecologically sound and economically viable is a totally different story!

Intensive production of agricultural and forestry monocultures is found in both advanced, developed countries (e.g., corn [*Zea mays* L.], soybean [*Glycine max* (L.) Merr.], pine [*Pinus* spp.], fruit and nut orchards, vineyards) and many tropical regions in the form of woody perennial tropical tree, shrub, and vine crops including oil palm (*Elaeis* spp.), rubber [*Hevea brasiliensis* (Willd. ex A. Juss.) Müll. Arg.], tea [*Camellia sinensis* (L.) Kuntze], coffee (*Coffea* spp.), pepper (*Piper nigrum* L.), and vanilla (*Vanilla planifolia* Jacks.) (Chambers, Pacey, & Thrupp, 1989; Jha et al., 2014; Liu, Kuchma, & Krutovsky, 2018; Pacheco, Gnych, Dermawan, Komarudin, & Okarda, 2017; Richards, 1985). On the other hand, agroforestry has remained the primary land use approach most common throughout the developing world (King, 1987; Mercer, 2004), where complex indigenous farming systems for food, fiber, and forage production have operated effectively for centuries (Nair, 1993). Not only have such agroforestry systems produced a variety of commodities for home use and/or sale, it is likely that they have offered a level of environmental protection unmatched by most modern land use technologies. Such dual features—production and protection—have become the basis for the concept of sustainability, which is now central to international development activities aimed at breaking the negative feedback relationship between intensive land use and progressive environmental degradation. Similarly, concepts such as "productive conservation" and "multifunctional agriculture," which combine production agriculture with conservation by introducing more sustainable agricultural practices, are increasingly being discussed as options in more developed countries and could easily incorporate agroforestry principles (Jordan, et. al., 2007). For example, the five principles of sustainable food and agriculture defined by the FAO (2018) include: (a) increase productivity, employment, and value addition in food systems, (b) protect and enhance natural resources, (c) improve livelihoods and foster inclusive economic growth, (d) enhance the resilience of people, communities, and ecosystems, and (e) adapt governance to the new challenges.

About four decades ago, agroforestry was "discovered" by the international scientific community as a practice in search of a science (Steppler, 1987). Since that time, an increasingly extensive research base has been developing to help understand, improve, and apply indigenous agroforestry practices in developing nations of the world (Nair, 1996; Garrity et al., 2010; van Noordwijk et al., 2019). Around the same time, academics started asking how such practices might be applied in more developed countries (e.g., Campbell et al., 1991; Gold & Hanover, 1987; Lassoie, Teel, & Davies, 1991). However, agroforestry practices were not new in the temperate context either. Native Americans, across what is now the United States and Canada, have been practicing indigenous forms of what could be termed landscape-scale agroforestry for millennia (Rossier & Lake, 2014). In the early decades of the 20th century, agroforestry plantings were done in the United States and Canada in the form of windbreaks and shelter belts as a response to the Dust Bowl of the 1930s. In the temperate zone, science-based agroforestry biophysical and socioeconomic research and practice gained attention in the1980s and has strongly increased in the past 40 yr. Interest in domestic agroforestry has continued to grow, particularly as the dual needs for enhanced environmental protection and new economic opportunity have increased in importance (Brown, Miller, Ordonez, & Baylis, 2018; Garrett et al., 1994; Jose, 2009; Jose, Gold, & Garrett, 2018). The realization that agroforestry systems are well suited for diversifying farm income while providing environmental services and ecosystem benefits has increased receptivity on the part of landowners (Rois-Díaz et al., 2018) Agroforestry systems offer great promise for the production of biomass for biofuel, specialty and organic crops, pasture-based dairy and beef, among others. Agroforestry also offers proven strategies for carbon sequestration, soil enrichment, biodiversity conservation, and air and water quality improvement not only for the landowners or farmers but for society at large (Dollinger & Jose, 2018; Holzmueller & Jose, 2012; Scherr & McNeely, 2007, 2008; Udawatta & Jose, 2012).

In this chapter, we demonstrate the linkages among emerging integrated management systems for agriculture and forestry and indicate possible roles that agroforestry could play in the continuing development of these new land use strategies. Opportunities for the development of domestic agroforestry practices are identified and progress toward meeting them highlighted. Possible approaches to overcoming constraints limiting the development of agroforestry in the United States are suggested. It is our purpose to provide a framework for the chapters that follow and to stimulate creative thinking and proactive behavior by scientists and management professionals responsible for developing and implementing new land use management strategies that are environmentally, socially, and economically sustainable.

Land use Management Systems in North America

Here we provide thoughts on new management systems that are emerging to help account for the complex demands currently being placed on the nation's rural lands. Specifically, agricultural and forestry land use practices are examined relative to certain biophysical and socioeconomic principles basic to natural resources management. We also provide a historical perspective for the evolution of forest management and agricultural production practices and for the development of domestic agroforestry activities.

Basic Principles Influencing Management Systems

Management can be considered as the planned intervention into natural processes to assure predictable outcomes of benefit to the health and welfare of humans. Hence, sociological factors often become the driving principles determining many land use decisions. For example, a stewardship ethic that places long-term social good above short-term personal gain can move people to spend time, effort, and money assuring the ecological integrity of land they currently own. In contrast, a pioneer ethic emphasizing the immediate needs of the individual can promote destructive activities that negatively impact future generations (Nash, 1982). This anthropocentric focus for management has been challenged for decades (Stone, 1996). Obviously, different user groups can hold very different views concerning the utilization, conservation, and preservation of our natural resources, often making the social context in which land use decisions are made highly contentious.

The social context for land use decision-making is also subject to increasingly rapid change as the pace of social evolution quickens in response to increased knowledge and technological advancements. For example, this century has witnessed major changes associated with the transition from a rural to an urban society, shifts in ethnic and age structures, a move to an information-based society, and periodic resurgence in the public's interest and concern about the environment and the use of the nation's farm and forest lands. Hence, management decisions socially acceptable in one generation may not be accepted in another (e.g., clear-cutting old-growth forests, eradicating predators, or indiscriminate pesticide use).

The United States is a capitalistic society, and the economic bottom line continues to drive many decisions concerning the production of food, forage, livestock, and fiber. We have been so successful in creating a higher order of socioeconomic organization through our effective harnessing of energy that subsistence living remains for only a few in North America. Agriculture and forestry are now big businesses operating in a dynamic world economy. Fortunately, there is a sound theory base supporting our understanding of the economic variables driving capitalism, such as cost/benefit ratios, supply–demand interrelationships, and marketplace dynamics. Unfortunately, much of this neoclassical theory simplifies or neglects critical issues, such as the long-term values associated with externalities arising from sound management practices, often making it inadequate for explaining the current realities of the land use and environmental decision-making process (Daly & Cobb, 1989; Tisdell, 1990).

Nonetheless, during the past two decades there has been increased interest in internalizing the environmental costs and benefits not necessarily reflected by our market system (Mann & Wustemann, 2008; Wang & Wolf, 2019). Payments for environmental or ecosystem services have entered the discussion of policymakers at both the federal and state levels in the United States (Mercer, Cooley, & Hamilton, 2011; Potter & Wolf, 2014). We have a voluntary market for carbon offsets in the United States and a developing market for water quality credits, both patterned after what has been considered to be a successful cap-and-trade system to control sulfur dioxide emissions (Börner et al., 2017; Gordon, 2007; Jack, Kousky, & Sims, 2008; Lowrance, 2007; Palma, Graves, Burgess, van der Werf, & Herzog, 2007b; Wang & Wolf, 2019).

Land use management is inherently interdisciplinary because of the multitude of interrelated factors that must be considered when deciding how best to optimize the use of land for realizing its multiple values (Ferraz-de-Oliveira, Azeda, & Pinto-Correia, 2016; Savory, 1988; Stankey, 1996). The extent to which scientific knowledge is useful in such a decision-making process depends on its ability to deepen managers' understanding of complex systems and how to adjust them to achieve specific objectives. An interdisciplinary approach is essential to the development of such knowledge (Chubin, Porter, Rossini, & Connolly, 1986). The study of interdisciplinary land use management systems, while previously overlooked (Stankey, 1996), has become a major topic of interest in the research and development community (LaCanne & Lundgren, 2018). The "tyranny of the disciplines," while still the norm in creating institutional obstacles to effective integration (Campbell, 1986), is no longer the only paradigm being promoted and is actively being superseded during the past decade by a shift toward increased diversification of landscapes and cropping systems (Geertsema et al., 2016; Liebman & Schulte, 2015). The theoretical base for the management of complex agroecosystems often does not meet the practical needs of the field-level manager (Wezel & Bellon, 2018). This can result in mismanagement by those owning land or controlling its use—unacceptable behavior in a society that is increasingly demanding sound ecological management of its natural resources.

Evolution of Management Systems

The United States inherited its forest management practices from Europe during the latter part of the 19th century and modified them to accommodate its large, sparsely populated country, which was rich in natural resources (Perlin, 1991; Williams, 1989). Prior to settlement by Europeans, Native Americans derived a variety of food, forage, and fiber products from forests while manipulating them primarily through the use of fire in what could be termed landscape-scale agroforestry (Carroll, 1973; Cronon, 1983; Rossier & Lake, 2014; Russell, 1982). European pioneers also derived most of their energy and construction materials from the forest (Carroll, 1973).

The Industrial Revolution brought with it new harvesting and milling technologies, which greatly enhanced the efficiency with which the nation's forest resources were exploited (Williams, 1989). Such forest practices accelerated as the population grew and became more urbanized. Around the turn of the 19th century, continuing over-exploitation stimulated public concern and the birth of America's conservation movement (Jordan, 1994), which included the development of professional forestry management agencies and academic institutions (Skok, 1996; Spencer, 1996). In 1905, the U.S. Forest Service was formally established to promote sustained-yield forestry, designed to provide wood fiber from the nation's forests forever (Steen, 1976). Conflicts over the single-purpose use of public forest lands led the U.S. Forest Service to develop its multiple-use approach to the management of national forests, which assured that, given a large enough and diverse enough land base, a full complement of forest uses could be enjoyed without conflict. Eventually, however, this approach also led to problems once the public began to question decisions being made about individual pieces of land, especially with respect to tradeoffs between wilderness preservation and timber production (Nash, 1982). Such concerns, together with a growing understanding of the impacts that plantation forestry has on biological diversity and the natural functioning of forest ecosystems, have stimulated the forestry profession to consider a new management strategy—ecosystem management—based on a holistic, integrative approach to land use (Coufal & Webster, 1996; Maser, 1994; Nunez-Mir, Iannonne, Curtis, & Fei, 2015; Probst & Crow, 1991; Stankey, 1996). Parallel to those efforts and because of the growing interest in preserving our national forests free from production activities, national forests are increasingly off limits to harvest, shifting production forestry and harvesting to private lands (Adams, Haynes, & Daigneault, 2006). Simultaneously, there is a growing public cry for less governmental regulation and a return to a conservation ethic embodied in the idea of sound stewardship (Jordan, 1994). Likewise, it took a century and a half for American agriculture to develop to the level of complexity that required an integrated management approach (National Research Council, 1989). Native Americans were hunter-gatherers, subsistence farmers, and also practiced indigenous forms of landscape-scale agroforestry (Rossier & Lake, 2014), while early immigrants were primarily hunter-gatherers and subsistence farmers (Russell, 1982). With population growth and industrial development came a growing need to improve food production capabilities and economic livelihoods of farmers to feed an ever-increasing urban society. The mid-1800s brought the development of the land grant university system and the initiation of an agricultural experiment station infrastructure that eventually built the world's greatest system for the intensive cultivation of commercial food products (National Research Council, 1996; Russell, 1982).

Domestic and global marketing uncertainties, high costs for equipment, seed, chemical and energy inputs, high interest rates, and regional identity and security issues are forcing many modern farmers to develop integrated farming systems involving the production of a variety of products. More recent public concerns about the environmental impacts of modern farming practices and food safety are prompting the development of a new management approach based on agroecology principles: alternative or sustainable agriculture (LaCanne & Lundgren, 2018; Liebman & Schulte, 2015; National Research Council, 1989, 1991, 1996) More recently, eco-agriculture and regenerative agriculture—integrating production and conservation at a landscape scale with the deliberate inclusion of perennial crops—have been put forth as new paradigms for linking production and conservation in our agricultural landscapes (Elevitch et al., 2018; Scherr & McNeely, 2007, 2008). Perennial trees and shrubs, and hence agroforestry practices, can serve important functions in such sustainable agricultural systems (Elevitch et al., 2018; Prinsley, 1992).

Evolution of North American Agroforestry

Although not defined as such until recently (Garrett et al., 1994; Gold & Hanover, 1987; Gordon & Newman, 1997; Rossier & Lake, 2014; Sinclair, 1999; Torquebiau, 2000), agroforestry-like practices have been part of North America's heritage. Native Americans and European pioneers practiced subsistence lifestyles based on integrated land use strategies that were similar in principle to the agroforestry being practiced by indigenous populations in today's developing countries (Carroll, 1973; King, 1987; Rossier & Lake, 2014; Russell, 1982). The widespread use of these strategies, however, largely disappeared during the last century with the concurrent development of separate agricultural and forestry research and management infrastructures. Today, an integrated, subsistence lifestyle is the chosen standard of living for a few independent, free-spirited individuals and an unfortunately necessary one for the economically marginalized rural poor. A few agroforestry practices survived into the mid-20th century associated with long-established organizations (e.g., the Northern Nut Growers Association) or as culturally acceptable complements to traditional farming enterprises (e.g., maple syrup production).

Periodic agricultural disasters have stimulated unique forestry activities that can also be considered agroforestry practices. In the 1930s, the Great Depression combined with the drought-induced Dust Bowl in the Great Plains caused severe economic and environmental perturbations throughout the agricultural community and the nation. The formation of the Civilian Conservation Corps promoted many conservation activities including the planting of millions of trees as windbreaks and plantations to help protect eroding farmlands (Hudson, 1981). Such ecological problems also stimulated interest in the use and genetic improvement of nut trees to reclaim and promote production from lands marginal for conventional farming practices (Smith, 1950). The farm crisis of the 1980s was less dramatic on a large scale, but it had devastating economic and social impacts on many rural communities (Fitchen, 1991). In response, congressional actions established alternative agricultural programs such as the Conservation Reserve Program, Low Input Sustainable Agriculture (renamed the Sustainable Agriculture Program), and the Integrated Pest Management Program.

In the first decade of the 21st century, there was an increased interest in the production of biofuels and a concerted government effort to develop the technologies to make biofuels a reality. One unintended impact of the interest and support for biofuels, and particularly corn-based ethanol, has been periodic increases in corn prices in the United States and around the world, igniting a "food versus fuel" debate. High commodity prices linked to the demand for biomass feedstocks for biofuels coupled with huge demand from China also resulted in farmers opting out of conservation programs and replacing conservation acres with commodity crops, with environmental consequences including increasing sediments and chemicals entering surface and ground waters (Jordan et al., 2007).

Simultaneously, spurts of environmental consciousness by the American public have promoted alternate land use practices, often involving unique mixes of trees, food crops, and livestock by non-traditional rural landowners. For example, the 1960s spawned a group of "back-to-the-land" environmentalists desiring low-impact communal lifestyles. Although most of these groups eventually disappeared, individuals committed to integrated land use practices remained to practice their more ecosystem-friendly forms of agriculture and to develop such organizations as the Land Institute, Rodale Research Center, and Wallace Center. The fact that the Northern Nut Growers Association was founded in 1910 is a testimony to the existence of such individuals for many years.

The past 40 yr have witnessed a growing understanding of the potential usefulness of agroforestry practices in addressing today's concerns over the economic and environmental sustainability of forest and farm lands. Gold and Hanover (1987) discussed two such practices: managing conifer sawlog with cattle grazing practices (silvopasture) and multi-cropping valuable hardwoods with agricultural crops (alley cropping). There are five widely recognized agroforestry practices in the United States today: (a) alley cropping, (b) riparian and upland buffers, (c) windbreaks, (d) silvopasture, and (e) forest farming. In addition to the five recognized practices, there is an emerging agroforestry practice called *urban food forests* that has gained considerable attention in the past decade (Bukowski & Munsell, 2018). The specifics concerning these six practices are examined later in this volume. In addition, with the growing interest in terrestrial carbon sequestration and alternative fuels provided by woody and herbaceous growth, there has been increasing interest in the role of agroforestry practices to sequester and store carbon as well as systems that produce biofuels. The National Agroforestry Center has continued to add support to those practices, providing publications and guidance to landowners and practitioners (Schoeneberger, 2005; USDA, 2015, 2019). Much progress has been attained toward the building of research, education, and application for domestic agroforestry that will foster the development of these practices (Gold, 2007; Gold, Hemmelgarn, & Mendelson, 2019; Gold & Jose, 2012). However, before discussing these concepts relative to the challenges still facing the development of agroforestry in the United States, we must first provide an appropriate context by considering opportunities for its development.

Opportunities for Agroforestry

There is a range of opportunities for the development of agroforestry in the United States, a topic first addressed relative to nut-tree crops in 1989 (Lassoie et al., 1991) and then more comprehensively by Lassoie and Buck (1991) and more recently by Jose et al. (2018). What follows is a consideration of the importance of agroforestry to the development of newly emerging (a) land use systems, (b) institutional arrangements, (c) scientific opportunities, and (d) knowledge systems. It is important to note that many of the following statements are in the process of ongoing validation and are areas under active evaluation by agroforestry professionals.

Opportunities for Practical Application

The possibilities for developing agroforestry for use in rural and urban areas of the United States is encouraging. It did function extensively at one time throughout North America (Carroll, 1973; Cronon, 1983; Russell, 1982), and refugia of such practices still exist today (e.g., see Campbell et al., 1991; Gold & Hanover, 1987; Schultz et al., 1995). Agroforestry is currently working effectively in many other developed (Burgess & Rosati, 2018; Gordon, Newman, & Coleman, 2018; Mead, 1995; Von Maydell, 1995) as well as developing (Hillbrand, Borelli, Conigliaro, & Olivier, 2017; Nair, 1989, 1993) countries. The political and social climate in the United States is rapidly changing, which is likely to allow the development of the new land use strategies that include agroforestry (Jose et al., 2018; Lovell et al., 2018; National Research Council, 1996; USDA, 2019). Opportunities for the development and application of domestic agroforestry practices can be separated into ecological, economic, and social components for discussion purposes; however, one needs to keep in mind that these components are interrelated, interlinked, and interactive.

Ecological

One of the primary advantages of agroforestry in the United States probably rests in its ecological benefits and resultant environmental protection characteristics (Garrett et al., 1994; Jose, 2009; Jose & Gordon, 2008; Jose, Walter, & Kumar, 2019; Kremen & Merenlender, 2018; Udawatta & Jose, 2012). As an ecologically based land management strategy, agroforestry practices help maintain ecosystem diversity and processes that are important to the long-term sustainability of any extractive land use practice such as agriculture and forestry. This approach offers the opportunity to maintain and possibly improve the quality of the soil resource by reducing erosion, enhancing nutrient capital, and improving water infiltration and retention rates (Dollinger & Jose, 2018; Udawatta, Gantzer, & Jose, 2017). Trees also moderate microclimatic extremes, assuring cooler summers and warmer winters. Such conditions are beneficial to the production of certain food crops and livestock under severe environmental conditions as well as to human comfort. Agroforestry practices can also result in decreases in chemical (e.g., fertilizers and pesticides) and energy inputs of farming systems, all important to an environmentally sensitive society (Jose, 2019; Lerch, Lin, Goyne, Kremer, & Anderson, 2017). There is a growing movement to value some of the environmental services provided by agroforestry practices, allowing their benefits to be translated into economic incentives for landowners.

Economic

Agroforestry offers financial opportunities associated with enhancing the profitability of traditional farming systems (Alavalapati & Mercer, 2004; Campbell et al., 1991; Nair, 1993; Van Vooren et al., 2016). One option is to reduce production costs by decreasing the need for external chemical, water, energy, and/or labor inputs. Another is that agroforestry might increase the net value of production from the nation's farmlands through optimizing mixtures of primary (e.g., corn, sawlogs, nuts, cattle, etc.) and secondary crops (e.g., mushrooms, Christmas trees, silvopasture-raised chickens, etc.) as well as ecological services (e.g., carbon sequestration and water quality credits) for which specialty markets have been or are being developed (Van Vooren et al., 2016; de Jalon et al., 2018). Total production might also be enhanced by increasing production from highly erodible or frequently flooded fragile lands without causing severe environmental degradation, for example through the use of tree crops as suggested by Smith (1950).

Social

The development of agroforestry in the United States has social ramifications that will be realized at the individual, community, and national levels. As a sustainable land use strategy, agroforestry practices can further the land stewardship concept (Jordan, 1994; Montambault & Alavalapati, 2005; Roesch-McNally, Arbuckle, & Tyndall, 2017; Udawatta et al., 2017; Weber, 1991) by providing assurance to landowners that they are meeting their ownership responsibilities to provide healthy ecosystems for future generations. If agroforestry proves to enhance the production capabilities of rural lands, such practices will help revitalize rural communities, which have become socially depressed because of recent economic problems (Jose et al., 2018). Farmers with limited land and immigrant populations interested in farming provide an important audience as well as a population that may be more interested in the more labor-intensive practices common to agroforestry (Faulkner, Owooh, & Idassi, 2014). Understanding the role that agroforestry might play in modern land use systems will also help individuals appreciate that people from developing countries have experiences, insights, and knowledge potentially helpful for solving many problems currently plaguing modern societies (Coulibaly, Chiputwa, Nakelse, & Kundhlande, 2016; Jose & Dollinger, 2019). Such an appreciation for the value of human capital and indigenous (i.e., local) knowledge (Rossier & Lake, 2014) will help reduce ethnocentric and educational biases that form barriers between individuals who must work together to successfully address today's environmental crises. Also important is the impression that the United States must make on the rest of the world with respect to its concern about the management of its own natural resources. Developing agroforestry practices and integrated agricultural and forestry land use systems will set international standards for ecologically sound management by example. Our concern about deforestation, desertification, and soil erosion in developing countries will gain more credibility once our own "ecological house" is in order.

Opportunities for Institutional Development

The continued development of a domestic focus on agroforestry offers unique opportunities for various organizations responsible for supporting the nation's food, forage, livestock, and fiber-producing networks. These include academic institutions, state and federal government agencies, non-governmental organizations (NGOs) that include conservation-oriented organizations, and the private sector. Such opportunities enhance the ability of these groups to provide support to rural communities while responding to public demand for more sustainable and environmentally benign land use practices (National Research Council, 1996; USDA, 2019). Previous experience in both developed and developing countries clearly demonstrates the need for institutional collaboration in developing effective agroforestry policies (Biggs, 1990). Domestic agroforestry has reached the point where new, interdisciplinary, and interagency approaches to integrated land use management are underway (USDA, 2019), thereby serving as a design methodology for reuniting the fields of agriculture and forestry in a common quest for sustainability.

Academic Institutions

The development of a domestic agroforestry program is especially important to the nation's land-grant institutions, and specifically their state land-grant colleges of agriculture and forestry. After successfully achieving the goal of enhanced scientific rigor (USDA, 1987), these institutions are now being criticized for moving away from their original applied missions, thus becoming less directly responsive to the needs of the public (National Research Council, 1996). In spite of this trend, the development of comprehensive agroforestry programs and multiple partnerships focused on helping private rural landowners is well underway across the United States (USDA, 2015).

Land-grant institutions are also responsible for educating future professionals. The teaching of agroforestry courses offers the opportunity to help meet the interests of students for an interdisciplinary, problem-solving education, which is difficult to provide due to the demands for scientific rigor within discipline-based curricula (Gold & Jose, 2012; Lassoie, 1990; Lassoie, Huxley, & Buck, 1994). More specifically, agroforestry can provide a model for teaching holistic approaches to land use management and may attract students from a wide variety of disciplines within the agricultural and natural resource sciences. Likewise, agroforestry provides an intersection between major fields of study, and reconnecting agriculture and forestry will certainly strengthen these fields as they move to develop a scientific basis for new management paradigms. In addition, new opportunities for funding and program development will arise as the importance of domestic agroforestry increases, thereby providing new areas for professional advancement by young academics.

Land-grant institutions are emphasizing interdisciplinary and transdisciplinary research to deal with real-world problems that cross disciplines. This approach also recognizes that many of the problems we face today require solutions that require an interdisciplinary and/or transdisciplinary approach (Stock & Burton, 2011). Agroforestry, with its roots as an applied science, provides ample opportunities for research across biophysical and social science disciplines to address applied problems.

Lastly, a domestic focus on agroforestry will further emphasize the importance of developing and maintaining a strong international component within the land-grant university system (Globalizing Agricultural Science and Education Programs in America Task Force, 1997) as many of the examples of successful agroforestry activities come from projects in developing countries (Nair, 1989, 1993; Garrity et al., 2010; Pinho, Miller, & Alfaia, 2012). Hence, a comprehensive agroforestry program from the nation's land-grant institutions could recommit and recharge the intellectual energy necessary to address the needs of the peoples of the world, a much broader mission than originally identified for these institutions (National Research Council, 1996). Online offering of degrees and certificates is another avenue by which land-grant institutions could offer much needed agroforestry training both nationally and globally. Such a program was initiated in 2013 at the University of Missouri, offering graduate degrees or certificates in agroforestry entirely online (Gold & Jose, 2012).

Government Agencies

State (e.g., departments of natural resources, environmental conservation, agriculture, and markets) and federal (e.g., U.S. Forest Service, USDA Natural Resources Conservation Service (NRCS), USDA Agricultural Marketing Service (AMS), USDA Agricultural Research Service (ARS), USDI Bureau of Land Management, and U.S. Environmental Protection Agency) agencies are gaining from the development of a domestic agroforestry program. The existing domestic program depends on, and promotes, interagency cooperation and effectiveness, areas always in need of improvement. Agroforestry provides a unique opportunity to foster new approaches to helping the farming and forestry communities with incentive programs, promoting needed rural development (Schoeneberger, Bentrup, & Patel-Weynand, 2017; USDA, 2019). This is especially important to small, independent farmers and nonindustrial forest landowners. While presently there is a trend toward rapid consolidation of small farms into larger corporate structures, with large and super-large farms controlling most of the fiber and food production in the country, there is growing recognition and appreciation for the important roles that small farms play in producing not only foodstuff but also a variety of economic, social, and environmental products and services (Schoeneberger et al., 2017). There is also a growing sector of beginning small farmers and immigrants entering rural areas interested in more diverse systems and organic options that adapt well to agroforestry management. This same population is often more open to the labor-intensive options associated with agroforestry production practices.

A report of the USDA National Commission on Small Farms (USDA, 1998) identified policy measures that are needed to enhance and preserve the important values of small farms; they recommended specifically that agroforestry offers small farm operators a means for economic diversification, windbreaks, biological diversity, and habitats for wildlife. The original publication was followed by another that offered specific legislation to support agroforestry (USDA, 2003). The report suggests that the USDA, through its extension, conservation, and forestry services, should make greater efforts to promote and support agroforestry as part of an economic and ecological strategy for a healthy agriculture. Of particular significance in establishing agroforestry policies and programs at the national level was the USDA Agroforestry Strategic Framework (USDA, 2011), which signified a major shift in the USDA's position on the value of agroforestry in today's agriculture. The strategic framework created a "road map" for advancing the science, practice, and application of agroforestry, broadening the USDA's role in agroforestry beyond that of just the National Agroforestry Center. For "buy-in" purposes, five USDA agencies and two non-USDA partners (The National Association of Conservation Districts and the National Association of State Foresters) were brought together to develop the framework in collaboration with stakeholders across the United States. In 2019, the USDA updated its Agroforestry Strategic Framework based on current agency needs and priorities, as well as additional input from partners and stakeholders (USDA, 2019). Within the farming and forestry sectors there is a growing trend toward the fragmentation of lands and expanded ownership by a larger group of small landowners, particularly in the eastern part of the country. Agroforestry can enhance the economic viability of owning and managing these units through the production and marketing of comparatively short-rotation, high-value specialty items in forest farming practices (e.g., see Chapter 9). Highly visible agroforestry programs that specifically address pressing environmental problems could greatly improve the public's image of agriculture and forestry as well as the agencies responsible for them. This could have important implications for their political futures.

Private Organizations

There is a wide variety of private groups—environmental organizations (e.g., Audubon, A Greener World, The Nature Conservancy, Trees Forever, the Sierra Club), foundations (e.g., Kellogg, McKnight, Pew Charitable Trust, Walton, Winrock), and institutions (e.g., the Land Institute, Rodale Research Center, Wallace Institute, Green Lands Blue Waters)—that are dedicated to finding alternative solutions to environmentally damaging land use practices and to rural development problems. Their diversity of interests often hinders collaboration as well as their meaningful interaction with governmental agencies and private individuals, organizations, and corporations. Because of its integrated approach, agroforestry might provide an opportunity for various audiences to develop a common agenda and approach for conservation and sustainable land use, particularly at a landscape level. Such cooperation could help everyone better understand divergent perspectives, thereby helping alleviate some of the constant pressures that exist between organizations with different concerns and goals. The Savanna Institute, formed in 2013, is a 501(c)(3) nonprofit organization working to lay the groundwork for widespread agroforestry in the U.S. Midwest. The Savanna Institute works in collaboration with farmers and scientists to develop perennial food and fodder crops within multifunctional polyculture systems grounded in ecology and inspired by the savanna biome. Private foundations also initiate new innovative programs that support agroforestry and enhance public agency competitive grant programs (e.g., Agroecology Fund, Cedar Tree Foundation).

Opportunities for the Scientific Community

The development of a domestic agroforestry program for the United States offers unique opportunities for the scientific community that embraces forest and agricultural sciences and can provide the opportunity for focusing issue-based science to address some of today's most complex problems. The scientific community currently is being challenged to search for viable solutions to complex environmental problems that are beyond its capabilities to address with customary precision and certainty (Burke et al., 2017). Consider the environmental and economic problems facing farmers and foresters today compared with the relatively simple production needs of the last century (National Research Council, 1996; Sampson & Hair, 1990). Agroforestry research experience in developing countries has shown it to be an effective means for interdisciplinary research teams to approach land use issues, in particular diagnosis and design methodologies (Murray & Bannister, 2004; Raintree, 1987, 1990). Similar work in Europe and North America now emphasizes the universality of ecological and socioeconomic issues, thereby blurring the distinction often made between domestic and international problems (Buck, 1995; Lovell et al., 2018). Agroforestry in developing countries is progressing with a combination of support from the research community (e.g., the World Agroforestry Center) and from the development assistance community (e.g., Heifer International) in promoting such practices. This is also beginning to unfold in the United States, in which both the underlying biophysical and socioeconomic science and broader knowledge infrastructure for agroforestry is beginning to reach critical mass, combining "bottom up, high touch" farmer-to-farmer approaches and "top down, high tech" scientific breakthroughs.

There is a growing interest in landscape level research on more sustainable land use systems that provide both income for farmers and ecological services for society (Lovell et al., 2010). Agroforestry concepts and applications provide ample opportunity to do just that (Brown et al., 2018; Palma et al., 2007a, 2007b). Research is now underway that demonstrates how those two objectives can be combined, providing opportunities for the scientific community to explore and identify new integrated land use options (Brown et al., 2018). Agroecology, eco-agriculture, and regenerative agriculture principles integrate biophysical, social, and economic factors at the landscape level and represent promise for moving agroforestry to the landscape level (Altieri, Nicholls, & Montalba, 2017; Geertsema et al., 2016; LaCanne & Lundgren, 2018; Liebman & Schulte, 2015; Scherr & McNeely, 2007).

Domestic agroforestry falls along the continuum of agroecology and regenerative agriculture, presenting a need for new types of information—a challenge that breeds creativity and vitality within the research community. Regardless of the scope, domestic agroforestry offers many opportunities for professional development arising from new research projects, education and training programs, and cooperative ventures with public agencies and private organizations.

Opportunities for the Development of New Knowledge Systems

In his review of the science of agroforestry, the director general of ICRAF argued that the key challenge posed by this field to the agricultural and forestry research communities is to develop a predictive understanding of the competition, complexity, profitability, and sustainability aspects of agroforestry practices (Sanchez, 1995). This would appear to hold true for the United States as well as developing, tropical countries. To evaluate these four key criteria for the performance of agroforestry, a sound understanding is needed of ecological processes (Ong & Huxley, 1996) as well as socioeconomic and policy conditions that affect agroforestry practices (Buck, 1995; Garrett & Buck, 1997) and how they can be optimized through management.

Innovative agroforestry practices in the United States encompass numerous characteristics and unique associations of component species as landowners experiment with various perennial and annual species, often in unconventional niches. Most of these associations have not been studied, thus significant knowledge gaps exist. A recently completed national assessment provides a science-based synthesis on the use of agroforestry for mitigation and adaptation services in the face of climatic variability and change. It serves as a framework for including agroforestry systems in agricultural strategies to improve productivity and food security and to build resilience in these landscapes. It also provides technical input on the need for innovative strategies to address significant climatic variability challenges faced by U.S. agriculture (Schoeneberger et al., 2017). Furthermore, the assessment reviews the social, cultural, and economic aspects of agroforestry and the capacity of agroforestry systems to provide multifunctional solutions. In addition, it presents a comprehensive North American perspective on the strengths and limitations of agroforestry through U.S.

regional overviews along with overviews for Canada and Mexico (Schoeneberger et al., 2017).

The challenges to generating practical, broadly useful knowledge about agroforestry are well documented (Sanchez, 1995), revolving around the comparative complexity and site specificity of various applications and thus the difficulty of generalizing from studies of particular practices. Each practice involves multiple components and processes, the dynamics of which change with time as the perennial components mature and assume different ecological and biological roles. Similarly, profitability, social acceptability, and regulatory incentives for practicing agroforestry vary and change as a function of complex interactions among a host of intended and unintended socioeconomic and policy factors (Van Vooren et al., 2016). These are exceptionally complex to untangle (Buck, 1995), but in recent years researchers have been working on new tools to deal with this complexity. For example, Hi-sAFe is a novel tool for exploring agroforestry designs, management strategies, and responses to environmental variation (Dupraz, Wolz, et al., 2019). Added to this are the institutional problems of dispersed, often uncoordinated resources that combine to influence the generation and use of new knowledge—mainly researchers, information, infrastructure, and financial support. Relevant and broadly encompassing scientific research in this context becomes prohibitively expensive—particularly in the current economic climate of the United States where agricultural research resources are increasingly scarce and often monopolized by "big business" interests whose central focus is on generating profitable products. Although there is some dedicated funding for integrated research on sustainable agriculture through the Sustainable Agriculture Research and Education (SARE) program of the USDA, and a variety of related funding opportunities through the USDA National Institute of Food and Agriculture (NIFA), funding is limited, extremely competitive, and relatively short term. Concerns about integrating conservation and sustainable development goals through agroforestry are likely to continue to receive limited priority.

We propose a complementary strategy for advancing understanding of the conditions under which the desired attributes of agroforestry practice may be achieved and how well various systems can be expected to perform. This involves harnessing the experience and learning processes of numerous, dispersed agroforestry practitioners into purposive knowledge networks. During the past decade, a series of regional agroforestry networks have been established (e.g., Mid-American Agroforestry Working Group [MAAWG], Northeast/Mid-Atlantic Working Group [NEMA], Pacific Northwest Agroforestry Working Group [PNAWG], Southwest Agroforestry Action Network [SWAAN], etc.). In addition, a number of regional specialty crop cooperatives have formed (e.g., Midwest Elderberry Cooperative, multiple chestnut cooperatives, etc.). What is still needed is to challenge these regional working groups to share and evaluate their experience with others about specific activities along integrative themes. Facilitators, who might come from universities, federal agencies such as the National Agroforestry Center, and/or NGOs including the Savanna Institute, would help to link landowners with one another and with other key actors from production, trade, NGOs, professional associations, land-grant universities, national agency research laboratories, the markets that are essential for viable systems, and various policy units. They would document individual and collective learning processes with an aim to move knowledge from the particular, context-specific state to a more global and predictive one integrating knowledge across landscapes.

Workshops and study tours designed to help participants recognize and evaluate the informal experimental design and evaluation processes in which landowners engage, and how they inform these processes through their respective learning networks, would serve to sharpen and focus the collective expert judgment that develops. In recent years, many such workshops have been established. For example, in 2013, the Center for Agroforestry, in conjunction with MAAWG and via initial funding from SARE, established an annual Agroforestry Academy to help address this need. As of 2019, 175 individuals (farmers and educators alike) have been trained across seven academies, and a longitudinal study is ongoing to extract lessons learned and to create a learning network among the trainees (Gold et al., 2019). The Savanna Institute is also very active in hosting workshops and study tours and linking farmers together in networks. These activities overlap with conventional extension roles in agriculture and forestry, helping to provide a dual purpose and justification for funding.

Numerous trainings, workshops, and study tours have been very successful in attracting agroforestry practitioners. These individuals, varying widely in age from their 20s to their 60s, are typically curious, open-minded landowners, many of whom come from an understanding of permaculture, who believe there may be a better or different way to manage agricultural and forestry resources than conventional land use approaches. They are also likely to have a multigenerational vision for the development of their production system, while at the same time adopting a willingness to compromise it in practical terms to the realities of today's transient society. Agroforestry attracts individuals who value hard work and understand the critical role of management in generating multiple outputs in as complementary and noncompetitive a manner as possible. They are likely to experiment with various components of their evolving production system and to have created a diverse network of information resources to assist their efforts to design new systems and informally test new hypotheses. Such people can be found in the membership of numerous organizations throughout the United States and Canada (e.g., the Association for Temperate Agroforestry [AFTA], the Appalachian Beginning Forest Farmer Coalition [ABFFC], etc.) that are concerned with the development and marketing of alternative crops and enterprises or the management of natural resources. In a highly connected world of social media, they can easily reach out to existing organizations, anticipating their role in satisfying their needs for learning, improving their practices, and addressing important social issues. Once they are part of such networks, they attract others to join.

Implementation of the proposed strategy is well underway, and critical perceptual and institutional barriers to improving the capacity for knowledge and information generation about agroforestry are being addressed. Scientific knowledge about agroforestry is rapidly being integrated into practice via the host of organizations previously mentioned (Gold, 2019).

The important implication is that landowners have now become an integral part of the knowledge generation process. This requires careful examination of the processes they use, the products they develop, and the various learning groups with whom they interact. In doing so, the research and development community now acknowledges and participates in the dense networks of informal learning about agroforestry that they understand and appreciate. As stated, numerous organizations are now playing important roles in developing generalizable knowledge if adequately recognized and organized to do so. Actions are being taken to link them. In this way, agroforestry now offers important opportunities fostering innovation in land use management.

Progress to Date and Challenges Ahead

The potential for domestic agroforestry and the constraints to its development that were first identified in 1989 (Lassoie et al., 1991) and then reexamined 2 yr later (Lassoie & Buck, 1991) are dramatically different from those facing us today (Gold, 2019). Agroforestry practices are becoming part of the repertoire of management strategies that are emerging from the research and development community to address complex land use sustainability issues within interdisciplinary forums.

As mentioned above, however, agroforestry is a hybrid of the established fields of agriculture and forestry, closely aligned with the science of agroecology and regenerative agriculture. Therefore, each new approach will face its own set of challenges as it moves from theory into practice. Practical application of these approaches also will face different challenges and offer different opportunities to the research and development community. These challenges and progress to date in meeting them are discussed here as well as specific recommendations to further advance agroforestry research, development, and practice in the United States.

Basic Challenges and Progress

Agroforestry in the United States has faced some unique challenges as an emerging land use strategy, many of which are being overcome. First, concepts and methodologies were originally obtained from international experiences primarily in developing, tropical countries with very different ecological and socioeconomic contexts. In recent decades, agroforestry in the United States and Canada (and Europe) has made huge strides to refine relevant concepts and methodologies that fit the temperate zone and the Western, industrialized realities in which we live. As such, domestic agroforestry has evolved at the intersection of the well-established fields of agriculture, horticulture, and forestry. As an emergent applied science, agroforestry has aligned with agroecology and established a research–education–development infrastructure that integrates across these well-established but separate disciplines.

Inherent Constraints Being Overcome

Because domestic agroforestry has evolved within a modern society primarily located in a temperate region of the world, it has faced inherent constraints not found in most developing countries. First, the climate in much of North America is not conducive to fast plant growth, especially by long-lived woody perennials. In addition, some of our indigenous tree species are naturally slow growing and yield only one primary product—usually timber. As a consequence, the use of trees for timber in many types of agroforestry practices do not directly yield useful or marketable products for many years—often after the life of the persons who planted them! Knowing this, domestic agroforestry has instead focused on overstory nut- and fruit-bearing trees and shrubs that come into economic production in 3-15 yr (e.g., elderberry [*Sambucus nigra* L. ssp. *canadensis* (L.) R. Bolli], aronia [*Aronia* sp.], eastern black walnut [*Juglans nigra* L.], pecan [*Carya illinoinensis* (Wangenh.) K. Koch], Chinese chestnut [*Castanea mollissima* Blume]). Furthermore, fast-growing species of the genus *Populus* (hybrid poplar, cottonwood, etc.) and *Salix* (clonal willow) (Robertson et al., 2017; Volk et al., 2006) are being used for biomass (MacPherson, 1995) and as woody florals (Gold, Godsey, & Josiah, 2004) and are integrated into riparian forest and upland buffer production systems to provide multiple products and environmental services. Finally, native perennial grasses (e.g., switchgrass [*Panicum virgatum* L.]) are also being used for biomass and other ecosystem services within a variety of agroforestry practices (Gamble, Johnson, Current, Wyse, & Sheaffer, 2016; Schulte et al., 2017)

The United States is a modern, industrialized nation with an increasingly large educated, urban population. Therefore, agroforestry practices are being developed to simultaneously address the market opportunities in urban areas while also meeting specific interests, needs, and problems of rural landowners. Currently, obstacles to agroforestry adoption exist but are in the process of being overcome (de Jalon et al., 2018; Wilson & Lovell, 2016). Barriers include the expense of establishment, landowners' lack of experience with trees (Faulkner et al., 2014), the time and knowledge required for management and marketing (Valdivia, Barbieri, & Gold, 2012), and a lack of understanding by extension and state and federal agency professionals.

Agroforestry practices also have to compete with commodity crops, which have well-developed government support systems providing insurance and price guarantees that significantly reduce landowner risk. Agroforestry practices do not, at present, have the same level of support, requiring that the landowner take on significant risk in adopting agroforestry practices. That said, the support structure and knowledge network for agroforestry is growing rapidly, addressing many of the issues constraining agroforestry adoption (Schoeneberger et al., 2017).

Evolving Infrastructure

The depth and breadth of the agroforestry research–education–application infrastructure has come a long way in the past 40 yr, developing most rapidly in the past decade. Coupled with an acceleration of biophysical and socioeconomic research, there are now positive changes in federal policy and positive market trends. The USDA–NRCS formally recognized temperate agroforestry practices in their cost-share Environmental Quality Incentives Program (EQIP), helping promote agroforestry through national policy. Further support for growth of the agroforestry sector comes from positive consumer and market trends: increased demand for and promoting of "buy local"; growth of direct-to-consumer farmers markets; continued growth in the organic sector; and strong interest in pasture-based livestock production.

Formally accredited online graduate certificate and master's degree programs have been established; numerous extended-duration training programs have been created and designed to train educators and landowners; NGOs (e.g., the Savanna Institute) and private sector (e.g., Iroquois Valley Farmland REIT, PBC Farms Beef) engage with landowners in agroforestry; multiple specialty crop and livestock cooperatives (e.g., elderberry, chestnut, hazelnut [*Corylus* spp.], aronia) have been formed; and robust financial decision support tools have been developed.

As it matures, this infrastructure must provide an interconnected feedback–feedforward knowledge system of researchers, teachers, extension personnel, and field practitioners to promote and support the development, refinement, and implementation of new ideas and practices (Gold, 2007).

Agroforestry as an Applied Science

The ongoing challenge is the continued development of domestic agroforestry practices along with the development of more discipline-based land use strategies of sustainable agriculture and forestry. Almost 30 yr ago, Lassoie and Buck (1991) called for a major national commitment similar to the one mounted near the turn of the 20th century for agricultural production and forest conservation. As we begin 2021, large-scale refocusing of the nation's resources and professional energies has yet to fully materialize owing to the strength and ingrained structure of our current institutions and steady stream of state budget tightening. Nonetheless, concerns about the environmental impacts of current land use practices and the deterioration of the land base and water are increasingly being recognized as important problems to address (e.g., hypoxia, soil health). In spite of institutional and fiscal limitations, steady development efforts are underway to move domestic agroforestry from concepts to practices. During the past decade, steady progress has been underway at many different levels toward building a research–education–practice infrastructure involving a unique partnership including academia, state and federal governments, NGOs, the private sector, and agroforestry practitioners.

Research and Development

Since the second North American Agroforestry Conference (NAAC) (Garrett, 1991), there has been a dramatic increase in the amount of biophysical and socioeconomic agroforestry research in the United States and Canada. This is directly reflected in the chapters within this third edition of *North American Agroforestry* along with other recently published works and edited volumes (Gordon et al., 2018; Mosquera-Losada & Prabhu, 2019; Schoeneberger et al., 2017). The 16th NAAC was held in 2019 and showcased a substantial amount of interdisciplinary research focused on specific opportunities where agroforestry practices can be applied. Temperate agroforestry research is regularly being reported at workshops and special sessions sponsored by professional societies, e.g., see recent abstracts of sessions at the American Society of Agronomy, Ecological Society of America, Society of American Foresters, and government agencies (e.g., USDA, 2019), along with active international conferences and symposia in Europe and elsewhere (Dupraz, Gosme, & Lawson, 2019). More scientific publications are appearing in a wider variety of scientific journals in addition to *Agroforestry Systems* (e.g., *Forest Ecology and Management*; *Society and Natural Resources*; *Agronomy Journal*; *Plant and Soil*; *Sustainability*; *Agriculture, Ecosystems, and Environment*; and the *Journal of Environmental Quality*).

Previously considered to be a new, interdisciplinary, applied science, agroforestry used to be equated with being professionally "vague" and "non-rigorous" by many working in more narrow scientific disciplines. However, the biennial NAAC, European Agroforestry conferences (EURAF), and the breadth and depth of the scientific literature are helping to change this situation by raising the professional recognition of those working in domestic agroforestry.

The volume of quality agroforestry research has increased dramatically in the past four decades, helping to support the application of agroforestry domestically. While the science of agroforestry lacks the full spectrum of understanding necessary to assure the successful widespread implementation of most agroforestry practices (e.g., information about specific species' responses to site characteristics, economics of production through time), a substantial body of research information has been developed and is increasing annually. Advances in both the biophysical and socioeconomic understanding of agroforestry practices is helping to reduce both biological and financial risks for producers.

Agroforestry scientists have found grant support through the many programs within the Agriculture and Food Research Initiative (AFRI), the nation's leading competitive grants program for the agricultural sciences. The NIFA awards AFRI research, education, and extension grants to improve rural economies, increase food production, stimulate the bioeconomy, mitigate the impacts of climate variability, address water availability issues, ensure food safety and security, enhance human nutrition, and train the next generation of the agricultural workforce. Multiple federal agencies and programs including NIFA, the USDA–AMS, USDA–ARS, Farm Service Agency, U.S. Forest Service, NRCS, SARE, U.S. Environmental Protection Agency, National Science Foundation, and National Institutes of Health all support facets of the science and application of agroforestry.

Agroforestry researchers have had particular funding success through USDA SARE grants and USDA–AMS Specialty Crop Block Grants. With rare exceptions, grant funding opportunities are competitive and, in light of constantly diminishing support for higher education, the competition for federal grant dollars is fierce—often funding <10% of submitted proposals. In this light, what is currently lacking is a dedicated research funding program specifically targeted to support agroforestry.

University Education

In 1997, 36 universities in 28 different states reported teaching at least one course dealing with agroforestry (Rietveld, 1997). As of 2017, 27 U.S. institutions reported current agroforestry course offerings (Wright, 2017). However, due to the presence of online agroforestry programs, educational access for those interested in studying agroforestry has increased (Gold, 2015; Gold & Jose, 2012). In addition, the breadth and availability of the relevant literature and up-to-date textbooks has continued to increase. In addition to this text, another recently updated text is dedicated to temperate agroforestry (Gordon et al., 2018), and other similar compendiums have been published (Mosquera-Losada & Prabhu,

2019). Agroforestry education is reviewed in detail in Chapter 19.

Within universities, agroforestry courses are most often offered through forestry, natural resources, or agriculture departments (Wright, 2017). In addition, agroforestry is often addressed within courses on sustainable agriculture, agroecology, integrated forest management, international agriculture, or sustainable development. Typically, courses dedicated solely to agroforestry consider both domestic and international aspects. Although many universities offer agroforestry courses, few offer comprehensive curricula, and most agroforestry courses are used to supplement disciplinary degree options at the undergraduate level and to help build interdisciplinary programs at the graduate level.

Few institutions possess the complement of faculty to offer the selection of courses believed necessary for a major in agroforestry or, if they have the faculty, it is difficult to bring them together to offer an integrated agroforestry curriculum (Gold & Jose, 2012; Lassoie, 1990; Lassoie et al., 1994). Agroforestry is not a discipline but rather an interdisciplinary field of study. Therefore, a comprehensive agroforestry curriculum (or even a single course) demands expertise from a wide variety of professionals, often from different academic units across campus. Such individuals are often fully committed to teaching responsibilities within their respective disciplines, making it difficult for them to engage in a new curriculum or team-taught course. This means that not only are their numbers relatively small, but there is also a widespread lack of extensive training and experience in agroforestry within the academic community, especially related to its application to North American conditions. Fortunately, this situation is changing as faculty gain relevant experience, more graduate students pursue agroforestry studies, and universities begin to hire those with such an education.

Agroforestry curricula tend to be carried by a limited number of faculty members (often one) and their graduate students working within either an agriculture or forestry academic unit (e.g., college, school, or department). Unfortunately, the decision typically is made by default: who has the interest and commitment to deal with an interdisciplinary topic like agroforestry, especially when considering its application to a modern, production-oriented society? This means that the administrative support for agroforestry can be quite weak, existing only at the margin of more commonly understood traditional teaching programs.

In the United States and Canada, notable exceptions to the general trend include agroforestry programs at the University of Missouri, Virginia Tech, and Laval University in Canada. These and a handful of other universities (e.g., the University of Florida, University of Minnesota, Cornell University) are actively training agroforestry professionals who are now filtering out to other schools in temperate North America, creating the human and applied research base that can be used to grow the discipline in the United States. The University of Missouri has had a sustained funding base for more than two decades and has developed increasingly robust agroforestry research, teaching, and outreach programs. In addition to its on-campus agroforestry graduate program, the University of Missouri established an online master of science program and an online graduate certificate in 2013. These fully online programs have provided access to agroforestry education regardless of geography (Gold & Jose, 2012). Between 2013 and 2018, more than 70 students have been admitted into these programs and 30 have received graduate credentials in agroforestry.

Despite the current limitations, agroforestry courses typically attract highly qualified students who often come with extensive international agroforestry experience, including the Peace Corps (Gold & Jose, 2012), or are familiar with permaculture, agroecology, and sustainable or regenerative agriculture. In the past, such interest was limited to graduate students seeking careers in international development. More recently, however, both undergraduates and graduate students have been attracted to agroforestry courses, probably reflecting their growing interest in courses dealing with issues of sustainability. Employment opportunities where agroforestry credentials are a definite plus are increasing. Many federal agencies (e.g., the NRCS), global, national, and regional conservation organizations (e.g., Heifer International, The Nature Conservancy, National Wild Turkey Federation, Trees Forever), along with NGOs specifically dedicated to agroforestry (e.g., Savanna Institute) are hiring individuals with agroforestry backgrounds.

Professional and Practitioner Training

The need for continuing education and training in agroforestry for both professionals and practitioners was recognized in the early 1990s. Specific needs for such training were identified for various regions of the United States (Merwin, 1997), and scattered regional trainings were held across a broad range of topics (Josiah, 1999); however, active professional training programs did not become commonplace until the early 2000s. Agroforestry training programs are reviewed in more detail in Chapter 19.

The first USDA agroforestry strategic framework (USDA, 2011) discussed the need for education and training of natural resource professionals, including training needs, methods, tools and certification, to effectively deliver agroforestry assistance. General recommendations included pursuing partnerships and cross-training opportunities with special interest groups and nontraditional partners and seeking training opportunities such as landowner-to-landowner, peer-to-peer, local organizations, and professional training of different audiences.

In 2019, the USDA released an updated Agroforestry Strategic Framework (USDA, 2019), revisiting priorities for professional education. Their primary objective was to increase the availability of information and tools that help natural resource professionals to provide technical, educational, financial, and marketing assistance. The outlined strategies included support for university efforts to develop agroforestry curricula and to offer a major, certificate, or area of expertise in agroforestry, providing natural resource professionals with an array of options for receiving and providing training and technical assistance in agroforestry technologies and landowner outreach, including professional meetings and conferences, stand-alone training activities, and online courses, and developing recognition mechanisms for professionals that have gained expertise in agroforestry through completion of a recommended set of agroforestry training requirements (e.g., agroforestry certification).

In spite of significant advances in both the science and practice of agroforestry during the past 35 yr, adoption has been limited. Up to about 2010, the situation persisted in which natural resource professionals and other educators were not well equipped to help landowners adopt agroforestry and benefit directly from an intensive immersion into agroforestry. Without being able to observe and understand the benefits of agroforestry, professionals lacked interest and, without interest, agroforestry practices were not being promoted or adopted. One of the most important contact points between landowners and natural resource professionals is the local county agent, often working for university extension, the USDA–NRCS, or a Soil and Water Conservation District. These are the professionals who help farmers as they adopt practices receiving local or federal government support. Although many of these professionals administer programs to which agroforestry practices might apply, the lack of knowledge or interest in those options by agents means that they are not suggesting agroforestry options to landowners, severely limiting the dissemination and demonstration of agroforestry practices.

One concrete step designed to help rectify this knowledge gap was created back in 2013. The University of Missouri and MAAWG collaborated to create a week-long intensive crash course in agroforestry planning and design: the Agroforestry Academy (Gold et al., 2019). The Agroforestry Academy, initially funded for 2 yr through an North Central Region–SARE Professional Development Program grant, was originally designed for professional development of natural resource professionals, extension agents, and other educators to advance the adoption of agroforestry as a cornerstone of productive land use in the Midwest. After the academy's second year, it was also opened up to landowners with a particular focus on opportunities for resource-limited farmers and military veteran farmers. Through other grant funding, scholarships have been provided to support military veterans. Up through 2021, the Agroforestry Academy has been offered for 7 yr with 175 educators and landowners trained in agroforestry. During the past decade (i.e., 2010–2020), many other spinoff trainings, offered throughout the United States, have evolved from or in parallel with the Agroforestry Academy.

Advanced training on the five agroforestry practices includes options for marketing, economic, social dimensions, and environmental services benefits and, coupled with practice in agroforestry planning and design, facilitates the development of an agroforestry knowledge network. In turn, this has helped to build the infrastructure needed to enhance landowner adoption of agroforestry, resulting in increased sustainability of rural communities and the food and agricultural system. As a result of the Agroforestry Academy and other training programs offered across the United States, educators and landowners are gaining an improved understanding of the design and implementation of agroforestry practices, including documented changes in awareness and knowledge and on-the-ground adoption (Gold et al., 2019).

Identification and Support of Practitioners

It is important to recognize the level of risk a practitioner takes on in adopting or practicing agroforestry. The predominant agricultural crops and, to a lesser extent, specialty nut and fruit crops often have extensive research bases that help reduce uncertainty. That research base, coupled with government-sponsored insurance and price support programs for many crops, significantly reduces the risk to landowners. The research base for agroforestry practices has grown substantially since 2000. While agroforestry practices do not enjoy the same level of support as commodity crops, both biophysical and socioeconomic research has been conducted to help reduce landowner risk. Government programs, especially USDA SARE grant programs directed to farmers, support the adoption and demonstration of sound agroforestry practices and are helping to address questions of risk for agroforestry adoption.

During the past decade (2010–2020), in addition to the National Agroforestry Center and the Association for Temperate Agroforestry, a number of regional agroforestry working groups have been established to bring agroforestry practitioners together. These informal networks are serving as venues for the exchange of knowledge and experiences among practitioners, cooperatives, researchers, outreach professionals, and NGOs. The growing list of key regional agroforestry working groups includes:

Northeast/Mid-Atlantic Agroforestry Working Group (NEMA) https://www.capitalrcd.org/nema-about-us.html
Mid-American Agroforestry Working Group (MAAWG) http://midamericanagroforestry.net/
Pacific Northwest Agroforestry Working Group (PNAWG) http://pnwagro.forestry.oregonstate.edu/
Southwest Agroforestry Action Network (SWAAN) https://aces.nmsu.edu/aes/agroforestry/southwest-agroforestry-w.html
Appalachian Beginning Forest Farmers Coalition (ABFFC) https://www.appalachianforestfarmers.org/
Savanna Institute http://www.savannainstitute.org/

In addition, many practitioners prefer to affiliate with associations or cooperatives involving like-minded individuals, and a number of these organizations support agroforestry specialty crop production (e.g., Northern Nut Growers Association, Chestnut Growers of America, Maple Producers Association, North American Ginseng Association, ABFFC, Northeast Organic Farmers Association, Nebraska Woody Florals). The Savanna Institute, as discussed above, is a nonprofit organization created to reap the full benefits of the experiences and knowledge emerging from the diversity of agroforestry practitioners.

At the federal level, the Cooperative Extension System has developed eXtension (eXtension.org), and within eXtension there are Communities of Practice. One such community of practice, created through a grant to Virginia Tech, is the Forest Farming Community (https://forest-farming.extension.org/). The Forest Farming Community includes forest farmers, university faculty, and agency personnel working together to provide useful farming information. The Forest Farming Community shares information about growing and selling high-value non-timber forest products. Members are from across the country and have experience farming and studying edible, medicinal, decorative, and craft-based products in woodlands. The community provides woodland owners and managers with information about startup, best practices, and markets and policies.

Because of the growing wealth of organizational resources and knowledge networks, agroforestry is becoming a more realistic and practical options for thousands of landowners.

Future Needs

In a short span of four decades, agroforestry in the United States has transitioned from a little-used name and practice to a science-based technology that is widely recognized. While the United States lacks a consistent national policy on agroforestry, the establishment of the USDA Agroforestry Strategic Framework for 2011–2016 (revised in 2019) has advanced agroforestry from a fragmented effort on the part of a few to an area of focus on the part of many. The question is no longer, do we need agroforestry, but rather what will agroforestry look like in the United States over the next four decades? The professional community continues to be focused on providing the biophysical and socioeconomic specifics needed to implement agroforestry on the ground. Such details comprise the rest of this volume, showing strong promise for the further development of agroforestry in the United States as well as for other developed, temperate regions of the world.

Although progress has been good, specific challenges still face the development of domestic agroforestry. First, we must continue to increase the amount of research being conducted, and this work must be interdisciplinary and focused on specific opportunities where agroforestry practices can be applied. Second, we must educate and train professionals who are capable of applying agroforestry research methodologies and results to real-world situations and, of course, assure employment opportunities for them at the end of their schooling. Third, we must cultivate and support a group of practitioners willing to work with researchers to test and evaluate new technologies. Fourth, we must educate the general public to understand the need to support the development of sustainable land use management systems like agroforestry and to appreciate the unique value of products from such systems. Lastly, collaboration must be stimulated among key individuals and organizations to further the ideals and practice of domestic agroforestry: extension personnel, researchers, and practitioners; different disciplines, departments, and colleges; and different public and private organizations, agencies, and institutions. In practice, the development of these technological and organizational components of a domestic agroforestry program should occur simultaneously and proceed in parallel rather than in series. Hopefully, an emerging national policy on agroforestry will provide the framework needed to address these challenges.

It is important that the research, extension, federal and state agencies, NGOs, associations, cooperatives, and the private sector continue to build and develop the knowledge network and infrastructure to support the growth of agroforestry. With dedicated, collaborative efforts, agroforestry practices will become increasingly important within North America's food, forage, and fiber production systems. It is hoped that this text provides support for innovative approaches to maintaining the long-term ecological integrity and productivity of the nation's farm and forest lands. Such is the essence of our society's quest for sustainability.

References

Adams, D. M., Haynes, R. W., & Daigneault, A. J. (2006). *Estimated timber harvest by U.S. region and ownership, 1950–2002* (Gen. Tech. Rep. PNW-GTR-659). Portland, OR: U.S. Forest Service, Pacific Northwest Research Station.

Alavalapati, J. R. R., & Mercer, D. E. (Ed.). (2004). Valuing agroforestry systems: Methods and applications. Dordrecht, the Netherlands: Kluwer.

Altieri, M. A., Nicholls, C. I., & Montalba, R. (2017). Technological approaches to sustainable agriculture at a crossroads: An agroecological approach. *Sustainability, 9,* 349. https://doi.org/10.3390/su9030349

Atangana, A., Khasa, D., Chang, S., & Degrande, A. (2013). *Tropical agroforestry*. Dordrecht, the Netherlands: Springer. https://doi.org/10.1007/978-94-007-7723-1

Biggs, S. D. (1990). A multiple source of innovation model of agricultural research and technology promotion. *World Development, 18,* 1481–1499. https://doi.org/10.1016/0305-750X(90)90038-Y

Börner, J., Baylis, K., Corbera, E., Ezzine-de-Blas, D., Honey-Rosés, J., Persson, U. M., & Wunder, S. (2017). The effectiveness of payments for environmental services. *World Development, 96,* 359–374. https://doi.org/10.1016/j.worlddev.2017.03.020

Brown, S. E., Miller, D. C., Ordonez, P. J., & Baylis, K. (2018). Evidence for the impacts of agroforestry on agricultural productivity, ecosystem services, and human well-being in high-income countries: A systematic map protocol. *Environmental Evidence, 7,* 24. https://doi.org/10.1186/s13750-018-0136-0

Buck, L. E. (1995). Agroforestry policy issues and research directions in the US and less developed countries: Insights and challenges from recent experience. *Agroforestry Systems, 30,* 57–73. https://doi.org/10.1007/BF00708913

Bukowski, C., & Munsell, J. (2018). *The community food forest handbook: How to plan, organize, and nurture edible gathering places*. White River Junction, VT: Chelsea Green.

Burgess, P., & Rosati, A. (2018). Advances in European agroforestry: Results from the AGFORWARD project. *Agroforestry Systems, 92,* 801–810.

Burke, T. A., Cascio, W. E., Costa, D. L., Deener, K., Fontaine, T. D., Fulk, F. A., . . . Zartarian, V. G. (2017). Rethinking environmental protection: Meeting the challenges of a changing world. *Environmental Health Perspectives, 125*(3), A43–A49. https://doi.org/10.1289/EHP1465

Campbell, D. T. (1986). Ethnocentrism of disciplines and the fish-scale model of omniscience. In D. E. Chubin, A. L. Porter, F. A. Rossini, & T. Connolly (Eds.), *Interdisciplinary analysis and research* (pp. 29–46). Mt. Airy, MD: Lomond.

Campbell, G. E., Lottes, G. J., & Dawson, J. O. (1991). Design and development of agroforestry systems for Illinois, USA: Silvicultural and economic considerations. *Agroforestry Systems, 13,* 203–224. https://doi.org/10.1007/BF00053579

Carroll, C. F. (1973). *The timber economy of Puritan New England*. Providence, RI: Brown University Press.

Chambers, R., Pacey, A., & Thrupp, L. A. (Eds.). (1989). *Farmer first: Farmer innovation and agricultural research*. London: Intermediate Technology Publications.

Chubin, D. E., Porter, A. L., Rossini, F. A., & Connolly, T. (Eds.). (1986). Interdisciplinary analysis and research. Mt. Airy, MD: Lomond.

Conway, G. R. (1987). The properties of agroecosystems. *Agricultural Systems, 24,* 95–117. https://doi.org/10.1016/0308-521X(87)90056-4

Coufal, J., & Webster, D. (1996). The emergence of sustainable forestry. In P. McDonald & J. Lassoie (Eds.), *The literature of forestry and agroforestry* (pp. 147–167). Ithaca, NY: Cornell University Press.

Coulibaly, J. Y., Chiputwa, B., Nakelse, T., & Kundhlande, G. (2016). *Adoption of agroforestry and its impact on household food security among farmers in Malawi* (ICRAF Working Paper 223). Nairobi, Kenya: World Agroforestry Centre. https://doi.org/10.5716/WP16013.PDF

Cronon, W. (1983). *Changes in the land: Indians, colonists and the ecology of New England*. New York: Hill and Wang.

Daly, H. E., & Cobb, J. B., Jr. (1989). *For the common good*. Boston, MA: Beacon Press.

de Jalon, S. G., Graves, A., Palma, J. H. N., Williams, A., Upson, M., & Burgess, P. J. (2018). Modelling and valuing the environmental impacts of arable, forestry and agroforestry systems: A case study. *Agroforestry Systems, 92,* 1059–1073. https://doi.org/10.1007/s10457-017-0128-z

Dollinger, J., & Jose, S. (2018). Agroforestry for soil health. *Agroforestry Systems, 92,* 213–219. https://doi.org/10.1007/s10457-018-0223-9

Downing, M., Volk, T. A., & Schmidt, D. A. (2005). Development of new generation cooperatives in agriculture for renewable energy research, development, and demonstration projects. *Biomass and Bioenergy, 28,* 425–434. https://doi.org/10.1016/j.biombioe.2004.09.004

Dupraz, C., Gosme, M., & Lawson, G. (Eds.). (2019). *Agroforestry: Strengthening links between science, society and policy: Book of Abstracts, 4th World Congress on Agroforestry.* Montpellier: CIRAD.

Dupraz, C., Wolz, K. J., Lecomte, I., Talbot, G., Vincent, G., Mulia, R., . . . van Noordwijk, M. (2019). Hi-sAFe: A 3D agroforestry model for integrating dynamic tree–crop interactions. *Sustainability, 11,* 2293. https://doi.org/10.3390/su11082293

Elevitch, C. R., Mazaroli, D. N., & Ragone, D. (2018). Agroforestry standards for regenerative agriculture. *Sustainability, 10*(9), 3337. https://doi.org/10.3390/su10093337

FAO. (2018). *Transforming food and agriculture to achieve the SDGs: 20 interconnected actions to guide decision-makers.* Rome: FAO.

Faulkner, P. A., Owooh, B., & Idassi, J. (2014). Assessment of the adoption of agroforestry technologies by limited-resource farmers in North Carolina. *Journal of Extension, 52*(5), 5rb7. Retrieved from https://joe.org/joe/2014october/rb7.php.

Feliciano, D., Ledo, A., Hiller, J., & Nayak, D. R. (2018). Which agroforestry options give the greatest soil and above ground carbon benefits in different world regions? *Agriculture, Ecosystems & Environment, 254,* 117–129. http://dx.doi.org/10.1016/j.agee.2017.11.032

Ferraz-de-Oliveira, M. I., Azeda, C., & Pinto-Correia, T. (2016). Management of montados and dehesas for high nature value: An interdisciplinary pathway. *Agroforestry Systems, 90,* 1–6. https://doi.org/10.1007/s10457-016-9900-8

Fitchen, J. M. (1991). *Endangered spaces, enduring places: Change, identity, and survival in rural America.* Boulder, CO: Westview Press.

Funabashi, M. (2018). Human augmentation of ecosystems: objectives for food production and science by 2045. *npj Science of Food, 2,* 16. https://doi.org/10.1038/s41538-018-0026-4

Gamble, J. D., Johnson, G., Current, D. A., Wyse, D. L., & Sheaffer, C. C. (2016). Species pairing and edge effects on biomass yield and nutrient update in perennial alley cropping systems. *Agronomy Journal, 108,* 1020–1029. https://doi.org/10.2134/agronj2015.0456

Garrett, H. E. (Ed.). (1991, 18–21 Aug.). *Proceedings of the Second Conference on Agroforestry in North America. Springfield, MO.* Columbia, MO: Association for Temperate Agroforestry.

Garrett, H. E., & Buck, L. E. (1997). Forest management practices in temperate zone agroforestry: Silvicultural and policy aspects in the United States. *Forest Ecology and Management, 91,* 5–15. https://doi.org/10.1016/S0378-1127(96)03884-4

Garrett, H. E., Jones, J. E., Kurtz, W. B., & Slusher, J.P. (1991). An evaluation of black walnut (*Juglans nigra* L.) agroforestry—Its design and potential as a land use alternative. *Forest Chronicles, 67,* 213–218. https://doi.org/10.5558/tfc67213-3

Garrett, H. E., Kurtz, W. B., Buck, L. E., Lassoie, J. P., Gold, M. A., Pearson, H. A., . . . Slusher, J. P. (1994). Agroforestry: An integrated land-use management system for production and farmland conservation. Washington, DC: USDA Soil Conservation Service.

Garrity, D. P., Akinnifesi, F. K., Ajayi, O. C., Weldesemayat, S. G., Mowo, J. G., Kalinganire, A., . . . Bayala, J. (2010). Evergreen Agriculture: A robust approach to sustainable food security in Africa. Food Security, 2(3), 197–214. https://doi.org/10.1007/s12571-010-0070-7

Geertsema, W., Rossing, W. A. H., Landis, D.A., Bianchi, F. J. J. A., van Rijn, P. C. J., Schaminée, J. H. J., . . . van der Werf, W. (2016). Actionable knowledge for ecological intensification of agriculture. *Frontiers in Ecology and the Environment, 14*(4), 209–216. https://doi.org/10.1002/fee.1258

Gillis, A. (1990). The new forestry. *BioScience, 40,* 558–562. https://doi.org/10.2307/1311294

Globalizing Agricultural Science and Education Programs in America Task Force. (1997). *An emerging agenda for sustainable agriculture, food, natural resources, rural and related human science programs.* Washington, DC: USDA–CSREES.

Gold, M. A. (2007). *Developing the infrastructure to stimulate agroforestry production in the U.S.* Presented at the 10th North American Agroforestry Conference, Quebec City, QC, Canada.

Gold, M. A. (2015). Evolution of U.S. agroforestry research and formalization of agroforestry education. *Inside Agroforestry,* 23(3). Retrieved from https://www.fs.usda.gov/nac/assets/documents/insideagroforestry/IA_Vol23Issue3.pdf.

Gold, M. A. (2019). Tracing 35 years of agroforestry development in the USA: Past, present, future. In C. Dupraz, M. Gosme, & G. Lawson (Eds.), *Agroforestry: Strengthening links between science, society and policy: Book of Abstracts, 4th World Congress on Agroforestry* (p. 865). Montpellier: CIRAD.

Gold, M. A., Godsey, L. D., & Josiah, S. J. (2004). Markets and marketing strategies for agroforestry specialty products in North America. *Agroforestry Systems, 61,* 371–382.

Gold, M. A., & Hanover, J.W. (1987). Agroforestry systems for the temperate zone. *Agroforestry Systems, 5,* 109–121. https://doi.org/10.1007/BF00047516

Gold, M. A., Hemmelgarn, H. L., & Mendelson, S. E. (2019). Academy offers professional development to boost agroforestry. *Forestry Source, 24*(3), 12–13.

Gold, M. A., & Jose, S. (2012). Developing an online certificate and master's degree program in agroforestry. *Agroforestry Systems, 86,* 379–385. https://doi.org/10.1007/s10457-012-9522-8

Gordon, A. M., & Newman, S.M. (Eds.). (1997). *Temperate agroforestry systems.* Wallingford, UK: CAB International.

Gordon, A. M., Newman, S. M., & Coleman, B. (Eds.). (2018). *Temperate agroforestry systems* (2nd ed.). Wallingford, UK: CABInternational.https://doi.org/10.1079/9781780644851.0000

Gordon, A. M. (2007). *Agroforestry systems and the invisible present: Ecological goods and services.* Keynote presented at the 10th North American Agroforestry Conference, Quebec City, QC, Canada.

Gruenewald, H., Brandt, B. K. V., Schneider, B. U., Bens, O., Kendzia, G., & Huttl, R.F. (2007). Agroforestry systems for the production of woody biomass for energy transformation purposes. *Ecological Engineering, 29,* 319–328. https://doi.org/10.1016/j.ecoleng.2006.09.012

Hillbrand, A., Borelli, S., Conigliaro, M., & Olivier, A. (2017). *Agroforestry for landscape restoration.* Rome: FAO.

Holzmueller, E. J., & Jose, S. (2012). Bioenergy crops in agroforestry systems: Potential for the U.S. North Central Region. *Agroforestry Systems, 85,* 305–314.

Hudson, N. (1981). *Soil conservation.* Ithaca, NY: Cornell University Press.

Jack, B. K., Kousky, C., & Sims, K. R. E. (2008). Designing payments for ecosystem services: Lessons from previous experience with incentive-based mechanisms. *Proceedings of the National Academy of Sciences, 105,* 9465–9470. https://doi.org/10.1073/pnas.0705503104

Jha, S., Bacon, C. M., Philpott, S. M., Méndez, V. E., Läderach, P., & Rice, R. A. (2014). Shade coffee: Update on a disappearing refuge for biodiversity. *BioScience, 64,* 416–428. https://doi.org/10.1093/biosci/biu038

Jordan, N., Boody, G., Broussard, W., Glover, J. D., Keeney, D., McCown, B. H., . . . Wyse, D. (2007). Sustainable development of the agricultural bio-economy. *Science, 316,* 1570–1571. https://doi.org/10.1126/science.1141700

Jordan, R. N. (1994). *Trees and people.* Washington, DC: Regnery Publishing.

Jose, S. (2009). Agroforestry for ecosystem services and environmental benefits: An overview. *Agroforestry Systems, 76*, 1–10. https://doi.org/10.1007/s10457-009-9229-7

Jose, S. (2019). Environmental impacts and benefits of agroforestry. In *Oxford research encyclopedia of environmental science*. Oxford, UK: Oxford University Press. https://doi.org/10.1093/acrefore/9780199389414.013.195

Jose, S., & Dollinger, J. (2019). Silvopasture: A sustainable livestock production system. *Agroforestry Systems, 93*, 1–9. https://doi.org/10.1007/s10457-019-00366-8

Jose, S., Gold, M. A., & Garrett, H. E. (2018). Temperate agroforestry in the United States: Current trends and future directions. In A. Gordon (Ed.), *Temperate agroforestry*. Wallingford, UK: CAB International. https://doi.org/10.1079/9781780644851.0050

Jose, S., & Gordon, A. M. (Eds.). (2008).*Toward agroforestry design: An ecological approach*. Dordrecht, the Netherlands: Springer. https://doi.org/10.1007/978-1-4020-6572-9

Jose, S., Walter, D., & Kumar, B. M. (2019). Ecological considerations in sustainable silvopasture design and management. *Agroforestry Systems, 93*, 317–331. https://doi.org/10.1007/s10457-016-0065-2

Josiah, S. (Ed.). (1999). *Proceedings of the North American Conference on Enterprise Development through Agroforestry: Farming the Agroforest for Specialty Products*. St. Paul, MN: Center for Integrated Natural Resources and Agriculture Management, University of Minnesota.

King, K. F. S. (1987). The history of agroforestry. In H. A. Steppler & P. K. R. Nair (Eds.), *Agroforestry: A decade of development* (pp. 3–11). Nairobi, Kenya: ICRAF.

Kremen, C., & Merenlender, A. M. (2018). Landscapes that work for biodiversity and people. *Science, 362*, eaau6020. https://doi.org/10.1126/science.aau6020

LaCanne, C. E., & Lundgren, J. G. (2018). Regenerative agriculture: Merging farming and natural resource conservation profitably. *PeerJ, 6*, e4428. https://doi.org/10.7717/peerj.4428

Lassoie, J. P. (1990). Towards a comprehensive education and training program in agroforestry. *Agroforestry Systems, 12*, 121–131. https://doi.org/10.1007/BF00055583

Lassoie, J. P., & Buck, L. E. (1991, 18–21 Aug.). Agroforestry in North America: New challenges and opportunities for integrated resource management. In H. E. Garrett (Ed.), *Proceedings of the 2nd Conference on Agroforestry in North America, Springfield, MO* (pp. 1–19). Columbia, MO: Association for Temperate Agroforestry.

Lassoie, J. P., Huxley, P., & Buck, L. E. (1994). Updating our ideas about agroforestry education and training. *Agroforestry Systems, 28*, 5–19. https://doi.org/10.1007/BF00711984

Lassoie, J. P., Teel, W. S., & Davies, K. M., Jr. (1991). Agroforestry research and extension needs for northeastern North America. *Forest Chronicles, 67*, 219–226. https://doi.org/10.5558/tfc67219-3

Laurence, J. R. (1987). Integrated natural resource management: Why? *Agriculture and Human Values, 4*, 94–99. https://doi.org/10.1007/BF01530645

Lerch, R. N., Lin, C.-H., Goyne, K. W., Kremer, R. J., & Anderson, S. H. (2017). Vegetative buffer strips for reducing herbicide transport in runoff: Effects of season, vegetation, and buffer width. *Journal of the American Water Resources Association, 53*, 667–683. https://doi.org/10.1111/1752-1688.12526

Liebman, M., & Schulte, L. A. (2015). Enhancing agroecosystem performance and resilience through increased diversification of landscapes and cropping systems. *Elementa: Science of the Anthropocene, 3*, 000041. https://doi.org/10.12952/journal.elementa.000041

Liu, C. L. C., Kuchma, O., & Krutovsky, K. V. (2018). Mixed-species versus monocultures in plantation forestry: Development, benefits, ecosystem services and perspectives for the future. *Global Ecology and Conservation, 15*, e00419. https://doi.org/10.1016/j.gecco.2018.e00419

Lovell, S. T., DeSantis, S., Nathan, C. A., Olson, M. B., Mendez, V. E., Kominami, H. C., . . . Morris, W. B. (2010). Integrating agroecology and landscape multifunctionality in Vermont: An evolving framework to evaluate the design of agroecosystems. *Agricultural Systems, 103*, 327–341. https://doi.org/10.1016/j.agsy.2010.03.003

Lovell, S. T., Dupraz, C., Gold, M., Jose, S., Revord, R., Stanek, E., & Wolz, K. (2018). Temperate agroforestry research: Considering multifunctional woody polycultures and the design of long-term field trials. *Agroforestry Systems*, 92:1397–1415. https://doi.org/10.1007/s10457-017-0087-4

Lowrance, R. (2007). *Ecological functions of riparian forest buffers*. Keynote presented at the 10th North American Agroforestry Conference, Quebec City, QC, Canada.

Lundgren, B. (1982). Introduction. *Agroforestry Systems, 1*, 3–6. https://doi.org/10.1007/BF00044324

MacPherson, G. (1995). *Homegrown energy from short-rotation coppice*. Ipswich, NY: Farming Press.

Mann, S., & Wustemann, H. (2008). Multifunctionality and a new focus on externalities. *The Journal of Socio-Economics, 37*, 293–307. https://doi.org/10.1016/j.socec.2006.12.031

Maser, C. (1994). *Sustainable forestry: Philosophy, science, and economics*. Delray Beach, FL: St. Lucie Press.

Mead, D. J. (1995). The role of agroforestry in industrial nations: The southern hemisphere perspective with special emphasis on Australia and New Zealand. *Agroforestry Systems, 31*, 143–156. https://doi.org/10.1007/BF00711722

Mercer, D. E. (2004). Adoption of agroforestry innovations in the tropics: A review. *Agroforestry Systems, 61*, 311–328 https://doi.org/10.1023/B:AGFO.0000029007.85754.70

Mercer, D. E., Cooley, D., & Hamilton, K. (2011). *Taking stock: Payments for forest ecosystem services in the United States*. Washington, DC: Forest Trends Association. Retrieved from https://www.forest-trends.org/wp-content/uploads/imported/ForestPES_Final.pdf

Merwin, M. (Ed.). (1997). *The status, opportunities and needs for agroforestry in the United States: A national report*. Columbia, MO: Association for Temperate Agroforestry. Retrieved from https://www.aftaweb.org/about/afta/2-uncategorised/35-agroforestry-opportunities.html.

Montambault, J. R., & Alavalapati, J. R. R. (2005). Socioeconomic research in agroforestry: A decade in review. *Agroforestry Systems, 65*, 151–161. https://doi.org/10.1007/s10457-005-0124-6

Mosquera-Losada, M. R., & Prabhu, R. (Eds.). (2019). *Agroforestry for sustainable agriculture*. Cambridge, UK: Burleigh Dodds Science. https://doi.org/10.19103/AS.2018.0041

Murray, G. F., & Bannister, M. E. (2004). Peasants, agroforesters, and anthropologists: A 20-year venture in income-generating trees and hedgerows in Haiti. *Agroforestry Systems, 61*, 383–397. https://doi.org/10.1023/B:AGFO.0000029012.28818.0c

Nair, P. K. R. (Ed.). (1989). *Agroforestry systems in the tropics*. Dordrecht, the Netherlands: Kluwer. https://doi.org/10.1007/978-94-009-2565-6

Nair, P. K. R. (1993). *Introduction to agroforestry*. Dordrecht, the Netherlands: Kluwer. https://doi.org/10.1007/978-94-011-1608-4

Nair, P. K. R. (1996). Agroforestry directions and literature trends. In P. McDonald & J. Lassoie (Eds.), *The literature of forestry and agroforestry* (pp. 74–95). Ithaca, NY: Cornell University Press.

Nair, P. K. R., Viswanath, S., & Lubina, P. A. (2017). Cinderella agroforestry systems. *Agroforestry Systems, 91*, 901–917. https://doi.org/10.1007/s10457-016-9966-3

Nash, R. (1982). *Wilderness and the American mind* (3rd ed.). New Haven, CT: Yale University Press.

National Research Council. (1989). *Alternative agriculture.* Washington, DC: National Academies Press.

National Research Council. (1991). *Sustainable agriculture research and education in the field.* Washington, DC: National Academies Press

National Research Council. (1996). *Colleges of agriculture and the land grant universities: Public service and public policy.* Washington, DC: National Academies Press.

Nunez-Mir, G. C., Iannonne, B.V., III, Curtis, K., & Fei, S. (2015). Evaluating the evolution of forest restoration research in a changing world: A "big literature" review. *New Forests, 46,* 669–682. https://doi.org/10.1007/s11056-015-9503-7

Ong, C. K., & Huxley, P. (1996). *Tree–crop interactions: A physiological approach.* Wallingford, UK: CAB International.

Pacheco, P., Gnych, S., Dermawan, A., Komarudin, H., & Okarda, B. (2017). *The palm oil global value chain: Implications for economic growth and social and environmental sustainability* (Working Paper 220). Bogor, Indonesia: Center for International Forestry Research. https://doi.org/10.17528/cifor/006405

Palma, J. H. N., Graves, A. R., Burgess, P. J., Keesman, K. J., van Keulen, H., Mayus, M., . . . Herzog, F. (2007a). Methodological approach for the assessment of environmental effects of agroforestry at the landscape scale. *Ecological Engineering, 29,* 450–462. https://doi.org/10.1016/j.ecoleng.2006.09.016

Palma, J., Graves, A. R., Burgess, P. J., van der Werf, W., & Herzog, F. (2007b). Integrating environmental and economic performance to assess modern silvoarable agroforestry in Europe. *Ecological Economics, 63,* 759–767. https://doi.org/10.1016/j.ecolecon.2007.01.011

Peichl, M., Thevathasan, N. V., Huss, J., & Gordon, A. M. (2006). Carbon sequestration potentials in temperate tree-based intercropping systems in southern Ontario, Canada. *Agroforestry Systems, 66,* 243–257. https://doi.org/10.1007/s10457-005-0361-8

Perlin, J. (1991). *A forest journey: The role of wood in the development of civilization.* Cambridge, MA: Harvard University Press.

Pinho, R. C., Miller, R. P., & Alfaia, S. S. (2012). Agroforestry and the improvement of soil fertility: A view from Amazonia. *Applied and Environmental Soil Science, 2012,* 616383. https://doi.org/10.1155/2012/616383

Potter, C. A., & Wolf, S. A. (2014). Payments for ecosystem services in relation to US and UK agri-environmental policy: Disruptive neoliberal innovation or hybrid policy adaptation? *Agriculture and Human Values, 31,* 397–408. https://doi.org/10.1007/s10460-014-9518-2

Prinsley, R. T. (1992). The role of trees in sustainable agriculture: An overview. *Agroforestry Systems, 20,* 87–115. https://doi.org/10.1007/BF00055306

Probst, J. R., & Crow, T. R. (1991). Integrating biological diversity and resource management. *Journal of Forestry, 89,* 12–17.

Raintree, J. B. (1987). The state of the art of agroforestry diagnosis and design. *Agroforestry Systems, 5,* 219–250. https://doi.org/10.1007/BF00119124

Raintree, J. B. (1990). Theory and practice of agroforestry diagnosis and design. In K. G. MacDicken & N. T. Vergara (Eds.), *Agroforestry: Classification and management* (pp. 58–97). New York: John Wiley & Sons.

Richards, P. (1985). *Indigenous agricultural revolution.* Boulder, CO: Westview Press.

Rietveld, W. J. (1997). Integrating agroforestry into USDA programs. USDA National Agroforestry Center. https://www.fs.usda.gov/nac/assets/documents/morepublications/agroforestry-usda-1997.pdf

Robertson, G. P., Hamilton, S. K., Barham, B. L., Dale, B. E., Izaurralde, R. C., Jackson, R. D., . . . Tiedje, J. M. (2017). Cellulosic biofuel contributions to a sustainable energy future: Choices and outcomes. *Science, 356,* eaal2324. https://doi.org/10.1126/science.aal2324

Roesch-McNally, G., Arbuckle, J. G,. & Tyndall, J. C. (2017). Soil as social-ecological feedback: Examining the "ethic" of soil stewardship among Corn Belt farmers. *Rural Sociology, 83*(1), 145–173. https://doi.org/10.1111/ruso.12167

Rois-Díaz, M., Lovric, N., Lovric, M., Ferreiro-Dominquez, N., Mosqueri-Losada, M. R., den Herder, M., . . . Burgess, P. (2018). Farmers' reasoning behind the update of agroforestry practices: Evidence from multiple case-studies across Europe. *Agroforestry Systems, 92,* 811–828. https://doi.org/10.1007/s10457-017-0139-9

Rossier, C., & Lake, F. (2014). *Indigenous traditional ecological knowledge in agroforestry* (Agroforestry Notes 44). Lincoln, NE: USDA National Agroforestry Center. Retrieved from https://www.fs.usda.gov/nac/assets/documents/agroforestrynotes/an44g14.pdf.

Russell, H. S. (1982). *A long deep furrow: Three centuries of farming in New England* (Abridged edition with forward by M. Lapping). Hanover, NH: University Press of New England.

Sampson, R. N., & Hair, D. (1990). *Natural resources for the 21st century.* Washington, DC: Island Press.

Sanchez, P.A. (1995). Science in agroforestry. *Agroforestry Systems, 30,* 5–55. https://doi.org/10.1007/BF00708912

Savory, A. (1988). *Holistic resource management.* Washington, DC: Island Press.

Scherr, S. J., & McNeely, J. A. (2007). *Farming with nature: The science and practice of ecoagriculture.* Washington, DC: Island Press.

Scherr, S. J., & McNeely, J. A. (2008). Biodiversity conservation and agricultural sustainability: Towards a new paradigm of 'ecoagriculture' landscapes. *Philosophical Transactions of the Royal Society B, 363,* 477–494. https://doi.org/10.1098/rstb.2007.2165

Schoeneberger, M. (2005, 12–15 June). Agroforestry: Working trees for sequestering carbon on ag-lands. In K. N. Brooks & P. F. Ffolliott (Eds.) Moving agroforestry into the mainstream: 9th North American Agroforestry Conference, Rochester, MN. Saint Paul, MN: University of Minnesota. Retrieved from https://www.cinram.umn.edu/sites/cinram.umn.edu/files/schoeneberger.pdf.

Schoeneberger, M. M., Bentrup, G., & Patel-Weynand, T. (Eds.). (2017). Agroforestry: Enhancing resiliency in U.S. agricultural landscapes under changing conditions (Gen. Tech. Rep. WO-96). Washington, DC: U.S. Forest Service. https://doi.org/10.2737/WO-GTR-96

Schulte, L. A., Niemi, J., Helmers, M. J., Liebman, M., Arbuckle, J. G., James, D. E., . . . Witte, C. (2017). Prairie strips improve biodiversity and the delivery of multiple ecosystem services from corn–soybean croplands. *Proceedings of the National Academy of Sciences, 114,* 11247–11252. https://doi.org/10.1073/pnas.1620229114

Schultz, R. C., Colletti, J. P., & Faltonson, R. R. (1995). Agroforestry opportunities for the United States of America. *Agroforestry Systems, 31,* 117–132. https://doi.org/10.1007/BF00711720

Searchinger, T., Waite, R., Hanson, C., Ranganathan, J., Dumas, P., & Matthews, E. (2018). *Creating a sustainable food future: A menu of solutions to feed nearly 10 billion people by 2050* (Final Report). Washington, DC: World Resources Institute. Retrieved from https://wrr-food.wri.org/sites/default/files/2019-07/WRR_Food_Full_Report_0.pdf.

Sinclair, F. L. (1999). A general classification of agroforestry practice. *Agroforestry Systems, 46,* 161–180. https://doi.org/10.1023/A:1006278928088

Skok, R. A. (1996). Forestry education in the United States. In P. McDonald and J. Lassoie (Eds.), *The literature of forestry and agroforestry* (pp. 168–197). Ithaca, NY: Cornell University Press.

Smith, J. R. (1950). *Tree crops: A permanent agriculture*. New York: Devin-Adair.

Spencer, J. S., Jr. (1996). The research publishing influence of the US Department of Agriculture Forest Service. In P. McDonald and J. Lassoie (Eds.), *The literature of forestry and agroforestry* (pp. 129–146). Ithaca, NY: Cornell University Press.

Stankey, G. H. (1996, 6–12 Aug.). Integrating natural resource planning and management: Social science perspectives. In E. Korpilahti, H. Mikkela, & T. Salonen (Eds.) Caring for the forest: Research in a changing world: Proceedings of the IUFRO 20th World Congress, Tampere, Finland (pp. 390–398). Vol. II. Jyvaskyla, Finland: Finnish IUFRO World Congress Organising Committee.

Steen, H. K. (1976). *The US Forest Service: A history*. Seattle, WA: University of Washington Press.

Steppler, H. A. (1987). ICRAF and a decade of agroforestry development. In H. A. Steppler and P. K. R. Nair (Eds.), *Agroforestry: A decade of development* (pp. 13–21). Nairobi, Kenya: ICRAF.

Stock, P., & Burton, R. J. F. (2011). Defining terms for integrated (multi-inter-trans-disciplinary) sustainability research. *Sustainability*, 3(8), 1090–1113. https://doi.org/10.3390/su3081090

Stone, C. D. (1996). *Should trees have standing?* (25th anniversary ed.). Dobbs Ferry, NY: Oceana Publications.

Tisdell, C. A. (1990). *Natural resources, growth, and development: Economics, ecology and resource scarcity*. Westport, CT: Greenwood Publishing.

Torquebiau, E. F. (2000). A renewed perspective on agroforestry concepts and classification. *Comptes Rendus de l'Académie des Sciences, Series III, Sciences de la Vie, 323*, 1009–1017. https://doi.org/10.1016/S0764-4469(00)01239-7

Turner, R. K. (Ed.). (1988). *Sustainable environmental management: Principles and practice*. Boulder, CO: Westview Press.

Udawatta, R. P., Gantzer, C. J., & Jose, S. (2017). Agroforestry practices and soil ecosystem services. In M.M. Al-Kaisi and B. Lowery (Eds.), *Soil health and intensification of agroecosystems* (pp. 305–333). London: Academic Press. https://doi.org/10.1016/B978-0-12-805317-1.00014-2

Udawatta, R., & Jose, S. (2012). Agroforestry strategies to sequester carbon in temperate North America. *Agroforestry Systems*, 86:225–242. https://doi.org/10.1007/s10457-012-9561-1

USDA. 1987. *Agricultural research for a better tomorrow*. Washington, DC: USDA.

USDA. 1998. *A time to act: A report of the USDA National Commission on Small Farms*. Washington, DC; USDA.

USDA. 2003. *Building on a time to act: A report by the USDA Advisory Committee on Small Farms*. Washington, DC: USDA.

USDA. 2011. *USDA agroforestry strategic framework, fiscal year 2011–2016*. Washington, DC: USDA. Retrieved from http://www.usda.gov/documents/AFStratFrame_FINAL-lr_6-3-11.pdf.

USDA. 2015. Twenty five years. *Inside Agroforestry* 23(3). Retrieved from https://www.fs.usda.gov/nac/assets/documents/insideagroforestry/IA_Vol23Issue3.pdf.

USDA. 2019. *Agroforestry strategic framework: Fiscal years 2019–2024* (Misc. Publ. 1615). Washington, DC: USDA. Retrieved from https://www.usda.gov/sites/default/files/documents/usda-agroforestry-strategic-framework.pdf.

Valdivia, C., Barbieri, C., & Gold, M. A. (2012). Between forestry and farming: Policy and environmental implications of the barriers to agroforestry adoption. *Canadian Journal of Agricultural Economics, 60*, 155–175. https://doi.org/10.1111/j.1744-7976.2012.01248.x

van Noordwijk, M., Rahayu, S., Gebrekirstos, A., Kindt, R., Tata, H. L., Muchugi, A., . . . Xu, J. (2019). Tree diversity as basis of agroforestry. In M. van Noordwijk (Ed.), *Sustainable development through trees on farms: Agroforestry in its fifth decade* (pp. 17–44). World Agroforestry (ICRAF) Southeast Asia Regional Program: Bogor, Indonesia.

Van Vooren, L., Reubens, B., Broekx, S., Pardon, P., Reheul, D., van Winsen, F., . . . Lauwers, L. (2016). Greening and producing: An economic assessment framework for integrating trees in cropping systems. *Agricultural Systems, 148*, 44–57. https://doi.org/10.1016/j.agsy.2016.06.007

Volk, T. A., Abrahamson, L. P., Nowak, C. A., Smart, L. B., Tharakan, P. J., & White, E. H. (2006). The development of short-rotation willow in the northeastern United States for bioenergy and bioproducts, agroforestry and phytoremediation. *Biomass and Bioenergy, 30*, 715–727. https://doi.org/10.1016/j.biombioe.2006.03.001

Von Maydell, H.-J. (1995). Agroforestry in central, northern, and eastern Europe. *Agroforestry Systems, 31*, 133–142. https://doi.org/10.1007/BF00711721

Wang, P., & Wolf, S. A. (2019). A targeted approach to payments for ecosystem services. *Global Ecology and Conservation*, 17, e00577. https://doi.org/10.1016/j.gecco.2019.e00577

Weber, L. J. (1991). The social responsibility of land ownership. *Journal of Forestry, 89*, 12–17.

Wezel, A., & Bellon, S. (2018). Mapping agroecology in Europe: New developments and applications. *Sustainability, 10*, 2751. https://doi.org/10.3390/su10082751

White, E. H., Abrahamson, L., Volk, T., Smart, L., Nakas, J., & Amidon, T. (2007). *Woody biomass feedstocks: Agroforestry and the energy crisis*. Keynote presented at the 10th North American Agroforestry Conference, Quebec City, QC, Canada.

Wiersum, K. F. (1990). Planning agroforestry for sustainable land use. In W. Budd, I. Duchart, L. H. Hardesty, & F. Steiner (Eds.), *Planning for agroforestry* (pp. 18–32). Amsterdam: Elsevier.

Williams, M. (1989). *Americans and their forests: A historical geography* (Studies in environment and history). Cambridge, UK. Cambridge University Press.

Wilson, M. H., & Lovell, S. T. (2016). Agroforestry: The next step in sustainable and resilient agriculture. *Sustainability, 8*, 574. https://doi.org/10.3390/su8060574

Wright, M. (2017). *Agroforestry education: The status and progress of agroforestry courses in the U.S.* (Master's thesis). Blacksburg, VA: Virginia Tech. Retrieved from https://vtechworks.lib.vt.edu/handle/10919/77521.

Study Questions

1. Throughout Chapter 1, the authors attempt to make a case for agroforestry's importance as a viable land use practice in North America. What are three major issues identified by the authors that agroforestry can be used to address in a cost effective manner?

2. To understand why agroforestry began in the United States, one must study the evolution of forest management. Of particular significance was a decision made by the U.S. Forest Service to manage public forest lands for multiple uses. What led to this decision?

3. Why has agroforestry always been the primary land use approach throughout the developing world, but is relatively new in developed nations?

4. In the late 1980s, Steppler (1987) suggested that agroforestry was "a practice in search of a science". What do you think was meant by this phrase? Has research in the past nearly four decades changed its validity?

5. Does agroforestry have a role in helping address global warming and dependence on foreign oil? Explain.

6. Do you agree that the importance of agroforestry in North America relates more to ecosystem services and resulting environmental protection than to production and economic gain? Justify your answer.

7. What role does state and federal policy play in the adoption of agroforestry? Has agroforestry policy development kept abreast of agroforestry technology development? Why or Why not? What do we need to do as agroforestry community to ensure the development of sound agroforestry policy?

2

Michael A. Gold and
Harold E. "Gene" Garrett

Agroforestry Nomenclature, Concepts and Practices

Application of agroforestry practices responds to economic (e.g., rural unemployment), environmental (e.g., soil erosion), and social (e.g., quality of life) issues common to all regions of the earth. However, differences exist between U.S. and Canadian agroforestry, tropical agroforestry, and agroforestry in other temperate regions of the world due to differences in ecosystems, their condition, and economic, social, cultural, and political realities.

Developing nations must deal with major issues that include inequitable land ownership and distribution (e.g., land and tree tenure), lack of access to credit, inability to purchase inputs (e.g., fertilizer, herbicides, pesticides, machinery), minimal rural infrastructure (e.g., roads, electricity, communications), and lack of information access (e.g., limited research and extension). Tropical agroforestry, long practiced and widely accepted by farmers, is viewed as an important alternative to traditional slash-and-burn agriculture and to conventional agriculture practiced on steep hillsides and marginal lands, practices that often result in overexploitation, massive erosion, and exhaustion of tropical soils. Whether highland or lowland tropics, wet or dry ecosystems, ecologically-based agroforestry practices help restore and maintain biodiversity, bring ecological stability to farms and watersheds, sustain production of basic needs, and create market opportunities for millions of rural poor (Garrity, 2005; Russell and Franzel, 2004; Nair et al., 2005; Garrity et al., 2010; Hillbrand et al., 2017).

In Europe, agroforestry applications are diverse and the development of agroforestry science parallels that in the United States and Canada (Palma et al., 2007; den Herder et al., 2017; Dupraz et al., 2018a; Mosquera-Losada and Prabhu, 2019). Differences arise due to Europe's patchwork of many countries, each with different land use practices and traditions in agriculture, forestry, and agroforestry (Eichhorn et al., 2006; Rois-Diaz et al., 2018; Gordon et al., 2018; Lovric et al., 2018). Agroforestry practices were widely utilized throughout Europe from Roman times until the post–World War II onset of agricultural industrialization (Lelle and Gold, 1994; Eichhorn et al., 2006). European Union subsidies have been largely directed to agriculture and forestry and as a result, acreage in traditional agroforestry practices declined dramatically during the latter half of the 20th century. Agroforestry systems have often been neglected in Europe because administrative structures within many national governments have considered that only agriculture or forestry are legitimate. This has resulted in the loss of agroforestry systems in European countries and a loss of the benefits that they provide (McAdam et al., 2009). The lack of recognition

North American Agroforestry, Third Edition. Edited by Harold E. "Gene" Garrett, Shibu Jose, and Michael A. Gold.

of agroforestry practices within the different sections of Europe's Common Agriculture Policy (CAP) has reduced the impact of CAP activities by overlooking land use practices that would optimize the use of agroforestry (Mosquera-Losada et al., 2018).

Native American Agroforestry

As is the case elsewhere throughout the world, agroforestry in the United States and Canada also has historic roots. Native Americans across what is now the United States and Canada have been practicing indigenous forms of what could be termed landscape-scale agroforestry for millennia (Rossier and Lake, 2014; Nelson, 2014; Anderson and Rosenthal, 2015). Some of these indigenous communities managed – and continue to manage – integrated systems of trees, plants, animals, and fungi in complex ways at multiple organizational scales (MacFarland et al., 2017).

Because indigenous peoples were forcibly removed from their aboriginal landscapes and/or their ability to manage, the long-standing indigenous agroforestry traditions of many Native peoples across the United States are unknown. United States fire suppression policies stopped Native American people from burning their agroforest landscapes in the complex and integrated ways they had developed over millennia to provide needed foods, fibers, fuels, and other resources as well as to manage the complex food and interaction webs inherent to the agroforest ecosystems with which they evolved (Norgaard, 2014; Anderson and Rosenthal, 2015).

Native Americans throughout much of California actively managed trees, understory plants, forages, and animal populations in such an integrated complex way that when John Muir arrived to Yosemite Valley and many other parts of California, he remarked upon their pristine, wild, garden-like quality, and stunning beauty. However, because they did not look like European agricultural systems, Muir did not fully understand the degree to which they had been managed by Native peoples (Anderson and Rosenthal, 2015). The Karuk Tribe in the Klamath Mountains of Northern California historically used fire, pruning, coppicing, and many other techniques to manage hundreds of plants, animals, and fungi in an integrated indigenous agroforestry system (Taylor and Skinner, 2003). This system includes tanoak and black oak acorn trees, tanoak mushrooms, elk, deer, evergreen huckleberries, blackcap raspberries, gooseberries, currants, hazel, willow, Indian potatoes, manzanita and madrone trees and their berries, elderberries, alder, yew, Douglas-fir, and so much more (Vinyeta et al., 2016).

Agroforestry as a Science

In spite of an increasing awareness of Native American traditional agroforestry practices over the past few decades, in the United States and Canada agroforestry has mainly been viewed as a new science and set of practices tailored to address numerous sustainability issues associated with production agriculture (Matson et al., 1997; Nair, 2007; Jose et al., 2018; USDA, 2019; Jose et al., 2022). Following an era of "efficient production" through the 1970's, U.S. and Canadian agriculture is slowly transitioning to an era of sustainable production and regenerative agriculture. Development of a sustainable, regenerative agriculture is stimulated by critical issues including long-term economic decline in rural America, need for crop diversification, and concern about soil erosion, environmental pollution, habitat loss, and climate change. These issues have been accentuated by repeated periods of massive flooding in the greater Mississippi River watershed (e.g., 1993, 2019) raising national awareness of problems stemming from increasingly large-scale monoculture production farming, especially excessive runoff and flooding, non-point source water pollution, and loss of critical wildlife habitat (Pimentel et al., 1995; Jordan et al., 2007; Porter et al., 2015; Lerch et al., 2017). Sustainable agricultural practices (e.g., use of cover crops, no till), organic agriculture, regenerative agriculture, are now recognized as viable additions to mainstream production agriculture (Liebman and Schulte, 2015; Geertsema et al., 2016; LaCanne and Lundgren, 2018; OTA, 2018).

While conservation programs were de-emphasized in the 1970's and early 1980's, the sustainable, regenerative agriculture, carbon farming movement has created a more receptive climate for agroforestry development (Liebman and Schulte, 2015; LaCanne and Lundgren, 2018; Feliciano et al., 2018). Agroforestry is directly relevant to sustainable and regenerative land use. The context is that agroforestry concepts and practices need to be incorporated into agricultural production systems to utilize productive niches, diversify products and income, and restore and/or enhance certain ecological services that are vital to achieve sustainable land use (Jordan et al., 2007; Udawatta et al., 2017;

Jose et al., 2018). Thus, agroforestry in the United States and Canada is not recognized as an end in itself, but rather an important ancillary discipline contributing to sustainable and regenerative agriculture and sustainable forestry.

It must be recognized that the present supporting infrastructure, including agricultural research, developed during the "efficient production era", remains strongly oriented toward commodity production (Sooby, 2003; IPES-Food, 2016). Moreover, in the United States and Canada, agriculture and forestry land uses are traditionally segregated on the land and in our institutions. Thus, many of the agroforestry concepts (i.e., integrating trees with crops and/or livestock) run counter to traditional thinking and existing infrastructure. To overcome these barriers, it is very important that agroforestry concepts and practices be relevant, pragmatic, and market-driven to foster interdisciplinary cooperation within our institutions, cooperation among businesses across the market value chain, and adoption on the land (Geertsema et al., 2016; LaCanne and Lundgren, 2018).

In summary, agroforestry in the U.S. and Canada is driven by sustainable development and growth in the "green" marketplace, which in turn will positively affect rural decline and the environmental impacts of agriculture. Agroforestry seeks to help bridge the gap between production agriculture and natural resource management. The development and definition of agroforestry reflects the commodity focus of agriculture, yet it is simultaneously and equally focused on conservation. Global awareness of tropical agroforestry has positively affected the development of agroforestry in the U.S. and Canada, and many of the current practices are adapted from the tropics. However, with foci and priorities differing from those in the tropics (Fig. 2–1), agroforestry over most of North America has evolved its own distinctive definition and nomenclature (Table 2–1). In the United States and Canada, agroforestry is in an active phase of development and, since 2010, has emerged over the past decade as both a science

North American – Temperate	**Tropical Regions**[1]
• Income Diversification o Market Driven, Value-Added	• Food/Energy Security – Basic Needs o Tree Fodder, Fuelwood o Security of Land and Tree Tenure
• Next Generation Farms/Farmers o Smaller Scale, Local Foods, Land Access, Social Justice	• Poverty Alleviation o Education, Income Generation, Markets, Value-Added
• Climate Change Mitigation o Ecosystem Resilience, Carbon Sequestration, Carbon Markets	• Build Human/Institutional Capacity o Advancement of Women, Rural Areas
• Environmental Protection o Air/Water Quality, Soil Erosion	• Health and Nutrition o Medicinal Plants, Healthy Diets
• Health and Nutrition o Healthy Diets: Nuts, Fruits	• Biodiversity Conservation
• Wildlife Habitat Restoration o Biodiversity Conservation	• Environmental Protection o Water Quality, Soil Erosion
• Aesthetics, Quality of Life	• Climate Change Mitigation o Carbon Markets, Desertification

[1]Adapted from Garrity (2004)

Fig. 2–1. Agroforestry priorities: Temperate and tropical.

Table 2–1. Global definitions of agroforestry.

Region	Definition	Citation
Canada	"An approach to land use that incorporates trees into farming systems, and allows for the production of trees and crops or livestock from the same piece of land in order to obtain economic, ecological, environmental and cultural benefits"	Gordon and Newman, 1997
Global, Tropics	"The set of land use practices involve the deliberate combination of trees (including shrubs, palms and bamboos) and agricultural crops and/or animals on the same land management unit in some form of spatial arrangement or temporal sequence such that there are significant ecological and economic interactions between tree and agricultural components"	Sinclair, 1999
Australia	"Agroforestry is the commitment of resources by farmers, alone or in partnerships, towards the establishment or management of trees and forest on their land"	Reid and Moore, 2018
Global, Tropics	"Agroforestry is a dynamic, ecologically based, natural resource management system that, through the integration of trees on farms and in the agricultural landscape, diversifies and sustains smallholder production for increased social, economic and environmental benefits"	Garrity, 2005
Tropics	"Agroforestry is any land-use system, practice or technology, where woody perennials are integrated with agricultural crops and/or animals in the same land management unit, in some form of spatial arrangement or temporal sequence"	Atangana et al., 2013
France	"The cultivation of the soil with a simultaneous or sequential association of trees and crops or animals to obtain products or services useful to man"	Torquebiau, 2000

and a defined set of practices, shaped and tailored to address urgent land use sustainability issues. Agroforestry completed the definition phase by the turn of the 21st century and is currently in a period of rapid scientific development (Nair, 2007; Feliciano et al., 2018: Jose et al., 2018; Garrett et al., 2022), outreach (Shelton et al., 2005; Gold et al., 2004; USDA, 2015; Elevitch et al., 2018) and farm-level application (Faulkner et al., 2014; Rios-Diaz et al., 2018; Van Noordwijk, 2019).

Definition and Practices

Definition

In the United States and Canada, agroforestry is defined as: *intensive land-use management that optimizes the benefits (physical, biological, ecological, economic, social) from biophysical interactions created when trees and/or shrubs are deliberately combined with crops and/or livestock.*

This definition is slightly modified from Garrett et al. (1994) to improve compatibility with traditional agriculture. The main difference is the removal of the word "systems" from the definition. In tropical agroforestry and throughout much of the temperate zone, the use of systems terminology has become the established norm (Nair, 1989; Sinclair, 1999; Newman and Gordon, 2018). However, many authors use the words "practice" and "system" interchangeably (Table 2–1) as in this volume. Agroforestry classification in the United States and Canada has evolved from agricultural traditions wherein an agricultural production system is an aggregation of various practices. Under this form of classification, agroforestry is most often recognized as a set of practices which are incorporated, along with other appropriate practices, into agricultural systems likes pieces of a puzzle fitting together at a variety of spatial scales (i.e., field, farm, watershed, landscape).

Criteria

Four key criteria characterize agroforestry practices in the United States and Canada and distinguish them from other practices (Merwin, 1997). To be called agroforestry, a land use practice must satisfy all of the following four criteria:

Intentional

Combinations of trees, crops, and/or livestock are *intentionally* designed, established and/or managed to work together and yield multiple products and benefits, rather than as individual elements which may occur together but are managed separately.

Intensive

Agroforestry practices are created and *intensively* managed to maintain their productive and protective functions, and often involve cultural operations such as cultivation, fertilization, irrigation, pruning and thinning.

Integrated

Components are structurally and functionally combined into a single, *integrated* management unit tailored to meet the objectives of the landowner. *Integration* may be horizontal or vertical, and above- or below-ground. *Integration* of multiple crops utilizes more of the productive capacity of the land and helps balance economic production with resource conservation.

Interactive

Agroforestry actively manipulates and utilizes the biophysical *interactions* among components to yield multiple harvestable products, while concurrently providing numerous conservation and ecological benefits.

Practices

Six categories of agroforestry practices (Table 2–2) that embody the above criteria are recognized in the United States and Canada by the Association for Temperate Agroforestry–AFTA (Merwin, 1997). The sixth agroforestry practice, urban food forests, has emerged over the past decade (Munsell et al., 2022).

Riparian and Upland Buffers

Riparian and upland buffers are strips of permanent vegetation, consisting of trees, shrubs, and grasses that are planted and managed together. Riparian buffers are placed between agricultural land (usually crop land or pastureland) and water bodies (rivers, streams, creeks, lakes, wetlands) to reduce runoff and non-point source pollution, stabilize streambanks, improve aquatic and terrestrial habitats, and provide harvestable products. Upland buffers are placed along the contour within agricultural crop lands to reduce runoff and non-point source pollution, improve internal drainage, enhance infiltration, create wildlife habitat and connective travel corridors and provide harvestable products.

Windbreaks

Trees or shrubs are planted as barriers to reduce wind speed. Windbreak practices include shelterbelts, timberbelts and hedgerows. Windbreaks are planted and managed as part of a crop or livestock operation. Field windbreaks are used to protect a variety of wind-sensitive row, forage, tree, and vine crops, to control wind erosion, and

Table 2–2. Six categories of agroforestry practices in the US and Canada.

Practice	Predominant Region (s)	Use(s)	Associated Technologies
Riparian and upland buffers	All Regions	Ameliorate nonpoint source pollution, abate soil erosion and nutrient loading, protect watersheds Modify microenvironments and protect aquatic habitats	Streambank bioengineering Constructed wetlands Green infrastructure
Windbreaks	Great Plains, All Regions	Protect and enhance production of crops and animals, control soil erosion, distribute snowfall. Trap snow.	Living snow fences
Alley cropping	Midwest, All Regions	Increases and diversifies farm crops and income	Plantation management
Silvopasture	West, Southeast, All Regions	Economic diversification, improve animal health, create wildlife habitat, fire protection, timber management	Pine straw harvest
Urban Food Forests	All Regions	Diverse urban food production, soil health and diversity, human health, nutrition and well-being, environmental and social justice benefits, education, community	Permaculture
Forest farming	All Regions	Income diversification	Forest management

to provide other benefits such as improved bee pollination of crops and wildlife habitat. Livestock windbreaks help reduce animal stress and mortality, feed and water consumption, and odor. Timberbelts are managed windbreaks designed to increase the value of the forestry component.

Alley Cropping

This practice combines trees planted in single or multiple rows with agricultural or horticultural crops cultivated in the alleyways between the tree rows. High-value hardwoods such as oak (*Quercus* sp.), walnut (*Juglans* sp.), chestnut (*Castanea* sp.) and pecan (*Carya illinoensis* (Wangenh.) K. Koch) are favored species in alley cropping practices, and many can provide high-value lumber or veneer logs. Crops grown in the alleys, and nuts from walnut, chestnut and pecan trees, provide annual income from the land while the longer-term wood crop matures (Gold, 2019). When specialty crops such as herbs, fruits, vegetables, nursery stock, or flowers are grown in the alleys, the microclimate created by the trees enables the economic production of these sensitive high-value crops in stressed environments.

Silvopasture

This practice combines trees with forage (pasture) and livestock production. Silvopasture can be established by adding trees to existing pasture, or by thinning an existing forest stand and adding (or improving) a forage component (Jose et al., 2017). The trees are managed for high-value sawlogs, and at the same time they provide shelter for livestock, protecting them from temperature stresses and reducing food and water consumption. Forage and livestock provide short-term income at the same time a crop of high-value sawlogs is being grown, providing a greater overall economic return from the land.

Forest Farming

High-value specialty crops are cultivated under the protection of a forest overstory that has been modified and managed to provide the appropriate microclimate conditions. Shade-tolerant specialty crops like ginseng (*Panax quinquefolium* L.), log-grown shiitake mushrooms (*Lentinula edodes* (Berkeley) Pegler), decorative ferns and spring ephemerals grown in the understory are sold for medicinal, botanical, food, decorative and handicraft, and landscaping products. Overstory trees are managed for high-value timber or veneer logs.

Urban Food Forests

In addition to the five recognized practices, there is an emerging, sixth agroforestry practice. Urban Food Forests, have gained considerable attention over the past decade (Lovell, 2010; Clark and Nicholas, 2013; Bukowski and Munsell, 2018; Park et al., 2019). Urban Food Forests are defined as: i) The intentional use of perennial food-producing plants to improve the sustainability and resilience of urban communities (Bukowski and Munsell, 2018); 2) A food forest is an edible, perennial, polyculture system that is designed and managed to mimic multistory forest structures and to function like a natural, self-sustaining forest (Park et al., 2018). The term food forest signifies an intentionally designed, highly integrated community of plants that has various vertical and horizontal plants and root layers that collectively provide edible products (Bukowski and Munsell, 2018). Urban food forestry is an emerging multifunctional and interdisciplinary approach to increasing urban sustainability and resilience, particularly where food security is concerned, and provides a starting point for bridging gaps in knowledge and practice between urban agriculture, urban forestry, and agroforestry. Also, as noted in Bukowksi and Munsell (2018), another commonly used term is community food forests.

Clark and Nicholas (2013) note that urban food forestry is a viable and important strategy to address multiple sustainability challenges (e.g., food security, climate change, and poverty), to contribute to human health by increasing affordable public access to and consumption of nutrient dense foods to combat hunger and obesity, and can be also used to promote sustainable urban development through providing enhanced ecosystem services.

Is it Agroforestry?

A key concern in developing agroforestry nomenclature for the United States and Canada is overlap and confusion with mainstream land use management disciplines, for example forestry, agriculture, and livestock production. A fundamental need was to develop a definition and criteria that would effectively distinguish practices that are agroforestry from those that are not (Garrett et al., 1994). Application of the four criteria defining agroforestry (intentional, intensive, integrative, and interactive) provides the key to determine what is and is not an agroforestry practice. Using these criteria, we have the basis to explain how variations of common land use practices can be properly described as agroforestry. The following examples are agroforestry because they satisfy all four of the criteria defining agroforestry.

Special Forest Products

Deliberate cultivation of an understory specialty crop beneath a forest canopy that has been modified and managed to provide the appropriate microclimate conditions in the understory is an example of forest farming (e.g., woods cultivated or wild simulated ginseng). The practice is created by design, is intensively managed, is integrated, and beneficial interactions are utilized. Thus, it is agroforestry, as contrasted to the gathering of naturally-occurring, unmanaged, specialty products (e.g., wild ginseng) from a natural forest stand.

Log-grown shiitake mushrooms, deliberately cultivated under the shade of the forest canopy is a legitimate forest farming practice. The practice is created by design, is intensively managed, is integrated, and beneficial interactions are utilized. Thus, it is agroforestry, in contrast to the production of shiitake on indoor sawdust bales which yields a similar (but not nutritionally identical) product but does not qualify as forest farming.

Nut Plantations and Fruit Orchards

When nut or fruit culture (i.e., planted in a plantation or orchard format) is combined with row crop or forage production, it is alley cropping (Gold, 2019). Crops grown in the between-row space can be changed over time to minimize competitive effects and/or adjust to changing market conditions. Windbreaks can be established to protect orchards in exposed, windy areas. They slow the wind, reduce water use, improve insect pollination, and increase pesticide use efficiency. In each case, the components are deliberately integrated and intensively managed within the plantation or orchard.

Managed Intensive Rotational Grazing

When trees are added to an existing pasture and the resulting area is managed for timber, forage and livestock, it is the agroforestry practice of silvopasture. The components are deliberately integrated and managed by design to enhance the biophysical interactions among components. Both the timber and forage components are designed to minimize competition and maximize complementarity. Neither opportunistic forest grazing nor grazing cattle without management within a plantation are agroforestry; both can be destructive to the forest, tree, and forage resources.

Agroforestry in the Landscape

A final issue to discuss is mosaics of monocultures in agricultural landscapes. Common features of agricultural landscapes throughout the United States and Canada are single-crop farm fields, woodlots and tree plantations, wetlands, and grazing lands. A physical proximity does not constitute agroforestry at the landscape level because there is no intentional integration and there is minimal interaction among components. In contrast, an agricultural landscape that contains windbreaks or riparian and upland buffers, that is, linear plantings including permanent woody vegetation strategically placed to maximize conservation benefits and create biophysical interactions with other components of the agricultural system, clearly demonstrates agroforestry at a landscape scale. The subject is discussed further in Table 2–3, which outlines agroforestry concepts.

Table 2–3. Agroforestry concepts.

Concepts/Comments
• Cascades of benefits are derived from beneficial process-level biophysical interactions created when trees and/or shrubs are deliberately combined with crops and/or livestock.
• Additional benefits are often derived from component integration when compared with traditional, segregated (agriculture and forestry) production. Through intensive and careful management, desired interactions are optimized and undesirable interactions are minimized. Agroforestry practices introduce, restore, and enhance biological diversity and agroecosystem resilience at field, farm, watershed and landscape levels.
• For many people, enhanced biodiversity is considered a benefit, e.g., increased wildlife; however, in agricultural regions, biodiversity must be managed to obtain desired effects like enhancement of selected fish, wildlife, and plant species. The challenge is to determine the type, amount and distribution of species that will provide an adequate level of desired benefits.
• Agroforestry contributes to integrated pest management by creating favorable habitats and microclimates to enhance the extent and effectiveness of natural pest controls.
Agroforestry is an essential component of effective conservation buffers creating positive impacts upon steep slopes, highly-erodible soils and collapsing streambanks.
• Agroforestry contributes to the maintenance of soil quality and productivity by keeping soil in place, enhancing nutrient absorption and cycling, intercepting water-borne pollutants, improving water filtration and retention capacity, and reducing flood damage.
Agroforestry increases the productive area of the farm by expanding use of vertical and horizontal space above- and below-ground and fully exploiting the diversity of useable niches.
• Agroforestry practices permit fuller use of the soil profile, maximize use of photosynthetic radiation, and lengthen the growing season.
• Agroforestry practices exploit additional field scale niches including border areas, marginal sites (rocky, infertile, too wet or dry) and steep slopes.

Departures from Traditional Agroforestry Nomenclature

There is obvious inconsistency in the nomenclature used to describe the six categories of agroforestry practices. In the United States, Canada and abroad, efforts have been made to clarify definitions and nomenclature in agroforestry (Table 2–1). Mantau et al. (2007) offer a thorough discussion of the concepts of classification and nomenclature with regard to non-timber forest products while Sinclair (1999) proposes a general classification of agroforestry practices. In the United States and Canadian nomenclature, two of the practices are named on the basis of function (windbreaks, and riparian and upland buffers), the names of two are based on the adoption of popularized names (forest farming, urban food forests).

As previously discussed, the nomenclature also departs from the systems terminology developed for tropical agroforestry and temperate agroforestry in other countries. Growing trees, crops, and animals in mixtures is a long-standing tradition of tropical farmers. Tropical agroforestry evolved from these age-old customs as well as more recent tropical agriculture paradigms of the 1960's and 1970's known as "cropping systems" and later as "farming systems" (Hildebrand, 1990). Subsequently, the nomenclature of tropical agroforestry tree, crop, and animal combinations was defined by the International Center for Research on Agroforestry (Lundgren and Raintree, 1982). During this definition phase for tropical agroforestry, a great deal of effort went into development of classification methodology. Classification and descriptive criteria were based on the situation and intended purpose to which agroforestry was being applied (Sinclair, 1999). The history of agroforestry classification has been reviewed and the five approaches to classify tropical agroforestry have been summarized (Nair, 1993; Atangana et al., 2013).

Nature of Components

Agrisilviculture describes crop–tree combinations, silvopasture describes tree–livestock combinations, and agrosilvopasture, describes crop–tree–livestock combinations.

Arrangement of Components

This criterion denotes whether the components exist simultaneously, overlap during part of a rotation, or follow in a prescribed sequence.

Functional Role

The primary use, production or conservation, is a common approach to classifying tropical agroforestry.

Agroecological Zone

The use of agroecological zones to classify agroforestry is based upon a characterization of climate, vegetation, and land-use capability, usually a region within a country, for example, humid lowlands, arid or semi-arid lands, or highlands.

Social and Economic Features

This approach uses scale of production and level of technology, for example, subsistence, intermediate, or commercial to classify agroforestry.

Perspectives on U.S. and Canadian Agroforestry

Finally, one must recognize that there are two distinct perspectives on agroforestry in the United States and Canada, and it is important to distinguish them from a nomenclature standpoint.

Agroforestry at the Practice Level

For field practitioners and landowners to understand, accept, and use agroforestry, it must be as pragmatic, market-focused, and adoptable as possible. Complex "systems terminology" is not acceptable. Consequently, a simple agroforestry nomenclature has been developed to make agroforestry practices compatible with, and complementary to, agricultural practices. The bottom line for agroforestry to succeed over most of North America is that it must be accepted and used within the agriculture community.

Agroforestry at the Science Level

Within the scientific community, agroforestry concepts have much in common with sustainable agriculture, agroecology, regenerative agriculture, and agroforestry in the rest of the world. Common goals are to conserve the natural resources upon which agriculture depends, minimize the environmental impacts of agriculture, maintain productivity and profitability, and provide for people's economic and social needs (Fig. 2–1). At the science level, it is more important to focus on agroforestry concepts and their underlying process level functions, and less important to debate nomenclature.

Aside from differences in nomenclature, the concepts of agroforestry over most of North America are not very different from those in the rest of the world. Agroforestry has emerged as a science-based practice and is increasingly finding its place in our agencies and educational institutions (Nair, 2007; USDA, 2017; Munsell and Chamberlain, 2019). However, the success of agroforestry science will ultimately be determined by the accomplishment of interdisciplinary research, development, and applications between forestry and natural resources and agriculture and livestock communities working in close cooperation with specialists in rural sociology, community development, applied economics, and marketing. The final measure of success will be agroforestry practices adapted to local conditions and seamlessly integrated into mainstream agriculture production systems in all regions of temperate North America.

Agroforestry Concepts

The effects of integrating trees into production agriculture systems are far-reaching, and address not only on-farm needs, but also numerous agriculturally-related problems causing increasing concern around the world. Growing trees in combination with crops and livestock has been shown to enhance crop yields (Kort, 1988; Dupraz et al., 2018b), improve animal health (Brunetti, 2006; Pent et al., 2022) and reduce losses, conserve soil and recycle nutrients, and reduce environmental impacts of agriculture (Udawatta et al., 2002; Blanco-Canqui et al., 2004; Dosskey et al., 2007; Lerch et al., 2017; Schulte et al., 2017), while producing various tree and specialty products (Gold et al., 2004; Mori et al., 2018). The postulated effects of agroforestry in the United States and Canada are presented in the form of verifiable agroforestry concepts (Table 2–3). Increasing amounts of data exist to support and prove these concepts. Current research and on-the-ground practices will continue to confirm and modify these concepts in the coming years.

Fundamentally, as the definition of agroforestry implies, the benefits from agroforestry are derived from the biophysical interactions created when trees and/or shrubs are combined with crops and/or livestock. Interactions refer to the influence of one component on the performance of the other components, and on the system as a whole. We seek to optimize these interactions to favor mutualism and commensalism and minimize competition and predation. Interactions include both above-ground and below-ground effects. Although the dynamics of component interactions is a complex research challenge (Jose and Holzmueller, 2022), the net effect of interactions is of practical significance, and creates the biophysical success or failure of an agroforestry practice. The observable net effects of component interactions are expressed by the terms complementary, supplementary, and competitive (Anderson and Sinclair, 1993; Ong et al., 2015; Jose and Holzmueller, 2022). Component interactions represent processes at the tree–crop interface and tree–crop–animal interface. These interactions can be positive (e.g., stress reduction, yield enhancement, soil retention, water capture) or negative (e.g., competition, allelopathy, pest enhancement) (Jose et al., 2004; Jose and Holzmueller, 2022). Consequently, it is imperative that agroforestry practices be properly designed and managed to optimize desired positive interactions and minimize the negative ones (Dupraz et al., 2019).

Our perception of agroforestry, its benefits, and its relative importance also depend on the scale of interest (e.g., field, farm, watershed, landscape). At the farm scale, benefits that accrue to the landowner are of primary importance while societal benefits are secondary. At larger landscape and watershed scales, societal benefits of conservation (e.g., water quality) are often valued equally with community viability (e.g., economic production) (Bentrup and Kellerman, 2004; Garrity, 2005). Thus, at the individual farm ownership scale, agroforestry focuses on utilizing the productive niches within the farm to meet the owner's conservation and income needs. At the landscape scale (Hillbrand et al., 2017; Kremen and Merenlender, 2018), agroforestry practices help create buffer zones (National Research Council, 1993; Schultz et al., 2022) within agricultural systems that enhance vital ecological services required for sustainability. At the watershed scale, agroforestry practices directly support community-based land stewardship by addressing both conservation and economic goals (Curtis et al., 1995; Jordan et al., 2007; USDA, 2017).

Agroforestry shares fundamental concepts and principles with sustainable agriculture, agroecology, and permaculture. These include judicious use of inputs, maintaining soil quality and productivity, minimizing the environmental impacts of agriculture, utilizing natural processes where possible and practical, and providing for human health, safety, and quality of life (Lovell et al., 2010; Ferguson and Lovell, 2014; Liebman and Schulte, 2015; Krebs and Bach, 2018). Agroforestry can contribute to the integrity of agroecosystems through the creation of buffer zones (National Research Council, 1993; Schoeneberger et al., 2006; USDA, 2017). These buffer zones expand the structural and spatial diversity of the system, and enhance certain ecological services generated from biodiversity and nutrient cycling (Edwards et al., 1993; Udawatta et al., 2017). Ultimately, management strategies must utilize agroforestry and other practices that generate biodiversity and nutrient cycling processes to provide ecological services like soil and nutrient retention, water capture and cycling, microclimate moderation, waste assimilation, and pest management (Costanza et al., 1997; Matson et al., 1997; Geertsema et al., 2016; LaCanne and Lundgren, 2018).

Windbreaks, the most understood and widely-used agroforestry practice, best illustrate the cascade of benefits from agroforestry practices at multiple scales (Fig. 2–2, also see Brandle et al., 2022). Windbreaks create farm scale buffer zones, which generate numerous benefits (physical, biological, ecological, and social). Physically, they slow the wind, create a microclimate more favorable to plant growth, decrease windborne soil erosion, reduce physical damage to emerging and sensitive crops, increase snow capture and distribution (Heavey and Volk, 2014), and protect crops and livestock from climatic extremes. Biologically, they provide habitat for natural enemies of crop pests (Altieri et al., 2017; Yang et al., 2019), enhancing biological controls

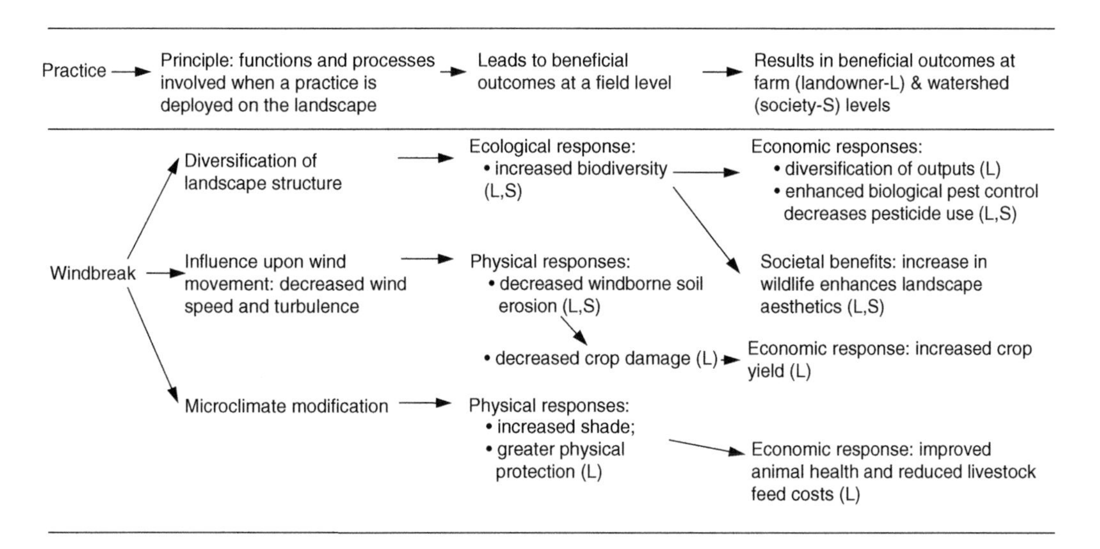

Fig. 2–2. Principles and benefits derived from the windbreak agroforestry practice.

and provide habitat for wildlife. Ecologically, windbreaks increase water capture and cycling, reduce runoff and flooding, and help maintain soil quality and productivity. Economically, windbreaks increase crop yields, reduce animal feed costs, increase the survival of newborns and can be designed to incorporate marketable products (Josiah et al., 2004; Baker et al., 2018). Social benefits consist of landscape diversity and protection of human environments from wind, dust, noise, and odors, as well as enhancement of wildlife and associated recreational opportunities. The outcome is multiple benefits that accrue to both the landowner and to society (Fig. 2–2).

Through the introduction of trees and the interactions they generate, agroforestry can significantly contribute to desirable ecosystem level services. Overall this can result in more structurally diverse (both above and below ground) agroecosystems that are richer in plant and animal biodiversity and have improved system self-maintenance and resistance to environmental stresses (Jose and Holzmueller, 2022; Udawatta et al., 2017; USDA, 2017). Based on these ecological principles, agroforestry practices can be an important tool to restore land use sustainability, overall ecosystem health, and be used to reclaim degraded lands. Thus, agroforestry is more than a set of practices; it is the incremental addition of trees to farming systems and farming landscapes, resulting in the generation and enhancement of desired ecological services considered vital for sustainable land use (Jordan et al., 2007; Geertsema et al., 2016; USDA, 2017).

To be effective, agroforestry must follow a grassroots approach tailored to: i) the individual landowner's special interests, problems, and needs (Rule et al., 2000); ii) the available productive niches; and iii) the local soil and climate conditions and existing and potential markets (Gold et al., 2004, 2013, 2018). This same approach also applies to community-based land stewardship (Curtis et al., 1995; Garrity 2005; Schoeneberger et al., 2006). Thus, although we now recognize six categories of agroforestry practices for nomenclature purposes, these practices are not rigid. Rather, the application of agroforestry should be seen as a common-sense approach tailored to local needs and conditions (Sobels et al., 2001; Rule et al., 2000; Rios-Diaz et al., 2018). The result is an approach that is similar to whole farm planning in that the emphasis is on the design of a system for a given field, farm, or watershed rather than the promotion of a particular land use option (Gold et al., 2013, 2018).

As found in the chapters within this volume, over the past decade, there has been rapid development of agroforestry science in support of agroforestry practices. Appropriate technologies, information, and tools have been developed to design agroforestry practices to achieve specific production and conservation objectives at the local level (Bentrup and Kellerman, 2004; Shelton et al., 2005; Gold et al., 2013, 2018; Wilson et al., 2018). Based on strong supporting data obtained from multiple studies accumulated at numerous sites over many years, agroforestry science is well on the way toward developing the principles (i.e., component interactions and ecosystem functions) that underlie these practices (Nair, 2007; Lovell et al., 2017; USDA, 2017; Munsell and Chamberlain, 2019).

As a science-based on interacting components within practices (i.e., combinations of trees, crops and livestock), agroforestry draws upon knowledge from many different disciplines. Beginning in 1990 and evolving rapidly in the past decade, a critical process-level, science-based approach to agroforestry research has gradually emerged. An understanding of component interactions is being assembled which will enable applications to be designed in a predictable manner. The bottom line for agroforestry is to be able to locally apply technologies that generate predictable and positive interactions, and optimize them for the benefit of the farmer and associated land resources, and for society as a whole. To achieve the bottom line, a fusion of top-down and bottom-up approaches are needed that are market-focused and result in the development of robust social networks (Rule et al., 2000; Valdivia et al., 2022) at multiple social and spatial scales (i.e., landowner, community, state, region, and nation).

References

Altieri, M.A., C.I. Nicholls, and R. Montalba. 2017. Technological approaches to sustainable agriculture at a crossroads: An agroecological approach. Sustainability 9:349. doi:10.3390/su9030349

Anderson, M.K., and J. Rosenthal. 2015. An ethnobiological approach to reconstructing indigenous fire regimes in the foothill chaparral of the Western Sierra Nevada. J. Ethnobiol. 35(1):4–36. doi:10.2993/0278-0771-35.1.4

Anderson, L.S., and F.L. Sinclair. 1993. Ecological interactions in agroforestry systems. Agroforestry Abstracts 6(2):57–91.

Atangana, A., D. Khasa, S. Chang, and A. Degrande. 2013. Definitions and classifications of agroforestry systems In: A. Degrande, D. Khasa, and S. Chang, (eds.), Tropical agroforestry. Dordrecht, The Netherlands: Springer. p. 35–47. doi:10.1007/978-94-007-7723-1.

Baker, T.P., M.T. Moroni, D.S. Mendham, R. Smith, and M.A. Hunt. 2018. Impacts of windbreak shelter on crop and livestock production. Crop Pasture Sci. 69(8):785–796. doi:10.1071/CP17242

Bentrup, G., and T. Kellerman. 2004. Where should buffers go? Modeling riparian habitat connectivity in northeast Kansas. J. Soil Water Conserv. 59:209–213.

Blanco-Canqui, H., C.J. Gantzer, S.H. Anderson, E.E. Alberts, and A.L. Thompson. 2004. Grass barrier and vegetative filter strip effectiveness in reducing runoff, sediment, nitrogen, and phosphorus loss. *Soil Sci. Soc. Am. J.* 68:1670–1678. doi:10.2136/sssaj2004.1670

Brandle, J. R., E. Takle and Z. Zhou. 2022. Windbreak practices. Chapter 5. In: H.E. Garrett, S. Jose, and M.A. Gold, (eds.), North American agroforestry. 3rd ed. Madison, WI: Agronomy Society of America.

Brunetti, J. 2006. Forage quality and livestock health: A nutritionist's view. In: T. Morris and M. Keilty, editors, Alternative health practices for livestock. Wiley-Blackwell Publishers, London. p. 85–103. doi:10.1002/9780470384978.ch8

Bukowski, C., and J. Munsell. 2018. The Community food forest handbook: How to plan, organize, and nurture edible gathering places. Chelsea Green Publishing, Hartford, VT.

Clark, K.H., and K.A. Nicholas. 2013. Introducing urban food forestry: A multifunctional approach to increase food security and provide ecosystem services. Landsc. Ecol. 28(9):1649–1669. doi:10.1007/s10980-013-9903-z

Costanza, R., R. d'Arge, R. de Groot, S. Farber, M. Grasso, B. Hannon, K. Limburg, S. Naeem, R.V. O'Neill, J. Paruelo, R.G. Raskin, P. Sutton, and M. van den Belt. 1997. The value of the world's ecosystem services and natural capital. Nature 387:253–260. doi:10.1038/387253a0

Curtis, A., J. Birkhead, and T. De Lacy. 1995. Community participation in landcare policy in Australia: the Victorian experience with regional landcare plans. Soc. Nat. Resour. 8:415–430. doi:10.1080/08941929509380933

den Herder, M., G. Moreno, R.M. Mosquera-Losada, J.H.N. Palma, A. Sidiropoulou, J.J. Santiago Freijanes, J. Crous-Duran, J.A. Paulo, M. Tomé, A. Pantera, V.P. Papanastasie, K. Mantzanas, P. Pachana, A. Papadopoulos, T. Plieninger, and P.J. Burgess. 2017. Current extent and stratification of agroforestry in the European Union. Agric. Ecosyst. Environ. 241:121–132. doi:10.1016/j.agee.2017.03.005

Dosskey, M.G., K.D. Hoagland, and J.R. Brandle. 2007. Change in filter strip performance over ten years. J. Soil Water Conserv. 62:21–32.

Dupraz, C., G.J. Lawson, N. Lamersdorf, V.P. Papanastasis, A. Rosati, and J. Ruiz-Mirazo. 2018a. Temperate agroforestry: The European way. p. 98–152. In: A.M. Gordon, S.M. Newman, and B. Coleman (ed.), Temperate agroforestry systems. 2nd Edition. Wallingford, U.K.: CABI.

Dupraz, C., C. Blitz-Frayret, I. Lecomte, Q. Molto, F. Reyes, and M. Gosme. 2018b. Influence of latitude on the light availability for intercrops in an agroforestry alley-cropping system. Agroforest Syst 92:1019–1033. doi:10.1007/s10457-018-0214-x

Dupraz, C., K.J. Wolz, I. Lecomte, G. Talbot, G. Vincent, R. Mulia, F. Bussière, H. Ozier-Lafontaine, S. Andrianarisoa, N. Jackson, G. Lawson, N. Dones, H. Sinoquet, B. Lusiana, D. Harja, S. Domenicano, F. Reyes, M. Gosme, and M. Van Noordwijk. 2019. Hi-sAFe: A 3D agroforestry model for integrating dynamic tree–crop interactions. Sustainability 11:2293. doi:10.3390/su11082293

Edwards, C.A., T.L. Grove, R.R. Harwood, and C.J.P. Colfer. 1993. The role of agroecology and integrated farming systems in agricultural sustainability. Agric. Ecosyst. Environ. 46:99–121. doi:10.1016/0167-8809(93)90017-J

Eichhorn, M.P., P. Paris, F. Herzog, L.D. Incoll, F. Liagre, K. Mantzanas, M. Mayus, G. Moreno, V.P. Papanastasis, D.J. Pilbeam, A. Pisanelli, and C. Dupraz. 2006. Silvoarable systems in Europe– past, present and future prospects. Agrofor. Syst. 67:29–50. doi:10.1007/s10457-005-1111-7

Elevitch, C.R., D.N. Mazaroli, and D. Ragone. 2018. Agroforestry standards for regenerative agriculture. Sustainability 10(9):3337. doi:10.3390/su10093337

Faulkner, P.A., B. Owooh, and J. Idassi. 2014. Assessment of the adoption of agroforestry technologies by limited-resource farmers in North Carolina. J. Ext. 52(5): 5RIB7. https://joe.org/joe/2014october/rb7.php.

Feliciano, D., A. Ledo, J. Hillier, and D.R. Nayak. 2018. Which agroforestry options give the greatest soil and above ground carbon benefits in different world regions? Agric. Ecosyst. Environ. 254:117–129 doi:10.1016/j.agee.2017.11.032

Ferguson, R.S., and S.T. Lovell. 2014. Permaculture for agroecology: Design, movement, practice and worldview. A Review. Agron. Sustain. Dev. 34:251–274. doi:10.1007/s13593-013-0181-6

Garrett, H.E., W.B. Kurtz, L.E. Buck, L.H. Hardesty, M.A. Gold, H.A. Pearson, J.P. Lassoie, and J.P. Slusher. 1994. Agroforestry: An integrated land use management system for production and farmland conservation. The Agroforestry component of the Resource Conservation Act appraisal for the Soil Conservation Service. NRCS, Washington, D.C.

Garrett, H.E., S. Jose, and M.A. Gold, editors. 2022. North American agroforestry. 3rd ed. American Society of Agronomy, Inc., Madison, WI.

Garrity, D.P. 2004. Agroforestry and the achievement of the Millennium Development Goals. Agrofor. Syst. 61:5–17.

Garrity, D.P. 2005. Forestry in agriculture: The vision of Landcare. p. 47–52. *In* A.G. Brown (ed.), Forests, wood and livelihoods: Finding a future for all. Record of a conference conducted by the ATSE Crawford Fund Parliament House, Canberra. 16 August 2005. Fyshwick, Australia; Crawford Fund. http://www.crawfordfund.org/publications/pdf/forestwoods.pdf#page=54

Garrity, D.P., F.K. Akinnifesi, O.C. Ajayi, S.G. Weldesemayat, J.G. Mowo, A. Kalinganire, M. Larwanou, and J. Bayala. 2010. Evergreen Agriculture: a robust approach to sustainable food security in Africa. Food Secur. 2(3):197–214. doi:10.1007/s12571-010-0070-7

Geertsema, W., W.A.H. Rossing, D.A. Landis, F.J.J.A. Bianchi, P.C.J. van Rijn, J.H.J. Schaminée, T. Tscharntke, and W. van der Werf. 2016. Actionable knowledge for ecological intensification of agriculture. Front. Ecol. Environ 14(4):209–216. doi:10.1002/fee.1258

Gold, M.A. 2019. Agroforestry for the cultivation of nuts. In: M.R. Mosquera-Losada and R. Prabhu, editors, Agroforestry for sustainable agriculture. Burleigh Dodds Science Publishing, Cambridge, UK. doi:10.19103/AS.2018.0041.17

Gold, M.A., L.D. Godsey, and S.J. Josiah. 2004. Markets and marketing strategy for agroforestry specialty products in North America. Agrofor. Syst. 61:371–382.

Gold, M.A., M.M. Cernusca, and M.M. Hall, editors. 2013. Handbook for agroforestry planning and design. Ann Arbor, MI: MU Center for Agroforestry. http://www.centerforagroforestry.org/pubs/training/HandbookP&D13.pdf

Gold, M.A., H.L. Hemmelgarn, G.O. Mori, and C. Todd, editors. 2018. Training manual for applied agroforestry practices. 2018 Edition. Ann Arbor, MI: MU Center for Agroforestry, http://www.centerforagroforestry.org/pubs/training/FullTrainingManual_2018.pdf

Gordon, A.M., S.M. Newman, and B. Coleman, editors. 2018. Temperate agroforestry systems. 2nd ed. Wallingford, U.K. CABI. doi:10.1079/9781780644851.0000

Heavey, J.P., and T.A. Volk. 2014. Living snow fences show potential for large storage capacity and reduced drift length shortly after planting. Agrofor. Syst. 88:803–814. doi:10.1007/s10457-014-9726-1

Hildebrand, P.E. 1990. Farming systems research-extension. In: J.G.W. Jones and P.R. Street, editors, Systems theory applied to agriculture and the food chain. Springer-Verlag, New York, p. 131–144.

Hillbrand, A., S. Borelli, M. Conigliaro, and A. Olivier. 2017. Agroforestry for landscape restoration. FAO, Rome.

IPES-Food. 2016. From uniformity to diversity: a paradigm shift from industrial agriculture to diversified agroecological systems. International Panel of Experts on Sustainable Food systems, Brussels, Belgium. http://www.ipes-food.org/_img/upload/files/UniformityToDiversity_FULL.pdf

Jordan, N., G. Boody, W. Broussard, J.D. Glover, D. Keeney, B.H. McGown, G. McIsaac, M. Muller, H. Murray, J. Neal, C. Pansing, R.E. Turner, K. Warner, and D. Wyse. 2007. Sustainable development of the agricultural bio-economy. Science 316: 1570–1571. http://www.sciencemag.org/cgi/content/full/316/5831/1570?ijkey=5fR9HWpeNY.HQ&keytype=ref&siteid=sci doi:10.1126/science.1141700

Jose, S., A.R. Gillespie, and S.G. Pallardy. 2004. Interspecific interactions in temperate agroforestry. Agrofor. Syst. 61:237–255.

Jose, S., M.A. Gold, and H.E. Garrett. 2018. Temperate agroforestry in the United States: Current trends and future directions. In: A. Gordon, S.M. Newman, and B. Coleman, editors, Temperate agroforestry systems. 2nd ed. CABI, Wallingford, UK. doi:10.1079/9781780644851.0050

Jose, S., B.M. Kumar, and D. Walter. 2017. Ecological considerations in sustainable silvopasture design and management. Agrofor. Syst. 93:317–331. doi:10.1007/s10457-016-0065-2

Jose, S., H.E. Garrett, M.A. Gold, J.P. Lassoie, L.E. Buck, and D. Current. 2022. Agroforestry as an integrated, multifunctional land use management strategy. Chapter 1. In: Garrett, H.E., S. Jose, and M.A. Gold, editors, North American agroforestry. 3rd ed. Agronomy Society of America, Madison, WI.

Jose, S., and E.J. Holzmueller. 2022. Tree-crop interactions in temperate agroforestry. In: H.E. Garrett, S. Jose, and M.A. Gold, editors, North American agroforestry. 3rd ed. Agronomy Society of America, Madison, WI.

Josiah, S.J., H. Brott, and J. Brandle. 2004. Producing woody floral products in an alleycropping system in Nebraska. Horttechnology 14(2):203–207. doi:10.21273/HORTTECH.14.2.0203

Kort, J. 1988. Benefits of windbreaks to field and forage crops. Agric. Ecosyst. Environ. 22–23:165–190.

Krebs, J., and S. Bach. 2018. Permaculture– Scientific evidence of principles for the agroecological design of farming systems. Sustainability 10(9):3218. doi:10.3390/su10093218

Kremen, C., and A.M. Merenlender. 2018. Landscapes that work for biodiversity and people. Science 362. doi:10.1126/science.aau6020

LaCanne, C.E., and J.G. Lundgren. 2018. Regenerative agriculture: merging farming and natural resource conservation profitably. PeerJ 6:e4428. doi:10.7717/peerj.4428

Lelle, M. and M.A. Gold. 1994. Agroforestry systems for temperate climates: Lessons from Roman Italy. Forest and Conservation History 38(3): 118–126.

Lerch, R.N., C.H. Lin, K.W. Goyne, R.J. Kremer, and S.H. Anderson. 2017. Vegetative buffer strips for reducing herbicide transport in runoff: Effects of buffer width, vegetation, and season. J. Am. Water Resour. Assoc. 53(3):1–17 (JAWRA). doi:10.1111/1752-1688.12526

Liebman, M., and L.A. Schulte. 2015. Enhancing agroecosystem performance and resilience through increased diversification of landscapes and cropping systems. Elementa. Science of the Anthropocene. 3:000041. doi:10.12952/journal.elementa.000041

Lovell, S.T. 2010. Multifunctional urban agriculture for sustainable land use planning in the United States. *Sustainability* 2:2499–2522. doi:10.3390/su2082499

Lovell, S.T., S. DeSantis, C.A. Nathan, M.B. Olson, V.E. Mendez, H.C. Kominami, D.L. Erickson, K.S. Morris, and W.B. Morris. 2010. Integrating agroecology and landscape multifunctionality in Vermont: An evolving framework to evaluate the design of agroecosystems. Agric. Syst. 103:327–341. doi:10.1016/j.agsy.2010.03.003

Lovell, S.T., C. Dupraz, M. Gold, S. Jose, R. Revord, E. Stanek, and K. Wolz. 2017. Temperate agroforestry research–Considering multifunctional woody polycultures and the design of long-term field trials. Agrofor. Syst. 92(5): 1397–1415. doi:10.1007/s10457-017-0087-4

Lovric, M., M. Rois-Diaz, M. den Herder, A. Pisanelli, N. Lovric, and P.J. Burgess. 2018. Driving forces for agroforestry update in Mediterranean Europe: Application of the analytic network process. Agroforest Syst 92: 863–876. doi:10.1007/s10457-018-0202-1

Lundgren, B.O., and J.B. Raintree. 1982. Sustained agroforestry. In: B. Nestel, editor, Agricultural research for development: Potentials and challenges in Asia. ISNAR, The Hague, The Netherlands. p. 37–49.

MacFarland, K., C. Elevitch, J.B. Friday, K. Friday, F.K. Lake, and D. Zamora. 2017. Chapter 5: Human dimensions of agroforestry systems. In: Schoeneberger, M.M.; Bentrup, G.; Patel-Weynand, T. eds. 2017. Agroforestry: Enhancing resiliency in U.S. agricultural landscapes under changing conditions. Gen. Tech. Report WO-96. Washington, DC: U.S. Department of Agriculture, Forest Service. p. 73–90.

Mantau, U., J.L.G. Wong, and S. Curl. 2007. Towards a taxonomy of forest goods and services. Small-scale For. 6:391–409. doi:10.1007/s11842-007-9033-z

Matson, P.A., W.J. Parton, A.G. Power, and M.J. Swift. 1997. Agricultural intensification and ecosystem properties. Science 277:504–509. doi:10.1126/science.277.5325.504

McAdam, J., P. Burgess, A. Graves, A. Riguero-Rodríquez, and M.R. Mosquera-Losada. 2009. Classifications and functions of agroforestry systems in Europe. In: A. Rigueiro-Rodríguez, J. McAdam, and M. Mosquera-Losada, editors, Agroforestry in Europe: Current status and future prospects. Springer Science + Business Media B.V., Dordrecht. p. 21–42.

Merwin, M.L. 1997. The status, opportunities and needs for agroforestry in the United States: A national report. Association for Temperate Agroforestry. Center for Agroforestry, University of Missouri, Columbia, MO. http://www.aftaweb.org/resources1.php?page=34#1

Mori, G.O., M.A. Gold, and S. Jose. 2018. Specialty crops in temperate agroforestry systems: Sustainable management, marketing and promotion for the Midwest region of the U.S.A. In: F. Montagnini, editor, Integrating landscapes: Agroforestry for biodiversity conservation and food sovereignty. Advances in agroforestry. Vol. 12. Springer, Dordrecht. p. 331–366. doi:10.1007/978-3-319-69371-2_14

Mosquera-Losada, M.R., J.J. Santiago-Freijanes, A. Pisanelli, M. Rois, J. Smith, M. den Herder, G. Moreno, N. Ferreiro-Domínguez, N. Malignier, N. Lamersdorf, F. Balaguer, A. Pantera, A. Rigueiro-Rodríguez, J.A. Aldrey, P. Gonzalez-Hernández, J.L. Fernández-Lorenzo, R. Romero-Franco, and P.J. Burgess. 2018. Agroforestry in the European Common Agricultural Policy. Agrofor. Syst. 92(4):1117–1127. doi:10.1007/s10457-018-0251-5

Mosquera-Losada, M.R., and R. Prabhu, editors. 2019. Agroforestry for sustainable agriculture. Burleigh Dodds Science Publishing, Cambridge, UK. doi:10.19103/AS.2018.0041

Munsell, J.F., C.J. Bukowski, M.M. Yanez, and J.A. Allen. 2022. Urban food forests and community agroforestry systems. In: H.E. Garrett, S. Jose, and M.A. Gold, editors, North American agroforestry. 3rd ed. Madison, WI: Agronomy Society of America.

Munsell, J.F., and J.L. Chamberlain. 2019. Agroforestry for a vibrant future: Connecting people, creating livelihoods, and sustaining places. Agroforest Syst 93(5):1605–1608. doi:10.1007/s10457-019-00433-0

Nair, P.K.R. 1989. Classification of agroforestry systems. In: P.K. Nair, editor, Agroforestry systems in the tropics. Kluwer Academic Publishers, Dordrecht, The Netherlands. doi:10.1007/978-94-009-2565-6_4

Nair, P.K.R. 1993. Classification of agroforestry systems In: An introduction to agroforestry. Kluwer Academic Publishers, Dordrecht, The Netherlands. doi:10.1007/978-94-011-1608-4_3

Nair, P.K.R., S.C. Allen, and M.E. Bannister. 2005. Agroforestry today: An analysis of the 750 presentations to the 1st World Congress of Agroforestry, 2004. J. For. 103(8):417–421.

Nair, P.K.R. 2007. The coming age of agroforestry. J. Sci. Food Agric. 87:1613–1619. doi:10.1002/jsfa.2897

National Research Council. 1993. Soil and water quality: An agenda for agriculture. National Academy Press, Washington, D.C.

Nelson, M. 2014. Indigenous Science and Traditional Ecological Knowledge. In: Warrior, R. *(Ed.), The World of Indigenous North America*. Routledge.

Newman, S.M., and A.M. Gordon. 2018. Temperate agroforestry: Key elements, current limits and opportunities for the future. In: Gordon, A.M., S.M. Newman, and B. Coleman (ed.), Temperate agroforestry systems. 2nd Edition. Wallingford, U.K. CABI.

Norgaard, K.M. 2014. The politics of fire and the social impacts of fire exclusion on the Klamath. Humboldt J. Soc. Relat. 36:77–101 https://www.jstor.org/stable/humjsocrel.36.77.

Ong, C.K., C.R. Black, and J. Wilson, editors. 2015. Tree-crop interactions: Agroforestry in a changing climate. 2nd ed. CABI, Wallingford, U.K. doi:10.1079/9781780645117.0000

OTA (Organic Trade Association). 2018. Maturing U.S. organic sector grows 6.4 percent in 2017. Organic Trade Association's 2018 Organic Industry Survey. Washington, D.C.: Organic Trade Association. https://ota.com/news/press-releases/20201

Palma, J.H.N., A.R. Graves, R.G.H. Bunce, P.J. Burgess, R. de Filippi, K.J. Keesman, H. van Keulen, F. Liagre, M. Mayus, G. Moreno, Y. Reisner, and F. Herzog. 2007. Modeling environmental benefits of silvoarable agroforestry in Europe. Ecol. Eng. 29:450–462. doi:10.1016/j.ecoleng.2006.09.016

Park, H., N. Turner, and E. Higgs. 2018. Exploring the potential of food forestry to assist in ecological restoration in North America and beyond. Restor. Ecol. 26(2):284–293. doi:10.1111/rec.12576

Park, H., M. Kramer, J.M. Rhemtulla, and C.C. Konijnendijk. 2019. Urban food systems that involve trees in North America and Europe: A scoping review. Urban Forestry and Urban Greening doi:10.1016/j.ufug.2019.06.003

Pent, G.J., J.H. Fike, J.N. Orefice, S.H. Sharrow, D. Brauer, and T.R. Clason. 2022. Silvopasture practices. In: H.E. Garrett, S. Jose, and M.A. Gold, editors, North American agroforestry. 3rd ed. Madison, WI: Agronomy Society of America.

Pimentel, D., C. Harvey, P. Resosudarmo, K. Sinclair, D. Kurz, M. McNair, S. Crist, L. Shpritz, L. Fitton, R. Saffouri, and R. Blair. 1995. Environmental and economic costs of soil erosion and conservation benefits. Science 267:1117–1123.

Porter, P.A., R.B. Mitchell, and K.J. Moore. 2015. Reducing hypoxia in the Gulf of Mexico: Reimagining a more resilient agricultural landscape in the Mississippi River Watershed. J. Soil Water Conserv. 70(3):63A–68A. doi:10.2489/jswc.70.3.63A

Reid, R., and R. Moore. 2018. Agroforestry systems in temperate Australia. Chapter 8. In: Gordon, A.M., S.M. Newman, and B. Coleman, (ed.). 2018. Temperate agroforestry systems. 2nd Edition. CABI. doi:10.1079/9781780644851.0195

Rios-Diaz, M., N. Lovric, M. Lovric, N. Ferreiro-Dominquez, M.R. Mosquera-Losada, M. den Herder, A. Graves, J.H.N. Palma, J.A. Paulo, A. Pisanelli, J. Smith, G. Moreno, S. Garcia, A. Varga, A. Pantera, J. Mirck, and P. Burgess. 2018. Farmers' reasoning behind the update of agroforestry practices: Evidence from multiple case-studies across Europe. Agrofor. Syst. 92(4):811–828. doi:10.1007/s10457-017-0139-9

Rossier, C., and F. Lake. 2014. Indigenous traditional ecological knowledge in agroforestry. Agroforestry Notes #44. USDA National Agroforestry Center, Lincoln, NE. https://www.fs.usda.gov/nac/assets/documents/agroforestrynotes/an44g14.pdf (accessed 5 Aug. 2020).

Rule, L.C., C.B. Flora, and S.S. Hodge. 2000. Social dimensions of agroforestry. p. 361–386. In H.E. Garrett, W.J. Rietveld and R.F. Fisher (ed.), North American agroforestry: An integrated science and practice. Agronomy Society of America, Madison, WI.

Russell, D., and S. Franzel. 2004. Trees of prosperity: Agroforestry, markets and the African smallholder. Agrofor. Syst. 61:345–355.

Schoeneberger, M.M., G. Bentrup, C.F. Francis, and R. Straight. 2006. Creating viable living linkages between farms and communities. In: C. Francis, R. Poincelot, and G. Bird, editors, Developing and extending sustainable agriculture. Haworth Press, Binghamton, NY. p. 225–246.

Schulte, L.A.J., Niemi, M.J. Helmers, M. Liebman, J.G. Arbuckle, D.E. James, R.K. Kolka, M.E. O'Neal, M.D. Tomer, J.C. Tyndall, H. Asbjornsen, P. Drobney, J. Neal, G. Van Ryswyk, and C. Witte. 2017. Prairie strips improve biodiversity and the delivery of multiple ecosystem services from corn–soybean croplands. PNAS 114(42):11247–11252. www.pnas.org/cgi/doi/10.1073/pnas.1620229114

Schultz, R.C., T.M. Isenhart, J.P. Colletti, W.W. Simpkins, R.P. Udawatta, and P.L. Schultz. 2022. Riparian and upland buffer practices. In: H.E. Garrett, S. Jose, and M.A. Gold, editors, North American agroforestry. 3rd ed. American Society of Agronomy, Madison, WI.

Shelton, D.P., R.A. Wilke, T.G. Franti, and S.J. Josiah. 2005. Farmlink- promoting conservation one-to-one. In: K.N. Brooks and P.F. Folliott, (ed.), Moving agroforestry into the mainstream. The 9th North American Agroforestry Conference Proceedings, June 12-15, 2005, St. Paul, MN. Dep. of Forest Resources, University of Minnesota, St. Paul, MN. 9 p. http://www.cinram.umn.edu/afta2005/pdf/Josiah.pdf

Sinclair, F.L. 1999. A general classification of agroforestry practice. Agrofor. Syst. 46:161–180. doi:10.1023/A:1006278928088

Sobels, J., A. Curtis, and S. Lockie. 2001. The role of Landcare group networks in rural Australia: exploring the contribution of social capital. J. Rural Stud. 17(3):265–276. doi:10.1016/S0743-0167(01)00003-1

Sooby, J. 2003. State of the states 2nd edition: Organic farming systems research at Land Grant Universities 2001 - 2003. Organic Farming Research Foundation, Santa Cruz, CA. http://ofrf.org/publications/pubs/sos2.pdf (accessed 30 Sept. 2007).

Taylor, A.H., and C.N. Skinner. 2003. Spatial patterns and controls on historical fire regimes and forest structure in the Klamath Mountains. Ecol. Appl. 13(3):704–719. doi:10.1890/1051-0761(2003)013[0704:SPACOH]2.0.CO;2

Torquebiau, E.F. 2000. A renewed perspective on agroforestry concepts and classification. C.R. Acad. Sci. Paris, Sciences de la vie. Life Sci. 323:1009–1017.

Udawatta, R.P., C.J. Gantzer, and S. Jose. 2017. Agroforestry practices and soil ecosystem services. In: M.M. Al-Kaisi and B. Lowery, editors, Soil health and intensification of agroecosystems. New York: Elsevier, p. 305–333. doi:10.1016/B978-0-12-805317-1.00014-2

Udawatta, R.P., J.J. Krstansky, G.S. Henderson, and H.E. Garrett. 2002. Agroforestry practices, runoff, and nutrient loss: a paired watershed comparison. J. Environ. Qual. 31: 1214–1225.

USDA (United States Department of Agriculture). 2015. Twenty-five years. Inside Agroforestry 23(3). USDA National Agroforestry Center. https://www.fs.usda.gov/nac/assets/documents/insideagroforestry/IA_Vol23Issue3.pdf

USDA (United States Department of Agriculture). 2017. Agroforestry: Enhancing resiliency in U.S. agricultural landscapes under changing conditions. In: M.M. Schoeneberger; G. Bentrup; T. Patel-Weynand (Eds.), Gen. Tech. Report WO-96. Washington, DC: U.S. Department of Agriculture, Forest Service. doi:10.2737/WO-GTR-96

USDA (United States Department of Agriculture). 2019. Agroforestry Strategic Framework 2019-2024. Miscellaneous Publication 1615. Washington, D.C.: USDA. https://www.usda.gov/sites/default/files/documents/usda-agroforestry-strategic-framework.pdf (accessed 12 July 2019).

Valdivia, C., M.A. Gold, C. Barbieri, L. Zabek, J. Arbuckle, and C. Flora. 2022. Human and institutional dimensions of agroforestry. In: H.E. Garrett, S. Jose, and M.A. Gold, editors, North American agroforestry. 3rd ed. Agronomy Society of America, Madison, WI.

van Noordwijk, M., ed. 2019. Sustainable development through trees on farms: Agroforestry in its fifth decade. Bogor, Indonesia: World Agroforestry (ICRAF) Southeast Asia Regional Program. http://old.worldagroforestry.org/downloads/Publications/PDFS/B19029.pdf

Vinyeta, K., F. Lake, and K. Norgaard. 2016. Chapter 3: Vulnerabilities of traditional foods and cultural use species. In The Karuk Tribe, (ed.), Karuk Tribe climate vulnerability assessment (p. 68–138). Karuk Tribe Council, Happy Camp, CA. https://karuktribeclimatechangeprojects.wordpress.com/chapter-3-vulnerabilities-of-traditional-foods-and-cultural-use-species/

Wilson, M., S.T. Lovell, and T. Carter. 2018. In: K. Gruley, and K. Keeley, editors, Perennial pathways. Planting tree crops: Designing & installing farm-scale edible agroforestry. Savanna Institute, Madison, WI. http://www.savannainstitute.org/resources.html.

Yang, L., B. Liu, Q. Zhang, Y. Pan, M. Li, and Y. Lu. 2019. Landscape structure alters the abundance and species composition of early-season aphid populations in wheat fields. Agric. Ecosyst. Environ. 269:167–173. doi:10.1016/j.agee.2018.07.028

Study Questions

1. Identify some of the key differences in priorities between temperate agroforestry in North America (US/Canada) and the tropics.

2. Define agroforestry as it is used in the US/Canada.

3. What are the 4 "I's" of agroforestry and why are they important?

4. Describe the "associated technologies" that are used in conjunction with riparian forest buffers and when and why they are used.

5. What are urban food forests and why have been recognized as a sixth recognized agroforestry practice?

6. How does agroforestry increase the productive area of the farm? Why would this be important to landowners practicing agroforestry?

7. Describe some of the main impacts that are realized when trees are integrated into agricultural production systems.

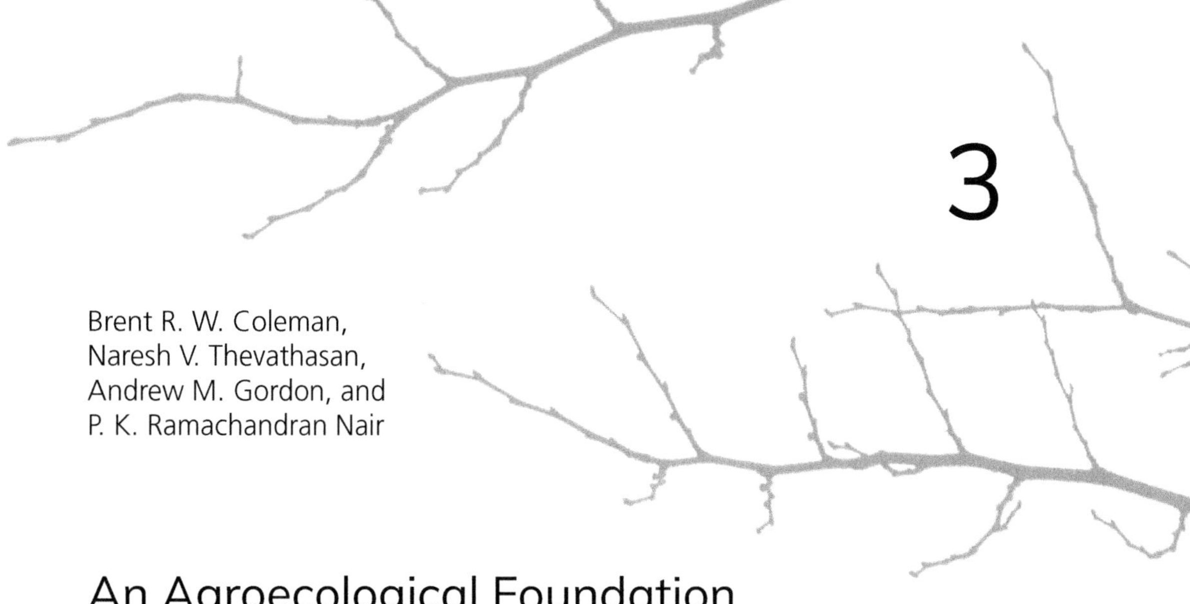

3

Brent R. W. Coleman,
Naresh V. Thevathasan,
Andrew M. Gordon, and
P. K. Ramachandran Nair

An Agroecological Foundation for Temperate Agroforestry

The primary natural ecosystems of North America are dominated by either perennial grasses or woody vegetation. Prior to European settlement, grasslands occupied 39% of the current United States, while forests and shrub-dominated systems covered most of the remainder (Sims, 1988). These highly diverse assemblages of species evolved during millions of years in response to major changes in climate and physiography. Powered almost exclusively by solar energy, they have sustained production and provided enormous ecosystem services, generating a legacy of rich soils and other biological wealth.

Because of this legacy, most of these ecosystems have been converted to agroecosystems (agricultural systems) through the substitution of annual food plants (e.g., corn [*Zea mays* L.], wheat [*Triticum aestivum* L.] and soybean [*Glycine max* (L.) Merr.]) for the original perennial vegetation. Where climate or other conditions preclude the planting of row crops, native grassland is exploited through the substitution of domestic livestock for native herbivores, and forests are intensively managed for timber production. The conversion has been extensive and thorough; at the extreme are states like Illinois that have seen a decrease of 99.9% of prairie acreage, with approximately 930 ha (2,300 acres) remaining statewide (Steinauer & Collins, 1996). Of the 930,000 ha (2.3 billion acres) of total land area in the United States, 17% is classified as cropland, 29% is classified as grassland pasture and rangeland, 28% is classified as forest-use land, 23% is classified as special use (parks and wildlife areas) or miscellaneous (wetlands, tundra, unproductive woodlands), and 3% as urban land (USDA, 2017). Approximately 43 million ha (106 million acres) are designated wilderness areas under the National Wilderness Preservation System, accounting for roughly 4.6% of the total land area of the United States (Watson, Matt, Knotek, Williams, & Yung, 2011). The landscape is now a "semi-natural matrix" (Roberts, 1988) within which humans and all other species must survive.

A similar situation exists in southern Canada, especially in southern Ontario, Quebec, and the Maritime Provinces. Herbaceous and woody biomass crops are being cultivated on increasing areas across North America as a means of producing bioenergy or for use as animal bedding. Warm-season grasses such as switchgrass (*Panicum virgatum* L.) and miscanthus (*Miscanthus* spp.) and fast-growing woody species such as hybrid willow (*Salix* spp.) and poplar (*Populus* spp.) are growing in popularity among producers as a result of their high yields, low nutrient requirements, broad environmental tolerances, and environmental benefits such as enhanced C sequestration potentials compared with conventional agricultural crops (Coleman et al., 2018; Graham et al., 2019). The ability of these crops to grow on

North American Agroforestry, Third Edition. Edited by Harold E. "Gene" Garrett, Shibu Jose, and Michael A. Gold.

marginal lands is likely to contribute to increasing popularity going forward. The ecological principles explored in this chapter would apply equally to temperate agroforestry and biomass crop production systems developed in these regions.

The goal of this land-use conversion has been to maximize the amount of net primary or secondary production from these systems that can be used by humans. In the short term, this goal has been met and a massive increase in food and wood supplies has been generated. However, the long-term consequences of these conversions bring into question the sustainability of this level of production. For example, in addition to solar energy, U.S. agroecosystems use large amounts of fossil fuels to power machinery or produce other inputs such as fertilizer and pesticides. Irrigated corn production in Nebraska, for example, requires an estimated average of nearly 86 million kJ ha^{-1} of fossil energy input (Pimentel, 2009), or 1 kJ input for each 1.65 kJ harvested. Conventional beef production requires 13 kg (29 lb) of grain and 30 kg (66 lb) of forage to produce 1 kg (2.2 lb) of beef, meaning fossil fuel energy inputs of 40 kJ kJ^{-1} (or 40 kcal energy input per 1 kcal) of beef protein, without considering the energy costs of processing or transportation (Pimentel et al., 2008). This energy profligacy occurs in a country that imports 10.14 million barrels per day of petroleum (U.S. Energy Information Administration, 2018a). Additionally, the United States also heavily relies on shale oil for its own domestic production, accounting for nearly 60% of the total U.S. crude oil production (U.S. Energy Information Administration, 2018b), with fracking posing devastating environmental impacts and requiring significantly greater amounts of energy to extract compared with conventional drilling.

Soil degradation caused by erosion, salinization, waterlogging, and such other processes are major environmental issues that seriously impact land use. For example, approximately 30% of U.S. cropland has been severely damaged because of erosion, salinization, or waterlogging (Pimentel et al., 1995). Soil loss by erosion continues at a rate of 1.54 billion Mg of soil per year, with water erosion causing annual soil losses of approximately 900 Mg annually, and wind erosion causing soil losses of nearly 640 Mg annually (Natural Resources Conservation Service, 2015).

The continuing decline of genetic diversity in agriculture is yet another issue of major concern. Instead of the 250–300 plant species found in an equivalent area of tall-grass prairie (Steiger, 1930), or the 100 species in a similar area of oak–hickory (*Quercus* spp.–*Carya* spp.) forest, a typical midwestern corn–soybean farm maintains only two species on a majority of the land area. In addition, genetic diversity within the major U.S. crops is quite low. Farmers who plant several hybrids or cultivars to increase their diversity are often planting essentially the same thing, under different names (National Academy of Science, 1972; Raeburn, 1995). These highly simplified single-species systems are at increased risk from pest outbreak or climate extremes.

Sustainability is "the concept about meeting today's needs without compromising the ability of future generations to satisfy their needs, and it strives to achieve a balance between ecological preservation, economic vitality, and social justice" (World Commission on Environment and Development, 1987). Our current farming system faces declining domestic energy reserves, soil loss in excess of regeneration, and a rapidly increasing human population with a concomitant increase in demand for agricultural products. Although farmers have adopted practices such as contour planting, no-till, and precision application of chemicals to reduce some of the negative effects of agriculture, farming systems based on monocultures or simple rotations of annuals are not sustainable without massive external inputs.

Further diversification of crops offers many advantages. The addition of herbaceous perennials such as alfalfa (*Medicago sativa* L.) and grasses increases the perennialism that is such a dominant feature of native ecosystems (Figure 3–1) (Van Andel, Bakker, & Grootjians, 1993). Reduced erosion, fixation of atmospheric N_2 by legumes, and reduced energy inputs (Heichel, 1978) are benefits of adding certain perennials to agroecosystems. Livestock offer further diversification and a mechanism for converting forages into higher value products (Bender, 1994).

Although forests are a major vegetation type in the United States, few farmers consciously integrate trees and other woody perennials into their farms as a way of increasing diversity and sustainability. This approach to farming is known as agroforestry and is defined generically as the integration of trees into agriculturally productive landscapes (Nair, 1993). Within a North American context, agroforestry can be more explicitly defined as: intensive land management that optimizes the benefits (physical, biological, ecological, economic, and social) from the biophysical interactions created when trees and/or shrubs are deliberately combined with crops and/or livestock (Garrett et al., 1994). A practice is deemed an agroforestry practice if it embraces four principal criteria as indicated by Gold and Garrett (Chapter 2) and shown in Table 3–1.

The key words in this definition are *interactions*, *benefits*, and *optimizes*. Biophysical interactions require a certain spatial and temporal proximity of the components. A woodlot on one corner of the farm, isolated from and not beneficially interacting with crops or livestock, does not constitute agroforestry by this definition (see discussion below). When trees, crops, and livestock are in close enough proximity to interact in a way that is significant to the farmer, agroforestry is created. The types of interactions depend on the species involved and their particular spatial and temporal relationships. Not all interactions are beneficial. For example, competition between trees and row crops for water, nutrients, and light can reduce row crop yields. Gray (2000) found that soybean yield and tree root quantity were negatively correlated, with 80% of tree roots being found in the A soil horizon. That study indicates, however, that soil cultivation suppresses tree root growth in the top portion of the soil, further establishing that appropriate tree management interventions can minimize competition with crops. Peng, Thevathasan, Gordon, Mohammed, & Gao (2015) also found a reduction in soybean yield when it was grown in a 26-year-old tree-based intercropping site with silver maple (*Acer saccharinum* Marsh.), hybrid poplar (*Populus deltoides* × *nigra*), and black walnut (*Juglans nigra* L.). In addition to belowground competition, the trees, now roughly 15–20 m tall, also reduced incident photosynthetically active radiation, thereby reducing net assimilation, growth, and soybean yield. Obtaining optimal benefits from agroforestry requires knowledgeable selection, placement, and management of the woody and non-woody components. Thinning and/or outright removal of some trees, as well as planting shade-tolerant crops, are management options to consider as the system ages. A random mixture is unlikely to perform well.

Unfortunately, there is no single optimal agroforestry design that interested farmers and ranchers can be encouraged to adopt. Differences in climate, topography, soils, crops, and livestock exist at scales that range from the local to the continental. Agroforestry practices must be designed to fit the particular ecological, social, and economic context of the farm in question. Component interactions in agroforestry practices have been investigated to a small extent (Ong & Huxley, 1996), and while the emphasis has been on tropical systems (e.g., Rao, Nair, & Ong, 1997), some information is also available for temperate agroforestry systems (Thevathasan & Gordon, 2004). Whether we are considering

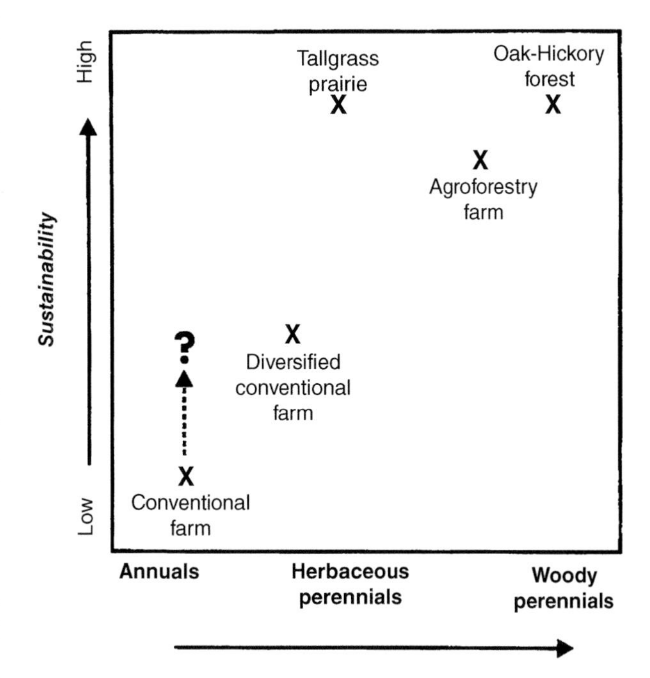

Fig. 3–1. Hypothetical relationship between perennialism and sustainability in selected natural ecosystems and agroecosystems (based on Van Andel et al., 1993).

Table 3–1. The four key criteria that characterize agroforestry practices (modified from University of Missouri Center for Agroforestry, 2018, pp. 9–10).

Criteria	Description
Intentional	Combinations of trees, crops, and/or livestock are intentionally designed, established, and/or managed to work together and yield multiple products and benefits, rather than as individual elements that may occur together but are managed separately. Agroforestry is neither monoculture farming nor is it a mixture of monocultures.
Intensive	Agroforestry practices are created and intensively managed to maintain their productive and protective functions and often involve cultural operations such as cultivation, fertilization, irrigation, pruning and thinning.
Integrated	Components are structurally and functionally combined into a single, integrated management unit tailored to meet the objectives of the landowner. Integration may be horizontal or vertical, above- or belowground, simultaneous or sequential. Integration of multiple crops utilizes more of the productive capacity of the land and helps to balance economic production with resource conservation.
Interactive	Agroforestry actively manipulates and utilizes the interactions among components to yield multiple harvestable products while concurrently providing numerous conservation and ecological benefits.

temperate or tropical agroforestry, Muschler, in *An Introduction to Agroforestry* (Nair, 1993), pointed out "that the complexity and lifespan of agroforestry makes investigations of mechanisms and processes extremely difficult." Leaving consideration of socioeconomic issues for later, how can we obtain the ecological knowledge necessary for the optimal design of a wide variety of temperate agroforestry practices?

The answer lies, at least in part, in the native ecosystems upon which U.S. agriculture is built. Highly sustainable, these systems were locally adapted to the environmental conditions under which they evolved. Natural ecosystems can provide models for the design of sustainable agroecosystems (Davies, 1994; Soule & Piper, 1992; Woodmansee, 1984). We believe that it is possible to identify structural and functional characteristics of natural ecosystems that contribute to their sustainability and then retain or incorporate these into agroecosystems while maintaining production. Regional and local differences in natural ecosystems can serve as guides for tailoring agroforestry practices that best fit a particular farm's environmental conditions. Our goal in the remainder of this chapter is to illustrate some of the structural and functional relationships among woody and herbaceous vegetation in natural ecosystems of the United States and to show how these relationships apply to agroforestry practices.

Ecological Interactions in Mixed Tree and Herb Systems

Categories of Systems

Ecosystem function is determined not just by species composition, but also by the spatial and temporal arrangement of the component species. Figure 3–2A identifies eight main types of North American temperate ecosystems based on the spatial and temporal relationships of their woody and herbaceous components. Figure 3–2B uses the same framework to identify structurally analogous agroforestry practices found in this region.

The main concern in agroforestry practices is performance or the relationship between structure and function. In each of the natural systems, there are certain processes that are most important in determining interactions among the woody and non-woody components and in turn the overall function of the system (Table 3–2). These same processes influence interactions among the components of analogous agroforestry practices.

Closed-Canopy Mesic Forests

Water, nutrients, and light (energy) are the main resources for which plants compete. In more mesic (wetter) sites with adequate nutrients, light is the limiting factor. Trees and shrubs invest heavily in structural components to lift leaves above competitors and capture light before it reaches the ground. Forests often have two or three canopy layers as trees, saplings, and shrubs capture light at different levels, and this vertical stratification may increase the total energy captured by the system. Total leaf area can be quantified as leaf area index (LAI), the ratio of total leaf surface area to unit ground surface area. Depending on leaf orientation, a canopy with an LAI of 3–4 can intercept 90% of the incident solar radiation (Loomis & Connor, 1992). Mature mesic forests generally have an LAI of 8–10 (Odum, 1971), so competition for light is intense, with only 1–5% of incident solar radiation reaching the forest floor in closed-canopy deciduous forests (Hicks & Chabot, 1985). For example, light penetration was 6% in a high-elevation fir forest with full crown density and 18% when crown density was 50% (Smith, 1985). Light quality as well as quantity is affected by tree canopies, with radiation below the canopy relatively enriched in red wavelengths (Atzet & Waring, 1970). The light environment under plant canopies is highly variable both spatially and temporally as sun flecks shift with changes in the angle of incident sunlight.

As a result of competition for light, mesic forests often have sparse ground-level vegetation. In a tulip tree–oak (*Liriodendron–Quercus*) forest in the southern Appalachians, only 2% of the total aboveground biomass consisted of herbaceous species (Harris, Sollins, Edwards, Dinger, & Shugart, 1975). However, some deciduous forests have rich herbaceous layers (Braun, 1967), and seasonal changes in LAI offer temporal niches for certain species. Spring ephemerals in forest understories are able to leaf out and capture substantial light energy before overstory canopy development occurs. Goldenseal (*Hydrastis canadensis* L.), an understory forb native to many eastern deciduous forests, produces >95% of its aboveground biomass within the first month of its growing season, well before the overstory canopy is developed (Eichenberger & Parker, 1976). Other species are adapted to full-shade conditions and experience peak development in late summer (Greller, 1988). The co-occurrence of these strategies results in increased capture of light energy as well as more efficient use of other resources. Nutrient uptake

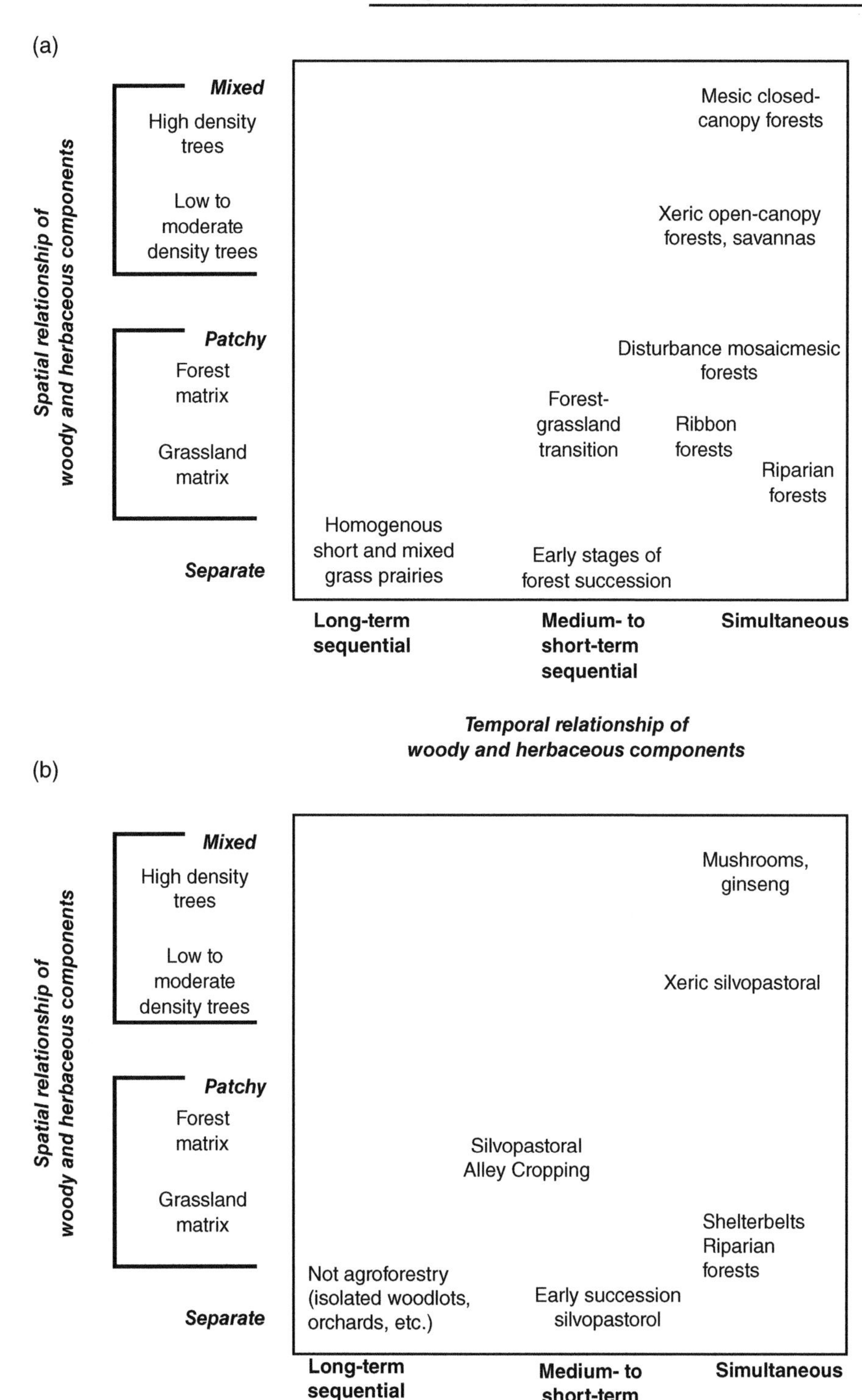

Fig. 3–2. (A) Categorization of ecosystems in terms of the spatial and temporal relationships of the woody and herbaceous components; and (B) categorization of temperate agroforestry practices in terms of the spatial and temporal relationships of the woody and herbaceous components.

Table 3–2. Summary of the most important processes in interactions between woody and herbaceous species in natural ecosystems of the United States and in analogous agroforestry practices.

Natural system category	Key processes in interactions among woody and herbaceous species	Analogous agroforestry practices in which these processes are important
Mesic forest, closed canopy	• canopy interception of solar radiation and modification of microclimate	• mushroom production • ginseng production
Disturbance patchiness in forest landscape	• gap-creating disturbances • edge effects • landscape processes	• none (in tropical areas this would be swidden agriculture)
Early successional systems	• progressive modification of microclimate as tree canopy closes	• black walnut alley cropping • silvopasture—grazing of early successional stages
Xeric forest, open canopy	• competition for water • localized interception of solar radiation	• silvopastoral practices
Mixture of forest and grass patches in transition zones	• topographic patterns often serve as template • chronic stress and disturbance	• silvopastoral practices
Ribbon forests	• windspeed reduction • snow distribution	• windbreaks
Riparian forests in grasslands	• corridors for movement of wildlife • specialized wildlife habitat • interception of sediment and nutrients	• riparian forests in cropland or pasture matrix
Isolated grasslands	• no interactions	• not agroforestry

by spring ephemerals may sequester up to 90% of the N and K that could potentially be leached during the spring from some Midwest forests (Blank, Olson, & Vitousek, 1980; Peterson & Roelf, 1982). Thus, temporal as well as spatial stratification plays a role in system function.

Forest canopies modify other aspects of microclimate in addition to radiation. During the day, interception of solar radiation by the tree canopy creates a temperature maximum at the height of maximum foliage density (Oke, 1987). This creates a temperature inversion that increases the atmospheric stability in the canopy relative to open terrain, partially decoupling the local atmosphere from the external environment. Windspeed decreases rapidly with distance into the canopy, while daytime humidity increases and CO_2 concentration decreases due to transpiration and photosynthesis by the foliage. At the forest floor, this altered environment affects seed germination, plant establishment, litter decomposition, and the population dynamics of microorganisms, insects, and other organisms (Belsky, 1994; Jackson, Strauss, Firestone, & Bartolome, 1990; Tiedemann & Klemmedson, 1973; Vetaas, 1992).

Agroforestry options for closed-canopy forests are limited to crops that are adapted to a low-light environment, such as shade-tolerant flowers. Shiitake mushrooms [*Lentinula edodes* (Berkeley) Pegler; Harris, 1986] and ginseng (*Panax quinquefolius* L.; Duke, 1989) fit perfectly in this situation, both requiring the protected environment of the forest floor. Shiitake is grown by inoculating logs with mushroom spawn and then stacking the logs under a hardwood or conifer canopy. If the site is a deciduous forest, shade cloth can be used to provide protection during leafless months. Ginseng, a medicinal herb, is cultivated in a variety of temperate deciduous forests, although most often associated with maple (*Acer saccharum* Marsh.) and beech (*Fagus grandifolia* Ehrh.). It grows well at light intensities from 5–30% and is sometimes intercropped with goldenseal to deter root rot (Duke, 1989).

Disturbance Patches and Early Successional Systems

Forest canopies are heterogenous. In regions where the climate is mesic enough to support closed-canopy forests, disturbances such as wind, avalanches, or fire create gaps that support herbaceous vegetation for a brief period of time. In old-growth forests of the eastern United States, 9.5% of the land area historically was in small gaps (created by the death of one to several trees) (Runkle, 1982). New gaps formed at a rate of 1% of the land each year while an equal area of gaps closed due to sapling growth, making this a landscape-level, steady-state process. Less frequently, larger areas are disturbed by hurricanes, fires, insect outbreaks (e.g., gypsy moth), and other large-scale events (Spies & Franklin, 1989). Since European settlement, most U.S. forests have been logged at least once.

The smaller the gap, the greater the edge effects on increasing competition for water and nutrients, shading, and reduction of windspeed. Edges are also zones of increased diversity and activity for many species of insects, birds, and mammals; at the landscape scale, the size and distribution of gaps is an important determinant of many forest functions, and edge "presence" in the landscape is often enhanced through the adoption of agroforestry systems. Swidden or slash-and-burn agriculture mimics the process of gap formation and succession in many tropical forests and has been called the most sustainable form of agriculture when practiced appropriately (Kleinman, Pimentel, & Bryant, 1995). An analogous form of temperate shifting agriculture was practiced by Native Americans in the New England region (Davies, 1994). In some temperate U.S. forests, logging of small patches to mimic the natural processes of gap formation offers a more sustainable alternative to large-scale clear-cuts (Maser, 1994).

Grasses and forbs dominate a gap immediately following disturbance but are soon replaced by trees or shrubs. This transition is known as *succession*, the "orderly process of community development that involves changes in species structure and community processes with time, and results from modification of the physical environment by the community" (Odum, 1971). Keever (1950) described a typical succession pattern for abandoned farmland in the North Carolina Piedmont with crabgrass (*Digitaria* spp.), asters (*Aster* spp.), and ragweed (*Ambrosia artemisiifolia* L.) dominating the first 2 yr, followed by broomsedge (*Agropogon virginicus* L.), which was gradually replaced in 10–15 yr by shortleaf (*Pinus echinata* Mill.) or loblolly (*Pinus taeda* L.) pines. A hardwood understory develops by 60 yr and forms the climax oak–hickory forest by 150 yr.

Succession results from the gradual modification of microclimate as the expanding canopy intercepts increased amounts of solar radiation each year. Enough light reaches the ground early in the successional process to support significant forage production. With proper management to limit damage to young trees, livestock can be rotationally grazed as part of a silvopastoral agroforestry practice. Stocking rates are reduced as tree growth reduces light levels until canopy closure eliminates forage production. Successional silvopastoral practices are particularly well developed in New Zealand and Australia (Anderson, Moore, & Jenkins, 1988) and in many parts of temperate Europe (Dupraz et al., 2018).

Successional principles can also be applied in cropping systems. Perhaps the best known temperate example is alley cropping with black walnut (Garrett & Kurtz, 1983; Garrett & Harper, 1998; Williams & Gordon, 1992; Thevathasan & Gordon, 2004). Black walnut is planted at wide spacings (e.g., 12 m), and row crops are grown in the alleys for up to 10 yr. When shading reduces row crop yields, forage crops are substituted either for haying or for direct grazing. By the time canopy closure ends profitable forage production, nut production provides income until the trees are cut for timber, and the process begins anew.

Xeric and Transitional Forests

On more xeric (drier) sites, moisture is limiting and competition for resources is greater belowground than aboveground. Forest canopies become more open as trees become more widely spaced, and a greater proportion of light reaches the ground. Higher light levels may allow the development of significant amounts of ground-level vegetation. Ponderosa pine (*Pinus ponderosa* Laws.) forests throughout the Rocky Mountains and longleaf pine (*Pinus palustris* Mill.) forests in the southeastern United States frequently have dense grass understories that are maintained in part by periodic fires (Daubenmire, 1978). In still drier areas, tree density decreases until scattered individuals in a grassland matrix form a savanna such as the blackjack oak (*Quercus marilandica* Muenchh.)–post oak (*Q. stellata* Wangenh.) savanna in eastern Texas, pinyon-juniper (*Pinus* sp.–*Juniperus* sp.) savanna in the southwestern United States, and the oak–hickory savanna in western Missouri. The oak savanna, characterized by a sparse overstory of oaks and an understory of herbs and grasses, is a transitional zone between the eastern forest and the grasslands (Packard, 1988). Oak savanna was once a major community across the Midwest—although it became severely diminished after the Euro-American settlement of the 1800s. Prior to settlement and overgrazing, large areas of sagebrush steppe in the Intermountain West also showed a co-dominance of shrubs (*Artemisia*) and perennial bunchgrasses (West, 1988).

As the preceding examples suggest, disturbance (e.g., grazing, browsing, drought, fire) is a critical mediator of the competition that occurs between trees and grasses (Belsky, 1994; Hamerlynck & Knapp, 1996; Jeltsch, Milton, Dean, & Van Rooyen, 1996). In the southeastern Coastal Plain, longleaf pine forests with a grassy understory are maintained by fires of 3–10-yr frequency that allow regeneration of the pines but prevent establishment of hardwoods, which have denser canopies than the pines and would inhibit grasses (Daubenmire, 1978). Most savannas are maintained by fire, and if fire is prevented or overgrazing leaves insufficient fuel to carry a fire, succession proceeds to a denser forest. Grasses are physiologically and morphologically adapted to burning. The ecological message is that a particular balance between grasses and trees can often be maintained only through regular disturbance.

As water becomes more limiting, trees disappear and grasses or shrubs dominate. Grasses have a high root/shoot mass ratio, which provides an advantage in competing for water and nutrients. However, a shrub such as mesquite

(*Prosopis* L.) also has an extensive lateral and vertical root system that allows it to compete effectively for water as well as nutrients against grasses in the arid grasslands in which it occurs (Tiedemann & Klemmedson, 1973). In other cases, competition for belowground resources is reduced by the exploitation of different soil layers by different species. In the blue oak (*Quercus douglasii* Hook. & Arn.) savanna, competition for water between trees and grasses is reduced by vertical stratification of the two root systems, with grasses occupying mainly the top meter of soil and oak roots penetrating >25 m (Jackson et al., 1990). This stratification also promotes more efficient cycling and retention of N in the ecosystem.

In addition to competition for resources, trees and grasses in these mixed systems may compete through direct interference. An example of this would be the allelopathic suppression of understory plants in oak forests in Oklahoma (McPherson & Thompson, 1972). Alternatively, some interactions may be positive. Survival of grass seedlings was three times greater within a California blue oak savanna than in adjacent open grassland (Jackson et al., 1990) due to the more favorable environment for seedling establishment (i.e., higher relative humidity, decreased evaporation, and increased near-surface soil moisture and nutrient levels).

Within a particular climatic region, topographic and soil patterns may have a strong influence on spatial patterns and interactions of woody and non-woody species. Throughout much of the Great Plains grasslands, trees and shrubs are restricted to riparian areas, rocky escarpments, mesic north-facing slopes, and other sites offering increased moisture availability and protection from fire. Rockier soils also provide better opportunity for tree seedling establishment in competition with the thick root mass of grasses (Wells, 1965). At the northern edge of the prairie, grasses on the uplands form a mosaic with groves of poplar (*Populus* sp.) located in depressions or on protected slopes (Daubenmire, 1978).

Significant grass production in a forest matrix allows timber production and grazing to coexist on >69 million ha (170 million acres) in the United States (U.S. Forest Service, 1981). The dual functions of these silvopastoral practices can be enhanced by management based on ecological principles. On mesic sites, thinning and pruning of trees maintains forage production while promoting high-quality timber. Prescribed burns can prevent invasion by undesirable species while maintaining an open and productive understory. In semiarid areas, avoiding overgrazing is the most effective means of preventing the replacement of grasses by shrubs.

Ribbon Forests and Windbreaks

When wind encounters the edge of a forest, some of the air is deflected over the canopy for a distance of up to 20 tree heights (Cionco, 1985; Fritschen, 1985). If the forest occurs as a narrow strip, this deflection of air creates a protected zone to the leeward in which wind speed is reduced, wind-related stresses such as desiccation are decreased, and snow deposition may increase.

This modification of microclimate is essential to the maintenance of ribbon forests (Billings, 1969; Peet, 1988) and is a fascinating feature of subalpine regions in the Rocky Mountains. Ribbon forests are arranged as alternating parallel strips of forest and moist alpine meadow oriented perpendicular to the prevailing winds. Snow accumulation to the lee of each forest strip inhibits seedling establishment, while tree growth rates at the far edge of each drift are increased by water from snowmelt and protection from desiccation by winter winds. Thus, the pattern and spacing of forest strips is determined by the effect of tree canopy structure on windspeed and snow deposition.

Ribbon forests are a classic model for one of the most common temperate agroforestry practices, windbreaks. Farm windbreaks are linear groups of trees that provide a sheltered microclimate for leeward fields. The extent and degree of shelter depends on the structural characteristics of the windbreak such as height, density, and orientation, and these can be manipulated to meet particular management goals (e.g., odor control). Dense windbreaks result in deposition of snow in drifts close to the leeward edge and act as living snow fences. More porous windbreaks cause snow to be distributed more evenly across the leeward field, a preferable situation if soil moisture conservation or protection of winter wheat from desiccation is the goal (Brandle & Finch, 1991; Mize, Brandle, Schoenberger, & Bentrup, 2008).

Riparian Forests

Particularly in arid and semiarid regions, riparian forests are often the only mesic vegetation type and serve a critical role as wildlife habitat. In Arizona and New Mexico, an estimated 80% of all vertebrates are dependent upon riparian habitat for at least part of their life cycle (Johnson, 1989). As linear features in the landscape, riparian forests may serve as corridors for the movement of many species between otherwise isolated patches of habitat (Forman & Godron, 1986). Woody vegetation also plays an integral role in the stabilization of streambanks (Smith, 1976), shading of streams reduces water temperature, and detritus inputs to the stream from the forest provide an energy source as well as habitat structure for aquatic organisms.

Because they occupy low spots in the landscape, riparian forests receive water and waterborne nutrients and sediment from upland areas, filtering and trapping many of these inputs before they reach the streambed (Lowrance et al., 1984). These forests interact not just with adjacent fields but with systems throughout the landscape, linked through the hydrologic pathways of the watershed. In agricultural regions, this landscape-level water quality function is particularly important. For example, despite large applications of N fertilizer to corn, peanut (*Arachis glabrata* Benth.), and other cropland in a Georgia Piedmont watershed, very little N left the watershed in streamflow due in part to accretion of N in the riparian forest biomass and denitrification in the saturated riparian soils (Lowrance, Leonard, Asmussen, & Todd, 1985). Maintenance of a young-age forest through selective logging can improve the water quality function of the stand by maintaining plant nutrient uptake at a high rate (Welsch, 1991). An excellent overview on the ecological impact of developing riparian forests can be found in Oelbermann, Gordon, and Kaushik (2008).

Isolated Grasslands

In large areas in the Great Plains, particularly in the more xeric short- and mixed-grass prairies, grasses and forbs exist largely independently of any woody species. The same situation exists in some of the larger high-elevation meadows in the Rockies and, pre-settlement, the grasslands of the Central Valley of California and the Palouse Prairie of eastern Washington.

How small a grassland can be before adjacent woodlands have a significant effect is, of course, a key question in terms of agroforestry design. Long-term—i.e., hundreds or thousands of years—these grasslands did not remain isolated, as evidenced for example by the presence of conifers throughout the Great Plains during the various glacial periods (Axelrod, 1985).

General Ecological Principles

We can highlight four ecological principles that will be of particular use in designing and evaluating agroforestry practices:

1. Ecosystems are distinguished by spatial and temporal heterogeneity. An ecosystem or landscape consists of a mosaic of patches and linear components. The boundaries between patches are often the site of increased rates of processes such as nutrient and energy exchange, competition, water flow, and movement of organisms (Ranney, Bruner, & Levenson, 1981; Holland, Risser, & Naiman, 1991). For example, most of the removal of NO_3 from water entering a Georgia riparian forest occurred in the first 10 m of a 55-m-wide forest (Lowrance, 1992). Designers of agroforestry practices should pay particular attention to the interfaces of woody and non-woody components within their systems (Dix et al., 1995).

   Temporal variability is also important. Some variability, such as diurnal and seasonal environmental change or longer term successional change, is predictable and can easily be considered in designing practices; an example would be a winter wheat alley cropping practice in which the wheat completes most of its growth before the tree crop leafs out each spring (Chirko, Gold, Nguyen, & Jiang, 1996; Thevathasan & Gordon, 2004). Other sources of variability (e.g., drought) are less predictable but no less important to practice design and function.

2. Disturbance is a primary determinant of ecosystem structure and function. Ecosystems have adapted to various degrees and combinations of fire, drought, wind, flood, pest outbreaks, and other disturbances. Much of the heterogeneity in landscape structure is due to patterns of disturbance. Removal of a critical disturbance, for instance fire from an oak savanna, is a major disruption of system function and may trigger a structural shift to a closed forest (Bragg, Knapp, & Briggs, 1993).

   Management of agroforestry practices requires the appropriate application of disturbance to maintain the state that best meets management goals. In conventional row-crop agriculture, tillage, a type of disturbance rarely seen in natural ecosystems, is required to inhibit normal successional processes and trajectories. Agroforestry managers need to consider the use of disturbance agents that mimic the natural disturbance patterns that maintain the agroecosystem at later stages of succession. Similarly, agroforestry practitioners should adopt management strategies such as pruning and root disking to reduce competitive interactions between trees and crops.

   Agroforestry practices must also be designed to handle sporadic, though inevitable, environmental stresses such as drought, high wind, intense rain, and extreme cold. Windbreaks, riparian forests, silvopasture, and other agroforestry practices that add perennial crops and groundcover to the farm will generally increase the system's resilience and resistance to these stresses.

3. Perennialism is the most common condition in natural ecosystems. Annual plants dominate only after certain disturbances and are quickly replaced in the successional process by perennials. Disturbances severe enough to provide an opening for annuals also provide a window for accelerated loss of soil and nutrients from the system. These windows are generally short in natural systems, but if the disturbance is repeated regularly, as in row-crop agroecosystems, the cumulative losses of soil and nutrients can greatly reduce the productive capacity of the system.

   Agroforestry practices provide one means of adding perennials to a conventional row-crop farming system. There are many non-woody perennials such as grasses, alfalfa, and clover (*Trifolium* spp.) that can also provide benefits. An optimal agroecosystem design will consider all potential perennial crops as well as appropriate annual crops.

4. Structural and functional diversity are important to ecosystem performance but are difficult to quantify. If an ecosystem includes species whose roots exploit different soil depths or whose leaves capture sunlight at different heights in the canopy, this structural and functional diversity may increase the efficiency of the system in using resources and maintaining production. However, many species function similarly, so species diversity alone is a poor measure of functional diversity (Olson & Francis, 1995). Agroforestry provides an obvious way to increase the structural diversity of a row-crop farm. While this change in diversity will undoubtedly have functional effects, the farm manager needs to carefully consider the relationship between structure and function as it applies to their management goals. Random additions of woody perennials will increase species diversity but are unlikely to produce optimal economic or ecological results.

Indicators of Agroecosystem Sustainability

In the remainder of this chapter, we examine the system-level effects of adding agroforestry practices to a conventional farm, such as those related to microclimate, as illustrated in Figure 3–3.

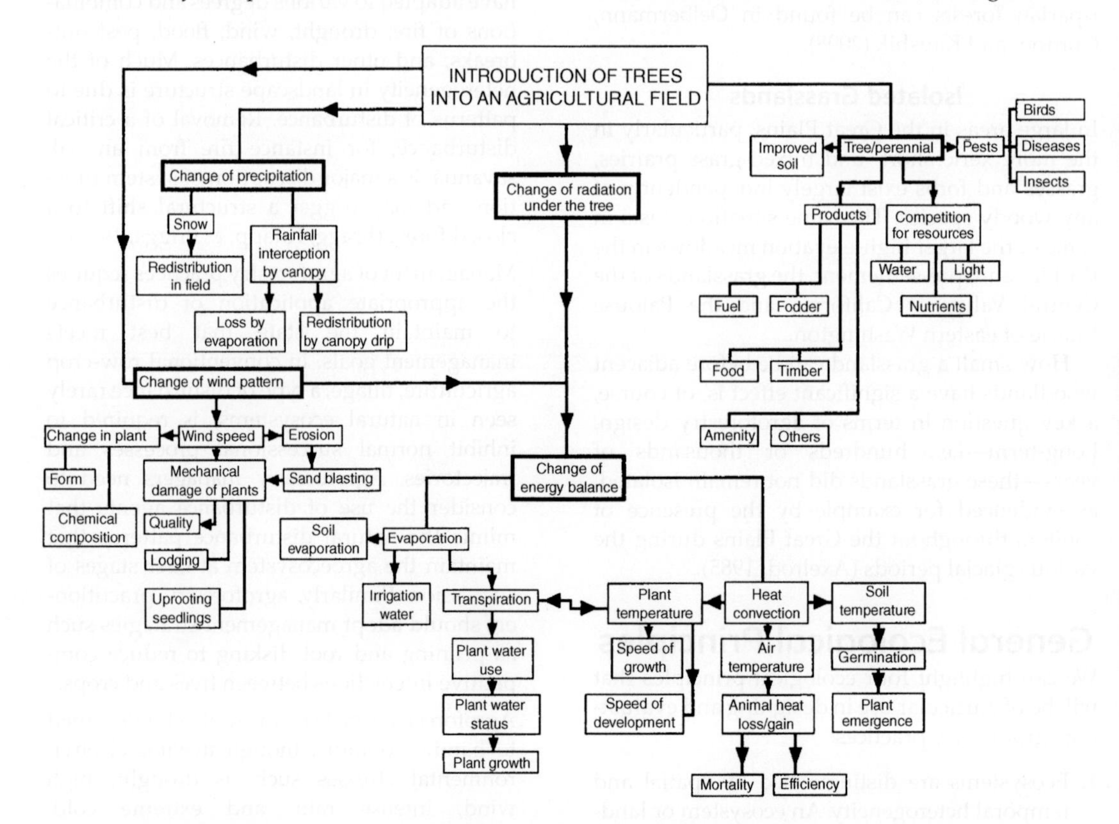

Fig. 3–3. Hypothetical changes in energy and nutrient fluxes, pools and conditions of existence, upon the introduction of trees via agroforestry systems into agricultural systems (from Brenner, 1996; reproduced with permission of CABI, Wallingford, UK).

However, to fully judge the effects, indicators of economic and social sustainability as well as of ecological sustainability must be considered. To be sustainable, an agroecosystem has to be profitable and it has to meet societal demands for food and fiber. If changes to a farm are made solely to improve the ecological trends illustrated in Table 3–3, the effect on the overall system may be negative. We should note that long-term gains may be justifiable reasons for introducing systems that in the short term may not be overly economically viable.

The issue of sustainability and the choice of indicators of agroecosystem condition have been considered frequently (Harrington, 1992; Lefroy & Hobbs, 1992; Stockle, Papendick, Saxton, Campbell, & van Evert, 1994; Campbell, Heck, Neher, Munster, & Hoag, 1995; Thevathasan et al., 2014). Although the debate continues about which group of indicators is most appropriate, there has been considerable convergence among the choices. We have compiled a suite of indicators (Table 3–4) based on our examination of Appendix 3-1, Odum (1985), (Table 3–3), and Francis, Aschmann, & Olson (1997), on indicators of functional sustainability of farms. This group of indicators reflects our summary view of agroecosystem sustainability, i.e., in an increasingly resource-poor world, farms that maintain a high rate of conversion of solar energy into marketable crops, minimize ancillary energy and material inputs, and preserve their natural capital (e.g., soil) will be the most sustainable.

Although it is fairly easy to determine which trend in an indicator favors sustainability, it is more difficult to quantify the particular values of an indicator that represent high or low sustainability. As indicated in the footnotes to Table 3–4, we set upper and lower bounds for our indicators based on benchmark farming systems in the region, such as irrigated continuous corn (e.g., high energy inputs), the properties of a particular soil (e.g., 11 Mg soil erosion per hectare is the tolerance limit for a Sharpsburg silty clay

Table 3–3. Trends expected in stressed ecosystems (Odum, 1985) and the evidence for these trends in a corn–soybean farm relative to a prairie or oak–hickory ecosystem (drawn from Appendix 3-1).

Trend	Farm characteristics in support
Energetics	
1. Community respiration increases	tillage increases decomposition of soil organic matter
2. P/R (production/respiration) becomes unbalanced (< or >1)	system production exceeds respiration due to export of net primary productivity (NPP) from system
3. P/B and R/B (maintenance/biomass structure) ratios increase	data not available
4. Importance of auxiliary energy increases	17.3×10^3 MJ ha^{-1} input (as fertilizer, fuel, labor, etc.)
5. Exported or unused primary production increases	450 g kg^{-1} (45%) of NPP exported as grain
Nutrient cycling	
6. Nutrient turnover increases	see no. 7
7. Horizontal transport increases and vertical cycling of nutrients decreases	internal N cycling decreases from 960 to 560 g kg^{-1} (96 to 56%) of total N flows
8. Nutrient loss increases (system becomes more "leaky")	loss of N from farm is 7 to 50 times greater than from natural ecosystems
Community structure	
9. Proportion of r-strategists increases	annual crops replace perennials
10. Size of organisms decreases	corn smaller than oak and soybean smaller than tall grasses
11. Lifespans of organisms or parts (e.g., leaves) decrease	crops are annuals
12. Food chains shorten	not shortened, but food web complexity likely reduced as one consumer (humans) co-opts almost half of NPP
13. Species diversity decreases and dominance increases	two species dominate
General system-level trends	
14. Ecosystem becomes more open (i.e., input and output environments become more important as internal cycling is reduced)	inputs of cultural energy and chemicals, and export of harvested crops are essential to system maintenance
15. Autogenic successional trends reverse (succession reverts to earlier stages)	system maintained at first year of secondary succession by annual tillage
16. Efficiency of resource use decreases	annual NPP reduced despite large inputs of external materials and energy
17. Parasitism and other negative interactions increase, and mutualism and other positive interactions decrease	chemical and energy inputs required to reduce specific pest populations on specific hosts
18. Functional properties (such as community metabolism) are more robust (homeostatic-resistant to stressors) than are species composition and other structural properties	despite drastic reduction in biodiversity and simplification of structure, system continues to be productive

Table 3–4. Selected indicators of sustainability for agroecosystems, and the indicator values for the conventional and agroforestry farms described in Table 3–5.

Indicator	Definition	Value indicating high sustainability	Value indicating low sustainability	Conventional farm	Agroforestry farm
Harvest [a]	weight of harvested crops and livestock, kg ha^{-1} (lb acre^{-1}) dry weight	7,952 (7,100)	0	3,805 (3,397)	3,923 (3,503)
Cultural energy input [b]	total non-solar energy inputs, MJ ha^{-1} (MJ acre^{-1})	0	59,259 (24,000)	17,264 (6,992)	14,091 (5,707)
Energy output/input [c]	ratio of energy in harvested crops to cultural energy inputs	5	<1	3.9	4.5
Energy capture efficiency [d]	energy in harvested crops as proportion of growing season PAR, %	1.0	0	0.38	0.35
Water use efficiency [e]	harvested biomass divided by AET (g m^{-1} mm^{-1})	1.15	0	0.61	0.61
Imported fertilizer [f]	N + P, kg ha^{-1} (lb acre^{-1})	0	151 (135)	44 (39)	26 (23)
N losses [g]	losses through erosion and leaching), kg ha^{-1} (lb acre^{-1})	0	45 (40)	28 (25)	20 (18)
Soil erosion [h]	wind + water, Mg ha^{-1} (tons acre^{-1})	0	11 (8)	11 (5)	7.8 (2.5)
N balance [i]	N inputs/N outputs	1	<0.8 or >1.2	0.6	0.64
P balance [j]	P inputs/P outputs	1	<0.8 or >1.2	0.7	0.46
Crop diversity [k]	no. of crops per farm	12	1	2	7
Hired labor [l]	h ha^{-1} (h acre^{-1})	0	5 (2)	1 (0.4)	5 (2)
Net income [m]	US$ ha^{-1} (US$ acre^{-1})	235 (95)	89 (36)	99 (40)	249 (101)
Capital borrowing [n]	debt/variable income	0	1	0.63	0.46
Farmer knowledge [o]	total skills and knowledge held by farm family	high	low	medium	high

[a] High value is dry weight of grain from Nebraska irrigated corn (9406 kg ha–1 (150 bu A–1)).

[b] The value indicating low sustainability is the energy input per hectare to produce irrigated corn in Nebraska (Pimentel, 1980).

[c] From Pimentel and Pimentel (1996), energy output/input ratio for U.S. soybean production is 4.15:1; Ohio alfalfa is 6.17:1; corn and wheat are around 2.5:1. So, 5:1 is a reasonable upper end to scale.

[d] Loomis & Connor (1992) showed that the theoretical maximum daily energy capture efficiency of a crop is 12% of photosynthetically active radiation (PAR). However, Tivy (1990, p. 109) wrote that only in exceptional cases do crop efficiencies exceed 2% PAR for an entire growing season, and efficiency in terms of economic yields is only 0.3 to 0.4%. If 2% capture of PAR is a high efficiency, then 1% PAR in harvest (50% of total net primary productivity harvested) is a high upper bound for energy capture efficiency.

[e] 1.15 is the water use efficiency for corn (grain only) on a central Iowa farm (Loomis & Connor, 1992).

[f] Irrigated corn yielding 9406 kg ha^{-1} (150 bu acre–1) would export 128 kg ha^{-1} (114 lb acre–1) N and 22 kg ha^{-1} (20 lb acre–1) P. g High value (45 kg ha^{-1} [40 lb acre–1]) is 2× the estimated N losses for corn on a central Iowa farm (Loomis & Connor, 1992).

[h] 11.2 Mg ha^{-1} (5 tons acre–1) is the soil loss tolerance (T-value) for a Sharpsburg silty clay loam with 4–6% slope.

[i] System outputs (harvest and losses) within ±20% of inputs (imported and N2 fixation) is considered close to balance (inputs/outputs = 1). Values greater than or less than 1 would indicate potential environmental problems or depletion of fertility.

[j] System outputs (harvest and losses) within ±20% of inputs (imported P) is considered close to balance (inputs/outputs = 1). Values greater than or equal to 1 would indicate potential environmental problems or a depletion of fertility.

[k] Bender (1994) grows 12 crops on his eastern Nebraska organic farm. Diversity of this magnitude is required to implement flexible rotations for weed control and fertility and to provide sod and pasture crops for grazing and erosion control.

[l] Irrigated corn in Nebraska requires 5 h labor ha–1 (2 h labor acre–1) (Selley, 1996).

[m] A 172-ha (425-acre) farm would have to generate $89 ha–1 ($36 acre–1`) in net income to keep a four-person family above the official poverty line ($15,141; U.S. Census Bureau, 1997, Table 732). An average size Nebraska cash grain farm (255 ha [630 acre]) generating $235 ha–1 ($95 acre–1) would be in the 90th percentile of net farm income for that type of farm (Johnson, 1995).

[n] A value of 1 indicates that the income remaining after fixed costs are covered is just sufficient to repay operating loans plus interest.

[o] This is very difficult to quantify, but it is assumed to be positively correlated with the number of crops and enterprises on the farm.

loam with 4–6% slope), or on economic benchmarks (e.g., the poverty level for a family of four). The goal is to ground the evaluations in a realistic assessment of the range of conditions in the region of interest.

Once indicators of sustainability have been defined, they can then be used to evaluate the effect of agroforestry practices on the sustainability of a farming system. Thevathasan et al. (2014) have suggested utilizing a common method for visualizing sustainability indices through the use of "amoeba diagrams," originally developed by Bell and Morse (2000). Amoeba diagrams are two-dimensional, multi-axis diagrams where the axis scale can be ordinal or relational (Figure 3–4). Using relational axes makes visual interpretation easier. In the absence of distinct values (or ranges of values) that are deemed thresholds of sustainability, data can be normalized against a reference state. The reference state may be determined by collecting information from a local site that reflects an ideal state of the ecosystem. This could be a site that has minimal disturbance and native vegetative cover, or it could be farmland that is currently managed under the best management practices.

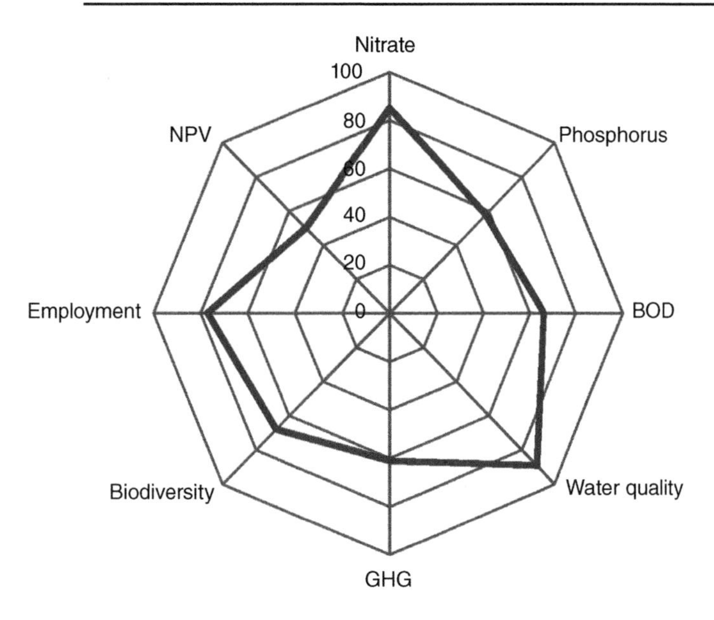

Fig. 3–4. Example of an amoeba diagram (NPV, net present value; BOD, biological oxygen demand; GHG, greenhouse gas) (adapted from Bell & Morse, 2000).

Table 3–5. Characteristics of two model farms in eastern Nebraska representing a conventional cash grain operation and an agroforestry alternative, both on a Sharpsburg silty clay loam with 4–6% slope.

Characteristic	Conventional farm	Agroforestry farm
Size, ha (acres)	264 (650)	172 (425)
Rented land, %	55	0
Crops, ha (acres)		
Corn	132 (325)	34 (83)
Soybean	132 (325)	61 (151)
Grain sorghum		34 (83)
Alfalfa		24 (60)
Christmas trees		4 (9)
Hazel nut production		6 (16)
Windbreaks		9 (23)
Area in perennials, %	0	25

Amoeba diagrams do not provide a composite value for sustainability. They are a visual representation that effectively gives equal weight to each index that will allow comparison and interpretation. Collecting the same set of data on the sustainable indicators with time, the user can see which areas are improving and which are declining while still getting a sense of the overall sustainability of the system.

Sustainability indices can also be assessed in more quantitative terms. We have undertaken a quantitative comparison of two synthetic farms modeled from regional data (Table 3–5). One of the synthetic farms is the conventional corn–soybean farm described in Appendix 3-1, while the other is a more diversified farm that incorporates windbreaks, an herbaceous perennial crop, and two woody perennial crops in block plantings.

The size and machinery complement of each synthetic farm was determined from a survey and analysis of Nebraska farms (Bernhardt, 1994), and a schedule of operations was developed for each farm based on best management practices for east-central Nebraska. The economic performance of the two systems was then quantified with a model developed by Olson (1998), and erosion and nutrient losses were evaluated with PLANETOR, a farm-scale environmental and economic model (Center for Farm Financial Management, University of Minnesota). Energy and nutrient budgets for each farm were compiled from published values of the embodied energy of farm inputs (Pimentel, 1980) and crop nutrient and energy contents (Church, 1984; Holland, Welch, et al., 1991). The values of each indicator for the two farms are given in Table 3–4.

A rapid appraisal of Table 3–4 suggests that the agroforestry farm is more sustainable than the conventional corn–soybean farm. Although the systems perform similarly as measured by production indicators (e.g., harvest, energy capture efficiency, water use efficiency), the agroforestry farm does better economically (net income, capital borrowing) and in some measures of resource conservation (e.g., erosion, N loss). Neither system has a sustainable nutrient balance in that each exports considerably more N and P than it imports.

Of course, there is no way to tell from system-level indicators how much of the improvement in the performance of the agroforestry farm is due to its woody perennial components. The underlying performance data (not shown) indicate that the tree components had a major impact on economic returns. Christmas trees and hazelnuts (*Corylus* L.) were very profitable, and windbreaks increased crop yields more than enough to compensate for the land taken out of production. Tree crops (with grassed alleys) eliminated water erosion on the land they occupied, although for the whole farm, alfalfa was equally important in reducing water erosion. Windbreaks provided no benefit in reducing wind erosion because soil loss by wind is insignificant on these soils when adequate residue is left each fall.

A final observation concerns the definition of agroforestry. The windbreaks on this model farm, by interacting with the field crops (biologically and physically), clearly meet the definition of agroforestry. The Christmas trees and hazelnut shrubs, although woody perennials, are planted in blocks and may have only minimal biophysical interaction with other components of the farming system. Does the inclusion of block plantings of trees on a farm necessarily constitute agroforestry? Not by the definition given earlier in this chapter (see Gold & Garrett, 2008), although other definitions of agroforestry would accept such a system on the landscape if it was developed in a temporal sense (Gordon, Newman, Coleman, & Thevathasan, 2018).

Without question, when the distribution of labor is considered (data not shown) on these two farms, there are advantages to having incorporated woody perennials into the farm system. The conventional farmer is very busy in the spring and early fall, with much less to do in-between. On the agroforestry farm, the hazelnuts require a great deal of labor for harvest in late July and early August, and Christmas tree sales provide work in late November and December. The inclusion of block plantings of tree crops represents both an economic and a social interaction with other components of the farm but not necessarily one of a biophysical nature. Agroforestry, in North America, is currently defined in terms of five individual practices, with a sixth one added recently (see Chapter 2); however, as it continues to evolve, a broader definition at farm and landscape scales may become appropriate.

Ecological Goods and Services

Ecological goods and services are defined as logical benefits resulting from the "normal" functioning of an ecosystem. Maximum production of such goods and services is associated with unstressed agroecosystems, which within the context of agroforestry would constitute a variety of temporal and spatial configurations of trees on the farming landscape. Humans benefit from the maintenance of these goods (e.g., fresh water) within the ecosystem, and the "flow" of these services (e.g., greenhouse gas mitigation) to other systems.

Agroforestry systems, regardless of type, are capable of providing numerous ecological goods and services, of a range of complexities, over long periods of time (Hunt, 2005; Jose, 2009; Nair, Gordon, & Mosquera-Losada, 2008). Indeed, agroforestry systems can be designed and engineered to provide specific quantities of particular goods and services. Nonetheless, the universal application of ecological principles to agroforestry system design and management is nearly impossible as a result of the many varied types of systems in existence—from riparian management systems that link terrestrial and aquatic systems to more traditional systems that integrate perennial plants with annual crops, with or without animals. The broad geographical range across which agroforestry systems may be successfully implemented and the scale at which interactions occur—from landscape to individual plant—also complicates the development of a universal understanding of nutrient and energy flows and the relationship of these to system productivity.

Although systems will differ in the nature and types of environmental services provided, some generalizations can be stated. Most agroforestry systems will tend to improve soils, including productivity, largely through the incorporation of organic matter and C into upper soil profiles from the production of annual litterfall from the tree component. As a result of the presence of perennial root systems, soil erosion relative to monocropped agroecosystems is often minimized.

With respect to the maintenance and proliferation of biodiversity, agroforestry systems often enhance the components of biodiversity, at scales ranging from the stand (farm level) to regional landscapes (Jose, 2009; Nair et al., 2008). Jose (2009) categorized the contributions of agroforestry systems toward biodiversity conservation into five major roles, which include: (a) species habitat provisioning; (b) germplasm preservation for sensitive species; (c) reduction of the rates of natural habitat loss through providing a more productive and sustainable alternative to traditional agricultural systems that may involve clearing natural habitats; (d) creation of corridors to connect habitat remnants for floral and faunal species; and (e) prevention of land degradation and habitat loss through provisioning additional ecosystem services such as erosion control and soil health enhancements. The potential for agroforestry systems to contribute to biodiversity conservation is especially high in the fragmented agricultural landscapes of North America, where widespread conventional agricultural practices are undoubtedly contributing to species decline across the continent. Gibbs et al. (2016) found that in a mature temperate tree-based intercropping system, avian species richness was nearly 1.5 times greater than in a conventionally managed sole-crop agricultural system (32 vs. 23 unique bird species). Additionally, avian diversity was more than twice as great in the tree-based intercropping system than in the sole-crop agricultural system (Shannon–Wiener diversity index values of 2.9 vs. 1.2). Tree-based intercropping systems may therefore be one method for slowing or even reversing the decline of migratory bird populations in North America, as seen during the last several decades.

Agroforestry systems are being promoted as a means of mitigating climate change as a result of their C sequestration potentials. In all systems, the storage of C is enhanced (Jose, 2009, 2019; Nair et al., 2008; Thevathasan & Gordon, 2004), not only through the perennial nature of the trees, but also through increased soil C storage. The C sequestration potential for agroforestry systems is dependent on the type of agroforestry system, in addition to species composition and age, geographic location, environmental factors, as well as system management practices (Jose, 2009). A 2006 study examining the C sequestration potentials in a 13-yr-old temperate tree-based intercropping system found that the carbon sequestration potential of systems incorporating barley (*Hordeum vulgare* L. 'OAC Kippen') and hybrid poplar (*Populus deltoids* × *Populus nigra* clone DN-177) were four times greater than in a barley and Norway spruce (*Picea abies* L.) system and five times greater than the examined sole-cropped barley system, with net C fluxes of 13.2, 1.1, and −2.9 Mg C ha^{-1} annually (Peichl, Thevathasan, Gordon, Huss, & Abohassan, 2006). Wotherspoon, Thevathasan, Gordon, and Voroney (2014), utilizing the same research site, also quantified the C sequestration potential of the then 25-yr-old temperate tree-based intercropping system and found net C fluxes for the soybean and hybrid poplar, soybean and Norway spruce, and sole-crop soybean systems of 2.1, 1.6, and −1.2 Mg C ha^{-1} yr^{-1}.

In addition to enhancing system-level C sequestration, agroforestry systems may also contribute to reduced greenhouse gas emissions (Thevathasan & Gordon, 2004). Through reduced fertilizer requirements and more efficient N cycling in tree-based intercropping systems, N_2O emissions reductions of nearly 1 kg ha^{-1} yr^{-1} compared with conventionally managed agricultural fields have been reported (Evers et al., 2010). Graungaard (2015) found that both tree species and proximity to trees influenced soil microbial communities. This study utilized a modified denitrification enzyme assay, which indicated a greater potential for N_2O production within tree-based intercropping systems comprised of hybrid poplar versus red oak (*Quercus rubra* L.). Tree species themselves are associated with unique microbial communities within agroforestry systems, which may play a role in ecosystem functioning, including N_2O and other greenhouse gas emissions.

Agroforestry systems also have the potential to reduce agricultural runoff, reducing sedimentation, nutrient runoff, and the leaching of pesticides into nearby waterbodies and beyond, contributing to eutrophication in the Great Lakes, the Gulf of Mexico, and elsewhere across the continent (Jose, 2009). Improving the quality of surface water that is adversely affected by runoff from heavily fertilized row-crop and pasture systems is an environmental benefit of agroforestry systems that is just beginning to be realized in quantitative terms (Michel, Nair, & Nair, 2007). Integrated riparian management systems address the interaction of terrestrial and aquatic environments in farming landscapes and can make major contributions to water quality at local scales and provide connectivity in agricultural landscapes at much larger scales (Schultz et al., 2000). Riparian buffers are able to reduce non-point-source pollution from agricultural fields through reduced runoff velocity and promotion of infiltration, increased nutrient retention through trees utilizing excess nutrients transported in runoff, and increased sediment deposition on land (Jose, 2009).

In intercropping systems, microclimate modification is common, and although the competition for water, light, and nutrient resources between the tree and crop components is complicated, improved and sustained crop yields have been noted (Thevathasan & Gordon, 2004). Due to enhanced structural diversity within agroforestry systems, microclimatic modifications and therefore plant growing conditions are not homogeneous as they are within conventional agricultural systems. A recent study by Coleman et al. (2020) found that both abiotic (light, soil moisture) and biotic (available soil nutrients) gradients within a 26-yr-old tree-based intercropping system intercropped with concentrated short-rotation willow (SV1; *Salix dasyclados* Wimm.) had significant influences on intraspecific variation in crop leaf traits, including increased specific leaf area and crop leaf N concentrations closest to the tree rows. These results contribute to an enhanced understanding of nutrient cycling within agroforestry systems and indicate that tree litter inputs may reduce the need for crop amendments, especially near the tree rows.

The presence of trees within agroforestry systems can also have more indirect influences on nutrient cycling and soil fertility. Price and Gordon (1998) examined the spatial and temporal distribution of earthworms in an 11-yr-old tree-based intercropping system planted with silver maple, white ash (*Fraxinus americana* L.), and hybrid poplar, in combination with soybean. The researchers found the greatest density of earthworms within the tree rows, with typically decreasing earthworm density towards the middle of the cropping alley. Earthworm density was drastically reduced in the summer, potentially tracking with reduced food availability (litterfall) and reduced soil moisture compared to the spring, and earthworm distribution tended to become more uniform during the summer. The authors found that earthworm density was highest near poplars, providing further evidence of the importance of tree species selection when considering soil fertility and other ecosystem functions.

Many additional goods and services can be provided by the suite of recognized agroforestry practices, including odor control (Tyndall & Grala, 2009), opportunities to embrace integrated pest management systems with reduced pesticide input (Diaz-Forestier, Gomez, & Montenegro, 2009), and the control of *Escherichia coli* outbreaks associated with manure application (Dougherty, 2007). Agroforestry systems can also enhance nutrient cycling and nutrient use efficiency with subsequent improvements in downstream water quality and reduced requirements for crop amendments (Jose, 2009). Thevathasan and Gordon (1997) utilized a 7–9-yr-old hybrid poplar tree based intercropping system planted with barley and found that mean nitrification rates, N availability, and C content were higher in soils closest to the poplar tree rows compared with the middle of the crop alley. It was also found that soil nitrification rates, soil C, and plant N uptake adjacent to the tree rows were influenced by the leaf biomass inputs of the preceding year, potentially contributing to increased aboveground biomass and greater grain N concentration in the barley intercrop.

In natural systems, a long-term ecological approach has proven useful to understanding the importance of (a) slow processes that occur on the scale of decades to centuries, (b) processes with high annual variability, (c) rare and unique events, (d) subtle processes, and (e) complex processes with many interacting factors. A long-term ecological research perspective also holds much potential for helping us understand agroforestry systems. The temporal context provided by engaging in such research can aid us greatly in understanding large-scale changes in ecosystem processes and thereby reveal the secrecy inherent in what has been termed "the invisible present" (Magnuson, 1990).

Such an approach to understanding the structure and function of agroforestry systems and the relationship of these parameters to net primary productivity is a strong foundation upon which to evaluate the production of ecological goods and services over long periods of time (Gordon & Jose, 2008).

Conclusions

Agroforestry offers a means of regaining some of the structural and functional characteristics that contribute to the sustainability of natural ecosystems that have been lost in the conversion of those ecosystems to homogeneous agroecosystems. An understanding of the structure and function of natural ecosystems is essential to the successful implementation of agroforestry if we wish to create more heterogeneous agroecosystems.

A complete knowledge of the many ecological processes and interactions responsible for a natural system's sustainability will always elude us—an ecosystem is just too complex. However, perennialism and a high proportion of area in mid- to late-successional states is the usual condition of natural ecosystems and an obvious goal in designing a sustainable agroecosystem. We are making progress in determining how to meet that goal, but there is much left to be learned in both basic and applied ecology. With respect to the inclusion of ecological principles within the broad field of agroforestry, some important areas warranting further research include: (a) the evaluation of net primary productivity (NPP = increment + litterfall + herbivory + mortality) for all types of agroforestry systems in different geographical regions; (b) the continued study of C sequestration and the emission of greenhouse gases such as CO_2, N_2O, and CH_4 in agroforestry systems; (c) the study of belowground interactions and processes in the realm of microbial ecology, root competition, and mycorrhizal associations; (d) the study of both positive and negative interactions among trees, shrubs, grasses, and forbs in agroforestry systems; (e) the implementation of long-term, system-level experiments and on-farm demonstrations including stronger and more specific economic analyses that include the value of all ecological goods and services; (f) the study of albedo—reflectance changes occurring at the landscape level as a result of agroforestry adoption that may have implications for global warming and C sequestration scenarios; (g) a comprehensive evaluation of the biology and economics of agroforestry on a variety of sites; (h) the study of all aspects of silvopastoral systems especially as they relate to potential lowering of wood quality and value, forest regeneration, greenhouse gas emissions, and the issue of animal welfare; (i) a continued screening of useful pharmaceutical and other chemical products from forest farming systems; and (j) a comprehensive evaluation of conservation biology principles—how can we incorporate information gleaned from natural forest systems that remain in our agricultural landscapes into agricultural systems that embrace woody perennials? A better appreciation of the ecosystem and environmental services provided through agroforestry would lead to the development of rigorous environmental and economic assessments and eventually to modification of community tax structures and environmental legislation.

The ecology of highly managed agroforestry systems is becoming better understood although much remains to be done. Two recent texts on the ecology of agroforestry systems (Batish, Kohli, Jose, & Singh, 2008; Jose & Gordon, 2008) present up-to-date research results on ecological interactions, belowground ecological processes, resource allocation and partitioning, and the modeling of these in both tropical and temperate agroforestry systems.

Acknowledgments

We wish to sincerely thank and acknowledge and are indebted to R.K. Olson, M.M. Schoeneberger, and S.G. Aschmann for their contributions to parts of this chapter. We have reproduced verbatim or with slight modifications of their writings in the first edition of this chapter. Similarly, we are also indebted to Caron et al. (2008), as some portions in this chapter have been adapted from their publication.

References

Anderson, G. W., Moore, R. W. & Jenkins, P. J. (1988). The integration of pasture, livestock and widely-spaced pine in south west Western Australia. *Agroforestry Systems, 6,* 195–211. https://doi.org/10.1007/BF02344759

Atzet, R., & Waring, R. H. (1970). Selective filtering of light by coniferous forests and minimum light energy requirements for regeneration. *Canadian Journal of Botany, 48,* 2163–2167. https://doi.org/10.1139/b70-312

Axelrod, D. I. (1985). Rise of the grassland biome, central North America. *The Botanical Review, 51,* 163–201. https://doi.org/10.1007/BF02861083

Batish, D. R., Kohli, R. K., Jose, S., & Singh, H. P. (Eds.). 2008. *Ecological basis of agroforestry.* Boca Raton, FL: CRC Press.

Bell, S., & Morse, S. (2000). *Sustainability indicators: Measuring the immeasurable.* London: Earthscan.

Belsky, A. J. (1994). Influences of trees on savanna productivity: Tests of shade, nutrients, and tree–grass competition. *Ecology, 75,* 922–932. https://doi.org/10.2307/1939416

Bender, J. (1994). *Future harvest.* Lincoln, NE: University of Nebraska Press.

Bernhardt, K. J. (1994). *Sustainable agriculture impact study.* 1994 ACE Project annual report. Lincoln, NE: Department of Agricultural Economics, University of Nebraska.

Billings, W. D. (1969). Vegetational pattern near alpine timberline as affected by fire–snowdrift interactions. *Vegetatio, 19,* 192–207. https://doi.org/10.1007/BF00259010

Blank, J. L., Olson, R. K., & Vitousek, P. M. (1980). Nutrient uptake by a diverse spring ephemeral community. *Oecologia, 47,* 96–98. https://doi.org/10.1007/BF00541781

Bragg, W. K., Knapp, A. K., & Briggs, J. M. (1993). Comparative water relations of seedling and adult *Quercus* species during gallery forest expansion in tallgrass prairie. *Forest Ecology and Management, 56,* 29–41. https://doi.org/10.1016/0378-1127(93)90101-R

Brandle, J. R., & Finch, S. (1991). *How windbreaks work* (EC 91-1763-B). Lincoln, NE: University of Nebraska Extension.

Braun, E. L. (1967). *Deciduous forests of eastern North America.* New York: Hafner Publishing Co.

Brenner, A. J. (1996). Microclimatic modifications in agroforestry. In C. K. Ong & P. Huxley (Eds.), *Tree-crop interactions: A physiological approach* (pp. 159–187). Wallingford, U.K.: CAB International.

Campbell, C., Heck, W., Neher, D., Munster, M., & Hoag, D. (1995). Biophysical measurement of the sustainability of temperate agriculture. In M. Munasinghe and W. Shearer (Eds.), Defining and measuring sustainability: The biophysical foundations (pp. 251–276). New York: United Nations University.

Caron, L., Duchesne, L., Gordon, A. M., Khasa, D., Kort, J., Olivier, A., . . . Vezina, A. (2008). Agroforestry. In R. Doucet (Ed.), *Manual of forestry for Quebec.* Quebec City, QC, Canada: University of Laval.

Chirko, C. P., Gold, M. A., Nguyen, P. V., & Jiang, J. P. (1996). Influence of direction and distance from trees on wheat yield and photosynthetic photon flux density (*Qp*) in a *Paulownia* and wheat intercropping system. *Forest Ecology and Management, 83,* 171–180. https://doi.org/10.1016/0378-1127(96)03721-8

Church, D. C. (1984). *Livestock feeds and feeding*. Corvallis, OR: O&B Books.

Cionco, R. M. (1985). Modeling windfields and surface layer wind profiles over complex terrain and within vegetative canopies. In B. A. Hutchinson & B. B. Hicks (Eds.), The forest–atmosphere interaction (pp. 501–520. Boston: D. Reidel. https://doi.org/10.1007/978-94-009-5305-5_30

Coleman, B., Bruce, K., Chang, Q., Frey, L., Guo, S., Tarannum, M.S., . . . Thevathasan, N. (2018). Quantifying C stocks in high-yield, short-rotation woody crop production systems for forest and bioenergy values and CO_2 emission reduction. *The Forestry Chronicle, 94*, 260–268. https://doi.org/10.5558/tfc2018-039

Coleman, B., Martin, A., Thevathasan, N., Gordon, A., and Isaac, M. (2020). Leaf trait variation and decomposition in short-rotation woody biomass crops under agroforestry management. *Agriculture, Ecosystems & Environment, 298*, 106971. https://doi.org/10.1016/j.agee.2020.106971.

Daubenmire, R. (1978). *Plant geography*. New York: Academic Press.

Davies, K. M., Jr. (1994). Some ecological aspects of northeastern American Indian agroforestry practices. *Northern Nut Growers Association Annual Report, 85*, 25–37.

Diaz-Forestier, J., Gomez, M., & Montenegro, G. (2009). Nectar volume and floral entomofauna as a tool for the implementation of sustainable apicultural management plans in *Quillaja saponaria* Mol. *Agroforestry Systems, 76*, 149–162.

Dix, M. E., Johnson, R. J., Harrell, M. O., Case, R. M., Wright, R. J., Hodges, L., . . . Hubbard, K. G. (1995). Influences of trees on abundance of natural enemies of insect pests: A review. *Agroforestry Systems, 29*, 303–311. https://doi.org/10.1007/BF00704876

Dougherty, M. C. (2007). Nitrate, ammonium and *Escherichia coli* NAR levels in mixed tree intercrop and monocrop systems (Doctoral dissertation). Retrieved from the Atrium repository (http://atrium.lib.uoguelph.ca/xmlui/handle/10214/19663).

Duke, J. A. (1989). *Ginseng: A concise handbook*. Algonac, MI: Reference Publications.

Dupraz, C., Lawson, G. J., Lamersdorf, N., Papanastasis, V. P., Rosati, A., & Ruiz-Mirazo, J. (2018). Temperate agroforestry: The European way. In A. M. Gordon, S. M. Newman, & B. R. W. Coleman (Ed.), *Temperate agroforestry systems* (2nd ed., pp. 98–152). Wallingford, U.K.: CAB International.

Eichenberger, M. D., & Parker, G. R. (1976). Goldenseal (*Hydrastis canadensis* L.) distribution, phenology and biomass in an oak–hickory forest. *Ohio Journal of Science, 76*, 204–210.

Evers, A. K., Bambrick, A., Lacombe, S., Dougherty, M. C., Peichl, M., Gordon, A. M., . . . Bradley, R. L. (2010). Potential greenhouse gas mitigation through temperate tree-based intercropping systems. *The Open Agriculture Journal, 4*, 49–57. https://doi.org/10.2174/1874331501004010049

Forman, R. T. T., & Godron, M. (1986). *Landscape ecology*. New York: John Wiley & Sons.

Francis, C. A., Aschmann, S. G., & Olson, R. K. (1997). Indicators of functional sustainability of farms and watersheds. In J. L. Steiner, A. J. Franzluebbers, C. B. Flora, & R. Janke (Ed.), Interactions: Investigating ecosystem dynamics at a watershed level: Proceedings, Athens, GA (pp. 64–66). Watkinsville, GA: USDA–ARS, Southern Piedmont Conservation Research Center.

Fritschen, L. J. (1985). Characterization of boundary conditions affecting forest environmental phenomena. In B. A. Hutchinson & B. B. Hicks (Eds.), *The forest–atmosphere interaction* (pp. 3–23). Boston: D. Reidel. https://doi.org/10.1007/978-94-009-5305-5_1

Garrett, H. E., & Harper, L. S. (1998). The science and practice of black walnut agroforestry in Missouri, USA: A temperate zone assessment. In L. E. Buck, J. P. Lassoie, & E. C. M. Fernandes (Ed.), *Agroforestry in sustainable agricultural systems* (pp. 97–110). Boca Raton, FL: CRC Press. https://doi.org/10.1201/9781420049473.ch5

Garrett, H. E., & Kurtz, W. B. (1983). Silvicultural and economic relationships of integrated forestry–farming with black walnut. *Agroforestry Systems, 1*, 245–256. https://doi.org/10.1007/BF00130610

Garrett, H. E., Kurtz, W. B., Buck, L. E., Lassoie, J. P., Gold, M. H., Pearson, H. A., . . . Slusher, J. P. (1994). *Agroforestry: An integrated land-use management system for production and farmland conservation*. Washington, DC: Natural Resource Conservation Service.

Gibbs, S., Koblents, H., Coleman, B., Gordon, A., Thevathasan, N., & Williams, P. (2016). Avian diversity in a temperate tree-based intercropping system from inception to now. *Agroforestry Systems, 90*, 905–916 [erratum: *90*, 917]. https://doi.org/10.1007/s10457-016-9901-7

Gold, M. A., & Garrett, H. E. (2008). Agroforestry nomenclature, concepts and practices. In H. E. Garrett (Ed.), *North American agroforestry: An integrated science and practice* (2nd ed.). Madison, WI: ASA. https://doi.org/10.2134/2009.northamericanagroforestry.2ed

Gordon, A. M., & Jose, S. (2008). Applying ecological knowledge to agroforestry design: A synthesis. In S. Jose and A. M. Gordon (Eds.), Towards agroforestry design: An ecological approach (pp. 301–306). Dordrecht, the Netherlands: Springer. https://doi.org/10.1007/978-1-4020-6572-9_18

Gordon, A. M., Newman, S. M., Coleman, B. R. W., & Thevathasan, N. V. (2018). Temperate agroforestry: An overview. In A. M. Gordon, S. M. Newman, & B. R. W. Coleman (Ed.), *Temperate agroforestry systems* (2nd ed., pp. 1–6). Wallingford, U.K.: CAB International.

Graham, J., Voroney, P., Coleman, B., Deen, B., Gordon, A., Thimmanagari, M., & Thevathasan, N. (2019). Quantifying soil organic carbon stocks in herbaceous biomass crops grown in Ontario, Canada. *Agroforestry Systems, 93*, 1627–1635. https://doi.org/10.1007/s10457-018-0272-0

Graungaard, K. C. (2015). *Bacterial communities associated with the cycling of nitrogen in a tree-based intercropping system* (Master's thesis). Retrieved from University of Guelph Atrium repository (http://hdl.handle.net/10214/9283).

Gray, R. (2000). *Root distribution of hybrid poplar in a temperate agroforestry intercropping system* (Master's thesis). Guelph, ON: Department of Environmental Biology, University of Guelph.

Greller, A. M. (1988). Deciduous forest. In M. G. Barbour & W. D. Billings (Eds.), *North American terrestrial vegetation* (pp. 287–316). New York: Cambridge Univ. Press.

Hamerlynck, E. P., & Knapp, A. K. (1996). Environmental and physiological factors influencing the distribution of oaks near the edge of their range. In D. H. Hartnett (Ed.), Proceedings of the 14th North American Prairie Conference: Prairie biodiversity (pp. 17–20). Manhattan, KS: Kansas State University.

Harrington, L. (1992). Measuring sustainability: Issues and alternatives. *Journal of Farming Systems Research–Extension, 3*, 1–19.

Harris, B. (1986). *Growing shiitake commercially*. Madison, WI: Science Tech.

Harris, W. F., Sollins, P., Edwards, N. T., Dinger, B. E., & Shugart, H. H. (1975). Analysis of carbon flow and productivity in a temperate deciduous forest ecosystem. In *Productivity of world ecosystems: Proceedings of a symposium* (pp. 116–122). Washington, DC: National Academies Press.

Heichel, G. H. (1978). Stabilizing agricultural energy needs: Role of forages, rotations, and nitrogen fixation. *Journal of Soil and Water Conservation, 33*(6), 279–282.

Hicks, D. J., & Chabot, B. F. (1985). Deciduous forest. In B. F. Chabot & H. A. Mooney (Ed.), *Physiological ecology of North American plant communities* (pp. 257–277). New York: Chapman and Hall.

Holland, M. M., Risser, P. G., & Naiman, R. J. (Ed.) (1991). *Ecotones: The role of landscape boundaries in the management and restoration of changing environments.* New York: Chapman and Hall.

Holland, B., Welch, A. A., Unwin, I. D., Buss, D. H., Paul, A. A., & Southgate, D. A. T. (1991). The composition of foods. Cambridge, U.K.: Royal Society of Chemistry.

Hunt, S. L. (2005). *Algorithm development for the quantification of the environmental impacts of agroforestry and the development of a framework for agroforestry evaluation.* Indian Head, SK: Agriculture and AgriFood Canada.

Jackson, L. E., Strauss, R. B., Firestone, M. K., & Bartolome, J.W. (1990). Influence of tree canopies on grassland productivity and nitrogen dynamics in deciduous oak savanna. *Agriculture, Ecosystems & Environment, 32*, 89–105. https://doi.org/10.1016/0167-8809(90)90126-X

Jeltsch, F., Milton, S., Dean, W. D. J., & Van Rooyen, N. (1996). Tree spacing and coexistence in semiarid savannas. *Journal of Ecology, 84*, 583–595. https://doi.org/10.2307/2261480

Johnson, A. S. (1989). The thin green line: Riparian corridors and endangered species in Arizona and New Mexico. In G. Mackintosh (Ed.), *Preserving communities and corridors* (pp. 35–46). Washington, DC: Defenders of Wildlife.

Johnson, B. (1995). A financial profile of Nebraska farm businesses (Report 140492). Lincoln, NE: Department of Agricultural Economics, University of Nebraska. Retrieved from http://ageconsearch.umn.edu/record/140492/files/un-l-rr-174.pdf

Jose, S. (2009). Agroforestry for ecosystem services and environmental benefits: An overview. *Agroforestry Systems, 76*, 1–10. https://doi.org/10.1007/s10457-009-9229-7

Jose, S. (2019). Environmental impacts and benefits of agroforestry. In *Oxford Research Encyclopedia of Environmental Science*. https://doi.org/10.1093/acrefore/9780199389414.013.195

Jose, S., & Gordon, A. M. (Eds.). 2008. *Toward agroforestry design: An ecological approach*. Dordrecht, the Netherlands: Springer. https://doi.org/10.1007/978-1-4020-6572-9

Keever, C. (1950). Causes of succession on old fields of the Piedmont, North Carolina. *Ecological Monographs, 20*, 229–250. https://doi.org/10.2307/1948582

Kleinman, P. J. A., Pimentel, D., & Bryant, R. B. (1995). The ecological sustainability of slash-and-burn agriculture. *Agriculture, Ecosystems & Environment, 52*, 235–249. https://doi.org/10.1016/0167-8809(94)00531-I

Lefroy, E., & Hobbs, R. (1992). Ecological indicators for sustainable agriculture. *Australian Journal of Soil and Water Conservation, 5*, 22–28.

Loomis, R. S., & Connor, D. J. (1992). *Crop ecology*. Cambridge, U.K.: Cambridge University Press. https://doi.org/10.1017/CBO9781139170161

Lowrance, R. (1992). Groundwater nitrate and denitrification in a Coastal Plain riparian forest. *Journal of Environmental Quality, 21*, 401–405. https://doi.org/10.2134/jeq1992.0047 2425002100030017x

Lowrance, R. R., Leonard, R. A., Asmussen, L. E., & Todd, R. L. (1985). Nutrient budgets for agricultural watersheds in the southeastern Coastal Plain. *Ecology, 66*, 287–296. https://doi.org/10.2307/1941330

Lowrance, R., Todd, R., Fail, J., Jr., Hendrickson, O., Jr., Leonard, R., & Asmussen, L. (1984). Riparian forests as nutrient filters in agricultural watersheds. *BioScience, 34*, 374–377. https://doi.org/10.2307/1309729

Magnuson, J. J. (1990). Long-term ecological research and the invisible present. *BioScience, 40*, 495–501. https://doi.org/10.2307/1311317

Maser, C. (1994). *Sustainable forestry*. Delray Beach, FL: St. Lucie Press.

McPherson, J. K., & Thompson, G. L. (1972). Competitive and allelopathic suppression of understory by Oklahoma oak forests. *Bulletin of the Torrey Botanical Club, 99*, 293–300. https://doi.org/10.2307/2997071

Michel, G. A., Nair, V. D., & Nair, P. K. R. (2007). Silvopasture for reducing phosphorus loss from subtropical sandy soils. *Plant and Soil, 297*, 267–276. https://doi.org/10.1007/s11104-007-9352-z

Mize, C. W., Brandle, J. R., Schoenberger, M. M., & Bentrup, G. (2008). Ecological development and function of shelterbelts in temperate North America. In S. Jose and A. M. Gordon (Eds.), *Towards agroforestry design: An ecological approach* (pp. 27–54). Dordrecht, the Netherlands: Springer. https://doi.org/10.1007/978-1-4020-6572-9_3

Nair, P. K. R. (1993). *An introduction to agroforestry*. Dordrecht, the Netherlands: Kluwer Academic. https://doi.org/10.1007/978-94-011-1608-4

Nair, P. K. R., Gordon, A. M., & Mosquera-Losada, M.-R. (2008). Agroforestry. In S. E. Jørgensen & B. D. Fath (Eds.), *Encyclopedia of ecology: Vol. 1. Ecological engineering* (pp. 101–110). Oxford, U.K.: Elsevier.

National Academy of Science. (1972). *Genetic vulnerability of major crops*. Washington, DC: National Academies Press.

Natural Resources Conservation Service. 2015. 2015 National Resources Inventory [Database]. Washington, DC.

Odum, E. P. (1971). *Fundamentals of ecology*. Philadelphia, PA: W.B. Saunders.

Odum, E. P. (1985). Trends expected in stressed ecosystems. *BioScience, 35*, 419–422. https://doi.org/10.2307/1310021

Oelbermann, M., Gordon, A. M., & Kaushik, N. K. (2008). Biophysical changes resulting from 16 years of riparian forest rehabilitation: An example from the southern Ontario agricultural landscape. In S. Jose and A. M. Gordon (Eds.), *Towards agroforestry design: An ecological approach* (pp. 13–26). Dordrecht, the Netherlands: Springer. https://doi.org/10.1007/978-1-4020-6572-9_2

Oke, T. R. (1987). *Boundary layer climates*. London: Methuen & Co.

Olson, R. K. (1998). Procedures for evaluating alternative farming systems: A case study for eastern Nebraska (Extension and education materials for sustainable agriculture 8). Lincoln, NE: Center for Sustainable Agricultural Systems, University of Nebraska.

Olson, R. K., & Francis, C. A. (1995). A hierarchical framework for evaluating diversity in agroecosystems. In R. K. Olson, C. A. Francis, and S. Kaffka (Eds.), *Exploring the role of diversity in sustainable agriculture* (pp. 5–34). Madison, WI: ASA, CSSA, & SSSA. https://doi.org/10.2134/1995.exploringroleofdiversity

Ong, C. K., & Huxley, P. (Eds.). (1996). Tree–crop interactions: a physiological approach. Wallingford, U.K.: CAB International.

Packard, S. (1988). Just a few oddball species: Restoration and the rediscovery of the tallgrass savanna. *Restoration & Management Notes, 6*, 13–20. https://doi.org/10.2307/43439280

Peet, R. K. (1988). Forests of the Rocky Mountains. In M. G. Barbour & W. D. Billings (Eds.), *North American terrestrial vegetation* (p. 63–101). New York: Cambridge University Press.

Peichl, M., Thevathasan, N. V., Gordon, A. M., Huss, J., & Abohassan, R. A. 2006. Carbon sequestration potentials in temperate tree-based intercropping systems, southern

Ontario, Canada. *Agroforestry Systems, 66,* 243–257. https://doi.org/10.1007/s10457-005-0361-8

Peng, X., Thevathasan, N. V., Gordon, A. M., Mohammed, I., & Gao, P. (2015). Photosynthetic response of soybean to microclimate in 26-year-old tree-based intercropping systems in southern Ontario, Canada. *PLOS ONE, 10*(6), e0129467. https://doi.org/10.1371/journal.pone.0129467

Peterson, D. L., & Roelf, GL. (1982). Nutrient dynamics of herbaceous vegetation in upland and floodplain forest communities. *The American Midland Naturalist, 107,* 325–339. https://doi.org/10.2307/2425383

Pimentel, D. (Ed.). (1980). *Handbook of energy utilization in agriculture.* Boca Raton, FL: CRC Press.

Pimentel, D. (2009). Energy inputs in food crop production in developing and developed nations. *Energies,* 2, 1–24. https://doi.org/10.3390/en20100001

Pimentel, D., Harvey, C., Resosudarmo, P., Sinclair, K., Kurz, D., McNair, M., . . . Blair, R. (1995). Environmental and economic costs of soil erosion and conservation benefits. *Science, 267,* 1117–1123. https://doi.org/10.1126/science.267.5201.1117

Pimentel, D., & Pimentel, M. (Eds.). 1996. *Food, energy, and society.* Niwot, CO: University Press of Colorado.

Pimentel, D., Williamson, S., Alexander, C. E., Gonzales-Pagan, O., Kontak, C., & Mulkey, S.E. (2008). Reducing energy inputs in the US food system. *Human Ecology, 36,* 459–471. https://doi.org/10.1007/s10745-008-9184-3

Price, G. W., & Gordon, A. M. (1998). Spatial and temporal distribution of earthworms in a temperate intercropping system in southern Ontario, Canada. *Agroforestry Systems, 44,* 141–149. https://doi.org/10.1023/A:1006213603150

Raeburn, P. (1995). *The last harvest.* New York: Simon & Schuster.

Ranney, J. W., Bruner, M. C., & Levenson, J. B. (1981). The importance of edge in the structure and dynamics of forest islands. In R. L. Burgess & D. M. Sharp (Eds.), *Forest island dynamics in man-dominated landscapes* (pp. 67–95). New York: Springer. https://doi.org/10.1007/978-1-4612-5936-7_6

Rao, M. R., Nair, P. K. R., & Ong, C. K. (1997). Biophysical interactions in tropical agroforestry systems. *Agroforestry Systems,38,*3–50.https://doi.org/10.1023/A:1005971525590

Roberts, L. (1988). Hard choices ahead on biodiversity. *Science, 241,* 1759–1761. https://doi.org/10.1126/science.241.4874.1759

Runkle, J. R. (1982). Patterns of disturbance in some old-growth mesic forests of eastern North America. *Ecology, 63,* 1533–1546. https://doi.org/10.2307/1938878

Schultz, R. C., Colletti, J. P., Isenhart, T. M., Marquez, C. O., Simpkins, W. W., & Ball, C. J. (2000). Riparian forest buffer practices. In H. E. Garrett, W. J. Rietveld and R. F. Fisher (Ed.), North American agroforestry: An integrated science and practice (pp. 189–281). Madison, WI: ASA.

Selley, R. A. (Ed.). (1996). Nebraska crop budgets (Nebraska Cooperative Extension EC96-872-S). Lincoln, NE: University of Nebraska.

Sims, P. L. (1988). Grasslands. In M. G. Barbour & W. D. Billings (Eds.), *North American terrestrial vegetation* (pp. 265–286). New York, Cambridge University Press.

Smith, D. G. (1976). Effect of vegetation on lateral migration of anastomosed channels of a glacier meltwater river. *Geological Society of America Bulletin, 87,* 857–860. https://doi.org/10.1130/0016-7606(1976)87<857:EOVOLM>2.0.CO;2

Smith, W. K. (1985). Western montane forests. In B. F. Chabot & H. A. Mooney (Eds.), *Physiological ecology of North American plant communities* (pp. 95–126). New York: Chapman and Hall. https://doi.org/10.1007/978-94-009-4830-3_5

Soule, J. D., & Piper, J. K. (1992). *Farming in nature's image.* Washington, DC: Island Press.

Spies, T. A., & Franklin, J. F. (1989). Gap characteristics and vegetation response in coniferous forests of the Pacific Northwest. *Ecology, 70,* 543–545. https://doi.org/10.2307/1940198

Steiger, T. L. (1930). Structure of prairie vegetation. *Ecology, 11,* 170–217. https://doi.org/10.2307/1930789

Steinauer, E. M., & Collins, S. L. (1996). Prairie ecology: The tallgrass prairie. In F. B. Samson & F. L. Knopf (Eds.), Prairie conservation: Preserving North America's most endangered ecosystem (pp. 39–52). Washington, DC: Island Press.

Stockle, C., Papendick, R., Saxton, K., Campbell, G., & van Evert, F. (1994). A framework for evaluating the sustainability of agricultural production systems. *American Journal of Alternative Agriculture, 9,* 45–50. https://doi.org/10.1017/S0889189300005555

Thevathasan, N. V., & Gordon, A. M. (1997). Poplar leaf biomass distribution and nitrogen dynamics in a poplar–barley intercropped system in southern Ontario, Canada. *Agroforestry Systems, 61,* 257–268.

Thevathasan, N. V., & Gordon, A. M. (2004). Ecology of tree intercropping systems in the north temperate region: Experiences from southern Ontario, Canada. *Agroforestry Systems, 61,* 257–268.

Thevathasan, N., Gordon, A., Simpson, J., Peng, X., Silim, S., Soolanayakanahally, R., & de Gooijer, H. (2014). Sustainability indicators of biomass production in agroforestry systems. *The Open Agriculture Journal, 8,* 1–11. https://doi.org/10.2174/1874331501408010001

Tiedemann, A. R., & Klemmedson, J. O. (1973). Effect of mesquite on physical and chemical properties of the soil. *Journal of Range Management, 26,* 27–29. doi:10.2307/3896877

Tivy, J. (1990). *Agricultural ecology.* New York: Longman.

Tyndall, J. C., & Grala, R. K. (2009). Financial feasibility of using shelterbelts for swine odor mitigation. *Agroforestry Systems, 76,* 237–250. https://doi.org/10.1007/s10457-008-9140-7

University of Missouri Center for Agroforestry. (2018). Training manual for applied agroforestry practices. Columbia, MO. Retrieved from http://www.centerforagroforestry.org/pubs/training/FullTrainingManual_2018.pdf

U.S. Census Bureau. (1997). Statistical abstract of the United States, 1996. Washington, DC. Retrieved from https://www2.census.gov/library/publications/1996/compendia/statab/116ed/tables/income.pdf

USDA. (2017). Major uses of land in the United States, 2012 (Economic Information Bulletin 178). Washington, DC: USDA Economic Research Service.

U.S. Forest Service. (1981). An assessment of the forest and range land situation in the United States (Forest Resource Rep. 22). Washington, DC.

U.S. Energy Information Administration. (2018a). Frequently asked questions: How much petroleum does the United States import and export. Washington, DC.

U.S. Energy Information Administration. (2018b). Frequently Asked Questions: How much shale (tight) oil is produced in the United States. Washington, DC.

Van Andel, J., Bakker, J. P., & Grootjians, A. P. (1993). Mechanisms of vegetation succession: A review of concepts and perspectives. *Acta Botanica Neerlandica, 42,* 413–433. http://dx.doi.org/10.1111/j.1438-8677.1993.tb00718.x

Vetaas, O. R. (1992). Micro-site effects of trees and shrubs in dry savannas. *Journal of Vegetation Science, 3,* 337–344. https://doi.org/10.2307/3235758

Watson, A., Matt, R., Knotek, K., Williams, D. R., & Yung, L. (2011). Traditional wisdom: Protecting relationships with wilderness as a cultural landscape. *Ecology and Society, 16*(1):36. https://doi.org/10.5751/ES-04003-160136

Wells, P. V. (1965). Scarp woodlands, transported grassland soils, and concept of grassland climate in the Great Plains region. *Science, 148,* 246–249. https://doi.org/10.1126/science.148.3667.246

Welsch, D. J. (1991). Riparian forest buffers (Northern Area State and Private Forestry Publ. Ser. NA-PR-07-91). Radnor, PA: U.S. Forest Service.

West, N. E. (1988). Intermountain deserts, shrub steppes, and woodlands. In M. G. Barbour & W. D. Billings(Eds.), North American terrestrial vegetation (pp. 209–230). New York: Cambridge University Press.

Williams, P. A., & Gordon, A. M. (1992). The potential of intercropping as an alternative land use system in temperate North America. *Agroforestry Systems*, *19*, 253–263. https://doi.org/10.1007/BF00118783

Woodmansee, R. G. (1984). Comparative nutrient cycles of natural and agricultural ecosystems: A step towards principles. In R. Lowrance, B. R. Stinner, & G. H. House (Eds.), Agricultural ecosystems: Unifying concepts (pp. 145–156). New York: John Wiley & Sons.

World Commission on Environment and Development. (1987). Our common future. Oxford, UK: Oxford University Press.

Wotherspoon, A., Thevathasan, N. V., Gordon, A. M., & Voroney, R. P. (2014). Carbon sequestration potential in five tree species in a 25-year-old temperate tree-based intercropping system in southern Ontario, Canada. *Agroforestry Systems*, *88*, 631–643. https://doi.org/10.1007/s10457-014-9719-0

Appendix 3-1

Comparative characteristics of a native tallgrass prairie, a conventional corn–soybean farm, and an oak–hickory forest.

Characteristic	Native tallgrass prairie	Conventional farm	Oak–hickory forest
Physiography	central lowlands of level to undulating terrain	central lowlands of level to undulating terrain	central lowlands and Ozark Plateaus
Climate	eastern Kansas	east-central Nebraska	north-central Missouri
Mean annual temperature,°C	12.8	10.0	12.2
Mean January temperature,°C	–2.3	–6.2	–2.5
Mean July temperature,°C	26.7	25.0	25.5
Avg. frost-free period, d	190	167	194
Mean annual precipitation, mm	800	695	1015
Quarterly precipitation, % of annual (Jan.–Mar./Apr.–June/July–Sept./Oct.–Dec.)	12/38/32/18	12/40/33/15	18/32/27/23
Annual precipitation variability, yr(1961–1990)			
<75% of avg.	4	5	3
>125% of avg.	4	7	2
Hydrology			
Actual evapotranspiration, mm	760	649	747
Potential evapotranspiration, mm	780	704	842
Percolation and runoff, mm	40	46	268
Soil			
Predominant soil order	Mollisol	Mollisol	Alfisol
Suborder	Udoll	Ustoll	Udalf
Total SOM, g m^{-2} (0–100 cm)	31,600	19,640	7,027–13,403
Soil N (0–100 cm)			
Inorganic N, g N m^{-2}	4.5	4.6–11.0	3.2–6.1
Organic N, g N m^{-2}	1,550–1,580	982	410–782
Community composition & structure			
Physiognomy	grasses dominant with a 0.5–3.3-m closed canopy	maize or soybean monocultures with uniform closed canopies (2.5–3 m maize; 0.7–1 m soybean)	broadleaf deciduous forest in which 50% or more of the trees are oak or hickory with a 5.5–8.3-m (avg. 6.3)
Dominant floristic type	perennial grasses	annual grass (maize) or annual broadleaf (soybean)	perennial broadleaf trees
Species richness (250 ha)	200–300 vascular plant species	dominance by two crop species	~94 vascular plant species
Species evenness	>50 % of aboveground biomass in 3–4 species	>90 % of aboveground biomass in 2 species	>90% of aboveground biomass in 3–4 species
Avg. annual max. aboveground biomass (includes litter), g m^{-2}	650–1,075	870 (live only)	10,934
Avg. annual max. belowground biomass, g m^{-2}	1,869–1,985	160	2,592
Belowground OM (includes living and SOM), % of total OM	>98	96	47–59
Leaf area index, m^2 m^{-2}	4	4–4.5	2.5–4.5
Root depth distribution	90% of root biomass in top 50 cm	>90% of root biomass in top 75 cm	>50% in upper 30 cm
Energy flow			
Solar radiation, growing-season PAR, kJ m^{-2}	2227 × 10^3	1772 × 10^3	2017 × 10^3
Non-solar energy inputs, MJ ha^{-1} (Mcal $acre^{-1}$)	0	17.3 × 10^3 (1670)	0
Annual aboveground NPP, g m^{-2}	300–700	983	597.6
Annual belowground NPP, g m^{-2}	500–1200	177	950
Proportion of total NPP as seed or grain, %	6	45	2

Table continued.

Characteristic	Native tallgrass prairie	Conventional farm	Oak–hickory forest
Energy capture efficiency (%)	0.72-1.09	0.74	1.08
Water use efficiency	1.8–2.18	1.87	2.07
Partitioning of aboveground NPP, %			
Herbivores	13–45	5	<0.2
Detritivores	55–87	50	99
Exported from the system	0	45	0
Nutrient cycling			
Avg. annual N inputs, g N m^{-2}	1.0	8.2	2
Avg. annual N losses, g N m^{-2}	0.26–0.6	2.2	2
Avg. annual N export in harvested crop, g N m^{-2}	0	10.9	0
Belowground N, % of total	>98	>98	92
Soil erosion, kg ha^{-1} yr^{-1}	minimal	11,200	minimal
Changes in soil SOM pool	stable	loss of ≥ 30% in first 30 yr of cultivation	stable

Note. SOM, soil organic matter; OM, organic matter; PAR, photosynthetically active radiation; NPP, net primary productivity.

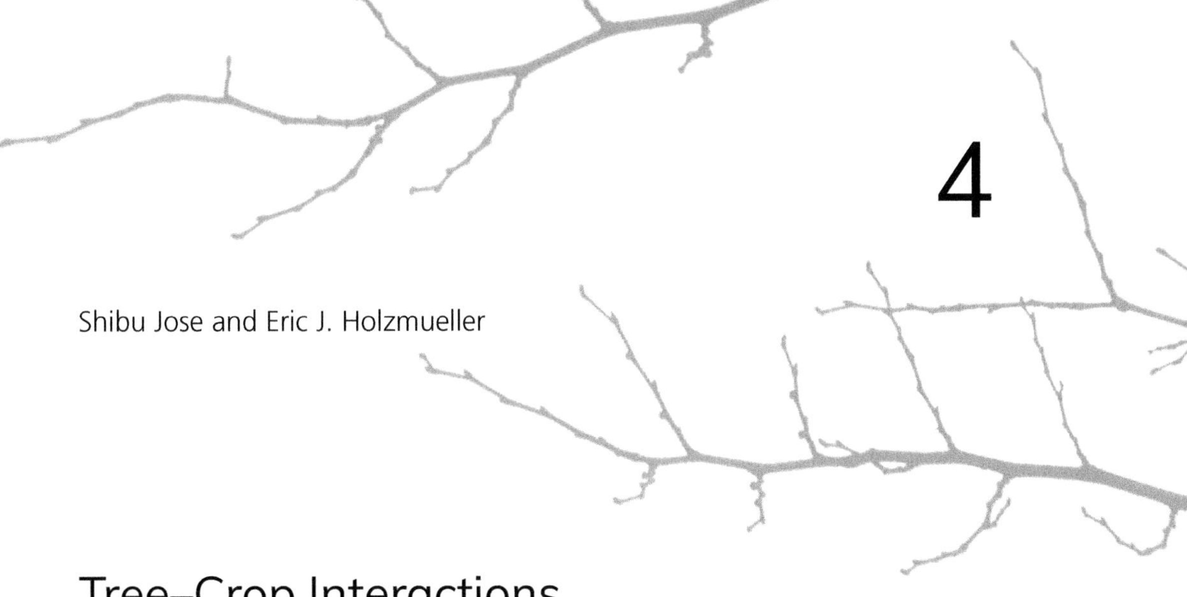

4

Shibu Jose and Eric J. Holzmueller

Tree–Crop Interactions in Temperate Agroforestry

Modern agricultural practices have allowed a dramatic increase in crop and livestock production during the past several decades; however, it has come at the expense of many environmental challenges and the loss of long-term agricultural sustainability (Foley et al., 2011; Funabashi, 2018; Poore & Nemecek, 2018; Tilman, Cassman, Matson, Naylor, & Polasky, 2002). Agroforestry, the intentional incorporation of trees, agricultural crops, and/or animals into a single land-use system, is one way to reduce the negative impacts of modern agriculture (Sanchez, 2002). By combining multiple components in the same system, there is potential to increase nutrient use efficiency; control subsurface water levels; improve soil, water, and air quality; provide favorable habitats for plant, insect or animal species; and create a more sustainable agricultural production system (Garrett, McGraw, & Walter, 2009; Garrity, 2004; Jose, 2009; Jose & Dollinger, 2019).

The incorporation of multiple species in a single ecological system or ecosystem such as agroforestry brings about a unique set of ecological interactions among the different species. An ecological interaction refers to the influence that one or more components of a system has on the performance of another component of the system and of the overall system itself (Nair, 1993). If two species are competing for the same resources, and do so equally, both species will likely exhibit lower productivity compared with their potential for independent growth. If an agroforestry system can be designed so that the physiological needs for particular resources are spatially or temporally different for the individual species growing in the system, then there is a possibility that the system may be more productive than the cumulative production of those species if they were grown separately on equal land area (Cannell, van Noordwijk, & Ong, 1996; Wojtkowski, 1998). With time, this advantage may disappear as competitive vectors overtake complementary interactions, but management intervention (e.g., thinning of trees, pruning of branches, disking to reduce root interactions, etc.) may bring the yield advantage back (Figure 4–1).

An understanding of both the biophysical processes and the mechanisms involved in the allocation of resources is essential for the development of ecologically sound agroforestry systems that are sustainable, economically viable, and socially acceptable. In this chapter, we examine the complex biophysical interactions that are central to the ecological sustainability of temperate agroforestry systems. Although our focus is on tree–crop interactions, we also review those interactions involving animals when appropriate. Reviews on the

North American Agroforestry, Third Edition. Edited by Harold E. "Gene" Garrett, Shibu Jose, and Michael A. Gold.

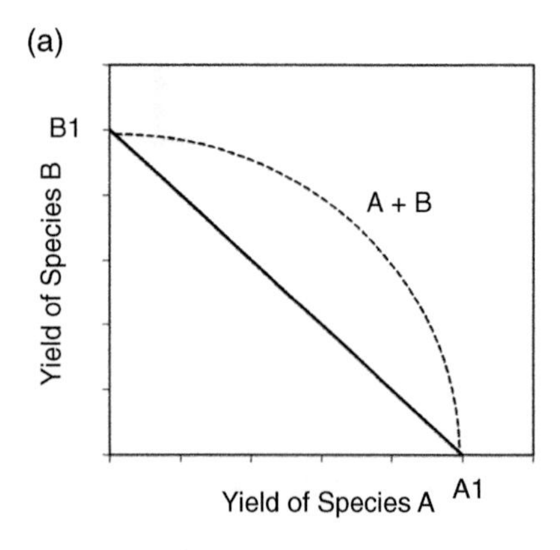

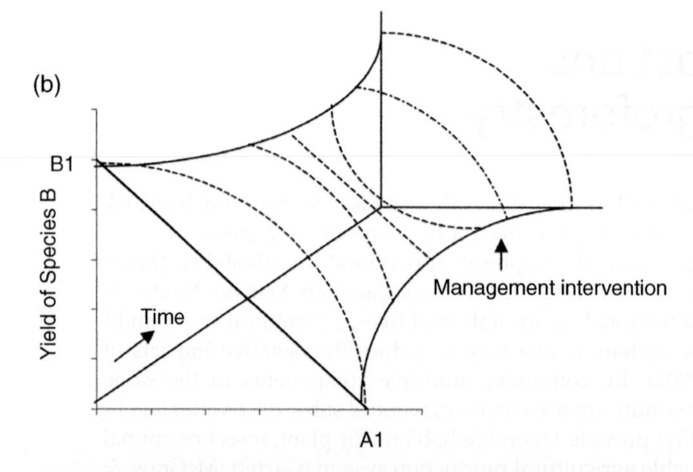

Fig. 4–1. The production possibility curves for two species, A and B: (a) Points A1 and B1 represent the maximum production potential if A and B were grown in monocultures, while Line A1–B1 represents the proportional yield of A and B when grown in mixtures, and the curve described by A+B represents overyielding (compared with either monoculture yield) of one possible mixture of A and B; (b) a hypothetical temporal production possibility surface for species A and B—as time progresses, overyielding gives away to underyielding, but a timely management intervention (e.g., root pruning of trees) alleviates competitive interactions, thereby resulting in overyielding again (reprinted with permission from Jose et al., 2004).

topic include Garcia-Barrios and Ong (2004), Jose, Gillespie, and Pallardy (2004), Thevathasan and Gordon (2004), Tsonkova, Böhm, Quinkenstein, and Freese (2012), and Atangana, Khasa, Chang, and Degrande (2014).

Species Coexistence and Ecological Interactions

It is important to review the theoretical basis for species coexistence before discussing the biophysical interactions among them. The competitive exclusion principle (CEP), termed by Hardin (1960), also known as the competitive displacement principle, Grinnell's axiom, the Volterra–Gause principle, or Gause's law, has been a cornerstone in ecological thinking regarding species coexistence for decades. The CEP is based on Gause's (1934) contention that two similar species competing for the same resources cannot stably coexist. Competition between species may lead to three outcomes: (a) if there is no trade-off between competitive abilities, competition will lead to competitive exclusion; however, (b) if there is a trade-off between competitive abilities for different resources, where the stronger competitor of one resource also obtains relatively more of another resource in which the other species is a stronger competitor and vice versa, competition will lead to stable coexistence; or (c) if there is a trade-off between competitive ability for resources and each species is a better competitor for a specific resource, competition will lead to alternative stable states (Passarge, Hol, Escher, & Huisman, 2006). In agroforestry systems, we deliberately mix tree and crop species so that they exert minimal or weak competition among themselves.

Even though there are examples of violations to the fundamental conditions of the CEP, using the concept as the initial inquiry of ecological thought has elucidated many other mechanisms that contribute to our knowledge of how species coexist in nature. For example, the resource-ratio hypothesis, proposed by Tilman (1980, 1982, 1990), has been used to explain species coexistence (for a review of examples in which the resource ratio theory has been tested, see Miller et al., 2004). According to this hypothesis, coexistence occurs where resource requirements differ among species. Greater capture of a limiting resource would be accompanied by an increased ability to utilize nonlimiting resources, which, by definition, are available but underutilized. In an agroforestry setting, based on the differences in physical or phenological characteristics of the component species, the interactions between tree and crop species may lead to an increased capture of a limiting growth resource. The system as a whole could then accrue greater total biomass than the cumulative production of those species if they were grown separately on an equivalent land area (Cannell et al., 1996).

Hubbell (2001) has challenged the notion that trade-offs are necessary for understanding broad patterns of species diversity and relative abundance. In contrast to trade-off-based theories, Hubbell developed a neutral model (united neutral theory) that explains plant species coexistence without any trade-offs. Neutral theories focus on "community drift" and explain the maintenance of biodiversity at large spatial and temporal scales by a balance between speciation and stochastic extinction events. These are caused by random drifts in population size in communities of ecologically identical (hence neutral) species, that is, without invoking any species-specific traits or interspecific trade-offs.

Finally, spatially explicit models of plant species coexistence have been developed (e.g., Gravel, Mouquet, Loreau, & Guichard, 2010; Isabelle, Damien, & Wilfried, 2014). They do not require trade-off or neutrality assumptions to explain plant species coexistence, and they predict coexistence if interactions among conspecifics (individuals of the same species) occur across larger distances than interactions among heterospecifics (individuals of different species). Moreover, they lend themselves to more direct experimental tests than the more general trade-off or neutral theories.

Analysis of ecological interactions has shown both competitive and facilitative (complementary) interactions in agroforestry systems (Jose et al., 2004), which occur both above- and belowground (Ong, Corlett, Singh, & Black, 1991; Singh, Ong, & Saharan, 1989). As stated by Shainsky and Radosevich (1992), mechanisms of competition for resources should at least include documentation of: (a) depletion of resources associated with the presence and abundance of plants; (b) changes in physiological and morphological growth responses associated with changes in the resource environment; and (c) correlations between the presence or abundance of neighbors, depression in resource availability, and physiological performance. In contrast, according to Kelty (2000), facilitative interactions are those in which one species benefits another and occur under four mechanisms: (a) increased nutrient cycling efficiency, e.g., increasing N availability by planting an N_2–fixing species with non-N_2–fixing species; (b) increased water and nutrient retention through improved soil structure; (c) increased water availability for understory species because of reduced evaporative demand or "hydraulic lift" of moisture from the lower levels in the soil by overstory species; and (d) decreases in productivity losses from insect pests, pathogens, and weeds.

Competition and facilitation are not necessarily independent of each other (Holmgren, Scheffer, & Huston, 1997); the balance between these factors may vary along a resource gradient (Brooker & Callaghan, 1998). Proper management of an agroforestry system that increases facilitative interactions and limits competitive interactions requires an understanding of the possible interactions in these systems. Therefore, an examination of both the effect that plants have on the shared resources and their response to the changed environment must occur in order for proper management to take place (Casper & Jackson, 1997; Goldberg, 1990).

Competitive Interactions—Aboveground

Competition for light

The incorporation of trees or shrubs in an agroforestry system can increase the amount of shading that plant species, primarily those in the understory, experience compared with growing in a monoculture. Green plants are photoautotrophs, and both the fraction of incident photosynthetically active radiation (PAR, 400–700-nm wavelength) that a species intercepts, and the ability of that species to convert radiation into energy (through photosynthesis) are important factors in plant biomass growth (Ong, Black, Marshall, & Corlett, 1996). Furthermore, these biomass growth factors are influenced by a number of additional factors including temperature, available water and nutrients, CO_2) level, aspect, time of day, photosynthetic pathway (C_3 vs. C_4), plant age and height, leaf area and angle, canopy structure, species combination, and transmission and reflectance traits of the canopy (Brenner, 1996; Kozlowski & Pallardy, 1997).

Numerous studies have examined shading and its effects on crop growth (Artru et al., 2017; Gillespie et al., 2000; Reynolds, Simpson, Thevathasan, & Gordon, 2007), and many of those studies have indicated that shading by tree species is a factor in reducing crop yield. For example, lower PAR levels resulting from overhead shading by hybrid poplar (*Populus* sp. clone DN-177) and silver maple (*Acer saccharinum* L.) significantly reduced the yield of maize (*Zea mays* L.) and soybean [*Glycine max* (L.) Merr.] in a temperate alley-cropping system in southern Ontario, Canada (Table 4–1) (Reynolds et al., 2007). The yields of soybean and maize were reduced by 49 and 51%, respectively, when PAR levels decreased by 29% at 2 m from silver maple tree rows. Similar results have also been reported for temperate silvopastoral systems. In Missouri, significant decreases in the mean dry weight of warm-season grasses was observed as

Table 4–1. Effects of tree (poplar and maple) competition on photosynthetically active radiation (PAR) and crop (soybean and maize) yield at two distances from tree rows for two growing seasons (modified from Reynolds et al., 2007).

Parameter (N = 6)	Crop	Control		Poplar		Maple	
		2 m	6 m	2 m	6 m	2 m	6 m
1997	soybean						
PAR, µmol s^{-1} m^{-2}		1,464.0 a	1,586.0 a	1,133.0 a	1,370.0 a	1,045.0 b	1,558.0 a
Yield, t ha^{-1}		2.51 a	2.59 a	1.04 b	1.97 a	1.29 b	2.00 a
1998	soybean						
PAR, µmol s^{-1} m^{-2}		1,405.0 a	1,158.0 a	746.0 b	1,296.0 a	670.0 b	1,336.0 a
Yield, t ha^{-1}		2.24 a	2.25 a	1.15 b	1.67 a	1.55 b	2.85 a
1997	maize						
PAR, µmol s^{-1} m^{-2}		1,528.0 a	1,579.0 a	952.0 b*	1,407.0 a*	1,075.0 b*	1,525.0 a*
Yield, t ha^{-1}		4.21 a	4.83 a	2.89 b	4.61 a	2.07 b	4.64 a
1998	maize						
PAR, µmol s^{-1} m^{-2}		1,422.0 a	1,200.0 a	794.0 b	1,117.0 a	481.0 b	1,420.0 a
Yield, t ha^{-1}		5.70 a	5.88 a	0.69 b	5.29 a	3.79 b	7.07 a

Note. Soybean and maize intercrops, July 1997 and July 1998. Within each treatment (control, poplar, maple), values in each row followed by the same letter are not significantly different (Tukey's HSD, $P < 0.05$).

* Significant at the 10% level.

the amount of available light declined (Lin, McGraw, George, & Garrett, 1999). In a silvopastoral aspen (*Populus tremuloides* Michx.) stand in Alberta, Canada, a decrease in canopy cover resulted in a significant increase in understory production, while understory production was only slightly affected by decreased belowground competition (Powell & Bork, 2006). When the canopy was removed, understory net primary production increased up to 275% compared with the control stands with full canopy.

The physiological basis of yield reduction due to shading has been investigated by several studies in temperate agroforestry systems (Albaugh et al., 2014; Ehret, Graß, & Wachendorf, 2015; Miller & Pallardy, 2001; Reynolds et al., 2007). Shading changes the quality of light reaching the understory canopy (Krueger, 1981). Since overhead canopies absorb both the longest and shortest wavelengths of the light spectrum (red and blue), diffuse radiation is primarily composed of medium-wavelength light (orange, yellow, and green). Growth regulating hormones and, therefore, growth are influenced by the interactions of the plant phytochrome system with red and infrared wavelengths (Baraldi, Bertazza, Bogino, Luna, & Bottini, 1995). Inadequate exposure to red light is known to influence stem production in clover (*Trifolium* sp.) (Robin, Hay, Newton, & Greer, 1994), tillering in grasses (Davis & Simmons, 1994a), flowering (Davis & Simmons, 1994b), and other basic plant growth processes (Sharrow, 1999).

Theoretically, one physiological response to shading depends on the pathway used by crop species to fix C (C_3 vs. C_4). In C_3 plants, as PAR increases from complete shade to approximately 25–50% of full sun there is a corresponding increase in the photosynthetic rate (P_{net}); however, as more light becomes available, P_{net} does not increase but rather levels off despite the additional increase in PAR (Figure 4–2). In contrast, in C_4 plants, P_{net} does not level off as PAR increases to full sunlight but rather continues to increase with increasing PAR (Figure 4–2). The difference between C_3 and C_4 plants is related to the pathway by which these two types of plants fix CO_2 (Kozlowski & Pallardy, 1997; Lambers, Chapin, & Pons, 1998). Theoretically, because of the ability of C_3 plants to maximize P_{net} growing under partial shade, they should be better suited for agroforestry practices than C_4 plants. However, field studies have produced mixed results.

In accordance with the theory that C_3 plants would not have reduced growth under shaded conditions, Wanvestraut, Jose, Nair, and Brecke (2004) reported no effects on the growth and yield of cotton (*Gossypium hirsutum* L.), a C_3 plant species, when grown under moderate shading in a temperate pecan [*Carya illinoinensis* (Wangenh.) K. Koch]–cotton alley-cropping system in Florida. Contrary to an anticipated yield decrease in maize, a C_4 species, in response to shading, Gillespie et al. (2000) reported no effect of shading in both black walnut (*Juglans nigra* L.)–maize and red oak (*Quercus rubra* L.)–maize alley-cropping systems in Indiana, which was not the expected result given the known strong positive correlation between PAR and P_{net} in C_4 plant species. Although these researchers found that, generally, the edge rows received lower PAR than the middle rows, particularly in the red oak–maize system because of the higher canopy leaf area, once competition for water and nutrients was removed through polyethylene root barriers and trenching, there was no indication of yield reduction because of reduced PAR (Figure 4–3), leading them to conclude that competition for light was not a factor for these two systems. Interestingly, however, Reynolds et al. (2007) reported that competition for light was a factor in both soybean (C_3 species) and maize yield reductions in a multispecies temperate agroforestry system in southern Ontario, Canada. In addition, they concluded that competition for light was more important than that for water during the study period. There are several reports from China showing reduced crop yield as a result of intercropping with trees. For example, Li, Meng, Dali, & Wang (2008) reported a 51% lower wheat (*Tritcum aestivum* L.) yield in a paulownia (*Paulownia* Siebold & Zucc.)–wheat intercropping system than sole cropping and attributed the reduction to shading. In a jujube (*Zizyphus jujuba* Mill.)–winter wheat–summer maize intercropping, Yang, Ding, Liu, Li, & Egrinya Eneji (2016) reported that the mean yield of winter wheat and summer maize was reduced by 35.6 and 35.2%, respectively, compared with monoculture. Zhang et al. (2017) also reported similar results from a jujube–winter wheat intercropping system.

Herbivory and physical damage

Damage to trees from associated animal species in certain agroforestry practices such as silvopasture can be substantial. Browsing animals such as goats, sheep, or deer are more likely to eat foliage, while large ruminants such as cattle are more likely to trample young trees. Generally, younger animals are more likely to damage trees than older, more experienced ones (Nowak, Blount, & Workman, 2002). Any browsing of the terminal shoot may result in deformity and loss of growth. Similarly, physical damage to bark or stem can

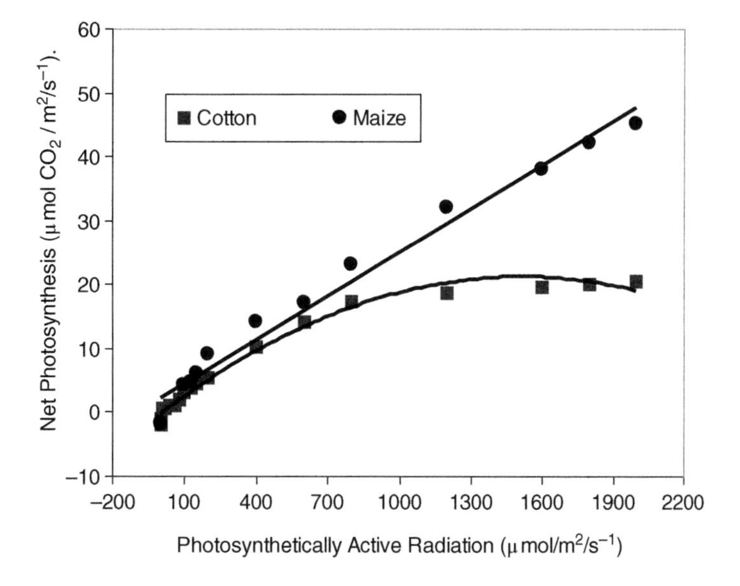

Fig. 4–2. Net photosynthesis as a function of photosynthetically active radiation in maize (C_4 plant) and cotton (C_3 plant) (based on data from Zamora et al. [2006] and Jose [1997])

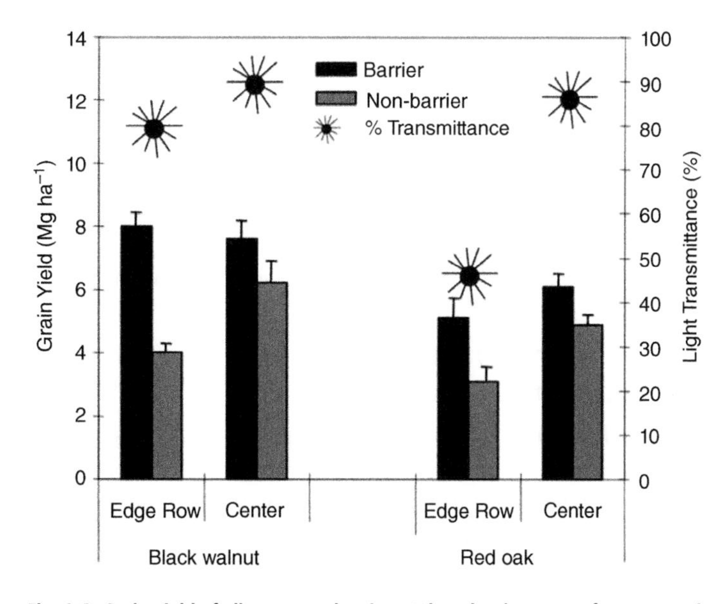

Fig. 4–3. Grain yield of alley-cropped maize at the edge (average of eastern and western rows closest to tree row) and alley center in two alley-cropping systems involving black walnut and red oak in southern Indiana. Light transmittance (as a fraction of full sunlight) reaching the top of edge and center-row plants is also shown (reprinted with permission from Jose et al., 2004).

result in loss of vigor and eventual death of saplings and young trees (Jose & Dollinger, 2019).

In a young silvopastoral system in Missouri, Lehmkuhler et al. (2003) reported significant damage to tree seedlings during the second year after planting when cattle were introduced. Seedlings that received protection using electric fencing were mostly undamaged (Figure 4–4). In a study in the Swiss Alps, Mayer, Stockli, Konold, and Kreuzer (2006) assessed cattle damage on naturally regenerated young Norway spruce [*Picea abies* (L.) Karst] following a summer grazing period. They observed that 4% of the young trees were browsed on the apical shoot, 10% were browsed on lateral shoots, and 13% of the trees showed other damage. The percentage of browsed or damaged (physical damage such as breaking seedlings or trampling) trees was positively correlated with the cattle stocking rate (livestock units per hectare) (Figure 4–5). This suggests that higher cattle stocking rates not only increase browsing pressure but also the risk of unintentional trampling of trees.

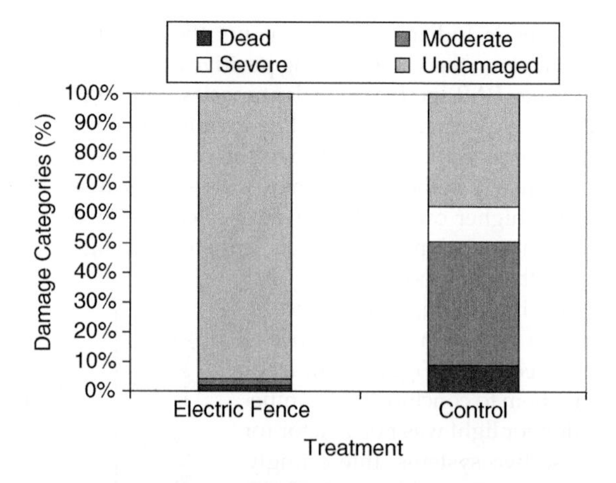

Fig. 4–4. Extent of damage to trees by cattle during second year after planting with and without electric fence protection in a silvopastoral system in Missouri (based on data from Lehmkuhler et al., 2003).

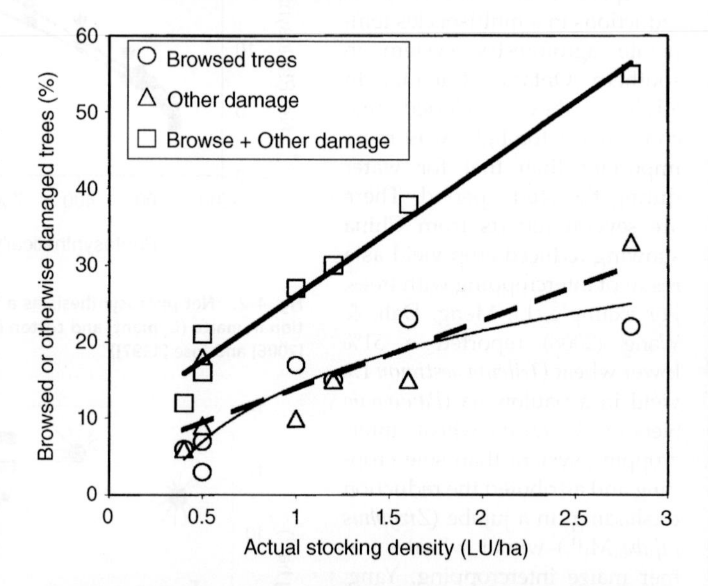

Fig. 4–5. Relationships between cattle stocking rate (livestock units, LU) and percentage of browsed trees, otherwise damaged trees, and the sum of browsed and otherwise damaged trees (modified from Mayer et al., 2006).

Damage or injury to animals as a result of trees can also occur in silvopastoral systems. In a recent survey of silvopastoral farmers in the northeastern United States, Orefice, Caroll, Conroy, & Ketner (2017) reported that farmers were concerned about falling tree branches as health risks for the animals. They also reported at least two forms of livestock injuries, one resulting from cows' tails being caught and torn off by woody vegetation and the other relating to hoof injury to pigs.

Facilitative Interactions—Aboveground

Modification of the microclimate

Trees can modify the microclimate of an agroforestry system, which, in turn, may benefit associated crop species. Despite the previous examples of competition for light, moderate shading can have a positive effect on crop growth. For example, Lin et al. (1999) found that because of shade tolerance, *Desmodium canescens* (L.) DC. and *D. paniculatum* (L.) DC., two warm-season legumes, had significantly higher dry weight under 50 and 80% shade than full sunlight in Missouri. Burner (2003) found that, across six harvesting periods, orchardgrass (*Dactylis glomerata* L.) yields did not differ among 8–10-yr-old loblolly pine (*Pinus taeda* L.) and shortleaf pine (*Pinus echinata* Mill.) silvopastures compared with yields in open pastures in Arkansas. Additionally, in the loblolly pine system, orchardgrass persistence was greater than in the open system (72 vs. 44% stand occupancy, respectively).

Shading can also have a positive effect on forage quality. Lin, McGraw, George, and Garrett (2001) reported that under an 80% shade treatment, the crude protein content of most of the cool-season forage grasses studied was greater

Table 4–2. Crude protein of selected introduced cool-season grasses when grown under three levels of shade during 1994 and 1995 in Missouri (modified from Lin et al., 2001).

Species	Crude protein		
	Full sun	50% Shade	80% Shade
	%		
Kentucky bluegrass	20.3 b	20.7 b	22.7 a
'Benchmark' orchardgrass	12.6 c	15.7 b	19.6 a
'Justus' orchardgrass	19.8 a	16.7 a	18.5 a
'Manhatten II' ryegrass	15.3 b	16.0 b	18.8 a
Smooth bromegrass	16.7 c	18.1 b	20.2 a
'KY31' tall fescue	14.0 b	15.0 b	18.1 a
'Martin' tall fescue	14.3 b	15.5 b	18.5 a
Timothy	15.4 c	17.6 b	20.4 a

Note. Means followed by the same letter within a row are not significantly different (Tukey's Studentized range test, $\alpha = 0.05$).

compared with the full sun treatment (Table 4–2). In a study of a 6–7-yr-old walnut–hybrid pine [pitch (*Pinus rigida* Mill.) × loblolly] and annual ryegrass (*Lolium multiflorum* Lam.) and cereal rye (*Secale cereale* L.) mixture silvopasture in Missouri, forage yield was slightly decreased in the silvopasture compared with forage yield in a nearby open pasture; however, forage quality was greater (Figure 4–6) and beef heifer average daily gain and gain per hectare were similar for the silvopasture and open pasture treatments (Kallenbach, Kerley, & Bishop-Hurley, 2006). Ford et al. (2019) observed similar results in Minnesota, where forage yield was lower but quality was greater in silvopastoral systems than open pastures. In a recent synthesis of information from several existing studies, Pang et al. (2019a, 2019b) showed that for a number of forage species (warm-season and cool-season grasses, forbs, and legumes), a moderate level of shading (45% of full sun) yielded the highest crude protein. Forage biomass yield also was either highest or similar to 100% sun for most of the studied species.

In addition to their effect on solar radiation, trees can also influence the microclimate of the surrounding area in terms of wind speed and humidity. Serving as windbreaks, trees slow the movement of air, thereby reducing evaporative stress. For example, in a silvopastoral system in Australia, wind speed was reduced up to 80% in a zone that extended 5*H* upwind and 25*H* downwind of the windbreak (where *H* is the height of the windbreak) (Cleugh, 2002). Windbreaks have also been shown to reduce evapotranspiration, improve the distribution and utilization of irrigation water, and improve crop water use efficiency (Davis & Norman, 1988). As shown in several studies, the wind reduction and improved microclimate resulting from planting windbreaks or shelterbelts in crop fields may translate into improved crop quality and yield within the sheltered areas (10–15*H*), (Brandle, Hodges, & Zhou, 2004; Kort, 1988). These effects, however, may vary with annual rainfall conditions (Rivest & Vézina, 2015).

Shading from trees can lower temperatures and reduce heat stress of crops in agroforestry systems. For example, in a silvopastoral system in west-central Spain, the presence of trees significantly lowered the air and soil temperature beneath the canopy on warm days and significantly increased both air and soil temperature beneath the canopy on cold days (Figure 4–7) (Moreno Marcos et al., 2007). Due to the air and temperature modifications caused by the tree shading, forage under the tree canopies began growing earlier in the growing season and continued growing later in the growing season in this system (Gómez-Gutierrez & Pérez-Fernández, 1996; Moreno Marcos et al., 2007). Similar results have been reported in other agroforestry systems. In their study of a pecan–cotton alley-cropping system in northwest Florida, Ramsey and Jose (2002) observed cotton plants germinating earlier in the growing season under pecan canopy cover compared with the cotton-only system, which was attributed to moister and cooler soil conditions. Tomato (*Lycopersicon esculentum* Mill.) and snap bean (*Phaseolus vulgaris* L.) showed earlier germination, accelerated growth, and increased yields under simulated narrow alleys than wider alleys in an alley-cropping study in Nebraska (Bagley, 1964; Garrett et al., 2009).

Enhancing beneficial insect populations

Variations in tree–crop combinations and spatial arrangements in agroforestry have been shown to have an effect on insect population density and species diversity (Altieri, 1991; Pardon et al., 2019). Agroforestry helps reduce pest problems because tree–crop combinations provide greater niche diversity and complexity than monoculture systems of annual crops (Martin-Chave, Béral, & Capowiez, 2019). This effect may be explained in one or more of the following ways: (a) wide spacing of host plants in the intercropping scheme may make the plants more difficult for herbivores to find; (b) one plant species may serve as a trap crop to detour herbivores from finding the other crop; (c) one plant species may serve as a repellent to the pest; (d) one plant species may serve to disrupt the ability of the pest to efficiently attack its intended host; and (e) the intercropping situation may attract more predators and parasites than monocultures,

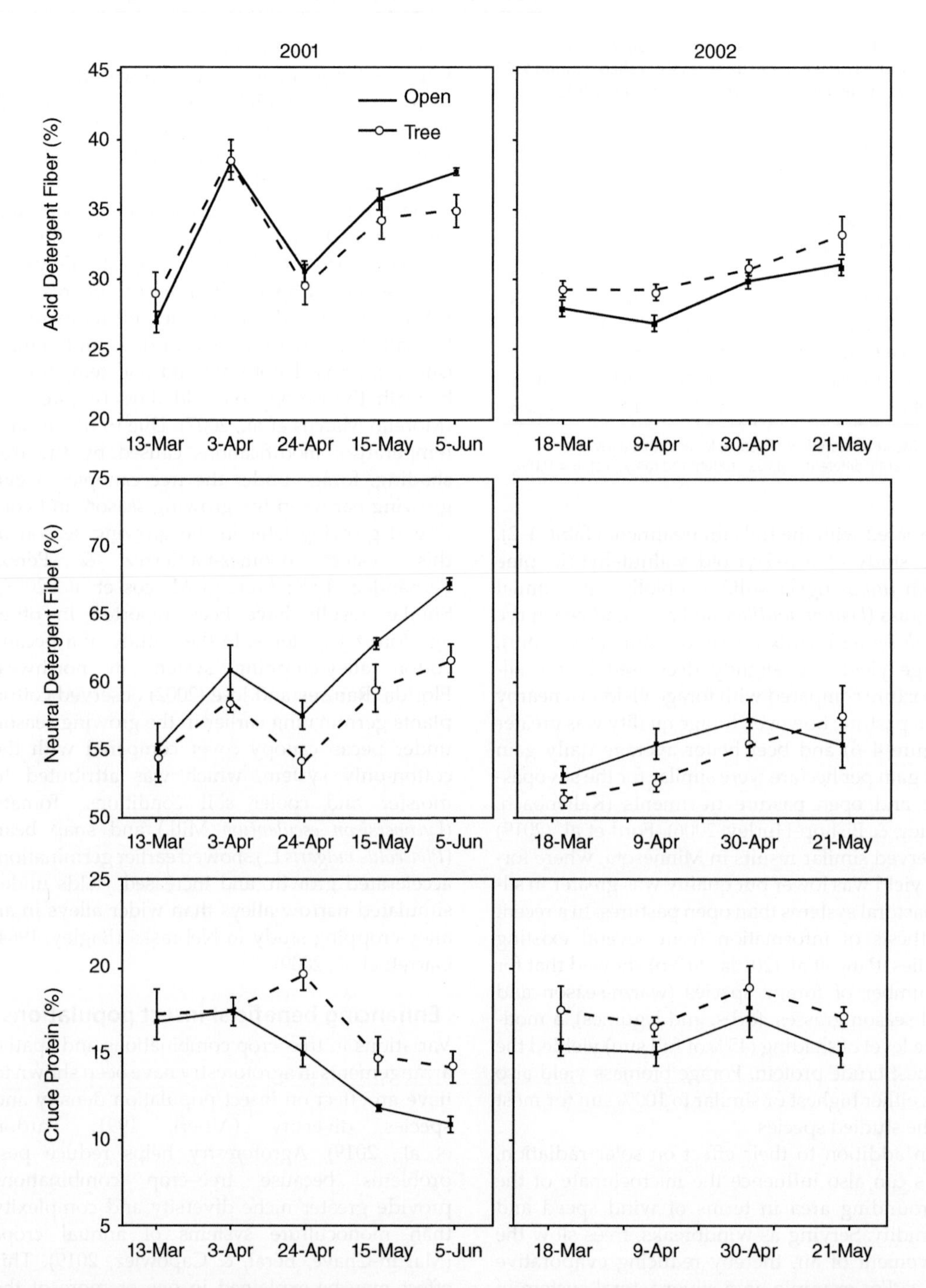

Fig. 4–6. Acid detergent fiber, neutral detergent fiber, and crude protein of annual ryegrass–cereal rye in Open and Tree pastures at the Horticulture and Agroforestry Research Center near New Franklin, MO. Bars indicate standard errors at each sampling (adapted from Kallenbach et al., 2006).

thus reducing pest density through predation and parasitism (Root, 1973; Jose et al., 2004).

Reports of agroforestry practices enhancing beneficial insects are limited in the temperate agroforestry literature. Studies with pecan, for example, have looked at the influence of ground covers on arthropod densities in tree–crop systems (Bugg, Sarrantonio, Dutcher, & Phatak, 1991; Smith et al., 1996). Bugg et al. (1991) observed that cover crops (e.g., annual legumes and grasses) sustained lady beetles (Coleoptera: Coccinellidae) and other arthropods that may be useful in the biological control of pests in pecan (Bugg et al., 1991; Garrett et al., 2009). However, Smith et al. (1996) found that ground cover had little influence on the type or density of arthropods present in pecan. Brandle et al. (2004) summarized the beneficial effects of windbreaks on natural enemies of crop pests. According to them, windbreaks influence the distributions of both predator and prey. Greater diversity of the edges provides numerous microhabitats for life-cycle activities and a variety of hosts, prey, pollen, and nectar sources. As windbreak structure becomes more complex, various microhabitats are created and insect populations increase in both number and diversity.

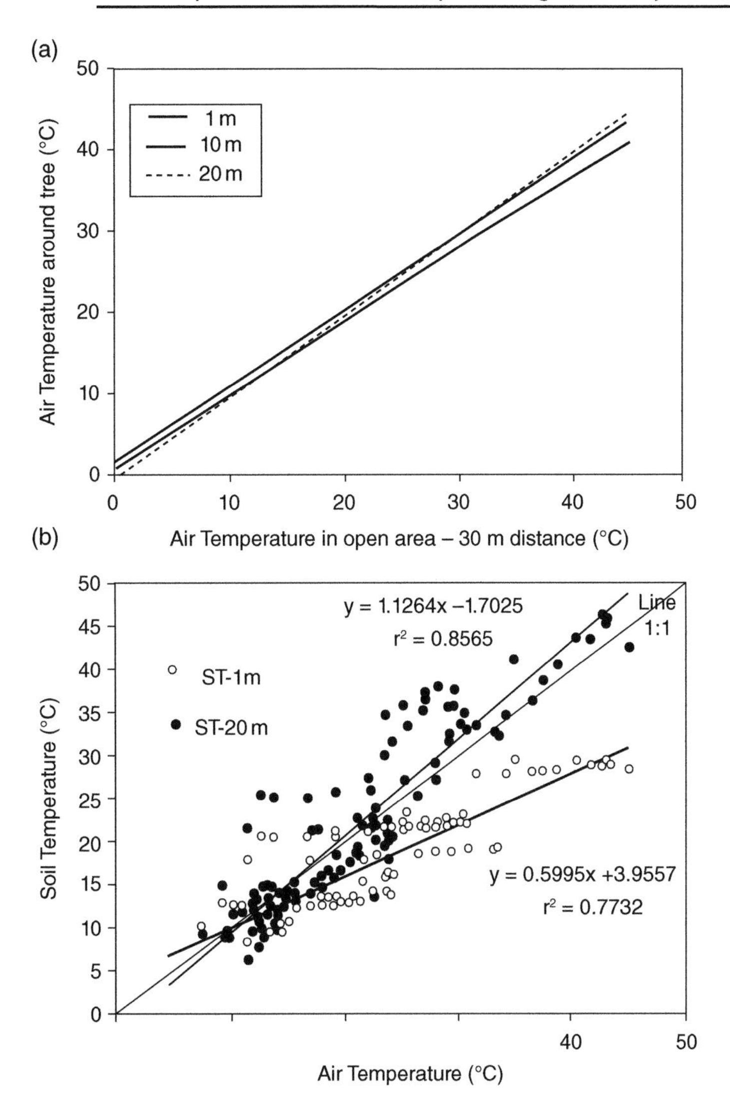

Fig. 4–7. Effect of trees on air and soil temperature of holm oak dehesas: (a) air temperature as a function of distance from the tree, and (b) soil temperature (ST) with respect to the air temperature both beneath and beyond the tree canopy (adapted from Moreno Marcos et al., 2007).

Improving Wildlife Habitat

Agroforestry practices, by increasing structural and compositional plant diversity on the landscape, provide improved wildlife habitat for many species. In some agriculture-dominated landscapes, windbreaks and riparian buffers offer the only woody habitat for wildlife (Johnson & Beck, 1988; Söderström, Svensson, Vessby, & Glimskär, 2001). Brandle et al. (2004) reported that in Nebraska, landowners identified wildlife as a primary reason for the establishment of windbreaks on agricultural land. In a comparison of maize monoculture with riparian buffer plantings of ladino clover (*Trifolium repens* L.) and orchardgrass with three tree species—green ash (*Fraxinus pennsylvanica* Marsh.), black walnut, and red oak—Gillespie, Miller, & Johnson (1995) showed that the riparian strips provided better habitat for birds than maize monoculture, with both higher bird densities and diversity.

Gibbs et al. (2016), in their study in Ontario, Canada, found that tree-based intercropping systems enhanced avian species diversity compared with old fields and monocropped areas. When a pasture is converted to agroforestry, birds normally associated with woodlands are being attracted to agroforestry and, along with birds normally found in open fields, create a unique assemblage of species. Mcadam, Sibbald, Teklehaimanot, and Eason (2007) examined the effect of temperate silvopastoral systems on certain invertebrate groups, including carabid beetles and spiders, and on the number of individuals and species of birds. They reported that the presence of trees on grasslands attracted invertebrates, which might have provided an enhanced food supply, which attracted birds. They concluded that, even at an early stage, silvopastoral systems had a positive impact on birds and could significantly enhance biodiversity.

Bobryk et al. (2016) demonstrated that the overall acoustic complexity index (ACI), a measure of species richness based on the sonic environment, was higher for a pecan alley-cropping system and a silvopasture system than a soybean monoculture in Missouri. There was a weak but significant relationship (R^2 = .30) between the ACI and overall structural complexity across different land-use systems. The sound sources identified included birds, amphibians, insects, and mammals. They concluded that habitat heterogeneity created by agroforestry was the reason for the acoustic and thereby species diversity observed in their study.

Competitive Interactions—Belowground

Despite the ability of many tree species to develop taproots that extend deep into the soil profile, typically most of the tree and crop roots in an agroforestry system are within the top 30 cm of the soil profile (Jose, Gillespie, Seifert, Mengel, & Pope, 2000; Jose, Williams, & Zamora, 2006; Zamora, Jose, & Nair, 2007). Consequently, there is intense competition for soil, water, and nutrients within the top 30 cm of the soil profile, which can affect crop production (Alley, Garrett, McGraw, & Blanche, 1999; Delate, Holzmueller, Mize, Frederick, & Brummer, 2005; Jose, Gillespie, Seifert, Mengel, & Pope, 2000; Zamora et al., 2007). For example, maize yields were 35 and 33% lower for black walnut and red oak alley-cropping systems, respectively, than for conventional crop systems (Jose, Gillespie, Seifert, Mengel, & Pope, 2000). Zamora et al. (2007) reported that belowground competition reduced the total root length in cotton plants by 33% when grown in a pecan–cotton alley-cropping system in Florida compared with cotton plants grown in a monoculture. Furthermore, they reported that cotton plants in the alley-cropped system had lower plant biomass production and root/shoot ratios than cotton plants grown in the monoculture system (Table 4–3). Competition between trees and crops can negatively affect tree growth as well. Delate et al. (2005) reported that, 4 yr after planting, trees planted with forage crops exhibited significantly less growth than the same tree species growing in a competition-free environment in a bottomland hardwood alley-cropping system in central Iowa.

Competition for water

Although belowground competition has been documented in many agroforestry systems, it is difficult to separate the belowground competition for water from that for nutrients. Competition for water, however, has been detected in some temperate agroforestry systems. In a silvopastoral study in the Canterbury region of New Zealand, Yanusa et al. (2005) reported a slightly lower water potential in radiata pine (*Pinus radiata* D. Don) trees planted in an alfalfa (*Medicago sativa* L.) pasture compared with trees growing in a vegetation-free control treatment. Competition for water led to a 44% decrease in instantaneous CO_2 assimilation, a 48% decrease in stomatal conductance, and a 64% decrease in growth between trees growing in the alfalfa pasture and trees growing in the alfalfa-free control treatment (Yanusa et al., 2005). In a recent study in a temperate silvopasture system in Patagonia, Argentina, Quinteros, Bava, Bernal, Gobbi, and Defosse (2017) studied the competition effects of herbaceous vegetation on the survival, growth, and plant water relations of planted *Nothofagus pumilio* seedlings and observed higher survival and growth and better plant water status where competition from herbaceous vegetation was controlled, indicating strong interspecific competition otherwise.

Water stress has also been found to affect crop plants. Multiple studies have reported large reductions in crop plant height (NeSmith & Ritchie, 1992a; Wanvestraut et al., 2004) and leaf area (Jose, Gillespie, Seifert, & Biehle, 2000; NeSmith & Ritchie, 1992b) and yield (Gillespie et al., 2000; Wanvestraut et al., 2004) when water is a limiting factor. Wanvestraut et al. (2004) also observed competition for water in a pecan–cotton alley-cropping system in Florida. By the end of the growing season, cotton plants in the barrier treatment were 26% taller and had 48% greater leaf area than non-barrier plants. Therefore, it is not surprising that cotton lint yield was 35% higher in the trenched barrier treatment than the

Table 4–3. Dry weight (DW) growth parameters of cotton grown in non-barrier, barrier, and monoculture treatments in the 2002 and 2003 growing seasons (modified from Zamora et al., 2007).

Treatment	Whole plant DW	Root DW	Shoot DW	Root/shoot ratio
	g plant^{-1}			
2002				
Non-barrier	29.78 b (2.67)	13.53 c (0.43)	26.25 b (1.81)	0.12 c (0.02)
Barrier	54.04 a (3.76)	17.60 b (0.23)	46.44 a (2.54)	0.16 b (0.01)
Monoculture	57.33 a (3.71)	18.70 a (0.69)	48.63 a (3.07)	0.19 a (0.01)
2003				
Non-barrier	25.13 c (1.88)	12.83 c (0.21)	22.29 c (1.71)	0.13 c (0.01)
Barrier	50.76 b (3.85)	16.88 b (0.41)	43.88 b (3.56)	0.16 b (0.01)
Monoculture	72.60 a (5.71)	12.19 a (0.96)	60.40 a (4.95)	0.21 a (0.01)

Note. In a given column for every growing season, means followed by a different letters are significantly different at α = 0.05. The standard error of the mean is given in parentheses.

control. Results similar to those of Wanvestraut et al. (2004) were reported in a black walnut–maize alley-cropping study conducted in Indiana (Gillespie et al., 2000; Jose, Gillespie, Seifert, & Biehle, 2000). Jose, Gillespie, Seifert, and Biehle (2000) reported a 21% increase in maize leaf area in the barrier treatment compared with the non-barrier treatment. In a companion study, Gillespie et al. (2000) reported a decreased maize yield when comparing alley-grown maize separated from the tree rows by a polyethylene root barrier versus alley-grown maize with no root barrier. By quantifying competition for water in the black walnut–maize alley-cropping system, Jose, Gillespie, Seifert, and Biehle (2000) concluded that severe competition for water was occurring between the trees and crops.

It is important to note that interspecific competition for water becomes increasingly intense when water levels decrease throughout the soil profile (Miller & Pallardy, 2001). Factors such as drought, the water holding capacity of the soil, and irrigation all play a role in the degree to which competition for water limits plant growth and productivity. Competition for water can be minimal given adequate levels of precipitation and/or irrigation in an agroforestry system.

Competition for nutrients

As in the case of conventional agriculture, nutrients can often be limiting in agroforestry systems. Therefore, many agroforestry systems are subject to fertilization, which is most commonly done at the level needed for the crop component to maintain high growth and productivity. Without fertilization, inter- and intraspecific competition for nutrients will be high and there will likely be a decrease in crop yields with time (Jose, Gillespie, Seifert, Mengel, & Pope, 2000).

Allen et al. (2004b) observed competition for N in a pecan–cotton alley-cropping system in northwestern Florida, where cotton plants in a barrier treatment had a 59% higher aboveground biomass than plants in a non-barrier treatment. Although a companion study indicated that competition for water was also a factor (Wanvestraut et al., 2004), the researchers hypothesized that because pecan trees leaf out earlier in the spring and have a high nutrient demand early in the growing season, the soil N was depleted before the cotton plants were established later in the growing season. Therefore, in this particular system, cotton plants are more likely to rely on fertilizer N to fulfill plant needs (Allen et al., 2004b).

Table 4–4. Comparison of height, diameter, and stem volume index of wild cherry and hybrid walnut trees 6 yr after planting for two treatments, intercropped or monoculture with weed control only (modified from Chifflot et al., 2006).

Parameter	Species	Treatment	
		Intercrop	Monoculture
Height (*H*), cm	wild cherry	522 a	470 b
	hybrid walnut	436 a	332 b
Diameter (*D*), cm	wild cherry	9.2 a	447.3 b
	hybrid walnut	47.8 a	444.7 b
Stem volume index (D^2H), 10^3 cm^3	wild cherry	447.2 a	431.6 b
	hybrid walnut	431.3 a	448.9 b

Note. Means followed by different letters are significantly different at $\alpha = 0.05$ for each row.

In addition, in a companion study, Allen et al. (2005) showed that N mineralization rates differed between barrier and non-barrier treatments in this pecan–cotton alley-cropping system. Higher rates of N mineralization were observed in the non-barrier treatment (26.05 mg kg^{-1} mo^{-1}) than the rates observed in the barrier treatment (19.78 mg kg^{-1} mo^{-1}), indicating that competitive interactions for water and N in the non-barrier treatment may have led to a decreased ability of the cotton plants to take up N (Allen et al., 2005).

If water is not a limiting factor, trees also benefit from increased nutrient availability in agroforestry systems. For example, Chifflot, Bertoni, Cabanettes, and Gavaland (2006) showed that 6 yr after planting, wild cherry (*Prunus avium* L.) and hybrid walnut trees (*J. nigra* × *J. regia*) grown in an intercropped agroforestry system with unirrigated cereal crops in France were significantly larger (Table 4–4) and had significantly greater N content and concentration (walnut hybrid only) than trees grown in a traditional plantation with weed control (Figure 4–8). They concluded that the trees benefited from N fertilization that was applied to the cropped alleyways.

Allelopathy

In addition to increased competition, the mixing of multiple species in a single system can potentially bring about other negative effects. One of those effects is allelopathy, which is caused by allelochemicals that are produced by some plant species and released into the soil by root exudation and aboveground litterfall. Allelochemicals have been documented to affect germination, growth, development, distribution, and reproduction of numerous plant species (Inderjit & Mallik, 2002). However, production rates of these chemicals in a given system depend on a variety of factors, including age of the species, density of the species, and the time of year. These factors, in combination with the

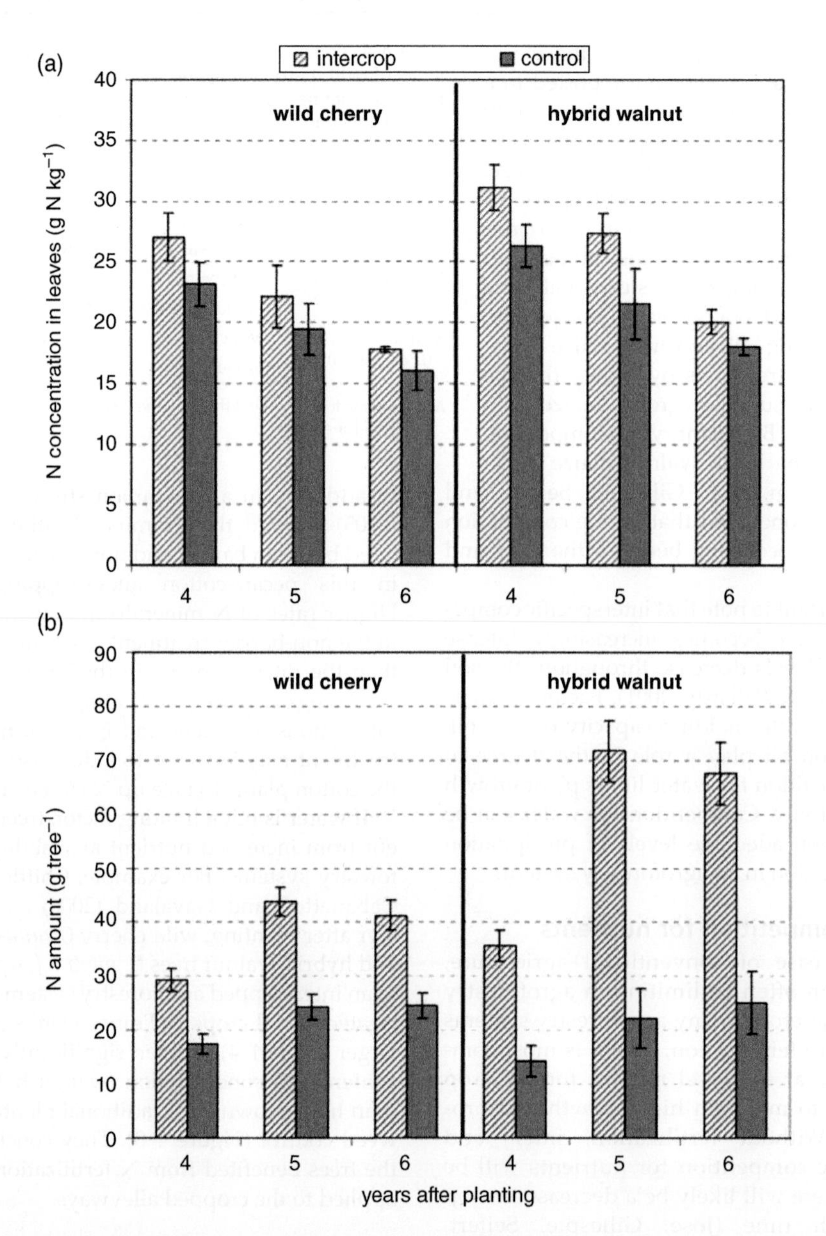

Fig. 4–8. Average leaf N (a) concentrations and (b) amounts in wild cherry and hybrid walnut trees 6 yr after planting with unirrigated cereal crops (intercropping) or in traditional monoculture plantations (control). The error bars show the 95% confidence intervals of the measured means (adapted from Chifflot et al., 2006).

residence time of the chemicals in the soil and plant resistance and/or tolerance to the chemicals, play a role in the degree to which these chemicals inhibit growth. It should also be noted that under certain soil conditions, such as low soil water regimes, these chemicals may be oxidized into a nontoxic form (Fisher, 1978).

There are several examples of allelopathy in temperate agroforestry systems (Geyer & Fick, 2015; Jose & Gillespie, 1998; Jose & Holzmueller, 2008; Thevathasan, Gordon, & Voroney, 1998) and most of them involve black walnut, a species that produces a phenolic compound called juglone (5-hydroxy-1, 4-naphthoquinone) that restricts the growth of some other species. Although Jose, Gillespie, Seifert, and Biehle (2000) documented that competition for water was the leading factor in reduced growth of alley-grown maize in a black walnut–maize alley-cropping system in Indiana, a companion study (Jose & Gillespie, 1998) also indicated the possibility of juglone phytotoxicity. Jose and Holzmueller (2007) reported sensitivity of cotton (*Gossypium* sp.) and peanut (*Arachis hypogaea* L.)

when exposed to juglone in hydroponic cultures. These are two common species with potential for alley cropping with pecan in the southern United States, and pecan also produces juglone (Jose, 2002). However, some species do not appear to be as susceptible to juglone (Geyer & Fick, 2015), and this should be taken into consideration when developing an agroforestry system.

It has also been reported that certain crop species may induce allelopathic effects on trees as well, including a decrease in development and growth (Smith, Wolf, Cheary, & Carroll, 2001; Todhunter & Beineke, 1979). In a study of containerized pecan trees, Smith et al. (2001) showed that allelochemical-containing leachates added to the containers from bermudagrass [*Cynodon dactylon* (L.) Pers.], cutleaf evening primrose (*Oenothera laciniata* Hill.), and tall fescue [*Schedonorus phoenix* (Scop.) Holub.] decreased pecan root weight by 17%, trunk weight by 22%, and total tree dry weight by 19% compared with the control treatment.

Management techniques to reduce the effects of allelopathy have also been examined in agroforestry systems. Jose, Gillespie, Seifert, and Biehle (2000) demonstrated that by separating the root systems of black walnut and maize using a polyethylene barrier, crop yield became similar to that of a monoculture. They further showed that the juglone concentration in the soil was negligible beyond the polyethylene barrier. Juglone concentration beyond the root barrier decreased to trace levels of 0.08 and 0.01 $\mu g\ g^{-1}$ soil (at distances of 2.45 and 4.25 m, respectively) in the barrier treatment compared with 0.42 and 0.32 $\mu g\ g^{-1}$ soil in the non-barrier control treatment.

A comprehensive study examining field soil juglone concentrations, sorption mechanisms, juglone production rates, and degradation rates and products (Von Kiparski, Lee, & Gillespie, 2007) showed that juglone can accumulate under field conditions, with release rates from black walnut being greater than abiotic and microbial transformation rates. In a 19-yr-old walnut plantation, surface soil pore water juglone concentrations approached but did not exceed the inhibition solution thresholds of typical intercrops. But substantially higher levels of juglone can be reversibly sorbed by soils, and true plant impacts may be a balance of responses to multiple stress conditions in the mixed systems. From greenhouse studies, it was determined that substantial quantities of juglone can be released into the rhizosphere, and so rooting patterns of intercrops will be of particular concern when judging allelopathic potential. However, soil chemistry will play a role in these intercrops, as this study showed that microbial activity will quickly degrade juglone and decrease persistence. Soils low in microbial activity, including subsurface horizons and acidic soils that are low in organic C and fertility, can accumulate juglone, and thus this negative interaction among interplanted species should continue to be considered in walnut and pecan agroforestry systems.

Facilitative Interactions—Belowground

Hydraulic lift

Hydraulic lift is the process by which deep-rooted plants transport or conduct water from deep within the soil and release it into the upper, drier regions of the soil. The process has been reported to be an appreciable water source for neighboring plants in some systems (Caldwell & Richards, 1989; Corak, Blevins, & Pallardy, 1987). This phenomenon can increase plant growth, in some cases, by increasing the availability of water for shallow-rooted plants and has important implications for ecosystem nutrient cycling and net primary productivity (Horton & Hart, 1998).

In a tropical agroforestry context, numerous studies have shown that trees can benefit associated crop plants through hydraulic lift by increasing water availability during dry periods when water would otherwise be unavailable (Burgess, Adams, Turner, & Ong, 1998; Dawson, 1993; Ong et al., 1999; van Noordwijk, Lawson, Soumaré, Groot, & Hairiah, 1996). In temperate agroforestry systems, however, research documenting the hydraulic lift phenomenon is limited. Hydraulic lift in temperate systems has been reported in *Quercus* sp. and *Pinus* sp. (Asbjornsen, Shepherd, Helmers, & Mora, 2008; Espeleta, West, & Donovan, 2004; Penuelas & Filella, 2003). These species are commonly used in temperate agroforestry systems, indicating a potential for these genera to be used in agroforestry to positively impact water relations. For example, Espeleta et al. (2004) reported hydraulic lift in longleaf pine (*Pinus palustris* Mill.), a species commonly used in silvopastoral systems in the southeastern United States. They reported hydraulic lift in two oak species (*Q. laevis* Walt. and *Q. incana* Bartr.) as well. They concluded that the ability of these species to redistribute water from the deep soil to the rapidly drying shallow soil has a strong positive effect on the water balance of understory plants.

Dinitrogen fixation

The incorporation of trees and crops that are able to biologically fix N_2 is fairly common and well researched in tropical agroforestry systems (Nair, Buresh, Mugendi, & Latt, 1999). In temperate systems, similar accounts of incorporating N_2–fixing trees into agroforestry are rare, perhaps because of the abundance and historically low cost of N fertilizer and the low value of N_2–fixing trees. Despite the infrequent use of biological N_2 fixation by trees in temperate agroforestry systems, there is potential for using N_2–fixing tree species native to temperate environments. Species from the genera *Robinia*, *Prosopis*, and *Alnus* have the potential to provide N_2 fixation benefits in temperate agroforestry systems (Boring & Swank, 1984; Seiter, Ingham, William, & Hibbs, 1995). Seiter et al. (1995) demonstrated this potential in a red alder (*Alnus rubra* Bong.)–maize alley-cropping system in Oregon. They observed, using a ^{15}N injection technique, that 32–58% of the total N in maize was obtained from N_2 fixed by red alder and that N transfer increased by shortening the distance between the trees and crops.

There are also several leguminous herbaceous plant species capable of fixing atmospheric N_2 in temperate agroforestry systems, including alfalfa, clover, hairy vetch (*Vicia villosa* Roth), and soybean (Troeh & Thompson, 1993). Although multiple studies have incorporated leguminous herbaceous species capable of biological N_2 fixation into temperate agroforestry systems (Alley et al., 1998; Delate et al., 2005; Gakis et al., 2004; Silva-Pando, Gonzalez-Hernandez, & Rozados-Lorenzo, 2002), few studies have actually quantified the effects that these species have on soil N (Dupraz et al., 1998; Waring & Snowdon, 1985). Nitrogen buildup in the soil is possible from leguminous herbaceous understory species; however, this is a slow process that does not occur immediately after herbaceous plant establishment. In a radiata pine–subterranean clover (*Trifolium subterraneum* L.) silvopasture in Australia, Waring and Snowdon (1985) observed a 36% increase in soil N at the end of seven growing seasons in the silvopasture, which corresponded to a 14% increase in tree diameter compared with pines growing in a monoculture without a subterranean clover understory.

Root plasticity

Many plant species show some degree of plasticity (the ability to respond to changes in local nutrient supplies or impervious soil layers) in their vertical (as well as lateral) root distribution (Kumar & Jose, 2018). Plants also exploit plasticity to avoid competition (Ong et al., 1996; Schroth, 1999). Belowground niche separation in response to competition can help component species in an agroforestry system to avoid competition. This can lead to complementary or facilitative interactions that help increase the production potential of the system.

It is possible to apply treatments such as repeated disking, knifing of fertilizer applications, or trenching, applied while trees are young, to force tree roots to grow deeper. Wanvestraut et al. (2004) observed pecan roots displaying plasticity by penetrating deeper soil strata, thereby avoiding a region of high cotton root density. This enhanced the overall water use efficiency of the system because the cotton plants were able to capitalize on the water available in the topsoil layer while the pecan trees exploited the moisture available in the deeper soil layers. Zamora et al. (2007) corroborated the findings of Wanvestraut et al. (2004) and confirmed the morphological plasticity of cotton roots in response to competition from pecan trees.

Dawson, Duff, Campbell, and Hirst (2001) demonstrated that cherry (*Prunus avium* L.) tree root distribution was influenced by grass competition in a silvopastoral system in Scotland. Cherry roots increased within the upper soil surface horizon after grass competition was removed with herbicides, and in areas where grass competition was not removed, the average depth of the tree roots increased with time.

Safety net role

In conventional agricultural systems, less than half of the applied N and P fertilizer is taken up by crops (Smil, 1999, 2000). Consequently, excess fertilizer is washed away from agricultural fields via surface runoff or leached into the subsurface water supply, thus contaminating water sources and decreasing water quality (Bonilla, Muñoz, & Vauclin, 1999; Ng, Drury, Serem, Tan, & Gaynor, 2000; Tilman et al., 2002). In an agroforestry system, however, trees with deep rooting systems potentially play the role of a "safety net" by retrieving excess nutrients that have been leached below the rooting zone of agronomic crops. These nutrients are then recycled back into the system through root turnover and litterfall, increasing the nutrient use efficiency of the system (van Noordwijk et al., 1996). Additionally, because trees have a longer growing season than most agronomic crops, tree roots occupying the same rooting zone as associated agronomic crops will increase nutrient use and use efficiency in an

Table 4–5. Percentage of sediment and nutrients removed by two riparian buffer systems in a study conducted in Iowa (adapted from Lee et al., 2003).

Sediment or nutrient	Switchgrass only buffer removal	Switchgrass and woody stem buffer removal
	%	
Sediment	95	97
Total N	80	94
NO_3–N	62	85
Total P	78	91
PO_4–P	58	80

agroforestry system by capturing nutrients before crops are planted and after crops are harvested.

Evidence supporting the safety net concept has been observed in field trials. In a pecan–cotton alley-cropping system in northwestern Florida, Allen et al. (2004a) reported a 245% NO_3–N increase at the 0.9-m depth when pecan roots were separated from cotton roots by a root barrier compared with the non-barrier treatment. These researchers suggested that this indicates the trees could potentially play the role of a N safety net by taking up N fertilizer from deep in the soil profile and redepositing it on the soil surface via litterfall (Allen et al., 2004a).

The safety net concept can be applied to other nutrients in agroforestry systems as well. In a silvopastoral system in Florida, Nair, Nair, Kalmbacher, and Ezenwa (2007) monitored soil P concentrations in pastures with and without 20-yr-old slash pine (*Pinus elliottii* Engelm.) trees. They found lower concentrations of P in the soil surface horizon and at the 1.0-m depth in pastures with trees, suggesting that silvopastoral associations enhance soil nutrient retention and limit nutrient transport in surface water. Lee, Isenhart, and Schultz (2003) documented increased nutrient removal efficiency when trees were incorporated into a riparian buffer strip placed on the border of agronomic field plots in their study in Iowa. They reported that a switchgrass (*Panicum virgatum* L.) and woody stem buffer removed similar amounts of sediment as a switchgrass-only buffer, but nutrient removal was increased by >20% in the switchgrass and woody stem buffer (Table 4–5).

The Future

It is true that we have made significant improvements in our understanding of ecological interactions in temperate agroforestry. As this chapter has revealed, we have information on above- and belowground interactions that define the ecological sustainability of some of the well-known agroforestry practices such as alley cropping and silvopasture in the United States and other temperate regions of the world. We also have information on the management techniques that may reduce competitive interactions while enhancing complementarity in those systems. However, our knowledge is still limited in several areas. For example, despite much research examining resource competition, we still lack a deeper understanding of the interactive effects of multiple resources on system productivity in several agroforestry systems. Modeling has helped us understand multiple resource interactions to a great extent (Lovell et al., 2017), but continued acquisition of information at a range of scales is urgently needed. Although information on specific components and their interactions are important, we also need to pay attention to interactions of agroforestry systems with the biotic and abiotic components of the surrounding landscape matrix. Watershed-level research and studies of agroforestry systems as wildlife habitats and corridors need to explore these relationships in detail.

Another area that needs immediate attention is the screening of species and germplasm for above- and belowground complementarity. Most of the improved germplasm currently used in agroforestry comes from breeding efforts for monoculture cropping systems. Breeding for crops and trees that can perform better under shade and under interspecific competition for water and nutrients needs to be initiated.

The available literature on facilitative interactions in temperate agroforestry is very limited. For example, the concept of hydraulic lift is yet to be experimentally proven in a temperate context. Information on the canopy and root architecture of many common agroforestry species is still not readily available. Dinitrogen fixation remains an unexplored and underutilized concept in many of the well-studied temperate agroforestry systems.

A number of agroforestry practices have received little attention from the scientific community despite their popularity. For example, forest farming is an attractive agroforestry practice in many parts of the United States. However, ecological sustainability or component interactions have seldom been investigated in these systems. Similarly, incorporation of high-value agronomic or horticultural crops into existing agroforestry practices or the design of new agroforestry systems also needs to be explored.

Conclusion

The ecological interaction among components in temperate agroforestry systems is a broad topic to cover in a single chapter; however, some generalizations can be made based on the available literature. For example, even though the impacts of competition for light, water, and nutrients can be difficult to separate for a given system due to temporal changes in climate, level of fertilization, and the varying effects that both species diversity and species size have on a system, crop production is typically decreased in areas immediately surrounding trees in agroforestry systems. In some cases, however, such as with shelterbelts, trees can have a positive impact on overall crop production or, in the case of silvopastoral systems, improve the quality of forage despite the decrease in forage production. Even though crop yields may be decreased compared with monocultural systems, by incorporating multiple species in a single system there is a possibility to increase the overall biomass production and economic value.

In addition to the potential for increased production from agroforestry systems compared with conventional agriculture, there are other potential benefits of agroforestry systems. For example, evidence supporting the safety net role of trees in temperate alley-cropping systems shows both a reduction in the contamination of surface and subsurface water and a reduced competition for nutrients. Furthermore, this safety net effect can also be seen in woody riparian buffer strips, which can be incorporated into any agricultural system, including conventional agriculture.

Several information gaps still exist that need immediate attention by the scientific community. These vary from quantifying the interactive effects of multiple resources on system productivity to studying broader landscape-level interactions of agroforestry systems within the landscape matrix. Despite these limitations, there is evidence that competition for light, water, and nutrients, in combination with other factors, such as allelopathy, can be managed through both the design and maintenance of agroforestry systems so that the competitive influence is minimized while the facilitative influence is maximized. This offers promise to the long-term ecological sustainability of agroforestry systems.

References

Albaugh, J. M., Albaugh, T. J., Heiderman, R. R., Leggett, Z., Stape, J. L., King, K., … King, J. S. (2014). Evaluating changes in switchgrass physiology, biomass, and light-use efficiency under artificial shade to estimate yields if intercropped with *Pinus taeda* L. *Agroforestry Systems, 88,* 489-503. https://doi.org/10.1007/s10457-014-9708-3

Allen, S. C., Jose, S., Nair, P. K. R., Brecke, B. J., Nair, V. D., Graetz, D. A., & Ramsey, C. L. (2005). Nitrogen mineralization in a pecan (*Carya illinoensis* K. Koch)–cotton (*Gossypium hirsutum* L.) alley cropping system in the southern United States. *Biology and Fertility of Soils, 41,* 28–37.

Allen, S. C., Jose, S., Nair, P. K. R., Brecke, B. J., Nkedi-Kizza, P., & Ramsey, C. L. (2004a). Safety-net role of tree roots: Evidence from a pecan (*Carya illinoensis* K. Koch)–cotton (*Gossypium hirsutum* L.) alley cropping system in the southern United States. *Forest Ecology and Management, 192,* 395–407.

Allen, S. C., Jose, S., Nair, P. K. R., Brecke, B. J., & Ramsey, C. L. (2004b). Competition for ^{15}N-labeled fertilizer in a pecan (*Carya illinoensis* K. Koch)–cotton (*Gossypium hirsutum* L.) alley cropping system in the southern United States. *Plant and Soil, 263,* 151–164.

Alley, J. L., Garrett, H. E., McGraw, R. L., & Blanche, C. A. (1998). Forage legumes as living mulches for trees in agroforestry practices: Preliminary results. *Agroforestry Systems, 44,* 281–291.

Altieri, M. A. (1991). Increasing biodiversity to improve pest management in agro-ecosystems. In D. L. Hawksworth (Ed.), *The biodiversity of microorganisms and invertebrates: Its role in sustainable agriculture* (pp. 165–182). Wallingford, UK: CAB International.

Artru, S., Garré, S., Dupraz, C., Hiel, M. P., Blitz-Frayret, C., & Lassois, L. (2017). Impact of spatio-temporal shade dynamics on wheat growth and yield, perspectives for temperate agroforestry. *European Journal of Agronomy, 82,* 60–70.

Asbjornsen, H., Shepherd, G., Helmers, M., & Mora, G. (2008). Seasonal patterns in depth of water uptake under contrasting annual and perennial systems in the Corn Belt region of the midwestern US. *Plant and Soil,* 308, 69–92.

Atangana, A., Khasa, D., Chang, S., & Degrande, A. (2014). *Tropical agroforestry.* Dordrecht, the Netherlands: Springer Nature.

Bagley, W. T. (1964). Responses of tomatoes and beans to windbreak shelter. *Journal of Soil and Water Conservation, 19,* 71–73.

Baraldi, R., Bertazza, G., Bogino, J., Luna, V., & Bottini, R. (1995). The effect of light quality on *Prunus cerasus*: II. Changes in hormone levels in plants grown under different light conditions. *Photochemistry and Photobiology, 62,* 800–803. https://doi.org/10.1111/j.1751-1097.1995.tb08732.x

Bobryk, C. W., Rega, C. C., Bardhan, S., Farina, A., He, H., & Jose, S. (2016). Assessing structural and compositional resources in temperate agroforestry systems using soundscape analysis. *Agroforestry Systems, 90,* 997–1008.

Bonilla, C. A., Muñoz, J. F., & Vauclin, M. (1999). Opus simulation of water dynamics and nitrate transport in a field plot. *Ecological Modelling, 122,* 69–80. https://doi.org/10.1016/S0304-3800(99)00119-2

Boring, L. R., & Swank, W. T. (1984). The role of black locust (*Robinia pseudoacacia*) in forest succession. *Journal of Ecology, 72,* 749–766.

Brandle, J. R., Hodges, L., & Zhou, X. (2004). Windbreaks in sustainable agriculture. *Agroforestry Systems,* 61, 65–78.

Brenner, A. J. (1996). Microclimatic modifications in agroforestry. In C.K. Ong and P. Huxley (Eds.), *Tree–crop interactions: A physiological approach* (pp. 159–187). Wallingford, UK: CAB International.

Brooker, R. W., & Callaghan, T. V. (1998). The balance between positive and negative plant interactions and its relationship to environmental gradients: A model. *Oikos, 81,* 196–207.

Bugg, R. L., Sarrantonio, M., Dutcher, J. D., & Phatak, S. C. (1991). Understory cover crops in pecan orchards: Possible management systems. *American Journal of Alternative Agriculture,* 6, 50–62.

Burgess, S. S. O., Adams, M. A., Turner, N. C., & Ong, C. K. (1998). The redistribution of soil water by tree root systems.

Oecologia, 115, 306–311. https://doi.org/10.1007/s004420050521

Burner, D. M. (2003). Influence of alley crop environment on orchardgrass and tall fescue herbage. *Agronomy Journal, 95,* 1163–1171. https://doi.org/10.2134/agronj2003.1163

Caldwell, M. M., & Richards, J. H. (1989). Hydraulic lift: Water efflux from upper roots improves effectiveness of water uptake by deep roots. *Oecologia, 79,* 1–5. https://doi.org/10.1007/BF00378231

Cannell, M. G. R., van Noordwijk, M., & Ong, C. K. (1996). The central agroforestry hypothesis: The trees must acquire resources that the crop would not otherwise acquire. *Agroforestry Systems, 34,* 27–31.

Casper, B. B., & Jackson, R. B. (1997). Plant competition underground. Annual Review of Ecology and Systematics, 28, 545–570. https://doi.org/10.1146/annurev.ecolsys.28.1.545

Chifflot, V., Bertoni, G., Cabanettes, A., & Gavaland, A. (2006). Beneficial effects of intercropping on the growth and nitrogen status of young wild cherry and hybrid walnut trees. *Agroforestry Systems,* 66, 13–21.

Cleugh, H. A. (2002). Field measurements of windbreak effects on airflow, turbulent exchanges and microclimates. *Australian Journal of Experimental Agriculture, 42,* 665–677.

Corak, S. J., Blevins, D. G., & Pallardy, S. G. (1987). Water transfer in an alfalfa–maize association: Survival of maize during drought. *Plant Physiology, 84,* 582–586.

Davis, J. E., & Norman, J. M. (1988). Effects of shelter on plant water use. *Agriculture, Ecosystems & Environment, 22–23,* 393–402. https://doi.org/10.1016/0167-8809(88)90034-5

Davis, M. H., & Simmons, S. R. (1994a). Tillering response of barley to shifts in light quality caused by neighboring plants. *Crop Science, 34,* 1604–1610. https://doi.org/10.2135/cropsci1994.0011183X003400060033x

Davis, M.H, & Simmons, S. R. (1994b). Far-red light reflected from neighboring vegetation promotes shoot elongation and accelerate flowering in spring barley plants. *Plant, Cell & Environment, 17,* 829–836. https://doi.org/10.1111/j.1365-3040.1994.tb00177.x

Dawson, L. A., Duff, E. I., Campbell, C. D., & Hirst, D. J. (2001). Depth distribution of cherry (*Prunus avium* L.) tree roots as influenced by grass root competition. *Plant and Soil, 231,* 11–19.

Dawson, T. E. (1993). Hydraulic lift and water use by plants: Implications for water balance, performance and plant–plant interactions. *Oecologia, 95,* 565–574.

Delate, K., Holzmueller, E., Mize, C., Frederick, D., & Brummer, C. (2005). Tree establishment and growth using forage ground covers in an alley-cropped system. *Agroforestry Systems, 65,* 43–52.

Dupraz, C., Simorte, V., Dauzat, M., Bertoni, G., Bernadac, A., & Masson, P. (1998). Growth and nitrogen status of young walnuts as affected by intercropped legumes in a Mediterranean climate. *Agroforestry Systems, 43,* 71–80.

Ehret, M., Graß, R., & Wachendorf, M. (2015). The effect of shade and shade material on white clover/perennial ryegrass mixtures for temperate agroforestry systems. *Agroforestry Systems,* 89, 557–570.

Espeleta, J. F., West, J. B., & Donovan, L. A. (2004). Species-specific patterns of hydraulic lift in co-occurring adult trees and grasses in a sandhill community. *Oecologia, 138,* 341–349.

Fisher, R. F. (1978). Juglone inhibits pine growth under certain moisture regimes. *Soil Science Society of America Journal, 42,* 801–803. https://doi.org/10.2136/sssaj1978.036159950042 00050030x

Foley, J. A., Ramankutty, N., Brauman, K., Cassidy, E. S., Gerber, J. S., Johnston, M., …Zaks, D. (2011). Solutions for a cultivated planet. *Nature, 478,* 337–342. https://doi.org/10.1038/nature10452pmid:21993620

Ford, M., Zamora, D., Current, D., Magner, J., Wyatt, G., Walter, D., & Vaughan, S. (2019). Impact of managed woodland grazing on forage quantity, quality and livestock performance: The potential for silvopasture in Central Minnesota, USA. *Agroforestry Systems, 93,* 67–79.

Funabashi, M. (2018). Human augmentation of ecosystems: Objectives for food production and science by 2045. *npj Science of Food,* 2, 16. https//doi.org/10.1038/s41538-018-0026-4

Gakis, S., Mantzanas, K., Alifragis, D., Papanastasis, V. P., Papaioannou, A., Seilopoulos, D., & Platis, P. (2004). Effects of understory vegetation on tree establishment and growth in a silvopastoral system in northern Greece. *Agroforestry Systems, 60,* 149–157.

Garcia-Barrios, L., & Ong, C. K. (2004). Ecological interactions, management lessons, and design tools in tropical agroforestry systems. *Agroforestry Systems, 61,* 221–236.

Garrett, H. E., McGraw, R. L., & Walter, W. D. (2009). Alley cropping practices. In H. E. Garrett (Ed.), North American agroforestry: An integrated science and practice (2nd ed., pp. 133–162). Madison, WI: ASA.

Garrity, D. P. (2004). Agroforestry and the achievement of the millennium development goals. *Agroforestry Systems, 61,* 5–17.

Gause, G.F. 1934. *The struggle for existence.* Baltimore, MD: Williams & Wilkins.

Geyer, W. A., & Fick, W. H. (2015). Yield and forage quality of smooth brome in a black walnut alley-cropping practice. *Agroforestry Systems, 89,* 107–112.

Gibbs, S., Koblents, H., Coleman, B., Gordon, A., Thevathasan, N., & Wiliams, P. (2016). Avian diversity in a temperate tree-based intercropping system from inception to now. *Agroforestry Systems, 90,* 905–916. https://doi.org/10.1007/s10457-016-9901-7

Gillespie, A. R., Jose, S., Mengel, D. B., Hoover, W. L., Pope, P. E., Seifert, J. R., …Benjamin, T. J. (2000). Defining competition vectors in a temperate alley cropping system in the midwestern USA: 1. Production physiology. *Agroforestry Systems, 48,* 25–40.

Gillespie, A. R., Miller, B. K., & Johnson, K. D. (1995). Effects of ground cover on tree survival and growth in filter strips of the Cornbelt region of the midwestern US. *Agriculture, Ecosystems & Environment, 53,* 263–270.

Goldberg, D. E. (1990). Components of resource competition in plant communities. In J. B. Grace & D. Timan (Ed.), *Perspectives on plant competition* (pp. 27–65). San Diego, CA: Academic Press.

Gómez-Gutierrez, J. M., & Pérez-Fernández, M. (1996). The "dehesas": Silvopastoral systems in semiarid Mediterranean regions with poor soils, seasonal climate and extensive utilisation. In M. Etienne (Ed.), *Western European silvopastoral systems* (pp. 55–70). Paris: INRA Editions.

Gravel, D., Mouquet, N., Loreau, M., & Guichard, F. (2010). Patch dynamics, persistence, and species coexistence in metaecosystems. *The American Naturalist, 176,* 289–302.

Hardin, G. (1960). The competitive exclusion principle. *Science, 131,* 1292–1297.

Holmgren, M., Scheffer, M., & Huston, M. A. (1997). The interplay of facilitation and competition in plant communities. *Ecology, 78,* 1966–1975.

Horton, J. L., & Hart, S. C. (1998). Hydraulic lift: A potentially important ecosystem process. *Trends in Ecology & Evolution, 13,* 232–235. https://doi.org/10.1016/s0169-5347(98)01328-7

Hubbell, S. P. (2001). *The unified neutral theory of species abundance and diversity.* Princeton, NJ: Princeton Univ. Press.

Inderjit, & Mallik, A. U. (2002). *Chemical ecology of plants: Allelopathy in aquatic and terrestrial ecosystems.* Basel, Switzerland: Birkhauser.

Isabelle, B., Damien, G., & Wilfried, T. (2014). FATE-HD: A spatially and temporally explicit integrated model for predicting vegetation structure and diversity at regional scale. *Global Change Biology, 20*, 2368–2378.

Johnson, R. J., & Beck, M. M. (1988). Influences of shelterbelts on wildlife management and biology. *Agriculture, Ecosystems & Environment, 22–23*, 301–335.

Jose, S. (1997). Interspecific interactions in alley cropping: The physiology and biogeochemistry (Doctoral dissertation). West Lafayette, IN: Purdue University.

Jose, S. (2002). Black walnut allelopathy: Current state of the science. In Inderjit & A. U. Mallik (Eds.), Chemical ecology of plants: Allelopathy in aquatic and terrestrial ecosystems (pp. 149–172). Basel, Switzerland: Birkhauser.

Jose, S. (2009). Agroforestry for ecosystem services and environmental benefits: An overview. *Agroforestry Systems, 76*, 1–10. https://doi.org/10.1007/s10457-009-9229-7

Jose, S., & Dollinger, J. (2019). Silvopasture: A sustainable livestock production system. *Agroforestry Systems, 93*, 1–9. https://doi.org/10.1007/s10457-019-00366-8

Jose, S., & Gillespie, A. R. (1998). Allelopathy in black walnut (*Juglans nigra* L.) alley cropping: I. Spatio-temporal variation in soil juglone in a black walnut–corn (*Zea mays* L.) alley cropping system in the mid-western USA. *Plant and Soil, 203*, 191–197.

Jose, S., Gillespie, A. R., & Pallardy, S. G. (2004). Interspecific interactions in temperate agroforestry. *Agroforestry Systems, 61*, 237–255.

Jose, S., Gillespie, A. R., Seifert, J. R., & Biehle, D. J. (2000). Defining competition vectors in a temperate alley cropping system in the mid-western USA: 2. Competition for water. *Agroforestry Systems, 48*, 41–59.

Jose, S., Gillespie, A. R., Seifert, J. R., Mengel, D. B., & Pope, P. E. (2000). Defining competition vectors in a temperate alley cropping system in the mid-western USA: 3. Competition for nitrogen and litter decomposition dynamics. *Agroforestry Systems, 48*, 61–77.

Jose, S., & Holzmueller, E. J. (2008). Black walnut allelopathy: Implications for intercropping. In R. S. Zeng, A. U. Mallik, & S. M. Luo (Eds.), Allelopathy in sustainable agriculture and forestry. New York: Springer. https://doi.org/10.1007/978-0-387-77337-7_16

Jose, S., Williams, R., & Zamora, D. (2006). Belowground ecological interactions in mixed-species forest plantations. *Forest Ecology and Management, 233*, 231–239. https://doi.org/10.1016/j.foreco.2006.05.014

Kallenbach, R. L., Kerley, M. S., & Bishop-Hurley, G. J. (2006). Cumulative forage production, forage quality and livestock performance from an annual ryegrass and cereal rye mixture in a pine–walnut silvopasture. *Agroforestry Systems, 66*, 43–53.

Kelty, M. J. (2000). Species interactions, stand structure, and productivity in agroforestry systems In M. S. Ashton & R. Montagnini (Eds.) *The silvicultural basis for agroforestry systems* (pp. 183–203). Boca Raton, FL: CRC Press.

Kort, J. (1988). Benefits of windbreaks to field and forage crops. *Agriculture, Ecosystems & Environment, 22–23*, 165–191.

Kozlowski, T. T., & Pallardy, S.G. (1997). Physiology of woody plants (2nd ed.). San Diego, CA: Academic Press.

Krueger, W. C. (1981). How a forest affects a forage crop. *Rangelands, 3*, 70–71.

Kumar, B. M., & Jose, S. (2018). Phenotypic plasticity of roots in mixed tree species agroforestry systems: Review with examples from peninsular India. *Agroforestry Systems, 92*, 59–69.

Lambers, H., Chapin, F. S., III, & Pons, T. L. (1998). *Plant physiological ecology*. New York: Springer.

Lee, K. H., Isenhart, T. M., & Schultz, R. C. (2003). Sediment and nutrient removal in an established multi-species riparian buffer. *Journal of Soil and Water Conservation, 58*, 1–8.

Lehmkuhler, J. W., Felton, E. E. D., Schmidt, D. A., Bader, K. J., Garrett, H. E., & Kerley, M. S. (2003). Tree protection methods during the silvopastoral-system establishment in midwestern USA: Cattle performance and tree damage. *Agroforestry Systems, 59*, 35–42.

Li, F. D., Meng, P., Dali, F., & Wang, B. P. (2008). Light distribution, photosynthetic rate and yield in a Paulownia–wheat intercropping system in China. *Agroforestry Systems, 74*, 163–172.

Lin, C. H., McGraw, R. L., George, M. F., & Garrett, H. E. (1999). Shade effects on forage crops with potential in temperate agroforestry practices. *Agroforestry Systems, 44*,109–119.

Lin, C. H., McGraw, R. L., George, M. F., & Garrett, H. E. (2001). Nutritive quality and morphological development under partial shade of some forage species with agroforestry potential. *Agroforestry Systems, 53*, 269–281.

Lovell, S. T., Dupraz, C., Gold, M., Jose, S., Revord, R., Stanek, E., & Wolz, K. J. (2017). Temperate agroforestry research: Considering multifunctional woody polycultures and the design of long-term field trials. *Agroforestry Systems, 92*, 1397–1415. https://doi.org/10.1007/s10457-017-0087-4

Martin-Chave, A., Béral, C., & Capowiez, Y. (2019). Agroforestry has an impact on nocturnal predation by ground beetles and Opiliones in a temperate organic alley cropping system. *Biological Control, 129*, 128–135. https://doi.org/10.1016/j.biocontrol.2018.10.009

Mayer, A. C., Stockli, V., Konold, W., & Kreuzer, M. (2006). Influence of cattle stocking rate on browsing of Norway spruce in subalpine wood pastures. *Agroforestry Systems*, 66, 143–149. https://doi.org/10.1007/s10457-005-5460-z

Mcadam, J. H., Sibbald, A. R., Teklehaimanot, Z., & Eason, W. R. (2007). Developing silvopastoral systems and their effects on diversity of fauna. *Agroforestry Systems, 70*, 81–89. https://doi.org/10.1007/s10457-007-9047-8

Miller, A. W., & Pallardy, S. G. (2001). Resource competition across the crop–tree interface in a maize–silver maple temperate alley cropping stand in Missouri. *Agroforestry Systems, 53*, 247–259.

Miller, T. E., Burns, J. H., Munguia, P., Walters, E. L., Kneitel, J. M., Richards, P. M., …Buckley, H. L. (2005). A critical review of 20 years' use of the resource ratio theory. *The American Naturalist, 165*, 439–448.

Moreno Marcos, G., Obrador, J. J., García, E., Cubera, E., Montero, M. J., Pulido, F., & Dupraz, C. (2007). Driving competitive and facilitative interactions in oak dehesas through management practices. *Agroforestry Systems, 70*, 25–40.

Nair, P. K. R. (1993). *An introduction to agroforestry*. Dordrecht, the Netherlands: Kluwer.

Nair, P. K. R., Buresh, R. J., Mugendi, D. N., & Latt, C. R. (1999). Nutrient cycling in tropical agroforestry systems: Myths and science. In L. E. Buck, J. P. Lassoie, & E. C. M. Fernandes (Eds.), *Agroforestry in sustainable agricultural systems* (pp. 1–31). Boca Raton, FL: CRC Press.

Nair, V. D., Nair, P. K. R., Kalmbacher, R. S., & Ezenwa, I. V. (2007). Reducing nutrient loss from farms through silvopastoral practices in coarse-textured soils of Florida, USA. *Ecological Engineering, 29*, 192–199.

NeSmith, D. S., & Ritchie, J. T. (1992a). Effects of soil water deficits during tassel emergence on development and yield component of maize (*Zea mays* L.). *Field Crops Research, 28*, 251–256.

NeSmith, D. S., & Ritchie, J. T. (1992b). Short- and long-term responses of corn to pre-anthesis soil water deficit. *Agronomy Journal, 84*,107–113. https://doi.org/10.2134/agronj1992.00021962008400010021x

Ng, H. Y. F., Drury, C. F., Serem, V. K., Tan, C. S., & Gaynor, J. D. (2000). Modeling and testing of the effect of tillage, cropping and water management practices on nitrate leaching in clay loam soil. *Agricultural Water Management, 43*, 111–131.

Nowak, J., Blount, A., & Workman, S. (2002). Integrated timber, forage and livestock production: Benefits of silvopasture (UF/IFAS Circ. 1430). Gainesville, FL: Florida Cooperative Extension Service, Institute of Food and Agricultural Sciences, University of Florida.

Ong, C. K., Black, C. R., Marshall, F. M., & Corlett, J. E. (1996). Principles of resource capture and utilization of light and water. In C. K. Ong & P. Huxley (Eds.), *Tree–crop interactions: A physiological approach* (pp. 73–158). Wallingford, UK: CAB International.

Ong, C. K., Corlett, J. E., Singh, R. P., & Black, C. R. (1991). Above and belowground interactions in agroforestry systems. *Forest Ecology and Management*, *45*, 45–57.

Ong, C. K., Deans, J. D., Wilson, J., Mutua, J., Khan, A. A. H., & Lawson, E. M. (1999). Exploring belowground complementarity in agroforestry using sap flow and root fractal techniques. *Agroforestry Systems*, *44*, 87–103.

Orefice, J., Caroll, J., Conroy, D., & Ketner, L. (2017). Silvopasture practices and perspectives in the northeastern United States. *Agroforestry Systems*, *91*, 149–160.

Pang, K., Van Sambeek, J. W., Navarrete-Tindall, N. E., Lin, C.-H., Jose, S., & Garrett, H. E. (2019a). Responses of legumes and grasses to non-, moderate, and dense shade in Missouri, USA: I. Forage yield and its species-level plasticity. *Agroforestry Systems*, 93, 11–24. https://doi.org/10.1007/s10457-017-0067-8

Pang, K., Van Sambeek, J. W., Navarrete-Tindall, N. E., Lin, C.-H., Jose, S., & Garrett, H. E. (2019b). Responses of legumes and grasses to non-, moderate, and dense shade in Missouri, USA: II. Forage quality and its species-level plasticity. *Agroforestry Systems*, 93:25–38. https://doi.org/10.1007/s10457-017-0068-7

Pardon, P., Reheul, D., Mertens, J., Reubens, B., De Frenne, P., De Smedt, P., …Verheyen, K. (2019). Gradients in abundance and diversity of ground dwelling arthropods as a function of distance to tree rows in temperate arable agroforestry systems. *Agriculture, Ecosystems & Environment*, *270*, 114–128.

Passarge, J., Hol, S., Escher, M., & Huisman, J. (2006). Competition for nutrients and light: Stale coexistence, alternative stable states, or competitive exclusion? *Ecological Monographs*, *76*, 57–72. https://doi.org/10.1890/04-1824

Penuelas, J., & Filella, I. (2003). Deuterium labeling of roots provides evidence of deep water access and hydraulic lift by *Pinus nigra* in a Mediterranean forest of NE Spain. *Environmental and Experimental Botany*, *49*, 201–208. https://doi.org/10.1016/S0098-8472(02)00070-9

Poore, J., & Nemecek, T. (2018). Reducing food's environmental impacts through producers and consumers. *Science*, *360*, 987–992.

Powell, G. W., & Bork, E. W. (2006). Aspen canopy removal and root trenching effects on understory vegetation. *Forest Ecology and Management*, *230*, 79–90.

Quinteros, C. P., Bava, J. O., Bernal, P. M. L., Gobbi, M. E., & Defosse, G. E. (2017). Competition effects of grazing-modified herbaceous vegetation on growth, survival and water relations of lenga (*Nothofagus pumilio*) seedlings in a temperate forest of Patagonia, Argentina. *Agroforestry Systems*, *91*, 597–611.

Ramsey, C. L., & Jose, S. (2002). Management challenges of pecan and pine based alley cropping systems of the Southern United States. In W. Schroder & J. Kort (Eds.), *Temperate agroforestry: Adaptive and mitigative roles* (pp. 158–163). Regina, SK, Canada: Plains and Prairie Forestry Association.

Reynolds, P. E., Simpson, J. A., Thevathasan, N. V., & Gordon, A. M. (2007). Effects of tree competition on corn and soybean photosynthesis, growth, and yield in a temperate tree-based agroforestry intercropping system in southern Ontario, Canada. *Ecological Engineering*, *29*, 362–371.

Rivest, D., & Vézina, A. (2015). Maize yield patterns on the leeward side of tree windbreaks are site-specific and depend on rainfall conditions in eastern Canada. *Agroforestry Systems*, *89*, 237–246.

Robin, C., Hay, M. J. M., Newton, P. C. D., & Greer, D. H. (1994). Effect of light quality (red:far-red ratio) at the apical bud of the main stolon on morphogenesis of *Trifolium repens* L. *Annals of Botany*, *74*, 119–123.

Root, R. (1973). Organization of a plant–arthropod association in simple and diverse habitats: The fauna of collards (*Brassica oleracea*). *Ecological Monographs*, *43*, 95–124.

Sanchez, P. A. (2002). Soil fertility and hunger in Africa. *Science*, *295*, 2019–2020.

Schroth, G. (1999). A review of belowground interactions in agroforestry, focusing on mechanisms and management options. *Agroforestry Systems*, 43, 5–34.

Seiter, S., Ingham, E. R., William, R. D., & Hibbs, D. E. (1995). Increase in soil microbial biomass and transfer of nitrogen from alder to sweet corn in an alley cropping system. In J. H. Ehrenreich, D. L. Ehrenreich, & H. W. Lee (Eds.), *Growing a sustainable future* (pp. 56–158). Boise, ID: University of Idaho.

Shainsky, L. J., & Radosevich, S. R. (1992). Mechanisms of competition between Douglas-fir and red alder seedlings. *Ecology*, *73*, 30–45.

Sharrow, S. H. (1999). Silvopastoralism: Competition and facilitation between trees, livestock, and improved grass–clover pastures on temperate rainfed lands. In L. E. Buck, J. Lassoie, & E. C. M. Fernandez (Eds.), *Agroforestry in sustainable agricultural systems* (pp. 111–130). Boca Raton, FL: CRC Press.

Silva-Pando, F. J., Gonzalez-Hernandez, M. P., & Rozados-Lorenzo, M. J. (2002). Pasture production in a silvopastoral system in relation with microclimate variables in the Atlantic coast of Spain. *Agroforestry Systems*, *56*, 203–211.

Singh, R. P., Ong, C. K., & Saharan, N. (1989). Above and below ground interactions in alley cropping in semiarid India. *Agroforestry Systems*, *9*, 259–274.

Smil, V. (1999). Nitrogen in crop production: An account of global flows. *Global Biogeochemical Cycles*, *13*, 647–662.

Smil, V. (2000). Phosphorus in the environment: Natural flows and human interferences. *Annual Review of Energy and the Environment*, *25*, 53–88.

Smith, M. W., Arnold, D. C., Eikenbary, R. D., Rice, N. R., Shiferaw, A., Cheary, B. S., & Carroll, B.L. (1996). Influence of ground cover on beneficial arthropods in pecan. *Biological Control*, *6*, 164–176.

Smith, M. W., Wolf, M. E., Cheary, B. S., & Carroll, B. L. (2001). Allelopathy of bermudagrass, tall fescue, redroot pigweed, and cutleaf evening primrose on pecan. *HortScience*, *36*, 1047–1048.

Söderström, B., Svensson, B., Vessby, K., & Glimskär, A. (2001). Plants, insects and birds in semi-natural pastures in relation to local habitat and landscape factors. *Biodiversity & Conservation*, *10*, 1839–1863. https://doi.org/10.1023/A:1013153427422

Thevathasan, N. V., & Gordon, A. M. (2004) Ecology of tree intercropping systems in the north temperate region: Experiences from southern Ontario, Canada. *Agroforestry Systems*, *61*, 257–268. https://doi.org/10.1023/B:AGFO.0000029003.00933.6d

Thevathasan, N. V., Gordon, A. M., & Voroney, R. P. (1998). Juglone (5-hydroxy-1,4 napthoquinone) and soil nitrogen transformation interactions under a walnut plantation in southern Ontario, Canada. *Agroforestry Systems*, 44, 151–162.

Tilman, D. (1980). A graphical-mechanistic approach to competition and predation. *The American Naturalist*, *116*, 362–393.

Tilman, D. (1982). *Resource competition and community structure*. Princeton, NJ: Princeton University Press.

Tilman, D. (1990). Constraints and tradeoffs: Toward a predictive theory of competition and succession. *Oikos, 58*, 3–15.

Tilman, D., Cassman, K. G., Matson, P. A., Naylor, R., & Polasky, S. (2002). Agricultural sustainability and intensive production practices. *Nature, 418*, 671–677.

Todhunter, M. N., & Beineke, W. F. (1979). Effect of fescue on black walnut growth. *Tree Planters' Notes, 30*, 20–23.

Troeh, F. R., & Thompson, L. M. (1993). Soils and soil fertility (5th ed.) New York: Oxford University Press.

Tsonkova, P., Böhm, C., Quinkenstein, A., & Freese, D. (2012). Ecological benefits provided by alley cropping systems for production of woody biomass in the temperate region: A review. *Agroforestry Systems, 85*, 133–152.

van Noordwijk, M., Lawson, G., Soumaré, A., Groot, J. J. R., & Hairiah, K. (1996). Root distribution of trees and crops: Competition and/or complementarity. In C. K. Ong & P. Huxley (Eds.), *Tree–crop interactions: A physiological approach* (pp. 319–364). Wallingford, UK: CAB International.

Von Kiparski, G. R., Lee, L. S., & Gillespie, A. R. (2007). Occurrence and fate of the phytotoxin juglone in alley soils under black walnut trees. *Journal of Environmental Quality, 36*, 709–717.

Wanvestraut, R., Jose, S., Nair, P. K. R., & Brecke, B. J. (2004). Competition for water in a pecan–cotton alley cropping system. *Agroforestry Systems, 60*, 167–179.

Waring, H. D., & Snowdon, P. (1985). Clover and urea as sources of nitrogen for the establishment of *Pinus radiata*. *Australian Forest Research, 15*, 115–121.

Wojtkowski, P. (1998). *The theory and practice of agroforestry design*. Enfield, NH: Science Publishers.

Yang, L. L., Ding, X. Q., Liu, X. J., Li, P. M., & Egrinya Eneji, A. (2016). Impacts of long-term jujube tree/winter wheat–summer maize intercropping on soil fertility and economic efficiency: A case study in the lower North China Plain. *European Journal of Agronomy, 75*, 105–117.

Zamora, D., Jose, S., & Nair, P. K. R. (2006). Interspecific interaction in a pecan–cotton alleycropping system in the southern United States: The production physiology. *Canadian Journal of Botany, 84*, 1686–1694.

Zamora, D., Jose, S., & Nair, P. K. R. (2007). Morphological plasticity of cotton roots in response to interspecific competition with pecan in an alleycropping system in the southern United States. *Agroforestry Systems, 69*, 107–116.

Zhang, W., Wang, B. J., Gan, Y. W., Duan, Z. P., Hao, X. D., Xu, W. L., & Li, L. H. (2018). Different tree age affects light competition and yield in wheat grown as a companion crop in jujube–wheat agroforestry. *Agroforestry Systems, 93*, 653–664.

Study Questions

1. Explain how the competitive exclusion principle and the unified neutral theory vary in explaining species coexistence.

2. How can we make use of physiological information such as photosynthetic pathway (e.g., C_3 vs. C_4) of component species in designing sustainable agroforestry systems?

3. Although shading is commonly associated with competition for light, shading can be beneficial in agroforestry systems. Describe the beneficial or facilitative role of shading with the help of examples.

4. What is meant by "safety-net hypothesis" in an agroforestry context? Provide an example each from (a) alley cropping, (b) silvopasture, and (c) riparian buffer.

5. Agroforestry systems seem to have fewer pest problems than their monoculture counterparts. Explain the ecological basis of this phenomenon.

6. Describe the concept of hydraulic lift and its importance in improving the water balance of forage grass in a longleaf pine (*Pinus palustris*) silvopasture in the southern United States.

7. Explain how plants utilize root plasticity in avoiding belowground competition when grown in species mixtures.

Section II

Agroforestry Practices

North American Agroforestry, Third Edition. Edited by Harold E. "Gene" Garrett, Shibu Jose, and Michael A. Gold.

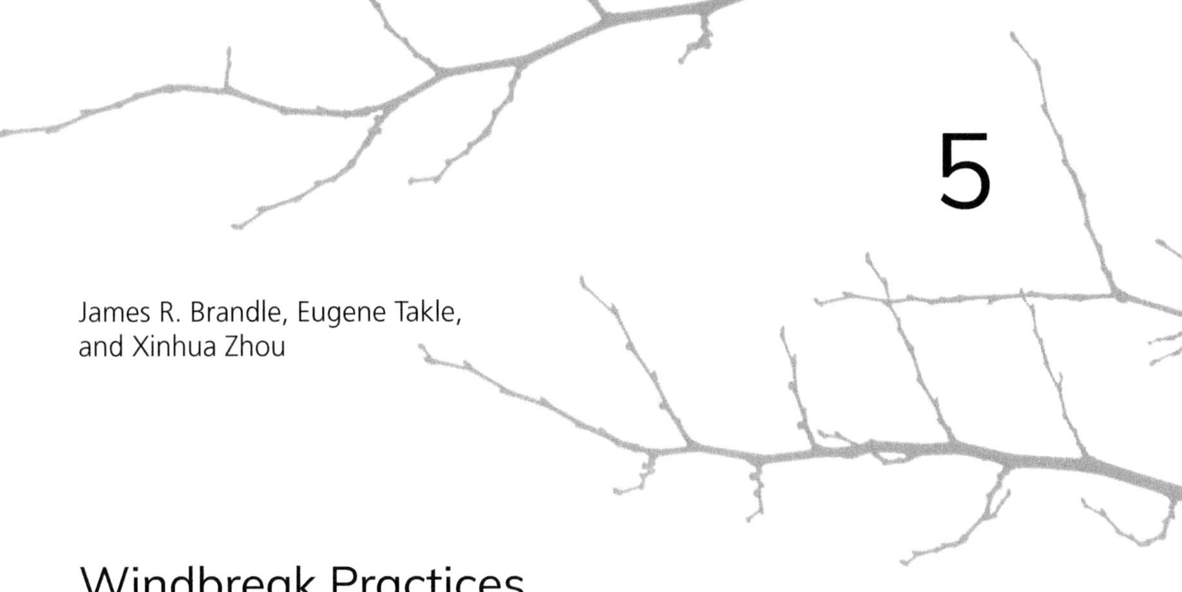

James R. Brandle, Eugene Takle,
and Xinhua Zhou

Windbreak Practices

Windbreaks and shelterbelts are barriers used to reduce wind speed. Usually consisting of trees and shrubs, they also may be perennial or annual crops, grasses, wooden fences, or other materials. They are used to protect crops and livestock, control erosion and blowing snow, define boundaries, provide habitat for wildlife, provide tree products, and improve landscape aesthetics.

The systematic use of windbreaks in agriculture is not a new concept. At least as early as the mid-1400s, the Scottish Parliament urged the planting of tree belts to protect agricultural production (Droze, 1977). From these beginnings, shelterbelts have been used extensively throughout the world to provide protection from the wind (Caborn, 1971). As settlement in North America moved west into the grasslands, homesteaders planted trees to protect their homes, farms, and ranches. In response to the 1930s Dust Bowl conditions, the U.S. Congress authorized the Prairie States Forestry Project. This conservation effort led to the establishment of 29,927 km (>96,000 ha) of shelterbelts in the Great Plains (Droze, 1977). In northern China, extensive planting of shelterbelts and forest blocks was initiated in the 1950s. Today the area is extensively protected, and studies have documented a modification in the regional climate (Wang, Zhang, Hasi, & Dong, 2010; Zhao, Xiao, Zhao, & Zhang, 1995; Zheng, Zhu, & Xing, 2016). Windbreak programs also have been established in Australia (Cleugh et al., 2002; Miller, Bohm, & Cleugh, 1995), New Zealand (Sturrock, 1984), Russia (Mattis, 1988), Argentina (Peri & Bloomberg, 2002), and other parts of the world. In policy circles, windbreak programs are considered an environmentally friendly technique for use by landowners. Although the value of protection is widely recognized, the inclusion of windbreaks as an integral component of sustainable agriculture in the United States remains limited.

The original goal of this chapter was to provide practical information for landowners, producers, conservation professionals, and students. In this third revision, we are grateful to the previous coauthors of this chapter (L. Hodges, J. Tyndall, R. Sudmeyer, & B. Wright) and acknowledge their contributions. It is our hope that this information will help others understand the value of windbreaks and encourage their inclusion as components of sustainable agricultural production systems. The chapter is divided into four main sections: (a) how windbreaks work, (b) how organisms respond to wind protection, (c) the design, management, and benefits of windbreaks, and (d) the overall role of windbreaks in the sustainable agricultural landscape. The emphasis is on temperate regions and, in most cases, on mechanized agriculture. This chapter will present only a summary of the wealth

North American Agroforestry, Third Edition. Edited by Harold E. "Gene" Garrett, Shibu Jose, and Michael A. Gold.

of information available on windbreaks. For more detail on any of the subjects covered here, see the original articles cited here, reviews by Brandle, Hintz, and Sturrock (1988), Miller et al. (1995), Cleugh et al. (2002), Gardiner, Berry, and Moulia (2016), and classical reviews by Bates (1911), van Eimern, Karschon, Razumava, and Robertson (1964), and Grace (1977). In addition, a searchable, comprehensive listing of published information is available at the University of Nebraska's website (https://snr.unl.edu/data/trees/research/windbreakbiblio.aspx).

How Windbreaks Work

Wind Flow in the Environment

Wind is defined as air in motion. It is caused by the differential heating of the Earth's surface, resulting in differences in pressure, and is influenced by Coriolis forces caused by the Earth's rotation. On a global scale, atmospheric circulation drives our daily weather patterns. On a microscale, there is a very thin layer of air (several millimeters or less) next to any surface within which transfer processes are controlled by the process of diffusion across the boundary layer. Between these two scales are the surface winds. They move in both vertical and horizontal directions and are affected by the conditions of the surfaces they encounter. Surface winds extend 50 to 100 m above the Earth's surface and are dominated by strong mixing or turbulence (Rosenberg, Blad, & Verma, 1983). These surface winds influence wind erosion, crop growth and development, animal health, and the general farm or ranch environment. They are also the winds that are affected by shelterbelts.

Although surface winds can be quite variable and the flows highly turbulent, the main component of the wind moves parallel to the ground. Wind speed at the soil surface approaches zero due to the frictional drag of the surface. The amount of drag is a function of surface roughness. In the case of vegetation, the height, uniformity, and flexibility of that vegetation determines the surface roughness and the amount of drag (Lowry, 1967). A rough surface such as wheat (*Triticum aestivum* L.) stubble has greater frictional drag, slower wind speeds, and greater turbulence near the surface than a relatively smooth surface, such as mown grass. A windbreak increases surface roughness and, when properly designed, reduces wind speed over large areas to the benefit of agriculture and other societal activities. Discussions of wind, wind profiles, turbulent transfer, and exchange coefficients may be found in McNaughton (1988) and Cleugh (2002). For our purposes, turbulent transfer rates are defined as the rates of exchange between the crop–soil continuum and the atmosphere for heat, water vapor, CO_2, and other air constituents caused by the vertical mixing of air. This is an area of current research as scientists attempt to model this mixing and its impact on surface vegetation (Helfer, Zhang, & Lemckert, 2009; Thuyet, Do, Santo, & Hung, 2014). A particularly noteworthy area of research is the merging of the new generation of crop models, such as APSIM, with agroforestry models (Luedeling et al., 2016) to capture the complexity of interactions between crops and trees for promoting sustainability, fostering food security, and reducing poverty in regions of fragile natural resource systems.

Wind Flow across a Barrier

A windbreak is a barrier on the land surface that obstructs the wind flow and alters flow patterns both upwind of the barrier (windward) and downwind of the barrier (leeward). As wind approaches a windbreak, some of the air passes through the barrier while the rest flows around the ends of the barrier or is forced up and over the barrier. These flow patterns are illustrated in Figure 5–1.

Shelterbelts can be considered as a collection of porous obstacles that create a series of pressure fields in the presence of wind in the atmospheric surface layer (Takle, Wang, Schmidt, Brandle, & Jairell, 1997). As air flow approaches the barrier, surface static pressure increases and reaches a maximum at the windward shelter zone near the edge of the barrier. This pressure drops as the wind passes through the barrier, reaching a minimum just to the lee of the barrier, then gradually increasing with increasing distance from the barrier (see Figure 5–2). The magnitude of the pressure difference between the windward and leeward sides of the windbreak is one factor determining the flow modification caused by the barrier and is a function of windbreak structure (Řeháček, Khel, Kucera, Vopravil, & Petera, 2017; Schmidt, Takle, Brandle, & Litvina, 1995; Takle et al., 1997).

As the air moves around or over the barrier, the streamlines of air are compressed (Patton, Shaw, Judd, & Raupach, 1998; van Eimern et al., 1964). This upward alteration of flow begins at some distance windward of the windbreak and creates a protected zone on the windward side of the windbreak where wind speed is reduced. This protected zone extends windward for a distance of 2–5*H*, where *H* is the windbreak height. A much larger region of reduced wind speed is created in the lee of the windbreak (Caborn, 1957;

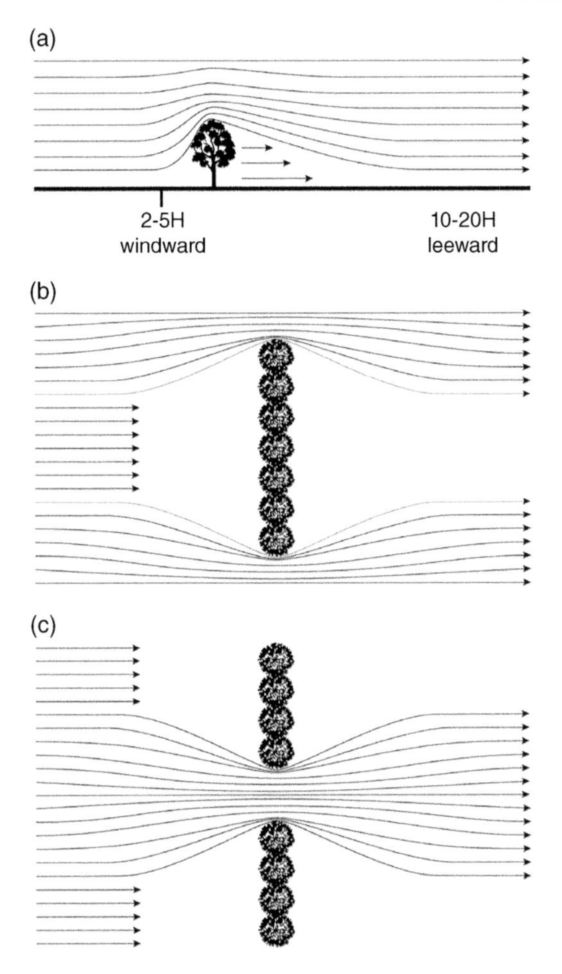

Fig. 5–1. Wind flow patterns (A) over, (B) around, and (C) through a field windbreak. Areas of increased wind flow are indicated by the close spacing of the lines; *H* is the height of the windbreak.

van Eimern et al., 1964). This zone typically extends for a distance of 10–30*H* (Brandle, 1990; Heisler & Dewalle, 1988; Středa, Malenová, Pokladniková, & Rožnovský, 2008; Wang & Takle, 1995b).

The magnitude of wind speed reduction at various locations within the protected zones is a function of the windbreak structure (Bird, Jackson, Kearney, & Roache, 2007; Brandle, 1990; Cornelis & Gabriels, 2005; Heisler & Dewalle, 1988; Wang & Takle, 1995a, 1996b). By adjusting the windbreak structure or density, different wind flow patterns and zones of protection are created. The best structure for a given windbreak depends on the objective of the windbreak. In general, the denser a windbreak the more wind speed is reduced. This general concept is illustrated in Figure 5–3, while specific design criteria are discussed below.

Windbreak Structure

The ability of a windbreak to reduce wind speed is a function of its *external structural features*—height, orientation, length, width, continuity or uniformity, and cross-sectional shape—and its *internal structural features*— the amount and arrangement of the solid and open portions and the surface area of the barrier components (Zhou, Brandle, Mize, & Takle, 2005; Zhou, Brandle, Takle, & Mize, 2002, 2008). The overall size of the protected zones, the extent of the wind speed reductions within the zones, and the resulting microclimate depend on these structural features (Wang and Takle, 1996b, 1997; Zhou et al., 2005, 2008). By manipulating windbreak structure through various management practices, a range

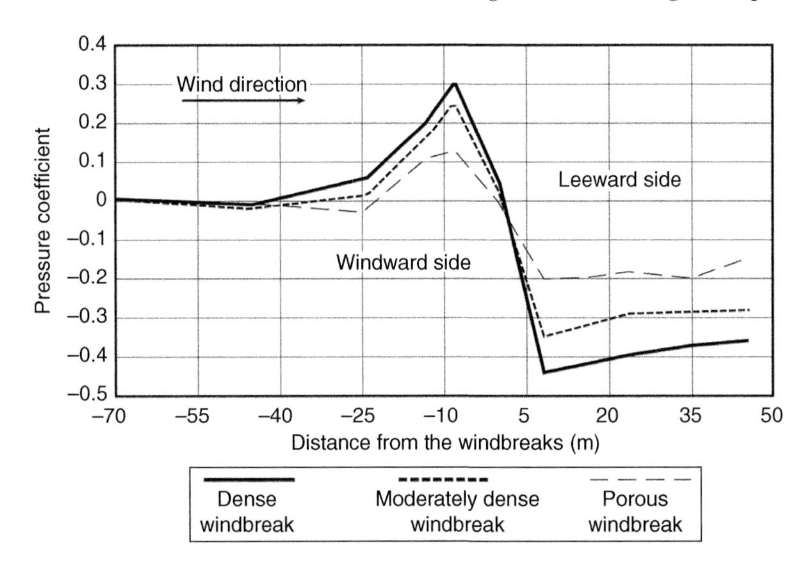

Fig. 5–2. Changes in the pressure coefficient at ground level windward and leeward of a two-row field windbreak with approximately 60% density. The leeward edge of the windbreak is designated as 0 *H*, where *H* is the height of the windbreak.

of conditions within the protected zone may be created that can be used to meet various design objectives (Brandle, 1990; Středa et al., 2008).

External Structure

Windbreak height (*H*) is the most important factor determining the extent of the protected zone and, combined with length, determines the total area protected. Windbreaks are most effective when they are oriented perpendicular to the wind. As winds become more oblique to the windbreak, the extent of the protected zone is reduced. The length of a windbreak should be at least 10 times its height to minimize the effect of wind flow around the ends of the windbreak. Windbreak width influences the windbreak effectiveness through its influence on density (Heisler & Dewalle, 1988). Windbreak continuity is also important. A gap or opening concentrates wind flow through the opening, creating a zone of increased wind speed to the lee. In both cases, increased flow around the ends or through a gap directly reduces the extent of the protected zone (Caborn, 1957; Jacobs, 1984) and reduces windbreak effectiveness in the area adjacent to the gap or end of the barrier (Figure 5–1C).

Internal Structure

Historically, windbreak structure was defined in terms of density or porosity, and these terms are still used in many situations (Thuyet et al., 2014). Windbreak *density* is the ratio of the solid portion of the windbreak to the total volume of the windbreak, while *porosity* is the ratio of the open portion of the windbreak to the total volume. The two terms are complementary in that their sum equals 1 or 100%. Wind flows through the open portions of a windbreak, thus as density increases, less wind passes through the barrier and wind speed reductions are greater.

The precise determination of windbreak density remains one of the problems facing researchers in windbreak technology. A solid barrier, such as a wall, will have a density of 100%. In the case of a slat fence or screen, the uniform size and distribution of the solid material and the relative "thinness" of the barrier make density easy both to determine and to manipulate.

For vegetative barriers, density is considerably more difficult to determine. There are a number of problems. First, it is impossible to have a vegetative barrier with 100% density. There will always be spaces between the various plant

Deciduous 25-35% Density
Open Wind Speed 10 m s^{-1}

H distance from windbreak	5H	10H	15H	20H	30H
m s^{-1}	5	6.5	8	8.5	10
% of open wind speed	50%	65%	80%	85%	100%

Conifer 40-60% Density
Open Wind Speed 10 m s^{-1}

H distance from windbreak	5H	10H	15H	20H	30H
m s^{-1}	3	5	6	7.5	9.5
% of open wind speed	30%	50%	60%	75%	95%

Multi-Row 60-80% Density
Open Wind Speed 10 m s^{-1}

H distance from windbreak	5H	10H	15H	20H	30H
m s^{-1}	2.5	3.5	6.5	8.5	9.5
% of open wind speed	25%	35%	65%	85%	95%

Solid Fence 100% Density
Open Wind Speed 10 m s^{-1}

H distance from windbreak	5H	10H	15H	20H	30H
m s^{-1}	2.5	7	9	9.5	10
% of open wind speed	25%	70%	90%	95%	100%

Fig. 5–3. Wind speed reductions at different distances to the lee of windbreaks with different densities, where *H* is the height of the windbreak.

elements. Unlike the slat fence, the size, shape, and arrangement of plant elements (stems, branches, and leaves) are not uniform. Similarly, the size and shape of the open spaces varies with season of the year (Koh, Park, Kang, & Lee, 2014) and branch or leaf movement. Furthermore, vegetative barriers have a significant width such that for any given transect through the barrier there is a unique arrangement of solid elements. Finally, as the angle of the approaching wind becomes more oblique to the barrier (less perpendicular), the length of the path of the wind through the barrier increases and the protected area decreases (Wang & Takle, 1996a). This is the same result as increasing the barrier width, which increases density and alters the effect of the windbreak on wind flow.

In the past, estimates of density or porosity were based on the relative abundance of solid or open areas as seen by an experienced observer or on how easy it was to see through the windbreak. We now refer to this as the *optical density* or *porosity*. With the advent of digital image processing techniques, the speed and accuracy of these estimations have improved significantly (Kenney, 1987; Loeffler, Gordon, & Gillespie, 1992; Středová, Podhrázská, Litschmann, Středa, & Rožnovský, 2012; Zhang, Brandle, Meyer, & Hodges, 1995). However, the issues related to the distribution of plant elements within the windbreak have not been resolved. While optical density is often used in applied situations, it is important to note that the path of the wind through the windbreak is not a straight line and that the wind flows over and around the various elements within the windbreak. To describe this flow, the three-dimensional or aerodynamic structure of the windbreak must be defined.

Heisler and Dewalle (1988) and others (Bird et al., 2007; Brandle, 1990; Caborn, 1957; Cornelis & Gabriels, 2005; Jensen, 1961; Read, 1964; van Eimern et al., 1964) suggested that the vertical distribution of density within the windbreak influences the wind flow response and windbreak effectiveness. Experience tells us that windbreaks that are porous in the lower levels will funnel wind through these areas, increasing wind speed in the lee of the windbreak and decreasing the level of leeward protection (Read, 1964). Thuyet et al. (2014) discussed the advantages of considering the three-dimensional structure of component trees of the shelterbelt in optimizing sheltering efficiency. Wind tunnel studies by Bitog et al. (2011) using actual black pine trees (*Pinus thunbergii* Parl.) produced porosity and resistance factors of various windbreak configurations for use in follow-on numerical simulations. They asserted that this method allows unique drag and porosity values to be determined for various tree types, density, and other characteristics.

Wang and Takle (1994, 1996b, 1997) developed numerical simulation models of the influence of shelterbelts on wind flow. Their results indicate that variation in the distribution of the surface area across the width of the windbreak may have minimal influence on shelter efficiency. Wang, Takle, and Shen (2001) attributed this to the fact that it is the overall structure of the barrier that creates the pressure fields driving the force equations in the model. However, field observations of the location of snowdrifts around windbreaks show that a dense windward tree row (i.e., conifer or shrub component) gives a different snowdrift pattern than a dense leeward tree row.

As these contradictions point out, the relationship between the internal structure of a windbreak and the resulting wind flow patterns will remain an active area of research, particularly with regard to field verification of numerical simulations of shelter effects. Discussions have led us to believe that the discrepancies between simulation results and field experience may be best resolved with better three-dimensional descriptions of windbreak structure. Most recently, Wu et al. (2018) studied shelter effects using a global dataset. They concluded that optical porosity is not a sufficient structural descriptor for shelter effects and recommended the development of three-dimensional structural descriptors for multiple-row windbreaks. Their findings reinforce those of Zhou et al. (2002, 2005, 2008), who described three-dimensional structural descriptors for various applications. Based on the earlier work of Wang and Takle (1996b), Zhou et al. (2002, 2005, 2008) developed two structural descriptors: *vegetative surface area density* (vegetative surface area per unit canopy volume) and *cubic density* (vegetative volume per unit of canopy volume), which were tested in numerical simulations of shelter effects. The first field tests of these descriptors indicated an improved ability to estimate the drag force term in the equations of motion used to predict boundary layer flows near windbreaks (Zhou, Brandle, & Takle, 2003; Brandle, Zhou, & Takle, 2003). Additional field verification of the rigor of these descriptors remains a research goal.

Microclimate Changes

While research on the aerodynamic structure continues, it is clear that dense windbreaks result in greater wind speed reductions, that the vertical distribution of structural components influences wind flow patterns, and that structure influences the amount of turbulence generated. All of these factors influence the microclimate changes that occur in the sheltered zones.

As a result of wind speed reduction and changes in turbulent transfer rates, the microclimate (temperature, precipitation, relative humidity, and CO_2) in the sheltered zone is altered (McNaughton, 1988, 1989). The magnitude of microclimate change for a given windbreak varies within the protected zone depending on existing atmospheric conditions, windbreak density and orientation, distance from the windbreak, time of day, height above the ground, and surface conditions (wet or dry) (Du, Ushiyama, & Maki, 2010). Most recently, Schmidt, Lischeid, and Nendel (2019) described the microclimate changes within the zone immediately adjacent to the boundary between a forested strip and arable land. They describe "S-shape microclimate gradients" within the transition zone of 50–80 m.

McNaughton (1988), following the terminology of Raine and Stevenson (1977), defined two zones in the lee of the windbreak:

1. The *quiet zone* extends from the top of the windbreak down to a point in the field located approximately 8*H* leeward, where both wind speed and turbulence are reduced.
2. The *wake zone* lies leeward of the quiet zone and extends approximately 20–25*H* from the barrier, where wind speed is reduced but turbulence is increased relative to open field conditions.

The boundary between these two zones is a function of windbreak structure and atmospheric stability and lies between 6 and 10*H* to the lee of the windbreak.

One useful concept explaining exchange rates between various surfaces and the atmosphere is the concept of *coupling* (Grace, Ford, & Jarvis, 1981). Monteith (1981) defined coupling as the capacity to exchange energy, momentum, or mass between two systems. Exchange processes between single leaves and the atmosphere or between plant canopies and the atmosphere are controlled by the gradients of temperature, humidity, and CO_2 that exist in the immediate environment above the surface. When these gradients are modified by shelter, the microclimate within the sheltered zone will be modified (Grace, 1981, 1988; McNaughton, 1988; Monteith, 1981).

In the quiet zone, the transfer coefficients are less and thus turbulent exchange is reduced. In the wake zone, transfer coefficients are greater and the rates of turbulent exchange are increased. As a result, the transport of heat, water vapor, and CO_2 within these two zones is different.

Radiation

Solar radiation provides essentially all of the energy received at the Earth's surface and influences most of the environmental conditions in which plants and animals live. On a regional scale, shelterbelts have minimal influence on the direct distribution of incoming radiation; however, they do influence radiant flux density (the amount of energy per unit surface area per unit time), primarily by shading and reflection in the area immediately adjacent to the windbreak. Schmidt et al. (2019) presented an excellent discussion of the interaction of orientation of a tree line and adjacent arable land.

Solar radiant flux density within and immediately adjacent to the windbreak is influenced by the sun angle (a function of location, season, and time of day) and by windbreak height, density, and orientation. Likewise, at any given location, the extent of the shaded zone is dependent on latitude, time of the day, season of the year, and the height of the windbreak. North–south-oriented windbreaks produce morning shade on the western side and afternoon shade on the eastern side. In the northern hemisphere, windbreaks oriented in an east–west direction produce a shaded area on the north side of the windbreak throughout the day while radiation is reflected off the south-facing surfaces, increasing radiant flux density adjacent to the windbreak. The amount of reflected radiant flux is dependent on the time of day, season of the year, and the reflectivity of the windbreak's vertical surface.

Air Temperature

In general, daytime temperatures within 8*H* of a medium-dense barrier can be several degrees warmer than temperatures in the open due to the reduction in turbulent mixing. This effect appears to be greater early in the growing season. Between 8 and 24*H*, daytime turbulence increases and air temperatures tend to be several degrees cooler than for unsheltered areas (McNaughton, 1988; Cleugh, 2002). Nighttime temperatures within 1 m of the ground are generally 1–2 °C warmer in the protected zone (up to 30*H*) than in the exposed areas (Hodges, Suratman, Brandle, & Hubbard, 2004; Read, 1964; Zhang, Brandle, Hodges, Daningsih, & Hubbard, 1999). In contrast, temperatures 2 m above the surface tend to be slightly cooler. On very calm nights, temperature inversions may occur and protected areas may be several degrees cooler at the surface than exposed areas due to the absence of even slight air movement (Argete & Wilson, 1989; McNaughton, 1988).

In warmer regions of the temperate zone, for example, southern Texas or Florida, temperature increases in shelter may exceed optimal temperatures for some crops or livestock. In these cases, the increase in sheltered temperature may increase plant or animal stress and decrease productivity. In more northern latitudes, temperature increases in the sheltered zones are generally beneficial to crop growth.

Soil Temperature

Average soil temperatures in shelter are slightly warmer than in unprotected areas (Hodges et al., 2004; McNaughton, 1988; Zhang et al., 1999). In most cases, this is due to the reduction in heat transfer away from the surface. In areas within the shadow of a windbreak, soil temperatures are lower due to shading of the surface. The magnitude of this effect is dependent on the height and orientation of the barrier and the angle of the sun (the size and duration of the shaded area). Conversely, soil temperatures may be slightly higher in areas receiving additional radiation reflected off the surface of the windbreak. These differences are greatest early in the season before the crop canopy closes (Caborn, 1957). Soil texture and soil moisture strongly affect soil heat retention and release and will influence the duration and magnitude of soil temperature differences between sheltered and unsheltered zones.

Frost

On clear, calm nights, infrared radiation emitted from the soil and vegetative surfaces is unimpeded. Under these conditions, surfaces may cool rapidly, resulting in decreased air temperature next to the surface. When this temperature decreases to the dew point, condensation forms on surfaces. If temperatures fall below freezing, this condensation freezes, resulting in radiation frost. In sheltered areas where wind speed is reduced, radiation frost may occur more frequently than in exposed areas, especially in sandy soils with low capacity to retain daytime solar gain. In contrast, advection frosts are generally associated with large-scale cold air masses. Strong winds are typically associated with the passage of the front and, while the radiative process contributes to heat loss, temperature inversions do not occur. Shelterbelts may offer some protection against advection frosts when episodes are of short duration and when windward temperatures are just below 0 °C. In sheltered areas, reductions in turbulent mixing (less mixing of the warm air near the surface with the colder air of the front) may reduce heat loss from the sheltered area and provide some degree of protection. The process may be influenced by evaporation from the soil surface and subsequent condensation of vapor on the leaves. If soil moisture is higher in shelter, then not only might there be less mixing and loss of water vapor, but sensible heat from the soil may be held in the crop canopy by the reduction in turbulent mixing, reducing the potential for frost. It is also possible that the increase in water vapor in the sheltered area will reduce the rate of radiative cooling (Rosenberg et al., 1983). At our research site in Nebraska, we have recorded frost occurring in sheltered areas when none has occurred in exposed areas and in exposed areas when none has occurred in the sheltered areas (Brandle & Hodges, unpublished data, 2004–2005). It should be noted that in all of these cases, temperatures were very close to freezing and may or may not have resulted in frost, depending on interacting microclimate conditions. A better understanding of the conditions leading to frost leeward of shelterbelts is needed if practical management recommendations are to be made for temperature-sensitive crops.

Precipitation

Rainfall over most of the sheltered zone is generally unaffected except in the area immediately adjacent to the windbreak. These areas may receive slightly more or less than the open field depending on wind direction, wind speed, and intensity of the rainfall. On the leeward side, there may be a small rain shadow where the amount of precipitation reaching the surface is slightly reduced. The converse is true on the windward side, as the windbreak may function as a barrier and lead to slightly higher levels of measured precipitation at or near the base of the trees due to increased stemflow or drip from the branches.

In contrast, the distribution of snow is greatly influenced by the presence of a windbreak and can be manipulated by managing windbreak density (Scholten, 1988; Shaw, 1988). A dense windbreak (>60% optical density) will lead to relatively short, deep snowdrifts on both the windward and leeward sides, while a more porous barrier (~35% optical density) will provide a long, relatively shallow drift primarily to the leeward side. In both cases, the distribution of snow and the resulting soil moisture will affect the microclimate of the site. In the case of field windbreaks, a uniform distribution of snow may provide moisture for significant increases in crop yield. This is especially true in areas where snowfall makes up a significant portion of the annual precipitation. In addition, fall-planted crops insulated by a blanket of snow are protected against desiccation by cold, dry winter winds (Brandle, Johnson, & Dearmont, 1984). The effect of density on snow distribution is illustrated in Figure 5–4.

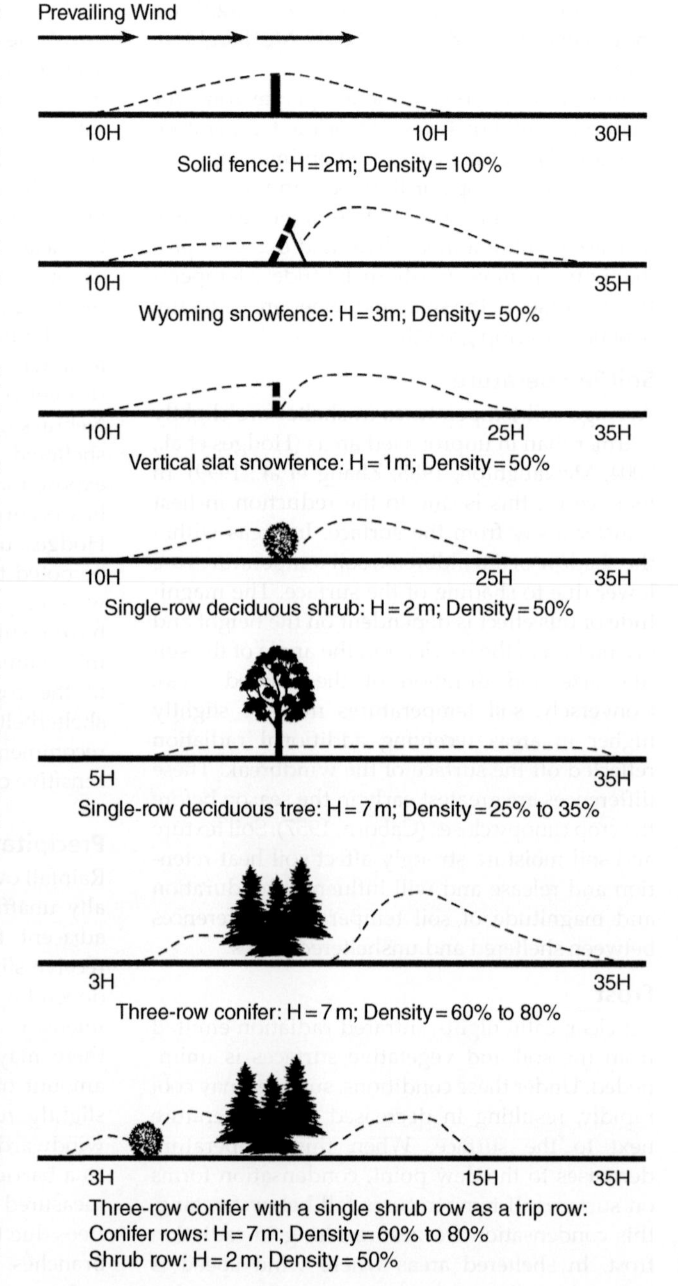

Fig. 5–4. The amount of snow storage windward and leeward of a snow fence or windbreak is determined by the height and density of the barrier; *H* is the height of the windbreak.

Humidity

Humidity, or the water vapor content of the air, is a major factor in the regulation of crop microclimate. Again, this is related to its role in the energy balance of the system (Rosenberg et al., 1983). Decreases in turbulent mixing reduce the amount of water vapor transported away from surfaces in the sheltered area. As a result, humidity and vapor pressure gradients are generally greater in shelter both during the day and at night (McNaughton, 1988). For example, Helfer et al. (2009) quantified the advantages of a transpiring shelterbelt located at the windward side of a small water body for reducing evaporative water losses. Because water vapor is a strong absorber of infrared radiation, higher humidity levels in shelter tend to protect the crop from radiative heat losses, reducing the potential for frost.

Evaporation

Evaporation from bare soil is reduced in shelter due to wind speed reductions and the reduction in transfer of water vapor away from the surface. In most cases, this is an advantage, conserving soil moisture for early season plant growth. In many cases, evaporation from leaf surfaces is also reduced in shelter. However, as plants get larger, with more complex canopy structures, sheltered crops may use more water than unsheltered crops (Rosenberg, 1966). In contrast, Sudmeyer, Crawford, Meinke, Poulton, & Robertson (2002) reported that greater leaf area did not increase soil water use. Under very limited moisture conditions, insufficient soil moisture may limit full development of the crop yield potential in larger plants found in sheltered areas.

Sugita (2018) provided a new look at the complex relationships between canopy layer transpiration and overall water use by a crop field. In his study, both increases and decreases in water consumption were reported for crop fields in the Nile Delta of Egypt. The magnitude of the change appears tied to canopy structure, most likely the height of the canopy. His theoretical discussion is far beyond the scope of this review, but his article offers a detailed discussion of the state of our knowledge of this important aspect of windbreak function. This remains a fertile area for research into water use in sheltered environments under a wide range of climatic conditions.

In most cases, increased humidity and reduced evaporation do not contribute to a higher incidence of disease. However, situations may occur where windbreak design, high humidity, rainfall, or irrigation contribute to abnormally high humidity levels in sheltered areas, leading to an increase in disease. Combined with lower nighttime temperatures in shelter, high humidity levels may cause more dew formation. In these cases, the added humidity and reduced evaporation in shelter may increase the possibility of disease. For example, to increase the incidence of white mold [*Sclerotinia sclerotiorum* (Lib.) de Bary] on dry edible bean (*Phaseolus vulgaris* L.) and identify resistant cultivars, Deshpande, Hubbard, Coyne, Steadman, and Parkhurst (1995) used closely spaced slat-fence windbreaks to increase humidity levels and dew formation in sheltered areas. In contrast, when windbreak systems are designed for optimal crop production, disease incidence is normally not a problem. During the past 40 yr of shelter research in eastern Nebraska, we have observed this phenomenon only twice, once in winter wheat (Brandle et al., 1984) and once in soybean [*Glycine max* (L.) Merr.] (Nieto & Brandle, unpublished data, 1996).

Windbreaks in Agricultural Production Systems

The goal of any system of windbreaks is to provide microclimate conditions that can be used for the benefit of the landowner. Two types of windbreaks have direct application to agricultural production systems: field windbreaks and livestock windbreaks. Two other types, farmstead windbreaks and living snow fences, provide indirect support to the agricultural operation and are significant components of any sustainable agricultural ecosystem. Here, we consider the effect of wind protection on individual plant growth and development, the effect of field windbreaks on crop production, and the benefits of protecting livestock under range and feedlot conditions from the adverse effects of wind.

Field Windbreaks

Agricultural producers frequently recognize the value of field windbreaks to reduce wind erosion (Tibke, 1988; Ticknor, 1988). In northern areas, the value of field windbreaks to harvest snow for crop production is also widely recognized (Scholten, 1988). Field windbreaks are often used to protect wind-sensitive crops such as fruits and vegetables (Baldwin, 1988; Norton, 1988). Unfortunately, their role in the protection of grain crops is less widely recognized (Brandle, 1990). In this review, the effect of wind on individual plant growth and development is considered first and then the overall benefits of field windbreaks at the farm scale are reviewed.

Physiological Response of Plants to Shelter

The effect of wind on plants is well studied and reviewed extensively (Cleugh, 1998; Coutts & Grace, 1995; Gardiner et al., 2016; Grace, 1977; Miller et al., 1995). Both photosynthesis and transpiration are driven in part by environmental conditions, particularly those within the leaf and canopy boundary layers. Because shelter modifies the microenvironment, it impacts plant productivity.

Plant temperature differences between sheltered and exposed sites are relatively small, on the order of 1–3 °C. In the quiet zone, where the rate of heat transfer from a plant is reduced, a slight increase in temperature can be an advantage, especially in cooler regions where even a small increase in plant temperature may have substantial positive effects on the rate of cellular processes and physiology (Grace, 1988; Van Gardingen & Grace, 1991). Lower nighttime temperatures in shelter may reduce the rate of respiration, resulting in higher rates of net photosynthesis and more vegetative growth or seed weight during grain-filling periods. Indeed, there are many examples of sheltered plants being taller and having more extensive leaf areas (Frank, Harris, & Willis, 1974; Grace, 1977; Ogbuehi & Brandle, 1981, 1982; Rosenberg, 1966). Higher soil temperatures in the sheltered zone may result in more rapid crop emergence and establishment, especially for crops with a high heat unit accumulation requirement for germination and establishment (Drew, 1982). In contrast, under hot conditions temperatures above the optimum for plant development may lead to periods of water stress if the plant is unable to adjust to the higher demands for moisture.

The overall influence of shelter on plant water relations is extremely complex and linked to the air temperature, soil moisture, and wind speed conditions found in shelter. Historically, the major effect of shelter and its influence on crop growth and yield was assumed to be due primarily to soil moisture conservation and a reduction in water stress of sheltered plants (Caborn, 1957; Grace, 1988; van Eimern et al., 1964). There is little question that evaporation rates are reduced in shelter (Grace, 1988; McNaughton, 1983, 1988); however, the effect on plant water status is less clear (Sugita, 2018). According to Grace (1988), transpiration rates may increase, decrease, or remain unaffected by shelter depending on wind speed, atmospheric resistance, and saturation vapor pressure deficit. Davis and Norman (1988) reviewed the concept of water use efficiency in shelter and concluded that under some conditions, sheltered plants make more efficient use of available water. Monteith (1993) suggested that water use efficiency in shelter was unlikely to increase except when there was a significant decrease in saturation vapor pressure deficit. Moreover, the increase in humidity in shelter would contribute to a decrease in saturation vapor pressure deficit and thus an increase in water use efficiency. However, sheltered plants tend to be taller and have larger leaf areas. Given an increase in biomass, sheltered plants have a greater demand for water and under conditions of limited soil moisture or high temperature may actually suffer greater water stress than exposed plants (Grace, 1988; Nuberg & Mylius, 2002; Rosenberg, 1966). Cleugh (2002) simulated a crop microclimate and found that shelter was more effective in reducing direct water loss from the soil than reducing transpiration. A negative consequence of increased temperatures in shelter may be an increased vapor pressure deficit at the end of the growing season (Sudmeyer, Crawford, et al., 2002). Overall, shelter improves water conservation and allows the crop to make better use of the available moisture over the course of a growing season. The magnitude of this response depends on the crop, stage of development, and environmental conditions. Additionally, species and ecotypes can vary in sensitivity to wind and response to wind protection (Emery, Reid, & Chinnappa, 1994; Van Gaal & Erwin, 2005).

Most recently, Dupraz et al. (2019) described a mechanistic, biophysical model, Hi-sAFe, that is designed to simulate interactions within agroforestry systems that mix trees with crops. This model combines a preexisting crop model (STICS) constructed as a simulation tool capable of working under agricultural conditions with several plasticity mechanisms responsive to tree–tree and tree–crop competition for light, water, and N. Understanding how plant water status affects the physiological and morphological aspects of crop response to wind and wind protection in production fields remains a fertile area for potential research.

Growth and Development Response of Plants to Shelter

As a result of favorable microclimate and the resulting physiological changes, the rate of growth and development of sheltered plants may increase. The increase in the rate of accumulation of heat units in shelter contributes to early maturity of many crops. For example, Ogbuehi and Brandle (1982) reported that flowering in soybean occurred 4 to 10 d earlier in sheltered fields than in unsheltered fields. Similar results have been reported for corn (*Zea mays* L.; Senaviratne, Udawatta, Nelson, Shannon, & Jose, 2012; Zohar & Brandle, 1978), cotton (*Gossypium hirsutum* L.; Barker, Hatfield, & Wanjura, 1989), and many vegetables (Baldwin, 1988; Hodges & Brandle, 1996). In previous studies at the University of Nebraska, earlier anthesis in muskmelon (*Cucumis melo* L.; Zhang et al., 1999) contributed to earlier harvest of sheltered plants (see Figure 5–5). In snap bean (*Phaseolus vulgaris* L.), an increase in soil temperature early in the season resulted in earlier maturity (Hodges et al., 2004). Similarly, in cultivar trials of cabbage [*Brassica oleracea* (L.) var. *capitate*] and pepper (*Capsicum annum* L.), most cultivars reached harvest maturity 3 to 10 d earlier in the sheltered fields (Hodges & Brandle, unpublished data, 1996). The ability to reach the early market with many of these perishable crops can mean sizable increases in economic returns to producers (Baldwin, 1988; Brandle, Hodges, & Stuthman, 1995; Hodges et al., 2004; Norton, 1988; Sturrock, 1984).

Vegetative growth or biomass is generally increased in sheltered environments (Baldwin, 1988; Bates, 1911; Caborn, 1957; Kort, 1988; Skidmore, Hagen, Naylor, & Teare, 1974; Stoeckeler, 1962; Sturrock, 1984; van Eimern et al., 1964) but not universally so. Nebraska research has demonstrated biomass and leaf area increases in sheltered soybean (Ogbuehi & Brandle, 1981, 1982), snap bean (Hodges et al., 2004), and muskmelon (Zhang et al., 1999) but not in corn (Zhang & Brandle, 1997) or alfalfa (*Medicago sativa* L.; Hans, 1987).

In many fruit and vegetable crops, reproductive growth is dependent on pollination by insects. In addition to the physical movement of the insect to the flower, the process has a number of critical aspects: attraction of the appropriate insect, receptivity of the stigmatic surface, pollen viability, rate of growth of the pollen tube, and fertilization of the ovule. All of these processes are partially dependent on the microclimate of the flower. In particular, they benefit from warm, moist, calm conditions similar to those found in sheltered areas during the spring (Norton, 1988). As a result, sheltered orchard and vineyard crops show significantly increased levels of fertilization and fruit formation that can be attributed to the improved microclimate in sheltered areas

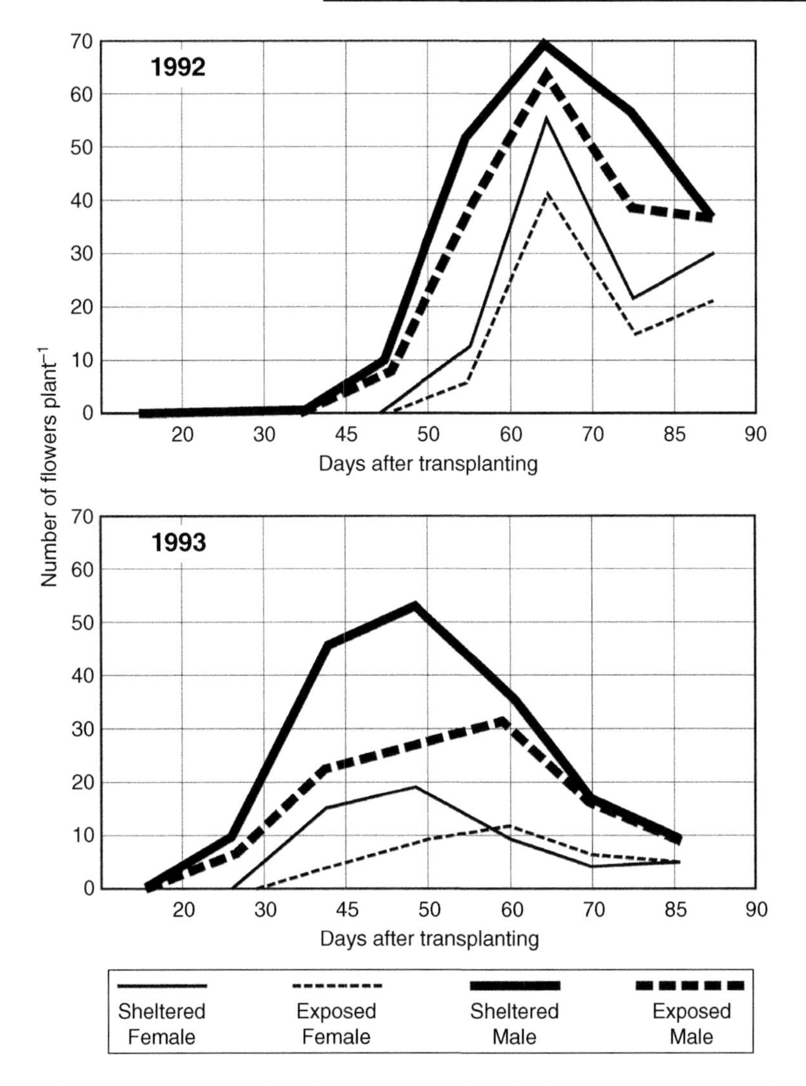

Fig. 5–5. Number and time of occurrence of male and female flowers of muskmelon grown under sheltered or exposed conditions in 1992 and 1993.

(Norton, 1988; Waister, 1972b) and/or enhanced pollinators from windbreak trees (Stanek, Lovell, & Reisner, 2019). Tamang, Andreu, and Rockwood (2010) described the value of single-row shelterbelts in reducing physical damage from tropical storms and potentially lowering canker infection in Florida citrus.

Wind also influences plant growth directly by the mechanical manipulation of plant parts (Miller et al., 1995). This movement may increase the radial enlargement of the stem, increase leaf thickness, reduce stem elongation and leaf area (Grace, 1988; Jaffe, 1976; Nobel, 1981), and affect cellular composition (Armbrust, 1982). On the whole-plant level, the interaction of ethylene and auxin (Biddington, 1986; Biro & Jaffe, 1984; Erner & Jaffe, 1982; Jaffe & Forbes, 1993) as well as possible inhibition of auxin transport (Mitchell, 1977) appear to be involved. The threshold wind speed and duration for these types of direct responses appears to be very low, perhaps as low as 1 m s^{-1} for <1 min (Garner & Bjorkman, 1996; Van Gaal & Erwin, 2005). As a result, these types of responses may be more indicative of a no-wind situation rather than an indicator of various wind speed differences as found in sheltered and non-sheltered conditions (Biddington, 1985; Miller et al., 1995; Van Gardingen & Grace, 1991).

Wind can cause direct physical damage to plants through abrasion and leaf tearing (Miller et al., 1995). Abrasion is caused when plant parts (leaves, stems, branches, or fruits) rub against each other. As tissue surfaces rub, the epicuticular waxes on the surfaces are abraded, increasing cuticular conductance and water loss (Pitcairn, Jeffree, & Grace, 1986;

Van Gardingen & Grace, 1991). The magnitude of the impact on transpiration is determined by the degree of abrasion and the relative importance of the epicuticular wax in controlling the total resistance of the cuticle to the diffusion of water vapor.

Tearing is common on leaves that are large, damaged by insects, or subjected to high wind speeds. Wind contributes to the abrasion of plant surfaces by wind-blown particulates (usually soil), often referred to as *sandblasting*. The extent of injury depends on the wind speed and degree of turbulence, the amount and type of abrasive material in the air stream, the duration of exposure, the plant species and its stage of development, and microclimatic conditions (Skidmore, 1966). Finch (1988) summarized the sensitivity of many crops to wind-blown soil based on estimates of crop tolerance to blowing soil (see Table 5–1). All three of these—abrasion, leaf tearing, and sandblasting—damage plant surfaces and can lead to uncontrolled water loss from the plant (Grace, 1977, 1981; Miller et al., 1995).

Plant lodging is another direct mechanical injury caused by wind. It takes two forms: stem lodging, where a lower internode permanently bends or breaks, and root lodging, where the soil or roots supporting the stem fail. Stem lodging is most common as crops approach maturity, while root lodging is more common on wet soils and during grain-filling periods (Easson, White, & Pickles, 1993; Miller et al., 1995; Pinthus, 1974). In both cases, heavy rainfall tends to increase the potential for lodging (Marshall, 1967).

Sheltered plants tend to be taller with heavier heads and reduced culm stiffness, characteristics that tend to contribute to lodging (Grace, 1977). Medium-dense shelterbelts tend to reduce crop lodging within the sheltered zone because of reduced wind speeds (Bates, 1944; Sturrock, 1981). As windbreak density increases, turbulence increases and the likelihood of lodging is greater (Kort, 1988).

Under extremely windy conditions, some plants may experience a phenomenon called *wind-snap*, *green-snap*, or *brittle-snap*, where the force of the wind breaks the stem. In 1993, eastern and central Nebraska experienced a severe windstorm in mid-July and a number of corn fields exhibited areas of brittle-snap (Elmore & Ferguson, 1996). It is interesting to note that we found areas of brittle-snap in several of our unsheltered corn fields but none in our sheltered fields. Meteorological measurements indicated that in the exposed areas, wind speeds exceeded 18 m s^{-1} while in sheltered fields wind speeds were generally <9 m s^{-1} (Brandle, unpublished data, 1993). The effect appeared to be related to stem characteristics of certain corn cultivars since not all cultivars exhibited damage.

Table 5–1. Estimated crop tolerances to damage by wind-blown soil (modified from Finch, 1988).

Crops grouped by tolerance	Species
Tolerant crops	
Barley	*Hordeum vulgare* L.
Buckwheat	*Fagopyrum* spp.
Flax	*Linum usitatissimum* L.
Millet	*Panicum miliaceum* L.
Oat	*Avena sativa* L.
Rye	*Secale cereal* L.
Wheat	*Triticum aestivum* L.
Moderately tolerant crops	
Corn	*Zea mays* L.
Grain sorghum	*Sorghum bicolor* (L.) Moench
Sunflower	*Helianthus annuus* L.
Very low tolerance crops	
Alfalfa seedlings	*Medicago sativa* L.
Cabbage and broccoli	*Brassica oleracea* L.
Cotton seedlings[a]	*Gossypium hirsutum* L.
Cucumber	*Cucumis sativus* L.
Flowers [a]	most species
Green, snap, or lima bean	*Phaseolus* spp. (all varieties)
Leafy vegetables[a]	All species
Muskmelon	*Cucumis melo* L.
Onion [a]	*Allium cepa* L.
Pea	*Pisum sativum* L. (all varieties)
Table and sugar beets	*Beta vulgaris* L.
Sugar beet seedlings [a]	
Soybean	*Glyine max* (L.) Merr.
Watermelon	*Citrullus lanatus* (Thumb) *Matsum. Ex Nakai*
Young orchards	most species

Note. Tolerant crops: estimated crop tolerance >5.4 Mt ha–1 yr–1; moderately tolerant crops: estimated crop tolerance between 2.7 and 5.4 Mt ha–1 yr–1; very low tolerance crops: estimated crop tolerance <2.7 Mt ha–1 yr–1.

[a] Tolerance <0.5 Mt ha–1 event–1.

Crop Yield Response to Shelter

While the influences of wind and shelter on individual plant processes are only partially understood and widely variable, the net effect of shelter on crop yield is generally positive (see Figure 5–6 and reviews by Baldwin [1988], Grace [1977], Kort [1988], and Norton [1988]). The reasons vary with crop, windbreak design, geographic location, moisture condition, and cultural practice. Here, we will focus on the benefits of shelter on the crop as a whole, field windbreak design, and their economics.

Stoeckeler (1962) conducted one of the most extensive studies on the effects of windbreaks on field crops in the northern Great Plains. Their survey of 184 cornfields and 94 fields of small grain

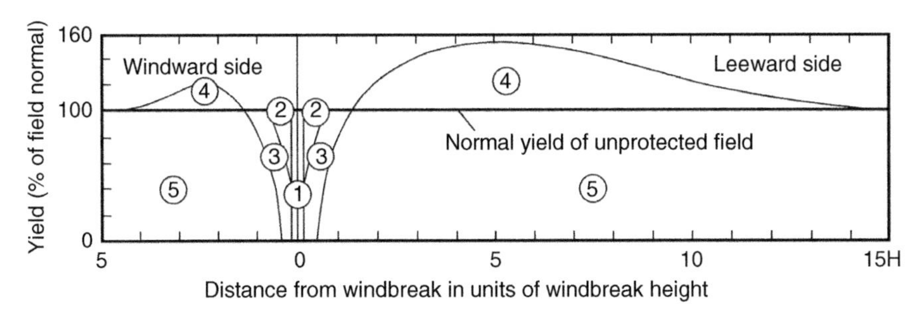

Fig. 5–6. The generalized case for crop yield responses for a field windbreak in the Great Plains.

indicated significant yield benefits in the sheltered zones of both east–west- and north–south-oriented field windbreaks. Also, Kort (1988) summarized yield responses for a number of field crops from temperate areas around the world. Average yield increases varied from 6 to 44% (see Table 5–2).

A close reading of the individual studies behind these averages indicates great variability in yield results. In most cases, the data indicate a strong positive response to shelter, while in others, the response is either neutral or negative. This is understandable because the final crop yield is the culmination of a series of interacting factors present throughout the growth and development of the crop. The possible combinations of growth response and microclimate conditions are unlimited, and the probability of a single combination and the corresponding crop response occurring on an annual basis is relatively small. It is likely that this variability in growing conditions and the stage of development at which a particular condition is present accounts for many of the contradictory responses reported in the literature. As Sturrock (1984) explained, the relationship between shelter and crop response is complex and dynamic, subject to continual change as a result of changes in microclimate, windbreak efficiency, and growth and development of the protected crop.

Table 5–2. Crop response to shelter (adapted from Baldwin, 1988; Brandle, Johnson, & Akeson, 1992; Kort, 1988).

Crop	Field years	Weighted mean yield increase
	no.	%
Spring wheat	190	8
Winter wheat	131	23
Barley	30	25
Oat	48	6
Rye	39	19
Millet	18	44
Corn	209	12
Soybean	17	15
Grass hay	14	20

Australian windbreak research demonstrates the complex interactions of climatic and edaphic influences where growing conditions are characterized by soils with low water holding capacity, terminal drought, and often dry conditions at the start of the growing season (Cleugh et al., 2002). While microclimate changes in the sheltered zone often increase crop yields, particularly of leguminous crops (Bennell & Verbyla, 2008; Bicknell, 1991; Nuberg, Mylius, Edwards, & Davey, 2002; Oliver, Lefroy, Stirzaker, & Davies, 2005; Sudmeyer, Adams, et al., 2002), these increases are often offset by yield declines in the zone of competition, where windbreak trees and crops compete for water (Nuberg et al., 2002; Oliver et al., 2005; Sudmeyer, Adams, et al., 2002; Unkovich et al., 2003). In the Australian example, the greatest benefits of windbreaks appear in dry, windy years when wind erosion and sandblast damage to establishing crops can cause significant losses to unprotected crops (Bennell, Leys, & Cleugh, 2007; Sudmeyer, Adams, et al., 2002). An economic evaluation of crops growing in windbreak systems in southwestern Australia found that the protection benefits they provided would offset all of the costs associated with establishment and competition if unprotected crops were damaged three to four times during the 30-yr life of the windbreak (Jones & Sudmeyer, 2002).

Another factor that may influence the crop response to shelter is the crop cultivar. Almost without exception, crops have been bred and selected under exposed conditions. As a result, most common cultivars represent those selections best able to perform under exposed conditions. To take full advantage of the microclimate conditions created by windbreaks, a producer should select crop cultivars best suited to sheltered conditions. For example, using shorter, thicker stemmed wheat cultivars will reduce the potential for lodging while taking advantage of the favorable growing conditions found in sheltered fields (Brandle et al., 1984).

In this review, we define horticultural crops primarily as fruits and vegetables. There are few studies on the production of floral crops in shelter, but we suspect that they would be extremely responsive to the microclimate of shelter due to their sensitivity to desiccation. In addition, high-quality factors, such as uniformity of size and shape, and the absence of physical defects, such as abrasion, are required for market acceptance.

Baldwin (1988) and Norton (1988) provided the most comprehensive reviews of horticultural crops and shelter. In horticultural crops, marketable yield, product quality, and earliness to market are of primary importance (Baldwin, 1988; Hodges et al., 2004; Hodges & Brandle, 1996; Hodges, Daningsih, & Brandle, 2006). Earliness is primarily a function of temperature and was discussed above under microclimate changes. Physiological and anatomical responses of snap bean to wind were found to interact with temperature, with plants being less responsive to wind when grown under cooler temperatures (Hunt & Jaffe, 1980). For horticultural crops grown under sheltered conditions, the moderation of temperature extremes, warmer soil and air temperatures, and improved plant water status contributed to yield increases in total marketable yield and individual fruit weight. The moderated microclimate in shelter contributes to longer flowering periods and increased bee activity and can result in improved fruit set and earlier maturity (Norton, 1988). Quality improvements have been reported for many crops including sugarbeet (*Beta vulgaris* L.; Bender, 1955, as cited in van Eimern et al., 1964), tobacco (*Nicotiana tabacum* L.) and French bean (*Phaseolus vulgaris* L.; Kreutz, 1952a, 1952b, as cited in van Eimern et al., 1964), strawberry (*Fragaria* sp.; Shah, 1970; Waister, 1972a, 1972b), lettuce (*Lactuca* sp.; Strupl, 1953, as cited in van Eimern et al., 1964), plum (*Prunus* sp.; de Preez, 1986), kiwifruit (*Actinidia chinensis* L.; McAneney & Judd, 1987), orange [*Citrus sinensis* (L.) Osbeck; Pohlan, Vazquez, & Garcia, 1986; Rodriquez, del Valle, Arango, Torres, & Fernandez, 1986], carrot (*Daucus carota* L.; Taksdal, 1992), potato (*Solanum tuberosum* L.; Sun & Dickinson, 1994), and others (for more details, see reviews by Baldwin [1988], Grace [1977], Norton [1988], Miller et al. [1995], and van Eimern et al. [1964]).

A recent study by Barbeau, Wilton, Oelbermann, Karagatzides, and Tsuji (2018) explored the interaction of willow (*Salix*) windbreaks in a subarctic community with changing climate conditions. Their research demonstrated positive yield benefits from windbreaks in the production system for potato and bush beans (*Phaseolus vulgaris* L.). The potential role of windbreaks in production systems impacted by changing climate conditions is another area for future research (Easterling, Hays, Easterling, & Brandle, 1997; Schoeneberger, Bentrup, & Patel-Weynand, 2017).

Wind-induced sandblasting and abrasion compound the direct effects of wind on the yield and quality of vegetable and specialty crops. As the amount of wind-blown soil, wind speed, or exposure time increases, crop survival, growth, yield, and quality decrease (Fryrear & Downes, 1975). Young plants tend to be more sensitive to damage (Liptay, 1987). Concern for damage by wind-blown soil is greatest during the early spring when stand establishment coincides with seasonally high winds and large areas of exposed soil during field preparation. Another critical time is during the flowering stage when rubbing and abrasion by wind-blown soil may result in damage to or loss of buds and flowers (Bubenzer & Weis, 1974). Vegetable producers need to be especially aware of the problems associated with wind erosion because the light-textured soils that favor vegetable production are most easily eroded.

Wind-blown soil and rain can carry inocula for bacterial and fungal diseases (Claflin, Stuteville, & Armbrust, 1973; Kahn, Conway, & Fisher, 1986; Pohronezny, Hewitt, Infante, & Datnoff, 1992), and wind-damaged plant tissues are potential entry points for pathogens, especially bacteria. For example, common blight of bean [*Xanthomonas phaseoli* (E.F. Sm) Dows] increased 120% when the duration of exposure to wind-blown, infected river sand increased from 3 to 5 min (Claflin et al., 1973). Similarly, bell pepper (Pohronezny et al., 1992) and prune (*Prunus domestica* L. 'French'; Michailides & Morgan, 1993) showed increased disease incidence when wind exposure increased. Windbreaks can also reduce the distribution and rate of spread of wind-blown, aphid-transmitted viruses (Simons, 1957). If windbreaks are too dense, higher humidity levels and slower drying can create conditions favorable for disease development. In some cases, windbreak vegetation or litter may serve as alternative hosts (Yang et al., 2019) or overwintering sites for various diseases or insect vectors of plant pathogens (Slosser & Boring, 1980). Insect populations may increase or decrease in the lee of windbreaks, with variable effects on the protected crops. A more complete discussion of insect distribution and movement in windbreak-protected crops is included below.

Field Windbreak Design

In designing any windbreak system, it is critical to have a good understanding of the objective of the planting. Windbreaks designed for snow management are different from those designed for wind erosion control or protection of summer crops. Field windbreaks should be designed to accommodate the cultural practices, equipment, and land situation of the individual farm operation. However, there are general principles that apply to the majority of situations (Finch, 1988). Here we consider the general principles of field windbreak design, looking first at individual windbreaks and then at windbreak systems. Later we will deal with the individual needs of wind erosion control (Tibke, 1988; Ticknor, 1988) and snow management (Scholten, 1988; Shaw, 1988).

Field windbreaks should be oriented perpendicular to the prevailing or problem winds to maximize the protected zone. If only a single windbreak is planted, it is usually located at the field edge such that the leeward zone extends into the crop field. This is not the most efficient location because all of the windward protection falls on non-crop ground. Locating the windbreak within the field at a distance of 2 to 5*H* from the field edge increases the amount of land protected by the windbreak and increases economic return. In most cases, a single windbreak will not protect the entire field and additional windbreaks, parallel to the first, will need to be established at intervals across the field. Typically, the distance between windbreaks should range from 10 to 20*H*, depending on the degree of protection desired and the size of the farm equipment. In many areas, problem winds will come from several directions. In these cases, additional windbreaks with different orientations may be required to achieve the desired level of protection (Finch, 1988).

The ideal field windbreak designed for maximum crop production should be one or two rows and composed of several tall, long-lived species with good rates of growth and similar growth forms. Individual species should tolerate local stress conditions and have good insect and disease resistance. Native species are usually a good choice. The overall windbreak should have an optical density during the growing season of approximately 40–60%, with a tall, narrow crown and a deep root system that minimizes the degree of competition with the adjacent crop (Cunningham, 1988).

With the new windbreak aerodynamic models relating windbreak flows to structural descriptors (Zhou et al., 2008), numerical analysis can simulate the shelter flows as influenced by windbreak structure that, in turn can then be managed by either the original planting design and/or management practices over the life of the windbreak. This modeling approach has allowed the evaluation of windbreaks in non-traditional situations. Ferreira (2011) used this approach to design a tree windbreak capable of moderating the uneven wind speed distribution across the width dimension of a high-level competition rowing channel. Avila-Sanchez, Pindado, Lopez-Garcia, & Sanz-Andres (2014) reported the advantages of shelter for the efficiency of rolling stock over railway embankments.

The Zone of Competition

One of the most commonly expressed concerns about field windbreaks is the impact of competition between the windbreak and adjacent crops. Reduced crop yields have been associated with less water in the soil due to exploitation by tree roots, shading close to trees, phytotoxins in the soil (allelopathy), rainfall interception, and competition for soil nutrients (Kort, 1988; Ong & Huxley, 1996). Competition is usually confined to the area occupied by tree roots (Greb & Black, 1961; Sudmeyer, Speijers, & Nicholas, 2004). North American studies have indicated that competition commonly extends between 0.5 and 2*H* (Kort, 1988) but can extend farther on occasion (Chaput & Tuskan, 1990; Greb & Black, 1961). The degree of competition varies with crop type, geographic location (Lyles, Tatarko, & Dickerson, 1984; Stoeckeler, 1962), tree species (Brandle & Kort, 1991; Greb & Black, 1961; Lyles et al., 1984), and soil or climate conditions (Sudmeyer, Adams, et al., 2002).

Because of the complex interactions of wind shelter and above- and belowground competition, it is often difficult to demonstrate the underlying mechanisms of tree–crop competition in the field. However, nutrient competition and shading probably play a relatively minor role compared with competition for water in most North American agroforestry systems (Hou et al., 2003; Jose, Gillespie, & Pallardy, 2004; Reynolds, Simpson, Thevathasan, & Gordon, 2007). In contrast, Ding and Su (2010) found that, in northwestern China, corn yields were reduced between

20 and 27% depending on aspect. There is no question that where crop growth is moisture limited, competition between the windbreak and crops has significant, negative impacts on yield (Jose et al., 2004; Jose, Gillespie, Seifert, & Biehle, 2000; Kowalchuk & de Jong, 1995; Reynolds et al., 2007; Rivest & Vezina, 2015; Sudmeyer, Hall, Eastham, & Adams, 2002). These conditions are often found in semiarid areas, on soils with low plant-available water capacity, such as sands or shallow soils, or in exceptionally dry years in more temperate climates.

Belowground competition can be minimized by using windbreak trees that are deep rooted and have limited lateral extent (Greb & Black, 1961). Where water is limiting and the lateral tree roots extending into the crop field are confined close to the soil surface, cutting the lateral tree roots (root pruning) can significantly improve crop yields (Chaput & Tuskan, 1990; Hou et al., 2003; Jose et al., 2000; Rasmussen & Shapiro, 1990; Sudmeyer et al., 2004; Sudmeyer & Flugge, 2005).

The effectiveness of root pruning depends on the rooting patterns of the windbreak trees and the depth of root pruning (Stoeckeler, 1962; Kort, 1988; Rasmussen & Shapiro, 1990). Where lateral roots are left uncut below the rip line, they will continue to grow and spread back through the soil with time, making subsequent ripping less effective (Sudmeyer et al., 2004). Root pruning must be repeated every 1 to 5 yr depending on tree species and local weather conditions (George, 1971; Lyles et al., 1984; Naughton & Capels, 1982; Stoeckeler, 1962; Sudmeyer & Flugge, 2005; Umland, 1979). North American studies have found that root pruning can improve crop yield in the competition zone by 10–44% (Chaput & Tuskan, 1990; Hou et al., 2003; Rasmussen & Shapiro, 1990).

The annual economic returns from root pruning depend on the magnitude and extent of competition losses, the ability to sever all or most tree lateral roots, the cost of ripping, the inherent productivity of the site, the costs associated with the root-pruning operation, and how often roots must be pruned. In Western Australia, the increase in annual returns from crops by root pruning windbreaks ranged between –AUS$14 and $309 km^{-1} (Sudmeyer & Flugge, 2005). Lyles et al. (1984) estimated an average economic return from root pruning a field windbreak protecting winter wheat in Kansas at US$205 km^{-1}.

Windbreak Economics

Field windbreaks have costs associated with establishment, maintenance, and removal. They occupy cropland, reducing the number of crop hectares, and compete with crops immediately adjacent to the windbreak (see above). From an economic perspective, the amount of land occupied by the windbreak and the degree of competition should be minimized to maximize the number of crop hectares available and the yield increases resulting from wind protection. Ideally, the windbreak should take advantage of both the windward and leeward protection zones. For a windbreak system to be profitable, the long-term average yield increase from the protected zones must be large enough to compensate for the land occupied by the windbreak, for the crop losses within the zone of competition, and for the costs associated with planting and maintaining the windbreak.

Using the general yield responses as described by Kort (1988), field windbreak systems that occupy between 5 and 6% of the crop field provide positive economic returns to producers based entirely on the increased yields found in sheltered areas (Brandle et al., 1984; Brandle, Johnson, & Akeson, 1992). Other benefits, such as wind erosion control, snow management, and wildlife habitat, provide additional returns to the landowner.

Using a net present value approach, Brandle and Kort (1991; also Kort & Brandle, 1991) developed an interactive computer model to evaluate the economic returns to grain producers when crops are protected by windbreaks. The analysis includes the costs of windbreak establishment and maintenance, the loss of crop yield due to hectares planted to trees, the loss of productivity associated with the zone of competition, the length of time required to grow the windbreak, and the cost of removal at some point in the future. These costs are offset by reduced input costs on those hectares removed from production and increased yields in the protected areas.

In 2003, Grala and Colletti (2003) and Helmers and Brandle (2005) completed economic analyses indicating positive economic benefits from field windbreaks. Grala and Colletti (2003) indicated that the magnitude of the economic response was dependent on the rate of growth of the windbreak and the total lifespan. Windbreaks that grew rapidly and lived longer were more economically beneficial. They emphasized the long-term nature of an investment in a windbreak system. Helmers and Brandle (2005) used integer programming techniques to determine the optimal spacing of field windbreaks for corn and soybean production. An optimal spacing of 13*H* increased net returns by 7.6% for corn and 9.2% for soybean on the windbreak investment above the net return for unprotected corn and soybean.

Wind Erosion Control

Of all the benefits of field windbreaks, wind erosion control is the most widely recognized and accepted. The link between wind speed and wind erosion is well established; when wind speed is reduced, the potential for wind erosion is reduced. This has a direct impact on both crop productivity and off-site costs.

As a soil erodes, its productivity is decreased due to the loss of fine soil particles containing organic matter and nutrients (Pimental et al., 1995; Williams, Lyles, & Langdale, 1981). In many cases, compensation for these losses is made by the addition of fertilizer, which increases crop production costs. In other cases, yields are reduced, resulting in lower economic returns. By controlling wind erosion, windbreaks limit long-term losses in soil productivity, reducing the need for added inputs. The reduction of these losses from wind erosion is an additional economic benefit flowing from the windbreak investment (Brandle, Johnson, & Akeson, 1992).

Off-site costs, which are more difficult to quantify, are also incurred by both the private and public sectors (Huszar & Piper, 1986; Piper, 1989) and include damage to water storage facilities, irrigation systems and road ditches (Ribaudo, 1986) and increased health care costs associated with blowing dust and asthma (Rutherford, Clark, McTainsh, Simpson, & Mitchell, 1999). In South Australia, Williams and Young (1999) estimated that the annual health care costs associated with wind erosion and asthma exceeded AUS$20 million. In extreme cases, wind-blown dust has contributed to highway accidents, resulting in traffic fatalities. Reducing off-site impacts and the associated costs are additional economic benefits from the windbreak investment and further justify public funding for wind erosion control.

Wind erosion is a natural process and its total control is neither practical nor desirable. What is of concern is *accelerated erosion* or erosion at rates in excess of the natural ability of the soil to replenish itself. Accelerated erosion occurs primarily on large open fields under dry conditions. It is enhanced when the soil is loose, dry, and finely granulated and when the soil lacks vegetative cover (Lyles, 1988).

For those soils most prone to erosion, wind speeds in excess of 3 to 5 m s^{-1} will cause the soil to move (Tibke, 1988; Woodruff, Lyles, Siddoway, & Fryrear, 1972; Zachar, 1982). It moves in three general ways (Lyles, 1988). The largest particles (500–1000 μm) are generally too large to be lifted above the surface by ordinary erosive winds and are either pushed, rolled, or driven along the surface in a process called *surface creep*. The smallest particles are generally <50 μm but may be as large as 100 μm. These are lifted into the air stream and may be carried for great distances. Certainly, the most dramatic of the three types of soil movement, *suspension*, generally accounts for <25% of wind erosion. Movement of soil particles in the range of 100 to 500 μm comprises the third and largest portion of soil erosion. In this process, called *saltation*, the individual particles are lifted from the soil surface to a height of 30 to 45 cm and then fall to the surface. As these particles strike the surface, they may break into smaller particles, dislodge other particles from the surface, or break down other surface particles, reducing them in size. Combined with the force of the wind, this process, known as *soil avalanching*, tends to increase the level of soil movement (Tibke, 1988). Because saltation initiates and sustains suspension and soil creep, control measures should focus on reducing the amount of saltation (Lyles, 1988).

Rates of wind erosion are determined by a number of factors: (a) the inherent erodibility of the soil, (b) the climatic conditions of the location, (c) ridge roughness, or the height and orientation of the crop rows, (d) the amount and type of vegetative or residue cover, and (e) the width of the field along the prevailing wind direction. From a management perspective, little can be done about either the soil properties or the climate of the area. In contrast, ridge roughness and vegetative cover can be manipulated by various cultural practices, and field windbreaks can be used to reduce the width of the field. Windbreaks mitigate wind erosion by reducing wind speed in the sheltered zone below the threshold for soil movement. By dividing the field into smaller units, windbreaks reduce the field width and interrupt soil avalanching.

The effectiveness of any barrier for wind protection depends in part on its shape, width, height, and density. Windbreaks designed to control wind erosion must have an optical density of at least 40% during the period when the soil is exposed to the erosive forces of the wind (Ticknor, 1988). Cornelis and Gabriels (2005) were more specific, recommending a uniform optical density of 65 to 80%, but cautioned that optimal design depends on the protection goals for the windbreak. Most often, protection is needed at the time of planting, when most deciduous trees are leafless. Typically, this means that the windbreak must contain either coniferous species or a dense shrub understory. However, Gonzales et al. (2018) demonstrated the effectiveness of single-row Osage-orange [*Maclura pomifera* (Raf.) C.K.

Schneid.] windbreaks located in Kansas. Spacing between field windbreaks designed for erosion control should be in the range of 10 to 20*H*. At spacings of 10*H* or less, the risk of wind erosion is negligible but economic returns are reduced. As windbreak spacings are increased to 15*H*, economic returns from crop protection increase while the risk of erosion, though increasing, remains low. As spacings approach 20*H*, the risk of erosion increases and economic returns from crop production decrease (Brandle, Johnson, & Akeson, 1992). The proper spacing for field windbreaks designed for wind erosion control depends on climatic conditions, soil properties, residue management practices, and the producer's willingness to accept the risk of erosion.

Snow Management

In many northern, semiarid areas, snow is a critical source of soil moisture for crop and forage production during the next growing season. Greb (1980) estimated that more than one-third of the snowfall in these northern areas is blown off the field. Much of this wind-blown snow is deposited in road ditches, gullies, or behind fence rows or other obstructions (Aase & Siddoway, 1976). Even more may simply evaporate (Schmidt, 1972; Tabler, 1975). Many factors influence snow distribution including: (a) the amount and specific gravity of the snow, (b) the topography and surface conditions, particularly the amount and type of vegetative cover or crop residue, (c) wind velocity and direction, and (d) the presence and characteristics of barriers to wind flow (Scholten, 1988).

Field windbreaks can help capture the moisture available in snow by slowing the wind and distributing the snow across the field. As a result, wheat yields on croplands protected by field windbreaks are increased 15–20% (Brandle et al., 1984; Kort, 1988; Lehane & Nielsen, 1961). These increases are a result of increased moisture due to snow capture and the protection of the wheat crop from desiccation.

Field windbreaks designed exclusively for the uniform distribution of snow across the field should have an optical density of no more than 40%. Planting a single row of a tall, deciduous tree species at a wide spacing (5–7 m between trees) perpendicular to the prevailing winter wind direction will provide good snow distribution across a field for a distance of 10 to 15*H*. Snow blowing over the tops of the trees falls out of the air stream on the relatively still, leeward side of the windbreak. Wind passing through the porous windbreak provides the mechanism to distribute the snow uniformly across the field. Very dense field windbreaks will cause snow to collect in narrow, deep drifts near the tree row (see Figure 5–4).

Areas or fields susceptible to wind erosion during winter present additional challenges because field windbreaks with densities <40%, which are ideal for uniform snow distribution, offer minimal wind erosion control. If the field is covered with snow, the soil is protected; however, many areas where snow is an important source of water do not have continuous winter snow cover. Increasing windbreak density will increase the size of the drift, and in more northern areas, may delay snowmelt and spring tillage operations due to wet conditions.

Integrated Pest Management and Windbreaks

Both crop pests and their natural enemies are influenced by the presence of windbreaks (Beecher, Johnson, Brandle, Case, & Young, 2002; Burel, 1996; Dix et al., 1995; Dix, Hodges, Brandle, Wright, & Harrell, 1997; Kahnonitch, Lubin, & Korine, 2018; Marshall, 1988; Morrison & Flores, 2013; Perkins, Johnson, & Blankenship, 2003; Pierce, Farrand, & Kurtz, 2001; Pisani Gareau & Shennan, 2010; Puckett, 2006; Quinkenstein et al., 2009; Shi & Gao, 1986; Solomon, 1981; Sreekar, Mohan, Das, Agarwal, & Vivek, 2013; Tremblay, Mineau, & Stewart, 2001; Tsitsilas, Hoffmann, Weeks, & Umina, 2010; Yang et al., 2019). This influence is reflected in the distribution of insects as a result of wind speed reductions in the sheltered zone (Lewis & Dibley, 1970; Heisler & Dix, 1988; Pasek, 1988) and as a function of additional foraging sites created both within the windbreak and in the sheltered zones (Corbett & Plant, 1993; Forman, 1995; Slosser & Boring, 1980; Southwood & Way, 1970). Recently, Nguyen and Nansen (2018) offered a review of the spatial distribution of insects related to edge environments. They reported that various mathematical models offer some understanding but that abiotic factors and crop vegetation traits probably bear some responsibility for this pattern of insect distribution. They emphasized that more research is needed as related to shelterbelts and insect distribution in crop fields.

In narrow vegetative or artificial windbreaks, insect distribution appears to be primarily a function of wind conditions (Pasek, 1988) (but see the recent review by Nguyen & Nansen, 2018). As windbreak structure becomes more complex, a variety of microhabitats are created, and insect and avian populations increase in both number and diversity. Greater vegetative diversity of the edges provides numerous microhabitats for life-cycle activities and a variety of hosts, prey, and pollen and nectar sources (Andow, 1991; Flint & Dreistadt, 1998). The addition of woody plants, particularly several rows of tall trees, increases the suitable habitat for numerous avian species (Hall, Nimmo, Watson, & Bennett, 2018; Igl, Kantrud, & Newton, 2018; Jobin, Choinière, & Bélanger, 2001; Pierce et al., 2001; Tremblay et al., 2001).

The impacts of the various insect distribution patterns are less clear (for more detail, see Dix et al., 1995; Pasek, 1988). Both positive and negative aspects are reported in the literature. For example, Slosser and Boring (1980) reported that in northern Texas the success of cotton boll weevils (*Anthonomus grandis* Boheman) overwintering in the litter of deciduous windbreaks was considerably greater than those overwintering in coniferous windbreaks. Danielson, Brandle, Hodges, and Srinivas (2000) reported mixed results for the presence of bean leaf beetle (*Cerotoma trifucata* Foster) in sheltered and unsheltered soybean fields in eastern Nebraska. In 70% of the cases, there were no differences in bean leaf beetle populations. Sheltered fields had significantly higher populations 20% of the time, and unsheltered fields had higher populations only 10% of the time. Corbett and Rosenheim (1996) found that French prune trees planted along the edges of vineyards in California provided significant overwintering habitat for *Anagrus*, an egg parasitoid of the grape leafhopper, *Erythroneura elegantula* (Kido, Flaherty, Bosch, & Valero, 1984). Riechert and Lockley (1984) reviewed the role of spiders as biological control agents and concluded that agricultural systems, with some type of perennial component where habitat structure, microclimate, and potential prey are maintained without annual disturbance, could benefit from spiders as biological control agents.

Pollinating insects are typically three times more abundant in sheltered fields than in exposed areas (Williams & Wilson, 1970), contribute to increased levels of pollination, and are dependent on the availability of non-crop habitat (Kremen, Williams, & Thorp, 2002). Bee flight is inhibited at wind speeds greater than 6.5–9 m s^{-1} and the increased levels of pollination that occur in sheltered areas have been attributed to the calmer, warmer conditions found in protected zones (Lewis & Smith, 1969; Norton, 1988).

Crops within the Windbreak

We have discussed the use of windbreaks to protect crops. Within the agroforestry concept, we should recognize the plant materials within the windbreak itself as potential products and as contributors to the total economic return from the agricultural system.

The management of existing multiple-row windbreaks (10 rows or more) for timber or fuelwood is similar to small woodlot management. Larger trees can provide lumber for crates and pallets. Various species of cedar (*Juniperus*) and Osage orange are resistant to decay and can be used for posts or poles. Cedar may be chipped or shaved for animal bedding and brings a premium when packaged for the small animal or pet market. There is growing interest in the utilization and processing of *Eucalyptus* species for a number of products (Rockwood & Bowman, 2017), especially in tropical areas but also in subtropical areas of North America. Wood chips of other species may be used for livestock bedding, landscape and garden mulches, and fuel. In areas near large urban markets, firewood can provide additional income. The production and marketing of alternative products from agroforestry has increased dramatically since 2004 (Gold, Godsey, & Josiah, 2004; Josiah, St-Pierre, Brott, & Brandle, 2004). More recently, an extensive review of the potential of non-timber products in the United States was provided by Chamberlain, Emery, and Patel-Weynand (2018). While their review focused on forestlands, the principles apply to windbreaks and shelterbelts, especially multiple-row belts. Potential products include small non-traditional fruits, hazelnuts (*Corylus* spp.), medicinal products, and woody florals. Incorporating these and other understory species into windbreaks provides additional density to the lower portions of the windbreak and may provide economic opportunities in local markets. The key to a successful agroforestry enterprise is the ability to recognize local market conditions and to supply products to that market (Brandle et al., 1995).

For those with a long-term outlook, new windbreaks can be designed to produce timber crops (Bagley, 1988; Sturrock, 1988). High-quality hardwoods, such as walnut (*Juglans*) and oak (*Quercus*), offer the best opportunities. In some cases, Christmas trees or nursery stock may be incorporated into a windbreak design. These types of crops require a little imagination, extensive management, a good understanding of windbreak ecology, and, in some cases, specialized equipment. Some are very labor intensive, and all require extensive business skills and a good understanding of marketing, but in each case, they may add considerable income to the overall economic return of a windbreak investment (Chamberlain et al., 2018; Gold et al., 2004; Josiah et al., 2004).

Livestock Windbreaks

Windbreaks play an important role in the protection of livestock, particularly young animals. In the northern Great Plains and the Canadian Prairie region, livestock protection is a vital part of successful operations. Producers in North and South Dakota have reported significant savings in feed costs, improved survival, and greater milk production when livestock are protected from winter storms (Stoeckeler & Williams, 1949). Livestock vary in their need for wind protection. Beef cattle are very hardy and require protection primarily during calving or during severe winter storms (Webster, 1970a, 1970b). Milk production is increased when dairy cattle are protected from cold, windy conditions (Johnson, 1965), and mortality is significantly decreased with protection of newborn lambs (Holmes & Sykes, 1984). Unfortunately, the literature on the effects of shelter on livestock production is not nearly as extensive as that pertaining to crop production. However, there does appear to be a consensus, especially among producers, that reducing wind speed in winter reduces animal stress, improves animal health, increases feed efficiency, and provides positive economic returns (Atchison & Strine, 1984; Quam, Johnson, Wight, & Brandle, 1994). Here we describe the responses of livestock to environmental conditions influenced by shelter, how shelter fits into livestock management systems, and the design and management of windbreaks for livestock protection.

Wind Chill Temperatures

The combined effect of low temperatures and high wind speeds is known as the *wind chill equivalent temperature* and is commonly referred to as the *wind chill factor*. It reflects the rate of sensible heat loss from the body. As wind speeds increase, the thickness of the boundary layer next to the body decreases and the rate of heat loss increases (Moran & Morgan, 1986). For example, when the air temperature is −18°C (~0 °F) and the wind speed is 12 m s^{-1} (~27 mph), the wind chill factor is −44°C (about −47°F). At this equivalent temperature, danger to animals increases, including freezing of exposed flesh. A windbreak would reduce wind speed by 50 to 60% and raise the equivalent temperature to −30°C (about −22°F), still stressful to young animals but of little consequence to healthy, mature animals.

Animal Response to Shelter

Livestock, like all warm-blooded animals, must maintain their body temperature within a very narrow range if they are to survive. Body temperatures outside this range induce either cold or heat stress and can cause death in a relatively short period. This temperature varies with species, breed, age, general health, animal weight, and season of the year. Fortunately, many types of livestock have excellent abilities to adapt to a wide range of low environmental temperatures (see Table 5–3) and maintain a constant body temperature (Primault, 1979).

Primault (1979) defined five thermal zones centered around a zone of thermal indifference. These zones vary with the species and age of the animal. Young animals tend to have high, narrow zones while older animals have lower and broader zones. Within the zone of thermal indifference, normal metabolism supplies the necessary energy to maintain body temperature. As air temperatures decrease, the animal must generate additional heat to maintain its critical body temperature and to survive. This requires the use of stored fat reserves or the ingestion of additional feed (Graham, Wainman, Blaxter, & Armstrong, 1959; Winchester, 1964; Young, 1983). In addition, long-term exposure to cold temperatures reduces the efficiency of feed utilization, meaning that not only must the animal eat more as it gets colder but also that the energy gained per unit of feed may decrease with continued exposure (Webster, 1970a, 1970b; Young & Christopherson, 1974). As air temperatures continue to decline, the ability to maintain body temperature is no longer sufficient to meet the animal's need and body temperature begins to fall, resulting in death.

Windbreaks for Livestock Operations

There are many benefits of windbreaks to the successful livestock operation. As in the case of crops, the goal is to use the microclimate

Table 5–3. Optimum temperature conditions for efficient livestock production systems.

Age or type of livestock	Temperature range
	°C
Calves for breeding	5–20
Calves while fattening	18–12 [a]
Young breeding cattle	5–20
Young cattle while fattening	10–20
Milk cows	0–15
Suckling pigs (newborn animals)	33–22[a]
Young pigs and pigs for slaughter	22–15 [a]
Pregnant and lactating sows	5–15
Lambs	12–16
Sheep for slaughter or wool	5–15
Horses	8–15
Newborn chicks	34–21[a]
Egg-laying hens	15–22

Note. Source: Primault (1979), reproduced with permission.

[a] Optimal temperature gradually decreases as animals age or gain weight.

conditions created by shelter to benefit the animal production system.

Animal behavior is influenced by cold, wet, and windy conditions, especially when combined with low temperatures (Graunke, Schuster, & Lidfors, 2011; Van laer, Ampe, Moons, Sonck, & Tuyttens, 2015). As minimum daily temperature decreases, cattle on rangeland spend less time grazing, reducing forage intake and weight gain (Adams, Nelsen, Reynolds, & Knapp, 1986; Kartchner, 1980; Malechek & Smith, 1976). In a pair of field studies of winter stalk grazing in east-central Nebraska (Jordon, Klopfenstein, Brandle, & Klemesrud, 1997; Morris et al., 1996), average winter temperatures were moderate and animals behaved similarly on both open and sheltered fields. However, on days with low temperatures (below − 20°C) and strong winds (>10 m s^{-1}), cattle sought any available shelter. In particular, it was noted that cattle on the sheltered fields were grazing in the sheltered zones, while cattle on the exposed fields were lying down in low areas to reduce stress associated with the cold, windy conditions. Even so, they concluded that shelter had little effect on weight gain from winter stalk grazing during mild winters in east-central Nebraska.

Bond and Laster (1974) investigated the impacts of providing shelter to livestock in confinement. Their results indicated that when given a choice of remaining in shelter or feeding in exposed areas, cattle spent more time in the sheltered zone than feeding. They concluded that shelter was not economical in feedlot situations in south-central Nebraska because animals spent less time feeding and gained less weight. In contrast, Anderson and Bird (1993) reported significant increases in average daily gain and daily feed intake in a North Dakota feeding study. Similarly, livestock feeders in South Dakota, Nebraska, and Kansas reported significant feed savings and increased weight gains (Atchison, 1976; Robbins, 1976). These differences emphasize the need to have properly designed windbreaks, with feeding areas well within the sheltered zone if the benefits of protection are to be realized. They also emphasize the need for long-term studies under various climatic conditions.

Properly designed livestock windbreaks provide additional benefits to the livestock producer. On rangeland, windbreaks located across the landscape will increase the amount of forage production on the sheltered areas (Kort, 1988) and provide protection for calving against early spring snowstorms. In a Kansas study, average calving success increased 2% when cows were protected by a windbreak (Quam et al., 1994). Furthermore, windbreaks can be designed to harvest snow and provide water to supplement stock ponds located in remote areas (Jairell & Schmidt, 1986, 1992; Tabler & Johnson, 1971).

Protecting confinement systems with multirow windbreaks can control snow drifting, enabling access to feedlots and other facilities such as grain and hay storage, and reducing costs associated with snow removal. Wind protection provides a more moderate working environment for feedlot workers, reducing their exposure to cold winds and increasing their efficiency. Windbreaks intercept dust, screen unsightly areas from the road or living area, and assist in control of odors.

Windbreak Design for Livestock Systems

As with other types of windbreaks, livestock windbreaks need to be designed for each specific operation. General principles are defined here, and a more complete discussion of design criteria is available in Dronen (1988).

Livestock protection requires that the windbreak system have an optical density of at least 60% during the winter months. To meet this need, livestock windbreaks should have from three to five rows of trees or shrubs, including at least one or two rows of dense conifers. Rows should extend at least 30 m past the area needing protection to prevent snow from drifting around the ends and into the livestock area. In areas with extreme winter conditions, such as the northern Great Plains and the Canadian Prairies, a minimum of five to seven rows is required for adequate protection.

Placement of the windbreak is critical. It should be located to provide protection from the prevailing winter winds and drifting snow. There should be sufficient distance (at least 50 m) between the windward row and the feeding or calving area to allow snow deposition. A shrub row located 10 to 15 m windward of the main windbreak will reduce snow deposition leeward of the main windbreak and allow greater flexibility in the livestock operation (Dronen, 1988). Loafing sheds should be located leeward of the drift zone (Jones, Friday, & DeForest, 1983). In areas with hot summers, particular attention must be paid to the distance between the leeward edge of the windbreak and feeding areas. Feed bunks should be located at least 25 m (typically 2 to 3*H*) leeward to prevent heat buildup in the feeding area, air stagnation, and animal stress.

In most cases, protection from two or three directions is best. For example, livestock facilities in most areas of the northern Great Plains should have protection on both the north and west exposures and, in some cases, on the east as well. Drainage for melting snow must be provided so that meltwater does not flow through the feeding area. Similarly, runoff from the feeding area should not drain through the windbreak, as high NO_3 levels can damage many tree species. All livestock windbreaks should be fenced to prevent damage by grazing livestock. Typical livestock windbreak systems are illustrated in Figure 5–7.

Windbreaks for Odor Mitigation

Odor mitigation from livestock operations remains a sensitive topic in many communities. As urban areas expand (Ashwood, Diamond, & Walker, 2019), agricultural communities have experienced an influx of people unfamiliar with the noise and odors of production agriculture. Livestock producers recognize the need for strategies to reduce potential conflicts, particularly with regard to odors associated with livestock production systems. Maurer et al. (2016) provided a review of the effectiveness of various technologies for mitigating odor and other factors associated with these systems.

The strategic use of shelterbelts for odor mitigation is gaining attention in livestock-producing regions. Research suggests that shelterbelts located near and within livestock facilities can play an important role in biophysically and sociopsychologically mitigating odor in an economically feasible way (Tyndall & Colletti, 2007; Tyndall & Randall, 2018; Ubeda, Lopez-Jimenez, Nicolas, & Calvet, 2013). Trees and shrubs alter air movement, helping to intercept, disperse, and/or dilute odors before they can accumulate and become a nuisance to downwind areas.

Because the livestock odor source is near the ground and the tendency of the odor is to travel along the ground, shelterbelts of modest heights (6–12 m) are ideal for plume interception, disruption, and dilution (Bottcher, 2001; Lin, Barrington, Nicell, Choiniere, & Vezina, 2006; Takle, 1983; Tyndall & Colletti, 2007). The majority of odorous chemicals and compounds are absorbed onto, concentrated by, and carried on particulates generated in animal facilities (e.g., animal houses and manure storage) or from land application (Bottcher, 2001). If particulate movement can be controlled, odor movement will be partially controlled (Hammond & Smith, 1981). Shelterbelts have been shown to mitigate odors by a combination of physical and social dynamics (Malone, van Wicklen, Collier, & Hansen, 2006; Tyndall & Colletti, 2007).

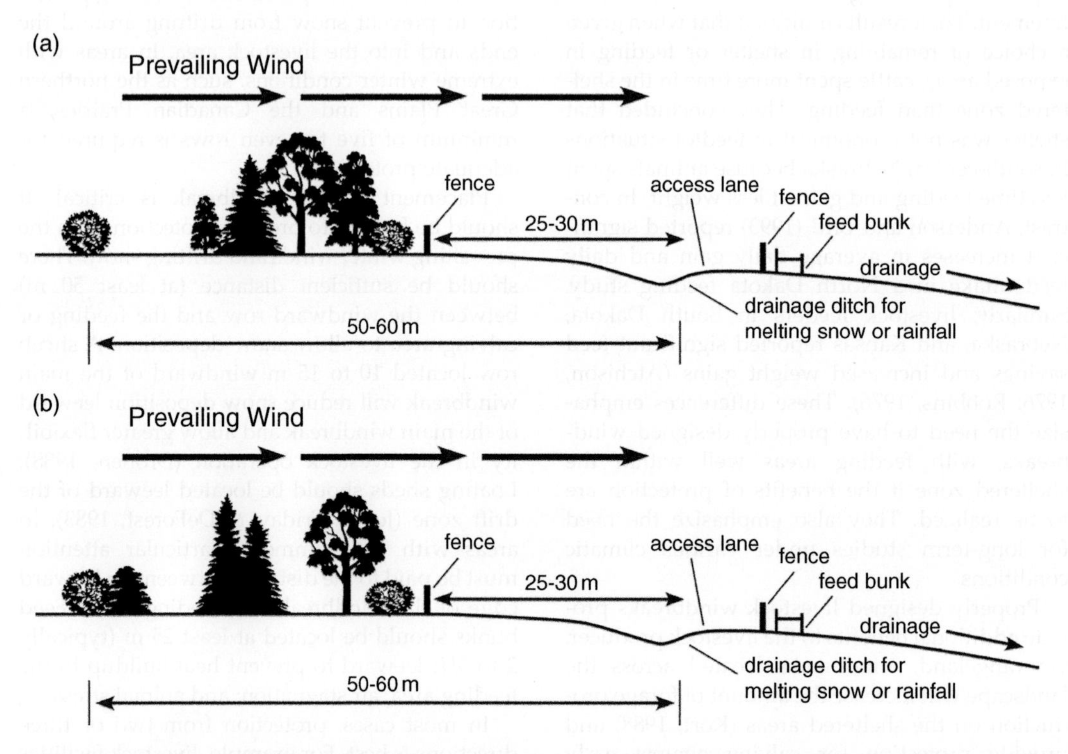

Fig. 5–7. Cross-section of a feedlot windbreak designed for wind and snow protection: (a) traditional multi-row windbreak with a row of shrubs on the windward side, and (b) modified twin-row, high-density windbreak including a double row of shrubs on the windward side to catch snow.

The most important odor mitigation dynamic provided by shelterbelts is vertical mixing through turbulence leading to enhanced dilution and dispersion of odor. Modeling the movement of odor up and over a shelterbelt, Lammers, Wallenfang, and Boeker (2001) observed that odor emissions from a livestock building would experience a significantly elevated airstream that is distributed by turbulent eddies, diluting the downwind stream of odor.

It is generally accepted that trees and other woody vegetation are among the most efficient natural filtering structures in a landscape, in part due to the very large total surface area of leafy plants (Bolund & Hunhammer, 1999). Field studies in Delaware (Malone et al., 2006) have quantified a 49 ± 27% ($p < .01$) reduction in particulate emissions from a working pullet facility with a 9.2-m-wide, three-row shelterbelt composed of bald cypress [*Taxodium distichum* (L.) Rich.], Leyland cypress (*Cupressocyparis leylandii* A.B. Jacks. & Dallim.), and eastern red cedar (*Juniperus virginiana* L.).

Laird (1997) and Thernelius (1997) modeled the potential of windbreaks to cause airstream fallout of odorous particulates by reducing wind speeds. Using an open-circuit wind tunnel, a small-scale model of an open-air ventilated hog confinement building, and a three-row simulated shelterbelt, mass transport of particulates was reduced by 35–56% (depending on wind angle and speed) as a result of reduced wind speeds (Laird, 1997).

Professionals involved with livestock agriculture generally accept that a well-landscaped operation that is visually pleasing or screened is more acceptable to the public than one that is not (Lorimor, 1998; National Pork Producers Council, 1995; Melvin, 1996). Focus groups in Iowa suggested that the general public is "highly appreciative" of more trees in agricultural landscapes and showed a "high level of agreement" that shelterbelts improve the site aesthetics of confinement livestock production. They also indicated a more positive view of the effectiveness of odor control practices when the sources of odor were hidden from view (Tyndall, 2006a). A general windbreak design for livestock odor control is illustrated in Figure 5–8.

The few studies that have attempted field quantification of reduced downwind odor concentration and movement have recorded reductions ranging from a low of 6% (Malone et al., 2006) to a high of 33% (Lin et al., 2006; Vezina, 2005). Financial analysis of shelterbelts used for odor mitigation across a series of hog production sites of varying scale and production types showed a range of costs from $0.03 to $0.33 per pig produced (Tyndall, 2006b), all well below producer-revealed expenses for odor management (Tyndall, 2006b). Shelterbelts are not a substitute for comprehensive odor management strategies. Rather, their use should be thought of as a complementary technology used within a "suite" of odor management strategies (Tyndall & Colletti, 2007).

Windbreak Technology at the Farm and Landscape Levels

Sustainable agriculture is a system of whole-farm resource use balanced with whole-farm productivity (Jackson & Jackson, 2002; Lefroy, Hobbs, O'Connor, & Pate, 1999). Agroforestry is one component of a successful sustainable agriculture system, and the use of field and livestock windbreaks within that system are specific management options. Mize et al. (2008) proposed a computer simulation model (SAMS) that used windbreak characteristics to calculate changes in wind speed and microclimate in shelter. It then used the climate data generated to grow a crop of corn and/or soybean using standard crop growth models. An economic analysis of the yield benefits indicated that windbreaks make economic sense from a production point of view but the value of other benefits including erosion control and wildlife need to be developed. This is another area for potential research.

According to some definitions, agroforestry must produce marketable products. In that sense, other types of windbreaks such as farmstead windbreaks or living snow fences are not agroforestry; however, if agroforestry is to be true to the basic ecological principles of sustainability, it must recognize the use of other types of windbreaks to support the whole-farm system and the agricultural ecosystem. To that end, here we identify other windbreak uses and their benefits and discuss very briefly the ecological implications of windbreak technology to support the farm operation. Those seeking a more detailed discussion of these concepts are referred to the Proceedings of the First International Symposium on Windbreak Technology (Brandle et al., 1988), the excellent text on landscape ecology by Forman (1995), the Proceedings from the Workshop on Agriculture as a Mimic of Natural Ecosystems (Lefroy et al., 1999), two texts—one by Jose and Gordon (2008) and the other by Batish, Kohli, Jose, and Singh (2008)—on the ecological basis of agroforestry, and most recently an overview by Schoeneberger et al. (2017).

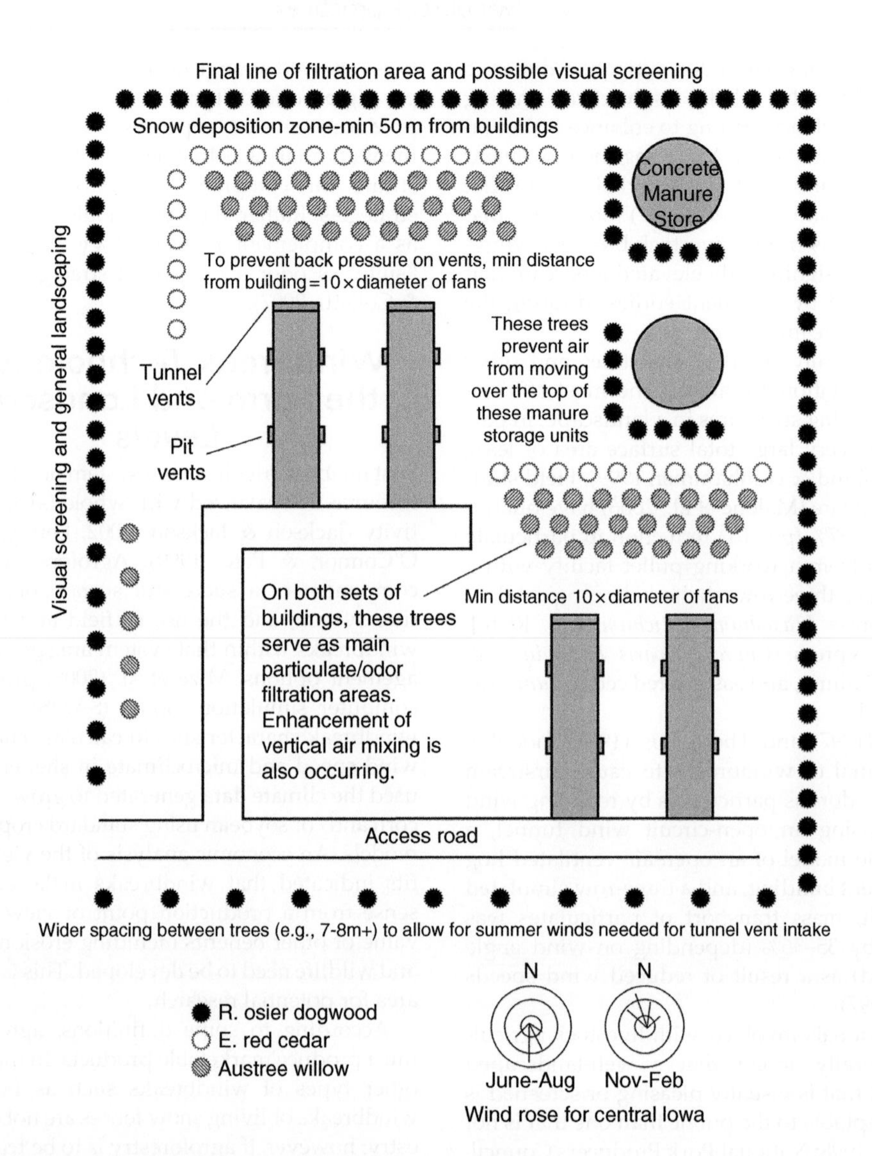

Fig. 5–8. Generalized windbreak design for odor mitigation in central Iowa.

Farmstead Windbreaks

The basic goal of a farmstead windbreak is to provide protection to the living and working area of a farm or ranch and thus to contribute to the overall well-being of the farm operation (Wight, 1988; Wight, Boes, & Brandle, 1991). The greatest economic benefit is derived from reducing the amount of energy needed to heat and cool the home. The amount of savings varies with climatic conditions (particularly wind and temperature), local site conditions, home construction, and the design and condition of the windbreak. Well-designed farmstead windbreaks can cut the average energy use of a typical farm or ranch home in the northern portions of the United States and Canada by 10–30% (Brandle, Wardle, & Bratton, 1992; Dewalle & Heisler, 1988).

Farmstead windbreaks improve living and working conditions by screening undesirable sights, sounds, smells, and dust crated by nearby agricultural activities or roads (Cook & van Haverbeke, 1971; Ferber, 1969; Wight, 1988). They reduce the effects of windchill and make outdoor activities less stressful. Properly located farmstead windbreaks can help in snow management, reducing the time and energy involved in snow removal from working areas and driveways. Locating the family garden within the sheltered zone improves yield and quality, and incorporating fruit and nut trees in the windbreak will give additional benefits. Multi-row farmstead windbreaks provide significant wildlife habitat in the form of nesting, feeding, singing, and breeding sites for many bird species and enrich the comfort and enjoyment of outdoor activities. Adding particular tree or shrub species to the windbreak can enhance the wildlife component and attract desirable species to the area (Johnson, Beck, & Brandle, 1991).

Designing a windbreak for a farmstead calls for considerations of seasonality of both the characteristics of the windbreak and the seasonal needs for specific kinds of sheltering—management of wind reduction, shade, snow deposition, temperature, human comfort, storm protection, and others. For instance, Koh et al. (2014) found that the sheltering effect in summer was higher by a factor of two than in autumn and a factor of 10 over winter.

Most important, designing a windbreak for a farmstead calls for consideration of the landowner's wishes. From a technical point of view, seasonality of the characteristics of the windbreak and the seasonal needs for specific kinds of sheltering are important. These include but are not limited to management of wind reduction, shade, snow deposition, temperature, human comfort, storm protection, and aesthetics.

Windbreaks for Snow Control

There are basically three objectives for snow management: (a) to spread snow across a crop field to protect the crop or to provide soil moisture for the next season, (b) to harvest snow for use in stock ponds, and (c) to prevent snow accumulation in undesirable locations, such as roadways or work areas (Scholten, 1988; Shaw, 1991). Each objective has specific design requirements. We have discussed briefly the use of field windbreaks to control snow on crop fields and have discussed the use of livestock windbreaks and farmstead windbreaks to provide snow control. In general, porous windbreaks spread snow across a large area, while dense windbreaks cause snow to accumulate in deep drifts near the windbreak.

Wildlife Windbreaks

In many agricultural areas, windbreak and riparian systems offer the only woody habitat for wildlife (Johnson, Beck, & Brandle, 1994). In Nebraska, foresters identify wildlife as a primary reason given by landowners for the establishment of windbreaks on agricultural land. Yahner (1982a, 1982b, 1982c) and others (see Johnson & Beck [1988] for an extensive review) have documented the critical nature of these habitats for various wildlife species. Johnson, Brandle, Fitzmaurice, and Poague (1993) reemphasized the potential role of these types of habitats in the control of crop pests in agricultural regions. Because of their linear nature, windbreaks are dominated by edge species (both plants and animals). As the width of a windbreak increases, species diversity increases as additional microhabitats are added (Forman, 1995; Forman & Baudry, 1984; Sreekar et al., 2013). In a Kansas study of habitat use within agricultural settings, these linear forests were favored by hunters because many game species are attracted to the cover provided by the woody vegetation. Cable and Cook (1990) estimated that annual economic returns associated with hunting linear forests in Kansas were in the range of $30 to 35 million.

More recently, interest has turned to the overall role of woody habitat in agricultural ecosystems (see discussion above on integrated pest management). Holland and Fahrig (2000) found that insect diversity in and adjacent to alfalfa fields was enhanced by adjacent woody borders. Yang et al. (2019) supported this finding. Perkins et al. (2003) indicated that management of the agricultural landscape for a diverse avifauna requires consideration of the amount and distribution of woody cover in the surrounding landscape. Similarly, the amount and distribution of grassland areas is critical to the success of grassland birds associated with agricultural landscapes (Hanson, 2007).

Windbreaks and Climate Change

Agroforestry practices are a critical component of our efforts to deal with climate change (Brandle & Schoeneberger, 2014; Schoeneberger et al., 2012, 2017). Brandle, Wardle, & Bratton (1992) reviewed the use of windbreaks as a means to reduce atmospheric CO_2 concentration. They identified not only the direct sequestration of C in the growing trees but also quantified the indirect benefits to agricultural production systems due to crop and livestock protection and energy savings. They estimated that a minimum windbreak planting program of 1.96 million ha would result in the storage of 22.2 million Mg of C. These numbers were estimated using forest tree biomass equations. Zhou et al. (2011, 2015) assessed the application of these equations to shelterbelt trees and suggested that a factor of 1.2 should be applied if windbreak biomass was to be estimated using forest tree biomass equations. Additional estimates of C storage by windbreak trees can be found in the work of Ballesteros Possu, Brandle, Domke, Schoeneberger, and Blankenship (2016), Ballesteros-Possu, Brandle, and Schoeneberger (2017), Hou, Young, Brandle, and Schoeneberger (2011), and Ziegler et al. (2016).

In addition, indirect benefits of windbreaks in the agricultural sector flowing from the reduction in the number of hectares farmed would reduce diesel fuel consumption by 1240 million L. Additional fuel savings from the protection of farmsteads and reduction in fertilizer use would save >5.4 billion m^3 of natural gas. These reductions in fossil fuel use could reduce CO_2 emissions by as much as 291 million Mg during the 50-yr life of the windbreak plantings (Brandle, Wardle, & Bratton, 1992).

Kort and Turnock (1998) conducted a study of the amount of C stored in shelterbelts of the Canadian prairie. They surveyed 11 sites and 12 major shelterbelt species. Based on their results, they estimated that a shelterbelt planting program of six million trees and shrubs in the Prairie Provinces could potentially sequester 0.4 million Mg C yr^{-1}.

Montagnini and Nair (2004) summarized the environmental benefits of agroforestry systems to the overall C sequestration issue. They estimated total annual potential C storage of >90 trillion Mg from five agroforestry practices, with windbreaks contributing 4 million Mg C yr^{-1}. Sánchez and McCollin (2015) have demonstrated multiple microclimate and environmental advantages of sheltering by hedgerows and dehesa in a Mediterranean, multipurpose crop and animal agroforestry system.

Windbreaks could also play a significant role in adaptation strategies as agricultural producers strive to adapt to changing climates. Easterling et al. (1997) reported that windbreaks could help maintain corn yields in eastern Nebraska under several climate scenarios. Using a crop modeling approach, they considered temperature increases up to 5 °C, precipitation levels of 70–130% of normal, and wind speed changes of ±30%. In all cases, sheltered crops continued to perform better than unsheltered crops. In all but the most extreme cases, windbreaks more than compensated for yield losses due to possible climate change, indicating the value of shelterbelts to ameliorate potential climate changes to the agricultural community.

Summary

In the context of agroforestry practices in temperate regions, windbreaks or shelterbelts are a major component of successful agricultural systems. By increasing crop production while reducing the level of inputs, they reduce the environmental costs associated with agriculture. They help control erosion, particularly wind erosion, and contribute to the long-term health of our agricultural systems. When various species are included in the design, they can contribute directly to the production of nuts, fruits, timber, and other wood products. When used in livestock production systems, they improve animal health, improve feed efficiency, mitigate odors, and contribute to the economic return of producers. When designed for snow management, they can capture snow for crop or livestock production.

As part of the overall agricultural enterprise, they reduce home energy consumption and improve working conditions within the farm area. When designed for snow control, they can reduce the costs of snow removal and improve access to livestock feeding areas. Windbreaks provide habitat for wildlife and a number of benefits to landowners and producers alike. The interspersion of woody wildlife habitat in agricultural areas contributes to a healthy and diverse wildlife population to the benefit of both hunters and non-hunters.

On a larger scale, windbreaks provide societal benefits both locally and on a larger, regional scale. Reductions in erosion not only benefit the landowner but reduce the off-site costs of erosion as well. Properly located, windbreaks reduce snow removal costs along highways, improve driving conditions, and reduce accidents. Windbreaks have the potential to assist with adapting to future changes in climate and may, in some cases, ease the economic burdens associated with change.

The integration of windbreaks and other agroforestry practices into sustainable agricultural systems can provide many rewards. It requires, however, careful consideration of all aspects of the agricultural system, an understanding of basic ecological principles, and a working knowledge of local conditions and markets.

References

Aase, J. K., & Siddoway, F. H. (1976). Influence of tall wheatgrass wind barriers on soil drying. *Agronomy Journal, 68,* 627–631. https://doi.org/10.2134/agronj1976.00021962006800040024x

Adams, D. C., Nelsen, T. C., Reynolds, W. L., & Knapp, B. W. (1986). Winter grazing activity and forage intake of range cows in the northern Great Plains. *Journal of Animal Science, 62,* 1240–1246. https://doi.org/10.2527/jas1986.6251240x

Anderson, V. L., & Bird, J. (1993). Effect of shelterbelt protection on performance of feedlot steers during a North Dakota winter. In *Beef Production Field Day Proceedings* (pp. 19–21). Vol. 16. Fargo, ND: Carrington Research Extension Center, North Dakota State University.

Andow, D. A. (1991). Vegetational diversity and arthropod population response. *Annual Review of Entomology, 36,* 561–586. https://doi.org/10.1146/annurev.en.36.010191.003021

Argete, J. C., & Wilson, J. D. (1989). The microclimate in the centre of small square sheltered plots. *Agricultural and Forest Meteorology, 48,* 185–199. https://doi.org/10.1016/0168-1923(89)90016-6

Armbrust, D. V. (1982). Physiological responses to wind and sandblast damage by grain sorghum plants. *Agronomy Journal, 74*, 133–135. https://doi.org/10.2134/agronj1982.00021962007400010034x

Ashwood, L., Diamond, D., & Walker, F. (2019). Property rights and rural justice: A study of U.S. right-to-farm laws. *Journal of Rural Studies, 67*, 120–129. https://doi.org/10.1016/j.jrurstud.2019.02.025

Atchison, F. D. (1976). Windbreaks for livestock protection in the central Great Plains. In R.W. Tinus (Ed.), *Windbreaks: What are they worth? Proceedings of the 34th Annual Meeting of the Forestry Committee, Great Plains Agricultural Council* (GPAC Publication no. 78, pp. 101–102).

Atchison, F. D., & Strine, J. H. (1984). *Windbreak protection for beef cattle* (Publication no. L-708). Manhattan, KS: Kansas State University Cooperative Extension Service.

Avila-Sanchez, S., Pindado, S., Lopez-Garcia, O., & Sanz-Andres, A. (2014). Wind tunnel analysis of the aerodynamic loads on rolling stock over railway embankments: The effect of shelter windbreaks. *Scientific World Journal, 2014*, 421829. https://doi.org/10.1155/2014/421829

Bagley, W. T. (1988). Agroforestry and windbreaks. *Agriculture, Ecosystems & Environment, 22–23*, 583–591. https://doi.org/10.1016/0167-8809(88)90047-3

Baldwin, C. S. (1988). The influence of field windbreaks on vegetable and specialty crops. *Agriculture, Ecosystems & Environment, 22–23*, 191–203. https://doi.org/10.1016/0167-8809(88)90018-7

Ballesteros Possu, W., Brandle, J. R., Domke, G. M., Schoeneberger, M., & Blankenship, E. (2016). Estimating carbon storage in windbreak trees on U.S. agricultural lands. *Agroforestry Systems, 90*, 889–904. https://doi.org/10.1007/s10457-016-9896-0

Ballesteros-Possu, W., Brandle, J. R., & Schoeneberger, M. (2017). Potential of windbreak trees to reduce carbon emissions by agricultural operations in the US. *Forests, 8*, 138. https://doi.org/10.3390/f8050138.

Barbeau, C. D., Wilton, M. J., Oelbermann, M., Karagatzides, J. D., & Tsuji, L. J. S. (2018). Local food production in a subarctic indigenous community: The use of willow (*Salix* spp.) windbreaks to increase the yield of intercropped potatoes (*Solanum tuberosum*) and bush beans (*Phaseolus vulgaris*). *International Journal of Agricultural Sustainability, 16*, 29–39. https://doi.org/10.1080/14735903.2017.1400713

Barker, G. L., Hatfield, J. L., & Wanjura, D. F. (1989). Influence of wind on cotton growth and yield. *Transactions of the ASAE, 32*, 97–104. https://doi.org/10.13031/2013.30968

Bates, C. G. (1911). *Windbreaks: Their influence and value* (USDA Forest Service Bulletin 86). Washington, DC: U.S. Government Printing Office.

Bates, C. G. (1944). *The windbreak as a farm asset* (USDA Farmers Bulletin 1405). Washington, DC: U.S. Government Printing Office.

Batish, D. R., Kohli, R. K., Jose, S., & Singh, H. P. (Eds.). (2008). *Ecological basis of agroforestry*. Boca Raton, FL: CRC Press.

Beecher, N. A., Johnson, R. J., Brandle, J. R., Case, R. M., & Young, L. J. (2002). Agroecology of birds in organic and nonorganic farmland. *Conservation Biology, 16*, 1620–1631. https://doi.org/10.1046/j.1523-1739.2002.01228.x

Bennell, M. R., Leys, J. F., & Cleugh, H. A. (2007). Sandblasting damage of narrow-leaf lupin (*Lupinus angustifolius* L): A field wind tunnel simulation. *Australian Journal of Soil Research, 45*, 119–128. https://doi.org/10.1071/SR06066

Bennell, M. R., & Verbyla, A. P. (2008). Quantifying the response of crops to shelter in the agricultural regions of South Australia. *Australian Journal of Agricultural Research, 59*, 950–957. https://doi.org/10.1071/AR08188

Bicknell, D. (1991, 30 Sept.–3 Oct.). The role of trees in providing shelter and controlling erosion in the dry temperate and semi-arid southern agricultural areas of Western Australia. *In* R. Prinsley (coordinator), *The role of trees in sustainable agriculture: Proceedings of a National Conference* (pp. 21–39). Canberra, ACT, Australia: Bureau of Rural Resources.

Biddington, N. L. (1985). A review of mechanically induced stress in plants. *Scientia Horticulturae, 36*, 12–20.

Biddington, N. L. (1986). The effects of mechanically-induced stress in plants: A review. *Plant Growth Regulation, 4*, 103–123. https://doi.org/10.1007/BF00025193

Bird, P. R., Jackson, T. T., Kearney, G. A., & Roache, A. (2007). Effects of windbreak structure on shelter characteristics. *Australian Journal of Experimental Agriculture, 47*, 727–737. https://doi.org/10.1071/EA06086

Biro, R. I., & Jaffe, J. J. (1984). Thigmomorphogenesis: Ethylene evolution and its role in the changes observed in mechanically perturbed bean plants. *Physiologia Plantarum, 62*, 289–296. https://doi.org/10.1111/j.1399-3054.1984.tb04575.x

Bitog, J. P., Lee, I.-B., Hwang, H.-S., Shin, M.-H., Hong, S.-W., Seo, I.-H., . . . Pang, Z. (2011). A wind tunnel study on aerodynamic porosity and windbreak drag. *Forest Science and Technology, 7*, 8–16. https://doi.org/10.1080/21580103.2011.559939

Bolund, P., & Hunhammer, S. (1999). Ecosystem services in urban areas. *Ecological Economics, 29*, 293–301. https://doi.org/10.1016/S0921-8009(99)00013-0

Bond, T. E., & Laster, D. B. (1974). Influence of windbreaks on feedlot cattle in the Midwest. *Transactions of the ASAE, 17*, 505–507, 512. https://doi.org/10.13031/2013.36893

Bottcher, R. E. (2001). An environmental nuisance: Odor concentrated and transported by dust. *Chemical Senses, 26*, 327–331. https://doi.org/10.1093/chemse/26.3.327

Brandle, J. R. (1990). Management of microclimate with windbreaks. In D. Sirois (Ed.), *Proceedings of the 17th Annual Meeting, Plant Growth Regulator Society of America* (pp. 61–69).

Brandle, J. R., Hintz, D. L., & Sturrock, J. W. (Eds.). (1988). *Windbreak technology*. Amsterdam: Elsevier.

Brandle, J. R., Hodges, L., & Stuthman, J. (1995). Windbreaks and specialty crops for greater profits. In W. J. Rietveld (Ed.), *Agroforestry and sustainable systems: Symposium Proceedings* (General Technical Report RM-GTR-261, p. 81–91). Fort Collins, CO: U.S. Forest Service, Rocky Mountain Forest & Range Experiment Station.

Brandle, J. R., Johnson, B. B., & Akeson, T. (1992). Field windbreaks: Are they economical? *Journal of Production Agriculture, 5*, 393–398. https://doi.org/10.2134/jpa1992.0393

Brandle, J. R., Johnson, B. B., & Dearmont, D. D. (1984). Windbreak economics: The case of winter wheat production in eastern Nebraska. *Journal of Soil and Water Conservation, 39*, 339–343.

Brandle, J. R., & Kort, J. (1991). WBECON: A windbreak evaluation model: 1. Comparison of windbreak characteristics. In S. Finch and C.S. Baldwin (Eds.), Windbreaks and agroforestry: 3rd International Symposium (p. 129–131).

Brandle, J. R., & Schoeneberger, M. M. (2014). Working trees: Supporting agriculture and healthy landscapes. *Journal of Tropical Forest Science, 26*, 305–308.

Brandle, J. R., Wardle, T. D., & Bratton, G. F. (1992)b. Opportunities to increase tree planting in shelterbelts and the potential impacts on carbon storage and conservation. *In* N. R. Sampson and D. Hair (Eds.), *Forests and global change. Vol. 1. Opportunities for increasing forest cover* (pp. 157–176). Washington, DC: American Forests.

Brandle, J. R., Zhou, X. H., & Takle, E. S. (2003). The influence of three dimensional structure of a tree shelterbelt on aerodynamic effectiveness. In *Land-use management for the future: Proceedings of the 6th Conference on Agroforestry in North America* (pp. 169–174).

Bubenzer, G. D., & Weis, G. G. (1974). Effect of wind erosion on production of snap beans and peas. *Journal of the American Society for Horticultural Science, 99*, 527–529.

Burel, F. (1996). Hedgerows and their roles in agricultural landscapes. *Critical Reviews in Plant Sciences, 15*(2), 169–190. https://doi.org/10.1080/07352689.1996.10393185

Cable, T. T., & Cook, P. S. (1990). The use of windbreaks by hunters in Kansas. *Journal of Soil and Water Conservation, 45*, 575–577.

Caborn, J. M. (1957). *Shelterbelts and microclimate* (Forestry Commission Bulletin 29). Edinburgh, Scotland, UK: Edinburgh University.

Caborn, J. M. (1971). The agronomic and biological significance of hedgerows. *Outlook on Agriculture, 6*, 279–284. https://doi.org/10.1177/003072707100600609

Chamberlain, J. L., Emery, M. R., & Patel-Weynand, T. (Eds.). (2018). Assessment of nontimber forest products in the United States under changing conditions (General Technical Report SRS-232). Asheville, NC: U.S. Forest Service, Southern Research Station. https://doi.org/10.2737/SRS-GTR-232)

Chaput, L. J., & Tuskan, G. A. (1990). Field windbreak management and its effect on adjacent crop yield. *North Dakota Farm Research, 48*(2), 6–28.

Claflin, L. E., Stuteville, D. L., & Armbrust, D. V. (1973). Windblown soil in the epidemiology of bacterial leaf spot of alfalfa and common blight of bean. *Phytopathology, 63*,1417–1419. https://doi.org/10.1094/Phyto-63-1417

Cleugh, H. A. (1998). Effects of windbreaks on airflow, microclimate and crop yields. *Agroforestry Systems, 41*, 55–84. https://doi.org/10.1023/A:1006019805109

Cleugh, H. A. (2002). Parameterising the impact of shelter on crop microclimates and evaporation fluxes. *Australian Journal of Experimental Agriculture, 42*, 859–874. https://doi.org/10.1071/EA02006

Cleugh, H. A., Prinsley, R., Bird, R.P., Brooks, S. J., Carberry, P. S., Crawford, M. C., . . . Wright, A. J. (2002). The Australian National Windbreaks Program: Overview and summary of results. *Australian Journal of Experimental Agriculture, 42*, 649–664. https://doi.org/10.1071/EA02003

Cook, D. I., & van Haverbeke, D. F. (1971). *Trees and shrubs for noise abatement* (Research Bulletin 246). Lincoln, NE: Nebraska Agricultural Experiment Station.

Corbett, A., & Plant, R. E. (1993). Role of movement in the response of natural enemies to agroecosystem diversification: A theoretical evaluation. *Environmental Entomology, 22*, 519–531. https://doi.org/10.1093/ee/22.3.519

Corbett, A., & Rosenheim, J. A. (1996). Impact of a natural enemy overwintering refuge and its interaction with the surrounding landscape. *Ecological Entomology, 21*, 155–164. https://doi.org/10.1111/j.1365-2311.1996.tb01182.x

Cornelis, W. M., & Gabriels, D. (2005). Optimal windbreak design for wind-erosion control. *Journal of Arid Environments, 61*, 315–332. https://doi.org/10.1016/j.jaridenv.2004.10.005

Coutts, M. P., & Grace, J. (Eds.). (1995). *Wind and trees*. Cambridge, UK: Cambridge University Press. https://doi.org/10.1017/CBO9780511600425

Cunningham, R. A. (1988). Genetic improvement of trees and shrubs used in windbreaks. *Agriculture, Ecosystems & Environment, 22–23*, 483–498. https://doi.org/10.1016/0167-8809(88)90040-0

Danielson, S. D., Brandle, J. R., Hodges, L., & Srinivas, P. (2000). Bean leaf beetle (Coleoptera: Chrysomelidae) abundance in soybean fields protected and unprotected by shelterbelts. *Journal of Entomological Science, 35*, 385–390. https://doi.org/10.18474/0749-8004-35.4.385

Davis, J. E., & Norman, J. M. (1988). Effects of shelter on plant water use. *Agriculture, Ecosystems & Environment, 22–23*, 393–402. https://doi.org/10.1016/0167-8809(88)90034-5

de Preez, N. D. (1986). The economic advantage of artificial windbreaks in plum cultivation on Tatura trellis system in a windy environment. *Deciduous Fruit Grower, 36*, 59–65.

Deshpande, R. Y., Hubbard, K. G., Coyne, D. P., Steadman, J. R., & Parkhurst, A. M. (1995). Estimating leaf wetness in dry bean canopies as a prerequisite to evaluating white mold disease. *Agronomy Journal, 87*, 613–619. https://doi.org/10.2134/agronj1995.00021962008700040002x

Dewalle, D. R., & Heisler, G. M. (1988). Use of windbreaks for home energy conservation. *Agriculture, Ecosystems & Environment, 22–23*, 243–260. https://doi.org/10.1016/0167-8809(88)90024-2

Ding, S., & Su, P. (2010). Effects of tree shading on maize crop within a Poplar-maize compound system in Hexi Corridor oasis, northwestern China. *Agroforestry Systems, 80*, 117–129. https://doi.org/10.1007/s10457-010-9287-x

Dix, M. E., Hodges, L., Brandle, J. R., Wright, R. J., & Harrell, M. O. (1997). Effects of shelterbelts on the aerial distribution of insect pests in muskmelon. *Journal of Sustainable Agriculture, 9*, 5–24. https://doi.org/10.1300/J064v09n02_03

Dix, M. E., Johnson, R. J., Harrell, M. O., Case, R. M., Wright, R. J., Hodges, L., . . . Hubbard, K. G. (1995). Influence of trees on abundance of natural enemies of insect pests: A review. *Agroforestry Systems, 29*, 303–311.

Drew, R. L. K. (1982). The effects of irrigation and of shelter from wind on emergence of carrot and cabbage seedlings. *Journal of Horticultural Science, 57*, 215–219. https://doi.org/10.1080/00221589.1982.11515043

Dronen, S. I. (1988). Layout and design criteria for livestock windbreaks. *Agriculture, Ecosystems & Environment, 22–23*, 231–240. https://doi.org/10.1016/0167-8809(88)90022-9

Droze, W. H. (1977). *Trees, prairies, and people: A history of tree planting in the Plains states*. Denton, TX: Texas Woman's University Press.

Du, M., Ushiyama, T., & Maki, T. (2010). Observations and numerical simulations of windbreak effects on surface temperature. In *HEAPFL'10: Proceedings of the 2010 international conference on theoretical and applied mechanics, and 2010 international conference on fluid mechanics and heat & mass transfer* (pp. 48–52). Stevens Point, WI: World Scientific and Engineering Academy and Society.

Dupraz, C., Wolz, K. J., Leconite, I., Talbot, G., Vincent, G., Mulia, R., . . . Van Noondwijk, M. (2019). Hi-sAFe: A 3D agroforestry model for integrating dynamic tree–crop interaction. *Sustainability, 11*, 2293. https://doi.org/10.3390/su11082293

Easson, D. L., White, E. M., & Pickles, S. J. (1993). The effects of weather, seed rate, and cultivar on lodging and yield in winter wheat. *Journal of Agricultural Science, 121*, 145–156. https://doi.org/10.1017/S0021859600077005

Easterling, W. E., Hays, C. J., Easterling, M. M., & Brandle, J. R. (1997). Modeling the effect of shelterbelts on maize productivity under climate change: An application of the EPIC model. *Agriculture, Ecosystems & Environment, 61*, 163–176. https://doi.org/10.1016/S0167-8809(96)01098-5

Elmore, R. W., & Ferguson, R. B. (1996). Brittle-snap in corn: Hybrid and environmental factors. In *Proceedings of the 51st Corn and Sorghum Research Conference, Chicago IL* (pp. 139–150). Alexandria, VA: American Seed Trade Association.

Emery, R. J. N., Reid, D. M., & Chinnappa, C. C. (1994). Phenotypic plasticity of stem elongation in two ecotypes of *Stellaria longipes*: The role of ethylene and response to wind. *Plant, Cell & Environment, 17*, 691–700. https://doi.org/10.1111/j.1365-3040.1994.tb00161.x

Erner, Y., & Jaffe, M. J. (1982). Thigmomorphogenesis: The involvement of auxin and abscisic acid in growth retardation due to mechanical perturbation. *Plant & Cell Physiology, 23*, 935–941. https://doi.org/10.1093/oxfordjournals.pcp.a076446

Ferber, A. E. (1969). *Windbreaks for conservation* (Agriculture Information Bulletin 339). Washington, DC: U.S. Government Printing Office.

Ferreira, A. D. (2011). Structural design of a natural windbreak using computational and experimental modeling. *Environmental Fluid Mechanics, 11*, 517–530. https://doi.org/10.1007/s10652-010-9203-y.

Finch, S. J. (1988). Field windbreaks: Design criteria. *Agriculture, Ecosystems & Environment, 22–23*, 215–228. https://doi.org/10.1016/0167-8809(88)90020-5

Flint, M. L., & Dreistadt, S. H. (1998). *Natural enemies handbook: The illustrated guide to biological pest control* (Publication 3386) Berkeley, CA: University of California, Division of Agriculture and Natural Resources.

Forman, R. T. T. (1995). *Land mosaics: The ecology of landscapes and regions*. Cambridge, UK: Cambridge University Press. https://doi.org/10.1017/9781107050327

Forman, R. T. T., & Baudry, J. (1984). Hedgerows and hedgerow networks in landscape ecology. *Environmental Management, 8*, 495–510. https://doi.org/10.1007/BF01871575

Frank, A. B., Harris, D. G., & Willis, W. O. (1974). Windbreak influence on water relations, growth, and yield of soybeans. *Crop Science, 14*, 761–765. https://doi.org/10.2135/cropsci1974.0011183X001400050044x

Fryrear, D. W., & Downes, J. D. (1975). Consider the plant in planning wind erosion control systems. *Transactions of the ASAE, 18*, 1070–1072. https://doi.org/10.13031/2013.36740

Gardiner, B., Berry, P., & Moulia, B. (2016). Review: Wind impact on plant growth, mechanics and damage. *Plant Science, 245*, 94–118. https://doi.org/10.1016/j.plantsci.2016.01.006

Garner, L. C., & Bjorkman, T. (1996). Mechanical conditioning for controlling excessive elongation in tomato transplants: Sensitivity to dose, frequency, and timing of brushing. *Journal of the American Society for Horticultural Science, 121*, 894–900. https://doi.org/10.21273/JASHS.121.5.894

George, E. J. (1971). *Effect of tree windbreaks and slat barriers on wind velocity and crop yields* (Production Research Report 121). Washington, DC: U.S. Government Printing Office.

Gold, M. A., Godsey, L. D., & Josiah, S. J. (2004). Markets and marketing strategies for agroforestry specialty products in North America. *Agroforestry Systems, 61*, 371–382.

Gonzales, H. B., Tatarko, J., Casada, M. E., Maghirang, R. G., Hagen, I. J., & Barden, C. J. (2018). Dust reduction efficiency of a single-row vegetative barrier (*Maclura pomifera*). *Transactions of the ASABE, 61*, 1907–1914. https://doi.org/10.13031/trans.12879

Grace, J. (1977). *Plant response to wind*. London: Academic Press.

Grace, J. (1981). Some effects of wind on plants. In J. Grace, E.D. Ford, & P.G. Jarvis (Eds.), *Plants and their atmospheric environment* (pp. 31–56). Oxford, UK: Blackwell Scientific.

Grace, J. (1988). Plant response to wind. *Agriculture, Ecosystems & Environment, 22–23*, 71–88. https://doi.org/10.1016/0167-8809(88)90008-4

Grace, J., Ford, E. D., & Jarvis, P. G. (Eds.). (1981). *Plants and their atmospheric environment*. Oxford, UK: Blackwell Scientific.

Graham, N. M., Wainman, F. W., Blaxter, K. L., & Armstrong, D. G. (1959). Environmental temperature, energy metabolism, and heat regulation in sheep: 1. Energy metabolism in closely clipped sheep. *Journal of Agricultural Science, 52*, 13–24. https://doi.org/10.1017/S0021859600035632

Grala, R. K., & Colletti, J. P. (2003). Estimates of additional maize (*Zea mays*) yields required to offset costs of tree-windbreaks in the midwestern USA. *Agroforestry Systems, 59*, 11–20. https://doi.org/10.1023/A:1026140208707

Graunke, K. L., Schuster, T., & Lidfors, L. M. (2011). Influence of weather on the behavior of outdoor-wintered beef cattle in Scandinavia. *Livestock Science, 136*, 247–255. https://doi.org/10.1016/j.livsci.2010.09.018

Greb, B. W. (1980). *Snowfall and its potential management in the semiarid central Great Plains* (Western Series no. 18). Oakland, CA: USDA–ARS.

Greb, B. W., & Black, A. L. (1961). Effects of windbreak plantings on adjacent crops. *Journal of Soil and Water Conservation, 16*(5), 223–227.

Hall, M., Nimmo, D., Watson, S., & Bennett, A. F. (2018). Linear habitats in rural landscapes have complementary roles in bird conservation. *Biodiversity and Conservation, 27*, 2605–2623. https://doi.org/10.1007/s10531-018-1557-3

Hammond, E. G., & Smith, R. J. (1981). Survey of some molecularly dispersed odorous constituents in swine house air. *Iowa State Journal of Research, 55*(4), 393–399.

Hans, T. G. (1987). *Effect of shelterbelts on growth, yield, and quality of alfalfa* (Medicago sativa *L.)* Master's thesis, University of Nebraska, Lincoln). Retrieved from https://digitalcommons.unl.edu/natresdiss/219/

Hanson, A. (2007). *Conservation and beneficial functions of grassland birds in agroecosystems* (Master's thesis, University of Nebraska, Lincoln). Retrieved from https://digitalcommons.unl.edu/natresdiss/245/

Heisler, G. M., & Dewalle, D. R. (1988). Effects of windbreak structure on wind flow. *Agriculture, Ecosystems & Environment, 22–23*, 41–69. https://doi.org/10.1016/0167-8809(88)90007-2

Heisler, G. M., & Dix, M. E. (1988). Effects of windbreaks on local distribution of airborne insects. *In* M.E. Dix and M. Harrell (Eds.), *Insects of windbreaks and related plantings: Distribution, importance, and management: Conference Proceedings, December 6, 1988, Louisville, KY* (General Technical Report RM-204, pp. 5–12). Fort Collins, CO: U.S. Forest Service Rocky Mountain Forest and Range Experiment Station.

Helfer, F., Zhang, H., & Lemckert, C. (2009). *Evaporation reduction by windbreaks: Overview, modelling and efficiency* (Technical Report 16). Queensland, Australia: Urban Water Security Research Alliance. Retrieved from http://www.urbanwateralliance.org.au/publications/UWSRA-tr16.pdf

Helmers, G., & Brandle, J. R. (2005). Optimum windbreak spacing in Great Plains agriculture. *Great Plains Research, 15*(2):179–198.

Hodges, L., & Brandle, J. R. (1996). Windbreaks: An important component in a plasticulture system. *HortTechnology, 6*, 177–181. https://doi.org/10.21273/HORTTECH.6.3.177

Hodges, L., Daningsih, E., & Brandle, J. R. (2006). Comparison of an antitranspirant spray, a polyacrylamide gel, and wind protection on early growth of muskmelon. *HortScience, 41*, 361–366. https://doi.org/10.21273/HORTSCI.41.2.361

Hodges, L., Suratman, M. N., Brandle, J. R., & Hubbard, K. G. (2004). Growth and yield of snap beans as affected by wind protection and microclimate changes due to shelterbelts and planting dates. *HortScience, 39*, 996–1004. https://doi.org/10.21273/HORTSCI.39.5.996

Holland, J., & Fahrig, L. (2000). Effect of woody borders on insect density and diversity in crop fields: A landscape-scale analysis. *Agriculture, Ecosystems & Environment, 78*, 115–122. https://doi.org/10.1016/S0167-8809(99)00123-1

Holmes, C. W., & Sykes, A. R. (1984). Shelter and climatic effects on livestock. In J. W. Sturrock (Ed.), Shelter research needs in relation to primary production (Water Soil Misc. Publ. 59, pp. 19–35). Wellington, New Zealand: National Shelter Working Party.

Hou, Q., Brandle, J., Hubbard, K., Schoeneberger, M., Nieto, C., & Francis, C. (2003). Alteration of soil water content consequent to root-pruning at a windbreak/crop interface in Nebraska, USA. *Agroforestry Systems, 57*, 137–147.

Hou, Q., Young, L. J., Brandle, J. R., & Schoeneberger, M. M. (2011). A spatial model approach for assessing windbreak growth and carbon stocks. *Journal of Environmental Quality, 40*, 842–852. https://doi.org/10.2134/jeq2010.0098

Hunt, E. R., & Jaffe, M. J. (1980). Thigmomorphogenesis: The interaction of wind and temperature in the field on the growth of *Phaseolus vulgaris* L. *Annals of Botany, 45*, 665–672. https://doi.org/10.1093/oxfordjournals.aob.a085875

Huszar, P. C., & Piper, S. L. (1986). Estimating the off-site costs of wind erosion in New Mexico. *Journal of Soil and Water Conservation, 41*, 414–416.

Igl, L. D., Kantrud, H. A., & Newton, W. E. (2018). Bird population changes following the establishment of a diverse stand of woody plants in a former crop field in North Dakota, 1975–2015. Great Plains Res. 28(1):73–90. https://doi.org/10.1353/gpr.2018.0006

Jackson, D. L., & Jackson, L. L. (Eds.). (2002). *The farm as natural habitat: Reconnecting food systems with ecosystems.* Island Press, Washington, DC.

Jacobs, A. (1984). The flow around a thin closed fence. *Boundary-Layer Meteorology, 28*, 317–328. https://doi.org/10.1007/BF00121311

Jaffe, M. J. (1976). Thigmomorphogenesis: A detailed characterization of the response of beans *(Phaseolus vulgaris* L.) to mechanical stimulation. *Zeitschrift für Pflanzenphysiologie, 77*, 437–453. https://doi.org/10.1016/S0044-328X(76)80017-7

Jaffe, M. J., & Forbes, S. (1993). Thigmomorphogenesis: The effect of mechanical perturbation on plants. *Plant Growth Regulation, 12*, 313–324. https://doi.org/10.1007/BF00027213

Jairell, R. L., & Schmidt, R.A. (1986). Scale model tests help optimize wind protection and water improvements for livestock. In D. L. Hintz and J. R. Brandle (Eds.), *International Symposium on Windbreak Technology, Proceedings: Lincoln, Nebraska, June 23–27, 1986* (GPAC Publ. 117, pp. 159-161).

Jairell, R. L., & Schmidt, R. A. (1992). Harvesting snow when water levels are low. In B. Shafer (Ed.), *Proceedings of the Western Snow Conference, Jackson, Wyoming, April 14–16, 1992, 60th Annual Meeting* (pp. 121–124). Portland, OR: Western Snow Conference.

Jensen, M. (1961). *Shelter effect: Investigation into the aerodynamics of shelter and its effects on climate and crops.* Copenhagen: Danish Technical Press.

Jobin, B., Choinière, L., & Bélanger, L. (2001). Bird use of three types of field margins in relation to intensive agriculture in Québec, Canada. *Agriculture, Ecosystems & Environment, 84*, 131–143. https://doi.org/10.1016/S0167-8809(00)00206-1

Johnson, H. D. (1965). Environmental temperature and lactation (with special reference to cattle). *International Journal of Biometeorology, 9*(2), 103–116. https://doi.org/10.1007/BF02188466

Johnson, R. J., & Beck, M. M. (1988). Influences of shelterbelts on wildlife management and biology. *Agriculture, Ecosystems & Environment, 22–23*, 301–335. https://doi.org/10.1016/0167-8809(88)90028-X

Johnson, R. J., Beck, M. M., & Brandle, J. R. (1991). *Windbreaks and wildlife* (EC 91-1771-B). Lincoln, NE: University of Nebraska Extension.

Johnson, R. J., Beck, M. M., & Brandle, J. R. (1994). Windbreaks for people: The wildlife connection. *Journal of Soil and Water Conservation, 49*(6), 546–550.

Johnson, R. J., Brandle, J. R., Fitzmaurice, R. L., & Poague, K. L. (1993). Vertebrates for biological control of insects in agroforestry systems. *In Biological control of forest pests in the Great Plains: Status and needs: Proceedings, 44th Annual Meeting, Great Plains Agricultural Council–Forestry Committee, June 28–July 1, 1993, Lubbock, TX* (GPAC Publ. 145, pp. 77–84).

Jones, D. D., Friday, W. H., & DeForest, S. S. (1983). Wind and snow control around the farm (NCR-191). West Lafayette, IN: Purdue University Cooperative Extension Service.

Jones, H. R., & Sudmeyer, R. A. (2002). Economic assessment of windbreaks on the south-eastern coast of Western Australia. *Australian Journal of Experimental Agriculture, 42*, 751–762. https://doi.org/10.1071/EA02010

Jordon, D. J., Klopfenstein, T., Brandle, J., & Klemesrud, M. (1997). Cornstalk grazing in protected and unprotected fields. In *1997 Nebraska beef report* (MP 67, p. 24). Lincoln, NE: University of Nebraska.

Jose, S., Gillespie, A. R., & Pallardy, S. G. (2004). Interspecific interactions in temperate agroforestry. *Agroforestry Systems, 61*, 237–255.

Jose, S., Gillespie, A. R., Seifert, J. R., & Biehle, D. J. (2000). Defining competition vectors in a temperate alley cropping system in the midwestern USA: 2. Competition for water. *Agroforestry Systems, 48*, 41–59. https://doi.org/10.1023/A:1006289322392

Jose, S., & Gordon, A. M. (Eds.). (2008). *Toward agroforestry design: An ecological approach.* Dordrecht, the Netherlands: Springer. https://doi.org/10.1007/978-1-4020-6572-9

Josiah, S. J., St-Pierre, R., Brott, H., & Brandle, J. (2004). Productive conservation: Diversifying farm enterprises by producing specialty woody products in agroforestry systems. *Journal of Sustainable Agriculture, 23*(3):93–108. https://doi.org/10.1300/J064v23n03_08

Kahn, B. A., Conway, K. E., & Fisher, C. G. (1986). Effects of wirestem, wind injury, and iprodione on yields of six broccoli cultivars. *HortScience, 21*, 1136–1139.

Kahnonitch, I., Lubin, Y., & Korine, C. (2018). Insectivorous bats in semi-arid agroecosystems: Effects on foraging activity and implications for insect pest control. *Agriculture, Ecosystems & Environment, 261*, 80–92. https://doi.org/10.1016/j.agee.2017.11.003

Kartchner, R. J. (1980). Effects of protein and energy supplementation of cows grazing native winter range forage on intake and digestibility. *Journal of Animal Science, 51*, 432–438. https://doi.org/10.2527/jas1980.512432x

Kenney, W. A. (1987). A method for estimating windbreak porosity using digitized photographic silhouettes. *Agricultural and Forest Meteorology, 39*, 91–94. https://doi.org/10.1016/0168-1923(87)90028-1

Kido, H., Flaherty, D. L., Bosch, D. F., & Valero, K. A. (1984). French prune trees as overwintering sites for the grape leafhopper egg parasite. *American Journal of Enology and Viticulture, 35*, 156–160.

Koh, I., Park, C. R., Kang, W., & Lee, D. (2014). Seasonal effectiveness of a Korean traditional deciduous windbreak in reducing wind speed. *Journal of Ecology and Environment, 37*, 91–97. https://doi.org/10.5141/ecoenv.2014.011

Kort, J. (1988). Benefits of windbreaks to field and forage crops. *Agriculture, Ecosystems & Environment, 22–23*, 165–190. https://doi.org/10.1016/0167-8809(88)90017-5

Kort, J., & Brandle, J. R. (1991). WBECON: A windbreak evaluation model: 2. Economic returns from a windbreak investment in the Great Plains. In S. Finch and C. S. Baldwin (Eds.), *Windbreaks and agroforestry: 3rd International Symposium on Windbreaks and Agroforestry, June 1991, Ridgetown College, Ridgetown, Ontario* (pp. 131–134).

Kort, J., & Turnock, R. (1998). Carbon reservoir and biomass in Canadian prairie shelterbelts. *Agroforestry Systems, 44*, 175–186. https://doi.org/10.1023/A:1006226006785

Kowalchuk, T. E., & de Jong, E. (1995). Shelterbelts and their effect on crop yield. *Canadian Journal of Soil Science, 75*, 543–550. https://doi.org/10.4141/cjss95-077

Kremen, C., Williams, N. M., & Thorp, R. W. (2002). Crop pollination from native bees at risk from agricultural intensification. *Proceedings of the National Academy of Sciences, 99*, 16812–16816. https://doi.org/10.1073/pnas.262413599

Laird, D. J. (1997). *Wind tunnel testing of shelterbelt effects on dust emissions from swine production facilities* (Master's thesis, Iowa State University). Retrieved from https://lib.dr.iastate.edu/rtd/18207/

Lammers, P. S., Wallenfang, O., & Boeker, P. (2001, July–August). *Computer modeling for assessing means to reduce odour emissions*. Paper 01-4042. Paper presented at the 2001 ASAE Annual International Meeting, Sacramento, CA.

Lefroy, E. C., Hobbs, R. J., O'Connor, M. H., & Pate, J. S. 1999. What can agriculture learn from natural ecosystems? *Agroforestry Systems, 45*, 423–436.

Lehane, J. J., & Nielsen, K. F. (1961). *The influence of field shelterbelts on climatic factors, soil drifting, snow accumulation, soil moisture and grain yields*. Swift Current, SK: Canada Department of Agriculture Experimental Farm.

Lewis, T., & Dibley, G. C. (1970). Air movement near windbreaks and a hypothesis of the mechanism of the accumulation of airborne insects. *Annals of Applied Biology, 66*, 477–484. https://doi.org/10.1111/j.1744-7348.1970.tb04627.x

Lewis, T., & Smith, B. D. (1969). The insect faunas of pear and apple orchards and the effect of windbreaks on their distribution. *Annals of Applied Biology, 64*, 11–20. https://doi.org/10.1111/j.1744-7348.1969.tb02850.x

Lin, X. J., Barrington, S., Nicell, J., Choiniere, D., & Vezina, A. (2006). Influence of windbreaks on livestock odor dispersion plumes in the field. *Agriculture, Ecosystems & Environment, 116*, 263–272. https://doi.org/10.1016/j.agee.2006.02.014

Liptay, A. (1987). Field survival and establishment of tomato transplants of various age and size. *Acta Horticulturae, 220*, 203–209. https://doi.org/10.17660/ActaHortic.1988.220.27

Loeffler, A. E., Gordon, A. M., & Gillespie, T. J. (1992). Optical porosity and windspeed reduction by coniferous windbreaks in southern Ontario. *Agroforestry Systems, 17*, 119–133. https://doi.org/10.1007/BF00053117

Lorimor, J. (1998). *Iowa Odor Control Demonstration Project: Landscaping* (Pm-1754h). Ames, IA: Iowa State University Cooperative Extension Service. Retrieved from http://www.extension.iastate.edu/Publications/PM1754h.pdf.

Lowry, W. P. (1967). *Weather and life: An introduction to biometeorology*. New York: Academic Press.

Luedeling, E., Smethurst, P. J., Baudron, F., Bayala, J., Huth, N. I., van Noordwijk, M., . . . Sinclair, F. L. (2016). Field-scale modeling of tree–crop interactions: Challenges and development needs. *Agricultural Systems, 142*, 51–69. https://doi.org/10.1016/j.agsy.2015.11.005

Lyles, L. (1988). Basic wind erosion processes. *Agriculture, Ecosystems & Environment, 22–23*, 91–101. https://doi.org/10.1016/0167-8809(88)90010-2

Lyles, L., Tatarko, J., & Dickerson, J. D. (1984). Windbreak effects on soil water and wheat yield. *Transactions of the ASAE, 20*, 69–72. https://doi.org/10.13031/2013.32737

Malechek, J. C., & Smith, B. M. (1976). Behavior of range cows in response to winter weather. *Journal of Range Management, 29*, 9–12. https://doi.org/10.2307/3897679

Malone, G. W., van Wicklen, G., Collier, S., & Hansen, D. (2006). Efficacy of vegetative environmental buffers to capture emissions from tunnel ventilated poultry houses. In V. P. Aneja, W. H. Schlesinger, R. Knighton, G. Jennings, D. Niyogi, W. Gilliam, & C. S. Duke (Eds.), *Proceedings: Workshop on Agricultural Air Quality: State of the Science, Potomac, Maryland, June 5–8, 2006* (pp. 875–878).

Marshall, E. J. P. (1967). The effect of shelter on the productivity of grasslands and field crops. *Field Crop Abstracts, 20*, 1–14.

Marshall, E. J. P. (1988). The ecology and management of field margin floras in England. *Outlook on Agriculture, 17*(4), 178–182. https://doi.org/10.1177/003072708801700408

Mattis, G. Y. (1988). Scientific achievements in agricultural afforestation in the USSR. In *Great Plains Agricultural Council Forestry Committee Proceedings, June 27–30, 1988, Regina, Saskatchewan* (GPAC Publ. 126, pp. 98–104).

Maurer, D. L., Koziel, J. A., Harmon, J. D., Hoff, S. J., Rieck-Hinz, A. M., & Andersen, D. S. (2016). Summary of performance data for technologies to control gaseous odor and particulate emissions from livestock operations: Air management practices assessment tool (AMPAT). *Data in Brief, 7*, 1413–1429. https://doi.org/10.1016/j.dib.2016.03.070

McAneney, K. J., & Judd, M. J. (1987). Comparative shelter strategies for kiwi fruit: A mechanistic interpretation of wind damage measurements. *Agricultural and Forest Meteorology, 39*, 225–240. https://doi.org/10.1016/0168-1923(87)90040-2

McNaughton, K. G. (1983). The direct effect of shelter on evaporation rates: Theory and experimental test. *Agricultural and Forest Meteorology, 34*, 315–322.

McNaughton, K. G. (1988). Effects of windbreaks on turbulent transport and microclimate. *Agriculture, Ecosystems & Environment, 22–23*, 17–39. https://doi.org/10.1016/0167-8809(88)90006-0

McNaughton, K. G. (1989). Micrometeorology of shelterbelts and forest edges. *Philosophical Transactions of the Royal Society of London, B. Biological Sciences, 324*, 351–368.

Melvin, S. W. (1996). Swine odor measurement and control issues reviewed. *Feedstuffs, 68*(22), 12–14.

Michailides, T. J., & Morgan, D. P. (1993). Wind scab of French prune: Symptomology and predisposition to preharvest and postharvest fungal decay. *Plant Disease, 77*, 90–93. https://doi.org/10.1094/PD-77-0090

Miller, J. M., Bohm, M., & Cleugh, H. A. (1995). *Direct mechanical effects of wind on selected crops: A review* (Technical Report 67). Canberra, ACT, Australia: Centre for Environmental Mechanics.

Mitchell, C. A. (1977). Influence of mechanical stress on auxin stimulated growth of excised pea stem sections. *Physiologia Plantarum, 41*, 129–134. https://doi.org/10.1111/j.1399-3054.1977.tb05543.x

Mize, C. W., Colletti, J., Batchelor, W., Kim, J. S., Takle, E. S., & Brandle, J. R. (2008). Modeling a field shelterbelt system with the Shelterbelt Agroforestry Modeling System. In D. R. Batish, R. K. Kohli, S. Jose, & H. P. Singh (Eds.), *Ecological basis of agroforestry* (pp. 287–300). Boca Raton, FL: CRC Press.

Montagnini, F., & Nair, P. K. R. (2004). Carbon sequestration: An underexploited environmental benefit of agroforestry systems. *Agroforestry Systems, 61*, 281–295.

Monteith, J. L. (1981). Coupling of plants to the atmosphere. In J. Grace, E. D. Ford, & P. G. Jarvis (Eds.), *Plants and their atmospheric environment* (pp. 1–29). Oxford, UK: Blackwell Scientific Publishers.

Monteith, J. L. (1993). The exchange of water and carbon by crops in a Mediterranean climate. *Irrigation Science, 14*, 85–91. https://doi.org/10.1007/BF00208401

Moran, J. M., & Morgan, M. D. (1986). *Meteorology: The atmosphere and the science of weather*. Edina, MN: Burgess Publishers.

Morris, C., Klopfenstein, T., Brandle, J., Stock, R., Shain, D., & Klemesrud, M. (1996). Winter calf grazing and field windbreaks. In *1996 Nebraska beef report* (MP 66-A, p. 44). Lincoln, NE: University of Nebraska.

Morrison, B. M. L., & Flores, S. A. (2013). Promoting biodiversity in agricultural landscapes: Native windbreaks support greater understory plant diversity in Monteverde, Costa Rica. *Journal of Young Investigators, 25*, 101–106.

National Pork Producers Council. (1995). A review of the literature on the nature and control of odors from pork production facilities [executive summary]. Des Moines, IA: National Pork Producers Council.

Naughton, G. C., & Capels, S. W. (1982). *Root-pruning Osage-orange windbreaks* (Contribution 82-419-A). Manhattan, KS: Kansas Agricultural Experiment Station.

Nguyen, H. D. D., & Nansen, C. (2018). Edge based distributions of insects: A review. *Agronomy for Sustainable Development, 38*, 11. https://doi.org/10.1007/s13593-018-0488-4

Nobel, P. S. (1981). Wind as an ecological factor. In O. L. Lange, P. S. Nobel, C. B. Osmond, & H. Ziegler (Eds.), Physiological plant ecology: I. Responses to the physical environment (pp. 475–500). Encyclopedia of Plant Physiology New Series Vol. 12A. Berlin: Springer.

Norton, R. L. (1988). Windbreaks: Benefits to orchard and vineyard crops. *Agriculture, Ecosystems & Environment, 22–23*, 205–213. https://doi.org/10.1016/0167-8809(88)90019-9

Nuberg, I. K., & Mylius, S. J. (2002). Effect of shelter on the yield and water use of wheat. *Australian Journal of Experimental Agriculture, 42*, 773–780. https://doi.org/10.1071/EA02013

Nuberg, I. K., Mylius, S. J., Edwards, J. M., & Davey, C. 2002. Windbreak research in a South Australian cropping system. *Australian Journal of Experimental Agriculture, 42*, 781–796. https://doi.org/10.1071/EA02014

Ogbuehi, S. N., & Brandle, J. R. (1981). Influence of windbreak shelter on light interception, stomatal conductance, and carbon dioxide exchange rate of soybean, *Glycine max* (L.) Merrill. *Transactions of the Nebraska Academy of Science, 9*, 49–54.

Ogbuehi, S. N., & Brandle, J. R. (1982). Influence of windbreak-shelter on soybean growth, canopy structure, and light relations. *Crop Science, 22*, 269–273. https://doi.org/10.2135/cropsci1982.0011183X002200020017x

Oliver, Y. M., Lefroy, E. C., Stirzaker, R., & Davies, C. L. (2005). Deep-drainage control and yield: The trade-off between trees and crops in agroforestry systems in the medium and low rainfall areas of Australia. *Australian Journal of Agricultural Research, 56*, 1011–1026. https://doi.org/10.1071/AR04213

Ong, C. K., & Huxley, P. (Eds.). (1996). *Tree–crop interactions: A physiological approach*. Wallingford, UK: CAB International.

Pasek, J. E. (1988). Influence of wind and windbreaks on local dispersal of insects. *Agriculture, Ecosystems & Environment, 22–23*, 539–554. https://doi.org/10.1016/0167-8809(88)90044-8

Patton, E. G., Shaw, R. H., Judd, M. J., & Raupach, M. R. (1998). Large-eddy simulation of windbreak flow. *Boundary-Layer Meteorology, 87*, 275–307. https://doi.org/10.1023/A:1000945626163

Peri, P.L., & Bloomberg, M. (2002). Windbreaks in southern Patagonia, Argentina: A review of research on growth models, wind speed reduction, and effects on crops. *Agroforestry Systems, 56*, 129–144. https://doi.org/10.1023/A:1021314927209

Perkins, M. W., Johnson, R. J., & Blankenship, E. E. (2003). Response of riparian avifauna to percentage and pattern of woody cover in an agricultural landscape. *Wildlife Society Bulletin, 31*, 642–660.

Pierce, R. A., Farrand, D. T., & Kurtz, W. B. (2001). Projecting the bird community response resulting from the adoption of shelterbelt agroforestry practices in eastern Nebraska. *Agroforestry Systems, 53*, 333–350. https://doi.org/10.1023/A:1013371325769

Pimental, D., Harvey, C., Resosudarmo, P., Sinclair, K., Kurz, D., McNair, M., . . . Blair, R. (1995). Environmental and economic costs of soil erosion and conservation benefits. *Science, 267*, 1117–1123.

Pinthus, M. J. (1974). Lodging in wheat, barley, and oats: The phenomenon, its causes, and preventive measures. *Advances in Agronomy, 25*, 209–263. https://doi.org/10.1016/S0065-2113(08)60782-8

Piper, S. (1989). Estimating the off-site benefits from a reduction in wind erosion and the optimal level of wind erosion control: An application in New Mexico. *Journal of Soil and Water Conservation, 44*(4):334–339.

Pisani Gareau, T., & Shennan, C. (2010). Can hedgerows attract beneficial insects and improve pest control? A study of hedgerows on Central Coast farms (Research Brief 13). Santa Cruz, CA: Center for Agroecology & Sustainable Food Systems, University of California. Retrieved from http://escholarship.org/uc/item/11d3v8p9

Pitcairn, C. E. R., Jeffree, C. E., & Grace, J. (1986). Influence of polishing and abrasion on the diffusive conductance of leaf surfaces of *Festuca arundinacea* Schreb. *Plant, Cell & Environment, 9*, 191–196. https://doi.org/10.1111/1365-3040.ep11611633

Pohlan, J. N., Vazquez, M., & Garcia, A. M. E. (1986). The influence of biotic and abiotic factors on external and internal fruit quality of Cuban orange. *Horticultural Abstracts, 56*, 3752.

Pohronezny, K., Hewitt, M., Infante, J., & Datnoff, L. (1992). Wind and wind-generated sand injury as factors in infection of pepper by *Xanthomonas campestris* pv *vesicatoria*. *Plant Disease, 76*, 1036–1039. https://doi.org/:10.1094/PD-76-1036

Primault, B. (1979). Optimum climate for animals. In J. Seemann, Y. I. Chirkov, J. Lomas, & B. Primault (Eds.), *Agrometeorology* (p. 182–189). Berlin: Springer. https://doi.org/10.1007/978-3-642-67288-0_26

Puckett, H. L. (2006). *Avian foraging use of the crop field-woody edge interface in agroecosystems in east central Nebraska* (Master's thesis, University of Nebraska).

Quam, V. C., Johnson, L., Wight, B., & Brandle, J. R. (1994). *Windbreaks for livestock production* (EC 94-1766-X). Lincoln, NE: University of Nebraska Cooperative Extension.

Quinkenstein, A., Wollecke, J., Bohm, C., Grunewald, H., Freese, D., Schneider, B.U., & Hüttl, R. F. (2009). Ecological benefits of the alley cropping in agroforestry system in sensitive regions of Europe. *Environmental Science & Policy, 12*, 1112–1121. https://doi.org/10.1016/j.envsci.2009.08.008

Raine, J. K., & Stevenson, D. C. (1977). Wind protection by model fences in simulated atmospheric boundary layer. *Journal of Wind Engineering and Industrial Aerodynamics, 2*, 159–180. https://doi.org/10.1016/0167-6105(77)90015-0

Rasmussen, S. D., & Shapiro, C. A. (1990). Effects of tree root-pruning adjacent to windbreaks on corn and soybeans. *Journal of Soil and Water Conservation, 45*, 571–575.

Read, R. A. (1964). Tree windbreaks for the central Great Plains (Agriculture Handbook 250). Washington, DC: U.S. Government Printing Office.

Řeháček, D., Khel, T., Kucera, J., Vopravil, J., & Petera, M. (2017). Effect of windbreaks on wind speed reduction and soil protection against wind erosion. *Soil & Water Research, 12*, 128–135. https://doi.org/10.17221/45/2016-SWR

Reynolds, P. E., Simpson, J. A., Thevathasan, N. V., & Gordon, A. M. (2007). Effects of tree competition on corn and soybean photosynthesis, growth, and yield in a temperate tree-based agroforestry intercropping system in southern Ontario, Canada. *Ecological Engineering, 29*, 362–371. https://doi.org/10.1016/j.ecoleng.2006.09.024

Ribaudo, M. O. (1986). Targeting soil conservation programs. *Land Economics, 62*, 402–411. https://doi.org/10.2307/3146472

Riechert, S. E., & Lockley, T. (1984). Spiders as biological control agents. *Annual Review of Entomology, 29*, 299–320. https://doi.org/10.1146/annurev.en.29.010184.001503

Rivest, D., & Vezina, A. (2015). Maize yield patterns on the leeward side of tree windbreaks are site-specific and dependent on rainfall conditions in eastern Canada. *Agroforestry Systems, 89*, 237–246. https://doi.org/10.1007/s10457-014-9758-6

Robbins, C. (1976). Economics of windbreaks and our cattle industry. In R. W. Tinus (Ed.), *Windbreaks: What are they worth? Proceedings of the Symposium, Great Plains Agricultural Council Forestry Committee, June 22–24, 1982, Dodge City, Kansas* (GPAC Publication 78, p. 107–108).

Rockwood, D. L., & Bowman, R. L. (2017). Medically related products obtainable from *Eucalyptus* trees. *International Biology Reviews, 1*(3), 1615. https://doi.org/10.18103/ibr.v1i3.1615

Rodriquez, R., del Valle, N., Arango, W., Torres, R., & Fernandez, M. (1986). Effect of shelterbelts on yields in Valencia late orange (*Citrus sinensis* (L.) Osbeck) plantations. *Centro Agricola, 12*, 71–80.

Rosenberg, N. J. (1966). Microclimate, air mixing, and physiological regulation of transpiration as influenced by wind shelter in an irrigated bean field. *Agricultural Meteorology, 3*, 197–224. https://doi.org/10.1016/0002-1571(66)90029-X

Rosenberg, N. J., Blad, B. L., & Verma, S. B. (1983). *Microclimate: The biological environment* (2nd ed.). New York: John Wiley & Sons.

Rutherford, S., Clark, E., McTainsh, G., Simpson, R., & Mitchell, C. (1999). Characteristics of rural dust events shown to impact on asthma severity in Brisbane, Australia. *International Journal of Biometeorology, 42*, 217–225. https://doi.org/10.1007/s004840050108

Sánchez, I. A., & McCollin, D. (2015). A comparison of microclimate and environmental modification produced by hedgerows and dehesa in the Mediterranean region: A study in the Guadarrama region, Spain. *Landscape and Urban Planning, 143*, 230–237. https://doi.org/10.1016/j.landurbplan.2015.07.002

Schmidt, M., Lischeid, G., & Nendel, C. (2019). Microclimate and matter dynamics in transition zones of forest to arable land. *Agricultural and Forest Meteorology, 268*, 1–10. https://doi.org/10.1016/j.agrformet.2019.01.001

Schmidt, R. A., Jr. (1972). *Sublimination of wind-transported snow: A model* (Res. Pap. RM-90). Fort Collins, CO: U.S. Forest Service, Rocky Mountain Forest and Range Experiment Station.

Schmidt, R. A., Takle, E. S., Brandle, J. R., & Litvina, I. V. (1995). Static pressure at the ground under atmospheric flow across a windbreak. In *Proceedings of the 11th Symposium on Boundary Layers and Turbulence, Charlotte, NC* (pp. 517–520). Boston, MA: American Meteorological Society.

Schoeneberger, M., Bentrup, G., de Gooijer, H., Soolanayakanahally, R., Sauer, T., Brandle, J., . . . Current, D. (2012). Branching out: Agroforestry as a climate change mitigation and adaptation tool for agriculture. *Journal of Soil and Water Conservation, 67*, 128A–136A. https://doi.org/10.2489/jswc.67.5.128A

Schoeneberger, M. M., Bentrup, G., & Patel-Weynand, T. (Eds.). (2017). *Agroforestry: Enhancing resiliency in U.S. agricultural landscapes under changing conditions.* (General Technical Report WO-96). Washington, DC: U.S. Forest Service.

Scholten, H. (1988). Snow distribution on crop fields. *Agriculture, Ecosystems & Environment, 22–23*, 363–380. https://doi.org/10.1016/0167-8809(88)90032-1

Senaviratne, G. M. M. M. A., Udawatta, R. P., Nelson, K. A., Shannon, K., & Jose, S. (2012). Temporal and spatial influence of perennial upland buffers on corn and soybean yields. *Agronomy Journal, 104*, 1356–1362. https://doi.org/10.2134/agronj2012.0081

Shah, S. R. H. (1970). The influence of windbreak on the development and yield of horticultural crop (genus *Fragaria*). *Agriculture Pakistan, 21*(2), 137–158.

Shaw, D. L. (1988). The design and use of living snowfences in North America. *Agriculture, Ecosystems & Environment, 22–23*, 351–362. https://doi.org/10.1016/0167-8809(88)90031-X

Shaw, D. L. (1991). Living snow fences: Protection that just keeps growing. Fort Collins, CO: Colorado Interagency Living Snow Fence Program, Colorado State University.

Shi, Z., & Gao, Z. (1986). On the ecological efficiency of shelterbelt network and its yield increasing effect in paddy fields. *Journal of Ecology, 5*(2), 10–14.

Simons, J. N. (1957). Effects of insecticides and physical barriers on field spread of pepper veinbanding mosaic virus. *Phytopathology, 47*, 139–145.

Skidmore, E. L. (1966). Wind and sandblast injury to seedling green beans. *Agronomy Journal, 58*, 311–315. https://doi.org/10.2134/agronj1966.00021962005800030020x

Skidmore, E. L., Hagen, L. J., Naylor, D. G., & Teare, I. D. (1974). Winter wheat response to barrier-induced microclimate. *Agronomy Journal, 66*, 501–505. https://doi.org/10.2134/agronj1974.00021962006600040008x

Slosser, J. E., & Boring, E. P., III. (1980). Shelterbelts and boll weevils: A control strategy based on management of overwintering habitat. *Environmental Entomology, 9*, 1–6. https://doi.org/10.1093/ee/9.1.1

Solomon, M. G. (1981). Windbreaks as a source of orchard pests and predators. In J.M. Thresh (Ed.), *Pests, pathogens, and diseases* (pp. 273–283). London: Pitman.

Southwood, T. R. E., & Way, M. J. (1970). Ecological background to pest management. In R. L. Rabb & F. E. Guthrie (Eds.), *Concepts of pest management, Proceedings of a conference* (pp. 6–29). Raleigh, NC: North Carolina State University.

Sreekar, R., Mohan, A., Das, S., Agarwal, P., & Vivek, R. (2013). Natural windbreaks sustain bird diversity in a tea-dominated landscape. *PLOS ONE, 8*(7), e70379. https://doi.org/10.1371/journal.pone.0070379

Stanek, E. C., Lovell, S. T,. & Reisner, A. (2019). Designing multifunctional woody polycultures according to landowner preferences in central Illinois. *Agroforestry Systems, 93*, 2294-2311. https://doi.org/10.1007/s10457-019-00350-2

Stoeckeler, J. H. (1962). *Shelterbelt influence on Great Plains field environment and crops: A guide for determining design and orientation* (U.S. Forest Service Production Research Report 62). Washington, DC: U.S. Government Printing Office.

Stoeckeler, J. H., & Williams, R. A. (1949). Windbreaks and shelterbelts. In A. Stefferud (Ed.) *Trees: The Yearbook of Agriculture, 1949* (pp. 191–199). Washington, DC: U.S. Government Printing Office.

Středa, T., Malenová, P., Pokladniková, H., & Rožnovský, J. (2008). The efficiency of windbreaks on the basis of wind field and optical porosity measurement. *Acta Universitatis Agriculturae et Silviculturae Mendelianae Brunensis, 56*(4), 281–288. https://doi.org/10.11118/actaun200856040281

Středová, H., Podhrázská, J., Litschmann, T., Středa, T., & Rožnovský, J. (2012). Aerodynamic parameters of windbreak based on its optical porosity. *Geophysics and Geodesy, 42*, 213–226.

Sturrock, J. W. (1981). Shelter boosts crop yield by 35 percent: Also prevents lodging. *New Zealand Journal of Agriculture, 143*, 18–19.

Sturrock, J. W. (1984). The role of shelter in irrigation and water use. In J.W. Sturrock (Ed.), Shelter research needs in relation to primary production (Miscellaneous Publication 94, pp. 79–86). Wellington, New Zealand: Ministry of Works and Development, Water and Soil.

Sturrock, J. W. (1988). Shelter: Its management and promotion. *Agriculture, Ecosystems & Environment, 22–23*, 1–13. https://doi.org/10.1016/0167-8809(88)90004-7

Sudmeyer, R. A., Adams, M., Eastham, J., Scott, P. R., Hawkins, W., & Rowland, I. (2002)a. Broad-acre crop yield in the lee of windbreaks in the medium and low rainfall areas of southwestern Australia. *Australian Journal of Experimental Agriculture, 42*, 739–750. https://doi.org/10.1071/EA02011

Sudmeyer, R. A., Crawford, M. C., Meinke, H., Poulton, P. L., & Robertson, M. J. (2002)b. Effect of artificial wind shelters on the growth and yield of rainfed crops. *Australian Journal of Experimental Agriculture, 42*, 841–858. https://doi.org/10.1071/EA02018

Sudmeyer, R. A., & Flugge, F. (2005). The economics of managing tree/crop competition in windbreak and alley systems. *Australian Journal of Experimental Agriculture, 45*, 1403–1414. https://doi.org/10.1071/EA04155

Sudmeyer, R. A., Hall, D. J. M., Eastham, J., & Adams, M. (2002)c. The tree-crop interface: The effects of root pruning in southwestern Australia. *Australian Journal of Experimental Agriculture, 42*, 763–772. https://doi.org/10.1071/EA02012

Sudmeyer, R. A., Speijers, J., & Nicholas, B. D. (2004). Root distribution of *Pinus pinaster, Eucalyptus globulus* and *E. kochii* and associated soil chemistry in agricultural land adjacent to tree lines. *Tree Physiology, 24,* 1333–1346. https://doi.org/10.1093/treephys/24.12.1333

Sugita, M. (2018). Do windbreaks reduce the water consumption of a crop field? *Agricultural and Forest Meteorology, 250–251,* 330–342. https://doi.org/10.1016/j.agrformet.2017.11.033

Sun, D., & Dickinson, G. R. (1994). A case study of shelterbelt effect on potato (*Solanum tuberosum*) yield on the Atherton Tablelands in tropical north Australia. *Agroforestry Systems, 25,* 141–151. https://doi.org/10.1007/BF00705674

Tabler, R. D. (1975). Estimating the transport and evaporation of blowing snow. In *Snow Management on the Great Plains, Symposium proceedings, Bismarck, North Dakota* (GPAC Publication 73, p. 85–104). Great Plains Agricultural Council, Research Committee.

Tabler, R. D., & Johnson, K. L. (1971). Snow fences for watershed management. In *Proceedings of the Symposium on snow and ice in relation to wildlife and recreation* (pp. 116–121). Ames, IA: University of Iowa Press.

Takle, E. S. (1983). Climatology of superadiabatic conditions for a rural area. *Journal of Applied Meteorology and Climatology, 22,* 1129–1132. https://doi.org/10.1175/1520-0450(1983)022<1129:COSCFA>2.0.CO;2

Takle, E. S., Wang, H., Schmidt, R. A., Brandle, J. R., & Jairell, R. L. (1997). Pressure perturbation around shelterbelts: Measurements and model results. In 12th Symposium on Boundary Layers and Turbulence, Vancouver, BC (pp. 563–564). Boston, MA: American Meteorological Society.

Taksdal, G. (1992). Windbreak effects on the carrot crop. *Acta Agriculturae Scandinavica, 42,* 177–183.

Tamang, B., Andreu, M. G., & Rockwood, D. L. (2010). Microclimate patterns on the leeside of single-row tree windbreaks during different weather conditions in Florida farms: Implication for improved crop production. *Agroforestry Systems, 79,* 111–122. https://doi.org/10.1007/s10457-010-9280-4

Thernelius, S. M. (1997). Wind tunnel testing of odor transportation from swine production facilities (Master's thesis, Iowa State University). https://doi.org/10.31274/rtd-180813-6974

Thuyet, D. V., Do, T. V., Santo, T., & Hung, T. T. (2014). Effects of species and shelterbelt structure on wind speed reduction in shelter. *Agroforestry Systems, 88,* 237–244. https://doi.org/10.1007/s10457-013-9671-4

Tibke, G. (1988). Basic principles of wind erosion control. *Agriculture, Ecosystems & Environment, 22–23,* 103–122. https://doi.org/10.1016/0167-8809(88)90011-4

Ticknor, K. A. (1988). Design and use of field windbreaks in wind erosion control systems. *Agriculture, Ecosystems & Environment, 22–23,* 123–132. https://doi.org/10.1016/0167-8809(88)90012-6

Tremblay, A. P., Mineau, P., & Stewart, R. K. (2001). Effects of bird predation on some pest insect populations in corn. *Agriculture, Ecosystems & Environment, 83,* 143–152. https://doi.org/10.1016/S0167-8809(00)00247-4

Tsitsilas, A., Hoffmann, A. A., Weeks, A. R., & Umina, P. A. (2010). Impact of groundcover manipulations on mite pests and their natural enemies. *Australian Journal of Entomology, 50,* 37–47. https://doi.org/10.1111/j.1440-6055.2010.00779.x

Tyndall, J. C. (2006a). Shelterbelts and livestock odor mitigation: A socio-economic assessment of pork producers and consumers. In V. P. Aneja, W. H. Schlesinger, R. Knighton, G. Jennings, D. Niyogi, W. Gilliam, and C. S. Duke (Eds.), Proceedings of a workshop on agricultural air quality: State of the science, Potomac, MD (pp. 341– 345).

Tyndall, J. C. (2006b). Financial feasibility of using shelterbelts for swine odor mitigation. In V. P. Aneja, W. H. Schlesinger, R. Knighton, G. Jennings, D. Niyogi, W. Gilliam, and C. S. Duke (Eds.), Proceedings of a workshop on agricultural air quality: State of the science, Potomac, MD (pp. 1170– 1174).

Tyndall, J. C., & Colletti, J. P. (2007). Mitigating swine odor with strategically designed shelterbelt systems: A review. *Agroforestry Systems, 69,* 45–65. https://doi.org/10.1007/s10457-006-9017-6

Tyndall, J. C., & Randall, J. (2018). VEB-Econ: A vegetative environmental buffer decision-support tool for environmental quality management. *Journal of Forestry, 116,* 573–580. https://doi.org/10.1093/jofore/fvy051

Ubeda, Y., Lopez-Jimenez, P. A., Nicolas, J., & Calvet, S. (2013). Strategies to control odours in livestock facilities: A critical review. *Spanish Journal of Agricultural Research, 11,* 1004–1015. https://doi.org/10.5424/sjar/2013114-4180

Umland, E. R. (1979). Root pruning as a management technique. In R. J. Gavit (Program Chair), *Windbreak management: A collection of papers presented at a workshop sponsored by the Great Plains Agricultural Council-Forestry Committee, October 23-25, 1979, Norfolk, NE* (GPAC Publication 92, pp. 101–111).

Unkovich, M., Blott, K., Knight, A., Mock, I., Rab, A., & Portelli, M. (2003). Water use, competition, and crop production in low rainfall, alley farming systems of south-eastern Australia. *Australian Journal of Agricultural Research, 54,* 751–762. https://doi.org/10.1071/AR03049

van Eimern, J., Karschon, R., Razumava, L. A., & Robertson, G. W. (1964). *Windbreaks and shelterbelts* (Technical Note no. 59). Geneva, Switzerland: World Meteorological Organization.

Van Gaal, T., & Erwin, J. E. (2005). Diurnal variation in thigmotropic inhibition of stem elongation. *HortTechnology, 15,* 291–294. https://doi.org/10.21273/HORTTECH.15.2.0291

Van Gardingen, P., & Grace, J. (1991). Plants and wind. *Advances in Botanical Research, 18,* 189–253. https://doi.org/:10.1016/S0065-2296(08)60023-3

Van Iaer, E., Ampe, B., Moons, C., Sonck, B., & Tuyttens, F. A. M. (2015). Wintertime use of natural versus artificial shelter by cattle in nature reserves in temperate areas. *Applied Animal Behaviour Science, 163,* 39–49. https://doi.org/10.1016/j.applanim.2014.12.004

Vezina, A. (2005). *Farmstead shelterbelts: Planning, planting and maintenance.* La Pocatiere, QC, Canada: Institut de Technologie Agrolimentaire.

Waister, P. D. (1972a). Wind as a limitation on the growth and yield of strawberries. *Journal of Horticultural Science, 47,* 411–418. https://doi.org/10.1080/00221589.1972.11514484

Waister, P. D. (1972b). Wind damage in horticultural crops. *Horticultural Abstracts, 42,* 609–615.

Wang, H., & Takle, E. S. (1994). Mesoscale and boundary-layer flows over inhomogeneous surfaces consisting of porous obstacles. In 6th Conference on Mesoscale Processes, Portland, OR (pp. 262–265). Boston, MA: American Meteorological Society.

Wang, H., & Takle, E. S. (1995a). A numerical simulation of boundary-layer flows near shelterbelts. *Boundary-Layer Meteorology, 75,* 141–173. https://doi.org/10.1007/BF00721047

Wang, H., & Takle, E. S. (1995b). Numerical simulations of shelterbelt effects on wind direction. *Journal of Applied Meteorology, 34,* 2206–2219. https://doi.org/10.1175/1520-0450(1995)034<2206:NSOSEO>2.0.CO;2

Wang, H., & Takle, E. S. (1996a). On shelter efficiency of shelterbelts in oblique winds. *Agricultural and Forest Meteorology, 81,* 95–117. https://doi.org/10.1016/0168-1923(95)02311-9

Wang, H., & Takle, E. S. (1996b). On three-dimensionality of shelterbelt structure and its influences on shelter effects. *Boundary-Layer Meteorology, 79,* 83–105. https://doi.org/10.1007/BF00120076

Wang, H., & Takle, E. S. (1997). Model simulated influences of shelterbelt shape on wind sheltering efficiency. *Journal of Applied Meteorology, 36*, 695–704. https://doi.org/10.1175/1520-0450-36.6.695

Wang, H., Takle, E. S., & Shen, J. (2001). Shelterbelts and windbreaks: Mathematical modeling and computer simulations of turbulent flows. *Annual Review of Fluid Mechanics, 33*, 549–586. https://doi.org/10.1146/annurev.fluid.33.1.549

Wang, X. M., Zhang, C. X., Hasi, E., & Dong, Z. B. (2010). Has the Three Norths Forest Shelterbelt Program solved the desertification and dust storm problems in arid and semiarid China? *Journal of Arid Environments, 74*, 13–22. https://doi.org/10.1016/j.jaridenv.2009.08.001

Webster, A. J. F. (1970a). Direct effects of cold weather on the energetic efficiency of beef production in different regions of Canada. Can. J. Anim. Sci. 50:563–573. https://doi.org/10.4141/cjas70-077

Webster, A. J. F. (1970b). Effects of cold outdoor environments on the energy exchanges and productive efficiency of beef cattle. In *Annual Feeder's Day report* (p. 30–38). Edmonton, AB, Canada: University of Alberta.

Wight, B. (1988). Farmstead windbreaks. *Agriculture, Ecosystems & Environment, 22–23*, 261–280. https://doi.org/10.1016/0167-8809(88)90025-4

Wight, B., Boes, T. K., & Brandle, J. R. (1991). Windbreaks for rural living (EC 91-1767-X). Lincoln, NE: University of Nebraska Extension.

Williams, J. R., Lyles, L., & Langdale, G. W. (1981). Soil erosion effects on soil productivity: A research perspective. *Journal of Soil and Water Conservation, 36*, 82–90.

Williams, P., & Young, M. (1999). Costing dust: How much does wind erosion cost the people of South Australia? Adelaide, SA, Australia: Policy and Economic Research Unit, CSIRO Land and Water. Retrieved from http://www.clw.csiro.au/publications/consultancy/1999/costing_dust.pdf

Williams, R. R., & Wilson, D. (1970). *Toward regulated cropping*. London: Grower Books.

Winchester, C. F. (1964). Symposium on growth: Environment and growth. *Journal of Animal Science, 23*, 254–264. https://doi.org/10.2527/jas1964.231254x

Woodruff, N. P., Lyles, L., Siddoway, L., & Fryrear, D. W. (1972). How to control wind erosion (USDA–ARS Agricultural Information Bulletin 354). Washington, DC: U.S. Government Printing Office.

Wu, T., Zhang, P., Zhang, L., Wang, J., Yu, M., Zhou, X. H., & Wang, G. (2018). Relationships between shelter effects and optical porosity: A meta-analysis for tree windbreaks. *Agricultural and Forest Meteorology, 259*, 75–81. https://doi.org/10.1016/j.agrformet.2018.04.013

Yahner, R. H. (1982a). Avian nest densities and nest-site selection in farmstead shelterbelts. *Wilson Bulletin, 94*, 156–175.

Yahner, R. H. (1982b). Avian use of vertical strata and plantings in farmstead shelterbelts. *Journal of Wildlife Management, 46*, 50–60. https://doi.org/10.2307/3808407

Yahner, R. H. (1982c). Microhabitat use by small mammals in farmstead shelterbelts. *Journal of Mammalogy, 63*, 440–445. https://doi.org/10.2307/1380441

Yang, L., Liu, B., Zhang, Q., Pan, Y., Li, M., & Lu, Y. (2019). Landscape structure alters the abundance and species composition of early-season aphid populations in wheat fields. *Agriculture, Ecosystems & Environment, 269*, 167–173. https://doi.org/10.1016/j.agee.2018.07.028

Young, B. A. (1983). Ruminant cold stress: Effect on production. *Journal of Animal Science, 57*, 1601–1607. https://doi.org/10.2527/jas1983.5761601x

Young, B. A., & Christopherson, R. J. (1974). Some effects of winter on cattle. In *Annual Feeder's Day report* (pp. 3–6). Edmonton, AB, Canada: University of Alberta.

Zachar, D. (1982). *Soil erosion*. Elsevier, Amsterdam.

Zhang, D. S., Brandle, J. R., Hodges, L., Daningsih, E., & Hubbard, K. G. (1999). The response of muskmelon growth and development to microclimate modification by shelterbelts. *HortScience, 34*, 64–68. https://doi.org/10.21273/HORTSCI.34.1.64

Zhang, H., & Brandle, J. R. (1997). Leaf area development of corn as affected by windbreak shelter. *Crop Science, 37*, 1253–1257. https://doi.org/10.2135/cropsci1997.0011183X003700040037x

Zhang, H., Brandle, J. R., Meyer, G. E., & Hodges, L. (1995). The relationship between open windspeed and windspeed reduction in shelter. *Agroforestry Systems*, 32:297–311. https://doi.org/10.1007/BF00711717

Zhao, Z., Xiao, L., Zhao, T., & Zhang, H. (1995). *Windbreaks for agriculture*. (In Chinese.) Beijing: China Forestry Publishing House.

Zheng, X., Zhu, J., & Xing, Z. (2016). Assessment of the effects of shelterbelts on crop yield at the regional scale in Northeast China. *Agricultural Systems, 143*, 49–60. https://doi.org/10.1016/j.agsy.2015.12.008

Zhou, X., Brandle, J. R., Awada, T. N., Schoeneberger, M. M., Martin, D. L., Xin, Y., & Tang, Z. 2011. The use of forest-derived specific gravity for the conversion of volume to biomass for open-grown trees on agricultural land. *Biomass and Bioenergy, 35*, 1721–1731.

Zhou, X. H., Brandle, J. R., Mize, C. W., & Takle, E. S. (2005). Three-dimensional aerodynamic structure of a tree shelterbelt: Definition, characterization and working models. *Agroforestry Systems, 63*, 133–147. https://doi.org/10.1007/s10457-004-3147-5

Zhou, X. H., Brandle, J. R., & Takle, E. S. (2003). Estimating the three dimensional structure of a green ash shelterbelt: Distribution of vegetative surface area. In T. R. Clason (Ed.), *Land-use management for the future: Proceedings of the 6th Conference on Agroforestry in North America, Hot Springs, AR. June 14–16, 1999* (pp. 182–186). Columbia, MO: Association for Temperate Agroforestry.

Zhou, X. H., Brandle, J. R., Takle, E. S., & Mize, C. W. (2002). Estimation of the three-dimensional aerodynamic structure of a green ash shelterbelt. *Agricultural and Forest Meteorology, 111*, 93–108. https://doi.org/10.1016/S0168-1923(02)00017-5

Zhou, X. H., Brandle, J. R., Takle, E. S., & Mize, C. W. (2008). Relationship of three-dimensional structure to shelterbelt function: A theoretical hypothesis. In D. R. Batish, R. K. Kohli, S. Jose, & H. P. Singh (Eds.), *Ecological basis of agroforestry* (pp. 273–285). Boca Raton, FL: CRC Press.

Zhou, X. H., Schoeneberger, M. M., Brandle, J. R., Awada, T. N., Martin, D. L., Chu, J. M., . . . Mize, C. W. (2015). Analyzing the uncertainties in use of forest-derived biomass equations for open-grown trees in agricultural land. *Forest Science, 61*, 144–161. https://doi.org/10.5849/forsci.13-071

Ziegler, J., Easter, M., Swan, A., Brandle, J., Ballesteros, W., Domke, G., . . . Paustian, K. (2016). A model for estimating windbreak carbon within COMET-Farm™. *Agroforestry Systems, 90*, 875–887. https://doi.org/10.1007/s10457-016-9977-0

Zohar, Y., & Brandle, J. R. (1978). Shelter effects on growth and yield of corn in Nebraska (in Hebrew with English abstract). La-Yaaran 28:1–4.

Study Questions

1. Explain the relationship between windbreak structure and the reduction of wind speed to the lee of the windbreak. What is the relationship between windbreak density and porosity?

2. How do optical density and aerodynamic density differ? From the standpoint of the field forester advising a landowner, which of these two terms is most useful? Why?

3. Different windbreak structures are used to accomplish different landowner objectives. Describe typical windbreaks (species and spacings) for crop protection, wind erosion control, snow management, and protection of livestock and buildings.

4. Working with a landowner, how would you explain the benefits of including windbreaks in their whole farm design and operation? What arguments would you make to convince the landowner that the investment in windbreaks was a good one? How would you counter the observation that yields are reduced adjacent to the windbreak? What factors influence the degree of competition between the windbreak and the adjacent crop?

5. What role(s) do windbreaks play in the overall health of the agricultural ecosystem?

6. Assuming that agricultural operations may be asked to be carbon neutral at some point in the future, what possible role(s) do windbreaks have in this effort? Identify at least two possible impacts of windbreaks on atmospheric CO_2 levels at the typical farm level.

7. Discuss the potential of windbreak systems to produce a marketable product. Identify and describe at least two challenges that a landowner might face if they chose to design a windbreak system capable of providing marketable products.

8. What are the economic factors that need to be considered in deciding if a windbreak investment provides positive economic returns to the landowner? Identify at least two market-based factors and at least two environmental factors.

9. Discuss the role of non-crop areas in biological control of crop pests.

10. Discuss the possible role(s) of field windbreaks in ameliorating spray drift and the spread of pollen from adjacent genetically modified organism (GMO) fields.

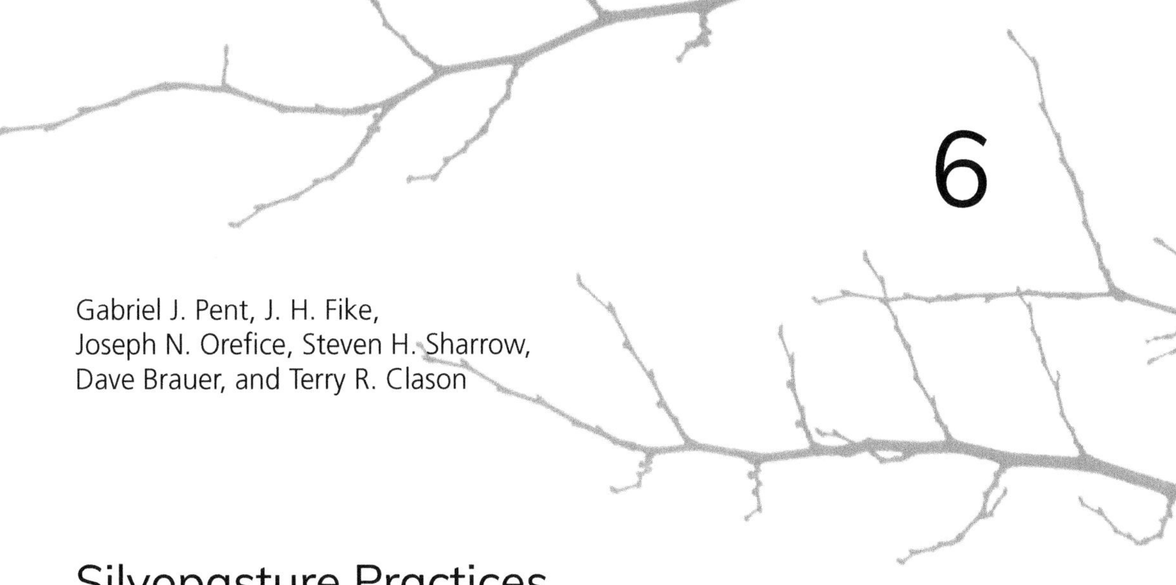

6

Gabriel J. Pent, J. H. Fike,
Joseph N. Orefice, Steven H. Sharrow,
Dave Brauer, and Terry R. Clason

Silvopasture Practices

The integration of livestock and trees takes many forms across North America. In some of these practices, such combinations receive little management, such as sparsely forested rangeland in the West and densely wooded livestock paddocks in the Northeast. In other practices, the interactions between livestock, forages, and trees are carefully planned, timed, and thoughtfully managed. Management complexity will intensify as farmers seek to optimize the interactions among trees, forages, and livestock, but such management of silvopastures—the integration and intentional production of trees, forages, and livestock on the same land—can result in a mutually beneficial framework for all components. This chapter begins by discussing the basis for livestock and tree integration in North America and then take a closer look at the interactions managed within silvopastures. Finally, we provide some examples of silvopastoral systems across particular regions of North America.

Livestock and Tree Integration in North America

Livestock grazing, woodlots, and low-intensity forestry are traditional uses for non-arable lands throughout temperate North America. These opportunistic uses of otherwise minimally productive lands occur across extensive areas of the western United States and western Canada, much of which is publicly owned. Private land holdings in the eastern and central portions of the United States often include woodlots, windbreaks, or non-industrial forests that are grazed by livestock. The largest blocks of grazed, privately owned industrial and non-industrial forests are found in the southern and southeastern United States. Given the widespread co-occurrence of grazing and forestry across North America, intensive and integrated production of livestock and tree products may be the most prevalent form of agroforestry found in the United States and Canada.

Statistics for the exact size of the land area producing joint livestock and tree products are difficult to obtain because the usual categories of forest, rangeland, and pasture imply a dominant use. They do not adequately deal with multiple product systems such as forested rangelands, grazed forests, and pastures with trees. Forests currently occupy approximately 256 million ha of land in the United States and 396 million ha in Canada, of which around 21% is grazed in the United States (Bigelow & Borchers, 2017; Canada's National Forest Inventory, 2013). The majority of grazed forestland in the United States Pacific and Mountain regions are public lands managed by the U.S. Forest Service, while grazed forestlands in the remaining regions are primarily private landholdings (Bigelow & Borchers, 2017). Forestry

North American Agroforestry, Third Edition. Edited by Harold E. "Gene" Garrett, Shibu Jose, and Michael A. Gold.

and grazing are competitive uses for lands that are too steep, rocky, wet, or infertile or consist of soils too thin to support crops or orchards. Forested lands and pastures or rangelands are common in all regions of the United States (Table 6–1), including highly agricultural regions such as the Midwest, Southern Plains, and Delta States. Grazing these non-arable lands is at times done simply because of the opportunity for additional forage utilization or the need for a place to keep livestock. Considerable amounts of forage may be available for grazing under trees in mature open-canopied forest stands, such as the semiarid conifer forests (Figure 6–1) and savannahs of the Northern Plains, Mountain, and Pacific states. Even closed-canopy forest sites may produce considerable amounts of grazeable ground vegetation following timber harvest or natural stand-opening events such as fire or windfall. Grazing by ruminant livestock is an obvious way to control vegetation that competes with trees while making beneficial use of a vegetation resource that would otherwise remain unexploited. However, unmanaged livestock grazing in closed-canopy forests may lead to serious soil degradation, reduction in tree vigor, and loss of some important ecological functions. Approximately one-fifth of all forestland in the United States is grazed by livestock, although it is not known how much of this grazing is managed to protect timber quality and ecosystem health. This amounts to about 54 million ha, or 17% of the total land grazed in the United States (Bigelow & Borchers, 2017).

Table 6–1. Area occupied by cropland, grassland or rangeland, and forestland in different regions of the 48 contiguous United States. Urban and special-use areas are not shown but are included in the total land area (based on Bigelow & Borchers, 2017).

Region	Cropland	Grassland pasture and range	Forest land	Total land
	10^6 ha			
Northeast	4.9	2.0	26.6	45.0
Lake states	16.3	3.4	21.0	49.3
Corn Belt	36.3	7.4	14.3	66.5
Northern plains	39.2	32.3	2.5	78.6
Appalachian states	8.1	5.6	28.6	49.9
Southeast	4.8	4.3	30.8	49.7
Delta states	7.3	3.7	21.3	36.8
Southern plains	16.4	50.3	8.9	85.4
Mountain states	16.3	132.3	40.0	221.6
Pacific states	8.8	23.4	24.0	82.4
Lower 48 states	158.4	264.7	217.9	765.3

Fig. 6–1. Grazed open canopied ponderosa pine forest, eastern Oregon.

In many regions, grazed forestlands occupy sites considered too marginal for agriculture or plantation forestry. In other regions, grazing occurs as a secondary use of lands whose primary purpose is for timber production. In both cases, grazing and forage management intensity tends to be low. Grazing of native vegetation by cattle (*Bos* spp.) is by far the most common form of forest grazing in North America. Although fencing, watering systems, and burning can be used to enhance grazing potential, forested range grazing remains an extensive approach to forage resource use and often lacks the planned interactions among trees, forage, and animals required to be true agroforestry. Silvopasture is an agroforestry practice that intentionally integrates trees, forage crops, and livestock into a structural system of mutually supportive planned interactions. These interactions are managed intensively to simultaneously produce wood or non-timber forest products, high-quality forage, and livestock on the same land unit in an environmentally sustainable manner. The forage component may be either native forage plants or introduced species managed as an improved pasture. This more intensive approach to land use provides the foundation for integrated commercial timber and livestock production systems. Both range and forestlands are increasingly being managed as integrated ecosystems that produce both saleable commodities, such as wood products, livestock, and hunting fees, as well as environmental services. Silvopastoralism reflects this ecosystem view. Environmental issues such as biodiversity, wildlife habitat, soil stabilization, watershed characteristics, pollution abatement, carbon sequestration, and aesthetic appeal are important design elements of silvopastoral systems.

The most commonly practiced forms of silvopastoralism in North America are integrated forest grazing and silvopastures. Integrated forest grazing uses livestock to harvest native forest plants as part of a planned forest ecosystem management system. Although often extensively (rather than intensively) managed, these systems can yield multiple goods and services with careful planning. Livestock become tools for managing forest trees and their understory plant communities for multiple outputs or outcomes, including timber, forage, wildlife habitat, fire fuel reduction, and improved water quantity and quality. Silvopasture is the intensive form of silvopastoralism. Trees and livestock are combined with improved pasture plants to form carefully designed systems that integrate intensive animal husbandry, silviculture, and forage agronomy (Figure 6–2). Silvopastures are highly productive and typically more complex to design and to manage than integrated forest grazing systems. Skillful selection of forage plants and manipulation of microenvironments within silvopastures can result in high-quality green feed available for livestock while providing them with shelter during inclement weather. Silvopastures compete economically for sites currently used for some high-value field crops when thoughtfully combined into an overall livestock, tree, and forage production system.

Fig. 6–2. Cattle grazing Coastal bermudagrass beneath southern pines in a southeastern silvopasture.

Silvopasture Concepts

To be successful, agroforestry systems must meet four requirements. They must be biologically possible, ecologically sustainable, socially acceptable, and economically feasible.

Silvopasture Biology and Ecology

The geographic range occupied by an organism is the result of its biological and ecological amplitudes. *Biological amplitude* refers to the environmental requirements of individual plants and animals. In contrast, *ecological amplitude* refers to the requirements of plants and animals as they interact with other organisms within communities. Silvopasture design involves selecting and managing components for desired interactions while respecting the biological limitations of each component. There is little point in promoting land use systems that do not meet the site requirements of the individual plant and animal components or that fail to integrate components into a properly functioning community. Placing plants in inappropriate site conditions, such as walnut (*Juglans* spp.) trees on thin, poorly drained soils or clovers (*Trifolium* spp.) on very acidic soils, is unlikely to be worthwhile. Likewise, combining fast growing trees with shade intolerant, slower growing trees or goats with young, highly palatable hardwood trees will present considerable problems to managers.

It is important, when considering site suitability, to recognize that trees will be growing in the same location for a long period of time. Their general health will reflect average conditions over time. However, they will most certainly also encounter unusual periods of weather, outbreaks of disease, and infestations of insects or other damaging organisms that may prove fatal. For instance, even apparently healthy conifer trees may cavitate (the water column breaks) and die very quickly in unusually hot and dry or cold and dry weather (Sharrow, 2004). The extreme as well as average site conditions will ultimately determine the success of plants in silvopastures.

Individual cohabitating organisms may help (facilitation) or may hinder (competition) each other. What we observe in an ecosystem is the net effect of these two processes. Organisms benefit each other when facilitation exceeds competition. When competition exceeds facilitation, negative consequences for one or both of the system components reduces their productivity. Plants may compete with each other directly through competition for site resources such as soil moisture and nutrients. A plant may also hinder other plants through interference by improving the habitat for diseases, herbivores, or other damaging organisms or by directly impeding the growth of other plants. For example, black walnut (*J. nigra* L.) trees produce an allelochemical, juglone, that inhibits the growth of plants in the Solanaceae family. Likewise, animals may compete with each other for access to food, shelter, and other site resources.

Animals and plants may also help each other. Hunter and Aarssen (1998) listed nine ways in which plants have been observed to facilitate the growth of other plants by modifying the physical and biotic environment, including: providing physical support, modifying the microclimate, changing soil nutrient availability and structure, directly transferring resources between interconnected plants, defending against predators, supporting a favorable soil microbe community, attracting pollinators, and attracting seed dispersal agents. Animals participate in many similar positive interactions. Animals may gather together to protect each other from inclement weather or for group protection from predators. Herds provide social opportunities for sharing knowledge and responsibilities for offspring rearing. Mixed species herds allow group defense, while different grazing habits of the individual species support efficient forage utilization and reduce the potential for the spread of weeds. The challenge of silvopasture system management is to successfully balance the facilitative and competitive interactions to meet the needs of individual components as well as to provide a framework within which plants and animals can successfully interact to form a productive community.

Silvopastoralism influences ecosystem processes of hydrology, nutrient cycling, energy flow, and succession primarily through manipulation of the community structure. The basic strategy is to combine trees, forages, and livestock in such a way that each component produces usable products (provisioning services) and contributes to the stability and conservation of land resources (supporting and regulating services). These systems also can contribute to enhanced aesthetics and outdoor activities (cultural services).

With good system design and management, both tree and livestock production may be enhanced because trees provide shelter for livestock and pasture plants, while livestock serve to control weeds, to recycle nutrients through their feces and urine, and to reduce escape cover for rodents that gnaw on trees. Supporting and regulating functions are often obtained at little cost as a result of production. For example, sheep (*Ovis aries*) may eat (and thus control) tansy ragwort (*Senecio jacobaea* L.), a weed toxic to cattle (Sharrow & Mosher, 1982). Dinitrogen fixation by plants such as subterranean clover (*Trifolium subterraneum* L.) or black locust trees (*Robinia pseudoacacia* L.) increase soil N when the plants shed their leaves and fine roots as a normal outcome of their growth. Greater nutrient capture and cycling also can occur with trees in pastures; e.g., soil NO_3 leaching is lower in silvopastures than in open pastures or pine plantations (Bambo, Nowak, Blount, Long, & Osiecka, 2009), probably due to the layered structure and dense rooting zone of a silvopasture. Although management intensity may be greater, silvopasture systems may require fewer external inputs such as nutrients or pesticides than open pastures. The presence of these service functions, together with efficient resource utilization among components in time and space, makes the cumulative production of well-designed and managed silvopasture systems greater than the isolated production of their individual components.

The reduced need for purchased inputs in silvopastures decreases dependency on outside suppliers, lowers operating costs, and increases potential profit margins of tree and livestock products. The proper unit of reference for the biological or economic productivity of intercrops, such as silvopastoral systems, is the entire system rather than the individual component. The efficiency of intercrops may be expressed by the land equivalency ratio (LER; Vandermeer, 1981), which represents the ratio of the land area managed in individual components to the intercrop-managed land area required to produce equivalent yields of each component. The LER of subterranean clover–conifer silvopastures ranges from 1.18 to 1.6 (Sharrow, Carlson, Emmingham, & Lavender, 1996), which compares favorably to values for other legume–nonlegume intercrops (Hiebsch & McCollum, 1987). In the Willamette Valley, near Corvallis, OR, it would take approximately 1.6 ha (0.96 ha of forest plus 0.64 ha of pasture) to equal the total aboveground productivity of 1 ha of silvopasture on a moderately productive commercial timber site (Sharrow et al., 1996). Similar animal performance per land area has been measured in open pastures and hardwood silvopastures (Fannon, Fike, Greiner, Feldhake, & Wahlberg, 2017; Kallenbach, Kerley, & Bishop-Hurley, 2006; Pent & Fike, 2016). Thus, any tree products produced in these silvopastures would represent the net productivity of silvopastures compared with open pastures.

This high biological productivity together with the large potential area suitable for silvopasture makes it a strong agent for C sequestration to help check global climate change

(Montagnini & Nair, 2004). Although most C offset programs have focused on forests' ability to store C in wood, the soil, rather than vegetation, is where most terrestrial C is stored. Grasslands may accrete as much C as forests (Corre, Schnabel, & Schaffer, 1999), but their contribution is often overlooked because it is predominately stored underground in soil organic matter (de Groot, 1990). The trees in silvopastures can also contribute to the C storage of soils, particularly at lower soil depths than grasses (Haile, Nair, & Nair, 2010). Silvopastures may outperform either pastures or forests as C sinks because they store C using both forest (wood) and grassland (soil organics) mechanisms. However, this dynamic depends on the rooting depth of the forages within a silvopasture and is not well studied across regions. Sharrow and Ismail (2004) estimated that cool-season pasture–Douglas-fir [*Pseudotsuga menziesii* (Mirb.) Franco] silvopasture in western Oregon accumulated approximately 740 kg ha^{-1} yr^{-1} more C than forest and 520 kg ha^{-1} yr^{-1} more C than pastures during its first 11 yr after planting. Potential C storage in U.S. silvopastures is substantial, estimated by Montagnini and Nair (2004) to be at least 9 Tg yr^{-1}, and considered by some as among the most effective means of drawing down atmospheric C (Hawken, 2017). Landowners would be encouraged to adopt silvopasture for environmental benefits if they could capture some of their financial value. Shrestha and Alavalapati (2004) estimated that Florida households might be willing to pay US$30–71 yr^{-1}, either directly or indirectly through subsidies, for environmental services from silvopastures. Cashing in this value has been difficult to achieve in practice. While C credits are traded in the United States, their value is inconsistent and varies wildly with public pressure on polluters and pollution regulations. Assuming that temperate silvopastures will sequester about 1.1 Mg C ha^{-1} yr^{-1}, this environmental service would have sold for around $14 on the California Air Resources Board auction in 2017. It is clear, however, that C trading markets will require more development before farmers accept them as incentives for implementing conservation practices such as silvopasture (Holderieath, Valdivia, Godsey, & Barbieri, 2012).

Silvopasture systems may be established following harvesting or thinning of existing tree stands, under open-canopied forests and woodlands, or by planting trees in pastures and other agricultural lands. In all cases, successional processes will begin based on existing vegetation and seeds present in the soil seed bank. Desired trees and forage plants are often planted to direct succession toward a desired plant community. The initial successional process favors existing established plants over establishing plants. Therefore, planting tree seedlings into an existing stand of perennial forage plants is generally less successful than planting trees and pasture plants at the same time so that they can establish together. Without controlling grass and other herbaceous vegetation, tree growth will be slowed relative to planting into bare ground (Houx et al., 2013).

Conceptually, there are three management periods in the life of a silvopasture system: tree or forage establishment, open-canopied forest, and closed-canopy forest. Whether a given silvopasture system passes through all these phases depends on the initial resource base, as well as design and implementation, and the tree stand's long-term management. The characteristics and length of the first stage depend on whether trees are planted into an established pasture (or cleared site) or whether a forest has been thinned and forages have been planted beneath the remaining trees.

In the tree establishment phase of pasture-to-silvopasture practices, managers couple standard pasture management practices with added efforts focused on tree protection and growth. Particular attention is paid to minimizing competition from resident forage plants, meeting the water requirements of tree seedlings, and preventing tree damage by wildlife and domestic animals. Seedling trees have relatively little direct impact on livestock production during the establishment phase (Lehmkuhler et al., 2003). However, within the first few years, moderately spaced hardwood saplings can alter the understory microclimate, increasing aboveground forage growth and altering understory competition (Buergler et al., 2005).

In forest- (or woodlot-) to-silvopasture practices, producers must select and thin trees to an appropriate density to permit acceptable forage seedling recruitment and subsequent forage growth. Often, soil nutrients and pH will need adjusting to support forage productivity. Site preparation such as mulching slash or stump grubbing can make the site clean enough that seed can be drilled into the soil to establish forages—but such clearing practices can be costly and may damage residual tree roots. Fire also can be used as a tool to remove debris and reduce the duff layer on the site, but care must be taken to minimize harm to the tree stand. Broadcasting seed or feeding hay with viable seeds underneath the residual trees may be a cost-effective means of establishing forages under trees, although seeding rates probably need to be higher than for drilling. Trampling and the addition of nutrients (particularly via manure and urine) can improve site suitability for new seedlings. The transition time to the open-canopied forest phase is much quicker when forages are seeded underneath mature trees and may be as short as 1 yr.

The open-canopied forest phase is the most interesting to agroforesters because during this period, competition and facilitation among trees, livestock, and pastures can greatly affect the productivity of each component. This is the period when LER is the greatest because both pasture and tree production contribute significantly to land productivity. The impacts of trees on understory forage production is now of special concern. If tree canopies are allowed to close, insufficient light reaches the ground to support forage for livestock grazing. The tree stand at this point essentially ceases to function as an agroforest and is managed as a late rotation forest. Managers more focused on livestock production often attempt to reduce or even prevent this phase by planting at low initial tree densities, thinning trees as soon as possible, or pruning off lower branches. For managers interested in long-term tree production, management is focused on getting some initial return from livestock and minimizing their potential negative effects on tree growth and quality.

Social Factors

The sociology and economics of silvopastoralism may be viewed at two scales, the societal regulatory scale and the individual practitioner scale. Silvopasture systems differ from agriculture, and even other forms of agroforestry, in their natural appearance. Many silvopasture systems are based on integration and management of naturally occurring combinations of native trees and understory plants or combinations of introduced trees and forages that resemble these native systems. Most silvopasture systems, therefore, are perceived as more similar to native forest or rangelands than agricultural lands and may be treated as such by governments and regulators. For example, forest land is often zoned in a separate category from agricultural land for taxation and regulatory purposes. While regulation of agriculture is mostly concerned with environmental pollution, soil erosion, and other off-site effects, regulation of forest practices are additionally aimed at maintaining the integrity of forest and woodland ecosystems. This places additional restrictions on silvopastures that would not be required for pastures or other agricultural lands. Requirements for unimpacted "set-back" zones along streams are often wider for forest than for agricultural lands. Agricultural and forest chemicals are labeled separately, such that a herbicide acceptable for use on agricultural lands may be illegal if applied to control the same weed on forest lands. Forest and agricultural lands may also have different tax systems. For example, agricultural lands are often assessed a yearly tax based on the value of the land. In some areas, forest lands are assessed a small yearly fire protection fee, and the main tax is paid as a "severance tax" when timber is cut and sold. Before converting pasture or agricultural lands to silvopastures, local regulators should be consulted to understand zoning and tax issues.

Societal desires for forest management are transmitted to land managers through economic and political means (Koch & Kennedy, 1991). At times, these two forces have come into conflict. Limited world timber supplies support high stumpage prices and encourage harvest, while public policy typically promotes sustainable multiple-use management of public and private forest lands through laws and other regulations that effectively reduce wood harvest. Economic incentives for afforestation are currently available through federal or state conservation reserve, wildlife habitat enhancement, or other reforestation programs. Federal incentives for silvopasture also exist, although not all states have opted to implement them. Within this regulatory environment, environmentally based multiproduct systems such as silvopasture systems have an advantage over plantation forestry. Social acceptability of silvopastoral systems may be increased by applying Brunson's (1993) principles of socially acceptable forestry. That is, in the context of silvopastoral design, natural-like systems are considered more acceptable, the perceived intents of management actions are as important as the actions taken, and all actions taken are judged relative to perceived alternatives. The intent to maintain natural processes within a healthy and productive ecosystem makes silvopastoralism potentially more socially acceptable than open pasture or plantation forestry systems.

As suburban growth encroaches on forests, and as people become increasingly sensitive to environmental issues, the visual appearance of forest view-sheds increasingly becomes a consideration for forest management. Forests and woodlands that were once quite remote are coming into full view of the public, which in turn is expressing definite opinions about the appropriateness of their management. Likewise, small but vocal local groups are beginning to discuss the desirability of preserving the visual "agricultural landscape" through the land use zoning process. For example, around the country, landowners can receive tax deferrals and even cash payments for surrendering the right to convert their rural land to suburban uses such as housing development. Silvopasture systems generally are visually acceptable within rural landscapes. In fact, there is some evidence that humans have an evolutionary predisposition to prefer savanna-type landscapes to forests (Balling & Falk, 1982). Integrated forest grazing systems often have a normal forest structure and an open grassy understory similar to early successional plant communities of native forests. The neat rows of pruned trees in silvopastures with a well-groomed pasture understory are open and park-like. Aesthetics may drive the adoption of silvopasture for some producers (Lawrence, Hardesty, Chapman, & Gill, 1992; Orefice, Carroll, Conroy, & Ketner, 2017; Workman, Bannister, & Nair, 2003), particularly in the urban fringe.

Silvopastures will only be adopted by practitioners if they perceive the benefits to outweigh the costs. Natural resource decisions often reflect both tangible benefits and intangible amenity values. Income considerations are an important factor in natural resource management (Zinkhan & Mercer, 1996), but the importance of amenity values should not be underestimated. Income implications can be readily evaluated by examining expected cash flows, internal rates of return on investment, and net present value. The intangible nature of amenity values such as beauty, fairness, desire to be a responsible neighbor, or leaving the land in good condition for the next generation are often overlooked because they are very difficult to quantify. They do influence individual decision-makers, however. Landowners often recognize these non-cash benefits and adjust their decisions to favor them. In a survey of southeastern landowners (Workman et al., 2003), >70% of respondents ranked aesthetics, wildlife habitat, soil conservation, biodiversity, and water quality as highly important benefits of agroforestry. In a survey in the state of Washington (Lawrence et al., 1992), non-industrial forest owners regarded aesthetic appeal (77%) and increased biodiversity (66%) as advantages of agroforestry. The two most frequently given reasons for owning land were to pass it on to children (80% of respondents) and to keep it natural (75% of respondents). Likewise, non-industrial forest landowners in Oregon ranked "good stewardship" and "leaving a legacy" above timber production, producing income, or grazing as the main objectives of their land management (Elwood, Hansen, & Oester, 2003).

Aesthetics, stewardship, and legacy are not always perceived as high priorities by those engaged in making a living from silvopastures, however. Of a group of 20 landowners practicing silvopasture in the U.S. Northeast, 80% listed shade for livestock and 30% listed animal welfare as reasons for their utilization of these systems (Orefice, Carroll, et al., 2017). In this case, only two landowners listed improved aesthetics as a reason for silvopasture implementation on their farms. There is growing interest in improving animal welfare, both at the societal regulatory scale and the individual practitioner scale. Although providing shade for livestock on pasture has been considered prohibitively expensive for typical extensive production systems (St-Pierre, Cobanov, & Schnitkey, 2003), trees in silvopasture provide shade for livestock while accruing value. The capacity of silvopastures to benefit animal welfare through shade and shelter while improving land productivity will probably be a central focus of future silvopasture research and adoption. Livestock producers have shown an interest in understanding more about these issues, with animal welfare being ranked in the top five priority management issues by beef cattle producers in the United States (USDA, 2016).

Economic Factors

Integrating trees, forage crops, and livestock creates a system that produces a constant flow of marketable products while maintaining long-term productivity. Cash flow is especially important for landholders who must support themselves while the pasture or cropland is being afforested. Economic timber rotations for conifers can be quite long, ranging from 20 yr in southern pine forests to 65 yr for coastal Douglas-fir in the Pacific Northwest and longer for eastern hardwoods. This places final timber harvest outside of the economic lifetime of many middle-aged landholders. Intensive silvopastoral practices of grazing, fertilizing, and thinning increases tree growth so less time is required to produce high-value timber products such as sawtimber and veneer logs. Early timber harvest speeds up cash flow and shortens the period of investment risk. Economic risks are reduced because livestock and forest components require different inputs, share few common diseases and pests, and sell into different markets. Risks can be further reduced by using tree and livestock components that have previously been successfully produced on similar sites and for which local markets and production infrastructure already exist, although timber markets may shift during the lifetime of a timber stand. Sharing costs between livestock and timber components reduces individual component production costs and enhances their marketing flexibility. Thus, silvopasture is often more profitable and less risky than either forest or livestock enterprises.

Long-term investments such as timber stands are very sensitive to the time value of money. Money made or saved now is much more valuable than money accrued later. This is because money invested in trees must compete with other potential investments. This "opportunity cost" accumulates each year much as a compound interest rate in a bank account does. So $1 invested in a timber stand now requires almost $24 of income at the end of a 65-yr rotation to equal money being put in the bank at 5% annual interest. This cost becomes even larger when one considers the possibility of money losing value over time (inflation). The buying power of our initial investment dollar was 100 cents. However, 65 yr later, 3% annual inflation would reduce its buying power to only about 8 cents. Reductions in the value of income due to opportunity costs and inflation can be minimized by reducing the initial investment and by emphasizing early income that can either be used to pay off the investment or to cover current expenses. Agroforestry can reduce investment costs by substituting service functions such as weed control by animal grazing for purchased inputs such as herbicides. It also generates income during the initial years of the timber rotation. This makes silvopastures very economically efficient compared with forest plantations. Although the amount of money saved or income generated during the early years of a timber rotation may seem small compared with the large amount of income when the final timber crop is sold, it has a huge effect on the profitability of the timber investment.

Increased economic diversity and higher total monetary returns were perceived as the primary tangible benefits of silvopastoral practices according to surveys of public land-use professionals in the Pacific Northwest and southern United States (Lawrence & Hardesty, 1992; Zinkhan, 1996). Seventy-four percent of private non-industrial forest managers surveyed in the state of Washington listed increased income as an advantage of agroforestry. Research supports these perceptions. A simulated loblolly pine (*Pinus taeda* L.)–forage–beef cattle system for the southern U.S. Coastal Plain found that a silvopasture net present value per unit area was 70% greater than a pure forestry operation (Dangerfield & Harwell, 1990). In Louisiana, a 'Coastal' bermudagrass [*Cynodon dactylon* (L.) Pers.]–loblolly pine agroforest produced 234 kg ha^{-1} yr^{-1} of meat and 3.3 m^3 ha^{-1} yr^{-1} of wood (Clason, 1995). Although establishment costs were $716 ha^{-1}, the internal rate of return for this silvopastoral system was 13%, which exceeded timber management and open pasture options by 4 and 7%, respectively. In western Oregon, KMX hybrid pine (*Pinus attenuata* Lemm. × *Pinus radiata* D. Don) silvopastures grazed with sheep were projected to yield a 22% internal rate of return after 22 yr (Sharrow & Fletcher, 1995). These results were attributed to comprehensive land utilization obtained by combining timber and livestock production, a reduction in time between cash flows by selling livestock, and synergies such as trees utilizing applied fertilizer and livestock manure.

Other land managers have different objectives, however. The majority of a group of 20 landowners utilizing silvopasture in the Northeast cited greater utilization of farm woodlots and expanded pasture acreage as their primary reasons for utilizing silvopasture (Orefice, Carroll, et al., 2017). In addition, trees can have a stabilizing effect on livestock physiology. Shelter and shade from trees can reduce the range in minimum to maximum daily body temperatures and help maintain lower body temperatures during periods of potential heat stress (Pent, Fike, & Kim, 2018). These benefits have the potential to improve animal productivity and feed use efficiency, which may provide a basis for accounting incurred tree management costs against livestock operation returns when determining tax deductions.

Historically, most silvopasture design has emphasized maximizing forage production and minimizing negative impact on trees. While the relatively high value of mature trees has made them a focus of planning, increasingly the shade provided to livestock is the predominant goal—although this is probably regionally specific. In the latter case, having a full stand of trees at rotation age may not be a primary production goal. Systems that emphasize tree production (agricultural production occurs until the forest closes) probably are more common with extensively managed landscapes.

Adoption of silvopasture technology has been more rapid by farmers and ranchers seeking to afforest agricultural lands or to manage non-industrial forest lands than by silviculturalists considering reforestation options for commercial forests. This may to some extent reflect the availability of investment capital following timber harvest. Reforestation of harvested lands can be financed from the proceeds of timber sales. The opportunity cost of this money is often lower than the interest paid on loans taken out by farmers or ranchers seeking to afforest pastures or cropland. Where forest practice regulations require replanting trees, reforestation may be considered a cost of harvest that is not carried forward as a cost of the next generation of trees. This greatly improves cash flow, increases the internal rate of return and net present value of the timber investment, and reduces the economic pressure for immediate income. Need for an income stream to support the costs of afforestation, on the other hand, encourages ranchers to continue to graze their lands.

Non-industrial forest managers tend to be balanced resource managers who value aesthetics and land conservation as well as income generation (Lawrence et al., 1992), making them natural clients for, or implementers of, agroforestry. This is especially true for the urban fringe lands where owners have other sources of cash income. For example, a survey of landowners found greater interest in agroforestry among those who had more recently acquired land and who managed their property more extensively; these owners had both greater discretionary income and multiple objectives in mind for the land's use (Trozzo, Munsell, & Chamberlain, 2014). Such findings suggest that largely untapped clientele and practitioners may exist for agroforestry systems that emphasize forage production, intangible benefits such as aesthetics, and independence or income through the production of non-timber forest products for sale or household subsistence needs. Clearly, in a silvopastoral context, successful integration of pastures with timber requires knowledge and managerial effort greater than managing either commodity separately. However, most agricultural and silvicultural information is specific to the technological context from which it came. The biological information necessary to produce such a hybrid system is just now being acquired. Silvopastures offer substantial rewards to knowledgeable managers but will be of limited use for those desiring a low level of management.

Integrating Silvopastoral Components

Silvopasture is one of the most complex forms of agroforestry. While alley cropping, forest farming, and urban food forests combine two components, trees and crops, silvopastoralism deliberately integrates three components: trees, understory vegetation, and domestic livestock. The many possible combinations of trees, livestock, and forage plants for a specific site provide a wide array of options to meet many different management goals. Considerations should be given to potential markets, to soil and climatic conditions, and to crop management compatibility when making tree, livestock, and forage crop selections. Most silvopastoral systems are designed to enhance the long-term value of the timber component while sustaining the short-term cash flow value of the livestock component.

Despite the added complexity attendant with having animals in the system, the base components about which decisions are made generally remain the system's trees and forages. Because a producer's choice of livestock most often is fixed, matching trees and forages for livestock (and not vice versa) becomes the primary consideration for implementation and management. General recommendations for tree selection are to choose marketable, high-quality timber species that grow rapidly and are deeply rooted. Coniferous trees are well-suited for silvopastures because they adapt to a variety of growing sites, respond rapidly to intensive management, have conical crowns that permit more light to reach the forest floor, are less likely to be browsed by livestock, and generally have well-defined markets. Several deciduous tree species also have desirable wood (and tree-crop products), along with morphological and phenological features that complement the production of cool-season forage species. However, difficulties with establishment, requirements for protection, and the length of the production cycle can present challenges to using hardwood species.

When choosing among the various forage species options for silvopastures, preference is given to plants that are nutritious; palatable to livestock; shade, drought, and grazing tolerant; and are responsive to intensive management. In parts of the United States with more extensive management (e.g., on federal lands in the Northwest or in southeastern plantations), the native herbaceous understory may be retained as the forage base. In contrast, most pasture systems of the humid East utilize naturalized (non-native) cool- and warm-season forage species as the primary herbaceous component given their productive potential. Reintroduction or management of native grasses into silvopastoral environments may be preferred by some, particularly among the conservation communities (Arbuckle, 2009), although matching native warm-season grass production with tree management may prove more challenging.

The capacity for trees and forages to share site resources well with each other is another key consideration when developing silvopastures. Resource sharing in time and space is a fundamental concept in selecting tree and forage components, and decisions are regionally (or site) specific. For the seasonal rainfall regimes of California and the Pacific Northwest, selecting annual forages that concentrate their growing cycle in the rainy season leaves all soil moisture in the dry season available to support tree growth. These species can also be shallow rooted, further minimizing competition with the tree crop for water and nutrients. Subterranean clover, for example, grows quickly in the spring, then dies. The plant works well in combination with Douglas-fir in the maritime zone of the Pacific Northwest because it completes most of its yearly growth before the spring rains end. The trees continue to grow until winter, using stored soil moisture and summer rainfall. Subterranean clover also restricts its root system to the top 15 cm of soil, leaving the soil moisture and nutrients below this shallow zone for other plants. Douglas-fir trees have deeper roots than clover that are generally able to penetrate lower into the soil profile. Similarly, in the U.S. Southeast, crimson clover (*T. incarnatum* L.), a cool-season annual, may provide winter grazing before giving way to a perennial, warm-season forage base.

Such resource partitioning in time and space provides design opportunities for the selection of tree and forage plants whose combined root systems are able to capture all site resources without undue competition in the zones of overlap. Most soil nutrients are associated with the organic matter zone near the soil surface. Both pasture plants and trees require soil nutrients to support their growth. Therefore, even deeply rooted trees will have fine roots in the upper soil layers where they will compete with pasture plants for nutrients. In alley cropping, managers may plow to prepare the land for cropping. This cuts off the shallow tree roots, reducing root competition with the annual crop within the plow zone. This is not possible with perennial crops such as native grasslands or permanent pastures, however. Thus, selection of tree and pasture plants that share soil resources well is critical in maintaining the health of both plant components.

In the humid, temperate eastern United States and Canada, where precipitation is more evenly distributed, light may be more limiting than water in some silvopastures. In this case, deciduous "warm-season" trees that leaf out relatively late in the spring and shed their leaves relatively early in the fall may be desirable. This phenological pattern allows more light to reach the forage understory during the peak periods of growth for cool-season forage species utilized in these regions. Several such tree species also have open-canopy architecture, as well as compound leaf shape, which supports light penetration during the summer growing season. These features are advantageous for cool-season forage plants by providing only limited competition during the most active growing phases (spring and fall) and buffering the herbaceous understory during summer. Understanding the phenology and morphology of tree and grass species, or their growth with time, can aid in the selection of complementary components of a silvopasture system and the proper management of competitive tree–forage interactions.

Animal Component

Most ruminant livestock species are viable candidates for silvopasture, but as with any agricultural enterprise, consideration should be given to existing markets and infrastructural requirements. If silvopastures are stocked with small ruminants, particularly goats (*Capra aegagrus hirucs*), then protection or tree maturity (crown height and bark thickness) may be more important considerations than if silvopastures are stocked with cattle. Utilizing rotational stocking and maintaining proper stocking rates are necessary steps in every silvopasture scenario to maintain the forage base and protect the trees and soil.

Animals in silvopastoral systems are both a product and a management tool. Techniques such as rotational stocking (grazing) allow producers to control the timing and frequency of grazing and provides an effective and inexpensive means of vegetation management. For many producers, livestock production is the only objective, but silvopastoral management offers the opportunity both to produce livestock and to use them to manipulate vegetation in creating wildlife habitat for desired game and non-game species. Sales of hunting rights and livestock products are a way of converting herbaceous vegetation into income. Wild animals and non-ruminants also can be used to manipulate silvopastoral ecosystems, but their activities are harder to direct in a precise manner. Large ungulates such as deer (*Odocoileus* spp.) and elk (*Cervus canadensis*) may damage trees by browsing or rubbing on young trees, and pigs (*Sus scrofa*) will cause significant harm to trees and ecosystems by rooting and wallowing if not managed. Wildlife and livestock may also benefit desired trees by reducing brush or consuming invasive vines and hardwood trees. In western Oregon, for example, deer have effectively controlled invasive bitter cherry [*Prunus emarginata* (Douglas ex Hook.) D. Dietr], which readily invades fenced areas and dense brush patches where the animals are excluded. Prescribed grazing with goats has been established as one technique for managing invasive shrubs such as autumn olive (*Elaeagnus umbellate* Thunb.) and vines such as kudzu [*Pueraria montana* (Lour.) Merr.] in forestlands, although multiple seasons of grazing are necessary for long-lasting control (Miller, Manning, & Enloe, 2010).

It is useful to note that grazing affects both the top (shoot) and the bottom (root) of a plant. The plant is a balanced system, with the root supplying water and nutrients to the top and the shoot supplying food to the roots. When the top is pruned back by grazing, the plant reduces the depth and spread of the root system to restore this balance. Therefore, what happens to the top of a plant is mirrored underground. Grasses have extensive fibrous root systems that make them fierce competitors for soil moisture and nutrients in the upper soil layers. Grazing can control competition between young trees and forage plants by removing forage leaf area prior to soil moisture depletion (Doescher, Tesch, Alejandro-Castro, 1987). Prescription grazing, which applies the proper amount, distribution, and season of grazing, has proven helpful in controlling competition between trees and forage plants (Doescher et al., 1987; Sharrow, 1994). If grazing is done properly, silvopasture trees may display less summer moisture stress and have greater diameter growth than trees in nearby ungrazed forest stands (Carlson, Sharrow, Emmingham, & Lavender, 1994). In a trial near Corvallis, OR, Douglas-fir trees in a silvopasture grew 14% faster in diameter than did adjacent forest trees during their first 4 yr (Sharrow, 1995). Approximately half of the additional diameter growth increase occurred during the dry summer–fall period when plant moisture stress limits tree growth.

The amount and quality of forage available for livestock in the agroforest varies with tree occupancy and the type of trees and forage crops present. Trees in young agroforests withdraw relatively few site resources, and understory forage yields approximate those of similar pastures. As trees grow, their demand for site resources increases and forage production declines accordingly (Sibbald, Griffith, & Elston, 1994). For example, Douglas-fir in western Oregon had relatively little effect on agroforest forage production until the trees were 9 yr old; then it declined rapidly to 54% of adjacent pasture yields when the trees were 11 yr old (Sharrow, 1991). Forage yields were greater under moderate densities of 7-yr-old black walnut and honeylocust (*Gleditsia triacanthos* L.) trees in an Appalachian silvopasture than where trees were spaced at higher or lower densities (Buergler et al., 2005). However, on an adjacent site, yields declined to about 85% of open pasture dry matter production as the trees matured and the canopies closed (Fannon et al., 2017). Livestock carrying capacity should be determined based on forage yields, similarly to open pastures. As a result, the stocking rate may be lower in silvopastures than in open pastures on similar sites, but individual animal productivity may compensate for this reduction because of benefits to animal welfare.

Although livestock and wildlife may play important roles in vegetation management, their potential to damage trees by browsing or trampling is often mentioned as a concern by prospective agroforesters (Lawrence & Hardesty, 1992; Zinkhan, 1996). Trampling damage is largely confined to very young trees and is more common with cattle than it is with sheep or goats. Livestock may chew or rub on trees, causing considerable damage to young or thin-barked species; even sizeable saplings can be at risk, however, as bulls have been observed walking them down to get a belly rub.

Livestock generally do not actively feed on conifers when other palatable forage is present (Doescher et al., 1987; Sharrow, 1994). Conifer foliage is most palatable when buds have just broken and the new growth is still light green in color (Sharrow, 1994). Animals do like variety in their diet and will eat a small amount of conifer foliage each day, especially when other sources of woody browse are not available. Goats tend to consume more browse than cattle or sheep and thus are generally more challenging to manage in silvopastures. Once the animals have decided that alternate forage is either unavailable or unattractive, active browsing on trees can quickly impact young trees. For this reason, some agroforesters prefer to either cut hay or to protect trees with fencing or individual tree tubes until the trees have grown above the reach of livestock (Sharrow & Fletcher, 1995). In some cases, protection from wildlife may be as important as protection from livestock (Bendfeldt, Feldhake, & Burger, 2001). This may be due to browsing, rubbing, or trampling by large animals or due to gnawing and chewing roots by small rodents.

Although browsing by livestock can interfere with the regeneration of hardwood stands, conifers are fairly tolerant of defoliation (Pearson & Cutshall, 1984) provided that the terminal bud remains intact (Sharrow, 1994; Sharrow, Leininger, & Osman, 1992). Young Douglas-fir, for instance, showed little reduction in either height or diameter growth following 50% defoliation in either spring or summer (Osman & Sharrow, 1993). Goats, horses, and to a lesser extent sheep or cattle will sometimes strip bark from young trees. Bark stripping is more likely to occur on young hardwoods than on older trees that have developed a thick, corky bark layer. Although the resins in conifer bark generally convey protection against bark stripping by livestock, goats have been known to peel bark off of conifers, including longleaf pine (*Pinus palustris* Mill.) (Karki, Karki, Khatri, & Tillman, 2018). Any break in the bark is undesirable because it provides an opportunity for disease or insects to enter the tree. The direct effect of bark removal interfering with water and nutrient flow through the stem usually does not reduce tree growth unless more than half of the circumference of the stem is debarked (Sharrow, 1994), but timber quality can be negatively affected. However, even when trees are temporarily damaged by grazing, the benefits of controlling competing vegetation may provide greater tree growth with time (Sharrow, Leininger, & Osman, 1992).

Along with the direct negative effects of livestock on trees, forest managers often mention soil compaction by livestock as a possible problem in grazing forestlands. In the southeastern pine zone, the combination of burning and moderate to heavy grazing increased bulk density, reduced pore space, and decreased percolation rates of the soil (Duvall & Linnartz, 1967). Bezkorowajnyj, Gordon, and McBride (1993) also observed an increase in soil bulk density from cattle grazing. Medium and high levels of soil compaction reduced water infiltration and N cycling, slowing seedling growth. Boyer (1967) reported that light cattle grazing on a longleaf pine site in southwestern Alabama reduced survival rates by 23% and diameter growth rates by 13% during the first 5 yr of the regeneration period. Negative effects of grazing on trees, often referred to as "soil compaction," actually includes both soil compaction and direct physical damage by trampling on shallow tree roots. Shallow-rooted trees such as Douglas-fir or western red cedar (*Thuja plicata* Donn ex D. Don) are particularly sensitive to direct root damage if livestock hooves actually penetrate the soil. Although most grazing, equipment movement, or even human foot traffic will compact soils to a measurable extent, only severe compaction generally reduces plant growth. Severe soil compaction reduces the water infiltration rate, soil aeration, and soil water holding capacity by collapsing both large and small soil pores. Lack of soil pore space to provide needed O_2 and water reduces plant growth. Not all soil compaction is undesirable, however. Mild to moderate compaction can benefit plants by increasing the soil water holding capacity in some soil types. During the early stages of compaction, large air-filled soil pores are collapsed to form smaller pores that hold soil water. Unless larger pores are so infrequent that poor soil aeration results, moderate compaction can actually increase plant production by increasing soil water storage. Sharrow (2007) reported that 11 yr of sheep grazing reduced the soil infiltration rate by reducing the volume of large soil pores and increased the volume of small pores, thus increasing soil water storage. Production of both Douglas-fir trees and pasture plants was higher on these moderately compacted silvopasture soils than on the uncompacted forest plantation sites nearby.

The compactness of a soil at any point in time is a balance of factors that compact soil and those that loosen it back up by forming new pore space. Physical un-compaction processes such as freezing–thawing or wetting–drying cycles and biological agents such as large and small burrowing organisms all are most active near the soil surface. This is significant because, unlike heavy machinery that compacts the soil to great depth, livestock hooves generally only compact the top 2–4 cm of soil. Such damage can be easily reversed by natural processes, while damage to tree roots may be irreversible. Sharrow (2007), for example, observed that the effects of 11 yr of sheep grazing on soil bulk density, total soil porosity, and air-filled pore space in the top 6 cm of soil in a silvopasture were erased by 2 yr of no grazing. It is unclear if high-producing silvopastures are more or less subject to soil compaction than are the generally lower producing grazed forest sites with native vegetation where stocking rates would be lower. The higher productivity of silvopastures should increase soil biological activity, while at the same time supporting more animal days of grazing with its associated greater foot traffic. Managing the intensity and length of time that animals have access to trees is critical for silvopastoral system health. Prescribed grazing management plans use water, salt, herding, and other tools to ensure that animals do not concentrate their use on small portions of the pasture. Proper animal distribution, together with avoiding grazing on heavy (clay) soils when they are saturated with water, minimizes the chance of severe soil compaction.

A primary factor for silvopasture adoption by farmers is the provision of shade and forage within the same pasture during the summer months (Orefice, Carroll, et al., 2017). Livestock find shade desirable and will actively seek it out. Unlike a solitary tree or a tree line along an open pasture, silvopastures promote greater landscape utilization and distribution by livestock through distributed shade (Karki & Goodman, 2010). This will have ramifications, not only for efficient forage utilization, but also for soil characteristics such as nutrient distribution and soil compaction. Without shade, livestock often seek water or mud sources as a means of cooling off, which may result in the degradation of sensitive environmental sites (Figure 6–3). Sheep in silvopastures spent around 97% of the entire daytime hours in the shade of hardwood trees during the summer months, compared with sheep in open pastures that had access to shade only for a small portion of the day (Pent & Fike, 2016; Table 6–2). The availability of shade alters animal behavior, and animals in shade display more restful activities (grazing, lying down) in the middle of the day compared with animals without shade. In addition to spending more time standing as a means of dissipating heat, stressed animals may display more agonistic behaviors (Améndola et al., 2016; Mitloehner & Laube, 2003). The provision of shade can have an impact on animal body temperatures, even for homeothermic mammals such as domestic livestock, which typically maintain body temperatures within a narrow range of 3°C. Ewe lambs in black walnut silvopastures in the mountains of Virginia were 0.4°C cooler than ewe lambs in open pastures during the summertime afternoon hours (Pent et al., 2018). Tree species selection may also have an impact on the cooling effect for livestock, as ewe lambs in honeylocust silvopastures were only 0.2°C cooler than ewe lambs in the open pastures at the hottest hour of the day. It is not clear whether this was a function of tree morphology, forage characteristics associated with different tree species, or both. Livestock may experience less nighttime heat loss underneath trees than directly underneath an open sky, which may be deleterious during periods of extreme heat stress. Nighttime heat loss may be as important as modulating the maximum daily temperature, but these conditions may be achieved by managing animals concurrently in a combination of open spaces and silvopastures.

Animal productivity in the establishment and open-canopied forest stages of a silvopasture generally is as good as or better than animal productivity in open pastures during the growing season. As noted above, during the establishment phase, trees have little impact on forage production and subsequent cattle production (Lehmkuhler et al., 2003). Slash pine (*Pinus elliottii* Engelm.) silvopastures with introduced warm-season grasses were stocked with cattle in Georgia 5 yr after establishment (Lewis, Burton, Monson, & McCormick, 1983). Cattle live weight gains were statistically similar 5 out of 9 yr following stocking in the open pastures and silvopastures but significantly lower in the

Fig. 6–3. Heifers in open pastures seeking relief from the heat by wallowing in mud, while heifers in loblolly pine silvopastures (background) rest in the shade.

Table 6–2. Time budgets and mean maximum core body temperatures of lambs in open pastures compared to lambs in black walnut or honeylocust silvopastures (time budgets source: Pent & Fike, 2016; core temperatures source: Pent et al., 2018).

Land use	Proportion of day spent in given activity				Max. core temperature
	Shade utilization	Lying down	Standing	Grazing	
	%				°C
Open pasture	34	6	33	61	40.3
Black walnut silvopasture	99	34	7	59	39.9
Honeylocust silvopasture	95	44	5	51	40.1

silvopastures every year following for 10 yr, mirroring the declining trends in grass productivity as a result of increasing tree shade. Cattle weight gains in pine–walnut silvopastures during the open-canopied forest stage in Missouri were similar to the gains of cattle in similar open pastures, despite nearly 20% less forage availability in the silvopastures (Kallenbach et al., 2006). Lamb weight gains have been similar whether they are stocked in open pastures or silvopastures or combinations of the two even in the mild climatic conditions of Appalachia (Neel & Belesky, 2017).Weight gains of lambs in hardwood silvopastures in Virginia were also as good as or better than weight gains of lambs in open pastures prior to thinning (Fannon et al., 2017) and again after a subsequent thinning 5 yr later (Pent & Fike, 2016). In all cases for these Appalachian systems, lower forage production in the silvopastures did not correspond with lower sheep performance. Cows and calves grazing in silvopastures created by thinning woodlots in Minnesota had similar weight gains compared with pairs grazing in open pastures (Ford et al., 2017). Similar or better weight gains for livestock in silvopastures compared with open pastures demonstrates the higher LER values possible for silvopastures.

The year-round productivity of forages and its effects on stocking rate and carrying capacity in silvopastures is less certain. Some reports have indicated that forages in silvopastures green up sooner in the spring than do those in open pastures (Kallenbach et al., 2006). This allows earlier grazing and more even forage distribution through the year. In an early-stage silvopasture, there was little difference in accumulated tall fescue [*Schedonorus arundinaceus* (Schreb.) Dumort.] forage mass in a fall stockpiled system within the alleys between tree rows (walnut between rows of pine) compared with an open pasture (Kallenbach, Venable, Kerley, & Bailey, 2010). However, the area excluded to protect the rows of trees limited the grazing area available to steers, which lowered animal output from the silvopasture by 35% relative to the open pasture. Lower forage yield in autumn under more mature (20-yr-old) walnut silvopastures (12.2- by 12.2-m tree spacing) dramatically reduced the carrying capacity for this system in a short-term (42-d) grazing experiment (Pent & Fike, 2018). There is little information available on the growth and welfare of livestock in silvopastures during the winter months, although the presence of trees may act as a windbreak, thereby reducing stress experienced by the livestock (Brandle, Hodges, & Zhou, 2004). This may improve reproductive performance and animal growth and reduce livestock mortality due to stress, particularly in extremely cold and windy climates.

Many native animals find the diverse structure of agroforests attractive. This may present both an opportunity and a challenge. Large herbivores, such as deer and elk, may make extensive use of silvopastures as feeding areas. The relatively early green-up and high nutritional value of forage in silvopastures and grazed forests makes them a better source of food for large herbivores than many native forest plant communities (Rhodes & Sharrow, 1990). Perching birds such as thrushes often use trees as observation posts from which to hunt worms and insects in the pasture areas between trees. In some cases, landowners can take advantage of such site use through the sale of hunting leases or rentals for birdwatching, but concentrated wildlife use can be a potential problem in small-sized plantations that draw animals from a large local area. Deer and elk may consume a considerable portion of the forage produced, browse young trees, and debark or break trees while rubbing the velvet from their antlers. This is of particular concern during the establishment phase of a silvopasture with young seedlings or sapling-sized trees. The terminal leader of conifer trees may be broken off when heavy bodied birds such as robins attempt to perch. Young trees may be protected from animal browsing damage by chemical deterrents or physical barriers (Bendfeldt et al., 2001; Sharrow & Fletcher, 1995). On smaller sites, three-dimensional fencing can be effective at excluding deer, but in some cases, the simplest solution may be to increase the agroforest size so that animal use does not exceed tolerable levels or to plant more trees with the expectation that damage losses over time will be less expensive than protection. In some cases, cattle may graze among newly established (and low-palatability) pines with little effect on short- or long-term tree production (Grelen, Pearson, & Thill, 1985; Pearson, Whitaker, & Duvall, 1971). In addition to physical or chemical barriers, it may prove possible to select and develop low-palatability trees (within a species) that discourage browsing (e.g., see Miller, O'Reilly-Wapstra, Potts, & McArthur, 2011). Altering animal behavior with aversion techniques may be useful in smaller, more intensive systems with small ruminants (e.g., see Burritt & Provenza, 1989; Manuelian, Albanell, Rovai, & Caja, 2016).

Forage Component

In the southeastern and northwestern United States, cool-season (C_3) forages such as clover species, perennial ryegrass (*Lolium perenne* L.), tall fescue, and orchardgrass (*Dactylis glomerata* L.) have been established as forage resources under forest canopies (Fribourg et al., 1989). Cool-season forage species are especially valued in the temperate regions of the United States because they provide winter and spring forage production as well as summer growth at higher elevations. In contrast, warm-season species are productive primarily in summer and require more light and higher temperatures than cool-season species. Bermudagrass, dallisgrass (*Paspalum dilitatum* Poir.), and bahiagrass (*Paspalum notatum* Flueggé), as well as other warm-season (C_4) grasses, have been used in silvopastures in the Southeastern, Appalachian and Delta states (Halls & Suman, 1954; Hughes, Hillmon, & Burton, 1965; Lewis et al., 1983). Good results have been achieved with Coastal bermudagrass in young southern pine agroforests (Clason, 1995), although the plant is sensitive to shading. Of the introduced warm-season grasses, bahiagrass, however, has proven to be the most successful and productive in growing beneath older trees (Burton, 1973; Lewis et al., 1983). Native warm-season grass species, including big bluestem (*Andropogon gerardii* Vitman), switchgrass (*Panicum virgatum* L.), indiangrass [*Sorghastrum nutans* (L.) Nash], and eastern gamagrass [*Tripsacum dactyloides* (L.) L.], have also been managed successfully in a silvopasture (Franzluebbers, Chappell, Shi, & Cubbage, 2017). These species may include some varieties that would have evolved within the silvopastoral-resembling pine savannas of the South.

Legumes such as subterranean clover, red clover (*Trifolium pratense* L.), and white clover (*Trifolium repens* L.) are often included in silvopastures (Figure 6–4). They are a highly nutritious food for livestock and may serve as the N source for the agroecosystem. *Rhizobium* bacteria live symbiotically in the roots of legumes. They derive their nutritional needs from the host plant and, in return, convert atmospheric N_2 gas into water-soluble forms that are available to the host. Clovers can "fix" considerable amounts of N. Although studies of N_2 fixation in silvopastures are limited, a healthy grass–clover pasture can fix >100 kg ha^{-1} yr^{-1} of atmospheric N_2 in open systems (Heichel, 1983). This is the equivalent of >200 kg ha^{-1} yr^{-1} of NH_4NO_3 fertilizer. Nitrogen present in clovers is transferred to associated grasses and trees when legume leaves, stems, and roots senesce and decompose. Grazing greatly speeds up this process. Livestock retain relatively little of the nutrients that they consume. Most nutrients pass through them and are quickly returned to the soil as urine and feces. As explained above, grazed plants generally reduce roots to re-establish an efficient root/shoot ratio. The senesced roots that are shed decompose, releasing their nutrients back into the soil nutrient pool where they can be reused. Livestock grazing, therefore, provides both an important nutrient transfer mechanism and a tool by which both aboveground and belowground organic matter dynamics may be managed.

Trees compete with forage plants for light, nutrients, and soil moisture. Competition for resources increases with decreasing distance to the tree. Water-soluble nutrients such as N and K are often taken up by plants along with water. It is very difficult, therefore, to separate their effects on plant growth. Burner and MacKown (2006) and Burner and Belesky (2008) attempted to assess competition for nutrients, light, and soil moisture independently. Tall fescue growing in the center of 2.5-m alleys between 8- to 9-yr-old loblolly pine trees had an N acquisition efficiency (kilograms of N accumulated per kilogram of N added) that was half that of fescue growing in the open when N was applied at <200 kg N ha^{-1} (Burner & MacKown, 2006), possibly because of reduced light or from tree root competition for soil moisture in silvopastures. Burner and Belesky (2008) compared the growth and physiology of tall fescue with and without irrigation in a later experiment at the same site. They concluded that differences in the photosynthetic assimilation rate had a greater effect on tall fescue growth than did soil moisture, implying that competition

Fig. 6–4. White clover forms a symbiotic relationship with rhizobial bacteria, which converts atmospheric N_2 into biologically available N for forages and trees.

for light was a significant determinant of forage productivity when a high canopy cover (approximately 70%) reduced understory light levels to <20% of that of open meadow. Silvopasture management often seeks to prevent understory light levels from declining below 50% of incident sunlight. In open-canopy stands, competition between trees and forage plants may be as much for moisture and soil nutrients as it is for light (Sharrow, 1999). This suggests that drought tolerance may be as important as tolerance to low light levels when selecting an understory forage plant for mid-rotation silvopastures in certain regions of the United States.

The degree of competition for light between trees and forages varies widely depending on tree and forage species age, morphology, phenology, and physiology, as well as on site and weather characteristics. Aboveground net primary production of cool-season forages was about 20% greater underneath medium-density, young black walnut and honeylocust silvopastures (approximately 32% full sun) compared with more open sites or approximately full sun (Buergler et al., 2005). This probably reflects changes in plant partitioning from roots to shoots for greater light capture (Belesky, 2005). However, it is unrealistic to expect forage yields under established trees to be as high as that of open pasture or rangeland. Cool-season plants using the C_3 photosynthetic pathway typically saturate with energy at about 50% of direct sunlight, although light saturation is rarely achieved in situ (Hopkins, 1999). Warm-season plants with the C_4 pathway can use much higher levels of light. Therefore, while C_3 plants such as ryegrass, bluegrass (*Poa* spp.), or orchardgrass grew as fast under 50% shade as they did in full sun, C_4 plants such as bermudagrass, big bluestem, and switchgrass produced only 66–75% as much forage under 50% shade as they did in full sun (Lin, McGraw, George, & Garrett, 2001). Once shade was increased to 80%, all plants suffered, but the C_3 grasses still produced 68–81% of full-sun yields while the C_4 grasses produced only 15–39% of full-sun yields. Studies in natural environments, however, have indicated that declining C_3 forage production begins at shade levels as low as 25% due to shading from neighboring plants or cloud cover. With C_3 forages, the negative relationship between shading and forage dry matter yield is generally a linear response (Feldhake, Neel, Belesky, & Mathias, 2005). Decreases in warm-season forage production have been reported when the tree canopy cover was minimal (Brauer, Burner, & Looper, 2004; Perry, Schacht, Ruark, & Brandle, 2009). Annual forage yields of a mixture of bermudagrass (C_4) and tall fescue (C_3) growing under 6–8-yr-old loblolly pine with measured tree canopies of 20–32% in central Arkansas were 50–60% less than those in open pastures, but bermudagrass yields declined at a greater rate than those of tall fescue (Brauer et al., 2004). Warm-season forage yields may drop off quickly as canopy cover and tree basal area increase beyond a critical threshold with decreasing light transmittance. For example, yields of 'Pensacola' bahiagrass, Coastal bermudagrass, and dallisgrass established in a 5-yr-old slash pine plantation (Hart, Hughes, Lewis, & Monson, 1970) declined by >80% as the canopy began to close 5 yr later. Data from Wolters, Martin, and Pearson (1982) relating declining forage production to increasing stand basal area found a declining curvilinear function. In Nebraska, smooth bromegrass (*Bromus inermis* Leyss.) yields decreased linearly while big bluestem yields fit an inverse polynomial model with decreasing light transmittance through Scots pine (*Pinus sylvestris* L.) (Perry et al., 2009). The critical threshold will vary with tree and forage species as well as with site characteristics. Canopy leaf area and geometry is also important in limiting understory plant access to light. The dense canopy of trees such as Douglas-fir and its tendency to retain multiple layers of branches near the ground make it a strong competitor with ground vegetation. Pruning the bottom branches from Douglas-fir can dramatically increase pasture production by allowing light to reach under the canopy as the sun angle changes during the day. Pines, in contrast, tend to have more open canopies with fewer but larger branches that project upward and that cast less shadow. Hardwood tree species differ considerably in morphology and phenology, with some species such as oaks providing a denser shade for a longer period of the year than compound-leafed or double-compound-leafed species such as black walnut and honeylocust, respectively. Besides tree species selection and managing the canopy with time for adequate light penetration, variation in forage species' light use efficiency or light saturation points could also be utilized to improve forage productivity or breed for shade tolerance in forage species. Light availability is a major factor in the tillering potential of grass species, but tillering in some species such as orchardgrass is less affected by tree shade than tillering in other species such as tall fescue (Belesky, Burner, & Ruckly, 2011). Studies indicating variation in forage species (and cultivar) light use efficiency or light saturation points suggest that these parameters could also be utilized to improve forage productivity or breed for shade tolerance in forage species.

As noted above, some studies have documented that cool-season forages begin growth earlier in the year in silvopastures than in open pastures (Kallenbach et al., 2006), probably due to the heat-trapping effects of conifers in the system. Others have seen slower spring green-up in deciduous-based silvopastures, probably a result of residual senesced leaf cover from oak trees during the autumn period (Feldhake, Neel, & Belesky, 2010). Leaf litter extracts from certain tree species have been shown to inhibit or delay grass and legume seed germination due to cations and allelochemicals present in the leachate (Halvorson, Belesky, & West, 2017). While allelochemicals can drastically affect the survival and growth of certain plant species, minerals from trees would be expected to have less of an effect on mature forage plants.

Forage nutritive value is also affected by trees. In general, shaded forages have greater crude protein and lower nonstructural carbohydrates (Buergler et al., 2006; Kephart & Buxton, 1993; Neel, Feldhake, & Belesky, 2008). Because plant crude protein levels typically are calculated as total N concentration (%) × 6.25, any NO_3 or other non-protein N present in forages would be counted in the calculation, resulting in an overestimation of actual protein (Neel et al., 2008). Forage NO_3 levels can rise with shading, although elevated concentrations probably would not be a concern for livestock except under extremely shaded environments (Neel et al., 2008) in combination with high fertility. It is thought that the maturity of forages in silvopasture is delayed relative to forages in unshaded sites and that cell size may be reduced in shaded environments, increasing the concentration of N (or calculated protein) within the cell (Kephart & Buxton, 1993). In the humid East, low energy, not low crude protein, more typically limits animal growth, particularly with cool-season forages. Thus, the increase often seen in crude protein concentrations under silvopastoral conditions may be described as an increase in nutritive value with little practical impact on animal performance. The decline in energy (often reported as total digestible nutrients) in forages from silvopastures compared with open pastures is a function of reduced total nonstructural carbohydrates and sometimes increased structural fiber content in shaded forages than in unshaded forages (Belesky, Chatterton, & Neel, 2006; Buergler et al., 2006; Neel et al., 2008; Neel, Felton, Singh, Sexstone, & Belesky, 2016). However, cooler environments (as might be expected under shade) are associated with greater total digestibility (Henderson & Robinson, 1982), which may partially offset the reductions in forage nonstructural carbohydrates. Forage chemical analyses may also be influenced by species within the sward, particularly when forages are analyzed from mixed grab samples of the most prevalent forages within the sward (Buergler et al., 2006). Time to maturity is often delayed in silvopasture systems relative to open pastures (Neel et al., 2016). Because maturity in forages largely drives forage nutritive value, forage characteristics may vary between silvopastures and open pastures at the same point in time. This phenomenon may be utilized strategically in a whole-farm pasture rotation as stocking may be delayed for silvopastures compared with open pastures without sacrificing forage quality.

Tree Component

Because forage yields decrease as tree canopies close, altering canopy closure patterns by reducing planting densities and changing spatial arrangement can substantially increase forage production in mid- to late-rotation agroforests. Planting fewer trees and aggregating trees into rows or clusters also facilitates agricultural operations (Sharrow, 1991). Trees can be planted in single widely spaced rows, in strips of double or triple rows with wider spacing between strips of trees, or in clusters. Cluster plantings may favor tree growth by quickly forming a forest microclimate within the cluster and by trees protecting each other from wind damage. However, a clustered arrangement, particularly in small fields, reduces the efficiency of fertilization, spraying, mowing, or other mechanical operations given the greater time required to maneuver between clusters. Planting trees in rows facilitates agricultural operations while greatly reducing tree impact on forage production with little immediate effect on tree growth (Sharrow, 1991). Sharrow et al. (1996) reported similar height and diameter growth of 8-yr-old Douglas-fir silvopasture trees planted 2.5 m apart in a rectangular grid as those planted in clusters of five trees each with 7.5 m between clusters.

In general, trees planted in single or double rows grow as fast as those planted at the same density (trees ha^{-1}) in conventional square grids until their canopies begin to overlap, provided that each tree has at least one side in full sun. It should be noted that stand density is averaged across the entire area planted. When trees are planted in single rows or in multiple-row sets, the density within the row or set will be higher than the average stand density. This can decrease tree performance within sets of rows unless pruning or early thinning is used to reduce intraspecific competition. Multiple row configurations such as three- or four-row sets separated by wide alleys reduce the performance of trees within the center rows once the tree canopies begin to coalesce, much as would happen in a plantation planting. Lewis, Tanner, and Terry (1985) reported that growth of slash and loblolly pine planted in double-row strips with 5.5-m spacing between strips was similar to a conventional 1.8- by 3.6-m planting. At age 13 yr, the mean height and diameter for the respective plantings were 10 and 9.5 m and 117 and 112 mm. Similar results have been reported for slash pine growing in central Florida (Ares, Brauer, & Burner, 2005). In young timber stands, interspecific competition between trees and ground vegetation is much more pronounced than is intraspecific competition with other trees. Neither tree density nor planting pattern affect tree growth until the trees' spheres of influence begin to touch. As might be expected, spatial pattern and density become more important as trees grow (Figure 6–5). At an initial tree density of 1,135 trees ha^{-1}, the basal area of 13-yr-old trees averaged 12 m^2 ha^{-1} in both single and double row configuration with alley widths of either 7.3 or 12–14 m (Ares et al., 2005). Trees in both single- and double-row plantings with similar alley widths were approximately the same size at age 18 yr. However, increasing the within-row density of either single- or double-row plantings to increase alley width (from 7.3 to 14.6 m in single-row or from 7.3 to 12.2 m in double-row plantings) reduced the tree basal area of 18-yr-old stands by approximately 20%. Ares and Brauer (2005) observed a 1.37-cm decrease in the diameter at breast height (DBH) for 19-yr-old loblolly pine trees grown in four-row configurations compared with double- or single-row configurations. This suggests the need for earlier or more aggressive thinning and pruning of stands with higher within-row density. Most of the changes in DBH were due to decreased DBH of the trees in the internal rows of the four-row configurations. Decreases in tree biomass and bole biomass were greater than changes in DBH. Reduced tree performance is not compensated for by the increased forage yield of multiple rows compared with single rows (Sharrow, 1991). Therefore, single- and double-tree rows are the most common patterns used in silvopastures.

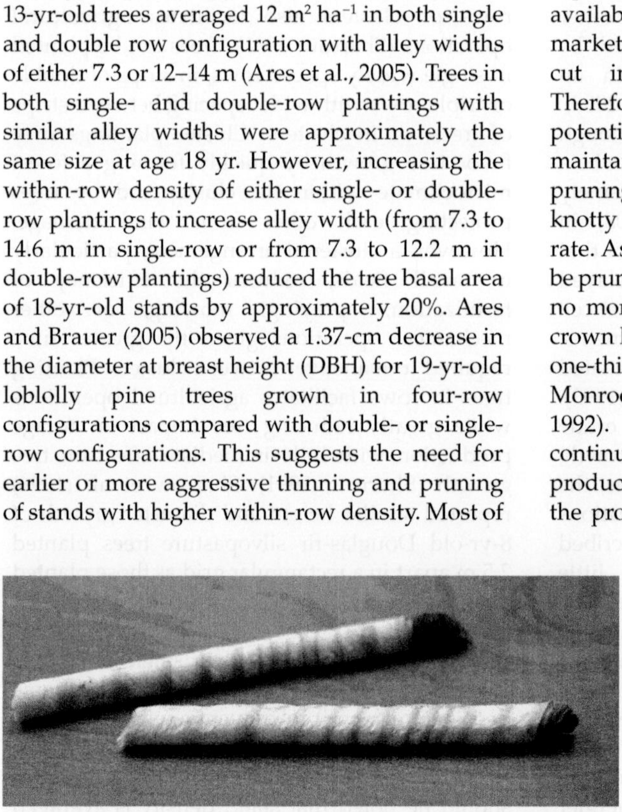

Although forage crops may benefit from low tree densities, the resulting "open grown" trees develop large branches extending close to the ground. The vascular system that supports these branches appears as knots in boards cut from these low-quality logs. It is best to prune lower limbs when they are still alive and of small diameter. When trees heal over the pruning wound, the knots associated with that former branch tend to be "tight knots" that will stay in the board. If canopies "lift" naturally by shading out lower branches that then die and partially decompose before being grown over, the resulting knots are loose and leave holes in boards cut from the tree. Historically, a premium price is paid for "clear" knot-free logs, but the capacity to capitalize on this premium will depend on the availability or proximity of these higher value markets. Clear logs can be peeled for veneer or cut into high-grade dimensional lumber. Therefore, pruning silvopasture trees has the potential both to increase log quality as well as to maintain forage production. The objective of pruning is to produce a log containing a small knotty core without reducing the tree growth rate. As silvopasture trees grow, lower limbs can be pruned in a series of canopy "lifts" that remove no more than one-third to one-half of the total crown length and maintains a live crown equal to one-third of the tree height (Fletcher, Logan, Monroe, Stephenson,& Withrow-Robinson, 1992). With conifers, pruning is generally continued to a final height of about 11 m. This produces two clear logs per tree. Log quality and the proportion of total tree biomass in the bole was greater 5 yr after pruning with 18-yr-old loblolly pine in central Arkansas (Ares & Brauer, 2005). However, less biomass was found in the stem and more in needles and branches in trees grown in silvopasture, even when pruned, compared with trees in an adjacent forest at a DBH of 25–26 cm. Sawtimber prices may or may not recover pruning costs, and the economic outcomes depend largely on the fluctuating value of knot-free logs on the world market at

Fig. 6–5. A thinning 3 yr prior to the collection of these loblolly pine cores resulted in increased growth following the development from forest to silvopasture.

the time of harvest (Fight et al., 1995). Because pruning is expensive and the removed branches pose a disposal problem, the practice often is limited to the "crop trees" that will be carried on to the final harvest age. Trees to be removed earlier during commercial thinning may not have enough growing years to produce sufficient clear wood to recover the cost of pruning and so are not pruned.

Silvopastoral trees often grow faster than trees under conventional forest management on the same site (Gibson, Clason, & Brozdits, 1994; Hughes et al., 1965; Sharrow, 1995). It is unclear how much of this increased growth can be attributed to management of competition between trees and ground vegetation and how much is the result of greater soil N status from N_2 fixation by silvopasture legumes. Silvopastures are also often fertilized with N, K, P, or S to promote herbage growth. Some of this fertilizer probably supports increased tree growth compared with unfertilized forests. Prescription grazing in the absence of fertilization has been reported to increase both the diameter and height for a number of conifer species (Sharrow, 1993, 1994), although livestock would probably speed the rate of nutrient cycling in these systems. Few studies are able to separate enhanced nutrient availability from reduced competition for soil moisture between trees and understory plants in agroforests. However, it is reasonable to assume that both factors combine to affect tree growth under grazing. Approximately half of the increased diameter growth of Douglas-fir in western Oregon silvopastures, compared with forests, occurs in the rainy spring period and half in the dry summer–fall period (Sharrow, 1995). This suggests that both increased soil fertility (in spring) and controlled competition for moisture (in summer–fall) contributed equally to the enhanced growth of agroforest trees. In more humid regions, soil nutrient availability may play a larger role in tree growth. Loblolly pine growth in silvopastures in the Coastal Plain of Louisiana responded positively to applications of poultry litter (Blazier et al., 2008).

Tree species are selected to provide specific production and service functions. Most silvopastoral research has focused on commercial indigenous conifer species grown for wood production. These include loblolly, longleaf, and slash pine in the southeastern region, and Douglas-fir and ponderosa pine (*Pinus ponderosa* Lawson & C. Lawson) in the northwestern region. Selection of populations of local trees that exhibit particularly rapid growth, good form, and adaptation to specific site conditions are a relatively cheap and effective way of improving tree performance. Widespread plants such as Douglas-fir, ponderosa pine, and loblolly pine often have genetically distinct populations (ecotypes) that are adapted to local site conditions. Moving these plants to other sites risks poor performance. For example, ponderosa pine in Oregon occurs both in the arid high desert zones of eastern Oregon and in the humid hills of western Oregon. Past attempts to use seed from eastern Oregon for afforestation of hill lands in western Oregon have largely failed because these desert trees are susceptible to disease and insect attack. Local ponderosa pine populations are more resistant to these attacks and have proven to be excellent commercial trees for hill land sites. Because of their drought tolerance, they do well on sites that are either too thin soiled or too wet for Douglas-fir. It is best if "mother trees" for seedlings be from local sites similar to where they will be used. Higher genetic potential trees such as improved selections of local cultivars or introduced new types of trees have many advantages for silvopastoral systems including a shorter establishment period (when trees may be damaged by big game and livestock), quicker harvest and return on investment, greater annual wood production increment, and increased income. However, fast growth must be sustained by rapid capture of site resources. Faster growing trees, logically, should be associated with increased demand on soil moisture and soil nutrients. This will make management of site resource partitioning between trees and forage plants in time and space more critical. Forest tree breeding programs have largely focused on gradual genetic improvements in forest tree populations instead of cultivar development for the sake of buffering against selection pressure over time (Neale & Kremer, 2011). These programs are relatively new and involve long generation times compared with agricultural crop breeding programs. However, the benefits include increased growth and improved disease resistance. Some fast-growing hybrid cultivars of pine, for example, have been rapidly realized in just several generations.

Hardwoods are used in many non-industrial silvopastoral systems. Hardwoods often require longer rotations to reach economic size. Hardwoods tend to be more palatable than conifers to both livestock and wild herbivores. This characteristic increases their risk of being damaged by livestock, although this may be manageable with appropriate grazing decisions. For example, Snell (1998) combined green ash (*Fraxinus pennsylvanica* Marsh.), American sycamore (*Platanus occidentalis* L.), and various red oak species with cool-season legumes and grasses. Initial hardwood growth and development was compatible with cattle without protection when grazing was limited to the tree crop's dormant growing season. The palatability of hardwood trees also makes them potentially valuable as a source of forage (Figure 6–6). Farmers have long used species such as willow (*Salix* spp.) and cottonwood (*Populus* spp.) as emergency feed resources, and some researchers have studied native and non-native trees for dedicated browse production (Addlestone, Mueller, & Luginbuhl, 1998; Animut et al., 2007; Burner, Pote, & Ares, 2005). Rapid establishment and production with time have been limitations in some of these studies, but bristly locust (*Robinia hispada* L.), a native, thornless, leguminous tree, was successfully established in an emulated silvopasture and provided nutritious browse for goats only 2 yr after planting (Burner & Burke, 2012). Although limited, producers have begun intentionally planting and managing species such as black locust and leucaena [*Leucaena leucocephala* (Lam.) de Wit] as browse. Some trees also produce fruits, nuts, or pods that may be utilized as additional forage resources. Honeylocust, for instance, produces large pods that can serve as a valuable source of livestock feed and have a nutritive value similar to whole ear corn (Wilson, 1991; Johnson et al., 2012, 2013). Other potential silvopasture trees include American and Chinese chestnut hybrids [*Castanea dentata* (Marsh) Borkh. × *mollissima* Blume], American elm (*Ulmus americana* L.), black walnut (*Juglans nigra* L.), cottonwood (*Populus deltoides* Bartr.), hickory (*Carya* sp.), Persian walnut (*Juglans regia* L.), persimmon (*Diospyros virginiana* L.), yellow poplar (*Liriodendron tulipifera* L.), red alder (*Alnus rubra* Bong.), pecan (*Carya illinoinensis* Wangenh), and some fruit trees (Bandolin & Fisher, 1991; Rule et al., 1994: Ares, Reid, & Brauer, 2006). Note that ash, American chestnut, American elm, and black walnut have existing pests or pathogens (emerald ash borer [*Agrilus planipennis*], Dutch elm disease [caused by *Ophiostoma* spp.], chestnut blight [caused by *Chryphonectria parasitica*], and thousand cankers disease [*Geosmithia morbida*] transmitted by the walnut twig beetle [*Pityophthorus juglandis*]) that currently limit or threaten their use, but as resistant trees are developed, silvopasture practices may offer part of a solution for bringing these trees back to their native landscapes. Climate, terrain, soils, markets, and social acceptance can influence crop tree selection, but a tree criterion unique to silvopasture design is how the tree will affect the light, water, and nutrient dynamics for the forage understory. The selected crop tree should yield high-value tree products, respond to management manipulation, and have an existing market infrastructure while maintaining favorable growing conditions for suitable forage production.

Fig. 6–6. Black locust trees planted into grass–clover pasture to fix N_2, provide forage, and produce rot-resistant fence posts.

Silvopastoral Regions

Silvopastures are intensively managed production systems. High productivity is generally required to justify the complex management of silvopastoral systems; therefore, the main silvopasture land base will probably be private, rural land. Extensive woodland grazing, on the other hand, is common on some federally owned lands. The continental United States has 737 million ha of rural land with distribution percentages for crop, range and pasture, and forested lands being 21, 36, and 30%, respectively (Bigelow & Borchers, 2017). In addition to an adequate land base, the eastern and northwestern regions of North America have mild, moist climates suited for commercial timber and livestock production. Their rich legacy of timber and livestock management practices make them an excellent choice for the development and use of silvopastures.

Southeast Region: History

In the southern United States, fire played an indispensable role in the growth and development of the forest forage resource. Low-intensity surface fires set by lightning were responsible for the open, grassy understory of longleaf pine forests (Franklin, 1997; Wright & Bailey, 1982). Native Americans managed these natural forage resources by burning to sustain forage growth and guide animal movement (MacCleery, 1992; Robbins & Wolf, 1994). European settlers introduced domesticated livestock, primarily cattle, into these fire-mediated ecosystems. Unlike the Native American spiritual sense of stewardship, early European forage utilization bordered on exploitation rather than sustainability. Fires set to "green up" the forest understory sometimes resulted in wildfire that destroyed the trees. Cattle were allowed to roam on an open-range basis, often grazing forestland not owned by the cattlemen (Healy, 1985). Uncontrolled livestock grazing created problems that included overgrazing, seedling trampling, and soil compaction. Rotational burning improved cattle performance, but heavy livestock concentration on burned forest range adversely impacted soils and reduced pine seedling growth. During the 1930s, landowners began fencing forest rangeland, equating good forestry with grazing prevention (Healy, 1985). This legacy of exploitative forestland grazing still flavors discussions between foresters and livestock graziers today. Many forest managers are reluctant to support grazing because they have had bad experiences in the past or have heard of such experiences from others.

Early range and forest scientists understood the importance of fire to southern pine woodland management. Many of the warm-season grasses native to southern range lands are quite coarse and stemmy when mature. Livestock and native herbivores such as deer tend to prefer grazing on recently burned areas because forage plants begin to grow there earlier in the spring and the new growth is not intermixed with coarse growth from past years. Prescribed burning proved useful in improving the nutritional value and maintaining the yields of native forages (Halls, 1957; Lewis & Hart, 1972). Low-intensity ground fires may also benefit southern pines by reducing disease problems and slowing the establishment of unwanted shrubs and hardwood trees (Wright & Bailey, 1982). Frequent burning, however, can deplete site productivity because much of the N in vegetation and litter is lost during combustion (Sharrow & Wright, 1977). Duvall and Whitaker (1964) found that cows and calves gained weight throughout the grazing season on rotationally burned forest rangeland. Rotational burning at 3–4-yr intervals maintained the nutritive content and palatability of native forages, removed pine litter, and suppressed competing brush. They concluded that rotational burning could be used to integrate range and timber management. However, the proper burning interval used to maintain soil fertility will vary among sites, with more productive sites tolerating shorter intervals between fires (Sharrow & Wright, 1977).

Southeast Region: Current Silvopastoral Situation

The Southeast region encompasses approximately 164 million ha across the southern and southeastern United States and includes all or a part of 14 states. The region extends from Virginia in the east, to Kentucky in the north, to the southern portion of Missouri and eastern portions of Oklahoma and Texas in the west, and to Florida in the south. More specifically, land use data (Bigelow & Borchers, 2017) from part of this region (Alabama, Florida, Georgia, and South Carolina) indicate that approximately 3% of total forest-use land is utilized for grazing. The majority of the resource occurs in the southern yellow pine, oak–pine, and oak–hickory forest cover types. Much of the region's arable land has been devoted to crop production. However, 9% of the total land currently is used for cattle and other livestock production (rangeland and pasture), while woodlands and forests (including grazed woodlands) occupy another 62% of the land. Much of the pasture and forestland is potentially suitable for silvopasture (Pearson, 1991). The region is still predominately rural, and the majority of this land is privately owned. Forestlands here are fairly productive. The majority (63%) of timber harvested in the United States in 2011 came from southern forests (Oswalt et al., 2014). This region has a long-standing tradition of low-intensity forestry and cattle grazing, which provides a firm basis for silvopastoral land use on pastures and non-industrial forest lands. Principal commercial tree species include slash pine, longleaf pine, and loblolly pine. However, many sites historically supported hardwood forests and mixed pine–hardwood stands.

Forest grazing is by far the most common form of livestock use in southern forests. This is usually a low-input, low-intensity management approach to land use. The use of planted and fertilized pasture with pine trees to form silvopastures is becoming more common throughout the region (Figure 6–7). Bandolin and Fisher (1991) cataloged numerous southern agroforestry systems that produced at least two of the following outputs: sawtimber, pulpwood, plywood, veneer, firewood, nuts, fruit, livestock, and human food. They concluded that pine–cattle grazing systems dominated southern agroforestry, with 40 million

Fig. 6–7. Recently pruned southern pine silvopasture planted in rows.

ha in the states of Alabama, Florida, Georgia, and Louisiana capable of supporting such systems. Southern land-use professionals and agroforestry producers listed a loblolly pine–grass–cattle mix as the most common silvopasture practice (Henderson & Mauer, 1993; Zinkhan, 1996). Cattle are by far the dominant livestock component (Zinkhan, 1996), although markets for small ruminants are expanding.

Under favorable climatic and soil conditions, a silvopasture is a biological, environmental, and economical approach to optimizing timber and livestock production. When the degree and timing of livestock grazing are properly controlled, southern pine woodlands may be grazed without endangering the trees (McKathen, 1980; Peebles, 1980). Pearson, Baldwin, and Barnett (1990) reported little or no damage to 1-yr loblolly and slash pines when the forage was intensively grazed for a short duration. However, high rates of mortality occurred when the forage adjoining 1-yr-old loblolly pine trees was continuously stocked. Clason and Oliver (1984) reported that Coastal bermudagrass growing in a properly managed, 30-yr-old, loblolly pine–shortleaf pine (*Pinus echinata* Mill.) forest can support a livestock grazing program while maintaining a high level of timber productivity. This system supported 7 mo of grazing (April–October) for 3.7 cow-calf grazing units ha^{-1} and produced 2 m^3 ha^{-1} of sawtimber annually for 16 yr. Longleaf and slash pine also respond positively to well-managed grazing. Lewis (1984) found that grazed longleaf pine survival was 15% less than ungrazed pine, but the grazed pines were 50% taller. Grazing reduced the level of plant competition, allowing full sunlight to reach the seedlings, thus enabling the seedlings to break out of the grass stage much earlier. Mills (1998) reported on a silvopasture in Florida where slash pine seedlings were planted in twin-row strips, 1.2 m between trees and 2.4 m between rows, with a 12.2-m open space between the strips. After the trees were planted, Pensacola bahiagrass was seeded on the open strips, and the area was cut twice for hay in the first growing season. Grazing was initiated when the trees were 3 yr old using 2.5 cow-calf grazing units ha^{-1}. The area was fertilized with broiler litter at 2 Mg ha^{-1} every 2 yr, overseeded annually with crimson clover, and limed every 3–4 yr according to soil test recommendations. This silvopasture has maintained annual cow herd conception rates at 90% and mean annual timber production at 3.4 m^3 ha^{-1}.

On the upper Coastal Plain of the southeastern United States, a silvopasture established in a 30-yr-old pine stand continuously produced timber and livestock for 25 yr (Clason & Oliver, 1984). Forage management practices combined with periodic timber harvest maintained a level of annual productivity that provided 168 d of warm-season grazing for 2.5 grazing units ha^{-1} while annually growing 2.5 m^3 ha^{-1} of sawtimber. The long-term production continuity of this silvopasture suggests that silvopastures of varying tree ages can be merged and managed on a landscape basis. There is a strong tradition of forestry and livestock production in southern woodlands. A large, well-managed contiguous forest and silvopasture land base would create a diversified commercial marketing system, which could stimulate rural economic development.

Northeast Region

The U.S. Northeast is dominated by second-growth forests that are a result of land clearing in the 18th and early 19th centuries followed by agricultural abandonment in the mid-19th and 20th centuries. The region is currently about 80% forest and predominantly hardwoods with a secondary component of conifers (Thompson, Carpenter, Cogbill, & Foster, 2013). Consistent annual rainfall and a temperate climate favors closed-canopy forests; areas that are not actively farmed typically convert back to woody vegetation. The modern forests of the region are crisscrossed with stone walls that serve as a testament to the glacial origin of its soils and its agricultural legacy.

Early European settlers in the region utilized livestock as an aid in clearing the land for agriculture. Small farms (around 50 ha) covered the landscape by the early 18th century, supplying food and clothing resources for the industrial, urban centers along the coast. The establishment of the Erie Canal and development of railroads led to a widespread abandonment of farms in the region as trade became possible with farmers in the fertile, arable lands of the Midwest. This first round of agricultural abandonment led to forests encroaching on all but the best agricultural soils. Dairies, however, remained scattered throughout the resulting agricultural landscape as refrigeration was not yet available for shipping milk great distances to urban areas. It became a common practice for farmers to fence in reforested lands and to turn dry cows, heifers, and other livestock loose into wooded paddocks without any rotational management. These areas held little forage for livestock. This practice continues to this day and results in poor forage production and degradation to tree and soil resources.

The mid-20th century witnessed a second round of agricultural abandonment in the Northeast during the second half of the century due to the boom of suburbs after World War II, high land tenure prices, and the development of convenient methods for transporting milk and other commodities from the Midwest. While deciduous forests successfully regrew in some of these abandoned agricultural fields, the invasion of alien shrubs, the foliage of which is often avoided by livestock and deer, led many of these former fields to develop into scrublands of non-native shrubs.

The working farms that remain in the region are typically on the best soils, and many are still utilizing woodlands as a place to keep livestock. Data from the 2012 USDA Census of Agriculture indicates that about one in every 2 ha of pastureland in the region is woodland pasture (Orefice & Carroll, 2017). It is unlikely that many of these woodland pastures are well managed. In fact, the same data set indicates that of 10,626 farms in New York and New England that utilize woodland pasture, only 456 of those farms also identify as practicing silvopasture or alley cropping (Orefice & Carroll, 2017). This suggests that the typical tree–livestock integration in the region is not silvopasture but wooded areas where livestock are stocked without regard for forage or tree productivity.

Increasing demand for local food and a new generation of farmers has spurred an interest in agroforestry, especially silvopasture, in the region. The dominance of woodland is a natural fit for conversion to silvopasture and there has been a growing effort to promote silvopasture as an alternative to the ecologically destructive practice of keeping livestock in closed-canopy forests (Chedzoy & Smallidge, 2011; Orefice & Carroll, 2017). The second-growth forests of the region are a natural target for silvopasture establishment because some of these areas are on fertile yet stony soils that can support high-quality, cool-season forage pastures. A study in New York (2012–2014) found that converting hardwood forest to silvopasture maintained soil quality better than converting the forest to treeless pasture (Orefice, Smith, Carroll, Asbjornsen, & Kelting, 2017). Orefice, Smith, Carroll, Asbjornsen, and Howard (2016) also contrasted the financial return of converting the forest to treeless pasture with leaving residual trees as a silvopasture, finding that the silvopasture establishment process was more financially viable than the establishment of the treeless pasture and yielded a similar amount of forage when orchardgrass was established underneath the residual trees.

Silvopasture practices in the region are diverse and driven by multiple goals, including shade for livestock, animal welfare, utilization of farm woodlots, and aesthetics (Orefice, Carroll, et al., 2017). Many of the diversified farms represented in this study were grazing sheep, cattle, and poultry under existing fruit orchards as a form of silvopasture (Figure 6–8). The same survey identified that some farmers were unfamiliar with silvopasture practices and easily confused them with any area where livestock are

Fig. 6–8. Cows grazing in an apple orchard in New England.

kept under trees. Specifically concerning were situations where farmers were keeping hogs in forested areas. Some landowners use hogs to root up invasive shrubs, yet hogs do not discriminate between desirable and undesirable woody species and can cause severe soil degradation. Other farmers attempt to mimic Mediterranean dehesa and montado silvopastures in which hogs are used to consume acorn mast, but unlike the Mediterranean systems, these northeastern U.S. farmers allow hogs access to woodlands throughout the year, even when acorns are not present, which results in significant physical damage to soils and trees.

Bringing the science of silvopasture to practitioners will be a challenge into the future as it involves, in most cases, breaking a long tradition of using farm woodlots as an unmanaged area to hold livestock without regard for potential tree or ecosystem outputs. Many misconceptions among farmers and foresters exist in the region regarding defining silvopasture and how to productively manage livestock, forage, and trees on the same unit of land (Orefice, Carroll, et al., 2017). Another challenge going forward for the Northeast will be how to effectively deal with lands dominated by invasive alien shrubs. Growing concerns over the water quality of lakes and streams in the region (Voigt, Lees, & Erickson, 2015) may begin to encourage the establishment of trees and pasture on currently tilled farmland. However, establishing trees into existing pastures is currently uncommon in the Northeast region.

Midwest Region: Pecan Practices

Silvopastoral practices are common within the floodplains of the midwestern United States, including eastern Kansas and Oklahoma and western Missouri. One such practice is an integration of native pecan trees with grasses and beef cattle. Currently in Oklahoma and Missouri, only 30% of pecan production is from improved cultivars (National Agricultural Statistics Service, 2017). This number has increased since 1991, when >90% of the pecan production in this region was from native stands (Reid & Hunt, 2000). Many of the stands in production in the past resulted from selective thinning of naturally generated forests in the years immediately after World War II. Future production will progressively come from planted and improved trees. The herbaceous understory in these systems prevents soil erosion. Grazing by cow-calf pairs typically begins in April when cool-season grass growth is vigorous and continues until forage is exhausted in late summer, about August. Understory forage composition is often a complex mixture of introduced and native grasses. Tall fescue can be a dominant plant in such systems (Ares et al., 2006). Grazing not only provides income from beef sales but reduces the need for mowing, plowing, or other forms of understory vegetation control. Reported forage yields range from 6,400 kg dry matter ha^{-1} in southern Oklahoma (Mitchell & Wright, 1991) to 2,000 kg dry matter ha^{-1} in a dry summer in southeastern Kansas (Ares et al., 2006). Native pecan trees often exhibit an alternate-year bearing pattern, with high nut yields 1 yr and lower nut yields the following year. Ares et al. (2006) reported that annual nut yields for 50–80-yr-old pecan trees were fairly predictable during a 20-yr period, averaging 360 kg ha^{-1} in an "off" year and 800 kg ha^{-1} in an "on" year. Despite a lack of change in nut yields, tree trunk diameter and stand basal area has increased with time and requires that 1–2 trees ha^{-1} be removed every 5–7 yr to prevent excessive canopy overlap. Timber from removed trees is seldom marketed. A variety of markets for this timber are available including low-value products such as fuel wood or hardwood pulpwood and high-value products such as sawlogs or veneer. The profitability of pecan silvopasture is largely determined by the income from nut sales. However, beef production provides more consistent income, thus reducing the magnitude of the effects of the alternate bearing pattern of nut production on annual income. The net present values of these silvopastures could be increased significantly by marketing wood from pruning and thinning trees to enhance nut production (Ares et al., 2006).

Northwest Region: Overview of Mixed Conifer Forest

The Northwest temperate zone includes Oregon, Washington, Idaho, Montana, and British Columbia. Natural forests range from the closed-canopy humid Douglas-fir–hemlock [*Tsugaheterophylla* (Raf.) Sarg.] coastal forests west of the Cascade mountain range to the more open-canopied interior mixed conifer forests and semiarid ponderosa pine–lodgepole pine (*Pinus contorta* Douglas ex Loudon) forests extending from the Cascades to the eastern edge of the Rocky Mountains. The two principal commercial timber trees of the region, Douglas-fir and ponderosa pine, together occupy over 28 million ha (280,000 km^2) in the Pacific Northwest and Rocky Mountain states. They are harvested for both solid wood and wood fiber products. Fuel wood has been an important subsistence product used by approximately 14% of households as a source of heat (U.S. Census, 1990), but this number has dropped to around 5% in recent years in the Northwest region (U.S. Census, 2016). Rangeland livestock production and forestry are major contributors to the largely natural-resource-based economy of the region. Approximately 5% of all privately owned non-urban land is pasture, 35% is rangeland, and 38% is forest. Understandably, by far the most prevalent agroforestry systems in the region are silvopastoral. More than half of all land in the U.S. portion of the region is federally owned, mostly managed by the U.S. Forest Service, the Bureau of Land Management, and to a lesser extent, the National Park Service. Private forest and rangelands are often intermingled with public lands. Rangeland grazing on private land is often coordinated with adjacent public lands through grazing permits or leases. Federal land management goals for multiple use and resource sustainability, including aesthetic as well as physical products, greatly impact management practices on private as well as public lands throughout the region. The vast majority (94%) of lands in British Columbia are Crown Lands and are governed through the Ministry of Forests, Lands, Natural Resource Operations and Rural Development (2011). Private lands in British Columbia are often clustered along the coastline or fertile valleys. Tenure to utilize provincial lands for agricultural use may be granted, but tenures granted for forestry use are predominant (69% of provincial land area) compared with tenures granted for agricultural use (<1% of the land area).

Cattle are the predominant livestock, with sheep being locally important, especially for use in herded bands for prescription grazing. In many areas, the distinction between range and forestland is confused by a considerable area of forested rangeland, which is managed for multiple uses including both livestock and tree production. Rangeland and pastures are frequently interspersed with forest. The rectangular land grants offered to settlers often included untillable portions, which were commonly used for livestock forage, farm woodlots, or persisted as relatively unmanaged forests. Homestead livestock were commonly allowed to forage in adjacent government forest lands. Privately owned hay meadows and rangelands are often integrated with publicly owned forests and rangelands to provide a year-round forage base for livestock. Currently, high prices being paid for timber are encouraging farmers and ranchers to reforest pastures and marginal croplands and to intensify the management of current woodlands. Low social acceptability of herbicides favors the use of non-chemical approaches to forest vegetation management, such as livestock grazing.

The most common types of integrated livestock–forestry systems in the Northwest region are: (a) integrated forest grazing in which trees are grown above native rangeland vegetation, (b) silvicultural prescription grazing in which livestock are used to facilitate tree establishment and growth, and (c) silvopastures in which livestock production is combined with commercial timber trees growing in introduced pasture. Although research experience with all three types of silvopastoralism has been accumulating since the 1950s, adoption of these systems has been slow. Considerable potential still exists for expansion of current silvopastoral technology to new users along with refinement of existing agroforestry approaches. In general, farmers and ranchers have been quicker to embrace agroforestry than have foresters, suggesting that systems that maintain forage and livestock production may be more rapidly adopted than those that focus more tightly on timber production.

When closed-canopy forests are opened up by timber harvesting, windthrow, or fire, they are capable of producing substantial amounts of ground vegetation to support both native and domestic herbivores. These early seral plant communities are referred to as "transitional ranges" by graziers, who graze them until the tree canopies again begin to exclude forage plants from the understory. Aggressive timber harvesting on both public forests and private commercial forests during the 1970s and 1980s produced a large inventory of early seral vegetation that competed with young tree regeneration for site resources. The relatively low biotic productivity of semiarid rangelands together with their highly seasonal forage production encourages large-scale livestock operations in which several different vegetation zones are integrated to provide a year-round forage base. Forested rangelands are an important link in this chain, providing green forage and shade during the summer and fall periods. In many ways, these systems mimic the migratory patterns of big game animals such as deer and elk, which follow seasonal changes of elevational vegetation zones to avoid unpleasant weather and to stay in green feed. Most interior forests and forested rangelands are managed under multiple use principles in which forage, wildlife, recreation, timber, and other natural resource values are harmonized. Whether current multiple use management is sufficiently aggressive in manipulating the interactions among components to classify it as agroforestry is a subject of debate. Examples of structurally integrated livestock–timber systems in which livestock, forage, and forest management are designed to facilitate each other are common in interior forests. Such systems clearly meet the systems perspective and purposefully managed interactions associated with agroforestry. Often, however, forested rangelands and grazed forests are primarily managed for either their forest or rangeland values. In this case, forest grazing would qualify as an agroforestry-like practice rather than as true agroforestry.

Increased areas of open-canopied conifer forests in the Interior West will increase forage production. Structurally, these native forests resemble silvopastures and should follow similar ecological principles. Stand-level investigations of tree–understory relationships on forested rangelands have shown a general reduction of forage production with increasing conifer tree basal area (Tapia, Ffolliott, & Garden, 1990) or canopy (Sibbald et al., 1994). Established trees and ground vegetation compete for both aboveground (light) and belowground (soil moisture and nutrients) site resources. Krueger (1981), however, noted that forest forage production generally does not correlate with conifer canopy cover until the average tree canopy cover exceeds 35%. Presumably, herbaceous production under dense tree canopies is limited by light, while that of younger or more open-canopied forest is reduced by competition with trees for other site resources. Several researchers have suggested that competition between large conifers and ground vegetation in open-canopied forest is primarily for soil resources (Krueger, 1981; McCune, 1986; Riegel, Miller, & Krueger, 1992). Thinning lodgepole pine stands improved forage production only when the stands were also fertilized (Lindgren & Sullivan, 2014). Soil nutrients are probably the most important factor in spring, while soil moisture dominates plant interactions in dry periods such as during summer or droughts (Riegel, Miller, & Krueger, 1991; Sharrow, 1995).

Soil resource sharing between grass and tree components is manipulated through both silvicultural and livestock management. Thinning or selectively harvesting trees to increase forage production is sometimes done, but the practice is not widespread. Most thinning is done for silvicultural reasons. Grazing is often timed to consume forage when it is green and nutritious. Most moisture use by plants is through evapotranspiration from living leaves. Grazing removes leaf area. In addition, grazed plants often shed roots to maintain an efficient root/shoot ratio (Motazedian & Sharrow, 1987). Timely grazing can reduce moisture withdrawal by grasses and shrubs, leaving more moisture for trees (Doescher et al., 1987). This is the basis for silvicultural prescription grazing in western conifer forests.

Silvicultural prescription grazing refers to grazing whose timing and intensity is designed to accomplish specific silvicultural objectives. Cattle, sheep, or goats are sometimes grazed for site preparation on harvested areas prior to replanting with trees to reduce vegetation that might impede planting crews or compete with new trees. Using cattle or sheep grazing to "release" young trees that are already on site from competing vegetation is more common. Specific recommendations and general principles for conifer release using cattle and sheep grazing have been reviewed by Sharrow (1993, 1994) and Doescher et al. (1987). Increased growth of young trees attributable to grazing has been reported for ponderosa pine, Douglas-fir, western white pine (*Pinus monticola* Douglas ex D. Don), western larch (*Larix occidentalis* Nutt.), and white spruce [*Picea glauca* (Moench) Voss]. However, reports of silvicultural grazing being ineffective in substantially increasing conifer growth are also common. Tree release by grazing is most likely to be successful when (a) livestock grazing is tightly controlled, (b) competing vegetation is reasonably palatable, (c) competing vegetation does not regrow quickly after grazing, (d) grazing occurs sufficiently early in the growing season that competing vegetation has not exhausted the soil moisture, (e) sufficient vigorous trees are present to benefit from release, and (f) the released trees are not palatable to livestock. When proper timing, intensity, duration, and class or type of livestock are applied to young conifer forests where grazable understory grasses or shrubs are competing with trees, increased tree growth of 5–10% can be achieved.

The more open-canopied inland and semiarid forests east of the crest of the Sierra Nevada–Cascade Range often produce grazable amounts of forage even when the trees are large. Periodic wildfire and fires set as a land management tool by indigenous people once consumed brush and killed conifer seedlings. This kept the forest open and park-like. Fire exclusion during the past century has allowed abundant tree reproduction, especially in the ponderosa pine and lodgepole pine zones. The resulting crowded, closed-canopy forests support little understory vegetation, and the weakly growing trees are susceptible to attack by insects. There is great interest in bringing these forests back to their former open-canopied structure through thinning and burning. Prescription grazing to stimulate the growth of grasses and to manage the growth of shrubs is a cost-effective and socially acceptable way to manage this understory vegetation for livestock production, big game habitat, and other multiple uses, particularly on productive sites. Greater species richness and diversity was evident in young, fertilized lodgepole pine rangelands where cattle were allowed to graze than in areas where cattle were excluded for 10 yr (Lindgren & Sullivan, 2012). Restored open-canopied forest should present substantial opportunities for agroforestry in both pine and mixed-conifer forests.

Northwest Region: Silvopastures

Silvopastures incorporating conifer trees with seeded grass–legume pastures are among the most intensively managed and most productive silvopastoral systems in the Northwest region (Figure 6–9). Although occasionally encountered east of the Cascade Mountains, these systems are currently most common in the valleys and foothills of western Oregon and western Washington. Most of the >1 million ha of hill land in the western Pacific Northwest is privately owned. Much of this land historically supported Oregon white oak (*Quercus garryana* Douglas ex Hook.) woodlands and savannahs. Hill lands are seldom used as croplands because of their steep slopes and shallow soils. Cattle and sheep grazing is the primary agricultural use. Large tracts of oak woodland were converted to improved pastures during the 1950s through early 1970s by felling oaks, burning, then seeding with forage legumes and perennial grasses. The resulting pastures are able to support one cow or five sheep per hectare without irrigation. Some of these lands are now being reforested as silvopastures. High current timber prices have dramatically increased landowner interest in planting conifers and justify a higher intensity of land management. Agroforestry presents opportunities to increase hill land productivity by producing both trees and livestock products, to increase the diversity of plants and animals present, and to improve cash flow by combining immediate income from grazing with later income from the sale of trees (Sharrow & Fletcher, 1995).

The original inhabitants of western Oregon were active land managers who used fire as a tool to produce grassy meadows and to keep oak woodlands open and park-like. Fire suppression in the last 150 yr has supported a successional process by which hardwood trees have invaded previously open grasslands and formerly open hardwood forests have become closed-canopy forests. Conifers, primarily Douglas-fir, grand fir [*Abies grandis* (Douglas ex D. Don) Lindl], and ponderosa pine are now beginning to overtop the canopy of hardwoods in many areas.

Apparently many hill lands will support conifer forests, but trees may be difficult to establish and growth rates are relatively slow compared with other commercial forest sites in western Oregon. Silvopastures may be successfully established in existing oak stands by thinning the oaks, then underplanting conifer trees and grass–clover pasture (Hedrick & Keniston, 1966). Young conifer trees have been observed to grow as fast under an open oak canopy as they do in clear-cut areas (Jaindl & Sharrow, 1988). Livestock grazing often increases the growth of young conifers by consuming vegetation that would otherwise compete with trees for stored soil moisture during summer droughts (Carlson et al., 1994). Retaining some large oak trees when establishing a silvopasture is an attractive option because many hill lands are near urban centers. Land use on the urban fringe must be especially sensitive

Fig. 6–9. Five-year-old Douglas-fir–subterranean clover–sheep silvopasture in western Oregon.

to environmental quality issues including environmental contamination, destruction of native plant or animal habitat, and visual appeal. Silvopastures are biologically more diverse than closed-canopy oak woodland and traditional forest or pasture monocultures. They are often park-like in appearance, and social acceptability may be higher than for traditional forest plantations or pastures.

Douglas-fir is the most common silvopasture timber tree in western Oregon and Washington, followed by ponderosa pine. Both trees are important commercial species native to the local area. Agroforests based on these components tend to be socially acceptable to local people, biologically feasible, and to have present the needed infrastructure to market and process its products. Douglas-fir and ponderosa pine are high-value products when mature, but plantations may require 50–70 yr to mature, and 30 yr or more before they produce significant income. Landowners wishing to plant trees into existing pastures or oak woodlands often lack the capital available to forest managers who have just sold a timber stand. Immediate income from livestock or hay is likely to be an important factor making agroforestry more widely acceptable than forestry, especially for lands that are currently occupied by pasture or noncommercial woodlands. Speeding up the timber crop cycle by either increasing the growth rate of native conifers or by using faster growing exotic trees is an attractive solution to problems of high initial investment followed by poor cash flow typical of timber rotations (Table 6–3). Fertilization with N, P, and S is a common practice on improved pastures but is rarely done for timber plantations. Silvopasture trees should benefit from access to both biologically fixed N_2 as well as the fertilizer nutrients applied to silvopastures.

Cool-season legumes, primarily subterranean clover or white clover, and grasses such as perennial ryegrass or orchardgrass are included as a forage crop. The extensive fibrous root system of pasture grasses makes them fierce competitors for soil moisture and nutrients in the upper soil layers. Herbaceous competition in established pastures is generally controlled during the initial 2 yr after tree planting by spraying a 1–2-m circle with herbicide around newly planted tree seedlings (Fletcher et al., 1992). In general, more vegetation control is needed to increase seedling growth than to just ensure survival of the seedling. Competition between young established trees and pastures is controlled by grazing to remove the forage canopy prior to soil moisture being exhausted (Doescher et al., 1987). When this is done, agroforest trees generally display less summer moisture stress and greater diameter growth than do nearby ungrazed forest stands (Carlson et al., 1994). In a trial near Corvallis, OR, for example, Douglas-fir silvopasture trees grew 14% faster in diameter than did adjacent forest trees during their first 4 yr (Sharrow, 1995).

The Future of Silvopastoral Systems

There is a long history of livestock grazing in North American forests. To some degree, livestock grazing under trees mimics the historic trend of native ungulates grazing vast savannah-like woodlands throughout North America (Noss, 2012). Managed grazing has evolved since European explorers and colonists introduced cattle, pigs, sheep, goats, and horses to North America. Early grazing was mostly unsupervised, with livestock having free range access to forests and woodlands. This uncontrolled grazing often resulted in damage to forest regeneration and gave livestock grazing a bad reputation with forest and woodland managers. Carefully controlled grazing now provides a tool by which livestock foraging can be used to further forest, as well as livestock, production goals. Silvopastoral systems seek to link the service and production functions of livestock, understory forage plants, and trees into a mutually supportive system of planned interactions. These functions sustainably

Table 6–3. Net cash flow per hectare and internal rate of return (IRR) of three alternative land uses for western Oregon oak woodlands. Analysis is based on a 20-yr KMX hybrid pine rotation and 10% discount rate. Costs and incomes are best estimates based on current market conditions and the experiences of local commercial agroforesters (modified from Sharrow & Fletcher, 1995).

Land use	Cash Income above Expenses					
	Years 1–5	Years 6-10	Years 11–15	Years 16-20	Total	IRR
	US$ ha^{-1}					%
KMX only	1,112	316	5,450	19,770	23,900	19
Sheep only	740	740	740	740	2,970	29
KMX + sheep	126	300	330	8,144	20,120	22

result in the marketable production of timber, fuel wood, livestock, and hunting as well as amenity values and environmental services. Environmental services such as clean air and water, scenic beauty, biodiversity, and C sequestration are already supported to some extent through tax abatement programs and cost sharing for sustainable practices. As these service values become more apparent to the general public, their willingness to pay for them should also increase. This new source of income should encourage increased application of ecologically based agricultural systems such as silvopastures, although it is clear that landowner adoption of silvopastoral practices is moving forward without strong external economic motivation due to an increasing sense of stewardship and land conservation. Independent of incentive programs, growing public and producer recognition of the challenges associated with climate change (Schoeneberger, Bentrup, & Patel-Weynand, 2017) is likely to spur development of these systems given their potential to increase system resiliency.

Approximately one-fourth of all forest-use land in the United States is currently grazed by livestock (Bigelow & Borchers, 2017). In addition, >35% of all land in the United States is managed as pasture or range for grazing. Silvopastoral management presents a sustainable framework for improving the productivity and services provided by these lands. The rising costs of land and external agricultural inputs are driving landowners to consider intensive approaches for maximizing productivity and efficiency. This may involve afforestation of pastures or expanding grazing acreage into managed woodlots. Although the area converted to date has been modest, silvopasture is a steadily increasing practice, particularly in the eastern and northwestern states. Pastures and woodlands often occur on lands that are marginal for crop production. These "secondary lands" have been under considerable development pressure as populations push out into the countryside seeking small farm and ranch residences. Silvopastures provide a socially acceptable mix of agricultural production, scenic beauty, and diversity of habitat for wildlife and livestock. Silvopastoral systems in North America will likely continue to be embraced most readily by ranchers and non-industrial forest landowners rather than by large private commercial or public land managers.

Adoption of silvopasture practices is currently hindered by landowner concern about livestock damaging trees, the economics of livestock and timber production, particularly during the establishment phase, and the complexity of managing joint production systems. Consulting foresters and other natural resource consultants often believed that grazing animals adversely affect forested ecosystems based on experiences with poorly designed livestock access to forests. Further research and technology transfer activities are needed to educate both landowners and consulting professionals that the financial risk and the risk to natural resources within a silvopasture can be quite small if such ecosystems are managed appropriately.

Agroforestry in general, and silvopasture systems in particular, are compatible with traditional agricultural practices and are favored by current economic, philosophical, and demographic trends in North America. Although the adoption of silvopasture systems has been slow, silvopastoral management has a bright future and will always find a place in modern agriculture.

References

Addlestone, B. J., Mueller, J. P., & Luginbuhl, J. M. (1998). The establishment and early growth of three leguminous tree species for use in silvopastoral systems of the southeastern USA. *Agroforestry Systems, 44,* 253–265. https://doi.org/10.1023/A:1006254812236

Améndola, L., Solorio, F. J., Ku-Vera, J. C., Améndola-Massiotti, R. D., Zarza, H., & Galindo, F. (2016). Social behaviour of cattle in tropical silvopastoral and monoculture systems. *Animal, 10,* 863–867. https://doi.org/10.1017/S1751731115002475

Animut, G., Goetsch, A. L., Aiken, G. E., Puchala, R., Detweiler, G., Krehbiel, C. R., . . . Dawson, L. J. (2007). Effects of pasture inclusion of mimosa on growth by sheep and goats co-grazing grass/forb pastures. *Journal of Applied Animal Research, 31,* 1–10. https://doi.org/10.1080/09712119.2007.9706619

Ares, A., & Brauer, D. (2005). Aboveground biomass partitioning in loblolly pine silvopastoral stands: Spatial configuration and pruning effects. *Forest Ecology and Management, 219,* 176–184. https://doi.org/10.1016/j.foreco.2005.08.042

Ares, A., Brauer, D. K., & Burner, D.M. (2005). Growth of southern pines at different stand configurations in silvopastoral practices. In K. N. Brooks & P.F. Ffolliott (Eds.), *Moving agroforestry into the mainstream: The 9th North American Agroforestry Conference Proceedings.* St. Paul, MN: Dep. of Forest Resources, University of Minnesota. Retrieved from https://www.cinram.umn.edu/sites/cinram.umn.edu/files/ares.pdf

Ares, A., Reid, W. C., & Brauer, D. (2006). Production and economics of native pecan silvopastures in central United States. *Agroforestry Systems, 66,* 205–215. https://doi.org/10.1007/s10457-005-8302-0

Arbuckle, J. G. (2009). Cattle and trees don't mix!?!: Competing agri-environmental paradigms and silvopasture agroforestry in the Missouri Ozarks. In A.J. Franzluebbers (Ed.), *Farming with grass: Achieving sustainable mixed agricultural landscapes* (pp. 116–133). Ames, IA: Soil and Water Conservation Society.

Balling, J. D., & Falk, J. H. (1982). Development of visual preference for natural environments. *Environment and Behavior, 14,* 5–28. https://doi.org/10.1177/0013916582141001

Bambo, S. K., Nowak, J., Blount, A. R., Long, A. J., & Osiecka, A. (2009). Soil nitrate leaching in silvopastures compared with open pasture and pine plantation. *Journal of Environmental Quality, 38*, 1870–1877. https://doi.org/10.2134/jeq2007.0634

Bandolin, T. H., & Fisher, R. F. (1991). Agroforestry systems in North America. *Agroforestry Systems, 16*, 95–118. https://doi.org/10.1007/BF00129742

Belesky, D. P. (2005). Growth of *Dactylis glomerata* along a light gradient in the central Appalachian region of the eastern USA: I. Dry matter production and partitioning. *Agroforestry Systems, 65*, 81–90. https://doi.org/10.1007/s10457-004-5725-y

Belesky, D. P., Chatterton, N. J., & Neel, J. P. S. (2006). *Dactylis glomerata* growing along a light gradient in the central Appalachian region of the eastern USA: III. Nonstructural carbohydrates and nutritive value. *Agroforestry Systems, 67*, 51–61. https://doi.org/10.1007/s10457-005-1112-6

Belesky, D. P., Burner, D. M., & Ruckly, J. M. (2011). Tiller production in cocksfoot (*Dactylis glomerata*) and tall fescue (*Festuca arundinacea*) growing along a light gradient. *Grass and Forage Science, 66*, 370–380. https://doi.org/10.1111/j.1365-2494.2011.00796.x

Bendfeldt, E. S., Feldhake, C. M., & Burger, J. A. (2001). Establishing trees in an Appalachian silvopasture: Response to shelters, grass control, mulch, and fertilization. *Agroforestry Systems, 53*, 291–295. https://doi.org/10.1023/A:1013367224860

Bezkorowajnyj, P. G., Gordon, A. M., & McBride, R. A. (1993). The effect of cattle foot traffic on soil compaction in a silvopastoral system. *Agroforestry Systems*, 21, 1–10. https://doi.org/10.1007/BF00704922

Bigelow, D. P., & Borchers, A. (2017). *Major uses of land in the United States, 2012* (Econ. Inf. Bull. 178). Washington, DC: USDA Economic Research Service.

Blazier, A. M., Gaston, L. A., Clason, T. R., Farrish, D. W., Oswald, B. P., & Evans, H. A. (2008). Nutrient dynamics and tree growth of silvopastoral systems: Impact of poultry litter. *Journal of Environmental Quality, 37*, 1546–1558. https://doi.org/10.2134/jeq2007.0343

Boyer, W. D. (1967). Grazing hampers development of longleaf pine seedlings in southwest Alabama. *Journal of Forestry, 65*, 336–338. https://doi.org/10.1093/jof/65.5.336

Brandle, J. R., Hodges, L., & Zhou, X. H. (2004). Windbreaks in North American agricultural systems. *Agroforestry Systems*, 61, 65–78.

Brauer, D., Burner, D., & Looper, M. (2004). Effects of tree configuration on the understory productivity of a loblolly pine–forage agroforestry practice. *American Forage Grassland Council Proceedings, 13*, 412–416.

Brunson, M. W. (1993). "Socially acceptable" forestry: What does it imply for ecosystem management? *Western Journal of Applied Forestry*, 8, 116–119. https://doi.org/10.1093/wjaf/8.4.116

Buergler, A. L., Fike, J. H., Burger, J. A., Feldhake, C. R., McKenna, J. R., & Teutsch, C. D. (2005). Botanical composition and forage production in an emulated silvopasture. *Agronomy Journal, 97*, 1141–1147. https://doi.org/10.2134/agronj2004.0308

Buergler, A. L., Fike, J. H., Burger, J. A., Feldhake, C. M., McKenna, J. R., & Teutsch, C. D. (2006). Forage nutritive value in an emulated silvopasture. *Agronomy Journal, 98*, 1265–1273. https://doi.org/10.2134/agronj2005.0199

Burner, D. M., & Belesky, D.P. (2008). Relative effects of irrigation and intense shade on productivity of alley-cropped tall fescue herbage. *Agroforestry Systems, 73*, 127–139. https://doi.org/10.1007/s10457-008-9118-5

Burner, D. M., & Burke, J. M. (2012). Survival of bristly locust (*Robinia hispada* L.) in an emulated organic silvopasture. *Native Plants Journal, 13*, 195–200. https://doi.org/10.3368/npj.13.3.195

Burner, D. M., & MacKown, C. T. (2006). Nitrogen effects on herbage nitrogen use and nutritive value in a meadow and loblolly pine alley. Crop Sci. 46:1149–1155. https://doi.org/10.2135/cropsci2005.08-0240

Burner, D. M., Pote, D. H., & Ares, A. (2005). Management effects on biomass and foliar nutritive value of *Robinia pseudoacacia* and *Gleditsia triacanthos* f. *inermis* in Arkansas, USA. *Agroforestry Systems, 65*, 207–214. https://doi.org/10.1007/s10457-005-0923-9

Burritt, E. A., & Provenza, F. D. (1989). Food aversion learning: Conditioning lambs to avoid a palatable shrub (*Cercocarpus montanus*). *Journal of Animal Science, 67*, 650–653. doi:10.2527/jas1989.673650x

Burton, G. W. (1973). Integrating forest trees with improved pastures. In R.S. Campbell and W.T. Keller (Eds.), Range resources of the southeastern United States (pp. 41–49). ASA Spec. Publ. 21. Madison, WI: ASA & CSSA.

Canada's National Forest Inventory. 2017, 1 May. Table 4.0. Area (1000 ha) of forest and non-forest land in Canada. Retrieved from https://nfi.nfis.org/resources/general/summaries/en/html/CA3_T4_FOR_AREA_en.html

Carlson, D. H., Sharrow, S. H., Emmingham, W. H., & Lavender, D. P. (1994). Plant–soil–water relations in forestry and silvopastoral systems in Oregon. *Agroforestry Systems, 25*, 1–12. https://doi.org/10.1007/BF00705702

Chedzoy, B. J., & Smallidge, P. J. (2011). *Silvopasturing in the Northeast: An introduction to opportunities and strategies for integrating livestock in private woodlands.* Ithaca, NY: Cornell Cooperative Extension.

Clason, T. R. (1995). Economic implications of silvipastures on southern pine plantations. *Agroforestry Systems, 29*, 227–238. https://doi.org/10.1007/BF00704870

Clason, T. R., & Oliver, W. M. (1984). Timber–pastures in loblolly pine stands. In M. K. Johnson (Ed.), *Proceedings of the 33rd Annual LSU Forestry Symposium* (pp. 127–137). Baton Rouge, LA: School of Forestry and Wildlife Management, Louisiana State University.

Corre, M. D., Schnabel, R. R., & Schaffer, J. A. (1999). Evaluation of soil organic carbon under forests, cool-season grasses and warm-season grasses in the northeastern U.S. *Soil Biology and Biochemistry, 31*, 1531–1539. https://doi.org/10.1016/S0038-0717(99)00074-7

Dangerfield, C. W., & Harwell, R. L. (1990). An analysis of a silvopastoral system for the marginal land in the Southeast United States. *Agroforestry Systems, 10*, 187–197. https://doi.org/10.1007/BF00122911

de Groot, P. (1990). Are we missing the grass for the trees? *New Science, 125*, 29–30.

Doescher, P. S., Tesch, S. D., & Alejandro-Castro, M. (1987). Livestock grazing: A silvicultural tool for plantation establishment. *Journal of Forestry, 85*, 29–37.

Duvall, V. L., & Linnartz, N. E. (1967). Influence of grazing and fire on vegetation and soil of longleaf–bluestem range. *Journal of Range Management, 20*, 241–247. https://doi.org/10.2307/3896259

Duvall, V. L., & Whitaker, L. B. (1964). Rotational burning: A forage management system for longleaf pine–bluestem ranges. *Journal of Range Management, 17*, 322–326. https://doi.org/10.2307/3895354

Elwood, N. E., Hansen, E. N., & Oester, P. (2003). Management plans and Oregon's NIPF owners: A survey of attitudes and practices. *Western Journal of Applied Forestry, 18*, 127–132. https://doi.org/10.1093/wjaf/18.2.127

Fannon, A. G., Fike, J. H., Greiner, S. P., Feldhake, C. M., & Wahlberg, M. A. (2017). Hair sheep performance in a mid-stage deciduous Appalachian silvopasture. *Agroforestry*

Systems, 93, 81–93. https://doi.org/10.1007/s10457-017-0154-x

Feldhake, C. M., Neel, J. P. S., & Belesky, D. P. (2010). Establishment and production from thinned mature deciduous-forest silvopastures in Appalachia. *Agroforestry Systems, 79,* 31–37. https://doi.org/10.1007/s10457-010-9289-8

Feldhake, C. M., Neel, J. P. S., Belesky, D. P., & Mathias, E. L. (2005). Light measurement methods related to forage yield in a grazed northern conifer silvopasture in the Appalachian region of eastern USA. *Agroforestry Systems, 65,* 231–239. https://doi.org/10.1007/s10457-005-1667-2

Fight, R. D., Johnson, S., Briggs, D. G., Fahey, T. D., Bolon, N. A., & Cahill, J. M. (1995). How much timber quality can we afford in coastal Douglas-fir stands? *Western Journal of Applied Forestry, 10,* 12–16. https://doi.org/10.1093/wjaf/10.1.12

Fletcher, R., Logan, R., Monroe, J., Stephenson, G., & Withrow-Robinson, B. (1992). Agroforestry in western Oregon (ORAF Rep. 12). Corvallis, OR: Benton County Extension.

Ford, M. M., Zamora, D. S., Current, D., Magner, J., Wyatt, G., Walter, W. D., & Vaughan, S. (2017). Impact of managed woodland grazing on forage quantity, quality and livestock performance: The potential for silvopasture in central Minnesota, USA. *Agroforestry Systems, 93,* 67–79. https://doi.org/10.1007/s10457-017-0098-1

Franklin, R. M. (1997). *Stewardship of longleaf pine forests: A guide for landowners.* Longleaf Alliance Rep. 2. Andalusia, AL: Longleaf Alliance, Solon Dixon Forestry Education Center.

Franzluebbers, A. J., Chappell, J. C., Shi, W., & Cubbage, F. W. (2017). Greenhouse gas emissions in an agroforestry system of the southeastern USA. *Nutrient Cycling in Agroecosystems, 108,* 85–100. https://doi.org/10.1007/s10705-016-9809-7

Fribourg, G. R., Wells, G. R., Calonne, H., Dujardin, E., Tyler, D. D., Ammons, J. T., . . . Percell, G. G. (1989). Forage and tree production on marginal soils in Tennessee. *Journal of Production Agriculture, 2,* 262–268. https://doi.org/10.2134/jpa1989.0262

Gibson, M. D., Clason, T. R., & Brozdits, G. A. (1994, November). Effects of silvipasture management on growth and wood quality of young loblolly pine. In M. B. Edwards (Ed.), Abstracts, 8th Biennial Southern Silvicultural Research Conference, Auburn AL (p. 48).

Grelen, H. E., Pearson, H. A., & Thill, R. E. (1985). Establishment and growth of slash pine on grazed cutover range in central Louisiana. *Southern Journal of Applied Forestry, 9,* 232–236.

Haile, S. G., Nair, V. D., & Nair, P. R. (2010). Contribution of trees to carbon storage in soils of silvopastoral systems in Florida, USA. *Global Change Biology, 16,* 427–438. https://doi.org/10.1111/j.1365-2486.2009.01981.x

Halls, L. K. (1957). Grazing capacity of wiregrass–pine ranges of Georgia. *Journal of Range Management, 10,* 1–5.

Halls, L. K., & Suman, R. F. (1954). Improved forage under southern pines. *Journal of Forestry, 52,* 848–851.

Halvorson, J. J., Belesky, D. P., & West, M. S. (2017). Inhibition of forage seed germination by leaf litter extracts of overstory hardwoods using in silvopastoral systems. *Agroforestry Systems, 91,* 69–83.

Hart, R. H., Hughes, R. H., Lewis, C. E., & Monson, W. C. (1970). Effect of nitrogen and shading on yield and quality of grasses grown under young slash pine. *Agronomy Journal, 62,* 285–287.

Hawken, P. (Ed.). (2017). *Drawdown: The most comprehensive plan ever proposed to reverse global warming.* New York: Penguin Random House.

Healy, R. G. (1985). *Competition for land in the American South.* Washington, DC: The Conservation Foundation.

Hedrick, D. W., & Keniston, R. F. (1966). Grazing and Douglas-fir growth in the Oregon white-oak type. *Journal of Forestry, 64,* 735–738.

Heichel, G. H. (1983). Nitrogen fixation of hay and pasture legumes. In D. B. Hannaway (Ed.), *Foothill for food and forests* (pp. 113–126). Oregon State University College of Agricultural Sciences Symposium Ser. 2. Beaverton, OR: Timber Press.

Henderson, D. R., & Mauer, T. A. (1993). Mid-South directory of agroforestry producers and researchers. Morrilton, AR: Winrock International.

Henderson, M. S., & Robinson, D. L. (1982). Environmental influences on yield and in vitro true digestibility of warm-season perennial grasses and the relationships to fiber components. *Agronomy Journal, 74,* 943–946. https://doi.org/10.2134/agronj1982.00021962007400060004x

Hiebsch, C. K., & McCollum, R. E. (1987). Area-×-equivalency ratio: A method for evaluating the productivity of intercrops. *Agronomy Journal, 79,* 15–22. https://doi.org/10.2134/agronj1987.00021962007900010004x

Holderieath, J., Valdivia, C., Godsey, L., & Barbieri, C. (2012). The potential for carbon offset trading to provide added incentive to adopt silvopasture and alley cropping in Missouri. *Agroforestry Systems, 86,* 345–353. https://doi.org/10.1007/s10457-012-9543-3

Hopkins, W. G. (1999). Introduction to plant physiology (2nd ed.). New York: John Wiley & Sons.

Houx, J. H., III, McGraw, R. L., Garrett, H. E., Kallenbach, R. L., Fritschi, F. B., & Rogers, W. (2013). Extent of vegetation-free zone necessary for silvopasture establishment of eastern black walnut seedlings in tall fescue. *Agroforestry Systems, 87,* 73–80.

Hughes, R. H., Hillmon, J. B., & Burton, G. W. (1965). Improving forage on southern pine woodlands (Paper 146). Asheville, NC: U.S. Forest Service, Southeastern Forest Experiment Station.

Hunter, A. F., & Aarssen, L. W. (1988). Plants helping plants. *BioScience, 38,* 34–40. https://doi.org/10.2307/1310644

Jaindl, R. G., & Sharrow, S. H. (1988). Oak/Douglas-fir and sheep: A three-crop silvopastoral system. *Agroforestry Systems, 6,* 147–152. https://doi.org/10.1007/BF02344751

Johnson, J. W., Fike, J. H., Fike, W. B., Burger, J. A., McKenna, J. R., Munsell, J. F., & Hodges, S. C. (2013). Millwood honeylocust trees: Seedpod nutritive value and yield characteristics. *Agroforestry Systems* 87(4):849–856. https://doi.org/10.1007/s10457-013-9601-5

Johnson, J. W., Fike, J. H., Fike, W. B., Burger, J. A., Munsell, J. F., McKenna, J. R., & Hodges, S. C. (2012). Millwood and wild-type honeylocust seedpod nutritive value changes over winter. *Crop Science, 52,* 2807–2816. https://doi.org/10.2135/cropsci2011.10.0542

Kallenbach, R. L., Kerley, M. S., & Bishop-Hurley, G. J. (2006). Cumulative forage production, forage quality and livestock performance from an annual ryegrass and cereal rye mixture in a pine walnut silvopasture. *Agroforestry Systems 66,* 43–53. https://doi.org/10.1007/s10457-005-6640-6

Kallenbach, R. L., Venable, E. B., Kerley, M. S., & Bailey, N. J. (2010). Stockpiled tall fescue and livestock performance in an early stage Midwest silvopasture system. *Agroforestry Systems, 80,* 379–384. https://doi.org/10.1007/s10457-010-9322-y

Karki, U., & Goodman, M. S. (2010). Cattle distribution and behavior in southern-pine silvopasture versus open-pasture. *Agroforestry Systems, 78,* 159–168. https://doi.org/10.1007/s10457-009-9250-x

Karki, U., Karki, Y., Khatri, R., & Tillman, A. (2018). Diurnal behavior and distribution patterns of Kiko wethers in southern-pine silvopastures during the cool-season grazing period. *Agroforestry Systems, 93*, 267–277. https://doi.org/10.1007/s10457-018-0229-3

Kephart, K. D., & Buxton, D. R. (1993). Forage quality responses of C_3 and C_4 perennial grasses to shade. *Crop Science, 33*, 831–837. https://doi.org/10.2135/cropsci1993.0011183X003300040040x

Koch, E. E., & Kennedy, J. J. (1991). Multiple-use forestry for social values. *Ambio, 20*, 330–333.

Krueger, W. C. (1981). How a forest affects a forage crop. *Rangelands, 3*, 70–71.

Lawrence, J. H., & Hardesty, L. H. (1992). Mapping the territory: Agroforestry awareness among Washington state land managers. *Agroforestry Systems, 19*, 27–36. https://doi.org/10.1007/BF00130092

Lawrence, J. H., Hardesty, L. H., Chapman, R. C., & Gill, S. J. (1992). Agroforestry practices of non-industrial private forest landowners in Washington state. *Agroforestry Systems, 19*, 37–55. https://doi.org/10.1007/BF00130093

Lehmkuhler, J. W., Felton, E. E. D., Schmidt, D. A., Bader, K. J., Garrett, H. E., & Kerley, M. S. (2003). Methods during the silvopastoral-system establishment in midwestern USA: Cattle performance and tree damage. *Agroforestry Systems 59*, 35–42. https://doi.org/10.1023/A:1026184902984

Lewis, C. E. (1984). Warm season forage under pine and related cattle damage to young pine. In M. K. Johnson (Ed.), *Proceedings of the 33rd Annual LSU Forestry Symposium* (pp. 66–78). Baton Rouge, LA: School of Forestry and Wildlife Management, Louisiana State University.

Lewis, C. E., Burton, G. W., Monson, W. G., & McCormick, W. C. (1983). Integration of pines, pastures, and cattle in south Georgia. *Agroforestry Systems, 1*, 277–297. https://doi.org/10.1007/BF00155936

Lewis, C. E., & Hart, R. H. (1972). Some herbage responses to fire on pine–wiregrass range. *Journal of Range Management, 25*, 209–213. https://doi.org/10.2307/3897057

Lewis, C. E., Tanner, G. W., & Terry, W. S. (1985). Double vs. single-row pine plantations for wood and forage production. *Southern Journal of Applied Forestry, 9*, 55–61. https://doi.org/10.1093/sjaf/9.1.55

Lin, C. H., McGraw, R. L., George, M. F., & Garrett, H. E. (2001). Nutritive quality and morphological development under partial shade of some forage species with agroforestry potential. *Agroforestry Systems, 53*, 269–281. https://doi.org/10.1023/A:1013323409839

Lindgren, P. M. F., & Sullivan, T. P. (2012). Response of plant community abundance and diversity during 10 years of cattle exclusion within silvopasture systems. *Canadian Journal of Forest Research, 42*, 451–462. https://doi.org/10.1139/x2012-003

Lindgren, P. M. F., & Sullivan, T. P. (2014). Response of forage yield and quality to thinning and fertilization of young forests: Implications for silvopasture management. *Canadian Journal of Forest Research, 44*, 281–289. https://doi.org/10.1139/cjfr-2013-0248

MacCleery, D. W. (1992). American forests: A history of resiliency and recovery (Publ. FS-540).:Washington, DC: U.S. Forest Service & Durham, NC: Forest History Society.

Manuelian, C. L., Albanell, E., Rovai, M., & Caja, G. (2016). How to create conditioned taste aversion for grazing ground covers in woody crops with small ruminants. *Journal of Visualized Experiments, 110*, e53887. https://doi.org/10.3791/53887

McCune, B. (1986). Root competition in a low-elevation grand fir forest in Montana: A trenching experiment. *Northwest Science, 60*, 52–54.

McKathen, G. (1980). The Spicer Field story. In: R. D. Child & E. K. Byington (Eds.), *Southern Forest Range and Pasture Resources: Proceedings of a Symposium, New Orleans, Louisiana* (pp. 208–211). Morrilton, AR: Winrock International.

Miller, A. M., O'Reilly-Wapstra, J. M., Potts, B. M., & McArthur, C. (2011). Field screening for genetic-based susceptibility to mammalian browsing. *Forest Ecology and Management, 262*, 1500–1506. https://doi.org/10.1016/j.foreco.2011.06.051

Miller, J. H., Manning, S. T,. & Enloe, S. F. (2010). *A management guide for invasive plants in southern forests* (Gen. Tech. Rep. SRS-131). Asheville, NC: U.S. Forest Service, Southern Research Station.

Mills, B. 1998. *Dynamic duo: Cattle and pine trees*. The Furrow, John Deere Agriculture.

Ministry of Forests, Lands, Natural Resource Operations, and Rural Development. 2010. *Crown land: Indicators & statistics report*. Victoria, BC: Province of British Columbia. Retrieved from http://www2.gov.bc.ca/assets/gov/farming-natural-resources-and-industry/natural-resource-use/land-water-use/crown-land/crown_land_indicators_statistics_report.pdf

Mitchell, R. L., & Wright, J. C. (1991). Experiences in pecan orchard floor vegetation management to stocker performances and evaluation of grazing management. *Annual Report, Northern Nut Growers Association*, 82, 72–79.

Mitloehner, F. M., & Laube, R. B. (2003). Chronobiological indicators of heat stress in *Bos indicus* cattle in the tropics. *Journal of Animal and Veterinary Advances, 2*, 654–659.

Montagnini, F., & Nair, P. K. R. (2004). Carbon sequestration: An underexploited environmental benefit of agroforestry systems. *Agroforestry Systems, 61*, 281–295.

Motazedian, I., & Sharrow, S. H. (1987). Persistence of a *Lolium perenne–Trifolium subterraneum* pasture under differing defoliation treatments. *Journal of Range Management, 40*, 232–236. https://doi.org/10.2307/3899085

National Agricultural Statistics Service. 2017, June. *Noncitrus fruits and nuts: 2016 summary*. Washington, DC: NASS. Retrieved from https://www.nass.usda.gov/Publications/Todays_Reports/reports/ncit0617.pdf

Neale, D. B., & Kremer, A. (2011). Forest tree genomics: Growing resources and applications. *National Review, 12*, 111–122.

Neel, J. P. S., & Belesky, D. P. (2017). Herbage production, nutritive value and animal productivity within hardwood silvopasture, open and mixed pasture systems in Appalachia, United States. *Grass and Forage Science, 72*, 137–153. https://doi.org/10.1111/gfs.12211

Neel, J. P. S., Feldhake, C. M., & Belesky, D. P. (2008). Influence of solar radiation on the productivity and nutritive value of herbage of cool-season species of an understorey sward in a mature conifer woodland. *Grass and Forage Science, 63*, 38–47. https://doi.org/10.1111/j.1365-2494.2007.00612.x

Neel, J. P. S., Felton, E. E., Singh, S., Sexstone, A. J., & Belesky, D. P. (2016). Open pasture, silvopasture and sward herbage maturity effects on nutritive value and fermentation characteristics of cool-season pasture. *Grass and Forage Science, 71*, 259–269. https://doi.org/10.1111/gfs.12172

Noss, R.F. 2012. *Forgotten grasslands of the South: Natural history and conservation*. Washington, DC: Island Press.

Orefice, J., & Carroll, J. (2017). Silvopasture—It's not a load of manure: Differentiating between silvopasture and wooded livestock paddocks in the northeastern United States. *Journal of Forestry, 115*, 71–72. https://doi.org/10.5849/jof.16-016

Orefice, J., Carroll, J., Conroy, D., & Ketner, L. (2017). Silvopasture practices and perspectives in the northeastern United States. *Agroforestry Systems, 91*, 149–160. https://doi.org/10.1007/s10457-016-9916-0

Orefice, J., Smith, R. G., Carroll, J., Asbjornsen, H., & Howard, T. (2016). Forage productivity and profitability in newly-established open pasture, silvopasture, and thinned forest production systems. *Agroforestry Systems, 93*, 51–65. https://doi.org/10.1007/s10457-016-0052-7

Orefice, J., Smith, R. G., Carroll, J., Asbjornsen, H., & Kelting, D. (2017). Soil and understory plant dynamics during conversion of forest to silvopasture, open pasture, and woodlot. *Agroforestry Systems, 91,* 729–739. https://doi.org/10.1007/s10457-016-0040-y

Osman, K. A., & Sharrow, S. H. (1993). Growth responses of Douglas-fir to defoliation. *Forest Ecology and Management, 60,* 105–117. https://doi.org/10.1016/0378-1127(93)90025-I

Oswalt, S. N., Smith, W. B., Miles, P. D., & Pugh, S.A. (2014). *Forest resources of the United States, 2012: A technical document supporting the Forest Service Update of the 2010 RPA Assessment* (Gen. Tech. Rep. WO-91). Washington, DC: U.S. Forest Service.

Pearson, H. A. (1991). Silvopasture: Forest grazing and agroforestry in the southern Coastal Plain. In D. R. Henderson (Ed.) *Mid-South Conference on Agroforestry Practices and Policies: Proceedings: West Memphis, Arkansas* (pp. 25–42). Morrilton, AR: Winrock International.

Pearson, H. A., Baldwin, V. C., & Barnett, J. P. (1990). Cattle grazing and pine survival and growth in subterranean clover pasture. *Agroforestry Systems, 10,* 161–168. doi:10.1007/BF00115364

Pearson, H. A., & Cutshall, J. R. (1984). Southern forest range management. In N. E. Linnartz & M. K. Johnson (Eds.), *Agroforestry in the southern United States: 33rd Annual Forestry Symposium* (pp. 36–52). Baton Rouge, LA: Louisiana Agricultural Experiment Station.

Pearson, H. A., Whitaker, L. B., & Duvall, V. L. (1971). Slash pine regeneration under regulated grazing. *Journal of Forestry, 69,* 745–746.

Peebles, H.A. 1980. Integrated forest and rangeland use. In R. D. Child & E. K. Byington (Eds.), *Southern Forest Range and Pasture Resources: Proceedings of a Symposium, New Orleans, Louisiana* (pp. 212–214). Morrilton, AR: Winrock International.

Pent, G. J., & Fike, J. H. (2016). Sheep performance, grazing behavior, and body temperatures in silvopasture systems. In American Forage & Grassland Council Annual Conference 2016, Baton Rouge, LA. Berea, KY: American Forage & Grassland Council. Retrieved from https://www.afgc.org/i4a/doclibrary/getfile.cfm?doc_id=457

Pent, G. J., & Fike, J. H. (2018). Lamb productivity on stockpiled fescue in honeylocust and black walnut silvopastures. *Agroforestry Systems, 93,* 113–121. https://doi.org/10.1007/s10457-018-0264-0

Pent, G. J., Fike, J. H., & Kim, I. (2018). Ewe lamb vaginal temperatures in hardwood silvopastures. *Agroforestry Systems.* 10.1007/s10457-018-0221-y

Perry, M. E. L., Schacht, W. H., Ruark, G. A., & Brandle, J. R. (2009). Tree canopy effect on grass and grass/legume mixtures in eastern Nebraska. *Agroforestry Systems, 77,* 23–35. https://doi.org/10.1007/s10457-009-9234-x

Reid, W., & Hunt, K. L. (2000). Pecan production in the northern United States. *HortTechnology, 10,* 298–301. https://doi.org/10.21273/HORTTECH.10.2.298

Rhodes, B. D., & Sharrow, S. H. (1990). Effect of grazing by sheep on the quantity and quality of forage available to big game in Oregon's Coast Range. *Journal of Range Management, 43,* 235–237. https://doi.org/10.2307/3898680

Riegel, G. M., Miller, R. F., & Krueger, W. C. (1991). Understory vegetation response to increasing water and nitrogen levels in a *Pinus ponderosa* forest in northeastern Oregon. *Northwest Science, 65,* 10–15.

Riegel, G. M., Miller, R. F., & Krueger, W. C. (1992). Competition for resources between understory vegetation and overstory *Pinus ponderosa* in northeastern Oregon. *Ecological Applications, 2,* 71–85. https://doi.org/10.2307/1941890

Robbins, W. G., & Wolf, D. W. (1994). *Landscape and the intermontane Northwest: An environmental history* (Gen. Tech. Rep. PNW-GTR-319). Portland, OR: U.S. Forest Service, Pacific Northwest Research Station.

Rule, L., Colletti, J., Liu, T., Jungst, S., Mize, C., & Schultz, R. (1994). Agroforestry in the midwestern United States. In R. C. Schultz & J. P. Colletti (Eds.), Opportunities for agroforestry in the temperate zone worldwide: Proceedings of the 3rd North American Agroforestry Conference, Ames IA (pp. 251–256). Ames, IA: Department of Forestry, Iowa State University.

Schoeneberger, M. M., Bentrup, G. & Patel-Weynand, T. (Eds.). 2017. *Agroforestry: Enhancing resiliency in U.S. agricultural landscapes under changing conditions* (Gen. Tech. Rep. WO-96). Washington, DC: U.S. Forest Service. https://doi.org/10.2737/WO-GTR-96

Sharrow, S. H. (1991). Tree planting pattern effects on forage production in a Douglas-fir agroforest. *Agroforestry Systems, 16,* 167–175. https://doi.org/10.1007/BF00129747

Sharrow, S. H. (1993). Animal grazing in forest vegetation management: A research synthesis. In T. B. Harrington & L. A. Parendes (Eds.), Forest vegetation management without herbicides: Proceedings of a workshop held February 18-19, 1992 (pp. 53–60). Corvallis, OR: Forest Research Laboratory.

Sharrow, S. H. (1994). Sheep as a silvicultural management tool on temperate conifer forest. *Sheep & Goat Research Journal, 10*(1), 97–104.

Sharrow, S. H. (1995). Agroforestry: Growth of trees planted in hill pasture. In *Abstracts, 48th Annual Meeting, Society for Range Management, 1995 January 14–20, Phoenix, AZ* (p. 58). Denver, CO: Society for Range Management.

Sharrow, S. H. (1999). Silvopastoralism: Competition and facilitation between trees, livestock, and improved grass–clover pastures on temperate rainfed lands. In L. E. Buck, J. P. Lassoie, & E. C. M. Fernandes (Eds.), *Agroforestry in sustainable agricultural systems* (pp. 111–130). Boca Raton, FL: CRC Press.

Sharrow, S. H. (2004). Cavitation may explain winter damage to rangeland vegetation. *The Grazier, 321,* 5–7 June. Retrieved from http://www.doctorrange.com/PDF/cavitation.pdf

Sharrow, S. H. (2007). Soil compaction by grazing livestock in silvopastures as evidenced by changes in soil physical properties. *Agroforestry Systems, 71,* 215–223. https://doi.org/10.1007/s10457-007-9083-4

Sharrow, S. H., Carlson, D. H., Emmingham, W. H., & Lavender, D. (1996). Productivity of two Douglas- fir/subclover/sheep agroforests compared to pasture and forest monocultures. *Agroforestry Systems, 34,* 305–313. https://doi.org/10.1007/BF00046930

Sharrow, S. H., & Fletcher, R. A. (1995). Trees and pastures: 40 years of agrosilvopastoral experience in western Oregon. In W. J. Rietveld (Ed.), Agroforestry and sustainable systems, Symposium proceedings (Gen. Tech. Rep. RM-GTR-261, pp. 47–52). Fort Collins, CO: U.S. Forest Service, Rocky Mountain Forest and Range Experiment Station.

Sharrow, S. H., & Ismail, S. (2004). Carbon and nitrogen storage in agroforests, tree plantations, and pastures in western Oregon USA. *Agroforestry Systems, 60,* 123–130. https://doi.org/10.1023/B:AGFO.0000013267.87896.41

Sharrow, S. H., Leininger, W. C., & Osman, K. A. (1992). Sheep grazing effects on coastal Douglas-fir growth: A ten-year perspective. *Forest Ecology and Management, 50,* 75–85. https://doi.org/10.1016/0378-1127(92)90315-Z

Sharrow, S. H., & Mosher, W. D. (1982). Sheep as a biological control agent for tansy ragwort. *Journal of Range Management, 35,* 480–482. https://doi.org/10.2307/3898610

Sharrow, S. H., & Wright, H. A. (1977). Proper burning intervals for tobosagrass in West Texas based upon nitrogen dynamics. *Journal of Range Management, 30,* 343–346. https://doi.org/10.2307/3897717

Shrestha, R. K., & Alavalapati, J. R. R. (2004). Valuing environmental benefits of silvopasture practice: A case study of Lake Okeechobee watershed in Florida. *Ecological Economics*, *49*, 349–359. https://doi.org/10.1016/j.ecolecon.2004.01.015

Sibbald, A. R., Griffith, J. H., & Elston, D. A. (1994). Herbage yield in agroforestry systems as a function of easily measured attributes of tree canopy. *Forest Ecology and Management*, *65*, 195–200. https://doi.org/10.1016/0378-1127(94)90170-8

Snell, T. K. (1998). Oklahoma projects combine timber production with grazing. *The Temperate Agroforester*, *6*(1), 1,6–8.

St-Pierre, N. R., Cobanov, B., & Schnitkey, G. (2003). Economic losses from heat stress by US livestock industries. *Journal of Dairy Science*, *86*, E52–E77. https://doi.org/10.3168/jds.S0022-0302(03)74040-5

Tapia, L. A. B., Ffolliott, P. F., & Garden, D. P. (1990). Herbage production–forest overstory relationships in two Arizona ponderosa pine forests. *Journal of Range Management*, *43*, 25–28. https://doi.org/10.2307/3899114

Thompson, J. R., Carpenter, D. N., Cogbill, C. V., & Foster, D. R. (2013). Four centuries of change in northeastern United States forests. *PLOS ONE*, *8*(9), e72540. https://doi.org/10.1371/journal.pone.0072540

Trozzo, K. E., Munsell, J. F., & Chamberlain, J. L. (2014). Landowner interest in multifunctional agroforestry riparian buffers. *Agroforestry Systems*, *88*, 619–629. https://doi.org/10.1007/s10457-014-9678-5

USDA. 2016. *Beef 2017: Information needs assessment survey results for the upcoming NAHMS beef 2017 study*. Washington, DC: USDA National Animal Health Monitoring System. Retrieved from https://www.aphis.usda.gov/animal_health/nahms/beefcowcalf/downloads/beef2017/Beef2017_NeedsAssess_1.pdf

U.S. Census. 1990. *Census of population, United States, social and economic characteristics*. Washington, DC: U.S. Department of Commerce.

U.S. Census. 2016. *American community survey 5-year estimates*. Washington, DC: U.S. Department of Commerce.

Vandermeer, J. (1981). The interference production principle: An ecological theory for agriculture. *BioScience*, *31*, 361–364. https://doi.org/10.2307/1308400

Voigt, B., Lees, J., & Erickson, J. (2015). *An assessment of the economic value of clean water in Lake Champlain* (Tech. Rep. 81). Grand Isle, VT: Lake Champlain Basin Program.

Wilson, A. A. (1991). Browse agroforestry using honeylocust. *The Forestry Chronicle*, *67*, 232–235.

Wolters, G., Martin, A., & Pearson, H. A. (1982). Forage response to overstory reduction on loblolly–shortleaf–hardwood forest range. *Journal of Range Management*, *35*, 443–446. https://doi.org/10.2307/3898601

Workman, S. W., Bannister, M. E., & Nair, P. K R. (2003). Agroforestry potential in the southwestern United States: Perceptions of landowners and extension professionals. *Agroforestry Systems*, 59, 73–83.

Wright, H. A., and A. W. Bailey. (1982). *Fire ecology: United States and southern Canada*. New York: John Wiley & Sons.

Zinkhan, F. C. (1996). Public land-use professionals' perceptions of agroforestry applications in the South. *Southern Journal of Applied Forestry*, *20*, 162–168. https://doi.org/10.1093/sjaf/20.3.162

Zinkhan, F. C., & Mercer, D. E. (1996). An assessment of agroforestry systems in the southern USA. *Agroforestry Systems*, *35*, 303–321. https://doi.org/10.1007/BF00044460

Study Questions

1. Why are silvopastoral systems the most common types of agroforestry in North America?

2. Define *silvopastoral system*, *integrated forest grazing*, and *silvopasture*. How do these practices differ?

3. Describe potential tree arrangements when establishing trees in pastures. What are the advantages and disadvantages of these planting configurations?

4. What are the three management stages of a silvopasture? How does management within each stage differ depending on the objectives for the site and existing site conditions?

5. How do the ecological concepts of *biological amplitude* and *ecological amplitude* apply to selecting tree, forage, and livestock components of silvopastoral systems?

6. Describe how trees and forages compete for resources in a silvopasture. How do these competitive interactions change seasonally or as the trees mature? How may livestock affect these interactions?

7. Based on surveys conducted throughout the United States, why do landowners adopt silvopasture practices?

8. Cool-season (C_3) plants are more often utilized in grazing systems in more northern regions of the United States, while warm-season (C_4) plants are more often utilized in southern pasture systems. Explain how tree–forage competition for light should be managed differently when growing these two types of forages in silvopastures. In what ways will managing trees together with these types of forages alter the expected production potential or suitability of these forages in a particular region?

9. How does soil compaction affect the rooting ability of plants? How can silvopasture managers reduce the risk of soil compaction in silvopastures?

10. What economic and social reasons may explain why farmers and ranchers are more likely to accept silvopastoral practices than are foresters?

11. Pruning trees may be a more common practice in silvopastoral systems than in commercial forests. What specific goals do silvopasture managers seek to achieve by pruning trees? What specific guidelines are followed to see that trees are not damaged by pruning?

12. Carbon sequestration is a type of ecosystem service provided by silvopastures. Explain why silvopastures may accrete (sequester) more C than either pastures or forests growing separately on the same site.

13. The major commercial timber trees used in the United States are conifers. Identify the two most common silvopasture trees utilized in the southeastern region and the two most common silvopasture trees utilized in the northwestern region.

14. There is growing interest in utilizing hardwood trees in silvopastures in certain regions of the country. Explain the advantages and disadvantages of managing hardwood trees in a silvopasture.

15. Livestock may be both a product and management tool for the silvopasture manager. Explain the benefits that silvopasture practices may provide to livestock, as well as the benefits or negative impacts that livestock may have on trees.

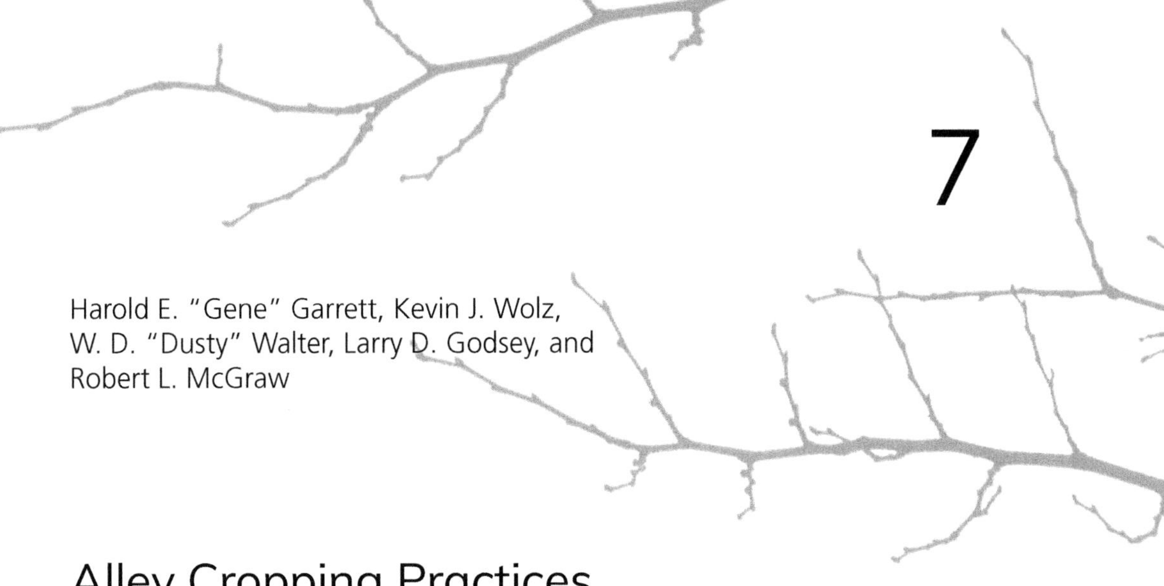

7

Harold E. "Gene" Garrett, Kevin J. Wolz,
W. D. "Dusty" Walter, Larry D. Godsey, and
Robert L. McGraw

Alley Cropping Practices

Agriculture across the temperate zone has resulted in many negative environmental consequences. From the loss of soil productivity, to the devastating effects of applying increasing rates of agrochemicals, such consequences indicate that innovative agricultural practices are needed (Amel, Manning, Scott, & Koger, 2017; Goucher, Bruce, Cameron, Koh, & Horton, 2017; Montgomery, 2007; Trudge, 2016). While alley cropping agroforestry practices may be only part of the solution, they can play a pivotal role in creating ecosystem services and production benefits that complement traditional agriculture.

Alley cropping is defined as the planting of crops between widely spaced rows of trees (Garrett et al., 1994; Gold & Garrett, 2009). While alley cropping is a common practice in parts of the tropics, is has only recently come into use in North America (Wolz & DeLucia, 2018). In contrast to its use in the tropics, temperate alley cropping has emphasized the production function of trees (e.g., timber, nuts, fruit) rather than their role in enhancing soil fertility. High-value hardwoods are the most commonly used trees in temperate alley cropping, in combination with row, forage, horticultural, and specialty crops. Tree–crop combinations in alley cropping are developed with both production and ecological benefits in mind (Smith, Pearce, & Wolfe, 2013) and can be highly diverse (Gordon et al., 2008; Morhart et al., 2014).

Tree and crop yields in alley cropping are typically, though not always, less than when the same plants are grown in monoculture (Beaudette et al., 2010; Nasielski et al., 2015). However, the combined value of trees and crops per unit area commonly increases relative to component monocultures. This results from synergistic interactions between trees and crops, leading to increased land-use efficiency and environmental benefits (Tsonkova, Böhm, Quinkenstein, & Freese, 2012). Furthermore, alley cropping provides the opportunity for landowners to enhance long-term profitability and diversify farm enterprises.

History of Alley Cropping

As with most agroforestry practices, various forms of alley cropping are ancient practices and were widely utilized around the world. However, these practices have declined with the trend to remove trees from agricultural landscapes, especially during the last century (Eichhorn et al., 2006; Nerlich, Graeff-Hönninger, & Claupein, 2013). A resurgence of alley cropping is thought to have its roots in the taungya system of forest management in Southeast Asia (Jordan, Gajaseni, & Watanabe, 1992). In exchange for tending teak

North American Agroforestry, Third Edition. Edited by Harold E. "Gene" Garrett, Shibu Jose, and Michael A. Gold.

plantations, workers had the opportunity to grow food crops between the trees. A major difference, however, between the taungya system and modern temperate alley cropping is that, in the latter, tree spacing is designed specifically to accommodate the alley crop's needs.

For much of the 19th and 20th centuries, a range of alley cropping practices were viewed primarily as an inexpensive means of establishing forests. However, beginning in the 1960s and 1970s, recognition of the benefits of growing trees and crops together to address hunger and ecological degradation expanded in many parts of the world. It was at this time, in the mid-1970s, that the Food and Agriculture Organization (FAO) of the United Nations decided to focus its attention in forest management development on the beneficial interactions that were being reported by researchers when trees and food crops were integrated into a single land-management system. Furthermore, with the establishment of the International Council for Research in Agroforestry (now the World Agroforestry Centre in Nairobi, Kenya), a broader movement began to advance the knowledge and the adoption of alley cropping.

During the next few decades, momentum around temperate agroforestry continued to grow, primarily in the form of experimental and demonstration plantings at universities around the world (Figure 7–1). Some of the most important temperate alley cropping research and demonstration sites during the last 30 yr include (Wolz & DeLucia, 2018):

- University of Missouri Center for Agroforestry & Hammons Products Co. (Stockton, MO, USA; established 1975): black walnut (*Juglans nigra* L.) with 12-m alleys within which row crops such as maize (*Zea mays* L.), soybean [*Glycine max* (L.) Merr.], and milo [*Sorghum bicolor* (L.) Moench] were double-cropped with wheat (*Triticum aestivum* L.) (Garrett, Walter, & Godsey, 2011)
- University of Guelph Agroforestry Research Station (Guelph, Canada; established 1988): a 30-ha trial alley cropping maize, soybean, and wheat between a diverse mix of 10 tree species (Thevathasan & Gordon, 2004)
- Domaine de Restinclieìres (Montpellier, France; established 1995): a wide range of alley cropping plots focusing primarily on alley cropping small grains between high-value hardwood trees (Dupraz et al., 2000; Lovell et al., 2018)
- University of Missouri Greenley Memorial Research Center (Novelty, MO, USA; established 1997): oak (*Quercus* spp.) alley cropped with maize and soybean as part of a paired watershed experiment comparing monoculture row crops and various perennial buffers (Udawatta, Krstansky, Henderson, & Garrett, 2002).

Most recently, interest in temperate alley cropping has continued to expand with both researchers and practitioners. Research and demonstration sites are being established at new locations, such as the University of Illinois Urbana-Champaign (Urbana, IL, USA; Lovell et al., 2018) and the Silva Tarouca Research Institute (Brno, Czech Republic; Houška, unpublished data, 2020), and farmer-led networks are driving alley cropping innovation and adoption, such as the Savanna Institute (www.savannainstitute.org) in the U.S. Midwest and the Association Française d'Agroforesterie (www.agroforesterie.fr) in France. Increasing emphasis is being placed on alley cropping systems with multiple tree species and strata (Lovell et al., 2018; Wolz, Lovell, et al., 2018), tree crops for food and fodder production (Revord, Lovell, Molnar, Wolz, & Mattia, 2019; Wolz, Lovell, et al., 2018), C sequestration potential (Chatterjee, Nair, Chakraborty, & Nair, 2018; Stefano & Jacobson, 2017), profitability at large scales (Wolz & DeLucia, 2019a), leveraging biophysical models to explore difficult questions (Dupraz et al., 2019;

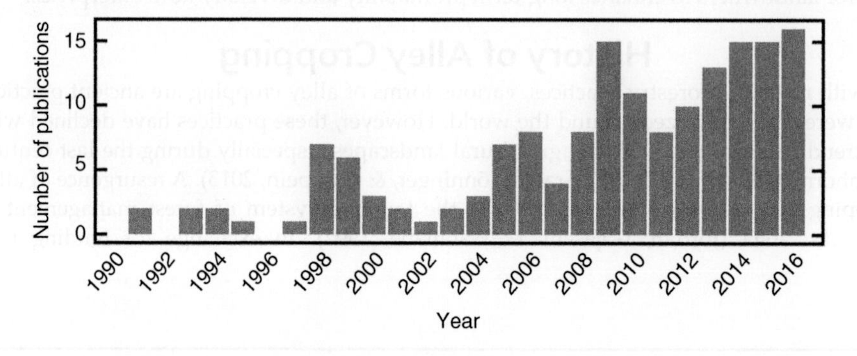

Fig. 7–1. Historical trend of peer-reviewed publications on temperate alley cropping field experiments (data from Wolz & DeLucia, 2018).

Luedeling et al., 2016), and social dimensions (Brown, Miller, Ordonez, & Baylis, 2018; Keeley et al., 2019; Mattia, Lovell, & Davis, 2018).

Tree Component

Tree Species Selection

The first and perhaps most important decision in designing an alley cropping practice is the selection of the tree species (Udawatta, Garrett, & Kallenbach, 2011). A combination of physical and economic criteria is typically used. Desirable tree characteristics vary depending on the goals and priorities of the user (Garrett, McGraw, & Walter, 2009; Molnar, Kahn, Ford, Funk, & Funk, 2013; Reisner, de Filippi, Herzog, & Palma, 2007). However, the species must (a) adapt to the site, (b) have high economic value, and (c) have morphological characteristics that create suitable microenvironments for the growth of alley crops (Jose, 2011; Udawatta, Nygren, & Garrett, 2005; Van Sambeek, Godsey, Walter, Garrett, & Dwyer, 2016). In some cases, a mixture of species can be considered (Wolz, Lovell, et al., 2018). While deciduous and coniferous species are suitable for alley cropping, research has emphasized deciduous species (Figures 7–2 and 7–3).

Desirable Characteristics of Trees in Alley Cropping

For many landowners, the financial benefits of alley cropping are often a top priority. Trees are perceived by many as a medium- or long-term investment, while the alley crop is viewed as a source of immediate income. Tree species that produce a marketable product (e.g., fruit, nut, floral green, chemical, etc.) other than wood, however, provide income within a shorter time and thus can be more financially appealing than species with timber value only (Mattia, Lovell, & Davis, 2018). However, tree species used in alley cropping are also often selected based on non-economic criteria, such as ecological and social factors (Alam et al., 2014; Tsonkova et al., 2012; Wilson & Lovell, 2016).

Meeting the physiological requirements of alley crops is critical in any alley cropping system (Ehret, Graß, & Wachendorf, 2015; Senaviratne, Udawatta, Nelson, Shannon, & Jose, 2012). The vigor and yields of the alley crop depend on the morphological characteristics of the tree species, so crown and foliage characteristics are important considerations in accommodating alley crop light requirements. Many grain and forage crops are shade intolerant and may not perform well in

Fig. 7–2. Photos of alley cropping with various trees and crop species: (a,b) how the complementary phenology of walnut and wheat can lead to minimal competition between these two species when mixed together (photos from Wolz & DeLucia, 2019b); (c) a wide alley used for hay production; and (d) a rare example of alley cropping with grape, a woody perennial crop (photo credit: Kevin Wolz).

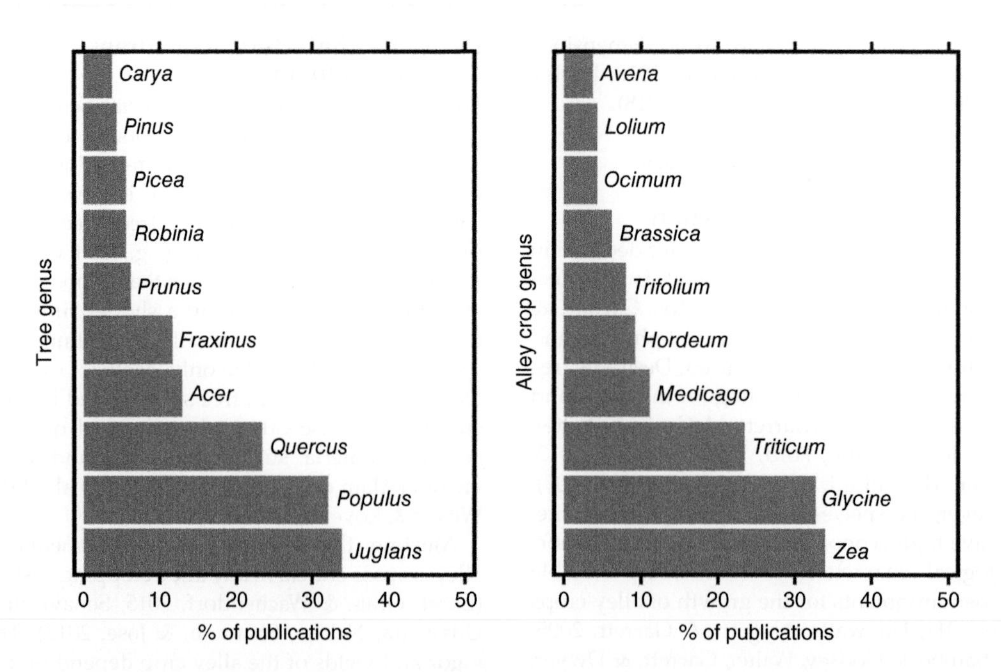

Fig. 7–3. Frequency of genera utilized in the tree component (left panel) and alley crop component (right panel) of temperate alley cropping field experiments (data from Wolz & DeLucia, 2018).

association with trees. While wide tree spacing will help, selecting species with characteristics that optimize the light environment under the canopy is also important. Some species, for example black walnut, by nature have small, sparse foliage and lend themselves well to an alley cropping regime (Andrianarisoa, Dufour, Bienaime, Zeller, & Dupraz, 2016; McGraw, Stamps, & Linit, 2005; Wolz & DeLucia, 2019a). However, most trees do not. If the agronomic crop component matures in the early spring, such as winter wheat, or heads out in the late fall such as grain sorghum [*Sorghum bicolor* (L.) Moench], a tree species should be incorporated that best accommodates the light needs of that specific crop, namely, a tree species that breaks dormancy late for winter wheat (Figures 7–2a and 7–2b) or a species that drops its leaves early for sorghum.

Other important tree foliage characteristics are the rate of leaf litter decomposition, acidity, and the presence of allelochemicals. Unless emphasis is on erosion control, a species whose foliage decomposes rapidly is desirable. Rapid decomposition avoids problems that can result from the alley crop being covered by leaves and enhances nutrient recycling (Zamora, Jose, & Napolitano, 2009). To maximize the health and vigor of the alley crop, tree species known to produce allelochemicals harmful to the alley crop of choice should be avoided. While some potential tree species have allelochemical properties (e.g., black walnut—Jose, 2002), an understanding of the sensitivity of the intercrop species to the allelochemical can still lead to a successful choice of alley crops (Cardinael, Mao, et al., 2015; Singh, Kohli, & Batish, 2001).

Competition for water between the tree and intercrop species not only affects the yields of the alley crop but also the growth of the trees (Jose, Gillespie, & Pallardy, 2004; Jose, Gillespie, Seifert, & Biehle, 2000; Wanvestraut, Jose, Nair, & Brecke, 2004). Because the vertical distribution of root systems varies among species, deep-rooted species that have a reduced volume of roots near the surface and yet a sufficient number to help with erosion control are desirable. As an example, Bouttier et al. (2014) found that when red oak (*Quercus rubra* L.) was intercropped with hay species, the oak allocated fewer fine roots to the top 10 cm of soil and more to depths between 10 and 30 cm, allowing the hay to have a greater proportion of its fine roots in the upper 10 cm. This arrangement provides the best opportunity to minimize competition for water between the tree and crop components and capture nutrients that might otherwise be lost to leaching (Allen, Jose, Nair, Brecke, & Ramsey, 2004) without using lateral root-pruning procedures (Bouttier et al., 2014; Wanvestraut et al., 2004).

Wildlife benefits are also an important consideration in the selection of a tree species. Linear plantings, as found in alley cropping, can serve as protective corridors for wildlife movement and the raising of young, as well as providing a food source. Incorporating fruit-bearing species can provide numerous wildlife benefits (Christoffel, 2013; Gibbs et al., 2016).

Non-Dinitrogen-Fixing Deciduous Species

Throughout the midwestern United States and parts of Canada (e.g., Ontario), many of the alley cropping tree species in use today are high-value hardwoods (Figure 7–3; Morhart et al., 2014). Perhaps the most widely planted species is black walnut (Wolz & DeLucia, 2018). Black walnut clearly represents an ideal alley cropping species (Garrett & Harper, 1999). Markets exist for both the wood and nuts. Moreover, because of the unique characteristics of the wood (i.e., color, workability, etc.) it is currently one of the most valuable tree species in North America.

Black walnut has growth and shade characteristics that embrace the production of many potential alley cropping species. It often has a foliage period of <135 d, being one of the last species to leaf out in the spring and one of the first to defoliate in the fall. The additional days of full sunlight in the spring are of great value to alley crops that ripen in spring, such as winter wheat or barley (*Hordeum vulgare* L.). Early foliage loss in late summer results in greater availability of sunlight and moisture for warm-season alley crop species. Even with full foliage, black walnut admits sufficient light to satisfy the needs of many forage legumes, cool-season grasses, and herbaceous specialty crops, including ginseng (*Panax quinquefolium* L.) and goldenseal (*Hydrastis canadensis* L.).

Black walnut's root system is ideally suited to alley cropping. While walnut grows best on deep, well-drained soils with a neutral pH, root characteristics are similar across a wide range of soil conditions (Pham, Yen, Cox, & Garrett, 1977). Young walnut typically produces a deep taproot that will penetrate more than 2 m in the absence of bedrock, claypan, or other physical barriers. The long lateral roots stay close to the surface, but most of the smaller roots turn down sharply into the soil (Pham et al., 1977; Yen, Pham, Cox, & Garrett, 1978). Feeder roots are normally concentrated in deeper soil layers, leaving a surface zone within which alley crops can develop their own root systems.

Unfortunately, black walnut is a known producer of an allelochemical (juglone), and some restrictions on the choice of companion species may result. Some undesirable (low-quality) native forage species, however, are unable to become established and grow under walnut, thus reducing competition and allowing other more desirable species to grow uninhibited (Brooks, 1951; Jose, 2002).

Pecan [*Carya illinoensis* (Wanganh.) K. Koch] is another tree species that is widely planted and well suited to alley cropping (Allen et al., 2005; Fletcher, Thetford, Sharma, & Jose, 2012). This species is especially popular throughout the southern states where long-established tree improvement programs have developed numerous high-yielding cultivars (Sparks, 1992; Thompson & Madden, 2003). While the natural range of pecan extends into northern Missouri and Illinois, the limited availability of improved selections has restricted its planting. Only in the recent past has in-depth research been undertaken to develop selections suitable for colder climates (Reid & Hunt, 2003; Thomas & Reid, 2006).

Pecan, like walnut, produces both wood and nut crops. Nut crops are the more valuable, however. Nut yields in an alley cropping system may be 2000 kg ha^{-1} or more (Reid & Olcott-Reid, 1985). While pecan produces more dense foliage than walnut, its canopy transmits sufficient light to support row crops or ground covers if trees are efficiently spaced and arranged. Pecan is not known to be a significant producer of allelochemicals; therefore, a greater variety of alley crops might be grown than with black walnut.

An unproven yet potentially good nut-bearing species for alley cropping is chestnut (*Castanea* sp.). Natural populations of the American chestnut [*Castanea* dentata (Marsh.) Borkh.] once grew throughout the eastern half of the United States but were eliminated by the chestnut blight [*Endothia parasitica* (Murr.) Barr.]. Reliable selections of the more resistant Chinese (*Castanea mollissima* Bl.) and Japanese chestnuts (*Castanea crenata* Sieb. & Zucc.) are available, together with some American hybrids (Miller, Miller, & Jaynes, 1996; Miller, 2003). Because disease-resistant trees of the pure American type are unavailable, although some promising hybrids have been developed, and the Asiatic (Chinese and Japanese) selections tend to produce logs of low value, the greatest current interest in this genus is for nut production (Aguilar, Cernusca, & Gold, 2009, 2010; Payne, Jaynes, & Kays, 1983). Chinese chestnut produces a nut that is low in fat, free of cholesterol and gluten, and high in vitamin C and complex carbohydrates. While numerous researchers have discussed the problems of establishing American markets for chestnut (Gold, Cernusca, & Godsey, 2005, 2006; Stebbins, 1990), the potential for market development is unquestionable and is being realized in some regions. The United States imports approximately 3200 Mg of chestnut each year but produces <1% of the total world production (Agricultural Marketing Resource Center, 2018). With the potential for producing more than 2200 kg ha^{-1} of nuts under alley cropping management, Chinese, Japanese, and American hybrid selections are all viable candidates. While great variance can be found in prices, the average retail price for chestnut in 2018 ranged from US$4.40 kg^{-1} to $11.00 kg^{-1} (Agricultural Marketing Resource Center, 2018).

The greatest single factor limiting the use of American chestnut in alley cropping is its susceptibility to the blight. A review by Miller (2003) suggested that field resistance to the blight exists within the American species even though there appears to be no complete immunity. Of the species available to North American growers, the Chinese chestnut is the most resistant, followed by the Japanese. The Japanese chestnut, however, is generally not as cold tolerant as the American and Chinese species (Miller, 2003). Choice of a cultivar for an alley cropping system should, in part, be determined by disease tolerance and adaptability to local climatic and soil conditions.

Of the non-nut-bearing hardwoods, honey locust (*Gleditsia triacanthos* L.) has received a great deal of attention for use in alley cropping (Gold & Hanover, 1993; Smith, 1914). Although Bryan, Berlyn, & Gordon (1996) suggested honey locust as a dinitrogen (N_2) fixer, it has not been confirmed as a significant contributor. However, as early as 1914, Smith (1929) advocated the planting of honey locust as a livestock food. Research conducted within the Tennessee Valley Authority in the 1930s (Tozer, 1980) clearly demonstrated that pods from select trees were valuable as a feed supplement for livestock; in alley cropping, pods could be raked and baled. Five-year-old grafted trees produce as much as 27 kg of pods. One hundred trees per hectare of this size and age could produce >2600 kg of pods with a nutritive value comparable to that of oat (*Avena sativa* L.) (Tozer, 1980) and whole-ear corn (Johnson et al., 2012, 2013). Historically, honey locust has not been in high demand as a timber species because of its thorniness and short-bole characteristics. Through selection, thornless cultivars with good stem form and heavy pod-bearing characteristics could potentially be made available for use. However, the presence of stray pollen from thorn-bearing varieties creates the potential for thornless trees to become the source for seedlings that bear thorns within a planting.

Both crown and root characteristics of honey locust are well suited for an alley cropping system. In general, the crown is broad, containing small-diameter branches covered with light to moderate foliage. While the period of leaf retention is long, only a light shade is produced. The root system develops from a central taproot that can extend downward 3–6 m, permitting the tree to absorb water and nutrients at great depths. Moreover, honey locust tends to have few lateral roots, thereby minimizing its competition with alley crops.

While producing greater shade and requiring more planning, many of the conventional hardwood timber species such as the oak (*Quercus* spp.) can also be adapted to alley cropping systems (Jose, Gold, & Garrett, 2012; Udawatta et al., 2005). Their slow initial growth has been somewhat overcome by recent advancements in the production of fast-growing planting stock (Mariotti, Maltoni, Jacobs, & Tani, 2015). Utilizing technology initiated by others (Huang, 1983), Forrest Keeling Nursery in Elsberry, MO, uses a root production method that produces containerized oak seedlings with the potential to grow ≥3 m in 5 yr. Recent research has demonstrated superior growth of some oak species throughout a 14-yr growing period when planted as a large containerized seedling rather than a bare-root seedling (Van Sambeek et al., 2016; Walter et al., 2013). Furthermore, swamp white (*Quercus bicolor* Willd.) and bur oak (*Q. macrocarpa* Michx.) seedlings grown using this procedure yield heavy mast crops within 5 yr, providing quick wildlife benefits to any alley cropping planting (Dey, Kabrick, & Gold, 2005; Kabrick, Dey, Van Sambeek, Wallendorf, & Gold, 2005). While many agroforesters incorporate tree species having only timber value, financial gain from such investments are likely to be less than for species that produce two or more crops (i.e., fruit, nuts, etc.). Early cash flows and short rotations are strongly correlated with the overall financial gain of any program (Godsey, Mercer, Grala, Grado, & Alavalapati, 2009). Therefore, planting the best stock available and enhancing growth through cultural treatments are important considerations when conventional hardwood timber species are used.

The use of hardwood species in making construction materials, and for conversion to bioenergy, increases the number of potential species for use in alley cropping. Many species with fast juvenile growth, such as poplar (*Populus* spp.), silver maple (*Acer saccharinum* L.), willow (*Salix* spp.), and black locust (*Robinia pseudoacacia* L.), can be placed in the linear plantings needed for alley cropping (Cardinael et al., 2012; Holzmueller & Jose, 2012; Jose & Bardhan, 2012). The rapid juvenile growth of these species yields high volumes per hectare (Gamble, Johnson, Sheaffer, Current, & Wyse, 2014; Gruenewald et al., 2007). With the projected high demand for hardwood chips from oriented strand board (OSB) manufacturers, the pulp and paper industry, and bioenergy producers, profitable alley cropping systems could be developed to meet these market needs (Howard, McKeever, & Liang, 2017; Spelter, McKeever, & Alderman, 2006).

Poplars have been widely planted in alley cropping in temperate zones around the world. Extensive plantings of poplar with row and vegetable crops have been reported in China (Dai, Sui, & Xie, 2006; Xiang, Shi, Baer, & Sturrock, 1990), Yugoslavia (FAO, 1980), Italy (Sekawin & Prevosto, 1978), and Turkey (Akbulut, Keten, & Stamps, 2003). In North America, research and demonstration has been underway since the mid-1960s to show the value of intercropping poplar with soybean, cotton (*Gossypium hirsutum* L.), biomass species, and a variety of other crops (Gamble et al., 2014; Rivest, Cogliastro, Vanasse, & Olivier, 2009). Early work was initiated in the Mississippi Delta by companies such as the Crown Zellerbach Corporation. Findings from this early work have been applied as far north as Ontario. At spacings of 2.7 by 5.5 m in eastern Ontario, maize and potato (*Solanum tuberosum* L.) have been successfully grown between rows of poplar during the very early years of a rotation before shade became a problem. The intercrop is used to offset the cost of establishing the trees (Raitanen, 1978). In Canada, maize and soybean grown in the alleys of fast-growing poplar have also been shown to increase farm profits (Mergen & Lai, 1982).

Dinitrogen-Fixing Deciduous Species

In the tropics, N_2–fixing tree species have long been important in alley cropping systems, especially under conditions of drought and poor-quality soils (Wolz & DeLucia, 2018; Makumba et al., 2006). However, their use in temperate regions has not been as extensive (Figure 7–3). In part, this is because of the greater accessibility and affordability of N fertilizers in North America and the general lack of research on N_2–fixing species in the temperate zone. This does not, however, suggest that they have limited potential. With the increased use of alley cropping on marginal lands, N_2–fixing trees and shrubs can play increasingly important roles in alley cropping systems (Gruenewald et al., 2007).

Dinitrogen-fixing woody species for use in temperate alley cropping may be either N_2–fixing leguminous (rhizobial-nodulated) or actinorhizal (Frankia-nodulated) plants. Within the temperate zone, perhaps the most widely planted and best known woody legume is black locust (Barrett, 1994; Gruenewald et al., 2007; Janke, 1990; Ntayombya & Gordon, 1995). Locust has been favorably compared to the N_2–fixing leucaena [*Leucaena leucocephala* (Lam.) deWit] tree of the tropics (Barrett & Hanover, 1991). Black locust is fast growing, produces a dense wood that is resistant to decay, and can accumulate 30–35 kg ha^{-1} of soil N annually (Boring & Swank, 1984). However, locust is attacked by a stem borer (*Megacyllene robiniae* Forster) capable of inflicting great damage. While infestations are more prevalent under conditions of poor to intermediate site quality, damage can occur anywhere. A comprehensive review of black locusts' potential in alley cropping can be found in Barrett (1994).

Other temperate legumes with potential application in alley cropping include species in these four genera:

- *Caragana* (e.g., the Siberian pea shrub used in the northern Great Plains)
- *Albizia* (e.g., *Albizia julibrissin* Durazz), commonly referred to as the silk tree or mimosa and widely planted as an ornamental in the southern United States
- *Lupinus* (some species may have value in the warmer climates of North America)
- *Prosopis* (mesquite)

The discovery of an erect, thornless, fast-growing mesquite in Peru offers opportunities for using this genus in the southwestern United States (Holden, 1996). Studies have demonstrated increases in wheat and oat when mesquite was used as a nurse crop. Furthermore, its wood has value for flooring and furniture, and the pod is consumed by livestock.

In contrast to leguminous plants, the value of actinorhizal plants is less understood and appreciated in North America, and yet their contribution in N_2 fixation worldwide may rival that of the legumes (Torrey, 1978). Within North America, several genera contain species of actinorhizal plants that may be of value in alley cropping. Included in this list are *Cercocarpus*, *Ceanothus*, *Datisca*, *Chamaebatia*, *Cowania*, *Comptonia*, *Sheperdia*, *Elaeagnus*, and *Alnus* (Dawson & Paschke, 1991). Species within most of these genera grow as shrubs and are found primarily in the western and southwestern areas of North America. Two genera, however, *Elaeagnus* and *Alnus*, are commonly grown in eastern North America. Both contribute significant quantities of N to the soil. Paschke, Dawson, and David (1989) demonstrated that through mineralization of organic material, a stand of autumn olive (*Elaeagnus umbellata* Thumb.) contributes as much as 236 kg

N ha^{-1} yr^{-1}. Daniere, Capellano, & Moirud (1986) showed contributions ranging from 48 to 185 kg ha^{-1} yr^{-1} for alder (*Alnus* sp.). However, many species of these genera, such as *Elaeagnus* and *Alnus*, can become invasive and the advantages and disadvantages must be weighed. See Dawson and Paschke (1991) for a comprehensive review of N_2–fixing species suitable for agroforestry.

Coniferous Species

Most of the information on alley cropping in North America addresses the use of deciduous, hardwood species (Figure 7–3; Wolz & DeLucia, 2018) because alley cropping research was initiated in deciduous rather than coniferous regions. Furthermore, because of the emphasis placed on conventional row crops as companion species, the greatest interest in alley cropping has been and continues to be in the midwestern United States and parts of Canada, where hardwoods are most prevalent. Nevertheless, as good markets exist for conifers in many regions, the opportunity to use conifers in alley cropping is excellent. Specialty crops, forages, horticultural species, and energy crops might be better suited than row crops as alley crops within a conifer alley cropping system (Haile, Palmer, & Otey, 2016; Minick et al., 2014).

Exotic Species

Even though native species are most commonly planted in alley cropping systems, the potential of exotic species that have been proven to be adaptable to a specific region should not be overlooked. The empress tree [*Paulownia tomentosa* (Thumb.) Siebold & Zucc. ex Steud.], for example, has proven to be an excellent species for alley cropping in China (Wei, 1986; Yin & He, 1997) and should be well suited for alley cropping in warmer climates in North America. Although valued only for its wood, *Paulownia*'s fast growth, high value, and adaptability to open, linear-shaped plantings make it a viable choice for use in alley cropping. In California, the English walnut (*Juglans regia* L.) is prized for its fruit. Large-scale, open plantings of this species could permit alley cropping while the trees are becoming established. Similarly, Chinese chestnut has great potential in the U.S. Midwest (Aguilar et al., 2009, 2010).

Tree Arrangement

Single Species versus Mixed Species

Temperate alley cropping systems have primarily utilized monoculture tree stands (Wolz & DeLucia, 2018). However, today's emphasis on maximizing biodiversity and creating more sustainable systems may encourage design innovation.

Evidence from both forestry and alley cropping systems suggest that planting multiple tree species in alley cropping could lead to many economic and ecological benefits (Lovell et al., 2018; Wolz, Lovell, et al., 2018). For example, meta-analyses of tree diversity in natural and managed forests have shown greater light capture (Sapijanskas, Paquette, Potvin, & Kunert, 2014), productivity (Zhang, Chen, & Reich, 2012), biomass accumulation (Piotto, 2008), and C storage (Hulvey et al., 2013) in polycultures than monocultures. Furthermore, a greater spatial and temporal distribution of roots in polyculture can improve nutrient cycling, N retention, water-use efficiency, and drought resilience (Jose, Williams, & Zamora, 2006; Lang et al., 2014; Pretzsch, Schütze, & Uhl, 2013; Schwarz et al., 2014). Intentional tree diversity can also stimulate an associated biodiversity of insects, pollinators, birds, mammals, and soil microbes (Malézieux et al., 2009; Perfecto, Mas, Dietsch, & Vandermeer, 2003). Finally, tree diversity can hedge financial risk by diversifying income over time (Cubbage et al., 2012).

In a survey of temperate alley cropping, Wolz and DeLucia (2018) found that the most common approach for mixing tree species in alley cropping was combining a fast-growing species like poplar with slow-growing, valuable hardwood species. This spreads the income potential across a longer time, and the fast-growing species force straight growth and natural pruning of the more valuable hardwood species. Mixing species, however, must be carefully planned (Wolz, Lovell, et al., 2018), as the juvenile growth patterns and management needs of different species can vary significantly. If species are not matched on the basis of early growth rates, one or more may soon dominate the site, neutralizing the benefits of a mixed planting. Studies with mixed plantings of walnut, European alder (*Alnus glutinosa* L.), and black locust have demonstrated that if they are not properly designed, any advantage that might be realized from N_2 fixation is lost through overtopping and shade problems (Van Sambeek, 1989).

Many studies have also demonstrated the benefits of mixing N_2–fixing and non-N_2–fixing trees (Morhart et al., 2014; Van Sambeek, 1989). In particular, black walnut has been extensively researched (Kessler, 1988; Ponder & Baines, 1985). The planting of N_2–fixing trees such as autumn olive, Russian olive (*Elaeagnus angustifolia* L.), European alder, and black locust in close association with walnut increases early growth (Van Sambeek, 1989). On sites with intermediate soil quality, growth increases may range from 100 to 200%. One study (Ponder & Baines, 1985) reported increases in walnut growth of >300%. Of the N_2–fixing species tested with walnut, autumn olive has been shown to result in the greatest and perhaps most consistent growth increase (Van Sambeek, 1989).

Additional difficulties and limitations of mixing tree species in alley cropping stem from the inherent complexity of managing polyculture systems. These include reconciling varying mechanical harvest requirements of different nut or fruit species, having farmers skilled in managing multiple species and monitoring multiple markets, and contrasting pesticide requirements and options across species (Wolz, Lovell, et al., 2018).

Single-Row versus Multiple-Row Configuration

Both single-row and multiple-row tree configurations are viable options in alley cropping. Multiple-row configurations group multiple tree rows closely together on either side of each alley, thereby increasing the number of trees per hectare while maintaining similar light transmittance to alleys. There are benefits and tradeoffs for each approach.

Tree species that show strong phototropic responses are not recommended for pure plantings using configurations of two or more rows. Under conditions of close spacings between rows, tops will diverge in two-row plantings and grow toward the open alleys, adversely affecting the value of the wood. Under conditions of three or more rows, only the outside rows will show such a response. If multiple-row configurations are adopted for species with strong phototropic responses, the outside rows should be planted with species that do not show strong phototropic responses.

In contrast, some tree species do not adapt well to single-row configurations under the open conditions of alley cropping. In meeting the needs of light-demanding species such as conventional row crops, spacing between tree rows can be wide while the within-row spacing is narrow. Because of the open conditions between rows, high variability exists among species to produce straight stems. One problem species is silver maple. A 12-yr study conducted at the University of Missouri (Garrett, unpublished data, 1995) demonstrated the difficulty of producing quality silver maple sawlogs without intensive management. Trees grow rapidly when planted on 1.5-m centers within the row, 19 m between rows, and intercropped with maize. However, because apical dominance is quickly lost in silver maple when a high incidence of light reaches the lower section of the stem, numerous sprouts develop at the base. Aggressive pruning is then required to maintain high-quality timber. For species with growth responses and market values similar to that of silver maple, a triple-row configuration using a "trainer" species in the outside rows that does not show a strong light response (e.g., pine) might be more economical.

Multiple-row configurations have, in some studies, also been shown to be better than single rows in optimizing wood and forage production. Sharrow (1991a) suggested that uniformly spaced (grid) tree plantings theoretically should minimize competition among trees while maximizing competition between trees and understory alley crops. Because of this, he concluded that a uniform pattern, commonly used in conventional forestry, might not be the most desirable for agroforestry designs such as silvopasture and alley cropping. Sharrow (1991a) used a computer simulation model incorporating data from a 10-yr-old Douglas-fir [*Pseudotsuga menziesii* (Mirb.) Franco]–grass–clover combination to show that forage production increased as the degree of tree aggregation increased from grids to triple rows. Furthermore, he concluded that pattern was as important as tree density in determining understory crop production. The analysis suggested that land planted with 900 trees ha^{-1} in double rows produced a similar amount of forage as land planted in a uniform grid with 450 trees ha^{-1}. Similarly, in a slash pine (*Pinus elliotti* Engelm.)–forage system over 13 yr, double-row configurations produced as much wood per hectare as the same number of trees in single-row configurations, but double-row configurations produced more forage than single rows (Lewis, Tanner, & Terry, 1985). To enhance the growth of trees in a multiple-row planting, staggering the trees between adjacent rows will permit maximum crown development.

Between- and Within-Row Spacing

Tree spacing between and within rows is a critical consideration in designing any alley cropping practice. Within-row spacing will vary with the primary objectives of the alley cropping system. If, for example, alley cropping is used to control an erosion problem, trees should be closely spaced to provide an immediate effect, or a shade-tolerant shrub with a shallow and highly diffuse root system should be strategically combined with the trees. If grafted nut trees (e.g., chestnut, walnut, pecan) are planted, within-row spacing will be higher to minimize costs and provide sufficient space for the grafted trees to fully develop their crowns for nut production.

In seedling plantings of wild or genetically diverse material, higher tree density may be necessary so that low-quality individuals can be thinned with time. Although the cost of planting extra trees and removing them later is greater than planting the trees initially at the desired spacing, in many instances this may be the only way landowners can develop a planting of high-quality trees (e.g., Jones, Haines, Garrett, & Loewenstein, 1994). Single, wide, between-row spacing is required in most alley cropping plantings; accommodating the extra trees is typically accomplished by reducing within-row tree spacing.

The overall success of any alley cropping system is often linked directly to the choice of between-row spacing (Gallagher, Mudge, Pritts, & DeGloria, 2015; McGraw et al., 2005). Between-row spacing depends on the emphasis placed on wood production versus some other tree-related crop (i.e., nuts, foliage production, C credits, etc.). For example, if the emphasis is on wood production, between-row spacing will be less than if nut production is emphasized.

Requirements for the growth of the intercrop species must also be met. Because many alley crops are shade intolerant (i.e., row crops, forages, small berry crops, etc.), alleys must be sufficiently wide to accommodate their light requirements. Furthermore, the spacing will be determined by the duration of the cropping regime selected. If shade-intolerant crops are to be grown in the alley for more than a few years (5–10 yr), then the spacing between rows must be wide enough to satisfy their needs (Figures 7–4 and 7–5). Studies in Missouri using black walnut and warm-season, conventional row crops—soybean and maize—have demonstrated that single rows spaced 12 m apart can accommodate the light and moisture needs for these crops for only 9–10 yr when tree lateral roots are not pruned to minimize competition between tree and crop (Garrett & Harper, 1999; Garrett & Kurtz, 1983). Similar findings were reported in Missouri for alfalfa (*Medicago sativa* L.) by McGraw et al. (2005). Regulation of surface tree roots could extend the cropping period. Winter wheat, which matures during late spring, can be grown for 10–15 yr in

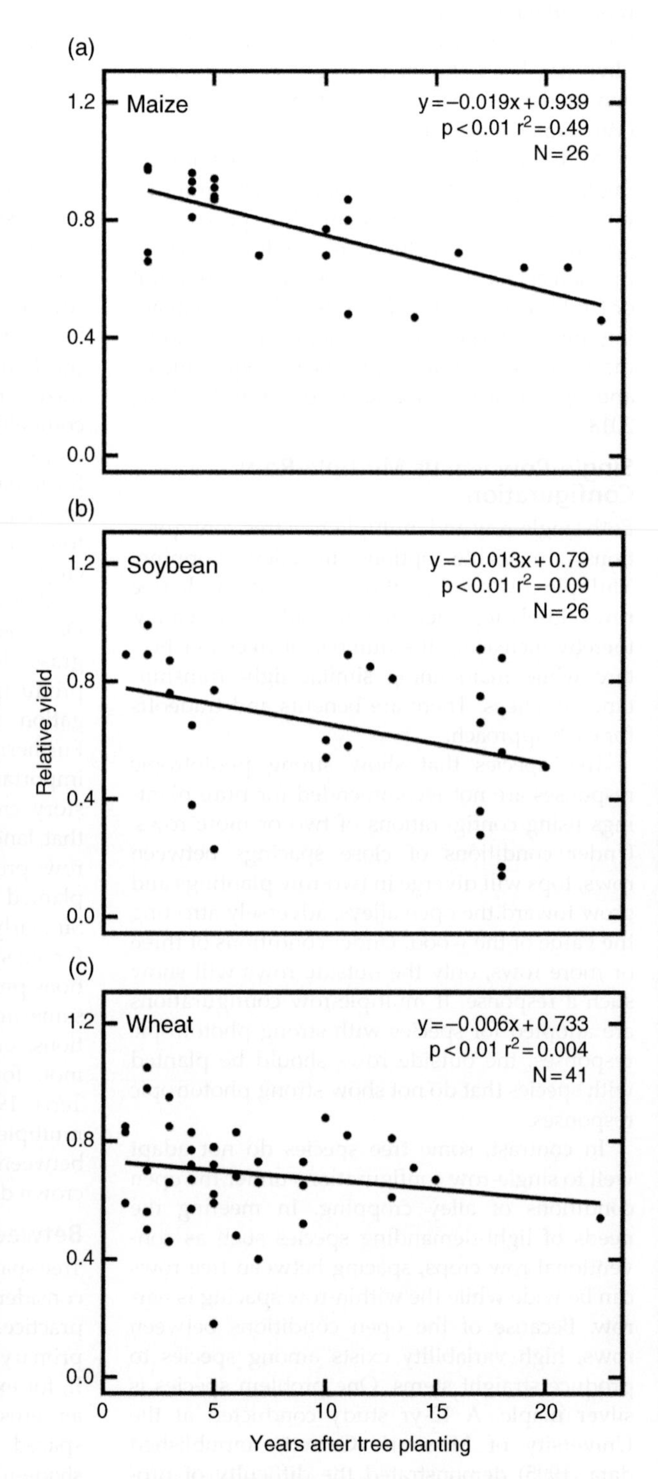

Fig. 7–4. Results from a literature survey of alley crop relative yields in temperate alley cropping field experiments with any tree species. The three most common temperate alley crop species shown are (a) maize, (b) soybean, and (c) wheat. Each point represents one site-year (reproduced from Wolz & DeLucia, 2019a).

Fig. 7–5. A black walnut site prepared for conversion from row crops to a hay crop at age 10 yr.

12-m alleys of walnut without significant losses in yield (Figure 7–4c). If row crops in the alleys are to be grown for 15–20 yr or longer, between-row spacing should be 24 m or more depending on the light needs of the alley crop and the management goal.

Concern about satisfying the light requirements of alley cropped species has resulted in research and much discussion of the subject. Talbot (2011), using findings from a computer simulation model, recommended that rows be spaced at twice the height of the mature tree to enable the yields of alley crops to remain sufficiently high to be profitable until the trees are harvested. Morhart et al. (2014) suggested that the required spacing, based on the final target diameter of the butt log, could be calculated from the projected crown diameter due to the close correlation existing between stem diameter at breast height and crown width.

Equipment available to the alley crop farmer can also be an important consideration for between-row spacing. For example, between-row spacing can be set as a multiple of the width of the equipment such as seeders, sprayers, and combines. In the U.S. Midwest, common row crop equipment widths are 30, 60, and even 120 ft.

Silviculture of Alley Cropped Trees

Weed Control

Weed control, which is critical to the success of any alley cropping practice, can be performed by either mechanical or chemical means (Dey, Gardiner, Kabrick, Stanture, & Jacobs, 2010; Dey et al., 2005; Garrett & Harper, 1999; Van Sambeek et al., 2016). When mechanical approaches are used, damage to the lower portion of the stem must be avoided. To avoid damage, weed control around the base of trees should be achieved chemically when conventional farm tilling equipment is used. Complete weed control by mechanical means is advocated only when some form of precision in-the-row tiller is used. Such equipment is specifically designed to minimize damage to young trees. It may also prove superior to chemical weed control as a result of the increased aeration and moisture penetration benefits derived from soil-surface scarification.

When conventional equipment is used between rows, deep tilling can be used to control competition for moisture and nutrients between the tree and alley crop components. Severing of the trees' surface roots reduces competition and helps force tree roots deeper into the soil to meet their moisture and nutrient needs (Miller & Pallardy, 2001). If this management approach is used, it should begin in the first year of the planting and be continued annually to keep lateral roots small and to train them to extend deep into the soil. Some research has also shown that competition with winter crops naturally induces deeper rooting of trees in alley cropping (Cardinael, Mao, et al., 2015).

Because of the ease of application and past success, chemical weed control is more popular than mechanical weed control. After planting, herbicides can be applied in a circle around each seedling or along both sides of a row of trees

Fig. 7–6. Chemical weed control along rows of 8-yr-old black walnut.

Fig. 7–7. A well-managed stand of black walnut alley cropped with a mixture of grasses and legumes. Remnants of a successful in-row Christmas tree planting are visible in the background.

(Figures 7–6 and 7–7). The area of the control zone around trees will vary with the landowner and the practice used. In all instances, it should be sufficient to reduce the competition for water and nutrients between the woody component, weeds, and the alley crop early in the rotation (Garrett & Harper, 1999; Houx et al., 2013). If sites are erodible, consideration should be given to planting on the contour using shallow-rooted "living mulches" in association with the trees, or using a commercial-grade, biodegradable mulch specifically designed for the purpose of weed control (Dey et al., 2005; Garrett, Jones, Kurtz, & Slusher, 1991; Van Sambeek et al., 2016).

Fertilization

Knowledge of the fertility requirements of the various tree species used in alley cropping practices is limited. The trees and crops grown together under alley cropping management compete directly for nutrients and are grown under microenvironmental conditions that may vary from their typical monoculture application, so it is likely that their fertility requirements will differ from those prescribed for conventional forestry or agricultural practices. Because substantial N is often removed when harvesting the alley crop, N supplements may be required in alley cropping systems to maintain favorable tree growth in addition to intercrop yields. For example, an English walnut–forage system in California required as much as 224 kg N ha^{-1} for nut production (Ramos, 1985). Nevertheless, for some trees species with low N requirements, the excess N not captured by the alley crop may be sufficient for growth. Indeed, the capturing of excess nutrients by tree roots has been demonstrated as an important benefit of alley cropping systems (Tsonkova et al., 2012; Wolz, Branham, & DeLucia, 2018).

Branch Pruning

Pruning requirements under alley cropping vary greatly depending on the short- and long-term objectives of the practice. In many situations, artificial pruning is recommended to increase the timber value of trees and enhance the microenvironment for the alley crop. In contrast, under certain alley cropping regimes, shorter boles with larger crowns may be favored to create the proper microenvironment for alley crops or to increase yields of nuts, fruit, or specialty products. In the U.S. Midwest, walnut alley cropping is more profitable if some clear-log length is sacrificed to increase the crown area for nut production (Kurtz, Garrett, & Kincaid, 1984). Furthermore, other fruit- and nut-bearing species (e.g., chestnut, pecan, pawpaw [*Asimina triloba* (L.) Dunal]) may provide greater economic gains with the development of larger crowns and shorter boles.

Tip pruning (the removal of the growing tip) of lower lateral branches is a practice unique to alley cropping management. Developing trees may be damaged as lower branches grow into the alleys and become entangled in farm equipment. Tip pruning prevents tree damage and enables the landowner to maintain larger live-crown ratios and faster growth early in the life of the trees than would be possible with complete branch removal (Garrett et al., 1991).

Root Pruning

Experiments on root pruning and the installation of root barriers strongly suggest that, even during the early years of tree development, competition for water and nutrients, or both, is the major reason for reduced yields. Competition can become so severe that row crop maturity is delayed and yields decreased (McGraw, unpublished data, 2000). Cultural treatments designed to reduce competition from tree roots near the soil surface can significantly increase yields of row crops and other crops established in the alleys without seriously affecting the growth of the trees (Jose, Gillespie, & Seifert, 1996; Miller & Pallardy, 2001). Kang (1993) emphasized the importance of root pruning in alley cropping in the tropics and indicated that early and repeated severing of lateral roots significantly reduces the number of roots in the plow zone and that roots can be trained to move deeper into the profile. Some research would even suggest that it is possible to create deeper rooting as a result of practicing alley cropping management. Cardinael, Mao, et al. (2015) demonstrated that competition with winter crops induces deeper rooting of walnut trees in a Mediterranean alley cropping agroforestry system. By managing root competition from the time of establishment, yields and profits can be significantly increased. Deep tilling adjacent to each tree row may reduce some competition. However, most studies have indicated that a deeper disturbance is needed to significantly reduce competition. A subsoiler tillage tool, specifically designed to sever roots to a depth of 0.6–0.9 m, might prove to be superior to tilling. Such an implement would quickly pay for itself on the basis of increased crop yields.

Allen et al. (2004) used a belowground polyethylene barrier to isolate pecan tree roots from cotton alleys in the southern United States and found increased yields and N concentrations in cotton compared with the non-barrier treatment. Miller & Pallardy (2001) studied the effects of trenching with a root barrier in a maize–silver maple alley cropping system in Missouri. They found that the grain yield of border-row maize plants lacking an adjacent barrier was depressed compared with that for maize plants with a root barrier present. Gillespie et al. (2000) examined alley cropping of maize grown with either black walnut or red oak in the U.S. Midwest. They found that during the course of 10 yr, maize yields in rows adjacent to tree rows declined by 50% or more. With the introduction of barriers to separate tree and crop root systems, yields in the rows near trees were equal to those of the center row and of yields in monoculture. Some research has shown that soybean may respond differently from maize and can even benefit from alley cropping without root control (Nasielski et al., 2015; Senaviratne et al., 2012).

Thinning

Both the timing of thinning and the number of trees to be removed in alley cropping varies greatly with species, site conditions, and management objectives. Forage, row crop, and specialty crop species used in alley cropping practices range from shade tolerant to shade intolerant. Semi-open crown conditions, maintained through regular thinning, are a prerequisite to maintaining the vigorous growth of shade-intolerant alley crops. Moreover, the quality and quantity of wood produced has a direct bearing on the expected economic gain when wood production is a goal.

Growth and Wood Quality

Research on the impact of alley cropping on tree growth and wood quality is limited. Cutter and Garrett (1993) found that height, diameter, and specific gravity are greater for alley cropped walnut trees than for trees grown in typical forestry monoculture. Furthermore, based on growth during the first 17 yr, 60-yr veneer log rotations may be possible for some alley cropped trees compared with a customary 80–100 yr or more for conventionally grown trees. Wolz and DeLucia (2019a) found similar results when modeling black walnut in alley cropping and monoculture forestry systems across the U.S. Midwest. Overall, results suggest that, if properly managed, the quality of trees grown under alley cropping will not vary significantly from that of forest-grown trees, but the time required to grow them to a marketable size can be greatly reduced (Cutter & Garrett, 1993).

Many landowners with alley cropping systems convert the alleys to forages and switch to a silvopasture practice when shade limits the growth of other crops. Such practices do not necessarily lower tree quality. In a study of the effects of >25 yr of cattle grazing on slash pine (*Pinus elliottii* Engelm.) in south-central Louisiana, regulated grazing (controlling the number of animals per hectare) was found to have no noticeable effect on either tree size or grade (Cutter, Hunt, & Haywood, 1998). Other wood quality indicators including earlywood–latewood ratios and specific gravity also showed no differences. Such findings have great significance in alley cropping, as many landowners are interested in transitioning from alley cropping to silvopasture in future years.

Crop Component

With proper planning, the light requirements of most crops, from open conditions to deep shade, can be accommodated in an alley cropping system (see Figures 7–2 and 7–3). This adaptability gives alley cropping great potential across the temperate zone. Nevertheless, crop selection depends on many factors, including site characteristics, tree species selection, and tree maturity.

Row Crops

In alley cropping systems with young trees, shade-intolerant row crops can be produced in the alleys until light, water, or nutrients become limiting from competition. Numerous studies have demonstrated the value of simultaneous production of trees and row crops (Figure 7–3). Examples are cotton(Allen, Nair, Graetz, Jose, & Nair, 2006; Zamora, Jose, & Nair, 2007), maize (Jose et al., 1996; Senaviratne et al., 2012; Udawatta, Motavalli, Jose, & Nelson, 2014), soybean (Garrett & Kurtz, 1983; Nasielski et al., 2015), milo (Garrett & Kurtz, 1983), wheat (Garrett et al., 1991; Inurreta-Aguirre, Lauri, Dupraz, & Gosme, 2018; Palma, Graves, Burgess, et al., 2007; Williams & Gordon, 1995), barley (Thevathasan & Gordon, 1996; Williams & Gordon, 1992), potato (Kapp, 1987), oat (Gordon & Williams, 1991), and pea (*Pisum* spp.) (Kapp, 1987).

Because many row crops are shade intolerant, their light requirements must be planned for. Depending on the tree species selected, shade can become a problem within 3 or 4 yr when using fast-growing species such as poplar. However, tree influence on alley crop yield can be minimized for a decade or more if slower growing species such as oak or walnut are planted with wide between-row spacing (Figure 7–4).

Research conducted on the growth of trees and conventional row crops in an integrated alley cropping configuration has proven the practice to be successful (Beaudette et al., 2010; Bradley,

Oliver, Thevathasan, & Whalen, 2008; Gordon et al., 2008). In particular, work on black walnut (Garrett & Harper, 1999; Wolz & DeLucia, 2019a) has been extensive (Figures 7–2a, 7–2b, and 7–3). Early in the rotation, wheat and other cool-season crops can be planted to within 0.9 m of the tree rows, as little or no weed maintenance is required during the winter and early spring. If tree rows are spaced 12 m apart, only 13% of the land is removed from crop production. With soybean and other warm-season crops, grass and herbaceous vegetation must be controlled within tree rows and along their edges. Space for equipment operation, approximately 2 m wide on each side of a row, should remain unplanted. With rows 12 m apart, approximately 38% of the land is used for trees and 62% is planted with the warm-season crop. By using a commercial mulch and planting the tree seedlings directly into it, some savings can be made on the percentage of land removed from cropping during the first few years. As the trees grow, the amount of lost space will increase because of the growth of branches over the alleys.

Studies conducted in Canada on alley cropping black walnut have demonstrated relationships similar to those observed in Missouri (Bradley et al., 2008; Nasielski et al., 2015; Williams & Gordon, 1992). Intercropping walnut with maize, soybean, and wheat has shown that all can be grown with no significant inhibition from chemicals produced by the trees. Differences in tree growth responses, however, have been found to be related to the intercrop used (Thevathasan & Gordon, 2004). Growth of trees during the first few years was found to be best with maize followed by soybean (Gordon & Williams, 1991; Williams & Gordon, 1992). Significant reductions in growth were associated with intercropping small cool-season grains. Differences in growth are believed to be related to differences in soil moisture available to the trees early in the growing season when trees typically do most of their growing. Soil moisture availability was always least in the spring when cool-season grain crops were used (Williams & Gordon, 1995).

Plantings in Indiana have also demonstrated that maize is a good choice as an alley crop for walnut. Under conditions of 1.5-m spacing within rows and only 6.8 m between rows, increased yields on a per-hectare basis early in the planting were observed because of an edge effect on maize rows planted adjacent to young trees. With six rows of maize planted in alleys, yields of 10,347 kg ha^{-1} were produced on land with a maize-production rating of 7838 kg ha^{-1} (Gogerty, 1994). Similar observations have been made in a separate planting in Indiana. At Purdue University, walnut was planted on a 1.2- by 8-m spacing and six rows of maize planted in the alley (Jose et al., 1996, 1997). Yields for the first 2 yr were 106 and 107%, respectively, of the expected farm average. Increased yields were again attributed to greater quantities produced in the outer rows (i.e., rows paralleling tree rows) that had access to more light and water. During Years 3–7, however, yields decreased, although they remained higher in the outer rows than in the middle rows.

Dinitrogen-fixing trees have long been favorites for alley cropping in the tropics, and, although in North America their wood value is not as great as that of many other species, the N made available to alley crops on intermediate and poor sites can be beneficial (Wolz & DeLucia, 2018). In the second year of a study in Ontario, yields of barley grown with black locust were 8% less than for barley grown alone even though the N content of the grain was 11% higher (Ntayombya & Gordon, 1995). Loss of yield was attributed to competition for moisture. In contrast, first-year responses of maize planted with 1-yr-old black locust in Ohio demonstrated that although the land area occupied by maize was reduced by 28%, maize yield was reduced by only 18%, indicating no reductions because of competition or chemical inhibition (Ssekabembe & Henderlong, 1991). A study in Michigan found that while an interplanting of maize did not affect the N_2–fixing capacity of black locust, black locust significantly inhibited the growth of maize. Either allelopathy or severe competition for nutrients reduced growth (Powers, Lantagne, Nguyen, & Gold, 1996). Seiter and William (1994) found that when sweet corn was planted in alley cropping designs with black locust and red alder (*Alnus rubra* Bong.), yields increased in close proximity to the tree rows.

Forage Crops

A common practice in North America is growing forage crops for harvest as hay. Alley cropping with forages should not be confused with silvopastoral management. In alley cropping, forage crops are grown for hay production between rows of planted trees, with no direct livestock integration (Figures 7–2c and 7–8). In contrast, silvopastoral management directly integrates livestock to harvest the forage that is produced under trees. The distinction between alley cropping with forages and silvopastoral management is important because of the vast differences created when livestock become part of the practice.

Fig. 7–8. A medium- to low-quality upland black walnut site managed primarily for walnut and hay production.

Published reports of alley cropping with forages in temperate climates are limited (Figure 7–3). Most of the literature on intercropping trees with forages pertains either to using forages as a cover crop or to silvopastoral management. Although much of this information is applicable to alley cropping, more research on hay production in alley cropping is needed. Recently, Lovell et al. (2018) described the establishment of a large-scale alley cropping research trial at the University of Illinois Urbana-Champaign that includes hay as an alley crop.

In the temperate zone, there is considerable potential for establishing alley cropping with forages (Pang et al., 2019a, 2019b). Land classified as grassland pasture is usually not suitable for row-crop production because it may be too sloping, droughty, or of poorer soil quality than land used for row crops. On many lands, forage production offers the best alternative for landowners. However, in the eastern-humid region of the United States, much of this grassland pasture would be suitable for tree production (Jose et al., 2012). Adding tree crops to these pastures could potentially provide additional income through nut or wood sales, improve biodiversity and wildlife habitat, and enhance the aesthetic value of the land without necessarily losing forage yield or quality. Ehret, Graß, & Wachendorf (2015) found that a clover (*Trifolium* spp.)–perennial ryegrass (*Lolium perenne* L.) mixture was highly suitable for alley cropping under moderate (50%) shade conditions.

Many forage species may be used in alley cropping practices, although most hay enterprises grow perennial forages because of their greater productivity and to avoid yearly establishment costs. Selecting a suitable forage species for alley cropping depends on many factors including tree species, site conditions, characteristics of the forage, and management objectives. Some general information about forages is given below. However, for explicit information on the characteristics of individual forage species, see Barnes, Nelson, Collins, and Moore (2003).

In general, legumes like alfalfa and clover have better forage quality (more protein and less fiber) than grasses. In association with *Rhizobium* bacteria, legumes fix atmospheric N_2 for themselves and contribute N to the soil, which can be used by the tree crop (Shults, 2017). Because of its high quality, legume hay is more marketable than grass hay and brings a better price when sold. Legumes, however, are more difficult to establish, harder to manage, and less persistent than grasses.

Cool-season perennial species are most productive in the spring and fall when temperatures are cool. During the hot summer months, growth slows and they often become dormant. Tall fescue

[*Schedonorus phoenix* (Scop.) Holub], orchardgrass (*Dactylis glomerata* L.), and timothy (*Phleum pratense* L.) are examples of cool-season perennial grasses. Alfalfa, red clover (*T. pratense* L.), white clover (*T. repens* L.), and birdsfoot trefoil (*Lotus corniculatus* L.) are examples of cool-season perennial legumes. Cool-season forages have a long growing season, are high yielding, and are the most commonly grown forage species for hay production. Because of their growth potential and long growing season, cool-season forages can be highly competitive with tree crops.

Warm-season perennial species grow most during the summer months. Bahiagrass (*Paspalum notatum* Flugge), big bluestem (*Andropogon gerardii* Vitman), and bermudagrass [*Cyndon dactylon* (L.) Pers.] are examples of warm-season perennial grasses. Bermudagrass is the most widely grown forage in southern pecan orchards (Holt, 1973). Sericea lespedeza [*Lespedeza cuneata* (Dum. Cours.) G. Don] and perennial peanut (*Arachis glabrata* Benth.) are examples of warm-season perennial legumes. Warm-season forages have a shorter growing season, although they are still very productive. In the early spring when some tree species are beginning growth, these forages are still dormant and are not competitive. These forages grow vigorously in the hot summer months; however, this is when competition with trees for water may be the greatest.

Winter annuals germinate and start their growth in the fall, become dormant or grow slowly during the coldest part of winter, then make their greatest growth in the early spring when they flower, set seed, and die. Although winter-annual grasses, such as cheatgrass (*Bromus tectorum* L.), often are a major component of pastures, few forage-type winter-annual grasses are grown for hay production in North America. Cereal-type winter-annual grasses, such as wheat, oat, or rye (*Secale cereale* L.), are sometimes grown and harvested for hay. Many winter-annual forage legumes are grown in North America: arrowleaf clover (*T. vesiculosum* Savi), berseem clover (*T. alexandrinum* L.), black medic (*Medicago lupulina* L.), crimson clover (*T. incarnatum* L.), hairy vetch (*Vicia villosa* Roth), rose clover (*T. hirtum* All.), and so on. These species are not as productive as perennials and usually produce only one cutting of hay. They do provide ground cover for erosion control during the winter and do not compete with the tree crop during the summer.

Summer-annual species germinate in the spring, grow during the summer, then flower, set seed, and die in the fall. Sudangrass [*Sorghum bicolor* (L.) Moench ssp. *drummondii* (Nees ex Steud.) de Wet & Harlan], foxtail millet [*Setaria italica* (L.) Beauv.], and crabgrass (*Erichloa polystachya* Kunth) are examples of summer-annual grass species. Korean and striate lespedeza (*Kummerowia* spp.) and cowpea [*Vigna unguiculata* (L.) Walp.] are examples of summer annual legume species.

Shading from trees can reduce light levels and temperature and increase relative humidity. Shade tolerance is an important characteristic for a forage crop grown in an alley cropping practice once the trees mature and begin to shade the alleys (Pang et al., 2019a). Most forage species are adapted to open fields and full sunlight. However, most individual leaves become saturated with light at levels less than full sunlight so they may tolerate some shading. The actual light intensity at which light saturation is reached is dependent on species, age, and prior light environment of the leaf, temperature, and CO_2 level (Beinhart, 1962; Hesketh, 1963; Scott & Menalda, 1970; Wolf & Blaser, 1972).

Neiderman and McGraw (unpublished data, 2006, University of Missouri Center for Agroforestry), compared the growth of alfalfa in full sun to alfalfa grown in artificial shade of 36 and 53% full sun. Shading was found to reduce yield, delay maturity, and reduce root nonstructural carbohydrates compared with full sun. McGraw et al. (2005) and McGraw, Stamps, and Linit (2008) grew alfalfa in open plots and in plots that were alley cropped between 20-yr-old black walnut trees planted in rows 24.4 and 12.2 m apart. Alfalfa yield was significantly reduced and maturity was delayed by the narrow 12.2-m tree spacing, but yield and maturity were not reduced in the centers of the wider 24.4-m alleys compared with the open plots. If alleys are sufficiently wide, the physiological needs of species like alfalfa could be accommodated.

While most forages are considered "sun" species and achieve maximum growth only with full sunlight, some forage species are productive in shade environments (Pang et al., 2019a, 2019b). Orchardgrass is often found growing under trees and is noted for being shade tolerant, hence the name "orchard" grass. When orchardgrass was grown at one-third incident sunlight, yield and persistence were not affected (Blake, Chamblee, & Woodhouse, 1966). Tall fescue also tolerates reduced light. Allard, Nelson, and Pallardy (1991) found that tall fescue shoot growth was similar whether grown at full sun or 60% full sun. Shoot growth was reduced at 30% full sun. Burner (2003) compared orchardgrass, tall fescue, and a binary mixture of orchardgrass and tall fescue in alleys of loblolly (*Pinus taeda* L.) and shortleaf pine (*Pinus echinata* Mill.). He found that orchardgrass and the binary mixture were superior to tall fescue in the pine alleys. Red clover is also considered a relatively shade-tolerant species whose leaves become light saturated at low light intensities (Bula, 1960).

Research on the shade tolerance of forage species for use in agroforestry is insufficient. During the past 30 yr, the University of Missouri Center for Agroforestry has been screening species for shade tolerance. Species have been grown initially in pots under artificial shade and full sun. Promising species are further tested in the field under trees. An initial screening of 27 forages identified some shade-tolerant species (Lin, McGraw, George, & Garrett, 2001; Lin, McGraw, et al., 2003; Lin et al., 1995). *Desmodium paniculatum* (L.) DC. and *D. canescens* (L.) DC., which are native legumes, had greater dry weights when grown under 50 and 80% shade than in full sun. Tall fescue produced more dry matter under 50% shade than in full sun. Kentucky bluegrass (*Poa pratensis* L.), orchardgrass, ryegrass, smooth bromegrass (*Bromus inermis* Leyss.), timothy, and white clover did not show a significant reduction in dry matter when grown at 50% shade compared with full sun. Comprehensive evaluations of approximately 30 yr of shade tolerance testing can be found in Pang et al. (2019a, 2019b).

Edaphic and climatic factors determine what forage species will persist and produce well at each location. Landowners must consider soil type, soil pH, soil fertility, soil drainage, and aspect, as well as rainfall and seasonal temperature fluctuations, when selecting a forage species. For example, bahiagrass and bermudagrass are adapted to parts of the southeastern United States but will not survive farther north. Some legumes, such as alfalfa, require a soil pH of 6.5–7.0 for maximum production, while others, such as birdsfoot trefoil or red clover, produce well at a lower soil pH. Also, a cultivar adapted to local conditions should be chosen. Alfalfa is grown from Florida to Canada; however, cultivars that perform well in Florida are not winter hardy and lack the dormancy response needed to survive farther north. Landowners should be careful to plant only species and cultivars that are adapted to their specific area and site.

Hay production can be a profitable enterprise (Moore & Nelson, 1995; Zulauf, 2018). Additionally, markets for hay are not just local; rather they are often driven by multiple factors that affect broad areas where livestock are raised. Factors such as higher maize prices, drought, flooding, and winter weather events have all contributed to broad markets for hay and quality baled forages, such as alfalfa, and its movement across the United States (McGinnis, 2018; Natzke, 2018). Good hay yields can be expected in an alley cropping system. In a walnut alley cropping practice in Missouri, hay yields in a red clover–orchardgrass mixture averaged 6.3 Mg ha^{-1} (Garrett, unpublished data, 1996). Hay yields were taken when trees grown on a 12.2- by 3.0-m spacing were 15–20 yr old. In another study in Texas, yields of Coastal and NK-37 bermudagrass exceeded 11 Mg ha^{-1} in a young irrigated pecan orchard (Holt, 1973). Sharrow, Carlson, Emmingham, and Lavender (1996) compared the productivity of two Douglas-fir–subterranean clover (*Trifolium subterraneum* L.) agroforestry practices with an improved pasture seeded with subterranean clover. They found that pasture production under agroforestry was 90% that of the open pastures.

Potential forage yields are influenced by tree management. Tree species, the pattern and density of the trees, and their age all affect the microenvironment in the alley where the forage crop is grown. Some tree species may be more adaptable to forage production than others. For example, black walnut has sparse foliage and a short growing season, leafing late in the spring and becoming dormant early in the fall (Garrett & Harper, 1999; Garrett & Kurtz, 1983; Mc Graw et al., 2008). These characteristics favor cool-season forages that produce most of their growth in spring and fall and are intermediate in shade tolerance.

Tree density and distribution pattern modify the microenvironment and may influence forage species selection. Sharrow (1991a) studied the effects of tree planting pattern on forage production in a Douglas-fir agroforest and found that forage production increased rapidly with increasing distance from the tree. Little effect was measured beyond 4.5 m or approximately two canopy diameters. He found that forage production increased as the degree of tree aggregation increased from grids to triple rows of trees (Sharrow, 1991b). In a hay alley cropping practice, the width of the forage harvesting equipment must also be considered when determining tree spacing. A single haying operation requires multiple trips across the field. The herbage must be cut, then raked into windrows to dry. The herbage is baled when it reaches the proper moisture level. Wider alley widths that are multiples of the width of the harvesting equipment make harvesting more efficient. Wider alleys also provide more direct sunlight and air circulation to enhance forage drying.

Specialty Crops

Many niche markets exist for a variety of specialty crops within any locale or region. While many specialty crops can be grown in forest farming practices (Chamberlain et al., 2009), alley cropping can also be adapted to include many of these crops (Figure 7–2d). Species that are light demanding can be established in the alleys, while those requiring some shade can be planted within the row as shade develops.

Landscaping and Christmas tree species can be grown either within the row (between the permanent trees) or between the rows in the alleys. One analysis that compared fuelwood, Christmas trees, and black walnut as the principal trees under alley cropping management demonstrated that Christmas tree alley cropping was the most profitable practice tested (Kurtz, Thurman, Monson, & Garrett, 1991). As the trees grow and begin to shade the alley, emphasis can switch from shade-intolerant to shade-tolerant species, such as redbud (*Cercis canadensis* L.), dogwood (*Cornus florida* L.), and so forth.

Plants that can be marketed for their medicinal, ornamental, or food values also provide unique marketing opportunities in alley cropping (Lovell et al., 2018). Small fruits, such as currant (*Ribes* spp.), blackberry (*Rubes* spp.), elderberry (*Sambucus* spp.), chokeberry (*Aronia* spp.), and serviceberry (*Amelanchier* spp.), can also be grown within tree rows and in alleys with the proper spacing of trees (e.g., Gallagher et al., 2015; Garrett, Kurtz, & Slusher, 1992). Many forest botanicals, such as floral greens, ginseng (*Panax* spp.), goldenseal (*Hydrastis canadensis* L.), mushrooms, and so forth, are highly valued and in demand (Carter, 1996; Persons & Davis, 2005). Worldwide use of herbal medicines rivals that of pharmaceuticals, with an estimated 80% of the worldwide population using herbals as a component of their primary healthcare (Chandler, 1996; Cohn, 1995; Ekor, 2014). However, this market is highly dynamic. Beginning in the late 1990s, the markets for many botanicals showed a significant decline (Blumenthal, 1999, 2002). The fastest growing markets are those for flowers, mushrooms, and medicinal herbs. This increase in demand has increased the destruction of natural ecosystems; consequently, restrictions have been placed on the quantities of native plants that may be harvested on many public lands. These restrictions provide opportunities for landowners adopting alley cropping. Landowners interested in supplying the specialty crop market should focus on high-value products and utilize alley cropping practices that provide the correct microclimatic conditions (Fletcher et al., 2012). While the quality of many special forest products (e.g., ginseng, goldenseal, etc.) decreases when they are produced under artificial shade, alley cropping, like forest farming, creates a natural shade environment permitting the maintenance of plant quality and value. Moreover, the microclimate requirements of species requiring higher light intensities (e.g., *Echinacea*), can be created in the alleys through the proper choice of tree species and their spacing (Gray, Garrett, et al., 2003; Gray, Pallardy, Garrett, & Rottinghaus, 2003).

Other specialty markets have yet to be fully developed. For example, the herbal tea market is growing and offers future opportunities for an alley cropping system (Global Industry Analysts, 2018). Similarly, products such as non-narcotic hemp (*Cannabis* spp.), which is being considered for many uses, including cosmetic products, may have a marketable future and could make an excellent alley crop species (Chandler, 1996).

Biomass Crops

Biomass has a wide range of uses as well as environmental, social, and economic benefits. Biomass supplies about 50% of the world's primary energy needs (Karekezi & Kithyoma, 2006) and is the fourth largest energy source, following coal, oil, and natural gas (Ladanai & Vinterbäck, 2009). The U.S. export markets for wood pellets have increased from <100,000 Mg in 2008 to >3 million Mg in 2015 (Howard et al., 2017). The potential exists in North America for even greater use of wood as an energy source because the worldwide market opportunity is high and it is renewable, with numerous advantages associated with burning wood compared with burning fossil fuels (Holzmueller & Jose 2012; Phillips, 1993).

Biomass may also be used in pulp and OSB manufacturing (Spelter, 1996; Spelter et al., 2006; Wiedenbeck & Araman, 1993). While in the long-term energy use may be a primary market for biomass produced in North America, in the near-term growers must look to the pulp, OSB, and similar markets. In particular, the OSB market offers opportunities in regions where manufacturing plants are located for establishing pine or fast-growing hardwoods. Such trees may be planted in linear alley cropping configurations as either the primary tree species or as the alley companion-crop species.

In particular, poplar and willow (*Salix* spp.) species have great potential as alley crops and have received the most research attention (Ehret, Buhle, Graß, Lamersdorf, & Wachendorf, 2015;

Gruenewald et al., 2007; Volk et al., 2006). Other fast-growing woody species include sycamore (*Platanus occidentalis* L.), some maple species (e.g., silver maple), and birches (*Betula* spp.). Depending on the availability of markets, many N_2–fixing species (i.e., black locust, alders, etc.) (Gruenewald et al., 2007) and coniferous species (Haile, Palmer, & Otey, 2016) could also be viable candidates. Growing woody biomass crops as alley crops has especially taken off in the U.S. Pacific Northwest (Stanton et al., 2002), although early results are promising in the Midwest as well (Moen, 1996).

Markets are also available for herbaceous biomass crops, although most markets are currently local. Production of herbaceous energy crops between dual rows of hybrid poplars is feasible (Colletti et al., 1994). Such combinations are complementary in that high volumes of both herbaceous and woody biomass are produced (Gamble et al., 2014; Haile et al., 2016). Potential dry weight yields of woody biomass may be as high as 9–13 Mg ha^{-1} yr^{-1}, while herbaceous species, such as switchgrass (*Panicum virgatum* L.), can yield as much as 8 Mg ha^{-1} yr^{-1} (Hall, Rosillo-Calle, Williams, & Woods, 1993). Haile et al. (2016) found that pure stands of pine or a monoculture of switchgrass would require 25–74% more land to produce the same amount of yield as an alley cropping system.

The use of herbaceous and woody biomass for energy production is not new (Hall & de Groot, 1988). Such use, however, has been slow to develop in North America. Today, because of the environmental problems associated with fossil fuels and nuclear reactors, as well as the availability of new technologies in the use of biomass, interest in converting biomass to energy is increasing (Gamble et al., 2014; Gruenewald et al., 2007; Holzmueller & Jose, 2012; Zerbe, 2006). Current trends suggest that available markets for biomass will increase during the next few decades, as will opportunities for combining biomass with permanent tree crops in alley cropping.

Economics of Alley Cropping

One of the greatest challenges for the adoption of alley cropping is overcoming the short-run economic perspective that is embedded in conventional agriculture. Nevertheless, alley cropping can be a viable land use option within that short-run economic perspective. Additionally, alley cropping also has the ability to impact long-run economic goals (Cubbage et al., 2012; Frey, Mercer, Cubbage, & Abt, 2010, 2013).

Economic Framework

The basic criteria for economic decision making is that the marginal benefit of any investment of time, money, or resources must exceed the marginal cost of those same investments. The term *marginal* refers to the additional satisfaction derived from consuming an additional unit of a good or service, or the maximum amount an individual is willing to pay for an additional unit of a good or service. From an accounting or financial perspective, the *marginal benefit* is measured in revenue or income and the *marginal cost* is measured in wages, rents, interest, and profit. Although many attempts have been made to put a price on the intangible benefits and costs, the value of the intangible aspects are based on the individual. For example, the value of a tree can be measured based on the potential or actual fruit or timber that the tree produces (value in exchange). Likewise, the value of a tree can be measured based on the shade, habitat, aesthetics, or nostalgia it provides (value in use). Value in exchange is easily measured, whereas value in use depends on the individual and is difficult to measure. For forestry concepts, it is difficult to combine these two values, especially since the ultimate financial value for a tree (e.g., timber harvest) may require that all other uses are stopped.

In comparing the marginal benefit to the marginal cost, it is also important to understand the concept of *opportunity cost*. Opportunity cost is simply the value of what must be given up when a choice is made. For example, in an alley cropping system, land must be converted from the production of one commodity to pursue the establishment of another commodity. From an accounting or financial perspective, if one commodity generates more revenue than another, then planting the lower revenue commodity would reflect a higher opportunity cost and would not be a financially feasible option unless it provided some other measurable benefit. A classic example of this is the standard maize–soybean rotations of the U.S. Midwest. Although maize probably generates the greatest revenue in most years, soybean provides essential N_2–fixing services that are beneficial to the land. High opportunity costs can be tolerated in short-run economic strategies as long as revenues exceed the variable costs of production. In all economic models, if costs exceed benefits, the opportunity cost of continuing in a practice is too high and other investments should be pursued.

Understanding the concepts of marginal benefit, marginal cost, and opportunity cost leads to the final economic concept, which is the *equi-*

marginal principle. The equi-marginal principle states that a person or decision maker will invest resources to the point where the marginal benefit divided by the marginal cost of all options are equal:

$$\frac{MB_a}{MC_a} = \frac{MB_b}{MB_b} = \frac{MB_c}{MC_c} = \ldots = \frac{MB_n}{MB_n}$$

where, the marginal benefit of option *a* (MB*a*) divided by the marginal cost of option *a* (MC_a) is equal to the marginal benefit of option *b* (MB_b) divided by the marginal cost of option *b* (MC_b), and so forth. More specifically, the marginal benefit of growing maize divided by the marginal cost of growing maize must equal the marginal benefit of growing trees divided by the marginal cost of growing trees or any other optional use of the land. If the equi-marginal principle is satisfied, then the landowner has achieved the optimal arrangement of land use. Additionally, if the marginal benefit divided by the marginal cost of one land use option is greater than the marginal benefit divided by the marginal cost of another land use option, then the prudent investor would invest more into the land use option that has the greatest marginal benefit/marginal cost ratio.

$$\frac{MB_a}{MC_a} > \frac{MB_b}{MC_b}$$

For example, if the marginal benefit divided by the marginal cost of planting maize is greater than the marginal benefit divided by the marginal cost of pasture, then the landowner should plant maize and forego establishing pasture.

These three economic concepts provide the basic economic framework in which all agroforestry practices should be designed and established. It should be noted that the discussion of the equi-marginal principle and the discussion of marginal benefit and marginal cost do not focus solely on marginal revenue or profit. Agroforestry, and alley cropping in particular, is clearly a practice with benefits that are not just reflected in revenues or profits.

Alley Cropping in the Short Run

In economics, the short run is a period of time in which at least one factor of production is fixed, meaning it cannot be changed. The factors of production include natural resources, human resources, human-made resources, and creative resources. In most cases, the natural resources are assumed to be fixed in the short run. More specifically, the short run would indicate that the parcel of land used in production cannot be increased, and all short-run production decisions are based on the limits of that natural resource asset. As a result, economic decisions seek to maximize benefits given a fixed natural resource.

An alley cropping system that is designed to provide complementary income or benefit streams can meet the demands of the short-run economic goals. Alley cropping systems that incorporate mutually beneficial species or alley crops can increase the total benefit derived from a fixed land resource. In the case of mutually beneficial alley crops, the total benefit derived from the whole may be greater than the sum of the individual parts. For example, wide alleys in row crops protected by tree rows have been shown to increase the overall yields from some row crops (Beaudette et al., 2010; Brandle, Hodges, & Zhou, 2004; Kort, 1988). Similarly, alley cropping of black walnut and alfalfa has been shown to improve pest management and alfalfa yields (Stamps, McGraw, Godsey, & Woods, 2009). Other examples of short-run economic benefits of alley cropping include chestnut and pumpkin (*Cucurbita* spp.), pecan and row crops, and almond (*Prunus amygdalus* Batsch). From a practical perspective, many orchards throughout the central United States unknowingly engage in alley cropping by mowing and selling hay from the alleys between their orchard trees. A short-run economic analysis of these practices would indicate that alley cropping provides the highest net benefit to the landowner given limited or fixed land resources.

Alley Cropping in the Long Run

The long run is defined as a period of time in which all factors of production are variable. In other words, land can be added or reduced, capital investments can be increased or decreased, and so on. From the alley cropping perspective, a long-run strategy may be to transition land from one productive use to another. For instance, as environmental concerns and production costs increase, landowners may choose to transition row crop production on highly erodible land into savannah management or recreational use. Alley cropping can provide the most economical transition method for this long-run strategy. More specifically, alley cropping allows the landowner to continue to derive an income from land that is being transitioned from one use to another. Although the cash benefit from the row crop and the non-cash benefit of the transitional land use may be difficult to compare directly, the equi-marginal principle would still be in effect, and the landowners would manage their land in a manner where the marginal benefit/marginal cost ratio of each land use would be equal in the long run.

From the perspective of long-term environmental transition strategies, alley cropping can reduce the short-run financial costs by providing offsetting revenue opportunities. A landowner who wishes to transition land from row crop production to timber or other permaculture opportunities can still earn revenue from crops grown in the alleys while the land is transitioning.

Economic Opportunity

Alley cropping provides the opportunity for landowners to diversify their production system or transition their production system from an annual crop to a perennial crop system. Several examples exist of perennial crop systems performing as well financially as annual crop systems. For example, pawpaw can produce up to 176 kg $tree^{-1}$ of fruit with market prices of up to US\$8.82 kg^{-1}. Average yields of pawpaw can be around 13.6 kg $tree^{-1}$, with full commercial production occurring in the seventh year after planting. Financially, 73 trees ha^{-1} could generate a gross income from \$8,698 to \$23,194 ha^{-1} (Byers, 2015). To put this into perspective, maize can yield 15.5 Mg ha^{-1} on the best cropland, generating a gross income of less than \$2,114 ha^{-1} at current market prices (\$137.79 Mg^{-1}). In addition to the revenue generated by the pawpaw, financial performance can be further enhanced by a hay crop or other annual crops grown in the alleys between the trees.

Pecan is a common tree nut that provides a great market opportunity throughout a large portion of the United States. About one-third of all domestically produced pecans are grown in Georgia, with Mitchell County, GA, being the top producer. In 2017, Mitchell County generated about \$40 million in revenue from pecan production (University of Georgia Extension, 2018). Improved cultivars of pecan can reach commercial maturity as early as 7 yr after planting (Nesbit, Stein, & Kamas, 2013). Low annual input costs and global markets enhance the opportunities from pecan.

A planting of Chinese chestnut trees in Missouri indicates that each tree can yield as much as 54 kg yr^{-1} of nuts at maturity, with a market value ranging from \$4.41 to \$11.03 kg^{-1}. Chestnut trees can reach productivity in <5 yr. Returns on investment can be as high as 12% over 40 yr.

These cases are just a few opportunities from alternative income sources in an alley cropping system. The biggest benefit of perennial crops is the absence of annual planting costs. For annual crops, these costs include the extensive use of fossil fuels. Alley cropping systems that include forage production along with perennial crop production have the lowest annual input costs and provide the greatest environmental benefits in terms of reduction of soil loss and increased wildlife habitat.

Economic Example

Comparing a conventional soybean rotation with a soybean–chestnut alley cropping system, the following assumptions are made:

> In the conventional dryland soybean system, expected yields, expected prices, and operating costs are based on the Crop Budget Generator model developed by the University of Missouri Food and Agriculture Policy Research Institute
>
> (Food & Agricultural Policy Research Institute, 2017).

> For the soybean–chestnut system, the same costs for dryland soybean are used; however, the costs are reduced by 25% to reflect the amount of land that is taken out of soybean production and planted with chestnut trees. The expected chestnut yields, expected chestnut prices and operating costs for chestnut are based on the Chestnut Decision Support Tool developed by the University of Missouri Center for Agroforestry using a fair site and \$4.41 kg^{-1} for the nuts
>
> (University of Missouri Center for Agroforestry, 2013).

Analyzing these two systems over 20 yr, the alley cropping system has higher financial returns than the conventional soybean system (Table 7–1).

Ecological Benefits

Overyielding

The most important ecological benefit of alley cropping is reducing the total area required for agricultural production via overyielding. Overyielding occurs when the trees and crops in alley cropping exhibit higher productivity, collectively, than tree and crop monocultures (Jose

Table 7–1. Comparison of conventional soybean production with soybean–chestnut alley cropping systems using the financial indicators of net present value (NPV), internal rate of return (IRR) and annual equivalent value (AEV) during a 20-yr period.

System	NPV at 6%	IRR	AEV
	US\$ ha^{-1}	%	US\$ ha^{-1}
Conventional soybean	6,761.85	6	589.53
Soybean–chestnut alley cropping	31,301.84	41	2,720.32

et al., 2004). Considerable research in temperate alley cropping has revealed that land-use efficiency can increase by 25–200% in alley cropping compared with tree and crop monocultures (Dubey, Sharma, Sharma, Sharma, & Kishore, 2016; Graves et al., 2007, 2010; Rivest, Cogliastro, Bradley, & Oliver, 2010). Dupraz and Liagre (2008) found this to be the case in an alley cropping study of walnut and wheat. They determined that a 140-ha farm where trees and crops were grown separately would be required to produce as much cereal and walnut timber as a 100-ha farm using alley cropping. Similarly, 25–74% more land was required in a monoculture to produce the same amount of yield when alley cropping switchgrass between loblolly pine (Haile et al., 2016).

The most common way to measure overyielding is the land equivalent ratio (Mead & Willey, 1980). This ratio compares the yields from growing two or more fully integrated plant species on a site with the yields of the same plants grown in a monoculture. This allows the ratio to be used in estimating the land base needed in a monocrop to equal the same production in an alley crop (Gruenewald et al., 2007).

A variety of mechanisms can drive overyielding. Niche differentiation occurs due to interspecific differences in resource use (e.g., of light, soil nutrients, pollinators). In alley cropping, this commonly occurs due to phenological differences between the tree and crop species (Dupraz et al., 2019). Positive, facilitative interactions among species can also drive overyielding. The most obvious example of facilitative interactions in alley cropping is N_2–fixing trees providing N for crops and crops, such as soybean, fixing N_2 that is made available to companion trees. However, other examples also exist. Deep tree roots can mine P from deep soil layers and return it to the soil surface via litter for the alley crop to utilize (Diemont et al., 2006). Trees can also moderate crop yield loss during drought (Nasielski et al., 2015) and move water from deep to shallow soil in a process called *hydraulic lift* (Burgess, Adams, Turner, White, & Ong, 2001). Finally, overyielding can also be driven by reductions in negative plant–soil feedbacks (Vandermeer, 1989).

Soil Stabilization

Many hectares of farmland have been abandoned in the past 200 yr due to soil erosion. In the United States, croplands lose soil at an average rate of 15 Mg ha^{-1} yr^{-1}, with much farmland losing at a rate above the sustainable level (Jose, Gold, & Garrett, 2018; Pimentel, 2006; Pimentel et al., 1995). The loss of productivity due to loss of arable land in the United States was approaching $40 billion yr^{-1} in the early 2000s (Pimentel, 2006). Farm consolidation has sped up the removal of shelterbelts and hedgerows that once served to protect soils from erosion. Furthermore, forests have been cleared from steep slopes and row crops planted on lands that, without the protection of a cover, erode at a level above tolerance (USDA, 2018).

Perhaps one of the most researched aspects of alley cropping is its value in controlling erosion. In the 1990s, the University of Missouri Center for Agroforestry established a paired watershed study consisting of a control, a watershed with grass buffers, and an alley cropped watershed consisting of oak, grass, and legume buffers. Early results demonstrated the advantages of alley cropping in reducing runoff and erosion under conditions of a maize–soybean rotation on a claypan soil (Udawatta, Anderson, et al., 2006; Udawatta Motavalli, et al., 2006; Udawatta et al., 2002, 2011). Improvements were attributed to changes in both physical and biological properties. Not only do alley crop buffers physically filter sediment that is moving in surface flow, it also reduces erosion by creating a zone for greater infiltration while altering the biological properties of the soil in the tree rows and alleys to create a soil surface that is less susceptible to erosion (Anderson, Udawatta, Seobi, & Garrett, 2009; Udawatta, Kremer, Garrett, & Anderson, 2009).

These early studies showed that the presence of tree–grass–legume buffers significantly increased the water-stable aggregates, soil C, soil pore size, functional diversity of enzyme activity, and other soil parameters that lead to the reduction of erosion. Of particular interest is the percentage of water-stable aggregates, which is a measure of resistance to the breakdown of soil by water and mechanical mechanisms. Kremer and Kussman (2011), working with a pecan–kura clover (*Trifolium ambiguum* M. Bieb.) alley cropping system found that water-stable aggregation improved by 50% and surface-soil shear strength improved significantly over that in a crop field. Studies in Europe also found that alley crop tree buffers placed on the contour could reduce erosion by as much as 70% (Palma, Graves, Bunce, et al., 2007).

Findings in the tropics have long suggested that a properly designed alley cropping practice helps regulate erosion by serving as a permeable barrier, developing a network of surface roots, increasing organic matter, creating a berm from soil accumulation on the upslope side of buffer strips, and by providing a source of ground cover (Kang, 1997). On moderate slopes (e.g., 15%), closely spaced trees and shrubs (i.e., 2–6 m) combined with grasses can slow the movement of water over a surface and filter out sediments and other contaminants (Jose et al., 2012, 2018; Senaviratne et al., 2012).

In designing alley cropping strips for soil erosion control, it may be undesirable to maximize the ground surface area covered by the upper canopy. High tree canopies provide little benefit in reducing the kinetic energy of raindrops and may actually increase the energy if the size of the drops increases on a leaf before being released. Soemarwoto (1987) reported a 135% increase in rainfall erosive capacity beneath a bamboo (*Bambusa* spp.) stand compared with the open. Wiersum (1985) found that by removing a high canopy of *Acacia* that had little understory, the erosive power of raindrops was reduced by 24%. Low-growing coffee (*Coffea arabica* L.) and tea [*Camellia sinensis* (L.) Kuntze] bushes beneath a plantation of shade trees significantly reduced the erosive power of raindrops compared with shade trees alone (Wiersum, 1984).

In contrast, species with low canopies and vertical stratification of canopies work the best to break the fall of raindrops (Kremer & Kussman, 2011). On steep slopes in Colombia, rows of *Gliricidia sepium* (Jacq.) Walp. reduced soil losses in maize plantings from 23 to >34.5 Mg ha^{-1} yr^{-1} to as low as 11.8 Mg ha^{-1} yr^{-1} (Van Eijk-Bos, Moreno, & Van Dijk, 1986). Similarly, in studies in Nigeria on *Leucaena* and *Gliricidia*, soil losses were reduced by >90%. Comparable reductions were observed in both total runoff and nutrient losses (Lal, 1984). While there is evidence that permeable runoff barriers alone are not as effective in reducing runoff as a heavy soil cover, a combination of trees and shrubs with grasses can effectively regulate runoff and is more effective than grass strips used alone (Maass, Jordan, & Sarakhan, 1988).

The ability to provide direct ground cover is the most important trait of an alley cropping system in maximizing soil stabilization (Young, 1989). In studies of *Acacia* plantations, 95% of the erosion control from trees on the site was attributed to the litter cover (Wiersum, 1985). Tree species with foliage that is slow to decompose can further maximize ground cover and erosion control.

Nutrient Retention and Water Quality Improvement

Since approximately two-thirds of all roots in annual and perennial herbaceous crops are found in the upper 50 cm of soil (Black, Masters, LeBauer, Anderson-Teixeria, & DeLucia, 2017), nutrients can leach from the system by getting past this point. Agricultural N leaching in North America accounts for about 80% of the N contributing to hypoxia conditions in the Gulf of Mexico (David, Drinkwater, & McIsaac, 2010; USEPA, 2007). The USEPA (2009) suggested that agricultural contaminants are the leading cause of water pollution in 44, 64, and 30% of evaluated river, lake, and estuary areas, respectively. Contaminants can be lost by surface movement and by seepage.

Research on riparian buffer strips in the United States has shown that strips 7 to 27 m wide can filter out 95% or more of the sediment moving from adjacent fields to streams. Reductions of 80% and more in contaminants, such as phosphate and nitrate that bond to sediment, have also been reported from the use of vegetative buffer strips along streams (Lee, Isenhart, & Schultz, 2003; Osborne & Kovacic, 1993; Schultz, Isenhart, Simpkins, & Colletti, 2004). In most instances, the riparian strips studied are significantly wider than the buffer areas that would normally be incorporated into an alley cropping practice. A riparian buffer, however, is designed to be the last line of defense against pollutants entering streamways. On uplands, which are the primary source of sediment and pollutants, alley cropping provides the first line of defense.

Alley cropping can reduce agricultural nonpoint-source pollution by increasing water-stable aggregates, soil C, microbial activity, functional diversity of enzyme activity, and deep roots (Lovell & Sullivan, 2006; Lowrance & Sheridan, 2005; Seobi, Anderson, Udawatta, & Gantzer, 2005; Udawatta et al., 2009). An experiment in northeastern Missouri on alley cropping in a "paired" watershed project demonstrated significant reductions in the loss of sediment, P, and N from a site maintained in a maize–soybean rotation (Udawatta et al., 2002). Tsonkova et al. (2012) found that agroforestry buffer strips reduced the loss of total N by 20–90%, NH_4–N by 20–100%, NO_3–N by 28–100%, total P by 8–91%, and PO_4–P by 58–100%. Allen et al. (2004) found that alley cropping reduced NO_3 at 0.3 m by 46% and at 0.9 m by 71% compared with a conventional agricultural monoculture crop. Wolz, Branham, and DeLucia (2018), in comparing a mixed nut–fruit tree alley cropping system to a maize–soybean conventional monoculture system, found an 82–91% reduction in NO_3 loss through leaching. In Belgium, Meiresonne, Schrijver, and Vos (2007) found that nutrient recycling in an alley cropping system with 18-yr-old poplar was very efficient, and only small quantities of nutrients were lost from the system. In combination with the natural deep-rooting characteristic of tree species, Mulia and Dupraz

(2006) demonstrated the plasticity of tree root systems that are displaced by root competition to deeper soil depths. To demonstrate just how effective this can be, Jorgensen and Hansen (1998), in fertilizing trees grown for biomass in an alley cropping practice, found that the average quantity of N leached from the tree buffer was 15 kg N ha^{-1} compared with 70–120 kg N ha^{-1} for a cultivated field planted to cereal crops. Similarly, Ryszkowski and Kedziora (2007) found that leaking NO_3 concentrations from cropland in Poland were reduced by 76–98% after passing under a shelterbelt, which had many of the characteristics of an alley cropping buffer.

Moreover, the environmental significance of having greater concentrations of roots deeper in the soil profile with alley cropping compared with a conventional crop monoculture has been clearly demonstrated (Beaudette et al., 2010; Dougherty, Thevathasan, Gordon, Lee, & Kort, 2009). Species whose roots are intermediate in depth as well as being deep rooted should also be established to capture nutrients and pesticides that might escape the more shallow-rooted species placed in the buffer to hold the surface soil (Allen et al., 2004; Jose et al., 2004).

In addition to reducing N leaching loss, trees also serve as a potential mitigation tool for climate change by significantly reducing N_2O, a known greenhouse gas involved in climate change (IPCC, 2014). Reductions in the emissions of N_2O of as high as 70% have been reported from incorporating woody perennials into monoculture annual crops (Amadi, Van Rees, & Farrell, 2016; Baah-Acheamfour, Carlyle, Lim, Bork, & Chang, 2016; Schoeneberger et al., 2012; Wolz, Branham, & DeLucia, 2018).

Bioremediation

Toxic agricultural chemicals, including herbicides and other pesticides, can accumulate in soil. Within the soil and litter layers, these toxic chemicals can be transformed to nontoxic forms by microbial decomposition and oxidation or reduction. Degradation or removal of these chemicals through biological processes before they move offsite is known as *bioremediation*. The literature on the bioremediation of agricultural contaminants is large. However, the value of alley cropping as a bioremediation tool has been less studied. The University of Missouri Center for Agroforestry began in-depth studies of this relationship in the early 2000s (Lin, Goyne, Lerch, & Garrett, 2010; Lin, Lerch, Garrett, & George, 2005, 2008; Lin, Lerch, Garrett, Jordan, & George, 2007; Lin, Lerch, Goyne, & Garrett, 2011; Lin, Lerch, et al., 2003; Lin et al., 2009).

To create a successful tree-based vegetative buffer zone for bioremediation, certain criteria must be met in species selection. The groundcover must be tolerant of the herbicides used and it must have some degree of shade tolerance. Moreover, it must possess the desired bioremediation capacity to adsorb and hold or further degrade the targeted agricultural chemicals (Lin, Lerch, Jordan, Garrett, & George, 2004; Lin et al., 2011).

One chemical that is of particular concern is atrazine. It has been in use for >40 yr and tens of millions of kilograms are applied annually to maize in the United States alone. Due to its persistence in the soil and its discovery in groundwaters, public concern has increased. Other chemicals that have been studied and are of interest are isoxaflutole, introduced as a possible substitute for atrazine in sensitive watersheds, and glyphosate. Studies by Lin et al. (2007, 2008, 2011) have concentrated on the bioremediation value of many grass species as candidates for incorporation into alley cropped tree buffers. The grass species found to be relatively tolerant to atrazine and isoxaflutole are switchgrass, tall fescue, and smooth bromegrass (Lin, Lerch, et al., 2003; Lin et al., 2004). While switchgrass proved in some ways to be superior to fescue and bromegrass in bioremediating atrazine, when isoxaflutole, NO_3, and glyphosate are included, fescue and bromegrass also proved superior to other grasses tested and are recommended.

Others have also demonstrated the benefits of vegetative buffer strips in reducing the movement of agricultural chemicals from croplands (Krutz, Senseman, Dozier, Hoffman, & Tierney, 2004; Schultz et al., 2000). In a study using a 9-m vegetative buffer strip, Hoffman, Gerik, and Richardson (1996) demonstrated a 44–50% reduction in atrazine levels. There are several physical, chemical, and biological mechanisms potentially involved in the bioremediation of soil-borne pesticides. They can be intercepted by roots and vegetation residue through physical adsorption and filtration. They can be metabolized by bacteria or taken up by plants. In addition, improvements in soil organic matter, porosity, and other soil attributes may result in increased adsorption and chemical hydrolysis within the buffer (Lin et al., 2004).

Other potentially serious contaminant types in alley cropping include veterinary antibiotics used in the livestock industry. Studies have found that 30–80% of an antibiotic dose can rapidly pass through the gastrointestinal tract of an animal and be deposited on the site in an unaltered state (Halling-Sørensen, Sengeløv, & Tjornelund, 2002; Sarmah, Meyer, & Boxall, 2006). Land application of animal manure in agriculture is believed to be the primary source of veterinary antibiotics released into the environment in North America (Boxall, 2008). These antibiotics may undergo biotic or abiotic degradation, sorption to soil minerals and organic matter, and/or movement to streams, rivers, and estuaries through leaching or surface runoff (Boxall, 2008). While few studies have investigated the value of buffers in removing antibiotics from surface runoff, Lin et al. (2010) and Chu, Goyne, Anderson, Udawatta, & Lin (2009) demonstrated plant differences in the capacity to dissipate two commonly administered veterinary antibiotics, sulfamethazine and tetracycline. Because of the highly unstable nature of tetracycline in a soil environment, it was difficult to determine the effects of the plant species studied. However, hybrid poplar trees showed highly increased dissipation of sulfamethazine compared with grass species or a control of soil only. The results indicated that the poplar stimulated microbial activity in the root zone more than the grass species. Increased microbial enzymatic activity was directly correlated with increases in sulfamethazine dissipation. Poplar has also been found by others to be a species worthy of incorporating into buffers. Burken and Schnoor (1997) studied its effects on the uptake and metabolism of atrazine, while Jordahl, Foster, Schnoor, and Alvarez (1997) reported on the effects of hybrid poplar on microbial populations important to hazardous waste bioremediation.

Wind Protection

Many findings on windbreak effects in the Great Plains have direct application in alley cropping. However, the effects of the trees in creating microclimates may be greater in alley cropping because of the closer spacing between tree rows. A study in Nebraska clearly demonstrated the effects of creating narrow alleys on crop yields. Two-meter-high, slat-type windbreaks (43% openings) were spaced 12.7 m apart and oriented east–west. Tomato (*Lycopersicum esculentum* Mill.) and snap bean (*Phaseolus vulgaris* L.) were planted in the alleys between the fences and in the open (controls). The microclimate created in these simulated alleys resulted in earlier germination, accelerated vegetative growth, earlier ripening of fruit, and 16–44% increases in the tomato and snap bean yields (Bagley, 1964). Except for accounting for competition between root systems if trees replaced the slat fences, the design used in this trial created conditions comparable to those found in the early years of an alley cropping system. If lateral roots of the trees were pruned, conditions would be even more similar.

Studies conducted on alley cropping *Paulownia* with wheat in temperate China demonstrated enhanced microclimatic conditions and increased wheat quality (Wang & Shogren, 1992). Wei (1986), studying the effects of alley cropping on microclimatic conditions, found that wind speeds were reduced by 45–50%, relative humidity was increased by 5–17%, evaporation was reduced by 15–30%, and water content in the tillage layer of the alley was increased by 5–15%. Similar findings were reported by Böhm, Kanzler, and Freese (2014) and Tsonkova et al. (2012). In Nebraska, Brandle et al. (2004) found that hedgerows altered wind movement patterns both upwind and downwind.

Carbon Sequestration

Global climate change is the preeminent crisis of our time (IPCC, 2014). Row-crop agriculture covers >1.28 billion ha of land globally (FAO, 2017), and it is estimated that the agricultural sector is the source of 10–12% of global anthropogenic greenhouse gas emissions (IPCC, 2014). Alley cropping has the potential to significantly reduce CO_2 and other greenhouse gas emissions in North America and worldwide (Dixon et al., 1994; Fargione et al., 2018; Mosquera-Losada, Freese, & Rigueiro-Rodroguez, 2011; Udawatta & Jose, 2012; Wolz, Lovell, et al., 2018). The amount of C sequestered can be large despite the small land area planted to trees due to the efficiency of C sequestration per unit area (Schoeneberger, 2009; Schoeneberger et al., 2012). This is due to the favorable conditions under which trees are grown in alley cropping. If merged with row crops, the trees benefit from being planted on quality sites, reaping the benefits from fertilization and weed control of the row crop and having increased light from growing under more open conditions than in a pure stand or woodlot setting. Although research is limited on C sequestration in alley cropping, existing results demonstrate substantial potential in alley cropping practices for capturing and storing C in above- and belowground biomass components (Chatterjee et al., 2018; Lorenz & Lal, 2014; Nair, Nair, Kumar, & Showalter, 2010; Stefano & Jacobson, 2017; Upson & Burgess, 2013).

Dixon, Winjum, & Schroeder (1993) estimated the aboveground storage of C by agroforestry practices to be about 63 Mg C ha^{-1} in temperate regions. Thevathasan and Gordon (2004), focusing on a 13-yr-old alley cropping trial in Canada, projected that it sequestered 14 Mg C ha^{-1} in aboveground biomass and another 25 Mg C ha^{-1} belowground. Differences in the sequestering potential among tree species are to be expected.

In a 25-yr-old alley cropping trial, hybrid poplar sequestered and stored 113.4 Mg C ha^{-1}, red oak stored 99.2 Mg C ha^{-1}, and black walnut stored 91.5 Mg C ha^{-1} (Wotherspoon, Thevathasan, Gordon, & Voroney, 2014). A soybean sole-cropping system stored only 71.1 Mg C ha^{-1}. Net C flux for poplar, oak, walnut, and soybean was 2.1, 0.8, 1.8, and −1.2 Mg C ha^{-1} yr^{-1}, respectively. Alley cropping proved superior to the conventional agriculture system in sequestering and storing C for all tree species tested.

In a comprehensive review of the effects of agroforestry practices in general on C sequestration, Dong-Gill, Kirschbaum, and Beedy (2016) estimated that agroforestry stands (at an average age of 14) having a tree–crop coexistence, such as in alley cropping, sequestered 7.2 ± 2.8 Mg C ha^{-1} yr^{-1}, with 70% in the aboveground biomass and 30% in the soil. To compare the potential of agroforestry strategies to sequester C in temperate North America, Udawatta and Jose (2012) estimated that 15.4 million ha (10% of the total cropland) under alley cropping would sequester 52.4 Tg C yr^{-1}. However, their estimate did not include the millions of hectares of marginal land that are going unused or are currently in forage crops for hay production or pastures.

Increased soil organic matter in alley cropping is probably due to an increase in root mass and the dropping of litter from trees and other perennial crops (Amadi et al., 2016; Paudel, Udawatta, & Anderson, 2011; Udawatta et al., 2009). In a comprehensive evaluation of the effects of soil enzyme activities and physical properties in a watershed managed under agroforestry and row cropping, Udawatta et al. (2009) reported that the C percentage in soil under row-crop management was 1.77%, while agroforestry buffers had 2.19%. Similarly, Amadi et al. (2016) found that the soil close to a windbreak buffer contained 27% greater organic C storage in the top 30 cm than did cropped fields.

Wildlife Benefits

The incorporation of alley cropping into an agricultural landscape increases the habitat diversity for wildlife (Gibbs et al., 2016; Millspaugh et al., 2009). Properly designed tree rows may have great value in enhancing wildlife numbers. This enhancement can result from an increase in the amount of edge, an increase in the diversity of vegetative types, or both.

For example, studies on bird use of alley cropped areas and farmstead shelterbelts have clearly shown greater diversity and density of bird species within the linear plantings of trees and shrubs than in cropped areas (Gibbs et al., 2016; Williams, Koblents, & Gordon, 1996; Yahner, 1982a, 1982b). Increased vertical complexity of managed grasslands in North Dakota has also been shown to increase bird usage (Renken, 1983). Studies of bird use of fencerows, which in many ways resemble a tree-row buffer strip in alley cropping, have shown that fencerows with more woody cover have greater total numbers of birds and more bird species (Best, 1983; Shalaway, 1985). Similarly, grass cover strips in crop fields have been found to contain nest densities of game and nongame birds that are one or two orders of magnitude greater than in adjacent crop fields (Wooley, Best & Clark, 1985).

Yet another advantage of linear tree–shrub–herbaceous plantings is found in their edge effect. Because of their extensive edge/volume ratios, alley cropping buffer strips can serve as a reservoir for many kinds of insects that attract bird species (Morandin & Kremen, 2013; Stamps, Woods, Linit, & Garrett, 2002). Strong correlations between the amount of edge and bird species abundance have been observed (Best, Whitmore, & Booth, 1990).

Alley cropping soils have also been shown to contain significantly greater living organisms (microbes, earthworms, etc.) than conventionally cropped fields (Bainard, Klironomos, & Gordon, 2011; Mungai, Motavalli, Kremer, & Nelson, 2005; Tsonkova et al., 2012; Weerasekara, Udawatta, Jose, Kremer, & Weerasekara, 2016). Significant increases in microbial activity in alley cropped sites have been reported (Weerasekara et al., 2016), including increases in the presence of mycorrhizae (Bainard et al., 2011; Unger, Goyne, Kremer, & Kennedy, 2013; Uri, Tullus, & Lohmus, 2002).

Insect Pest Control

Crop protection from insect pests can result from any one of five mechanisms (Stamps & Linit, 1997; Stamps et al., 2002):

- Reduced visibility of pest target
- Diluted number of potential hosts with increased plant diversity
- Physical interference with pest movement
- Creation of environments that are less favorable to building pest populations
- Creation of environments that are more favorable to parasites and predators

Each of these factors can come into play in alley cropping systems, thereby enhancing crop and/or tree growth and yield. Spatial patterns of vegetation strongly influence the biology of arthropods both directly and indirectly (Graham & Nassauer, 2019). In particular, buffer areas of linear tree plantings, characteristic of alley cropping, that are distinctly different from the crops on either side can play significant roles in plant–insect interactions (Akbulut et al., 2003). Bugg, Sarrantonio, Dutcher, and Phatak (1991) found that annual legumes and mixtures of annual legumes and grasses used as cover crops in pecan orchards sustained lady beetles (Coleoptera: Coccinellidae) and other arthropods that may aid the biological control of pecan pests.

Because of the edge/volume ratio found in corridors associated with alley cropping, organisms found within the corridors tend to have strong interactions with adjacent areas found on both sides. Tree–shrub–groundcover strips may increase or decrease pest density in adjacent crops, depending on multiple factors (Kranz, Wolz, & Miller, 2018). These natural habitats may serve as a reservoir of pest species, as overwintering sites for predators, or as a source of food. Alternatively, fields in association with wooded edge habitats have been reported to harbor fewer pests than fields without wooded edge. An increased abundance of natural enemies and more effective biological control of pests are often found where uncultivated habitats, especially corridors, occur with crops (Linit & Stamps, 1996; Stamps et al., 2002).

Some forest-dwelling carabids— predators of insect pests—are typically never found more than about 45 m from a forest edge. With narrow linear plantings, such as with alley cropping, carabids can move deeper into crop fields to help control pests. One study demonstrated that tree corridor plantings resulted in carabids being able to move as far as 20 km into unforested agricultural landscapes (Burel, 1989). Moreover, studies in Michigan have shown significantly higher ichneumonid wasp parasitism of European corn borer larvae near wooded edges than near non-wooded edges or field interiors (Landis & Haas, 1992). Similarly, Girma, Rao, and Sithanantham (2000) found a significant reduction in maize stalk borers with tree intercropping compared with a monoculture.

Biological pest control in alley cropping could lead to a reduction of pesticides used and their potential for buildup on site. Linit and Stamps (1996) and Stamps et al. (2002) found that growing alfalfa between rows of walnut trees compared with in the open significantly reduced the population of alfalfa weevils to the extent that only one, instead of two, application of insecticide was required for weevil control. They attributed this to the increase in predatory insects found within the alleys.

Potential for Alley Cropping in the United States

Existing Alley Cropping Area

As a conservation practice, alley cropping use has seemingly been declining in the United States in recent years. In 2002, approximately 29,000 ha of cultivated and uncultivated cropland were established with alley cropping (USDA, 2015). By 2015, this number had declined to 21,974 ha (USDA, 2018). However, this may not present a clear picture of the land area on which alley cropping is being adopted. The definition of the conservation practice of *contour orchard* is the planting of orchards, vineyards, or other perennial crops so that all cultural operations are done on or near the contour. While this definition provides leeway to include other practices besides alley cropping, the contour orchard practice meets the alley cropping definition in many instances. Therefore, the 44,596 ha enrolled in 2015 with the USDA may include land uses that meet the definition of an alley cropping practice (USDA, 2018). If the specific alley cropping practice of the USDA–NRCS is broadened to include those enrolled as contour orcharding, then the total hectares that can be identified as alley cropping significantly increase.

Target Adoption Areas

Alley cropping adoption may initially have the most success on "marginal" lands (i.e., lands that are marginal in productivity for conventional row crop production). For example, a vast area of land within the generally highly productive U.S. Midwest can be unprofitable in many years (Brandes et al., 2016). These marginal row crop areas can easily be identified using GIS, and targeted adoption in these areas could lead to disproportionately higher ecological benefits (Brandes et al., 2018; Mattia, Lovell, & Fraterrigo, 2018).

Agricultural areas with erosion concerns (i.e., Land Capability Class II or higher and Capability Subclass e) are also prime candidates for conservation practices that include alley cropping. In 2015, >75.9 million ha were designated this way in the United States; this constitutes approximately 51% of all U.S. cropland (USDA, 2018).

Across the United States, >47 million ha of pastureland exist and another 164 million ha of rangeland. Approximately 7 million ha of pastureland have high potential for conversion to cropland and could be alley cropped. Another

16 million ha have medium potential for conversion. More than 9 and 16 million ha of the rangeland have high or medium potential, respectively, for conversion to cropland. These would probably accommodate some form of alley cropping, if only for hay and wood production (USDA, 2018).

While these ecological considerations can help identify target adoption areas for alley cropping, they have so far failed to drive adoption due to the lack of robust market mechanisms to monetize their value. Instead, profitability remains the primary driver of alley cropping and other alternative agricultural practices. From this perspective of ignoring ecological drivers for alley cropping adoption, Wolz and DeLucia (2019a) conducted a comprehensive economic analysis comparing a pure maize–soybean rotation to an alley cropping rotation with black walnut. Their results indicated higher profitability with alley cropping on >23% of cropland in the Midwest (Figure 7–9). Furthermore, the economic competitiveness of alley cropping was not correlated to how marginal the soil was for crop productivity, indicating that alley cropping can be economically favorable even on highly productive cropland.

Timber Markets

Since the early 1990s, dramatic changes have created unique opportunities for nonindustrial private-forest landowners and landowners with unforested lands who are interested in planting trees. Many domestic sources of raw wood products that have been relied on during the past in North America have greatly diminished. This is especially true in the United States, where a reduced emphasis on wood production on federal lands has decreased harvesting from national forests. Moreover, changes in management strategies on many state-owned lands have also decreased the volume of wood harvested and general management conducted. In the United States, public lands have never been the biggest source of wood—in 2012, public forestland represented 42% of U.S. forests but accounted for only 11% of the 2016 annual volume of growing stock removed (Oswalt, Miles, Pugh, & Smith, 2018).

Private forestry contributes significantly toward meeting U.S. wood needs. In 2017, farmers and other private landowners represented approximately 38% of all the forested lands in the United States (Oswalt et al., 2018). Because they—in combination with the forest products

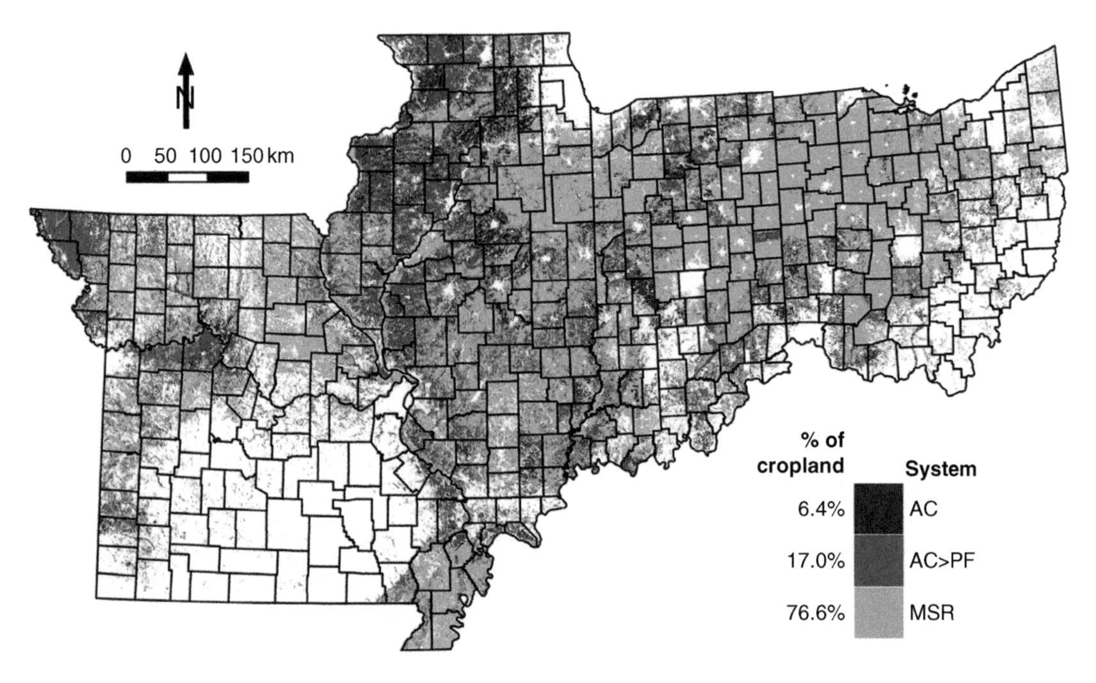

Fig. 7–9. A map of four midwestern states (Missouri, Illinois, Indiana, and Ohio) indicating the distribution of currently cultivated land on which black walnut alley cropping (AC; blue and red areas) or black walnut production forestry (PF; red areas) are more profitable to the landowner over the long-run, assuming a 5% discount rate. Gray areas represent currently cultivated land that is more profitable remaining in the maize–soybean rotation (MSR) than converting to black walnut alley cropping, although this analysis did not include alley cropping with any other tree species; white areas are currently not cultivated land (reproduced from Wolz & DeLucia, 2019a).

industry—satisfied 89% of the total timber needs in the United States in 2016, one or both sources must make up the loss from public lands. While some potential exists for increasing yields on industry land, the greatest potential is found in the nonindustrial private sector. The loss of raw wood material from federal lands is especially significant to the small family-farm landowner looking for new financial opportunities. Because projected increases in U.S. wood consumption come at a time of decreased wood availability from public lands, the owners of small farms have a unique opportunity. Consumption has been projected to rise approximately 140% from 1996 to 2050 (Haynes, 2003). With an abundance of idle or near-idle farmland available for tree planting, wood production could become part of a strategy to put these lands back into production while providing a source of income for the owner.

There were just over 2million farms in the United States in 2016, making up slightly more than 368 million ha, most of which produced few food crops (National Agricultural Statistics Service, 2017). If alley cropping was adopted, the landowners would have excellent markets for their wood products (Smith, Miles, Vissage, & Pugh, 2003). By selecting intercropping species for which known markets exist and diversifying so that several markets could be explored simultaneously, they could improve financially (Cubbage et al., 2012; Haile et al., 2016).

Fruit and Nut Markets

While fruit and nuts markets are manifold and diverse, some important general trends shed positive light on the incorporation of fruit and nut trees into alley cropping systems. Most strikingly, global production of the top four most produced temperate fruit and nut crops substantially increased in 2010–2014 relative to 2000–2004 (FAO, 2017): apple (*Malus pumila* Mill.) (46%), grape (*Vitis* spp.) (20%), pear (*Pyrus* spp.) (55%), peach [*Prunus persica* (L.) Batsch], nectarine [*Prunus persica* (L.) Batsch var. *nucipersica* (Suckow) C.K. Schneid.], and plum (*Prunus* spp.) (49%), almond (71%), chestnut (109%), hazelnut (*Corylus* spp.) (13%), and walnut (180%). However, opportunities for fruits and nuts in alley cropping are not limited to these already common species. Other species, such as pawpaw, persimmon (*Diospyros virginiana* L.), elderberry, chokeberry, black currant (*Ribes nigrum* L.), heartnut (*Juglans ailantifolia* Carrière), hickory (*Carya* spp.), and many more already have substantial regional markets that are growing rapidly (e.g., Mori, Gold & Jose, 2017).

Conclusions

Never in history has there been greater concern for the consequences of human land use. Agriculture is a leading cause of climate change and ecological degradation. Alley cropping is a transformative solution to the problems of agriculture across the temperate zone (Fischer et al., 2017; DeLonge, Miles, & Carlisle, 2016; Wilson & Lovell, 2016; Van Rees, 2008). These systems leverage strategic combinations of trees and crops, providing many economic and ecological advantages. Economic drivers of alley cropping include overyielding, robust timber and fruit or nut markets, and resilience via product diversification. Ecological benefits include enhanced land-use efficiency, C sequestration, soil and nutrient stabilization, wildlife habitat, and resilience to ecological pressures.

References

Agricultural Marketing Resource Center. (2018). *Chestnuts.* Ames, IA: AgMRC. Retrieved from https://www.agmrc.org/commodities-products/nuts/chestnuts

Aguilar, F. X., Cernusca, M. M., & Gold, M.A. (2009). Conjoint analysis of consumer preferences for chestnut attributes. *HortTechnology, 19,* 216–223. https://doi.org/10.21273/HORTSCI.19.1.216

Aguilar, F. X., Cernusca, M. M., & Gold, M. A. (2010). Frequency of consumption, familiarity and preferences for chestnuts in Missouri. *Agroforestry Systems, 79,* 19–29. https://doi.org/10.1007/s10457-009-9266-2

Akbulut, S. K., Keten, A., & Stamps, W.T. (2003). Effect of alley cropping on crops and orthropod diversity in Duzce, Turkey. *Journal of Agronomy and Crop Science, 189,* 261–269. https://doi.org/10.1046/j.1439-037X.2003.00042.x

Alam, M., Olivier, A., Paquette, A., Dupraz, J., Reveret, J. P., & Messier, C. (2014). A general framework for the quantification and valuation of ecosystem services of tree-based intercropping systems. *Agroforestry Systems, 88,* 679–691. https://doi.org/10.1007/s10457-014-9681-x

Allard, G., Nelson, C. J., & Pallardy, S. G. (1991). Shade effects on growth of tall fescue: I. Leaf anatomy and dry matter partitioning. *Crop Science, 31,* 163–167. https://doi.org/10.2135/cropsci1991.0011183X003100010037x

Allen, S. C., Jose, S., Nair, P. K. R., Brecke, B. J., Nair, V. D., Graetz, D. A., & Ramsey, C. L. (2005). Nitrogen mineralization in a pecan (*Carya illinoensis* K. Koch)–cotton (*Gossypium hirsutum* L.) alley cropping system in the southern United States. *Biology and Fertility of Soils, 41,* 28–37. https://doi.org/10.1007/s00374-004-0799-2

Allen, S. C., Jose, S., Nair, P. K. R., Brecke, B. J., & Ramsey, C. L. (2004). Competition for ^{15}N-labeled fertilizer in a pecan (*Carya illinoensis* K. Koch)–cotton (*Gossypium hirsutum* L.) alley cropping system in the southern United States. *Plant and Soil, 263,* 151–164. https://doi.org/10.1023/B:PLSO.0000047732.95283.ac

Allen, S. C., Nair, V. D., Graetz, D. A., Jose, S., & Nair, P. K. R. (2006). Phosphorus loss from organic versus inorganic fertilizers used in alleycropping on a Florida Ultisol. *Agriculture, Ecosystems & Environment, 117,* 290–298. https://doi.org/10.1016/j.agee.2006.04.010

Amadi, C. C., Van Rees, K. C. J., & Farrell, R. E. (2016). Soil–atmosphere exchange of carbon dioxide, methane and nitrous oxide in shelterbelts, compared with adjacent cropped fields. *Agriculture, Ecosystems & Environment*, 223:123–134. https://doi.org/10.1016/j.agee.2016.02.026

Amel, E., Manning, C., Scott, B., & Koger, S. (2017). Beyond the roots of human inaction: Fostering collective effort toward ecosystem conservation. *Science*, *356*, 275–279. https://doi.org/10.1126/science.aal1931

Anderson, S. H., Udawatta, R. P., Seobi, T., & Garrett, H. E. (2009). Soil water content and infiltration in agroforestry buffer strips. *Agroforestry Systems*, *75*, 5–16. https://doi.org/10.1007/s10457-008-9128-3

Andrianarisoa, K. S., Dufour, L., Bienaime, S., Zeller, B., & Dupraz, C. (2016). The introduction of hybrid walnut trees (*Juglans nigra × regia* cv. NG23) into cropland reduces soil mineral N content in autumn in southern France. *Agroforestry Systems*, *90*, 193–205.

Baah-Acheamfour, M., Carlyle, C. N., Lim, S. S., Bork, E. W., & Chang, S. X. (2016). Forest and grassland cover types reduce net greenhouse gas emissions from agricultural soils. *Science of the Total Environment*, *571*, 1115–1127. https://doi.org/10.1016/j.scitotenv.2016.07.106

Bagley, W. T. (1964). Responses of tomatoes and beans to windbreak shelter. *Journal of Soil and Water Conservation*, *19*, 71–73.

Bainard, L. D., Klironomos, J. N., & Gordon, A. M. (2011). Arbuscular mycorrhizal fungi in tree-based intercropping systems: A review of their abundance and diversity. *Pedobiologia*, *54*, 57–61. https://doi.org/10.1016/j.pedobi.2010.11.001

Barnes, R. F, Nelson, C. J., Collins, M., & Moore, K. J. (Eds.). (2003). *Forages. Vol. 1: An introduction to grassland agriculture* (6th ed.) Ames, IA: Iowa State University Press.

Barrett, R. P. (1994). Evaluating black locust for alley cropping. In R. C. Schultz & J. P. Colletti (Eds.) *Opportunities for agroforestry in the temperate zone worldwide: Proceedings of the 3rd North American Agroforestry Conference* (pp. 159–162). Ames, IA: Department of Forestry, Iowa State University.

Barrett, R. P., & Hanover, J. W. (1991). *Robinia pseudoacacia*: A possible temperate zone counterpart to *Leucaena*? In H. E. Garrett (Ed.), *Proceedings of the 2nd North American Agroforestry Conference* (pp. 27–41). Columbia, MO: School of Natural Resources, University of Missouri.

Beaudette, C., Bradley, R. L., Whalen, J. K., McVetty, P. B. E., Vessey, K., & Smith, D. L. (2010). Tree-based intercropping does not compromise canola (*Brassica napus* L.) seed oil yield and reduces soil nitrous oxide emissions. *Agriculture, Ecosystems & Environment*, *139*, 33–39. https://doi.org/10.1016/j.agee.2010.06.014

Beinhart, G. (1962). Effects of temperature and light intensity on CO_2 uptake, respiration, and growth of white clover. *Plant Physiology*, *37*, 709–715. https://doi.org/10.1104/pp.37.6.709

Best, L.B. (1983). Bird use of fencerows: Implications of contemporary fencerow management practices. *Wildlife Society Bulletin*, *11*, 343–347.

Best, L. B., Whitmore, R. C., & Booth, G. M. (1990). Use of cornfields by birds during the breeding season: The importance of edge habitat. *The American Midland Naturalist*, *123*, 84–99. https://doi.org/10.2307/2425762

Black, C. K., Masters, M. D., LeBauer, D.S., Anderson-Teixeria, K.J., & DeLucia, E.H. (2017). Root volume distribution of maturing perennial grasses revealed by correcting for minirhizotron surface effects. *Plant and Soil*, 419, 391–404. https://doi.org/10.1007/s11104-017-3333-7

Blake, C. T., Chamblee, D. S., & Woodhouse, W. W., Jr. (1966). Influence of some environmental and management factors on the persistence of ladino clover in association with orchardgrass. *Agronomy Journal*, *58*, 487–489. https://doi.org/10.2134/agronj1966.00021962005800050009x

Blumenthal, M. (1999). Herb market levels after five years of boom: 1999 sales in mainstream market up only 11% in first half of 1999 after 55% increase in 1998. *HerbalGram*, *47*, 64–65.

Blumenthal, M. (2002). Herb sales down in mainstream market, up in natural food stores. *HerbalGram*, *55*, 60.

Böhm, C., Kanzler, M., & Freese, D. (2014). Wind speed reductions as influenced by woody hedgerows grown for biomass in short rotation alley cropping systems in Germany. *Agroforestry Systems*, *88*, 579–591. https://doi.org/10.1007/s10457-014-9700-y

Boring, L. R., & Swank, W. T. (1984). The role of black locust (*Robinia pseudoacacia*) in forest succession. *Journal of Ecology*, *72*, 749–766. https://doi.org/10.2307/2259529

Bouttier, L., Paquette, A., Messier, C., Rivest, D., Olivier, A., & Cogliastro, A. (2014). Vertical root separation and light interception in a temperate tree-based intercropping system of eastern Canada. *Agroforestry Systems*, *88*, 693–706. https://doi.org/10.1007/s10457-014-9721-6

Boxall, A. B. A. (2008). Fate and transport of veterinary medicines in the soil environment. In D. S. Aga (Ed.), *Fate of pharmaceuticals in the environment and in water treatment systems* (pp. 123–137). Boca Raton, FL: CRC Press.

Bradley, R., Oliver, A., Thevathasan, N., & Whalen, J. (2008). Environmental and economic benefits of tree-based intercropping systems. *Policy Options*, *29*, 46–49.

Brandes, E., McNunn, G. S., Schulte, L. A., Bonner, I. J., Muth, D. J., Babcock, B. A., . . . Heaton, E. A. (2016). Subfield profitability analysis reveals an economic case for cropland diversification. *Environmental Research Letters*, *11*, 014009. https://doi.org/10.1088/1748-9326/11/1/014009

Brandes, E., McNunn, G. S., Schulte, L. A., Muth, D. J., Van Locke, A., & Heaton, E. A. (2018). Targeted subfield switchgrass integration could improve the farm economy water quality and bioenergy feedstock production. *GCB Bioenergy*, *10*, 199–212. https://doi.org/10.1111/gcbb.12481

Brandle, J. R., Hodges, L., & Zhou, X. H. (2004). Windbreaks in North American agricultural systems. *Agroforestry Systems*, 61:65–78.

Brooks, M. B. (1951). *Effects of black walnut trees and their products on other vegetation* (Bulletin 347). Morgantown, WV: West Virginia Agricultural Experiment Station. http://dx.doi.org/10.33915/agnic.347

Brown, S. E., Miller, D. C., Ordonez, P. J., & Baylis, K. (2018). Evidence for the impacts of agroforestry on agricultural productivity, ecosystem services, and human well-being in high-income countries: A systematic map protocol. *Environmental Evidence*, *7*, 24. https://doi.org/10.1186/s13750-018-0136-0

Bryan, J. A., Berlyn, G. P., & Gordon, J. C. (1996). Towards a new concept of the evolution of symbiotic nitrogen fixation in the Leguminosae. *Plant and Soil*, *186*, 151–159. https://doi.org/10.1007/BF00035069

Bugg, R. L., Sarrantonio, M., Dutcher, J. D., & Phatak, S.C. (1991). Understory cover crops in pecan orchards: Possible management systems. *American Journal of Alternative Agriculture*, *6*, 50–62. https://doi.org/10.1017/S0889189300003854

Bula, R. J. (1960). Vegetative and floral development in red clover as affected by duration and intensity of illumination. *Agronomy Journal*, *52*, 74–77. https://doi.org/10.2134/agronj1960.00021962005200020005x

Burel, F. (1989). Landscape structure effects on carabid beetles spatial patterns in western France. Landscape Ecology, 2, 215–226. https://doi.org/10.1007/BF00125092

Burgess, S. O., Adams, M. A., Turner, N. C., White, D. A., & Ong, C.K. (2001). Tree roots: Conduits for deep recharge of soil water. *Oecologia*, *126*, 158–165. https://doi.org/10.1007/s004420000501

Burken, J. G., & Schnoor, J. L. (1997). Uptake and metabolism of atrazine by poplar trees. *Environmental Science & Technology*, *31*, 1399–1406. https://doi.org/10.1021/es960629v

Burner, D. M. (2003). Influence of alley crop environment on orchardgrass and tall fescue herbage. *Agronomy Journal, 95,* 1163–1171. https://doi.org/10.2134/agronj2003.1163

Byers, P. L. (2015). *Pawpaw: Unique native fruit.* Columbia, MO: University of Missouri Extension. Retrieved from http://extension.missouri.edu/greene/documents/Horticulture/Presentations/PawpawMOA2_7_15.pdf

Cardinael, R., Chevallier, T., Barthes, B. G., Saby, N. P. A., Parent, T., Dupraz, C., . . . Chenu, C. (2015). Impact of alley cropping agroforestry on stocks, forms and spatial distribution of soil organic carbon: A case study in a Mediterranean context. *Geoderma, 259–260,* 288–299. https://doi.org/10.1016/j.geoderma.2015.06.015

Cardinael, R., Mao, Z., Prieto, I., Stokes, A., Dupraz, C., Kim, J.H., & Jourdan, C. (2015). Competition with winter crops induces deeper rooting of walnut trees in a Mediterranean alley cropping agroforestry system. *Plant and Soil, 391,* 219–235. https://doi.org/10.1007/s11104-015-2422-8

Cardinael, R., Thevathasan, N., Gordon, A., Clinch, R., Mohammed, I., & Sidders, D. (2012). Growing woody biomass for bioenergy in a tree-based intercropping system in southern Ontario, Canada. *Agroforestry Systems, 86,* 279–286. https://doi.org/10.1007/s10457-012-9572-y

Carter, Z. (1996). Truffling among the trees. *Tree Farmer,* Sept./Oct., pp. 12–14.

Chamberlain, J. L., Mitchell, D., Brigham, T., Hobby, T., Zabek, L., & Davis, J. (2009). Forest farming practices. In H. E. Garrett (Ed.), *North American agroforestry: An integrated science and practice* (2nd ed., p. 219–255). Madison, WI: ASA.

Chandler, H. (1996). Seeing the forest instead of the trees. *Tree Farmer,* Sept./Oct., pp. 6–9.

Chatterjee, N., Nair, P. K. R., Chakraborty, S., & Nair, V. D. (2018). Changes in soil carbon stocks across the Forest–Agroforest–Agriculture/pasture continuum in various agroecological regions: A meta-analysis. *Agriculture, Ecosystems & Environment, 266,* 55–67. https://doi.org/10.1016/j.agee.2018.07.014

Christoffel, R. (2013). Agroforestry and wildlife. In M. A. Gold, M. Cernusca, & M. Hall (Eds.), *Training manual for applied agroforestry practices* (pp. 127–137). Columbia, MO: University of Missouri Center for Agroforestry.

Chu, B., Goyne, K. W., Anderson, S. H., Udawatta, R. P., & Lin, C. H. (2009). Veterinary antibiotic sorption to agroforestry buffer, grass buffer and cropland soils. In M. A. Gold & M. M. Hall (Eds.), Agroforestry comes of age: Putting science into practice. Proceedings of the 11th North American Agroforestry Conference (p. 31–40). Columbia, MO: Center for Agroforestry, University of Missouri.

Cohn, L. (1995). The growing market for special forest products: An opportunity and challenge. *Forest Perspectives, 4,* 4–7.

Colletti, J., Mize, C., Schultz, R., Faltonson, R., Skadberg, A., Mattila, J., . . . Brown, R. (1994). An alley cropping biofuel system: Operation and economics. In R. C. Schultz & J. P. Colletti (Eds.), Opportunities for agroforestry in the temperate zone worldwide: Proceedings of the 3rd North American Agroforestry Conference (pp. 303–310). Ames, IA: Department of Forestry, Iowa State University.

Cubbage, F., Glenn, V., Mueller, J. P., Robison, D., Myers, R., Luginbuhl, J. M., & Myers, R. (2012). Early tree growth, crop yields and estimated returns for an agroforestry trial in Goldsboro, North Carolina. *Agroforestry Systems, 86,* 323–334. https://doi.org/10.1007/s10457-012-9481-0

Cutter, B. E., & Garrett, H. E. (1993). Wood quality in alley cropped eastern black walnut. *Agroforestry Systems, 22,* 25–32. https://doi.org/10.1007/BF00707467

Cutter, B. E., Hunt, K., & Haywood, J. D. (1998). Tree/wood quality in slash pine following long-term cattle grazing. *Agroforestry Systems, 44,* 305–312. https://doi.org/10.1023/A:1006219231801

Dai, X. Q., Sui, P., & Xie, G. H. (2006). Water use and nitrate nitrogen changes in intensive farmlands following introduction of poplar (*Populus × euramericana*) in a semi-arid region. *Arid Land Research and Management, 20,* 281–294. https://doi.org/10.1080/15324980600904734

Daniere, C., Capellano, A., & Moirud, A. (1986). Dynamique de l'azote dans un peuplement natural d'*Alnus incana* L. Moench. *Acta Oecologica, Oecologia Plantarum, 7,* 165–175.

David, M. B., Drinkwater, L. E., & McIsaac, G. F. (2010). Sources of nitrate yields in the Mississippi River basin. *Journal of Environmental Quality, 39,* 1657–1667. https://doi.org/10.2134/jeq2010.0115

Dawson, J. O., & Paschke, M. W. (1991). Current and potential uses of nitrogen-fixing trees and shrubs in temperate agroforestry systems. In H. E. Garrett (Ed.), Proceedings of the 2nd North American Agroforestry Conference (pp. 183–209). Columbia, MO: School of Natural Resources, University of Missouri.

DeLonge, M. S., Miles, A., & Carlisle, L. (2016). Investing in the transition to sustainable agriculture. *Environmental Science & Policy, 55,* 266–273. https://doi.org/10.1016/j.envsci.2015.09.013

Dey, D. C., Gardiner, E. S., Kabrick, J. M., Stanture, J. A., & Jacobs, D. F. (2010). Innovations in afforestation of agricultural bottomlands to restore native forests in the eastern USA. *Scandinavian Journal of Forest Research, 25*(Suppl. 8), 31–42. https://doi.org/10.1080/02827581.2010.485822

Dey, D. C., Kabrick, J. M., & Gold, M. A. (2005). Evaluation of RPM oak seedlings in afforesting floodplain crop fields along the Missouri River. In D. R. Weigel, J. W. Van Sambeek, & C. H. Michler (Eds.) Ninth workshop on seedling physiology and problems in oak plantings (abstracts) (Gen. Tech. Rep. NC-262). St. Paul, MN: U.S. Forest Service, North Central Research Station.

Diemont, S. A. W., Martin, J. F., Levy-Tacher, S. I., Nigh, R. B., Lopez Ramirez, P., & Duncan, J. G. (2006). Lacandon Maya forest management: Restoration of soil fertility using native tree species. *Ecological Engineering, 28,* 205–212. https://doi.org/10.1016/j.ecoleng.2005.10.012

Dixon, R. K., Brown, S., Houghton, R. A., Solomon, A. M., Trexler, M. C., & Wisniewski, J. (1994). Carbon pools and flux of global forest ecosystems. *Science, 263,* 185–190. https://doi.org/10.1126/science.263.5144.185

Dixon, R. K., Winjum, J. K., & Schroeder, P. E. (1993). Conservation and sequestration of carbon: The potential of forest and agroforestry management practices. *Global Environmental Change, 3,* 159–173. https://doi.org/10.1016/0959-3780(93)90004-5

Dong-Gill, K., Kirschbaum, M. U. F., & Beedy, T. L. (2016). Carbon sequestration and net emissions of CH_4 and N_2O under agroforestry: Synthesizing available data and suggestions for future studies. *Agriculture, Ecosystems & Environment, 226,* 65–78. https://doi.org/10.1016/j.agee.2016.04.011

Dougherty, M. C., Thevathasan, N. V., Gordon, A. M., Lee, H., & Kort, J. (2009). Nitrate and *Escherichia coli* NAR analysis in tile drain effluent from a mixed tree intercrop and monocrop system. *Agriculture, Ecosystems & Environment, 131,* 77–84. https://doi.org/10.1016/j.agee.2008.09.011

Dubey, S. K., Sharma, N., Sharma, J. P., Sharma, A., & Kishore, N. (2016). Assessing citrus (lemon) based intercropping in the irrigated areas of northern plains of Haryana. *Indian Journal of Horticulture, 73,* 441–444. https://doi.org/10.5958/0974-0112.2016.00094.3

Dupraz, C., Auclair, D., Barthelemy, D., Caraglio, Y., Sabatier, S., Bariteau, M., . . . Maillet, J. (2000). *Programme intégré de recherches en agroforesterie á restincliéres (PIRAT).* Montpellier, France: INRA.

Dupraz, C., & Liagre, F. (2008). *Agroforesterie: Des arbres et des cultures.* Paris: Editions France Agricole.

Dupraz, C., Wolz, K. J., Lecomte, I., Talbot, G., Vincent, G., Mulia, R., . . . Van Noordwijk, M. (2019). Hi-sAFe: A 3D Agroforestry model for integrating dynamic tree–crop interactions. *Sustainability, 11*(8), 2293. https://doi.org/10.3390/su11082293

Eichhorn, M. P., Paris, P., Herzog, F., Incoll, L.D., Liagre, F., Mantzanas, K., . . . Dupraz, C. (2006). Silvoarable systems in Europe: Past, present and future prospects. *Agroforestry Systems, 67*, 29–50. https://doi.org/10.1007/s10457-005-1111-7

Ehret, M., Buhle, L., Graß, R., Lamersdorf, N., & Wachendorf, M. (2015). Bioenergy provision by an alley cropping system of grassland and shrub willow hybrids: Biomass, fuel characteristics and net energy yields. *Agroforestry Systems, 89*, 365–381. https://doi.org/10.1007/s10457-014-9773-7

Ehret, M., Graß, R., & Wachendorf, M. (2015). The effect of shade and shade material on white clover/perennial ryegrass mixtures for temperate agroforestry systems. *Agroforestry Systems, 89*, 557–570. https://doi.org/10.1007/s10457-015-9791-0

Ekor, M. (2014). The growing use of herbal medicines: Issues relating to adverse reactions and challenges in monitoring safety. *Frontiers in Pharmacology, 4*, 177. https://doi.org/10.3389/fphar.2013.00177

FAO. (1980). *Poplars and willows in wood production and land use* (FAO Forestry Series 10). Rome: FAO.

FAO. (2017). *FAOSTAT* [Database]. Rome: FAO.

Fargione, J. E., Bassett, S., Boucher, T., Bridgham, S. D., Conant, R. T., Cook-Patton, S. C., . . . Griscom, B. W. (2018). Natural climate solutions for the United States. *Science Advances, 4*, eaat1869. https://doi.org/10.1126/sciadv.aat1869

Fischer, J., Abson, D. J., Bergsten, A., Collier, N. F., Dorresteign, I., Hanspach, J., & Senbeta, F. (2017). Reframing the food-biodiversity challenge. *Trends in Ecology & Evolution, 32*, 335–345. https://doi.org/10.1016/j.tree.2017.02.009

Fletcher, E. H., III, Thetford, M., Sharma, J., & Jose, S. (2012). Effect of root competition and shade on survival and growth of nine woody plant taxa within a pecan (*Carya illinoinensis* (Wangenh.) C. Koch) alley cropping system. *Agroforestry Systems, 86*, 49–60. https://doi.org/10.1007/s10457-012-9507-7

Food & Agricultural Policy Research Institute. (2017). *Crop budget generator, Version 11.7.17*. Columbia, MO: FAPRI, University of Missouri. Retrieved from https://www.fapri.missouri.edu/wp-content/uploads/2018/02/CBG_11-7-17.xls

Frey, G. E., Mercer, D. E,. Cubbage, F. W., & Abt, R.C. (2010). Economic potential of agroforestry and forestry in the Lower Mississippi alluvial valley with incentive programs and carbon payments. *Southern Journal of Applied Forestry, 34*, 176–185. https://doi.org/10.1093/sjaf/34.4.176

Frey, G. E., Mercer, D. E., Cubbage, F. W., & Abt, R. C. (2013). A real options model to assess the role of flexibility in forestry and agroforestry adoption and disadoption in the Lower Mississippi alluvial valley. *Agricultural Economics, 44*, 73–91. https://doi.org/10.1111/j.1574-0862.2012.00633.x

Gallagher, E. J., Mudge, K. W., Pritts, M. P., & DeGloria, S. D. (2015). Growth and development of 'Illini Hardy' blackberry (*Rubus subgenus* Eubatus Focke) under shaded systems. *Agroforestry Systems, 89*, 1–17. https://doi.org/10.1007/s10457-014-9738-x

Gamble, J. D., Johnson, G., Sheaffer, C. C., Current, D. A., & Wyse, D. L. (2014). Establishment and early productivity of perennial biomass alley cropping systems in Minnesota, USA. *Agroforestry Systems, 88*, 75–85. https://doi.org/10.1007/s10457-013-9657-2

Garrett, H. E., Buck, L. E., Gold, M. A., Hardesty, L. H., Kurtz, W. B., Lassoie, J. P., . . . Slusher, J. P. (1994). Agroforestry: An integrated land-use management system for production and farmland conservation. Washington, DC: Soil Conservation Service.

Garrett, H. E., & Harper, L. S. (1999). The science and practice of black walnut agroforestry in Missouri, USA: A temperate zone assessment. In L. E. Buck, J. P. Lassoie, & E. C. M. Fernandes (Eds.), Agroforestry in sustainable agricultural systems (pp. 97–110). Boca Raton, FL: CRC Press.

Garrett, H. E., Jones, J. E., Kurtz, W. B., & Slusher, J. P. (1991). Black walnut (*Juglans nigra* L.) agroforestry: Its design and potential as a land-use alternative. *The Forestry Chronicle, 67*, 213–218. https://doi.org/10.5558/tfc67213-3

Garrett, H. E., & Kurtz, W. B. (1983). Silvicultural and economic relationships of integrated forestry farming with black walnut. *Agroforestry Systems, 1*, 245–256. https://doi.org/10.1007/BF00130610

Garrett, H. E., Kurtz, W. B., & Slusher, J. P. (1992). Walnut agroforestry (Guide Sheet G5020). Columbia, MO: College of Agriculture, University of Missouri.

Garrett, H. E., McGraw, R. L., & Walter, W. D. (2009). Alley cropping practices. In H. E. Garrett (Ed.), North American Agroforestry: An integrated science and practice (2nd ed., pp. 133–162). Madison, WI, ASA. https://doi.org/10.2134/2009.northamericanagroforestry.2ed

Garrett, H. E., Walter, W. D., & Godsey, L. D. (2011). Alley cropping: A relic from the past or a bridge to the future? *Inside Agroforestry, 19*(2), 1, 11.

Gibbs, S., Koblents, H., Coleman, B., Gordon, A., Thevathasan, N., & Williams, P. (2016). Avian diversity in a temperate tree-based intercropping system from inception to now. *Agroforestry Systems, 90*, 905–916. https://doi.org/10.1007/s10457-016-9901-7

Gillespie, A. R., Jose, S., Mengel, D. B., Hoover, W. L., Pope, P. E., Seifert, . . . Benjamin, T. J. (2000). Defining competition vectors in a temperate alley cropping system in the midwestern USA. *Agroforestry Systems, 48*, 25–40. https://doi.org/10.1023/A:1006285205553

Girma, H., Rao, M. R., & Sithanantham, S. (2000). Insect pests and beneficial arthropods' population under different hedgerow intercropping systems in semiarid Kenya. *Agroforestry Systems, 50*, 279–292. https://doi.org/10.1023/A:1006447813882

Global Industry Analysts. (2018, May). Herbal supplements and remedies: Market analysis, trends, and forecasts (MCP-1081). San Jose, CA. Retrieved from https://www.strategyr.com/MCP-1081.asp

Godsey, L. D., Mercer, D. E., Grala, R. K., Grado, S. C., & Alavalapati, J. R. R. (2009). Agroforestry economics and policy. In H. E. Garrett (Ed.), North American Agroforestry: An integrated science and practice (2nd ed., pp. 315–337). Madison, WI: ASA.

Gogerty, R. (1994). Companion crops for corn. *Furrow, 99*, 7–8.

Gold, M. A., Cernusca, M. M., & Godsey, L. D. (2005). Update on consumers' preferences on chestnuts. *HortTechnology, 15*, 904–906. https://doi.org/10.21273/HORTTECH.15.4.0904

Gold, M. A., Cernusca, M. M., & Godsey, L. D. (2006). Competitive market analysis: Chestnut producers. *HortTechnology, 16*, 360–369. https://doi.org/10.21273/HORTTECH.16.2.0360

Gold, M. A., & Garrett, H. E. (2009). Agroforestry nomenclature, concepts, and practices. In H. E. Garrett (Ed.), North American Agroforestry: An integrated science and practice (2nd ed., pp. 45–56). Madison, WI: ASA.

Gold, M. A., & Hanover, J. W. (1993). Honeylocust (*Gleditsia triacanthos*), a multipurpose tree for the temperate zone. *International Tree Crops Journal, 7*, 189–207. https://doi.org/10.1080/01435698.1993.9752919

Gordon, A. M., Thevathasan, N., Klironomos, J., Bradley, R., Shipley, B., Cogliastro, A., . . . Whalen, J. (2008). Agroforestry in the world: Lessons for Canada. *Policy Options, 29*, 50–53.

Gordon, A. M., & Williams, P. A. (1991). Intercropping valuable hardwood tree species and agricultural crops in southern

Ontario. *The Forest Chronicles*, *67*, 200–208. https://doi.org/10.5558/tfc67200-3

Goucher, L., Bruce, R., Cameron, D. D., Koh, S. C. L., & Horton, P. (2017). Environmental impact of fertiliser embodied in a wheat-to-bread supply chain. *Nature Plants*, *3*, 17012. https://doi.org/10.1038/nplants.2017.12

Graham, J. B., & Nassauer, J. I. (2019). Wild bee abundance in temperate agroforestry landscapes: Assessing effects of alley crop composition, landscape configuration, and agroforestry area. *Agroforestry Systems*, *93*, 837–850. https://doi.org/10.1007/s10457-017-0179-1

Graves, A. R., Burgess, P. J., Keesman, K. J., van der Werf, W., Dupraz, C., van Keulen, H., . . . Mayus, M. (2010). Implementation and calibration of the parameter-sparse Yield-SAFE model to predict production and land equivalent ratio in mixed tree and crop systems under two contrasting production situations in Europe. *Ecological Modelling*, *221*, 1744–1756. https://doi.org/10.1016/j.ecolmodel.2010.03.008

Graves, A. R., Burgess, P. J., Palma, J. H. N., Herzog, F., Moreno, G., Bertomeu, M., . . . van den Briel, J. P. (2007). Development and application of bio-economic modeling to compare silvoarable, arable, and forestry systems in three European countries. *Ecological Engineering*, *29*, 434–449. https://doi.org/10.1016/j.ecoleng.2006.09.018

Gray, D. E., Garrett, H. E., Pallardy, S. G., & Rottinghaus, G.E. (2003). An introduction to incorporating botanicals into alley cropping practices. In T. R. Clason (Ed.), Land-use management for the future: Proceedings of the 6th North American Conference on Agroforestry (pp. 203–212). Columbia, MO: Association for Temperate Agroforestry.

Gray, D. E., Pallardy, S. G., Garrett, H. E., & Rottinghaus, G. E. (2003). Acute drought stress and root age effects on alkamide and phenolic acid content in purple coneflower roots. *Planta Medica*, *69*, 50–55. doi:10.1055/s-2003-37026

Gruenewald, H., Brandt, B. K. Y., Schneider, B. U., Oliver, B., Kendzia, G., & Huttl, R. F. (2007). Agroforestry systems for the production of woody biomass for energy transformation purposes. *Ecological Engineering*, *29*, 319–328. https://doi.org/10.1016/j.ecoleng.2006.09.012

Haile, S., Palmer, M., & Otey, A. (2016). Potential of loblolly pine: Switchgrass alley cropping for provision of biofuel feedstock. *Agroforestry Systems*, *90*, 763–771. https://doi.org/10.1007/s10457-016-9921-3

Hall, D. O., & de Groot, P. J. (1988). Biomass for fuel and food: A parallel necessity. *Advances in Solar Energy*, *4*, 1–90. https://doi.org/10.1007/978-1-4613-9945-2_1

Hall, D. O., Rosillo-Calle, F., Williams, R. H., & Woods, J. (1993). Biomass for energy: Supply prospects. In T. B. Johansson, H. Kelly, A. K. N. Reddy, & R. H. Williams (Eds.), Renewable energy: Sources for fuels and electricity (pp. 593–652). Washington, DC: Island Press.

Halling-Sørensen, B., Sengeløv, G., & Tjornelund, J. (2002). Toxicity of tetracyclines and tetracycline degradation products to environmentally relevant bacteria, including selected tetracycline-resistant bacteria. *Archives of Environmental Contamination and Toxicology*, *42*, 263–271. https://doi.org/10.1007/s00244-001-0017-2

Haynes, R.W. (Tech. Coord.) (2003). *An analysis of the timber situation in the United States: 1952 to 2050* (Gen. Tech. Rep. PNW-GTR-560). Portland, OR: U.S. Forest Service, Pacific Northwest Research Station.

Hesketh, J. D. (1963). Limitations to photosynthesis responsible for differences among species. *Crop Science*, *3*, 493–496. https://doi.org/10.2135/cropsci1963.0011183X000300060011x

Hoffman, D. W., Gerik, T. J., & Richardson, C. W. (1996). Use of contour strip cropping as a best management practice to reduce atrazine contamination of surface water. In M. Straškraba (Ed.), *Diffuse Pollution '95: Selected Proceedings of the 2nd IAWQ International Specialized Conference and Symposia on Diffuse Pollution* (p. 595–596). Oxford, UK: Pergamon Press.

Holden, C. (1996). Random samples. *Science*, *271*, 1813.

Holt, E. C. (1973). *Forage production in pecan orchards* (Texas Agricultural Experiment Station Bull. 1131). College Station, TX: Texas A&M University.

Holzmueller, E. J., & Jose, S. (2012). Biomass production for biofuels using agroforestry: Potential for the North Central Region of the United States. *Agroforestry Systems*, *85*, 305–314. https://doi.org/10.1007/s10457-012-9502-z

Houx, J. H., III, McGraw, R. L., Garrett, H. E., Kallenbach, R. L., Fritschi, F. B., & Rogers, W. (2013). Extent of vegetation-free zone necessary for silvopasture establishment of eastern black walnut seedlings in tall fescue. *Agroforestry Systems*, *87*, 73–80. https://doi.org/10.1007/s10457-012-9523-7

Howard, L., McKeever, D. B., & Liang, S. (2017). *U.S. forest products annual market review and prospects, 2013–2017* (Research Note FPL-RN-0348). Madison, WI: U.S. Forest Service, Forest Products Laboratory.

Huang, B. K. (1983). System engineering in precision automatic transplanting. *AMA–Agricultural Mechanization in Asia, Africa, and Latin America*, *14*, 11–19.

Hulvey, K. B., Hobbs, R. J., Standish, R. J., Lindenmayer, D. B., Lach, L., & Perring, M. P. (2013). Benefits of tree mixes in carbon plantings. *Nature Climate Change*, *3*, 869–874. https://doi.org/10.1038/nclimate1862

Inurreta-Aguirre, H. D., Lauri, P.-E., Dupraz, C., & Gosme, M. (2018). Yield components and phenology of durum wheat in a Mediterranean alley-cropping system. *Agroforestry Systems*, *92*, 961–974. https://doi.org/10.1007/s10457-018-0201-2

IPCC. (2014). *Climate Change 2014: Mitigation of climate change.* Contribution of Working Group III to the Fifth Assessment Report to the Intergovernmental Panel on climate change. Geneva, Switzerland: IPCC.

Janke, R. R. (1990). Alley cropping with nitrogen-fixing trees and vegetables in the northeast USA. In P. Williams (Ed.), *Agroforestry in North America: Proceedings of the First Conference on Agroforestry in North America, Guelph, ON, Canada* (p. 254). Ontario Ministry of Agriculture and Food.

Johnson, J. W., Fike, J. H., Fike, W. B., Burger, J. A., McKenna, J. R., Munsell, J. F., & Hodges, S. C. (2013). Millwood honeylocust trees: Seedpod nutritive value and yield characteristics. *Agroforestry Systems*, *87*, 849–856. https://doi.org/10.1007/s10457-013-9601-5

Johnson, J. W., Fike, J. H., Fike, W. B., Burger, J. A., Munsell, J. F., McKenna, J. R., & Hodges, S. C. (2012). Millwood and wild-type honeylocust seedpod nutritive value changes over winter. *Crop Science*, *52*, 2807–2016. https://doi.org/10.2135/cropsci2011.10.0542

Jones, J. E., Haines, J., Garrett, H. E., & Loewenstein, E. F. (1994). Genetic selection and fertilization provide increased nut production under walnut–agroforestry management. In R. C. Schultz & J. P. Colletti (Eds.), *Opportunities for agroforestry in the temperate zone worldwide: Proceedings of the 3rd North American Agroforestry Conference* (pp. 39–42). Ames, IA: Department of Forestry, Iowa State University.

Jordahl, J. L., Foster, L., Schnoor, J. L., & Alvarez, P. J. J. (1997). Effect of hybrid poplar trees on microbial populations important to hazardous waste bioremediation. *Environmental Toxicology and Chemistry*, *16*, 1318–1321. https://doi.org/10.1002/etc.5620160630

Jordan, C. F., Gajaseni, J., & Watanabe, H. (1992). *Taungya: Forest plantations with agriculture in Southeast Asia*. Wallingford, UK: CAB International.

Jorgensen, U., & Hansen, E. M. (1998). Nitrate leaching from Miscanthus, willow, grain crops and rape. In M. Worgetter (Ed.), Proceedings of the International Workshop on

Environmental Aspects of Energy Crop Production, Brasimone, Italy (pp. 207–218). Wieselburg, Austria: BLT.

Jose, S. (2002). Black walnut allelopathy: Current state of the science. In Inderjit & A. U. Malik (Eds.), *Chemical ecology of plants: Allelopathy in aquatic and terrestrial ecosystems* (pp. 149–172). Berlin: Birkhauser.

Jose, S. (2011). Managing native and non-native plants in agroforestry systems. *Agroforestry Systems, 83,* 101. https://doi.org/10.1007/s10457-011-9440-1

Jose, S., & Bardhan, S. (2012). Agroforestry for biomass production and carbon sequestration: An overview. *Agroforestry Systems, 86,* 105–111. https://doi.org/10.1007/s10457-012-9573-x

Jose, S., Benjamin, T., Stall, T., Gillespie, A. R., Hoover, W. L., Mengel, D. B., . . . Biehle, D. J. (1997). Biology and economics of a black walnut–corn alley cropping system. In J. Van Sambeek (Ed.), Knowledge for the future of black walnut: Proceedings of the 5th Black Walnut Symposium, Springfield, MO (Gen. Tech. Rep. NC-191, pp. 203–208). St. Paul, MN: U.S. Forest Service North Central Forest Experiment Station.

Jose, S., Gillespie, A. R., & Pallardy, S. G. (2004). Interspecific interactions in temperate agroforestry. *Agroforestry Systems, 61,* 237–255.

Jose, S., Gillespie, A. R., & Seifert, J. (1996). The microenvironmental and physiological basis for temporal reductions in crop production in an Indiana alley cropping system. In J. H. Ehrenreich, D. L. Ehrenreich, & H. W. Lee (Eds.), Growing a sustainable future: Proceedings of the 4th North American Agroforestry Conference, Boise, ID (pp. 54–55). Moscow, ID: University of Idaho.

Jose, S., Gillespie, A. R., Seifert, J., & Biehle, D. J. (2000). Defining competition vectors in a temperate alley cropping system in the midwestern USA: 2. Competition for water. *Agroforestry Systems, 48,* 41–59. https://doi.org/10.1023/A:1006289322392

Jose, S., Gold, M. A., & Garrett, H. E. (2012). The future of temperate agroforestry in the United States. In P. K. R. Nair & D. Garrity (Eds.), Agroforestry: The future of global land use (pp. 217-245). Advances in Agroforestry 9. Dordrecht, the Netherlands: Springer. https://doi.org/10.1007/978-94-007-4676-3_14

Jose, S., Gold, M. A., & Garrett, H. E. (2018). Temperate agroforestry in the United States: Current trends and future directions. In A. M. Gordon, S. M. Newman, & B. R. W. Coleman (Eds.), Temperate agroforestry systems (2nd ed., pp. 50–71). Wallingford, UK: CAB International. https://doi.org/10.1079/9781780644851.0050

Jose, S., Williams, R., & Zamora, D. (2006). Belowground ecological interactions in mixed-species forest plantations. *Forest Ecology and Management, 233,* 231–239. https://doi.org/10.1016/j.foreco.2006.05.014

Kabrick, J. M., Dey, D. C., Van Sambeek, J. W., Wallendorf, M., & Gold, M. A. (2005). Soil properties and growth of swamp white oak and pin oak on bedded soils in the Lower Missouri River floodplain. *Forest Ecology and Management, 204,* 315–327. https://doi.org/10.1016/j.foreco.2004.09.014

Kang, B. T. (1993). Alley cropping: Past achievements and future directions. *Agroforestry Systems, 23,* 141–155. https://doi.org/10.1007/BF00704912

Kang, B. T. (1997). Alley cropping: Soil productivity and nutrient recycling. *Forest Ecology and Management, 91,* 75–82. https://doi.org/10.1016/S0378-1127(96)03886-8

Kapp, G. 1987. Agrisilviculture: A system of agroforestry in 18th and 19th-century Germany. *Plant Research and Development, 26,* 36–45.

Karekezi, S., & Kithyoma, W. (2006). *Bioenergy and agriculture: Promises and challenges: Bioenergy and the poor* (2020 Visions focus brief: Focus 14, Brief 11). Washington, DC: International Food Policy Research Institute.

Keeley, K. O., Wolz, K. J., Adams, K. I., Richards, J. H., Hannum, E., von Tscharner, F. S., & Ventura, S. J. (2019). Multi-party agroforestry: Emergent approaches to trees and tenure on farms in the Midwest USA. *Sustainability, 11,* 2449. https://doi.org/10.3390/su11082449

Kessler, K. J., Jr. (1988). Companion plantings of black walnut with autumn olive to control walnut anthracnose. *Phytopathology, 78,* 1606.

Kort, J. (1988). Benefits of windbreaks to field and forage crops. *Agriculture, Ecosystems & Environment, 22–23,* 165–190. https://doi.org/10.1016/0167-8809(88)90017-5

Kranz, A. J., Wolz, K. J., & Miller, J. R. (2018). Effects of shrub crop interplanting on apple pest ecology in a temperate agroforestry system. *Agroforestry Systems, 93,* 1179–1189. https://doi.org/10.1007/s10457-018-0224-8

Kremer, R. J., & Kussman, R. D. (2011). Soil quality in a pecan–kura clover alley cropping system in the midwestern USA. *Agroforestry Systems, 83,* 213–223. https://doi.org/10.1007/s10457-011-9370-y

Krutz, L. J., Senseman, S. A., Dozier, M. C., Hoffman, D. W., & Tierney, D. P. (2004). Infiltration and adsorption of dissolved metolachlor, metolachlor oxanilic acid, and metolachlor ethanesulfonic acid by buffalograss (*Buchloe dactyloides*) filter strips. *Weed Science, 52,* 166–171. https://doi.org/10.1614/WS-03-033R

Kurtz, W. B., Garrett, H. E., & Kincaid, W. H. (1984). Investment alternatives for black walnut plantation management. *Journal of Forestry, 82,* 604–608. https://doi.org/10.1093/jof/82.10.604

Kurtz, W. B., Thurman, S. E., Monson, M. J., & Garrett, H. E. (1991). The use of agroforestry to control erosion: Financial aspects. *The Forest Chronicles, 67,* 254–257. https://doi.org/10.5558/tfc67254-3

Ladanai, S., & Vinterbäck, J. (2009). *Global potential of sustainable biomass for energy*. Uppsala, Sweden: Department of Energy and Technology, Swedish University of Agricultural Sciences.

Lal, R. (1984). Soil losses from tropical arable lands and its control. *Advances in Agronomy, 37,* 183–248. https://doi.org/10.1016/S0065-2113(08)60455-1

Landis, D. A., & Haas, M. J. (1992). Influence of landscape structure abundance and within-field distribution of European corn borer larvae parasitoids in Michigan. *Environmental Entomology, 21,* 409–416. https://doi.org/10.1093/ee/21.2.409

Lang, A. C., von Oheimb, G., M. Scherer-Lorenzen, B. Yang, S. Trogisch, H. Bruelheide, . . . W. Härdtle. (2014). Mixed afforestation of young subtropical trees promotes nitrogen acquisition and retention. *Journal of Applied Ecology, 51,* 224–233. https://doi.org/10.1111/1365-2664.12157

Lee, K.-H., Isenhart, T. M., & Schultz, R. C. (2003). Sediment and nutrient removal in an established multi-species riparian buffer. *Journal of Soil and Water Conservation, 58,* 1–8.

Luedeling, E., Smethurst, P. J., Baudron, F., Bayala, J., Huth, N. I., van Noordwijk, M., . . . Sinclair, F. (2016). Field-scale modeling of tree–crop interactions: Challenges and development needs. *Agricultural Systems, 142,* 51–69. https://doi.org/10.1016/j.agsy.2015.11.005

Lewis, C. E., Tanner, G. W., & Terry, W. S. (1985). Double vs. single-row pine plantations for wood and forage production. *Southern Journal of Applied Forestry, 9,* 55–61. https://doi.org/10.1093/sjaf/9.1.55

Lin, C. H., Goyne, K. W., Lerch, R. N., & Garrett, H. E. (2010). Dissipation of sulfamethazine and tetracycline in the root zone of grass and tree species. *Journal of Environmental Quality, 39,* 1269–1278. https://doi.org/10.2134/jeq2009.0346

Lin, C. H., Lerch, R. N., Garrett, H. E., & George, M. F. (2005). Incorporating forage grasses in riparian buffers for bioremediation of atrazine, isoxaflutole and nitrate in Missouri. *Agroforestry Systems, 63,* 87–95.

Lin, C. H., Lerch, R. N., Garrett, H. E., & George, M. F. (2008). Phytoremediation of atrazine by five forage grass species: Transformation, uptake, and detoxification. *Journal of Environmental Quality*, *37*, 196–206. https://doi.org/10.2134/jeq2006.0503

Lin, C. H., Lerch, R. N., Garrett, H. E., Jordan, D., & George, M. F. (2007). Ability of forage grasses exposed to atrazine and isoxaflutole to reduce nutrient levels in soils and shallow groundwater. *Communications in Soil Science and Plant Analysis*, *38*, 1119–1136. https://doi.org/10.1080/00103620701327976

Lin, C. H., Lerch, R. N., Goyne, K. W., & Garrett, H. E. (2011). Reducing herbicide and veterinary antibiotic losses from agroecosystems using vegetative buffers. *Journal of Environmental Quality*, *40*, 791–799. https://doi.org/10.2134/jeq2010.0141

Lin, C. H., Lerch, R. N., Johnson, W. G., Jordan, D., Garrett, H. E., & George, M. F. (2003). The effect of five forage species on transport and transformation of atrazine and Balance (isoxaflutole) in lysimeter leachate. *Journal of Environmental Quality*, *32*, 1992–2000. https://doi.org/10.2134/jeq2003.1992

Lin, C. H., Lerch, R. N., Jordan, D., Garrett, H. E., & George, M. (2004). The effects of herbicides (atrazine and Balance) and ground covers on microbial biomass carbon and nitrate reduction. In Proceedings of the 8th North American Agroforestry Conference, Corvallis, OR (pp. 182–195).

Lin, C. H., McGraw, R. L., George, M. F., & Garrett, H. E. (2001). Nutritive quality and morphological development under partial shade of some forage species with agroforestry potential. *Agroforestry Systems*, *53*, 269–281. https://doi.org/10.1023/A:1013323409839

Lin, C. H., McGraw, R. L., George, M. F., & Garrett, H. E. (2003). Shade effects on quality and morphological development of forage crops with potential in temperate agroforestry practice. In T. R. Clason (Ed.), Land-use management for the future: Proceedings of the 6th North American Conference on Agroforestry, Hot Springs, AR (pp. 127-139).

Lin, C. H., McGraw, R. L., George, M. F., Garrett, H. E., Piotter, B. J., & Alley, J. L. (1995). Effects of shade on forage crops that have potential use in agroforestry. In *1995 Agronomy abstracts* (p. 53). Madison WI: ASA.

Lin, C. H., Thompson, B. M., Hsieh, H. Y., Lerch, R. N., Kremer, R. J., & Garrett, H. E. (2009). Introduction of atrazine-degrading *Pseudomonas* sp. strain ADP to enhance rhizodegradation of atrazine. In M. A. Gold and M. M. Hall (Eds.), Agroforestry comes of age: Putting science into practice: Proceedings of the 11th North American Agroforestry Conference, Columbia, MO (pp. 183–190).

Linit, M.J., and W.T. Stamps. (1996). Plant diversity and arthropod communities: Implications for agroforestry. In J. H. Ehrenreich, D. L. Ehrenreich, & H. W. Lee (Eds.), Growing a sustainable future: Proceedings of the 4th North American Agroforestry Conference (pp. 56–58). Moscow, ID: University of Idaho.

Lorenz, K., & Lal, R. (2014). Soil organic carbon sequestration in agroforestry systems. A review. *Agronomy for Sustainable Development*, *34*, 443–454. https://doi.org/10.1007/s13593-014-0212-y

Lovell, S. T., & Sullivan, W. C. (2006). Environmental benefits of conservation buffers in the United States: Evidence, promise and open questions. *Agriculture, Ecosystems & Environment*, *112*, 249–260. https://doi.org/10.1016/j.agee.2005.08.002

Lovell, S. T., Dupraz, C., Gold, M., Jose, S., Revord, R., Stanek, E., & Wolz, K. J. (2018). Temperate agroforestry research: Considering multifunctional woody polycultures and the design of long-term field trials. *Agroforestry Systems*, *92*, 1397–1415. https://doi.org/10.1007/s10457-017-0087-4

Lowrance, R., & Sheridan, J. M. (2005). Surface runoff water quality in a managed three zone riparian buffer. *Journal of Environmental Quality*, *34*, 1851–1859. https://doi.org/10.2134/jeq2004.0291

Maass, J. M., Jordan, C. F., & Sarakhan, J. (1988). Soil erosion and nutrient losses in seasonal tropical agroecosystems under various management techniques. *Journal of Applied Ecology*, *25*, 595–607. https://doi.org/10.2307/2403847

Makumba, W., Janssen, B., Oenema, O., Akinnifesi, F. K., Mweta, D., & Kwesiga, F. (2006). The long-term effects of a gliricidia–maize intercropping system in southern Malawi, on gliricidia and maize yields, and soil properties. *Agriculture, Ecosystems & Environment*, *116*, 85–92. https://doi.org/10.1016/j.agee.2006.03.012

Malézieux, E., Crozat, Y., Dupraz, C., Laurans, M., Makowski, D., Ozier-Lafontaine, H., . . . Valantin-Morison, M. (2009). Mixing plant species in cropping systems: Concepts, tools and models: A review. *Agronomy for Sustainable Development*, *29*, 43–62. https://doi.org/10.1051/agro:2007057

Mariotti, B., Maltoni, A., Jacobs, D. F., & Tani, A. (2015). Container effects on growth and biomass allocation in *Quercus robur* and *Juglans regia* seedlings. *Scandinavian Journal of Forest Research*, *30*, 401–415.

Mattia, C. M., Lovell, S. T., & Davis, A. (2018). Identifying barriers and motivators for adoption of multifunctional perennial cropping systems by landowners in the Upper Sangamon River Watershed, Illinois. *Agroforestry Systems*, *92*, 1155–1169. https://doi.org/10.1007/s10457-016-0053-6

Mattia, C., Lovell, S. T., & Fraterrigo, J. (2018). Identifying marginal land for multifunctional perennial cropping systems in the Upper Sangamon River Watershed, Illinois. *Journal of Soil and Water Conservation*, *73*, 669–681. https://doi.org/10.2489/jswc.73.6.669

McGinnis, M. (2018). Hay shortage grows, prices nearly double. *Successful Farming*, 18 April. Retrieved from https://www.agriculture.com/news/crops/hay-shortage-grows-prices-nearly-double

McGraw, R. L., Stamps, W. T., Houx, J. H., & Linit, M. J. (2008). Yield, maturation, and forage quality of alfalfa in a black walnut alley-cropping practice. *Agroforestry Systems*, *74*, 155–161.

McGraw, R. L., Stamps, W. T. & Linit, M. (2005). Yield and maturation of alfalfa in a black walnut alley cropping practice. In K. Brooks & P. Ffolliott (Eds.), Moving agroforestry into the mainstream: Proceedings of the 9th North American Agroforestry Conference Proceedings. Columbia, MO: Association for Temperate Agroforestry.

Mead, R., & Willey, R. W. (1980). The concept of a "land equivalent ratio" and advantages in yields from intercropping. *Experimental Agriculture*, *16*, 217–228. https://doi.org/10.1017/S0014479700010978

Meiresonne, L., Schrijver, A. D., & Vos, B. (2007). Nutrient cycling in a poplar plantation (*Populus trichocarpa* × *Populus deltoides* 'Beaupre') on former agricultural land in northern Belgium. *Canadian Journal of Forest Research*, *37*, 141–155. https://doi.org/10.1139/x06-205

Mergen, F., & Lai, C. K. (1982). Professional education in agroforestry in North America. In E. Zulberti (Ed.), *International workshop on professional education in agroforestry* (p. 39–55). Nairobi, Kenya: ICRAF.

Miller, A. W., & Pallardy, S. G. (2001). Resource competition across the crop–tree interface in a maize–silver maple temperate alley cropping stand in Missouri. *Agroforestry Systems*, *53*, 247–259. https://doi.org/10.1023/A:1013327510748

Miller, G. (2003). Chestnuts. In D. W. Fulbright (Ed.), *A guide to nut tree culture in North America* (Vol. 1, pp. 167-181). Northern Nut Growers Association.

Miller, G., Miller, D. D., & Jaynes, R. A. (1996). Chestnuts. In J. Janick and J. N. Moore (Eds.), *Fruit breeding* (pp. 99–123). New York: John Wiley & Sons.

Millspaugh, J. J., Schulz, J. H., Mong, T. W., Burhans, D., Walter, W. D., Bredesen, R., . . . Dey, D. C. (2009). Agroforestry wildlife benefits. In H. E. Garrett (Ed.), North American agroforestry: An integrated science and practice (2nd ed., pp. 257–286). Madison, WI: ASA.

Minick, K. J., Strahm, B. D., Fox, T. R., Suere, E. B., Leggett, Z. H., & Zerpa, J. L. (2014). Switchgrass intercropping reduces soil inorganic nitrogen in a young loblolly pine plantation located in coastal North Carolina. *Forest Ecology and Management, 319,* 161–168. https://doi.org/10.1016/j.foreco.2014.02.013

Moen, M. (1996). Agroforestry grows in Minnesota. *Spectrum* (Minneapolis, MN), 2, 8–11.

Molnar, T. J., Kahn, P. C., Ford, T. M., Funk, C. J., & Funk, C.R. (2013). Tree crops, a permanent agriculture: Concepts from the past for a sustainable future. Resources, 2, 457–488. https://doi.org/10.3390/resources2040457

Montgomery, D. R. (2007). Soil erosion and agricultural sustainability. *Proceedings of the National Academy of Science, 104,* 13268–13272. https://doi.org/10.1073/pnas.0611508104

Moore, K. C., & Nelson, C. J. (1995). Economics of forage production and utilization. In R. F Barnes, D. A. Miller, and C. J. Nelson (Eds.), *Forages. Vol. I: An introduction to grassland agriculture* (pp. 189–202). Ames, IA: Iowa State University Press.

Morandin, L. A., & Kremen, C. (2013). Hedgerow restoration promotes pollinator populations and exports native bees to adjacent fields. *Ecological Applications, 23,* 829–839. https://doi.org/10.1890/12-1051.1

Morhart, C. D., Douglas, G. C., Dupraz, C., Graves, A. R., Nahm, M., Paris, P., . . . Spiecker, H. (2014). Alley coppice: A new system with ancient roots. *Annals of Forest Science, 71,* 527–542.

Mori, G. O., Gold, M., & Jose, S. (2017). Specialty crops in temperate agroforestry systems: Sustainable management, marketing and promotion for the Midwest Region of the U.S.A. In F. Montagnini (Ed.), *Integrating landscapes: Agroforestry for biodiversity conservation and food sovereignty* (pp. 331–366). Advances in Agroforestry 12. Cham, Switzerland: Springer. https://doi.org/10.1007/978-3-319-69371-2_14

Mosquera-Losada, M. R., Freese, D., & Rigueiro-Rodroguez, A. (2011). Carbon sequestration in European agroforestry systems. In B. Kumar and P. Nair (Eds.), Carbon sequestration potential of agroforestry systems. (Vol. 8, pp. 43–59). Dordrecht, the Netherlands: Springer Netherlands.

Mulia, R., & Dupraz, C. (2006). Unusual fine root distributions of two deciduous tree species in southern France: What consequences for modeling of tree root dynamics? *Plant Science, 281,* 71–85. https://doi.org/10.1007/s11104-005-3770-6

Mungai, N. W., Motavalli, P. P., Kremer, R. J., & Nelson, K. A. (2005). Spatial variation of soil enzyme activities and microbial functional diversity in temperate alley cropping systems. *Biology and Fertility of Soils, 42,* 129–136. https://doi.org/10.1007/s00374-005-0005-1

Nair, P. K. R., Nair, V. D., Kumar, B. M., & Showalter, J. M. (2010). Carbon sequestration in agroforestry systems. *Advances in Agronomy, 108,* 237–307. https://doi.org/10.1016/S0065-2113(10)08005-3

Nasielski, J., Furze, J. R., Tan, J., Bargaz, A., Thevathasan, N.V., & Isaac, M. E. (2015). Agroforestry promotes soybean yield stability and N_2-fixation under water stress. *Agronomy for Sustainable Development, 35,* 1541–1549. https://doi.org/10.1007/s13593-015-0330-1

National Agricultural Statistics Service. 2017. 2017 Census of agriculture: United States summary and state data (AC-02-A-51). Washington, DC: USDA–NASS. Retrieved from http://agcensus.mannlib.cornell.edu/AgCensus/censusParts.do?year=2002

Natzke, D. (2018). Hay market insights: Average prices move to three-year high. *Progressive Forage,* 8 June.

Nerlich, K., Graeff-Hönninger, S., & Claupein, W. (2013). Agroforestry in Europe: A review of the disappearance of traditional systems and development of modern agroforestry practices, with emphasis on experiences in Germany. *Agroforestry Systems, 87,* 475–492.

Nesbit, M., Stein, L., & Kamas, J. (2013). Improved pecans. In *Texas fruit and nut production* (E-609, p. 12). College Station, TX: Texas A&M AgriLife Communications.

Ntayombya, P., & Gordon, A. M. (1995). Effects of black locust on productivity and nitrogen nutrition of intercropped barley. *Agroforestry Systems, 29,* 239–254. https://doi.org/10.1007/BF00704871

Osborne, L. L., & Kovacic, D. A. (1993). Riparian vegetated buffer strips in water quality restoration and stream management. *Freshwater Biology, 29,* 243–258.

Oswalt, S. N., Miles, P. D., Pugh, S. A., & Smith, W. B. (2018). *Forest Resources of the United States, 2017: A technical document supporting the Forest Service 2020 update of the RPA Assessment* (Gen. Tech. Rep. WO-97). Washington, DC: U.S. Forest Service.

Palma, J. H. N., Graves, A. R., Bunce, R. G. H., Burgess, P. J., de Flippi, R., Keesman, K. J., . . . Herzog, F. (2007). Modeling environmental benefits of silvoarable agroforestry in Europe. *Agriculture, Ecosystems & Environment, 119,* 320–334. https://doi.org/10.1016/j.agee.2006.07.021

Palma, J. H. N., Graves, A. R., Burgess, P. J., Keesman, K. J., van Keulen, H., Mayus, M., . . . Herzog, F. (2007). Methodological approach for the assessment of environmental effects of agroforestry at the landscape scale. *Ecological Engineering, 29,* 450–462. https://doi.org/10.1016/j.ecoleng.2006.09.016

Pang, K., Van Sambeek, J. W., Navarrete-Tindall, N. E., Lin, C. H., Jose, S., & Garrett, H. E. (2019a). Responses of legumes and grasses to non-, moderate, and dense shade in Missouri, USA: I. Forage yield and its species-level plasticity. *Agroforestry Systems, 93,* 11–24. https://doi.org/10.1007/s10457-017-0067-8

Pang, K., Van Sambeek, J. W., Navarrete-Tindall, N. E., Lin, C.H., Jose, S., & Garrett, H. E. (2019b). Responses of legumes and grasses to non-, moderate, and dense shade in Missouri, USA: II. Forage quality and its species-level plasticity. *Agroforestry Systems, 93,* 25–38. https://doi.org/10.1007/s10457-017-0068-7

Paschke, M. W., Dawson, J. O., & David, M. B. (1989). Soil nitrogen mineralization in plantations of *Juglans nigra* interplanted with actinorhizal *Elaeagnus umbellata* or *Alnus glutinosa. Plant and Soil, 118,* 33–42. https://doi.org/10.1007/BF02232788

Paudel, B. R., Udawatta, R. P., & Anderson, S. H. (2011). Agroforestry and grass buffer effects on soil quality parameters for grazed pasture and row-crop systems. *Applied Soil Ecology, 48,* 125–132. https://doi.org/10.1016/j.apsoil.2011.04.004

Payne, J. A., Jaynes, R. A., & Kays, S. J. (1983). Chinese chestnut production in the United States: Practice, problems, and possible solutions. *Economic Botany, 37,* 187–200. https://doi.org/10.1007/BF02858784

Perfecto, I., Mas, A., Dietsch, T., & Vandermeer, J. (2003). Conservation of biodiversity in coffee agroecosystems: A tri-taxa comparison in southern Mexico. *Biodiversity & Conservation, 12,* 1239–1252. https://doi.org/10.1023/A:1023039921916

Persons, W. S., & Davis, J. M. (2005). Growing and marketing ginseng, goldenseal, and other woodland medicinals. Fairview, NC: Bright Mountain Books.

Pham, C. H., Yen, C. P., Cox, G. S., & Garrett, H. E. (1977). Slope position, soil water storage capacity and black walnut root development. In W.E. Balmer (Ed.), Proceedings: Soil moisture . . . site productivity symposium, Myrtle Beach,

SC (pp. 326–335). Atlanta, GA: U.S. Forest Service, Southern Area.

Phillips, J. (1993). Utilities (re)discover electricity from biomass. *Biologue, 11*, 23–29.

Pimentel, D. (2006). Soil erosion: A food and environment threat. *Environment, Development and Sustainability, 8*, 119–137.

Pimentel, D., Harvey, C., Resosudarmo, P., Sinclair, K., Kurz, D., McNair, M., . . . Blair, R. (1995). Environmental and economic costs of soil erosion and conservation benefits. *Science, 267*, 1117–1123. https://doi.org/10.1126/science.267.5201.1117

Piotto, D. (2008). A meta-analysis comparing tree growth in monocultures and mixed plantations. *Forest Ecology and Management, 255*, 781–786. https://doi.org/10.1016/j.foreco.2007.09.065

Ponder, F., Jr., & Baines, D. M. (1985). Growth and nutrition of planted black walnut in response to several cultural treatments. In J. O. Dawson & K. A. Majerus (Eds.), *Fifth Central Hardwood Forest Conference: proceedings of a meeting held at the University of Illinois at Urbana-Champaign, Illinois, April 15–17, 1985* (pp. 15–18). Urbana-Champaign: Dep. of Forestry, University of Illinois.

Powers, M. P., Lantagne, D. O., Nguyen, P. V., & Gold, M. A. (1996). Fertilizer and maize intercropping effects on nitrogen-fixation in black locust (*Robinia pseudoacacia* L.). In J. H. Ehrenreich, D. L. Ehrenreich, & H. W. Lee (Eds.), *Growing a sustainable future: Proceedings of the 4th North American Agroforestry Conference* (pp. 62–64). Moscow, ID: University of Idaho.

Pretzsch, H., Schütze, G., & Uhl, E. (2013). Resistance of European tree species to drought stress in mixed versus pure forests: Evidence of stress release by inter-specific facilitation. *Plant Biology, 15*, 483–495. https://doi.org/10.1111/j.1438-8677.2012.00670.x

Raitanen, W. E. (1978). Energy, fibre and food: Agroforestry in eastern Ontario. In *Proceedings of the Eighth World Forestry Congress, Jakarta, Indonesia*.

Ramos, D. E. (Ed.) (1985). *Walnut orchard management* (Publ. 21410). Davis, CA: Cooperative Extension, University of California.

Reid, W., & Hunt, K. L. (2003). Pecan production in the Midwest. In D. W. Fulbright (Ed.), *A guide to nut tree culture in North America* (Vol. 1, pp. 107-115). Northern Nut Growers Association.

Reid, W., & Olcott-Reid, B. (1985). Profits from native pecans. *Annual Report of the Northern Nut Growers Association, 76*, 77–83.

Reisner, Y., de Filippi, R., Herzog, F., & Palma, J. (2007). Target regions for silvoarable agroforestry in Europe. *Ecological Engineering, 29*, 401–418. https://doi.org/10.1016/j.ecoleng.2006.09.020

Renken, R. B. (1983). *Breeding bird communities and bird habitat associations on North Dakota waterfowl production areas of three habitat types* (Master's thesis). Retrieved from Iowa State University digital repository. https://lib.dr.iastate.edu/cgi/viewcontent.cgi?article=19685&context=rtd

Revord, R., Lovell, S., Molnar, T., Wolz, K. J., & Mattia, C. (2019). Germplasm development of underutilized temperate U.S. tree crops. *Sustainability, 11*, 1546. https://doi.org/10.3390/su11061546

Rivest, D., Cogliastro, A., Bradley, R. L., & Oliver, A. (2010). Inter-cropping hybrid poplar with soybean increases microbial biomass, mineral N supply and tree growth. *Agroforestry Systems, 80*, 33–40. https://doi.org/10.1007/s10457-010-9342-7

Rivest, D., Cogliastro, A., Vanasse, A., & Olivier, A. (2009). Production of soybean associated with different hybrid poplar clones in a tree-based intercropping system in southwestern Quebec, Canada. *Agriculture, Ecosystems & Environment, 131*, 51–60. https://doi.org/10.1016/j.agee.2008.08.011

Ryszkowski, L., & Kedziora, A. (2007). Modification of water flows and nitrogen fluxes by shelterbelts. *Ecological Engineering, 29*, 388–400. https://doi.org/10.1016/j.ecoleng.2006.09.023

Sapijanskas, J., Paquette, A., Potvin, C., & Kunert, N. (2014). Tropical tree diversity enhances light capture through crown plasticity and spatial and temporal niche differences. *Ecology, 95*, 2479–2492. https://doi.org/10.1890/13-1366.1

Sarmah, A. K., Meyer, M. T., & Boxall, A. B. A. (2006). A global perspective on the use, sales, exposure pathways, occurrence, fate, and effects of veterinary antibiotics (VAs) in the environment. *Chemosphere, 65*, 725–759. https://doi.org/10.1016/j.chemosphere.2006.03.026

Schoeneberger, M. M. (2009). Agroforestry: Working trees for sequestering carbon on agricultural lands. *Agroforestry Systems, 75*, 27–37. https://doi.org/10.1007/s10457-008-9123-8

Schoeneberger, M., Bentrup, G., Gooijer, H. D., Soolanayakanahally, R., Sauer, T., Brandle, J., . . . Current, D. (2012). Branching out: Agroforestry as a climate change mitigation and adaptation tool for agriculture. *Journal of Soil and Water Conservation, 67*, 128A–136A. https://doi.org/10.2489/jswc.67.5.128A

Schultz, R. C., Colletti, J. P., Isenhart, T. M., Marquez, C. O., Simpkins, W. W., & Ball, C. J. (2000). Riparian forest buffer practices. In H. E. Garrett (Ed.), *North American agroforestry: An integrated science and practice* (pp. 189–281). Madison, WI: ASA.

Schultz, R. C., Isenhart, T. M., Simpkins, W. W., & Colletti, J. P. (2004). Riparian forest buffers in agroecosystems: Lessons learned from the Bear Creek Watershed, central Iowa, USA. *Agroforestry Systems, 61*, 35–50. https://doi.org/10.1023/B:AGFO.0000028988.67721.4d

Schwarz, M. T., Bischoff, S., Blaser, S., Boch, S., Schmitt, B., Thieme, L., . . . Wilcke, W. (2014). More efficient aboveground nitrogen use in more diverse Central European forest canopies. *Forest Ecology and Management, 313*, 274–282. https://doi.org/10.1016/j.foreco.2013.11.021

Scott, D., & Menalda, P. H. (1970). CO exchange of plants: 2. Response of six species to temperature and light intensity. *New Zealand Journal of Botany, 8*, 361–368. https://doi.org/10.1080/0028825X.1970.10429136

Seiter, S., & William, R. (1994). Interplanting trees as a green manure crop in a vegetable production system in Oregon. In R. C. Schultz & J. P. Colletti (Eds.), Opportunities for agroforestry in the temperate zone worldwide: Proceedings of the 3rd North American Agroforestry Conference (pp. 159–162). Ames, IA: Department of Forestry, Iowa State University.

Sekawin, M., & Prevosto, M. (1978). Technical and economic analysis of the influence of management system and intercropping in a poplar plantation located in Piacentro. (In Italian.) *Cellulosa e Carta*, 8.

Senaviratne, G. M. M. A., Udawatta, R. P., Nelson, K. A., Shannon, K., & Jose, S. (2012). Temporal and spatial influence of perennial upland buffers on corn and soybean yields. *Agronomy Journal, 104*, 1356–1362. https://doi.org/10.2134/agronj2012.0081

Seobi, T., Anderson, S. H., Udawatta, R. P., & Gantzer, C. J. (2005). Influence of grass and agroforestry buffer strips on soil hydraulic properties for an Albaqualf. *Soil Science Society of America Journal, 69*, 893–901. https://doi.org/10.2136/sssaj2004.0280

Shalaway, S. D. (1985). Fencerow management for nesting birds in Michigan. *Wildlife Society Bulletin, 13*, 302–306.

Sharrow, S. H. (1991a). Tree planting pattern effects on forage production in a Douglas-fir agroforest. *Agroforestry Systems, 16*, 167–175. https://doi.org/10.1007/BF00129747

Sharrow, S. H. (1991b). Tree density and pattern as factors in agrosilvopastoral system design. In H. E. Garrett (Ed.), Proceedings of the 2nd North American Agroforestry Conference (pp. 242–246). Columbia, MO: School of Natural Resources, University of Missouri.

Sharrow, S. H., Carlson, D. H., Emmingham, W. H., & Lavender, D. (1996). Productivity of two Douglas fir–subclover–sheep agroforests compared to pasture and forest monocultures. *Agroforestry Systems, 34*, 305–313. https://doi.org/10.1007/BF00046930

Shults, P. (2017). *Exploring the benefits of cover crops to agroforestry tree plantations: An analysis of direct and indirect nitrogen transfer in alley cropping systems* (Master's thesis). Retrieved from Michigan State University Libraries digital repository. https://d.lib.msu.edu/etd/4678

Singh, H. P., Kohli, R. K., & Batish, D. R. (2001). Allelopathic interference of *Populus deltoides* with some winter season crops. *Agronomie, 21*, 139–146. https://doi.org/10.1051/agro:2001114

Smith, J., Pearce, B. D., & Wolfe, M. S. (2013). Reconciling productivity with protection of the environment: Is temperate agroforestry the answer? *Renewable Agriculture and Food Systems, 28*, 80–92. https://doi.org/10.1017/S1742170511000585

Smith, J. R. (1914). Soil erosion and its remedy by terracing and tree planting. *Science, 39*:858–862. https://doi.org/10.1126/science.39.1015.858

Smith, J. R. (1929). *Tree crops: A permanent agriculture*. New York: Harper and Row.

Smith, W. B., Miles, P. D., Vissage, J. S., & Pugh, S. A. (2003). *Forest resources of the United States, 2002* (Gen. Tech. Rep. NC-241). St. Paul, MN: U.S. Forest Service North Central Research Station.

Soemarwoto, O. (1987). Homegardens: A traditional agroforestry system with a promising future. In H. A. Steppler and P. K. R. Nair (Eds.), *Agroforestry: A decade of development* (pp. 157–170). Nairobi, Kenya: ICRAF.

Sparks, D. (1992). *Pecan cultivars: The orchard's foundation*. Watkinsville, GA: Pecan Production Innovations.

Spelter, H. (1996). *Capacity, production, and manufacture of wood-based panels in the United States and Canada* (Gen. Tech. Rep. FPL-GTR-90). Madison, WI: U.S. Forest Service, Forest Products Laboratory.

Spelter, H., McKeever, D., & Alderman, M. (2006). *Status and trends: Profile of structural panels in the United States and Canada* (Res. Note FPL-RP-636). Madison, WI: U.S. Forest Service, Forest Products Laboratory.

Ssekabembe, C. K., & Henderlong, P. R. (1991). Belowground interactions in alley cropping: Appraisal of first-year observations on maize grown in black locust alleys. In H. E. Garrett (Ed.), Proceedings of the 2nd North American Agroforestry Conference (pp. 58–73). Columbia, MO: School of Natural Resources, University of Missouri.

Stamps, W. T., & Linit, M. J. (1997). Plant diversity and arthropod communities: Implications for temperate agroforestry. *Agroforestry Systems, 39*, 73–89. https://doi.org/10.1023/A:1005972025089

Stamps, W. T., McGraw, R. L., Godsey, L. D., & Woods, T. L. (2009). The ecology and economics of insect pest management in nut tree alley cropping systems in the midwestern United States. *Agriculture, Ecosystems & Environment, 131*, 4–8. doi:10.1016/j.agee.2008.06.012

Stamps, W. T., Woods, T. W., Linit, M. J., & Garrett, H. E. (2002). Arthropod diversity in alley cropped black walnut (*Juglans nigra* L.) stands in eastern Missouri, USA. *Agroforestry Systems, 56*, 167–175. https://doi.org/10.1023/A:1021319628004

Stanton, B., Eaton, J., Johnson, J., Rice, D., Schuette, B., & Moser, B. (2002). Hybrid poplar in the Pacific Northwest: The effects of market-driven management. *Journal of Forestry, 100*, 28–33. https://doi.org/10.1093/jof/100.4.28

Stebbins, R. L. (1990). Requirements for a United States chestnut industry. In J. Janick and J. E. Simon (Eds.), *Advances in new crops* (pp. 324–327). Portland, OR: Timber Press.

Stefano, A., & Jacobson, M. G. (2017). Soil carbon sequestration in agroforestry systems: A meta-analysis. *Agroforestry Systems, 92*, 285–299.

Talbot, G. 2011. *L'integration spatiale et temporelle des competitions pour l'eau et la lumiere dans un systeme agroforestiers noyers-cereales permet-elle d'en comprende la productivite?* (Doctoral dissertation, Université de Montpellier, France). Retrieved from thèses-EN-ligne (tel-00664530, version 1). https://tel.archives-ouvertes.fr/tel-00664530

Thevathasan, N., & Gordon, A. M. (1996). Poplar leaf biomass distribution and nitrogen dynamics in a poplar–barley intercropped system. In J. H. Ehrenreich, D. L. Ehrenreich, & H. W. Lee (Eds.), *Growing a sustainable future: Proceedings of the 4th North American Agroforestry Conference* (pp. 65–69). Moscow, ID: University of Idaho.

Thevathasan, N. V., & Gordon, A. M. (2004). Ecology of tree intercropping systems in the north temperate region: Experiences from southern Ontario, Canada. *Agroforestry Systems, 61*, 257–268.

Thomas, A. L., & Reid, W. R. (2006). Hardiness of black walnut and pecan cultivars in response to an early hard freeze. *Journal of the American Pomological Society, 60*, 90–94.

Thompson, T. E., & Madden, G. (2003). Pecans. In D. W. Fulbright (Ed.), A guide to nut tree culture in North America (Vol. 1, pp. 79–104). Northern Nut Growers Association.

Torrey, J. G. (1978). Nitrogen fixation by actinomycete-nodulated angiosperms. *BioScience, 28*, 586–592. https://doi.org/10.2307/1307515

Tozer, E. (1980). These trees, which yield more cattle feed per acre than oats may lead to a more permanent agriculture. *Horticulture, 58*, 49–52.

Trudge, C. (2016). *Six steps back to the land: Why we need small mixed farms and millions more farmers*. Cambridge: Green Books.

Tsonkova, P., Böhm, C., Quinkenstein, A., & Freese, D. (2012). Ecological benefits provided by alley cropping systems for production of woody biomass in the temperate region: A review. *Agroforestry Systems, 85*, 133–152. https://doi.org/10.1007/s10457-012-9494-8

Udawatta, R. P., Anderson, S. H., Gantzer, C. J., & Garrett, H. E. (2006). Agroforestry and grass buffer influence on macropore characteristics: A computed tomography analysis. *Soil Science Society of America Journal, 70*, 1763–1773. https://doi.org/10.2136/sssaj2006.0307

Udawatta, R. P., Garrett, H. E., & Kallenbach, R. (2011). Agroforestry buffers for nonpoint source reductions from agricultural watersheds. *Journal of Environmental Quality, 40*, 800–806. https://doi.org/10.2134/jeq2010.0168

Udawatta, R. P., & Jose, S. (2012). Agroforestry strategies to sequester carbon in temperate North America. *Agroforestry Systems, 86*, 225–242. https://doi.org/10.1007/s10457-012-9561-1

Udawatta, R. P., Kremer, R. J., Garrett, H. E,. & Anderson, S. J. (2009). Soil enzyme activities and physical properties in a watershed managed under agroforestry and row-crop systems. *Agriculture, Ecosystems & Environment, 131*, 98–104.

Udawatta, R. P., Krstansky, J. J., Henderson, G. S., & Garrett, H. E. (2002). Agroforestry practices, runoff, and nutrient loss: A paired watershed comparison. *Journal of Environmental Quality, 31*, 1214–1225. https://doi.org/10.2134/jeq2002.1214

Udawatta, R. P., Motavalli, P. P., Garrett, H. E., & Krstansky, J. J. (2006). Nitrogen and nitrate losses in runoff from three adjacent corn-soybean watersheds. *Agriculture, Ecosystems & Environment, 117*, 39–48. https://doi.org/10.1016/j.agee.2006.03.002

Udawatta, R. P., Motavalli, P. P., Jose, S., & Nelson, K. A. (2014). Temporal and spatial differences in crop yields of a mature silver maple alley cropping system. *Agronomy Journal, 106,* 407–415. https://doi.org/10.2134/agronj2013.0429

Udawatta, R. P., Nygren, P., & Garrett, H. E. (2005). Growth of three oak species during establishment of an agroforestry practice for watershed protection. *Canadian Journal of Forest Research, 35,* 602–609. https://doi.org/10.1139/x04-206

University of Missouri Center for Agroforestry. (2013). Chestnut Decision Support Tool, Version 2.0. Retrieved from http://www.centerforagroforestry.org/profit/ChestnutDecisionSupportTool.xlsm

Unger, I. M., Goyne, K. W., Kremer, R. J., & Kennedy, A. C. (2013). Microbial community diversity in agroforestry and grass vegetative filter strips. *Agroforestry Systems, 87,* 395–402. https://doi.org/10.1007/s10457-012-9559-8

University of Georgia Extension (2018, 28 Sept.). *Pecans.* Retrieved from http://extension.uga.edu/topic-areas/fruit-vegetable-ornamentals-production/pecans.html

Upson, M. A., & Burgess, P. J. (2013). Soil organic carbon and root distribution in a temperate arable agroforestry system. *Plant and Soil, 373,* 43–58. https://doi.org/10.1007/s11104-013-1733-x

Uri, V., Tullus, H., & Lohmus, K. (2002). Biomass production and nutrient accumulation in short-rotation grey alder (*Alnus incana* (L.) Moench) plantation on abandoned agricultural land. *Forest Ecology and Management, 161,* 169–179. https://doi.org/10.1016/S0378-1127(01)00478-9

USDA. 2015. *Summary Report: 2012 National Resources Inventory,* Washington, DC: Natural Resources Conservation Service. Retrieved from https://www.nrcs.usda.gov/Internet/FSE_DOCUMENTS/nrcseprd396218.pdf

USDA. 2018. *Summary Report: 2015 National Resources Inventory,* Washington, DC: Natural Resources Conservation Service. Retrieved from https://www.nrcs.usda.gov/Internet/FSE_DOCUMENTS/nrcseprd1422028.pdf.

USEPA. (2007). *Hypoxia in the northern Gulf of Mexico: An update by the EPA Science Advisory Board* (EPA.SAB 08-003). Washington, DC: USEPA.

USEPA. (2009). *National water quality inventory: Report to Congress, 2004 reporting cycle.* Washington, DC: USEPA.

Vandermeer, J. (1989). *The ecology of intercropping.* Cambridge, UK: Cambridge University Press. https://doi.org/10.1017/CBO9780511623523

Van Eijk-Bos, C., Moreno, L.A., & Van Dijk, K. (1986). *Barreras vivas de Gliricidia sepium (Jacq.) Steud. (Matarraton) y su efecto sobre la perdida de suelo en terrenos de colinas bajas, Uraba, Colombia.* Bogota, Colombia: CONIF-Informa.

Van Rees, K. (2008). Developing a national agroforestry and afforestation network for Canada. *Policy Options, 29,* 54–57.

Van Sambeek, J. W. (1989). Vegetation management in established stands. In J. E. Phelps and D. R. McCurdy (Eds.), *Proceedings, 4th Black Walnut Symposium,* Carbondale, IL (p. 125). Indianapolis, IN: Walnut Council.

Van Sambeek, J. W., Godsey, L. D., Walter, W. D., Garrett, H. E., & Dwyer, J. P. (2016). Field performance of *Quercus bicolor* established as repeatedly air-root-pruned container and bareroot planting stock. *Open Journal of Forestry, 6,* 163–176. https://doi.org/10.4236/ojf.2016.63014

Volk, T. A., Abrahamson, L. P., Nowak, C. A., Smart, L. B., Thorakan, P. J., & White, E. H. (2006). The development of short-rotation willow in the northeastern United States for bioenergy and bioproducts, agroforestry, and phytoremediation. *Biomass and Bioenergy, 30,* 715–727. https://doi.org/10.1016/j.biombioe.2006.03.001

Walter, W. D., Godsey, L. D., Garrett, H. E., Dwyer, J. P., Van Sambeek, J. W., & Ellersieck, M. R. (2013). Survival and 14-year growth of black, white, and swamp white oaks established as bareroot and RPM-containerized planting stock. *Northern Journal of Applied Forestry, 30,* 43–46. https://doi.org/10.5849/njaf.11-047

Wang, Q., & Shogren, J. F. (1992). Characteristics of the crop–paulownia system in China. *Agriculture, Ecosystems & Environment,* 39, 145–152. https://doi.org/10.1016/0167-8809(92)90050-L

Wanvestraut, R., Jose, S., Nair, P. K. R., & Brecke, B. J. (2004). Competition for water in a pecan–cotton alley cropping system in the southern United States. *Agroforestry Systems, 60,* 167–179.

Weerasekara, C., Udawatta, R. P., Jose, S., Kremer, R. J., & Weerasekara, C. (2016). Soil quality differences in a row-crop watershed with agroforestry and grass buffers. *Agroforestry Systems, 90,* 829–838.

Wei, J. (1986). A study on the structure and economic return of the crop–paulownia system. In *Proceedings, Symposium of Agroforestry Systems in China* (pp. 33–39). Nanjing, China: Nanjing Forestry University.

Wiedenbeck, J. K., & Araman, P. A. (1993). Possible demands for eastern hardwoods resulting from harvest restrictions in the Pacific Northwest. *Forest Products Journal, 43,* 51–57.

Wiersum, K. F. (1984). Surface erosion under various tropical agroforestry systems. In C. L. O'Loughlin & A. G. Pearce (Eds.), *Symposium on Effects of Forest Land Use on Erosion and Slope Stability* (Workshop Rep. 9, pp. 231–239). Honolulu, HI: East–West Environment and Policy Institute.

Wiersum, K.F. (1985). Effects of various vegetation layers in an *Acacia auriculiformis* forest plantation on surface erosion in Java, Indonesia. In S. A. El-Swaify, W. C. Moldenhauer, & A. Lo (Eds.), *Soil erosion and conservation* (pp. 79–89). Ankeny, IA: Soil Conservation Society of America.

Williams, P. A., & Gordon, A. M. (1992). The potential of intercropping as an alternative land use system in temperate North America. *Agroforestry Systems, 19,* 253–263. https://doi.org/10.1007/BF00118783

Williams, P. A., & Gordon, A. M. (1995). Microclimate and soil moisture effects of three intercrops on the tree rows of a newly planted intercropped plantation. *Agroforestry Systems, 29,* 285–302. https://doi.org/10.1007/BF00704875

Williams, P. A., Koblents, H., & Gordon, A. M. (1996). Bird use of an intercropped, corn, and old field in southern Ontario, Canada. In J. H. Ehrenreich, D. L. Ehrenreich, & H. W. Lee (Eds.), *Growing a sustainable future: Proceedings of the 4th North American Agroforestry Conference* (pp. 158–162). Moscow, ID: University of Idaho.

Wilson, M. H., & Lovell, S. T. (2016). Agroforestry: The next step in sustainable and resilient agriculture. *Sustainability, 8,* 574. https://doi.org/10.3390/su8060574

Wolf, D. D., & Blaser, R. E. (1972). Growth rate and physiology of alfalfa as influenced by canopy and light. *Crop Science, 12,* 23–26. https://doi.org/10.2135/cropsci1972.0011183X001200010008x

Wolz, K. J., Branham, B. E., & DeLucia, E. H. (2018). Reduced nitrogen losses after conversion of row crop agriculture to alley cropping with mixed fruit and nut trees. *Agriculture, Ecosystems & Environment,* 258:172–181. https://doi.org/10.1016/j.agee.2018.02.024

Wolz, K. J., & DeLucia, E. H. (2018). Alley cropping: Global patterns of species composition and function. *Agriculture, Ecosystems & Environment, 252,* 61–68. https://doi.org/10.1016/j.agee.2017.10.005

Wolz, K. J., & DeLucia, E. H. (2019a). Black walnut alley cropping is economically competitive with row crops in the Midwest USA. *Ecological Applications, 29*(1), e01829. https://doi.org/10.1002/eap.1829

Wolz, K. J., & DeLucia, E. H. (2019b). Black walnut alley cropping is economically competitive with row crops in the Midwest USA. *Bulletin of the Ecological Society of America, 100*(1), e01500. https://doi.org/10.1002/bes2.1500

Wolz, K. J., Lovell, S. T., Branham, B. E., Eddy, W. C., Keeley, K., Revord, R. S., . . . DeLucia, E. H. (2018). Frontiers in alley cropping: Transformative solutions for temperate agriculture. *Global Change Biology, 24*, 883–894.

Wooley, J. B., Jr., Best, L. B., & Clark, W. R. (1985). Impacts of no-till row cropping on upland wildlife. *Transactions of the North American Wildlife and Natural Resources Conference, 50*, 156–168.

Wotherspoon, A., Thevathasan, N. V., Gordon, A. M., & Voroney, R. P. (2014). Carbon sequestration potential of five tree species in a 25-yr-old temperate tree-based intercropping system in southern Ontario, Canada. *Agroforestry Systems, 88*, 631–643. https://doi.org/10.1007/s10457-014-9719-0

Xiang, K., Shi, J., Baer, N. W., & Sturrock, J. W. (1990). *Protective plantation technology*. Harbin, China: Publishing House of Northeast Forestry University.

Yahner, R. H. (1982a). Avian nest densities and nest-site selection in farmstead shelterbelts. *Wilson Bulletin, 94*, 156–175.

Yahner, R. H. (1982b). Avian use of vertical strata and plantings in farmstead shelterbelts. *Journal of Wildlife Management, 46*, 50–60. https://doi.org/10.2307/3808407

Yen, C. P., Pham, C. H., Cox, G. S., & Garrett, H. E. (1978). Soil depth and root development patterns of Missouri black walnut and certain Taiwan hardwoods. In E. V. Eerden and J. M. Kinghorn (Eds.), Proceedings, Symposium on root form of planted trees, Victoria, BC, Canada (pp. 36–43). Victoria, BC: Canadian Forest Service, Pacific Forest Research Centre.

Yin, R., & He, Q. (1997). The spatial and temporal effects of paulownia intercropping: The case of northern China. *Agroforestry Systems, 37*, 91–109. doi:10.1023/A:1005837729528

Young, A. (1989). *Agroforestry for soil conservation*. Wallingford, UK: CAB International.

Zamora, D. S., Jose, S., & Nair, P. K. R. (2007). Morphological plasticity of cotton roots in response to interspecific competition with pecan in one alleycropping system in the southern United States. *Agroforestry Systems, 69*, 107–116. https://doi.org/10.1007/s10457-006-9022-9

Zamora, D. S., Jose, S., & Napolitano, K. (2009). Competition for ^{15}N labeled nitrogen in a loblolly pine–cotton alley cropping system in the southeastern United States. *Agriculture, Ecosystems & Environment, 131*, 40–50. https://doi.org/10.1016/j.agee.2008.08.012

Zerbe, J. I. (2006). Thermal energy, electricity, and transportation fuels from wood. *Forest Products Journal, 56*, 6–14.

Zhang, Y., Chen, H. Y. H., & Reich, P. B. (2012). Forest productivity increases with evenness, species richness and trait variation: A global meta-analysis. *Journal of Ecology, 100*, 742–749. https://doi.org/10.1111/j.1365-2745.2011.01944.x

Zulauf, C. (2018). A look back at the U.S. hay market over the last 100 years. *Dairy Herd Management*, 21 November. Retrieved from https://www.dairyherd.com/article/look-back-us-hay-market-over-last-100-years

Study Questions

1. How does a projected increase in U.S. wood consumption benefit family farms and what role might alley cropping play?

2. Desirable characteristics of a tree species vary depending on the goals and objectives of an agroforestry practice. What are the desirable characteristics for trees used in a temperate zone alley cropping practice?

3. Under what circumstances in designing an alley cropping practice is a multiple-tree row design superior to a single-tree row design?

4. What factors must be considered when establishing the width of alleys in an alley cropping practice?

5. What are the potential ecological benefits of alley cropping? How might different design aspects impact the strength of these ecological benefits?

6. What are the major opportunities in and hurdles to expanding alley cropping adoption in the temperate zone?

7. By definition, when is the growing of forages in tree alleys alley cropping and when is it silvopasture?

8. In designing alley cropping strips for erosion control, why might it be undesirable to maximize the ground surface area covered by the upper tree canopy? If you want to minimize erosion where you have tall trees, what other plant dimension must be accounted for in your design?

9. Considering the equi-marginal principle, is it possible that a land-use practice would show negative financial returns and still be equi-marginally balanced? Explain.

10. A few examples of alternative perennial crops were given in the text that were options for alley cropping production. What other alternative crops could be used in alley cropping and how do they compare economically to conventional crops?

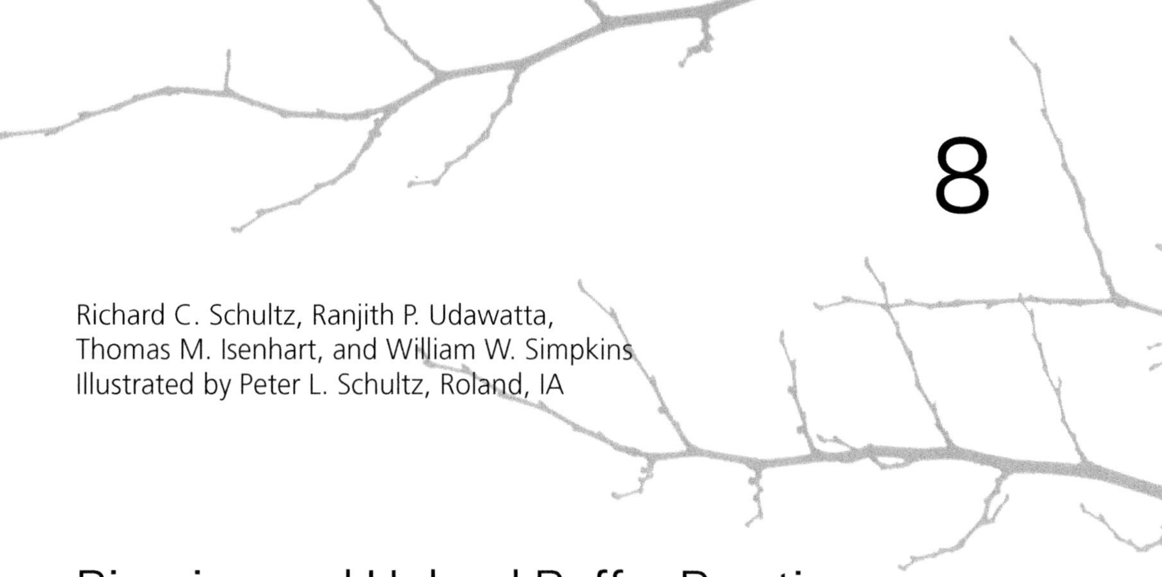

8

Richard C. Schultz, Ranjith P. Udawatta,
Thomas M. Isenhart, and William W. Simpkins
Illustrated by Peter L. Schultz, Roland, IA

Riparian and Upland Buffer Practices

As an agroforestry practice, buffers and filter strips play an important role in the movement of water and nonpoint-source pollutants through the watersheds where they are applied. Because of their positions in the landscape, they provide effective connections between the upland and aquatic ecosystems in a watershed and have been described as "one of the most effective tools for coping with nonpoint-source pollution" (Palone & Todd, 1997; Schultz, Isenhart, Beck, Groh, & Davis, 2019).

A riparian area can be defined in numerous ways. The U.S. Forest Service (Welsch, 1991) defines it as:

> *the aquatic ecosystem and the portions of the adjacent terrestrial ecosystem that directly affect or are affected by the aquatic environment. This includes streams, rivers, lakes, and bays and their adjacent side channels, flood plain, and wetlands. In specific cases, the riparian area may also include a portion of the hillslope that directly serves as streamside habitat for wildlife.*

Lowrance, Leonard, & Sheridan (1985) defined the riparian ecosystem as:

> *a complex assemblage of plants and other organisms in an environment adjacent to water. Without definite boundaries, it may include streambanks, flood plain, and wetlands, forming a transitional zone between upland and aquatic habitat. Mainly linear in shape and extent, they are characterized by laterally flowing water that rises and falls at least once within a growing season.*

The Coastal Zone Management Handbook (Palone & Todd, 1997) defines riparian areas as:

> *vegetated ecosystems along a water body through which energy, materials, and water pass. Riparian areas characteristically have a high water table and are subject to periodic flooding and influence from the adjacent water body. These systems encompass wetlands, uplands, or some combination of these two land forms. They will not in all cases have all the characteristics necessary for them to be classified as wetlands.*

Finally, the Bureau of Land Management (1993) defines riparian areas as

> *a form of wetland transition between permanently saturated wetlands and upland areas. These areas exhibit vegetation or physical characteristics reflective of permanent surface or subsurface water influence. Lands along, adjacent to, or contiguous with perennially and intermittently*

North American Agroforestry, Third Edition. Edited by Harold E. "Gene" Garrett, Shibu Jose, and Michael A. Gold.

flowing rivers and streams, glacial potholes, and the shores of lakes and reservoirs with stable water levels are typical riparian areas. Excluded are such sites as ephemeral streams or washes that do not exhibit the presence of vegetation dependent on free water in the soil

(Prichard, 1993).

The common threads among these and other definitions are that a riparian area (a) is adjacent to a body of water, (b) has no clearly defined boundaries, (c) is a transition between aquatic and upland environments, and (d) is linear in nature.

A riparian forest buffer is specifically defined as:

an area of trees, usually accompanied by shrubs and other vegetation, that is adjacent to a body of water and which is managed to maintain the integrity of stream channels and shorelines, to reduce the impact of upland sources of pollution by trapping, filtering and converting sediment, nutrients, and other chemicals, and to supply food, cover, and thermal protection to fish and other wildlife

(Palone & Todd, 1997).

Riparian forest buffers differ from vegetative or grassed filter strips in that they are dominated by a functional forest ecosystem.

Native grass filters can also be designed to intercept and treat nonpoint-source pollutants and serve as functional ecosystems. However, many non-native, cool-season grass filter strips are designed specifically for pollutant removal and are not multifunctional ecosystems by themselves. These filter strips often are not designed for large runoff events, which may inundate them with sediment after only a few events (Dillaha, Reneau, Mostaghimi, & Lee, 1989). However, they are critical to the design of successful riparian forest buffers that intercept surface runoff as concentrated flow. Where concentrated flow enters forest buffers without grass filters, gullies may develop and continue through the forest buffer, thus providing no effective treatment of the surface runoff (Knight, 2007). Cool-season grass filter strips do serve an excellent purpose when used as grass waterways designed to carry surface runoff without soil erosion.

While non-native plant species may work as well as native species in terms of controlling nonpoint-source pollutants, native plants are important for the habitat functions of vegetative buffers (Henningsen & Best, 2005). These multispecies buffers and filter strip plant communities also help enhance numerous other ecosystem services including pollination, soil health, and the aesthetic value of the landscape. Upland buffers with cool- or warm-season grass alone or combined with shrubs and/or trees are also used to reduce nonpoint-source pollutants and prevent gully formation in agricultural watersheds (Blanco-Canqui, Gantzer, Anderson, Alberts, & Thompson, 2004; Dosskey, Hoagland, & Brandle, 2007; Udawatta, Garrett, & Kallenbach, 2011; Udawatta, Krstansky, Henderson, & Garrett, 2002). These vegetative buffers, filter strips, or midslope contour buffers are narrower than riparian forest buffers, usually less than 15 m (50 ft) in width, and are located in the upland areas of the watersheds (Daniels & Gilliam, 1996; Dosskey, 2001). Although some of the functionalities are common for both upland and riparian buffer practices, they should not be confused with riparian forest buffers or grass filter strips. One major difference between these two systems is their location and associated functions. Upland buffers designed as crosswind trap strips can trap snow, providing in-field protection against wind erosion and improved soil moisture and keeping eroded soil closer to the source compared with the final-defense line of a streamside riparian buffer (USDA Farm Service Agency, 2007). Similar to riparian buffers, these multispecies upland buffers show enhanced soil microbial diversity, enzyme activities, C, nutrients, soil water recharge, soil physical properties, and microclimatic differences (Akdemir, Anderson, & Udawatta, 2016; Alagele, Anderson, & Udawatta, 2019; Alagele, Anderson, Udawatta, Veum, & Rankoth, 2019; Sahin, Anderson, & Udawatta, 2016; Seobi, Anderson, Udawatta, & Gantzer, 2005; Udawatta et al., 2006; Udawatta, Kremer, Nelson, Jose, & Bardhan, 2014; Udawatta, Rankoth, & Jose, 2019).

Riparian forest buffers can vary in design in response to management objectives. Their tree, shrub, and grass structure may differ from one location to the next, but they are usually dominated by the woody tree and shrub component. They can consist of existing riparian forests or be established on previously cultivated or grazed land. While this chapter focuses primarily on upland and riparian forest buffers in the agricultural landscape, these buffers may also be found in forested, suburban, and urban landscapes.

Palone and Todd (1997) summarized the differences among riparian forest buffer uses in different landscapes. In a forest landscape, riparian forest buffers are usually referred to as streamside management zones. Their purpose is to establish natural vegetated filters, control erosion, and increase both surface and subsurface filtration of sediment and chemicals adjacent to natural or constructed water bodies. Management activities are usually restricted within this zone. Harvesting is limited to the removal of individual trees or specific basal areas (Verry et al., 1999). Some organizations suggest leaving a minimum of 4.6 m^2 (50 ft^2) of basal area after harvest. A rule of thumb is to leave 50% of the original canopy cover after a harvest. Machine access is often restricted, and applications of fertilizers and pesticides are strictly controlled. Streamside management zones vary between 7.5 and 90 m (25 and 300 ft) in width. In a suburban landscape, riparian forest buffers usually consist of existing forests that border on residential or commercial developments. If these forests are protected and managed, they can provide the natural functions of filtering and processing of suburban pollutants associated with runoff from roads, lawns, and construction sites. In addition, they provide noise control and screening along with their wildlife habitat and aesthetic benefits. In a more urbanized environment, riparian forest buffers come under more pressure from development (Pouyat, Groffman, Yesilonis, & Hernandez, 2002; Pouyat, Szlavecz, & Yesilonis, 2010). Water tables in urban areas are usually lower than in similar locations in non-urbanized areas and soils usually have lower N content; therefore, buffer denitrification activity is not as essential a process as it is in agricultural landscapes (Gift, Groffman, Kaushal, & Mayer, 2010). Urban buffers are also very narrow and fragmented to the point that they may not be completely functional forest ecosystems. However, with proper planning and zoning, these forests can play an important role in stormwater management. These forests are often the largest and most continuous forests in the urban environment and therefore provide significant wildlife habitat and recreational opportunities. However, as forests are encroached on by urban development and fragmentation, their pollution control ability is often reduced. Because these buffers often are located on highly prized land, the goal of suburban and urban planning is to protect them from development and overuse by the public.

Riparian forest buffers in the agricultural landscape can take many forms. In the more humid parts of the eastern and midwestern United States, they typically exist as continuous narrow bands or irregular patches of remnant forest along portions of meandering streams. Most of these are natural remnants of previous forests. Most have been harvested numerous times and/or grazed by livestock. In some cases, riparian forest buffers had to be reestablished from scratch along previously cropped, grazed, and/or channelized reaches of streams. In the arid and semiarid West, riparian forest buffers may consist of narrow tree and shrub zones in a vast expanse of grazed, dry, upland shrubs and grasses.

Buffers in the agricultural landscape serve to reduce losses of soil and nutrients that typically remain within an ecosystem of undisturbed or perennial vegetation. By definition these buffers are placed at the edges of these "leaky" ecosystems to keep the soil and materials from completely escaping. In the ideal agricultural landscape, the goal is to keep the soil and materials close to their location of origin. Riparian forest buffers should serve as the last buffer in a system of buffers that keep the potential pollutants out of surface waters. Contour buffer strips, filter strips, grassed waterways, and vegetative barriers (USDA–NRCS Conservation Practice Standards 332, 393 [USDA–NRCS, 1997b], 412 [USDA–NRCS, 1999], and 601, respectively) have been designed to provide the infield buffering required to minimize the amount of materials that riparian buffers must handle (USDA–NRCS, n.d.). Several of these conservation buffers are narrow strips (2–3 m) of trees and grass–legume combinations planted along the field contours (Udawatta et al., 2002). Analyzing two large studies (de la Crétaz & Barten, 2007, pp. 105–130; Omernik, Abernathy, & Gale, 1981) concluded that the land use pattern of the entire watershed may be as important as or more important than the presence or absence of riparian buffers in determining water quality. While impaired stream conditions were associated with impaired riparian buffers, heavy nutrients, and suspended sediment loads (Meador & Goldstein, 2003), Omernik et al. (1981) found that the proximity of agriculture or forest to streams had no relationship with stream water quality but the proportion of land use under each management was highly correlated with stream water quality. Yu, Xu, Wu, and Zuo (2016) found that slope and the proximity of agriculture and urban lands to the stream had a significant impact on water quality, especially during the dry season. Other studies have acknowledged that a combination of best management practices are needed for the protection of our nation's water bodies. Mayer, Reynolds, McCutchen, and Canfield (2007) stated that to be successful in removing excess nutrients from surface and subsurface waters, comprehensive management approaches are required that involve a network of buffers, conservation practices, and nutrient management plans.

Wider riparian buffers along larger rivers reduce levee breaks and protect agricultural lands from floods and sand deposition. A study conducted by Allen, Dwyer, Wallace, and Cook (2003) along a 565-km (351-mile) stretch of the Missouri River showed that buffers >100 m (328 ft) in width were effective in protecting levees. Forty-one percent of the levee failure was associated with no woody corridors. The study further emphasized the importance of a continuous and intact buffer and the maintenance of buffer vegetation.

In urban areas within subdivisions, riparian buffers can be found along streams and lakes. Some of these buffers are within 5–10 m (16–32 ft) of homes. These buffers help reduce the pollutants including lawn nutrients, chemicals, sediment, and yard waste entering waterbodies. These buffers usually have larger older trees, mid-canopy trees, and an understory. These buffers also provide soil stability on sloping lands closer to streams. Since nutrients are received from yards, the understory and vegetation near the lawn and buffer edge are denser compared with a riparian buffer in a typical agricultural setting.

The main purpose of upland buffer practices is to reduce nonpoint-source pollutants from agricultural watersheds and to improve water quality. Design factors of these buffers vary with soil type, slope, precipitation, and management, and establishment must comply with local, state, and federal practice standards. However, these design factors change over a period of years due to changes in the vegetation and the buffer's influence on soil properties (Dosskey et al., 2007). The width of the cropped strip must be designed to accommodate some multiple of full equipment width, while the width of permanent vegetation should ensure sufficient grade to avoid ponding and to provide significant processing of nonpoint-source pollutants. In general, contour buffers, agroforestry buffers, and grass barriers should parallel the contour lines as closely as possible. Filter strips are typically the widest (>5 m) of these various practices and established between field borders and waterways (Blanco-Canqui et al., 2004). Grass, grass–legume combinations, shrubs, or trees established in these buffers should be compatible with site conditions. In areas where filter strips are established for wildlife habitat development, plants may be selected to satisfy the target wildlife.

Management practices within these zones include the removal of undesired vegetation, branch removal, replanting, harvesting, mowing, root pruning, pest control, and the removal of trapped soil. Mowing and harvesting helps to maintain stem density and height for maximum filtration of nonpoint-source pollutants. Harvesting of nuts from nut-bearing species, ornamental and woody plants, as well as the presence of wildlife may generate income in addition to that from the agricultural crops (Gold, Godsey, & Josiah, 2004). Sediment that comes from the crop areas may accumulate along the field edge of the buffer and may reduce or even smother the density of grasses and forbs in the filter strip edge of the buffer. Therefore, accumulated sediment should be removed and spread evenly in the adjacent crop areas. Landowners must inspect buffers for possible damages and ephemeral gully erosion, and these should be repaired before they worsen. Although fertilizer application is strictly controlled in riparian buffers, limited and targeted fertilizer application is recommended to enhance initial buffer growth. Because compaction of soils in the buffer strips may not serve the intended purpose, buffer strips should not be used as travel paths for livestock or equipment. In contrast, grass waterways can provide livestock and vehicular crossings when soil moisture conditions are favorable. However, heavy equipment movement should not be permitted, especially when soil moisture conditions are unfavorable.

Studies conducted in Missouri with alley cropping and riparian buffers have shown that mature buffer vegetation competes for resources and reduces crop yields (Senaviratne, Udawatta, Nelson, Shannon, & Jose, 2012; Udawatta, Gantzer, Reinbott, Wright, & Pierce, 2016; Udawatta, Nelson, Jose, & Motavalli, 2014). Therefore, root pruning, branch thinning, and tree removal could be practiced to minimize crop yield reductions by riparian and upland buffers. These management practices help reduce competition for resources including light, water, nutrients, and soil volume. These studies also have shown less impact on crop yields with Conservation Reserve Program filter strips or other shorter vegetation between the crop area and trees because the short vegetation has no or less shading effects. Another method of reducing competition between a riparian forest buffer and adjacent crop fields is to add a grass filter of anywhere between 9 and 45 m (30–150 ft) in width, the width depending on whether the edge of the forest buffer consists of trees or shrubs.

The movement of water in the hydrologic cycle has been extensively modified by agricultural and grazing practices. Without riparian buffers, surface runoff with its associated sediment, nutrient, and chemical loads can reach open bodies of water rapidly during and shortly after a rainfall event. Nutrients and chemicals also can find their way into streams, lakes, and coastal waters through the groundwater and through field tile drainage systems. As mentioned above, the combination of upland and riparian forest buffers in these landscapes serves extremely important functions for improving water quality, providing wildlife habitat, and stabilizing channels that have often been extensively modified. These functions include trapping, filtering, and converting sediment and agricultural chemicals before they reach the stream, providing stream bank and channel stability, and providing habitat for both terrestrial and aquatic organisms. By slowing the downstream flow of floodwaters, they can also play a role in reducing downstream floods during storm events. Their ability to sequester large amounts of plant and soil C can play a role in mitigating global climate change. Buffers also improve the biotic integrity of both the aquatic and terrestrial landscape by providing a diverse and ecologically functioning community. Finally, they provide wind and visual barriers, improve the aesthetics of the landscape, and can provide both market and non-market benefits to the landowner and society in general. Studies have shown that the establishment of specifically selected species and diverse vegetation in these buffers enhances degradation, adsorption, storage, utilization, and volatilization of various agrochemicals within a buffer, thus helping to improve water quality in rivers, streams, lakes, and ponds (Chu, Anderson, Goyne, Lin, & Lerch, 2013; Chu, Goyne, Anderson, Lin, & Lerch, 2013; Chu, Goyne, Anderson, Lin, & Udawatta, 2010; Lin, Goyne, Kremer, Lerch, & Garrett, 2010; Lin, Lerch, Garrett, & George, 2004; Lin, Lerch, Garrett, Li, & George, 2007; Lin, Lerch, Goyne, & Garrett, 2011). These benefits have been attributed to diverse microbial communities, various exudates, and soil reactions. For example, Lin et al. (2004, 2007, 2010, 2011) found shorter half-lives and faster degradation of herbicides, antibiotics, and degradation products in multispecies buffers. Various organic compounds in root exudates and microbial chemicals and their activities result in less harmful chemicals and CO_2.

As mentioned above, the position of buffers in the landscape provides effective connections between the upland and aquatic ecosystems in a watershed. Where they have been removed and their functions lost, surface water quality and the integrity of aquatic ecosystems have been compromised. Because riparian forest communities evolved in the most fertile and moist position of the landscape, they can often be reestablished rather easily. However, in many agricultural landscapes, land uses have so dramatically changed the hydrology that these communities cannot be restored to their original natural condition. In many cases, water table depths have been lowered to the point where restored buffers require a different community structure to function properly. However, with proper planning and design, the functions of a healthy riparian forest community or upland buffer can be reestablished. To appreciate the role of the riparian forest and upland buffer communities in the landscape and to understand how they function, it is necessary to review the movement of water through the landscape.

Hillslope and Channel Processes in Agricultural Landscapes

Water, in a landscape, moves both above and below the soil surface, from higher to lower elevations, concentrating in depressions, gullies, creeks and streams of ever-increasing size. The landscape surface, which concentrates that water into a specific stream, is called a *drainage basin* or *watershed*. A drainage basin is defined by the *watershed divide*, a continuous ridge surrounding the area from which water drains into a given channel system (Figure 8–1). As can be seen in Figure 8–1, watersheds are nested within other watersheds, giving rise to streams of different orders. A first-order stream is a perennial stream (one that flows year-round) with no perennial stream tributaries. A second-order stream arises when two first-order streams flow into each other. A third-order stream results from two second-order streams, and so forth (Strahler,1957).

The shape of a drainage basin and the flow, timing, and chemistry of the stream(s) in the basin are functions of the climate, geomorphology, and land use of the region. Climate provides the energy for modifying the shape of the landscape by regulating temperatures, including freeze–thaw cycles, and the timing, quantity, and type of precipitation. These directly influence the weathering of rock materials, the composition of plant communities, and the production of soil.

Vegetation is especially important in this context because it provides the organic matter for building and protecting the soil and the mechanism for evapotranspiration. These influences on soil formation and evapotranspiration are important in regulating streamflow. The water balance equation demonstrates that streamflow is the difference between precipitation and evapotranspiration:

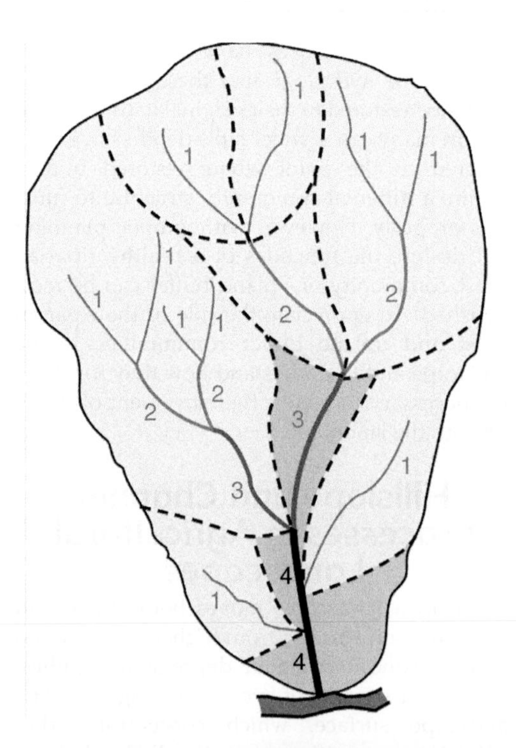

Fig. 8–1. Watershed and sub-watersheds of a fourth-order stream. The watershed divide surrounds all the land that drains surface water into the stream. The dotted lines outline various sub-watersheds for lower order streams. Numbers indicate stream order according to the Horton–Strahler method of classification. First-order streams are either intermittent or perennial, with no other intermittent or perennial channels entering them. Second-order streams are formed by the junction of two first-order streams, and third-order streams by two second-order streams, and so on. Perennial streams are those that essentially flow continuously throughout the year and are generally associated with a water table. Intermittent streams are those that flow only during the wet times of the year when they receive water from springs, rising water tables, or surface runoff. Ephemeral channels are always above the water table and flow only during and immediately after a rain, fed by surface runoff. Shaded areas on the diagram are the only areas that drain directly into the third- and fourth-order channels (after Leopold, Wolman, & Miller, 1964).

$$Q = P - \mathrm{ET} \pm \Delta S$$

where Q is streamflow, P is precipitation, ET is evapotranspiration, and S is the storage of water in the watershed.

In watersheds of the humid and subhumid temperate region, evapotranspiration from perennial plant communities can account for up to 66% of the water loss, leaving only 33% as streamflow (Hewlett, 1982). When perennial vegetation is converted to annual crops, evapotranspiration is reduced. Through canopy and litter layer interception of precipitation, vegetation protects the soil surface against erosion. Vegetation is a sink for nutrients that might otherwise leach or move with eroded soil particles from the landscape. Vegetation provides stability to stream banks and diverse habitat for aquatic and terrestrial wildlife. While plant communities help minimize excessive movement of water and materials in a drainage basin, some movement of organic plant and animal materials as well as inorganic nutrients and soil naturally takes place.

Organic and inorganic materials are transported through a drainage basin by several physical processes. Strictly speaking, material transfer involves erosion, leaching, transport, and deposition. These processes are important in determining the morphology of the drainage basin and the redistribution and export of nutrients from the basin. Erosion, leaching, transport, and depositional processes can create specific land forms that dictate the movement of water through the landscape and that offer contrasting habitats for terrestrial and aquatic organisms. These processes also influence the rates and patterns of succession that follow disturbances such as landslides and deposition of alluvial material following flood events.

Material transfer processes are broadly grouped into those affecting hillslopes and those operating in stream channels. Hillslope processes supply dissolved and particulate organic and inorganic material to the channel, where channel processes take over to break down and transport the material from the drainage basin. Hillslope processes can be divided into four broad classes (Figure 8–2). These are litterfall, surface erosion, solution transport, and mass movements.

Litterfall transfers organic matter in the form of leaves and stems to the soil surface or into an adjacent stream channel. The accumulated surface organic matter provides a protective litter layer to reduce raindrop impact. As this organic matter decomposes and becomes incorporated into the soil, it helps build and maintain soil structure and fertility and provides a C source for soil organisms. In riparian forest ecosystems, vegetation provides large woody debris to the channel. This woody debris regulates channel morphology through debris dams and provides a food source for aquatic organisms.

In humid and subhumid climates where vegetation and litter cover is continuous, surface runoff is almost nonexistent because high soil organic matter helps provide good soil structure and high infiltration rates. As a result, surface erosion is minimal, measured in tens of kilograms per hectare rather than tonnes. The protective layer of litter and the high infiltration rates of undisturbed surface soils allow most of the precipita-

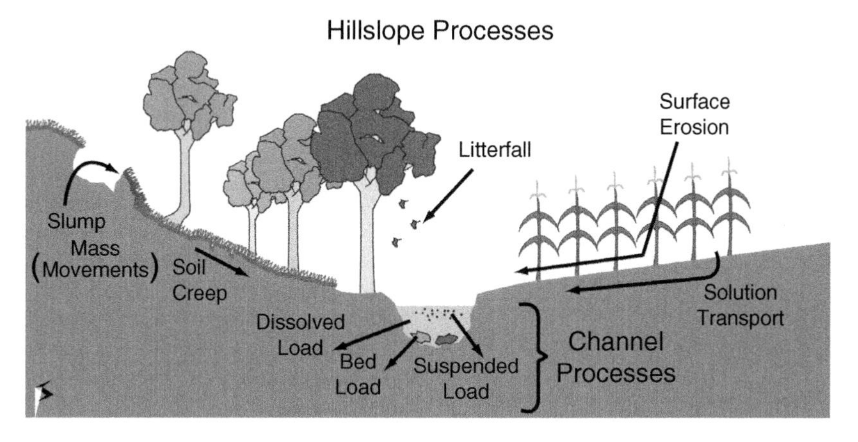

Fig. 8–2. Hillslope processes that transfer materials from upland and riparian ecosystems to the aquatic system. Channel processes transfer material from upstream to downstream watersheds.

tion that reaches the ground to enter the soil and move slowly through the profile to streams or through plants back to the atmosphere. In dry climates with incomplete canopy cover and a reduced or absent litter layer, much of the precipitation may naturally run over the surface, resulting in high erosion rates.

Solution transport carries dissolved materials that originate from the leaching of vegetation or soil, the weathering of bedrock, and the inputs of atmospheric deposition. In humid and subhumid ecosystems with continuous plant cover, these solutions follow torturous pathways through the soil, while in dry climates a significant portion may travel over the surface directly into stream channels. In the soil, the dissolved materials interact with living plant roots, microbial organisms, clay minerals, and humus before entering the groundwater or a stream channel. Materials may be removed from the solution by adsorption to the clay and humus or may be absorbed by plant roots and microbial organisms that create a "living filter" that reduces the concentration of solutes in the soil solution. In humid and subhumid climates with continuous plant cover, the landscape tends to keep most of its dissolved materials in the ecosystem by rapidly immobilizing them in the living biomass or adsorbing them in soil rich in organic and clay fractions. In drier climates with less vegetation cover or in agricultural landscapes where plant cover is primarily annual with long fallow periods, the ecosystem tends to "leak" nutrients. Once in the streams, they may be carried completely out of the watershed. In such cases, fertility losses may limit plant growth and site productivity.

Finally, mass movements may cause large quantities of material to move rapidly, as in landslides and debris avalanches, or slowly, as in slumps and soil creep. When these occur near streams, large quantities of material may rapidly enter the channel system, often resulting in changes in channel morphology. Vegetation tends to stabilize landscapes by removing excess moisture through evapotranspiration and providing soil strength by deep rooting.

The result of these hillslope transport processes is that upland watershed material is ultimately transported and deposited along and in streams. Along ephemeral channels (channels that are not connected to the water table and carry water only during and shortly after precipitation events), deposition is primarily the result of downslope movement of material that was directly upslope. Such deposits are called *colluvial deposits*. Along intermittent (channels that carry water during the wet part of the year) and perennial (channels that carry water most of the year) channels, there is enough energy that the colluvial material is eroded and transported in the channel and deposited farther downstream. Such deposits are called *alluvial deposits* and are responsible for forming the floodplain along streams and *point bars*, the sand and gravel deposits on the inside of channel bends.

Channel erosion and deposition are indicative of the channel processes of material transport. These include *dissolved transport*, the movement of materials in solution in the channel; *suspended sediment transport*, the movement of material in colloidal to sand-size fractions in suspension; and *bed load transport*, the movement of material larger than coarse sand that is rolled and bounced along the channel bottom. Lane (1955) developed the qualitative stream balance equation to show the relationship between stream discharge, channel gradient, sediment load, and sediment size:

$$Q_w S = Q_s D_{50}$$

where Q_w is stream discharge, S is the channel gradient, Q_s is sediment discharge or load, and D_{50} is sediment size. As can be seen in this equation, action that increases discharge may increase the gradient of the channel as well as increase the sediment load and/or sediment size. In young or very dry landscapes with little plant cover and soil development, much of the precipitation runs over the surface, reaching the stream channels rapidly and providing high discharges. Under such conditions, the channel cuts downward, creating a steep gradient that is capable of producing discharge with high sediment loads. Downcutting slows when the channel gradient approaches equilibrium with channel control points. *Channel control points* are features that cannot be easily eroded, thus controlling the depth of the channel above them. Permanent control points include exposed bedrock, while less permanent controls may be beaver (*Castor canadensis* Kuhl) or natural debris dams created from large woody debris. Control points create the pools and riffles that are important for healthy aquatic ecosystems. As the channel gradient is reduced, velocities are reduced and sediment is dropped out of the water, especially during flood events, creating point bars, natural levees, and floodplains.

A *floodplain* is the area along a stream that carries out-of-channel water from storm events

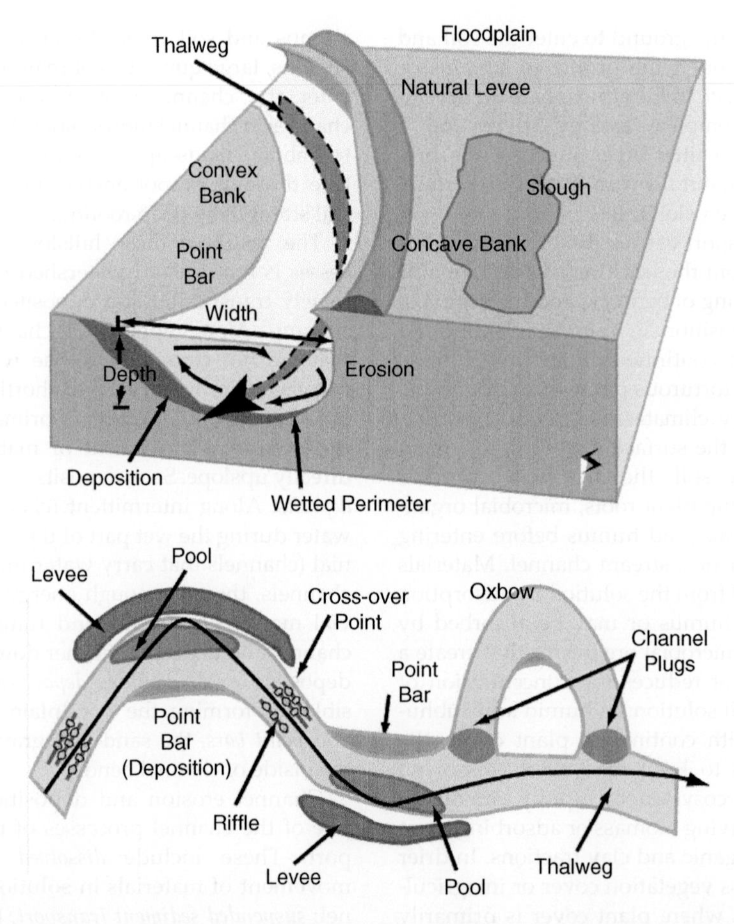

Fig. 8–3. Upper diagram shows floodplain and channel dimensional features. A natural levee forms from coarse sediment dropped from floodwater as its velocity rapidly slows on coming out of the channel. A slough consists of low wet spots that form behind levees because water is trapped or from oxbows that were cut off. The outside bank of the channel is concave because it is actively eroded by the thalweg, the highest current in the channel. The inside bank is convex because deposition of sediment from the slower current forms a point bar on that side of the channel. The wetted perimeter is the linear length of the bottom and sides of the channel that hold water. The lower diagram shows two-dimensional channel features. Pools and riffles are formed in response to the location of the thalweg in the meandering channel. Pools are associated with the scouring of outside banks or below riffles, and riffles are associated with the crossover points where the thalweg moves from one side of the channel to the other. Oxbows are portions of the channel that have been cut off because the meandering bend could no longer hold all of the flow during floods.

(Figure 8–3). The channel is said to be in contact with its floodplain when it can only carry storms of a 1–2-yr frequency (Leopold, Wolman, & Miller, 1964). Under these conditions the floodplain is used to dissipate the high energy associated with floods. Floodplains tend to be narrow along small streams because of the limited volume of water available for flooding. They increase in size as the watershed and associated channel gets larger and flooding becomes potentially greater.

If the channel is deeper and carries storms with a frequency greater than every 2–4 yr, the volume and velocity of discharge in the channel is increased, producing greater channel erosion. Downcutting is part of channel evolution that may lead to unstable bank conditions in young or degraded landscapes. The channel evolution model of Schumm (1977) can be used to assess the condition of stream banks and channels (Figure 8–4). Streams in contact with their floodplain are shown in Stage I. When an action such as channelization results in rapidly accelerating discharge or increasing channel gradient, downcutting and incision occur, constituting Stage II. The *headcut* is the point of active downcutting that moves up the main channel and all its tributaries. As the headcut passes, the channel deepens and stream banks become vertical. As the bank heights increase, they approach the critical height at which they begin to collapse. When banks begin to collapse, Stage III has begun. Collapse will continue as long as channel discharge is large enough to move the sediment associated with the collapsing banks downstream. Once the channel is wide enough that the energy of the discharge is spread across a wide enough channel so that all the collapsed sediment is not removed, sediment will remain at the toe of the banks long enough to allow plants to become reestablished, thus stabilizing the toe. When this occurs, the channel will be at Stage IV and, given enough time, may move on to Stage V.

Once the channel has reached Stage V, it may begin to move in a sinuous pattern from side to side, creating meanders. As it does, the material on the outside of the bends is eroded and deposited on the inside of the bend producing a relatively flat surface known as the point bar (Figure 8–3). This flat surface is the beginning of the floodplain. It is composed of the coarse material deposited by bedload from the scouring of the outsides of bends and then is covered by finer silt and clay from suspended load that is deposited by the frequent flooding of the point bar. As the channel system continues to develop, the floodplain grows in size until it can be many kilometers wide, as seen on rivers such as the Mississippi. Other dominant features of the floodplain include natural levees that form along the edge of outside bends as a result of rapid deposition of sediment when floodwaters that over-

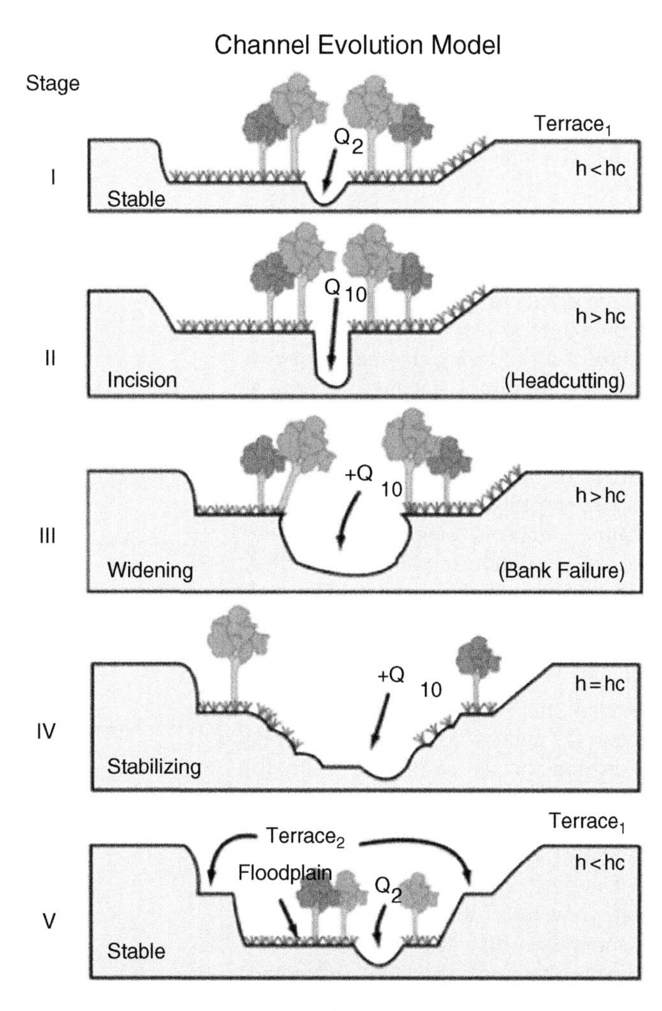

Fig. 8–4. The channel evolution model demonstrates the cycle of events that occur when a channel is out of equilibrium with its discharge or sediment load or its slope has been modified downstream. The *Q* values show the size of storm event volume that the channel can carry, with the subscripts denoting the average recurrence interval in years. Incised channels carry larger storm flows, increasing the velocity of flow which, in stable channels, would be reduced because much of the flow would be spread over the floodplain. The *h* is the present bank height and h_c is the critical bank height at which failure occurs (adapted from Schumm, Harvey, and Watson, 1984).

top the banks slow and drop sediment. Behind these levees, wet depressions called *sloughs* may develop because water cannot easily return to the channel. Finally, *oxbows* may develop when meander bends become so tight that they are cut off by floodwaters. Streams and their floodplains are so intimately linked that they should be managed as integral parts of a single ecosystem (National Research Council, 1993).

Stream channels are long linear systems that generally increase in width, depth, and volume from their source to their mouth. The stream channel of a healthy stream usually consists of a series of riffles, pools, and runs that reoccur at an interval of about five to seven times the width of the channel (Figure 8–3). Riffles are areas of rock and gravel where turbulence provides opportunities for aeration and habitat for macroinvertebrates. They usually occur at the crossover point in the channel where the *thalweg* (the portion of the channel with the most rapid current) moves from one side of the stream to the other. *Riffles* are control points that control downcutting and undercutting in the channel. *Pools* are depressions located along the outside bends of meandering channels that also provide important habitat for aquatic organisms. *Runs* are areas of flat, shallow water often located just downstream of riffles where the current is moving rapidly. During the process of channel incision, Stages II–III of the channel evolution model, pools, riffles, and runs may not be present. The relationship between the stream and the landscape changes along its length (Figure 8–5). A stream system can be divided into three major geomorphic zones: (a) the *production zone*—the zone of erosion and sediment production, (b) the *transport zone*—the zone of storage and transport, and (c) the *deposition zone* (Schumm, 1977).

The production zone is in the headwater reaches of the fluvial system and encompasses first- through third-order streams that are intimately in contact with their watershed. This zone may be at high elevations in the mountainous regions of the eastern and western United States or may be found in the gently undulating landscapes of the central and southern United States. The streams in this zone have the steepest gradients in the fluvial system. In the steeper regions of the country, these channels may be relatively straight and V-shaped with steep banks unless the channel bottom consists of bedrock, which serves as a channel control. In flatter landscapes, natural channels may be shallow with low banks. In this zone, the distance between the hilltops and channels is the shortest of any landscape position, providing ample opportunities for sediment from surface erosion to run off into the channel system. The floodplain in this zone is narrow and floods only for short periods of time immediately following precipitation events. The riparian vegetation in this zone has strong control over the aquatic environment of the stream (Karr, 1991). Because the low-order streams of this zone reach out farthest into the landscape, agricultural processes and urban development are closest to the channel (distance from divide to channel is shortest in this

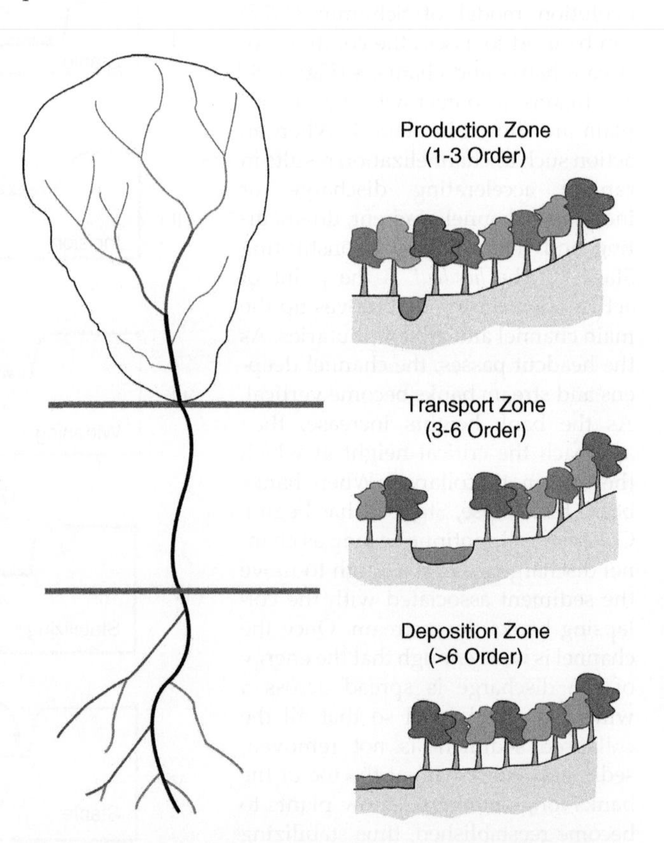

Fig. 8–5. The geomorphic zones of a fluvial system. In the production zone, surface sediment is detached and transported by erosion and runoff. In this zone, channels are in the closest contact with the uplands of the watershed because the distance from any upland point to a channel is relatively short, increasing the potential that detached soil particles can reach the surface channel. The transport zone is found along midsized streams and is the zone of transport and sediment storage. Sediment is stored in the channel as point and channel bars. The deposition zone is found along the largest streams where the gradient is small and the channel makes a confluence with an even larger stream or a lake or ocean. Here sediment is deposited in alluvial fans.

zone) and have the potential to provide large amounts of input to the channel.

The transport zone lies downstream of the zone of erosion and generally consists of the fourth- through sixth-order streams. These streams accumulate the sediment and nutrient loads that are produced by the low-order streams and transport them farther downstream. The actual land area that drains directly into these streams is a relatively small portion of the whole drainage basin area (Figure 8–1). These channels may still have relatively steep gradients but may have either a V- or U-shaped channel depending on topography and geology. The floodplain in this zone is still relatively narrow and composed of a high percentage of coarse-textured alluvial material.

The deposition zone is found along the highest order streams of the fluvial system. These are low-gradient streams that move through gently sloping, wide valleys. The rivers are wide and move slowly, depositing sediment in large, broad floodplains through which they meander. Alluvial soils grade from coarse textured near the stream channel to fine textured at the far edges of the floodplain. The flooding of the broad riparian zone tends to be seasonal following spring or rainy season floods.

As can be expected, riparian zone functions differ depending on in which functional stream zone they are located (Figure 8–5). The narrowest riparian zone would be found in the zone of production and would increase in width and complexity as it proceeds through the other two zones. The wider the zone becomes, the more it begins to function as a sink and source for materials from both the uplands and the channel. As the channels get wider in the lower two zones, the riparian vegetation also has less direct impact on the aquatic environment in the channel.

Burt (1997) noted that Schumm's scheme of zonation fits the river continuum concept proposed by Vannote, Minshall, Cummins, Sedell, & Cushing (1980) (Figure 8–6). That scheme couples continual downstream changes in channel morphology, energy, and organic matter with aquatic biota. In the production zone, streams are small, often highly oxygenated, and have inputs of organic matter from the riparian ecosystem as their primary source of energy. In this zone, in-stream photosynthesis is minimal and most of the invertebrate populations of organisms belong to a functional group of shredders of coarse particulate organic debris that has fallen into the stream

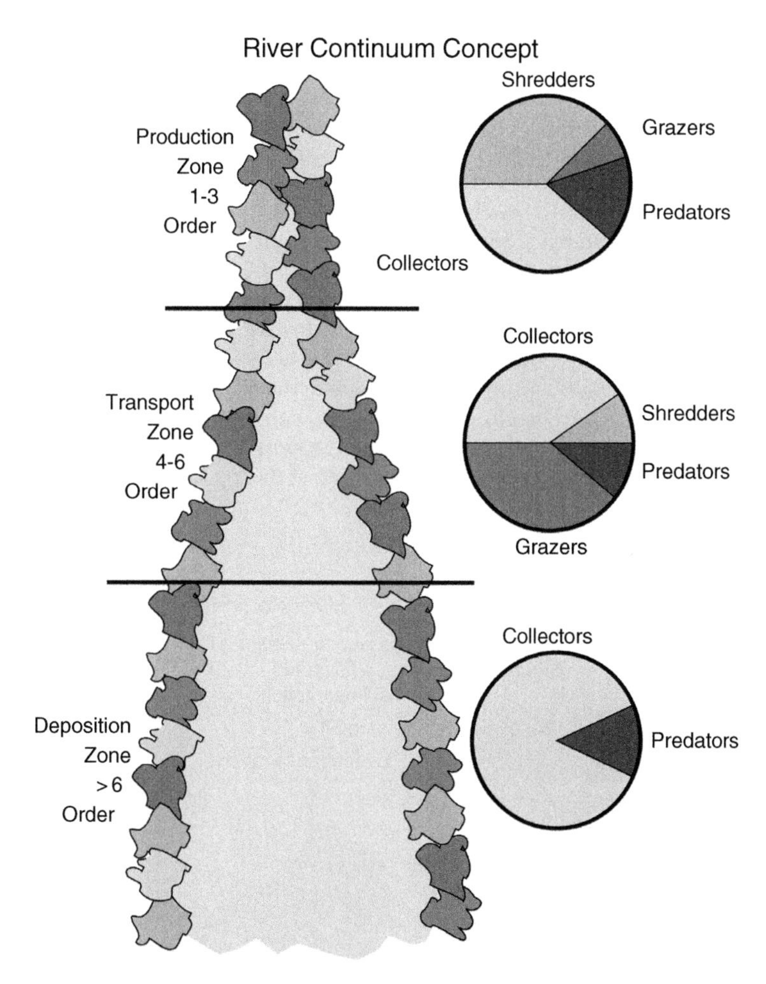

Fig. 8–6. The river continuum concept depicts the relationship between the geomorphic zones of a fluvial system, the impact of riparian vegetation on in-stream processes, and the invertebrates in the stream. In the zone of production, the riparian vegetation may shade much of the channel. The large and fine organic matter debris provided to the channel serves as the primary source of carbon for the aquatic ecosystem. In the zone of transport, the organic carbon input from the riparian vegetation is replaced by in-stream production of carbon from aquatic plant life that benefits from increased solar inputs to the channel. In the zone of deposition, increased turbidity may reduce in-stream production of carbon, and invertebrates may rely primarily on suspended carbon fragments moving in the channel from upstream (adapted from Vannote et al., 1980).

from the adjacent riparian vegetation. The shredders themselves either eat the shredded material or leave some suspended in the water column to be extracted and eaten by the functional group of collector invertebrates.

In the transfer zone, the channel is wider and the influence of the riparian ecosystem is reduced. In this zone, energy is produced in the stream by aquatic plants or transferred from upstream in the form of fine particulate organic matter that was not extracted by the collectors. The major functional group of invertebrates in this zone are the grazers of aquatic plants, which also may release fine particulate matter into the water column that can then be extracted by collectors. Finally, in the deposition zone, sediment and solute loads are great enough to reduce light penetration and therefore in-stream energy production by plants, with the biota relying on the transfer of energy from upstream. The major functional group of invertebrates in this reach is the collector group. Thus, activities in the headwater production zone can have important impacts on the downstream ecosystem.

The important concept to remember is that energy and matter are continually moving downstream through the fluvial system. Solutes usually spend a shorter time in the system than sediments that can be temporarily stored many times as they move through the channel system. It is also important to remember that most of the land area in a river basin is located in the headwaters zone of erosion and sediment production, where the influence of agricultural practices can be rapidly coupled with the channel system. In fact, nearly three-fourths of the total stream length in the United States is made up of first- and second-order streams (Dunne & Leopold, 1978). Riparian management practices, whose primary goals are to reduce nonpoint-source pollutant inputs from cultivated fields and urban and suburban landscapes, are most effective in the production and upper transport zones of the landscape. In the zone of deposition, riparian management practices are most effective at attenuating flood flows. Large expansive forests in these wide, flat bottoms provide the friction and storage needed to slow the velocity of flood flows (Verry et al., 1999).

The Hydrologic Cycle: Paths within an Ecosystem

In landscapes covered with native vegetation, precipitation is first intercepted by vegetation surfaces, where it can be temporarily stored (interception storage) and evaporated, or it can run down branches and stems as stemflow, or accumulate on leaf surfaces and fall as throughfall (Figures 8–7 and 8–8). Raindrops associated with throughfall are often larger than those not intercepted and have a different chemistry

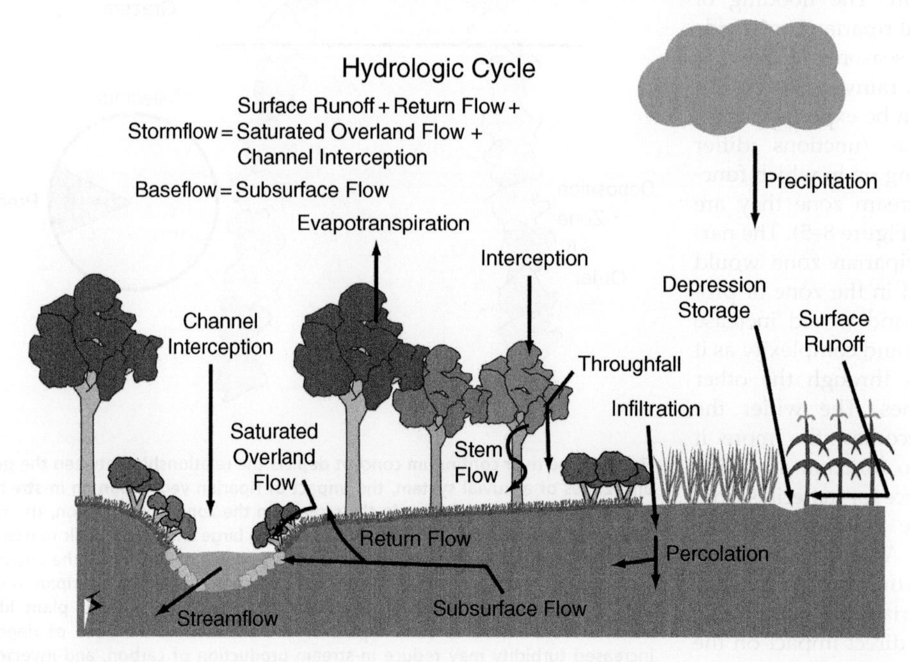

Fig. 8–7. The hydrologic cycle showing the paths of water movement through a riparian zone. Similar paths are present in all watersheds.

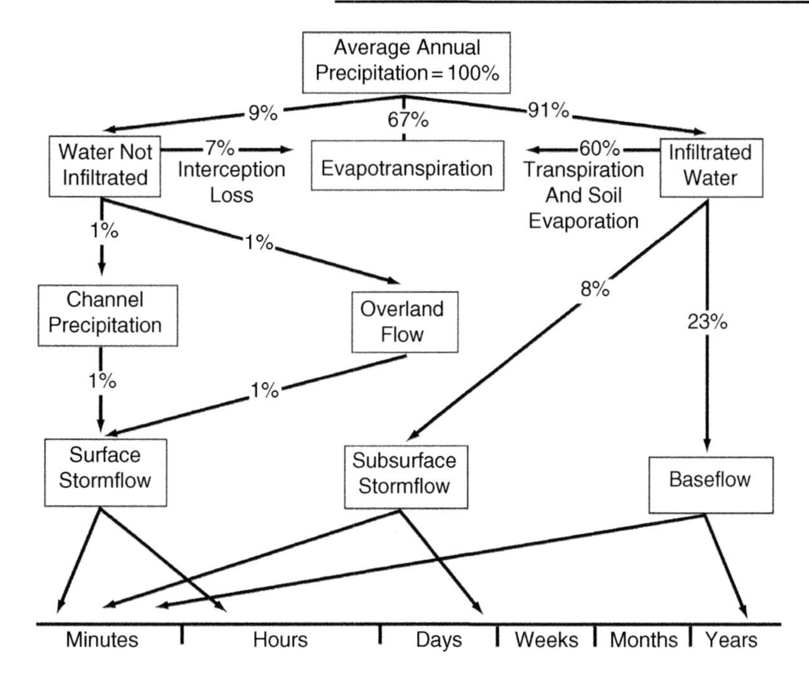

Fig. 8–8. Fate and temporal scale of precipitation in undisturbed watersheds in the eastern United States (adapted from Hewlett, 1982).

resulting from their contact with plant surfaces and associated dry deposition. If the larger drops reach their terminal velocity before reaching the soil surface, they will have more erosive force than non-intercepted raindrops. This situation can be found in some old-growth forests and in some forest plantations where the distance between the lowest branches in the canopy and the ground is greater than 10 m, the distance needed for the largest raindrop to reach terminal velocity. At terminal velocity, a falling large raindrop has 2.01 mJ of kinetic energy, which can easily dislodge soil particles and promote erosion (Gantzer, Buyanovsky, Alberts, & Remley, 1987). The potential of greater erosion under these conditions stresses the crucial protective role of the understory vegetation and the surface organic matter in protecting the soil. Upon hitting the litter layer, water may come to rest in depressions within the litter or between soil particles and become part of depression storage, which is temporarily held and evaporated. Once depression storage has been filled, water infiltrates into the soil surface. *Infiltration* refers to the entrance of water into the mineral soil. Soil texture and structure, along with antecedent moisture conditions, determine the rate of infiltration. In undisturbed soils under continuous native vegetation cover, infiltration rates usually far exceed even the highest precipitation rates. If mineral soils have been exposed or compacted, or if they are covered with impervious materials, infiltration rates can drop dramatically or cease to exist. Under these conditions, precipitation rates often exceed infiltration rates, and overland flow or surface runoff begins. *Overland flow* and *surface runoff* refer to water that never infiltrates the soil from where it is intercepted to where it finally enters an ephemeral, intermittent, or perennial channel. Surface runoff moves rapidly to a stream, producing *storm flow* (stream discharge flowing during and immediately after a heavy rain).

Once in the soil, water percolates vertically in response to gravity or moves laterally in response to slope or restrictive soil horizons, parent material, or bedrock. Subsurface flow, which moves slowly downslope through the macro- and micropores of the soil, may saturate the floodplain soil near the stream, producing *return flow,* which is water that exfiltrates from the soil to the surface. Return flow and rainfall falling on the saturated soil produce *saturated overland flow* (saturated surface runoff). Subsurface flow from this saturated zone also contributes to storm flow and is called *subsurface storm flow* (interflow). Subsurface flow is also the source of *base flow*, the flow that supplies water to channels between storm events. Because subsurface flow from soil near the channel can contribute to both storm flow and base flow, differentiating between the two near the end of a storm event often becomes arbitrary.

Rain falling near the boundary of an undisturbed watershed will not make it to the stream during a storm event. The saturated zone of the soil will produce storm flow in the stream when return flow, saturated overland flow, and subsurface flow reach the channel during the storm.

The saturated zone is dynamic, expanding and contracting depending on the intensity and duration of the rain and the antecedent moisture conditions prior to the rain. This zone is called the *variable source area* and is responsible for generating both storm flow and base flow in naturally vegetated drainage basins (Hewlett, 1982) (Figure 8–9). Prior to a rain and assuming a long period without prior rain, the saturated zone of the soil is limited to areas adjacent to perennial channels. As rain begins to fall on the watershed, areas of shallow soil or areas that are already moist are the first to become saturated. These areas are usually at the bottom of slopes along stream channels. As the rain continues, channels expand upslope into ever increasingly wet areas that can produce storm flow. Not only do channels elongate up into the watershed but they also begin to fill locally from return flow, saturated overland flow, subsurface storm flow, and surface runoff. In disturbed basins, storm flow can be generated by a much larger surface area because of high surface runoff from reduced infiltration rates. Under these conditions, storm flow occurs rapidly and extensively. After the storm has ended, the variable source area begins to shrink as subsurface water provides base flow to the channel. The residence time of the subsurface flow depends on the slope of the landscape, the porosity of the soil, and the evapotranspirational demand of the plant community. Thus, the variable source expands and contracts in response to storm events. The actual extent of the change depends on the physical and biological characteristics of the watershed and the rainfall and land-use patterns.

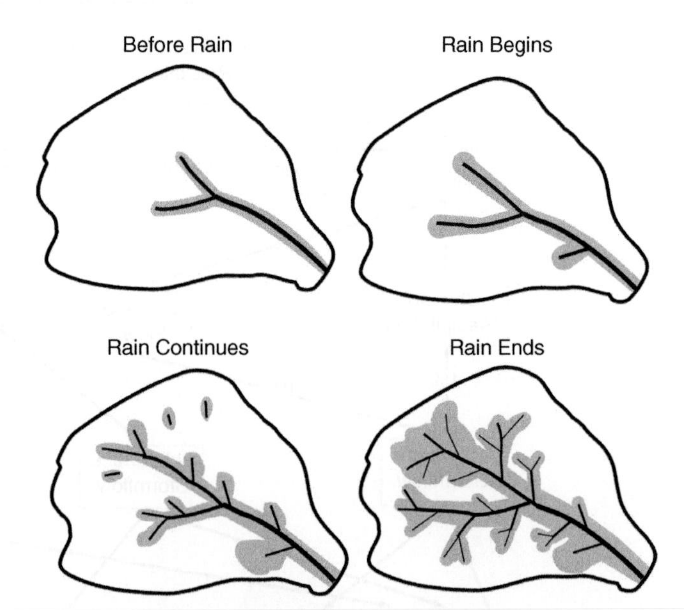
Before Rain
Rain Begins
Rain Continues
Rain Ends
1. Channel system elongates into intermittent and ephemeral channels
2. Saturated source area along channels increase in width

Fig. 8–9. The variable source area that contributes stormflow and base flow to a dynamic channel system. After rain has begun, the channel system elongates into intermittent and ephemeral channels as soils adjacent to the channels become saturated. The saturated source area along channels also increases in width as rainfall continues. After rain ends, the channel system again begins to contract. Ephemeral channels receive most of their water from surface runoff, while perennial channels receive water from surface runoff, subsurface soil water, and groundwater (adapted from Hewlett, 1982).

In summary, while some of the water flowing in streams is the direct result of precipitation on the channel, most is derived from surface and subsurface runoff originating in its drainage basin. The actual amount and rate of movement of surface and subsurface runoff is determined by (a) the geomorphology of the landscape, including topography and parent materials, (b) the soils and antecedent soil moisture conditions, (c) the climate, especially the frequency, intensity, and duration of precipitation events, (d) the vegetation and animals, and (e) land-use practices. These factors influence water movement through their impacts on soil formation and soil quality.

Soil quality is "the capacity of the soil to promote the growth of plants, protect watersheds by regulating the infiltration and partitioning of precipitation, and prevent water and air pollution by buffering potential pollutants" (National Research Council, 1993). This definition of soil quality implies that a combination of physical and biological processes that have developed over time at a specific location allow a soil to (a) act as a medium for plant growth, (b) regulate and partition water flow, and (c) serve as an environmental filter (Wagenet & Hutson, 1997). The factors that influence soil quality are: texture; aggregation; depth of topsoil and rooting; infiltration rates and bulk density; water holding capacity; soil organic matter; pH; electrical conductivity; extractable N, P, and K; microbial biomass C and N; potentially mineralizable N; soil respiration; water content; and temperature (Doran & Parkin, 1996; Karlen et al., 1997). Naturally vegetated soils in the floodplain often have high quality because of their depositional nature and their soil moisture. This high quality is especially important in maintaining stream water quality. Numerous studies have shown significantly greater amounts of C, N, enzyme activities, microbial diversity, and other soil physical properties within riparian and upland buffers. These effects have increased as the system matured. Many studies have shown lower bulk density, greater infiltration, and more water storage in upland buffers than with row crop management and grazing practices (Alagele, Anderson, & Udawatta, 2019; Alagele, Anderson, Udawatta, et al., 2019; Kumar, Anderson, Bricknell, & Udawatta, 2008; Kumar, Anderson, Udawatta, & Kallenbach, 2012; Sahin et al., 2016). At the Greenley Center paired watershed study, Seobi et al. (2005) showed a 2.3% reduction in bulk density after 6 yr of buffers, and Akdemir et al. (2016) noted an 8.5% lower bulk density on the same watersheds after 17 yr. These buffer practices also improve geometrical pore parameters measured by computed tomography (Kumar, Anderson, & Udawatta, 2010; Udawatta & Anderson, 2008). Buffers contained a greater number of pores, larger pores, and greater porosity, fractal, and circularity parameters. Comparing soil thermal properties of agroforestry buffers, grass buffers, and prairies with a corn (*Zea mays* L.)–soybean [*Glycine max* (L.) Merr.] rotation, Adhikari, Udawatta, Anderson, and Gantzer (2014) demonstrated better buffering capacity of soils under buffers and prairies than row crops. These beneficial effects help promote the persistence of diverse soil microbial communities by reducing stress conditions and preserving soil moisture and C.

The floodplain and its associated riparian zone is the location of the saturated variable source area that provides base flow to the stream. By nature of its lack of slope and nearness to the channel, the riparian zone has a high water table with a long residence time and slow discharge. As the floodplain gets wider, water has a longer residence time, especially if the alluvial sediments are fine textured. This means that the soil solution is in contact with the living filter of plant roots and microbial organisms for a longer time, thereby providing an important biological influence on the quality of water in solution transport. The potential effectiveness of the living filter is a function of soil quality.

The connections between the riparian zone and subsurface flow from the uplands are very dependent on the geomorphology of the site. Where the alluvial deposits are deep and the water table lies below the rooting zone or where the alluvial soil is underlain by a coarse-textured material, water may percolate below the rooting zone of the vegetation, bypassing the living filter (Figure 8–10). It is very important to recognize the complexity of water movement through floodplains or riparian zones (Correll, 1997). Not only can upslope water move through or below the riparian community but stream channel water may provide input to groundwater or surface flooding. Groundwater flow paths may vary throughout the year depending on water table depth, precipitation, and the sequence and structure of the sediments underlying the riparian

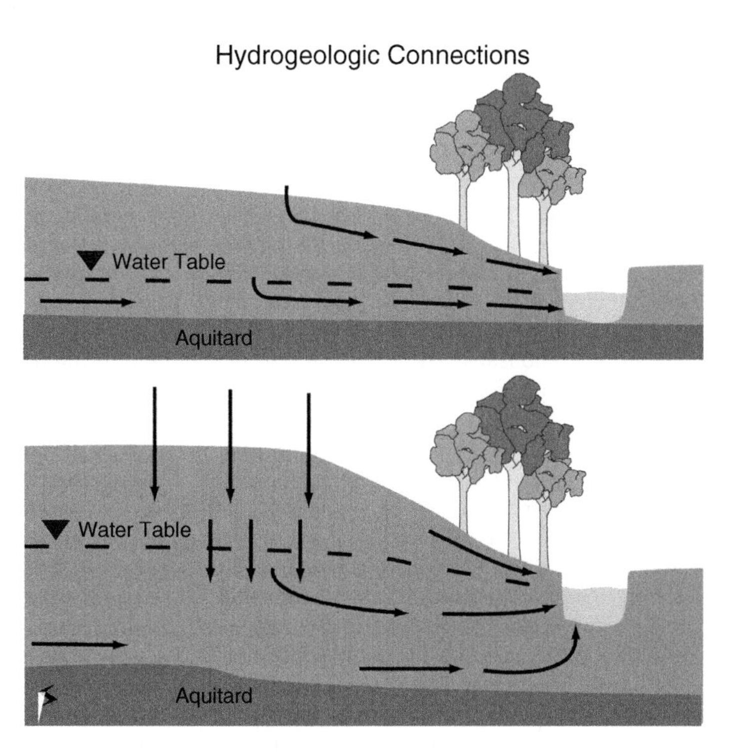

Fig. 8–10. Hydrogeologic connections as they influence water movement through or below the living filter of the buffer. Where aquitards are shallow, the living filter can significantly reduce nonpoint source pollutants in solution. If the aquitards are deep, significant quantities of water may pass below the living filter, bypassing any of the biological processes of the filter.

zone (Simpkins et al., 2002; Vidon & Smith, 2007). To fully understand the complexity of water movement through a riparian zone, it is imperative that the hydrogeological characteristics of the floodplain be established (Schultz, Isenhart, Simpkins, & Colletti, 2004).

Functions of Riparian Buffers

The active floodplain along a stream is also referred to as the *riparian zone*. A floodplain may contain several inactive terraces that are elevated above the active floodplain. Because of landscape position, the riparian zone is intimately linked to the stream channel and its aquatic ecosystems as well as to the upland ecosystems. As a result, the active flood plain and riparian zone play a critical role in the hydrology of watersheds. Riparian zones may contain sharp biological and physical gradients because they receive frequent disturbances from floods. This results in a plant community that often contains a mosaic of age classes of upland species as well as riparian or bottomland species adapted to abundant water (Gregory, Swanson, McKee, & Cummins, 1991). Their typically long and narrow nature, along with their unique physical and biological processes, allow riparian zones to act as strategic buffers between upland and aquatic ecosystems. Although riparian zones may occupy as little as 1% of the land area of a watershed, these ecosystems are among the most productive in the landscape (Chaney, Elmore, Platts, 1990). Riparian zones provide important links between terrestrial upland ecosystems and aquatic stream or lake ecosystems (Correll, 2005; Elmore, 1992; Lee, Smyth, & Boutin, 2004; Osborne & Kovacic, 1993). Some of their most important functions are: (a) filtering and retaining sediment; (b) immobilizing, storing, and transforming chemical inputs from uplands; (c) controlling stream environments and morphology; (d) controlling aquatic and terrestrial habitats; (e) providing water storage and recharge of subsurface aquifers; (f) reducing floods and/or moderating their impacts on adjacent land; and (g) protecting levees. The importance of each of these functions is influenced by the riparian zone location within the whole stream system. While each of the functions may apply in any setting, riparian ecosystems may exert their greatest influence on water quality along first- to third-order streams because this is the zone of sediment and solute production in a watershed and most of this production passes through the riparian community. Because more than 90% of stream lengths in a watershed are in first- to third-order streams, these riparian plant communities have a major impact on water quality throughout the whole basin. Riparian ecosystems may have their greatest impact on aquatic habitat along the mid-order streams (third–sixth order), which make up the zone of transport because these channels have sufficient flow and woody debris to support an active aquatic community. Along the highest order streams (higher than sixth), which make up the zone of deposition, the wide riparian forests and wetlands may provide their greatest influence on flood attenuation.

Filtering and Retaining Sediment

Riparian communities and other buffer systems can remove large amounts of suspended sediments and associated nutrients from overland flow entering from the uplands and from floodwaters entering from the stream channel (Broadmeadow & Nisbet, 2004; Correll, 2005; Nair & Graetz, 2004; Schultz et al., 2004). Government-supported funding programs in North America and Europe have been made available to restore and protect these buffers for environmental and economic concerns. According to Palone and Todd (1997), high levels of sediment removal for 30-m-wide mature forest buffers are 85–95% of the incoming sediment while low levels are between 40 and 64%. Vegetative filter strips with grass and grass–shrub–tree buffers also are an effective best management practice for controlling sediment and sediment-bound nutrients, with removal efficiencies ranging from 53 to 98% of sediment in runoff (Blanco-Canqui et al., 2004; Dillaha et al., 1989; Lee, Isenhart, & Schultz, 2003; Patty, Réal, & Gril, 1997; Robinson, Ghaffarzadeh, & Cruse, 1996; Schmitt, Dosskey, & Hoagland, 1999). Research also has shown that most of the deposits occur within the first 4 to 7.5 m of the filter strip, and thereafter, sediment deposition through the strip diminishes exponentially with increasing width (Foster, 1982; Schmitt et al., 1999). Surveys of narrow forest buffers without adjacent grass filters show numerous gullies moving through them from concentrated flow that originates in the adjacent crop field, suggesting that multiple zones of different vegetation are needed (Schultz et al., 2004).

The largest sediment deposition generally occurs at the upslope edge of the buffer where coarse sediment is deposited first, with finer sediments being trapped within the buffer. A secondary deposit may occur at the stream edge from sediment dropped from floodwater. The most effective sediment trapping occurs when concentrated flow from crop fields is slowed and spread at the edge of the buffer, dropping most of the large sediment and organic debris in the crop field before entering the buffer itself (Knight, 2007). The vertical structure and density of the standing plants and the organic litter on the soil surface provide the frictional surfaces that slow surface runoff, causing the sediment to be deposited (Cooper, Gilliam, Daniels, & Robarge, 1987; Dillaha et al., 1989; Lee, Isenhart, Schultz, & Mickelson, 1999, 2000; Lowrance, McIntyre, & Lance, 1988; Magette, Brinsfield, Palmer, & Wood, 1989). The benefit of properly designed and maintained grass filters adjacent to the forest buffers is that they provide the resistance needed to spread channelized flow entering them. Because of the density of overstory canopies, many forest buffers cannot maintain the density of living ground cover plants to provide enough resistance to spread concentrated flow. In many deciduous forest buffers, the forest floor is of minimal thickness because of rapid decomposition and incorporation of litter into the surface soil (Knight, 2007). In these situations, concentrated flow can easily move through the buffer, creating gullies and negating the potential water quality benefits of properly designed buffers.

In addition to sediments, dissolved and adsorbed nutrients and pesticides can also be removed from surface runoff within the riparian buffer and filter communities (Lee et al., 2003; Pankau, Schoonover, Williard, & Edwards, 2012). Mechanisms for these reductions include infiltration of runoff, dilution by precipitation, and adsorption and desorption reactions within the buffer system. In a study conducted in the Bear Creek watershed in central Iowa, average cumulative 30- and 60-min infiltration rates were greater under 5-yr-old restored buffers than under cultivated corn and soybean fields and intensively grazed pastures (Bharati, Lee, Isenhart, & Schultz, 2002). Corresponding changes in bulk density and porosity were also significant. Bulk densities in the same soil mapping unit under the buffers were significantly lower than under the other land uses. Improvements in infiltration, saturated hydraulic conductivity, pore parameter, and water storage as a system matures have been reported in a series of studies conducted at agroforestry watersheds in Missouri (Kumar et al., 2008; Sahin et al., 2016; Seobi et al., 2005). These studies showed 46 and 14 times greater saturated hydraulic conductivity in agroforestry buffers than under pasture and row crop management practices (Kumar et al., 2008; Seobi et al., 2005).

Forest- and grass-dominated riparian communities differ in their ability to trap and assimilate sediment and associated chemicals. The capacity of a grass-dominated community to trap sediment depends on its hydraulic resistance, the particle size of the sediment, the flow rate of the runoff, and the topography of the area (Dabney, McGregor, Meyer, Grissinger, & Foster, 1994; Dabney, Yoder, & Vieira, 2012). The longevity of sediment trapping ability varies between forest and grass communities. Cooper et al. (1987) and Lowrance et al. (1988) suggested that forest riparian buffers can filter sediments over long periods whereas Dillaha et al. (1989) and Magette et al. (1989) indicated that cool-season grass-dominated buffers may have short sediment filtering lives because they may become compromised by too much sediment. Shortly after becoming established, grass buffers may remove up to 98% of the sediment and bound N and P (Dillaha et al., 1989; Lee et al., 2003). If short, cool-season grasses are replaced by tall warm-season prairie grasses, grass buffer strips have a longer sediment-trapping life span (Dabney et al., 1994; Schultz, Isenhart, & Colletti, 2005). The taller grasses are capable of backing up water at the grass strip edge, causing the sediment to drop out before entering the grass zone. Dosskey et al. (2007) showed that most of the changes in grass buffers occur within three growing seasons after establishment, and changes in soil infiltration accounted for most changes. Comparing tree versus grass for 10 yr, they further concluded that both grass and forest vegetation are equally good as filter strips for at least 10 yr after establishment. However, filtration of sediment from flood flows will become different between the forests and grasses due to improvements in soil bulk density, hydraulic properties, and pore parameters with tree maturity (Seobi et al., 2005; Udawatta, Anderson, Gantzer, & Garrett, 2006).

Buildup of natural levees on stream banks along higher order streams will depend on the frictional resistance of buffer vegetation and flood flows. Well-developed natural levees may create wet meadows or floodplain forest wetlands that could enhance nutrient reduction in water moving through the buffers (Chaney et al., 1990).

Nutrient and Chemical Processing

A large body of evidence indicates that riparian and other buffers are effective at immobilizing, storing, and transforming chemical inputs from uplands (Correll, 2005; Dosskey et al., 2010; Lovell & Sullivan, 2006; Márquez, Garcia, Schultz, & Isenhart, 2017). According to Palone and Todd (1997), high levels of N and P removal for 30-m-wide mature forest buffers range from 68–92 and 70–81%, respectively, of the incoming nutrients while low levels range from 15–45 and 24–50%, respectively. Lee et al. (2003) found that adding a 16-m-wide forest buffer to a 7-m-wide native grass filter increased nutrient removal by more than 20–94% of the total N, 85% of the NO_3–N, 91% of the total P, and 80% of the PO_4–P in the runoff. Many other studies have shown that riparian forests and grass communities reduce N by 40–100 and 10–60%, respectively (Osborne & Kovacic, 1993; Sweeney & Newbold, 2014), and most of the removal of NO_3 has been shown to occur within the first 10–20 m of the forest–crop boundary (Balestrini, Arese, Delconte, Lotti, & Salerno, 2011; Jacobs & Gilliam, 1985; Peterjohn & Correll, 1984). A comprehensive review on N removal by riparian buffers has concluded that surface N removal is related to buffer width while subsurface removal is not (Mayer et al., 2007).

Two long-term upland buffer studies in Missouri using the paired watershed approach with a corn–soybean rotation and a replicated study with cattle grazing have shown reductions in runoff, sediment, and nutrient losses in runoff from these watersheds (Udawatta et al., 2002; Udawatta, Garrett, & Kallenbach, 2011). The paired and replicated watershed studies were established in 1991 and 2000, respectively. These studies have shown improvements in filtration efficiencies as the system matured. These benefits have been attributed to perennial vegetation and improvements in soil parameters. Cottonwood (*Populus deltoides* W. Bartram ex Marshall) was the buffer tree species on the grazing watersheds. Corn–soybean watersheds have grass only and grass with pin (*Quercus palustris* Münchh.), bur (*Quercus macrocarpa* Michx.), and swamp white (*Quercus bicolor* Willd.) oak trees. On these two watershed studies, APEX watershed model simulations demonstrated long-term benefits of these buffers (Kumar, Udawatta, Anderson, & Mudgal, 2011; Senaviratne et al., 2018). In a 30-yr simulation study on the corn–soybean watersheds, Senaviratne et al. (2018) compared six different buffer scenarios, with each scenario occupying 20% of the watershed, on reduction of runoff and sediments to streams. The simulation study showed that buffers next to the outlet and a buffer in the upland area of the steepest slopes were the most effective in reducing runoff, sediment, and P from these watersheds.

Annual P discharges reviewed by Beaulac and Reckhow (1982) averaged around 0.2 kg ha^{-1} for forests, 1 kg ha^{-1} for pasture, and 2 kg ha^{-1} for row crops. Udawatta et al. (2002) found that a watershed with an agroforestry treatment of trees planted in contour grass strips reduced total P losses by 17% compared with contour grass strips without trees, which reduced total P losses by 8%. While these differences suggest that forests and agroforestry practices may be more efficient in reducing N and P within the watershed and in the riparian zone, it must be remembered that the grass sites in many studies consist of short cool-season grasses that lie upslope of a forest site. Thus, positional differences in water table depth, soil drainage, or organic C may account for the differences rather than vegetation type (Schilling, Zhongwei, & Zhang, 2006; Verchot, Franklin, & Gilliam, 1997). On the other hand, grass, grass–shrub, and grass–tree buffers have also been shown to be effective in removing nutrients, sediment, and pesticides from surface and subsurface flow (Abu-Zreig, Rudra, Whiteley, Lalonde, & Kaushik, 2003; Dillaha et al., 1989; Olilo et al, 2016; Schultz et al., 2004; Simpkins et al., 2002). To be successful in removing excess nutrients from surface and subsurface waters, comprehensive management approaches are required that involve a network of buffers, conservation practices, nutrient management plans, maintenance of a good ground cover, and government regulations and standards (Mayer et al., 2007).

The mechanisms of chemical removal include plant and microbial uptake and immobilization, microbial transformations in surface water and groundwater, and adsorption to soil and organic matter particles. The effectiveness of these processes will depend on the age and condition of the vegetation, soil characteristics such as porosity, aeration, and organic matter content, the depth to shallow groundwater, and the rate at which surface and subsurface waters move through the buffer (Groffman, Gold, & Simmons, 1992; Lowrance, 1992; Puckett, 2004; Schultz et al., 2019; Simpkins et al., 2002). Uptake of nutrients by grass can be further enhanced by stimulating grass growth (Edwards, Hutchens, Rhodes, Larson, & Dunn, 2000). The majority of P moving in the watershed is adsorbed to soil and organic particles that are moving as a result of surface erosion (Figure 8–11). In a 7-yr study with three adjacent corn–soybean watersheds, Udawatta, Motavalli, and Garrett (2004) showed a strong relationship between sediment and P loss in runoff. The study also showed that the largest five runoff events removed greater than 80% of the total loss during the 7-yr study period, emphasizing the importance of conservation efforts for reduction of nonpoint-source pollutants from these larger events.

There are some studies showing significant dissolved reactive phosphate moving in groundwater through buffers (Blattel, Williard, Baer, & Zaczek, 2005). The groundwater P reductions shown in these studies are less than for N. Nitrogen moves most readily in solution as NO_3. As it moves, it may be assimilated by plants and microbes, denitrified, or leached from the watershed (Figure 8–12). The importance of any one of the removal methods is difficult to assess because very few studies have accurately measured the amount of nutrients removed at a given site by each of these mechanisms (Correll, 1997).

Plants can assimilate and immobilize nutrients such as N and P as well as heavy metals and pesticides. However, to be effective at removing these chemicals, plants must have access to high water tables or the capillary fringe, or there must be sufficient unsaturated subsurface flow through the rooting zone. Many biogeochemical reactions in the riparian zone require anaerobic or low-oxidation soil status at least part of the year (Correll, Jordan, & Weller, 1994; Naiman, Décamps, McClain, & Likens, 2005, pp. 125–158). While many plants cannot maintain root systems under continuously saturated soil conditions, many are able to function in the zone of fluctuating water tables during the growing season. Most plants are also not able to remove chemicals from water that is moving too rapidly over the surface or as preferential flow through macropores. Correll et al. (1994) observed that NO_3 is not effectively reduced in coarse-textured soils under high flow events when much of the

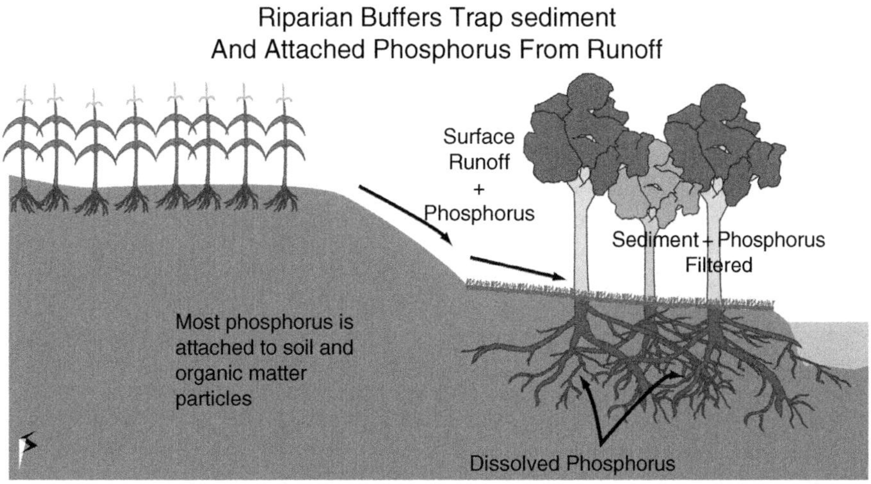

Riparian Buffers Trap sediment
And Attached Phosphorus From Runoff
Surface
Runoff
+
Phosphorus
Sediment + Phosphorus
Filtered
Most phosphorus is
attached to soil and
organic matter
particles
Dissolved Phosphorus

Fig. 8–11. Phosphorus movement through a riparian forest buffer. Riparian forest buffers filter sediment and attached phosphorus from surface runoff, and dissolved phosphorus may be taken up by the biota of the living filter (adapted from Welsch, 1991).

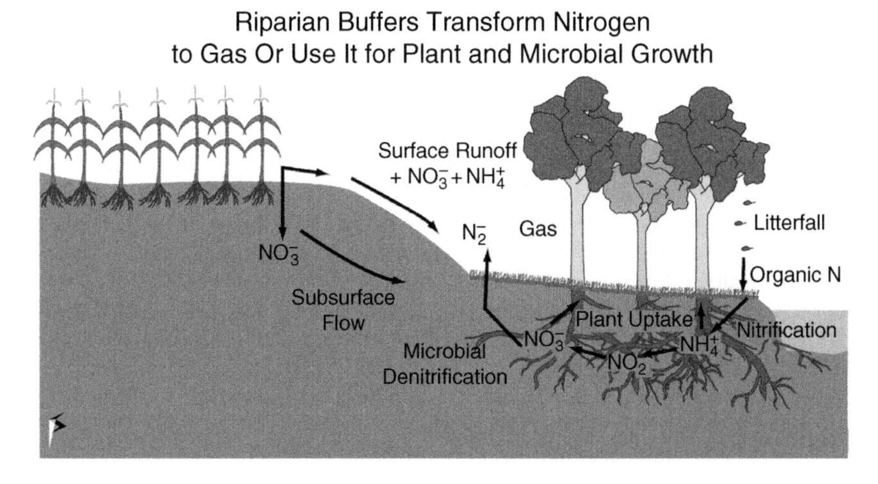

Fig. 8–12. Nitrogen movement through a riparian forest buffer. Riparian forest buffers provide carbon for the numerous microbial processes involved in nitrogen transformation. In a healthy riparian ecosystem, microbial and plant processes can immobilize or transform most of the nitrogen moving through the system (adapted from Welsch, 1991).

annual N loading of the buffer might be taking place.

Riparian vegetation will be an effective sink only as long as the plants are actively accumulating biomass. Once annual biomass production is equal to or less than litterfall and senescence of suppressed trees occurs, there will be no net addition of nutrients to the standing biomass sink. Plants must be harvested before that time if they are to add annual increments of nutrients to the standing biomass sink. However, vegetation whose standing biomass has reached equilibrium may still play an important role as a nutrient sink. For example, the release of N by decomposition of their litter on or just below the soil surface may be beneficial if the vegetation removed the nutrients from the groundwater, where the potential for transformation to harmless byproducts is often quite low (Groffman et al., 1992; Lowrance, 1992). Harvesting must be done with care, as it can result in soil compaction, which may reduce soil infiltration. Plant nutrient removal is also a seasonal phenomenon. Deciduous forests actively accumulate nutrients only during the growing season. Much of the nonpoint-source pollutant movement through the buffer may occur in late winter and early spring following the winter thaw or associated with early spring rains. Cool-season grasses and forbs in buffers and filter strips are very important in accumulating nutrients during this period while most of the woody vegetation is still inactive. The need to provide a living plant system to take up these early-season nutrients suggests that tree and shrub densities near the grass filter–forest buffer edge be less dense than those near the stream edge to provide the necessary sunlight to maintain a grass-covered soil surface not only in spring but throughout the year. Reviews of the literature by Mayer et al. (2007) and Schultz et al. (2019) suggested that wider buffers and buffers with herbaceous and forest–herbaceous vegetation combinations are most effective at removing N moving through the buffers.

Microbial processes of immobilization, degradation of organic compounds, and denitrification are also important in reducing nonpoint-source pollutants in the landscape (Dosskey et al., 2010; Puckett, 2004). Microbes will assimilate and immobilize nonpoint-source pollutants similar to plants. In soils under native forest or prairie ecosystems, high organic matter contents support large microbial populations that can act as a significant nutrient sink. Soils under restored buffers show rapid increases in organic matter and soil aggregation, both of which can stimulate microbial activity (Márquez et al., 2004; Márquez, Garcia, Schultz, & Isenhart, 2017, 2019). In low-organic-matter soils, the small microbial biomass and rapid turnover make them a minor nutrient sink. Microbes may also degrade many organic compounds through aerobic, anaerobic, chemoautotrophic, and heterotrophic pathways. Little is known about these processes in riparian buffers but, as with immobilization and denitrification, the amount of metabolic breakdown of organic compounds is often dependent on readily available organic matter in the soil (Dosskey et al., 2010; Martin, Kaushik, Trevors, & Whiteley, 1999; Mayer et al., 2007; National Research Council, 1993).

Under anaerobic conditions, microbes can reduce NO_2 and NO_3 into N_2 gas, the major atmospheric gas that we breathe. This process of denitrification occurs under anaerobic conditions in surface soils of riparian forests and is controlled by the availability of O_2, NO_3, and C (Burgin & Groffman, 2012; Hefting et al., 2004; Lowrance et al., 1995; Martin et al., 1999). Generally, a reduction in denitrification is found with depth and is related to a reduction in C availability. The small amounts of C that do exist may result from dissolved organic C transport from root leakage or may be related to buried A horizons from paleosols (Schoonover & Williard, 2003). The low soil C levels in the deeper soils are sometimes used to suggest that plant uptake is the primary method of NO_3 reduction. Even though denitrification may be slow deeper in the profile, the high residence time of water moving through the soil can account for significant NO_3 removal. Parkin (1987) and Legout, Molenat, Lefebvre, Marmonier, & Aquilna (2005) found denitrification occurring in localized micro-sites in unsaturated soils. Lowrance, Vellidis, and Hubbard (1995) reported higher denitrification rates under cool-season grass than under adjacent forests zones, while Verchot et al. (1997) found higher rates under the forest zone than under the grass zone. In landscapes where forest vegetation does not provide deep roots for nutrient uptake, native warm-season grasses, such as switchgrass (*Panicum virgatum* L.), can be used to remove NO_3 that has leached below the normal rooting zone of most crops (Huang, Rickerl, & Kephart, 1996). A review of 13 studies by Puckett (2004) suggested that denitrification takes place in any hydrogeological setting where anaerobic conditions are combined with sufficient organic C to create reducing conditions and residence times are sufficient to remove the NO_3.

Much of the research that has been conducted on riparian buffers has been conducted in the Piedmont and Coastal Plain of the northeastern and southeastern United States. Because of the uncertainty of mechanisms that account for the loss of nutrients and the complex relationships between local geomorphology, soils, vegetation, and hydrology, it is important that further research be conducted in other regions of the United States. While wider vegetated buffers are usually more efficient at removing nutrients, buffers will have little effect on groundwater and chemicals that move below the active buffer soil–plant–microbe zone. Hickey and Doran (2004) suggested that much of the data on the efficiency of riparian buffers has been collected from buffers that are significantly wider than what most farmers would be willing to remove from active crop production. In addition, the long-term nutrient removal effectiveness of reconstructed buffers is not known in any region (Osborne & Kovacic, 1993; Puckett, 2004).

Recent deep well studies in Missouri and Pennsylvania have shown that deep rooted riparian tree buffers effectively remove nutrients from ground and deep waters by plant uptake, denitrification, and other soil–water reactions (Figure 8–13). Deep wells below the tree–grass buffer had lower NO_3 and total N concentrations in a 2-yr study with weekly sampling in Missouri than the grass-only buffer (Wickramaratne 2017). These mini-watersheds with cottonwood trees were established in 2000, and water samples were collected in 2014–2016 under cattle grazing management. The soils of the study region are deep loess. The study also showed that these concentrations varied with rain events. Similar to these findings, a riparian buffer study in Pennsylvania with deep well monitoring reported significantly lower NO_3 concentrations beneath the buffer than in upland areas (Newbold, Herbert, Sweeney, Kiry, & Alberts, 2010). The riparian buffer system removed 90 kg NO_3 ha^{-1} yr^{-1} or 26% of upslope subsurface inputs.

The influence of riparian buffers on stream water quality must also be assessed in the context of stream order. Nonpoint source pollutants in a first-order stream must all pass through or below a riparian buffer before getting into the channel, assuming that a buffer exists along the full length of the channel. However, in higher order streams, the influence of the riparian buffer on stream water quality is confounded by the fact that upstream inputs of nonpoint-source pollutants may already be in the water. The higher the stream order, the less impact the local riparian buffer has. As a result, the most important buffers for maintaining water quality are those in the first- to third-order streams, the zone of

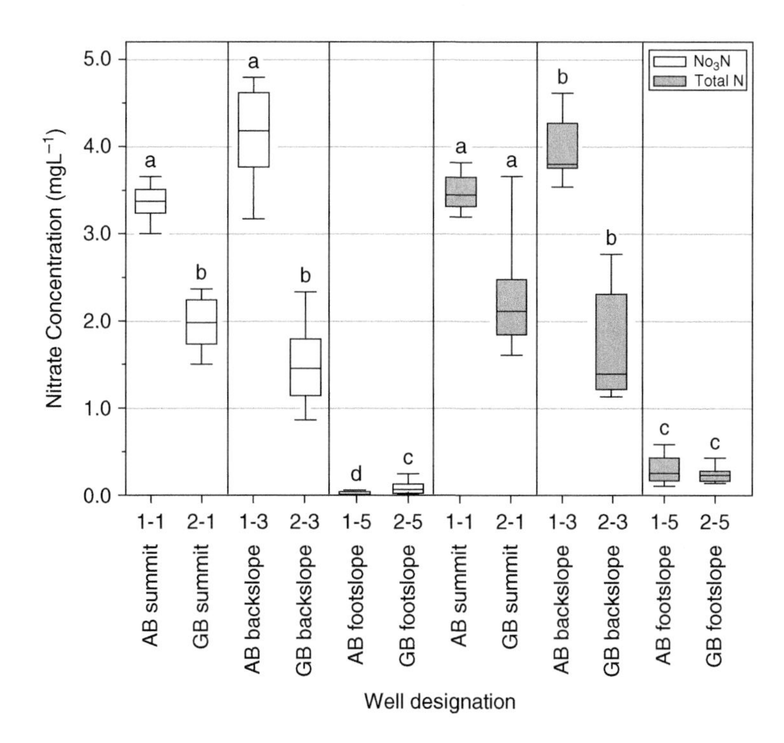

Fig. 8–13. The median and interquartile ranges of NO_3–N (N = 58–97) and total N (TN, N = 41–87) concentrations in groundwater monitoring wells of AB (agroforestry buffer) summit, AB backslope, AB footslope, GB (grass buffer) summit, GB backslope, and GB footslope at the Horticulture and Agroforestry Research Center (HARC), New Franklin, MO, in two adjacent grazing watersheds. For each well, letters indicate significant differences in mean rank (α =.05).

production in the river system (Correll, 2005; Schultz et al., 2019). Riparian buffers along higher order streams have other functions that are as much or more important than reducing chemical and sediment loads to the channel.

Controlling Stream Environments and Morphology

Riparian vegetation can control available light and the size and shape of the channel. It can also provide organic matter inputs that can serve as a C source for aquatic organisms. Riparian vegetation controls the quantity and quality of solar radiation reaching the water surface in lower order streams where tree crowns may close over the stream and thus influence in-stream production, water temperature, and O_2 content (Gregory et al., 1991; Naiman et al., 2005, pp. 125–158; Sweeney, 1992). In the eastern United States, openings in the canopy over the stream occur when stream width exceeds 20 m, widths found in fourth- to fifth-order streams (Feld et al., 2018; Palone & Todd, 1997). Organic matter input into the stream from riparian vegetation is an important energy source for aquatic organisms. Differences in the quality and quantity of organic matter inputs between conifer and deciduous forests and between forests and grasslands often determine the structure of the invertebrate populations in the stream (Broadmeadow & Nisbet, 2004; Gregory et al., 1991; Gurtz, Marzolf, Killingbeck, Smith, & McArthur, 1988; Murphy & Giller, 2000). Mature, diverse forest buffers provide a wide variety of C sources to the stream. Under the dense canopies of these forests, the highest proportion of in-stream C comes from litter inputs. In-stream plant life consists primarily of diatoms growing on rocks. If these forests are cut and large amounts of light reach the channel, the in-stream plant life becomes dominated by filamentous algae. These changes in food quantity and quality result in major changes in the macroinvertebrates and fish species present in the stream.

Large woody debris in the stream channel influences the physical structure of the channel, diversifying the habitat by controlling the distribution of pools, which store and detain sediment, and riffles, which help oxygenate the water (Bisson et al., 1987; Gregory et al., 1991; Naiman et al., 2005, pp. 125–158; Sweeney, 1992; Verry et al., 1999). Large woody debris consists of branches and roots that are at least 10 cm in diameter. This material can accumulate in the outside bends of larger streams, providing habitat for fish and other organisms. In smaller order streams, large woody debris forms debris dams that reduce the stream gradient, reducing the erosive potential and thereby stabilizing the channel. Debris dams also trap litter that provides a C source for aquatic organisms. Large woody debris decomposes more slowly than leaf litter, thus providing longer term nutrients to the channel. Ideally, 40–100 pieces of large woody debris per kilometer of stream should be available to provide good fish habitat in the northeastern United States (Marcus, Marston, Colvard, & Gray, 2002; Palone & Todd, 1997). Most of that large woody debris is produced within 18 m of the stream.

Forest buffers also provide woody roots along the bank, which stabilize the banks and provide habitat for organisms. Woody and fibrous roots of plants growing on the stream bank provide strength to hold the stream bank in place (Zaimes, Tufekcioglu, & Schultz, 2019). Plant roots increase soil stability by mechanically reinforcing the soil and by reducing the weight of the soil through evapotranspiration (Gyssels, Poesen, Bochet, & Li, 2005; Pohl, Alig, Korner, & Rixen, 2009; Waldron & Dakessian, 1982). Woody plant roots provide superior soil stabilization compared with those of herbaceous plants because of their deeper rooting habit and their larger permanent roots (Waldron, Dakessian, & Nemson, 1983; Wynn et al., 2004). Woody roots provide protection against the hydraulic pressures of high flows, while fibrous roots bind the finer soil particles (Elmore, 1992; Verry et al., 1999; Wynn et al., 2004). In the agricultural regions of the midwestern United States, grass filters may be as or more effective at reducing bank erosion as woody root systems, especially if channels are sloped and not vertically incised (Lyons, Trimble, & Paine, 2000; Schultz et al., 2004, 2019).

If forests are cleared and large woody debris is lost from the channel, the channel gradients may be increased and channel erosion will be accelerated, thus moving the channel through the various stages of the channel evolution model (Figure 8–4) and ultimately resulting in a shallower and wider channel. If the banks are converted from forest cover to grass, the channel may narrow along Stage I channels because the grass develops a sod that slowly encroaches on the channel. In summary, opening the channel to more sunlight and narrowing the channel will have dramatic effects on habitat diversity and aquatic organisms (Lyons et al., 2000).

Water Storage, Groundwater Recharge, Flood Attenuation, and Hydrogeological Considerations

Vegetated riparian zones slow flood flows by allowing water to spread and soak into the soil, thereby recharging the local groundwater and extending the base flow through the summer season (Elmore, 1992; Naiman et al., 2005, pp. 125–158; Wissmar & Swanson, 1990). Palone and Todd (1997) pointed out that forests can capture and absorb 40 times more rainfall than disturbed agricultural and construction site soils and 15 times more than turfgrass- or pasture-covered soils. They described the riparian buffer community as the "right-of-way" for the stream, including its floodwater. The multiple stems of forest buffers provide a frictional surface that slows floodwaters and causes sediment and debris to drop out. Grass-covered floodplains, when submerged, do not slow floodwaters as effectively. The sediment trapped in riparian forest buffers reduces downstream sedimentation of reservoirs.

Riparian forests protect stream banks during floods and provide large woody debris that can produce debris dams that reduce down cutting, trap sediment, and dissipate energy (Dosskey et al., 2010; Verry et al., 1999; Zaimes, Schultz, & Isenhart, 2008). Floods are important for the health of a stream system in that they rearrange channel materials through scour and deposition, deposit new fertile soil from upstream, and may uproot or break down riparian vegetation. The activities associated with most floods may act to revitalize the channel and riparian system and are nature's way of maintaining diversity in the ecosystem (Naiman et al., 2005, pp. 1–18). Many of the woody plants in the riparian zone are able to regenerate vegetatively through root or stump sprouts in response to floods. The resprouting plants assure a continued frictional surface. Channels that are in contact with their floodplain will have frequent small floods, whereas channels that have lost that contact will have fewer but more extreme floods.

Water storage in upland buffer soils varies with the time of year, precipitation conditions, and vegetation type (Alagele, Anderson, & Udawatta, 2020; Sahin et al., 2016; Udawatta, Anderson, Motavalli, & Garrett, 2011). In these studies in Missouri, soil moisture contents were lower throughout the growing season in the perennial vegetative buffer (grass only and grass plus tree) areas than under corn and soybean (Figure 8–14). Soil moisture was monitored at 15-min intervals for a 0–50-cm soil profile using soil moisture sensors at 5-, 10-, 20-, and 40-cm depths from 2003 to 2017. Large rain events during the growing season showed greater infiltration rates and soil water storage for perennial vegetation areas than crop areas. Soil water contents were the lowest for perennial vegetation before the rain and highest after rain events (Figure 8–15). These studies further demonstrated greater soil water recharge during the recharge period within the buffers than the crop areas. The improved infiltration and storage have been attributed to improvements in soil C and physical properties by the perennial vegetation and no disturbance. Greater dewatering during the growing season, improved storage during rain events, and the recharge period help reduce surface runoff and losses of nutrients and sediment from agricultural watersheds. Dewatering also helps reduce deep leaching of water and nutrients and thus better protect groundwater. However, site-suitable buffer design criteria and tree species should be used to maintain suitable soil-water conditions for crop production. Either low-water-demanding trees, narrower buffers, a reduced number of buffers, or scattered trees can be adopted for water-limited site conditions.

Terrestrial Habitat

Riparian ecosystems are important travel corridors for both animals and plants. They provide lush and diverse habitat for wildlife and, because of their rich, moist microenvironments, they are often the part of the landscape with the highest plant species diversity, especially in watersheds with high proportions of agricultural land use (Gregory et al., 1991; Harvey et al., 2008; Naiman, Décamps, & Pollock, 1993; Naiman et al., 2005, pp. 125–158; Verry et al., 1999). The rich plant species diversity can provide a multistoried canopy with many diverse habitats. Management of riparian forest buffer corridors to maintain high species diversity can also provide the opportunity for alternative income for landowners (Peters & Hodge, 2000; USDA National Agroforestry Center, 2004). In bird surveys conducted in 5-yr-old established riparian forest buffers in the Bear Creek watershed in central Iowa, species numbers were four to six times greater than in intensively grazed pastures or narrow strips of annual and perennial weeds (Taylor, Schuster, Verdon, Pease, & Isenhart, 2000). Surveys in the same buffers 5 yr later showed a further increase of 15 additional species in the reestablished buffers (Berges, Schulte, Isenhart, & Schultz, 2010). Another bird survey study in the same watershed conducted in May–July found an average of eight species in the crop fields, 23 species in a 2-yr-old buffer, and 42 in a 9-yr-old buffer, each buffer consisting of the same plant community design (Berges et al., 2010). Because riparian ecosystems are long and narrow, they also provide a high proportion of edge habitat, which is often multistoried. The two edges of riparian communities

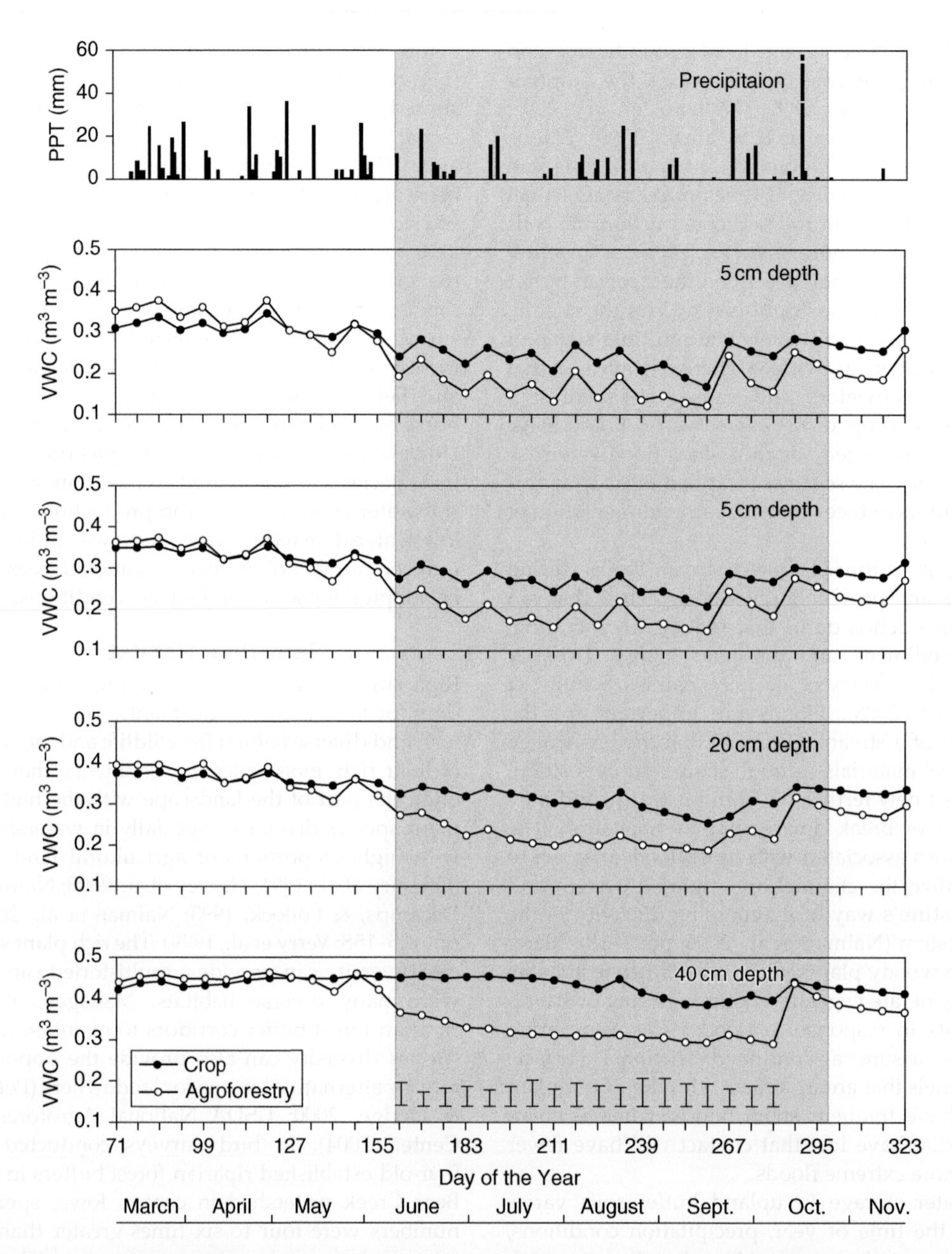

Fig. 8–14. Daily precipitation (PPT) and volumetric soil water content (VWC) at 5-, 10-, 20-, and 40-cm soil depths for crop and agroforestry treatments during 2007 at the Greenley Memorial Research Center, Novelty, MO. The shaded area represents the crop period for soybean, and bars on the 40-cm graph indicate LSD values for significant differences at the α =.05 level (source: Udawatta, Anderson, et al., 2011).

are different because one borders on water and the other on upland communities. These long, narrow communities also serve as travel corridors, connecting different upland and aquatic habitats. In this role they may serve species needing large interior habitats by simply providing critical connections between blocks of habitat that may have large interior to edge ratios. Because they are often the only extensive perennial plant community, riparian forest buffers may harbor animals that may be considered pests to local farmers and landowners. In some parts of the country, for example, deer (*Odocoileus virginianus* Zimm.) and beaver (*Castor canadensis*) may be seen as pests. Beavers build dams that may back water into drainage tiles, while deer may use the cover of the buffer from which to feed on adjacent crops. Work is presently underway looking at riparian forest buffers as a refuge for predators of insect pests.

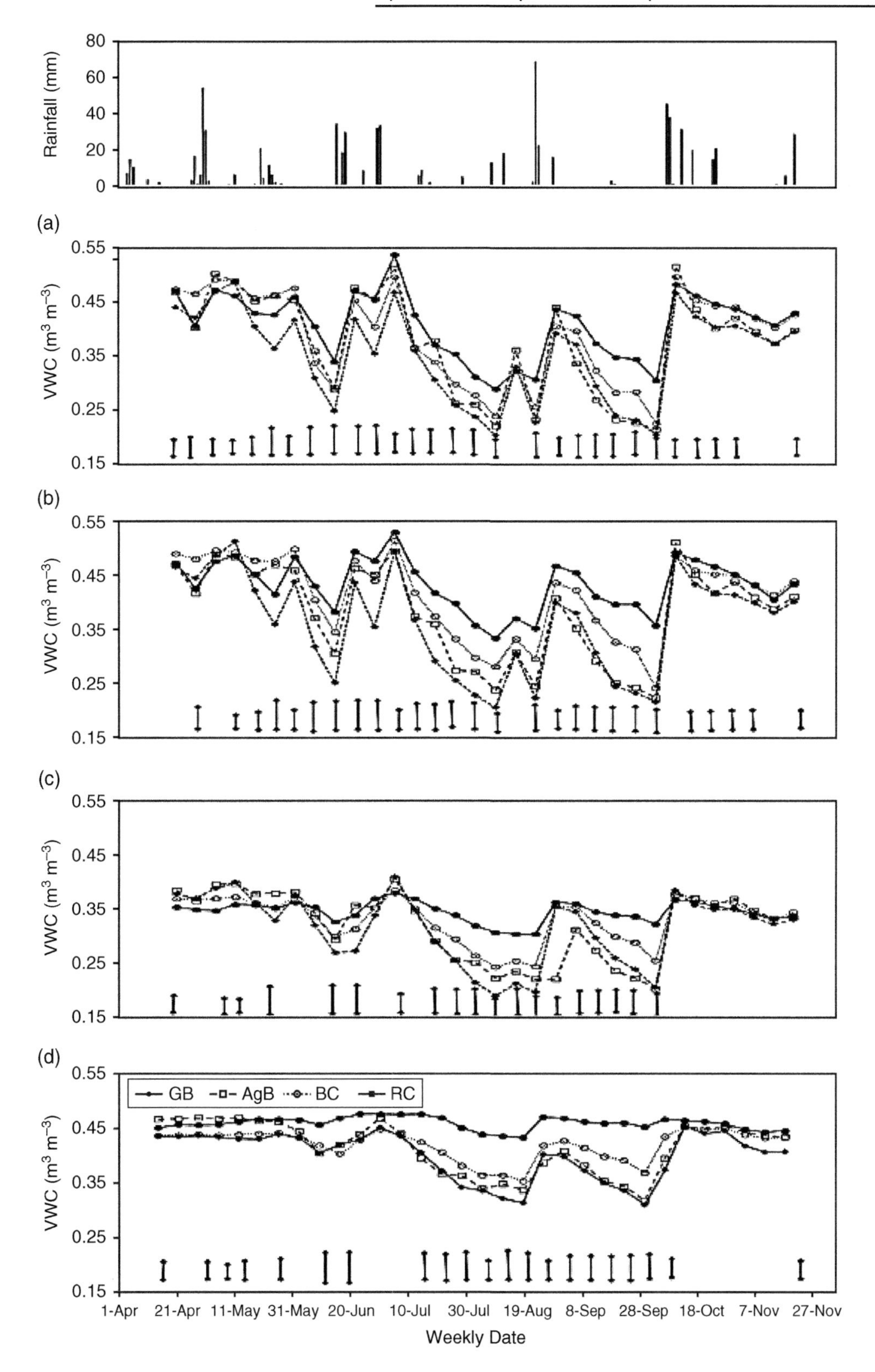

Fig. 8–15. Daily precipitation and volumetric soil water content (VWC) at 5-, 10-, 20-, and 40-cm soil depths for grass buffer (GB), agroforestry buffer (AB), biomass crop (BC), and row crop (RC) treatments at the Greenley Memorial Research Center, Novelty, MO. Bars on the 40-cm graph indicate LSD values for significant differences at the α =.05 level (source: Alagele et al., 2020).

Buffers also help soil microbial diversity and diversity islands (Alagele, Anderson, Udawatta, et al., 2019; Udawatta et al., 2019). Studies have shown greater enzyme activities and microbial diversity under trees than on cropped areas. In a recent study, Alagele, Anderson, Udawatta, et al. (2019) noticed greater enzyme activities at 50 cm from the tree base than 150 cm for four major enzymes involved in mineralization and nutrient transformation. These differences have diminished as the system matured and due to the uniform distribution of roots and litter material. Similar observations have been reported in Canada, suggesting improved soil biodiversity and associated degradation of various chemicals.

Economic and Social Benefits

Riparian ecosystems can produce many market and non-market products. With careful harvesting, high-quality wood and fiber products may be removed. Careful management may produce high-quality forage for livestock producers. Numerous recreational activities are enhanced by these communities, including fishing and hunting, canoeing and boating, hiking, camping, and picnicking. The use of carefully constructed and maintained trails can provide activities ranging from running to cycling and motorbiking (Bentrup, 2008; Palone & Todd, 1997). Finally, nature appreciation and relaxation can be enjoyed in riparian communities.

The role of riparian communities in providing high-quality water, reducing flooding, and sequestering of C for improved global climate gives us non-market products whose value is difficult to assess. Growing concerns about global climate change and ecosystem services that can mitigate it is creating more interest in trying to assign monetary values to restored riparian communities (Lal, Delgado, Groffman, & Millar, 2011; Smith et al., 2013).

Present Condition of Agricultural and Grazing Landscapes

Highly productive agricultural regions are a mosaic of crop and grazing lands, human habitations, and small remnants of native prairie, wetland, and forest ecosystems. The Midwest Corn Belt region can be used to demonstrate the human impacts of agriculture on the landscape because this region has had the greatest loss of native perennial vegetation of any region in the United States. Large portions of natural ecosystems have been converted to intensively managed agroecosystems in the 12 states ranging from Ohio to the eastern portions of the Dakotas, Nebraska, and Kansas, and from the southern portions of the Lake States to the northern half of Missouri. In Iowa, for example, 99% of the original prairie and wetland area and more than 80% of the original forest area have been converted to other uses (Bishop & van der Valk, 1982; Thomson & Hertel, 1981) (Figure 8–16). Ohio, Indiana, Illinois, and Missouri drained more than 85% of their wetlands by the mid 1980s (Dahl, Johnson, & Frayer, 1991). In most of the region, less than 20% of the natural prairie, forest, wetland, and riparian ecosystems still exist (Burkart, Oberle, Hewitt, & Pickus, 1994).

In a typical watershed in central Iowa, about 50% of the total length of the stream channel may be cultivated to the bank edge (Figure 8–17). Another 30% of the length may be in pasture, most of which is overgrazed (Bercovici, 1994). Most of the wetlands have been converted to cropland by the installation of drainage ditches and/or subsurface drainage tiles (Figure 8–18). Many kilometers of stream channels have been straightened to increase the area of the riparian zone that can be cultivated and to speed water transport from the landscape (Figure 8–19).

In areas with higher rainfall, narrow corridors of unmanaged woody vegetation may be left between the channel and the crop fields, suggesting abundant riparian buffers. However, in a study in northeastern Missouri (Herring, Schultz, & Isenhart, 2006) it was found that the narrow tree strips adjacent to first- and second-order streams were typically not wide enough to provide adequate stream protection and many of the wider ones did not have a grass filter adjacent to them, allowing concentrated flow from adjacent crop fields to continue through the narrow forest strips directly into the channel. Replacement of native perennial ecosystems with annual row-crop agriculture has major impacts on soil quality, erosion, and moisture (Chaplot, Saleh, Jaynes, & Arnold, 2004; David & Gentry, 2000; Magner & Alexander, 2002). Soils lie fallow for more than half of the year, during which time they are not protected by a crop canopy or a significant organic litter layer. This results in surface compaction and erosion. Annual cultivation and crop removal also promotes compaction, erosion, and a loss of organic matter (Helmers et al., 2012; Hernandez-Santana, Zhou, Helmers, Asbjornsen, & Kolka, 2013). This results in a loss of soil structure and reduces infiltration rates, thereby accelerating surface runoff (Bharati, 1997; Márquez et al., 2004, 2017). The loss of a perennial plant community also reduces annual evapotranspiration from the landscape, reducing the amount of plant uptake of water and nutrients. Because of the lack of a perennial plant community and

Fig. 8–16. A corn and soybean field on land that was once covered by native prairie. The loss of perennial plant cover and the introduction of mechanical tillage have resulted in major changes in soil and hydrologic conditions. These include reductions in surface organic residue, soil organic matter, soil structure, and infiltration and in higher surface soil temperatures and greater surface runoff. These changes have increased stormflow and reduced base flow of streams in many watersheds. Higher concentrations of sediment and agrichemicals can also be expected (Wickramaratne, 2017, with permission of University of Missouri Press).

Fig. 8–17. Farming through the riparian zone to the edge of the stream modifies the hydrologic cycle. The loss of stabilizing perennial plant roots allows surface runoff to move quickly to the stream and has resulted in unstable stream banks prone to failure. In some agricultural watersheds, it is estimated that more than half of the sediment in streams comes from bank failure.

Fig. 8–18. A field drainage tile passes below the riparian buffer and carries large quantities of upland soil water and soluble chemicals to the stream. The living filter of the riparian zone has no effect on tile water. In some stream systems, drainage tiles provide the major proportion of base flow.

Fig. 8–19. A stream channel that once meandered has been straightened, reducing channel storage and increasing streamflow velocity. Similar channels have been dug in areas that previously had no channels, increasing the overall flow of water in surface channels. Increased flooding can result from these straightened channels.

reduced microbial activity stemming from reduced aeration and soil C, nutrients are more readily lost by leaching and surface runoff (Haake, 2003; Nelson, 2003; Tufekcioglu, Raich, Isenhart, & Schultz, 2003). The landscape is therefore more "leaky" than under natural conditions (Figure 8–20). One of the major problems associated with agricultural production in the United States is movement of sediment, fertilizers, and pesticides from the uplands into surface waters. Croplands contribute 43 and 40% of the annual inputs of N and P, respectively, to surface waters, while pasture and rangelands contribute 25% and 28%, respectively (Welsch, 1991). Average annual soil erosion is greater than 6.7 Mg ha^{-1} in much of the central part of this region, and in some areas is greater than 11.2 Mg ha^{-1} despite the fact that many of these same areas have over 50% of the land in upland conservation practices (Burkart et al., 1994).

In large areas of the Midwest, the glacially young landscape with its poor internal drainage and numerous pothole wetlands has been crisscrossed by networks of drainage tiles to increase the amount of tillable land. These porous clay or plastic pipes rapidly remove water from the uplands and deliver them to the streams. This rapid movement of water has reduced the antecedent moisture content of the soils. The loss of residence time has reduced or eliminated contact with the living filter of plant roots and microbial populations that could remove dissolved chemicals from the water. In many first- to third-order streams, tile flow is sufficient to provide as much as 90% of the base flow of the stream during the summer months, resulting in a significant lowering of the water table. The lower water table produces dry conditions later in the growing season. The same drainage systems that allow crops to grow in areas that may be too wet in the spring may also reduce growth because of near-drought conditions later in the year.

Chaney et al. (1990), Chaney, Elmore, and Platts (1993), Elmore (1992), Fleischner (1994), and Robertson and Rowling (2000) identified livestock grazing as having dramatically changed riparian zones in the rangelands of the West. The changes due to livestock have been so great and cover so much of the western landscape that it is even difficult to determine what the natural vegetation was or what the effects of livestock grazing have been (Fleischner, 1994). These riparian areas are the most biodiverse ecosystems in the landscape, even though they occupy less than 1%

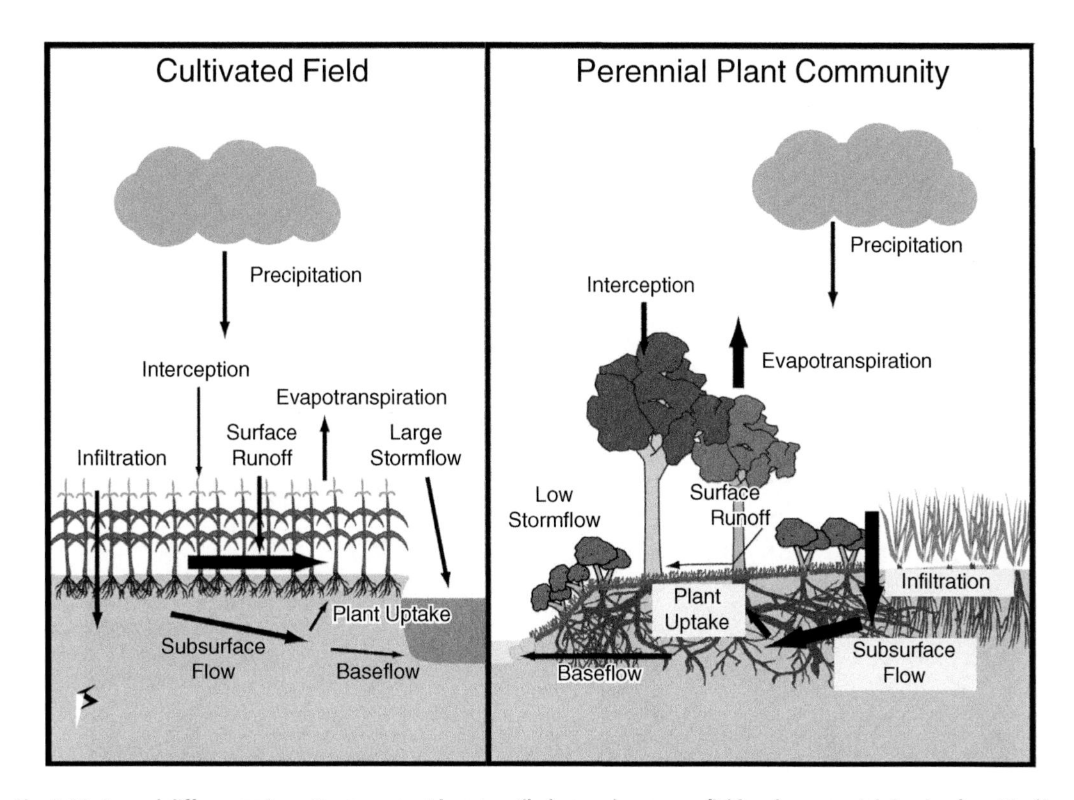

Fig. 8–20. Annual differences in water movement in a non-tiled annual row-crop field and a perennial riparian forest buffer. More surface runoff and less total evapotranspiration result in larger stormflow in the row-crop field. while higher rates of infiltration and annual evapotranspiration reduce stormflow and increase base flow.

of the landscape. More than 75% of the wildlife in many of these watersheds depends on the riparian zone for existence (Chaney et al., 1990). Riparian ecosystems in the arid and semiarid West also function to filter sediment, stabilize stream banks, store water, and recharge subsurface aquifers (Chaney et al., 1990; Elmore, 1992; Fleischner, 1994). In the midwestern United States, Tufekcioglu, Isenhart, et al. (2012) and Tufekcioglu, Schultz, Zaimes, Isenhart, and Tufekcioglu (2012) found that riparian grazing had significant impacts on stream bank erosion and sediment and P input to streams.

In cropland landscapes of the Midwest and eastern United States, riparian areas along meandering streams with their irregular shapes are difficult to cultivate with large equipment, resulting in many kilometers of streams being fenced for grazing (Figure 8–21). Under these conditions, livestock grazing can do extensive damage to stream banks and channels (Tufekcioglu, 2006; Zaimes, Schultz, & Isenhart, 2004, 2006) Livestock walking along the edge of the stream bank, loafing under stream-side trees, and entering into the channel create conditions of bank instability and erosion. Livestock will also rub, trample, and browse the vegetation, and urinate or defecate directly in the stream. Stream channels along intensively grazed pastures may have more than three times the length of severely eroding banks compared with channels with riparian forest buffers or grass filters (Zaimes et al., 2006). In a study of pastures in three regions of Iowa, Tufekcioglu (2006) found that loafing areas and channel access points accounted for only 2.7% of the channel bank length but supplied more than 70% of the bank-contributed sediment and 55% of the total P to the channel. Stream bank erosion can contribute 25–80% or more of the total sediment load to the channel, and grazing systems are usually at the upper end of these contributions (Amiri-Tokaldany, Darby, & Tosswell, 2003; Schilling & Wolter, 2000; Simon, Rinaldi, & Hadish, 1996; Wilkin & Hebel, 1982; Zaimes et al., 2004, 2008). Row crop cultivation to near the edge of the stream bank can result in stream bank sediment losses as great as those along intensively grazed pastures (Zaimes et al., 2004, 2006).

To gain more cropland and remove water more quickly from the landscape, many streams have been channelized or straightened. The loss of meanders from a channel reduces channel water storage, increasing the velocity of flow and therefore the erosive potential. Removal of large woody debris results in a loss of grade control and an increase in flood velocities. The building of dikes and levees also increases flood peaks downstream. The result of all of these actions is an increase in velocity and volume, with eventual

Fig. 8–21. Cattle grazing in riparian zones damage stream banks and add pollutants directly to the water. Reduced infiltration due to soil compaction increases surface runoff in fenced riparian pastures like the one shown.

downcutting of channels from storm events that occur with frequencies of 10 or more years. When these events come out of their banks, they can produce more serious flooding damage than in channels that are in contact with their floodplain and able to carry only flows from storms of 2–3-yr frequency. The wholesale conversion of perennial vegetation to row-crop agriculture, the draining of wetlands, and the manipulation of stream channels and their riparian vegetation can dramatically reduce wildlife habitat. In some agricultural landscapes with fencerow-to-fencerow cultivation, there is almost no wildlife habitat available (Schultz et al., 2004). Loss of riparian vegetation reduces insect populations on which many birds depend, requiring them to move over larger areas, in smaller numbers, and with greater chances for predation. In a study in Iowa, Stauffer and Best (1980) predicted that the removal of woody vegetation from a riparian community would eliminate 78% of 41 species of breeding birds even though herbaceous cover remained. Another 12% of the species would decrease in number, while 10% would increase. Similar predictions were developed by Geier and Best (1980) for nine species of small mammals faced with the same reduction in woody vegetation. Pease and Isenhart (1996) found that a 4-yr-old multispecies buffer composed of trees, shrubs, and native grasses supported almost four times the number of bird species during the summer than did an adjacent channelized reach with about 4 m of annual weeds and cool-season grasses. Even where riparian communities are left intact in a largely row-crop landscape, concern exists that these narrow ecosystems may act as biological traps for some species. Berges et al. (2010) found almost twice as many birds, as well as higher richness and diversity, in established riparian buffers than in adjacent crop fields of corn and soybean.

The large-scale modification of regional hydrology and terrestrial and aquatic ecosystems, especially in response to urbanization, has had a profound impact on the biological integrity of the surface waters (Snyder, Young, Villella, & Lemarie, 2003). Menzel (1983) reviewed the natural structure and function of stream ecosystems of the Corn Belt region with special reference to the impacts of past and present agricultural management practices. He concluded that impacts on water quality were not the sole problem, and that aspects of water quantity, habitat structure, and energy transfer are often profoundly affected by agricultural land-use practices (Figure 8–22). This alteration of the physical, chemical, and biological processes associated with the water resource has dramatically reduced the biological integrity of the aquatic systems in the region (Karr, 1991).

Modern product-oriented agriculture has put agroecosystems at risk. The production- oriented function of this landscape has produced unintended and undesirable environmental consequences that include loss of biodiversity, detrimental alteration of waterways and groundwater aquifers, and loss of significant portions of the productive topsoil, resulting in greater need for fertilizer and energy inputs (Scanlon, Jolly,

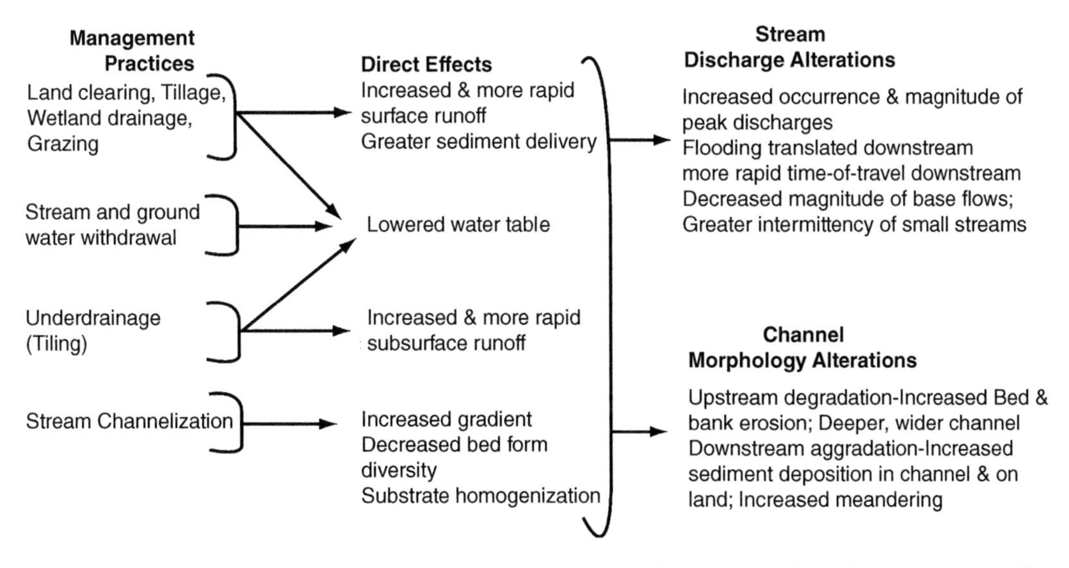

Fig. 8–22. Alterations in stream discharge and morphology brought about by agricultural land-use practices (from Menzel, 1983).

Scophocleous, & Zhang, 2007; Schulte, Liebman, Asbjornsen, & Crow, 2006). Nonpoint-source pollution has become pervasive because of rapid surface and subsurface water movement and reduced soil residence time for agrichemicals. It is now apparent that upland conservation practices alone are not effective in reducing nonpoint-source pollutants (Burkart et al., 1994; Cho, Vellidis, Bosch, Lowrance, & Strickland, 2010; National Research Council, 1993; Tomer & Locke, 2011). A challenge for resource managers in these modified landscapes is the development and implementation of restoration-based management approaches that build on traditional soil and water conservation and pollution control efforts. One promising approach to increase the effectiveness of efforts to protect soil and water quality, while also enhancing the physical, chemical, and biological integrity of the terrestrial and aquatic ecosystems, is the creation or restoration of landscape buffers, especially riparian buffers (Castelle, Johnson, & Conolly, 1994; Cho et al., 2010; National Research Council, 1993). However, few landscape restoration models exist for use in intensively modified agricultural ecosystems, and significant issues about their efficiency and design must be clarified before they can be effectively implemented across all temperate agroecosystems (Schulte et al., 2006). These regional landscape issues include plant species selection, optimal buffer widths for various purposes, quantification of process rates that reduce nonpoint-source pollutants, longevity of the buffers as nutrient and sediment sinks, criteria for identifying riparian zones in need of buffers, and criteria for long-term management of buffers (Castelle et al., 1994; National Research Council, 1993; Osborne & Kovacic, 1993; Schultz et al., 2004).

Additional questions about riparian management must be asked in landscapes where stream banks are unstable, field tile drains are prevalent, and riparian zones are grazed. In landscapes with deeply incised and unstable banks, riparian buffers may not become established rapidly enough to stabilize the stream banks before they become victims of bank collapse. In such cases, stream bank bioengineering must be an integral part of the riparian management system. Another option might be to plant the first tree row far enough back from the bank that the mature trees would not shade the banks. Shrubs could be planted as the woody plant species between the trees and the bank. Such a planting design would allow the banks to slump, and in many cases grass and other herbaceous species would have enough light to become established to stabilize the slumped bank. If the slumped bank were shaded by trees, little vegetation would get established, allowing further bank collapse to continue.

Where field drainage tiles drain the uplands, riparian buffers would effectively capture sediment moving with surface runoff but would only effectively reduce the subsurface nonpoint-source pollutants from a relatively narrow zone adjacent to the channel. Field tiles effectively carry water and pollutants below the buffer without treatment. In such cases, saturated buffers or constructed wetlands that capture field tile water from large collection tiles may be required to provide an effective riparian management system. Saturated buffers are designed by installing a perforated field tile along and parallel to the field edge of the buffer. Field tile from the crop field are intercepted prior to entering the buffer with a control box that allows flow from the intercepted tile to be discharged into the tile parallel to the buffer. Water flowing out of that tile then moves through the riparian buffer. Up to as much as 60% of the tile flow from the fields can be moved through the buffer, dramatically reducing the amount of tile water flowing under the buffer and directly into the stream. Water passing through the buffer from the parallel field tile loses most of its N before becoming a part of the soil water released by the buffer into the stream. Finally, in those riparian zones where grazing is practiced, modifications of the buffer may be needed to address both the grazing pressures on nonpoint-source pollutants and the need for access to water by livestock.

Superimposed on the physical aspects of landscape restoration efforts are the social, political, and economic questions associated with land ownership patterns (Schulte et al., 2006). Large percentages of the land are no longer owned by those who farm it. In many cases, tenant farmers have very different land management objectives than the landowners. Some farmers see buffers as taking land out of production, something they cannot afford to do at the low prices they receive for many of their crops. Most landowners seem to agree that landscape buffers would function to improve water quality, but most also indicate a need for shared private and public responsibility associated with the voluntary establishment of such buffers (Colletti, Premachandra, Schultz, Rule, & Gan, 1993). Thus, for implementation to occur on a watershed scale, restoration models must remain flexible enough to fit the objectives of landowners and yet be scientifically robust enough to attain governmental and public acceptance.

Restoration of Riparian Zone Functions

General Considerations

Before discussion of specific restoration models and methods is possible, it is important to review some general restoration principles. Riparian buffers should be viewed as one tool in a box full of best management practices (BMPs) that can be used to address nonpoint-source pollution problems in the agricultural and grazing landscape. In most cases, they should be viewed as the last line of defense against pollutants reaching streams. They should be part of a land-use plan that contains numerous upland BMPs that keep pollutants close to their source. In the production zone of a fluvial system where first- to third-order streams are located, the slope distance between the top of the watershed and the nearest stream channel can be very short. In this zone, riparian buffers are probably the most important BMP for protecting stream water quality and biological integrity. It is therefore important to buffer these streams before moving downstream into the zones of transport or deposition, especially since the vast majority of stream miles in a watershed consist of first- to third-order streams (Correll, 2005).

When developing a watershed buffer plan, it is important to consider the connectivity of the buffers both along the stream corridor and between the stream corridor and upland buffers or other perennial plant communities. It does little good for stream water quality or biotic integrity if buffers are installed randomly along the corridor with large gaps of unprotected channel left in contact with nonpoint-source pollutant sources. Unfortunately, the voluntary, first-come, first-served approach of riparian buffer installation that has occurred with government programs such as the Conservation Reserve Program leaves a patchwork of unconnected buffers spread throughout the watershed, with many large gaps. Connectivity is imperative for maintaining healthy in-stream aquatic habitat and travel corridors for wildlife. Because it is usually not possible to install riparian buffers along entire lengths of channels, land-use practices and pollutant travel pathways must be evaluated and buffers placed at the most problematic sites. Placing well-designed riparian buffers at sites that provide the highest percentage of pollutant inputs is probably more effective than placing narrower and simpler continuous buffers along the whole channel system including at sites that produce very little pollutant input. By integrating the riparian buffers with upland buffers, nonpoint-source pollutants can be kept close to their source, reducing the need for the riparian buffer to capture the brunt of the pollutant load.

In designing buffers, plant species and types and combinations are very important. If the goal is to restore riparian buffers to near-original plant communities, studying the composition and structure of a reference stream buffer community would be very useful. *Reference streams* are local streams with intact, or as near to intact as possible, riparian buffers in an undisturbed condition. Evaluation of the horizontal and vertical structure and diversity of these communities can be used to develop guidelines for providing buffers with optimal sediment, chemical, and flood debris trapping abilities and optimal habitat for local fauna. In areas where patches of native plant communities exist within an otherwise poorly developed riparian buffer community, it is important to preserve the native communities and work them into the new buffer design. Where possible, it is always best to use native plant species in a buffer design, as these species are well adapted to the landscape. However, there may be times when introduced species may provide better function because of irreparable changes that have taken place in the hydrology of the landscape.

Improvement of soil quality should be a paramount consideration in the design of a riparian buffer. A major priority should be the development of an effective living filter that can function as both an above- and belowground filter and sink for nonpoint-source pollution and can regulate and partition the flow of water through the buffer. However, it is also necessary at the time of buffer design and establishment to develop the long-term management practices that will maintain the living filter functions for the target nonpoint-source pollutants. In designing the living filter capability for the target dissolved chemicals, it is important to establish their pathway through the riparian buffer so that realistic pollutant reductions can be expected. To do this, it is necessary to establish a clear picture of the hydrogeological characteristics of the site to determine what proportion of the water moves through the living filter and what proportion may move below it. Any design features that can extend the depth of the living filter should then be considered.

Finally, all of these considerations have to be made in the context of the landowner's objectives and the present government or non-governmental organization (NGO) cost-share assistance opportunities that exist (Schultz et al., 2004). To get landowners to establish riparian buffers requires that they see benefits for themselves and their families. This may mean using species combinations and buffer widths that may not be optimal from a biological or physical perspective but which will be acceptable to the landowner. Cost-share assistance programs may put additional constraints on buffer designs but may also help landowners accept riparian practices.

Schultz et al. (2004) have provided a series of landowner concerns and site inventory questions that should be considered before designing a riparian or upland buffer on a specific site. It is important that the research and professional communities continue to study and monitor riparian buffers to help policy and government officials accept more flexibility in buffer designs than exist at the present time if they are to be nearly universally accepted by landowners.

In the following sections, riparian and upland buffers in the row-crop agricultural landscapes of the eastern and midwestern United States and in the predominantly grazing landscapes of the western United States are discussed. General forest riparian buffers are compared with grass filter strips, upland tree–grass buffers, and various grazing strategies. Design, installation, and maintenance are discussed. Much of the discussion is focused on studies conducted and experiences gained in the Bear Creek National Research and Demonstration Watershed in central Iowa and at the University of Missouri Center for Agroforestry, Greenley Memorial Research Center at Novelty, MO. Benefits and costs of various options are considered.

Riparian Buffer Models for the Agricultural Landscape

Two general riparian zone hydrologic conditions exist in U.S. agricultural landscapes. The hydrology of the agricultural landscape in the central and eastern United States has not been as dramatically modified as that of the recently glaciated Upper Midwest because the landscape is geologically older and has a higher channel density to carry water naturally from agricultural fields. In this landscape, many intact forest buffers still exist, and others, in various stages of development, can be restored to naturally functioning conditions. In the recently glaciated landscape of the Upper Midwest, less time has passed to develop a stream channel network. As a result, large numbers of natural pothole wetlands existed at the time of European settlement in the area called the Prairie Pothole Region. Major development of agricultural land uses have accelerated surface and subsurface (field tile drainage networks) runoff and resulted in deeply incised stream channels that have significantly lowered water tables. Thus, riparian soil moisture conditions have been dramatically modified and with them the potential for reestablishing true riparian plant communities. As a result, complete restoration of the riparian community is often not possible, but reestablishment of riparian zone function is possible with a variety of buffer practices.

Two separate, but similar, riparian management models have developed in row-crop regions of the country. The Riparian Forest Buffer (RFB) model was developed in the originally forested eastern United States, while the Riparian Management Practice model containing a multispecies riparian buffer (MRFB), a saturated riparian buffer component that intercepts field tile water, a stream bank bioengineering component, a constructed wetland component, and a controlled grazing component was developed in the more hydrologically modified Midwest that includes the recently glaciated young Des Moines Lobe landscape. Both address restoration of similar functions with a multi-zone buffer system. However, the zones vary depending on the landscape and landowner objectives and the concerns of the region in which they were developed. In addition, the Riparian Management Practice model consists of components that address the problems of stream bank erosion of deeply incised channels and field tile drainage, which are prevalent in much of the midwestern agricultural landscape.

Both of these riparian buffer models were developed as alternatives to cool-season grass filter strips, which provide perennial vegetation cover in the riparian zone but may not be as effective at stabilizing the stream banks of deeply incised channels, may be overwhelmed by high sediment loads in surface runoff, and provide less wildlife habitat (Dillaha et al., 1989). Some of these shortcomings of cool-season grass filters can be overcome by using native warm-season grasses and forbs as grass filters. Both models were used to develop the USDA Forest Riparian Buffer Conservation Practice (392-1) (USDA–NRCS, 1997a).

Riparian Forest Buffer Model

The U.S. Forest Service and the USDA–NRCS developed the initial guidelines for a Streamside Forest Buffer (Welsch, 1991). This buffer had three distinct zones (Figure 8–23). Zone 1 is a minimum 5-m-wide strip of undisturbed mature trees that begins at the edge of the stream bank and provides the final filter for materials moving through the buffer. It directly influences the in-stream ecosystem by providing shade and large and small organic matter inputs. The purpose of this zone is to create an undisturbed, stable ecosystem that

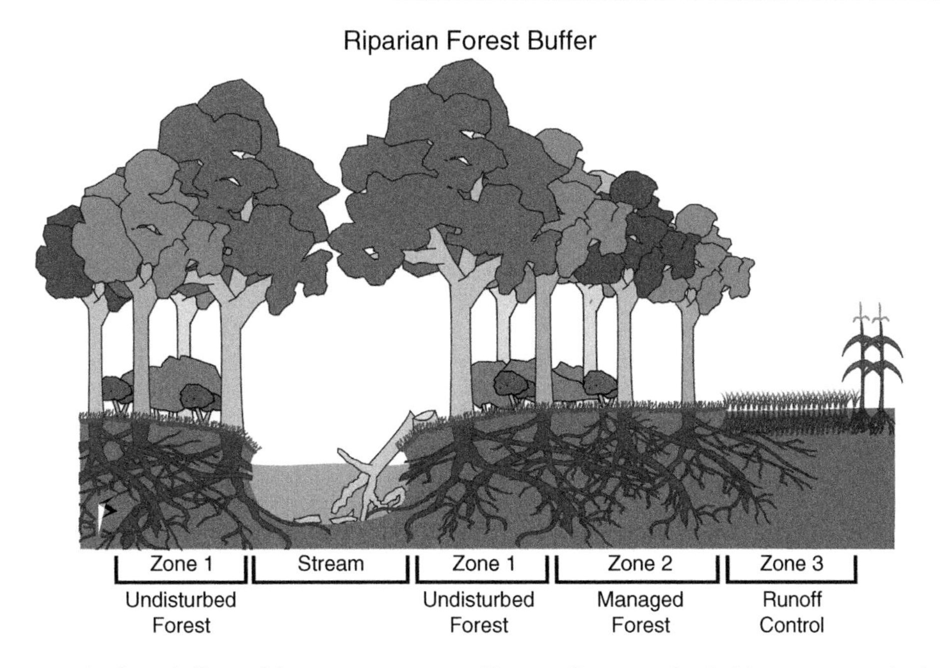

Fig. 8–23. A riparian forest buffer model. In Zone 1, tree removal is generally not permitted. This assures stream bank protection and provides large woody debris to the stream. In Zone 2, periodic harvesting is necessary to remove the woody plant nutrient sink and maintain nutrient uptake through vigorous plant growth. In Zone 3, controlled grazing or haying can be permitted under certain conditions (adapted from Welsch, 1991).

provides bank stability, an environment for dissolved soil water nutrients to interact with the living filter including plants that shade the stream, stable water temperatures, and both fine particulate organic matter and large woody debris to the stream. For this zone to be effective, flow must be either as sheet or subsurface. Concentrated flows must be converted to sheet flow before they enter the zone. Drainage from subsurface tiles must not be allowed to pass under the buffer if their water is to be filtered. Instead, a major portion of the discharge should flow into a tile that runs parallel to the RFB on the field side of the buffer. This perforated tile allows water to seep out of the tile, flowing through the soil under the MRFB before entering the stream. The buffer zone should consist of a mixture of species of native trees and shrubs to provide a variety of organic debris for microbial and invertebrate use. Management of this zone is restricted to removal of problem trees or an occasional removal of an extremely valuable tree. Over-mature trees are valued because they provide large woody debris for the stream. Logging equipment is excluded except at designated stream crossings. Likewise, grazing is excluded from this zone.

Zone 2 is upslope of Zone 1 and is a zone of trees that is at least 18 m wide and managed to provide maximum infiltration of surface runoff and nutrient uptake and storage while also providing organic matter for microbial processing of agrichemicals. The purpose of this zone is to provide the necessary contact time for biological processes associated with microbial activity and to provide plant uptake to remove nonpoint-source pollutants from soil water. Long-term storage of the plant-removed pollutants is provided by the trees, and periodic harvesting is used to remove them from the site and maintain viability for continued plant uptake. Once again, native tree and shrub species on the site should be managed, with planting done to stabilize the site if necessary. Dinitrogen-fixing plants should not be encouraged if NO_3 removal is an important activity of the buffer. Multiple-use management for timber and wildlife can be compatible with nonpoint-source pollutant removal. As in Zone 1, subsurface tile drainage should not be allowed to pass under the buffer, and livestock should not be allowed to graze in the buffer.

Zone 3 is a zone of grazed or ungrazed grass, a minimum of 6 m wide, that converts concentrated flow from the upland to sheet flow, either naturally or by the use of structures. This zone filters sediment from the sheet flow and causes the water and agrichemicals to infiltrate into the biologically active rooting zone where nutrient uptake and microbial processing occurs. The zone is composed of grasses and forbs that must be removed to provide effective nutrient sequestering. Removal may be in the form of grazing or mowing but is imperative for maintaining the vigor of the plant community.

The RFB model was developed after extensive reviews of forested riparian zones research in the eastern United States. However, this model is not as well suited to the agroecosystems of the Midwest and Great Plains, where many deeply incised, small-order streams drain highly modified agricultural and grazed landscapes. The native riparian vegetation of many of these first- and second-order watersheds would have been prairie communities with riparian forests along third and higher order streams often located on the east side of the stream or along steep valleys.

Farmers in this landscape view the value of trees along streams differently than many of their counterparts in the eastern United States. To make much of the midwestern landscape suitable for row-crop agriculture, the landscape had to be drained by lowering the water table with subsurface drainage tiles or drainage ditches. Many natural streams were channelized or drainage ditches were dug to accelerate the flow from the landscape and to produce rectangular fields. Landowners spent many hours and much money clearing woody vegetation from along the channels to increase the rate of flow of the system. Removal of perennial vegetation accelerated surface and subsurface runoff, increasing stream flow and decreasing the residence time of contact between the water and vegetation, ultimately increasing the amount of nonpoint-source pollutants lost. To landowners in this region, large woody debris in the stream is seen as a hindrance to drainage.

Development of the Riparian Management System model was first begun by the Department of Forestry, now the Department of Natural Resource Ecology and Management, at Iowa State University in 1989 and later by the Leopold Center for Sustainable Agriculture to function within this modified landscape and within the context of landowner concerns (Schultz et al., 1995). The Riparian Management System is an integrated management system that includes a MRFB, stream bank bioengineering to stabilize stream banks, small constructed wetlands placed within the buffer to intercept field tiles before they empty into the stream, and controlled grazing of riparian zones planted to cool- or warm-season grasses (Schultz et al., 1995) (Figure 8–24).

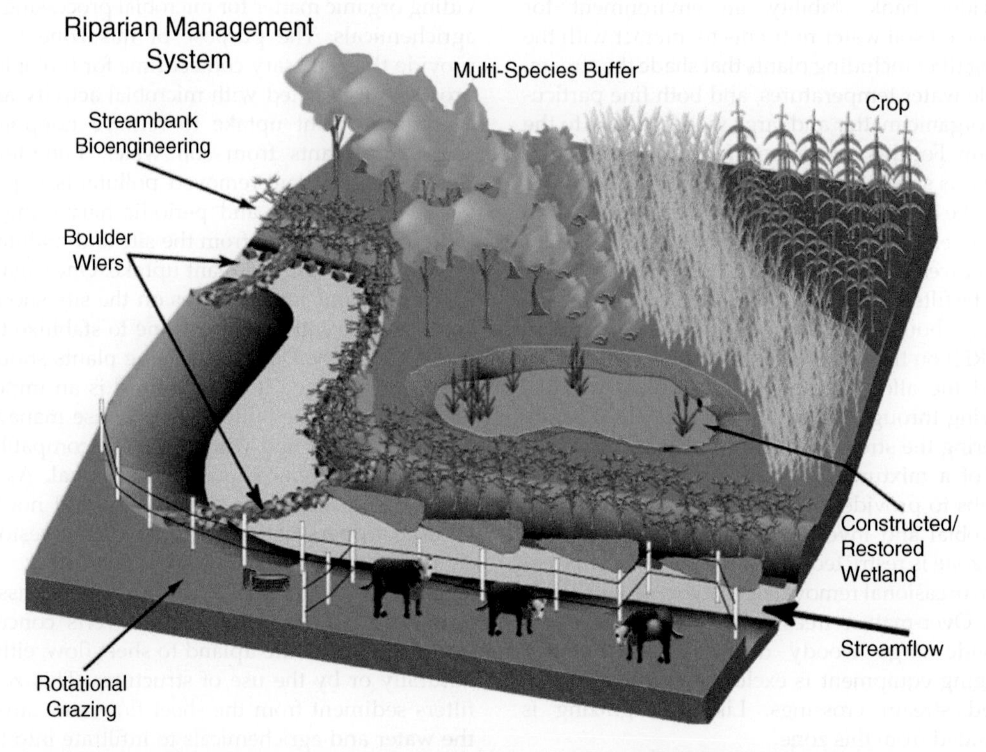

Fig. 8–24. A riparian management system model. This model consists of four practices: the multispecies riparian forest buffer (MRFB) with woody and warm-season grass zones that intercept nonpoint-source pollutants from adjacent land; stream bank bioengineering that provides bank stability, and constructed or restored wetlands that intercept and process subsurface drainage tile water before it enters the stream. Rotational grazing may be used where a cool-season or warm-season filter strip replaces the MRFB.

In many of the landscapes where the practice is used, it is imperative that several of the components be used together to address problems. In areas such as the Midwest and Great Plains, streams may be so deeply incised that installing a MRFB without also including stream bank bioengineering may mean failure of the relatively narrow buffers unless those buffers are made substantially wider to anticipate losses of plants during establishment of the buffers. Likewise, many of these areas are also laced with subsurface drainage tiles that completely bypass the MRFB. As many as 10 tiles km^{-1} may be found and may be the major source of base flow in the first-order streams of the region. Without some method of intercepting and removing this chemical load, stream water quality improvement is limited even with an MRFB. Therefore, constructed wetlands are also a component of the Riparian Management System. Several states in the region have now developed programs to specifically locate larger wetlands in major tile networks to capture and biologically treat high nutrient loads (Conservation Reserve Enhancement Program [CREP], part of the Conservation Reserve Program [CRP] of the USDA Farm Service Agency, https://www.fsa.usda.gov/programs-and-services/conservation-programs/conservation-reserve-enhancement/).

Multi-Species Riparian Forest Buffer Model

The MRFB contains a woody plant zone and a native warm-season grass zone of various widths and species compositions depending on landowner objectives, the adjacent upland land-use practices, and the characteristics of the riparian zone. The basic model consists of a 20-m-wide strip of trees, shrubs, and warm-season grasses and forbs on either side or on both sides of the stream channel but may vary from as little as a 10-m width on a side to as much as 45-m width (Figure 8–25).

Zone 1, adjacent to the streambank, may vary from one-third to two-thirds of the total width of the buffer and contains a mixture of rapid- and slow-growing trees and shrubs. The goal of the woody zone is to provide (a) stream bank stability, (b) a long-term nutrient sink, (c) a frictional surface to reduce the velocity of flood flows that come out of the banks, (d) improved soil quality, (e) wildlife habitat, and (f) shade and fine particulate organic matter for aquatic ecosystems in the stream. The order and number of the tree and shrub rows within the woody zone may vary depending on landowner objectives and the width of the buffer. Wider spacing of trees and shrubs is recommended to help maintain an actively growing ground cover of grasses and forbs to provide cover and resistance to surface water movement and raindrop impact on bare soil. It is important to provide enough sunlight to the stream bank edge to allow grasses and forbs to cover as much of the bank as possible. To accomplish this, the first tree row should not be planted too close to the bank. The actual distance from the bank to the first row of trees will depend

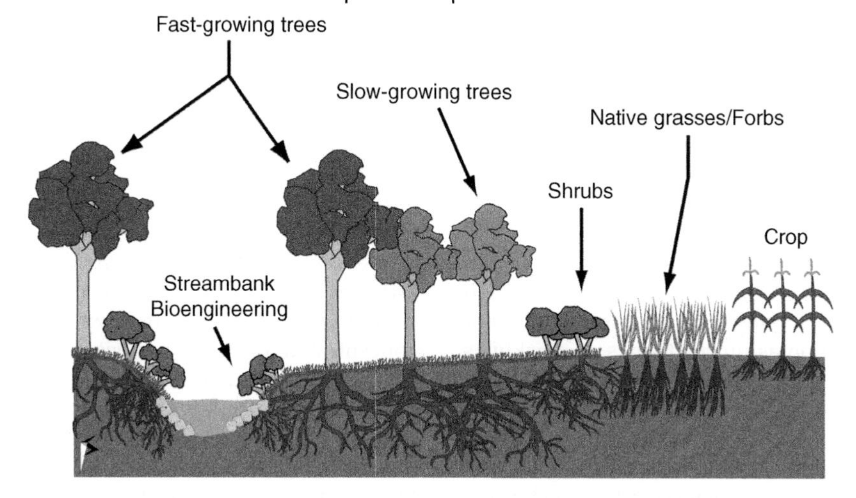

Fig. 8–25. The multispecies riparian forest buffer model. This buffer consists of a managed woody zone of fast-growing and slower growing trees and shrubs that provide a nutrient and sediment sink, wildlife habitat, bank stability, and modification of the aquatic environment. Production of large woody debris is discouraged by harvesting stream-edge trees before they fall. Species used along the bank will regenerate from stump or root sprouts. The native grass and forb zone intercepts and processes surface runoff from adjacent land and provides high infiltration.

on the width and orientation of the channel in relation to the daily movement of the sun. As a channel gets narrower, the banks are more easily shaded by trees planted directly adjacent to the channel. Shaded banks often support little plant growth, making them more prone to bank erosion. One or two shrub rows can replace the first and second rows of the tree strip if woody roots are desired along the bank to increase stability. A grass strip that is continuous with the grass and/or forbs planted on the bank could also be used up to the first tree row.

One of the major differences between this model and the U.S. Forest Service and NRCS model is the goal of providing large woody debris for the stream channel. In some areas where the MRFB is applied, landowners are concerned that large woody debris in the channel will slow water flow enough to impede subsurface tile drainage and keep the crop ground too wet for most crops. As a result, the whole woody zone is considered a managed zone and trees should be removed if they threaten to fall into the channel. The fast-growing riparian species that are planted along the bank will regenerate from stump or root sprouts, reducing the disturbance associated with harvesting and replanting of trees. As suggested above, if landowners are so concerned about the possibility of fallen trees clogging the channel, the best design would be to place a row or two of shrubs nearest the channel, with trees in the subsequent rows. In situations where a landowner desires large in-stream woody debris, the design can reflect that desire by planting trees right along the edge and not removing them before they fall in.

Fast-growing riparian tree species are frequently planted closest to the stream bank to rapidly develop bank stability and provide evidence of rapid woody plant development to landowners who are skeptical of trees. If, throughout the year, the rooting zone along the stream bank is more than 1.2 m above the growing season water table and soils are well drained, then upland deciduous trees and shrub species can be planted in the innermost three to four rows. Although these slower growing upland species will not begin to function as nutrient sinks as quickly as faster growing species, they will provide diversity and a potentially higher quality product to the landowner at harvest. While conifers are not recommended for wide-scale planting in the riparian zone because they catch too much flood debris and create too much shade on the soil surface, small patches of densely planted conifers in the innermost rows of the tree zone can provide excellent winter cover for birds.

Machine or hand planting of seedlings or cuttings is the major method of establishing MRFBs because little native riparian forest or native grass vegetation remains. Tree species should be varied both between and within rows to provide maximum diversity. Attention should be given to placing species in suitable microsite conditions, especially with respect to moisture. Planting in rows provides access for maintenance of trees and the grass cover between the tree and shrub rows. More naturalized buffers can be produced by planting seedlings in meandering rows or using direct seeding techniques at the time of establishment.

Shrubs can play an important role in the MRFB design. Their multiple stems can be very important in slowing flood flows and trapping large woody debris that may be floating downstream during a flood event. A major goal of watershed management is often to reduce downstream flood flows by desynchronizing the delivery of water from small-order streams to the larger order streams. The MRFB can slow floodwaters that move through the multi-stemmed trees and shrubs and the woody shoots in stream bank bioengineering. Not only can they help to desynchronize downstream flooding but they may also reduce the area of crop fields that get flooded or covered with flood-carried debris. Most first- and second-order streams have flashy flood flows that minimize long-term flooding even with riparian vegetation that may slow the flood flows. Shrubs can be placed along the edge of the bank to reduce the shade on stream banks where shade-intolerant grasses are being grown to provide soil cover and stability. Shrubs also are useful for diversifying wildlife habitat and providing alternative berry and nut crops for interested landowners (Peters & Hodge, 2000). It is even possible to plant fruit trees in place of the shrubs to provide an alternative crop for the landowner.

While the present USDA forest buffer standards do not include a grass zone at the field edge of the buffer, the standards do suggest that a grass–forb filter strip (NRCS 393 practice standard, USDA–NRCS, 1997b) can be planted adjacent to the forest buffer where substantial concentrated or sheet surface flow exists. We strongly believe that a filter strip of either cool-season or, even better, a zone of warm-season grasses and forbs should be planted. That zone should be at least 7 m or more in width and function to (a) intercept and dissipate the energy of surface runoff by producing sheet flow, (b) trap sediment and agricultural chemicals in the surface runoff, and (c) improve soil quality by increasing infiltration capacity and microbial activity as a result of their annually high turnover of roots. This zone is imperative if the MRFB is to function properly. Ideally, the leading edge of the grass filter should be dense enough to slow and spread any incoming concentrated flow into sheet flow before it enters the filter. This results in much of the larger sediment being deposited in the crop field before it enters the filter. Warm-season grasses are better suited to the filter strip in a MRFB than most of the cool-season grasses that are used for grassed waterways (USDA–NRCS, 1999). The warm-season grasses are taller and have stiffer stems and significantly deeper roots. The warm-season grasses have eight times greater root mass extending more than three times as deep as cool-season grasses (Schultz et al., 1995; Tufekcioglu et al., 2003). Switchgrass is preferred where surface runoff is a major problem. Unlike many warm-season bunch grasses, switchgrass is a stiff, single-stemmed grass that grows very dense. Dabney et al. (1994) have demonstrated that as little as a 12-cm-wide strip of switchgrass can dam water as high as 10 cm deep. These characteristics help get sediment dropped in the field before it enters the filter rather than in the filter, as is the case for cool-season grasses. One concern of landowners is that a berm of sediment may form at the edge of the grass filter and cause water to flow along it and then move through as concentrated flow. Several practices may be used to reduce that likelihood. Dillaha and Inamdar (1996) suggested placing water bars perpendicular to the filter at intervals of 15–30 m to route the water through the grass. They also suggest that if moldboard plowing is done, the last plow pass should turn the soil toward the buffer and then be carefully disked to minimize gully formation. Annual disking along the edge can also pull the sediment back up into the field.

A warm-season grass filter may function longer to trap sediment than a cool-season grass filter because the latter may lose its effectiveness at slowing overland flow after only a few surface runoff events because the grass becomes buried by sediment (Dillaha & Inamdar, 1996). If that occurs, the grass strip must be completely reestablished to develop the plant structural conditions that will encourage sheet flow.

One of the concerns of using pure switchgrass is that it does not provide the best nesting habitat for game birds. Ground-nesting birds will be more successful in a mixed planting of other warm-season bunch grasses and forbs. Switchgrass does provide excellent winter cover in conditions of high snow and wind because it is not easily bent over under high snow loads or by wind. Wildlife habitat and aesthetics can be further improved with the addition of forbs to the mixture of grasses. One word of caution is that most switchgrass that can be purchased at local farmer cooperatives are selected cultivars that are very aggressive and should not be planted as part of a warm-season grass mixture. If switchgrass is desired as part of a mixture, local ecotypes should be used. Where a mixture of grasses is desired in an area where surface runoff can be a problem, a minimum of 2.5 m of switchgrass can be planted at the crop field edge of the strip, with the mixed grasses planted downslope of that strip. Management of a native grass filter is critical to maintaining the filter's function and species mix. This narrow strip of native perennials is inundated annually with annual and perennial weed seeds and if not maintained may succumb. A burning, harvesting, or flash grazing scheme will help remove biomass that can smother regenerating plants. These maintenance activities also help remove the nutrients sequestered in the standing biomass.

As mentioned above, riparian buffers and filters are the last conservation practice in the landscape that helps to prevent nonpoint-source pollutants from getting into the stream. For them to function properly, especially in complex landscapes with long and steep slopes, upland buffers and filters must also be installed. These buffers could be either grass or grasses and forbs only or a combination of grass and trees, or grass, shrubs, and trees similar to the designs that have just been discussed. In most cases, however, these upland buffer systems will be narrower than the riparian systems. In a study in northeastern Missouri with three adjacent watersheds, Udawatta et al. (2002, 2006) showed that tree–grass and grass buffers removed significant amounts of sediment and nutrients from runoff. The study design consisted of 3–4-m-wide buffers established at 22–36-m spacing on contours. Grass strip and agroforestry watersheds were established by planting redtop (*Agrostis gigantea* Roth), brome (*Bromus* sp.), and birdsfoot trefoil (*Lotus corniculatus* L.). Pin oak, swamp white oak, and bur oak trees were alternately planted at 3-m spacing at the center of the grass buffers on the agroforestry watershed. This tight spacing allows thinning at a later stage after identifying the most suitable tree species for the site and environmental benefits.

Several factors should be considered when establishing agroforestry buffers. Buffer strip spacing is determined by soil erodibility, slope degree, slope length, rainfall erosivity, and management factors. The Universal Soil Loss Equation can be used to determine spacing between buffers. Wider spacing between buffers can be allowed on flat areas with less potentially erodible soils, while highly erodible soils on steep slopes require narrow spacing between buffers. The second factor that should be considered in buffer spacing is the size of the farm machinery that is to be used. Crop area should be wide enough to allow multiple passes during planting, harvesting, and other operations. Similarly, buffer strips should be wide enough for mowing or hay harvesting paths on either side of tree rows. For example, if a hay harvester is 2 m wide, then the buffer should be at least 5 m wide to make harvesting paths and not damage trees. Buffers established in an east–west direction may result in the least shading effect on crops when trees are larger. Selection of grass, legume, and tree species depend on landowner objectives. To improve the quality of harvested hay, a legume can be planted with the grass in the buffer. Cool-season grass species help reduce runoff early in the season when spring and early summer rains occur. Warm-season grass species can be effective during the summer and fall months. Selection of the proper grass species may help to draw additional income if the land owner is interested in supporting wildlife habitat development. Habitat Buffers for Upland Birds (CP 33) is one of the USDA Farm Service Agency, Conservation Reserve Program practices that pays landowners on a per-hectare basis for 10–15 yr. Trees can be selected to generate additional income through selling nuts or timber. Several tree species should be alternately planted at 3-m spacing, with no one species making up more than 25% of the total trees in the buffer. Trees should be protected from deer damage by installing wire mesh fences (1-m height and 1-m diameter). Also, application of slow-release fertilizers may enhance early growth. Weed mats of approximately 1.5 by 1.5 m around the base of the trees reduce early competition for resources and help improve early growth. Buffers should be inspected after large rainfall events for possible damages, and these should be corrected before significant erosion occurs. Harvesting of hay and pruning of trees should be conducted to reduce insect damage and for additional income.

Species Selection for RFBs, MRFBs, and Upland Buffers

Tree species that are commonly recommended for planting in both the Chesapeake Bay watershed and in the Midwest include species that have a high tolerance to flooding, such as black willow (*Salix nigra* Marsh.), boxelder (*Acer negundo* L.), river birch (*Betula nigra* L.), and silver maple (*Acer saccharinum* L.); species with a moderate tolerance to flooding, such as green ash (*Fraxinus pennsylvanica* Marsh.), hackberry (*Celtis occidentalis* L.), red maple (*Acer rubrum* L.), and sycamore (*Platanus occidentalis* L.); and species with a low tolerance to flooding, such as red oak (*Quercus rubra* L.), white oak (*Quercus alba* L.), pin oak, swamp white oak, and bur oak (Palone & Todd, 1997; Schultz et al., 2004; Udawatta et al., 2002). Common shrubs that are recommended include silky dogwood (*Cornus amomum* Mill.), gray dogwood (*Cornus racemosa* Lam.), elderberry (*Sambucus canadensis* L.), and ninebark [*Physocarpus opulifolius* (L.) Maxim.].

Common warm-season grass and forb mixes for somewhat poorly to well drained sites may include switchgrass, eastern gamagrass [*Tripsacum dactyloides* (L.) L.], purple coneflower [*Echinacea purpurea* (L.) Moench], blazing star [*Liatris spicata* (L.) Willd.], and bee balm (*Monarda didyma* L.). On moderately well to well drained sites, species may include big bluestem (*Andropogon gerardii* Vitman), switchgrass, Indiangrass [*Sorghastrum nutans* (L.) Nash], butterfly milkweed (*Asclepias syriaca* L.), and black-eyed Susan (*Rudbeckia hirta* L.). Milkweed is especially important for monarch bufferflies. Cool-season grass–forb mixtures can include redtop, brome, and birdsfoot trefoil.

Numerous other local trees, shrubs, and warm-season grasses and forbs are available to provide maximum function and diversity to the buffers. Native species and native seed sources usually give the best results. Contact local county extension, district forester, and NRCS offices for other species recommendations.

Site Evaluation and Buffer Design

There are four basic steps that should be followed when designing a riparian buffer (Dosskey, Schultz, & Isenhart, 1997; Iowa State University Extension, 2002). They are (a) determine what benefits are needed and what the landowner's objectives and concerns are, (b) identify the best types of vegetation to provide those benefits, (c) determine the minimum acceptable buffer width, and (d) develop an installation and maintenance plan.

Determine What Benefits Are Needed

Every site evaluation should begin with communication with the landowner. If landowner objectives and concerns are not addressed as part of the design process, even the best design to meet the physical and biological conditions of the site may not be accepted or maintained by a landowner who sees problems with it. Schultz et al. (2004)

developed a list of 17 concerns that landowners may have with riparian and upland buffers and filters. They have also developed a list of 17 questions to help guide the field assessment of a potential buffer site.

The field assessment to determine the benefits that are needed for the site should begin with an evaluation of its physical characteristics. Major soil features that should be evaluated include texture and structure, pH, and depth to the water table, especially during the growing season. Texture and structure are indicators of the potential soil moisture holding capacity and fertility, important factors for plant species selection. Many plant species are sensitive to extremes in soil pH, as it influences nutrient availability, among other things. Finally, depth to the water table helps define the depth of the rooting zone, as few species have roots that can live in saturated soils for any length of time. Wet soils exist where the growing season water table height reaches within 0.5 m of the surface. If the average water table only reaches a height of 1.5 m below the surface during most of the growing season, the site has moist soils and is well suited for a wide variety of buffer species. Where stream channels have been incised into the landscape, water tables have dropped and effective rooting depths have increased, increasing the number of species that may grow on the site, including species that are only found in upland landscape positions.

Evaluation of macro- and microtopography are also important to successful design and establishment of buffers. The slope and aspect of a site determine solar energy inputs that influence temperatures and soil moisture. South- and west-facing slopes will be drier and warmer than east- and north-facing ones. North- and east-facing stream banks will be shaded for more of the day, and that could influence species selection, especially along narrow streams where mature trees from one side of the stream cast shade on the plants on the opposite side of the stream. Microtopographic variations within the riparian zone are also important to identify because subtle differences of less than 0.5 m can influence the successful survival and growth of certain species on a site. Natural levees are usually deep sand deposits that can provide relatively dry conditions compared with depressional old abandoned channels or sloughs that may hold standing water. The stream bank condition should be described as part of the topographic evaluation. The stage of the channel development, according to the channel evolution model, will identify the restoration options that may be available. In addition to stream bank condition, field tiles entering the channel should be located and marked, as they influence buffer design and may require special design modification.

As part of the initial site evaluation, it is also important to identify the vegetation that is presently on the site. Is the site presently in row-crop agriculture, grazed pasture, or forest cover? Are there high numbers of exotic species that are often major weeds in crop fields? An inventory of the vegetation will not only identify what plant species are needed but also what site preparation and planting techniques will be required to control potentially competing and noxious species.

One of the best ways for making these initial site evaluations is to use NRCS soil survey maps and USGS quadrangle maps. Digitized maps using GIS can be used with a series of overlays to improve the planning process.

In the process of making the site evaluations, problems such as cropping to the edge of the stream bank, accelerated bank erosion, livestock grazing on the stream banks with uncontrolled access to the channel, high sediment load in the channel, lack of shade or woody debris in the channel, minimal wildlife habitat along the riparian zone or the adjacent upland should be identified. After the major site problems have been identified, the objectives and the concerns of the landowner should be revisited. His or her concerns for stopping surface runoff or bank erosion or wanting wildlife habitat should be considered. Concerns about such things as in-stream large woody debris and beaver dams blocking field tile drainage or dislike of trees and shrubs of specific species should be respected. Where landowners will be using available government or NGO cost-share programs, concerns about reclaiming the land after the program has lapsed should also be addressed. Once all of these problems have been identified, they should be prioritized with the help of the landowner.

Identify the Best Vegetation for Providing the Needed Benefits

There are three general types of plants that can be included in a buffer, namely, grasses and forbs, shrubs, and trees, and each is able to provide different benefits. Each buffer design should be flexible enough to use these different plant types to the best of their ability for addressing the major prioritized problems of the site. Constraints may exist because of landowner concerns or certain cost-share programs that have specific requirements. Table 8–1 compares the relative benefits of each plant type in agricultural riparian buffers. The basic designs of the RFB and MRFB have already been presented.

Table 8–1. Relative effectiveness of different vegetation types for providing specific benefits.

Benefit	Vegetation Type		
	Grass	Shrub	Tree
Stabilize bank erosion	low	high	high
Filter sediment	high	medium	medium
Filter nutrients, pesticides, microbes			
Sediment-bound	high	medium	medium
Soluble in surface runoff	medium	low	low
Soluble in subsurface flow	high	medium	high
Aquatic habitat	low	medium	high
Wildlife habitat			
Range, pasture, prairie wildlife	high	medium	low
Forest wildlife	low	medium	high
Economic products	medium	low	medium
Visual diversity	low	medium	high
Flood protection	low	medium	high

Note. Source: Dosskey, Schultz, & Isenhart (1997).

While the RFB and MRFB are two buffer designs that have been shown to be effective, it is very important to provide as much flexibility in design as possible to improve landowner acceptance. Although the species combinations of these two model buffers provide very effective riparian buffer plant communities, there are other plant combinations that also can be effective. These may include combinations with narrower grass zones and more trees or shrubs in cases where timber products or wildlife habitat with more vertical structure are major objectives. Ideally the tree zone should occupy at least one-third of the total width of the MRFB plus the adjacent grass and/or grass–forb filter strip. Wider woody plant zones are recommended for buffers planted into grazed pastures. It is also possible that shrubs could be substituted for trees in the tree zone or that the first few rows near the stream could be planted to shrubs instead of trees to address the concerns of landowners who want to minimize the risk of large woody debris falling into the stream channel. The NRCS specific design criteria may differ among states and regions. Substitution of shrubs in place of trees and the use of conifers in buffers may vary. Restrictions on substitutions may impact landowner willingness to accept the buffer practice. It is therefore important to check local species recommendations before getting deeply involved in the planning process.

In some cases, only warm-season grasses and forbs might be planted in headwater reaches where landowners want to try to recreate a more typical presettlement landscape. If stream banks are not deeply incised, such an option might be well founded. If the banks need stabilization, bioengineering techniques could use warm-season grasses that continue on across the buffer. An all warm-season grass and forb planting would be considered a filter strip under Farm Bill programs.

Several variations may be developed for the grass zone. Switchgrass is frequently recommended for this zone because it is a strong, single-stemmed grass that is very effective at slowing water and trapping sediment and associated nutrients. Where the width of the grass zone is greater than 7 m and the adjacent slope is less than 15%, a combination of other warm-season grasses may be used. In these cases, the outer 3–4 m of the zone could be planted to pure switchgrass and the rest planted to combinations of bunch grasses such as big bluestem and Indiangrass. These bunch grasses are not as effective at slowing water because of the bunching nature, but in combination with the switchgrass strip and forbs they can be effective. They also provide better wildlife habitat than pure switchgrass.

Landowners might opt completely for a cool-season filter strip instead of a forest buffer. In that case, some state NRCS programs specify that the strip has to be twice as wide as a warm-season grass strip because of the reduced sediment and nutrient trapping capability of the cool-season grass filter. Carefully managed, these filter strips could be grazed using rotational management practices. Ultimately, site conditions, major buffer biological and physical functions, owner objectives, and cost-share assistance program requirements should be considered in specifying species combinations.

In a tree–grass upland buffer system, legumes, cool-season grasses, warm-season grasses, or shrubs can be planted in the buffer areas. Careful selection of plants for the buffer is important. Successful agroforestry practices optimize resource utilization among trees, grasses, and crops, which can be accomplished by proper species selection and management (Sanchez, 1995). Species selection should be guided by species compatibility with site, climatic, and management options. Based on seedling and sapling survival, initial growth rate, and rooting patterns of an agroforestry practice in the claypan region of northern Missouri, pin oak and swamp white oak appeared to be better suited than bur oak (Udawatta, Nygren, & Garrett, 2005). Some trees with more vigorous horizontal roots may compete for resources with the accompanying crop. Trees with a greater proportion of vertical roots are preferred, as they tend to produce more roots deeper into the profile that are less competitive with adjacent crops (Udawatta et al., 2005). Trees that have fewer surface roots, similar to *Paulownia elongata* S.Y. Hu, which has less than 2% of its roots in the surface 0–20-cm soil depth, are ideal for these buffers (van Noordwijk & van de Geijn, 1996). However, even tree species that generally produce deep roots may be confined to producing shallower, flatter root systems in restrictive claypan soils. Finally, planting seedlings with well-established root systems such as those produced under the Root Production Method (RPM) helps to improve survival, but these seedlings require large 30–40-cm-diameter holes for planting.

Determine the Minimum Acceptable Buffer or Filter Width

The minimum acceptable buffer or filter width is one that provides acceptable levels of all needed benefits at a reasonable cost (Dosskey et al., 1997). Minimum width is determined by the specific benefit that requires the greatest width. Figure 8–26 presents a general comparison of buffer widths for specific benefits. For most of these benefits, research information is limited, so the widths indicated represent best estimates. The widths may vary depending on the value of the resource being protected, e.g., whether it is a high-value trout stream or a channelized agricultural stream. The width may vary depending on the site and watershed traits, the desired benefits of the buffer, the landowners objectives, and the intensity of adjacent land use—the more intense the use the wider the buffer should be (Dabney, Moore, & Locke, 2006; Palone & Todd, 1997).

Buffers may be narrower if they are maintained in a healthy condition. For example, if they are to function as sediment traps and nutrient sinks, they must be designed and maintained so that they are not inundated by sediment, and biomass must be periodically removed to renew the storage potential of the plant sinks. Although there are several vegetation types that can be used in buffers, trees provide the greatest range and number of benefits (Palone & Todd, 1997). Some of the benefits that are enhanced by trees include protection against stream bank erosion; increased N, P, and sediment removal; reduction of downstream flooding; thermal protection of stream channel water; food and habitat for terrestrial and aquatic organisms; foundation for present and future greenways; increased property values; enhanced potential for stream restoration; reduced watershed imperviousness; and protection of associated riverine wetlands.

The widths shown in Figure 8–26 provide ranges for various conditions that may be encountered. Along channels that are in contact with

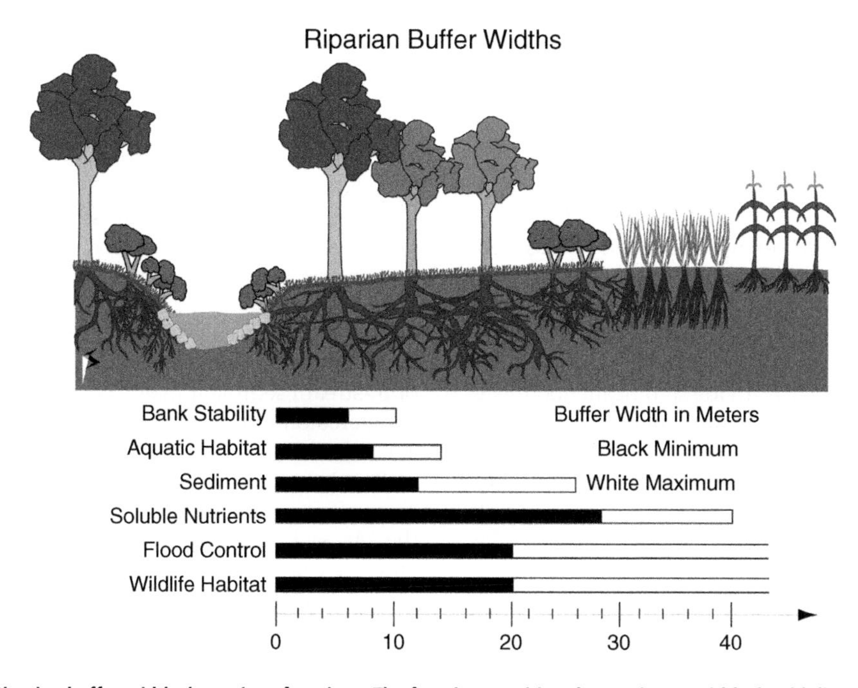

Fig. 8–26. Riparian buffer widths by various functions. The function requiring the maximum width should dictate the actual width of the buffer. These widths are destined to change as more research and field monitoring is done.

their floodplain, a very narrow band of vegetation, located primarily on the bank itself, is all that is needed. However, along deeply incised channels with vertical banks (Stages 2 or 3 of the Channel Evolution Model), widths of at least three times the vertical height of the bank may be needed to stabilize the banks. Under the worst-case scenarios, complex soil bioengineering solutions may be needed (see below).

Relatively narrow buffers are able to provide adequate aquatic habitat unless temperature control is needed. Most of the litter that provides an organic food source for aquatic organisms comes from within a few meters of the bank. Likewise, the source of large woody debris in the stream originates within 5–10 m of the bank (Welsch, 1991). However, if shade is needed to improve habitat, then stream width, canopy height, and sun angle at solar noon become important and may require widths of up to 30 m for streams in the zone of production and transport.

For upland slopes of less than 15%, up to 80% of the sediment and adsorbed chemicals will settle out within the first 10 m of properly designed grass filters (Lee et al., 2003). If the finer suspended sediments are to be trapped, significantly wider buffers are needed. Widths for soluble subsurface nutrient removal are the most difficult to predict because many of the controlling variables are difficult to measure. The hydrogeology of the site, which controls the pathway and residence time of water moving into and through or below the living filter, microbial processes, and plant nutrient uptake as well as soil quality must all be assessed to establish a treatment width (Dabney et al., 2006; Simpkins et al., 2002). Thirty-meter widths of efficient forest riparian buffers in soils where an impermeable zone (aquitard) is within 4–5 m of the surface can remove up to 90% of the NO_3 moving through them (Simpkins et al., 2002). Widths for flood attenuation must be based on the roughness of the buffer, the infiltration and moisture holding capacity of the buffer soil, and the size of the watershed above the point of interest. In many cases, buffers as wide as the floodplain are needed to provide significant floodwater storage and downstream attenuation. Finally, buffer widths for wildlife habitat may vary depending on the wildlife in question. For most wildlife, the buffer provides a corridor with extensive edge in relation to interior habitat. For larger wildlife species, corridor widths of 100–150 m may be needed. Welsch (1991) summarized the work of others and suggested that buffer widths could be 20% of the total nonpoint-source pollutant area, or widths for Land Capability Classes I, II, and V of 29 m; III and IV of 36 m; and VI and VII of 52 m. The U.S. Forest Service RFB model has a width of at least 29 m. More recently, the NRCS model of the Riparian Forest Buffer Practice (391) generally includes only Zones 1 and 2 consisting of unmanaged and managed forest. The total width of this buffer is suggested to be 30% of the geomorphic floodplain or a minimum of 10 m (35 ft). These widths may vary by upslope conditions and by state (NRCS Field Office Technical Guide, http://www.nrcs.usda.gov/technical/efotg/). Instead of a Zone 3 as proposed by Welsch (1991), the NRCS model suggests that a grass filter (NRCS 393 practice standard, USDA–NRCS, 1997b) be added between Zone 2 and the upland if erosion activities warrant it. Dabney et al. (2006) suggested that some buffer is better than none and that buffers of less than 1-m width can trap significant amounts of sediment and soluble nutrients. They pointed out that the leading edge does the majority of the work and if that edge is dense enough, much of the deposition takes place upslope of the edge. This work points out the importance of the grass filter at the edge of a riparian forest buffer. Many narrow forest strips in agricultural landscapes that are assumed to function as buffers do not because they lack the grass filter edge. Personal observations have shown that these forest strips frequently have classic gullies through them carrying concentrated flow from surface runoff from adjacent agricultural fields. In fact, frequently, runoff from grass waterways that exit into forested strips result in the development of major gullies that dramatically extend the surface drainage network and may become first-order channels with sufficient incision. This is a major issue in row-crop areas where grass waterways are common. While sheet and rill erosion in fields may be controlled by infield practices such as terraces, in-field vegetation strips, and grass waterways, accelerated erosion from gully formation provides significant sediment to the expanding drainage network. This problem has not been well quantified, but preliminary indications in some areas suggest that as much as 80% of in-stream sediment may come from gully and bank erosion in these actively developing drainage networks (Zaimes et al., 2019).

Recommended riparian forest buffer dimensions often vary from state to state. For example, the Massachusetts Department of Conservation and Recreation (2003) specifies that buffer width should be increased by 12 m (40 ft) for every 10% slope increment on poorly and moderately drained soils for up to 20% slopes. On well-drained soils, a 12-m (40-ft) increase for every 10% slope increment is recommended for slopes greater than 20%. In Pennsylvania, the BMP requires continuous buffers along perennial and intermittent stream channels (Lynch & Corbert, 1990).

As mentioned above, when designing buffers for upland areas, attention should be paid to the spacing between buffers. Crop areas or crop alleys should be wide enough to allow multiple paths for farm equipment, and buffers should not affect regular agricultural operations. Buffers should be as close to parallel to field contour lines as possible. Buffer widths should also satisfy buffer maintenance requirements such as mowing. For example, a buffer should be wide enough so mowing can be performed on both sides of the tree line in a tree–grass buffer. Landowners should contact federal, state, and county agencies for establishment guidelines and other regulations.

Upland contour buffer strip dimensions also vary from state to state and with land management and landscape features according to NRCS guidelines (www.nrcs.usda.gov/technical/efotg). These buffers are suitable for slopes ranging from 4 to 8% and in regions with low to moderate rainfall intensities. The purpose of buffers is to reduce the transport of sediment and other water-borne contaminants. The NRCS does not recommend contour buffers for long slopes (lengths greater than 1.5 times the critical length for contouring) unless terraces are installed. The buffer width at the narrowest point shall be no less than 4.5 m (15 ft) for grass and 9 m (30 ft) for a grass–legume combination. Contour buffers in Texas shall be no less than 4.5 m (15 ft) for grass or a grass–legume combination and 9 m (30 ft) for legume alone. Some other states recommend 4.5-m buffers for slopes greater than 3%. Landowners who are interested in wildlife species are required to maintain a 9 m or wider buffer for nesting and escape cover. In some states, the maximum spacing between buffers shall not exceed 45–60 m (150–200 ft).

Filter strips are often established along the field edge and shall not be less than 6 m (20 ft) wide and established on the approximate contour. The drainage area/filter strip area ratio shall be less than 70:1, 60:1, and 50:1 for areas with Revised Universal Soil Loss Equation rainfall erosivity (*R*) factor values of 0–35, 35–175, and >175, respectively. The NRCS also recommends 1 ha of filter area with a 6-m filter strip for every 15 ha of contributing land with <3% slope and 9 m for slopes >10%. Filter strip dimensions vary with the objective. For example, for <2% slopes with Hydrologic Group A soils, 6 m is recommended to reduce sediment, particulate organic matter, and pathogens, while an 18-m strip is recommended for animal wastewater systems. Filter strip dimensions vary with slope length and percentage, erosion rate, runoff, and vegetation parameters. While edge-of-field filter strips may take up space in upland fields, they help keep eroded sediment closer to the source than if the only buffer in the farming landscape is a riparian buffer or filter. Grassed waterways should be able to convey the peak runoff expected from a storm of 10-yr frequency and 24-h duration. They can have either a trapezoidal or parabolic shape. The bottom width of the waterway shall not exceed 15 m (50 ft). The waterway should have the capacity to keep the water level at or below the water level in a terrace, diversion, or other tributary.

Field borders are another form of buffer that can be established around the perimeter of the field, or as a minimum on slopes >5%, or on long slopes, or where runoff enters and leaves the field. A minimum width of 9 m has been proposed for wildlife improvement. Establishment of wider field borders can serve several purposes including space for turning farm equipment and improved wildlife habitat.

This discussion has provided some of the basics and some of the government program recommendations for determining average buffer widths. Actual widths of the buffers should be adjusted to fit the site. The buffer boundary should be as gentle and smooth along a crop field border as possible to facilitate cultivation and harvesting, even though this may not be the best design for wildlife. Along meandering streams, this may mean that the actual buffer width may be wider or narrower than the average at some points. If the width needs adjustment, the plant community of most importance should remain the widest. For example, if the cropland adjacent to the buffer has a slope of >5%, then the woody plant strips of the buffer should be narrowed in favor of the grass strip. If, on the other hand, the slope is <5% and the channel is deeply incised and widening, then the grass strip should be narrowed in favor of the woody strip.

Develop an Installation and Maintenance Plan

Once the site evaluation has been completed, the basic buffer design selected, and the average width determined, an installation and maintenance plan must be developed. Several general considerations will be mentioned here before a more complete discussion is pursued. First, rely on local knowledge to select native species that are best suited for the specific site. Incorporate existing perennial vegetation into the buffer design where possible, as this vegetation has already stabilized the site and can act as a local seedbank to diversify the rest of the buffer with time. It may take 10 or more years for a planted buffer to reach its full functioning potential. The goal of a buffer planting is to provide a rough surface of vegetation that covers the whole buffer site to reduce raindrop impact as well as slow surface runoff. It may therefore be necessary to plant a ground cover such as less-competitive grasses between tree and shrub rows even if this reduces the growth rate of the woody plants. Spacing of the trees and shrubs should be wide enough to maintain the ground cover throughout the rotation, which may require some thinning of the trees. Maintenance is necessary to control weed growth in both the woody and grass components of buffers, especially in native grass buffers where these narrow strips are constantly under pressure from annual and perennial invasive weeds. Not only are weeds competitive with the desired plants, but landowners want assurance that their buffers are not a weed seed bank for their adjacent crop fields. Maintenance should be continued throughout the life of the buffer, and lack of maintenance often spells failure for the system. Finally, a harvest rotation should be determined at the time of establishment because harvesting is necessary to maintain a viable biomass nutrient sink.

East–west directional buffers may have the least shading effects on crop areas. If shading is a concern, woody plantings should consist of shrubs or small trees and/or a wider filter strip or grass–forb zone should be established. In some cases where landowners are concerned about competition between tree roots and crops, trenching may be performed to reduce the competition. This may be important where narrow field buffers are planted but may not be of concern with riparian forest buffers that have a grass–forb filter strip between the woody strip and the crop field.

Specifics of Site Preparation, Layout, and Maintenance of Buffers

Site preparation should begin the fall prior to planting. Four different types of sites may be encountered. They are pasture, crop field, old field (abandoned crop field), and existing forest sites. If the site has been pastured, preparation for woody plants is accomplished by eliminating competing perennial vegetation in 1-m-wide strips or circles where trees or shrubs will be planted. This can be accomplished with fall tillage and/or herbicide application (such as glyphosate). If the proposed grass zone is to be replanted with warm-season grasses, the cool-season sod can be killed with herbicide in the fall, which can be repeated, if needed, in the spring. Use care, however, when applying herbicides in the riparian zone, as the chemicals may kill needed vegetation growing on the bank or go directly into the stream. If cultivation is used in the grass zone, several light diskings and packing will have to be completed in the spring before planting to eliminate some of the potential weed seeds in the seed bank and to prepare the bed for the small native plant seeds. Where concentrated flow is a problem, cultivation should be minimized to reduce erosion.

If an area has been used for row crops, the ground should be cultivated with a disk in spring, and the area where trees and shrubs will be planted should be seeded with a mixture of perennial cool-season grasses, such as perennial rye (*Lolium perenne* L.) and timothy (*Phleum pratense* L.). These cool-season grasses are less competitive with trees and shrubs than other species such as brome (*Bromus inermis* Leyss.) and fescue (*Festuca* sp.). Local natural resource conservationists are sources for other recommended grass species. The grass planting is very important because the goal of the buffer is to provide a rough frictional surface for overland flow and to protect the soil from direct raindrop impact. Clean cultivating between the trees and shrubs may maximize woody plant growth but will not protect the exposed soil from raindrop impact or slow overland concentrated flow, and the cultivated zone may become a source rather than sink of sediment. Most raindrops reach terminal velocity in 10 m (35 ft) or less, and raindrops falling from the leaves or trees are often larger than those in the free-falling rain event. Previously cultivated areas to be planted to warm-season grasses and forbs should also be disked and packed in preparation for broadcasting or drilling the seed. Use local recommendations for planting density.

Abandoned crop fields or old field sites are often covered with tree saplings, shrubs, and vines. The goal in preparing these sites is to release those tree saplings that can be used and kill the other vegetation. This may take more than one season to accomplish because most of the perennial plants will have well-established root systems and can readily resprout from roots or cut stumps. The best approach is to use selective herbicides to kill the undesirable plants. Cutting the shrubs and vines and treating their stumps in the spring right after they have leafed out usually results in a successful kill. Basally treating undesired saplings can accomplish the same results. Once the undesired plants have been killed, the site can be cultivated as needed or left for hand planting.

On sites with existing forest buffers, timber stand improvement practices can be used to revitalize the stand. This may mean thinning the stand to allow the remaining trees to grow more rapidly or to allow enough light to reach the forest floor to stimulate regeneration of desired species and develop a ground cover. Thinning of the stand may also provide opportunities for underplanting with desired species. Timber stand improvement requires knowledge of succession and species competition and should be conducted under the guidance of a professional forester. Assistance may be obtained from local state or county foresters. They can develop a plan, recommend consultants to conduct the work, and inspect the work to make sure it was properly done.

In agricultural regions where field drainage tiles are present, any clay or perforated drainage tiles running through the buffer should be replaced with solid polyvinyl chloride (PVC) tile because tree and shrub roots can plug them. If tiles cannot be replaced, plant a 4-m-wide strip of cool-season grasses directly over the tile (2 m [6–7 ft] on each side of the tile) (Figure 8–27). The shallow roots of the grass will not interfere with the tile. Adjacent to both sides of the cool-season grass, plant a 3-m (10-ft) wide strip of warm-season grass. While the roots of these grasses may extend 1 m (3.3 ft) or more into the soil, they do not extend very far laterally. Next to the warm-season grass, plant one or more rows of shrubs and then plant trees. Trees should be planted a minimum of one mature tree height away from the center line of the tile, as their roots can be expected to grow laterally at least one tree height from the base of the tree. With this planting design, tile flow will not become impeded.

Planting density and species selection will depend on landowner objectives and the size of the plant material used. Densities in mature riparian forests are roughly 360 stems ha^{-1} (150 stems acre^{-1}), corresponding to a spacing of approximately 5.3 by 5.3 m (17 by 17 ft) (Palone & Todd, 1997). Depending on the size of the plants used, planting densities can vary from this density to higher densities if thinning can be used once the seedlings have become saplings. Nursery stock can be obtained as balled-and-burlapped stock or container stock but is expensive to buy and expensive to plant. The only place where such plant material may be justified is in an urban buffer planting. Larger bare-root saplings and whips can also be planted. These 1.5-m (5-ft) tall plants are less expensive to purchase and plant than balled-and-burlapped or container stock but still may exceed the budget available for most plantings in an agricultural setting. They usually require auger planting to assure good long-term survival and growth. While these are large plants to begin with, they are also quite unbalanced in terms of their root/shoot ratio and may require a number of years devoted to root system development before they begin rapid height growth.

Bare-root seedlings are the least expensive plants to purchase and the easiest to plant. If good quality seedlings are planted carefully and maintained properly, they can grow to almost the same heights as bare-root saplings in an equal time period. The key to success is to plant

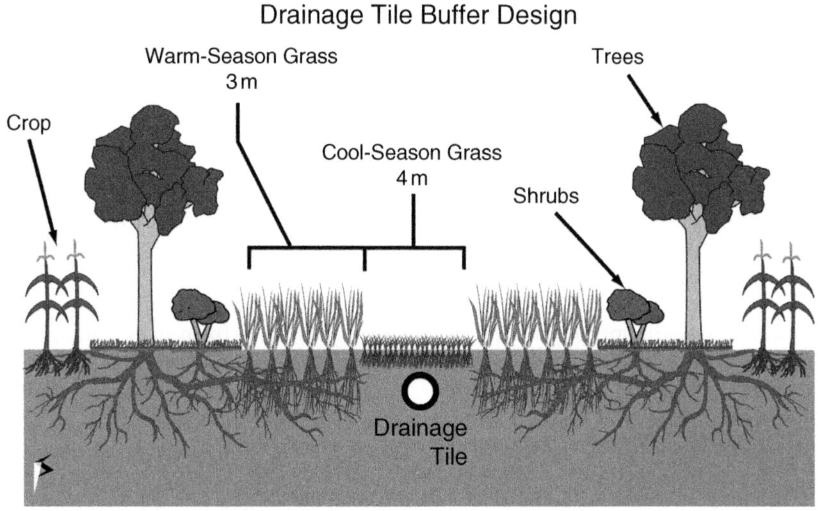

Drainage Tile Buffer Design

Fig. 8–27. Buffer design modification to accommodate subsurface drainage tiles. Trees and shrubs should not be planted closer than one mature plant height from the tile.

seedlings with a minimum of six large lateral roots and a large caliper diameter, keeping them moist throughout the handling and planting process and taking care during planting. Consider ordering 10–15% more seedlings than needed. The additional plants can be planted in a nearby "holding" area and used for replacement plantings in the following year.

Seedlings may be planted in the fall or spring depending on the planting window that exists. If there is sufficient time for lifting of seedlings and field planting with several weeks of soil temperatures warm enough for root growth, then fall planting can be successful. Where these conditions do not exist, it is best to plant in the spring. Before planting, soak seedlings and rooted cuttings in water for 2 to 4 h and unrooted cuttings for 24 h. This will ensure well-hydrated plants that can survive the initial shock of planting. Seedlings should be planted with root collars slightly below the soil surface. Planting holes should be large enough to accommodate fully extended roots. If roots are longer than the radius of the hole, they should be pruned. Holes should be closed with no air pockets around any of the roots. For unrooted cuttings, such as willow cuttings, plant deep enough to leave only one or two buds above the ground. If more than one or two buds break in the spring before root primordia have been initiated and grown, the cutting will desiccate and die. Seed for broadcast planting can be collected and planted in the fall or stratified for spring planting (contact your local service forester for details). If bare-root seedlings are used, rows should be between 3 and 4 m (10–13 ft) apart with 2.5–3 m (6–10 ft) between trees within the rows. Shrubs can be planted at densities of 1–1.5 m (3–5 ft) within the row. If production of biomass for energy is a goal, use closer spacing within rows. For timber production, wider spacing between and within rows can be used. The wider the initial spacing is, the longer weed control practices will have to be applied, especially if a good cover of cool-season grass isn't established. However, wider spacing is recommended to develop and maintain a ground cover for slowing and dispersing concentrated flow and protecting the soil against sheet flow. Actual row widths should also be wide enough to allow the landowner to use available equipment to mow for the first 3–5 yr.

Cool-season grasses for interrow planting or as a filter strip can be sown in the fall prior to planting woody plants in the spring. Typical rates for seeding are between 10 and 15 kg ha^{-1} (9–13 lb acre^{-1}) but vary by species. Warm-season grasses can be drilled in late spring with a grass seed drill or broadcast seeded. Rates of 8–10 kg ha^{-1} (7–9 lb acre^{-1}) sown by mid-June are typical, but actual timing and rates should be modified according to local recommendations. Because of the small size of many of these seeds, it is important that the seedbed be packed prior to drilling or after broadcast sowing. If the grass zone is to be planted with cool-season rather than warm-season grasses, some cost-share programs require that the zone be twice as wide.

Weed control is essential for the survival and rapid growth of all established plants. Non-chemical weed control techniques such as organic mulch, weed control fabrics, or mowing can be used. However, several of these techniques can be expensive so careful use of pre-emergent herbicides will be more typical. Chemical weed control should be continued in strips for the first 2–3 yr or until the woody plants are above the height of most of the competing weeds.

Grass between the tree and shrub rows should be mowed once or twice during the growing season for the first 2–3 yr to help mark the rows and keep weeds from flowering. Late fall mowing also removes rodent habitat and helps minimize plant damage during the winter months. Mow the warm-season grass two to three times per year at a height of about 35 cm to control broadleaf weeds. Native grasses allocate much of their first-year energy to root growth. In agricultural areas where weed control is necessary to produce high-quality crop yields, the buffer should not serve as a source of weed seed. Mowing controls the production of weed seeds as well as insects and diseases that may be detrimental to either crops or trees and shrubs in the buffer planting. Under some conditions, mowed grass can be baled for other uses. If possible, a warm-season grass zone should be burned in early spring for the first 3–5 yr until establishment is complete. Subsequently, a burning rotation of 3–5 yr can be used or the grass biomass may be removed by mowing and baling. This baled biomass generally cannot be used or sold if the buffer is under a Farm Bill contract. Burning frequency and methods is another practice that should be checked with local experts.

In areas where considerable deer populations exist, a 1-m-diameter wire mesh can be installed to protect trees from deer damage and improve initial growth. Tree growth can be drastically reduced by deer browsing. For example, 1.2-m-tall trees were planted at the Greenley paired watershed study in northeastern Missouri in 1997, and the average height was 0.8 m in 1999. All trees recovered and grew after wire mesh cages were installed. Trees with wire cages approached an 8-m average height in 2018, while the average height of trees without protection was only 4 m. However, the impact of the initial damage on young saplings during the 1997–1999 period could not be estimated, thus protection of valuable trees is important especially during early years of growth.

Buffers must be monitored and managed to maintain their maximum nonpoint-source pollution removal abilities. They should be inspected at least once a year and always within a few days after severe storms for evidence of sediment or debris deposits that may concentrate rather than disperse concentrated flow from the adjacent fields. Repairs should be made as soon as possible. If a berm from sediment deposits or tillage develops along the field edge of the grass zone or filter strip, a disk may be needed to pull soil back into the crop field.

After the first 5 yr, the native grass zone or filter strip should be harvested or burned on a biannual basis. Periodic or regular removal of biomass promotes dense upper plant and root growth, which is needed to provide the sink for nutrient removal from subsurface water. If the warm-season grass zone cannot be harvested or burned, some of the grass may be removed by short periods of controlled grazing, using electric fences to keep livestock away from the stream. Grazing should not be allowed when the soil is wet, as compaction rapidly reduces the effectiveness of the buffer. Furthermore, wildlife habitat and nesting issues should be considered in all management decisions.

The use of fast-growing tree species (e.g., willow, cottonwood, poplar hybrids (*Populus* sp.), silver maple, and green ash) ensures rapid growth and effective use of nutrients and other excess chemicals in the soil. To remove nutrients and chemicals stored in the woody biomass sink, it is necessary to harvest trees based on their planting density. Trees should be harvested when the stand begins to naturally thin itself. As long as the number of living stems remains constant and growing, the nutrient sink is actively increasing. However, if some of the smaller trees in the canopy begin to die, the biomass of the stand has reached a relatively constant level that will now be continually redistributed on fewer and fewer trees. This self-thinning phase of stand development signals the time at which harvesting for nutrient sink maintenance should begin. If an active ground cover is being maintained, then harvesting may be required even earlier, with the timing being determined by the condition of the ground cover.

If harvested in winter, most of the fast-growing species listed above will regenerate from stump sprouts, thereby maintaining root system integrity and continued protection of stream banks. Trees can be harvested in whole rows, blocks within rows, or small groups of trees between and within rows. Harvesting also promotes more diverse habitat by introducing various age classes of woody plants with different heights and crown forms. High-value species such as walnut or oak can be managed for sawlog production. Tree selection and thinning promote faster growth and higher quality material than when trees are allowed to grow without management. Local service foresters can provide the necessary expertise to determine the proper management for these trees.

If problems with beaver develop, such as the loss of large numbers of trees or unwanted beaver dams, a controlled trapping program may be needed. Increased diversity attracts many kinds of wildlife to an area, including some that may be perceived as a nuisance. Beaver can be trapped during regular trapping seasons or with special permits. Beaver, however, provide benefits to a stream system by developing pools for trapping organic matter and sediment behind their dams. The dams act as grade-control structures, reducing stream gradients, providing riffle-like aeration, and reducing bank erosion by raising the water level in the channel and thereby reducing bank height. The pools also provide additional storage during flood events.

The effectiveness of existing riparian forest buffers has already been discussed, but a case study of the effectiveness of established buffers on previously cultivated and grazed land will show how quickly such a plant community can reestablish some of the important hydrologic functions of a riparian zone. A riparian management practice was established on a farm along Bear Creek, near Roland, IA, in 1990 (Isenhart et al., 1997). At that time, a portion of the riparian zone was being grazed while another was cropped down to the stream edge. Multi-species riparian forest buffers were established in both kinds of land-use areas. Figure 8–28 shows before and after pictures demonstrating the kind of changes that took place after establishment of the MRFB. The first picture was taken in early 1990, prior to planting. The second shows the same site at the beginning of the fifth growing season.

Research conducted on this and adjacent farms showed that the sediment in surface runoff was reduced by 70–80% in the first 3 m (10 ft) of the native grass–forb strip, with another 10%

Fig. 8–28. A before-and-after sequence showing the results of riparian buffer establishment on a farm in central Iowa. The top picture was taken in March of 1990 and shows a field cultivated to the stream edge on the right and a grazed riparian zone on the left of the channel. A multispecies buffer was established in April 1990. The lower picture shows the same site in June 1994 at the beginning of the fourth growing season.

reduction in the next 1 m (3.3 ft) of native grass–forb filter strip. Infiltration rates in previously cropped soils have gone from 1.5–4 cm h^{-1} (0.6–1.6 inches h^{-1}) to as high as 12–15 cm h^{-1} (4.7–6 inches h^{-1}) after 5 yr of buffer community growth. Soil organic matter under the buffer has increased by an average of 50% in the top 35 cm (14 inches) during a 7-yr period. Fine-root biomass in the top 35 cm (14 inches) of soil under switchgrass was five times greater than under corn in August when maximum corn fine-root growth occurs. Fine-root biomass under trees and switchgrass varies between 7.5 and 14.2 Mg ha^{-1} (6.7–12.7 lb $acre^{-1}$), while corn fine-root biomass peaks at 3 Mg ha^{-1} (2.7 lb $acre^{-1}$). The residence time for groundwater moving below the buffer varies from 2 to 3 mo, providing time for nonpoint-source pollutants to be in contact with the living filter. The MRFB reduced NO_3 and atrazine agrichemical pollutants moving in the soil solution of the rooting zone or in the shallow groundwater by >90%, with resulting concentrations well below the maximum contaminant levels allowed by the USEPA. Because of the higher soil organic matter, denitrification rates also increased under the buffer. In summary, the MRFB concept was capable of significantly reducing nonpoint-source pollutants after only a few years of growth. Continued improvement of the MRFB was expected based on comparisons with established natural forest buffers in the area. While MRFBs are very effective at mitigating surface-derived pollutants, their impact on soil solution pollutants is only effective for those pollutants that are derived from the immediate upslope source area but not those carried in solution in tile drains that bring soil water from greater distances.

The sites where the monitoring was conducted are underlain by a dense glacial till, which forces water to move laterally through the unsaturated zone and shallow groundwater. This provides a tight connection between the upland crop fields, the buffer community, and the stream. Under these geophysical conditions, a 20-m (65-ft) wide MRFB on each side of a stream is sufficient to effectively reduce chemical and sediment loading of the stream. In other landscape settings, however, a wider buffer may be necessary to obtain the same kind of reductions. It is therefore imperative that a careful site evaluation be conducted prior to designing the buffer. As the slope, depth to bedrock or confining layer, intensity of land use, or total area of the land producing nonpoint-source pollutants increases, or as soil permeability decreases, a wider MRFB is required.

Riparian buffers cannot and should not be expected to intercept and process all nonpoint-source pollutants coming from the adjacent crop fields but rather should be part of a conservation plan that includes upland buffers as well. Numerous studies support the idea that establishment of upland buffers improves surface and subsurface water quality and soil quality. A long-term study is being conducted in northeastern Missouri on claypan soils in three adjacent watersheds. The watersheds are primarily managed on a corn–soybean rotation, and a paired watershed approach has been used to examine contour grass buffer and agroforestry (trees and grass) effects on nonpoint-source pollutant reduction. In addition, changes in soil moisture, soil physical properties, and microbial diversity have been examined (Udawatta et al., 2002).

Results from this work show that the establishment of upland agroforestry and grass buffers reduce total N, NO_3, and total P in runoff (Udawatta et al., 2002). Soil bulk density, hydraulic conductivity, porosity, and soil water storage in the buffers also have improved significantly (Seobi et al., 2005; Udawatta et al., 2006). Microbial diversity, enzyme activities, soil aggregate stability, and soil C and N are higher in the buffer areas than in the crop areas (Udawatta, Kremer, Garrett, & Anderson, 2007). It could be speculated that improved soil physical properties, microbial diversity, and enzyme activities may imply greater bioremediation potential of upland buffers helping to reduce nonpoint-source pollutants from these agricultural watersheds. In another study, Lin, Lerch, Kremer, Garrett, Udawatta, & George (2005) found that higher degradation activity of metolachlor in vegetative buffers than crop areas coincided with higher dehydrogenase and fluorescein diacetate activities. Adoption of these in-field buffer practices not only improves soil and water quality but watersheds become more diverse in plant communities and aesthetically pleasing. Nut-bearing trees can be incorporated into the field buffers to provide extra income. Buffers also can be planted with ornamental and flowering woody species instead of grass species for additional income.

Thus, both upland and riparian buffers are needed to effectively reduce nonpoint-source pollutants of surface waters. While riparian buffers provide the last protection against nonpoint-source pollutants entering surface water, in certain instances they may not be able to stabilize the stream bank, which may be a major source of sediment to the stream. As a result, stream bank bioengineering practices may be needed as an integral part of a well-buffered agricultural landscape.

Stream Bank Bioengineering as Part of the Riparian Buffer Practice

In many agricultural landscapes, riparian buffers, by themselves, cannot stabilize stream banks rapidly enough to contain the accelerated movement of modified channels. These are usually in areas with deeply incised channels, where banks are too steep to seed grasses and too high for woody root systems to penetrate to the bottom of the bank. In such areas, undercutting continues to undermine any vegetation along the banks, and banks can easily collapse. In such cases, stream bank bioengineering must become an integral part of riparian zone restoration, or the buffer itself must be made wide enough to accommodate the expected movement of the stream.

Streams naturally move across a floodplain. This movement is caused by turbulent water eroding the outside of bends and depositing on point bars on the inside of bends. This pattern of movement results in a sinuous channel of meanders. In many agricultural landscapes, eroding stream banks may be responsible for a major portion of the sediment load in a stream (Sekely, Mulla, & Bauer, 2002; Schilling & Wolter, 2000). Because bank erosion is a natural process, not all eroding banks should be protected unless the goal is to stop meandering in natural streams or in channelized reaches where meanders are becoming reestablished. Some stream bank erosion is tolerable. Simonson, Lyons, and Kanehl (1994) suggested that healthy streams in Wisconsin can have up to 20% of their total bank length eroding. Stream banks along row-cropped fields and pastures with direct livestock access to the channel had 30–40% of their bank lengths severely eroding in three regions of Iowa (Zaimes et al., 2004, 2006). Because of the vast lengths of stream banks that are eroded, it is not feasible to consider stabilizing most of them except with the indirect methods of planting riparian forest buffers. The woody plant roots can provide reinforcement of the banks if the channels are not incised to depths below the root zone. If incisions extend below the root zone, undercutting can result in large trees falling into the stream often creating large debris dams that can result in more channel movement. This is especially true on low-order streams. If stream bank erosion threatens structures or other high-value property, two general approaches can be taken to stabilize the banks.

One is the "hard engineering" approach of armoring banks with various kinds of rip-rap that may consist of rock, concrete structures, or other inert materials. The second approach, which is often cheaper and more environmentally friendly, is the "soft engineering" or bioengineering approach, which uses plant materials. The root systems of living plants hold the soil together and strengthen the stream bank while the stems increase the roughness of the bank surface, slowing water flow and causing sediment deposition along the bank. The stems also buffer the bank against large floating debris and provide improved in-stream and near-stream aquatic and wildlife habitat. In severely eroded situations, this plant material is often used in combination with bank reshaping and hard engineering practices along the toe of the bank. Stream bank bioengineering systems are often installed using unrooted plant material that is established during the dormant season.

The goals of stream bank bioengineering systems are:

- to change the steep angle on actively eroding banks to a gradual slope on which plants may naturally seed and become established or on which cuttings or seedlings of woody plants or native grasses may be planted
- to provide as much frictional material as possible to slow water movement and reduce erosion
- to keep the bottom of the bank (toe of the bank) stable so no undercutting of the upper bank can occur
- to carefully tie the bioengineering system into the bank on both the upstream and downstream ends of the restoration practice
- to reduce stream sediment loads and improve water quality
- to improve both the in-stream and stream bank aquatic and wildlife habitat

Stream bank bioengineering does not work under all stream conditions. If a stream is still actively downcutting, stream bank bioengineering structures may be ineffective until the downcutting is stopped. Under these conditions, channel grade control measures such as boulder weirs or Newbury weirs (Newbury & Gaboury, 1993) can be installed to slow the downcutting. Bioengineering becomes less effective on the larger order streams in the zones of production and transport (higher than fifth order). Stream bank bioengineering requires a professional understanding of channel dynamics because incorrect installation of bioengineering practices can actually result in more problems than existed before.

There are numerous kinds of bioengineering designs that can be used for stream bank stabilization. Some are more time consuming and expensive to install than others. A detailed discussion of a wide variety of techniques can be found in USDA–NRCS (1996) and in Hoag and Fripp (2002).

Several techniques that have broad application on headwater streams in agricultural landscapes and that can be installed with a minimum of people at a reasonable cost are described here. Examples of these structures include live stakes and posts, joint plantings, live fascines, and brush mattresses. These can be used in combination with each other and in combination with nonliving structures such as dead tree revetments, fiber mats, or rock rip-rap. The practices that are used depend on the severity of bank instability and the objective of the stabilization activity. The living plant material provides a porous, frictional surface that slows water, causing sediment deposition. It also provides habitat for the aquatic community and develops roots that stabilize the bank and are self-repairing after beaver, flood, or ice damage has occurred.

Living Structures

Live stakes and posts are used extensively in bioengineering designs because of their ability to develop roots and shoots from the dormant woody material (Figures 8–29, 8-30, 8-31, 8-32 and 8–33). They are also one of the quickest designs to install and one of the least expensive. Almost any willow species can be used. Most are tolerant of flooding and sediment deposition, but most have a low tolerance to shade. Willow (*Salix* sp.), boxelder, eastern cottonwood, red-stemmed (*Cornus stolonifera* Michx.) and silky dogwood, nannyberry (*Viburnum lentago* L.), arrowwood (*Viburnum dentatum* L.), and American elderberry are examples of tree and shrub species that can be planted as unrooted cuttings in bioengineering applications. Willow, boxelder, and American elderberry have a high tolerance to flooding and sediment deposition, while the other species have medium to low tolerance to both. All of the shrubs are moderately tolerant of shade. These tolerances suggest that willow, boxelder, and American elderberry can be planted closer to the bottom of the bank, while the others should be planted from the middle to the top of the bank depending on the dynamics of the channel. More detailed information on these and other species can be found in USDA–NRCS (1996) and from a local service forester or NRCS District Conservationist and in Hoag and Fripp (2002).

Stakes vary in size from 1.3 to 3.8 cm in diameter and from 0.5 to 1.2 m in length. Posts can be 3.8–15 cm (1.5–6 inches) in diameter and 1.5–3 m (5–10 ft) in height. In the simplest installations, the larger posts are placed along the toe of the slope, in the base-flow channel, and the smaller stakes are placed in the stream bank above the posts. If the bank is less than 2–2.5 m (6.5–8.2 ft) high and the slope has an angle of 80° or less, posts and stakes can be planted directly into the bank. Under these conditions, three rows of posts should be installed, with the first row in the stream channel and the second row at the toe of the slope. Spacing should be 0.6–1.2 m (2–4 ft) between posts within and between the rows using a triangular planting design with posts offset by rows. Posts should be planted to a depth of one-half to two-thirds their length. In these plantings, the bedrock should be more than 1.2 m below the bed surface, and posts should be planted as deep as possible. It is imperative that the posts be planted right-side up and that the butt end be tapered for easier installation. Mechanical driving rams or stingers may be used to install these long posts.

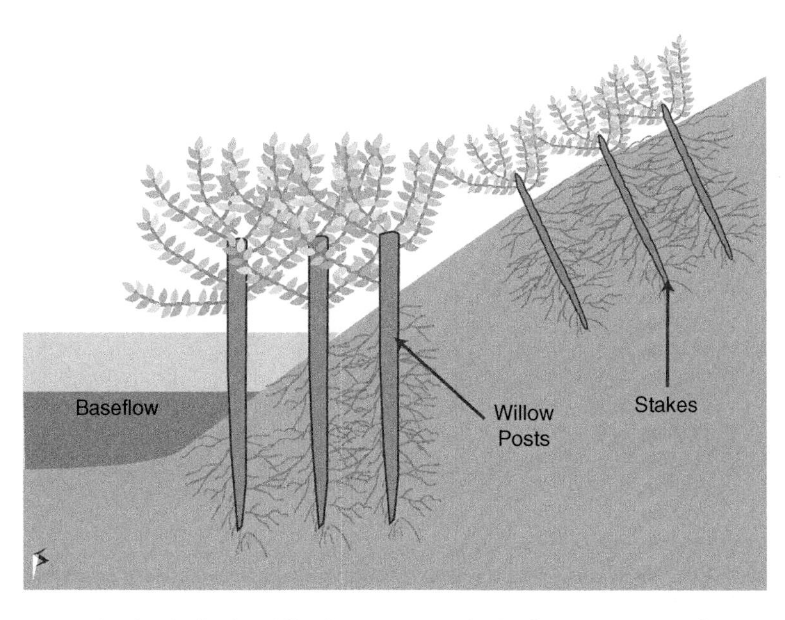

Fig. 8–29. Willow posts and stakes for bank stabilization. Posts range in size from 4 to 15 cm in diameter and 1.5 to 4 m in length. Stakes range in size from 1.5 to 4 cm in diameter and 0.5 to 1.2 m in length. Both are harvested and planted when they are dormant (modified from USDA–NRCS, 1996).

Fig. 8–30. A 5-m-wide stream bank "blowout" that occurred within the first year after establishment of a riparian buffer. Note the two rows of trees in the distance that had been continuous throughout the blowout area when first planted.

Fig. 8–31. Manual installation of small willow posts and stakes. Notice that several rows of posts are planted at the toe of the bank in the streambed to control the toe of the bank from undercutting.

Fig. 8–32. Growth of the willow posts and stakes after 4 yr. The bank is now protected against high-velocity water by multiple stems.

Fig. 8–33. A willow post porous dam placed in a gully. Willow posts are pounded into the bottom of the gully as tightly together as possible. Tying them together with twine reinforces their strength. A cut eastern redcedar or other conifer can be tied to the willows on the uphill side of the gully to improve initial sediment trapping.

Above the three rows of posts, two or three more rows of stakes should be installed at similar spacing. Stakes should be planted so that at least three-quarters of their length is below ground. Only two or three buds should be exposed above ground. Stakes should be a minimum of 60 cm (24 inches) long and should be installed perpendicular to the slope of the bank. Spacing should be between 0.6 and 1.2 m (2–4 ft) using a triangular spacing. Stakes can be pounded into the bank with rubber mallets or installed with a water jet system (Drake & Langel, 1998). Seedlings of a variety of shrub species can be planted along the top third of most stream banks in place of unrooted stakes. As can be seen in Figure 8–30, 8-31 and 8–32, this simple method can effectively cover a bank with a dense frictional surface that slows water and captures sediment. If individual stems are buried or broken, more will sprout from either the roots or stems depending on the species. This method is not effective on banks that are vertical and more than 1.8 m high. In those situations, the bank needs to be reshaped to a 2:1 or 3:1 slope or other practices must be used.

Posts and stakes can be used as grade control structures in surface gullies that empty into streams. In this case, two rows of posts are planted across the gully as close together as possible with the posts above ground as tall as the depth of the gully. If the posts are very tall, they should be tied together with untreated twine. Dead conifers such as cedar (*Juniperus* sp.) trees can be tied to the post structure on the upstream side of these live dams to help increase sediment deposition (Figure 8–33). Gullies fill quickly with sediment that serves as a seedbed for other plants and spreads concentrated flow into sheet flow.

A *joint planting* is a specialized use of live stakes or posts when rip-rap is used in the system (Figure 8–34). Rip-rap can be placed to the bankfull height of the channel to stabilize the toe of the bank in systems where undercutting may be a significant problem. Joint planting can be done after the rip-rap has been put in place, but it may be very difficult to work the stakes or posts through the cracks between the rocks. Another way to install a joint planting is to place the posts or larger stakes in the streambed and along the toe of the slope before adding the rip-rap. Careful addition of rip-rap will scar some posts and stakes but will not break most of them. The posts and stakes in the rip-rap produce a root mat in the soil below the rock that reduces the erosion of fine material from below the rip rap, helping to further stabilize the bank. The shoots above the rip-rap help dissipate more energy than the rock alone and provide additional streamside habitat.

Live fascines can be used in combination with other structures to help make the stream bank more stable against erosion (Figure 8–35 and 8–36).

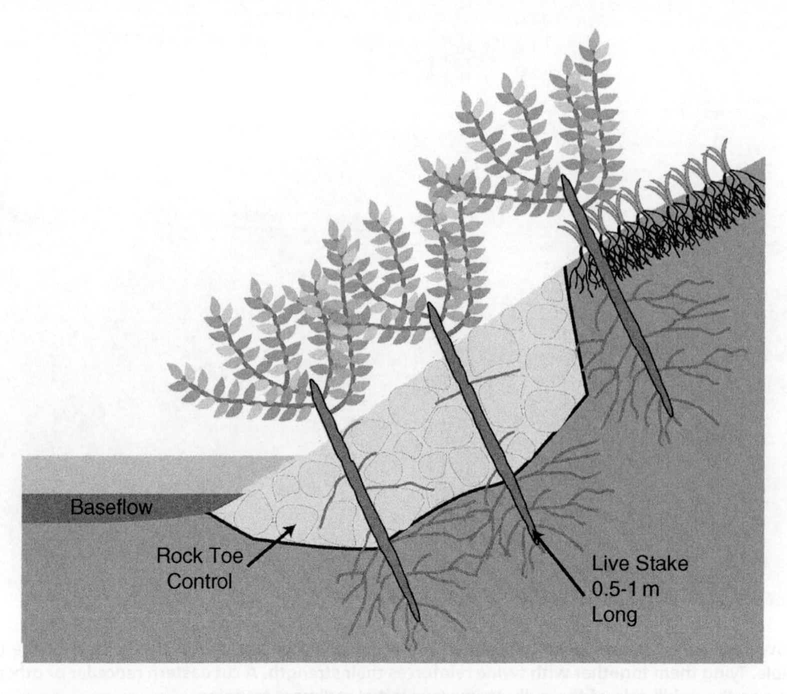

Fig. 8–34. A joint planting where willow posts are used in conjunction with rock rip-rap. Posts are usually installed before the rock is put in place. Damage to the willows is usually not enough to kill them. Stakes are planted within cool- or warm-season grasses above the joint planting (modified from USDA–NRCS, 1996).

Live fascines consist of long bundles of dormant, live branch cuttings that are tied together with twine to resemble long sausages. This material may consist of the small-diameter branches and tops of the posts and stakes that are installed on the lower portion of the banks. Bundles of 1.5–6 m (5–20 ft) in length and 15–20 cm (6–8 inches) in diameter, with overlapping branches, buds all

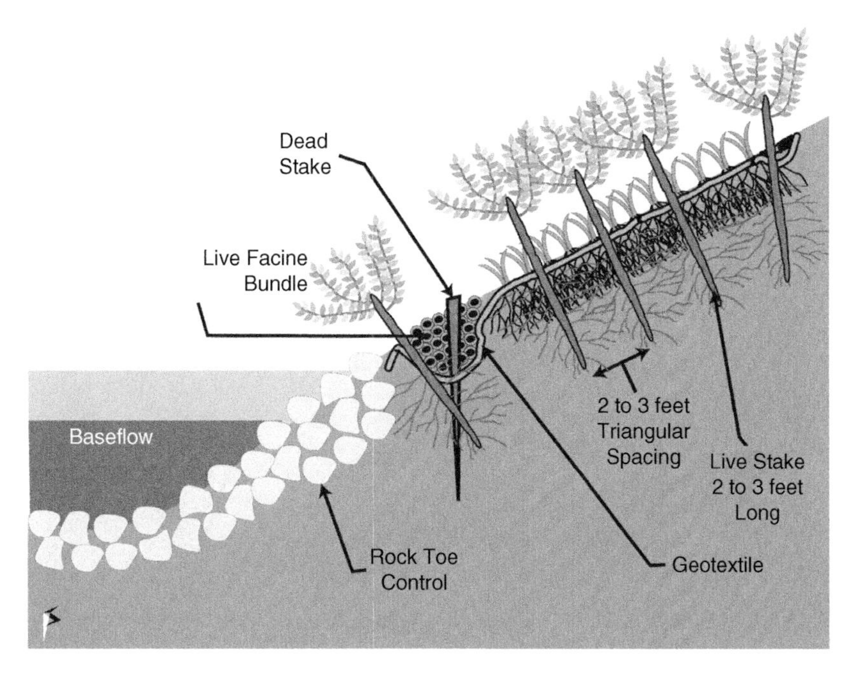

Fig. 8–35. A live fascine used to anchor the bottom of geotextile. Grass is planted under the geotextile through which stakes are planted. Both live and dead stakes are used to anchor the live fascine (after USDA–NRCS, 1996).

Fig. 8–36. A brush mattress installation. The base of the stems are buried in a trench, and the branches are held in contact with the soil of the bank by stakes and wire. This contact allows the willow to develop roots and shoots across the whole slope. A live fascine bundle can be placed in the trench to help anchor the branches.

facing in the same direction (butt ends all pointing the same direction), can be produced just before use. Branches with 1.3–5-cm (0.6–2-inch) diameters are tied together with untreated baling twine. Bundles are placed in shallow trenches and covered with soil, leaving the top branches barely uncovered so exposed buds will develop branches while buried adventitious buds will

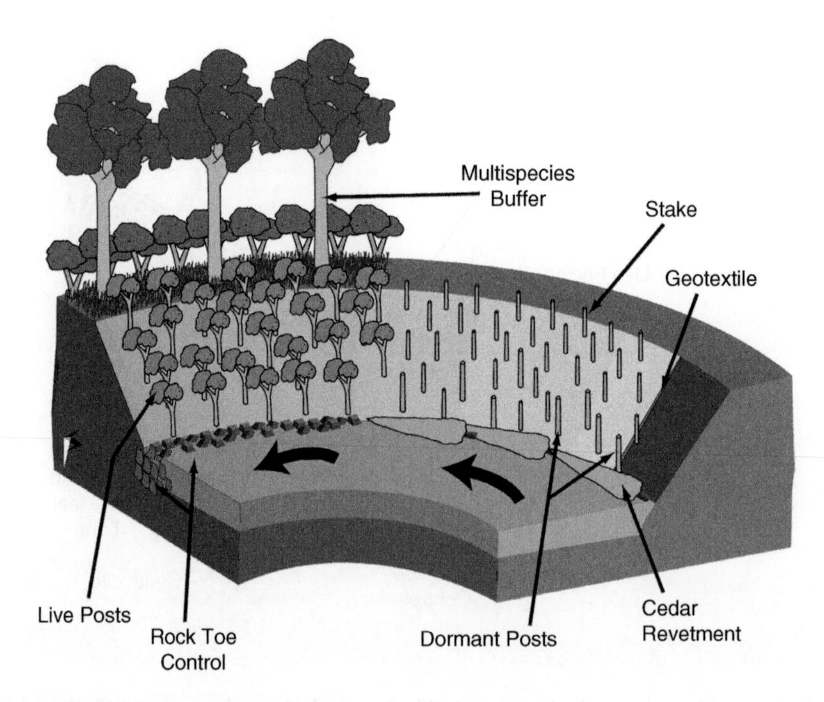

Fig. 8–37. A combination of live and dead materials used in the stream bank bioengineering practice. Fresh-cut eastern redcedar used as a tree revetment and rock rip-rap can provide bank toe control, and geotextile can reduce bank erosion until grasses, shrubs, and stakes become established.

Fig. 8–38. Backhoe sloping a 4.5-m vertical stream bank to a 2:1 slope as part of the soil bioengineering practice.

Fig. 8–39. Eastern redcedar being cabled to posts to protect the toe of the bank from undercutting. The base of the redcedars are pointing upstream and each tree overlaps the downstream tree.

Fig. 8–40. Coconut fiber geotextile being installed up and down the slope, with the upstream sheet overlapping the downstream sheet. The geotextile can also be installed parallel to the bank. Grass seed was spread on the bank before the geotextile was installed.

Fig. 8–41. Willow planting after installation of the redcedar revetment, the geotextile, and the fascines.

develop roots. Live and dead stakes are used to help hold the fascines in place after they have been covered with soil but before they have rooted. Dead stakes should be at least 0.8 m (2.5 ft) long and pounded through the center of the fascine. Live stakes should be placed on the downhill side of the fascines.

Live fascines can be placed along the contours of gently sloping banks to slow bank collapse and trap any moving soil behind the branch dams, thus producing a slope with small terraces. If they are used as the primary bank treatment, they should be placed at intervals of 1–2 m (3.2–6.5 ft) between bundles, with the narrower spacing used for banks that are steeper than 3:1 slope. Live fascines can also be effectively used with geotextile erosion fabrics to anchor the borders of the fabric to the slope. They also can be used to secure the base ends of brush mattresses to the bank. Live fascines are very effective for producing a long, strongly rooted line of branches on the bank. They can be constructed of the branch material that is left over from the preparation of the live posts. However, their construction is time consuming so they are best used in conjunction with other structures.

Brush mattresses use live branch material similar in size to that used in the live fascine (Figure 8–35). A dense mat of branches is placed vertically across the stream bank to provide immediate protection against erosion. Shallow trenches 30–45 cm (12–18 inches) across and 15–20 cm (6–8 inches) deep are dug along the contour at the base of the slope, and basal ends of the branches are placed in them. Live fascines can be used with a brush mattress by placing them in the trench and sticking the brush mattress branches into them. A layer one to two branches thick is placed over the bank so that it covers an area of up to 3 m (10 ft) upslope of the trench. Both live and dead stakes are pounded into the bank at a spacing of 0.5–1 m^2 (1.6–3.2 ft^2) over the slope, and wire is stretched diagonally between the dead stakes. Approximately 15 cm (6 inches) of the dead stakes is left exposed before wire is crisscrossed over the bank and attached about 5 cm (2 inches) from the tops of the stakes. Once the wire has been installed, the stakes are pounded farther into the ground to ensure good contact between the branches and the soil. Live stakes can be used for wiring but will ultimately be girdled by the wire. Their initial rooting will help hold the system anchored to the slope before the branches of the mattress become rooted. Finally, a light covering of soil is placed over the branches, leaving the tops of some of the branches exposed. This step is imperative if branches are to produce roots and stems that will protect the slope against further erosion.

Brush mattress covered slopes are vulnerable to erosion until the branches have rooted and produced new branches. However, once this happens, the slope is well covered with a dense mat of living plant material. The brush mattress restores riparian habitat and effectively traps sediment that can provide the conditions for further colonization of native vegetation. They are best used on 2:1 or gentler slopes. As with live fascines, they are time consuming to prepare and install.

Collecting and Handling Live Material

The structures discussed above are constructed of live plant material. Cuttings can be collected from vigorous trees near the construction site or from established cutting orchards. Dormant material should be collected during the winter or very early spring. If the material is to be used for posts or stakes, all side branches should be removed at harvest time. Cuttings should be stored in a cool (0°C) (32°F), moist, and dark location until planting. If the butt ends have dried substantially by planting time, a portion should be cut off to expose fresh wood. Stakes should be cut to the sizes needed as close to the time of planting as possible. To help seal the top ends and to make it easier to plant the stakes right side up, the top 2–5 cm (0.8–2 inches) should be dipped in a 50:50 mix of light-colored latex paint and water. Soak the bottom third or more of the stakes and posts at least 24 h and as much as 5–6 d before planting to help initiate root growth. Rooted seedlings of shrubs can be successfully planted on the upper two-thirds of a bank if flooding is not expected to be severe before the root systems begin to grow or if the seedlings are used with geotextile mats (see discussion below).

Landowners with serious erosion problems may find a cutting orchard the most sensible source for a ready supply of cuttings. Cutting orchards can be established from 45-cm (18-inch) stakes with a spacing of 4 m (12 ft) between rows and 1.5 m (5 ft) within the rows. If the orchard is planted on the floodplain, the first harvest can often be made after the second or third growing season. After the first harvest, most stumps will produce at least four viable stems for subsequent harvests, which may occur as frequently as every 2–3 yr. Many willow cuttings will produce up to 4.5 m (15 inches) of growth by the end of the third growing season, yielding a 2–14-cm (0.8–5.5-inch) diameter post, 1–2 m (3.3–6.6 ft) long, with several additional stakes in the top. In subsequent cuttings, four times that number will be produced. Shrubs of other species can also be planted and cuttings collected for use in bioengineering applications or as other agroforestry products such as ornamental branches and fruits for decoration.

Whole trees of eastern redcedar (*Juniperus virginiana* L.) or other small conifers or hardwoods used for toe slope revetments need not be alive. These trees commonly are found in ditches, unused areas around farmsteads, and small forested lots. Trees should be selected with branches that are at least 1.5 cm in diameter and distributed as close to the base of the tree as possible.

Nonliving Structures

Living plant material requires time to get established. On banks where erosion is severe enough that living material may not get established fast enough, nonliving structures may be needed in conjunction with bank reshaping and living material to stabilize the bank (Figures 8–37 and 8–38).

Tree revetments involve staking dead trees along eroded stream banks to protect them until other living woody vegetation can regrow (Figures 8–39, 8-40, 8-41, 8-42, and 8–43). Eastern redcedar trees are ideal for this purpose because they have dense branches and, if cut alive, will hold their leaves for extended periods after they have been installed. The leaves and branches catch sediment that provides a medium for other plant species to grow in. The branches also provide shade and habitat for the aquatic community. Bundles of dead, densely branched hardwoods such as silver maple, oak, and ash can also be used.

Tree revetments can be used on vertical banks of up to 4.5 m (15 ft) high along streams that do not drain more than 80,000 ha (198,000 acres) above the revetment. They should not be used on channel reaches that are actively downcutting or in straight channelized reaches. Selected trees should have dense branches to the base and be large enough to cover two-thirds the height of the bank. Any trunk (basal stem) that does not have branches should be cut off. Open-grown trees usually have the best shape for use in this practice. Two rows of trees can be used—more than two rows becomes very difficult to install. The trees are anchored to the banks using earth anchors, fence posts, or living willow posts. In Missouri, recommendations are for anchors that withstand a pullout force of at least 500 kg (1100 lb). A minimum of 0.5-cm (2-inch) diameter cable and cable clamps should be used. Laconia arrowhead anchors (10–15 cm or 4–6 inches), Duckbill 88 anchors, 14-cm (5.5-inch) steel T-posts, or 10-cm (4-inch) helical anchors can be used. Bedrock or a stable bottom should be within 1.2 m (4 ft) of the toe of the slope, and the soil should be dense enough to hold the anchors.

As with all stream bank bioengineering, it is imperative that this practice starts and ends at stable bank positions. The first trees should be

Fig. 8–42. An installation similar to the one in Figures 8–35 through 8–39 but with rock installed for toe control.

Fig. 8–43. Bioengineering system shown in Figure 8–39 2 mo after installation.

placed at a stable stream bank location at the downstream end of the project site. The butt end of the tree should point upstream and be cabled to anchors, leaving enough cable to attach to the next tree. The next tree should be placed upstream, overlapping the first by one-third and cabled to it. No large gaps should be left between the overlapping trees, as these are locations where scour could begin to wash out the whole structure. A second row may be placed on top of the first row so that at least two-thirds of the bank height is covered by the tree revetment. Care should be taken that the presence of the trees along the slope does not reduce the width of the channel by more than 20%. To hasten the reestablishment of living trees, willow posts or stakes can be planted into the revetment either at the time of establishment or after the revetment has captured a significant amount of sediment. Tree revetments have a life expectancy of about 5 yr, during which time other woody vegetation can become established. Tree revetments can be used alone or with other stream bank bioengineering techniques. For example, tree revetments can be used to stabilize the toe of the bank while other bioengineering methods protect the upper portion of the bank.

Rock rip-rap may be used for toe slope stabilization where severe undercutting is likely (Figure 8–37, 8-42, and 8–43). The rock should be placed at least 30–60 cm (12–24 inches) above the base flow water line or to the bank-full height if that can be easily identified. Sizing of the rock is important and related to stream size and the volume and velocity of water. Gabion rock with average diameters of 15–22 cm (6–8.7 inches) may be mixed with larger rock to provide a good mix of sizes. The technical staff of the NRCS can provide assistance with selecting the proper size rock. The rock should be tied into the bank at stable upstream and downstream locations for the treated area. Proper installation may require that a trench be prepared to get the rock in contact with stable channel bottom material. Living posts or stakes (see joint plantings above) may be planted among the rock to provide a continuous tie with the upslope vegetation and to stabilize the fine sediments below the rock, which may be eroded by turbulent flow within the rip-rap. It is usually best to install the posts and then fill in with rock. Careful placement of rock does minimal damage to the willow posts.

Geotextile fabrics constructed of jute, coconut, or other fibers can be used in conjunction with stakes and live fascines as well as with other structures (Figure 8–37, 8-40, 8-41, 8-42 and 8–43). These fabrics are meant to provide temporary bank protection while the plant material is getting established. Fabrics come in various widths and lengths. They can be installed with the long axis vertical or horizontal to the bank. If they are to be installed horizontal to the channel, the width of the fabric should cover the total height of the stream bank. If it does not, turbulent flow from the stream may get under the fabric at the overlapping seams and tear it from the bank.

Where the width of the fabric is narrow, it should be placed up and down the slope, with the upstream sheet overlapping the downstream sheet. This is best accomplished by beginning installation at the downstream end of the project and working upstream, overlapping the top sheet over the previous one. Before the fabric is laid down, a shallow trench 45–60 cm (18–24 inches) wide and 15–20 cm (6–8 inches) deep should be dug just above the base flow height of the water. The bottom of the fabric should be placed in this trench, with a live fascine to stabilize it.

Another shallow trench should be prepared at the top of the slope and the top end of the fiber mat buried in it. In addition, a cool-season or warm-season grass mixture should be spread at a rate of 5–7.5 kg ha^{-1} (4.5–7 lb $acre^{-1}$) under the fabric. The grass germinates quickly, grows through the fabric, and secures it to the soil before any live stakes or posts, placed through the mat or fascines anchoring the base, begin to grow. Willow posts are installed through the fascines, and soil is added to fill the trench and cover all but the top of the fascines. Stakes of willow or other woody species can be placed at regular intervals throughout the fabric. Where rock rip-rap is used for a toe control, it is possible to install the fabric first and then cover the bottom with rip-rap.

Project Design

The goals of stream bank stabilization are to change the steep angle of actively eroding banks to a more gradual slope, move the rapid current out away from the stream bank by providing a dense frictional surface of woody stems, trap enough sediment to provide sites for native plants to seed into, and improve terrestrial and aquatic wildlife habitat.

The project design depends on the size of the stream, the stability of the channel bottom and stream bank, and the severity of the stream bank erosion. In small streams in which the top of the bank is 1.2–1.8 m (4–6 ft) above the base flow water surface, stabilization can often be accomplished simply by installing tree revetments or live posts, stakes, and cuttings along the toe and in the vertical wall of the stream bank.

In a larger stream, additional material must be used to slow down the current, cut down erosion, and allow the cuttings to root. Figures 8–38 through 8–43 show the sequence of several approaches. In this example, the channel was in Stage 3 of the channel evolution model and the banks were higher than their critical stable height. As a result, a backhoe was used to slope them to a 2:1 (horizontal/vertical) angle. The toes of the slopes were stabilized by wiring eastern redcedar to posts. On sites where toe slope scour may be very severe, rock rip-rap should be used in place of the tree revetment. A shallow trench should be dug above the cedar revetment and a geotextile fiber installed. Willow posts are installed in the revetment through the fascine and in the lower part of the geotextile in three rows at a spacing of 1 m between trees and 1–1.2 m (3.3–4 ft) between rows. Stakes are planted throughout the fabric-covered bank. Finally, a buffer is planted along the top of the slope. The buffer is very important to the success of the stream bank stabilization work, as it reduces any overland flow that may compromise the stream bank stabilization structure.

The best time to install a stream bank stabilization planting is early spring, soon after ice breakup and before buds begin to swell on the native trees. Once established, the planting should be monitored monthly during the first 2 yr for pest damage and after each major storm event for water damage. Damaged areas should be repaired as soon as possible.

The design of stream bank bioengineering projects should have professional input. The professional help is critical in making an assessment of the flow dynamics and condition of the channel and stream bank. A plan should include a design and materials list and should consider such items as who will install the system (landowner, natural resource consultant, or citizen action groups) and whether a cost-share is available. Locating native willow for stakes and posts and ordering live plants from nurseries should be planned for early in the fall prior to the spring installation.

Constructed Wetlands to Mitigate Subsurface Drainage Tile Flow

In areas of artificial tile drainage, small wetlands can be constructed at the end of field tiles to interrupt and process nonpoint-source pollutants before they enter water bodies. These 0.5–1-m (1.5–3.3-ft) deep depressions are constructed at a ratio of 1:100 (1 ha [2.5 acres]) of wetland for 100 ha [247 acres] of drainage) to provide significant NO_3 reduction. A berm should be constructed along the stream (Figure 8–44). It can be stabilized on the stream side with willow cuttings and seeded with a mixture of prairie grasses and forbs. If a coarse-textured soil is encountered, the bottom of the wetland can be sealed with clay and topped with the original soil. A gated control structure for controlling the water level should be installed at the outflow into the stream. In designing the wetland, it is important to remember that most of the chemical transformation and retention occurs at or near substrates (sediments or plant litter). Wetlands containing large amounts of vegetation and decaying plant litter will thus have a much greater capacity for pollutant removal. Any management technique that accelerates vegetation establishment (active regeneration) and litter buildup (addition of organic substrate) will improve chemical retention or reduction.

Fig. 8–44. A 1-yr-old constructed wetland showing the outflow structure.

Because of the large number of small drainage tiles in some agricultural landscapes, it is unlikely that constructed wetlands will be economically, socially, or physically possible on a significant proportion of them. Many tiles are at depths of 1.5 m (5 ft) or greater, and the process of bringing them to the surface would back water up for significant distances, rewetting areas designed to be drained. Preliminary work suggests that the few large collector tiles (>45-cm [18-inch] diameter) present in many watersheds contribute the largest proportion of tile flow to streams. These large tiles are designed to collect the flow from many small tiles in the uplands and carry their water to the nearby stream or drainage ditch. Many of these large tiles are also located beneath grass waterways that carry surface runoff to the stream. The state of Iowa has targeted these larger tiles for remediation through a Conservation Reserve Enhancement Program (CREP). This program is a state and federal program that provides incentives to private landowners to develop and restore wetlands that intercept tile drains that drain more than 202 ha (500 acres). Landowners receive annual land payments over 15 yr and reimbursements for costs of wetland and buffer establishment. Easements to maintain the wetlands and buffers are required for a minimum of 15 yr beyond the CREP payments, for a total of 30 yr. Additional one-time, upfront incentive payments are used to encourage participating landowners to enter into perpetual easements. This kind of a program can result in reductions of 40–90% of the NO_3 and up to 90% of the herbicide in tile water.

While small constructed wetlands can be designed to fit within a buffer zone, there are few incentive programs to support their construction. It would be a major challenge to convince enough landowners to install the number of small wetlands required to provide the benefits of one large one. Constructed and restored wetlands are one more tool that can improve the water quality in some agricultural watersheds.

Riparian Grazing Practices

The semiarid and arid western rangelands cover a wide latitudinal and elevational topographic range with many potential plant communities. This broad landscape distribution requires that prescriptions of riparian grazing practices be made on a site-by-site basis. In most cases, managing access of livestock to the riparian zone may be the best way to bring about vegetational changes in the plant community. Little emphasis is placed on mechanically manipulating the site as is the case with the MRFB practice. In most cases, western riparian grazing practices do not completely exclude livestock from the riparian zone but rather control access based on site conditions and season. While fencing and complete exclusion are often recommended in the narrow riparian pasture corridors in the Midwest and the East, this is not feasible in the West. In many areas of the West, the forage in the riparian zone is very important to the overall grazing allotment because of the productivity in this moist environment. Restoration of riparian pastures in the West is primarily a matter of managing access and relying on the seedbank to restore the vegetation.

In developing a grazing plan for a given riparian zone, several general principles should be followed (Chaney et al., 1993). First, grazing access to the riparian zone should be limited during those times when stream bank soils are moist and most susceptible to compaction and collapse. This condition frequently exists during the early spring following snowmelt and early spring rains. However, there are areas where the critical season may be in midsummer when intense thunderstorm activity and rapid runoff may occur. Second, enough living plant material should be left on the stream bank to ensure protection of the banks (Clary & Webster, 1990). Plant material is often measured as residual stubble height or utilization as a percentage by weight of herbaceous forage. Stubble heights of 10–15 cm (4–6 inches) and utilization of 30–50%, depending on grazing practice, are often recommended. The lower values apply to riparian areas that do not meet vegetation standards or are sensitive fish production areas, while the higher values apply to areas that meet standards or that are rotationally grazed. Third, grazing pressure should be sufficiently controlled to allow desirable plants time to regrow and store enough carbohydrates for overwinter dormancy and competition with other undesirable species. Many native plant species have been replaced by exotics in western riparian meadows because of compaction, loss of litter, and exposure of bare soil. Many exotics are opportunists that can take advantage of these degraded conditions. Various grazing strategies are available to meet these criteria.

Before specific grazing practices can be developed for a site, the present condition of the site must be evaluated so the desired plant community can be determined based on management objectives. The Bureau of Land Management (1993), which manages 112 million ha (276 million acres) of public land, has developed a method for evaluating the proper functioning condition of the landscape. Even though only 9% of the land managed by the BLM is riparian-wetland area, it comprises the most productive land. The goal of the BLM is to restore and maintain 75% of these areas in their proper functioning condition (Bureau of Land Management, 1993):

Riparian-wetland areas are functioning properly when adequate vegetation, land form, or large woody debris is present to:

1. *dissipate stream energy associated with high waterflow, thereby reducing erosion and improving water quality;*
2. *filter sediment, capture bedload, and aid floodplain development;*
3. *improve flood-water retention and groundwater recharge;*
4. *develop root masses that stabilize stream banks against cutting action;*
5. *develop diverse ponding and channel characteristics to provide the habitat and the water depth, duration, and temperature necessary for fish production, waterfowl breeding, and other uses; and*
6. *support greater biodiversity (Bureau of Land Management,* 1993).

Riparian areas are functioning properly when there is adequate structure present to provide the listed benefits that are applicable to a particular area.

Chaney et al. (1993), Clary and Webster (1990), and Elmore (1992) provided the following summary of grazing strategies for western riparian zones to help maintain their proper functioning conditions. If possible, riparian areas should be maintained separate from upland pasture so that separate management strategies can be exercised. Grazing should coincide with the physiological needs of the target plant species. Livestock should be kept off stream banks when they are most susceptible to damage. Lengthening rest times between grazing and controlling the intensity of grazing increases plant vigor, encourages more desirable plant species composition, and allows banks to heal. If riparian areas are to be restored, livestock should be fenced out for as long as it takes for the vegetation and stream banks to recover. Riparian areas with poor recovery potential should be permanently excluded from grazing.

In those areas where fencing is used to exclude livestock from the riparian zone, alternative watering sources must be developed. Pasture nose pumps and wind- and solar- powered pumps provide opportunities for getting water to sites outside the riparian zone. Gravity-feed watering systems may also be possible if springs are located in the area. An alternative to fencing may be the planting of a hedgerow of dense native shrubs. One plan under consideration would provide three rows of shrubs, with each row offset from the other, to provide a dense visual and physical barrier for the livestock. The row farthest from the stream bank would be planted to a shrub with thorns to help dissuade livestock from entering the hedgerow. Armored access points to the stream channel could be provided for crossing and water. An electric fence could be used during the establishment period. The hedgerow would have many of the same benefits that woody plants have in MSFBs. Finally, Chaney et al. (1993) suggested that no one grazing system applies to all riparian locations and that any grazing strategy is only as good as the management applied.

Final Thoughts on Riparian Forest and Upland Buffers

Riparian forest buffers are an agroforestry practice that, when properly installed, can reduce the movement of nonpoint-source pollutants to surface and groundwaters. Limited research from sites around the world has demonstrated that in the right location, riparian forest buffers can remove significant amounts of sediment, nutrients, antibiotics, and pesticides from both surface and subsurface waters. Other potential benefits of riparian forest buffers include slowing floodwaters, stabilizing stream banks against erosion, reducing levee breaks, and providing wildlife and aquatic habitat and harvestable products for landowners.

While many studies have demonstrated the benefits of both riparian forest and upland buffers, there are many landscapes in which buffers have not been studied. In these locations, there is a great need to study the factors controlling buffer efficacy. For example, the geomorphology of some landscapes would suggest that flow paths of groundwater might move soluble chemicals below the living filter of the buffer. While it is possible that there is sufficient C to fuel denitrification in some of these landscapes, little has been done to study this or other mechanisms in these settings.

While buffer widths required to trap sediment in surface runoff are reasonably well established, the widths needed to biologically reduce the chemical load of shallow groundwater in landscapes where groundwater has access to the living filter have not been established. Additionally, issues such as the combination of plant species providing the optimum root systems, the most rapid improvement of soil quality, and the ideal surface friction for surface runoff need to be studied along with the value of buffers for wildlife habitat and commercial products for landowners. The effectiveness of narrow strips of existing forests along streams should be evaluated, as some recent surveys would suggest that without the grass strip to intercept and spread concentrated flow from the adjacent crop fields, gullies can develop and carry pollutants directly through the forests. A management option in these forests might be to create more savanna-like conditions that would allow more perennial ground cover to establish. However, such a savanna system would require significant management to maintain the reduced density of woody plants to allow the ground cover to flourish. Establishing a grass filter adjacent to such naturally occurring narrow forest strips would be a more effective method of providing the protection needed, as maintenance of the grass strip is easier to apply and more familiar to many landowners. These grass strips reduce the shading impact of trees and crop yields.

Acceptance of riparian forest buffers by farmers in many regions dominated by intensive row-crop agriculture has been slow for several reasons. First, farmers are more familiar with grass filter strips than they are with buffers incorporating woody plants. Second, they are concerned that they are committing valuable cropland to a practice that is not familiar to them and from which they do not see much direct benefit. Many farmers see the greatest benefit of buffers accruing to those downstream, with their own benefit being primarily in the form of wildlife habitat (see Chapter 13). Third, many farmers who establish riparian forest buffers do so under government programs such as the Conservation Reserve Program. Many of them are concerned that after these programs are ended, they will have to reclaim these riparian areas for crop production and the cost of tree removal will be great. Fourth, many farmers accept new practices after seeing them demonstrated in their local area. Because the application of riparian forest buffers has expanded more rapidly than the research and demonstration to support them, there are still relatively few demonstration sites available for farmers to see a functioning buffer. Supporting coalitions of public and private organizations to establish demonstrations such as the Iowa Buffer Initiative and the Illinois Buffer Partnership organized by the NGO Trees Forever can help spread the word to willing landowners.

While the science has not kept up with the application of riparian forest buffers, there is sufficient evidence to support the concept that riparian buffers are crucial landscape elements for improving water quality and the biotic integrity of streams. Their placement is critical if stream water quality improvements are to be achieved. Placement of riparian buffers should begin in headwater streams and proceed downstream. Random placement of short segments of buffers throughout many watersheds will provide field-level reductions in nonpoint-source pollutants but will not make significant contributions to stream water improvements. To optimize riparian forest buffer benefits, it is also necessary to properly buffer the rest of the landscape to keep nonpoint-source pollutants as close to their sources as possible. The living filters of riparian forest buffers cannot, on their own, provide the improvement that is needed and desired for our streams and waterways, but they can go a long way toward protecting them.

References

Abu-Zreig, M., Rudra, R. P., Whiteley, H. R., Lalonde, M. N., & Kaushik, N. K. (2003). Phosphorus removal in vegetated filter strips. *Journal of Environmental Quality, 32*, 613–619. https://doi.org/10.2134/jeq2003.6130

Adhikari, P., Udawatta, R. P., Anderson, S. H., & Gantzer, C. J. (2014). Soil thermal properties under prairies, conservation buffers, and corn–soybean land use systems. *Soil Science Society of America Journal, 78*, 1977–1986. https://doi.org/10.2136/sssaj2014.02.0074

Akdemir, E., Anderson, S. H., & Udawatta, R. P. (2016). Influence of agroforestry buffers on soil hydraulic properties relative to row crop management. *Soil Science, 181*, 368–376. https://doi.org/10.1097/SS.0000000000000170

Alagele, S. M., Anderson, S. H., & Udawatta,. R. P. (2019). Biomass and buffer management practice effects on soil hydraulic properties compared to grain crops for claypan landscapes. *Agroforestry Systems, 93*, 1609–1625. https://doi.org/10.1007/s10457-018-0255-1

Alagele, S.M., Anderson, S. H., & Udawatta, R. P. (2020). Agroforestry, grass, biofuel crop, and row-crop management effects on soil water dynamics for claypan landscapes. *Soil Science Society of America Journal, 84*, 203–219. https://doi.org/10.1002/saj2.20026

Alagele, S. M., Anderson, S. H., Udawatta, R. P., Veum, K. S., & Rankoth, L. M. (2019). Effects of conservation practices on soil quality compared with a corn–soybean rotation on a claypan soil. *Journal of Environmental Quality, 48*, 1694–1702. https://doi.org/10.2134/jeq2019.03.0121

Allen, S. B., Dwyer, J. P., Wallace, D. C., & Cook, E. A. (2003). Missouri river flood of 1993: Role of woody corridor width in levee protection. *Journal of the American Water Resources Association, 39*, 923–933. https://doi.org/10.1111/j.1752-1688.2003.tb04416.x

Amiri-Tokaldany, E., Darby, S. E., & Tosswell, P. (2003). Bank stability: Analysis for predicting reach scale land loss and sediment yield. *Journal of the American Water Resources Association, 39*, 897–909. https://doi.org/10.1111/j.1752-1688.2003.tb04414.x

Balestrini, R., Arese, C., Delconte, C. A., Lotti, A., & Salerno, F. (2011). Nitrogen removal in subsurface water by narrow buffer strips in the intensive farming landscape of the Po River watershed, Italy. *Ecological Engineering*, *37*, 148–157.

Beaulac, M. N., & Reckhow, K. H. (1982). An examination of land use–nutrient export relationships. *Water Resources Bulletin*, *18*, 1013–1022. https://doi.org/10.1111/j.1752-1688.1982.tb00109.x

Bentrup, G. 2008. *Conservation buffers: Design guidelines for buffers, corridors, and greenways* (Gen. Tech. Rep. SRS-109). Asheville, NC: U.S. Forest Service, Southern Research Station.

Berges, S. A., Schulte, L. A., Isenhart, T. M., & Schultz, R. C. (2010). Bird species diversity in riparian buffers, row crop fields, and grazed pastures within agriculturally dominated watersheds. *Agroforestry Systems*, *79*, 97–110. https://doi.org/10.1007/s10457-009-9270-6

Bercovici, M. M. (1994). *Riparian buffer strips in the Bear Creek Watershed* (Master's thesis). Iowa State University. https://doi.org/10.31274/rtd-180813-7293

Bharati, L. 1997. *Infiltration in a Coland clay loam under a six-year old multi-species riparian buffer strip, cultivated row crops and continuously grazed pasture* (Master's thesis). Iowa State University. https://doi.org/10.31274/rtd-180813-7530

Bharati, L., Lee, K. H., Isenhart, T. M., & Schultz, R. C. (2002). Soil-water infiltration under crops, pasture, and established riparian buffer in midwestern USA. *Agroforestry Systems*, *56*, 249–257. https://doi.org/10.1023/A:1021344807285

Bishop, R. A., & van der Valk, A. G. (1982). Wetlands. In T. C. Cooper (Ed.), *Iowa's natural heritage* (pp. 208–229). Des Moines, IA: Iowa Academy of Science.

Bisson, P. A., Bilby, R. E., Bryant, M. D., Doloff, C. A., Grette, G. B., House, R. A., . . . Sedell, J. R. (1987). Large woody debris in forested streams in the Pacific Northwest: Past present, and future. In E. O. Salo & T. W. Cundy (Eds.), *Streamside management: Forestry and fishery interactions* (pp. 143–190). Seattle, WA: Institute of Forest Resources, University of Washington.

Blanco-Canqui, H., Gantzer, C. J., Anderson, S. H., Alberts, E. E., & Thompson, A. L. (2004). Grass barrier and vegetative filter strip effectiveness in reducing runoff, sediment, nitrogen, and phosphorus loss. *Soil Science Society of America Journal*, *68*, 1670–1678. https://doi.org/10.2136/sssaj2004.1670

Blattel, C. R., Williard, K. W. J., Baer, S. G., & Zaczek, J. J. (2005). Abatement of ground water phosphate in giant cane and forest riparian buffers. *Journal of the American Water Resources Association*, *41*, 301–307. https://doi.org/10.1111/j.1752-1688.2005.tb03736.x

Broadmeadow, S., & Nisbet, T. R. (2004). The effects of riparian forest management on the freshwater environment: A literature review of best management practice. *Hydrology and Earth System Sciences*, *8*, 286–305. https://doi.org/10.5194/hess-8-286-2004

Bureau of Land Management. (1993). Riparian area management: Process for assessing proper functioning condition (TR 1737-9). Washington, DC: Bureau of Land Management.

Burgin, A. J., & Groffman, P. M. (2012). Soil O_2 controls denitrification rates and N_2O yield in riparian wetland. *Journal of Geophysical Research*, *117*, G01010. https://doi.org/10.1029/2011JG001799

Burkart, M. R., Oberle, S. L., Hewitt, M. J., & Pickus, J. (1994). A framework for regional agroecosystems characterization using the National Resource Inventory. *Journal of Environmental Quality*, *23*, 866–874. https://doi.org/10.2134/jeq1994.00472425002300050002x

Burt, T. P. (1997). The hydrological role of floodplains within the drainage basin system. In N. E. Haycock, T. P. Burt, K. W. T. Goulding, & G. Pinay (Eds.), *Buffer zones: Their processes and potential in water protection* (pp. 21–32). Harpenden, UK: Quest Environmental.

Castelle, A. J., Johnson, A. W., & Conolly, C. (1994). Wetland and stream buffer size requirements: A review. *Journal of Environmental Quality*, *23*, 878–882. https://doi.org/10.2134/jeq1994.00472425002300050004x

Chaney, E., Elmore, W., & Platts, W. S. (1990). *Livestock grazing on western riparian areas*. Washington, DC: USEPA.

Chaney, E., Elmore, W., & Platts, W. S. (1993). *Livestock grazing on western riparian areas: Managing change*. Washington, DC: USEPA.

Chaplot, V., Saleh, A., Jaynes, D. B., & Arnold, J. (2004). Predicting water, sediment and NO_3–N loads under scenarios of land-use and management practices in a flat watershed. *Water, Air, and Soil Pollution*, *154*, 271–293. https://doi.org/10.1023/B:WATE.0000022973.60928.30

Cho, J., Vellidis, G., Bosch, D. D., Lowrance, R., & Strickland, T. (2010). Water quality effects of simulated conservation practice scenarios in the Little River experimental watershed. *Journal of Soil and Water Conservation*, *65*, 463–473. https://doi.org/10.2489/jswc.65.6.463

Chu, B., Anderson, S. H., Goyne, K. W., Lin, C. H., & Lerch, R. N. (2013). Sulfamethazine transport in agroforestry and cropland soils. *Vadose Zone Journal*, *12*(2). https://doi.org/10.2136/vzj2012.0124

Chu, B., Goyne, K. W., Anderson, S. H., Lin, C. H., & Lerch, R.N. (2013). Sulfamethazine sorption to soil: Vegetative management, pH, and dissolved organic matter effects. *Journal of Environmental Quality*, *42*, 794–805. https://doi.org/10.2134/jeq2012.0222

Chu, B., Goyne, K. W., Anderson, S. H., Lin, C. H., & Udawatta, R. P. (2010). Veterinary antibiotic sorption to agroforestry buffer, grass buffer and cropland soils. *Agroforestry Systems*, *79*, 67–80. https://doi.org/10.1007/s10457-009-9273-3

Clary, W. P., & Webster, B. F. (1990). Riparian grazing guidelines for the intermountain region. *Rangelands*, *12*, 209–212.

Colletti, J. P., Premachandra, W., Schultz, R. C., Rule, L. C., & Gan, J. (1993, 15–18 Mar.). A socio-economic assessment of an Iowa watershed. In *Riparian Ecosystems in the Humid U.S., Proceedings of the Conference, Atlanta, GA* (pp. 358–372). Washington, DC: National Association of Conservation Districts.

Cooper, J. R., Gilliam, J. W., Daniels, R. B., & Robarge, W. P. (1987). Riparian areas as filters for agricultural sediment. *Soil Science Society of America Journal*, *51*, 416–420. https://doi.org/10.2136/sssaj1987.03615995005100020029x

Correll, D. L. (1997). Buffer zones and water quality protection: General principles. In N. E. Haycock, T. P. Burt, K. W. T. Goulding, & G. Pinay (Eds.), *Buffer zones: Their processes and potential in water protection* (pp. 7–20). Harpenden, UK: Quest Environmental.

Correll, D. L. (2005). Principles of planning and establishment of buffer zones. *Ecological Engineering*, *24*, 433–439. https://doi.org/10.1016/j.ecoleng.2005.01.007

Correll, D. L., Jordan, T. E., & Weller, D. E. (1994). Failure of agricultural riparian buffers to protect surface waters from groundwater contamination. In J. Gibert, J. Mathieu, & F. Fournier (Eds.), *Groundwater/surface water ecotones: Biological and hydrological interactions and management options* (pp. 162–165). Cambridge, UK: Cambridge University Press.

Dabney, S. M., McGregor, K. C., Meyer, L. D., Grissinger, E. H., & Foster, G. R. (1994). Vegetative barriers for runoff and sediment control. In J. K. Mitchell (Ed.), *Integrated resource management and landscape modifications for environmental protection* (pp. 60–70). St. Joseph, MI: ASAE.

Dabney, S. M., Moore, M. T., & Locke, M. A. (2006). Integrated management of in-field, edge-of-field, and after-field buffers. *Journal of the American Water Resources Association*, *42*, 15–24. https://doi.org/10.1111/j.1752-1688.2006.tb03819.x

Dabney, S. M., Yoder, D. C., & Vieira, D. A. N. (2012). The application of the Revised Universal Soil Loss Equation,

Version 2, to evaluate the impacts of alternative climate change scenarios on runoff and sediment yield. *Journal of Soil and Water Conservation*, *67*, 343–353. https://doi.org/10.2489/jswc.67.5.343

Dahl, T. E., Johnson, C. E., & Frayer, W. E. (1991). *Status and trends of wetlands in the conterminous United States, mid-1970's to mid-1980's*. Washington, DC: U.S. Fish and Wildlife Service.

Daniels, R. B., & Gilliam, J. W. (1996). Sediment and chemical load reduction by grass and riparian filters. *Soil Science Society of America Journal*, *60*, 246–251. https://doi.org/10.2136/sssaj1996.03615995006000010037x

David, M. B., & Gentry, L. E. (2000). Anthropogenic inputs of nitrogen and phosphorus and riverine export for Illinois, USA. *Journal of Environmental Quality*, *29*, 494–508. https://doi.org/10.2134/jeq2000.00472425002900020018x

de la Crétaz, A. L., & Barten, P. K. (2007). *Land use effects on streamflow and water quality in the northeastern United States*. Boca Raton, FL: CRC Press. https://doi.org/10.1201/9781420008722

Dillaha, T. A., III, & Inamdar, S. P. 1996. Buffer zones as sediment traps or sources. N. E. Haycock, T. P. Burt, K. W. T. Goulding, & G. Pinay (Eds.), *Buffer zones: Their processes and potential in water protection* (pp. 33–42). Harpenden, UK: Quest Environmental.

Dillaha, T. A., Reneau, R. B., Mostaghimi, S., & Lee, D. (1989). Vegetative filter strips for agricultural nonpoint source pollution control. *Transactions of the ASAE*, *32*, 513–519. https://doi.org/10.13031/2013.31033

Doran, J. W., & Parkin, T. B. (1996). Quantitative indicators of soil quality: A minimum set. In J. W. Doran & A. J. Jones (Ed.), *Methods for assessing soil quality* (pp. 25–37). SSSA Special Publication 49. Madison, WI: SSSA.

Dosskey, M. G. (2001). Toward quantifying water pollution abatement in response to installing buffers on crop land. *Environmental Management*, *28*, 577–598. https://doi.org/10.1007/s002670010245

Dosskey, M. G., Hoagland, K. D., & Brandle, J. R. (2007). Change in filter strip performance over ten years. *Journal of Soil and Water Conservation*, 62, 21–32.

Dosskey, M., Schultz, R., & Isenhart, T. (1997). *How to design a riparian buffer for agricultural land* (Agroforestry Note 4). Lincoln, NE: National Agroforestry Center.

Dosskey, M. G., Vidon, P., Gurwick, N. P., Allan, C. J., Duval, T. P., & Lowrance, R. (2010). The role of riparian vegetation in protecting and improving chemical water quality in streams. *Journal of the American Water Resources Association*, *46*, 261–277.

Drake, L., & Langel, R. (1998, 22–27 Mar.). Deep-planting willow cuttings via water jetting. In D. F. Hayes (Ed.), *Engineering approaches to ecosystem restoration: ASCE Wetlands and Engineering & River Restoration Conference, Denver, CO* (pp. 1046–1051). Reston, VA: American Society of Civil Engineers. https://doi.org/10.1061/40382(1998)171

Dunne, T., & Leopold, L. B. (1978). *Water in environmental planning*. San Francisco, CA: W.H. Freeman and Co.

Edwards, D. R., Hutchens, T. K., Rhodes, R. W., Larson, B. T., & Dunn, L. (2000). Quality of runoff from plots with simulated grazing. *Journal of the American Water Resources Association*, *36*, 1063–1073. https://doi.org/10.1111/j.1752-1688.2000.tb05710.x

Elmore, W. (1992). Riparian responses to grazing practices. In R. J. Naiman (Ed.), *Watershed management* (pp. 442–457). New York, Springer. https://doi.org/10.1007/978-1-4612-4382-3_17

Feld, C. K., Fernandes, M. R., Ferreira, M. T., Hering, D., Ormerod, S. J., Venohr, M., & Gutierrez-Canovas, C. (2018). Evaluating riparian solutions to multiple stressor problems in river ecosystems: A conceptual study. *Water Research*, *139*, 381–394. https://doi.org/10.1016/j.watres.2018.04.014

Fleischner, T. L. (1994). Ecological costs of livestock grazing in western North America. *Conservation Biology*, *8*, 629–644. https://doi.org/10.1046/j.1523-1739.1994.08030629.x

Foster, G. R. (1982). Modeling the erosion process. In C. T. Hann, H. P. Johnson, & D. L. Brakensiek (Eds.), *Hydrological modeling of small watersheds* (pp. 297–380). ASAE Monograph 5. St. Joseph, MI: ASAE.

Gantzer, C. J., Buyanovsky, G. A., Alberts, E. E., & Remley, P. A. (1987). Effects of soybean and corn residue decomposition on soil strength and splash detachment. *Soil Science Society of America Journal*, *51*, 202–206. https://doi.org/10.2136/sssaj1987.03615995005100010042x

Geier, A. R., & Best, L. B. (1980). Habitat selection by small mammals of riparian communities: Evaluating effects of habitat alterations. *Journal of Wildlife Management*, *44*, 16–24. https://doi.org/10.2307/3808346

Gift, D. M., Groffman, P. M., Kaushal, S. S., & Mayer, P.M. (2010). Denitrification potential, root biomass, and organic matter in degraded and restored urban riparian zones. *Restoration Ecology*, *18*, 113–120. https://doi.org/10.1111/j.1526-100X.2008.00438.x

Gold, M. A., Godsey, L. D., & Josiah, S. J. (2004). Market and marketing strategies for agroforestry specialty products in North America. *Agroforestry Systems*, *61*, 371–382.

Gregory, S. V., Swanson, F. J., McKee, W. A., & Cummins, K. W. (1991). An ecosystem perspective of riparian zones. *BioScience*, *41*, 540–551.

Groffman, P. M., Gold, A. J., & Simmons, R. C. (1992). Nitrate dynamics in riparian forests: Microbial studies. *Journal of Environmental Quality*, *21*, 666–671. https://doi.org/10.2134/jeq1992.00472425002100040022x

Gurtz, M. E., Marzolf, G. R., Killingbeck, K. T., Smith, D. L., & McArthur, J. V. (1988). Hydrologic and riparian influences on the import and storage of coarse particulate organic matter in a prairie stream. *Canadian Journal of Fisheries and Aquatic Sciences*, *45*, 655–665. https://doi.org/10.1139/f88-079

Gyssels, G., Poesen, J., Bochet, E., & Li, Y. (2005). Impact of plant roots on the resistance of soils to erosion by water: A review. *Progress in Physical Geography*, *29*, 189–217. https://doi.org/10.1191/0309133305pp443ra

Haake, D. M. (2003). *Land use effects on soil microbial carbon and nitrogen in riparian zones of northeast Missouri* (Master's thesis). Iowa State University. https://doi.org/10.31274/rtd-20200803-205

Harvey, C.A., Komar, O., Chazdon, R., Ferguson, B. G., Finegan, B., Griffith, D. M., . . . Whishnie, M. (2008). Integrating agricultural landscapes with biodiversity conservation in the Mesoamerican hotspot. *Conservation Biology*, *22*, 8–15. https://doi.org/10.1111/j.1523-1739.2007.00863.x

Hefting, M. J., Clement, C., Dowrick, D., Cosandey, A. C., Bernal, S., Cimpian, C., . . . Pinay, G. (2004). Water table elevation controls on soil nitrogen cycling in riparian wetlands along a European climatic gradient. *Biogeochemistry*, *67*, 113–134. https://doi.org/10.1023/B:BIOG.0000015320.69868.33

Helmers, M. J., Zhou, X., Asbjornsen, H., Kolka, R., Tomer, M. D., & Cruse, R. M. (2012). Sediment removal by prairie filter strips in row-cropped ephemeral watersheds. *Journal of Environmental Quality*, *41*, 1521–1539. https://doi.org/10.2134/jeq2011.0473

Henningsen, J. C., & Best, L. B. (2005). Grassland bird use of riparian filter strips in southeast Iowa. *Journal of Wildlife Management*, *69*, 198–210. https://doi.org/10.2193/0022-541X(2005)069<0198:GBUORF>2.0.CO;2

Hernandez-Santana, V., Zhou, X., Helmers, M. J., Asbjornsen, H., & Kolka, R. (2013). Native prairie filter strips reduce runoff from hillslopes under annual row-crop systems in Iowa, USA. *Journal of Hydrology*, *477*, 94–103. https://doi.org/10.1016/j.jhydrol.2012.11.013

Herring, J. P., Schultz, R. C., & Isenhart, T. M. (2006). Watershed scale inventory of existing riparian buffers in northeast Missouri using GIS. *Journal of the American Water Resources Association, 42*, 145–155. https://doi.org/10.1111/j.1752-1688.2006.tb03830.x

Hewlett, J. D. (1982). *Principles of forest hydrology*. Athens, GA: University of Georgia Press.

Hickey, M. B. C., & Doran, B. (2004). A review of the efficiency of buffer strips for the maintenance and enhancement of riparian ecosystems. *Water Quality Research Journal, 39*, 311–317. https://doi.org/10.2166/wqrj.2004.042

Hoag, J. C., & Fripp, J. (2002). *Streambank soil bioengineering field guide for low precipitation areas*. Aberdeen, ID: USDA–NRCS Aberdeen Plant Materials Center.

Huang, Y., Rickerl, D. H., & Kephart, K. D. (1996). Recovery of deep-point injected soil nitrogen-15 by switchgrass, alfalfa, ineffective alfalfa, and corn. *Journal of Environmental Quality, 25*, 1394–1400. https://doi.org/10.2134/jeq1996.00472425002500060033x

Iowa State University Extension. (2002). *Stewards of our streams: Assessing the need for a riparian management system* (Pm-1626d). Ames, IA: Iowa State University Extension.

Isenhart, T. M., Schultz, R. C., & Colletti J. P. (1997). Watershed restoration and agricultural practices in the Midwest Bear Creek of Iowa. In J. E. Williams, C. A. Wood, & M. P. Dombeck (Eds.), *Watershed restoration: Principles and practices* (pp. 318–334). Bethesda, MD: American Fisheries Society.

Jacobs, T. C., & Gilliam, J. W. (1985). Riparian losses of nitrate from agricultural waters. *Journal of Environmental Quality, 14*, 472–478. https://doi.org/10.2134/jeq1985.00472425001400040004x

Karlen, D. L., Mausbach, M. J., Doran, J. W., Cline, R. G., Harris, R. F., & Schuman, G. E. (1997). Soil quality: A definition, and framework for evaluation. *Soil Science Society of America Journal, 61*, 4–10. https://doi.org/10.2136/sssaj1997.03615995006100010001x

Karr, J. R. (1991). Biological integrity: A long-neglected aspect of water resource management. *Ecological Applications, 1*, 66–84. https://doi.org/10.2307/1941848

Knight, K. (2007). *Effectiveness of naturally-occurring riparian forest buffers and grass filter strips at buffering concentrated flow from row crop fields to streams in northeast Missouri* (Master's thesis). Iowa State University. https://doi.org/10.31274/rtd-180813-7422

Kumar, S., Anderson, S. H., Bricknell, L. G., & Udawatta, R. P. (2008). Soil hydraulic properties influenced by agroforestry and grass buffers for grazed pasture systems. *Journal of Soil and Water Conservation, 63*, 224–232. https://doi.org/10.2489/jswc.63.4.224

Kumar, S., Anderson, S. H., & Udawatta, R. P. (2010). Agroforestry and grass buffer influences on macropores measured by computed tomography under grazed pasture systems. *Soil Science Society of America Journal, 74*, 203–212. https://doi.org/10.2136/sssaj2008.0409

Kumar, S., Anderson, S. H., Udawatta, R. P., & Kallenbach, R. L. (2012). Water infiltration influenced by agroforestry and grass buffers for a grazed pasture system. *Agroforestry Systems, 84*, 325–335. https://doi.org/10.1007/s10457-011-9474-4

Kumar, S., Udawatta, R. P., Anderson, S. H., & Mudgal, A. (2011). APEX model simulation of runoff and sediment losses for grazed pasture watersheds with agroforestry buffers. *Agroforestry Systems, 83*, 51–62. https://doi.org/10.1007/s10457-010-9350-7

Lal, R., Delgado, J. A., Groffman, P. M., & Millar, N. (2011). Management to mitigate and adapt to climate change. *Journal of Soil and Water Conservation, 66*, 276–285. https://doi.org/10.2489/jswc.66.4.276

Lane, W. W. (1955). Design of stable channels. *Transactions of the American Society of Civil Engineers, 120*, 1234–1279.

Lee, K., Isenhart, T. M., & Schultz, R. C. (2003). Sediment and nutrient removal in an established multi-species riparian buffer. *Journal of Soil and Water Conservation*, 58:1–8.

Lee, K., Isenhart, T. M., Schultz, R. C., & Mickelson, S. K. (1999). Nutrient and sediment removal by switchgrass and cool-season filter strips in central Iowa, USA. *Agroforestry Systems, 44*, 121–132. https://doi.org/10.1023/A:1006201302242

Lee, K., Isenhart, T. M., Schultz, R. C., & Mickelson, S. K. (2000). Multi-species riparian buffer system in central Iowa for controlling sediment and nutrient losses during simulated rain. *Journal of Environmental Quality, 29*, 1200–1205. https://doi.org/10.2134/jeq2000.00472425002900040025x

Lee, P., Smyth, C., & Boutin, S. (2004). Quantitative review of riparian buffer width guidelines from Canada and the United States. *Journal of Environmental Management, 70*, 165–180. https://doi.org/10.1016/j.jenvman.2003.11.009

Legout, C., Molenat, J., Lefebvre, S., Marmonier, P., & Aquilna, L. (2005). Investigation of biogeochemical activities in the soil and unsaturated zone of weathered granite. *Biogeochemistry, 75*, 329–350. https://doi.org/10.1007/s10533-005-0110-0

Leopold, L. B., Wolman, M. G., & Miller, J. P. (1964). Fluvial processes in geomorphology. San Francisco, CA: W.H. Freeman.

Lin, C. H., Goyne, K. W., Kremer, R. J., Lerch, R. N., & Garrett, H. E. (2010). Dissipation of sulfamethazine and tetracycline in the root zone of grass and tree species. *Journal of Environmental Quality, 39*, 1269–1278. https://doi.org/10.2134/jeq2009.0346

Lin, C. H., Lerch, R. N., Garrett, H. E., & George, M. F. (2004). Incorporating forage grasses in riparian buffers for bioremediation of atrazine, isoxaflutole and nitrate in Missouri. *Agroforestry Systems, 63*, 91–99. https://doi.org/10.1023/B:AGFO.0000049437.70313.ef

Lin, C. H., Lerch, R. N., Garrett, H. E., Li, Y. Z., & George, M. F. (2007). An improved HPLC-MS/MS method for determination of isoxaflutole (Balance) and its metabolites in soils and forage plants. *Journal of Agricultural and Food Chemistry, 55*, 3805–3815. https://doi.org/10.1021/jf063322g

Lin, C. H., Lerch, R. N., Goyne, K. W., & Garrett, H. E. (2011). Reducing herbicides and veterinary antibiotic losses from agroecosystems using vegetative buffers. *Journal of Environmental Quality, 40*, 791–799. https://doi.org/10.2134/jeq2010.0141

Lin, C. H., Lerch, R. N., Kremer, R. J., Garrett, H. E., Udawatta, R. P., & George, M. F. (2005, 12–15 June). Soil microbiological activities in vegetative buffer strips and their association with herbicide degradation. In K. N. Brooks & P. F. Ffolliott (Eds.), Moving agroforestry into the mainstream: 9th North American Agroforestry Conference Proceedings, St. Paul, MN [CD-ROM]. https://www.cinram.umn.edu/sites/cinram.umn.edu/files/lin.pdf.

Lovell, S. T., & Sullivan, W. C. (2006). Environmental benefits of conservation buffers in the United States: Evidence, promise, and open questions. *Agriculture, Ecosystems & Environment, 112*, 249–260. https://doi.org/10.1016/j.agee.2005.08.002

Lowrance, R. (1992). Groundwater nitrate and denitrification in a coastal plain riparian soil. *Journal of Environmental Quality, 21*, 401–405. https://doi.org/10.2134/jeq1992.00472425002100030017x

Lowrance, R., Leonard, R., & Sheridan, J. (1985). Managing riparian ecosystems to control nonpoint pollution. *Journal of Soil and Water Conservation, 40*, 87–91.

Lowrance, R., McIntyre, S., & Lance, J. C. (1988). Erosion and deposition in a coastal plain watershed measured using Cs-137. *Journal of Soil and Water Conservation, 43*, 195–198.

Lowrance, R., Vellidis, G., & Hubbard, R. K. (1995). Denitrification in a restored riparian forest wetland. *Journal of Environmental Quality, 24*, 808–815. https://doi.org/10.2134/jeq1995.00472425002400050003x

Lynch, J. A., & Corbert, E. S. (1990). Evaluation of best management practices for controlling nonpoint pollution from silvicultural operations. *Water Resources Bulletin, 29*, 369–382.

Lyons, J., Trimble, S. W., & Paine, L. K. (2000). Grass versus trees: Managing riparian areas to benefit streams of central North America. *Journal of the American Water Resources Association, 36*, 919–930.

Magette, W. L., Brinsfield, R. B., Palmer, R. E., & Wood, J. D. (1989). Nutrient and sediment removal by vegetated filter strips. *Transactions of the ASAE, 32*, 663–667. https://doi.org/10.13031/2013.31054

Magner, J. A., & Alexander, S. C. (2002). Geochemical and isotopic tracing of water in nested southern Minnesota cornbelt watersheds. *Water Science & Technology, 45*, 37–42. https://doi.org/10.2166/wst.2002.0199

Marcus, W. A., Marston, R. A., Colvard, C.R., Jr., & Gray, R. D. (2002). Mapping the spatial and temporal distributions of wood debris in streams of the Greater Yellowstone ecosystem, USA. *Geomorphology, 44*, 323–335. https://doi.org/10.1016/S0169-555X(01)00181-7

Márquez, C. O., Garcia, V. J., Cambardella, C. A., Schultz, R. C., & Isenhart, T. M. (2004). Aggregate-size stability distribution and soil stability. *Soil Science Society of America Journal, 68*, 725–735. https://doi.org/10.2136/sssaj2004.7250

Márquez, C. O., Garcia, V. J., Schultz, R. C., & Isenhart, T. M. (2017). Assessment of soil degradation through soil aggregation and particulate organic matter following conversion of a riparian buffer to continuous cultivation. *European Journal of Soil Science, 68*, 295–304. https://doi.org/10.1111/ejss.12422

Márquez, C. O., Garcia, V. J., Schultz, R. C., & Isenhart, T. M. (2019). A conceptual framework to study soil aggregate dynamics. *European Journal of Soil Science, 70*, 466–479. https://doi.org/10.1111/ejss.12775

Martin, T. L., Kaushik, N. K., Trevors, J. T., & Whiteley, H. R. (1999). Review: Denitrification in temperate climate riparian zones. *Water, Air, and Soil Pollution, 111*, 171–186. https://doi.org/10.1023/A:1005015400607

Massachusetts Department of Conservation and Recreation. 2003. *Ware River watershed land management plan 2003–2012*. Boston: Massachusetts Dep. of Conservation and Recreation.

Mayer, P. M., Reynolds, S.K., Jr., McCutchen, M. D., & Canfield, T. J. (2007). Meta-analysis of nitrogen removal in riparian buffers. *Journal of Environmental Quality, 36*, 1172–1180. https://doi.org/10.2134/jeq2006.0462

Meador, M. R., & Goldstein, R. M. (2003). Assessing water quality at large geographical scales: Relations among landuse, water physicochemistry, riparian condition, and fis community structure. *Environmental Management, 31*, 504–517. https://doi.org/10.1007/s00267-002-2805-5

Menzel, B. W. (1983). Agricultural management practices and the integrity of instream biological habitat. In F. W. Schaller & G. W. Bailey (Eds.), *Agricultural management and water quality* (pp. 305–329). Ames, IA: Iowa State University Press.

Murphy, J. F., & Giller, P. S. (2000). Seasonal dynamics of macroinvertebrate assemblages in the benthos and associated with detritus packs in two low-order streams with different riparian vegetation. *Freshwater Biology, 43*, 617–631. https://doi.org/10.1046/j.1365-2427.2000.t01-1-00548.x

Naiman, R. J., Décamps, H., McClain, M. E., & Likens, G. E. (2005). *Riparia: Ecology, conservation, and management of streamside communities*. New York: Academic Press. https://doi.org/10.1016/B978-012663315-3/50006-X

Naiman, R. J., Décamps, H., & Pollock, M. (1993). The role of riparian corridors in maintaining regional biodiversity. *Ecological Applications, 3*, 209–212. https://doi.org/10.2307/1941822

Nair, V. D., & Graetz, D. A. (2004). Agroforestry as an approach to minimizing nutrient loss from heavily fertilized soils: The Florida experience. *Agroforestry Systems, 61*, 269–279. https://doi.org/10.1023/B:AGFO.0000029004.03475.1d

National Research Council. 1993. *Soil and water quality: An agenda for agriculture*. Washington, DC: National Academies Press.

Nelson, J. L. (2003). *Denitrification in riparian soils in three northeast Missouri watersheds* (Master's thesis). Iowa State University.

Newbold, J. D., Herbert, S., Sweeney, B. W., Kiry, P., & Alberts, S. J. (2010). Water quality functions of a 15-year-old riparian forest buffer system. *Journal of the American Water Resources Association, 46*, 299–310. https://doi.org/10.1111/j.1752-1688.2010.00421.x

Newbury, R. W., & Gaboury, M. N. (1993). *Stream analysis and fish habitat design: A field manual*. Gibsons, BC, Canada: Newbury Hydraulics.

Olilo, C. O., Onyndo, J. O., Moturi, W. N., Muia, A. W., Ombui, P., Shivoga, W. A., & Roegner, A. F. (2016). Effect of vegetated filter strips on transport and deposition rates of *Escherichia coli* in overland flow in the eastern escarpments of the Mau Forest, Njoro River watershed, Kenya. *Energy, Ecology and Environment, 1*, 157–182. https://doi.org/10.1007/s40974-016-0006-y

Omernik, J. M., Abernathy, A. R., & Gale, L. M. (1981). Stream nutrient level and proximity of agriculture and forest lands to streams: Some relationships. *Journal of Soil and Water Conservation, 36*, 227–231.

Osborne, L. L., & Kovacic, D. A. (1993). Riparian vegetated buffer strips in water-quality restoration and stream management. *Freshwater Biology, 29*, 243–258. https://doi.org/10.1111/j.1365-2427.1993.tb00761.x

Palone, R. S., & Todd, A. H. (Eds.). (1997). *Chesapeake Bay riparian handbook: A guide for establishing and maintaining riparian forest buffers* (NA-TP-02-97). Radnor, PA: U.S. Forest Service, Northern Forest Experiment Station.

Pankau, R. C., Schoonover, J. E., Williard, K. W. J., & Edwards, P. J. (2012). Concentrated flow paths in riparian buffer zones of southern Illinois. *Agroforestry Systems, 84*, 191–205. https://doi.org/10.1007/s10457-011-9457-5

Parkin, T. B. (1987). Soil microsites as a source of denitrification variability. *Soil Science Society of America Journal, 51*, 1194–1199. https://doi.org/10.2136/sssaj1987.03615995005100050019x

Patty, L., Réal, B., & Gril, J. J. (1997). The use of grassed buffer strips to remove pesticides, nitrate and soluble phosphorus compounds from runoff water. *Pesticide Science, 49*, 243–251. https://doi.org/10.1002/(SICI)1096-9063(199703)49:3<243::AID-PS510>3.0.CO;2-8

Pease, J., & Isenhart, T. M. (1996). Riparian management systems for enhanced wildlife habitat. In R. C. Schultz & T. M. Isenhart (Eds.), *Riparian management systems (RiMS) design, function and location: A progress report of the Agroecology Issue Team of the Leopold Center for Sustainable Agriculture* (pp. 148–156). Ames, IA: Iowa State University.

Peterjohn, W. T., & Correll, D. L. (1984). Nutrient dynamics in an agricultural watershed: Observations on the role of a riparian forest. *Ecology, 65*, 1466–1475. https://doi.org/10.2307/1939127

Peters, S. M., & Hodge, S. S. (2000). *Agroforestry: An integration of land use practices* (UMCA-2000-1). Columbia, MO: University of Missouri Center for Agroforestry.

Pohl, M., Alig, D., Korner, C., & Rixen, C. (2009). Higher plant diversity enhances soil stability in disturbed alpine ecosystems. *Plant and Soil, 324*, 91–102.

Pouyat, R., Groffman, P., Yesilonis, I., & Hernandez, L. (2002). Soil carbon pools and fluxes in urban ecosystems.

Environmental Pollution, *116*(Suppl. 1), S107–S118. https://doi.org/10.1016/S0269-7491(01)00263-9

Pouyat, R. V., Szlavecz, K., & Yesilonis, I. D. (2010). Chemical, physical, and biological characteristics of urban soils. In J. Aitkenhead-Peterson and A. Volder (Eds.), Urban ecosystem ecology (pp. 119–152). Agronomy Monograph 55. Madison, WI: ASA, CSSA, and SSSA. https://doi.org/10.2134/agronmonogr55.c7

Prichard, D. (1993). Riparian area management (Tech. Ref. 1737-9). Washington, DC: U.S. Bureau of Land Management.

Puckett, L. J. (2004). Hydrogeologic controls on the transport and fate of nitrate in ground water beneath riparian buffer zones: Results from thirteen studies across the United States. *Water Science & Technology*, *49*, 47–53. https://doi.org/10.2166/wst.2004.0160

Robertson, A. L., & Rowling, R. W. (2000). Effects of livestock on riparian zone vegetation in an Australian dryland river. *Regulated Rivers: Research & Management*, *16*, 527–541. https://doi.org/10.1002/1099-1646(200009/10)16:5<527::AID-RRR602>3.0.CO;2-W

Robinson, C. A., Ghaffarzadeh, M., & Cruse, R. M. (1996). Vegetative filter strip effects on sediment concentration in cropland runoff. *Journal of Soil and Water Conservation*, *50*, 227–230.

Sahin, H., Anderson, S. H., & Udawatta, R. P. (2016). Water infiltration and soil water content in claypan soils influenced by agroforestry and grass buffers compared to row crop management. *Agroforestry Systems*, *90*, 839–860. https://doi.org/10.1007/s10457-016-9899-x

Sanchez, P. A. (1995). Science in agroforestry. *Agroforestry Systems*, *30*, 5–55. https://doi.org/10.1007/BF00708912

Scanlon, B. R., Jolly, I., Scophocleous, M., & Zhang, L. (2007). Global impacts of conversions from natural to agricultural ecosystems on water resources: Quantity vs quality. *Water Resources Research*, *43*, W03437. https://doi.org/10.1029/2006WR005486

Schilling, K. E., & Wolter, C. F. (2000). Applications of GPS and GIS to map channel features in Walnut Creek, Iowa. *Journal of the American Water Resources Association*, *36*, 1423–1434. https://doi.org/10.1111/j.1752-1688.2000.tb05737.x

Schilling, K. E., Zhongwei, L., & Zhang, Y. K. (2006). Groundwater–surface interaction in the riparian zone of an incised channel, Walnut Creek, Iowa. *Journal of Hydrology*, *327*, 140–150. https://doi.org/10.1016/j.jhydrol.2005.11.014

Schmitt, T. J., Dosskey, M. G., & Hoagland, K. D. (1999). Filter strip performance and processes for different vegetation, widths and contaminants. *Journal of Environmental Quality*, *28*, 1479–1489. https://doi.org/10.2134/jeq1999.00472425002800050013x

Schoonover, J. E., & Williard, K. W. J. (2003). Groundwater nitrate reduction in giant cane and forest riparian buffer zones. *Journal of the American Water Resources Association*, *39*, 347–354. https://doi.org/10.1111/j.1752-1688.2003.tb04389.x

Schulte, L. A., Liebman, M., Asbjornsen, H., & Crow, T. R. (2006). Agroecosystem restoration through strategic integration of perennials. *Journal of Soil and Water Conservation*, 61, 164A–169A.

Schultz, R. C., Colletti, J. P., Isenhart, T. M., Simpkins, W. W., Mize, C. W., & Thompson, M. L. (1995). Design and placement of a multi-species riparian buffer strip system. *Agroforestry Systems*, *29*, 201–226. https://doi.org/10.1007/BF00704869

Schultz, R. C., Isenhart, T. M., Beck, W., Groh, T., & Davis, M. (2019). Agroforestry practices: Riparian forest buffers and filter strips. In M. R. Mosquera-Losada & R. Prabhu (Eds.), Agroforestry for sustainable agriculture. Cambridge, UK: Burleigh Dodds. https://doi.org/10.19103/AS.2018.0041.01

Schultz, R. C., Isenhart, T. M., & Colletti, J. P. (2005). Riparian buffer systems in crop and rangelands. In R. J. Rietveld (Tech. Coord.), Agroforestry and Sustainable Systems, Symposium Proceedings (Gen. Tech. Rep. RM-GTR-261, pp. 13–28). Fort Collins, CO: U.S. Forest Service Rocky Mountain Forest and Range Experiment Station.

Schultz, R. C., Isenhart, T. M., Simpkins, W. W., & Colletti, J. P. (2004). Riparian forest buffers in agroecosystems: Lessons learned from the Bear Creek Watershed, central Iowa, U.S.A. *Agroforestry Systems*, *61*, 35–50. https://doi.org/10.1023/B:AGFO.0000028988.67721.4d

Schumm, S. A. (1977). *The fluvial system*. New York: John Wiley & Sons.

Schumm, S. A., Harvey, M. D., & Watson, C. C. (1984) *Incised channels: Morphology, dynamics, and control*. Highlands Ranch, CO: Water Resources Publications.

Sekely, A. C., Mulla, D. J., & Bauer, D. W. (2002). Streambank slumping and its contribution to the phosphorus and suspended sediment loads of the Blue Earth River, Minnesota. *Journal of Soil and Water Conservation*, *57*(8), 243–250.

Seobi, T., Anderson, S. H., Udawatta, R. P., & Gantzer, C. J. (2005). Influence of grass and agroforestry buffer strips on soil hydraulic properties for an Albaqualf. *Soil Science Society of America Journal*, *69*, 893–901. https://doi.org/10.2136/sssaj2004.0280

Senaviratne, G. M. M. A., Udawatta, R. P., Nelson, K. A., Shannon, K., & Jose, S. (2012). Temporal and spatial influence of perennial upland buffers on corn and soybean yields. *Agronomy Journal*, *104*, 1356–1362. https://doi.org/10.2134/agronj2012.0081

Senaviratne, G. M. M. A., Baffaut, C., Lory, J. A., Udawatta, R. P., Nelson, N. O., Williams, J. R., & Anderson, S. H. (2018). Improved APEX model simulation of buffer water quality benefits at field scale. *Transactions of the ASABE*, *61*, 603–616.

Simon, A., Rinaldi, M., & Hadish, G. (1996). Channel evolution in the loess area of the midwestern United States. In Proceeding of the Sixth Federal Interagency Sedimentation Conference, Las Vegas, NV (pp. 86–93). Vol. 3. Washington, DC: U.S. Government Printing Office.

Simonson, T. D., Lyons, J., & Kanehl, P. D. (1994). Guidelines for evaluating fish habitat in Wisconsin streams (Gen. Tech. Rep. NC-164). St. Paul, MN: U.S. Forest Service, North Central Forest Experiment Station.

Simpkins, W. W., Wineland, T. R., Andress, R. J., Johnston, D. A., Isenhart, T. M., & Schultz, R. C. (2002). Hydrogeological constraints on riparian buffers for reduction of diffuse pollution: Examples from the Bear Creek Watershed in Iowa. *Water Science and Technology*, *45*, 61–68. https://doi.org/10.2166/wst.2002.0205

Smith, P., Ashmore, M. R., Black, H. I. J., Burgess, P. J., Evans, C. D., Quine, T. A, . . . Orr, H.G. (2013). The role of ecosystems and their management in regulating climate, and soil, water and air quality. *Journal of Applied Ecology*, *50*, 812–829. https://doi.org/10.1111/1365-2664.12016

Snyder, C. D., Young, J. A., Villella, R., & Lemarie, D. P. (2003). Influences of upland and riparian land use patterns on stream biotic integrity. *Landscape Ecology*, *18*, 647–664. https://doi.org/10.1023/B:LAND.0000004178.41511.da

Stauffer, D. F., & Best, L. B. (1980). Habitat selection by birds of riparian communities: Evaluating effects of habitat alterations. *Journal of Wildlife Management*, 44. https://doi.org/10.2307/3808345

Strahler, A. N. (1957). Quantitative analysis of watershed geomorphology. *Transactions of the American Geophysical Union*, *38*, 913–920. https://doi.org/10.1029/TR038i006p00913

Sweeney, B. W. (1992). Streamside forests and the physical, chemical, and trophic characteristics of Piedmont streams in eastern North America. *Proceedings of the Academy of Natural Science of Philadelphia*, *144*, 291–340.

Sweeney, B. W., & Newbold, J. D. (2014). Streamside forest buffer width needed to protect stream water quality, habitat, and organisms: A literature review. *Journal of the American Water Resources Association, 50*, 560–584.

Taylor, K., Schuster, J., Verdon, J., Pease, J., & Isenhart, T. (2000). Bird diversity within re-established riparian buffers in an agricultural watershed. In R. C. Schultz & T. M. Isenhart (Eds.), *Riparian management systems (RiMS) design, function and location: A progress report of the Agroecology Issue Team of the Leopold Center for Sustainable Agriculture* (pp. 69–75). Ames, IA: Iowa State University.

Thomson, G. W., & Hertel, H. G. (1981). The forest resources of Iowa in 1980. *Proceedings of the Iowa Academy of Science, 88*, 2–6.

Tomer, M. D., & Locke, M. A. (2011). The challenge of documenting water quality benefits of conservation practices: A review of USDA–ARS's conservation effects assessment project watershed studies. *Water Science &Technology, 64*, 300–310. https://doi.org/10.2166/wst.2011.555

Tufekcioglu, A., Raich, J. W., Isenhart, T. M., & Schultz, R. C. (2003). Biomass, carbon and nitrogen dynamics of multispecies riparian buffer zones within an agricultural watershed in Iowa, USA. *Agroforestry Systems, 57*, 187–198. https://doi.org/10.1023/A:1024898615284

Tufekcioglu, M. (2006). *Riparian land-use impacts on stream bank soil and phosphorus losses from grazed pastures* (Master's thesis). Iowa State University. https://doi.org/10.31274/rtd-180816-174

Tufekcioglu, M., Isenhart, T. M., Schultz, R. C., Bear, D. A., Kovar, J. L., & Russell, J. R. (2012). Stream bank erosion as a source of sediment and phosphorus in grazed pastures of the Rathbun Lake Watershed in southern Iowa, USA. *Journal of Soil and Water Conservation, 67*, 545–555. https://doi.org/10.2489/jswc.67.6.545

Tufekcioglu, M., Schultz, R. C., Zaimes, G. N., Isenhart, T. M., & Tufekcioglu, A. (2012). Riparian grazing impacts on streambank erosion and phosphorus loss via surface runoff. *Journal of the American Water Resources Association*, 49(1):103–113.

Udawatta, R. P., & Anderson, S. H. (2008). CT-measured pore characteristics of surface and subsurface soils as influenced by agroforestry and grass buffers. *Geoderma, 145*, 381–389. https://doi.org/10.1016/j.geoderma.2008.04.004

Udawatta, R. P., Anderson, S. H., Gantzer, C. J., & Garrett, H. E. (2006). Agroforestry and grass buffer influence on macropore characteristics: A computed tomography analysis. *Soil Science Society of America Journal, 70*, 1763–1773. https://doi.org/10.2136/sssaj2006.0307

Udawatta, R. P., Anderson, S. H., Motavalli, P. P., & Garrett, H. E. (2011). Calibration of a water content reflectometer and soil water dynamics for an agroforestry practice. *Agroforestry Systems, 82*, 61–75. https://doi.org/10.1007/s10457-010-9362-3

Udawatta, R. P., Gantzer, C. J., Reinbott, T. M., Wright, R. L., & Pierce, R. A., II. (2016). Yield differences influenced by distance from riparian buffers and CRP. *Agronomy Journal, 108*, 647–655. https://doi.org/10.2134/agronj2015.0273

Udawatta, R. P., Garrett, H. E., & Kallenbach, R. L. (2011). Agroforestry buffers for non-point source pollution reductions from agricultural watersheds. *Journal of Environmental Quality, 40*, 800–806. https://doi.org/10.2134/jeq2010.0168

Udawatta, R. P., Kremer, R. J., Garrett, H. E., & Anderson, S. H. (2007). Soil aggregate stability and enzyme activity in agroforestry and row-cropping systems. In A. Olivier (Ed.), *Economic opportunities and environmental benefits of agroforestry when trees and crops get together: 10th North American Agroforestry Conference Proceedings, Québec City, Canada* (pp. 165–175).

Udawatta, R. P., Kremer, R. J., Nelson, K. A., Jose, S., & Bardhan, S. (2014). Soil quality of a mature alley cropping agroforestry system in temperate North America. *Communications in Soil Science and Plant Analysis, 45*, 2539–2551. https://doi.org/10.1080/00103624.2014.932376

Udawatta, R. P., Krstansky, J. J., Henderson, G. S., & Garrett, H. E. (2002). Agroforestry practices, runoff, and nutrient loss: A paired watershed comparison. *Journal of Environmental Quality, 31*, 1214–1225. https://doi.org/10.2134/jeq2002.1214

Udawatta, R. P., Motavalli, P. P., & Garrett, H. E. (2004). Phosphorus loss and runoff characteristics in three adjacent agricultural watersheds with claypan soils. *Journal of Environmental Quality, 33*, 1709–1719. https://doi.org/10.2134/jeq2004.1709

Udawatta, R. P., Nelson, K. A., Jose, S., & Motavalli, P. P. (2014). Temporal and spatial differences in crop yields of a mature silver maple alley cropping system. *Agronomy Journal, 106*, 407–415. https://doi.org/10.2134/agronj2013.0429

Udawatta, R. P., Nygren, P. O., & Garrett, H. E. (2005). Growth of three oak species during establishment in an agroforestry practice for watershed protection. *Canadian Journal of Forest Research, 35*, 602–609. https://doi.org/10.1139/x04-206

Udawatta, R. P., Rankoth, L. M., & Jose, S. (2019). Agroforestry and biodiversity. *Sustainability, 11*, 02879. https://doi.org/10.3390/su11102879

USDA Farm Service Agency. (2007). FSA handbook: Agricultural Resource Conservation Program. Washington, DC: USDA–FSA. https://www.fsa.usda.gov/Internet/FSA_File/2-crp_r06_a06.pdf.

USDA National Agroforestry Center. (2004). Working trees for water quality. Lincoln, NE: National Agroforestry Center, University of Nebraska.

USDA–NRCS. (n.d.). National conservation practice standards. Washington, DC: USDA–NRCS. https://www.nrcs.usda.gov/wps/portal/nrcs/detail/national/technical/?cid=nrcsdev11_001020

USDA–NRCS. (1996). Chapter 16: Streambank and shoreline protection (2nd ed.). In Engineering field handbook. Washington, DC: USDA–NRCS. https://efotg.sc.egov.usda.gov/references/public/IA/Chapter-16_Streambank_and_Shoreline_Protection.pdf.

USDA–NRCS. (1999). Conservation practice standard: Grassed waterway (Code 412). In Field office technical guide: Practice standards and specifications. Washington, DC: USDA–NRCS.

USDA–NRCS. (1997a). Conservation practice standard: Riparian forest buffer (Code 392). In Field office technical guide: Practice standards and specifications. Washington, DC: USDA–NRCS.

USDA–NRCS. (1997b). Conservation practice standard: Filter strip (Code 393). In Field office technical guide: Practice standards and specifications. Washington, DC: USDA–NRCS.

van Noordwijk, M., & van de Geijn, S. C. (1996). Root, shoot, and soil parameters required for process oriented model of crop growth limited by water and nutrients. *Plant and Soil, 183*, 1–25. https://doi.org/10.1007/BF02185562

Vannote, R. L., Minshall, G. W., Cummins, K. W., Sedell, J. R., & Cushing, C. E. (1980). The river continuum concept. *Canadian Journal of Fisheries and Aquatic Sciences, 37*, 130–137. https://doi.org/10.1139/f80-017

Verchot, L. V., Franklin, E. C., & Gilliam, J. W. (1997). Nitrogen cycling in Piedmont vegetated filter zones: II. Subsurface nitrate removal. *Journal of Environmental Quality, 26*, 337–347. https://doi.org/10.2134/jeq1997.00472425002600020003x

Verry, E. S., Hornbeck, J. W., & Dolloff, C. A. (Eds.). (1999). *Riparian management in forests of the continental eastern United States.* Boca Raton, FL: CRC Press.

Vidon, P., & Smith, A. P. (2007). Upland controls on the hydrological functioning of riparian zones in glacial till valleys of the Midwest. *Journal of the American Water Resources Association, 43*, 1524–1539.

Wagenet, R. J., & Hutson, J. L. (1997). Soil quality and its dependence on dynamic physical processes. *Journal of Environmental Quality*, 26, 41–48. https://doi.org/10.2134/jeq1997.00472425002600010007x

Waldron, L. J., & Dakessian, S. (1982). Effect of grass, legume, and tree roots on soil shearing resistance. *Soil Science Society of America Journal, 46*, 894–899. https://doi.org/10.2136/sssaj1982.03615995004600050002x

Waldron, L. J., Dakessian, S., & Nemson, J. A. (1983). Shear resistance enhancement of 1.22 meter diameter soil cross sections by pine and alfalfa roots. *Soil Science Society of America Journal, 47*, 9–14. https://doi.org/10.2136/sssaj1983.03615995004700010002x

Welsch, D. J. (1991). *Riparian forest buffers: Function and design for protection and enhancement of water resources* (NA-PR-07-91). Radnor, PA: U.S. Forest Service, Northern Forest Experiment Station.

Wickramaratne, N. (2017). *Groundwater nitrogen and phosphorus dynamics under cattle grazing and row crop management in two contrasting soils in Missouri* (Master's thesis). University of Missouri. https://mospace.umsystem.edu/xmlui/handle/10355/63364

Wilkin, D. C., & Hebel, S. J. (1982). Erosion, redeposition, and delivery of sediment to Midwestern streams. *Water Resources Research, 18*, 1278–1282. https://doi.org/10.1029/WR018i004p01278

Wissmar, R. C., & Swanson, F. J. (1990). Landscape disturbances and lotic ecotones. In R. J. Naiman & H. Décamps (Eds.), *The ecology and management of aquatic-terrestrial ecotones* (pp. 65–89). Paris: UNESCO.

Wynn, R. M., Mostaghimi, S., Burger, J. A., Harpold, A. A., Henderson, M. B., & Henry, L. A. (2004). Variation in root density along stream banks. *Journal of Environmental Quality, 33*, 2030–2039. https://doi.org/10.2134/jeq2004.2030

Yu, S., Xu, Z., Wu, W., & Zuo, D. (2016). Effect of land use types on stream water quality under seasonal variation and topographic characteristics in the Wei River basin, China. *Ecological Indicators, 60*, 202–212. https://doi.org/10.1016/j.ecolind.2015.06.029

Zaimes, G. N., Schultz, R. C., & Isenhart, T. M. (2004). Stream bank erosion adjacent to riparian forest buffers, row-crop fields, and continuously-grazed pastures along Bear Creek in central Iowa. *Journal of Soil and Water Conservation, 59*, 19–27.

Zaimes, G. N., Schultz, R. C., & Isenhart, T. M. (2006). Riparian land uses and precipitation influences on stream bank erosion in central Iowa. *Journal of the American Water Resources Association, 42*, 83–97. https://doi.org/10.1111/j.1752-1688.2006.tb03825.x

Zaimes, G. N., Schultz, R. C., & Isenhart, T. M. (2008). Streambank soil and phosphorus losses under different riparian land-uses in Iowa. *Journal of the American Water Resources Association, 44*, 935–947. https://doi.org/10.1111/j.1752-1688.2008.00210.x

Zaimes, G. N., Tufekcioglu, M., & Schultz, R. C. (2019). Riparian land-use impacts on stream bank and gully erosion in agricultural watersheds: What we have learned. *Water, 11*(7), 1343. https://doi.org/10.3390/w11071343

Study Questions

1. Explain similarities and differences in riparian and upland buffer systems and their roles in terms of water quality protection.

2. Describe the major functions and benefits of a riparian buffer system.

3. Define major zones within a riparian buffer system and explain the processes and mechanisms that occur within each zone.

4. What factors are important when determining the width of a riparian buffer for the protection of stream habitat?

5. Why is it important to consider the land-use pattern of an entire watershed when designing practices to address water quality issues?

6. Can you identify soil–site–climatic conditions that enhance denitrification in riparian and upland buffer practices?

7. Why are multispecies buffers potentially more effective than single-species buffers for water quality protection and wildlife improvements?

8. How might a riparian buffer design (width, species, zones, etc.) and a management plan (establishment, maintenance, and protection) for a row-cropped watershed and a watershed with intensive grazing differ? Assume soil, site, landscape, and precipitation conditions are similar for both watersheds.

9. What are the potential roles of riparian and upland buffers in addressing climate change? Consider carbon sequestration, nutrient cycling, mineralization, immobilization, exchange of atmospheric gases, and species diversity when answering this question.

9

J. L. Chamberlain, John Davis, and
John F. Munsell

Forest Farming Practices

People have informally farmed forests for nontimber products for generations. In recent years, formal cultivation using forest farming practices has grown in popularity. Forest farming consists of growing and harvesting nontimber forest products in natural woodland habitats using agroforestry principles and practices. This agroforestry practice produces high-quality and sustainable raw nontimber materials that can be sold as primary products or used in value-added processing. It has gained a great deal of traction, leading to increased interest in studying ways to successfully forest farm and build profitable markets. The contemporary practice of forest farming in temperate North America has increased in large part because it provides landowners and tenant farmers income opportunities and diversification, a foundation for improving forest resources management, and nontimber production systems that increase biological diversity.

The purpose of this chapter is to present historical and modern perspectives of North American forest farming as well as examples of dynamic contemporary practices that showcase the many opportunities associated with temperate forest farming. The chapter is designed to encourage the reader to think about and to look beyond these perspectives and examples and continue their study of forest farming in temperate North America and beyond.

During the last several decades, forest farming has evolved into an agroforestry or complementary land-use practice to cutting timber that leads to the comprehensive and sustainable management of forest resources. Early interest in forest farming focused largely on developing countries, where people often directly depend on nontimber forest products for basic needs. Forest farming in North America is relatively new and still evolving, but interest is growing and the potential to diversify and stabilize income, improve forest health, and promote new "green" enterprise in this forest type is tremendous.

A student of temperate forest farming in North America will benefit from exposure to its historical evolution, the nuances of its terminology, historical and contemporary market dynamics, and the diverse range of products and production systems. This chapter was designed to be a primer that covers the basics of forest farming in North America. It also addresses associated opportunities and challenges across the continent and offers thoughts about the future of forest farming in this largely temperate region.

North American Agroforestry, Third Edition. Edited by Harold E. "Gene" Garrett, Shibu Jose, and Michael A. Gold.

What is Forest Farming?

In the earliest stages of development, forest farming was championed as a way to increase and diversify woodland production. Instead of management solely for timber and other wood products, forest farming was viewed as a way to increase the range of forest products beyond timber into nontimber products such as food, medicine, and additional salable materials. This concept integrated forestry with farming, animal husbandry, and horticulture to maximize output and optimize conservation (Douglas & Hart, 1976).

Some of the first forest farming visionaries maintained that a fully applied forest farm integrated three main components—trees, livestock, and forage. Trees would provide timber and associated wood products and help conserve soil and manage microclimates. Livestock would benefit from the shade, forage, and protection produced by the trees. When fully integrated, each component would become essential elements of an ecosystem. In its earliest agroforestry form, forest farming resembled more of what is considered silvopasture today (see Chapter 6). As it evolved, forest farming came to exclude livestock and focus more on cultivation of naturally growing and marketable understory nontimber crops.

According to Douglas and Hart (1976), forest farming combines the ecological stability of natural forests and the productivity of agricultural systems. Their vision of forest farming was considered most appropriate for marginal lands that are not well suited for intensive agriculture. Subsistence, low-income, and underserved owners and tenants have historically farmed marginal land. Douglas and Hart (1976) thus submitted that forest farming was relevant to a large segment of America's landowners. It was accordingly considered the "tool" with the greatest potential to feed people and animals, to regenerate soils and restore aquifers, to control floods and drought, and to create beneficial microclimates for production. As this vision evolved, so did the field of agroforestry and its new concept of forest farming, which focused on what can be undertaken in forests to manage for timber and nontimber products simultaneously.

Forest farming involves the cultivation or management of understory crops in an established or developing forest (Center for Subtropical Agroforestry, 2007; Cornell Cooperative Extension, 2007; National Agroforestry Center, 1997; University of Missouri Center for Agroforestry, 2006). It is a type of agroforestry that involves co-management of multiple crops (e.g., fungi, timber, plants) in a woodland setting. Forest farming may take place in a natural forest or in a plantation and is a production system that can improve forest health by managing forests for diverse ecosystem products and services. Management may range from intensive cultivated systems (Figure 9–1) in which plants, fungi, and other nontimber crops are planted in beds in the understory of a forest and grown in high populations with extensive inputs, to much less intensive approaches in which wild nontimber natural stands of desired understory plants are managed to enhance production, yields, and marketability of existing species.

Historical Perspective

The principles and practices of forest farming have origins deep in history, traditions, and culture. Humans have long gathered nontimber forest products, stewarded plants and fungi, and nurtured these resources for their health and well-being. Long before the technology existed to cut timber on a large scale, these forest products were gathered for personal consumption, and forests were tended to guarantee the long-term availability of understory crops. To ensure the future availability of these products, humans managed their patches by planting seeds, pulling weeds, and protecting crops from thieves and herbivores. The earliest people to inhabit North America's forests understood the importance of conserving these resources and took actions to help sustain them. Later, "hunters" of American ginseng (*Panax quinquefolius* L.), for example, would purportedly collect and sow seed from the plants before harvesting the prized roots. Today, forest farming is a formal practice that carries this stewardship tradition forward and renders a range of forest products, and is characterized by various levels of systematic intensity and a holistic forest management perspective.

Forest farming is not exclusive to temperate North America. In fact, the tropical regions and developing countries may have a longer history of traditional forest farming. For example, Asia provides the largest collection of historical information regarding forest farming. In western Asia, tree crops have been regarded as vital components of permanent agricultural production systems. The carob tree (*Ceratonia siliqua* L.) has been cultivated on marginal forest and farm sites in Cyprus and Syria for generations. In Southeast Asia and the Pacific Islands, trees of the genus *Pithecelobium* have been farmed in forests for production of feedstock as well as a staple human food. White mulberry (*Morus alba* L.) trees are farmed in Afghanistan to provide a flour substitute for traditional breads. In Bangladesh, the traditional homestead agroforestry systems are a diverse array of trees and understory plants grown in small holdings to provide for household consumption and sale.

Fig. 9–1. Forest farming can be intensive on small plots with multiple species of understory plants (photo credit: John Munsell, Virginia Tech).

Fig. 9–2. Maple syrup production, now a sophisticated forest farming practice, is attributed to Native Americans, who harvested sap from sugar maple trees (photo credit: Keith Otto - Cornell University, Uihlein Maple Research Forest).

North American Context

Native American people have a long history of farming forests that includes propagation, pruning, tending, weeding, selective harvest, and habitat modification (Turner, 2001; Turner & Cocksedge, 2001). Numerous examples of their use and management of nontimber resources by Native American peoples of the United States, Canada, and Mexico can be found throughout the forested regions. In northern climates, maple syrup production (Figure 9–2) is perhaps one of the most well-known examples of forest farming with origins prior to European settlement. Forest farming systems in Mexico have been developed and refined over many centuries (Goméz-Pompa, Kaus, Jimenez-Ozornio, Bainbridge, & Rorive, 1993). Native people stewarded the forests to ensure long-term availability of the plants that they used. They would manage plant populations by burning the forest floor, planting seeds, and nurturing growth. Many of these practices continue today based on the transfer of traditional knowledge through generations. The integration and management of crops in forested landscapes is an important complement to the integration of trees into agricultural landscapes.

Throughout the United States there is a long history of farming the forest for native plants. In the southwest, pods of *Prosopis* (aka, mesquite) were farmed from the forest to provide a staple used to make flour. Early farmers in the southern United States planted honey locust (*Gleditsia triacanthos* L.) in crop fields to supply forage for winter feed. The leaves and fruit of the mulberry tree were recognized a long time ago by farmers in the south as a livestock feed source. Farmers would manage mulberry trees in their forests with the knowledge that "one good fruiting mulberry tree could supply enough food for one pig for two months" (Douglas and Hart, 1976, p. 29). Early forest farmers would manage stands for mulberry while allowing pigs to forage under the canopy. Today, that practice is called silvopasture, but it retains elements of forest farming.

In 1929, J. Russell Smith, one of the early proponents of forest farming, described how "certain crop-yielding trees could provide useful substitutes for cereals in animal feeds as well as conserve the environment" in the rural United States. His early discussion about forest farming focused on harvesting products from trees to supplement farm production. Today, forest farming has progressed beyond this model. It has expanded to use the space under trees in an agroforestry model to produce crops and other products. Trees have expanded from singular sources of timber to multifunctional providers of services (e.g., shade and protection) that enhance the production of understory crops.

Smith (1929) proposed "progressive establishment of massive complexes of tree farms." He envisioned "hills green with crop-yielding trees" in place of poor pastures, eroding gullies, and abandoned farm lots. His ideal farm consisted of level and gently sloped lands that were protected by terraces. Other areas were planted with trees under which was planted high quality forage grasses. Smith advocated "two-storied" agriculture that allowed farmers to grow trees while raising livestock under their shade. Now, more than 80 yr after these visionary words were posed, silvopasture is well accepted in many parts of North America, and forest farming has emerged as the practice of growing nontimber crops in forests.

Agroforestry and forest farming in Mexico have been developed and refined over centuries (Goméz-Pompa et al., 1993). Growing food, medicine, and other nontimber forest products under natural forest canopy has provided the basic needs of people throughout Mexico. Many farmers have detailed knowledge of plants, soils, and regeneration that they use in the management of trees. According to Goméz-Pompa et al. (1993), forest farming practices developed over time include protection, cultivation, selection, and introduction of trees into the milpas (traditional farming system), fallows, plantations, natural forests, and forest gardens.

Forest farms in Mexico can be found on family plots, village land, and surrounded by natural forests. The abundance of trees in forest gardens—60–80 species in a family plot and 100–200 species in a village—is outstanding (Herrera, 1990, in Goméz-Pompa et al., 1993). Forest farms with more than 75 crop species and fruit trees can be found in the midst of natural forests (Nations & Komer, 1983 in Goméz-Pompa et al., 1993). These are managed and harvested until the forest canopy closes enough to reduce crop production. Many of the trees are common in natural forests, but new species, such as papaya, guava, and banana, have been introduced. Indigenous and exotic species of herbs, shrubs, vines, and epiphytes are grown in the shade. Recognizing the usefulness of wild species, forest farmers allow wildlings to become established when they germinate under the trees. Selection and protection of useful plants extends beyond forest gardens to the management of traditional milpas.

Advantages and Disadvantages

Forest farming has advantages and disadvantages compared with conventional forestry or farming (University of Missouri Center for Agroforestry, 2006). Forest farming can improve forest health by increasing biological diversity, prompting the removal of damaged and infected vegetation, and actively managing forest resources (University of Missouri Center for Agroforestry, 2006). It can result in additional and diversified income opportunities by producing raw material for different markets and value-added products.

At the same time, forest farming requires more intensive management and longer time horizons, which demand greater skills and increased efforts. The markets for many forest farmed products may be less than adequately understood, which increases the need for more research and assistance. Often the task of learning about and entering new markets is daunting, and the integration of forestry and farming with new plants requires broad knowledge that spans techniques and strategies for growing and managing trees, understory crops, and their interactions.

What are Nontimber Forest Products?

A variety of terms have been used to describe products other than timber that come from forests. Words commonly used to describe these products include, but are not limited to, secondary, minor, special or specialty, non-wood, and non-traditional. Students of forest farming must be aware, and have a clear understanding, of the many terms used to describe nontimber forest farmed products.

In many cases, the term does not accurately or adequately describe the products. Often, the nontimber forest products are neither minor nor secondary but may be major components of rural household economies. Frequently, they are commodities, marketed in large volumes and at low prices, as opposed to specialty products that are typically sold in small quantities at premium prices. The collection and use of some products have a longer tradition in human society than commercial timber harvesting. People collected berries and other edible products from the forests long before they had the technology to harvest timber. In some cases, they are produced by trees and may even be from wood collected from the forest. Regardless of terminology, it is important to understand the nuances of each.

The U.S. Forest Service defines "special forest products" in the national strategy as products derived from biological resources collected in forests, grasslands, and prairies for personal, educational, commercial, and scientific uses (U.S. Forest Service, 2001). In the strategy, special forest products exclude saw timber, pulpwood, cull logs, small round wood, house logs, utility poles, minerals, animal parts, rocks, water, and soil (U.S. Forest Service, 2001).

In 1999, the U.S. Congress passed legislation recognizing the importance of "forest botanicals" and requiring that better efforts be made to manage them for their associated products. For the purpose of Public Law 106-113, Congress defined "forest botanical products" as naturally occurring mushrooms, fungi, flowers, seeds, roots, barks, leaves, and other vegetation (or portions thereof) that grow on National Forest System lands. The term does not include trees, except as provided in regulations issued under this section by the Secretary of Agriculture (Department of the Interior and Related Agencies Appropriations Act of 2000).

A more common and widespread term is *nontimber forest products* (NTFPs). This term relates to plants, parts of plants, fungi, and other biological materials that are harvested from within and on the edges of natural, manipulated, or disturbed forests. Nontimber forest products come from natural forests as well as from plantations. They originate from fungi, moss, lichen, herbs, vines, shrubs, or trees. Many organs are harvested, including the roots, tubers, leaves, bark, twigs and branches, fruit, sap and resin, as well as the wood. In some cases, such as wild onion (ramps), the entire plant is removed and consumed. They may be processed into finished products, such as carvings, walking sticks, jams, jellies, tinctures, or teas.

In Canada, the term *nontimber forest products* may take on a broader definition. The Centre for Non-Timber Resources defines them as products from all botanical (plant) and mycological (mushroom) species in the forest other than those that produce timber, pulpwood, shakes, or other wood products. The designation includes services such as tourism and education relating to the products. Some groups in Canada, particularly First Nations (indigenous peoples), consider NTFPs to include forest animals. Animal products are commonly considered NTFPs in other parts of the world (for example "bush meat" in Africa) as well.

There are five broad market segments of commercially traded NTFPs: (a) culinary; (b) medicinal and dietary supplements; (c) decorative floral; (d) nursery and landscaping; and (e) fine arts and crafts (Chamberlain, Bush, & Hammett, 1998; Chamberlain et al., 2018).

- **Culinary products**: Food harvested from the forests include mushrooms, ferns, and the fruits, leaves, and roots of many plant species. Perhaps the most commonly collected culinary forest products are assorted berries and mushrooms. A commonly collected culinary forest product in the early spring is ramps (*Allium tricoccum* Ait.), wild onions that are available for only a few weeks until the forest canopy closes and their leaves senesce. Another important culinary species, black walnut (*Juglans nigra* L.), which is native to the eastern United States, is used in the medicinal and dietary supplement industry as well. Honey, from bees raised in a forest setting, also is considered an edible NTFP.
- **Medicinal and dietary supplements**: Forest plants used for their therapeutic value are marketed as medicines or as dietary supplements.

Plants that have been tested for safety and efficacy and meet strict U.S. Food and Drug Administration (FDA) standards can be marketed as medicines or drugs. More than 25% of all prescriptions dispensed in the United States contain active ingredients extracted from higher order plants (Farnsworth & Morris, 1976). Plants and plant products that do not meet the strictest FDA standards are marketed as dietary supplements in the United States. These products are legally considered food items, and product labels can make no claims about their medical benefits.

- **Floral decoratives**: Many forest plants and parts of plants are used in decorative arrangements, to complement and furnish the backdrop for flowers, as well as for the main component of dried ornaments. The end uses for many forest harvested floral greens include fresh and dried flowers, aromatic oils, greenery, basket fillers, wreaths, and roping, as well as craft items. Floral products from the oak ecosystems of southern Appalachia include vines of various species of grape, kudzu (*Pueraria lobata* Willd.), and smokevine (*Aristolochia macrophylla* Lam.) for wreaths and baskets, leaves of galax (*Galax urceolata* Poir.) for floral decorations, and twigs from several tree species. Several genera of moss are harvested from forests and used domestically and exported to the European floral industry.
- **Nursery and landscaping**: Live forest plants are collected for the nursery and landscaping industry and may be farmed in the forest. In North Carolina, Fraser fir [*Abies fraseri* (Pursh) Poir.] seedlings are pulled from the forest floor to be transplanted to Christmas tree farms. Rhododendron (*Rhododendron* spp.), azalea (*Rhododendron* spp.), and mountain laurel (*Kalmia latifolia* L.) as well as cactus (from the U.S. Southwest) are dug from forests and sold for landscaping. Understory forest plants, such as lady's slipper orchids (*Cypripedium* spp.), trillium (*Trillium* spp.), and wildflowers, also are collected for the landscaping industry. Included in this segment of the NTFP industry are native plants used in horticulture and ecosystem restoration (whole-plant extraction).
- **Fine arts and crafts**: The number of NTFPs used to make fine art and crafts is limited only by the creativity of the artisan. Some of the more important wood-based NTFPs used for arts and crafts include the stems of sassafras (*Sassafras albidum* Nutt.) for walking sticks and willow (*Salix* spp.) stems for furniture. Vines, particularly grapevine (*Vitis* spp.) and smokevine, are used to make specialty wood-based products as well as floral decorative products. A variety of hardwoods are used for carving. Baskets are crafted from the bark of paper birch (*Betula papyrifera* Marshall) that is stripped from trees and splits of wood from white oak (*Quercus alba* L.) and ash (*Fraxinus* spp.) trees.

While categorizing the products in this way is useful and convenient, many have multiple uses and may end up serving more than one market. For example, black cohosh (*Actaea racemosa* L.) is a popular medicinal plant and is used in natural landscapes (Figure 9–3). Black walnut has many

Fig. 9–3. Black cohosh is valued as a medicinal forest product and in some cases used in landscaping. Its roots can be divided for propagation in a forest farming system (photo credit: James Chamberlain, U.S. Forest Service).

uses: the nut is edible, the shell can be used as an abrasive, and the extract from the shells and husks is medicinal. In the Pacific Northwest and British Columbia, salal (*Gaultheria shallon* Pursh) is used for its foliage (as a floral green) and for its berries as an edible forest product. Oregon grape [*Mahonia aquifolium* (Pursh) Nutt.] is used in landscaping and as a floral green, while the berries are harvested for jellies and wines and the roots for natural dyes and medicinal applications (Pojar & Mackinnon, 1994). Organizing NTFPs into market segments does not preclude having specific species in several categories but recognizes that species may have multiple uses and therefore the potential for multiple products.

Mexico has a diversity of plants that could be used in forest farming, yet very few of the NTFPs from Mexico are marketed or regulated. Most are used for personal consumption at the household level. According to Blancas, Caballero, and Beltrán (2017), more than half of all NTFPs come from tropical forests and about a quarter from mountain mesophyll forests. Mosses in the genera *Campylopus*, *Hypnum*, *Leptodontium*, *Polytrichum*, and *Thuidium* are harvested in humid, high-mountain forest areas of Mexico (CONABIO, 2012), mostly to supply urban areas of the State of Mexico and Mexico City during the Christmas season. Some, like candelilla (*Euphorbia antisyphilitica* Zucc.), which is used for wax production, grow mostly in the Chihuahuan Desert.

Some of the more prominent NTFPs include *Agave* spp. for cosmetics and beverages, bamboos in the *Olmeca*, *Otatea*, and *Alonemia* genera, medicinal plants [*Arctostaphylus* spp., *Caesalpinia cacalao*, *Mimosa tenuiflora* (Willd.) Poir., *Turnera diffusa* Willd. ex Schult., and *Yucca schidigera* Roezl ex Ortgies], mosses, nuts, ornamental palms, pinions, sotol (a distilled beverage), spices (laurel, oregano), and *Tillandsia* leaves. *Aechmea magdalenae* (André) André ex Baker (known as ixtle or pita), a forest understory bromeliad, is harvested by indigenous communities to extract fibers, which have been used to make ropes, nets, fishing lines, sandals, and strings for musical instruments (Schultes, 1941). Indigenous people have traditionally consumed forest mushrooms such as *Amanita caesarea*, *Cantharellus cibarius*, *Morchella esculenta*, *M. conica*, *M. elata*, and *Tricholoma magnivelare* (Guzmán, 1999).

Who Might Forest Farm NTFPS?

Many of the reasons why someone may begin forest farming NTFPs are the same as, or similar to, the reasons why they might invest in any agroforestry system. Choices depend on the person's interests and expertise as well as the land and financial resources. In some cases, the system chosen will be designed to address a specific ecological function or to mitigate a perceived concern. Practices may be incorporated by farmers seeking to diversify their operation or better utilize their woodlands. Or they may fit well into lifestyle, land use, and holistic environmental problem solving. By diversifying crops, products, production cycles, and land management systems, forest farmers can reduce financial risk and generate environmental, cultural, and recreational benefits.

Forest farmers can realize many economic, ecological, and social benefits. Economic benefits range from added income from new crops that provide interim income while longer term crops mature, to improved revenues from marginal lands that are otherwise questionable for more traditional agricultural production. Labor requirements may be diversified, as new crops can have different growth and production time frames. Ecological benefits may be realized as a greater diversity of plants are grown or nurtured under the forest canopy. Water quality and quantity may be enhanced, which improves resources for aquatic fauna. Overall, forest farming can provide economic and conservation incentives, thereby enhancing environmental stewardship and community development.

Timber harvesting is not a primary objective for the vast majority of forestland owners in North America (Butler et al., 2016). Goals such as wildlife, aesthetics, and other forms of recreation typically are more important for many forest owners. Nontimber forest products are of interest to a sizable segment of forestland owners (Strong & Jacobson, 2005; Trozzo, Munsell, & Chamberlain, 2014; Workman, Bannister, & Nair, 2003). The United States, for instance, is in the midst of the largest intergenerational transfer of land, and a growing number of private forest landowners are faced with degraded forests that have been exploited for timber (Munsell, Germain, Luzadis, & Bevilacqua, 2009). Thus, nontraditional, comprehensive, and diversified approaches to managing woodlands are ever more necessary as forest landowners deal with rehabilitation and shift toward longer term holistic management. Progressive forest landowners may desire to manage for timber and nontimber products, and thus forest farming may be attractive.

Forest landowner objectives vary, but in general most balance economic and amenity goals. They want to keep the forest healthy and encourage a diversity of plants, animals, and fungi, while also leveraging salable goods within their forest to help offset expenses such as taxes and maintenance.

They may have little interest in cutting timber, but most typically do at some point (Jones, Luloff, & Finley, 1995; Turner, Finely, & Kingsley, 1977). Alternative activities, such as growing their food, making medicine, and living and working close to the forest, are increasingly of interest.

Landowners who practice permaculture are often interested in creating food forests or forest gardens. Permaculture has many definitions, but in general it is a permanent, sustainable system incorporating principles of natural ecosystems, agriculture, and social design. A permaculturist will strive to design and maintain a healthy forest that supports diverse populations of vegetative, fungal, microbial, and animal life while at the same time providing food, energy, and other resources for the people who tend that forest. The management approach is gentle but deliberate. Other basic tenets of permaculture design are integration, diversity, no waste production, the usefulness of everything, and patterns. Mudge and Gabriel (2014) described in detail how to design and manage a forest farm based on permaculture principles.

Landowners interested in pursuing forest farming need to examine internal and external factors that could influence their success. Many new enterprises require additional skills and expertise, and there may be sizable capital or labor investments to consider and budget. The competition in some markets, such as edible mushrooms, Christmas trees, and honey and related products, may have profit margins that make these alternatives less attractive. Interested landowners need to examine the markets and fully understand the potential and pitfalls of each possible venture. Although there are many challenges in developing forest farming, a diversified land use and management strategy can be economically rewarding to landowners willing to invest the time and energy.

Establishing a Forest Farm

In general, the new forest farmer should think about starting small, keep careful notes, and expect to make many modifications along the way. It takes many years for a farmer to master a new crop, and forest farmers can expect a similar time frame. The first few years are spent learning how to grow and meet the buyer's needs. A few more years are needed to acquire equipment and to refine processes to increase efficiency and profitability. And then, with time, the forest farmer will build confidence and experience.

Forest farms can be set up in natural or planted forests. A general recommendation is to progress cautiously. One of the first considerations is selecting the best site that is appropriate for the desired crop species. The site then must be prepared by removing competing vegetation and debris. Before the site is ready, planting stock must be produced or procured. Maintenance of the site after planting and until harvest requires reoccurring labor inputs. Post-harvest activities are needed to create the environment for future production.

Site evaluation is the first step to setting up a forest farm. The forest farmer needs to assess the location for planting and determine if it is appropriate: slope, soils, shade, water, and species. Each situation is different and depends on ecoregion, geography, forest type, and microclimatic variables. For example, forest farmers with deciduous or mixed forests want 75–80% shade provided by deeply rooted trees such as beech (*Fagus grandiflora* Ehrh.), birch, maple (*Acer* spp.), poplar (*Populus* spp.), and basswood (*Tilia americana* L.) to grow more native medicinal herbs. Solid stands of conifers or other shallow-rooted trees compete too much for water and nutrients with most of the forest herbs that growers may want to produce. Most forest botanicals require a moist, well-drained soil and may not do well in areas with prolonged standing water. Wooded, sloped hillsides that drain well are preferred by many forest herbs. For example, American ginseng and many of the forest herbs associated with it thrive on well-drained, north-or east-facing slopes. Look for other woodland botanicals, or similar plants such as Solomon's seal (*Polygonatum biflorum* Walt.), growing in the area. If there are none, it might not be a good place to plant, although a lack of companion or indicator species is by no means a guarantee that forest farming is not possible.

A general recommendation is to study the habitat characteristics of new plants. If you can find growing guidelines for the species of interest, use them. If not, start with any existing production practices recommended for American ginseng and goldenseal (*Hydrastis canadensis* L.) and modify as necessary (Figure 9–4). Indeed, much of what we know about forest farming represents the accumulation of producer experience given minimal historical investment in research related to woodland species cultivation.

Some forest farming species may be threatened, endangered, or of special concern. As such, they may be protected by state, federal, and/or international laws. As an example, American ginseng and goldenseal are regulated by international treaties. In undertaking an assessment, the forest farmer should check with local and federal departments of agriculture about the status of the plant of interest. At the same time, forest farmers need to be especially concerned about introducing aggressive or non-native invasive species that

Fig. 9–4. Wild-simulated American ginseng (left) and goldenseal (right) take about 7 yr before the roots are large enough to be harvested (photo credit: Jeanine Davis, North Carolina State University).

can displace or destroy native plant populations and habitats.

Many forest farming species tolerate a variety of soil types, although in general, heavy clays and very sandy soils are not desired by woodland botanicals that usually grow in hardwood forests. These species prefer loamy soil with high organic matter. In doing a site assessment, the forest farmer should collect soil samples from prospective sites and have them analyzed for nutritional status. For example, American ginseng grows best with a soil pH adjusted to about 5.5 and with phosphorus and calcium readily available. Depending on soil type, this may require the addition of lime, phosphate, and/or gypsum. In the case of ginseng, changing the soil pH and adding fertilizer may increase root growth and yield but may decrease the value of the root or make it more prone to disease. For most of the other woodland herbs, little or no soil amendments are needed if a good site is selected. The soil analysis information will indicate if soil amendments are necessary (Davis & Persons, 2014).

There are basically two types of forest farming systems: woods cultivated and wild simulated. In a woods-cultivated system, the objective is to produce the highest yields feasible by growing intensively. Contemporary production practices and pest control are used. The objective of a wild-simulated production system is to mimic the way the plants grow naturally. It is simple, inexpensive, and requires few inputs. Generally, the plants grow more slowly and yields are lower than in the other system, but since inputs are lower, profits may be similar or even higher. The final product in woods-cultivated systems may be less valuable than that produced in a wild-simulated system if there is a discernable difference in quality.

In a wild-simulated production system, site preparation requires raking aside leaves to expose the soil in the production area. Some forest farmers sprinkle gypsum or lime on the soil and rake it into the top few centimeters of soil. Seeds are scattered across the soil surface, or shallow holes or trenches are dug to accommodate rootstock. Whether using seeds or rootstock, the planting density is fairly thin so that plants have plenty of room to grow and will not compete with each other. The seeds or rootstock are covered with soil, and the leaves are redistributed across the planted area. In general, seeds are planted about twice the depth of the thickness of the seed, while rootstock is planted deep enough to stay moist. The plants are allowed to grow naturally. Most forest farmers do not spray fungicides or add any additional fertilizer. Most will try to pull big weeds and control any animals that may feed on the plants.

For the woods-cultivated system, all obstructions such as stumps, rocks, and big roots are removed (Figure 9–5). The soil is tilled, and soil amendments such as lime and phosphate are incorporated. Raised beds are often built to improve soil drainage. Seed or root stock is planted more densely than in the wild-simulated system. A thick layer of mulch is added, often straw, but in some areas composted sawdust, hardwood bark, or leaves are used. The plants are monitored closely and may be fertilized and sprayed with fungicides or insecticides on a regular basis. Weed and animal control are practiced.

Fig. 9–5. Woods-cultivated forest farming entails building raised beds, clearing competing understory vegetation, and tending plots as if they were a garden (photo credit: John Munsell, Virginia Tech).

For many forest farmed species, the root is the part of economic value. Depending on the species, plants must be allowed to grow 3–8 yr before the roots are large enough to harvest. During this time, seed and sometimes foliage can be harvested. Seeds of forest botanicals usually have unusual requirements that must be met for them to germinate. For example, American ginseng, goldenseal, and bloodroot (*Sanguinaria canadensis* L.) seeds must never be allowed to dry out. Ginseng seed must be stratified for a year before sowing. In contrast, goldenseal seed usually best germinates if fresh seed are sown. Each species has its own requirements for germination.

When roots are large enough for harvest, they must be carefully dug to minimize injury. In a wild-simulated system, this is done manually with a spade or fork. In a woods-cultivated system, this process may be partially mechanized by using a tool, similar to a potato digger, to bring the roots to the surface for collection. Roots must be washed and then, for most markets, dried. The washing and drying process must be done properly to maintain quality and should be practiced before the main crop is harvested. Practice sessions can create samples to be sent to prospective buyers for evaluation or testing for heavy metals, pesticides, bioactive constituents, and microbial contamination. Dried roots and herbs must be carefully packed into the kind of containers specified by the buyer. These are usually cardboard barrels or poly sacks. Packaged herbs should be stored in a cool, dry atmosphere and protected from rodents and insects.

Since these medicinal herb products will be consumed by people or animals, it is important to follow good agricultural practices and abide by all relevant food safety rules as outlined in the Food Safety and Modernization Act (U.S. Food and Drug Administration, 2018a). Manufacturers are required to abide by federally mandated good manufacturing practices (U.S. Food and Drug Administration, 2018b). These do not directly affect the forest farmer, although they indirectly affect them because the manufacturer must provide positive identification for all ingredients and certificates of analyses that may include microbial, heavy metal, pesticide, and constituent testing. Forest farmers can make their products more attractive to buyers by providing a voucher specimen or other evidence of product identity; growing in areas that are not contaminated with heavy metals, pesticides, or manure; washing roots with clean water; drying properly; and packaging and storing in a manner that will protect the integrity of the product (Davis, 2016).

Forest farmers cultivate nontimber crops in a forest and must manage the overstory to create and maintain canopy conditions suitable for production. Silviculture is the practice of establishing and sustainably managing biological conditions in a forest that support landowner objectives (Nyland, 2002). Most silviculturists manage forests for sustained yields of timber using science-based metrics related to tree density and composition. The principles and practices of silviculture, however, broadly relate to multiple production and amenity objectives associated with forests. Tending trees using associated guidelines such as relative density, particular thinning and species selection strategies, and age class management regimes can help forest farmers maintain productive and sustainable agroforests. Of particular concern to forest farmers is the management of available site resources to ensure appropriate growing conditions for nontimber crops.

In a diverse forested setting, there are many demands for water, light, and nutrients. The basis for silvicultural decision-making is forest inventory, which typically involves plot sampling distributed throughout a stand of trees. Plot data are used to study central tendencies and outliers related to key management variables and are useful for guiding management prescriptions to balance competition for resources and maximize site utilization. Important silvicultural variables include tree density, tree diameter, species composition, tree health and vigor, tree age, uniformity of tree distribution, advanced tree regeneration, herbaceous and fungal density and distribution, downed woody debris, and other habitat and ecological parameters.

In the long run, successful forest farmers must focus attention on the understory crops and management of a dynamic forest agroecosystem. Inventory data are used to estimate the overall condition of the forest and develop management prescriptions that change conditions to meet owner objectives. As the forest grows, trees and plant density can alter growing conditions and limit resource availability to the extent that timber and nontimber productivity significantly decline and become increasingly impacted by mortality.

Compatible management, or co-management, using silvicultural prescriptions and harvesting operations is often necessary to adjust growing conditions, ensure healthy and vigorous tree, plant, and fungal communities, and provide additional economic returns from a well-managed forest (Haynes, Monserud, & Johnson, 2003). Co-management (Hobby, Dow, & MacKenzie, 2006), relative to the trees and understory crops, reflects a continuum of activities and intensities from wild harvest to extensive and intensive forest farming approaches (Munsell, Davis, & Chamberlain, 2018). Activities range from passive, such as planting seeds from plants as they are harvested and allowing them to grow, to highly intensive management, such as cultivating the forest floor, removing competing vegetation, and planting improved stock of NTFP species, along with management of tree stem and canopy conditions (Cocksedge & Hobby, 2006; Davis & Persons, 2014; Mudge & Gabriel, 2014). Thus, the more active phases of co-management are consistent with extensive to intensive forest farming systems, as trees and herbaceous plants are considered in management.

Co-management that might be regarded as extensive forest farming activities include integrating conifer foliage collection with pruning and juvenile spacing, managing and studying the effects of silviculture and/or zoning on mushroom productivity, controlled burns for specific species regeneration, riparian area restoration with economic species, thinning and spacing to enhance the understory, targeted brushing to enhance noncompetitive brush species, partial harvests, and longer tree rotations for mushroom production.

Market Perspective for Forest Farmed Products

Although there is a long history of using NTFPs in North America, they resurfaced as potential income sources in the late 20th century. In the early 1990s, bumper crops of edible mushrooms appeared in many national forests in Oregon and Washington (Freed, 1994) as a result of major forest fires and spurred an increased interest in these alternative forest products. The findings of medical research helped to increase market demand for medicinal forest products (Eisenberg et al., 1993; Le Bars et al., 1997; Stix, 1998). Other social trends, including interest in foods foraged from the forests (e.g., Urban Food Forests, see Chapter 10) and natural products, have helped to spur demand for NTFPs.

Although no formal estimates have been made of the total value of the NTFP markets in North America, available data illustrate the economic importance of some individual products. Also, as these products are not fully integrated into systems to track and monitor production, studies that quantify their values are dated. Nonetheless evidence indicates that large quantities of products with high values are sold each year.

To illustrate, estimates of the volumes of NTFPs harvested from U.S. forests have been prepared for a variety of product categories. Based on permits to harvest from national forests and Bureau of Land Management forests, Chamberlain, Teets, and Kruger (2018) estimated that more than 2.5 million kg (5.6 million lb) of plant material were harvested for fine arts and crafts across the country in 2013. That same year, they estimated that more than 303,906 kg (670,000 lb) and 1.3 million l (300,000 gal) of food were harvested on public forestlands. In that same year, permits were issued for the harvest of more than 18,143 kg (40,000 lb) of medicinal plants from public forestlands.

The value of these, and other NTFPs, from public forestlands is substantial. For the 10-yr period of 2004–2013, the average annual wholesale value of NTFPs was estimated at US$907 million (Chamberlain et al., 2018). Wholesale value ranged from $793 million to $1,017 million during that period. In 2013, the estimated wholesale value of all NTFPs harvested from public forestlands exceeded $950 million (Chamberlain et al., 2018).

Evidence of the value of culinary forest products at national and local levels is significant as well. The wholesale value of food items alone harvested from public forests in 2013 was estimated at $77 million. Schlosser and Blatner (1995) estimated the wholesale value of wild edible mushrooms in Washington, Oregon, and Idaho at $41.1 million. They estimated that in 1992 buyers of mushrooms in the Pacific Northwest purchased $20.3 million of products from more than 10,000 harvesters. Collectors of the fruit of eastern black walnut were paid more than $2.5 million in 1995 (J. Jones, formerly of Hammons Products Co., Stockton, MO, personal communication, 1996). Volunteer fire departments in western North Carolina generate from 30–90% of their budgets from annual festivals that celebrate ramps (wild onions).

In Canada, Wetzel, Duchesne, and Laporte (2006) estimated the output of forest-based foods to be CDN$1.33 billion, with potential to expand to CDN$5.4 billion, including wild harvested and forest farmed products. The value of wild mushrooms collected from British Columbia during the decade ending in 2006 ranged from CDN$10–42 million, with an average annual value of CDN$29 million (Cocksedge & Hobby, 2006). The export value of pine mushrooms during the decade of 1995–2005 ranged from CDN$6–32 million, while the value of chanterelle mushrooms ranged annually from CDN$1–6 million.

Herbal and medicinal forest products contribute value to rural, regional, and national economies. In 1998, the total retail market for medicinal herbs in the United States was estimated at US$3.97 billion, more than double the estimate for North America in 1996 (Brevoort, 1998; Yuan & Grunwald, 1997). Twenty years later, in 2016, retail sales of all herbal products (forest harvested and cultivated) contributed more than US$6.5 billion to the dietary supplement industry that was valued at $36.7 billion (Smith, Kawa, Eckl, & Johnson, 2017). This segment of the industry is growing at an estimated annual rate of more than 7% (American Botanical Council, 2017).

Several medicinal forest products illustrate the significant value of these products. In 2016, retail sales, through mainstream multi-outlet channels, of black cohosh, saw palmetto (*Sereno repens* Bartram), and elderberry (*Sambucus* spp.), exceeded US$68 million (Smith, Kawa, Eckl, Morton, & Stredney, 2017). Retail sales of black cohosh alone was greater than $36 million, although this represented a 15% decrease from the previous year. Saw palmetto and black cohosh retail sales through U.S. natural product channels were $7.5 million and $3.5 million, respectively, in 2016. Each year, the sale of American ginseng root generates, on average, about $27 million for rural harvesters (Chamberlain, Prisley, & McGuffin, 2013).

The floral industry relies heavily on products gathered from the forests and contributes significantly to the economies of Canada and the United States. An early study of the floral greens and decoratives segment revealed that these businesses contributed more than US$128.5 million to the economy of the Pacific Northwest (Schlosser & Blatner, 1994). In 1989, buyers in western Washington and Oregon and southeastern British Columbia purchased US$38 million worth of floral greens and US$9.6 million of boughs. In Canada, the export value of the floral greens sector was estimated to range from CDN$27 million to $65 million over 5 yr (2000–2005), with an average annual value of approximately CDN$40 million (Cocksedge & Hobby, 2006). Bough collection for holiday wreaths is a major economic activity in other places, particularly the northern states of Wisconsin and Michigan.

Other studies also indicate that the floral decorative segment has tremendous value. In 1995, the United States exported moss and lichen, most of which was from southern or northwestern forests, valued at more than US$14 million (Goldberg, 1996). Muir, Norman, and Sikes (2006) reported that from 1998 to 2003 the United States exported between 4,808 dry Mg (5,300 U.S. tons) and 18,416 dry Mg (20,300 U.S. tons) of moss and lichen, and overall sales (domestic and exports) were between US$6 million and $165 million each year.

North Carolina, Florida, and Georgia are the leading pine straw producers. In 1996, the estimated market value for pine straw in North Carolina was about $50 million (Rowland, 2003).

In 2003, the market value for pine straw in Florida was about $79 million (Hodges, Mulkey, Alavalapati, Carter, & Kiker, 2005), and in 2012 pine straw was almost 10% of Georgia's forest products market, at $59 million (University of Georgia Center for Agribusiness & Economic Development, 2013, pp. 109–110). Finally, one company in southwest Virginia specializing in pine roping had retail sales in excess of $1.5 million in 1997 (Hauslohner, 1997).

Other forest plants are used in the floral industry. Galax, an evergreen groundcover, has been harvested from the forests of western North Carolina and southern Virginia since before the 20th century. The primary source of this important floral product is seven counties in western North Carolina, where millions of leaves are harvested annually. Pickers are paid by the leaf and can earn $50–$120 for a box of 5,000 leaves (Predny & Chamberlain, 2005). By some estimates, galax harvesting contributes many millions of dollars to the local economy (Greenfield & Davis, 2003).

Estimates of NTFP value are not based on consistent or reliable collection of production and value data. Few NTFPs are tracked regularly, and the values of those that may have more regular reporting are probably underreported because most wild harvesting is documented poorly or not at all.

Forest Farming Culinary Products

Most people do not think of the forest as a source of food, but there are a surprising number of plants that have been consumed traditionally by people who live in or near the woods. Edible and culinary products that can be farmed in the forests of North America include mushrooms, ferns, and plants harvested for their fruits, leaves, and roots. Every year, people forage these plants from the forests and share them with their friends and family. The wild foods and local foods movements, however, have created commercial demand for many of these unusual and tasty forest plants. Forest foods can now be found at farmers' markets, roadside stands, specialty markets, natural foods stores, and even some supermarkets. Some have been highlighted on cooking shows and in gourmet magazines. Many are sold over the Internet; a search with the term *wild foods* will reveal a tremendous number of sources.

There is potential to make a profit from cultivating some forest foods, but understanding the markets is critical. Only a small percentage of the public will buy these products, but those who do are devoted customers who will have a tendency to return year after year. This group includes wild food enthusiasts, people who grew up eating these foods, "foodies," and gourmet chefs at high-end restaurants. The average consumer will stop and look at these with interest but probably will not buy. To be successful, the forest farmer will have to find a way to tap into the market. Local food campaigns and the Internet are obvious places to start.

Vegetables

There are many forest vegetables that have potential for forest farming. Appendix 9-1 provides an illustrative list of edible forest products that can be forest farmed. Most are wild harvested for personal use or local sales. One of the more popular forest vegetables in the eastern United States is the ramp, a member of the onion family related to leeks and garlic. There is an increasing number of enterprising landowners growing ramps for commercial sale (Figure 9–6).

Ramps are native to the eastern North American mountains and can be found growing in patches in rich, moist, deciduous forests from Canada, west to Missouri and Minnesota, and south to North Carolina and Tennessee. In early spring, ramps send up smooth, broad, lily-of-the-valley-like leaves that senesce by mid-May before the white flowers appear. The bulbs have the pleasant taste of sweet spring onions with a strong garlic-like aroma.

As one of the first plants to emerge in the spring, ramps traditionally were consumed as the season's first "greens." They were considered a tonic because they provided vitamins and minerals following long winter months without any fresh vegetables. Traditions evolved around the annual gathering and preparation of this pungent plant, and in the early 20th century rural communities started hosting annual spring ramp festivals. These festivals have become major tourist attractions and are promoted by the communities in which they are held. The tremendous volumes of ramps that are consumed at these festivals are gathered from nearby forests.

Many high-end restaurants serve ramps in a variety of dishes, increasing the demand for large, consistent supplies of the wild forest plant. In an effort to conserve native populations and meet the rising demand, forest farming of ramps is strongly encouraged. Harvesting ramps from easily accessible, concentrated plantings would benefit festival participants, chefs, and consumers and create a new marketable product for forest farmers.

Fig. 9–6. Forest farming ramps, an edible forest product, can be a productive use of small forest holdings (photo credit: John Munsell, Virginia Tech).

In the southeastern United States, ramps begin emerging in March and grow rapidly through April before the leaves senesce in early May. They grow best in cool, shady areas with damp soil and an abundance of decomposed leaf litter or other organic matter. In late May and early June, after the leaves senesce, a flower stalk fully emerges. The flower blooms and the seeds develop in late summer. The seeds mature atop a leafless stalk and eventually fall to the ground to germinate near the mother plant. The timing of these events is usually delayed at high elevations and more northern locations.

Ramps grow naturally under a forest canopy of beech, birch, sugar maple (*Acer saccharum* Marsh.), and/or poplar. Other forest trees under which ramps will grow include buckeye (*Aesculus* spp.), American basswood, hickory (*Carya* spp.), and oak (*Quercus* spp.). A forested area with any of these trees provides a possible location for planting a ramp crop. Areas that host understory species, such as trillium, toothwort (*Cardamine* spp.), nettle (*Urtica dioica* L.), black cohosh, American ginseng, bloodroot, trout lily (*Erythronium umbilicatum* Parks & Hardin), bellwort (*Uvularia* spp.), and mayapple (*Podophyllum peltatum* L.) should be suitable for growing ramps.

Greenfield and Davis (2001) recommended growing ramps in well-drained sites that have rich, moist soil with high amounts of organic matter. Soil moisture is an important variable influencing the seed germination, seedling emergence rate, survival, and growth rate of the plant. Adequate soil moisture must be maintained throughout the year, not just during the active growing season. The growth period for ramps is limited to about 10 wk in the spring, during which time the plant is dependent on adequate light, moisture, and nutrients.

Although ramp seeds can be sown any time the soil is not frozen, early to late fall is usually considered the best time for seeding ramps. Fresh ramp seeds have a dormant, underdeveloped embryo. The seed requires a warm, moist period to break root dormancy and a subsequent cold period to break shoot dormancy. With enough warm weather after sowing, root dormancy will be broken. During winter, cold weather is sufficient to break shoot dormancy, and the plants emerge in spring. If there is not an adequate warm period after sowing, the seed will not germinate until the second spring. Thus, ramp seeds can take 6–18 months to germinate. For example, in Mills River, NC, ramp seeds that were sown in the fall of 1999 and spring of 2000 all germinated in April 2001 (Davis & Persons, 2014). Providing adequate soil moisture and protection from wildlife are key factors in determining where and when to sow seeds. Production from seeds to root harvest can take 5–7 yr.

To plant under a forested canopy, rake back the leaves on the forest floor, remove any unwanted weeds, tree sprouts, or roots. If the soil is not high in organic matter, incorporate organic materials such as composted leaves and other decaying plant material from the forest. Loosen the soil and rake to prepare a fine seed bed. Sow seeds thinly on top of the ground, pressing them gently into the soil. Cover the seeds with several inches of leaves to retain soil moisture and to protect the seeds from wildlife.

Many growers prefer planting bulbs or young plants instead of sowing seeds. Since germination of the seed can take up to 18 months, transplants and bulbs may be a good alternative for the beginning ramp grower. Planting large bulbs can provide harvestable ramps within 2–3 yr. Bulbs can be purchased in February and March and transplanted as soon as the soil is thawed and free of snow. If bulbs are to be dug for transplanting, once the ground has thawed, gently dig the ramps, taking great care not to damage the rhizomes, roots, or bulbs. In a prepared planting bed, transplant the bulbs approximately 7.6 cm (3 inches) deep, and 10.2–15.2 cm (4–6 inches) apart, allowing all the roots to be buried and keeping just the very tip of the bulb above the surface. Planting bulbs at the proper depth is important for survival. Transplant leafed-out plants at the same depth they had been growing and space 10.2–15.2 cm (4–6 inches) apart. If space is limited, clumps of four or five plants can be grouped together. Mulch the planting bed with at least 5–7.6 cm (2–3 inches) of leaf litter.

Hardwood leaves provide the best mulch for ramps. Poor results have been obtained with pine bark and commercial mulches, and they should be avoided until further research is done. The effects of mulching are numerous: decaying organic matter provides essential elements like nitrogen, much needed moisture is retained within the mulched area, and the mulch acts as an insulator to protect the plants from freezing temperatures. In addition, mulching helps to suppress weeds as well as protect newly sown seeds and seedlings from wildlife.

In native populations, ramps usually form large colonies or patches. Often the bulbs are so densely spaced that they outcompete other vegetation. In a forest farming system, do not harvest plants until they have filled the planting area, have large bulbs, and have flowered. If the whole plot is to be harvested at one time, a general recommendation is to have enough patches to allow for a 5–7- yr rotation. That is, to have an annual harvest, harvest only one-fifth to one-seventh of your production area each year. When harvesting a portion of a plot, no more than 15% of the ramps should be removed. If the thinning method is used, great care should be taken not to damage plants that are not harvested.

Digging methods are the same as those described for transplanting. Again, great care should be taken not to damage the bulbs. While harvesting, keep the dug ramps cool and moist. When harvesting is complete, wash the ramps thoroughly and trim off the roots. Pack in waxed cardboard produce boxes and store in a cool place, preferably a walk-in cooler. Do not store in airtight containers.

Conservationists and ecologists across the native ramp range are encouraging ramp growers and wild harvesters to harvest only the leaves of the ramp plant, leaving the bulbs intact to regrow the following year. There is some evidence that this will help keep ramp populations healthy, although more research needs to be done to better understand at what stage the leaves can be cut, whether this can be done every year, and how close to the bulb the cut can be made.

Very little information is available on pests and diseases that may affect ramps. In North Carolina and Tennessee, Septoria leaf spot has been observed in wild and cultivated ramps. Although the spot was unsightly on the foliage, it did not appear to adversely affect plant yields in 2001 (Davis & Persons, 2014). The long-term effects of the disease are unknown. The allium leaf miner (*Phytomyza gymnostoma*) and the leek moth (*Acrolepiopis assectella* Zeller) are emerging insect pests affecting ramps in the Northeast (Barringer, Fleischer, Roberts, Spichiger, & Elkner, 2018; Davis & Persons, 2014). New ramp plantings do not compete successfully with weeds, thus weeds should be controlled until the plants are well established.

Mushrooms and Fungi

Many edible mushrooms, such as shiitake (*Lentinula edodes* Berk.), maitake (*Grifola frondosa* Dicks.), lion's mane (*Hericium erinaceus* Bull.), and oyster (*Pleurotus* spp.) can be grown in a forest farming setting. The shiitake mushroom is the most popular for small-scale cultivation. Production of shiitake in the United States started about three decades ago, when demand exceeded the ability of importers to fulfill orders and the technology for landowner production became readily available and simple. Rural development agencies began promoting shiitake mushroom production as an alternative income source for forest landowners. Many landowners started producing this valuable mushroom, and today it is well accepted in gourmet markets.

Shiitake mushrooms grow best on hardwood logs cut from live trees (Mudge & Gabriel, 2014).

The first step in producing shiitake mushrooms is to select the best tree species for the logs. Shiitake mushrooms grow best on white oak and red oak (*Quercus rubra* L.), although hard maple also works well. They also can be produced, with varying degrees of success, on ironwood (*Carpinus caroliniana* Walt.), alder (*Alnus* spp.), cottonwood (*Populus deltoides* W. Bartram ex Marshall), poplar, beech, and sweetgum (*Liguidambar styraciflua* L.).

Logs should be cut from living trees that have no decay. The best time to harvest the logs is during the dormant months, when the wood has the greatest amount of stored carbohydrates. The bark also will remain intact longer if the logs are cut during this time of year. The logs should be inoculated within 3 wk after felling. The longer cut logs are left un-inoculated, the greater the chance that native wood-decay fungi will invade the logs and compete with the shiitake mushroom mycelium to reduce yields.

Log length is not as critical as log diameter. The length should be determined by what is most easily managed by the person involved in moving them. In general, logs 0.9–1.2 m (3–4 ft) long and larger than 7.6 cm (3 inches) in diameter work well. Smaller logs will dry out more quickly. Logs greater than 15.2 cm (6 inches) in diameter may produce longer but need more inoculum to compensate for the greater diameter. It is important to retain the moisture content of the wood while keeping the bark relatively dry. If cut logs will not be inoculated for several weeks after cutting, they should be covered with a porous material (e.g., burlap or muslin) and watered regularly.

Logs are prepared to receive the spawn (a substrate that contains active mycelium) by drilling a diamond pattern of holes through the bark and into the sapwood. Holes should be drilled 15.2 cm (6 inches) apart within rows along the length of the log, with 5.1–10.2 cm (2–4 inches) between rows. The number of rows along the length of the log depends on the diameter of the log. In general, there should be one row less than the diameter in inches of the log. So, if a log is 15.2 cm (6 inches) in diameter, there would be five rows of holes. The diameter and depth of the holes depend on the type and variety of spawn used.

The fungus is introduced by inoculating the logs with mycelium in the form of spawn. Suppliers may recommend cold, warm, or wide-ranging weather spawn, depending on local growing conditions. Spawn can be refrigerated for several weeks but should be kept at room temperature for a few days before inoculation. Spawn is available in three forms: sawdust plugs, dowels, or a pre-sealed plug that requires no wax at the inoculation site. Spawn should be inserted into the holes immediately after the holes are drilled to reduce possible contamination. Once the holes are filled, they should be sealed with hot wax (paraffin) or impermeable plugs to prevent drying. Some growers also seal the ends of logs with wax to reduce contamination and moisture loss.

After inoculation, the logs are stacked and protected from moisture loss to allow the shiitake mycelium to spread throughout the log. This incubation, the *spawn run*, period takes from 6 to 18 months and depends on the type of spawn, log size, moisture content of the log, and temperature. Protect the logs during the incubation period by providing sufficient shade (60–80%). Log moisture content should be monitored to prevent the logs from drying out. Logs should be stacked in a particular pattern—the lean-to and criss-cross are two common designs (Figure 9–7). Logs should not be placed directly on bare soil but should be raised off the ground.

The spawn run is complete when white mycelia appear on the end of the logs. The fungi will fruit when the weather conditions are favorable, which typically occurs in the spring and fall. Forcing production to meet a set schedule is possible by soaking the logs in water for 48–72 h; fruiting should begin within a few days. If left alone (not soaked), logs will produce for a longer time period and eventually produce about the same amount. As the optimal harvest time lasts only about 12 h, it is important to check daily for mushrooms. Once the logs begin to fruit, they will produce mushrooms a few times a year for up to 3 yr. After each harvest, the logs need 8–12 wk of rest to allow the mycelia to reproduce.

To create a weekly market for mushrooms, you should fruit 1/12 of the logs weekly. Ideally, logs should be moved to a production house. Mushrooms will not fruit when temperatures exceed 29.4 °C (85 °F) or go below 10 °C (50 °F). To fulfill a weekly market throughout most of the year (not the dead of winter), you should have a controlled-environment building that can be heated and cooled to fruit the most desirable mushrooms.

Farming a forest for mushrooms can be lucrative, but successful commercial producers are those who market them well. The final decision to grow mushrooms as an alternative forest product should be based on economics. As markets develop and more people begin to grow mushrooms, profit margins decrease. The successful producer will figure out how to compete with established and experienced firms by finding niche markets and producing high-quality, low-cost products. One option is to cut and sell

Fig. 9–7. Shiitake mushroom growing on logs stacked in the teepee formation (photo credit: Jeanine Davis, North Carolina State University).

logs to mushroom producers. Another is to market logs that are inoculated and allowed to age for 4–5 months.

A search of the internet can help identify buyers of edible mushrooms. Large grocery chains rely on wholesale distributors for products, and tapping into that segment may be difficult for small producers. It may be possible to sell to wholesaler distributors, but direct sales to local restaurants and consumers through farmer's markets may prove more profitable for small entrepreneurs. To succeed in the edible mushroom business, the entrepreneur must find niche markets and differentiate their products from the many other mushroom producers.

Forest Farming Medicinal Products

For as long as people have wandered the forests, they have gathered and made medicines from plants. With time, people in North America began cultivating herbs, most notably ginseng. By the early 1990s commercial production of ginseng under artificial shade was big business in many places including Wisconsin, Ontario, and British Columbia. Cultivating herbs in the forest, however, was done on a relatively small scale by a few growers scattered across North America. This has changed dramatically in the past 30 yr. There are now many commercial plantings of a wide variety of medicinal herbs in the forests. Individual plantings range in size from a few hundred square meters to more than 20 ha (50 acres).

American Ginseng

An examination of the history of growing medicinal plants in forests is important to better understand the potential of forest farming in North America. In the early 1700s in Canada and in the mid-1800s in the United States, there were "ginseng gold rushes" during which tens of thousands of kilograms of American ginseng were harvested from the forests for export to China. It was not long before the wild populations were severely depleted by overharvesting, and people began experimenting with cultivating the plant.

The first attempts at farming ginseng in North America failed miserably, but in the 1870s, Abraham Whisman of Virginia learned to successfully cultivate it. About that same time, George Stanton in New York started growing it in large quantities under wood lath shade structures. He is generally recognized as the first commercial ginseng grower in America. Prices paid for cultivated ginseng at that time were high, and soon "garden culture" of ginseng became popular enough that the USDA published its first ginseng production publication in 1895. In 1905, Cornell University released a bulletin on diseases of ginseng (Van Hook, 1904). Unfortunately, Alternaria blight soon became a serious problem in many commercial ginseng gardens, and with no means to effectively control the disease, the industry dwindled to about 9 ha (23 acres) by 1909. The discovery that Bordeaux mixture, a copper fungicide, worked well in controlling Alternaria blight caused an upsurge in production, and by 1929, there were approximately 175.5 ha (434 acres) of ginseng under cultivation in the United States.

After World War II, there was a slow increase in ginseng farming in the United States, with the bulk of the acreage being in Wisconsin. According to Davis and Persons (2014), production in the United States peaked in 1996, with more than 1 million kg (2.3 million lb) of dried root produced, of which approximately 998,000 kg (2.2 million lb) were field grown. This coincided with a significant increase in production of American ginseng in China and Canada. As expected, this resulted in an oversupply and a decline in prices.

In southern Ontario, Canada, Clarence Hellyer successfully grew ginseng in 1896 under a wood lath structure. His sons, Audrey and Russell, formed Hellyer Brothers in 1918 and grew ginseng commercially until 1970 (Davis & Persons, 2014). In 1962, there were only eight ginseng farms in Ontario, but by 1983 this number had grown to about 60 farms. By the middle of the 1990s, there was about 2041 ha (4,500 acres) of ginseng under cultivation in Ontario and almost 1,361 ha (3,000 acres) in British Columbia (British Columbia Ministry of Agriculture, Food and Fisheries, 2003a). All of this production was under artificial shade.

Similar to the earlier "ginseng gold rushes" in both countries, the boom was followed by bust. Supply grew more rapidly than demand, and prices fell. By 2003, the area under cultivation in British Columbia had declined to about 688 ha (1,700 acres) (British Columbia Ministry of Agriculture, Food and Fisheries, 2003a), representing 162 ha (400 acres) harvested and 590,000 kg (1.3 million lb) of root sold (British Columbia Ministry of Agriculture, Food and Fisheries, 2003b). The number of growers had declined from approximately 130 in the mid-1990s to 40 in 2003 (British Columbia Ministry of Agriculture, Food and Fisheries, 2003a).

The history of forest farming ginseng is difficult to trace because no records were kept for sale of this type of ginseng. Forest-grown ginseng roots were usually mixed in with wild roots and no distinction was made between the two. There is one paragraph devoted to "forest plantings" in the 1913 USDA Farmers' Bulletin (Van Fleet, 1913) on cultivation of ginseng. The 1921 USDA Farmer's Bulletin (Stockberger, 1921) explained that forest plantings were less expensive to establish but yielded about half as much as those under artificial shade. It also explained that growers on the Pacific coast could not grow ginseng in the woods. Stockberger (1921) noted that, "There is always a ready sale for the cultivated roots which closely resemble the wild in quality and conditions, and prudent growers will not fail to adopt the wild root as the standard of future production."

Growing ginseng in the forest produces "wilder looking" roots than growing in artificially shaded open fields (Figure 9–8). An early ginseng manual states that, "Ginseng grown in

Fig. 9–8. American ginseng has been forest farmed in the United States since the mid-1800s, although wild-harvested ginseng is still preferred (photo credit: Gary Kauffman, U.S. Forest Service).

the natural forest bed will command much greater, more attractive prices than its cultivation under artificial shade" (Bryant, 1949). That manual strongly recommended growing ginseng in the natural forest. Yet, there was not widespread interest in growing ginseng in beds in the woods until the early 1990s. It was even more recently that growers seriously tried to produce the most natural looking ginseng by growing in the forest using a "wild simulated" method.

Following the crash in the ginseng industry in the mid 1990s, many of the artificial shade growers who did not quit growing ginseng completely converted to forest farming. In 2000, there were more than 4,000 growers in the United States forest farming ginseng, representing more than 809 ha (2,000 acres) (Davis & Persons, 2014). At the same time in Quebec, there were an estimated 283 ha (700 acres) of ginseng being forest farmed (Nadeau, Ginseng Boreal, Plessisville, QC, Canada, personal communication, 2000). Market predictions indicate that the demand for woods-cultivated and wild-simulated ginseng will continue to increase.

Goldenseal

There are reports from the mid-1700s of European settlers in North America harvesting goldenseal from the forests. By 1860, the herb was in high demand, and by the 1880s there was concern about the impact of overharvesting on native populations (Lloyd, 1912; Van Fleet, 1914). The USDA began experimenting with goldenseal cultivation under artificial shade in 1899 and published the first bulletin on it in 1905 (Henkel & Klugh, 1905). In the fall of 1903, the rising popularity of goldenseal as a medicinal tea pushed prices to US$2.20 per kilogram. This caught the interest of many ginseng farmers, who then began cultivating goldenseal. In contrast to ginseng, however, much of the early goldenseal cultivation appears to have taken place in the forest. The first large-scale commercial producer of goldenseal was the Skagit Valley Golden Seal Farm in the state of Washington in 1905 (Veninga & Zaricor, 1976).

In the early 20th century, goldenseal was planted as an ornamental and grown on a large scale for pharmaceutical purposes (Hus, 1907). The author's concern was that propagation of goldenseal was almost exclusively by division of rootstock, which was not a very efficient method. Seed propagation, however, was extremely difficult and only a few growers attempted it. Hus (1907) stated, "The importance of sowing fresh seed cannot be over emphasized; it is one of the essentials of success." Throughout his short article, Hus dispensed information that would benefit any modern-day goldenseal grower.

A 1912 article by John Uri Lloyd described experiments by Dr. H.T. Grime in Indiana from 1908 in which goldenseal was grown in gardens shaded by beans grown on poles, fruit trees, and grapevines and occasionally sprayed with Bordeaux mixture. Lloyd (1912) reported that the plants grew rapidly, had leaves 30 cm (12 inches) in diameter, and exhausted the soil "worse than tobacco." He concluded that goldenseal was easy to cultivate, and the greatest threat to natural woodland cultivation would come from the poacher.

Goldenseal cultivation, under artificial shade and in the forest, rose and fell in step with supply and demand for the next few decades. Some farmers, including the Skagit Valley operation, experienced severe disease problems that caused them to cease production or move their operations completely. By 1960, the USDA estimated that there were fewer than 2 ha (5 acres) of goldenseal under cultivation in the country. Veninga and Zaricor (1976) estimated that there were 40.5 ha (100 acres) or less in cultivation by the mid-1970s. In the early 1990s, demand for goldenseal rose once again. Pressure on wild populations was at dangerous levels, prompting action on the part of the government and nonprofit agencies to protect the plant. Growers across North America began producing the crop once again.

This latest increase in demand for goldenseal coincided with the drop in demand for ginseng. As a result, many ginseng growers in Wisconsin and Ontario began growing goldenseal. Much of the goldenseal was planted under the same shade structures the ginseng had been grown under. In other areas, forest farming of goldenseal was more popular. According to the American Herbal Products Association (2007), in 1998 there were 42 ha (104 acres) of goldenseal cultivated under artificial shade and 14.6 ha (36 acres) under forest farming. In 2004–2005, there were 25.5 ha (63 acres) under artificial shade and 22.6 ha (56 acres) in the forests. The American Herbal Products Association reported that in 1998, 117,409 dried kg (258,843 lb) of wild goldenseal root and rhizome were harvested compared with only 2,923 kg (6,445 lb) of cultivated material.

Small commercial gardens of ginseng and goldenseal are being planted on private forestland across North America. Growth in this segment of the industry is driven by several factors. Many small landowners want to make extra income from their woodlands without cutting timber. There is a rising demand for domestically produced, high-quality, forest-grown, certified organic material. There is more information and support for growers wishing to produce these crops. People enjoy the connection to their land and heritage and the sense of sustainability that they get from growing their own native medicinal plants.

Other Medicinals

American ginseng and goldenseal are not the only medicinal plants that can be forest farmed (Appendix 9-2). Interest in growing forest botanicals is driven by a variety of forces. As the natural products industry grows, demand for raw materials increases. Recent food safety and quality issues with inexpensive foreign imports have convinced some companies to purchase more domestically produced herbs. Consumers are driving the demand for certified organic products. Concerns about the conservation of wild-harvested herbs are putting pressure on manufacturers to buy cultivated herbs. For many herbs, this has elevated the prices paid for cultivated material. At the same time, increasing numbers of forest landowners are looking for alternative crops, and since many of the herbs in demand are native to North American forests, growers often assume they should be easy to grow.

Forest Farming Floral and Landscaping Products

Creative and entrepreneurial landowners can farm their forests for plants used in the floral industry. Many forest plants and parts of plants are used in arrangements, to complement and furnish the backdrop for flowers, as well as for the main component of dried ornaments. The end uses for many floral greens include fresh or dried flowers, aromatic oils, greenery, basket filler, wreaths, and roping. Floral products include various species of grapevine, kudzu, and smokevine for wreaths and baskets, galax for floral decorations, and twigs from several tree species. Several genera of moss are harvested from forests and used domestically or exported to the European floral industry. These and others can be farmed from the forests by private landowners.

An array of plant species can be forest farmed for the decorative market. Baskets made from branches, needles, and wood splits have ready markets. The tips of pruned evergreen trees can be fashioned into wreaths, roping, and garlands. Christmas trees, from native or exotic evergreens, are a well-established segment of the decorative market. In Vancouver, BC, Canada, the United Flower Growers Cooperative has recorded commercial sales of more than 60 wild-harvested forest species. Some of the top-selling products included: beargrass [*Xerophyllum tenax* (Pursh) Nutt.], pussy willow (*Salix discolor* Muhl.), boxwood (*Buxus* spp.), moss, rose hips (*Rosa* spp.), huckleberry (*Vaccimium* spp.), horsetail (*Equisetum* spp.), ferns, dogwood (*Cornus* spp.), and scotch broom (*Cytisus scoparius* L.) (Cocksedge & Hobby, 2006).

In a forest farming system, it may be necessary or desirable to thin trees to decrease shade or increase spacing. Understory forest trees and plants can be nurtured and harvested to produce useful products. Many native plants are valued for landscaping and could be farmed in a forest setting. Trees, shrubs, and perennials are commonly planted around homes, offices, stores, and streets for both beauty and function. Landowners can supplement their income by forest farming desirable plants to sell as transplants to nurseries, landscape businesses, or homeowners.

A landowner interested in supplying the landscape industry can grow desired plants under the forest canopy. As transplanting causes stress that may weaken or kill the plant, only strong thriving plants that are free from pests and disease should be harvested. Transplanting is best done in late fall and early spring while plants are dormant and the soil is not frozen. The less the root system is disturbed, the better chance a plant has of surviving. After digging, transplants should be watered and placed in an area that is protected from wind and direct sunlight until the landowner is ready to plant or sell them.

Herbaceous perennials may be dug and transferred to a pot for transport. Use a spade or garden fork to gently loosen the soil around the roots. Pulling a plant by the stem or crown can cause damage; the plant should be lifted by supporting the roots instead. Extra soil may be needed to fill in around the roots to stabilize the transplant in the container.

Larger trees and shrubs will need to be balled and wrapped in burlap. Since most of the active water- and nutrient-absorbing roots are at the periphery of the root system, it is necessary to prune the roots to encourage rooting closer to the trunk. New roots will form at the cut edge, inside the area that will be balled.

Galax is a low-growing plant with glossy, attractive, heart-shaped leaves that last a long time after being picked. Depending on what time of year they are harvested, galax leaves range in color from bright green to dark red. They are very desirable in the florist industry. Most of the galax leaves on the market are wild harvested, but galax can be cultivated. It is such a slow-growing plant, however, that few people in the nursery industry have shown interest in it, but it fits well in a diversified forest farming operation. Galax can be grown from seed sown directly in a nursery bed in the woods. Seeds can be sown in flats of potting media in a greenhouse and set out in the woods when the plants are 2 yr old. Galax also can be propagated by dividing the rhizomes in early spring, soaking the rhizome pieces in water at room temperature overnight, and planting them out as you would most woodland botanicals. The plants should be allowed to grow for a minimum of 4 yr before cutting foliage for market (Davis & Persons, 2014).

Trees and Related Products Integral to Forest Farming

Trees are essential to forest farming, and the products from trees can be integral as well. Trees provide services to the understory crops, such as shade, moisture retention and nutrients. There is a wide range of NTFPs that come from trees. These include medicinal products such as Balm of Gilead (*Populus balsamifera* L.) buds, black walnut hulls, and wild cherry (*Prunus serotina* Ehrh.) fruit. There are also a number of barks from tree roots and trunks that are important in the medicinal herb industry, including slippery elm (*Ulmus rubra* Muhl.), black haw (*Viburnum prunifolium* L.), fringe tree (*Chionanthus virginicus* L.), mountain maple (*Acer spicatum* Lam.), prickly ash (*Zanthoxylum americanum* Mill. and *Z. clava-herculis* Mill.), sassafras, eastern white pine (*Pinus strobus* L.), white willow (*Salix alba* L.), and wild cherry. Leaves of some trees are also harvested for the herb and natural dye industry, including black walnut, butternut (*Juglans cinerea* L.), witch hazel (*Hamamelis virginiana* L.), and New Jersey tea (*Ceanothus americanus* L.). To date, wild populations of these trees are usually managed in a forest farming situation, but plantings of slippery elm and witch hazel are becoming more common (Davis & Persons, 2014). Other trees are tapped for their sap, and growing understory plants under these trees can be a viable forest farming system. Understanding the potential for various tree species and related products in forest farming is important in developing dynamic and resilient systems. Presented below are a few examples of tree products that can be integrated in forest farming.

Tapping Trees for Syrup

In North America, tree sap products have increased in economic and cultural importance and can provide additional forest farming inputs. Native people traditionally harvested the sugar maple for its sap to produce sugar as a condiment in foods. Early European settlers learned to make sugar from the Native people and with time developed syrup-making methods. Sap collection entailed boring holes in trees with augers and using wooden spikes to tap the trees and collect the sap in wooden buckets. The sap was then boiled in iron kettles and, later, in flat-bottomed tin pans (Ramlal, Fox, Pate, & Guerra, 2007).

While the use of sugar maple is well established in eastern North America, there is also a history of Native people using sap from big leaf maple (*Acer macrophylum* Pursh.) and paper birch (Turner, 1998). A specialty syrup industry associated with these species (Hobby, Maher, & Keller, 2007) is emerging, and capturing this opportunity has potential for forest farming.

Sugar maple is widespread as a dominant or co-dominant tree species in mixed hardwood forests of eastern North America (Godman, Yawney, & Tubbs, 1990; Tirmenstein, 1991; USDA, 1997). Sugar maple forms pure stands but also grows mixed with other hardwoods and scattered conifers. Common co-dominants include beech, birch, and American basswood. Sugar maples are very tolerant of shade (Ontario Ministry of Natural Resources, 1995; Tirmenstein, 1991; USDA, 1997). They can survive in the shade of other species for years until an opening in the canopy occurs and they are released to grow in partial or full sunlight. This factor makes sugar maple a good candidate for compatible management with other timber crops, where it could be intercropped with coniferous species and/or other mixed hardwoods to promote biodiversity while providing income possibilities between timber rotations.

As sugar maple can be managed either as pure or mixed stands, forest farmers can plan multiple objectives of timber and nontimber opportunities and develop steady cash flows. *Sugar bush*, a term used to describe a stand of sugar maple trees managed for sap production, can be intercropped with longer lived forest perennials and other understory species to create dynamic forest farming systems (Small Woodlands Program of BC, 2001). Sugar bush growth is improved by the presence of nurse trees and logs that provide weed control, reduce wind, and improve tree form and can be harvested once the maple trees mature. Sugar bush stands are less compatible with other forest values such as crops that require flooding, or that require substantial sunlight, as well as the grazing of animals.

Pine Needles for Mulch and Landscaping

Harvesting pine needles, also known as *pine straw*, offers an interim income stream while timber or pulpwood stands are maturing. Pine straw makes attractive landscape mulch and protects the roots of plants from extreme temperatures, supplies nutrients upon decomposition, and reduces weed growth, erosion, and evaporation of water from the soil. The low pH of the resin on the needles creates a preferred environment for acid-loving landscape plants such as azalea, rhododendron, camellia (*Camellia japonica* L.), gardenia (*Gardenia* spp.), and blueberry (*Vaccinium* ssp.). With this in mind, it may be possible to farm the forest for multiple crops—pine straw and live plants for landscaping. Compared with other mulches, pine straw may last longer and cover more area per cost of materials. It has become a preferred mulch throughout much of North America. Pine straw is flammable and should be integrated into homeowner fire plans.

A good site to establish a pine straw operation should be relatively flat with minimal soil erosion potential. The species that produce the most desirable straw are longleaf (*Pinus palustris* P. Mill.) and slash (*Pinus elliottii* Engelm.) pine. Loblolly pine (*Pinus taeda* L.) also may be used, though the needles are shorter and more difficult to bale. Stands with basal areas of 6.96–11.6 m^2 ha^{-1} (75–125 ft^2 $acre^{-1}$) can produce approximately 125–175 bales per raking, each weighing about 13.6 kg (30 lb). Pine straw is the secondary crop to timber and spacing should be determined by the primary objective of growing wood. The first harvest can begin as early as 8–12 yr in plantations, later in natural stands.

Honey Bee Products

The land under forest trees can be used to raise honey bees (*Apis mellifera* L.). These beneficial insects provide valuable products and services, especially pollination of adjacent crops. A single hive can produce 176–265 kg (80–120 lb) of honey each year. Sourwood [*Oxydendrum arboreum* (L.) DC] honey is a particularly desirable product, especially in the southern United States. Specialty honey products such as flavored honeys, packaged honey gifts, creamed honey, honey wine, and mead may command higher prices. Pollen, which contains high levels of protein and other nutrients, is used as a food additive, medicine, and in cosmetics. Beeswax is used in candles, cosmetics, foundation sheets for frames, and other assorted products. Wax is harvested from the cappings that are removed during honey extraction and from other broken combs in the hive. Propolis is a mixture of beeswax and resins from plants and is used in the hive to reduce the entrance, repair cracks, cap brood, and seal off intruders. Antibacterial properties of propolis make it useful in medicines, particularly for wound healing. Further, hives may be rented out to crop growers for pollination services.

Marketing Forest Farmed Products

Marketing is the process of planning and implementing a strategy that includes everything from concept development, pricing, promotion, and distribution of what is being offered through to the exchange of product for money (Ambus, Davis-Case, Mitchell, & Tyler, 2007). Marketing and the economics of production are major challenges.

Effectively marketing forest farming products makes the difference between success and failure for an operation. Prior to undertaking a new initiative with significant investments, forest farmers should develop detailed marketing and business plans. A well thought-out marketing strategy will help focus objectives, distinguish products from others, and improve bargaining positions.

New forest farmers need to learn all they can about the industry and how it works. They should inquire of the agencies (university extension services; state, provincial, and federal departments of agriculture; and agricultural nonprofits) and search the Internet for support programs. Beginning forest farmers should attend conferences and workshops and visit experienced forest farmers to learn about the industry. While most farmers do not want to give away their secrets, most are proud of their successes and enjoy sharing what they are doing with other interested farmers. Offering to pay for consulting time makes the sharing of information a business transaction that most farmers are willing to engage in. There are many "tricks" to dealing with this industry that only experience, yours and theirs, will teach you.

An essential part of the process of entering the market is to develop a clear picture of the industry (Ambus et al., 2007), which includes developing a clear understanding of a number of factors. First, knowledge of product standards is essential. To effectively market forest farmed products, the form in which products are sold, requirements for product handling, and the minimum or preferred purchasing volumes are critical to clearly understand. Second, forest farmers must be aware of external influences within the industry, including fluctuations in markets, effects of climate on local supplies, and seasonal purchasing trends. Finally, the successful forest farmer will be aware of and understand the implications of industry trends. This includes expected long-term growth or declines in markets, development of new product areas, pricing directions, shifts and swings, and new cultivars for production (Ambus et al., 2007).

Refining the marketing plan as knowledge develops may be wise because new options may have better potential in the marketplace. Consideration should be given to a number of factors, including the proximity of buyers, demand and supply of the raw materials, the timing of labor needs and potential competition for labor needs, and the potential for an adequate return on investment. All of these should be compared with other product options (Ambus et al., 2007).

The natural products industry deals with raw materials differently than the average agricultural commodity. The majority of the raw materials for the industry have been obtained by harvesting from natural populations, and that practice remains the norm today. There are many manufacturers of natural products that use NTFPs, although there are relatively few companies who supply the raw materials. Most manufacturers do not purchase directly from the harvester or the forest farmer. They buy raw materials from dealers, who buy from other dealers or directly from harvesters. Beginning forest farmers may be challenged to find buyers or to convince a buyer to purchase their products. As with any commodity, relationships and trust between harvesters and buyers is essential.

There are three main ways to generate revenue from forest farming products and services: marketing commodities, marketing value-added products, and marketing services and experiences.

Commodities

There is a wide variety of products that can be potentially marketed as commodities from forest farms. Medicinal herbs, edible products, floral decoratives, and live plants for the nursery landscaping segment are just a few examples of products that can be sold into "commodity" or raw material markets. Pricing for commodities can range from relatively stable to significantly fluctuating depending on supply and demand, including supply from other parts of the world where similar species grow. In general, commodities are marketed in large volumes at low prices per unit.

Selling commodities may be the simplest way to enter the market—and is potentially less risky than other approaches—but may not always provide the best returns. In areas where there are buyers in reasonable proximity, selling raw product is an easy way to become familiar with the sector and with the demands of the market for product quality. Production information will be plant and/or region specific, offering opportunities to test and adapt available information to new regions. Similarly, marketing principles are highly portable, thus producers may learn as much from another producer or resource professional across the country as from someone located within their province or state.

In the end, forest farmers may determine that selling commodities does not provide the returns they are looking for or need to stay in the production of a particular good or service. Rather than abandoning the idea of producing a crop, it may be worthwhile to take the time to explore potential value-added opportunities that may provide better returns on investment.

Value-Added Products

When making decisions about the choice of products and how to market them, forest farmers should consider ways to add value. Adding value to forest farmed products is a way to receive higher prices for the same amount of raw material. For forest farmers, adding value to crops can make a significant difference to the bottom line. Adding value to products takes time, resources, creativity, and skill. Some forest farmers may prefer to market commodities, and this is a personal decision that must be considered.

Value can be added to raw products by changing the form of the product or by selling further along the marketing chain (Small Woodlands Program of BC, 2001). Changing the form of the product includes grading, drying, processing, canning, and freezing. Selling further along the market chain may include through retail outlets as opposed to a wholesaler. Apart from gaining an increase in price for the same volume of raw materials, processing raw materials can provide other benefits. Processing allows the forest farmer to sell some products out of season or over a longer period of time. Processing also allows for differentiating products using the same primary material to create a "niche" market.

Greater profits often are achieved by adding value, but there are also greater costs and risks. These include increased requirements for investment in materials and equipment, training, and operating capital as well as market research. Prices increase further down the marketing chain, but it may also take a good deal of time to find and sell products to a large number of retail enterprises rather than to a single broker or processing plant.

Useful information for producing and marketing value-added products is available from many sources that deal with general industry categories, such as food processing and landscaping. Other resources to consider for valuable information include: business planning resources; services and programs available from provincial or state, federal, and local agencies; nonprofits; and university extension programs. As with any business venture, the forest farmer will have to do the research, consider resources, and "crunch the numbers" to determine if a specific idea for adding value is worthwhile.

Services and Experiences

Forest farmers can learn from the agritourism industry about the potential for marketing services and experiences. For a growing population of urban, educated, and affluent people, rural life, rural skills, and rural products can exert a powerful pull to experience these activities. Depending on location, forest farms may have the potential to provide services to local communities and visitors. For example, mushroom and berry-picking outings and festivals, visits to a birch or maple forest for "sugaring off," or cut-your-own Christmas trees with hot cider and perhaps a wagon ride are experiences and services that consumers are willing to pay for. Classes and instruction in making wreathes, rustic furniture, floral design with wild plants, gourmet cooking with wild and local foods, or other crafts are also popular in many places.

Guided experiences on a forest farm are ways of adding value to the resources. Festivals also hold potential for capturing more value at the local level by combining a number of activities to draw people in. The organization required to run a successful festival is significant but so too can be the returns. Berry festivals, wild-mushroom festivals, and herb gatherings have met with success in various locations. There is also the possibility of organizing a fall harvest festival where forest farm products could play important roles. Often one specific product is the focus of a festival, but the spin-offs for the community are much greater than created by the single resource.

Sustainability

Sustainable production of NTFPs through forest farming is one of the basic premises of this agroforestry practice. The concept of forest farming is that it is a way to produce these products without long-term detriment (i.e., sustainably) to the plant populations. Realizing this goal is not a given, however; the forest farmer needs to be aware of the potential for unsustainable production. This is not of particular concern in woods-cultivated forest farming, where the area is planted, tended, and harvested when the product is ready. In wild-simulated forest farming, conversely, sustainable production is an issue that needs consideration. Without planning and managing for sustainable production, it is possible for plant populations to be exploited and decline with time.

Simply, sustainable production with forest farming requires that harvest volumes do not exceed growth volumes. Fundamentally, the amount harvested cannot surpass the volume of product produced. If 45.5 kg (100 lb) of product is generated each year, then the amount harvested cannot exceed that amount. If it does, then the plant population will decline with time, although the decline may not be readily apparent.

The challenge of meeting the goal of sustainable production is that determining how much product is produced each year is challenging at best. With timber, determining how much biomass is added annually simply requires measuring the height and diameter of the tree and calculating the increase in volume. In producing a nontimber product that does not require extraction of the entire plant, or does not increase mortality of the plant, such as harvesting fruits, seeds, or leaves, the potential for unsustainable production is not a major concern. When producing a product that cannot be seen or measured, such as roots, determining how much biomass is added each year is not easily possible. To ensure that production is sustainable, forest farmers need methods to estimate the annual mean increase in volume of the desired product. Research has been done to estimate the amount of black cohosh roots in a patch (Chamberlain et al., 2013) but not to determine how much is produced annually. In general, methods to estimate sustainable production are lacking.

To address this issue, forest farmers need to inventory the plant population to estimate its size and then monitor the population with time. An inventory of the plant population provides information on the number of plants before harvesting begins. Estimating the age class distribution of the population, the proportion of adults and juvenile plants, is important as it allows the forest farmer to determine the stability of the plant population. New cohorts should be added to the population regularly. If with time new plants are not being added to the population, or if juvenile plants are not becoming adults, there should be concern that production is not sustainable. A forest farmer confronted with this scenario has serious choices to make. The forest farmer could cease production in the declining patch and move to another patch. This may assuage the immediate impact of declining production but it does not mitigate the problem. The astute forest farmer would continue to monitor the plant population in the fallowed patch to see if new plants were being added and if juveniles were becoming adults. If this was not happening, the wise forest farmer would start planting seeds or seedlings to help increase the population. Similar actions would be taken in new patches as the forest farmer moved to other populations. In this way, production of NTFPs through wild-simulated forest farming can be sustainable.

Conclusions

Douglas and Hart (1976) recognized that farm forests were ecosystems in themselves. They proposed designing them to conform to ecological principles and practices. The systems should at least preserve, and ideally improve, ecological functions. Modern forest farming, whether extensive or intensive and well planned, holds to these principles. The fundamental purpose of forest farming as suggested by Douglas and Hart (1976) is to integrate the components into a complete stable dynamic system that supports the productive function of them all.

Most contemporary discussions of forest farming include only medicinal plants and other crops such as mushrooms and shade-tolerant plants, yet there is a broader range of opportunities for landowners to farm in their forests for various product markets. A critical feature that distinguishes forest farming from other agroforestry systems is that forest farming incorporates shade-tolerant, nontimber forest resources with trees that form a closed canopy and may be grown for timber using co-management strategies based in silviculture and other holistic ecosystem approaches to forest management. These nontimber forest resources are a main component of forest farming, and the co-management of overstory trees with shade-tolerant understory plants is a major objective and challenge of forest farming.

There are advantages and disadvantages to forest farming of which landowners need to be aware. When well designed and implemented, forest farming can help improve forest health by increasing plant diversity as well as removing injured or contaminated vegetation. Forest farming can lead to additional income opportunities as well. Conversely, it may require more and different skills and expertise. Compared with traditional agricultural or forest commodities, there may be few marketing structures in place and relatively sparse information regarding crop management.

Integrating forestry and farming activities requires broader knowledge to successfully manage the trees, understory, and their interactions, but it is also an opportunity for rehabilitating degraded stands and finding holistic management possibilities for next-generation owners. Forest farming can take more time and energy, which may be limiting factors. In general, landowners may lack the knowledge and expertise to understand and enter the markets for many forest farming products. The task of learning about these new opportunities may be overwhelming for many, but for the patient entrepreneur, forest farming can be rewarding.

Public policies and programs seldom provide support or encouragement for forest farming. From the perspective of agricultural policy, forest farmed products may not be considered "crops" for purposes such as crop insurance, marketing assistance, or other government support. In the long term, therefore, forest farmers should work with other individuals and organizations with similar aims to influence policy and programs that are more supportive of multiple uses of forest lands.

Despite difficulties and challenges, there are many forest farming opportunities for landowners to pursue to generate extra income. Those with large holdings of pine forests might think about managing for pine straw or coinciding pruning activities with evergreen bough markets. A forest landowner with a propensity for growing or digging native plants might do well to investigate starting a landscaping business. Those that are partial to gardening or animal husbandry could grow mushrooms or raise bees for additional income. Landowners with an inclination for making crafts could farm a variety of forest species that are used for decorative products.

Whichever forest farming alternative is most appealing, the interested landowner should always investigate the costs, benefits, and potential pitfalls of each option. If there is a ready market, what barriers might impede a landowner from entering the market? The landowner must have a clear picture of the economics before committing resources to any venture. Interested landowners need to evaluate whether they have the skills needed to undertake the desired alternative. One of the first things a landowner should do is inventory and identify resources, habitats, opportunities, and constraints. If possible, invite a forestry consultant, state forestry agency professional, or extension agent to walk the property and discuss possible options. The decision to grow a specific crop or embark on a forest farming alternative, however, is ultimately the producer's decision. Thus, potential forest farmers need to identify where the information and resource gaps occur and to assess them against their degree of acceptable risk in pursuing alternatives of interest.

Acknowledgments

This chapter builds on the excellent contributions of all co-authors of the chapter presented in the second edition of *North American agroforestry: An integrated science and practice*. We expressly acknowledge that the second edition had tremendous contributions from Canadian colleagues. Unfortunately, we were unable to secure their contribution for this edition, but we want to make sure that their previous inputs are acknowledged. Much of the text devoted to forest farming in Canada should be credited to contributions of Darcy Mitchell, Tim Brigham, Tom Hobby, and Lisa Zabek. Thank you!

References

Ambus, L, Davis-Case, D., Mitchell, D., & Tyler, S. (2007). Strength in diversity: Market opportunities and benefits from Small Forest Tenures British Columbia. *Journal of Ecosystems & Management*, *8*(2), 88–89.

American Botanical Council. (2017, 7 Sept.). *US sales of herbal supplements increase by 7.7% in 2016: Total retail sales surpass $7 billion in 2016, the 13th consecutive year of sales growth* [Press release]. http://cms.herbalgram.org/press/2017/USSalesofHerbalSupplementsIncreaseby77percentin2016.html?t=1504734295&ts=1504734303&signature=d764fdbd5fcd4e7a55eda008335e0c73.

American Herbal Products Association. (2007). *Tonnage survey of select North American wild-harvested plants, 2004–2005*. Silver Spring, MD: American Herbal Products Association.

Barringer, L. E., Fleischer, S. J., Roberts, D., Spichiger, S.-E., & Elkner, T. (2018). The first North American record of the allium leafminer. *Journal of Integrated Pest Management*, *9*(1), 8. https://doi.org/10.1093/jipm/pmx034

Blancas Vázquez, J., Caballero, J., & Beltrán Rodríguez, L. (2017). *Los productos forestales no maderables de México* (CONACYT Project no. 280901). Mexico City, Mexico: Red Temática Productos Forestales No Maderables y Consejo Nacional de Ciencia y Tecnología.

Brevoort, P. (1998). The booming U.S. botanical market: A new overview. *HerbalGram*, *4*, 33–45.

British Columbia Ministry of Agriculture, Food and Fisheries. (2003a). *Ginseng production guide for commercial growers*. Victoria, BC, Canada: BCMAFF.

British Columbia Ministry of Agriculture, Food and Fisheries. (2003b). *Annual B.C. horticultural statistics*. Victoria, BC, Canada: BCMAFF.

Bryant, G. (1949). *The Wildcrafters ginseng manual: A guide to American ginseng* (4th ed.). Rockville, IN: Laurence Barcus.

Butler, B., Hewes, J. H., Dickinson, B. J., Andrejczyk, K., Butler, S. M., & Markowiski-Lindsay, M. (2016). *USDA Forest Service national woodland owner survey: National, regional, and state statistics for family forest and woodland ownerships with 10+ acres, 2001–2013* (Res. Bull. NRS-99). Newtown Square, PA: U.S. Forest Service, Northern Research Station. https://doi.org/10.2737/NRS-RB-99

Center for Subtropical Agroforestry. (2007). *Forest farming*. Gainesville, FL: University of Florida.

Chamberlain, J., Bush, R., & Hammett, A. L. 1998. Non-timber forest products: The other forest products. *Forest Products Journal*, *48*(10), 10–19.

Chamberlain, J., Davis, J., Duguid, M., Ellum, D., Farrell, M., Friday, J. B., . . . Zasada, J. (2018). Non-timber forest products & production. In J. L. Chamberlain, M. Emery, & T. Patel-Weynand (Eds.), *Assessment of non-timber forest products in the United States under changing conditions: A technical synthesis report for the 2017 National Climate Assessment* (Gen. Tech. Rep. SRS-232, pp. 9–58). Asheville, NC: U.S. Forest Service, Southern Research Station.

Chamberlain, J., Teets, A., & Kruger, S. (2018). *Nontimber forest products in the United States: An analysis for the 2015 National Sustainable Forest Report* (Gen. Tech. Rep. SRS-229). Asheville, NC: U.S. Forest Service, Southern Research Station.

Chamberlain, J. L., Ness, G., Small, C. J., Bonner, S. J., & Hiebert, E. B. (2013). Modeling below-ground biomass to improve sustainable management of *Actaea racemosa*, a globally important medicinal forest product. *Forest Ecology and Management*, *293*, 1–8.

Chamberlain, J. L., Prisley, S., & McGuffin, M. (2013). Understanding the relationship between American ginseng harvest and hardwood forests inventory and timber harvest to improve co-management of the forests of eastern United States. *Journal of Sustainable Forestry*, *32*, 605–624.

Cocksedge, W., & Hobby, T. (2006). *Critical information for policy development and management of non-timber forest products in British Columbia: Baseline studies on economic value and compatible management* (Forest Science Program Project Y061065) [Executive summary]. Victoria, BC, Canada: Royal Roads University. Retrieved from https://www.for.gov.bc.ca/hfd/library/FIA/2006/FSP_Y061065a.pdf

CONABIO. 2012. Musgos, hepáticas y antoceros [Data set]. Retrieved from https://enciclovida.mx/especies/135296-bryophyta.

Cornell Cooperative Extension. 2007. *Forest farming in New York's southern tier*. Ithaca, NY: Cornell University.

Davis, J. M. (2016). *Suggested good agricultural and collection practices for North Carolina herbs* (AG-810). Raleigh, NC: NC State Extension.

Davis, J. M. & Persons, W. S. (2014). *Growing and marketing ginseng, goldenseal and other woodland medicinals*. Gabriola Island, BC, Canada: New Society Publishers.

Department of the Interior and Related Agencies Appropriations Act of 2000, Pub. L. 106-113, 113 Stat.1501A-199 (2000). https://www.govinfo.gov/content/pkg/PLAW-106publ113/pdf/PLAW-106publ113.pdf.

Douglas, J. S. & Hart, R. A. J. (1976). *Forest farming: Towards a solution to problems of world hunger and conservation*. London: Intermediate Technology Publications.

Eisenberg, D. M., Kessler, R. C., Foster, C., Norlock, F. E., Calkins, D. R., & Delbanco, T. L. (1993). Unconventional medicine in the United States. *The New England Journal of Medicine*, *328*(4), 246–252.

Farnsworth, N. R., & Morris, R. W. (1976). Higher plants: The sleeping giant of drug development. *American Journal of Pharmacy and the Sciences Supporting Public Health*, *148*(2), 46–52.

Freed, J. (1994). Special forest products: Past, present, future. In C. Schnept (Ed.), *Dancing with an elephant: Proceedings of the conference, the business and science of special forest products* (pp. 1–11). Portland, OR: Western Forestry and Conservation Association.

Godman, R. M, Yawney, H. W., & Tubbs, C. H. (1990). Sugar maple. In R. M. Burns & B. H Honkala (Tech. Coord.), *Silvics of North America (Hardwoods)* (Timber management research, agricultural handbook 654, pp. 78–91) Washington, DC: U.S. Forest Service.

Goldberg, C. (1996). From necessity, new forest industry rises. *New York Times, National Report Section, 24 Mar.*, p. 1.

Gomez-Pompa, A., Kaus, A., Jimenez-Ozornio, J., Bainbridge, D., & Rorive, V. M. (1993). Mexico. In Sustainable agriculture and the environment in the humid tropics (pp. 483–548). Washington, DC: National Academies Press. https://doi.org/10.17226/1985

Greenfield, J., & Davis, J. M. (2001). *Cultivation of ramps* (Allium tricoccum *and* A. burdickii). Horticulture information leaflet 133. Raleigh, NC: Department of Horticultural Science, North Carolina State University.

Greenfield, J., & Davis, J. (2003). *Collection to Commerce: Western North Carolina non-timber forest products and their markets*. Retrieved from https://ntfpinfo.us/docs/other/GreenfieldDavis2003-NCNTFPandMarkets.pdf.

Guzmán, G. (1999, 25 Oct.). Los hongos en México. *La Jornada Ecológica*. Retrieved from https://www.jornada.com.mx/1999/11/14/eco-textos.html.

Hauslohner, A. W. (1997, 6 Dec.). Couple builds green empire with pine-roping outfit. *Roanoke Times, Metro Edition*, A1.

Haynes, R. W., Monserud, R. A., & Johnson, A. C. (2003). Compatible forest management: Background and context. In R. A. Monserud, R. W. Haynes, & A. C. Johnson (Eds.), *Compatible forest management* (pp. 3–32). Dordrect, the Netherlands: Kluwer Academic Publishers. http://www.fs.fed.us/pnw/pubs/journals/pnw_2003_haynes001.pdf

Henkel, A., & Klugh, F. G. (1905). *Golden seal* (Bull. 51, Misc. Papers, Part 6, p. 35–46). Washington, DC: U.S. Government Printing Office.

Herrera Castro, N. (1990). *Estudios ecológicos en los huertos familiares Mayas*. Riverside, CA: Maya Sustainability Project.

Hobby, T., Dow, K., & MacKenzie, S. (2006). *Commercial development of NTFPs and forest bio-products: Critical factors for success—A salal* (Gaultheria shallon) *case study*. Victoria, BC, Canada: Centre for Non-Timber Resources, Royal Roads University.

Hobby, T., Maher, K., & Keller, E. (2007). *A case study of bigleaf maple sap harvesting on Vancouver Island*. Victoria, BC, Canada: Centre for Non-Timber Resources, Royal Roads University.

Hodges, A.W., Mulkey, W. D., Alavalapati, J. R. R., Carter, D. R., & Kiker, C. F. (2005). *Economic impacts of the forest industry in Florida, 2003*. Gainesville, FL: University of Florida, Institute of Food & Agricultural Sciences.

Hus, H. (1907). The germination of *Hydrastis canadensis*. In *Eighteenth Annual Report of the Missouri Botanical Garden* (p. 85–95). St. Louis, MO: Missouri Botanical Garden.

Jones, S. B., Luloff, A. E., &Finley, J. C. (1995). Another look at NIPFs: Facing our "myths". *Journal of Forestry, 93*(9), 41–44. https://doi.org/10.1093/jof/93.9.41

Le Bars, P. L., Katz, M. M., Berman, N., Itil, T. M., Freedman, A. M., & Schatzberg, A. F. (1997). A placebo-controlled, double-blind randomized trial of an extract of *Gingko biloba* for dementia. *The Journal of the American Medical Association, 278*, 1327–1332.

Lloyd, J. U. (1912). The cultivation of hydrastis. *Journal of the American Pharmaceutical Association, 1*, 5–12. https://doi.org/10.1002/jps.3080010103

Mudge, K., & Gabriel, S. (2014). *Farming the woods: An integrated permaculture approach to growing food and medicinals in temperate forests*. White River Junction, VT: Chelsea Green Publishing.

Muir, P., Norman, K., & Sikes, K. (2006). Quantity and value of commercial moss harvest from forests of the Pacific Northwest and Appalachian regions of the US. *The Bryologist, 109*(2), 197–214.

Munsell, J. F., Davis, J. M., & Chamberlain, J. L. (2018). Forest farming. In M. Gold, H. Hemmelgarn, G. Ormsby-Mori, & C. Todd (Eds.), *Training manual for applied agroforestry practices* (pp. 115–126). Columbia, MO: University of Missouri Center for Agroforestry.

Munsell, J. F., Germain, R. H., Luzadis, V. A., & Bevilacqua, E. (2009). Owner intentions, previous harvests, and future timber yield. *Northern Journal of Applied Forestry, 60*, 374–381.

National Agroforestry Center. 1997. Forest farming: An agroforestry practice (Agroforestry Note 7). Lincoln, NE: USDA National Agroforestry Center.

Nations, J. D., & Komer, D. I. (1983). Central America's tropical forests: Positive steps for survival. *Ambio, 12*(5), 233–239.

Nyland, R. D. (2002). *Silviculture: concepts and applications* (2nd ed.). New York: McGraw-Hill.

Ontario Ministry of Natural Resources. (1995). Sugar maple. *Ontario Extension Notes*. Retrieved from http://www.lrconline.com/Extension_Notes_English/pdf/sgr_mpl.pdf.

Pojar, J., & Mackinnon, A. (1994). *Plants of coastal British Columbia*. Vancouver, BC, Canada: Lone Pine Publishing.

Predny, M. L., & Chamberlain, J. L. (2005). *Galax* (Galax urceolata): *An annotated bibliography* (Gen. Tech. Rep. SRS-87). Asheville, NC: U.S. Forest Service, Southern Research Station.

Ramlal, E., Fox, G., Pate, G., & Guerra, P. (2007). *Commercial development of non-timber resources: A case study of maple syrup in Ontario*. Guelph, ON, Canada: University of Guelph.

Rowland, G. (2003). *Pine straw market*. New Albany, MS: North Central Mississippi Resource Conservation and Development Council.

Schlosser, W. E., & Blatner, K. A. (1994). An economic overview of the special forest products industry. In C. Schnept (Ed.), *Dancing with an elephant: Proceedings of the conference, the business and science of special forest products* (pp. 11–23). Portland, OR: Western Forestry and Conservation Association.

Schlosser, W. E., & Blatner, K. A. (1995). Economic and marketing implications of the wild edible mushroom harvest in Washington, Oregon and Idaho. *Journal of Forestry, 93*(3), 31–36.

Schultes, R. E. (1941). Plantae Mexicanae IX. *Aechmea magdalenae* and its utilization as a fiber plant. *Botanical Museum Leaflets, Harvard University, 9*(7), 117–122.

Small Woodlands Program of BC. (2001). *A guide to agroforestry in BC*. Victoria, BC, Canada: Forest Renewal BC.

Smith, J. R. (1929). *Tree crops: A permanent agriculture*. New York: Harcourt, Brace and Company.

Smith, T., Kawa, K., Eckl, V., & Johnson, J. (2017). Sales of herbal dietary supplements in US increased 7.5% in 2015: Consumers spent $6.92 billion on herbal supplements in 2015, marking the 12th consecutive year of growth. *HerbalGram, 11*, 67–73.

Smith, T., Kawa, K., Eckl, V., Morton, C., & Stredney, R. (2017). Market report: Herbal supplement sales in US increase 7.7% in 2016. *HerbalGram, 115*, 56–65.

Stix, G. (1998). Plant matters. *Scientific American, 278*, 301.

Stockberger, W. W. (1921). Ginseng culture (Farmers' Bull. 1184). Washington, DC: U.S. Government Printing Office.

Strong, N. A., & Jacobson, M. G. (2005). Assessing agroforestry adoption potential utilizing market segmentation: A case study in Pennsylvania. *Small-Scale Forest Economics, Management and Policy, 4*, 215–228.

Tirmenstein, D. A. (1991). *Acer saccharum*. In Fire effects information system [Database]. Missoula, MT: U.S. Forest Service, Rocky Mountain Research Station, Fire Sciences Laboratory. Retrieved from https://www.fs.fed.us/database/feis/plants/tree/acesac/all.html.

Trozzo, K., Munsell, J. F., & Chamberlain, J. L. (2014). Landowner interest in multifunctional agroforestry riparian buffers. *Agroforestry Systems, 8*, 619–629.

Turner, J. B., Finely, J. C., & Kingsley, N. P. (1977). How reliable are woodland owners' intentions? *Journal of Forestry, 75*, 498–499.

Turner, N. J. (1998). *Plant technology of British Columbia First Peoples*. Vancouver, BC: University of British Columbia Press.

Turner, N. J. (2001). Doing it right: Issues and practices of sustainable harvesting of non-timber forest products relating to First Peoples in British Columbia. *BC Journal of Ecosystems and Management, 1*(11), 44–53.

Turner, N. J., & Cocksedge, W. (2001). Aboriginal use of non-timber forest products in northwestern North America: Applications and issues. In M. R. Emery & R. J. McLain (Eds.), *Non-timber forest products: Medicinal herbs, fungi, edible fruits and nuts and other natural products from the forest* (pp. 31–57). Binghamton, NY: Haworth Press.

University of Georgia Center for Agribusiness & Economic Development. (2013). 2012 Georgia farm gate value report (AR-13-01). Retrieved from https://caed.uga.edu/content/dam/caes-subsite/caed/publications/annual-reports-farm-gate-value-reports/2012-farm-gate-value-report.pdf.

University of Missouri Center for Agroforestry. 2006. *Agroforestry practices: Forest farming*. Retrieved from http://www.centerforagroforestry.org/practices/ff.php.

USDA. (1997). *Plant guide for sugar maple, ACSA3,* Acer saccharum *Marsh*. In PLANTS [Database]. Washington, DC: Natural Resources Conservation Service. Retrieved from https://plants.usda.gov/plantguide/pdf/pg_acsa3.pdf

U.S. Forest Service. (2001). *National strategy for special forest products*. Washington, DC: U.S. Forest Service.

U.S. Food and Drug Administration. (2018a, 27 Feb.). *Current good manufacturing practices (CGMPs) for dietary supplements*. Retrieved from https://www.fda.gov/Food/GuidanceRegulation/CGMP/ucm079496.htm.

U.S. Food and Drug Administration. 2018b. *Food Safety Modernization Act (FSMA)*. Retrieved from https://www.fda.gov/Food/GuidanceRegulation/FSMA.

Van Fleet, W. (1913). *The cultivation of American ginseng* (Farmers' Bull. 551). Washington, DC: U.S. Government Printing Office.

Van Fleet, W. (1914). *Goldenseal under cultivation* (Farmers' Bull. 613). Washington, DC: U.S. Government Printing Office.

Van Hook, J. M. (1904). *Diseases of ginseng* (Bull. 219). Ithaca, NY: Cornell University Agricultural Experiment Station of the College of Agriculture, Botanical Department (Extension Work).

Veninga, L., & Zaricor, B. R. (1976). The cultivation of goldenseal. In R. M. Smartt (Ed.), *Goldenseal/Etc.: A pharmacognosy of wild herbs* (pp. 53–72). Santa Cruz, CA: Ruka Publications.

Wetzel, S., Duchesne, L. C., & Laporte, M. F. (2006). *Bioproducts in Canada's forests: New partnerships in the bioeconomy*. Dordrecht, the Netherlands: Springer.

Workman, S. W., Bannister, M. E., & Nair, P. K. E. (2003). Agroforestry potential in the southeastern United States: Perceptions of landowners and extension professionals. *Agroforestry Systems, 59*, 73–83.

Yuan, R., & Grunwald, J. (1997). Germany moves to the forefront of the European herbal medicine industry. *Genetic Engineering News, 17*(8), 14.

Appendix 9-1—Culinary Plants that can be Forest Farmed

There is an abundance of edible plants that may have a market. Many plant identification books include brief descriptions of or references to historical usage (e.g., Pojar & Mackinnon, 1994). Other sources provide more detailed information on edibility and uses. Some of the plants described below may be obscure, but they illustrate the vast potential of vegetables that can be farmed in the forest. The fact that a plant is discussed below does not mean that it has a ready market. Also, forest farmers need to be concerned about introducing invasive plants and of exploiting rare plants, as both have serious implications. Much of the information provided here is derived from two online databases: the USDA Plants Database (https://plants.usda.gov/java) and Plants for a Future (https://www.pfaf.org/user/Default.aspx).

Bean salad, rosy twistedstalk, scootberry [*Streptopus lanceolatus* Aiton var. *roseus* (Michx.) Reveal]: This native perennial can be found growing in moist wooded areas throughout much of eastern North America. The young leaves and shoots are used in salads or cooked as greens. The small, edible fruit has a melon flavor but can be toxic if eaten in large quantities. The plant is easily propagated by seed or root division. This plant is listed as threatened, endangered, or of special concern in five states. Check on local regulations concerning cultivation and sale of the plant.

Bear grass spiderwort, Virginia spiderwort (*Tradescantia virginiana* L.): This native perennial grows naturally in moist, shaded areas throughout the eastern United States and California. The young leaves and shoots are eaten raw in salads or cooked as greens. The attractive flowers are also edible. The plant also has a number of medicinal uses. Seeds can be started indoors and set out in the spring. Shoot cuttings root easily.

Branch lettuce, mountain lettuce, lettuceleaf saxifrage [*Saxifraga micranthidifolia* (Haw.) Steud.]: This native perennial can be found in the mid-Atlantic states. It is listed as threatened or endangered in two states, so check on local regulations before cultivating or selling it. Branch lettuce grows in moist soils in light shade to full sunlight. In some areas, this is a common springtime food. It is used as a salad green or cooked vegetable. A traditional meal would consist of branch lettuce fried in bacon grease with ramps or wild onions, pinto beans, and corn bread. It can be propagated by dividing the plants in early spring.

Burdock, gobo (*Arctium lappa* L.): Introduced from Europe and Asia, burdock is a biennial that can be found throughout North America. It grows in moist soil in light shade to full sun. Burdock is commercially produced in some areas for its roots. Leaves, stems, and roots are edible. Very young roots can be eaten raw, but usually the roots are boiled, steamed, or sliced for a stir-fry. Young stalks and leaves are eaten raw or cooked. Burdock is also a very important medicinal herb. The plant grows easily from seed. Harvest first-year plants because roots of 2-yr-old plants are woody. There may be some toxicity issues with burdock; the most commonly reported problem is skin sensitivity to the hairs on the seeds.

Dandelion (*Taraxacum* spp., esp. *T. officinale* F.H. Wigg): This common perennial plant grows throughout North America in full sun and partial shade. Some species are native and some are introduced. It is produced commercially as a salad crop. The raw leaves are tastefully bitter. The root can also be eaten raw or cooked. The flowers are eaten raw or fried as fritters. Dandelion wine is popular, and a tea is made from roasted roots. Dandelion is a potent medicinal herb. This plant is easy to grow from seed but definitely grows better in some areas than others. Dandelion can be weedy or invasive.

Dock, yellow dock, curly dock (*Rumex* spp., esp. *R. crispus* L.): This perennial, from Europe, Asia, and Africa, now grows throughout North America. It prefers partial shade, such as found at the forest edge. It is considered seriously weedy or invasive in some states. The young leaves can be eaten raw in salads or cooked like spinach. Also like spinach, it is high in oxalic acid, so eating large quantities may cause problems. Dock is also an important medicinal herb. This plant is very easily grown from seed.

Miner's lettuce (*Claytonia perfoliata* Donn ex Willd.): This annual native plant is widespread over western North America and is considered a weed in some areas. It grows in full shade to full sun. The leaves are eaten raw or cooked. It grows easily from seed and will readily self-seed.

Stinging nettle (*Urtica dioica* L.): This native perennial grows throughout North America. It is weedy or invasive in some states, and many people consider it an uncomfortable nuisance to have around. The plant has stinging hairs on the leaves that cause pain and irritation when touched, yet this plant has a reputation as being an ideal source of vitamins and minerals, especially iron. It is a well-known medicinal herb. The plant grows in the shade and in the forest edge. The young leaves are cooked and served as a green or added to soups. The dried leaves make a pleasant-tasting tea. Start seedlings in the greenhouse and set out in the spring.

Poke sallet, poke, American pokeweed (*Phytolacca americana* L.): This tall, distinctive native perennial prefers to grow on the edges of the woods. Birds love the berries, but the raw plant is highly toxic to livestock and humans. It is, however, a highly desired traditional food in the South. Young leaves and shoots are gathered in the spring and boiled two times, discarding the water each time to get rid of the toxins. It is then boiled a third time till tender and seasoned with salt and fat back. The plant also has medicinal purposes. Similar to rhubarb, this vegetable should only be sold to people who understand its toxicity and how to prepare it. This plant can be found growing throughout most of North America. It can be very weedy and invasive. Although it can be easily grown from seeds or divisions, poke is so prevalent that it can probably just be wild harvested and managed as such.

Sweet salad, Solomon's seal [*Polygonatum biflorum* (Walter) Elliott]: This native perennial can be found growing in moist, rich woods throughout most of North America. It can be weedy or invasive in some areas. Young shoots are boiled and eaten as a vegetable. It is easiest to propagate by dividing large plants in early spring or fall.

Upland cress, creasy greens, creasy sallet, early yellowrocket, or early watercress [*Barbarea verna* (Mill.) Asch.]: This low-growing plant from Europe resembles watercress. It prefers to grow on the edge of the forest, in a moist but well-drained site where it gets some sun and shade. It is a perennial, but when grown as a salad plant it is usually treated as an annual. The young leaves have a hot, spicy flavor. They are usually served cooked, but increasingly are being used raw in salads. In northern locations, seed can be sown in succession from spring through early fall for harvest over an extended season. In the south, seeds are usually sown during the fall for harvest in late winter and early spring. At harvest, leaves may be cut for a"cut-and-come-again crop" or the entire plant may be cut.

Watercress (*Nasturtium officinale* W.T. Aiton): This creeping, small-leafed perennial is native to Eurasia but can now be found throughout most of North America. It is a peppery-flavored salad herb that is also used on sandwiches and in soups. Rich in vitamins and minerals, it has a long history of use as a medicinal herb. It likes to grow in clean, flowing streams or in very wet soil in shaded areas. Young seedlings or cuttings can be planted in the spring. If grown in water, care must be taken to avoid infestation with parasites or other human pathogens. This plant is considered invasive in some areas and may be banned in some states.

Appendix 9-2—Medicinal Plants that can be Forest Farmed

Most of the herbs described have many traditional and modern uses. Only a few uses are mentioned to give a sense of the plant's usefulness. There is an established market for most of the following herbs, and some commercial forest production is underway in North America. Readers interested in pursuing these alternatives are encouraged to do more research and to learn as much as possible about the plant and its production and markets. The information contained here is derived from Davis and Persons (2014).

American ginseng (*Panax quinquefolius* L.) has a long history of cultivation in North America, and information on how to grow it is readily available. It is, however, probably the most difficult to grow of all the herbs listed here. It is prone to disease and takes a very long time to reach harvestable size. If you live in an area where people harvest wild ginseng, your ginseng is also at risk from poaching. Ginseng is propagated by seed and by transplanting 1- or 2-yr-old roots. Ginseng is used as a general tonic, to improve fertility, to reduce stress, and to treat certain diseases.

Bethroot, red trillium, purple trillium (*Trillium erectum* L.) grows to about 0.5 m in height and has a single brown or greenish-purple flower. The roots of this plant are harvested in the fall and used to treat hemorrhages, skin infections, and heart palpitations, among other disorders. This plant is challenging to cultivate and there is no forest-farmed material on the market.

Black cohosh (*Actaea racemosa* L.) is a very popular woman's herb used for the treatment of unpleasant menopausal symptoms. The popularity of this herb has caused concern for the species well-being, and cultivation is strongly encouraged. Fortunately, it is easy to grow. It is an attractive shrub, which can get quite tall, with delicate, lacy-type foliage. It thrives in rich, moist soil. The roots are harvested in early fall for medicinal use. It can be propagated by dividing the rootstock in the spring or fall. Seed propagation can be difficult. Demand for cultivated material is increasing.

Black root, Culver's root [*Veronicastrum virginicum* (L.) Farw.] is a tall perennial that grows in moist forests throughout North America. The dried root is used as a mild laxative. The fresh root is toxic.

Bloodroot (*Sanguinaria canadensis* L.) is an attractive, low-growing plant that puts on pretty white flowers in very early spring. It can often be found growing in deep woods. It spreads naturally by rhizomes and seed and can be easily propagated by both. The root has a long history of use by Native Americans and has proven antimicrobial activities. It can be toxic, however, and should not be used casually. In the past decade, bloodroot was used in a toothpaste and livestock feed.

Blue cohosh [*Caulophyllum thalictroides* (L.) Michx.] is another beautiful plant that is commonly found in hardwood forests in southern Appalachia. The plant grows to about 1 m in height and can have very blue foliage. The rhizomes can be divided in spring or fall. Seeds are difficult and slow to germinate. Blue cohosh is another woman's herb and was traditionally used to aid in childbirth.

Boneset (*Eupatorium perfoliatum* L.) is a shrub that grows in woods throughout eastern North America. The top of the plant is harvested and dried and used as a laxative and to treat coughs, fevers, and chest illnesses. Boneset is grown in a wild-simulated system by a few growers, particularly in the southeastern United States.

False unicorn, star root, devil's bit, fairywand [*Chamaelirium luteum* (L.) A. Gray] is a rather unobtrusive plant that starts out as a very low rosette of leaves. It has separate male and female plants, which send up very tall flower stalks. The plant is slow to grow, and mortality can be high in some years. False unicorn can be grown from seed or rootstock cuttings. It is considered an important woman's herb but also is used to treat a wide range of disorders such as pain, poor appetite, and cough.

Ginkgo (*Ginkgo biloba* L.) is an ancient tree that is used extensively as an ornamental in cities across North America. The leaves are harvested and used to enhance memory and treat circulatory problems. Ginkgo is usually grown in a monoculture system, and there are several very large plantations in the United States (observations by the authors). Growing it in a forest of mixed species is rare.

Goldenseal (*Hydrastis canadensis* L.) is used for many purposes including as a treatment for AIDS, cancer, various digestive disorders, and to boost the immune system. It once grew abundantly in forests across eastern North America. Native populations have been seriously reduced by overcollection and it is now an endangered species in some states and regulated under the Convention on International Trade in Endangered Species of Wild Fauna and Flora. Goldenseal is propagated by divisions and seeds.

Lady's slipper (*Cypripedium* spp.) is a highly desirable ornamental and medicinal plant of which many species grow across North America. Most buyers stopped buying lady's slippers because wild populations were being so heavily damaged. Some cultivated material is now on the market and small amounts are again in trade. Lady's slippers have distinctive pouch-like flowers of a variety of colors. They prefer rich woods and wet areas. Some of the species are very difficult to propagate. The roots are used as a mild sedative, to treat headaches and depression, and for menstrual difficulties.

Mayapple (*Podophyllum peltatum* L.) grows on the edge of the woods. It emerges in early spring and produces a tall plant (0.3–0.61 m [1–2 ft]) with an umbrella-like leaf. It is easy to propagate by dividing the rhizomes. There is interest in its use for the treatment of cancer, liver problems, and constipation.

Maypop, purple passionflower (*Passiflora incarnata* L.) is a perennial vine that grows on the forest edge and in open woods, mostly in the eastern United States. The top of the plant, with flowers and fruit, is harvested in the summer. It is used as a mild sedative and to treat skin problems. Maypop is produced on a small scale by at least a few farmers.

Oregon grape [*Mahonia aquifolium* (Pursh) Nutt.] is an evergreen perennial shrub. Different species grow throughout North America. The roots contain berberine and are sometimes used as a goldenseal substitute. It has antimicrobial properties and is a liver stimulant.

Partridgeberry, squaw vine, squaw berry (*Mitchella repens* L.) is a low-growing perennial vine found in eastern North America. The top of the plant is harvested in the fall and used to treat diarrhea, as a diuretic, and to aid in childbirth. There are some small-scale producers of the plant.

Pinkroot, Carolina pinkroot, Indian pink [*Spigelia marilandica* (L.) L.] is a beautiful plant with elongated flowers that are red on the outside and yellow on the inside. It is easy to propagate by root divisions. It is commonly used to aid digestion.

Skullcap (*Scutellaria lateriflora* L.) is a low-growing perennial that grows in open woods and on the edge of woods. The roots are harvested in the fall and used as a laxative and to treat respiratory and uterine problems. Skullcap is grown commercially in some areas in the woods and open fields.

Slippery elm (*Ulmus rubra* Muhl.) is a tall tree that grows in the woods throughout eastern North America. The inner bark is harvested in the spring and fall. It is used as a laxative and to treat skin conditions, sore throats, stomach ulcers, and wounds. Cultivation is steadily increasing.

Spikenard (*Aralia racemosa* L.) grows 0.3–3 m (1–10 ft) tall and bares elongated flower stalks covered with yellow-green flowers that develop into purple berries. It grows in rich woods and on riverbanks. Spikenard is easy to cultivate and can be propagated by division. The roots are harvested in the fall and used to treat many ailments, including backaches.

Stargrass, true unicorn root, devil's bit, blazing star (*Aletris farinosa* L.) is a perennial that grows in moist woods and meadows. The roots are harvested in the fall. It is used as a sedative and tonic. There is little trade of cultivated or wild-harvested material at this time.

Stone root, Canada horse-balm (*Collinsonia canadensis* L.) is a perennial that grows in moist woods. The roots are harvested in the fall and used as a sedative and tonic. It is known to be grown on a very small scale by a few growers.

Virginia snakeroot (*Aristolochia serpentaria* L.) is a perennial that grows in moist woods in eastern North America. Roots are harvested in the fall and used to treat wounds and skin ulcers.

Wild ginger (*Asarum canadense* L.) is a low-growing plant with heart-shaped leaves and brown bell-like flowers. It grows in cool, shaded, moist woods and is propagated by root division. It is used to treat "gas" and motion sickness and as a stimulant.

Wild yam (*Dioscorea villosa* L.) is a perennial vine that grows in forests and on the forest edge in the eastern United States. Roots are harvested in the fall and used to treat menopausal symptoms, asthma, and gastrointestinal problems. The amount of cultivated wild yam is small but increasing.

Yellow indigo, wild indigo (*Baptisia tinctoria* L.) grows to about 1.2 m (4 ft) in height and has small leaves and a large number of small yellow flowers. It prefers a little more light and a drier soil than the other plants discussed here. Propagate by cuttings or seed. It has a long history of use for treatment of sore throats, typhus, wounds, and to enhance the immune system.

Yellowroot (*Xanthorhiza simplicissima* Marshall) is a common shrub that is often found growing in damp woods and on stream banks in the mountains. It has a bright yellow root that is used as a tonic and to treat many of the disorders for which goldenseal is used. It is propagated by division.

Study Questions

1. Define and provide examples of forest farming in North America.

2. What are three major benefits a landowner can expect to realize through forest farming?

3. Discuss some major challenges that a landowner may face before and following adoption of forest farming as an alternative land use practice.

4. The early visionaries of forest farming maintained that this integrated land practice had three main components. Name and discuss the three components and how the forest farming concept has changed with time.

5. What are some of the major advantages and disadvantages of forest farming?

6. Identify the three primary methods of generating revenue from forest farming products and services.

7. Describe the evolution of ginseng cultivation in the United States. What were the major challenges in the early production of this forest medicinal herb?

8. What factors should a landowner consider when considering forest farming as an alternative land-use practice?

9. Identify the major challenge with sustainable wild-simulated forest farming, and explain what a forest farmer can do to address the challenge.

10. What is the first thing a forest landowner should do to determine the feasibility of establishing a forest farm?

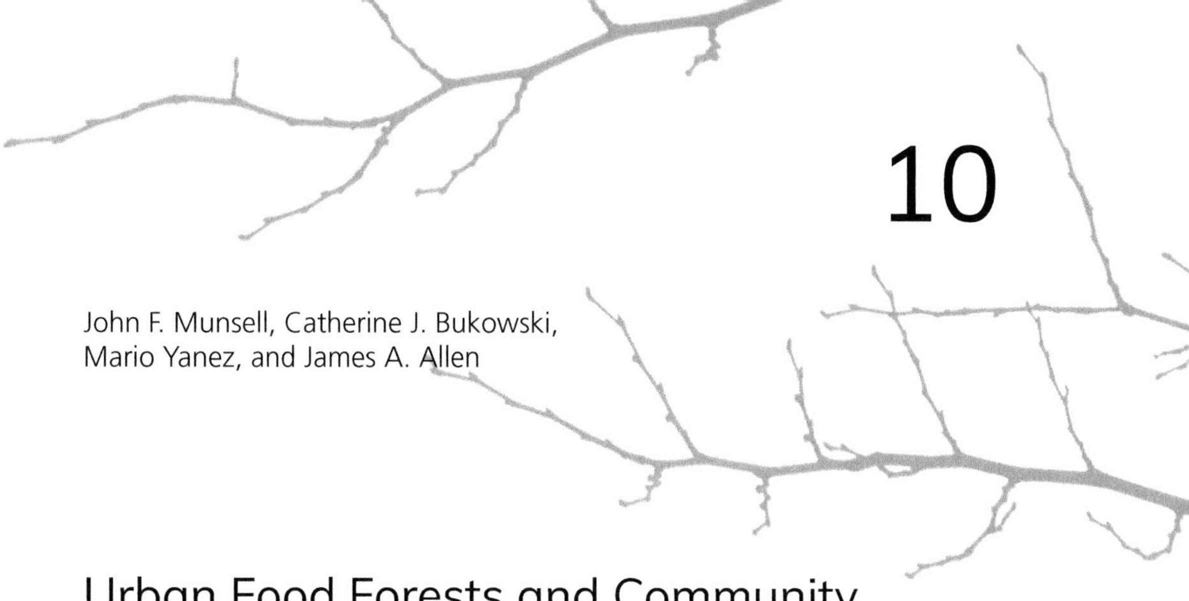

10

John F. Munsell, Catherine J. Bukowski,
Mario Yanez, and James A. Allen

Urban Food Forests and Community Agroforestry Systems

The expansion of agroforestry applications in North American cities and towns constitutes a fundamental shift in the use of private property and public green space in built environments. This change is amplified by unprecedented rates of migration from rural areas to population centers and the fact that more than half of the world's population now lives in cities (United Nations, 2017). For example, more than 80% of U.S. residents live in urban areas (American Community Survey, 2016). At the same time, the world's population is rapidly growing, with more than nine billion people expected to inhabit the Earth by 2050 (United Nations, 2017). Where and how humans produce food, timber, fiber, and other basic materials is transforming as humans navigate these crossroads, and the relevance of agroforestry in built environments is increasing in the midst of this sea change.

Agroforestry in towns and cities largely involves multistoried systems comprised of diverse food-producing annual and perennial species planted together in spaces such as abandoned lots, roadside swales, or rooftops but also in larger landscapes such as church campuses and parks. Food is central to these projects, but many other benefits are derivable, including wood, craft materials, medicinal products, pollinator habitat, ecosystem services, and space for community gatherings. In this context, the flexibility inherent in agroforestry is leveraged to create systems that help combat food insecurity (Nytofte & Henriksen, 2019), restore and retain resilient ecosystems (Lovell, 2010), and supplement civic activities and welfare (Bukowski & Munsell, 2018) in the places where people are concentrated.

Urban food forests are the cornerstone of agroforestry applications in population centers. They are on the frontline of activity and presented as a sixth and emerging temperate zone agroforestry practice in Chapter 2. Urban food forests typically include combinations of perennial and annual food-producing species that are appropriate for a particular place such as a backyard, swale, or rooftop. The function and structure of an urban food forest can vary, but the design generally involves integrating annual and perennial species into the same space and partitioning resource competition and creating complementary relationships between them when possible (Figure 10–1). Stacking and spacing plants and fungi in time and space is intended to mimic interactions in a forest ecosystem and beget an integrated, largely self-sustaining and multifunctional food system (Park, Turner, & Higgs, 2018).

Urban food forests are championed as a way to address some of the world's most pressing human health and equity issues (Clark & Nicholas, 2013). This has been an important driver of grassroots agroforestry awareness and adoption of urban food forests and a

North American Agroforestry, Third Edition. Edited by Harold E. "Gene" Garrett, Shibu Jose, and Michael A. Gold.

Fig. 10–1. A simplified cross-sectional schematic depicting vertical stacking of annual and perennial species in a multistory urban food forest (top left). Various strata are labeled, as well as climbing species such as vines. An image from Trozzo, Munsell, and Chamberlain (2013) demonstrates horizontal spacing in an agroforestry food forest to maximize site occupancy of stacked perennial and annual species (top right). Below are two images from the same multistory urban food forest in Iowa City, IA, that demonstrate vertical stacking and horizontal spacing of diverse species.

fundamental factor in the rising rates of community participation. However, urban food forests also are increasingly wrapped into formal municipal-scale planning and development policies that aim to interconnect applications and rehabilitate damaged ecosystems, repurpose abandoned and blighted areas, provide resident learning grounds, and improve civic activity and well-being across towns and cities (e.g., Colinas, Bush, & Manaugh, 2019; Lehmann, Lysák, Schafer, & Henriksen, 2019; Riolo, 2019). Scaling connectedness between urban food forest initiatives involves systems thinking, where particular projects form nodes of activity that are connected and managed within an urban food forestry network that sustains production in corridors across built environments.

The number of agroforestry projects in cities and towns has grown in Europe, North America, and other industrialized regions (Hübner, Künstle, Munsell, & Pauleit, 2018), and by the late 2000s scores of urban food forest initiatives were underway on vacant lots, marginal sites, public land, and private property. Many were the result of private residents acting individually or in small informal groups, but others were led by government agencies, educational institutions, nongovernmental community groups, and applied and scientific agroforestry stakeholders (Bukowski & Munsell, 2018). The overarching focus is on practical public and private property applications that improve access to nutritious and diverse food. However, many also marry production with outputs such as civic engagement and public participation, ecosystem rehabilitation and stewardship, and community education (Bukowski & Munsell, 2018; Hemmelgarn & Munsell, 2021). In the meantime, policies and overarching community initiatives that support agroforestry applications in towns and cities likewise have increased (Salbitano, Borelli, Conigliaro, & Chen, 2016).

The emergence of agroforestry in the built environment has captured student and stakeholder attention, and recent applications build on the field's rural tradition. This interest in towns and cities follows, in many ways, the same simple premise that underpins rural applications, namely that designing with nature instead of against it is necessary for success. This is especially true when connecting agroforestry projects in the complex constructed places where humans live into one system because the spectrum of production includes raw material ranging from food to medicine to wood, as well as eco-social goals like air quality, community education, and civic well-being. The adaptability and creativity inherent to agroforestry is compelling in ever-growing built environments as the importance of whole ecosystem health increases. Nowhere else is this more relevant than on the often irregular and disjointed arable spaces in and around the areas where people are concentrated.

This chapter focuses on urban and community agroforestry strategies that connect and expand urban food forestry systems. It offers insights into how these systems are designed and scaled to provide food, timber, fiber, and other materials in addition to eco-social benefits. Productive placemaking using agroforestry is a chief theme and is described along with beneficial primary and secondary outputs from practices such as urban food forests and other food forestry applications. Also addressed are the various forms of community assets that constitute needs and strengths and how residents use this knowledge to organize and create productive places. As towns and cities synergize and leverage their strengths and address critical needs through agroforestry, they fashion places that connect communities defined by and embedded in landscape ecosystem function and structure. Case studies drive home system concepts where the past and present intersect, existing perspectives are reinforced, unique features are highlighted, and outlooks on the future of agroforestry in towns and cities are provided.

Multifunctional Greenspace in Urban Planning

Agroforestry is largely seen in terms of rural land use practices that integrate and thereby diversify farm and forest products such as crops, livestock, timber, and non-timber forest products (Mercer & Pattanayak, 2003). It is a model for optimizing the agroecological production of raw material while conserving natural resources. When agroforestry is practiced in towns or cities, production is optimized with ecosystem services like soil and water quality conservation, air quality, carbon sequestration, habitat, and pollinator forage. However, also explicitly targeted are whole-ecosystem benefits that involve the intentional integration of humans and physical spaces into a single eco-social matrix (Figure 10–2).

Agroforestry systems in towns and cities typically depend on a team of residents working together in collaborative processes to design, install, and manage projects that serve a wide variety of purposes for community members and the places they live. Projects largely focus on how to maximize use-benefit for a community rather than a sole individual, but the intent also typically includes expectations that onsite visitation and programming will pay forward and result in applications on private property and shared

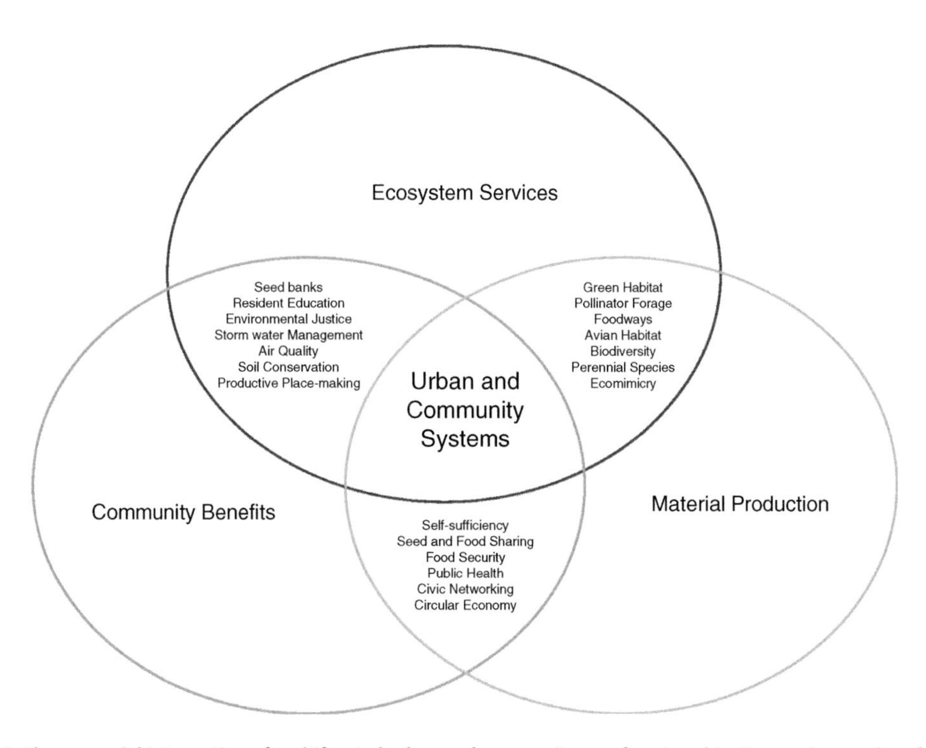

Fig. 10–2. The eco-social intersection of multifaceted urban and community agroforestry objectives and examples of aligned site and design goals (adapted from Bukowski & Munsell, 2018).

commons such as apartment complex grounds and homeowner's association landscapes (Bukowski & Munsell, 2018). Overall, urban and community agroforestry offers a pathway, whether on public or private land, for aggregating perennial and annual species to achieve a broad spectrum of eco-social outcomes.

Increasing populations pose new planning challenges for agroforestry systems where open space is at a premium and often under threat of development. Whether on public or private land, they can provide important benefits such as human health and community well-being, ecological habitat, ecosystem services, as well as food, wood products, herbs, and natural materials for arts, crafts, and traditional uses (Douglas, 2012; Hansen & Pauleit, 2014; Schewenius, McPhearson, & Elmqvist, 2014; Shackleton, Hurley, Dahlberg, Emery, & Nagendra, 2017; Synk et al., 2017; Tzoulas et al., 2007). This is needed more than ever as many urban and suburban communities merge in metro megaregions, where surrounding suburban sprawl connects population centers through conversion of fields and forests into shopping centers, housing, highways, business parks, and recreation areas. The result is that the locus of production has been pushed farther away, which runs counter to interests in local materials and community connections. Thus, multifunctional agroforestry systems that reclaim and regenerate community ecosystems are timely, and the field offers proven solutions that can be scaled and shaped to meet community goals.

Historical Precedent

Towns and cities in modern-day North America have largely excluded food-producing perennial species in public spaces (Coffey, 2019; Coffey, Munsell, Hübner, & Friedel, 2021). Producing food and fiber using combinations of trees, crops, and even livestock has generally not been part of a concerted and permanent contemporary strategy for managing greenspace in built environments. This is often due to management and liability concerns such as fruit decay and rodents, falling matter, and infestation (Bukowski & Munsell, 2018). Nevertheless, there have been notable periods of intentional food production, such as wartime Victory Gardens in the United States. However, virtually all were limited to annual crops, and gardens often were abandoned once the need for national unity or economic hardship subsided. Public interest in urban and community agroforestry systems and the increasing number of projects and publications, as well as a growing base of governmental and nongovernmental advocates, are shaping a new era focused on long-term polyculture solutions. This has led to an uptick in planning, establishing, and managing agroforestry projects in urban and community ecosystems. Central to this is the idea that these applications should reflect the eco-social fabric of a particular place and thereby benefit local communities while also helping to address some of the planet's most pressing problems.

Agroforestry is a combination of land use science, traditional farming practices, and local ecological knowledge. It reflects early forms of indigenous land use, like those in North America, where forests and fields were managed intensively using fixed polyculture practices in and around concentrated centers of community activity (Doolittle, 2004). In the mid-20th century, scientists the world over began studying methods of multi-cropping to develop verifiable and reproducible systems based on traditional diversified land use (Bene, Beall, & Côté, 1977). The result was the basis for contemporary models of agroforestry that systematically blend and layer plants, shrubs, and trees to minimize competition, facilitate enhanced growth, optimize multispecies production, and conserve natural capital. Tropical home gardens are perhaps one of the oldest agroforestry practices and a primary basis for developing early agroforestry concepts and practices. They are multilayered mixed forest gardens around homesteads and throughout communities that integrate a wide range of species within production plots and corridors.

Modern-day agroforestry in towns and cities reflects many facets of the traditional tropical home garden, where multifunctional shrubs and tree species are combined with crops and livestock in and around the places where people live. Like the home garden, these systems are integrated into the eco-social context of a particular community and are thus fundamental to the health and well-being of built environments. This renaissance of sorts is growing, and a number of municipalities are bucking trends of limited applications of agroforestry (Park et al., 2018). For instance, some government agencies have added urban agroforestry practices to their lexicon and offer support through education, outreach, and grants. The Browns Mill Food Forest in Atlanta, GA, is a good example (https://www.conservationfund.org/projects/food-forest-at-browns-mill). This project acquired land through a combination of public and private sector support, such as the U.S. Forest Service's Urban and Community Forestry Program, The Conservation

Fund, the City of Atlanta's Office of Resilience and Department of Parks and Recreation, and Trees Atlanta. In many ways, the surge of urban and community initiatives in the past 15 yr is shaping a new era for agroforestry in North America, where projects optimize production locally as much as they bridge the divide between urban and rural production and consumption.

Primary Production through Urban and Community Systems

Residents in towns and cities depend largely on raw materials that are produced far from population centers, often with negative consequences to the communities and ecosystems from which they are supplied. Environmental sociologist John Bellamy Foster publicized the term *metabolic rift* to describe the separation of large concentrations of humans from the rural ecosystems that provide many of the goods and services on which they depend (Thackara, 2019). This concept is a useful framework for studying, quantifying, and mapping what can be thought of as the metabolism of population centers, the premise of which is that the gulf between where most people live and where their food and materials are produced is wide. When it comes to optimizing the production of food, timber, fiber, and other material in proximal urban and community spaces, agroforestry systems provide a degree of connectedness and functionality by which provisioning, regeneration, conservation, and community benefits can be achieved, thereby decreasing the divide between residents and the locus of production.

Farming and forestry in rural areas has long played a vital role in feeding and sheltering growing urban and suburban populations, but as cities and neighborhoods continue to sprawl and transportation distances increase, the nature of community self-sufficiency and energy efficiency has come into question (Madlener & Sunak, 2011). Efforts to improve collective self-sufficiency in cities and towns reflects a growing consciousness regarding the negative consequences of metabolic rift but also the desire to reclaim physical connections to the places and spaces where people live (e.g., Askerlund & Almers, 2016). This is especially true in the face of challenges such as natural disasters, changing climate, and pest and disease outbreaks, all of which affect the sources of imported food, timber, fiber, and other important materials. Thus, ownership of the locus of production is as much about mitigating problems as it is about reclaiming meaningful corridors of concentrated community activity where raw materials are produced. Underpinning this balance is that people in population centers must work together to determine what they can do because a singular outcome (e.g., crop to market) is insufficient. This notion is key to the relevance and impact of agroforestry practices in towns and cities, and undergirds the whole-ecosystem productive placemaking that characterizes them.

Residents and governmental and nongovernmental agroforestry allies alike are enacting agroforestry through local food movement initiatives, such as community-supported agriculture, farmers' markets, and community gardens (Gold, Godsey, & Josiah, 2004; Scherr, 2004). Additionally, many communities are entertaining agroforestry-inspired perennial solutions that offer a wide range of products, as well as community benefits and ecosystem rehabilitation. As a result, there are numerous guidebooks and how-to manuals that promote urban and community systems (e.g., Hemenway, 2017). Thus, it is an important time for the field of agroforestry as the local food movement has now grown to include patterned systems that demonstrate how food and other raw materials can be produced using combinations of perennial and annual crops in places and at scales not previously thought possible in modern-day North America.

Increased land use diversity and creativity are generally found in places where ecosystems overlap. This also is the case with agroforestry in urban and community settings, where innovation is observed most often at the intersections of various fields, such as urban forestry, regional planning, urban agriculture, landscape architecture, permaculture, and economic development (Bukowski & Munsell, 2018). At its core, primary production involves provisioning of basic material needs—food, fuel, fiber, fodder, and natural medicine. To close the metabolic gap, residents, community groups, religious organizations, town planners and resource managers, and elected officials across North America are actively leading urban food forest projects on underutilized urban, suburban, and peri-urban spaces. The same is true on private property, although agroforestry applications in this context come with different sets of opportunities and challenges that are typically but not always confined to families or individuals. The overall implication is that intensive and intentional production of food, fiber, timber, and other important raw materials like natural medicines and decorative materials is no longer solely a rural proposition. What is more, well-designed and productive urban practices form nodes of activity that can be scaled by connecting structure and leveraging relevant functions, which ultimately can be woven into a community's eco-social fabric and complement rather than conflict with rural production through localized agroecological farming and forestry.

Agroforestry systems in towns and cities, like those in rural areas, are constrained by climate, moisture, and nutrients. However, in some ways amending urban and community applications to support production may not be as energy intensive because waste streams with as much as 40% organic composition are concentrated and can be redirected and used to regenerate soil and enhance urban primary production (Cofie, Bradford, & Drechsel, 2006). In addition, cities in North America, for example, have an abundance of underutilized space, much of which is considered arable using agroforestry. A shift toward primary production through the use of agroforestry can help close metabolic rifts in cities and towns and spur secondary production comprised of proximal economies where residents and groups and organizations make, serve, sell, and distribute food, timber, fiber, and other natural resources products locally and regionally. The result is that humans are able to connect more intimately to the materials they consume and thus better discern the negative consequences of displacing sources and modes of production. Just as important, urban and community agroforestry systems stand to benefit local economies by building capacity through meaningful and authentic production that satisfies human, community, and ecological needs.

Productive Placemaking and Community Strategies

Civic considerations and community benefits take on greater importance when designing, implementing, and managing agroforestry systems in places where humans are concentrated. This distinction is important for conceptualizing how urban community systems fit into the field of agroforestry. Human communities evolve culturally and make meaning within the places where they live and then pass that meaning and way of knowing to generations that follow (Wilson, 1997). The increasing number of people in metropolitan regions gives rise to myriad cultures stretched across socioeconomic classes, as well as magnifies the challenges of community health and well-being in densely populated environments. Agroforestry is a blended approach to farming and forestry. However, applications in population centers also must function in unique and confined spaces, in most cases without maximum yield for product sales as a chief aim. Rather, the exigency is to create systems that address complex eco-social challenges, such as food insecurity, urban land regeneration, economic injustice, poor water and air quality, climate instability, soil loss, accumulation of greenhouse gases, and possibly and most importantly, the well-being of future eco-social systems. The benefits of doing so means communities use these spaces for multiple benefits that greatly increase civic engagement, help revitalize economies, and regenerate ecosystems.

Agroforestry projects in towns and cities provide a place where residents can visit, meet, share, and enjoy themselves in the midst of primary production. These social "products" provide knowledge and experience that can be passed on to neighbors and family. Oftentimes, they are the primary output. Many have been established not only to amplify production but also to accomplish objectives such as introducing residents to agroforestry practices in the hope they will apply them elsewhere. Other examples include food access and nutrition, community rallies, youth engagement and education, preserving genetic material for vegetative propagation, and providing or specifying plant species and cultivars that are appropriate for or acclimated to local conditions (Bukowski & Munsell, 2018). Agroforestry applications in towns and cities can vary widely in scale, form, and function (Park & Higgs, 2018). Regardless of size, the hallmark of successful urban and community agroforestry systems is an equitable mixture of production, regeneration, conservation, and community objectives, and successful projects in these settings are designed to achieve all according to local needs and preferences.

The intersection of production, ecological stewardship, and community planning through agroforestry can be thought of as *productive placemaking*, the implications of which are significant in terms of land use policy in cities and towns because it is a multifaceted and nuanced approach to the planning, design, and management of public spaces that contribute to human health, happiness, and well-being. The features of productive placemaking are best defined by residents who are embedded in their local eco-social system and consequently able to design meaningful systems. Roles and responsibilities are naturally selected and shared in this way, which sets in motion a community approach rooted in local ecology. Even when practices are exported from a community project, then established on and confined to private property, contributions to the well-being of a town or city are evident through productive base points. Hence, placemaking in the context of urban and community agroforestry is a process and outcome that depends upon particular procedures, or action items, that a given community sees fit to follow when designing and implementing practices. As such, the nuances affecting scale, form, and function are best defined by residents and the local ecology because eco-social systems are the most appropriate frames for community change, making modified places not only productive but also relevant.

The skills and needs of a community that can be leveraged to achieve ecological and social benefits using agroforestry are often codified in terms of an array of capital. It is important to note that these attributes are not confined to monetary concepts or forms of quantitative accounting but rather constitute a holistic decision-making framework that can help people organize and apply their skills to address and leverage the interests, needs, and opportunities relevant to particular eco-social systems. In terms of agroforestry, skills and needs are balanced across material production (primary production), benefits tied to the resulting materials (secondary production), and the regeneration of ecosystems and conservation of natural resources (Figure 10–3). This diversified assemblage empowers residents and optimizes the extent to which people and places are invested across a spectrum of goals and objectives. Although there are others, a common framework for defining clear lines of community development planning and action is the *community capitals* model. The diverse attributes in this model represent a range of potential concrete action items, community strengths, and eco-social needs that residents can account for when determining how to work together to make productive places using agroforestry in towns and cities.

Flora, Flora, and Gasteyer (2016) defined seven types of capital as: (a) social, (b) human, (c) natural, (d) cultural, (e) political (sometimes referred to as institutional), (f) built (sometimes referred to as produced or manufactured), and (g) financial. A systems perspective is critical in terms of agroforestry applications in the built environment, even if confined to private property, and this framework lays out a broad array of complementary assets that are important to consider in urban and community contexts (Table 10–1). Bukowski and Munsell (2018) studied the ways food forest projects invest in these seven types of community capital to identify the strengths and weaknesses related to the design, implementation, and long-term management of agroforestry systems in towns and cities. Their findings demonstrate that the nature of implementation in population-rich settings influences not only normative land use values but also the effect that agroforestry has on those values. Determining the extent to which a community focuses on any given type of capital is therefore critical to the balance that helps or hurts the stability of applications in the built environment. Said differently,

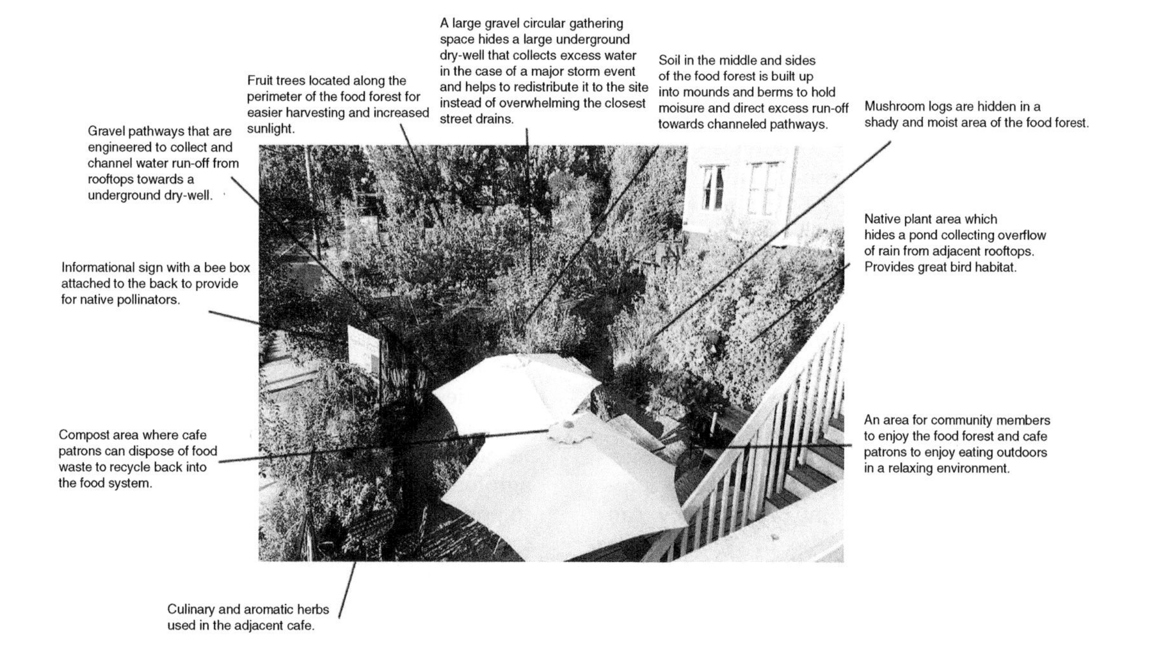

Fig. 10–3. Image of a small urban and community agroforestry project in Portland, OR. Annotations emphasize food and fiber production, community and civic benefits, ecological regeneration, and natural resources conservation.

Table 10–1. Descriptions of seven kinds of community capital representing strategic attributes on which community members may decide to focus when designing, implementing, and managing urban and community systems across three key objectives: (a) food and fiber production, (b) environmental conservation, and (c) community and civic benefits.

Capital	Description	Example assets	Example outflows
Social	strength, direction, and nature of direct and indirect associations among involved people	networks with various backgrounds, interests, and experience; inclusive events, teamwork, and diversity	bonding over similarities and bridging differences; collaborative greening and food production activities
Human	beliefs, energy, skills, education, and ideas that may be reflected in urban and community systems	leveraging collective skill sets to enhance supporting projects, as well as organizing and defining governance	Learning about alternative food sources, planting methods, community change, and environmental stewardship
Natural	living and nonliving components that support production, conservation, and community	planting diverse annual and perennial edible species; amending soil; bioremediation; riparian projects and habitat	relaxation and recreation; soil conservation, food production, vegetation, and habitat; water and air quality; aesthetics
Cultural	preferences and perceptions of people regarding their community in terms of norms and values	engaging concepts of production; improve access, exchange new and innovative methods; share and build	new forms of diversified and responsive management; creative ideas that enhance experience and productivity
Political	formal and informal resource and skillset allocations that drive policies and civic decisions	awareness among formal and informal community leaders; collective effort and structures that enable local production	affecting decisions, policies, and private and public short- and long-term land use change; advocacy and publicity
Built	infrastructure such as roads, buildings, fences, sheds, and drainage systems	brick and mortar that supports education, conservation, and community activities; artwork, trail systems, and benches	covered meeting space and protection of equipment; local art; memorials and celebratory events
Financial	monetary assets that affect site capacity and support future programming and development	secure external donors or investors; additional support and spin-off private community enterprise and cost savings	advertising and incentives; growth in infrastructure and means for improving capacity and incentivizing commitment

Note. Example inflows and outflows are provided for each type of capital. Adapted from Bukowski and Munsell (2018).

thoughtful and equitable project decisions lead to the accumulation of meaningful accomplishments that seed benefits extending well beyond the boundary of any given place where agroforestry is implemented to include reducing the metabolic rift between urban and rural communities.

As Table 10–1 shows, Flora, Flora, and Gasteyer (2016) community capital constitutes a framework for organizing and acting to create productive and impactful urban and community agroforestry systems. Short summaries of the relevance of community capital provide a conceptual basis for design, implementation, and long-term management.

1. *Social capital* includes assets such as resident networking, friendships, cooperation, and trust that define and underpin agroforestry in towns and cities. Agroforestry helps strengthen the bonds between people that work by providing a common ground where they can grow, harvest, and share food. When people work together on agroforestry, they create and nurture multifunctional places and experiences that can bridge social divides.
2. *Human capital* is the capacity that community residents bring to an agroforestry project in a neighborhood, town, or city. Attributes such as community experience, personal skills, formal and informal education, and passion are relevant. Human assets in the pluralistic sense often are a key driver of agroforestry design.
3. *Natural capital* is a critical investment that most urban and community agroforestry projects place at or near the top in terms of importance. Goals are to decrease the metabolic rift, increase primary and secondary production, and set in motion productive placemaking that focuses on healthy soil, unpolluted and abundant water, diverse habitat, carbon stocks, and clean air.
4. *Cultural capital* is the set of values and norms tied to traditions, mores, religions, and other identities and beliefs among the people that live together in, and see themselves as, a town, neighborhood, suburb, or city. Investments in agroforestry that include cultural exchange are important for enhancing shared learning and growing experiences among the people working together.
5. *Political capital* often involves working constructively with civil servants, elected officials, and political appointees. Some agroforestry projects explicitly make political investments a

priority, while others rank political capital much lower in terms of priorities. Another strategy is self-governance and community-based organizing focused on local or regional decision-making within the context of agroforestry.

6. *Built capital* consists of constructed or crafted infrastructure. Agroforestry projects generally plan for efficient maintenance and meaningful use, but structures like fences, pavilions, benches and tables, pathways, and storage sheds often require short- and long-term planning and management. Interest and community involvement can increase through installations such as artwork and meeting and working places. Infrastructure also may be belowground, such as dry wells or drainage systems.
7. *Financial capital* is the monetary and credit investments that underwrite agroforestry projects. Grant funding often is secured early, but other forms of investment are possible and often materialize in concert with or because of grants. Some projects tie into municipal support, whereas others secure private donations or are largely self-sufficient, and in some cases resources are secured through fundraising.

Scalability and Adaptability

Capital investments in agroforestry on individual lots is but one strategy. When agroforestry projects are connected, they form *foodways*—a network of connected urban food forestry projects that constitute a multifunctional agroforestry corridor (Figure 10–4). Connectivity in this regard reflects a cultural concept that typically refers to the intersection of food in history with how certain cultural practices are formed and for what reasons. These intertwined agroforestry practices are an element of productive placemaking shaped by cultural expressions in a corridor comprised of edible and useful species that address community priorities. Through scale and change across an interconnected pattern, urban and community agroforestry systems can positively affect the livability of towns and cities by forming eco-social pathways that provide multiple benefits for residents, wherein community engagement, informal learning, arts and culture, and livelihoods are chief among them.

Agroforestry practices arranged and interconnected within foodways can magnify community and ecological impacts (Borelli, Conigliaro, Quaglia, & Salbitano, 2017). Integrated systems are able to form part of a larger strategy for human communities that seek to meet some or all of their material needs while growing circular economies, reducing carbon footprints, decreasing waste, and enhancing community and neighborhood health and well-being. These goals can be achieved through intentional agroforestry design and management, investment in the community capital deemed critical for a project, and on-site maintenance, repair, reuse, remanufacturing, refurbishing, and recycling. In this way,

Fig. 10–4. Foodways are interconnected agroforestry projects in a town or city. They represent the intersection of small-scale productive placemaking that extends across sections of a built environment to form a larger network of cultural, ecological, and agricultural assemblages. Urban and community agroforestry systems that are interconnected provide multiple primary and secondary production benefits.

agroforestry systems in the built environment together with rural counterparts compose a science-based framework that transcends traditional notions of scale and metabolic delineations. Applications are possible on large farms, in vast woodlands, in multifunctional foodways, and on abandoned quarter-acre lots, all of which can be interconnected. In this way, there is potential to bond people and places through agroforestry bioregions that link cities, suburbs, towns, and rural areas into one shared production system.

At its core, agroforestry design mimics ecosystems and combines species that produce more food and fiber than if they were grown alone. As in rural areas, ecosystem scale in built environments can vary widely, depending on physical constraints and the nature and diversity of interspecies relationships, and many production practices follow guidelines and structured practices but generally stem from observed qualitative patterns in natural ecosystems (Hemenway, 2017). Agroforestry practices also benefit from decades of scientific research, and because of this, applications in towns and cities profit from systematic design, establishment, and management that can be used at different intensities in various settings to complement localized primary and secondary production. The benefit is that ecosystem scale varies in size and complexity in multistory configurations ranging from matters such as soil microbial management (e.g., Lorenz & Lal, 2014) to climate change mitigation (e.g., Verchot et al., 2007). This ultimately enriches planning for urban and community applications, which better translates the value of multifunctional design and management of agroforestry systems focused on the health and well-being of community residents and places.

Working beyond single crop monocultures or annual-only gardens is the aim of urban and community agroforestry systems. Typical applications include combinations of crops in an agroforestry polyculture defined by a mixture of perennial and annual species laid out in complementary and functional spatial patterns (Figure 10–5). Sometimes this involves rows of trees and shrubs but also can be somewhat random or fixed in particular locations for production as well as education and community activities. However, projects of any scale are largely purposive and, as Figure 10–5 depicts, usually are intended to provide equitable environmental and civic benefits for residents and communities. This requires going about species selection in a different way, where diversity and density are determined not only by goals of raw material production but also by the ways in which civic and environmental benefits are enhanced.

The scalability of agroforestry practices and associated benefits are a useful strategy for community and urban initiatives. The value-added proposition is that whole health is combined with production, regeneration, and conservation along with civic goals such as environmental literacy, community collaboration, and human health. Existing projects demonstrate that the facets of agroforestry in built environments are broad and applications generally involve a unique combination of fundamental agroecological principles that form a set of science-based regenerative practices wherein food, timber, fiber, and other important materials are produced. As in all agroforestry systems, planning and adaptation are necessary with time, involving both biophysical and civic goals. Both are critical because projects are circumscribed by how well plants and trees work together but also the extent to which citizens and communities are engaged and rewarded. Given longstanding and emergent networks of sustainable land uses, connecting agroforestry projects across urban, suburban, and even rural contexts in multifunctional foodways could pay huge dividends when it comes to sustainable communities and global environmental health.

Fig. 10–5. Images indicating the balance of scope and scale of urban and community systems, cutting across eco-social aspects and underpinned by unique, place-based design.

Urban and Community Systems in Practice

Unlike cropping and contour farming on open lands, agroforestry applications in towns and cities typically involve the complex integration of multiple species that are "stacked" or "assembled" to maximize production in confined and often irregularly shaped spaces. This is due to the size of most open areas in built environments and the ways in which buildings, streets, bridges, and other infrastructure have been designed and integrated. The aim often is to create resilient multifunctional projects, whether stand-alone or connected, that benefit cities, neighborhoods, towns, and suburbs and effectively occupy built spaces and efficiently augment the supply of raw materials from outlying rural areas. In some rare cases, such as intentional independent communities and neighborhoods, a vast amount of food, timber, fiber, and other valuable materials are grown and harvested in urban and community systems that are fully integrated into the spaces and places people live.

The following case studies highlight key aspects of urban and community agroforestry principles and practices across an array of contexts. Regardless of the setting, they demonstrate that human culture is as important as material production, whether agroforestry is practiced in a neighborhood, town, or city. The lines of historical and space-appropriate design, productive placemaking, primary and secondary production, community capital investments, and the varying nature of urban and community design and management that meet ecological, production, and civic needs are found in each example. The first case study is an illustration of productive placemaking within a connected agroforestry corridor—or foodway—that occurred through the power of awareness building and demonstration by one family in an Arizona neighborhood. The second represents scaled efforts to plan and create a sizable agroforestry foodway that supports primary and secondary production along a corridor in need of physical reclamation and social equity. The third is a set of short vignettes demonstrating the power of productive placemaking even in the smallest of urban and community confines.

The Dunbar/Spring neighborhood in Tucson, Arizona: Agroforestry at the intersection of placemaking and neighborhood enlivening

Dunbar/Spring is a small, culturally diverse neighborhood not far from downtown Tucson, AZ. The area experienced a gradual decline during much of the 20th century, resulting in increases in crime and property abandonment. Like the rest of Tucson, the neighborhood has a hot, desert climate, with only about 28 cm of annual rainfall and temperatures in the summer that frequently exceed 38 °C.

In 1994, Brad Lancaster and his brother Rodd purchased an old house on a 1/20-ha lot in the neighborhood. Both had a strong interest in water conservation and began to install rainwater collection cisterns, swales, and small infiltration basins (i.e., rain gardens). In addition to work on their own property, they also turned their attention to the approximately 6-m-wide right-of-way between their property and the street. Rights-of-way were common throughout the neighborhood and typically barren areas of compacted soils where trash accumulated and cars often parked, despite the neighborhood's exceptionally wide streets.

The transformation of their property was dramatic and included planting dozens of food- and medicine-producing trees, shrubs, and other plants, with heavy but not exclusive emphasis on species native to the Sonoran Desert (Figure 10–6). The plants are irrigated with rainwater or gray water, using a variety of innovative techniques such as curb cuts. Curb cuts, and a related technique called curb cores, allow small infiltration basins in the rights-of-way to be periodically filled with rainwater that would otherwise run down the street to the nearest stormwater drain.

While the transformation of the Lancaster property is interesting in its own right, the real value is in how widely the associated techniques have been adopted in the neighborhood, leading to productive placemaking and foodway connections. In a community of small lots, trees and shrubs numbering in the thousands have been planted on numerous properties, in public rights-of-way, and even within traffic circles and chicanes built in the streets (Figure 10–7). The diffusion of agroforestry from a locus of origin and design to scores of other sites throughout the neighborhood is a prime example of what can be accomplished at the intersection of agroforestry, urban forestry, and community development, particularly when investments in demonstrable community capital assets build momentum that translate into meaningful and observable benefits.

From an agroforestry perspective, the neighborhood is a relatively large, dispersed foodway that forms a multifunctional system providing numerous ecological, material, and civic outputs (Figure 10–8). Major overstory species are mostly

Fig. 10–6. Transformation of properties in the Dunbar/Spring neighborhood in Tucson, AZ, using multifunctional and multi-species agroforestry plantings largely of native species. Plantings were supported by water management techniques such as curb core, displayed in the lower left corner. The nested image in the upper right is of the neighborhood before transformation.

Fig. 10–7. Productive and diverse species stacked in open spaces such as traffic circles and chicanes in the Dunbar/Spring neighborhood in Tucson, AZ. Multifunctional species planted in small agroforestry configurations connect across built structures such as sidewalks, rights-of-way, and private property.

natives such as mesquite (*Prosopis* spp.) and desert ironwood (*Olneya tesota* A. Gray), which produce edible seeds and pods. Lower strata include non-native fruit trees such as pomegranate (*Punica granatum* L.) and various citrus and native plants such as wolfberry (*Lycium* spp.) and various species of cacti. Some of the food and other products are harvested by individual property owners, but there also are community gathering and processing events, some of which center on the harvest of mesquite pods, thus enhancing returns on community investments in capital such as social, cultural, and human attributes.

Key results, from the standpoint of urban forestry, include significant improvement to the neighborhood aesthetics, as well as a cooling effect that is very noticeable when walking from a section of right-of-way without any trees into one that has been transformed. The overall effect on the community is also readily apparent. Among other things, there is art in the traffic circles and rights-of-way, places designed specifically for community members to gather, and community events such as the mesquite pod harvests and mesquite pancake breakfasts (Figure 10–9).

Progress in the Dunbar/Spring neighborhood has not come easily, however. It involved (and continues to involve) a door-to-door campaign to engage property owners, as well as many hours spent by Brad Lancaster and others at community and city government meetings. In addition,

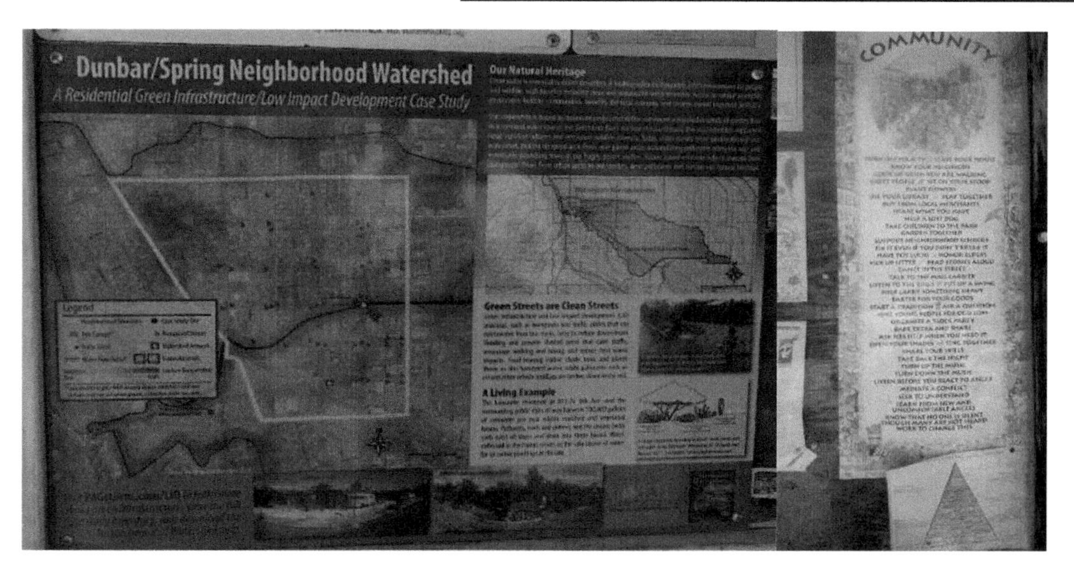

Fig. 10–8. Interpretive signs conveying the multifunctional nature of the foodway transformation in the Dunbar/Spring neighborhood in Tucson, AZ. Natural heritage underpins the design, as well as productive placemaking and civic goals.

Fig. 10–9. Artwork complements agroforestry infrastructure in the Dunbar/Spring neighborhood in Tucson, AZ. Installations represents an important investment in the experiential facet of an urban and community application.

Brad and others have submitted numerous grant proposals to the City of Tucson and other funding sources, which has helped pay for workshops (e.g., on pruning), much of the work in the rights-of-way, and construction of traffic circles and chicanes.

Transformations in this neighborhood are profiled in other ways, but much more could be learned by the systematic attention of researchers. Possible areas of inquiry include the effectiveness of tree management and water conservation practices, amounts and types of food produced, and effects on the local microclimate. Behavioral researchers also could study the work that led to the adoption of agroforestry practices in this and other neighborhoods, the challenges that limit adoption or even result in resistance by some property owners, and socioeconomic impacts such as the influence on crime, community cohesion, and property values.

The Bronx River Foodway: Agroforestry at the intersection of placemaking and urban planning

Once heavily polluted and endangered, the Bronx River has made a dramatic comeback during a 17-yr period, thanks to the more than 120 community groups that form the Bronx River Alliance. This network is currently implementing an agroforestry system foodway along the river's edge, which will constitute a 37-km-long ribbon of open spaces with a multiuse paths extending along the full length of the river in Westchester County and The Bronx. It ties together new open spaces, existing parks, and reclaimed waterfronts. The productive landscape with community-operated spaces provides opportunities for diverse cultural expression (Figure 10–10). As for most urban and community agroforestry systems in North American cities, a project of this magnitude would not be possible without collaboration.

The primary partner involved is New York City Parks, dedicating staff time and funding. Inhabit Earth pitched the concept, brings funding to the project, and is leading the design effort. Many other community partners are participating, including Youth Ministries for Peace and Justice, The Point CDC, and Swale Project, as well as a diverse team of designers.

During the summer of 2017, several components of design were implemented on a 0.8-ha portion of Concrete Plant Park, a former industrial site in the South Bronx reclaimed by community groups (Figure 10–11). The public park, almost 3 ha in size, is located on the west bank of the Bronx River. The surrounding community is ethnically diverse, highly industrialized, and economically distressed. Local obesity and malnutrition rates are high, as are rates of diabetes-related mortality. Ironically, it is adjacent to the largest food distribution facility in the world. In advance of implementation, designers met with dozens of stakeholder groups and more than 90 community residents to design an urban and community system. As the design unfolded, 0.4 ha of the park was dedicated to a forest garden with a winding path through various patches of fruits, nuts, perennial vegetables, pollinator and medicinal plants, and herbs (Figure 10–12).

A second 0.4 ha of the design is dedicated to production using an alley cropping system to demonstrate agroforestry techniques. An orchard and several supporting perennial strips are due to be planted on contour, leaving wide alleys for the production of crops. In between the forest garden and primary production area, a planned village area will demonstrate various styles of home and community permaculture gardening using shipping containers, hoop houses, and other temporary structures. This area represents the blending of agroforestry technologies, as well as space for public art, performances, storytelling, and educational activities. There also are a dozen or so interpretive discovery stations that inform visitors of urban and community systems concepts.

Large-scale urban and community agroforestry projects like these can enable allied professions such as planners, architects, urban foresters, public works personnel, decision-makers, researchers, activists, nongovernmental organizations, and developers to join forces and leverage skill sets. The result is that local governments, community

Fig. 10–10. The Bronx River foodway is a 37-km-long investment in productive placemaking that includes agroforestry applications interspersed with other complementary landscape patterns that enhance secondary production. The site includes community-owned and -defined spaces that provide opportunities for cultural and civic growth.

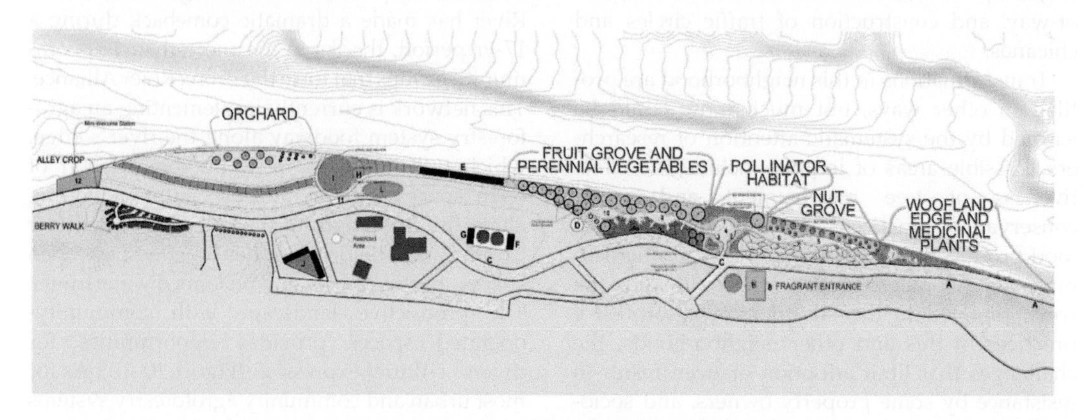

Fig. 10–11. First segment of the Bronx River foodway. The 0.8-ha installation was integrated on 3 ha and included perennial species and fruits, nuts, perennial vegetables, pollinator and medicinal plants, and herbs.

Fig. 10–12. Aerial image of early agroforestry establishment along the Bronx River foodway. Urban food forest plantings have occurred in the green lawn space to the right, and ultimately will tie into additional projects across a 9.3-ha integrated urban and community agroforestry foodway.

groups, and residents across North America are implementing agroforestry via civic collaboration and regenerative design that transforms places and optimizes strategies, which brings out the best in our towns and cities. Add individual applications on private properties, apartment roofs and commons, and businesses and the connectivity of urban and community agroforestry at scale forms a mosaic of production systems where practices are stacked, tailored to particular needs and goals, but always integrated and interactive in terms of production, conservation, and community benefit.

Edible enlivening vignettes: Agroforestry and productive placemaking on street corners and abandoned lots

Urban food forests are created by planting a variety of annual and perennial plants together to form diverse food-producing agroforestry ecosystems. These living assemblies of edible plants including fruit trees, vegetables, herbs, and other products are biologically complex and systematically modeled after a young forest where diverse plant species live together, flowering and fruiting at different times of the year. From today's intentional community food agroforests to the ancient gardens of the Maya, these systems have a long historical food production legacy. In the modern-day spirit of urban and community agroforestry, they also increasingly serve as places of community renewal by merging recreation, socializing, art, and other civic activities where food and fiber are produced and environmental services are provided.

Sometimes referred to as forest gardens, urban food forests, as defined here and in Chapter 3, are agroforestry projects in the built environment that can be small enough to grow in a yard or large enough to cover a city block or multiple city blocks. Many emerging civic renewal initiatives have implemented the small-scale, community-oriented versions typically around 0.4 ha in size. These urban and community systems are found in municipalities large and small and, as of 2019, there were more than 60 of these initiatives underway in more than 35 states. Each of these applications has a unique cultural, socioeconomic, and political context that shapes productive placemaking. Vignettes of three systems in different cities in the United States (Troy, NY; Syracuse, NY; and Providence, RI) illustrate how green urban re-enlivenment strategies that use agroforestry practices and principles are ultimately grounded in the context of community.

Troy, New York

Troy was once a booming producer of textiles and steel in the era of waterway transportation. Like many industrial cities in the northeastern United States, it experienced significant urban decline in the years following World War II. The legacy of this decay was clearly visible in the abandoned buildings and vacant properties that peppered its city center by the end of the 20th century. North Troy's Sixth Avenue neighborhood exemplifies this deindustrialization, with high rates of vacant homes, crime, violence, food insecurity, and poverty. More than 65% of neighborhood children live in households below the poverty line.

A local nongovernmental organization called the Sanctuary for Independent Media supports a wide range of community renewal activities such as food literacy and access, social and environmental justice, and youth education. The organization sponsors multiple programs that practice and demonstrate urban and community agroforestry production, environmental restoration, and the ways in which science and art can be blended. Ultimately, the Sanctuary provides the community, specifically the youth, with practical tools for expressing themselves creatively in places shaped by urban food forest systems. The Sanctuary's work involves transforming open areas into art and garden zones, which include annual and perennial species that are stacked in multistory agroforestry configurations and serve as the backdrop for educational programs and community food production.

The Collard City Growers is a small urban and community demonstration and composting group that leads programs where local children connect with food by growing it. A second site, the L-lot, is a phyto- and myco-remediation experimental site intended to study and ultimately help inform property rehabilitation using agroforestry vegetation and polyculture configurations. The third site, Freedom Square, is a highly visible corner lot that was collaboratively designed in the spirit of productive placemaking. What was once a vacant space has been transformed into a community plaza where neighbors recreate and socialize with a beautiful mosaic mural stage, youth bike path, an educational moth garden, and a young urban and community agroforestry system.

The food-producing agroforestry zones occupy the edges of Freedom Square in the spirit of a foodway, and community members pick the fruits, berries, and vegetables that the system produces (Figure 10–13). A unique component of Freedom Square is a moth garden that attracts insects for pollination. The garden is used as a creative educational feature at night called "Moth Cinema," where lighted moths cast amplified shadows on an adjacent white building. The Sanctuary demonstrates that creative, artistic components integrated into urban and community systems adds another element that invites community members to begin using the space quickly during early stages of development when the plants are still young.

Fig. 10–13. A very small, young, urban and community agroforestry planting juxtaposed with the mosaic mural crafted by community members.

Syracuse, New York

Similar to Troy, Syracuse experienced a decline in infrastructure at the end of the 20th century and likewise initiated several urban revitalization projects. One project involved a community food forest based on the principles of urban and community agroforestry systems that aim to create productive places and functional spaces. The Rahma Free Health Clinic Edible Snack Garden is on a corner lot adjacent to a gas station convenience store that is the primary food source in an impoverished, high-crime southside neighborhood. The lot was vacant and overgrown and located between the clinic and convenience store.

The Alchemical Nursery, a local organization that promotes eco-social landscape development and community food forestry, collaborated with the clinic to transform the lot using urban and community design (Figure 10–14). The group raised money for materials through crowd funding, organized volunteers to establish the site, and established protocols to assist with maintenance. The clinic brings produce from the garden into the office for patients and encourages them to use the agroforestry garden to glean food and spend time outdoors.

Along the path that traverses the urban food forest, one can find edible fruits from tree species such as cherry (*Prunus avium* L.), plum (*Prunus domestica* L.), and apple (*Malus domestica* Borkh.) that form a canopy. The more adventuresome can forage in the understory filled with low-growing bushes of golden raspberry (*Rubus ellipticus* Sm.), blackberry (*Rubus* sp.), currant (*Ribes* sp.), seaberry (*Hippophae rhamnoides* L.), and groundcovers such as strawberry (*Fragaria* sp.) and perennial sunchoke (*Helianthus tuberosus* L.). Like most other newly established sites, it resembles a young diverse forest that has been designed to produce food, fiber, wood, and other important products, provide ecosystem services, and enhance civic well-being. Birds and pollinator insects frequent the site as habitat and layers of vegetation help with water retention and soil conservation. There are multiple narrow paths weaving through the site that invite people to explore the natural ecology and landscape aesthetics. One path leads to the front corner of the Snack Garden, where there is an annual garden bed near the sidewalk displaying easily recognizable and quick-growing produce to passersby, which is intended to entice them to enter and forage. The Snack Garden site is a useful example of small-scale urban and community agroforestry, where primary and secondary production obscure the lines between raw material production health care, and community engagement.

Providence, Rhode Island

Agroforestry in Providence, RI, represents another instance of a diverse, uniquely designed urban and community system geared to enliven

Fig. 10–14. Simple seating and an interpretive sign along a footpath that guides community members to multifunctional agroforestry features in a densely populated southside neighborhood in Syracuse, NY.

what once was an underutilized and underappreciated open space in a densely populated metropolitan landscape. The project in Providence is located near a community garden in Roger Williams Park adjacent to the park's botanical center. The land borders an artificial lake with a thin buffer of trees lining the banks, and the agroforestry plantings are a continuation of the lake's forest edge. A primary objective in this case is water quality, and plantings are intended to serve as a natural filter for surface water runoff to the water body.

The location of the agroforestry enlivenment project was previously an open grassy area, and numerous environmental and community benefits have been documented following establishment. Master Gardeners helping to maintain the space have found turtles nesting in the riparian areas, and community gardeners have reported benefits from the increase in pollinators that are attracted to the food forest. Another interesting feature is that visitors to the site can learn about historically important plants such as New Jersey tea (*Ceanothus americanus* L.), which was used as a substitute during the boycott of British tea in the late 1700s, and the American chestnut [*Castanea dentata* (Marshall) Borkh.], which once grew abundantly in eastern North America. By intentionally choosing historically significant plants as part of the system composition, this food forest offers an array of educational opportunities in additional to urban and community ecosystem services and production.

Conclusion

Agroforestry is a scientific land use as well as a long-standing indigenous knowledge system and an important facet of traditional diversified farming and forestry practices. In North America, agroforestry embodies this legacy, which is manifested in the form and function of codified temperate zone practices. The systematic application of agroforestry in North American towns and cities is relatively new, but they too reflect this tradition. Conceptual linkages to tropical home gardens, which concentrate polyculture production in and around the places where people live, mean agroforestry is as relevant in neighborhoods and boroughs as it is on rural farms. This chapter provided a definition of urban food forests and highlighted urban food forestry as a popular framework for designing and managing agroforestry systems in built environments. It also highlighted the multifunctional and adaptive value in interconnecting practices at scale across built environments, even potentially bridging production systems and reducing the metabolic rift between cities and towns and rural environments.

An important outcome of agroforestry system connectivity is the creation of productive places that involve a mixture of biological parameters, relevant community objectives, and resident participation. This is true on open public space and on private property and is generally necessary as a basis for adopting foodways more broadly that tie together urban food forest practices in places like a historic neighborhood, a street corner, along a riverside, or on a college campus. The separation of urban, suburban, and even small towns from the locus of food, fiber, and wood production forms the premise for reclaiming local production through interconnected agroforestry systems across built environments. It is there that projects complement rather than conflict with rural production, particularly in light of population growth and expanding megaregions.

Primary and secondary production represent two important aspects of urban and community systems that define and demarcate productive placemaking. Primary outputs are specific to onsite productivity and largely reflect agroforestry design based on the biophysical attributes of a place. Secondary benefits are those that represent the ways in which people make use of intentional urban and community systems through market or amenity benefits. Because various forms and tangents of production are possible, diverse agroforestry systems in towns and cities like those in rural areas are a strength rather than a liability. Also important is the range in agroforestry practices in urban and community settings—from small-scale private and reclaimed properties to extensive plantings in public spaces networked through foodway design. To achieve meaningful placemaking outcomes, whether based on primary or secondary production, investments in diverse community assets, or capital, are necessary. Strategic directions determined by communities in this regard balance community stocks and flows that achieve a comprehensive level of production.

Agroforestry applications in built environments are productive and creative placemaking farming and forestry strategies where people produce tangible and intangible benefits in and around the places and spaces where they live. They are shaping systems that supplement the evolution of temperate zone agroforestry, and what makes them unique is that this is underscored, and only possible, by the commitment and sustained involvement of local residents, organizations, technical experts, advocates, and elected and appointed officials. Urban and community systems, like those in rural farming and forestry contexts, are not a short-term proposition. They typically require multiple growing seasons before rendering the intended multiple benefits. Yet in their infancy and into maturity, the payoff can be substantial because the majority of the world's population can take ownership of and shape raw material production as well as realize civic benefits such as education, nutrition, and aesthetics to name just a few.

References

American Community Survey. 2016, 8 Dec. *New census data show differences between urban and rural populations.* Washington, DC: U.S. Census Bureau. Retrieved from https://www.census.gov/newsroom/press-releases/2016/cb16-210.html.

Askerlund, P., & Almers, E. (2016). Forest gardens: New opportunities for urban children to understand and develop relationships with other organisms. *Urban Forestry & Urban Greening, 20,* 187–197.

Bene, J. G., Beall, H. W., & Côté, A. (1977). *Trees, food and people: Land management in the tropics.* Ottawa, ON, Canada: International Development Research Center.

Borelli, S., Conigliaro, M., Quaglia, S., & Salbitano, F. (2017). Urban and peri-urban agroforestry as multifunctional land use. In J. Dagar & V. Tewari (Eds.), *Agroforestry* (pp. 705–724). Singapore: Springer.

Bukowski, C. J., & Munsell, J. F. (2018). *The community food forest handbook.* White River Junction, VT: Chelsea Green.

Clark, K. H., & Nicholas, K. A. (2013). Introducing urban food forestry: A multifunctional approach to increase food security and provide ecosystem services. *Landscape Ecology, 28,* 1649–1669. https://doi.org/10.1007/s10980-013-9903-z

Coffey, S. E. (2020). *Edible green infrastructure in the United States: Policy at the municipal level* [Unpublished master's thesis]. Virginia Tech.

Coffey, S. E., Munsell, J. F., Hübner, R., & Friedel, C. R. (2021). Public food forest opportunities and challenges in small municipalities. *Urban Agriculture & Regional Food Systems.* https://doi.org/10.1002/uar2.20011

Cofie, O., Bradford, A. A., & Drechsel, P. (2006). Recycling of urban organic waste for urban agriculture. In R. van Veenhuizen (Ed.), *Cities farming for the future: Urban agriculture for green and productive cities* (pp. 207–229). Leusden, the Netherlands: ETC–Urban Agriculture.

Colinas, J., Bush, P., & Manaugh, K. (2019). The socio-environmental impacts of public urban fruit trees: A Montreal case-study. *Urban Forestry & Urban Greening, 45,* 126132. https://doi.org/10.1016/j.ufug.2018.05.002

Doolittle, W. E. (2004). Permanent vs. shifting cultivation in the eastern woodlands of North America prior to European contact. *Agriculture and Human Values, 21,* 181–189.

Douglas, I. (2012). Urban ecology and urban ecosystems: Understanding the links to human health and wellbeing. *Current Opinion in Environmental Sustainability, 4,* 385–392.

Flora, C. B., Flora, J. L., & Gasteyer, S. P. (2016). *Rural communities: legacy and change* (5th ed.). Boulder, CO: Westview Press.

Gold, M. A., Godsey, L. D., & Josiah, S. J. (2004). Markets and marketing strategies for agroforestry specialty products in North America. In P. K. R. Nair, M.R. Rao, & L. E. Buck (Eds.), *New vistas in agroforestry: A compendium for 1st World Congress of agroforestry, 2004* (pp. 371-382). Dordrecht, the Netherlands: Springer.

Hansen, R., & Pauleit, S. (2014). From multifunctionality to multiple ecosystem services? A conceptual framework for multifunctionality in green infrastructure planning for urban areas. *Ambio, 43,* 516–529.

Hemenway, T. (2017). *The permaculture city: Regenerative design for urban, suburban, and town resilience.* White River Junction, VT: Chelsea Green.

Hemmelgarn, H. L., & Munsell, J. F. (2021). Exploring "beyond food" opportunities for biocultural conservation in urban forest gardens. *Urban Agriculture & Regional Food Systems.* https://doi.org/10.1002/uar2.20009

Hübner, R., Künstle, S., Munsell, J. F., & Pauleit, S. (2018, 27 Nov.–1 Dec.). *The rise of urban agroforestry systems: A comparative analysis of the United States/Canada and Germany.* Paper presented at World Forum on Urban Forests, Mantova, Italy.

Lehmann, L. M., Lysák, M., Schafer, L., & Henriksen, C. B. (2019). Quantification of the understorey contribution to carbon storage in a peri-urban temperate food forest. *Urban Forestry & Urban Greening, 45,* 126359.

Lorenz, K. & Lal, R. (2014). Soil organic carbon sequestration in agroforestry systems: A review. *Agronomy for Sustainable Development, 34,* 443–454.

Lovell, S. T. (2010). Multifunctional urban agriculture for sustainable land use planning in the United States. *Sustainability, 2,* 2499–2522. https://doi.org/10.3390/su2082499

Madlener, R., & Sunak, Y. (2011). Impacts of urbanization on urban structures and energy demand: What can we learn for urban energy planning and urbanization management? *Sustainable Cities and Society, 1*(1), 45–53.

Mercer, D. E., & Pattanayak, S. K. (2003). Agroforestry adoption by smallholders. In E.O. Sills & K. L. Abt (Eds.), *Forests in a market economy* (pp. 283–299). Dordrecht, the Netherlands: Springer.

Nytofte, J. L. S., & Henriksen, C. B. (2019). Sustainable food production in a temperate climate: A case study analysis of the nutritional yield in a peri-urban food forest. *Urban Forestry & Urban Greening, 45,* 126326.

Park, H., & Higgs, E. (2018). A criteria and indicators monitoring framework for food forestry embedded in the principles of ecological restoration. *Environmental Monitoring and Assessment, 190,* 113. https://doi.org/10.1007/s10661-018-6494-9

Park, H., Turner, N., & Higgs, E. (2018). Exploring the potential of food forestry to assist in ecological restoration in North America and beyond. *Restoration Ecology, 26,* 284–293.

Riolo, F. (2019). The social and environmental value of public urban food forests: The case study of the Picasso Food Forest in Parma, Italy. *Urban Forestry & Urban Greening, 45,* 126225.

Salbitano, F., Borelli, S., Conigliaro, M., & Chen, Y. (2016). *Guidelines on urban and peri-urban forestry.* Forestry Paper 178. Rome: FAO.

Scherr, S.J. (2004). Building opportunities for small-farm agroforestry to supply domestic wood markets in developing countries. *Agroforestry Systems, 61,* 357–370.

Schewenius, M., McPhearson, T., & Elmqvist, T. (2014). Opportunities for increasing resilience and sustainability of urban social–ecological systems: Insights from the URBES and the Cities and Biodiversity Outlook projects. *Ambio, 43,* 434–444.

Shackleton, C., Hurley, P., Dahlberg, A., Emery, M., & Nagendra, H. (2017). Urban foraging: A ubiquitous human practice overlooked by urban planners, policy, and research. *Sustainability, 9,* 1884.

Synk, C. M., Kim, B. F., Davis, C. A., Harding, J., Rogers, V., Hurley, P. T., & Nachman, K. E. (2017). Gathering Baltimore's bounty: Characterizing behaviors, motivations, and barriers of foragers in an urban ecosystem. *Urban Forestry & Urban Greening, 28,* 97–102.

Thackara, J. (2019). Bioregioning: Pathways to urban-rural reconnection. *She Ji: The Journal of Design, Economics, and Innovation, 5*(1), 15–28.

Trozzo, K. E., Munsell, J. F., & Chamberlain, J. L. (2013). *How to plan for and plant streamside conservation buffers with native fruit and nut trees and woody floral shrubs* (ANR-69P). Blacksburg, VA: Virginia Cooperative Extension.

Tzoulas, K., Korpela, K., Venn, S., Yli-Pelkonen, V., Kaźmierczak, A., Niemela, J., & James, P. (2007). Promoting ecosystem and human health in urban areas using green infrastructure: A literature review. *Landscape and Urban Planning, 81*(3), 167–178.

United Nations. 2017. *World population projected to reach 9.8 billion in 2050, and 11.2 billion in 2100.* New York: UN Department of Economic and Social Affairs. Retrieved from https://www.un.org/development/desa/en/news/population/world-population-prospects-2017.html.

Verchot, L. V., Van Noordwijk, M., Kandji, S., Tomich, T., Ong, C., Albrecht, A., . . . Palm, C. (2007). Climate change: Linking adaptation and mitigation through agroforestry. *Mitigation and Adaptation Strategies for Global Change, 12,* 901–918.

Wilson, R. (1997). A sense of place. *Early Childhood Education Journal, 24,* 191–194.

Study Questions

1. What are some potential barriers to the establishment of food forests in built environments?

2. What is the dominant form of agroforestry in the built environment? Describe the basic design principles of this new temperate zone practice and list what is supplied.

3. Imagine that you work for a city government or nongovernmental organization interested in urban food forests. How would you go about promoting the establishment of a food forest, especially if the project is set to occur in a neighborhood where you do not live?

4. Explain how and why outcomes and aims listed in the eco-social Venn diagram are together multifunctional and why that is important for a community agroforestry project.

5. Do you think the opportunities and constraints for community agroforestry systems relate more to the biophysical ecosystem or to social and economic conditions? Why did you answer the way you did?

6. How might the implementation of urban agroforestry projects contribute to meeting ecosystem and social justice goals?

7. Suppose you have been asked to plan a community agroforestry project. What would you want to know about a community to design a project that aligns with their assets and addresses their needs?

8. How do community agroforestry systems contribute to productive placemaking, and how do they enhance primary production in the built environment?

9. What is the difference between an urban food forest and a foodway?

10. What are some of the potential economic benefits and implications of urban food forests and community agroforestry systems?

11. Explain how urban and community agroforestry can address pressing social issues listed in the chapter and use specific examples.

12. Which of the other temperate zone agroforestry practices besides food forests have the most potential in built environments? Why did you choose that answer?

Section III

Agroforestry Benefits

North American Agroforestry, Third Edition. Edited by Harold E. "Gene" Garrett, Shibu Jose, and Michael A. Gold.

11

Chung-Ho Lin*, Eric E. Weber,
W. D. "Dusty" Walter,
Teng Teeh Lim, and
Harold E. "Gene" Garrett

Vegetative Environmental Buffers for Air Quality Benefits

Abbreviations:
CAFO concentrated animal feeding operations
LAD leaf area density
PM particulate matter
VEB vegetative environmental buffers
VOC volatile organic compounds

Emissions of malodor from concentrated animal feeding operation (CAFO) facilities are an increasing environmental concern for producers and nearby local communities. Odorous chemicals including ammonia (NH_3), hydrogen sulfide (H_2S) and a wide range of volatile organic compounds (VOCs) have been identified as a result of emission tests from CAFO waste decomposition. The compounds are often characterized based on their chemical functional groups. A wide range of malodors that are characteristic of CAFO emissions exist. Studies suggest that long-term exposure to these odors has an effect on secretory immune function which directly or indirectly triggers a physiologic effect among neighbors of industrial CAFO facilities.

Use of vegetative environmental buffers (VEB) for odor abatement is new and the science in support of using VEBs for this purpose is limited. However, reports in the literature strongly suggest that significant quantities of odor compounds are highly correlated with malodors and can be effectively removed through the use of VEB technology. The VEB effectiveness is known to be related to its physical location, species composition, density, and geometric configuration. Odor reduction by VEB occurs via physical interception, dilution, and chemical adsorption. The objectives of this chapter are to summarize the current research efforts and scientific findings involving VEB-mitigation technologies for reducing emissions of malodors and other health hazardous materials, such as bioaerosals from CAFOs. This chapter has synthesized scientific findings on: i) impacts of odor and particulate matter emission on environmental quality, human health and social economy, ii) mechanisms involved in malodor abatement by VEBs, and iii) design recommendations and management strategies.

*Corresponding author (LinChu@missouri.edu)

North American Agroforestry, Third Edition. Edited by Harold E. "Gene" Garrett, Shibu Jose, and Michael A. Gold.

Impact of Odor and Particulate Matter Emission on Environmental Quality and Human Health

Financial considerations have motivated producers to use CAFOs as the preferred approach to livestock production, especially in swine and poultry industries. Concerns associated with potential environmental and health effects of odor emissions and particulate matter (PM) from CAFOs has risen with their increasing operation scale, geographic concentration, and suburbanization (National Research Council, 2002). In an effort to reduce odor emissions from swine CAFOs, 44 of the 50 states in the United States have enacted air emission policies to directly or indirectly reduce odors from these operations (Vander, 2001). Recently, CAFOs have faced the major challenge of reducing emissions of malodorous compounds.

Most of the odorous compounds that are emitted from CAFO operations result from anaerobic decomposition of wastes by microorganisms. The decomposition of this complex organic mixture is mediated by a diverse group of obligate and facultative anaerobic microorganisms (Zahn et al., 1997). In the United States, more than 75% of CAFO waste is processed anaerobically and the manure waste is often stored for as long as 13 mo before land application (Zahn et al., 1997; Zahn et al., 2001). O'Neill and Phillips (1992) developed and compiled a comprehensive list of 168 different malodorous compounds emitted from swine wastes. A wide range of malodors that are characteristic of swine CAFO emissions have been identified, including NH_3, H_2S, amines, carboxylic acids, ethers, volatile fatty acids, indoles, skatole, phenols, mercaptans, alcohols, and carbonyls (Cai et al., 2006; O'Neill and Phillips, 1992). Several classes of these odorous compounds have been linked to degraded environmental quality and human health. For example, findings suggest that H_2S is a potent neurotoxin, and that chronic exposure to even low ambient levels can lead to irreversible damage to the brain and central nervous system (Kilburn, 1997; Kilburn, 1999; Kilburn and Warshaw, 1995). Children are among the most susceptible to this poisonous gas. Several VOCs emitted from swine facilities, particularly indoles and skatole, have also been demonstrated to have distinct toxic effects on humans (Spoelstra, 1977; Yokoyama and Carlson, 1981). The common characterization of the malodorous VOCs pollutants and their concentrations measured in a field swine facility in northern Missouri are illustrated in Table 11–1 and Table 11–2, respectively (Lin et al., 2011; Yasuhara et al., 1984). Figure 11–1 and 11–2 demonstrate the representative dispersion of NH_3 and H_2S near the facility (Lin et al., 2011).

Airborne dust and PM play a major role in the spreading of odor. Dust and airborne PM emitted from swine confinement originates from feces, animal skin, feathers, and feed particles. Studies have shown that there are approximately 70 to 100 volatile compounds adsorbed on dust (Mackie et al., 1998; Wang et al., 1998). Past studies have reported aerial dust concentrations ranging from 212,000 to 73,550,000 particles m^{-3}, with 71% of the particles less than 5 µm in size (Honey and McQuitty, 1979; Rosentrater, 2003). In 1986, more than 700,000 people in the United States were exposed to hazardous levels of CAFO dust (Mutel et al., 1986), and the number has increased as more CAFOs are now in operation. Approximately 70% of those exposed, including workers, family members of workers, and veterinarians, suffered from various respiratory ailments, such as organic toxic dust syndrome, chronic bronchitis, hypersensitive pneumonitis, and occupational asthma (Donham and Gustafson, 1982).

Table 11–1 Characterization of the malodor VOCs compounds (based on Yasuhara et al., 1984).

VOCs compound	Smells Like	Odor threshold (µg m^{-3})
dichloromethane		3000.00
Propionic acid	fecal	2.50
methylbutanol		80.00
Methyl disulfide	nauseating, fecal	0.00
Toluene		80.00
Butyric acid	Fecal, stench	0.30
capronaldehyde		28.00
Isovaleric acid	fecal	0.26
P-xylene		350.00
Valeric acid	fecal	0.26
O-xylene		350.00
Octyl aldehyde		0.00
O-cresol	fecal	0.40
P-cresol	fecal	0.05
3-methylphenol	fecal	0.40
Nonyl aldehyde		0.00
4-ethylphenol	pungent	3.50
3-ethylphenol	pungent	3.50
Indole	Fecal, nauseating	0.64
2-methylindole	Fecal, nauseating	0.35
3-methylindole	Fecal, nauseating	0.35
butylphenol		0.00
Dimethyldisulfide	Decayed vegetable	1.00
butanedione		0.00

Table 11–2 . Concentrations of VOCs (μg m^{-3}) at source, 15 m from the source and along the VEBs established near a cooperating deep-pit CAFO swine facility in Missouri. The standard deviations are expressed with the values in in the parentheses (*n*=10).

Compounds	Source	15 m from source	30 m from source (along the tree line)
		μg m^{-3}	
Dichloromethane	NA	NA	NA
Propionic acid	1.81 (1.31)	15.22 (12.98)	0.00 (0.00)
Methylbutanol	2.38 (1.54)	0.02 (0.02)	2.23 (1.81)
Methyl Disulfide	0.003 (0.00)	0.00 (0.00)	0.01 (0.01)
Toluene	8.85 (7.80)	2.91 (2.25)	0.00 (0.00)
Butyric acid	0.68 (0.39)	1.28 (1.06)	0.22 (0.14)
Capronaldehyde	6.64 (4.27)	0.18 (0.18)	0.00 (0.00)
Isovaleric acid	1.38 (1.1)	0.00 (0.00)	0.00 (0.00)
P-Xylene	0.44 (0.14)	0.68 (0.42)	0.11 (0.11)
Valeric acid	3.08 (2.95)	0.45 (0.45)	0.00 (0.00)
O-Xylene	1.04 (0.38)	0.51 (0.34)	1.21 (0.96)
Octyl Aldehyde	1.92 (0.91)	1.20 (0.59)	0.82 (0.67)
O-cresol	0.001 (0.00)	0.00 (0.00)	0.00 (0.00)
P-cresol	43.65 (22.6)	2.75 (2.34)	3.04 (2.22)
3-methylphenol	29.55 (13.95)	5.22 (3.90)	1.45 (1.05)
Indole	0.04 (0.03)	0.03 (0.03)	0.00 (0.00)
2-methylindole	0.15 (0.11)	0.00 (0.00)	0.00 (0.00)
3-methylindole	0.21 (0.11)	0.03 (0.03)	0.00 (0.00)
Butyphenol	0.00 (0.00)	0.00 (0.00)	0.00 (0.00)

The emitted dusts can also carry other biological and biochemical agents (Heber et al., 1988; Wang et al., 1998). Airborne culturable and nonculturable bacteria, fungi, parasites, viruses, algae, and fragments of microbial agents (endotoxins) may be contained within bioaerosols (Liebers et al., 2006). Antibiotic residues like sulfamethazine, oxytetracycline and chloramphenicol have also been found in airborne dust originating from swine CAFOs (Hamscher et al., 2003). These bioaerosols pose potential health risks for humans via inhalation of the contaminated dust. One survey of swine CAFO workers revealed that at airborne dust concentrations of around 2.2 mg m^{-3}, workers inhaled approximately 6.3 mg of dust per day, putting them at risk for exposure to various classes of antibiotics and biological agents (Donham, 1998; Mutel et al., 1986; Rosentrater, 2003). These airborne biological agents and sub-therapeutic concentrations of antibiotics can result in greater risks for respiratory ailments among exposed workers and family members living near swine CAFOs.

The dispersion of airborne emissions from CAFOs is affected by factors such as topography, air temperature, humidity, ventilation system, flow rate, diet, prevailing winds, animal activity, and building orientation (Chapin et al., 1998; Heber et al., 1988). The transport of odor plumes often decreases exponentially with distance (Bremberg, 1994). Dispersion modeling is typically used to estimate the concentration of pollutants, such as NH_3 and H_2S, at specified distances from an emission source (US Environmental Protection Agency, 2004). A number of steady state plume models incorporating air dispersion based on planetary boundary layer turbulence structure, and scaling concepts (for example, the American Meteorological Society and Environmental Protection Agency Regulatory Model, AERMOD) are being developed to more accurately predict air pollutant transport from livestock operations (U.S. Environmental Protection Agency, 2019). Other non-steady state dispersion models, such as the California Puff Model (CALPUFF), were developed to simulate the effects of time- and space-varying meteorological conditions on pollution transport, transformation, and removal. They can be applied for long-range transport and for complex terrain (US Environmental Protection Agency, 2004).

Impact of Odor Emission on Psychological Health

Overall, research on odor effects on mood and psychological functioning is still in its infancy. Most odor research has examined the positive or stress-reducing effects of positive odors (i.e., jasmine, lemon, orange, lavender, aromatherapy), but very few studies have examined negative odors. One study suggested that long-term exposure to negative odor has an effect on secretory immune function that directly or indirectly triggers a physiologic effect among neighbors of industrial hog operations (Avery, 2003). As might be expected, malodors produce a strong negative effect that motivates individuals to escape the setting (Asmus and Bell, 1999). In a field study, Steinheider et al. (1998) showed that close proximity to a swine CAFO was associated with increased gastric distress, as well as greater subjective annoyance. In a more recent field study, Horton et al. (2008) showed that greater perceived swine CAFO malodors were associated with greater stress and negative emotions among nearby residents. Specifically, Herz et al. (2004) showed that associating an odor with frustration at Time 1 produced reduced performance of important tasks at Time 2 when the odor was reintroduced. As shown by Knasko et al. (1990), even the mere suggestion that a negative odor is present (whether or not one is) is enough to affect

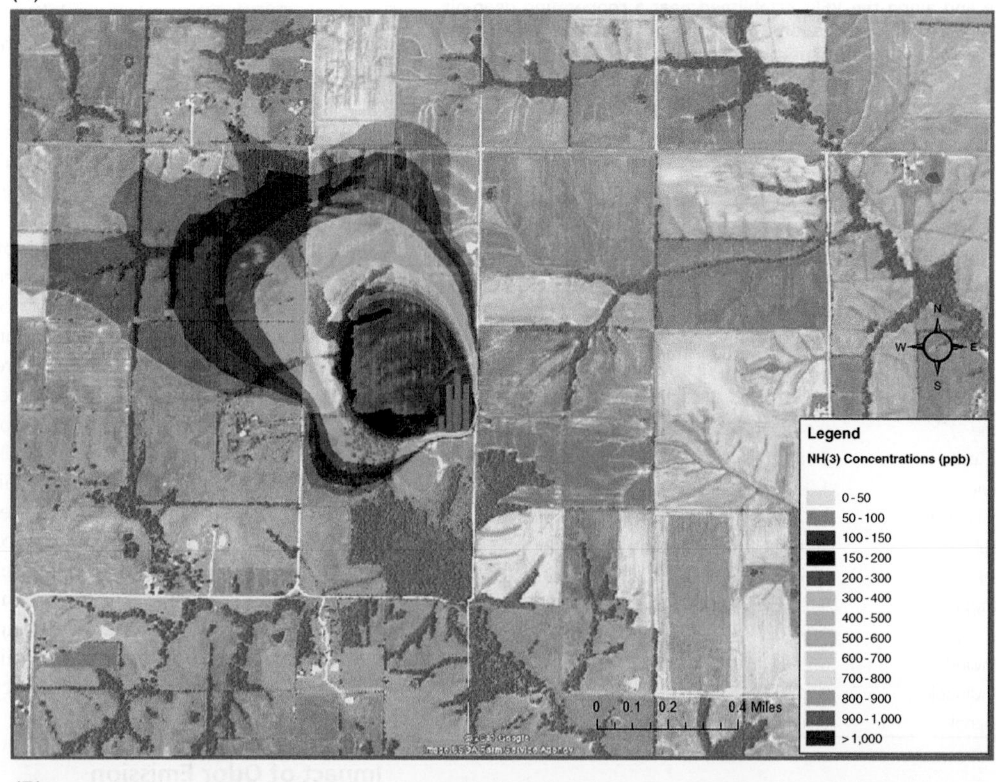

(B)

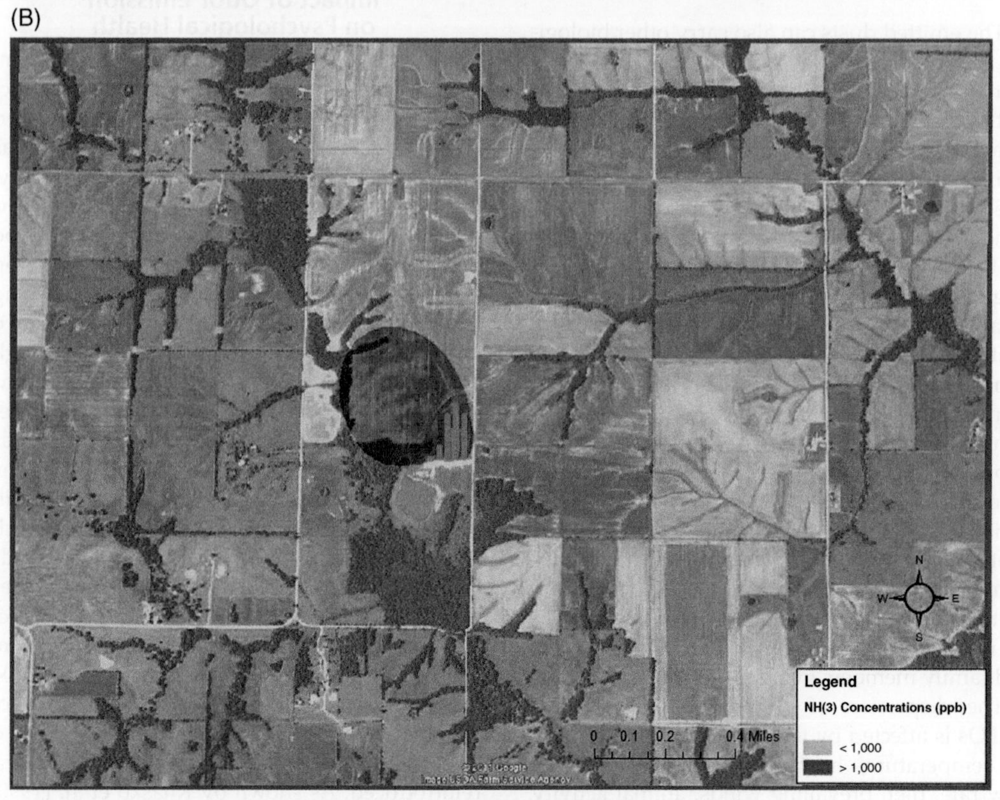

Fig. 11–1. Concentration contour maps of NH_3 sampled on 7/26/2010 (A) and contour maps of concentration values range > smell threshold concentration 1000 ppb (B) near a swine CAFO facility in northeastern Missouri, U.S.

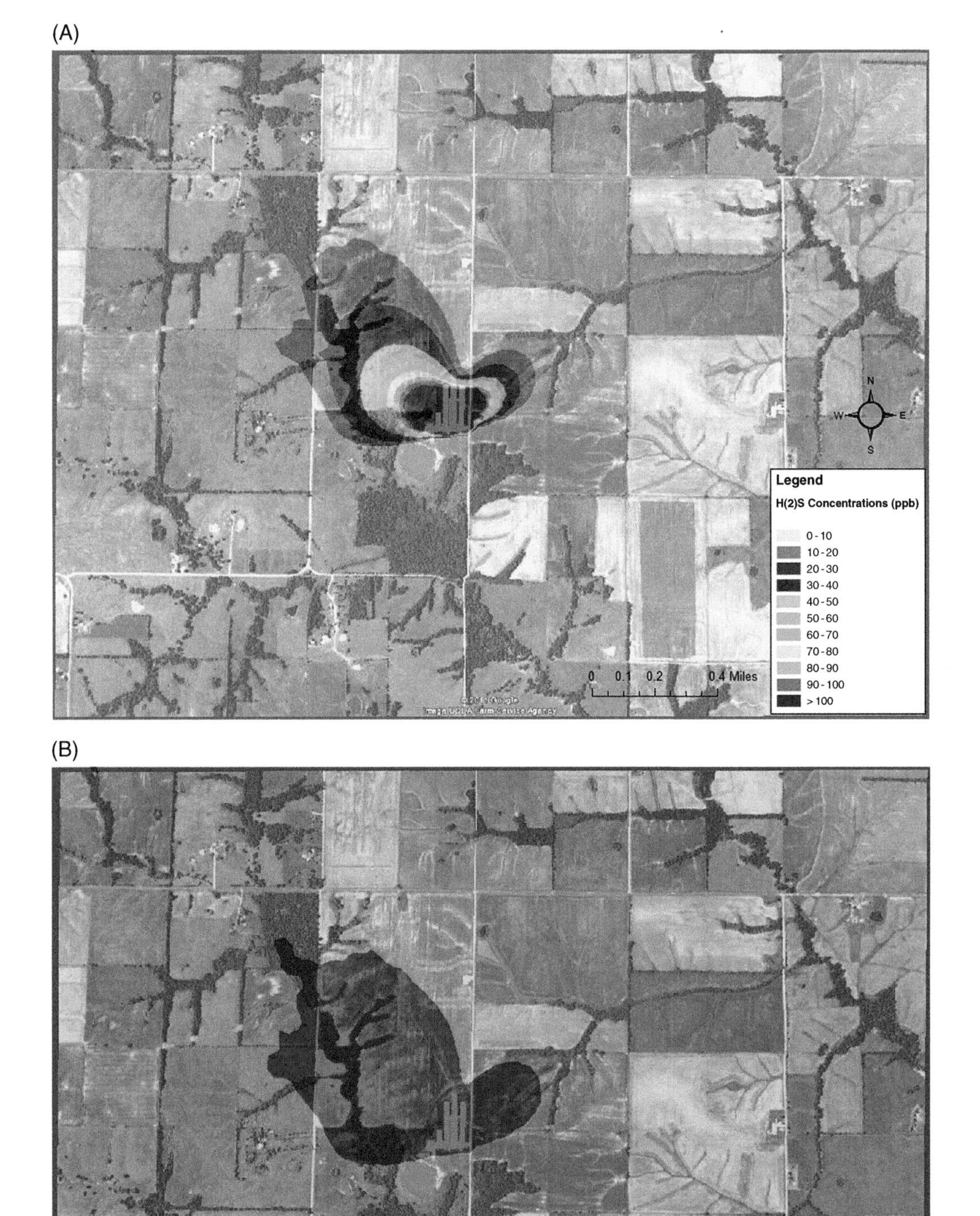

Fig. 11–2. Concentration contour maps of H_2S (A) and contour maps of concentration values range > smell threshold concentration 8 ppb (B) near a swine CAFO facility in northeastern Missouri, U.S.

peoples' mood and well-being. Donham (1998) referred to an "environmental stress syndrome" that might produce psychological symptoms even when environmental stressors are objectively undetectable.

Impact of Odor Emission on Real Estate Values

Odor emissions from CAFOs may have significant economic impacts on the surrounding area. The most common impact is reduced real estate property values in the vicinity of the CAFO (Park et al., 2004; Ready and Abdalla, 2005; Saphores and Aguilar-Benitez, 2005). Studies conducted throughout the United States by these authors have shown that home prices can decline by at least 3.4 to 6.4%. In addition to spatial proximity, Herriges et al. (2005) showed that the building of new livestock facilities resulted in homes located directly downwind to lose 14 to 16% of their value. Isakson and Ecker (2008) report that in Iowa, a house located within 2.5 miles of and directly downwind from a swine CAFO would lose about 15.3% of its value. Milla et al. (2005) suggest that an increase in the number of hogs on a farm at a fixed distance from a property has a direct and negative effect on the property's market value. Based on a regression of home sale prices, Milla et al. (2005) reported that for every 1% increase in hog density (number of hogs in the nearest hog farm divided by linear distance from house), property values declined by 0.03% (Milla et al., 2005). Using a combination of geographic information systems–derived variables (including distance, wind angle, and house descriptors), Isakson and Ecker (2008) concluded that prevailing winds have a significant negative effect on home prices when located within three miles of a CAFO, but the size of the operation (number of animals) is the primary factor affecting the price of homes located more than three miles from the CAFO. Clearly, the research findings from a broad geographic area of the United States show that swine CAFOs have a significant negative economic impact on surrounding properties.

Mitigation of Odor Using Vegetative Environmental Buffers

The use of vegetative environmental buffers (VEBs) for odor abatement is a new management practice, and the science in support of using VEBs for this purpose is still limited. While reports in the literature strongly suggest that significant quantities of compounds known to correlate highly with malodors can be removed through the use of VEB technology (e.g., 47% and 50% reduction in NH_3 and dust emissions, respectively), the overall effect on reducing odor, based on the literature, appears to be highly variable ranging from 6% to 49% (Malone et al., 2006b; Parker et al., 2012). The VEB effectiveness is known to be related to its physical location, species composition, density, and geometric configuration. Odor reduction by VEB occurs via physical interception, dilution, and chemical adsorption (Tyndall and Colletti, 2007). The VEB canopy encourages the interception of odor carriers, such as dust and organic particulates. In addition, VEBs reduce wind speed, facilitating the deposition of PM and bioaerosols (Tyndall and Colletti, 2007). The turbulence created by VEBs can also dilute the odor by forcing the mixing of odor with clean air.

Reviews conducted by Tyndall and Colletti and others, (Baldauf, 2017; Malone et al., 2006a; McDonald et al., 2016; Tyndall and Colletti, 2007; Hewitt et al., 2019), show that odors can be ameliorated by VEBs through the following mechanisms: i) physical dilution and diffusion, ii) encouraged dust and aerosol deposition by reducing wind speeds, iii) physical interception of dust and aerosols, and iv) creating sinks for chemical constituents of odor. In this chapter, we have concluded the recommended design configurations, species consideration and management strategies based on the results of available field studies and wind tunnel simulations.

1. Optimize the crown porosity and configuration to facilitate the physical dilution and vertical diffusion.

Vegetative environmental buffers provide semipermeable physical barriers to deflect the flow of the air, introducing surface turbulence that intercepts and disrupts odor plumes, promotes vertical dispersion, and therefore, the VEBs enhance the physical dilution of the pollutants by mixing with clean air and vertically extending the effective path-length of air from source to receptor (Hewitt et al., 2019).

The surface characteristics of a VEB play a critical role in determining the turbulence capability of the nearby atmosphere (Weber, 2012). The characteristic variables, such as surface roughness length, albedo, and Bowen ratio, significantly affect the turbulent characteristics of the flow. These variables have been used to calculate the structure and turbulence of the boundary layer for a VEB system in several simulation studies (Weber, 2012). In general, the canopy of a coniferous plantation has 30% to 160% more surface roughness lengths, numerical values to express the roughness of the surface, during the spring (30%), autumn (63%) and winter (160%), compared with a canopy of a deciduous plantation (US Environmental Protection Agency, 2004). Therefore, the introduced roughness of the coniferous VEB plantation tends to more favorably promote turbulence in the boundary layer (Weber, 2012).

The reduced physical transport, increased dilution processes and enhanced vertical dispersion of pollutants by VEBs, have been quantified through simulation studies (Weber, 2012). Using the collected baseline ground-measured data, and the AERMOD with integration of the Landscape Management System (LMS) tree growth model, the effects of a VEB on the dispersion and vertical movement of the odorous chemicals can be successfully simulated, predicted and validated (Lin et al., 2011; Weber, 2012). For example, the 2D AERMOD simulations for dispersion of NH_3 demonstrate that the surface concentration maximum to the Northeast of a facility is reduced by approximately 27% when the VEB is fully developed around the facility (Figure 11–3A and B). The results from 3D AERMOD simulations show a vertical dispersion plume that is primarily centered over the facility. Before a VEB is implemented, a concentration maximum is found over the facility and vertical dispersion of the pollutant is limited (Figure 11–4**A**). When a VEB is fully established around a facility, the upper-air concentration maximum over the facility becomes more concentrated and vertical dispersion in general increases around the facility (Figure 11–4B).

In a simulated tunnel experiment, it was demonstrated that more than 40% of ammonia can be reduced in the downwind (100 m; at 1.5 m elevation from the ground, close to the nose position) by a 3 m high VEB barrier through its vertical dispersion, increased turbulence and initial mixing, and redirected vertical dispersion (Weber, 2012). Additional height did not result in improved effectiveness (Figure 11–5). The simulation study also suggested that the effectiveness of VEBs in controlling odor is much better when the VEBs are established closer to the source (Weber, 2012). Similar findings reported by a field study concluded that odor dispersion was improved when the source was located 15 m upwind from the windbreak, rather than 60 m (Lin et al., 2006).

A tree canopy sharply reduces the wind velocity and increases turbulence, and therefore reduces transport of odor constituents and particulates from the source and promotes physical dilution. Several structural factors, such as crown porosity, plant height, thickness, crown geometry and foliage distribution, influence how the VEBs vegetation interacts with flow (Baldauf, 2017; Hewitt et al., 2019). With crown porosities above 50%, a considerable reduction in the turbulence flow can occur and therefore no recirculating region forms behind the VEB (Baltaxe, 1967; Bradley and Mulhearn, 1983). Conifers have been found to offer more wind resistance and produce more odor dispersion, as compared with deciduous trees (Lin et al., 2006). A three-row configuration (first row of willow trees plus two rows of jack pine or eastern red cedar trees) or a single row of willow trees has exhibited significantly higher turbulence (mixing) intensity as compared with a single row of other hardwood deciduous trees (Sauer et al., 2008). The effectiveness of a single willow tree row in enhancing turbulence and decreasing wind speed appears to be similar to the three-tree-row (willow, cedar and pine) configuration perhaps because of its unique branch characteristics (Sauer et al., 2008). A wedge-shaped design with multiple rows of different heights tends to generate higher air-flow over the shelterbelt (Tyndall and Colletti, 2007). The wedge-shaped VEBs facing the prevailing winds often create more turbulence as well as promote better vertical dispersion (Tyndall and Colletti, 2007). A wedge- shaped configuration also helps protect shrub and tree seedlings in a three–row, shrub-evergreen-deciduous tree design during the establishment phase.

2. Species selection, crown management, and optimized configuration encourage dust and aerosol deposition.

The encouraged deposition process is one of the critical mitigation processes to reduce pollutant transport. It is well known that the deposition of gaseous pollutants and particulates, and the interception of aerosols are significantly greater with woodlands than grasslands (Beckett et al., 2001).

The effects of a species traits on PM deposition on the plant surface in woody species have been investigated by Litschke and Kuttler (2008). The filtration performance of plants is determined by the shape of the plant, crown architecture, thickness and porosity, spatial structure of branches and twigs, and the chemical and physical properties of the leaves or needles. The findings from a wind tunnel simulation have shown that selected conifers (pine and cypress) have a much higher deposition velocity of particles than deciduous trees (maple and poplar) (Beckett et al., 2001). Field studies also show that coniferous species capture more particles than species with broadleaves (Beckett et al., 1999). For particles between 5 and 10 microns, Arizona cypress, American holly, and eastern red cedar trapped significantly more per surface area compared to other plant

(A)

(B)

Fig. 11–3. Twelve-hour AERMOD model simulation showing 2D spatial dispersion of NH_3 without the presence of a VEB (A), and with a fully developed VEB (B). The 2D AERMOD simulations for dispersion of NH_3 demonstrated that the surface concentration maximum to the Northeast of the facility is reduced by approximately 27% when the VEB is fully developed around a swine CAFO facility in northeastern Missouri, U.S.

species tested (Jerez and Bray, 2016). In a study comparing the effectiveness of five trees species, Pine (*Pinus* spp.) captured significantly more particles than cypress (*Cupressus* spp.) and broad-leaved species. Among the broad-leaved species, whitebeam [*Sorbus aria* (L.) Crantz] captured the most and poplar (*Populus* spp.) the least weight of particles (Beckett et al., 1999). In a similar study, the finer more complex structure of the foliage of two conifers (P. nigra and C. leylandii) demon-

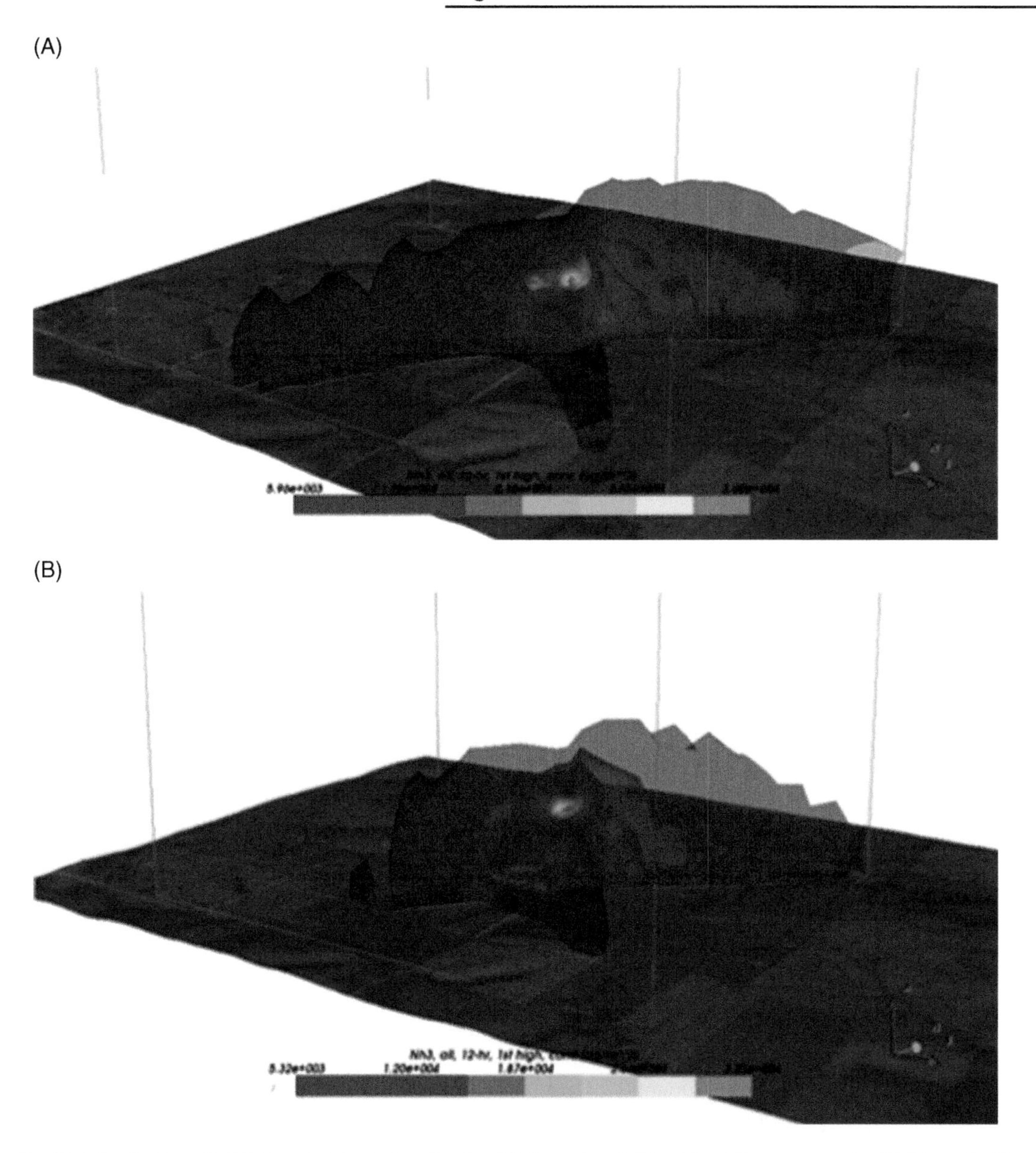

Fig. 11–4. Twelve-hour AERMOD model simulation showing 3D dispersion of NH_3 without the presence of a VEB (A), and with a fully developed VEB (B) around a swine CAFO facility in northeastern Missouri, U.S. The VEB encourages the vertical dispersion of the air pollutants.

strated much greater effectiveness at capturing pollutant particles as compared with the broad-leaved species, including maple (*Acer campestre* L.), whitebeam (*Sorbus intermedia* (Ehrh.) Pers.), and poplar (*Populus deltoides* W. Bartram ex Marshall) (Beckett et al., 2001). Furthermore, Beckett et al., found that among the broad-leaved species studied; those with rough leaf surfaces are most effective at capturing particles.

The complexity of the leaf morphology and surface chemistry often dictate the success of the biofiltration process. The presence of hairs or waxes on the leaf surface, which differs considerably between species, has a major influence on how the leaves capture particles. The waxy coated cuticle often helps plants like conifers protect their leaves by preventing water loss and makes them more durable and less palatable than deciduous species. Waxes have been found to significantly increase the PM deposition by almost two fold in *Tilia spp.* compared with *Platanus* (Dzierzanowski et al., 2011; Grote et al., 2016).

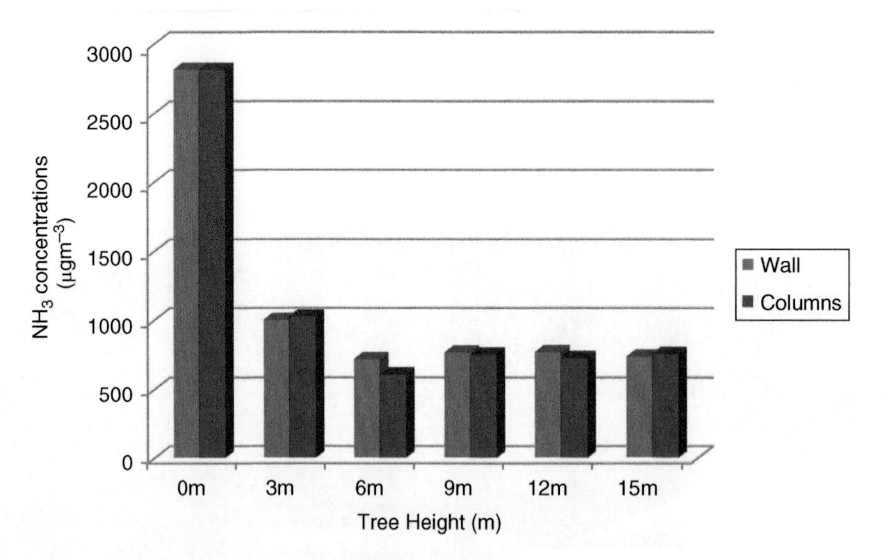

Fig. 11–5. Average NH_3 concentrations measured 100 m downwind of a CAFO facility as a function of VEB height. Concentrations shown in μgm⁻³. Blue columns: 15 m wall VEB; Red columns: 15 m column line VEB.

Another study showed a possible encapsulation of PM deposition in the wax layer during the growing season (Hofman et al., 2014). The waxy, lipophilic cuticle enhances the adsorption of hydrophobic pollutants like VOCs into plant tissues through the hydrophobic partition process (Taiz et al., 2014). The waxy, lipid-based and lipophilic polymers like cutin, the predominant biomolecules of the cuticle on the epidermis of conifer species, are excellent accumulators of lipophilic pollutants such as VOCs and hydrophobic PM, such as dusts, skin, hair, feathers, feces, and excrement particles released from CAFO facilities. In a study examining PM accumulation by 47 woody species commonly cultivated in Europe, Sæbø et al. (2012) reported a positive relationship between PM accumulation, hair density on the leaves and the quantity of leaf waxes. Contrary to previous studies, no correlations were found between PM accumulation and leaf surface roughness and leaf size of tested species (Sæbø et al., 2012).

Leaf Area Density (LAD) has been used as an important parameter for assessing the capacity of VEBs to deposit pollutants within the canopy as well as to reduce air flow turbulence (Tong et al., 2016). The LAD is determined by the ratio of the leaf surface area to the total volume occupied by the vegetative element. A study reported that increased LAD can significantly enhance the deposition process, especially for particles smaller than 50 nm, however, the effect is nonlinear (Tong et al., 2016). The LAD could be used as an excellent tool for crown management to maintain a VEB's performance.

The width of a VEB barrier is another important factor in enhancing the deposition process within the tree canopy. A design with a wide VEB barrier and a high LAD has been recommended to maximize the capacity to deposit pollutants (Tong et al., 2016). Tong et al., also suggested a design of vegetation–solid barrier combinations, that is, planting trees next to a solid barrier, like a building, could substantially increase the effect to capture PM; however, the effects of VEBs are particle size-dependent. Modeling and wind tunnel analyses using LAD to estimate the effects of porosity on pollutants captured, suggests that a minimum thickness of approximately 5 m, extending up to 10 m or more is required to effectively filter aerosols (Neft et al., 2016).

About 50% porosity of the canopy has been recommended for filtering pollutants (Baldauf, 2017). A porosity higher than 50% often allows the pollutants to easily pass through the VEB canopy. In contrast, a high density canopy will limit the deposition process within the canopy since the air movement is forced around the VEB. Thus, the vegetation porosity should be optimized in VEB systems to balance the effects of vertical dilution and deposition processes. Tyndall and Larsen (2013) concluded that VEB porosities of 35 to 50% might have the greatest impact on the degree of turbulent transfer, extent of reduced wind speed, and extent of odor plume exposure to foliar surface area (Hofer, 2009; Lin et al., 2007; Tyndall and Colletti, 2007; Tyndall and Larsen, 2013).

3. Enhanced adsorption, absorption, biodegradation and uptake of pollutants.

Leaves are known to uptake both gaseous and dissolved NH_3 (Adrizal et al., 2008; Yin et al., 1998). Studies have demonstrated a significant lowering of NH_3 concentrations with the presence of tree canopies suggesting that a portion of the atmospheric NH_3 was being absorbed and adsorbed by the plants (Adrizal et al., 2008; Patterson et al., 2008). Plant species located in front of exhaust fans have often been found to have higher growth rates compared with control plants due to higher NH_3 exposure (Patterson et al., 2006). However, plant responses to NH_3 exposure are often species-dependent. A VEB study to assess plant tolerance to poultry farm emissions found that elm, switchgrass, and giant cane show increased growth rates when they are exposed to air with higher ammonia concentrations (Belt, 2015). A greater N retention in streamco willow and lilac than in juniper was reported (Adrizal et al., 2008). In a similar study, Patterson et al. (2008) reported the potential of poplar to retain greater foliar N than spruce. However, when the concentration of NH_3 vapors exceed the tolerance threshold level for sensitive species, the excessive NH_3 can lead to immediate tissue necrosis, retarded growth, and greater frost sensitivity (Belt, 2015; Van der Eerden et al., 1998).

Plant leaves have also been shown to absorb atmospheric VOC pollutants. A study in a deciduous forest confirmed that the VOC (PAHs), including phenanthrene, anthracene, and pyrene were absorbed within and above the forest canopy during the early spring (Choi et al., 2008). While some studies have reported that urban forests near roads do not significantly reduce gaseous air VOC pollutants such as BTEX, VEBs established near CAFOs significantly reduce VOCs by capturing particles, such as dust and feather and fecal debris that carry the VOC pollutants (Parker et al., 2012). Wind tunnel VOC flux studies have also shown that the foliage of VEBs can significantly lower the levels of VOCs (78% to 98%) emitted from CAFOs by capturing the PM (Parker et al., 2012). The findings suggest that a large proportion of VOCs move with PM released by the CAFO facilities, and that the vegetation either adsorbs or absorbs the VOCs not attached to the PM. High concentrations of VOCs often found during the summer are rapidly attenuated as they travel through a VEB canopy (Figure 11–6).

Odorous pollutants are washed off VEBs by precipitation, or drop to the ground with leaf and twig fall. The permanent removal of the odorous PM, and gaseous pollutants, such as VOCs, NH_3 and H_2S can then be achieved by plant uptake and biodegradation in the rhizosphere. Several soil fungi strains have been found to be able to utilize VOCs, including benzene, toluene, and xylene, as sole sources of carbon and catalyze them in enzymatic degradation processes (Jin et al., 2006; Prenafeta-Boldú et al., 2002).

The microbes colonized on leaf surfaces and in leaves (endophytes) have also been shown to be able to biodegrade pollutants, such as VOCs and NH_3. The leaf-associated microbes (e.g., *Pseudomonas putida)* can rapidly break down several VOCs, such as xylenes, toluene, and benzene (De Kempeneer et al., 2004; Wei et al., 2017). The endophytic *Burkholderia cepacia* harbored inside leaves has also shown the capacity to degrade toluene (Barac et al., 2004). *Methylobacterium* sp., in poplar leaves, were found to be able to degrade and transform several classes of xenobiotic compounds (Van Aken et al., 2004). And, a comprehensive study reported the enhanced degradation of VOCs by the presence of phyllosphere (leaf-associated microbes) communities. This study suggested that the VOC pollutants, such as phenols can be accumulated by leaves and subsequently made available to bacteria in the phyllosphere for degradation. Community diversity of the leaf-associated microbial degraders on 10 ornamental plant species and their bioactivity toward degradation of VOCs has been characterized (Yutthammo et al., 2010). The phyllosphere harbors a wide range of bacterial degraders including Acinetobacter, Pseudomonas, Pseudoxanthomonas, and Mycobacterium. The study found that the VOC-degrading bacteria accounted for about 1 to 10% of the total heterotrophic phyllosphere populations depending on plant species. Additional desired characteristics and considerations for VEBs are identified in Table 11–3.

From the limited studies conducted, one must conclude that VEBs, when properly designed and positioned, offer great potential for improving air quality. While most research has focused on more highly adverse and toxic conditions, such as those found in the vicinity of CAFO facilities, air pollution which lowers the quality of life and can lead to health issues, is a common problem throughout North America and beyond. Agroforestry practices that advocate for "working trees", such as found in VEBs, can have a lasting effect in improving air quality and the environment in which we live.

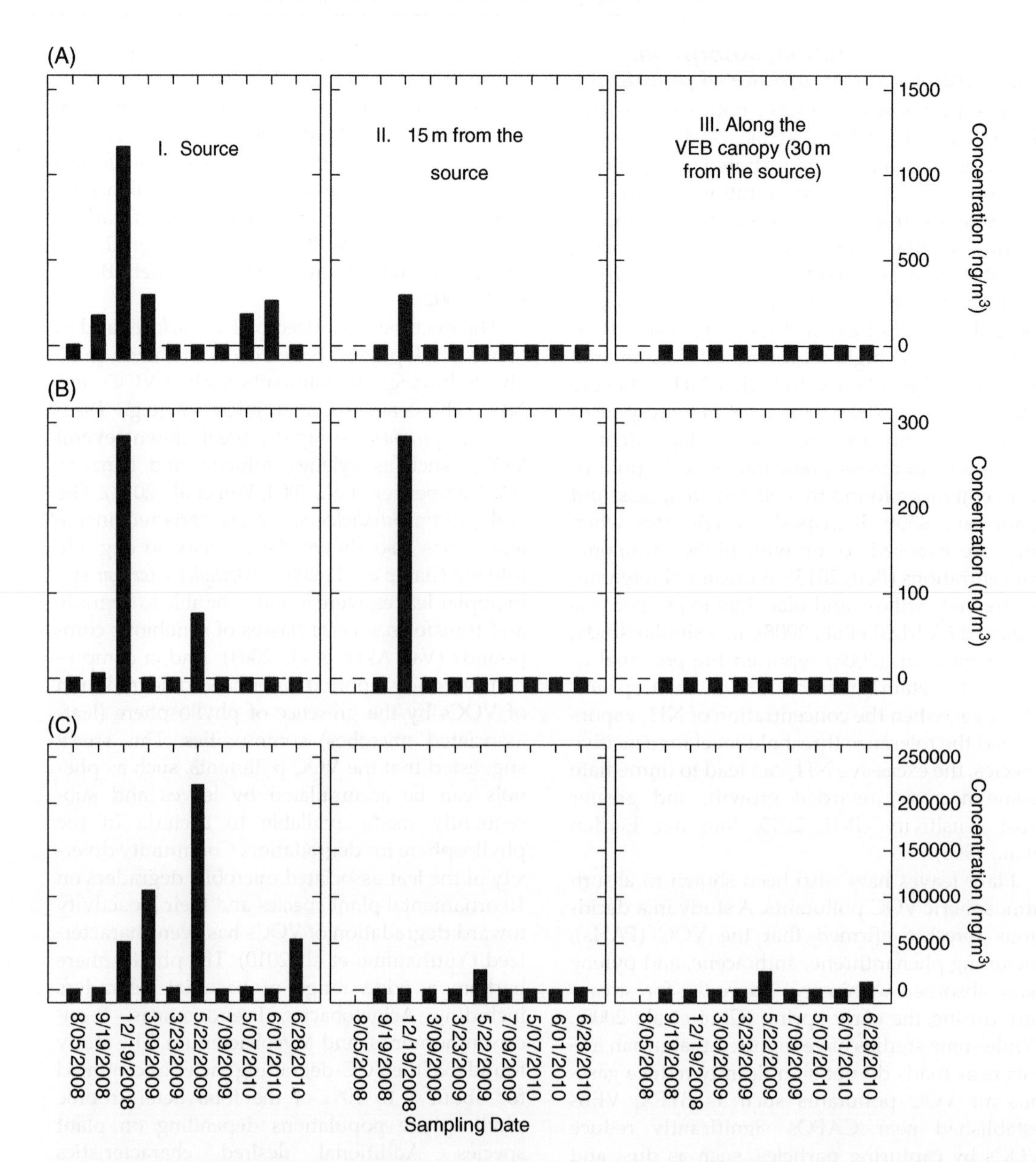

Fig. 11–6. Concentrations of (A) 3-methylindole, (B) Indole and (C) *p*-cresol near an exhaust (source), 15 m away from the source (between), and at tree line (tree), approximately 30 m away along the VEBs (between 6 Aug. 2008 and 28 June 2010).

Table 11–3. Additional desired characteristics and considerations for VEBs.

I. Desired Characteristics for Species Selection	References
Resistant to high concentration of dusts	(USDA NRCS, 2007)
Resistant to high concentration of ammonia	(USDA NRCS, 2007)
Resistant to high concentration of VOCs	(USDA NRCS, 2007)
Resistant to deer browsing	(USDA NRCS, 2007)
Medium to fast growing rates	(USDA NRCS, 2007)
Native (USDA hardiness range)	(USDA NRCS, 2007)
Plants with smaller leaves trap more dust and odors than plants with larger leaves	(Belt, 2015)
Waxy leaves	(Belt, 2015)

Table continued.

I. Desired Characteristics for Species Selection	References
Rough leaf surfaces (leaf hairs, veins)	(Belt, 2015)
Complex waxy and/or hairy surfaces with high surface area	(Belt, 2015)
Evergreens is preferred for year-round protection	(Belt, 2015)
Low branching (shrubs and trees) or leaves extend to the ground	(Belt, 2015)
II. Desired Configuration and Design	
Height maintained with minimum 3 to 5 m (< 3 m from the source) and 6 to 10 m (3 to 6 m from the source)	(Baldauf, 2017)
Thickness maintained with 10 m or more if it is possible	(Baldauf, 2017)
Canopy with minimum 35-50% porosity	(Tyndall and Larsen, 2013)
Wedge shaped shelterbelts create more turbulence as well as push air streams higher into the lower atmosphere	(Tyndall and Colletti, 2007)
Spacing within the row	
Small Shrubs (1.2 to 3.6 m tall) – 1.2 to 1.8 m apart	(USDA NRCS, 2007)
Large Shrubs and Small Trees (3.6 to 9 m tall) – 1.8 to 3 m apart	(USDA NRCS, 2007)
Large Trees (over 9 m tall) – 3 to 4.2 m apart	(USDA NRCS, 2007)
Evergreen Trees – 2.4 to 4.2 m apart depending on the species	(USDA NRCS, 2007)
Spacing between rows	
5 to 7 m	(USDA NRCS, 2007)
Time for planting	
Evergreen-early spring	
Deciduous-either in the spring or autumn	(USDA NRCS, 2007)
Weed control	
Plastic weed barrier with organic mulch is preferred (also help retain moisture and promote plant growth)	(USDA NRCS, 2007)

References

Adrizal, P., Patterson, H., Hulet, R.M., Bates, R.M., Despot, D.A., Wheeler, E.F., Topper, P.A., Anderson, D.A., and Thompson, J.R. (2008). The potential for plants to trap emissions from farms with laying hens: 2. ammonia and dust. *J. Appl. Poult. Res.* 17, 398–411. doi:10.3382/japr.2007-00104

Asmus, C.L., and Bell, P.A. (1999). Effects of environmental odor and coping style on negative affect, anger, arousal, and escape. *J. Appl. Soc. Psychol.* 29, 245–260. doi:10.1111/j.1559-1816.1999.tb01384.x

Avery, R. (2003). Health effects associated with exposure to airborne emissions from industrial hog operations in Eastern North Carolina. MS Thesis. Chapel Hill, NC: University of North Carolina at Chapel Hill.

Baldauf, R. (2017). Roadside vegetation design characteristics that can improve local, near-road air quality. *Transp. Res. Part D Transp. Environ.* 52, 354–361. doi:10.1016/j.trd.2017.03.013

Baltaxe, R. (1967). Air flow patterns in the lee of model windbreaks. *Archiv für Meteorologie, Geophysik und Bioklimatologie, Serie B* 15, 287. doi:10.1007/BF02243857

Barac, T., Taghavi, S., Borremans, B., Provoost,A., Oeyen, L. Colpaert, J.V., Vangronsveld, J., van der Lelie, D. (2004). Engineered endophytic bacteria improve phytoremediation of water-soluble, volatile, organic pollutants. *Nature Biotechnology* 22, 583–588.

Beckett, K.P., Freer-Smith, P., and Taylor, G. (1999). Effective tree species for local air quality management. *Tree Species and Air Quality* 26, 12–19.

Beckett, K.P., Freer-Smith, P.H., and Taylor, G. (2001). Particulate pollution capture by urban trees: Effect of species and windspeed. *Glob. Change Biol.* 6, 995–1003. doi:10.1046/j.1365-2486.2000.00376.x

Belt, S.V. (2015). Plants tolerant of poultry farm emissions in the Chesapeake Bay watershed. *Study Report-Norman A.* Beltsville, MD: *Berg National Plant Materials Center*. p. 1–25.

Bradley, E.F., and Mulhearn, P.J. (1983). Development of velocity and shear stress distribution in the wake of a porous shelter fence. *J. Wind Eng. Ind. Aerodyn.* 151, 45–156.

Bremberg, B. (1994). Livestock and the environment: Law and policy aspects of odor. Texas Institute for Applied Environmental Research. Stephensville, TX: Tarleton State University.

Cai, L., Koziel, J.A., Lo, Y.-C., and Hoff, S.J. (2006). Characterization of volatile organic compounds and odorants associated with swine barn particulate matter using solid-phase microextraction and gas chromatography–mass spectrometry–olfactometry. *J. Chromatogr. A* 1102, 60–72. doi:10.1016/j.chroma.2005.10.040

Chapin, A., Boulind, C., and Moore, A. (1998). Controlling odor and gaseous emission problems from industrial swine facilities. New Haven, CT: Yale Environmental Protection Clinic.

Choi, S.-D., Staebler, R.M., Li, H., Su, Y., Gevao, B., Harner, T., and Wania, F. (2008). Depletion of gaseous polycyclic aromatic hydrocarbons by a forest canopy. *Atmos. Chem. Phys.* 8, 4105–4113. doi:10.5194/acp-8-4105-2008

De Kempeneer, L., Sercu, B., Vanbrabant, W., Van Langenhove, H., and Verstraete, W. (2004). Bioaugmentation of the phyllosphere for the removal of toluene from indoor air. *Appl. Microbiol. Biotechnol.* 64, 284–288. doi:10.1007/s00253-003-1415-3

Donham, K.J. (1998). Occupational health risks for swine producers: Inferences for public health risks for people living in the vicinity of swine production units. Pollution Prevention Info House. Des Moines: Iowa Department of Natural Resources.

Donham, K.J., and Gustafson, K.E. (1982). Human occupational hazards from swine confinements. *In* W. D. Kelley, (ed.), Agricultural respiratory hazards. Cincinnati, OH: American Conference of Governmental Industrial Hygienists.

Dzierżanowski, K., Popek, R., Gawrońska, H., Saebø, A., and Gawroński, S.W. (2011). Deposition of particulate matter of different size fractions on leaf surfaces and in waxes of urban forest species. *Int. J. Phytoremediation* 13, 1037–1046. doi:10.1080/15226514.2011.552929

Grote, R., Samson, R., Alonso, R., Amorim, J.H., Cariñanos, P., Churkina, G., Fares, S., Thiec, D.L., Niinemets, Ü., and Mikkelsen, T.N. (2016). Functional traits of urban trees: Air pollution mitigation potential. *Front. Ecol. Environ.* doi:10.1002/fee.1426

Hamscher, G., Pawelzick, H.T., Sczesny, S., Nau, H., and Hartung, J. (2003). Antibiotics in dust originating from a pig-fattening farm: A new source of health hazard for farmers. *Environ. Health Perspect.* 111, 1590–1594. doi:10.1289/ehp.6288

Heber, A., Stroik, J.M., Faubion, J.M., and Willard, L.H. (1988). Size distribution and identification of aerial dust particles in swine finishing buildings. *Trans. ASAE* 31, 882–887. doi:10.13031/2013.30794

Herriges, J. A., Secchi, S., and Babcock, B. A. (2005). Living with Hogs in Iowa: The Impact of Livestock Facilities on Rural Residential Property Values. *Land Economics* **81**, 530–545.

Herz, R. S., Schankler, C., and Beland, S. (2004). Olfaction, emotion, and associative learning: Effects on motivated behavior. *Motivation and Emotion* **28**, 363–383.

Horton, R. A. (2008). "Malodor from industrial hog operations, stress, negative mood, and secretory immune function in nearby residents.Dissertation Abstracts International: Section B: The Sciences and Engineering."

Hewitt, C.N., Ashworth, K., and MacKenzie, A.R. (2019). Using green infrastructure to improve urban air quality (GI4AQ). *Ambio* 49, 62–73. doi:10.1007/s13280-019-01164-3

Hofer, B. (2009). Effect of a shelterbelt on H2S concentrations from swine barns. M.S. Thesis. Brookings, SD: South Dakota State University.

Hofman, J., Wuyts, K., Wittenberghe, S.V., and Samson, R. (2014). On the temporal variation of leaf magnetic parameters: Seasonal accumulation of leaf-deposited and leaf-encapsulated particles of a roadside tree crown. *Sci. Total Environ.* 493, 766–772. doi:10.1016/j.scitotenv.2014.06.074

Honey, L.F., and McQuitty, J.B. (1979). Some physical factors affecting dust concentrations in a pig facility. *Canadian Agricultural Engineering* 21, 9–14.

Isakson, H., and Ecker, M. (2008). An analysis of the impact of swine CAFOs on the value of nearby houses. *Agricultural Economics* **39**, 365–372.

Jerez, S.B., and J. Bray. (2016). Demonstrating the effect of trees for controlling particulate matter, ammonia, and odor from poultry buildings-Final report. Washington, D.C.: United States Department of Agriculture- Natural Resources Conservation Service. p. 1–50.

Jin, Y., Veiga, M., and Kennes, C. (2006). Performance optimization of the fungal biodegradation of a-pinene in gas-phase biofilter. *Process Biochem.* 41, 1722–1728. doi:10.1016/j.procbio.2006.03.020

Knasko, S.C., Gilbert, A.N., and Sabini, J. (1990). Emotional state, physical well-being, and performance in the presence of feigned ambient odor. *Journal of Applied Social Psychology* **20**, 1345–1357.

Kilburn, K. (1997). Civil Action File No.: 970-CV-238. Los Angeles, CA: University of Southern California School of Medicine Environmental Sciences Laboratory.

Kilburn, K.H. (1999). Evaluating health effects from exposure to hydrogen sulfide: Central nervous system dysfunction. *Environ Epidemiol Toxico* 1, 207–217.

Kilburn, K.H., and Warshaw, R.H. (1995). Hydrogen sulfide and reduced-sulfur gases adversely affect neurophysiological functions. *Toxicol. Ind. Health* 11, 185–197. doi:10.1177/074823379501100206

Liebers, V., Brüning, T., and Raulf-Heimsoth, M. (2006). Occupational endotoxin-exposure and possible health effects on humans. *Am. J. Industr. Med.* 49, 474–491. doi:10.1002/ajim.20310

Lin, C.-H., Garrett, H.E., and Walter, D.D. (2011). Windbreak odor abatement-The small business development authority. Jefferson City, MO: Missouri Department of Agriculture.

Lin, X.-J., Barrington, S., Nicell, J., Choinie're, D., and Ve'zina, A. 2006. Influence of windbreaks on livestock odour dispersion plume in the field. *Agric. Ecosyst. Environ.* 116, 263–272. doi:10.1016/j.agee.2006.02.014

Lin, X.J., Barrington, S., Nicell, J., and Choiniere, D. (2007). Effect of natural windbreaks on maximum odour dispersion distance (MODD). *Canadian Biosystems Engineering* 49, 21–32.

Litschke, T., and Kuttler, W. (2008). On the reduction of urban particle concentration by vegetation- A review *Meteorologische Zeitschrift* 17, 229–240.

Mackie, R.I., Stroot, P.G., and Varel, V.H. (1998). Biochemical identification and biological origin of key odor components in livestock waste. *J. Anim. Sci.* 76, 1331–1342. doi:10.2527/1998.7651331x

Malone, G., VanWicklen, G., and Collier, S. (2006a). Efficacy of vegetative environmental buffers to mitigate emissions from tunnel-ventilated poultry houses. Newark, DE: University of Delaware.

Malone, G.W., VanWicklen, G., Collier, S., and Hansen, D. (2006b). Efficacy of vegetative environmental buffers to capture emissions from tunnel ventilated poultry houses. In Aneja, V.P., Schlesinger, W.H., Knigton, R., Jennings, G., Niyogi, D., Gilliam, W., and Duke, C.S., (eds.), Workshop on agricultural air quality: State of the Science, Potomac, MD. 5–8 June 2006.

McDonald, R., Kroeger, T., Boucher, T., Longzhu, W., and Salem, R. (2016). Planting healthy air-A global analysis of the role of urban trees in addressing particulate matter pollution and extreme heat. Arlington, VA: The Nature Conservancy.

Milla, K., Thomas, M.H., and Ansine, W. (2005). Evaluating the effect of proximity to hog farms on residential property values: A GIS-based hedonic price model approach. *URISA Journal* 17, 27–32.

Mutel, C.F., Donham, K.J., Merchant, J.A., Redshaw, C.P., and Starr, S.D. (1986). Unit 4: Livestock confinement dusts and gases In Kay, B.J., (ed.), Agricultural respiratory hazards education series. Des Moines, IA: American Lung Association of Iowa.

National Research Council. (2002). The scientific basis for estimating emissions from animal feeding operations interim report. Washington, D.C.: National Academy Press.

Neft, I., M. Scungio, Culver, N., and Singh, S. (2016). Simulations of aerosol filtration by vegetation: Validation of existing models with available lab data and application to near-roadway scenario. *Aerosol Sci. Technol.* 50, 937–946. doi:10.1080/02786826.2016.1206653

O'Neill, D., and Phillips, V. (1992). A review of the control of odour nuisance from livestock buildings. 3. Properties of the odorous substances which have identified in livestock wastes or in the air around them. *J. Agric. Eng. Res.* 53, 23–50. doi:10.1016/0021-8634(92)80072-Z

Park, D., Seidl, A.F., and Davies, S.P. (2004). The effect of livestock industry location on rural residential property values. Colorado State University Extension EDR 04-12. Fort Collins, CO: Colorado State University.

Parker, D.B., Malone, G.W., and Walter, W.D. (2012). Vegetative environmental buffers and exhaust fan deflectors for reducing downwind odor and VOCs from tunnel-ventilated swine barns. *Trans. ASABE* 55, 227–240. doi:10.13031/2013.41250

Patterson, P., Adrizal, R., Bates, C., Myers, G., Martin, R., Shockey, R., and Grinten, M.V.D. (2006). Plant foliar nitrogen and temperature on commercial poultry farms in

Pennsylvania. *Proceedings of the Workshop on Agricultural Air Quality*, p. 453-457.

Patterson, P.H., Adrizal, A., Hulet, R.M., Bates, R.M., Myers, C.A., Martin, G.P., Shockey, R.L., and van der Grinten, M. (2008). Vegetative buffers for fan emissions from poultry farms: 1. temperature and foliar nitrogen. *J. Environ. Sci. Health Part B* 43, 199–204. doi:10.1080/03601230801890179

Prenafeta-Boldú, F.X., Vervoort, J., Grotenhuis, J.T.C., and van-Groenestijn, J.W. (2002). Substrate interactions during the biodegradation of benzene, toluene, ethylbenzene, and xylene (BTEX) hydrocarbons by the fungus Cladophialophora sp. strain T1. *Appl. Environ. Microbiol.* 68, 2660–2665.

Ready, R., and Abdalla, C. (2005). The amenity and disamenity impacts of agriculture: Estimates from a hedonic pricing model. *Am. J. Agric. Econ.* 87, 314–326. doi:10.1111/j.1467-8276.2005.00724.x

Rosentrater, K. A. (2003). Performance of an electrostatic dust collection system in swine facilities. *Agricultural Engineering International: The CIGR EJournal* V.

Sæbø, A., Popek, R., Nawrot, B., Hanslin, H.M., Gawronska, H., and Gawronski, S.W. (2012). Plant species differences in particulate matter accumulation on leaf surfaces. *Sci. Total Environ.* 427-428, 347–354. doi:10.1016/j.scitotenv.2012.0,.084

Saphores, J., and Aguilar-Benitez, I. (2005). Smelly local polluters and residential property values: A hedonic analysis of four orange county cities. *Estud. Econ.* 20, 197–218.

Sauer, T., Haan, J.F., Tyndall, J., Hernandez-Ramirez, G., Trabue, S., Pfeiffer, R., and Singer, J. (2008). Vegetative buffers for swine odor mitigation- Wind tunnel evaluation of air flow dynamics. In National Conference on mitigating air emissions from animal feeding operations: Exploring the advantages, limitations, and economics of mitigation technologies, Des Moines, IA, 19-21 May 2008. Ames, IA: Iowa State University. p. 30–34.

Spoelstra, S.F. (1977). Simple phenols and indoles in anaerobically stored piggery wastes. *Journal of Science and Agriculture* 28, 415–423.

Steinheider, B., Both, R., and Winneke, G. (1998). Field studies on environmental odors inducing annoyance as well as gastric and general health-related symptoms. *J. Psychophysiol.* 12, 64–79.

Taiz, L., Zeiger, E., Møller, I.M., and Murphy, A. (2014). *Plant physiology and development*. 6th edition, Sinauer: Sunderland, MA.

Tong, Z., Baldauf, R., Isakov, V., Deshmukhd, P., and Zhang, K.M. (2016). Roadside vegetation barrier designs to mitigate near-road air pollution impacts. *Sci. Total Environ.* 541, 920–927. doi:10.1016/j.scitotenv.2015.09.067

Tyndall, J., and Colletti, J. (2007). Mitigating swine odor with strategically designed shelterbelt systems: A review. *Agrofor. Syst.* 69, 45–65. doi:10.1007/s10457-006-9017-6

Tyndall, J.C., and Larsen, G.L.D. (2013). Vegetative environmental buffers for odor mitigation. Cliva, IA: Pork Information Gateway. p. 1–7.

USEPA. 2019. Air quality dispersion modeling. Washington, D.C.: U.S. Environmental Protection Agency Support Center for Regulatory Atmospheric Modeling (SCRAM). http://www.epa.gov/scram001/dispersionindex.htm. *[2019 is year accessed]*.

US Environmental Protection Agency. (2004). User's Guide for the AERMOD Meteorological Preprocessor (AERMET). EPA-454/B-03-002. Research Triangle Park, NC: U.S. Environmental Protection Agency.

Van Aken, B., Yoon, J.M., and Schnoor, J.L. (2004). Biodegradation of nitro-substituted explosives 2,4,6-trinitrotoluene, hexahydro-1,3,5-trinitro-1,3,5-triazine, and octahydro-1,3,5,7-tetranitro-1,3,5-tetrazocine by a phytosymbiotic Methylobacterium sp. associated with poplar tissues (Populus deltoides × nigra N34). *Appl. Environ. Microbiol.* 70, 508–517. doi:10.1128/AEM.70.1.508-517.2004

Vander, W. G. (2001). 44 states regulate odors on hog farms. *National Hog Farmer, 15 March.*

Van der Eerden, L.J.M., de Visser H.B., van Dijk C.J. (1998). Risk of damage to crops in the direct neighbourhood of ammonia sources. *Environmental Pollution* **102**, 49–53.

Wang, X., Stroot, P.G., Zhang, Y., and Riskowski, G.L. (1998). Odor carrying characteristics of dust from swine facilities. *Proceeding for 1998 ASAE Annual International Meeting*, Orlando, FL, 12-16 July, 1998. (1998), 1–11.

Weber, E.E. (2012). The use of dispersion modeling to determine the feasibility of vegetative environmental buffers (VEBs) at controlling odor dispersion. PhD dissertation, Columbia, MO: University of Missouri.

Wei, X., Lyu, S.,Yu, Y., Wang, Z., Liu, H., Pan, D., and Chen, J. (2017). Phylloremediation of air pollutants: Exploiting the potential of plant leaves and leaf-associated microbes. *Front. Plant Sci.* 8. doi:10.3389/fpls.2017.01318

Yasuhara, A., Fuwa, K. and Jimbu, M. (1984). Identification of odorous compounds in fresh and rotten swine manure. *Agric. Biol. Chem.* 48, 3001–3010.

Yin, Z.-H., Kaiser, W., Hebera, U., and Raven, J.A. (1998). Effects of gaseous ammonia on intracellular pH values in leaves of C3- and C4-plants. *Atmos. Environ.* 32, 539–544. doi:10.1016/S1352-2310(97)00165-9

Yokoyama, M.T., and Carlson, J.R. (1981). Production of skatole and para-cresol by rumen:*Lactobacillus* sp. *Appl. Environ. Microbiol.* 41, 71–76. doi:10.1128/AEM.41.1.71-76.1981

Yutthammo, C., Thongthammachat, N., Pinphanichakarn, P., and Luepromchai, E. (2010). Diversity and activity of PAH-degrading bacteria in the phyllosphere of ornamental plants. *Microb. Ecol.* 59, 357–368. doi:10.1007/s00248-009-9631-8

Zahn, J.A., Hatfield, J.L., Do, Y.S., DiSpirito, A.A., Laird, D.A., and Pfeiffer, R.L. (1997). Characterization of volatile organic emissions and wastes from a swine production facility. *J. Environ. Qual.* 26, 1687–1696. doi:10.2134/jeq1997.00472425002600060032x

Zahn, J.A., Hatfield, J.L., Laird, D.A., Hart, T.T., Do, Y.S., and DiSpirito, A.A. (2001). Functional classification of swine manure management systems based on effluent and gas emission characteristics. *J. Environ. Qual.* 30, 635–647. doi:10.2134/jeq2001.302635x

Study Questions

1. While air quality is a major problem in many areas worldwide, it is especially serious in close location to Confined Animal Feeding Operations (CAFOs). What is the major cause (source) of most odorous compounds in the vicinity of swine CAFOs?

2. Identify 5 factors that are known to play significant roles in the dispersion of airborne emissions from CAFOs.

3. What is the relationship between CAFOs and real estate values? Is it as simple as the distance between the CAFO facility and the real estate in question? Explain.

4. The literature suggests that while VEBs can remove significant quantities of compounds known to be carriers of odors, the overall effect on reducing odor is variable and less than would be expected. Explain how this is possible.

5. How does one determine Leaf Area Density (LAD) and why is it considered an important parameter for measuring the effectiveness of a VEB in reducing air flow, pollutant removal, etc.?

6. With the knowledge currently available relating to the characteristics for an effective VEB (i.e., evergreen vs. deciduous species, porosity, height, depth, width, location, etc.), design, what might be considered as an effective VEB for a CAFO facility?

12

Ranjith P. Udawatta, Stephen H. Anderson,
Robert J. Kremer, and
Harold E. "Gene" Garrett*

Agroforestry for Soil Health

Abbreviations

AF	agroforestry
CS	carbon sequestration
CT	computed tomography
ES	ecosystem services
PLFA	phospholipid fatty acids
SBD	soil biodiversity
SC	soil carbon
SH	soil health
SOM	soil organic matter

In the United States, corn (*Zea mays* L.) yields have increased by 0.12 Mg ha^{-1} annually since 1955 due to advances in breeding, management, technology, and input intensification (Nielson, 2012). However, it has come at an expense of ecosystem services (ES) including degradation of environmental benefits, soil health (SH), and long-term agricultural sustainability (Montgomery, 2007). Soil degradation destroys soil properties and functions thus reducing SH and land productivity (Al-Kaisi, 2008; Montgomery, 2007). Soil health, according to the Natural Resources Conservation Service, is "the continued capacity of a soil to function as a vital living ecosystem that sustains plants, animals, and humans." The term implies enhanced land productivity and environmental benefits from maintenance and improvements in soil properties while sustaining plants, animals, and humans.

At the 21st World Congress of Soil Science in Brazil, the FAO Director stated, "Soil degradation affects food production, causing hunger and malnutrition, amplifying food price volatility, forcing land abandonment and involuntary migration, leading millions into poverty." And according to Dent (2007), "Land degradation is a long-term loss of ecosystem functions and services, caused by disturbances from which the system cannot recover unaided." According to Montgomery (2007), there is three times more land degradation due to conventional agriculture than due to conservation agriculture and >75 times greater than that due to native vegetation. As degraded land is abandoned, land degradation causes a shrinkage in the available agricultural land (Bakker et al., 2005; Boardman, 2006; Lal, 1996). The issue is further exasperated due to the expansion of urban, industrial, and commercial areas. In the United States, cropland acreage decreased by 22% between 1982 and 2007 (USDA Economic Research Service, 2011).

*Corresponding author (UdawattaR@missouri.edu)

North American Agroforestry, Third Edition. Edited by Harold E. "Gene" Garrett, Shibu Jose, and Michael A. Gold.

In recent years, studies have shown relationships between SH and human health (Antunes et al., 2013; Brevik & Sauer, 2015; Oliver & Gregory, 2015). Human health depends on nutritious food and clean air and water, and therefore long-term sustainable soil use is vital for human health (Wall & Six, 2015). Others have found relationships between little green space and increasing chronic inflammatory disorders (Hanski et al., 2012). Numerous soil and human health studies have emphasized the importance of SH for human health, allergies, toxicities, and others, although science is just beginning to link SH to human health (Antunes et al., 2013).

Soil erosion may account for a 10% increase in the total U.S. agricultural energy use to offset losses of nutrients, water, and soil productivity (Pimentel et al., 1995). In many regions, soil loss exceeds soil formation (FAO, 2015). Projected increases in rainfall intensity of 51% during the next century could further destroy soil resources, water, and SH. However, conservation practices could help reduce the damage and conserve soil resources. For example, in the United States, sheet and rill soil erosion on cropland estimated with the Revised Universal Soil Loss Equation (RUSLE) decreased from 1.52×10^9 Mg yr^{-1} in 1982 to 0.87×10^9 Mg yr^{-1} in 2007, a 43% reduction due to conservation practices (USDA–NRCS, 2010).

Increasing demand for food to meet the demands of the projected 9 billion people, decreasing land resources, and potential consequences of climate change have emphasized the importance of soil conservation and SH (Howden et al., 2007; Lal, 2010; Tilman, Cassman, Matson, Naylor, & Polasky, 2002). Recent concerns about soil, water, and environmental degradation demand sustainable soil management practices that are less dependent on external inputs but with efficient nutrient cycles, increased productivity, and reduced degradation of natural resources (Swift, Izac, & van Noordwijk, 2004).

Conventional agricultural practices impose unsustainable stresses on soil, water, and biodiversity, and the world needs highly productive farming systems that also have a lower environmental footprint (Ponisio et al., 2015). The role of agroforestry (AF) in improving soil properties and ES has been reviewed by Buresh and Tian (1998), Udawatta, Gantzer, and Jose (2017), and Young (1997) among others.

Agroforestry is an intensive land management practice where trees and shrubs are intentionally integrated into crop and livestock management practices to optimize numerous benefits arising from biophysical interactions among the components (Gold & Garrett, 2009). There are six main AF practices: riparian buffers, alley cropping, windbreaks, silvopasture, forest farming, and urban food forests. Riparian buffers exist around water bodies, while upland buffers are mostly located on contours to create alley cropping. Windbreaks protect crops, livestock, and farm structures from wind and snow. Silvopasture is the intentional integration of trees, forage, and livestock, which are managed intensively. In forest farming, economically valuable crops like ginseng (*Panax* spp.) are grown under forest canopies. Urban food forests, as the name implies, are forested urban areas managed for food production (see Chapter 10). Traditional AF of the tropics closely resembles natural rainforests and therefore has been suggested for promoting wildlife and conserving biodiversity while providing other ES. A recent study by Zomer et al. (2016) has shown that nearly half of the global agricultural land, about 1 billion ha, has >10% tree cover.

Agroforestry can contribute to sustainable production and ES through improvements in SH (Gold & Hanover, 1987). Integration of AF practices into agroecosystems enhances SH by its influence on C sequestration (CS; Nair, Kumar, & Nair, 2009; Nair, Nair, et al., 2009; Stefano & Jacobson, 2018; Udawatta & Jose, 2012), soil properties (Adhikari, Udawatta, Anderson, & Gantzer, 2014; Akdemir, Anderson, & Udawatta, 2016; Chu, Goyne, Anderson, Lin, & Udawatta, 2010; Kumar, Anderson, Bricknell, Udawatta, & Gantzer, 2008; Kumar, Anderson, Udawatta, & Kallenbach, 2012; Seobi, Anderson, Udawatta, & Gantzer, 2005; Tufekcioglu, Raich, Isenhart, & Schultz, 2003; Udawatta, Anderson, Gantzer, & Garrett, 2008; Udawatta, Gantzer, Anderson, & Garrett, 2008; Udawatta, Kremer, Garrett, & Anderson, 2009), removing contaminants (Kaur, Singh, Kaur, & Singh, 2018; Lin, Goyne, Kremer, Lerch, & Garrett, 2010; Lin, Lerch, Goyne, & Garrett, 2011), and increasing biodiversity (Bhagwat, Willis, Birks, & Whittaker, 2008; Jose, 2012; Polglase et al., 2008; Sistla et al., 2016; Udawatta, Rankoth, & Jose, 2019). Recent research has shown improved soil water dynamics, soil hydraulics, CS, and soil biodiversity (SBD) when riparian and upland buffers and windbreaks are integrated within row crop agriculture (Schultz et al., 2009; Udawatta et al., 2017). Silvopasture has been shown to improve SH and animal welfare as well as to extend the grazing period (Kallenbach, 2009; Kumar et al., 2008; Kumar, Anderson, & Udawatta, 2010; Kumar, Udawatta, & Anderson, 2010; Paudel, Udawatta, Kremer, & Anderson, 2011, 2012). Multispecies AF practices promote greater SBD and thereby influence nutrient cycling, decontamination, and soil reactions compared with less diverse management practices (Bhagwat et al., 2008; Jose, 2012; Polglase et al., 2008; Sistla et al., 2016; Udawatta et al., 2019). A meta-analysis in Europe has shown positive effects of AF on SH, SBD, nutrient supply, erosion control, and many other ES (Torralba, Fagerholm, Burgess, Moreno, & Plieninger, 2016). Although AF's potential for SH benefits has been widely recognized, the current literature is limited on information for temperate regions and does not have a systematic evaluation of all the SH benefits of AF, thus restricting AF integration for the development of sustainable management plans. This chapter highlights the benefits of AF practices on SH parameters including soil C (SC), physical properties, biological properties, chemical properties, degradation of harmful chemicals, SBD, and ES.

Agroforestry and Soil Carbon

Declining soil C and biodiversity leads to soil degradation, and this can lead to food insecurity and declining ecosystem sustainability (Godfray et al., 2010; Montgomery 2010). Poor soil management and land-use practices have resulted in 42×10^9 to 78×10^9 Mg of C lost during the past century (Lal, 2004). Agroforestry could be a partial solution to increase SC and was approved by both the afforestation and reforestation programs under the Clean Development Mechanisms of the Kyoto Protocol for enhanced CS (IPCC, 2007; Smith et al., 2007; Watson et al., 2000). According to Sanchez (2000), 35% of the forest C that was lost due to slash and burn can be recovered by AF. Agroforestry practices store C in trees for extended periods and potentially slow deforestation (Schroeder, 1993). In a meta-analysis, conversion of agriculture, pasture, grazing, and uncultivated lands to AF showed a significant increase in SC (Stefano & Jacobson, 2018). These increases ranged from 9 to 34% within the surface 1-m soil depth. Nair, Kumar, and Nair (2009) and Nair, Nair, et al. (2009) ranked SC stocks from the greatest to the smallest in terms of plant associations: forests > AF > tree plantations > arable crops.

On a global scale, 43% of the 22.2 million km^2 of cropland has >10% tree cover, and the tree cover has increased by 3.7% during the last 10 yr (Zomer et al., 2016). Using 1-km resolution geospatial remote sensing data, this ICRAF-sponsored project estimated 1 billion ha of AF-type agricultural lands worldwide. Global croplands sequester 54.3 Pg of C annually, and 75% of this C is contributed by the trees. In the United States, 10% of the cropland or 18 million ha under alley cropping and 78 million ha under silvopasture have the potential to sequester 535 Tg C yr^{-1} (Udawatta & Jose, 2012). The potential to sequester C by windbreaks is about 8.8 Tg C yr^{-1}. Riparian buffers, alley cropping, and silvopasture could sequester a significant amount of C. Udawatta and Jose (2012) have estimated a 642-Tg C sequestration potential by improved silvopasture, alley cropping, windbreaks, and riparian buffers in the United States (Figure 12–1).

Carbon sequestration varies with tree species, density, maintenance, soil, site, age, and climatic factors. In general, AF reduces tillage and soil disturbance and thereby further enhances maintenance and increases C pools. Many studies have reported increased CS with increasing tree density. These improvements have been

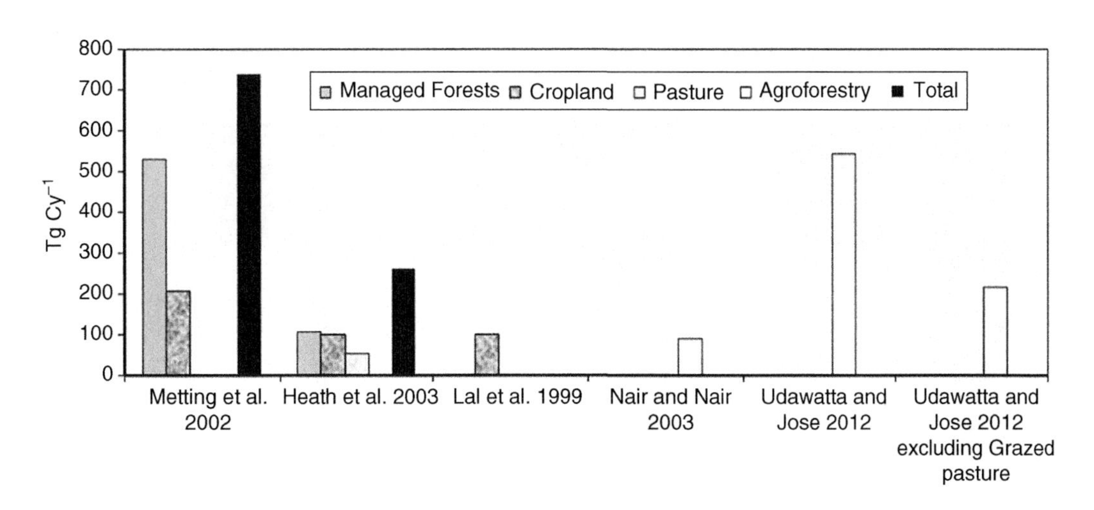

Fig. 12–1. Carbon sequestration potential for various management systems in the United States (source: Udawatta & Jose, 2012).

attributed to enhanced root development, deep roots, perennial vegetation, and reduced disturbance (Kumar, Udawatta, & Anderson, 2010; Udawatta & Jose, 2012). In Iowa, poplar (*Populus* spp.) trees have sequestered 5, 8, and 2.5 times more below- and aboveground C than corn, soybean [*Glycine max* (L.) Merr.], and pasture management practices (Tufekcioglu et al., 2003; Figure 12–2). Mature (60 yr) riparian buffers sequestered >100 Mg C ha^{-1} in South Carolina (Giese, Aust, Kolka, & Trettin, 2003). Studying soil C in a grazing system, Kumar, Udawatta, and Anderson (2010) reported 4.5 times greater root length, 3% greater root C, and 115% greater soil C in a 1-m soil depth of AF than grazing areas. In Alberta, Canada, Baah-Acheamfour, Carlyle, Bork, and Chang (2014) and Banerjee et al. (2016) observed that AF plots with trees had two to three times more soil organic and dissolved C than adjacent agricultural plots. The amount of C in the soil changes with the distance from the tree base and with soil depth (Figure 12–3; Sauer, Cambardella, & Brandle, 2007; Seiter, Ingham, William, & Hibbs, 1996; Thevathasan & Gordon, 1997).

In recent years, many plot, field, and watershed scale studies and reviews have shown improvements in soil C due to the integration of AF (Stefano & Jacobson, 2018; Udawatta & Jose, 2012). However, proper maintenance is required for enhanced CS. As trees and their root systems mature, growth rates decline and the accumulation of above- and belowground biomass plateaus (Naiman, Décamps, McClain, &

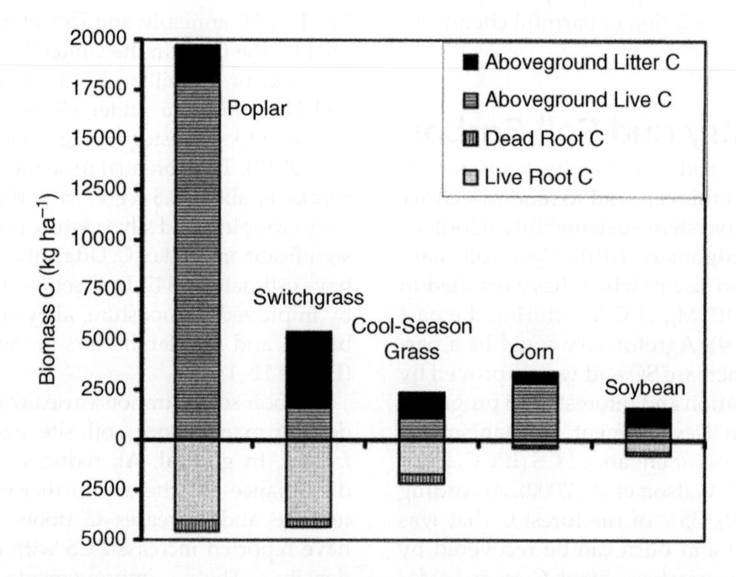

Fig. 12–2. Above- and belowground biomass C in a riparian zone with trees, grass, and crops in Iowa, USA (Tufekcioglu et al., 2003).

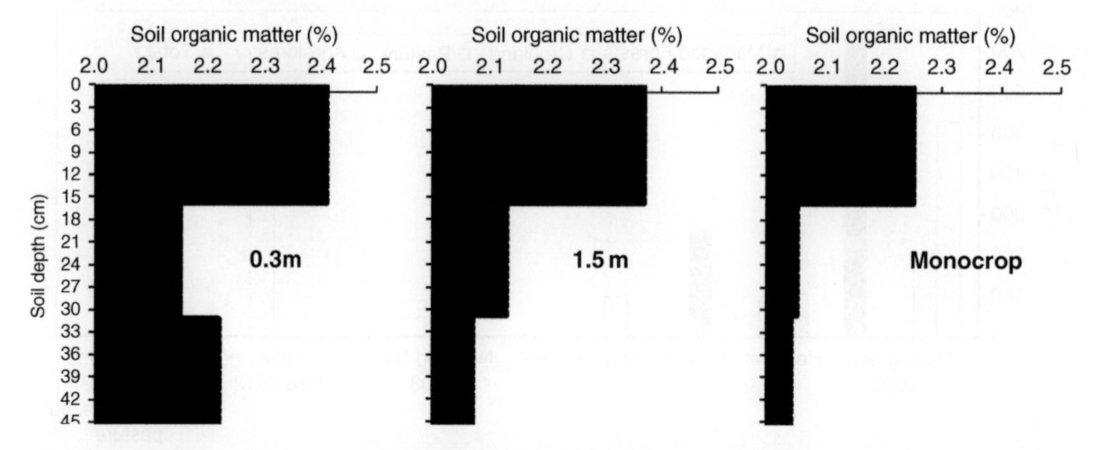

Fig. 12–3. Soil organic matter percentage decreased with increasing distance and depth from tree rows for a 4-yr-old red alder–corn alley cropping system in western Oregon, USA (adapted from Seiter, William, & Hibbs, 1999).

Likens, 2005). Therefore, thinning of older trees and shrubs and regeneration or replanting are required for maintenance and enhanced CS. Additionally, selection of fast-growing species could further enhance CS in AF soils.

Three regional-scale windbreak projects in the United States, Canada, and the former Soviet Union have demonstrated enhanced CS potential in addition to expected improvements in soil conservation, soil properties, and land productivity and addressing of famines. President Franklin D. Roosevelt initiated the Prairie States Forestry Project to stabilize blowing soil and improve land productivity. From North Dakota to Texas, 223 million trees were planted on 30 million ha (114,700 mi^2) of shelterbelts throughout the Great Plains states. In a recent 50-yr model simulation study, Possu, Brandle, Domke, Schoeneberger, and Blankenship (2016) estimated mean C storage of 2.45 ± 0.42 and 4.39 ± 1.74 Mg ha^{-1} for conifer and broad-leaved species of these windbreaks. Agriculture Canada has estimated that $>218 \times 10^6$ Mg of C was sequestered by planting >610 million shelterbelt trees in Canada. Other goods and services from these shelterbelts had a value greater than Can$600 million in Canada. The "Great Plan for the Transformation of Nature" of Joseph Stalin in the USSR resulted in the planting of about 5.7 million ha of windbreaks to defeat drought and improve crop yields and soils. A comprehensive regional study has estimated C storage of 8.16 Mg C ha^{-1} in semiarid, 19.05 Mg C ha^{-1} in subhumid, 45.36 Mg C ha^{-1} in humid, and 57.15 Mg C ha^{-1} in temperate ecoregions (Nair, Kumar, & Nair, 2009; Nair, Nair, et al., 2009).

Other SH benefits including improvements in soil water relationships, soil conservation, heat transfer, nutrient supply, water quality, soil microbial activity, and biodiversity occur in all regions and at all scales due to enhanced SC (Adhikari et al., 2014; Chirwa, 2006; Oelbermann, Voroney, Kass, & Schlönvoigt, 2006; Oelbermann, Voroney, Thevathasan, et al., 2006; Udawatta et al., 2009; Unger, Goyne, Kremer, & Kennedy, 2013). This is because integration of AF practices enhances soil physical, chemical, and biological properties, thus further enhancing ES. Deep roots bring mineral nutrients to the topsoil, improve soil nutrient status, and make available these nutrients for cash crops and AF vegetation (Chirwa, 2006). Decomposition of pruned material and litter from AF and legumes enriches soil nutrient status and enhances the CS potential (Oelbermann & Voroney, 2007; Oelbermann, Voroney, Kass, & Schlönvoigt, 2006; Oelbermann, Voroney, Thevathasan, et al., 2006). Improved soil moisture conditions in agroforestry practices such as alley cropping and silvopasture further enhance biomass growth and soil activities (Balandier, de Montard, & Curt, 2008; Udawatta, Anderson, Motavalli, & Garrett, 2011).

Perennial vegetation in AF practices has many advantages over annual monoculture plantings for storing C in the above- and belowground biomass, soil, living and dead organisms, and root exudates (Albrecht & Kandji, 2003; Cairns & Meganck, 1994; Pinho, Miller, & Alfaia, 2012). Because forest and grassland sequestration and storage patterns are active under AF, a higher percentage of C is allocated to the belowground biomass through an extended growing season (Kort & Turnock, 1999; Morgan et al., 2010; Schroeder, 1994; Sharrow & Ismail, 2004). Carbon sequestration under AF practices is more efficient because, structurally and functionally, diverse vegetation uses resources more efficiently (Sanchez, 2000; Sharrow & Ismail, 2004; Steinbeiss et al., 2008; Thevathasan & Gordon, 2004). Trees, grasses, and shrubs in a corn–soybean alley cropping practice begin photosynthesis before the cash crop is established and continue after the cash crop is harvested. In a multispecies AF practice, resource needs vary by species in the amount, time, and availability during the year (Jose, Gillespie, & Pallardy, 2004). Diverse vegetation promotes diverse soil communities, further enhancing CS (Paudel et al., 2011; Udawatta et al., 2009). The sequestration potential varies by ecosystem region, soil, climate, management practice, species composition, and age of an individual species (Jose, 2009; Pinho et al., 2012). However, a properly designed AF practice can enhance CS without sacrificing crop yields when appropriate management practices are implemented.

Soil Physical Properties

The annual economic cost of water and wind erosion in the United States is $37.6 billion (Uri, 2001). Soil erosion by rainfall is projected to increase between 16 and 58% in the 21st century (Nearing, 2001; Nearing, Pruski, & O'Neal, 2004), which emphasizes the importance of soil conservation and improved SH. Soil conservation will be a strong priority due to more frequent and extreme events associated with climate change. The soil conservation role of AF has been reviewed by Buresh and Tian (1998) and Young (1997), among others. The simplest explanation of how AF reduces erosion is that litter material and other inputs on the soil surface reduce the dislodging of soil particles and protect soils from raindrop impact and surface flow. At the terminal velocity, a single raindrop has 2.01 mJ of kinetic energy, which can dislodge soil to produce splash erosion (Gantzer, Buyanovsky, Alberts, & Remley, 1987). On the other hand, during moisture-limited or drought periods, effective soil conservation approaches are required to conserve soil and water and to reduce wind erosion while controlling damage to crops, livestock, and structures. Trees, shrubs, and grasses under AF in an agriculture system reduce soil bulk density and improve aggregate stability, porosity, water holding capacity, infiltration, soil water dynamics, soil thermal properties, and other soil physical properties (Bharati, Lee, Isenhart, & Schultz, 2002; Kumar, Anderson, & Udawatta, 2010; Sauer et al., 2007; Seobi et al., 2005; Udawatta et al., 2009; Udawatta, Anderson, et al., 2011; Udawatta, Anderson, Gantzer, & Garrett, 2006).

Aggregate Stability

Aggregate stability is an indicator of resistance to disruption when outside forces are applied to soil structural units. There is a strong relationship between aggregate stability and soil organic matter (SOM). Aggregate stability is determined by SOM, biota, ionic bonds, clay, and carbonates (Six, Elliott, & Paustian, 2000). At the microscopic level, clay and organic matter act as a bridge between sand and silt particles and form microaggregates. Macroaggregates are formed by the bonding of fungal hypae, plant roots, and other stabilizing agents (McLaren & Cameron, 1996; Oades, 1984). Soil aggregation potentially enhances organic matter stabilization and residence time in soils (Novara, Armstrong, Gristina, Semple, & Quinton, 2012).

In general, AF areas are usually managed with very little disturbance, and thus aggregate stability is greater than in disturbed areas and cropped areas (Gupta, Kukal, Bawa, & Dhaliwal, 2009; Paudel et al., 2011; Udawatta et al., 2009; Udawatta, Gantzer, et al., 2008). Tree-based systems with potentially greater biomass production help increase aggregate stability as well as SOM buildup. Significantly greater aggregate stability, compared with row crops and grazing, has been reported in AF systems (Paudel et al., 2011, 2012; Udawatta et al., 2009). According to Paudel et al. (2011), AF and grass areas were found to have three times more aggregates than row crop areas. They attributed the increased aggregate stability to SOM accumulation, increased tree roots, and reduced soil disturbance. As SOM increases, a concomitant increase in aggregate stability occurs, reducing soil erodibility (Al-Kaisi, Douelle, & Kwaw-Mensah, 2014). Root exudates, faunal excretion, and detritus add to organic matter and improve soil aggregate stability (Kremer & Kussman, 2011). Therefore, soils under AF have better structure and health, thus providing better soil resiliency.

Bulk Density

Incorporation of AF within row crop and grass management practices reduces soil bulk density (Akdemir et al., 2016; Kumar et al., 2008; Seobi et al., 2005; Udawatta et al., 2006). Studying the effects of AF on soil bulk density in row-crop watersheds, Seobi et al. (2005) found a 2.3% reduction after 6 yr of AF in the watersheds. Bulk density of the surface 10 cm of soil under AF was 1.13 g cm^{-3} compared with 1.45 g cm^{-3} in the cropped areas of the same watersheds (Udawatta et al., 2006). Again, Akdemir et al. (2016) observed significant reductions in bulk density 17 yr after establishment of AF. In their study, buffers had 8.5% lower bulk density than cropped areas.

Studying deep loess soils in central Missouri, Kumar et al. (2008) noticed a 12.6% higher bulk density for rotationally grazed and continuously grazed treatments (1.41 and 1.45 g cm^{-3}) compared with a grass buffer and AF buffer treatment (1.25 and 1.29 g cm^{-3}). These differences were reported 6 yr after the establishment of AF. Reduced bulk density helps increase infiltration, soil porosity, water storage, microbial activities, root growth, and SH and thereby helps control nonpoint-source pollution and improve land productivity.

Porosity

Porosity can be measured using traditional methods such as the soil core method (Anderson, Gantzer, & Brown, 1990), Boyle's Law porosimetry (American Petroleum Institute, 1960), and thin section analysis (Van Golf-Recht, 1982). Improved porosity can help increase water infiltration, improve water storage in soils, and enhance heat and gas transmission. Among the three main pore categories, macropores (>1000-μm diameter) and mesopores (10–1000--μm diameter) are important for water movement, while micropores (<10-μm diameter) are important for plant water storage.

Porosity measured by conventional methods in AF areas was found to be 3% greater 6 yr after the establishment of AF than in cropped areas (Figure 12–4; Seobi et al., 2005). On the same watersheds, Akdemir et al. (2016) observed 4.7% greater porosity in AF soils than cropped areas. In another study under grazing management, Kumar et al. (2008) found significantly greater porosity in AF areas than in conventional grazing

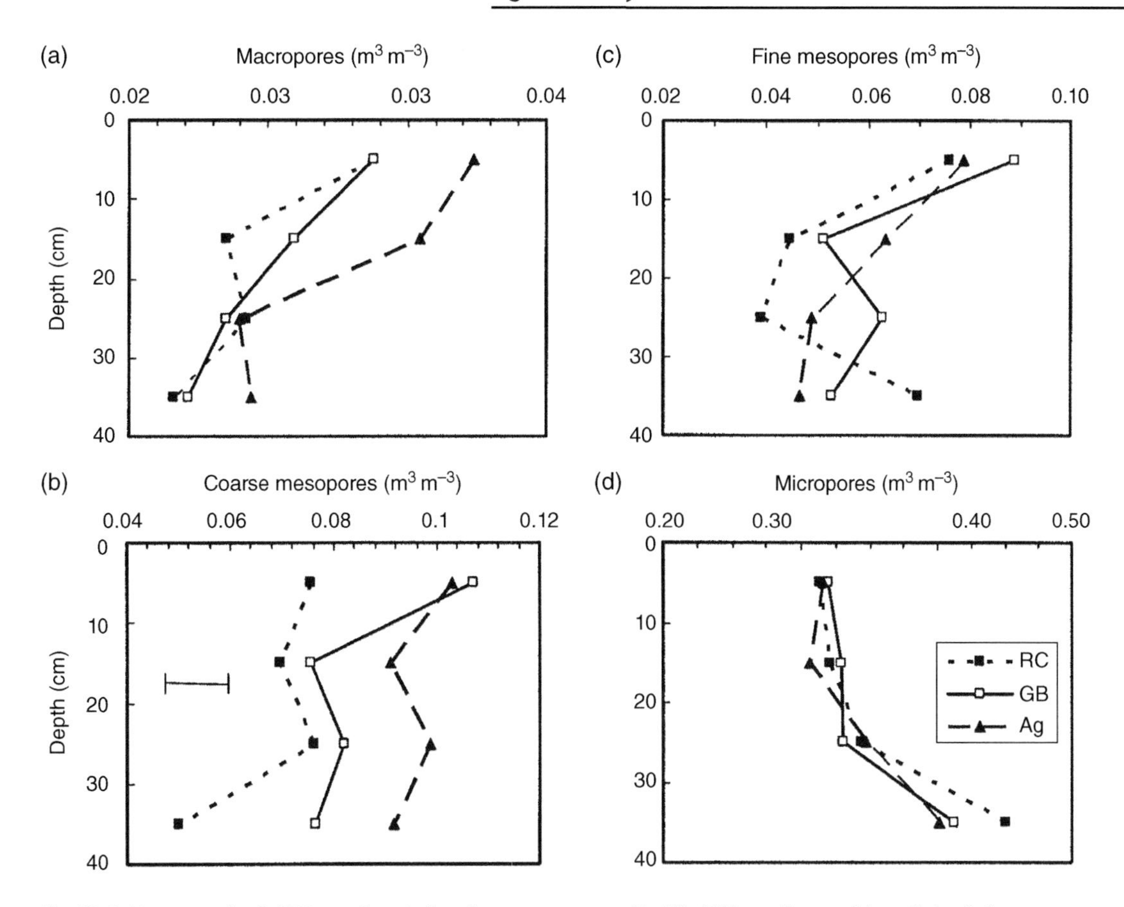

Fig. 12–4. Macroporosity (>1000-μm diameter) and coarse mesoporosity (60–1000-μm diameter) by soil depth for row crop (RC), grass buffer (GB), and agroforestry (Ag) at Greenley Memorial Research Center, University of Missouri (source: Seobi et al., 2005).

areas in Missouri. In their study, AF, grass buffers, and rotationally grazed pastures had 5.7, 4.5, and 3.9 times higher soil macroporosity than the continuously grazed pasture areas. The grazed areas had lower macroporosity and mesoporosity than AF and grass buffers. However, greater macroporosity can help release more fecal coliform bacteria and nutrients to other soil regions and groundwaters (Boyer & Neel, 2010). Therefore, livestock and crops should not be established near sensitive water resources.

Pores measured by computed tomography (CT) at 200-μm resolution for AF, grass, and cropped areas were 207, 87, and 44 pores in a 3632-mm^2 soil area (Udawatta et al., 2006). Two and a half and 3.6 times greater numbers of macropores were observed in the AF than in the grass and cropped areas. On the same watershed, Udawatta and Anderson (2008) evaluated CT-measured pore parameters among AF, grass buffers, and row-crop areas. These two studies of AF with row-crop management explained 64% of the variation in saturated hydraulic conductivity (K_{sat}) by the CT-measured macropores. For a grazing system with AF, Kumar, Anderson, and Udawatta (2010) showed that the number of pores, macropores, macroporosity, and coarse mesoporosity were greater under AF than grazing areas (Figure 12–5). They reported 13 times greater macroporosity for AF buffers (0.053 m^3 m^{-3}) than grazing areas (0.004 m^3 m^{-3}). Computed-tomography studies demonstrate that integration of AF in row-crop and grazing systems improves the number of pores, soil porosity, macroporosity, and mesoporosity, which are important for the movement and storage of water and air within soils.

Udawatta, Gantzer, et al. (2008) reported five and three times greater flow paths in AF and grass buffers than in row crops. Their study used 76-mm-long and 76-mm-diameter cores imaged at 84-μm resolution and 73- by 73- by 84-μm voxel size to quantify pore connectivity, tortuosity, and other geometrical pore parameters. The row-crop treatment had 112% greater pore path tortuosity than AF and lower characteristic pore path length

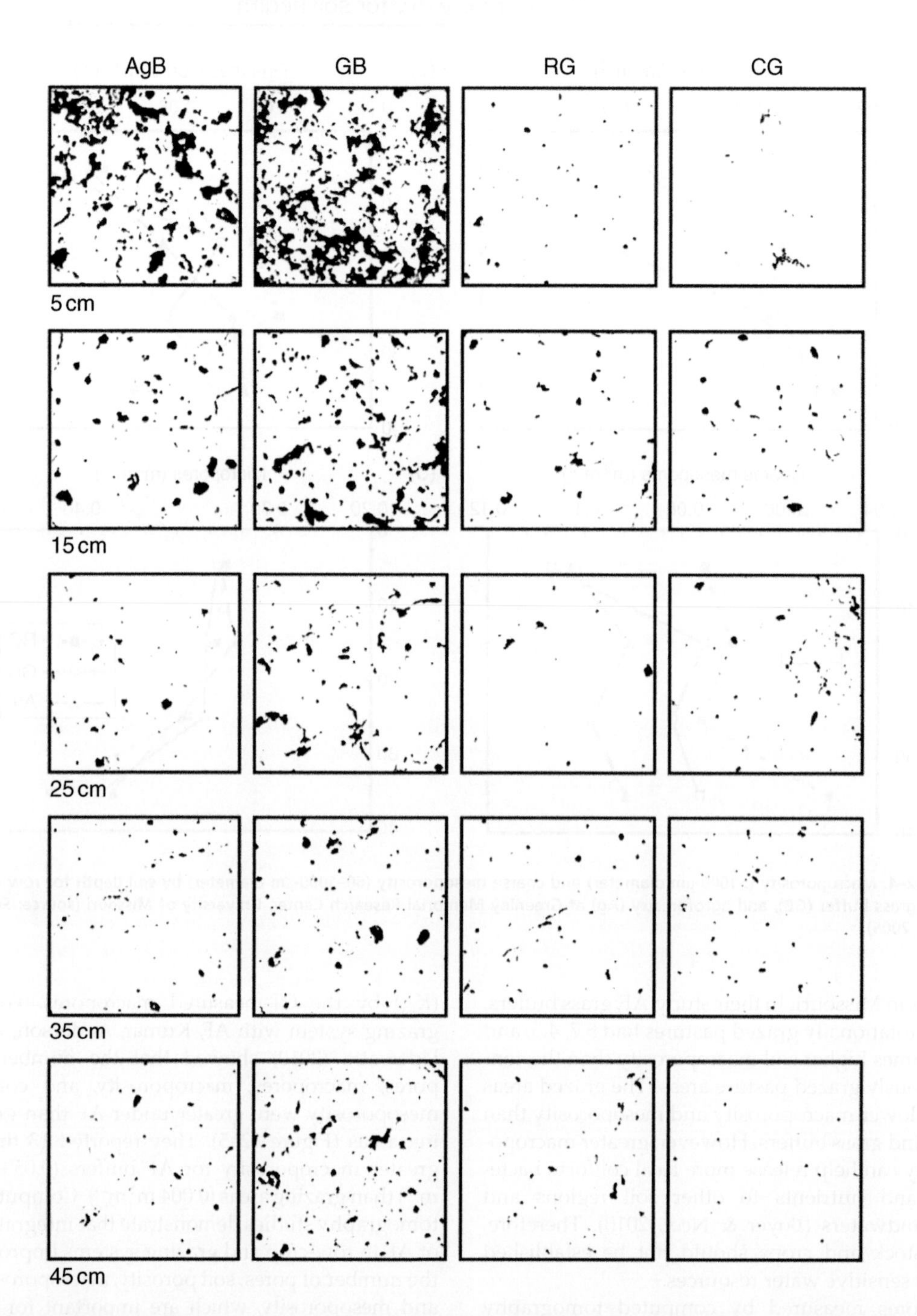

Fig. 12–5. Air-filled pores (black) in a 2500-mm² area of agroforestry buffer (AgB), grass buffer (GB), rotationally grazed pasture (RG), and continuously grazed pasture (CG) as determined by computer tomography for 5-, 15-, 25-, 35- and 45-cm depths from the surface at the Horticulture and Agroforestry Research Center, University of Missouri (source: Kumar, Anderson, & Udawatta, 2010).

constants. The differences among management were attributed to SC and roots of the perennial vegetation.

Hydraulic Conductivity and Infiltration

Shallow, compacted, and eroded soils, or soils with low organic matter, hold less water. Agroforestry practices have been shown to improve soil water holding capacity, K_{sat}, and infiltration. Furthermore, AF reduces erosion potential, runoff, and compaction (Kumar et al., 2008; Seobi et al., 2005; Udawatta, Garrett, & Kallenbach, 2011). In Missouri, Kumar et al. (2012) observed 31 and 46 times greater quasi-steady-state infiltration and K_{sat} for AF buffers of a silvopasture study than for pasture treatments. For an AF alley cropping practice in Missouri, Seobi et al. (2005) observed 3 and 14 times greater K_{sat} in grass and AF buffers compared with a corn–soybean rotation. Recent studies on the same watersheds have shown greater improvements in K_{sat} in AF areas than cropped areas (Akdemir et al., 2016).

In an AF study in Iowa, Bharati et al. (2002) showed five times greater infiltration in AF soil than soil cultivated with corn, soybean, and pasture. In Missouri loess soils, infiltration was improved by a factor of 80 compared with grazing areas (Kumar et al., 2008). Another study in Missouri under a corn–soybean rotation reported 40 times greater infiltration under AF than in cropped claypan soils (Seobi et al., 2005). Improved infiltration of AF favors more soil water movement through the profile and reduced runoff potential. These improvements can also benefit crop growth by providing greater access to water and nutrients.

Soil Moisture

Soil moisture is critical for agricultural production and 3/4 of human fresh water use is on rainfed agriculture. The impact of water stress during extended droughts can be minimized by integration of AF, resulting in improvements in soil water holding capacity, retention, recharge, storage, redistribution, and hydraulic lift (Alagele, Anderson, & Udawatta, 2018; Anderson, Udawatta, Seobi, & Garrett, 2009; Sahin, Anderson, & Udawatta, 2016; Udawatta & Anderson, 2008). Soil water dynamics of an AF alley cropping practice in Missouri showed lower soil water storage during the growing season and greater soil water storage during recharge periods (Figure 12–6). Anderson et al. (2009) found 0.9 and 1.1 cm greater water storage in a 30-cm soil profile of 6-yr-old grass and AF buffers, respectively, than in cropped areas. They attributed these differences to improvements in soil physical properties and C storage. Integration of AF thereby helps reduce runoff potential during rainy periods and stores more water during recharge periods for future crop use. Additionally, with changes in microclimate including reductions in soil temperature, wind speed, air temperature, and evaporation, AF practices enhance soil water storage and plant water-use efficiency (Brandle, Hodges, & Zhou, 2004).

Agroforestry also helps reduce drought stress by redistribution of soil water and hydraulic uplift (Bayala, Heng, van Noordwijk, & Ouedraogo, 2008; Dimitriou, Busch, Jacobs, Schmidt-Walter, & Lambersdorf, 2009; van Noordwijk, Lawson, Soumare, Groot, & Hairiah, 1996). Deep roots of the perennial vegetation can extract soil water from deeper horizons and bring it to the surface for redistribution. This water is made available for other plants with shallow roots. Similar findings have been reported by Balandier et al. (2008) in South America. The hydraulic uplift and redistribution can be further improved by selection of site-suitable trees and management practices (Udawatta, Gantzer, Reinbott, Wright, & Pierce, 2016). Reduction of heat stress, redistribution of water, and hydraulic lift help annual crops especially during drought periods.

Soil Thermal Properties

Soil thermal properties are one of the least monitored soil physical parameters, and their effects are not well documented in terms of land management and SH. Adhikari et al. (2014) examined soil heat capacity and thermal conductivity among AF, prairie, and row-crop management practices in long-term studies. The results of their study indicated that thermal conductivity was significantly higher and heat capacity was significantly lower for row-crop management than for AF and prairie systems. The lower thermal conductance of soils under AF and prairies implies increased longevity of C and conservation of soil water. In their study, SC was positively correlated with heat capacity and negatively correlated with thermal conductivity. Soils with a vegetation cover and crop residue reduce thermal conductivity and heat transfer into the soil and tend to have higher water content, lower temperatures, and lower soil heat fluxes. Their results also imply that the reduced heat flow and greater buffer capacity of soils under AF can help climate mitigation and enhance ecosystem resilience. These favorable conditions also support greater diversity of soil fauna.

Deep roots of trees, the addition of litter material and C, greater diversity, and reduced disturbance contribute to these positive SH benefits. Improvements in soil physical properties help control nonpoint-source pollution and enhance soil biological activities, CS, and land productivity. Similarly, redistribution of water by the deep roots of the perennial vegetation and reductions in water losses enhance crop productivity (Brandle, Hodges, Tyndall, & Sudmeyer, 2009). Agroforestry-induced soil thermal properties also help improve SH.

Soil Biological Properties

Soil biology involves all living organisms found in the soil environment; their various

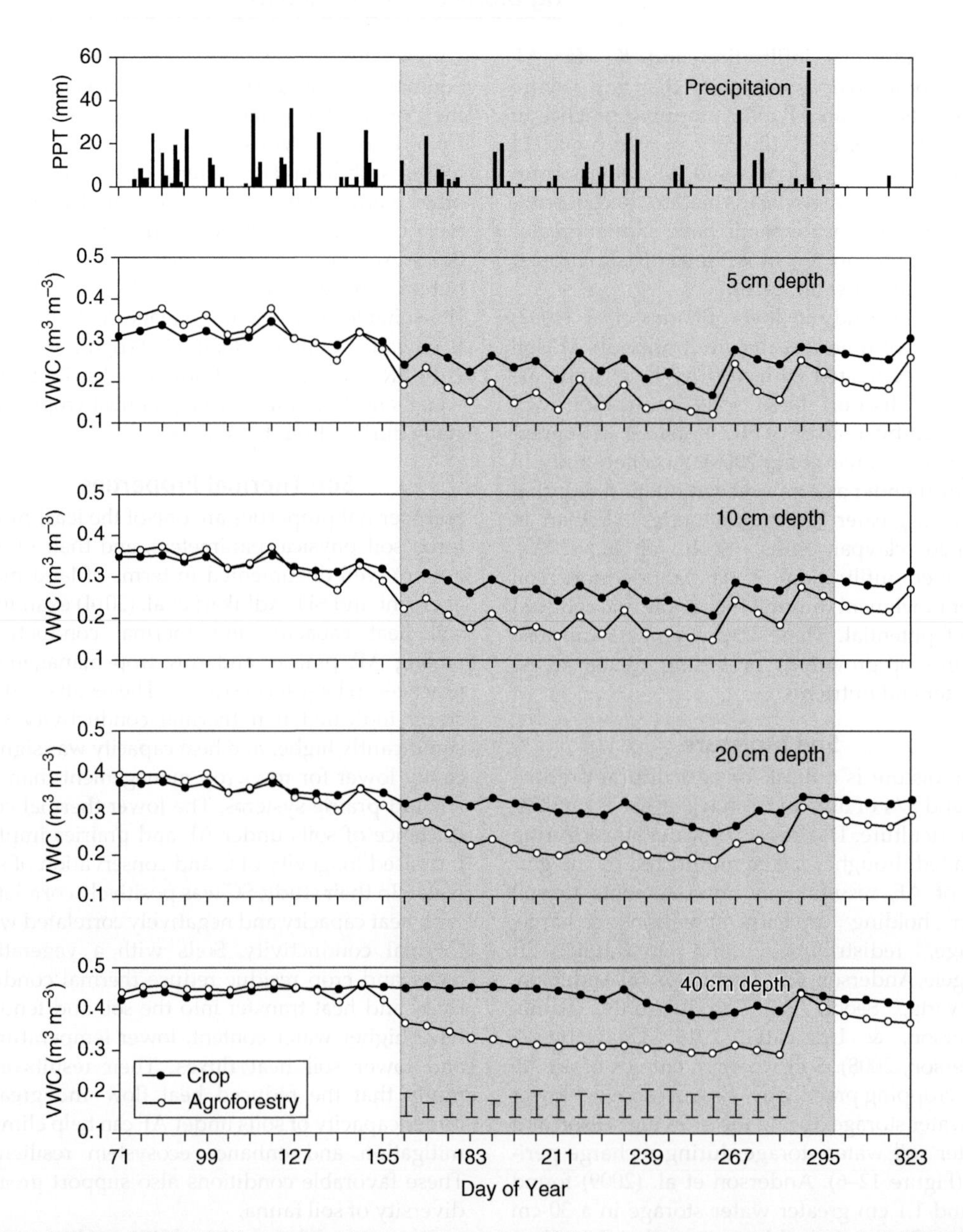

Fig. 12–6. Daily precipitation (PPT) and volumetric soil water content (VWC) for soybean (cropping period shaded light gray) and agroforestry treatments at 5-, 10-, 20-, and 40-cm soil depths for 2007 at the paired watershed study, Greenley Research Center, University of Missouri. Bars indicate LSD at α =.05 for differences in water content between treatments (source: Udawatta, Anderson, et al., 2011).

characteristics including morphology, physiology, and behavior or activities; ecological relationships among biological communities including plants; distribution within specific habitats; and their relationship to soil management, plant and animal productivity, and environmental quality. Living organisms mediate numerous simple and complex soil ecological processes (e.g., nutrient cycling) that are critical for SH, sustainability, land productivity, and restoration of degraded ecosystems. Thus, soil biological processes affect soil chemical and physical properties, nutrient dynamics, water availability, maintenance of soil condition (limiting soil degradation including erosion), and the impacts of pests and diseases.

Various groups of soil organisms are vital for a wide range of SH and ES (Wardle, Giller, & Barker, 1999; Wall, 2004). Soil microfauna (body width <100 μm), mesofauna (100 μm–2 mm), and macrofauna (>2 mm) contribute to many essential soil functions including decomposition, nutrient cycling, disease suppression, regulation of plant growth, and primary productivity. Bacteria and fungi are largely responsible for these functions. Micro- and mesofauna contribute to nutrient cycling, soil aggregate formation, and suppression of pests and diseases. They also affect nutrient availability due to preferential feeding patterns on microflora at particular times and soil microsites. Soil macrofauna, classed as large-bodied herbivores, decomposers, and predators, contribute to diverse ecosystem functions such as physically processing organic materials in preparation for further decomposition and nutrient cycling, contributing to optimum soil structure, suppressing pests and diseases, and enhancing plant growth. As the various components of the soil organisms complete their life cycles, their dead biomass contributes to SOM during decomposition via the process of microbial turnover, in which a significant amount of C enters soil organic C pools and N, P, and S are mineralized for uptake by living microbiota and plant roots. Overall, soil microbiota composition, microbial activity, and microbial biomass are good indicators of soil quality for assessing land management effects on the environment (Anderson & Domsch, 1990; Boerner, Decker, & Sutherland, 2000; Schloter, Dilly, & Munch, 2003).

Soil Biological Communities in Agroforestry Systems

The soil biological community comprises the collective groups of all organisms in the soil environment, each inhabiting particular niches or "biomes." Among all soil communities, soil microbial communities can be characterized by various methods including physiological profiling (phenotyping) based on utilization patterns of a battery of substrates (Zak, Willig, Moorhead, & Wildman, 1994); molecular techniques based on detecting genomic sequences of DNA extracted from soil (Drenovsky, Feris, Batten, & Hristova, 2008); or biochemical analysis of cell membrane phospholipid fatty acids (PLFA) extracted from soil (Zelles, 1999). Phospholipid fatty acids analysis detects lipids of microbial membranes as "biomarkers" for specific groups of microorganisms, producing a profile or "fingerprint" of the community structure. Biomarkers specific to functional groups of microorganisms include bacteria, actinobacteria, fungi, arbuscular mycorrhizal fungi, and protists. Thus, rapid changes in the microbial community structure are directly related to changes in PLFA patterns. Also, the total PLFA concentration is a measure of viable microbial biomass (Zelles, 1999).

The PLFA characterization approach used in a long-term study of an alley cropping AF system comprised of fruit trees with alleys of native grasses and forbs demonstrated that alley vegetation supported higher total soil microbial biomass than alleys of tall fescue (*Festuca arundinacea* Schreb.) or adjacent unmanaged pasture and row-cropped fields (Figure 12–7; Kremer & Hezel, 2013). Although total PLFA does not provide information on microbial community composition, a higher biomass suggests a more complex community structure relative to sites with lower total PLFA. Microbial biomass is constantly reused through a cycle of biogenesis to active growth and to decomposition ("microbial turnover"), eventually contributing to SOM formation and N mineralization, important for SH and plant nutrition, respectively. Thus, increased microbial biomass (higher total PLFA) resulting from increased rhizodeposition of labile C contents in the alleys of native plants demonstrates that diverse vegetation supplements overall AF productivity, similar to that suggested by LaCanne and Lundgren (2018) for regenerative agriculture systems.

Agroforestry practices show a more robust microbial community composition, as illustrated by a fruit tree–native vegetation alley cropping practice compared with a monoculture grass and row-crop practice (Figure 12–7). Total soil bacteria was increased; however, Gram-negative bacteria were particularly abundant. Gram-negative bacteria rapidly metabolize readily available C released via rhizodeposition while other members of the soil food web metabolize more complex C, leading to the development of more diverse microbial communities (Bardgett & Wardle, 2010). Many Gram-negative rhizobacteria are involved in nutrient cycling, including N mineralization, plant growth stimulation, and antibiotics that suppress soil and root pathogens, enhancing overall SH benefits. The long-term fruit tree AF practice with perennial native vegetation also increased arbuscular mycorrhizal fungi compared with tree fruit and single-species alleys of tall fescue (Kremer, Hezel, & Veum, 2015).

Soil bacterial diversity in a 21-yr-old AF system in Missouri was found to not differ significantly between silver maple (*Acer saccharinum* L.) tree rows and alley-cropped soils (Bardhan, Jose, Udawatta, & Fritschi, 2013). Mungai, Motavalli, Kremer, and Nelson (2005), previously working on the same site, reported similar results. Alleys were established with a no-till corn–wheat (*Triticum aestivum* L.) rotation; however, tree roots overlapped into soils of alleys after 20 yr, causing

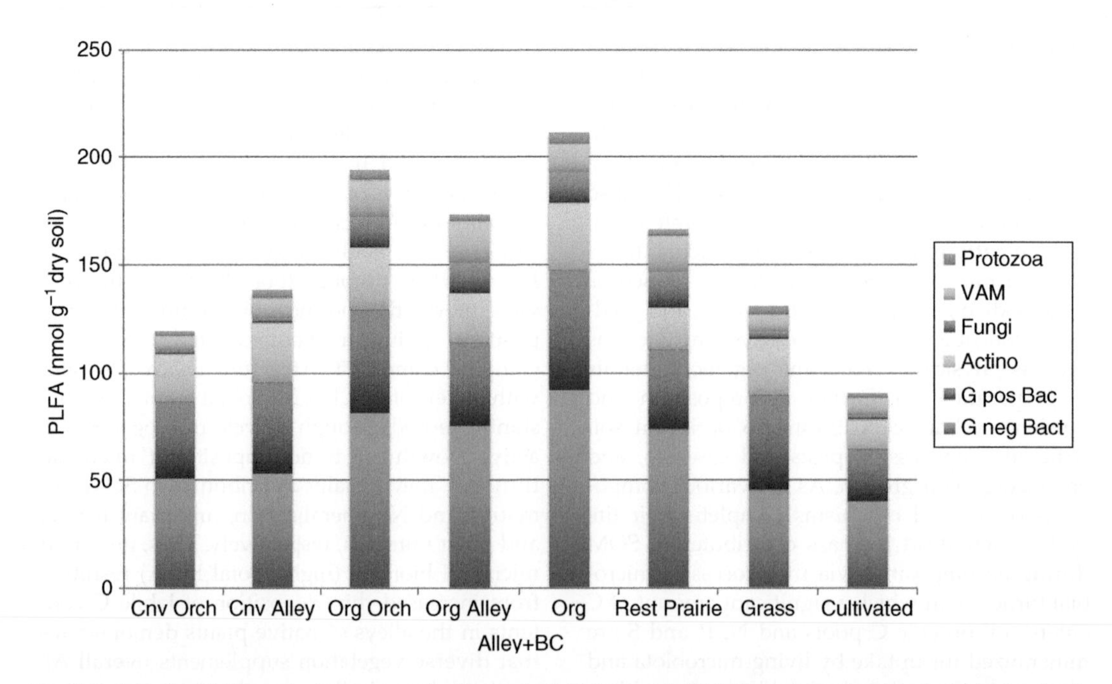

Fig. 12–7. Soil microbial biomass shown as total soil phospholipid fatty acids (PLFA) (cumulative height of each bar) and soil microbial community structure by major PLFA microbial groups (individual segments within each bar) under various production practices including a fruit tree (orchard)–native plants alley system at Prairie Birthday Farm, Clay County, Missouri, in 2014. VAM, vesicular arbuscular mycorrhizae; Actino, actinobacteria; G pos Bac, Gram-positive bacteria; G neg Bact, Gram-negative bacteria; Cnv, conventional system; Org, organic system; BC, biochar amendment; Rest Prairie, restored prairie; Grass, mixed cool-season pasture; Cultivated, corn–soybean row crop rotation (source: Kremer et al., 2015).

a more uniform soil microbiome, possibly due to increasing the variety of C exudates from both trees and crops.

Studying three AF types across a 270-km soil climate gradient in Alberta, Canada, Banerjee et al. (2016) observed significantly greater bacterial abundance measured by ^{16}S rRNA gene copies in hedgerows and woodlands than in agricultural lands without trees. Bacterial community composition differed among land-use practices, and the sites with trees increased all soil C fractions. In their study, plots with trees in AF showed greater bacterial abundance and species richness than agricultural plots. The number of individual bacterial species was also different among land-use types. Proteobacteria (37–40%), acidobacteria (14–19%), and actinobacteria (14–26%) were abundant in AF systems, while Gammaproteobacteria and Bacteroidetes were higher in agricultural lands. Supporting these findings, Zak, Holmes, White, Peacock, and Tilman (2003) observed increasing microbial abundance with an increasing number of plants. Another study in northeastern Canada showed significantly greater microbial resilience in windbreak AF soils 30 d after disturbance than in agricultural soils (Rivest, Lorente, Olivier, & Messier, 2013). Similar results were found in a laboratory study 24 and 48 h following disturbance. Because litter species diversity increases the stability of organisms (Keith et al., 2008), the greater resilience was attributed to greater diversity of the litter and rhizodeposition products of trees.

Higher microbial biomass and N mineralization rates were also observed in temperate windbreaks than under conventional agriculture (Kaur, Gupta, & Singh, 2000; Wojewoda & Russel, 2003). These benefits have been attributed to the quantity and quality of litter, rhizodeposition products, and changes in the soil. According to Bainard, Klironomos, and Gordon (2011), AF alters the composition of microbial communities as soil physio-chemical properties are different from those under conventional agriculture. Differences in microbial parameters can also be attributed to the absence of disturbance, greater aggregate stability, complex organic compounds, and microclimate (Amador, Glucksman, Lyons, & Gorres, 1997; Boerner et al., 2000; Dornbush, 2007; Helgason, Walley, & Germida, 2010; Mungai et al., 2005; Udawatta, Anderson, et al., 2008). Additionally, better geometrical soil pore parameters and water dynamics may also help promote microbial parameters within the perennial management practices (Anderson, Udawatta, Kumar, Gantzer, & Rachman, 2010; Kumar, Anderson, & Udawatta, 2010; Udawatta, Anderson, et al., 2011; Udawatta, Ganter, et al., 2008). Within the perennial vegetation, favorable soil thermal parameters reduce heat and other stress conditions and may favor greater microbial presence (Adhikari et al., 2014). The absence of vegetation (i.e., cover crops) between cropping seasons and the use of inorganic fertilizers and pesticides during crop production likely suppress the development of soil bacterial communities in crop areas, resulting in a difference from the tree row.

Fungi

Differences in fungal densities and community structures also have been reported between AF and monocropping practices. In Germany, a greater abundance of saprotrophic and ectomycorrhizal fungi was found under silvopasture than pastures without trees (Beuschel, Piepho, Joergensen, & Wachendorf, 2019). That study showed an increasing fungal C/bacterial C ratio, which indicates a potential increase of fungal C in tree rows of the study. They also found a fungal community structure shift toward saprotrophic and ectomycorrhizal fungi, which was attributed to more complex and diverse organic material, reduced soil disturbance, and soil pore geometry.

The abundance of fungi measured by PLFA profile analysis was also found to be greater in tree-based systems than adjacent cropping systems in Canada (Lacombe, Bradley, Hamel, & Beaulieu, 2009). The spatial difference and diversity of fungi pores were significant on tree-based intercropping sites compared with a monoculture (Chifflot, Rivest, Olivier, Cogliastro, & Khasa, 2009). A significant effect on the distribution of pores in a poplar tree intercropped AF system was reported. Zhang, Zhang, Zhou, Fang, and Ji (2018) investigated plant–tree associations and the soil fungal community using molecular techniques. In three practices—barley (*Hordeum vulgare* L.) alone, barley with *Populus euramericana* Guinier, or barley with *Taxodium distichum* (L.) Rich.—fungal diversity was greater in the AF rhizospheres than the bulk soil but did not differ among the practices.

Arbuscular mycorrhizal and ectomycorrhizal fungi are important to SH and plant growth. They improve soil structure, nutrient status, and microbial community structure and suppress weed populations (Bainard, Koch, Gordon, & Klironomos, 2012). Arbuscular mycorrhizal richness was found to be greater in tree-based cropping systems than conventional management at the University of Guelph Research Station, Canada (Bainard et al., 2012; Bainard, Koch, Gordon, & Klironomos, 2013). Furthermore, fungal community composition was significantly different between AF and conventional cropping systems, with several more taxa found in the AF systems. Similarly, mycorrhizae, detected using PLFA analysis of soils in an oak (*Quercus* spp.) AF practice, accumulated similar biomass densities in both the grass alleys and within the tree rows, but both were significantly higher than in the adjacent cultivated row-crop field (Unger et al., 2013). Soil microbial components (bacteria, fungi, protozoa) were also greater for the tree rows and grass alleys relative to the row-crop field.

Because of the high density of roots of multiple vegetation types including trees and crops, inevitable belowground interactions occur within the soil (Jose et al., 2004; Jose, Gillespie, Seifert, & Biehle, 2000). According to a review by Bainard, Klironomos, and Gordon (2011), the limited research on SH relationships with fungi from AF suggests that integration of AF increases fungal diversity and abundance more than monocrop systems. However, research on the relationships between microbial communities and AF has yielded limited, inconclusive, and sometimes even contradictory results (Bardhan et al., 2013; Lacombe et al., 2009; Mungai et al., 2005; Rivest et al., 2013; Saggar, Hedley, & Salt, 2001). A very complex interaction of factors comes into play. For example, mycorrhizal fungi that were more abundant in an orchard AF practice with native vegetation in alleys relative to pure fescue alleys were further increased simply with the addition of biochar in the tree rows (Kremer et al., 2015). All of the above emphasizes the importance and need for long-term studies with standard management, soil amendment, and data collection protocols.

Earthworms and Other Macrofauna

Earthworm biomass and population density was found to increase near trees in an intercropping AF planting in Canada compared with various distances from the trees (Price & Gordon, 1998). Even species differences were observed. The density was greatest for poplar (182 m^{-2}), medium for white ash (*Fraxinus americana* L.; 90 m^{-2}) and lowest for silver maple (71 m^{-2}). Another study evaluating the relationships among earthworm casts, tree distribution, and mulch showed that the spatial distribution of earthworm casts was closely related to the distribution of trees (Pauli, Oberthur, Barrios, & Conacher, 2010). A survey of 13 AF practices in France consisting of various tree species, all with alleys managed under row-crop rotation, showed that soil tillage and inorganic fertilization significantly increased total earthworm abundance and biomass in tree rows due to increased soil organic C and the lack of disturbance compared with alleys and treeless control plots (Cardinael et al., 2018). This study demonstrated how earthworm activity and associated ES are modified by AF systems and suggests the need for proper planning for optimum soil biological benefits.

Many studies have indicated greater beneficial effects from greater diversity, abundance, and functions of soil fauna within AF compared with monocrops. However, a limited number of studies in the tropics have suggested the need for caution about making generalizations regarding other soil organisms (Barrios, Sileshi, Shepherd, & Sinclair, 2012). Studies in the tropics have shown increases of some organisms with certain tree species and other organisms with other tree species. For example, Sileshi and Mafongoya (2007) found an abundance of earthworms and beetles with legume species producing high-quality biomass in Zambia. The same study showed increases of millipedes and beetles with legume species producing low-quality biomass while spiders and centipedes were not influenced by the quality of the biomass.

Most soil organisms obtain their energy from C. With a few exceptions, living and decomposing plant parts provide almost all of the C and energy. Therefore, changes in plant communities, plant litter, litter composition and forms, as well as dead and live soil fauna can change soil functions. Agroforestry's diverse plant communities and structure modify important soil functions, with possible feedback to the above- and belowground components of the AF practice.

Soil Biological Activities

Incorporation of trees and other perennial vegetation directly influences the microclimate, soil environment, and ecosystem processes (Jose, 2009) and thus the activities of soil organisms and soil properties. The spatial distribution of crops, pastures, shrubs, and trees, and the species composition are different in AF relative to monocrop management, and these factors influence the quality and quantity of organic matter and many other soil properties (Banerjee et al., 2016). Research has shown changes in soil water dynamics, soil hydraulics, physical parameters, and chemical properties due to the integration of tree- and plant-induced heterogeneity (Berg, 2013; Chu et al., 2010; Seobi et al., 2005; Udawatta, Anderson, et al., 2011). The activities of soil organisms and soil functions are influenced by these physiochemical changes and thus influence soil biological properties and activities mediated by soil fauna and flora. Soil microbial communities and their activities are highly dependent on the quantity and quality of above- and belowground inputs of plant-derived organic matter (Myers, Zak, White, & Peacock, 2001; Zak et al., 2003). Soil enzyme activities are related to SC (Mungai et al., 2005; Paudel et al., 2011; Udawatta et al., 2009), microbial community structure (Bardhan et al., 2013), soil physiochemical characteristics (Amador et al., 1997), vegetation (Sinsabaugh, Carreiro, & Repert, 2002), and disturbance (Boerner et al., 2000). Additionally, trees within AF can influence the abundance, diversity, and functions of soil organisms. Information on the composition of these communities, their diversity, and the factors influencing their activities can help us understand many soil processes and manage AF more efficiently (Banerjee et al., 2016).

Among the microbial parameters, enzyme activities are promising indicators of SH because of their rapid response to changes in soil management (Bandick & Dick, 1999; Schloter et al., 2003). Soil enzymes play key biochemical functions in the overall processes of organic matter decomposition, nutrient mineralization and cycling, nutrient availability, biodegradation of synthetic compounds, and synthesis of plant-growth-regulating substances, thereby mediating critical roles in most biochemical and ecological processes in the soil ecosystem (Bardgett & van der Putten, 2014; Sinsabaugh, Antibus, & Linkins, 1991). Therefore, assessment of enzymatic activities in the ecosystem aid in quantifying and evaluating specific biological processes in the soil.

Soil enzyme activities are sensitive SH indicators for microbial activity used to differentiate various soil and crop management regimes and to quantify specific soil biological processes (Bandick & Dick, 1999). In various AF practices, selected soil enzymes are significantly higher under both grass and grass-plus-tree strips than in continuously cropped alleys (Mungai et al., 2005; Myers et al., 2001; Udawatta et al., 2009). Soil physical properties, including bulk density, are improved under AF buffer management, demonstrating the linkage of biological, chemical, and physical indicators in overall SH (Udawatta et al., 2009). Paudel et al. (2011, 2012) conducted SH assessments in an alley cropping practice with eastern cottonwood trees (*Populus deltoides* W. Bartram ex Marshall) plus tall fescue grass buffers, tall fescue grass buffers alone, and permanent pasture alleys of tall fescue plus forage legumes. The pasture alleys were subjected to intermittent grazing by cattle, and all treatments were compared with a row-cropping (corn–soybean rotation) system. All SH indicators and biological activities were highest in the alleys with perennial vegetation compared with the row crop, including enzymatic activities in the grazed pastures, demonstrating the benefits of including livestock in the silvopasture component of this AF practice.

In cropping watersheds with AF, Weerasekara, Udawatta, Jose, Kremer, & Weerasekara (2016) noticed significantly greater enzyme activities between AF and conventional crop areas 10 yr after the establishment of buffers. Evaluation of a mature silver maple alley cropping practice by Mungai et al. (2005) revealed greater microbial biomass and activity in perennial buffer soils compared with crop areas. In a pecan [*Carya illinoinensis* (Wangenh.) K. Koch] AF practice with alleys intercropped with the perennial legume kura clover (*Trifolium ambiguum* M. Bieb.), SH indicators including soil organic C content, soil enzyme activities, and stable soil aggregates improved compared with no or unmanaged vegetation in alleys of adjacent trees (Kremer & Kussman, 2011). Soil enzyme activity at this site gradually increased in alleys and tree rows relative to pasture and annual crop sites during the 7-yr monitoring period and was also affected by the landscape position of the alleys (Figure 12–8).

Although soil glucosidase activity is considered a good indicator of management-induced changes in SH (Bandick & Dick, 1999; Stott, Andrews, Liebig, Wienhold, & Karlen, 2010), it may not be sensitive enough for differentiating the effects of management variations on microbial activity in soils and rhizospheres of AF practices. The substrate for glucosidase, cellobiose, is a product of cellulose decomposition, and ~80% of some plant root exudates are water-soluble carbohydrates, organic acids, and amino acids (Hütsch, Augustin, & Merbach, 2002). Because of the wide C/N ratio of these root exudates, rapid metabolism by the proliferating rhizosphere microbial community results in temporary N immobilization before synthesis of extracellular enzymes can mineralize N from organic materials and the dead microbial biomass (Paterson, 2003; Schenck zu Schweinsberg-Mickan, Jörgensen, & Müller, 2012). Thus, more sensitive SH indicators for microbial activity are required that include enzymes that degrade soluble carbohydrates such as sucrose or glucose and/or amino acids present in the rhizosphere.

Increased enzyme activities indicate enhanced potential to degrade cellulose, hemicellulose, chitin, peptidoglycan, and proteins, which leads to subsequent improved mineralization and nutrient cycling. A considerable number of studies have reported significant differences between AF and cropped areas for SC and some biological parameters, while the influence on microbial parameters has been inconsistent

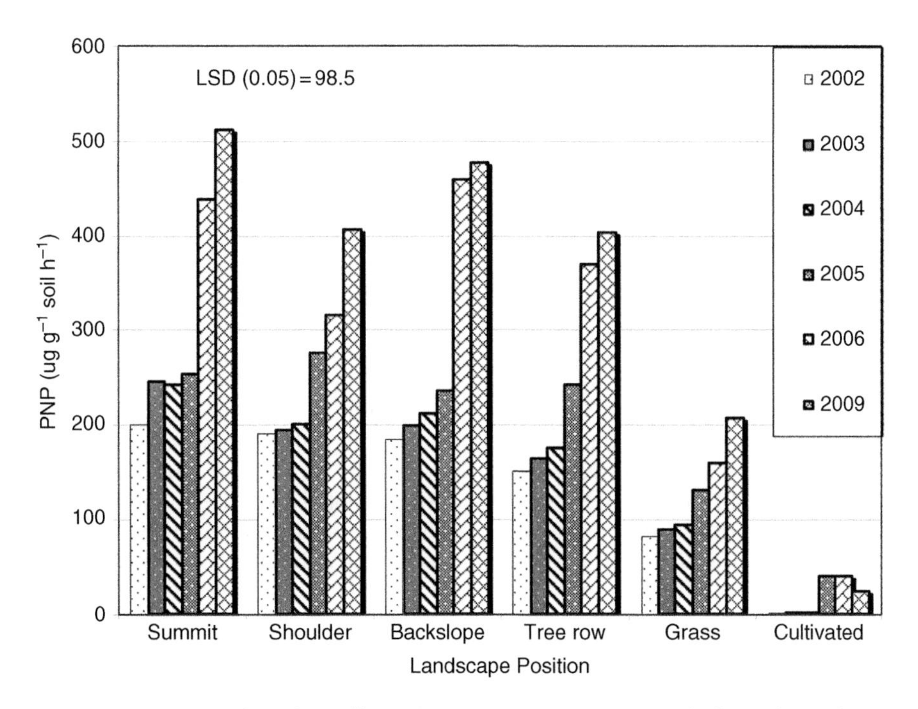

Fig. 12–8. Soil glucosidase activity in kura clover alleys, pecan tree row, grass pasture, and cultivated annual crop sites across various landscape positions in a pecan agroforestry practice, 2002–2009; PNP is *p*-nitrophenol, a product for quantifying glucosidase activity (from Kremer and Kussman, 2011; used with permission).

(Bardhan et al., 2013; Cardinael et al., 2017; Paudel et al., 2011, 2012; Mungai et al., 2005; Udawatta et al., 2009). Various studies have reported either greater biological activity and microbiological properties or no difference between AF and monocropped areas. Bambrick et al. (2010) found no significant differences in C and microbiological properties between 4-, 8-, and 21-yr-old tree-based oat (*Avena sativa* L.)–corn–corn rotational systems in Canada. At the Greenley paired watersheds in Missouri, Weerasekara et al. (2016) also observed no differences in enzyme activities at 30, 90, and 150 cm from the tree base. In Germany, Nii-Annang, Grünewald, Freese, Hüttl, and Dilly (2009) found no significant difference for microbial indicators at 0- to 3- and 3- to 10-cm soil depths 9 yr after tree establishment.

Some studies have reported decreasing differences in microbial parameters between AF and crop areas as the system matured and tree and shrub vegetation occupied more soil and aboveground space (Bambrick et al., 2010; Bardhan et al., 2013). Findings from mature AF systems may show a better distribution of leaves, litter, and roots within the crop and tree areas. Therefore, differences in soil parameters may diminish in these systems due to a tree-induced, uniform increase in soil quality indices. However, direct comparisons are limited due to differences in spatial arrangements of trees, soils, climate, management, and landscape parameters among studies.

Soil Biodiversity

The Millennium Ecosystem Assessment initiated by the United Nations Convention on Biological Diversity explained how human well-being depends on the state and use of the Earth system and its biodiversity. Soil biodiversity can be defined as the "variation in soil life, from genes to communities, and the variation in soil habitats, from micro-aggregates to entire landscapes" (Turbé et al., 2010). Biodiversity enhances agricultural sustainability as it contributes to water and nutrient efficiencies, improves SH, and protects against soil-borne diseases (Brussaard, de Ruiter, & Brown, 2007). A series of reviews and a meta-analysis have emphasized the importance of SBD for ES and of declining SBD to reduce ES, which can magnify with time (Balvanera, Kremen, & Martinez-Ramos, 2005; Cardinale et al., 2007; Hillebrand & Matthiessen, 2009; Hooper et al., 2005; Stachowicz, Bruno, & Duffy, 2007). Agroforestry can be used for biodiversity conservation because AF naturally has significantly greater SBD than monocropping management practices (Bardhan, Jose, Biswas, Kabir, & Rogers, 2012; Burgess, 1999; Stamps, Woods, Linit, & Garrett, 2002; Thevathasan & Gordon 2004; Torralba et al., 2016; Udawatta et al., 2019).

Increased use of AF for SBD conservation has been reported in both temperate and tropical regions (Noble & Dirzo, 1997; Huang et al., 2002). The diverse vegetation of AF has indirect effects on SBD by providing C and energy, regulating soil water dynamics, and changing nutrient status. Leguminous trees and shrubs can affect the SBD in unique ways, although the number of species is limited in temperate regions relative to the tropics. According to a recent study (Prober et al., 2015), plant diversity is positively correlated with microbial β diversity, which indicates greater microbial diversity with more diverse species communities (Banerjee et al., 2016).

Significantly greater diversity has been reported in AF than crop, pasture, forest, and tree monoculture management practices (Bainard, Koch, et al., 2011; Huang et al., 2002; Sistla et al., 2016; Stamps & Linit, 1997; Torralba et al., 2016; Unger et al., 2013; Zhang et al., 2018). In a meta-analysis, Bhagwat et al. (2008) reported 60% greater mean richness of taxa in AF compared with forests. Others have observed more diverse and greater microbial communities and functions under AF than cropped areas (Bainard et al., 2012; Mungai et al., 2005; Paudel et al., 2011, 2012; Udawatta et al., 2009). A Canadian study by Lacombe et al. (2009) with PLFA profiling showed significantly greater arbuscular mycorrhizae in tree-based systems than an adjacent conventional monocropping system. Chifflot et al. (2009) recorded significantly greater fungal diversity and spores in tree-based alley cropping than in monocrop forests. Using data for a larger region, Torralba et al. (2016) showed an overall positive effect of AF on biodiversity and ES in Europe.

In Canada, arbuscular mycorrhizal communities were more diverse in alley cropping systems near trees, and it was concluded that integration of trees helped improve diversity (Bainard, Klironomos, & Gordon, 2011). In a model run, Polglase et al. (2008) found that the SBD of AF is profitable in sequestering about 8 to 10 times more C in soils than the actual average rates for Australia. A plot study in Canada showed that AF promoted bacterial abundance and richness (Banerjee et al., 2016). In that study, microbial community structure was predictable in AF systems. Others have observed significantly greater phylotypic richness of mycorrhizae in intercropped alleys (Bainard, Klironomos, & Gordon, 2011). Agroforestry-induced SBD improvement can be attributed to food, shelter, habitat, and other resources produced by the multispecies vegetation of AF.

Conversion of AF to monocultures has been shown to reduce SBD (Lawton et al., 1998; Perfecto, Rice, Greenberg, & Van Der Voort, 1996; Schroth et al., 2004). In contrast, conversion of monocultures to AF increases SBD because AF harbors greater species richness and diversity (Jose, 2012; Varah, Jones, Smith, & Potts, 2013). In spite of greater diversity in AF compared with adjacent forests, AF usually has a lower number of endemic species due to intensive management (Bhagwat et al., 2008; Noble & Dirzo, 1997).

Because of the diversity of plant communities, AF usually has spatially concentrated soil communities associated with specific tree species and shrubs (Bainard et al., 2012; Pauli et al., 2010). Bainard, Koch, et al. (2011) found that white ash and poplar had a greater richness than Norway spruce [*Picea abies* (L.) Karst.] in Canada. Specific arbuscular mycorrhizal communities were observed in corn roots growing in a Norway spruce AF association compared with silver maple (Bainard et al., 2012). Another study in eastern Canada also showed greater arbuscular mycorrhizal fungal diversity in an AF system compared with a poplar plantation (Chifflot et al., 2009). Agroforestry systems have the potential to modify the soil community structure because trees and soil physiochemical properties are highly correlated (Bainard, Koch, et al., 2011). According to Rivest et al. (2013) the lower microbial metabolic quotient (respiration/biomass ratio or qCO_2) values of AF compared with conventional cropping suggests greater microbial substrate use under AF. These findings imply positive effects on soil biochemical processes and microbial resilience, ultimately leading to greater productivity and system resilience to water stress or climate change (Rivest et al., 2013). Greater diversity and activities are vital for SH and ES because most soil activities are mediated by soil communities. Research may be needed in this discipline to quantify the SBD of AF in temperate regions by using newer metagenomic, metabolomic, and proteomic approaches to understand synergies among these organisms in enhancing SH.

Agroforestry reduces the impact of disturbance, synthetic chemical use, and monocropping, and therefore strategic planning and integration of AF can further improve SBD. Under AF, fertilizer use can be reduced due to the provision of nutrients made available through litter input and leguminous plants placed at strategic locations within the system, thereby enhancing biodiversity. The use of synthetic inorganic fertilizers reduces SBD (Helgason, Daniell, Husband, Fitter, & Young, 1998; Kleijn et al., 2009). Others have noticed less effective fungal communities in fertilized agricultural systems. Reduced tillage also favors greater diversity as tillage destroys fungal mycelial networks (Mathimaran et al., 2007; Simmons & Coleman, 2008). Recent studies and findings from other parts of the world suggest that rationally selected species combinations placed at strategic locations within landscapes combined with suitable management can further enhance SBD as AF practices inherently reduce disturbance and chemical use.

Soil Chemical Properties

Soil Enrichment

Soil enrichment functions of AF include improving soil quality and maintaining long-term productivity and land sustainability while reducing water pollution and hypoxia conditions (Jose, 2009; Udawatta et al., 2009; Udawatta, Garrett, et al., 2011). Improvements of SH by trees occur through increased supply and availability of nutrients, reduction of losses of soil and nutrients, recycling of nutrients and improving nutrient cycling (Buresh & Tian, 1998; Udawatta et al., 2017). Agroforestry retains nutrients and C through their filtration and reduction of water erosion (Broadmeadow & Nisbet, 2004; Nair & Graetz, 2004; Schultz et al., 2009; Udawatta, Garrett, et al., 2011; Udawatta, Krstansky, Henderson, & Garrett, 2002). Trees in AF practices enhance the nutrient pool as evidenced by higher extractable P, total N, and mineralization compared with conventional agriculture (Kaur et al., 2000; Rivest et al., 2013). In Canada, N release from annual litter fall of a hybrid poplar was 7 kg N ha^{-1} yr^{-1} in an intercropped system (Thevathasan & Gordon, 2004). The addition of organic matter from diverse vegetation promotes diversity within soils. These materials usually have lower decomposition rates than the residue of associated crops (Rivest et al., 2013). In Florida, sandy loam soils under silvopasture showed greater P storage capacity and less buildup of nutrients than pastures without trees (Michel, Nair, & Nair, 2007). This also indicates reduced losses to groundwater and water contamination. Zuo and Zhang (2009) were able to show that intercropping dicots and graminaceous species can effectively increase the Zn and Fe in crops.

Deep roots of AF trees and shrubs capture nutrients from deeper horizons that are lost from the crop root zone and from the weathering zones and makes them available to shallow-rooted crops (Allen et al., 2004; Chirwa, 2006; Jose et al., 2000; van Noordwijk et al., 1996). Reduction of leaching and capturing of nutrients from deep soils are explained by "safety net" and "nutrient pump mechanisms." These two functions and SH can be further improved by integrating leguminous species and selecting trees and shrubs with proportionately greater vertical rooting than horizontal rooting, thus reducing the competition for resources (Udawatta et al., 2006; Udawatta, Nelson, Jose, & Motavalli, 2014; Table 12–1). In support of these concepts, a study in Denmark found that N removal below the 1-m root depth was strongly correlated with root density levels (Figure 12–9; Thorup-Kristensen, 2001; Thorup-Kristensen & Rasmussen, 2015). These results imply that N removal by deep-rooted perennial vegetation can be improved by selection of plants with deep roots and implementing cultural practices such as trenching to limit tree roots in the upper horizons (Udawatta et al., 2014, 2016). These mechanisms within AF soils also prevent contamination of the groundwater because they capture nutrients lost from shallow-rooted pastures and crops (Jose, 2009). For example, the "safety net" in an Appalachian silvopasture practice reduced NO_3 leaching losses (Boyer & Neel, 2010). Furthermore, these two mechanisms also recycle synthetic fertilizers applied to crops and pastures, thus improving nutrient use efficiency and economics.

Table 12–1. The percentage of cross-sectional area of horizontal (vertical insertion angle <45°) and vertical (>45°) roots for pin, swamp white, and bur oak trees at the Paired watershed study, Greenley Research Center of University of Missouri.

Species	Horizontal roots (%)	Vertical roots (%)
Pin oak	51	49
Swamp white oak	60	40
Bur oak	44	56

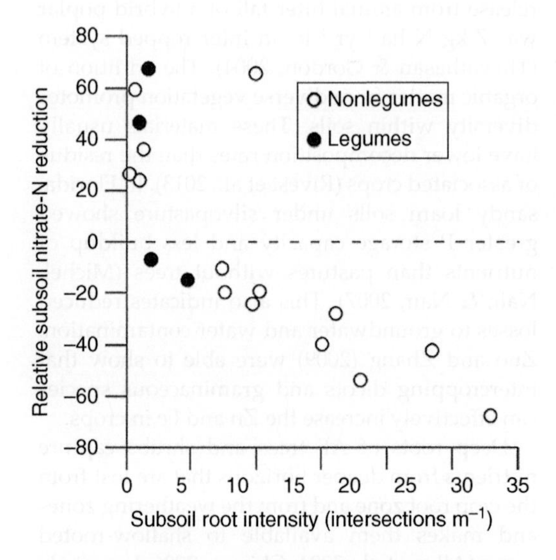

Fig. 12–9. Linear relationship of subsoil nitrate reduction by catch crops and root growth (source: Thorup-Kristensen & Rasmussen, 2015).

Perennial vegetative buffers along water bodies and upland crop areas reduce nutrient losses from row-crop and grazing practices, retain them within the cropped and grazing areas, and enrich soil nutrient status and soil properties, thereby improving land productivity (Schultz et al., 2009; Udawatta, Garrett, & Kallenbach, 2011). The four-zone buffer system, as described by Schultz et al. (2009), includes grasses next to the crop area in the fourth zone and shrubs in between the grasses and high-value, slow-growing tree zone in the third zone. The vegetation in these two zones helps reduce concentrated flow, enhance sedimentation, and retain sediment-bound nutrients on the soil surface. The second zone consists of high-value timber species, while the zone next to the stream is planted with fast-growing, water-tolerant species. These two zones also improve retention of nutrients while stabilizing the stream bank and retaining soil. Additionally, roots of the vegetation of all four zones remove nutrients from the subsurface and thereby retain nutrients within the buffers. In Canada, Banerjee et al. (2016) observed greater concentrations of various forms of N (NH_4, NO_3, total) and C (total organic C, dissolved, and total) within shelterbelt, hedgerow, and silvopasture practices than in agriculture and grasslands.

Two metadata analyses and numerous other studies have shown relationships between nutrient filtration and sedimentation with buffer width (Dillaha, Reneau, Mostaghimi, & Lee, 1989; Lee & Jose, 2003; Liu, Zhang, & Zhang, 2008; Mayer, Reynolds, McCutchen, & Canfield, 2007; Patty, Réal, & Gril, 1997; Rhoades, Nissen, & Kettler, 1997; Robinson, Ghaffarzadeh, & Cruse, 1996; Schmitt et al., 1999). Wider buffers are more effective, although productive land is taken out of crop production or grazing. Greater N than P removal has been reported in many studies (Dosskey, 2001; Schmitt, Dosskey, & Hoagland, 1999). Moreover, nutrient retention efficiency and soil enrichment can be improved by periodic sediment removal and thinning of the vegetation.

Although legumes are used for N_2 fixation, soil enrichment, production of green manure, fuel wood, and fodder in many tropical AF practices (Nygren & Leblanc, 2009; Jose, 2009; Rao, Nair, & Ong, 1998), the number of legume species used in temperate AF is limited. However, using ^{15}N, Seiter et al. (1996) in Oregon demonstrated that 32–58% of the total N for corn could be obtained from N_2 fixed by strategically positioned red alder (*Alnus rubra* Bong.). Decomposition of pruned material and litter from AF species and legumes also improves the nutrient status and enhances the CS potential of soils (Oelbermann, Voroney, Thevathasan, et al., 2006). Furthermore, studies have shown other improvements including soil and microbial C, residual N after legume-enriched fallow, long-term accumulation of C and N, and increased N mineralization and nitrification rates in soils with legumes (Dulormne, Sierra, Nygren, & Cruz, 2003; Mazzarino, Szott, & Jimenez, 1993; Sierra, Dulormne, & Desfontaines, 2002; Sierra & Nygren, 2005; Ståhl, Nyberg, Högberg, & Buresh, 2002).

Arbuscular mycorrhizae and ectomycorrhizae can also reduce leaching and complement the nutrient capture of deep-rooted trees. They increase the uptake of less mobile nutrients and thus reduce leaching losses. Mycorrhizal fungi improve soil nutrient availability by increasing the solubility, transformation, and translocation of nutrients (Bainard, Klironomos, & Gordon, 2011). They also greatly increase the ability of plants to take up water and certain nutrients by increasing their effective root surface area (Smith & Read, 2008). These associations often protect plants from pests and diseases and enhance plant health (Newsham, Fitter, & Watkinson, 1995). In addition, certain plant species are mycorrhizal obligatory (they need the fungal association for their survival).

Decontamination

Mining, industrialization, and rapid urbanization in many parts of the world have caused significant soil contamination (Gomes, 2012; Hundal et al., 2006; Rattan, Datta, Chandra, & Sharma, 2002). Some soils are contaminated with large concentrations of heavy metals including Cr, Zn, Pb, Cd, Cu, Hg, and Ni (Hundal, Kumar, Singh, & Manchanda, 2006; Rattan et al., 2002). Large concentrations of other chemicals such as herbicides, pesticides, antibiotics, personal care products, and other toxic compounds also accumulate in soil and water bodies (Hayes et al., 2002; Kim & Carlson, 2007; Lin et al., 2011; USEPA, 2007). Mineralized per- and polyfluoroalkyl substances, a soil contaminant that is highly resistant to degradation and extremely toxic, has been found in the drinking water of 27 states (Ahrens, 2011; Conder, Hoke, DeWolf, Russell, & Buck, 2008; Lau et al., 2007). In the United States, there are 1858 Superfund sites and more than 450,000 brownfields (USEPA, 2018a, 2018b).

The combination of soils, crops, and trees within AF is especially beneficial to the removal of hazardous chemicals such as herbicides, antibiotics, salinity, and heavy metals. The effectiveness of the processes is largely determined by climate, management, tree species, and plant density. Selection of highly effective tree species or combinations and silvicultural management practices will help enhance the benefits arising from the integration of AF (Rockwood et al., 2004).

Healthy soils play a paramount role in the overall cleaning of the environment. Riparian buffers can serve as an effective decontamination technique to remove contaminants before they enter water bodies (Chu et al., 2010; Lin et al., 2010, 2011; Rockwood et al., 2004). Commonly used buffer species like poplars and forage grasses (e.g., *Panicum virgatum* L.) phytoremediate atrazine (6-chloro-N^2-ethyl-N^4-isopropyl-1,3,5-triazine-2,4-diamine) by plant uptake, degradation, and detoxification (Chang, Le, & Je, 2005; Lin, Lerch, Garrett, Li, & George, 2007). Evaluating 11 different willow (*Salix* spp.) cultivars in Quebec, Canada, Bell et al. (2014) found that the abundance of the fungal class Pezizomycetes in hydrocarbon-contaminated soils was directly related to willow phylogeny. Although specific bacterial groups are highly correlated with hydrocarbon degradation, fungi provide translocation paths for hydrocarbon-degrading bacteria through their hypae to the substrate (Bell et al., 2013; Wick, Furuno, & Harms, 2010).

Soils in AF practices have the capacity to degrade or reduce toxic or hazardous compounds through various mechanisms (Andrews, Karlen, & Cambardella, 2004; Dominati, Patterson, & Mackay, 2010). These mechanisms include soil microbial activities (degradation), phytostabilization, phytoextraction, and rhizodegradation (Lin et al., 2007, 2010). Agroforestry systems with greater biodiversity promote greater degradation and stronger binding of contaminants (Andrews et al., 2004; Chu et al., 2010; Dominati et al., 2010; Lin et al., 2010). Root exudates and root decomposition products including phenolic and carboxyl groups, N-heterocyclic compounds, lignin, and decomposition products serve as binding sites (Cheng & Kuzyakov, 2005; Chu et al., 2010; Lin et al., 2010; Thiele-Bruhn, Seibicke, Schulten, & Leinweber, 2004). Chu et al. (2010) showed that oxytetracycline and sulfadimethoxine were strongly sorbed into AF soils, compared with soils from crop areas, and thereby the concentrations were reduced in runoff waters. They attributed these differences to organic compounds within the AF soils. Additionally, some tree root exudates in the rhizosphere promote degradation by soil fauna and the bonding of chemical compounds to soil particles (Chu et al., 2010).

Shorter half-lives and quicker degradation of herbicides, antibiotics, and degradation products were found by Lin, Lerch, Garrett, and George (2004) and Lin et al., (2007, 2010, 2011) in multispecies AF buffers than in non-tree practices. Yang, Wang, Chen, and Li (2011) and Yang, Wang, Cai, and Li (2012) attributed these benefits to enzymatic activities, root exudates, and physical parameters of the soils. The first phase of reduction of pollutants moving to water is brought about by improved infiltration, which reduces the losses in runoff water (Krutz, Senseman, Dozier, Hoffman, & Tierney, 2004; Lin et al., 2004). Subsequent steps involve microbial and biochemical processes that produce less harmful degradation products and CO_2 (Lin et al., 2004, 2010, 2011; Mandelbaum, Wackett, & Allan, 1993). For example, Lin et al. (2010) demonstrated that the half-life of sulphamethazine was 2.1 d in the root zone of poplar and 7.6 d in control treatments.

Phytoremediation—the use of plants to bioremediate contaminated soil, water, and air—has emerged as a more cost-effective, noninvasive, and publicly preferred approach to the removal of environmental contaminants than other options (Boyajian & Carreira, 1997). Phytoremediation includes phytofiltration, rhizofiltration, phytoextraction, phytoimmobilization, phytostabilization, phytodegradation, and rhizodegradation (Arthur et al., 2005). Although the phytoremediation prospective is governed mainly by the plants and trees used, soil communities, and various exudates of soil organisms, the genetics of the selected plants also play a valuable role. Therefore, selection of the proper tree species and cultivar can enhance the decontamination process. Fast-growing tree species produce high biomass and deep roots that tolerate and accumulate large amounts of contaminants (Dhillon, Dhillon, & Thind, 2008; Gomes 2012). Such species are preferred because larger above- and belowground biomass can accumulate more contaminants than slow-growing trees or annual crops (Dhillon et al., 2008). Contaminants can remain within the tree biomass for extended periods of time, or the trees can be harvested and used for bioenergy or other products with environmentally sound contaminant disposal protocols. Rockwood et al. (2004) have summarized tree species used in the United States for phytoremediation of heavy metals and organic compounds. Among many species commonly used in AF buffers, tree species such as poplar and willow have been shown to be very effective in bioremediation of contaminated soils (Gomes, 2012; Rockwood et al., 2004). These trees within riparian buffers can serve as an effective phytoremediation barrier to remove heavy metals, chemical compounds, herbicides, antibiotics, and personal care products before they enter water bodies (Chu et al., 2010; Lin et al., 2010, 2011; Rockwood et al., 2004).

Phytoextraction and accumulation potential varies by tree species and the number of trees planted. For example, Cd extraction was greater in *Melia azedarach* L. than *Eucalyptus teretocornis* Sm. (Kaur et al., 2018). Short-rotation woody crop species of *Eucalyptus*, *Populus*, and *Salix*, primarily grown for fuelwood and established at close spacing for 10- to 15-yr rotations, have been shown to have greater potential than slow-growing trees (Rockwood et al., 2004). Furthermore, a 5-yr study conducted in the United Kingdom demonstrated that phytoremediation was enhanced when chemical leachates were applied during the growing season (Rockwood et al., 2004).

Soil Health and Ecosystem Services

"An ecosystem service is a benefit to society derived from a healthy ecosystem property or process" (SSSA Science Policy Office, n.d.). Soils are essential in providing a wide range of ES including plant production, air quality, water quality, recycling of animal and plant products, removal of contaminants from the environment, and climate mitigation. How we manage the soils determines how well these functions are performed. The perennial nature of trees when introduced into agricultural systems has a profound effect on soil properties, nutrient dynamics, soil, water, biodiversity, CS, and soil functions (Barrios et al., 2012). A variety of spatial and temporal configurations of trees, shrubs, and grasses with crop and/or livestock in a landscape of AF contributes to short- and long-term ES. Agroforestry-induced SH improvements enhance all four categories of ES (Table 12–2), and these benefits can range from farm to global scales. Accumulation of SC and greater biodiversity in healthy soils under diverse vegetation can contribute to numerous ES including climate mitigation and supplying food, fiber, energy, and a good environment. In a recent review, Dollinger and Jose (2018) summarized the SH and ES of AF in great detail.

Supporting Services

Supporting services for SH include soil formation, nutrient cycling, and primary production. Soil biodiversity has been shown to be greater in

Table 12–2. Improved soil health of agroforestry and related ecosystem services.

Supporting	Provisional	Regulation	Cultural
Soil formation	Food supply	Water quality	Aesthetic value
Nutrient cycling	Water supply	Water supply	
Primary production	Fiber and fuel supply	Soil conservation	
	Genetic resource	Air quality	
		Climate regulation	

AF soils than conventional agricultural soils, forest, and barren lands (Udawatta et al., 2019). Improved SH provides a better habitat for soil organisms and thus helps soil formation and nutrient cycling. Changes in physical parameters and accumulation of organic matter also enhances soil formation. Diverse plant species growing in soils use nutrients at different times and at different rates, and therefore replenishment of cations and anions may benefit in such healthy soils. Diverse communities living in healthy soils contribute to efficient nutrient cycling and availability of limited nutrients, thus providing an improved environment for enhanced weathering of minerals—a process that occurs over a long period of time.

Primary production is enhanced in healthy AF soils by providing an improved medium for seed germination, survival, and growth. Healthy soils provide the habitat and life-support necessities for survival and growth of aboveground and belowground communities. These include annuals, perennials, and micro- and macrofauna and -flora. Improvements in SH through increased SC provides a better growing environment with greater moisture, nutrients, and air for flora and fauna. Dinitrogen-fixing trees and shrubs, as well as N_2–fixing bacteria and mycorrhizae in AF practices can provide increased N and other nutrients for greater productivity and sustainability of the land. Maintenance of a suitable status of soil water, nutrients, and aeration, along with decontamination, helps provide a better plant environment. These soil changes may favor some plants and organisms over others. It is also important to note that soil conditions can be modified to improve certain soil functions. For example, acidification may enhance the growth and survival of some selective plants and soil organisms while decreasing the growth of others. Enhanced growth of living organisms further helps to improve SH parameters.

Provisional Services

Improvements in SH by AF practices enhances provisional services provided by the soil. These include the supply of food, fiber, fuel/energy, water, and genetic resources and materials. Improved SH increases the survival, growth, and production of biomass. Improvements in plant growth conditions are attributed to better soil density and porosity and the availability of nutrients and water. Soil nutrient enrichment and soil conservation also improve crop growth and yields. Improved soil physical parameters and water holding capacity can help enhance plant biomass accumulation and crop yields while protecting crops from water and wind erosion as well as drought conditions or other adverse climatic conditions. Water infiltration, storage, recharge, and flood control also can contribute to improved SH and thus production of food, fiber, energy, and other materials. Soil biology contributing to nutrient cycling and the availability of nutrients also improves crop yield.

Regulatory Services

Regulatory services of healthy soils include regulation of water quality, water supply, air quality, erosion control, and climate. Healthy soils filter nutrients, contaminants, and harmful organisms and regulate water quality. Soil biodiversity, various plant and other exudates, and the formation of various organic compounds help reduce contaminants, thus regulating water quality, soil quality, and air quality. Better soils that hold more nutrients, water, and chemicals can reduce the degradation of water quality and regulate water supply. All of these benefits can be attributed to improvement in soil physical properties and an increase of SC and soil exudates. Healthy soils also help regulate pest and disease populations through competition, predation, and parasitism (Susilo et al., 2004).

Soil conservation is another significant benefit of healthy soils. Improved soil physical, chemical, and biological parameters reduce soil erosion. Improved soil helps maintain better air quality by reducing dust and particulate material. As a soil is drying, it will be more susceptible to wind erosion and the generation of dust and larger particles. Healthy soils reduce wind erosion and the development of dust and other particles. Windbreaks established to combat famines and droughts in the United States, Canada, and the former USSR are good examples of AF improving almost all ES, including air quality.

Climate regulation is enhanced by healthy soils. Agroforestry soils have a greater potential to sequester and store C than monocropped soils. The higher above- and belowground C inputs by trees in AF practices contribute heavily to climate regulation. Furthermore, improvements in other soil properties and the enhancement of the growth of the vegetation further increases C storage and helps climate regulation. Benefits such as nutrient and water conservation also provide a better environment for the germination, survival, and growth of plants.

Cultural Services

Aesthetic values of the soil are improved by SH. Agroforestry contributes to a pleasing landscape that is distinctly superior to conventional agriculture, forest, and barren lands. These improvements also help improve domestic animal welfare in silvopasture and wildlife.

Soil carbon is a factor that has influence on almost all soil parameters. Ecosystem services provided by SH that improve supporting, regulating, provisional, and cultural ES are interrelated and therefore it is difficult to assign a single parameter that is responsible for a particular service (e.g., increased SC is related to almost all benefits arising from improved SH).

Practical Implications of Soil Health and Sustainability

The mean farm size in the United States almost doubled between 1982 (238 ha) and 2007 (448 ha) (MacDonald, Korb, & Hoppe, 2013). Modern agricultural practices most often use continuous monocropping, large machinery, and inputs including fertilizers and agrochemicals. These practices have affected many social, ecological, and environmental aspects of our lives. For example, the Hypoxia Zone of the Gulf of Mexico expanded from 10,000 km^2 in 1985 to 21,000 km^2 in 2007, and the change is strongly correlated with the nutrient inputs and cropland acreage in the Midwest. Such negative effects, including the degradation of SH and ES, have resulted in a recent trend toward the adoption of more sustainable and diverse agricultural practices that closely mimic AF's multispecies and diverse management of agricultural lands.

A wide array of conservation practices within AF can be integrated into agricultural management practices to conserve and enhance the quality of soil, water, and the environment and help sustain food and fiber production. Existing data from the temperate region and many years of data gathering from tropical areas have shown that integration of AF principles into conventional agriculture improves SH, land productivity, and aesthetic values of the land. Soil health improvements with AF practices can contribute to sustainable production and ES. In addition to increased crop and pasture production, SH provides improved habitat for diverse soil communities.

Integration of AF into agroecosystems enhances SH by its influence on SC, other soil properties, nutrient cycling, contaminant removal, and increased biodiversity. Many studies in different regions of the temperate zone have shown that AF practices such as riparian and upland buffers, windbreaks, alley cropping, and silvopasture improve SC. Efficient nutrient cycling, decontamination, and soil reactions are improved by multispecies AF because of greater soil biodiversity and improved soil physical, chemical, and biological properties. Moreover, a study in Australia reported enhanced climate mitigation potential from improved SH.

These benefits can be further enhanced by selection of site-suitable species and proper species combinations and by implementing appropriate management practices. For example, if CS is a priority, cottonwood trees in a buffer can sequester significantly more C than more commonly used species. Direct and indirect benefits of improved water quality, soil biodiversity, CS, and decontamination have long-term health benefits for human beings in addition to enhanced SH. Numerous other ecosystem benefits including reductions in nutrient losses and incidents of pests and diseases, and degradation of environmental quality have been identified as benefits from AF-induced SH improvements. These changes boost supporting, provisional, regulatory, and cultural ES.

Improvements in soil physical properties like porosity may enhance water infiltration and thus mitigate potential groundwater contamination and loss of nutrients. With the integration of AF, SBD may not be as diverse as under forested conditions because AF favors certain groups of flora and fauna over others, but diversity is still greatly enhanced. Selected trees, shrubs, pasture, crops, animals, and their combinations may enhance the growth of certain pests and pathogens, so one must be aware of possibly creating such problems within AF practices. Therefore, AF planning needs to consider short- and long-term interactions and outcomes before implementation.

Strategic placement of AF on agroecosystem landscapes can contribute to additional environmental, land production, and landowner economic benefits due to improved SH. Landowners and farmers need not use their most productive land for AF. They can identify the most vulnerable areas of the farm or landscape and less productive areas for establishment.

Establishment of upland buffers improves retention of soil and nutrients in upland areas. These selected areas can be planted with income-generating trees, shrubs, grasses, and pollinators for additional short- and long-term benefits. Like conventional agriculture, AF practices also face threats from climate change and management decisions. Therefore, all available information must be used to develop management plans because AF is very complex compared with monocrop agriculture. Future research may focus on the selection of the most critical parameters for SH indices that can be used to design management guidelines for AF. Landowners and farmers who adopt AF practices will provide multiple services and benefits to our society at local, regional, continental, and ecoregional global scales.

References

Adhikari, P., Udawatta, R. P., Anderson, S. H., & Gantzer, C. J. (2014). Soil thermal properties under prairies, conservation buffers, and corn/soybean land use systems. *Soil Science Society of America Journal, 78,* 1977–1986. https://doi.org/10.2136/sssaj2014.02.0074

Ahrens, L. (2011). Polyfluoroalkyl compounds in the aquatic environment: A review of their occurrence and fate. *Journal of Environmental Monitoring, 13,* 20–31. https://doi.org/10.1039/C0EM00373E

Akdemir, E., Anderson, S. H., & Udawatta, R. P. (2016). Influence of agroforestry buffers on soil hydraulic properties relative to row crop management. *Soil Science, 181,* 368–376. https://doi.org/10.1097/SS.0000000000000170

Alagele, S. M., Anderson, S. H., & Udawatta, R. P. (2018). Biomass and buffer management practice effects on soil hydraulic properties compared to grain crops for claypan landscapes. *Agroforestry Systems, 93,* 1609–1625. https://doi.org/10.1007/s10457-018-0255-1

Albrecht, A., & Kandji, S. T. (2003). Carbon sequestration in tropical agroforestry systems. *Agriculture, Ecosystems & Environment, 99,*15–27. https://doi.org/10.1016/S0167-8809(03)00138-5

Al-Kaisi, M. (2008). *Soil erosion, crop productivity and cultural practices* (PM 1870). Ames, IA: Iowa State University Extension.

Al-Kaisi, M. M., Douelle, A., & Kwaw-Mensah, D. (2014). Soil microaggregate and macroaggregate decay over time and soil carbon change as influenced by different tillage systems. *Journal of Soil and Water Conservation, 69,* 574–580. https://doi.org/10.2489/jswc.69.6.574

Allen, S. C., Jose, S., Nair, P. K. R., Brecke, B. J., Nkedi-Kizza, P., & Ramsey, C. L. (2004). Safety-net role of tree roots: Evidence from a pecan (Carya illinoensis K. Koch)–cotton (Gossypium hirsutum L.) alley cropping system in the southern United States. *Forest Ecology and Management, 192,* 395–407. https://doi.org/10.1016/j.foreco.2004.02.009

Amador, J. A., Glucksman, A. M., Lyons, J. B., & Gorres, J. H. (1997). Spatial distribution of soil phosphatase activity within a riparian forest. *Soil Science, 162,* 808–825. https://doi.org/10.1097/00010694-199711000-00005

American Petroleum Institute. (1960). API recommended practice for core analysis procedures (Report 40). Dallas, TX.

Anderson, S. H., Gantzer, C. J., & Brown, J. R. (1990). Soil physical properties after 100 years of continuous cultivation. *Journal of Soil and Water Conservation, 45,* 117–121.

Anderson, S. H., Udawatta, R. P., Kumar, S., Gantzer, C. J., & Rachman, A. (2010). CT-measured macropore parameters for estimating saturated hydraulic conductivity at four study sites. In R. J. Gilkes and N. Prakongkep (Eds.), *Soil solutions for a changing world: 19th World Congress of Soil Science.* Brisbane, Australia.

Anderson, S. H., Udawatta, R. P., Seobi, T., & Garrett, H. E. (2009). Soil water content and infiltration in agroforestry buffer strips. *Agroforestry Systems, 75,* 5–16. https://doi.org/10.1007/s10457-008-9128-3

Anderson, T. H., & Domsch, K. H. (1990). Application of eco-physiological quotients (qCO_2 and qD) on microbial biomass from soils of different cropping histories. *Soil Biology and Biochemistry, 22,* 251–255. https://doi.org/10.1016/0038-0717(90)90094-G

Andrews, S. S., Karlen, D. L., & Cambardella, C. A. (2004). The soil management assessment framework: A quantitative soil quality evaluation method. *Soil Science Society of America Journal, 68,* 1945–1962. https://doi.org/10.2136/sssaj2004.1945

Antunes, P. M., Franken, P., Schwarz, D., Rillig, M. C., Cosme, M., Scott, M., & Hart, M. M. (2013). Linking soil biodiversity and human health: Do arbuscular mycorrhizal fungi contribute to food nutrition? In D. W. Hall et al. (Eds.), *Soil ecology and ecosystem services* (pp. 153–175). Oxford, UK: Oxford University Press. https://doi.org/10.1093/acprof:oso/9780199575923.003.0015

Arthur, E. L., Rice, P. J., Rice, P. J., Anderson, T. A., Baladi, S. M., Henderson, K. L. D., & Coats, J. R. (2005). Phytoremediation: An overview. *Critical Reviews in Plant Sciences, 24,* 109–122. https://doi.org/10.1080/07352680590952496

Baah-Acheamfour, M., Carlyle, C. N., Bork, E. W., & Chang, S. X. (2014). Trees increase soil carbon and its stability in three agroforestry systems in central Alberta, Canada. *Forest Ecology and Management, 328,* 131–139. https://doi.org/10.1016/j.foreco.2014.05.031

Bainard, L. D., Klironomos, J. N., & Gordon, A. M. (2011). Arbuscular mycorrhizal fungi in a tree-based intercropping system: A review of their abundance and diversity. *Pedobiologia, 54,* 57–61. https://doi.org/10.1016/j.pedobi.2010.11.001

Bainard, L. D., Koch, A. M., Gordon, A. M., & Klironomos, J. N. (2012). Temporal and compositional differences of arbuscular mycorrhizal fungal communities in conventional monocropping and tree-based intercropping systems. *Soil Biology and Biochemistry, 45,* 172–180. https://doi.org/10.1016/j.soilbio.2011.10.008

Bainard, L. D., Koch, A. M., Gordon, A. M., & Klironomos, J. N. (2013). Growth response of crops to soil microbial communities from conventional monocropping and tree-based intercropping systems. *Plant and Soil, 363,* 345–356. https://doi.org/10.1007/s11104-012-1321-5

Bainard, L. D., Koch, A. M., Gordon, A. M., Newmaster, S. G., Thevathasan, N. V., & Klironomos, J. N. (2011). Influence of trees on the spatial structure of arbuscular mycorrhizal communities in a temperate tree-based intercropping system. *Agriculture, Ecosystems & Environment, 144,* 13–20. https://doi.org/10.1016/j.agee.2011.07.014

Bakker, M. M., Govers, G., Kosmas, C., Vanacker, V., van Oost, K., & Rounsevell, M. (2005). Soil erosion as a driver of land-use change. *Agriculture, Ecosystems & Environment, 105,* 467–481. https://doi.org/10.1016/j.agee.2004.07.009

Balandier, P., de Montard, F., & Curt, T. (2008). Root competition for water between trees and grass in silvopastoral plots of 10 year old *Prunus avium.* In D. R. Batish, R. K. Kohli, S. Jose, and H. P. Singh, (Eds.), *Ecological basis of agroforestry* (pp. 253–270). Boca Raton, FL: CRC Press.

Balvanera, P., Kremen, C., & Martinez-Ramos, M. (2005). Applying community structure analysis to ecosystem function: Examples from pollination and carbon storage.

Ecological Applications, 15, 360–375. https://doi.org/10.1890/03-5192

Bambrick, A. D., Whallen, J. K., Bradley, R. L., Cogliastro, A., Gordon, A. M., Olivier, A., & Thevathasan, N. V. (2010). Spatial heterogeneity of organic carbon in tree-based intercropping systems in Quebec and Ontario, Canada. *Agroforestry Systems, 79,* 343–353. https://doi.org/10.1007/s10457-010-9305-z

Bandick, A. K., & Dick, R. P. (1999). Field management effects on soil enzyme activities. *Soil Biology and Biochemistry, 31,* 1471–1479. https://doi.org/10.1016/S0038-0717(99)00051-6

Banerjee, S., Baah-Acheamfour, M., Carlyle, C. M., Bissett, A., Richardon, A. E., Siddique, T., . . . Chang, S. X. (2016). Determinants of bacterial communities in Canadian agroforestry systems. *Environmental Microbiology, 18,* 1805–1816. https://doi.org/10.1111/1462-2920.12986

Bardhan, S., Jose, S., Biswas, S., Kabir, K., & Rogers, W. (2012). Biodiversity in homegardens in Bangladesh: A comparison. *Agroforestry Systems, 85,* 29–34. https://doi.org/10.1007/s10457-012-9515-7

Bardhan, S., Jose, S., Udawatta, R. P., & Fritschi, F. (2013). Microbial community diversity in a 21-year old temperate alley cropping system. *Agroforestry Systems, 87,* 1031–1041. https://doi.org/10.1007/s10457-013-9617-x

Bardgett, R. D., & van der Putten, W. H. (2014). Belowground biodiversity and ecosystem functions. *Nature, 515,* 505–511. https://doi.org/10.1038/nature13855

Bardgett, R. D., & Wardle, D. A. (2010). *Aboveground–belowground linkages: Biotic interactions, ecosystem processes, and global change.* Oxford, UK: Oxford University Press.

Barrios, E., Sileshi, G. W., Shepherd, K., & Sinclair, F. (2012). Agroforestry and soil health: Linking trees, soil biota, and ecosystem services. In D. H. Wall et al. (Eds.), *Soil ecology and ecosystem functions* (pp. 315–330). Oxford, UK: Oxford University Press.

Bayala, J., Heng, L. K., van Noordwijk, M., & Ouedraogo, S. J. (2008). Hydraulic redistribution study in two native tree species of agroforestry parklands of West African dry savanna. *Acta Oecologica, 34,* 370–378.

Bell, T. H., Hassan, S. E., Lauron-Moreau, A., Al-Otaibi, F., Hijri, M., Yergeau, E., & St-Arnaud, M. (2014). Linkage between bacterial and fungal rhizosphere communities in hydrocarbon-contaminated soils is related to plant phylogeny. *The ISME Journal, 8,* 331–343. https://doi.org/10.1038/ismej.2013.149

Bell, T. H., Yergeau, E., Maynard, C., Juck, D., Whyte, L. G., & Greer, C. W. (2013). Predictable bacterial composition and hydrocarbon degradation in Arctic soils following diesel and nutrient disturbance. *The ISME Journal, 7,* 1200–1210. https://doi.org/10.1038/ismej.2013.1

Berg, M. P. (2013). Patterns of biodiversity at fine and small spatial scales. In D. W. Hall et al. (Eds.), *Soil ecology and ecosystem functions* (pp. 136–152). Oxford, UK: Oxford University Press.

Beuschel, R., Piepho, H., Joergensen, R. G., & Wachendorf, C. (2019). Similar spatial patterns of soil quality indicators in three poplar-based silvo-arable alley cropping systems in Germany. *Biology and Fertility of Soils, 55,* 1–14. https://doi.org/10.1007/s00374-018-1324-3

Bhagwat, S. A., Willis, K. J., Birks, H. J. B., & Whittaker, R. J. (2008). Agroforestry: A refuge for tropical biodiversity? *Trends in Ecology and Evolution, 23,* 261–267. https://doi.org/10.1016/j.tree.2008.01.005

Bharati, L., Lee, K.-H., Isenhart, T. M., & Schultz, R. C. (2002). Soil water infiltration under crops, pasture, and established riparian buffer in midwestern USA. *Agroforestry Systems, 56,* 249–257. https://doi.org/10.1023/A:1021344807285

Boardman, J. (2006). Soil erosion science: Reflections on the limitations of current approaches. *Catena, 68,* 73–86. https://doi.org/10.1016/j.catena.2006.03.007

Boerner, R. E. J., Decker, K. L. M., & Sutherland, E. K. (2000). Prescribed burning effects on soil enzyme activity in a southern Ohio hardwood forest: A landscape-scale analysis. *Soil Biology and Biochemistry, 32,* 899–908. https://doi.org/10.1016/S0038-0717(99)00208-4

Boyajian, G. E., & Carreira, L. H. (1997). Phytoremediation: A clean transition from laboratory to marketplace? *Nature Biotechnology, 15,* 127–128. https://doi.org/10.1038/nbt0297-127

Boyer, D. G., & Neel, J. P. S. (2010). Nitrate and fecal coliform concentration differences at the soil/bedrock interface in Appalachian silvopasture, pasture, and forest. *Agroforestry Systems, 79,* 89–96. https://doi.org/10.1007/s10457-009-9272-4

Brandle, J. R., Hodges, L., Tyndall, J., & Sudmeyer, R. A. (2009). Windbreak practices. In H. E Garrett (Ed.), *North American agroforestry: An integrated science and practice* (2nd ed., pp. 75–104). Madison, WI: ASA. https://doi.org/10.2134/2009.northamericanagroforestry.2ed.c5

Brandle, J. R., Hodges, L., & Zhou, X. H. (2004). Windbreaks in North American agricultural systems. *Agroforestry Systems, 61,* 65–78.

Brevik, E. C., & Sauer, T. J. (2015). The past, present, and future of soils and human health studies. *SOIL, 1,* 35–46. https://doi.org/10.5194/soil-1-35-2015

Broadmeadow, S., & Nisbet, T. R. (2004). The effects of riparian forest management on the freshwater environment: A literature review of best management practice. *Hydrology and Earth System Sciences, 8,* 286–305. https://doi.org/10.5194/hess-8-286-2004

Brussaard, L., de Ruiter, P. C., & Brown, G. G. (2007). Soil biodiversity for agricultural sustainability. *Agriculture, Ecosystems & Environment, 121,* 233–244. https://doi.org/10.1016/j.agee.2006.12.013

Buresh, R. J., & Tian, G. (1998). Soil improvement by trees in sub-Saharan Africa. *Agroforestry Systems, 38,* 51–76. doi:10.1023/A:1005948326499

Burgess, P. J. (1999). Effects of agroforestry on farm diversity in the UK. *Scottish Forestry, 53,* 24–27.

Cairns, M. A., & Meganck, R. A. (1994). Carbon sequestration, biological diversity, and sustainable development: Integrated forest management. *Environmental Management, 18,* 13–22. https://doi.org/10.1007/BF02393746

Cardinael, R., Chevallier, T., Cambou, A., Béral, C., Barthès, B. G., Dupraz, C., . . . Chenu, C. (2017). Increased soil organic carbon stocks under agroforestry: A survey of six different sites in France. *Agriculture, Ecosystems & Environment, 236,* 243–255. https://doi.org/10.1016/j.agee.2016.12.011

Cardinael, R., Hoeffner, K., Chenu, C., Chevallier, T., Béral, C., Dewisme, A., & Cluzeau, D. (2018). Spatial variation of earthworm communities and soil organic carbon in temperate agroforestry. *Biology and Fertility of Soils, 55,* 171–183. https://doi.org/10.1007/s00374-018-1332-3

Cardinale, B. J., Wright, J. P., Cadotte, M. W., Carroll, I. T., Hector, A., Srivastava, D. S., . . . Weis, J. J. (2007). Impacts of plant diversity on biomass production increase through time because of species complementarity. *Proceedings of the National Academy of Sciences, 104,* 18123–18128. https://doi.org/10.1073/pnas.0709069104

Chang, S. W., Lee, S. J., & Je, C. H. (2005). Phytoremediation of atrazine by poplar trees: Toxicity, uptake and transformation. *Journal of Environmental Science and Health, Part B, 40,* 801–811. https://doi.org/10.1080/03601230500227483

Cheng, W., & Kuzyakov, Y. (2005). Root effects on soil organic matter decomposition. In R. W. Zobel and S. F. Wright (Eds.), *Roots and soil management: Interactions between roots and the soil* (pp. 119–143). Madison, WI: ASA, CSSA, and SSSA. https://doi.org/10.2134/agronmonogr48.c7

Chifflot, V., Rivest, D., Olivier, A., Cogliastro, A., & Khasa, D. (2009). Molecular analysis of arbuscular mycorrhizal community structure and spores distribution in tree-based intercropping and forest systems. *Agriculture, Ecosystems & Environment, 131,* 32–39. https://doi.org/10.1016/j.agee.2008.11.010

Chirwa, P. W. (2006). Nitrogen dynamics in cropping systems in southern Malawi containing Gliricidia sepium, pigionpea and maize. *Agroforestry Systems, 67,* 93–106. https://doi.org/10.1007/s10457-005-0949-z

Chu, B., Goyne, K. W., Anderson, S. H., Lin, C. H., & Udawatta, R. P. (2010). Veterinary antibiotic sorption to agroforestry buffer, grass buffer, and cropland soils. *Agroforestry Systems, 79,* 67–80. https://doi.org/10.1007/s10457-009-9273-3

Conder, J. M., Hoke, R. A., DeWolf, W., Russell, M. H., & Buck, R. C. (2008). Are PFCAs bioaccumulative? A critical review and comparison with regulatory criteria and persistent lipophilic compounds. *Environmental Science & Technology, 42,* 995–1003. https://doi.org/10.1021/es070895g

Dent, D. (2007). Land. In R. Mnatsakanian (Ed.), *Global Environment Outlook 4: Environment for development* (pp. 81–114). Nairobi, Kenya: United Nations Environment Programme.

Dhillon, K. S., Dhillon, S. K., & Thind, H. S. (2008). Evaluation of different agroforestry tree species for their suitability in the phytoremediation of seleniferous soils. *Soil Use and Management, 24,* 208–2016. https://doi.org/10.1111/j.1475-2743.2008.00143.x

Dillaha, T. A., Reneau, R. B., Mostaghimi, S., & Lee, D. (1989). Vegetative filter strips for agricultural non-point-source pollution control. *Transactions of the ASAE, 32,* 513–519. https://doi.org/10.13031/2013.31033

Dimitriou, I., Busch, G., Jacobs, S., Schmidt-Walter, P., & Lambersdorf, N. (2009). A review of the impact of short rotation coppice cultivation on water issues. *Landbauforschung Volkenrode, 59,* 197–206.

Dollinger, J., & Jose, S. (2018). Agroforestry for soil health. *Agroforestry Systems,* 92, 213–219. https://doi.org/10.1007/s10457-018-0223-9

Dominati, E., Patterson, M., & Mackay, A. (2010). A framework for classifying and quantifying the natural capital and ecosystem services of soils. *Ecological Economics, 69,* 1858–1868. https://doi.org/10.1016/j.ecolecon.2010.05.002

Dornbush, M. E. (2007). Grasses, litter and their interaction affects microbial biomass and soil enzyme activity. *Soil Biology and Biochemistry, 39,* 2241–2249. https://doi.org/10.1016/j.soilbio.2007.03.018

Dosskey, M. G. (2001). Toward quantifying water pollution abatement in response to installing buffers on crop land. *Environmental Management, 28,* 577–598. https://doi.org/10.1007/s002670010245

Drenovsky, R. E., Feris, K. P., Batten, K. M., & Hristova, K. (2008). New and current microbiological tools for ecosystem ecologists: Towards a goal of linking structure and function. *The American Midland Naturalist, 160,* 140–159. https://doi.org/10.1674/0003-0031(2008)160[140:NACMTF]2.0.CO;2

Dulormne, M., Sierra, J., Nygren, P., & Cruz, P. (2003). Nitrogen fixation dynamics in a cut-and-carry silvopastoral system in the subhumid conditions of Guadeloupe, French Antilles. *Agroforestry Systems, 59,* 121–129. https://doi.org/10.1023/A:1026387711571

FAO. (2015). Status of the world's soil resources. Rome: FAO.

Gantzer, C. J., Buyanovsky, G. A., Alberts, E. E., & Remley, P. A. (1987). Effects of soybean and corn residue decomposition on soil strength and splash detachment. *Soil Science Society of America Journal, 51,* 202–206. https://doi.org/10.2136/sssaj1987.03615995005100010042x

Giese, L. A. B., Aust, W. M., Kolka, R. K., & Trettin, C. C. (2003). Biomass and carbon pools of disturbed riparian forests. *Forest Ecology and Management, 180,* 493–508. https://doi.org/10.1016/S0378-1127(02)00644-8

Godfray, H. C. J., Beddington, J. R., Crute, I. R., Haddad, L., Lawrence, D., Muir, J. F., . . . Toulmin, C. (2010). Food security: The challenge of feeding 9 billion people. *Science, 327,* 812–818. https://doi.org/10.1126/science.1185383

Gold, M. A., & Garrett, H. E. (2009). Agroforestry nomenclature, concepts, and practices. In H. E. Garrett (Ed.), *North American agroforestry: An integrated science and practice* (2nd ed., pp. 45–55). Madison, WI: ASA.

Gold, M. A., & Hanover, J. W. (1987). Agroforestry systems for the temperate zone. *Agroforestry Systems, 5,* 109–121. https://doi.org/10.1007/BF00047516

Gomes, H. I. (2012). Phytoremediation for bioenergy: Challenges and opportunities. *Environmental Technology Reviews, 1,* 59–66. https://doi.org/10.1080/09593330.2012.696715

Gupta, N., Kukal, S. S., Bawa, S. S., & Dhaliwal, G. S. (2009). Soil organic carbon and aggregation under poplar based agroforestry system in relation to tree age and soil type. *Agroforestry Systems, 76,* 27–35. https://doi.org/10.1007/s10457-009-9219-9

Hanski, I., von Hertzen, L., Fyhrquist, N., Koskinen, K., Torppa, K., Laatikainen, T., . . . Haahtela, T. (2012). Environmental biodiversity, human microbiota, and allergy are interrelated. *Proceedings of the National Academy of Sciences, 9,* 8334–8339. https://doi.org/10.1073/pnas.1205624109

Hayes, T., Haston, K., Tsui, M., Hoang, A., Haeffele, C., & Vonk, A. (2002). Herbicides: Feminization of male frogs in the wild. *Nature, 419,* 895–896. https://doi.org/10.1038/419895a

Heath, L. S., Kimble, J. M., Birdsey, R. A, & Lal, R. (2003). The potential of U.S. forest soils to sequester carbon. In J. M. Kimble, L. S. Heath, R. A. Birdsey, & R. Lal (Eds.) The potential of US forest soils to sequester carbon and mitigate the greenhouse effect (pp. 385–394). Boca Raton, FL: CRC Press.

Helgason, T., Daniell, T. J., Husband, R., Fitter, A. H., & Young, J. P. W. (1998). Ploughing up the wood-wide web? *Nature, 394,* 431–431. https://doi.org/10.1038/28764

Helgason, B. L., Walley, F. L., & Germida, J. J. (2010). No-till and soil management increases microbial biomass and alters community profiles in soil aggregates. *Applied Soil Ecology, 46,* 390–397. https://doi.org/10.1016/j.apsoil.2010.10.002

Hillebrand, H., & Matthiessen, B. (2009). Biodiversity in a complex world: Consolidation and progress in functional biodiversity research. *Ecology Letters, 12,* 1405–1419. https://doi.org/10.1111/j.1461-0248.2009.01388.x

Hooper, D. U., Chapin, F. S. I., Ewel, J. J., Hector, A., Inchausti, P., Lavorel, S., . . . Wardle, D.A. (2005). Effects of biodiversity on ecosystem functioning: A consensus of current knowledge and needs for future research. *Ecological Monographs, 75,* 3–35. https://doi.org/10.1890/04-0922

Howden, S. M., Soussana, J. F., Tubiello, F. N., Chhetri, N., Dunlop, M., & Meinke, H. (2007). Adapting agriculture to climate change. *Proceedings of the National Academy of Sciences, 104,* 19691–19696. https://doi.org/10.1073/pnas.0701890104

Huang, W., Luukkanen, O., Johanson, S., Kaarakka, V., Räisänen, S., & Vihemäki, H. (2002). Agroforestry for biodiversity conservation of nature reserves: Functional group identification and analysis. *Agroforestry Systems, 55,* 65–72. https://doi.org/10.1023/A:1020284225155

Hundal, H. S., Kumar, R., Singh, D., & Manchanda, J. S. (2006). Available nutrients and heavy metal status of soils of Punjab, northwest India. *Journal of the Indian Society of Soil Science, 54,* 50–56.

Hütsch, B. W., Augustin, J., & Merbach, W. (2002). Plant rhizodeposition: An important source for carbon turnover in soils. *Journal of Plant Nutrition and Soil Science, 165,* 397–407.https://doi.org/10.1002/1522-2624(200208)165:4<397::AID-JPLN397>3.0.CO;2-C

IPCC. (2007). Climate change 2007: Synthesis report. Contribution of Working Groups I, II and III to the Fourth

Assessment Report of the Intergovernmental Panel on Climate Change. Geneva, Switzerland: IPCC. Retrieved from http://www.ipcc.ch/pdf/assessment-report/ar4/syr/ar4_syr.pdf.

Jose, S. (2009). Agroforestry for ecosystem services and environmental benefits: An overview. *Agroforestry Systems, 76*, 1–10. https://doi.org/10.1007/s10457-009-9229-7

Jose, S. (2012). Agroforestry for conserving and enhancing biodiversity. *Agroforestry Systems, 85*, 1–8. https://doi.org/10.1007/s10457-012-9517-5

Jose, S., Gillespie, A. R., & Pallardy, S. G. (2004). Interspecific interactions in temperate agroforestry. *Agroforestry Systems*, 61:237–255. https://doi.org/10.1023/B:AGFO.0000029002.85273.9b

Jose, S., Gillespie, A. R., Seifert, J. R., & Biehle, D. J. (2000). Defining competition vectors in a temperate alleycropping system in the midwestern USA: 2. Competition for water. *Agroforestry Systems, 48*, 41–59. https://doi.org/10.1023/A:1006289322392

Kallenbach, R. (2009). Integrating silvopastures into current forage–livestock systems. In M. A. Gold and M. M. Hill (Eds.), *Agroforestry comes of age: Putting science into practice. Proceedings of the 11th North American Agroforestry Conference* (pp. 455–461). http://www.centerforagroforestry.org/pubs/proceedings.pdf.

Kaur, B., Gupta, S. R., & Singh, G. (2000). Soil carbon, microbial activity, and nitrogen availability in a agroforestry systems on moderately alkaline soils in northern India. *Applied Soil Ecology, 15*, 283–294. https://doi.org/10.1016/S0929-1393(00)00079-2

Kaur, B., Singh, B., Kaur, N., & Singh, D. (2018). Phytoremediation of cadmium-contaminated soil through multipurpose tree species. *Agroforestry Systems*, 92, 473–483. https://doi.org/10.1007/s10457-017-0141-2

Keith, A. M., van der Wal, R., Brooker, R. W., Osler, G. H. R., Chapman, S. J., Burslem, D. F. R. P., & Elston, D. A. (2008). Increasing litter species richness reduces variability in a terrestrial decomposer system. *Ecology, 89*, 2657–2664. https://doi.org/10.1890/07-1364.1

Kim, S.-C., & Carlson, K. (2007). Temporal and spatial trends in the occurrence of human and veterinary antibiotics in aqueous and river sediment matrices. *Environmental Science & Technology, 41*, 50–57. https://doi.org/10.1021/es060737+

Kleijn, D., Kohler, F., Baldi, A., Batáry, P., Concepción, E. D., Clough, Y., . . . Verhulst, J. (2009). On the relationship between farmland biodiversity and land-use intensity in Europe. *Proceedings of the Royal Society B, 276*, 903–909. https://doi.org/10.1098/rspb.2008.1509

Kort, J., & Turnock, R. (1999). Carbon reservoir and biomass in Canadian Prairie shelterbelts. *Agroforestry Systems, 44*, 175–186. https://doi.org/10.1023/A:1006226006785

Kremer, R. J., & Hezel, L. F. (2013). Soil quality improvement under an ecologically based farming system in northwest Missouri. *Renewable Agriculture and Food Systems, 28*, 245–254. https://doi.org/10.1017/S174217051200018X

Kremer, R. J., Hezel, L. F., & Veum, K. S. (2015). Soil health improvement in an organic orchard production system in northwest Missouri. In B. Baker (Ed.) *Proceedings of the organic agriculture research symposium.* Retrieved from https://eorganic.info/sites/eorganic.info/files/u27/07-Kremer,Hezel,Veum-Soil_Health-Final.pdf; https://www.youtube.com/watch?v=I1Br5r5tcPQ&index=7&list=PLZMuQJAj6rOqB8rDwoVdQi4kNtcgmicbv

Kremer, R. J., & Kussman, R. D. (2011). Soil quality in a pecan–kura clover alley cropping system in the midwestern USA. *Agroforestry Systems, 83*, 213–223. https://doi.org/10.1007/s10457-011-9370-y

Krutz, L. J., Senseman, S. A., Dozier, M. C., Hoffman, D. W., & Tierney, D. P. (2004). Infiltration and adsorption of dissolved metolachlor, metolachlor oxanilic acid, and metolachlor ethanesulfonic acid by buffalograss (*Buchloe dactyloides*) filter strips. *Weed Science, 52*, 166–171.

Kumar, S., Anderson, S. H., Bricknell, L. G., Udawatta, R. P., & Gantzer, C. J. (2008). Soil hydraulic properties influenced by agroforestry and grass buffers for grazed pasture systems. *Journal of Soil and Water Conservation, 63*, 224–232. https://doi.org/10.2489/jswc.63.4.224

Kumar, S., Anderson, S. H., & Udawatta, R. P. (2010). Agroforestry and grass buffer influences on macropores measured by computed tomography under grazed pasture systems. *Soil Science Society of America Journal, 74*, 203–212. https://doi.org/10.2136/sssaj2008.0409

Kumar, S., Anderson, S. H., Udawatta, R. P., & Kallenbach, R. L. (2012). Water infiltration influenced by agroforestry and grass buffers for a grazed pasture system. *Agroforestry Systems, 84*, 325–335. https://doi.org/10.1007/s10457-011-9474-4

Kumar, S., Udawatta, R. P., & Anderson, S. H. (2010). Root length density and carbon content of agroforestry and grass buffers under grazed pasture systems in a Hapludalf. *Agroforestry Systems, 80*, 85–96.

LaCanne, C. E., & Lundgren, J. G. (2018). Regenerative agriculture: Merging farming and natural resource conservation profitably. *PeerJ, 6*, e4428. https://doi.org/10.7717/peerj.4428

Lacombe, S., Bradley, R. L., Hamel, C., & Beaulieu, C. (2009). Do tree-based intercropping systems increase the diversity and stability of soil microbial communities? *Agriculture, Ecosystems & Environment, 131*, 25–31. https://doi.org/10.1016/j.agee.2008.08.010

Lal, R. (1996). Deforestation and land-use effects on soil degradation and rehabilitation in western Nigeria: III. Runoff, soil erosion and nutrient loss. *Land Degradation & Development, 7*, 87–98. https://doi.org/10.1002/(SICI)1099-145X(199606)7:2<87::AID-LDR219>3.0.CO;2-X

Lal, R. (2004). Soil carbon sequestration impacts on global climate change and food security. *Science, 304*, 1623–1627.

Lal, R. (2010). Managing soil and ecosystems for mitigating anthropogenic carbon emissions and advancing global food security. *BioScience, 60*, 708–721. https://doi.org/10.1525/bio.2010.60.9.8

Lal, R., Kimble, J. M., Follett, R. F., & Cole, C. V. (1999). The potential of U.S. cropland to sequester carbon and mitigate the greenhouse effect. Boca Raton, FL: Lewis Publishers.

Lau, C., Anitole, K., Hodes, C., Lai, D., Pfahles-Hutchens, A., & Seed, J. (2007). Perfluoroalkyl acids: A review of monitoring and toxicological findings. *Toxicological Sciences, 99*, 366–394.

Lawton, J. H., Bignell, D. E., Bolton, B., Bloemers, G. F., Eggleton, P., Hammond, P. M., . . . Watt, A. D. (1998). Biodiversity inventories, indicator taxa and effects of habitat modification in tropical forest. *Nature, 391*, 72–76. https://doi.org/10.1038/34166

Lee, K. H., & Jose, S. (2003). Soil respiration and microbial biomass in soils under alley cropping and monoculture cropping systems in southern USA. *Agroforestry Systems, 58*, 45–54. https://doi.org/10.1023/A:1025404019211

Lin, C. H., Lerch, R. N., Garrett, H. E., & George, M. F. (2004). Incorporating forage grasses in riparian buffers for bioremediation of atrazine, isoxaflutole and nitrate in Missouri. *Agroforestry Systems, 63*, 91–99. https://doi.org/10.1023/B:AGFO.0000049437.70313.ef

Lin, C. H., Lerch, R. N., Garrett, H. E., Li, Y. Z., & George, M. F. (2007). An improved HPLC-MS/MS method for determination of isoxaflutole (Balance) and its metabolites in soils and forage plants. *Journal of Agricultural and Food Chemistry, 55*, 3805–3815. https://doi.org/10.1021/jf063322g

Lin, C.-H., Goyne, K. W., Kremer, R. J., Lerch, R. N., & Garrett, H. E. (2010). Dissipation of sulfamethazine and tetracycline in the root zone of grass and tree species. *Journal of Environmental Quality, 39*, 1269–1278. https://doi.org/10.2134/jeq2009.0346

Lin, C.-H., Lerch, R. N., Goyne, K. W., & Garrett, H. E. (2011). Reducing herbicides and veterinary antibiotics losses from agroecosystems using vegetative buffers. *Journal of Environmental Quality, 40*, 791–799. https://doi.org/10.2134/jeq2010.0141

Liu, X., Zhang, X., & Zhang, M. (2008). Major factors influencing the efficacy of vegetated buffers on sediment trapping: A review and analysis. *Journal of Environmental Quality, 37*, 1667–1674. https://doi.org/10.2134/jeq2007.0437

MacDonald, J. M., Korb, P., & Hoppe, R. A. (2013). *Farm size and the organization of U.S. crop farming* (Economic Research Report no. 152). Washington, DC: USDA Economic Research Service.

Mandelbaum, R. T., Wackett, L. P., & Allan, D. L. (1993). Rapid hydrolysis of atrazine to hydroxyatrazine by soil bacteria. *Environmental Science & Technology, 27*, 1943–1944.

Mathimaran, N., Ruh, R., Jama, B., Verchot, L., Frossard, E., & Jansa, J. (2007). Impact of agricultural management on arbuscular mycorrhizal fungal communities in Kenyan ferralsol. *Agriculture, Ecosystems & Environment, 119*, 22–32. https://doi.org/10.1016/j.agee.2006.06.004

Mayer, P. M., Reynolds, S. K., Jr., McCutchen, M. D., & Canfield, T. J. (2007). Meta-analysis of nitrogen removal in riparian buffers. *Journal of Environmental Quality, 36*, 1172–1180. https://doi.org/10.2134/jeq2006.0462

Mazzarino, M. J., Szott, L., & Jimenez, M. (1993). Dynamics of soil total C and N, microbial biomass, and water soluble C in tropical agroecosystems. *Soil Biology and Biochemistry, 25*, 205–214. https://doi.org/10.1016/0038-0717(93)90028-A

McLaren, R. G., & Cameron, K. C. (1996). *Soil science: Sustainable production and environmental protection* (2nd ed.). New York: Oxford University Press.

Metting, F. B., Jacobs, G. K., Amthor, J. S., & Dahlman, R. (2002). Terrestrial carbon sequestration potential. *ACS Division of Fuel Chemistry, Preprints, 47*, 5–6.

Michel, G.-A., Nair, V. D., & Nair, P. K. R. (2007). Silvopasture for reducing phosphorus loss from subtropical sandy soils. *Plant and Soil, 297*, 267–276. https://doi.org/10.1007/s11104-007-9352-z

Montgomery, D. R. (2007). Soil erosion and agricultural sustainability. *Proceedings of the National Academy of Sciences, 104*, 13268–13272. https://doi.org/10.1073/pnas.0611508104

Montgomery, H. L. (2010). *How is soil made?* New York: Crabtree Publishing.

Morgan, J. A., Follett, R. F., Allen, L. H., Grosso, S. D., Derner, J. D., Dijkstra, F., . . . Schoeneberger, M. M. (2010). Carbon sequestration in agricultural land of the United States. *Journal of Soil and Water Conservation, 65*, 6A–13A. https://doi.org/10.2489/jswc.65.1.6A

Mungai, N. W., Motavalli, P. P., Kremer, R. J., & Nelson, K. A. (2005). Spatial variation of soil enzyme activities and microbial functional diversity in temperate alley cropping practices. *Biology and Fertility of Soils, 42*, 129–136. https://doi.org/10.1007/s00374-005-0005-1

Myers, R. T., Zak, D. R., White, D. C., & Peacock, A. (2001). Landscape-level patterns of microbial community composition and substrate use in upland forest ecosystems. *Soil Science Society of America Journal, 65*, 359–367. https://doi.org/10.2136/sssaj2001.652359x

Naiman, R. J., Décamps, H., McClain, M. E., & Likens, G.E. (2005). Structural pattern. In R. J. Naiman, H. Décamps, & M. E. McClain (Eds.), *Riparia: Ecology, conservation, and management of streamside communities* (pp. 79–123). New York: Academic Press. https://doi.org/10.1016/B978-012663315-3/50005-8

Nair, P. K. R., Kumar, B. M., & Nair, V. D. (2009). Agroforestry as a strategy for carbon sequestration. *Journal of Plant Nutrition and Soil Science, 172*, 10–23. https://doi.org/10.1002/jpln.200800030

Nair, P. K. R., & Nair, V. D. (2003). Carbon storage in North American agroforestry systems. In J. M. Kimble, L. S. Heath, R. A. Birdsey, & R. Lal (Eds.) The potential of US forest soils to sequester carbon and mitigate the greenhouse effect (pp. 333–346). Boca Raton, FL: CRC Press.

Nair, P. K. R., Nair, V. D., Gama-Rodriguez, E. F., Garcia, R., Haile, S. G., Howlett, D. S., . . . Tonucci, R. G. (2009). Soil carbon in agroforestry systems: An unexplored treasure? *Nature Precedings*. 10.1038/npre.2009.4061.1

Nair, V. D., & Graetz, D. A. (2004). Agroforestry as an approach to minimizing nutrient loss from heavily fertilized soils: The Florida experience. *Agroforestry Systems, 61*, 269–279.

Nearing, M. A. (2001). Potential changes in rainfall erosivity in the U.S. with climate change during the 21st century. *Journal of Soil and Water Conservation, 56*, 229–232.

Nearing, M. A., Pruski, F. F., & O'Neal, M. R. (2004). Expected climate change impact on soil erosion rates: A review. *Journal of Soil and Water Conservation, 59*, 43–50.

Newsham, K. K., Fitter, A. H., & Watkinson, A. R. (1995). Arbuscular mycorrhiza protect an annual grass from root pathogenic fungi in the field. *Journal of Ecology, 83*, 991–1000. https://doi.org/10.2307/2261180

Nielson, R. L. (2012). Historical corn grain yields for Indiana and the U.S. *Corny News*. Retrieved 1 June 2018 from https://www.agry.purdue.edu/ext/corn/news/timeless/yieldtrends.html.

Nii-Annang, S., Grünewald, H., Freese, D., Hüttl, R. F., & Dilly, O. (2009). Microbial activity, organic C accumulation and ^{13}C abundance in soils under alley cropping systems after 9 years of recultivation of quaternary deposits. *Biology and Fertility of Soils, 45*, 531–538. https://doi.org/10.1007/s00374-009-0360-4

Noble, I. R., & Dirzo, R. (1997). Forests as human-dominated ecosystems. *Science, 277*, 522–525. https://doi.org/10.1126/science.277.5325.522

Novara, A., Armstrong, A., Gristina, L., Semple, K. T., & Quinton, J. N. (2012). Effects of soil compaction, rain exposure and their interaction on soil carbon dioxide emission. *Earth Surface Processes and Landforms, 37*, 994–999. https://doi.org/10.1002/esp.3224

Nygren, P., & Leblanc, H. A. (2009). Natural abundance of ^{15}N in two cacao plantations with legume and non-legume shade trees. *Agroforestry Systems, 76*, 303–315. https://doi.org/10.1007/s10457-008-9160-3

Oades, J. M. (1984). Soil organic matter and structural stability: Mechanism and implication for management. *Plant and Soil, 76*, 319–337. https://doi.org/10.1007/BF02205590

Oelbermann, M., & Voroney, R. P. (2007). Carbon and nitrogen in a temperate agroforestry system: Using stable isotopes as a tool to understand soil dynamics. *Ecological Engineering, 29*, 342–349. https://doi.org/10.1016/j.ecoleng.2006.09.014

Oelbermann, M., Voroney, R. P., Kass, D. C. L., & Schlönvoigt, A. M. (2006). Soil carbon and nitrogen dynamics using stable isotopes in 19- and 10-yr old tropical agroforestry systems. *Geoderma, 130*, 356–367. https://doi.org/10.1016/j.geoderma.2005.02.009

Oelbermann, M., Voroney, R. P., Thevathasan, N. V., Gordon, A. M., Kass, D. C. L., & Schlönvoigt, A. M. (2006). Soil carbon dynamics and residue stabilization in a Costa Rican and southern Canadian alley cropping system. *Agroforestry Systems, 68*, 27–36. https://doi.org/10.1007/s10457-005-5963-7

Oliver, M. A., & Gregory, P. J. (2015). Soil, food security, and human health: A review. *European Journal of Soil Science, 66*, 257–276.

Paterson, E. (2003). Importance of rhizodeposition in the coupling of plant and microbial productivity. *European Journal of Soil Science, 54*, 741–750. https://doi.org/10.1046/j.1351-0754.2003.0557.x

Patty, L., Réal, B., & Gril, J. J. (1997). The use of grassed buffer strips to remove pesticides, nitrate and soluble phosphorus compounds from runoff water. *Pesticide Science*, *49*, 243–251. https://doi.org/10.1002/(SICI)1096-9063(199703)49:3<243::AID-PS510>3.0.CO;2-8

Paudel, B. R., Udawatta, R. P., Kremer, R. J., & Anderson, S. H. (2011). Agroforestry and grass buffer effects on soil quality parameters for grazed pasture and row-crop systems. *Applied Soil Ecology*, *48*, 125–132. https://doi.org/10.1016/j.apsoil.2011.04.004

Paudel, B. R., Udawatta, R. P., Kremer, R. J., & Anderson, S. H. (2012). Soil quality indicator responses to crop, grazed pasture, and agroforestry buffer management. *Agroforestry Systems*, *84*, 311–323. https://doi.org/10.1007/s10457-011-9454-8

Pauli, N., Oberthur, T., Barrios, E., & Conacher, A. J. (2010). Fine-scale spatial and temporal variation in earthworm surface casting activity in agroforestry fields, western Honduras. *Pedobiologia*, *53*, 127–139. https://doi.org/10.1016/j.pedobi.2009.08.001

Perfecto, I., Rice, R. A., Greenberg, R., & Van Der Voort, M. E. (1996). Shade coffee: A disappearing refuge for biodiversity. *BioScience*, *46*, 598–608. https://doi.org/10.2307/1312989

Pimentel, D., Harvey, C., Resosudarmo, P., Sinclair, K., Kurz, D., McNair, M., . . . Blair, R. (1995). Environmental and economic costs of soil erosion and conservation benefits. *Science*, *267*, 1117–1123. https://doi.org/10.1126/science.267.5201.1117

Pinho, R. C., Miller, R. P., & Alfaia, S. S. (2012). Agroforestry and the improvement of soil fertility: A view from Amazonia. *Applied and Environmental Soil Science*, *2012*, 616383. https://doi.org/10.1155/2012/616383

Polglase, P., Paul, K., Hawkins, C., Siggins, A., Turner, J., Booth, T., . . . Carter, J. (2008). Regional opportunities for agroforestry systems in Australia (RIRDC Publication no. 08/176). Kingston, ACT, Australia: Rural Industries Research and Development Corporation.

Ponisio, L. C., M'Gonigle, L. K., Mace, K. C., Palomino, J., de Valpine, P., & Kremen, C. (2015). Diversification practices reduce organic to conventional yield gap. *Proceedings of the Royal Society B*, *282*, 20141396. https://doi.org/10.1098/rspb.2014.1396

Possu, W. B., Brandle, J. R., Domke, G. M., Schoeneberger, M., & Blankenship, E. (2016). Estimating carbon storage in windbreak trees on U.S. agricultural lands. *Agroforestry Systems*, *90*, 889–904. https://doi.org/10.1007/s10457-016-9896-0

Price, G. W., & Gordon, A. M. (1998). Spatial and temporal distribution of earthworms in a temperate intercropping system in southern Ontario, Canada. *Agroforestry Systems*, *44*, 141–149. https://doi.org/10.1023/A:1006213603150

Prober, S. M., Left, J. W., Bates, S. T., Borer, E. T., Firn, J., Harpole, W. S., . . . Fierer, N. (2015). Plant diversity predicts beta but not alpha diversity of soil microbes across grasslands worldwide. *Ecology Letters*, *18*, 85–95. https://doi.org/10.1111/ele.12381

Rao, M. R., Nair, P. K. R., & Ong, C. K. (1998). Biophysical interactions in tropical agroforestry systems. *Agroforestry Systems*,*38*,3–50.https://doi.org/10.1023/A:1005971525590

Rattan, R. K., Datta, S. P., Chandra, S., & Sharma, N. (2002). Heavy metals and environmental quality. *Fertiliser News*, *47*, 21–40.

Rhoades, C. C., Nissen, T. M., & Kettler, J. S. (1997). Soil nitrogen dynamics in alley cropping and no-till systems on Ultisols of Georgia Piedmont, USA. *Agroforestry Systems*, *39*, 31–44. https://doi.org/10.1023/A:1005995201216

Rivest, D., Lorente, M., Olivier, A., & Messier, C. (2013). Soil biochemical properties and microbial resilience in agroforestry systems: Effects on wheat growth under controlled drought and flooding conditions. *Science of the Total Environment*, *463–464*, 51–60. https://doi.org/10.1016/j.scitotenv.2013.05.071

Robinson, C. A., Ghaffarzadeh, M., & Cruse, R. M. (1996). Vegetative filter strip effects on sediment concentration in cropland runoff. *Journal of Soil and Water Conservation*, *50*, 227–230.

Rockwood, D. L., Naidu, C. V., Carter, D. R., Rahmani, M., Spriggs, T. A., Lin, C., . . . Segrest, S. A. (2004). Short-rotation woody crops and phytoremediation: Opportunities for agroforestry? *Agroforestry Systems*, *61*, 51–63.

Saggar, S., Hedley, C. B., & Salt, G. J. (2001). Soil microbial biomass, metabolic quotient, and carbon and nitrogen mineralization in 25-year old Pinus radiata agroforestry regimes. *Australian Journal of Soil Research*, *39*, 491–504. https://doi.org/10.1071/SR00012

Sahin, H., Anderson, S. H., & Udawatta, R. P. (2016). Water infiltration and soil water content in claypan soils influenced by agroforestry and grass buffers compared to row crop management. *Agroforestry Systems*, *90*, 839–860. https://doi.org/10.1007/s10457-016-9899-x

Sanchez, P. A. (2000. Linking climate change research with food security and poverty reduction in the tropics. *Agriculture, Ecosystems & Environment*, *82*, 371–383. https://doi.org/10.1016/S0167-8809(00)00238-3

Sauer, T. J., Cambardella, C. A., & Brandle, J. R. (2007). Soil carbon and litter dynamics in a red cedar–scotch pine shelterbelt. *Agroforestry Systems*, *71*, 163–174. https://doi.org/10.1007/s10457-007-9072-7

Schenck zu Schweinsberg-Mickan, M., Jörgensen, R. G., & Müller, T. (2012). Rhizodeposition: Its contribution to microbial growth and carbon and nitrogen turnover within the rhizosphere. *Journal of Plant Nutrition and Soil Science*, *175*, 750–760. https://doi.org/10.1002/jpln.201100300

Schloter, M., Dilly, O., & Munch, J. C. (2003). Indicators for evaluating soil quality. *Agriculture, Ecosystems & Environment*, *98*, 255–262. https://doi.org/10.1016/S0167-8809(03)00085-9

Schmitt, T. J., Dosskey, M. G., & Hoagland, K. D. (1999). Filter strip performance and processes for different widths and contaminants. *Journal of Environmental Quality*, *28*, 1479–1489. https://doi.org/10.2134/jeq1999.00472425002800050013x

Schroeder, P. (1993). Agroforestry systems: Integrated land use to store and conserve carbon. *Climate Research*, *3*, 53–60. https://doi.org/10.3354/cr003053

Schroeder, P. (1994). Carbon storage benefits of agroforestry systems. *Agroforestry Systems*, *27*, 89–97. https://doi.org/10.1007/BF00704837

Schroth, G., da Fonseca, G. A. B., Harvey, C. A., Gascon, C., Vasconcelos, H. L., & A.-M.N. Izac. (2004). *Agroforestry and biodiversity conservation in tropical landscapes*. Washington, DC: Island Press.

Schultz, R. C., Isenhart, T. M., Colletti, J. P., Simpkins, W. W., Udawatta, R. P., & Schultz, P. L. (2009). Riparian and upland buffer practices. In H. E. Garrett (Ed.), *North American agroforestry: An integrated science and practice* (2nd ed., pp. 163–218). Madison, WI: ASA.

Seiter, S., Ingham, E. R., William, R. D., & Hibbs, D. E. (1996). Increase in soil microbial biomass and transfer of nitrogen from alder to sweet corn in an alley cropping system. In J. H. Ehrenreich, D. L. Ehrenreich, & H. W. Lee (Eds.), *Growing a sustainable future: Proceedings of the 4th North American Agroforestry Conference* (pp. 56–158). Boise, ID: University of Idaho.

Seiter, S., William, R. D., & Hibbs, D. E. (1999). Crop yield and tree-leaf production in three planting patterns of temperate-zone alley cropping in Oregon, USA. *Agroforestry Systems*, *46*, 273–288.

Seobi, T., Anderson, S. H., Udawatta, R. P., & Gantzer, C. J. (2005). Influences of grass and agroforestry buffer strips on soil hydraulic properties. *Soil Science Society of America Journal, 69*, 893–901. https://doi.org/10.2136/sssaj2004.0280

Sharrow, S. H., & Ismail, S. (2004). Carbon and nitrogen storage in agroforests, tree plantations, and pastures in western Oregon, USA. *Agroforestry Systems, 60*, 123–130. https://doi.org/10.1023/B:AGFO.0000013267.87896.41

Sierra, J., Dulormne, M., & Desfontaines, L. (2002). Soil nitrogen as affected by Gliricidia sepium in a silvopastoral system in Guadeloupe, French Antilles. *Agroforestry Systems, 54*, 87–97. https://doi.org/10.1023/A:1015025401946

Sierra, J., & Nygren, P. (2005). Role of root inputs from a dinitrogen-fixing tree in soil carbon and nitrogen sequestration in a tropical silvopastoral system. *Australian Journal of Soil Research, 43*, 667–675. https://doi.org/10.1071/SR04167

Sileshi, G., & Mafongoya, P. L. (2007). Quantity and quality of organic inputs from coppicing leguminous trees influence abundance of soil macrofauna in maize crops in eastern Zambia. *Biology and Fertility of Soils, 43*, 333–340. https://doi.org/10.1007/s00374-006-0111-8

Simmons, B. L., & Coleman, D. C. (2008). Microbial community response to transition from conventional to conservation tillage in cotton fields. *Applied Soil Ecology, 40*, 518–528. https://doi.org/10.1016/j.apsoil.2008.08.003

Sinsabaugh, R. L., Antibus, R. K., & Linkins, A. E. (1991). An enzymic approach to the analysis of microbial activity during plant litter decomposition. *Agriculture, Ecosystems & Environment, 34*, 43–54. https://doi.org/10.1016/0167-8809(91)90092-C

Sinsabaugh, R. L., Carreiro, M. M., & Repert, D. A. (2002). Allocation of extracellular enzymatic activity in relation to litter composition, N deposition and mass loss. *Biogeochemistry, 60*, 1–24. https://doi.org/10.1023/A:1016541114786

Sistla, S. A., Roddy, A. B., Williams, N. E., Kramer, D. B., Stevens, K., & Allison, S. D. (2016). Agroforestry practices promote biodiversity and natural resource diversity in Atlantic Nicaragua. *PLOS ONE, 11*(9), e0162529. https://doi.org/10.1371/journal.pone.0162529

Six, J., Elliott, E. T., & Paustian, K. (2000). Soil macroaggregate turnover and microaggregate formation: A mechanism for C sequestration under no tillage agriculture. *Soil Biology and Biochemistry, 32*, 2099–2103. https://doi.org/10.1016/S0038-0717(00)00179-6

Smith, P., Martono, D., Cai, Z., Gwary, D., Janzen, H., Kumar, P., . . . Sirotenko, O. (2007). Agriculture. In B. Metz, O. Davidson, P. Bosch, R. Dave, & L. Meyer (Eds.), *Climate change 2007: Mitigation of climate change* (pp. 497–540). Cambridge, UK: Cambridge University Press.

Smith, S. E., & Read, D. J. (2008). *Mycorrhizal symbiosis* (3rd ed.). New York: Academic Press.

SSSA Science Policy Office. (n.d.). *Soil ecosystem services*. Madison, WI: SSSA. Retrieved 8 Nov. 2018 from https://www.soils.org/files/science-policy/issues/reports/sssa-soils-eco-serv.pdf.

Stachowicz, J. J., Bruno, J. F., & Duffy, J. E. (2007). Understanding the effects of marine biodiversity on communities and ecosystems. *Annual Review of Ecology, Evolution, and Systematics, 38*, 739–766. https://doi.org/10.1146/annurev.ecolsys.38.091206.095659

Ståhl, L., Nyberg, G., Högberg, P., & Buresh, R. L. (2002). Effects of planted tree fallows on soil nitrogen dynamics, above-ground and root biomass, N_2-fixation and subsequent maize crop productivity in Kenya. *Plant and Soil, 243*, 103–117. https://doi.org/10.1023/A:1019937408919

Stamps, W. T., & Linit, M. J. (1997). Plant diversity and arthropod communities: Implications for temperate agroforestry. *Agroforestry Systems, 39*, 73–89. https://doi.org/10.1023/A:1005972025089

Stamps, W. T., Woods, T. W., Linit, M. J., & Garrett, H. E. (2002). Arthropod diversity in alley cropped black walnut (Juglans nigra L.) stands in eastern Missouri, USA. *Agroforestry Systems, 56*, 167–175. https://doi.org/10.1023/A:1021319628004

Stefano, A. D., & Jacobson, M. G. (2018). Soil carbon sequestration in agroforestry systems: A meta-analysis. *Agroforestry Systems, 92*, 285–299.

Steinbeiss, S., Beßler, H., Engels, C., Temperton, V. M., Buchmanns, N., Roscher, C., . . . Gleixner, G. (2008). Plant diversity positively affects short-term soil carbon storage in experimental grasslands. *Global Change Biology, 14*, 2937–2949. https://doi.org/10.1111/j.1365-2486.2008.01697.x

Stott, D. E., Andrews, S. S., Liebig, M. A., Wienhold, B. J., & Karlen, D. L. (2010). Evaluation of β-glucosidase activity as a soil quality indicator for the soil management assessment framework. *Soil Science Society of America Journal, 74*, 107–119. https://doi.org/10.2136/sssaj2009.0029

Susilo, F. X., Neutel, A. M., van Noordwijk, M., Hairiah, K., Brown, G., & Swift, M. J. (2004). Soil biodiversity and food webs. In M. van Noordwijk, G. Cadisch, & C. K. Ong. (Eds.) Below-ground interactions in tropical agroecosystems: Concepts and models with multiple plant components (pp. 285–302). Wallingford, UK: CAB International.

Swift, M. J., Izac, A.-M. N., & van Noordwijk, M. (2004). Biodiversity and ecosystem services in agricultural landscapes—Are we asking the right questions? *Agriculture, Ecosystems & Environment, 104*, 113–134.

Thevathasan, N. V., & Gordon, A. M. (1997). Poplar leaf biomass distribution and nitrogen dynamics in a poplar–barley intercropped system in southern Ontario, Canada. *Agroforestry Systems, 37*, 79–90. https://doi.org/10.1023/A:1005853811781

Thevathasan, N. V., & Gordon, A. M. (2004). Ecology of tree intercropping systems in the north temperate region: Experiences from southern Ontario, Canada. *Agroforestry Systems, 61–62*, 257–268. https://doi.org/10.1023/B:AGFO.0000029003.00933.6d

Thiele-Bruhn, S., Seibicke, T., Schulten, H.-R., & Leinweber, P. (2004). Sorption of sulfonamide pharmaceutical antibiotics on whole soils and particle-size fractions. *Journal of Environmental Quality, 33*, 1331–1342. https://doi.org/10.2134/jeq2004.1331

Thorup-Kristensen, K. (2001). Effect of deep and shallow root systems on the dynamics of soil inorganic N during 3-year crop rotations. *Plant and Soil, 288*, 233–248. https://doi.org/10.1007/s11104-006-9110-7

Thorup-Kristensen, K., & Rasmussen, C. R. (2015). Identifying new deep rooted plant species suitable as undersown nitrogen catch crops. *Journal of Soil and Water Conservation, 70*, 399–409. https://doi.org/10.2489/jswc.70.6.399

Tilman, D., Cassman, K. G., Matson, P. A., Naylor, R., & Polasky, S. (2002). Agricultural sustainability and intensive production practices. *Nature, 418*, 671–677. https://doi.org/10.1038/nature01014

Torralba, M., Fagerholm, N., Burgess, P. J., Moreno, G., & Plieninger, T. (2016). Do European agroforestry systems enhance biodiversity and ecosystem services? *Agriculture, Ecosystems & Environment, 230*, 150–161. https://doi.org/10.1016/j.agee.2016.06.002

Tufekcioglu, A., Raich, J. W., Isenhart, T. M., & Schultz, R. C. (2003). Biomass, carbon and nitrogen dynamics of multi-species riparian buffers within an agricultural watershed in Iowa, USA. *Agroforestry Systems, 57*, 187–198. https://doi.org/10.1023/A:1024898615284

Turbé, A., DeToni, A., Benito, P., Lavelle, P., Lavelle, P., Ruiz, N., . . . Mudgal, S. (2010). *Soil biodiversity: Functions, threats and tools for policymakers* (Technical Report 2010-049). Paris: Bio Intelligence Service.

Udawatta, R. P., & Anderson, S. H. (2008). CT-measured pore characteristics of surface and subsurface soils as influenced by agroforestry and grass buffers. *Geoderma, 145*, 381–389. https://doi.org/10.1016/j.geoderma.2008.04.004

Udawatta, R. P., Anderson, S. H., Gantzer, C. J., & Garrett, H. E. (2006). Agroforestry and grass buffer influence on macropore characteristics: A computed tomography analysis. *Soil Science Society of America Journal, 70*, 1763–1773. https://doi.org/10.2136/sssaj2006.0307

Udawatta, R. P., Anderson, S. H., Gantzer, C. J., & Garrett, H. E. (2008). Influence of prairie restoration on CT-measured soil pore characteristics. *Journal of Environmental Quality, 37*, 219–228. https://doi.org/10.2134/jeq2007.0227

Udawatta, R. P., Anderson, S. H., Motavalli, P. P., & Garrett, H. E. (2011). Clay and temperature influences on sensor measured volumetric soil water content. *Agroforestry Systems, 82*, 61–75. https://doi.org/10.1007/s10457-010-9362-3

Udawatta, R. P., Gantzer, C. J., Anderson, S. H., & Garrett, H. E. (2008). Agroforestry and grass buffer effects on high resolution X-ray CT-measured pore characteristics. *Soil Science Society of America Journal, 72*, 295–304. https://doi.org/10.2136/sssaj2007.0057

Udawatta, R. P., Gantzer, C. J., & Jose, S. (2017). Agroforestry practices and soil ecosystem services. In M. M. Al-Kaisi and B. Lowery (Eds.), Soil health and intensification of agroecosystems (pp. 305–334). San Diego, CA: Academic Press. https://doi.org/10.1016/B978-0-12-805317-1.00014-2

Udawatta, R. P., Gantzer, C. J., Reinbott, T. M., Wright, R. L., & Pierce, R. A. (2016). Temporal and spatial yield differences influenced by riparian buffers and CRP. *Agronomy Journal, 108*, 647–655. doi:10.2134/agronj2015.0273

Udawatta, R. P., Garrett, H. E., & Kallenbach, R. L. (2011). Agroforestry buffers for nonpoint source pollution reductions from agricultural watersheds. *Journal of Environmental Quality, 40*, 800–806. https://doi.org/10.2134/jeq2010.0168

Udawatta, R. P., & Jose, S. (2012). Agroforestry strategies to sequester carbon in temperate North America. *Agroforestry Systems, 86*, 225–242. https://doi.org/10.1007/s10457-012-9561-1

Udawatta, R. P., Kremer, R. J., Garrett, H. E., & Anderson, S. H. (2009). Soil enzyme activities and physical properties in a watershed managed under agroforestry and row-crop system. *Agriculture, Ecosystems & Environment, 131*, 98–104. https://doi.org/10.1016/j.agee.2008.06.001

Udawatta, R. P., Krstansky, J. J., Henderson, G. S., & Garrett, H. E. (2002). Agroforestry practices, runoff, and nutrient loss: A paired watershed comparison. *Journal of Environmental Quality, 31*, 1214–1225. https://doi.org/10.2134/jeq2002.1214

Udawatta, R. P., Nelson, K. A., Jose, S., & Motavalli, P. P. (2014). Temporal and spatial differences in crop yields of a mature silver maple alley cropping system. *Agronomy Journal, 106*, 407–415. https://doi.org/10.2134/agronj2013.0429

Udawatta, R. P., Rankoth, L. M., & Jose, S. (2019). Agroforestry and biodiversity. *Sustainability, 11*, 02879. https://doi.org/10.3390/su11102879

Unger, I. M., Goyne, K. W., Kremer, R. J., & Kennedy, A.C. (2013). Microbial community diversity in agroforestry and grass vegetative filter strips. *Agroforestry Systems, 87*, 395–402. https://doi.org/10.1007/s10457-012-9559-8

Uri, N. D. (2001). The environmental implications of soil erosion in the United States. *Environmental Monitoring and Assessment, 66*, 293–312. https://doi.org/10.1023/A:1006333329653

USDA Economic Research Service. (2011). *Major uses of land in the United States, 2007* (Economic Information Bulletin no. EIB-89). Retrieved from https://www.ers.usda.gov/publications/pub-details/?pubid=44630.

USDA-NRCS. (2010). Soil erosion on cropland 2007. Retrieved from https://www.nrcs.usda.gov/wps/portal/nrcs/detail/national/technical/?cid=stelprdb1041887.

USEPA. (2007). *Preliminary interpretation of the ecological significance of atrazine stream-water concentrations using a statistically-designed monitoring program*. Alexandria, VA: USEPA Office of Prevention, Pesticides, and Toxic Substances.

USEPA. (2018a). Search for superfund sites where you live. Retrieved 10 Apr. 2018 from https://www.epa.gov/superfund/search-superfund-sites-where-you-live.

USEPA. (2018b). Overview of EPA's Brownfields Program. Retrieved 10 Apr. 2018 from https://www.epa.gov/brownfields/overview-brownfields-program.

Van Golf-Recht, T. D. (1982). *Fundamentals of fractured reservoir engineering*. Amsterdam: Elsevier.

van Noordwijk, M., Lawson, G., Soumare, A., Groot, J. J. R., & Hairiah, K. (1996). Root distribution of trees and crops: Competition and/or complementarity. In C. K. Ong & P. Huxley (Eds.), *Tree–crop interactions* (pp. 319–364). Wallingford, UK: CAB International.

Varah, A., Jones, H., Smith, J., & Potts, S. G. (2013). Enhanced biodiversity and pollination in UK agroforestry systems. *Journal of the Science of Food and Agriculture, 93*, 2073–2075. https://doi.org/10.1002/jsfa.6148

Wall, D. H. (2004). *Sustaining biodiversity and ecosystem services in soils and sediments* (SCOPE 64). Washington, DC: Island Press.

Wall, D. H., & Six, J. (2015). Give soils their due. *Science, 347*, 694–695. https://doi.org/10.1126/science.aaa8493

Wardle, D. A., Giller, K. E., & Barker, G. M. (1999). The regulation and functional significance of soil biodiversity in agroecosystems. In D. Wood and J. M. Lenné (Eds.), *Agrobiodiversity: Characterization, utilization and management* (pp. 87–121). Wallingford, UK: CAB International.

Watson, R. T., Noble, I. R., Bolin, B., Ravindranathan, N. R., Verardo, D. J., & Dokken, D. J. (Eds.) (2000). *Land use, land-use change, and forestry* (IPCC special report). Cambridge, UK: Cambridge University Press. Retrieved from https://archive.ipcc.ch/ipccreports/sres/land_use/index.php?idp=501.

Weerasekara, C., Udawatta, R. P., Jose, S., Kremer, R. J., & Weerasekara, C. (2016). Soil quality differences in a row-crop watershed with agroforestry and grass buffers. *Agroforestry Systems, 90*, 829–838. https://doi.org/10.1007/s10457-016-9903-5

Wick, L. Y., Furuno, S., & Harms, H. (2010). Fungi as transport vectors for contaminants and contaminant-degrading bacteria. In K. N. Timmis (Ed.), *Handbook of hydrocarbon and lipid microbiology* (p. 1555–1561). Berlin: Springer. https://doi.org/10.1007/978-3-540-77587-4_107

Wojewoda, D., & Russel, S. (2003). The impact of a shelter belt on soil properties and microbial activity in an adjacent crop field. *Polish Journal of Ecology, 51*, 291–307.

Yang, C., Wang, M., Cai, W., & Li, J. (2012). Bensulfuron-methyl biodegradation and microbial parameters in a riparian soil as affected by simulated saltwater incursion. Clean Soil Air *Water, 40*, 348–355.

Yang, C., Wang, M., Chen, H., & Li, J. (2011). Responses of butachlor degradation and microbial properties in a riparian soil to the cultivation of three different plants. *Journal of Environmental Sciences, 23*, 1437–1444. https://doi.org/10.1016/S1001-0742(10)60604-3

Young, A. (1997). *Agroforestry for soil conservation*. Wallingford, UK: CAB International.

Zak, D. R., Holmes, W. E., White, D. C., Peacock, A. D., & Tilman, D. (2003). Plant diversity, soil microbial communities, and ecosystem function: Are there any links? *Ecology, 84*, 2042–2050. https://doi.org/10.1890/02-0433

Zak, J. C., Willig, M. R., Moorhead, D. L., & Wildman, H. G. (1994). Functional diversity of microbial communities: A quantitative approach. *Soil Biology and Biochemistry, 26*, 1101–1108. https://doi.org/10.1016/0038-0717(94)90131-7

Zelles, L. (1999). Fatty acid patterns of phospholipids and lipopolysaccharides in the characterization of microbial communities in soil: A review. *Biology and Fertility of Soils, 29*, 111–129. https://doi.org/10.1007/s003740050533

Zhang, Q., Zhang, M., Zhou, P., Fang, Y., & Ji, Y. (2018). Impact of tree species on barley rhizosphere-associated fungi in an agroforestry ecosystem as revealed by ^{18}S rDNA PCR-DGGE. *Agroforestry Systems, 92*, 541–554.

Zomer, R. J., Neufeldt, H., Xu, J., Ahrends, A., Bossio, D., Trabucco, A., . . . Wang, M. (2016). Global tree cover and biomass carbon on agricultural land: The contribution of agroforestry to global and national carbon budgets. *Scientific Reports, 6*, 29987. https://doi.org/10.1038/srep29987

Zuo, Y., & Zhang, F. (2009). Iron and zinc biofortification strategies in dicot plants by intercropping with gramineous species: A review. *Agronomy for Sustainable Development, 29*, 63–71. https://doi.org/10.1051/agro:2008055

Study Questions

1. Explain why soil health is important for agricultural sustainability, food security, and environmental quality.

2. Explain why agroforestry is a better option for carbon sequestration and other ecosystem services compared with monocrop management or forestry.

3. Which soil pore size classes are most important for water transmission and what are their size ranges? Which soil pore size classes are most important for plant-available water storage and what are their size ranges? How do land management and agroforestry practices change the volume fractions of these pore size classes?

4. Describe what is meant by soil heat capacity and soil thermal conductivity. How do land management and agroforestry practices affect these soil thermal properties?

5. Discuss the difference between soil hydraulic conductivity and water infiltration. How do land management and agroforestry practices change these soil hydraulic properties?

6. Identify the major functions of the various soil microbial and meso- and macrofaunal communities in a well-managed agroforestry practice.

7. List some soil biological activities that can be measured and used as indicators for soil health assessment of agroforestry systems.

8. Discuss how management practices implemented in agroforestry influence soil biodiversity, ecosystem services, and overall soil health. Consider the comparative impacts of "conventionally managed" systems (i.e., crop production fields, pastures, timber plantation monocultures, etc.) in your discussion.

9. Describe how agroforestry can help improve soil chemical health.

10. Discuss the advantages and disadvantages of agroforestry for satisfying crop nutrient demand for food security.

13

Thomas W. Bonnot, Joshua J. Millspaugh,
John H. Schulz, Dirk Burhans,
Daniel C. Dey, and
W. D. "Dusty" Walter

Managing for Wildlife in Agroforestry

Agroforestry provides an opportunity to link timber production with benefits for wildlife, including improved habitat and more connected landscapes. Many modern agricultural practices such as monocultural production (Soule, Carre, & Jackson, 1990), the use of pesticides and herbicides, and increases in field size have proven detrimental to many wildlife species (Warner & Etter, 1985). Because of the increasing sentiment for diverse ecosystems that are economically viable, agroforestry may be an option for some private landowners interested in linking commodity production with wildlife benefits (Husak, 2001). Many landowners view the presence of wildlife as an important byproduct to the production of wood products on their land, especially in the southeastern United States (Allen, Bernal, & Moulton, 1996). In addition to wildlife benefits, agroforestry offers diverse environmental, aesthetic, and recreational opportunities over many other modern agricultural practices through diversity in plantings, both structurally and spatially (Jose, 2009).

The types of wildlife species that benefit from agroforestry practices vary among regions and ecotypes and are dependent on the landscape context, size of the agroforestry area, and the types, spatial configuration, and age of plantings. For example, wildlife species that are sensitive to habitat fragmentation (e.g., spotted owls [*Strix occidentalis*] and the pileated woodpecker [*Dryocopus pileatus*]) will not benefit in the same capacity as those adapted to fragmented habitats (e.g., blue jay [*Cyanocitta cristata*] and white-tailed deer [*Odocoileus virginianus*]). The wildlife benefits derived from agroforestry practices are driven by the goals and investment of the landowner and constrained by the habitat and landscape features of the proposed project. To maximize the feasibility of meeting those goals, traditional agroforestry plantings can be slightly modified and selected to meet the needs of wildlife species with little impact to the production of wood products or field management. By altering traditional agroforestry plantings and selecting tree and shrub species carefully, landowners can also develop a new wildlife product and diversify their returns. The wildlife production gained from the conversion to an agroforestry practice can allow landowners to better balance the compromise between the competing interests of agriculture, wildlife, and potential wood products. That balance can be achieved through possibilities such as cost-share from national conservation programs and earning potential through conservation easements and lease hunting opportunities, not to mention the aesthetic benefits perceived by the landowner.

There is an inherent compromise when balancing the production of wood products, crop income, and wildlife abundance in agroforestry practices. It is often easier for landowners

North American Agroforestry, Third Edition. Edited by Harold E. "Gene" Garrett, Shibu Jose, and Michael A. Gold.

to see the direct benefit of a traditional agriculture practice through a practical cost versus income comparison compared with an agroforestry practice where the aesthetic, ecosystem health, and alternative income benefits may be important considerations. For example, when converting traditional agricultural land into agroforestry plantations, there may be a temporary loss of productive land with associated losses of income. However, these short-term losses can be reduced or offset by considering the short- and long-term earning potential of wildlife hunt leases and long-term investment of wood products. If the landowner resides in a region where lease hunting is popular or where a highly desirable huntable species is common, the short-term loss of agricultural income can be at least partially reclaimed. If a goal of the conversion to agroforestry is to create an alternative revenue stream through hunt leases, landowners must create a plan that accounts for not only increasing the population of the main species or group of species of interest (e.g., male deer with large antlers) but considers how wildlife diversity and abundance changes with time.

This chapter is focused on wildlife benefits in agroforestry settings. Wildlife production is a valuable byproduct that can be complementary to the goals of tree and crop production. Wildlife is also an important consideration for landowners because it may diversify income opportunities and offset costs, particularly early in the initiation of agroforestry practices, with only modest alterations to plantings and management techniques. This chapter describes the response of wildlife to a diversity of agroforestry settings, but it also discusses general ecological issues, such as scale and habitat fragmentation, that should be considered when attempting to maximize wildlife benefits. Finally, it concludes with some general recommendations to improve wildlife benefits in agroforestry settings.

Ecological Considerations for Wildlife in Agroforestry Settings

Having realistic expectations when managing for wildlife in agroforestry settings necessitates an understanding of ecological concepts such as habitat and species interactions and how they relate at different spatial scales. This section provides an overview of these issues to provide context for our recommendations and conclusions about improving wildlife benefits in agroforestry settings.

Scale, Patch Size, and Fragmentation

The effectiveness of management for wildlife populations will be dependent in part on considerations of scale (Donovan et al., 2002). Often the scale of agroforestry practices is small enough (e.g., 4–8 ha) that benefits to many wildlife species are not attainable or practical. For some forest and grassland songbirds, even large habitat patches will have lower benefits if not part of a larger forested or grassland landscape (Fitzgerald, Herkert, & Brawn, 2000; Fitzgerald & Pashley, 2000). Birds may not be present in small patches, or if present, may not exhibit a positive population growth rate due to lowered reproductive success from brown-headed cowbird (*Molothrus ater*) parasitism and nest predation that is often experienced in landscapes that are fragmented (Robinson, Thompson, Donovan, Whitehead, & Faaborg, 1995). Mammalian populations may experience similar fates if habitats are isolated. In west-central Indiana, species richness of small mammals was highest in continuous forest sites and increased with area (Nupp & Swihart, 2000). Furthermore, one needs to consider the normal density of wildlife within the context of the size of the agroforestry area. For example, if our goal for northern bobwhite quail (*Colinus virginianus*) density was one per 0.8 ha, this density implies that one covey of bobwhite might be produced per 8.1 ha, which might be acceptable to some landowners. Given a 44% harvest rate, including crippling (Roseberry & Klimstra, 1984), the additional covey would only result in approximately three additional quail in the bag during an average hunting season. Within the context of quail population and habitat management, the size and location of habitat management efforts are serious considerations if increased hunting opportunity is one of the desired outcomes (Schulz, Millspaugh, Zekor, & Washburn, 2003).

Benefits to wildlife at the level of the agroforestry plot may even be negated by factors operating at larger scales. Elliott and Root (2006) believed that the linear and fragmented makeup of their riparian forest sites in northeastern Missouri resulted in a small mammal species assemblage dominated by habitat generalists. Also in northeastern Missouri, Peak, Thompson, and Shaffer (2004) concluded that buffer strips did not provide better songbird breeding opportunities because of the overriding influence of agricultural landscapes on increased nest predation. Similarly, Davros, Debinski, Reeder, and Hohman (2006) reported that landscape context was a critical determinant of butterfly use of filter strips in southwestern Minnesota. Such studies demonstrate the importance of landscape context; the wildlife benefit that may be derived from agroforestry is directly related to the surrounding habitat matrix.

Woodlots within a traditional agricultural setting have fewer wildlife benefits. For example, in agroecosystems dominated by intensive monoculture row-crop farming (e.g., corn [*Zea mays* L.] or soybean [*Glycine max* (L.) Merr.]), wildlife benefits will probably be reduced (Tewksbury et al., 2006). In particular, there is an increase in nest predation in areas within an agricultural matrix (Tewksbury et al., 2006) when compared with other environments. Using a meta-analysis, Chalfoun, Thompson, and Ratnaswamy (2002) found that small-scale edge and patch effects were most common in forests where agriculture was the dominant land use. Higher predation rates might relate to increased predator densities in these areas because additional food is available in agricultural settings (Dijak & Thompson, 2000; Marzluff, Gehlbach, & Manuwal, 1998). Thus, landscape context is important in determining the overall benefits that may be derived from agroforestry practices.

Where possible, the use of larger or more connected restoration fragments is more desirable than the creation of small, unconnected fragments. Twedt and Cooper (2005) indicated that edge effects on nest survival for some forest birds were less severe where reforestation was widespread in the landscape. Because cowbird parasitism typically decreases with distance from forest edges, they recommended reforestation for forest birds near large preexisting forest tracts rather than near small preexisting patches (Twedt & Cooper, 2005). In a survey of the Mississippi Alluvial Valley, researchers found that most remaining forest patches were small (<1,012 ha; Twedt & Loesch, 1999). They suggested that, at least for forest songbird species, reforestation efforts should be concentrated on large tracts, either by adding or linking to existing forested patches of land. For amphibians, fragmentation of natural habitats limits dispersal while decreasing opportunities for wetland colonization (Semlitsch, 2000). These studies collectively underscore the importance of habitat connectivity within a larger spatial framework. Agroforestry offers an opportunity to minimize the negative consequences of fragmentation by reducing habitat isolation (Allen, 1990) provided plantings are well thought out and well connected with other habitats.

Minimizing habitat isolation also helps prevent predator traps, which result when prey species are attracted to isolated patches of habitat, which in turn increases predator use of those sites (Fretwell & Shipley, 1981), and increased amount of edge, which makes prey more vulnerable (Wilcove, 1985). Agroforestry plantations could result in predator traps because sometimes these areas are isolated and maintain high edge to interior ratios, which favor predation (Tewksbury et al., 2006). Such areas are sometimes attractive to prey species, such as migrating birds. For example, Kelly, Dyer, and Chesemore (1990) described agroforestry sites in the San Joaquin Valley in California as "biological magnets" for birds. More research is needed to better understand whether agroforestry sites that attract wildlife are actually predator traps.

Source and Sink Populations

Wildlife populations residing in agricultural landscapes often exhibit source–sink population dynamics. In such cases, sinks (e.g., small marginal patches of habitat) are supported and sustained by immigration from sources (e.g., larger high-quality patches of habitat; Pulliam, 1988). In this context, agroforestry fields might be habitat sinks for wildlife species. The characteristics of agroforestry sites, such as the presence of vegetation edges, might contribute to the presence of predators. Increasing numbers of predators along with increased nest predation (Hoover, Tear, & Baltz, 2006) and cowbird parasitism can negate the reproductive success of animals within agroforestry environments. Thus, wildlife abundance in agroforestry settings might be determined more by the availability of surrounding sources than production of wildlife from within the area.

Although more research is needed to address the hypothesis that agroforestry settings may act as population sinks, work by the authors (Millspaugh and Dey, unpublished data, 2004) with eastern cottontail rabbits (*Sylvilagus floridanus*) supports the hypothesis. During the 3-yr period from 2001 to 2004, we used a combination of mark–recapture and telemetry studies to assess cottontail rabbit demographics (e.g., density, survival rates) and movements within 4–16.2-ha agroforestry fields. The bottomland sites were located in central Missouri and planted with swamp white oak (*Quercus bicolor* Willd.) and pin oak (*Quercus palustris* Münchh.) seedlings (Grossman, Gold, & Dey, 2003; Shaw, Dey, Kabrick, Grabner, & Muzika, 2003). When compared with control sites, we observed similar rabbit densities; however, we noted that rabbit survival rates in agroforestry plots were only one-third those of rabbits in control plots (Millspaugh, unpublished data, 2004). Predation was the major source of mortality of rabbits in agroforestry plots. Population growth rates were negative in agroforestry plots and positive in control plots. Our telemetry work demonstrated that rabbits from surrounding sites dispersed to agroforestry plots; once in agroforestry plots, there was no indication of further movement beyond the sites.

Wildlife Damage

Because of the food and habitat provided by agroforestry systems, they are susceptible to crop and tree damage from wildlife, both desired and invasive species. For example, ecosystem damage by feral swine (*Sus scrofa*) has become an issue for landowners in recent decades. Among agricultural producers, rooting by feral swine is the most widespread source of conflict (Campbell & Long, 2009). Campbell and Long (2009) summarized multiple studies on how feral swine can cause extensive damage to row crops and forest plantations, rooting up seedlings and consuming the roots of recently planted trees. However, eastern cottontail rabbits and white-tailed deer are probably the most common cause of damage in agroforestry practices. The two species routinely forage on tree and

Fig. 13–1. (A) Damage from deer rubbing antlers on tree; (B) evidence of deer browsing on tree; (C) damage from rabbit herbivory; (D) tree guard to protect against rabbit herbivory; and (E) chicken wire used to protect tree from wildlife damage. All photos taken within agroforestry stands from Plowboy Bend Conservation Area, Missouri (Shaw et al., 2003).

crop plantings (Figure 13–1); deer also rub their antlers on trees during the fall mating season, causing additional sapling mortality (Figure 13–1). Conover (2002) estimated that white-tailed deer losses to commercial timber throughout the midwestern and eastern United States were $1.6 billion annually. In Nebraska, Johnson and Timm (1987) reported that rabbits caused an estimated $2.2 million in loss to major field crops and destroyed nearly 200 ha of new forest plantations. Swihart and Yahner (1983) observed significant damage by eastern cottontails on trees planted for shelterbelts and windbreaks in Minnesota. In a southern Alabama silvopasture, Moore (2012) concluded that wildlife browse damage slowed the growth of loblolly pine (*Pinus taeda* L.) seedlings, ultimately delaying when livestock grazing could be implemented. Gordon (2015) noted that soybean was difficult to cultivate in their silvopasture due to white-tailed deer herbivory.

Damage caused by rabbits most often occurs in the winter, when rabbits routinely feed on buds, bark, shoots, and twigs of vines, shrubs, and trees, especially when the ground is snow covered (Haugen, 1942; Schwartz & Schwartz, 1995). During the summer, rabbit damage to woody plants is minimal because of the abundance of preferred grasses, legumes, and forbs. In winter, rabbits can cause severe damage to tree reproduction by pruning, barking, and girdling stems and shoots, which often increases seedling mortality (Geis, 1954; Meiners & Martinkovic, 2002). Trees with thin bark, young stump sprouts, and seedlings are particularly vulnerable. In Missouri, herbivory to oak (pin and swamp white) seedlings primarily consisted of barking, girdling, and shoot clipping by rabbits.

There are various options for minimizing wildlife damage to agroforestry plantings beyond obvious efforts to trap, remove, or cull problem individuals. One option is through habitat management. In Missouri, scientists identified that creating an unfavorable understory condition for rabbits in agroforestry plots promote regeneration efforts. Dugger, Dey, and Millspaugh (2003) observed that damage to planted oaks was substantially greater in plots that contained natural vegetation than in those that were planted in a redtop grass (*Agrostis gigantea* Roth). Because redtop grass cover was not good rabbit habitat in the winter, planted seedlings were more likely to survive the winter with little herbivory damage from rabbits. Even subtle differences in vegetation structure were important determinants of rabbit density and herbivore damage to tree seedlings (Dugger et al., 2003). In areas where rabbit herbivory is severe, fall mowing of natural vegetation to eliminate winter cover may be an alternative, but this needs to be evaluated in future studies. Similarly, others have also suggested planting non-target crops to divert browsing pressure (Swihart & Yahner, 1983).

Other methods for deterring browse damage relate to directly excluding wildlife from browsing crops or trees. Rigid plastic mesh tubes have been used with success to limit herbivory to young trees (Black, 1992). Similarly, shelter tubes have reduced rabbit and deer herbivory on a variety of hardwood tree species (Potter, 1988). However, one issue that requires consideration is the microenvironment within the tube (Potter, 1988; Sharrow, 2001). The area within the tube is warmer and more humid during the day (Potter, 1988), which may alter tree performance; compared with mesh tubes, Sharrow (2001) observed higher survival and increased growth of honey locust (*Gleditsia triacanthos* L.) and black locust (*Robinia pseudoacacia* L.) that had shelter tubes. In this case, 60-cm tubes were sufficient to protect against small herbivores and 180-cm tubes protected trees from deer. Repellents are a popular method for preventing mammalian herbivory because they are perceived as humane, easy to apply, and effective. For example, Williams and Short (2014) compared eight different commercial repellents against physical exclusion on levels of rabbit browsing of garden vegetation. Their results indicate that repellent usage could be a practical solution for deterring rabbit herbivory in agroforestry fields, but this has yet to be tested.

Implications for Management

When landowners are interested in increasing wildlife on their property, success improves if they clearly identify their goals (e.g., increased hunting for family and friends, additional income, and/or viewing new wildlife species). It is also necessary to identify short- and long-term wildlife objectives. By addressing these questions, it becomes possible to identify appropriate management strategies and select appropriate tree and shrub plantings. It also becomes possible to arrange plantings to maximize benefits to the wildlife of interest. Without explicit planning, a landowner may later be disappointed by a lack of return on their investment.

It is important to remember that not all wildlife benefit from agroforestry practices. Landowners are more likely to achieve success if habitat development and management activities focus on meeting the needs associated with a more specific group of wildlife. Landowners should consider the intended use of wildlife, such as viewing versus hunter harvest, when planning agroforestry practices. For example, landowners should recognize that as forests mature (i.e., succession), the wildlife species that occupy these forests will change. Early in the development of agroforestry habitats (0–15 yr), we can expect that early successional species (e.g., bobwhite, rabbits, field sparrow [*Spizella pusilla*]) will be benefited most (Figure 13–2). As the plantings mature (15–30 yr), there will be a transitional period as the trees mature and the ground cover starts to diminish. During this time period, species such as northern cardinals (*Cardinalis cardinalis*) and brown thrashers (*Toxostoma rufum*) will benefit most. Finally, as the trees mature (30–60 yr), we can expect late successional species such as wild turkey (*Meleagris gallopavo*) and gray squirrel (*Sciurus carolinensis*) to benefit most due to the production of mast, for example. Recognition that wildlife benefits and opportunities will change as plantings mature allows landowners to set and attain realistic goals and avoid later disappointment.

In addition to the issues discussed above about landscape context, other physical characteristics such as field size should be considered when setting goals for any agroforestry practice. Such factors inherently limit what can be accomplished and what can be expected at the site. For example, our discussion of scale issues and quail indicated that we might expect 0.8 quail per hectare in an optimally managed site. In the case of an 8-ha agroforestry site, this might mean an additional covey of quail. Although this benefit might be attractive to a landowner, there would be few corresponding economic benefits. In contrast, western Missouri agroforestry fields planted with sunflower (*Helianthus annuus* L.) as a cover crop allowed harvest of more than 50 mourning doves (*Zenaida macroura*) per hectare (Bonnot, Schulz, & Millspaugh, 2011). Therefore, if there is interest in economic returns through lease hunting opportunities, mourning doves might be a more appropriate option. However, it depends on the objectives of the landowner and the value placed on different wildlife species.

Landowners have multiple options for investing in wildlife benefits, and decisions should be directly tied to their objectives. Such objectives should be clearly stated at the outset because they will determine decisions about types of trees, distance between plantings, cover crops, and even field configuration. Some options might prove economically viable (e.g., a mourning dove lease program) whereas others might offer more aesthetic opportunities (e.g., wildlife viewing in general). It is important for landowners to recognize that agroforestry plantings are not static environments; setting of objectives must take into account changes in the vegetative structure of the area as the trees mature. For example, a mourning dove lease program may not be viable once trees mature and the area that may be planted to sunflower is diminished. As with any investment, both short- and long-term goals must be considered along with changing expectations with time. Lastly, there are also often undesirable consequences of attracting wildlife, including damage concerns.

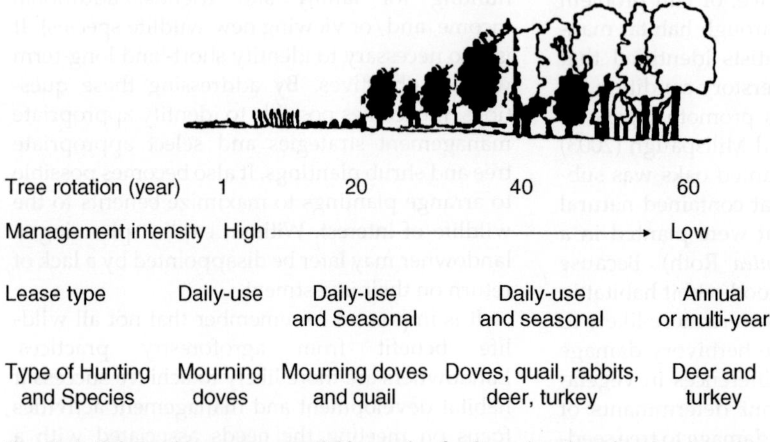

Fig. 13–2. Relationships among available species for lease hunting showing how tree growth through an agroforestry tree rotation changes the corresponding wildlife habitat and resulting lease hunting options.

Wildlife Responses to Agroforestry Practices

Knowledge of a species' ecology and habitat enables landowners to begin to integrate wildlife considerations (such as habitat management) to benefit the species along with their ongoing land management objectives. For this reason, many extension organizations produce documents that describe the habitat needs of many wildlife species of interest to landowners. For example, the University of Missouri's Center for Agroforestry and the USDA's National Agroforestry Center provide guides on managing habitats for species such as white-tailed deer, eastern wild turkey, honey bee (*Apis mellifera*), and songbirds. Although we do not describe them here, we recommend that landowners interested in managing for specific species contact their local extension office to inquire about such information. Many scientific studies have evaluated wildlife response to various agroforestry practices directly, and we summarize those findings below. In addition, researchers and managers have also explored the effects of certain restoration and conservation approaches that, while agroforestry was not the primary goal, are similar and may yield insights for wildlife in agroforestry applications. Thus, we include this information when relevant.

Alley Cropping

Alley cropping systems are designed to grow an annual or perennial crop between rows of high-value trees, like oak (*Quercus* spp.), pecan [*Carya illinoinensis* (Wangenh.) K. Koch], or walnut (*Juglans* spp.), until the trees are harvested or the alley crops are shaded out. Key attributes of these systems for wildlife include the diversification of vegetation structure and the forage and habitat provided by the trees and crops. Studies of alley cropping practices have found increased use by invertebrates, mammals, and birds. In Mississippi, greater numbers of white-tailed deer and mourning doves used a loblolly pine system than surrounding plots (Gordon, 2015; Manning, 2005). In Missouri, Bobryk et al. (2016) measured levels of diversity in invertebrates across various agroforestry systems. Although they found diversity was less in a pecan alley cropping system than in an adjacent forest, the alley cropping system had greater diversity than the silvopasture and soybean monoculture practices. In other studies, variations in tree–crop combinations and spatial arrangements have been shown to affect insect population density and species diversity (Jose, 2009).

The wide spacing between trees in these systems creates a system of corridors and protective cover important for movements of wildlife such as bobwhite and deer. Once trees begin maturing, precluding crops and reducing suitability for grassland species, the remaining understory vegetation, along with shade and mast from the trees, provides habitat and protective cover for woodland-associated wildlife, including species of salamanders, reptiles, and birds. For example, although eastern collared lizards (*Crotaphytus collaris*) occupy glade habitats, their populations depend on adjacent woodlands for dispersal (Templeton, Brazeal, & Neuwald, 2011).

Cover Crop

The choice of cover crop in alley cropping and other agroforestry systems greatly affects wildlife response. A variety of cover crops are available (e.g., shrub vs. forb vs. warm-season vs. cool-season grass) that have demonstrated benefits for different groups of wildlife. More traditional agricultural crops can attract popular game species. Deer and mourning doves showed increased use of alley cropping systems planted with corn and soybean (Gordon, 2015). Sunflower is an important crop for mourning doves across much of their range. For example, in a Missouri agroforestry study, Bonnot et al. (2011) found that increasing the amount of hectarage planted in sunflower resulted in increased mourning dove harvest by hunters. Similarly, sunflower plots in North Dakota were used by a greater diversity of migratory birds than other crops (Hagy, Linz, & Bleier, 2010).

In the last two decades, an increasing amount of research has shown the benefit of native warm-season grasses as habitat for birds. Common species used in planting include big bluestem (*Andropogon gerardii* Vitman) and switchgrass (*Panicum virgatum* L.). In southwestern Pennsylvania, Giuliano and Daves (2002) observed greater bird abundance, species richness for several birds, including field sparrow, grasshopper sparrow (*Ammodramus savannarum*), song sparrow (*Melospiza melodia*), chipping sparrow (*Spizella passerina*), and vesper sparrow (*Pooecetes gramineus*), within native warm-season grass fields than cool-season grasses. Similarly, Washburn, Barnes, and Sole (2000) discussed the benefits of these grasses to bobwhite and other wildlife in Kentucky, and Blank (2013) concluded that conservation plantings of native warm-season grasses created habitat for bobwhite in Maryland and Delaware.

Including legumes and forbs in grass planting mixtures can also benefit soils, wildlife, and even trees. Increased invertebrate biomass occurs in cover containing these plants (Bugg, Sarrantonio, Dutcher, & Phatak, 1991; Burger, Kurzejeski, Dailey, & Ryan, 1993). This effect provides an important food source for bobwhite during brood rearing (Burger et al., 1993). In addition, using legumes as cover can benefit tree plantings. Van Sambeek and Garrett (2004) found that the use of legumes led to better growth of hardwood seedlings and saplings compared with plantings with a ground cover of mowed or unmowed weeds. They also found that black walnut (*Juglans nigra* L.) grew better with some woody cover crops than with managed or unmanaged ground cover alone.

An important consideration when deciding cover for alley cropping will be any direct impacts to wildlife associated with crop harvest. Activities such as haying or mowing can harm individuals, destroy nests, or cause them to be abandoned (Ryan, Pierce, Suedkamp-Wells, & Kerns, 2002). For example in their study of bobolinks (*Dolichonyx oryzivorus*), Bollinger, Bollinger, and Gavin (1990) found that 85% of nests were lost to actions related to crop harvest. Even individuals that survive can face increased predation from decreased cover after haying. The use of warm-season grasses over cool-season species can help reduce these risks. Both Washburn et al. (2000) and Giuliano and Daves (2002) reported greater breeding success by wildlife in these fields because haying of warm-season grasses occurs later in the summer, allowing more nesting to be complete before harvest. Additionally, warm-season species are mowed at taller heights, which allows more cover from predators than cool-season grasses.

Silvopasture

A silvopasture combines trees, forage, and livestock in an intensively managed system. Silvopastures are typically less diverse than a natural forest understory but, similar to alley cropping, can provide quality habitat for wild turkey and other animals by incorporating native grasses and forbs. Unlike alley cropping, however, a reduction of vertical cover may affect use by wildlife, although few have studied wildlife responses to this practice. When comparing acoustic samples from a silvopasture, alley cropping, forest, and soybean field, Bobryk et al. (2016) observed fewer invertebrates in the silvopasture than in the alley cropping or forest but more than in the soybean field. They also noted no frog or toad species in any of the managed systems, only the forest. In their review of grazing impacts on grassland birds, Ryan et al. (2002) noted a range of tolerances for bird species to grazing that varied with factors such as the timing, frequency, and intensity of grazing. Species that tolerated grazing include mourning dove, burrowing owl (*Athene cunicularia*), horned lark (*Eremophila alpestris*), and brown-headed cowbird. Other species, such as Le Conte's sparrow (*Ammospiza leconteii*), common yellowthroat (*Geothlypis trichas*), sedge wren (*Cistothorus stellaris*), and mallard (*Anas platyrhynchos*), were noted as intolerant. The effects were both direct (e.g., nest mortality) and indirect given the vegetation structure, which determined habitat suitability. However, steps could be taken to reduce these risks (Ryan et al., 2002). Silvopastures may be suitable for arboreal species that require little to no understory. For example, Stainback and Alavalapati (2004) suggested that grazing cattle in a longleaf pine (*Pinus palustris* Mill.) plantation could create and enhance habitat for red-cockaded woodpeckers (*Leuconotopicus borealis*) in the South.

Riparian Buffer Strips

Riparian forest buffer strips are streamside wildlife plantings of trees, accompanied by a mixture of grasses, shrubs, and forbs, to maintain the integrity of a stream or shoreline and to reduce the impact of upland pollution and erosion (see Chapter 8). However, the unique habitat features of riparian buffers (proximity to water, food, structure, thermal cover, and so forth) are important to both aquatic and terrestrial wildlife. These areas are also used as movement corridors by birds, reptiles, and amphibians and presumably by small mammals (Clark & Reeder, 2007). For example, in an area dominated by extensive crop fields, the threatened Louisiana black bear (*Ursus americanus luteolus*) used riparian forest buffers to travel between hardwood patches (Anderson, 1997). Johnson and Buffler (2008) suggested that riparian buffers are the most critical habitat component for the greatest number of wildlife and fish species in the Intermountain West.

By reducing bank erosion and in-stream sedimentation and helping to maintain water temperatures, riparian forest buffers can improve habitat for fisheries. These areas provide coarse woody debris and detritus inputs that serve as food and habitat structure for aquatic species (Warrington et al., 2017). While detailing the benefits of agroforestry, Bentrup (2014) described a case study in the Tucannon River in Washington, where spring Chinook salmon (*Oncorhynchus tshawytscha*) runs hit a low of 54 fish in 1995 and juvenile salmonids were absent in lower reaches of the river. However, since 1999, more than 400 ha of riparian forest buffers and other restoration measures have been implemented, reducing summer water temperatures by about 12 °C. Young salmon are returning and are now using areas of the river that were previously too warm for them. Chinook adults have increased in number to 1239 in 2012. Because temperature is known to affect fish, amphibian, and invertebrate life histories, the benefits of riparian buffers are significant; stream biota could provide benefits in terms of mitigating some of the ecological effects of climate change on water temperature (Bowler, Mant, Orr, Hannah, & Pullin, 2012).

As an option in national conservation funding programs such as the Conservation Reserve

Program, implementation of riparian buffer strips has increased greatly in the last few decades. In their review of wildlife response to riparian buffers, Marczak et al. (2010) pointed out that for some taxa, buffers fail to provide the same quality of habitat as intact riparian forests. However, these practices have improved habitat for birds, amphibians, and even fish in many agricultural settings. For example, in an Iowa study of riparian forest buffers, Berges, Schulte Moore, Isenhart, and Schultz (2010) found that reestablishing native riparian vegetation in areas of intensive agriculture will provide habitat for a broad suite of bird species and that specific species will reflect successional stage, horizontal and vertical vegetative structure, and compositional diversity of the buffer vegetation. A 14-yr-old buffer had the most vertical and horizontal stratification, with a greater variety of tree and shrub sizes, and thus attracted a greater number of forest bird species known to be associated with forest and edge (i.e., Baltimore oriole [*Icterus galbula*], eastern phoebe [*Sayornis phoebe*], and red-eyed vireo [*Vireo olivaceus*]). The lack of large trees in the 9-yr-old buffer attracted species known to respond to shrub and edge habitat, such as the common yellowthroat, and American goldfinch (*Carduelis tristis*). And finally, savanna sparrows (*Passerculus sandwichensis*), and dickcissels (*Spiza americana*) were found within the 2-yr-old buffer that was dominated by grasses. Similarly, Peak and Thompson (2006) concluded that buffers in northern Missouri can increase bird diversity in narrow riparian forests through the greater microhabitat diversity found when combining buffer strips with adjacent forest. Often the diversity of birds is greater along forest corridors because of the interspersion of deciduous and evergreen species (Clark & Reeder, 2005). However, Clark and Reeder (2007) warned of riparian forest practices that create "hard edges" so that edge effects are often more pronounced, leading to avoidance by some species such as regal fritillaries (*Speyeria idalia*), bobolinks, dickcissels, and red-winged blackbirds (*Agelaius phoeniceus*).

In some cases, the benefits of buffer strips are constrained by the same landscape fragmentation processes affecting shelter belts. Tewksbury et al. (2006) reported that nest predation increased in forest buffers with more agriculture in the landscape in Idaho and Montana. In Maine, Vander Haegen and DeGraaf (1996) found higher predation in riparian buffer strips compared with riparian forests and recommended that buffer strips be >150 m to minimize detrimental effects. However, Clark and Reeder (2005) suggested that common forest wildlife species are often better adapted to the edge effects of riparian corridors embedded in forested landscapes (Peak et al., 2004).

Stream buffer strips also are beneficial to amphibians. In comparing total salamander abundance and amphibian species richness in western Oregon, Vesely and McComb (2002) reported positive benefits from riparian buffer strips. Peterman and Semlitsch (2009) found similar abundances of larval salamanders in buffers and controls in the Appalachia region and indicated that these areas can help address the impacts of roads. Other studies have reported benefits of riparian buffers to amphibians (Hawkes & Gregory, 2012; Perkins & Hunter, 2006). However, Marczak et al. (2010) concluded that buffers do not replace unaltered riparian forests for salamander habitat. They point to insufficient size and edge effects across buffer applications for inconsistencies in responses. Typically when buffer zones are determined to mitigate edge effects, they are based on criteria that protect aquatic resources alone and do not include movements and habitat for wildlife (Crawford & Semlitsch, 2007). For example, in southern Appalachian streams, Crawford and Semlitsch (2007) identified an overall buffer width near 100 m to protect stream amphibians. In their review of riparian forestry impacts on amphibian populations, Olson, Anderson, Frissell, Welsh, and Bradford (2007) suggested buffers of 40–100 m. These buffers are sufficient to preserve the majority of aquatic–riparian-dependent species such as salamanders (Peterman & Semlitsch, 2009).

Windbreaks and Shelter Belts

Shelter belts have a long and complicated history of use as a habitat management tool. Shelter belts (also called windbreaks) are planted perpendicular to the prevailing winds to help reduce field, building, and livestock exposure to winds and thereby reduce wind erosion and evapotranspiration of soil moisture. Shelter belts can provide cover, foraging sites, and travel corridors for wildlife, but their benefits for some taxa can depend on the buffer's characteristics in addition to varying with season, succession, and the surrounding landscape.

Generally, many wildlife species respond favorably to shelter belts in agricultural settings. These practices offer more constant habitat in the shifting landscapes, and the woody structure and associated understory increase niche diversity (Bentrup, 2014). For example, generalists, such as eastern cottontail rabbits and white-tailed deer, are likely to benefit from shelter belts because of the interspersion of habitats that provide food (e.g., row crops) and cover (e.g., tree plantings; Allen et al., 1996). These mammals will feed on fruits or stems and twigs of shrubs and trees, use the shrubs for shelter and to protect their young during rearing seasons, and find warmth from cold winter winds inside the plantings (Janke, 2016). Brandle, Hodges, & Zhou (2004) observed greater density and diversity of insect populations in windbreaks because they provided varied microhabitats for life-cycle activities and a variety of hosts, prey, pollen, and nectar sources. Bentrup (2014) highlighted examples where native bees and insect predators were greater in hedgerows and provided increases in crop pollination and pest control in adjacent crop fields. Birds make extensive use of windbreaks, nesting in trees and shrubs, eating soft or hard mast (fruits and nuts) produced by shrubs during fall and winter, seeking shelter from temperature extremes during winter or summer, and escaping predators throughout the year. Johnson and Beck (1988) identified many species of birds using shelter belts from North Carolina to Iowa.

Several researchers (Cable, 1991; Capel, 1988; Janke, 2016; Johnson & Beck, 1988) have discussed wildlife benefits from various shelter belt designs (Figure 13–3). The size of the shelter belt is often considered most important to bird diversity (Cassel & Wiehe, 1980; Schroeder, Cable, & Haire, 1992). Johnson and Beck (1988) reported that the densities and diversity of wildlife may be found in shelter belts that are larger with a diversity of structure, including deciduous and coniferous trees, shrubs, and a diverse understory of grasses and forbs (Figure 13–3). Johnson and Beck (1988) suggested that snags, a well-developed canopy layer, and limited overgrazing would also promote wildlife benefits in shelter belts.

The benefit of shelter belts for bird conservation has received considerable attention since the establishment of modern conservation initiatives. In all, research suggests that the response of birds to these linear patches of habitat is mixed. In an extensive review of the effects of various buffer applications on wildlife, Clark and Reeder (2005) summarized how these practices may have limited benefits for birds during breeding. For example, studies have shown that grassland birds such as dickcissels, grasshopper sparrows and bobolinks avoid habitats with woody vegetation (e.g., Adams, Burger, & Riffell, 2015; Ribic & Sample, 2001; West et al., 2016). In eastern Nebraska, Pierce, Farrand, and Kurtz (2001) reported that shelter belts favor forest-edge and generalist guild bird species, with negative impacts to grassland birds. In addition to generalists, shrub nesting and woodland species nest at greater densities in these areas than in surrounding row crops. Northern bobwhite, in particular, has been found in greater densities on farms with shelter belts (Bromley, Wellendorf, Palmer, & Marcus, 2002). However, a meta-analysis by Schlossberg and King (2008) in the Northeast found that shrubland birds are not "edge birds" as they actually avoid edges in favor of interior shrublands.

A more important issue for birds nesting in or near shelter belts is that reproductive success is often impaired because many of the generalist species that use these habitats also happen to be predators like raccoons (*Procyon lotor*), snakes,

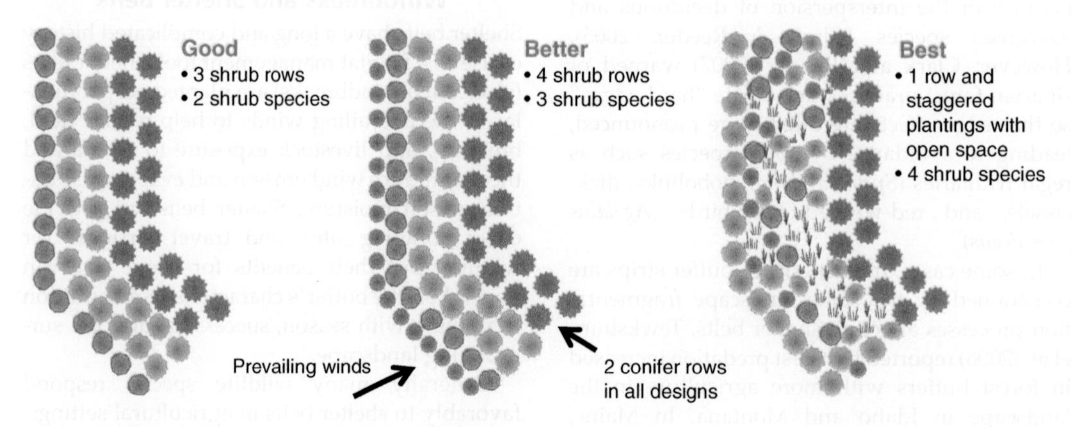

Fig. 13–3. Suggested designs of shelterbelts for wildlife use from Iowa State Extension and Outreach. The quality of a windbreak for wildlife increases with the diversity of plants, like shrubs, wildflowers, and grasses, and the size of the windbreak. Windbreaks with at least five or more rows provide the best habitat. Breaking up the rigid straight lines in typical plantings can also help protect wildlife from predators that tend to search for prey down linear features. Thinning and planting inside established rows of windbreaks is another way to increase diversity while retaining the function of outside rows of the windbreak (from Janke, 2016).

and raptors. Many studies have reported lower productivity in or near woody edges (Clark & Reeder, 2005). In southwestern Missouri, dickcissel and Henslow's sparrow (*Ammodramus henslowii*) nest success was lower near woody edges (Winter, Johnson, & Faaborg, 2000). In addition to raptors, the presence of mature trees provides perches for brown-headed cowbirds to parasitize nests. Winter et al. (2000) found that parasitism of dickcissel nests in southwestern Missouri was greater within 50 m of a woody edge. Despite these impacts during breeding, Evans et al. (2014) observed considerable use by birds in other seasons, concluding that by providing a direct source of winter food and cover resources, buffers such as windbreaks are an important nonbreeding habitat for wintering farmland bird populations. Janke and Gates (2013) identified that early succession woody cover is necessary for bobwhites during the nonbreeding season throughout their range.

These studies point to complex issues related to the utility of shelter belts for wildlife. Grassland birds may be negatively impacted during breeding but rely on them during winter. While the presence of predators may impact some species, they are still a valuable addition to the diversity of wildlife on the agricultural landscape. Furthermore, in row-crop settings, these areas are valuable to wildlife because they create areas of perennial vegetation that are less disturbed relative to surrounding annually changing crop fields.

Restoration of Bottomland and Floodplain Forests

Within the discipline of agroforestry there exist many potential benefits for wildlife in bottomland forests. Floodplain forests occur in the low, flood-prone areas along rivers. In their natural state, they were sinks for sediments and nutrients, provided temporary storage of floodwaters, stored significant amounts of carbon in tree biomass and soils, and provided extensive habitat for flora and fauna (Faulkner, Barrow, Keeland, Walls, & Telesco, 2011). Millions of hectares of historically forested floodplain have been lost or converted to agriculture in the 19th and 20th centuries (Abernethy & Turner, 1987; Twedt, 2006, and references therein). The capability of the remaining forests to provide ecosystem services has also declined because of forest fragmentation, altered hydrology, sedimentation and water pollution, invasive exotic plants, and indiscriminate timber harvesting (Gardiner & Oliver, 2005). Government incentives and other opportunities have resulted in widespread restoration potential for bottomland forests (Haynes, 2004). It is estimated that 300,000 ha of cleared land has been restored to forest cover during the 30 yr through 2005 (King, Twedt, & Wilson, 2006). More than 12,000 additional hectares have been restored since 2004 under the Ivory-Billed Woodpecker Recovery Plan (U.S. Fish and Wildlife Service, 2010). Agroforestry plantings can be located and designed to offer key ecological attributes normally provided by these forests (Dosskey, Bentrup, & Schoeneberger, 2012).

Bottomland forests can be incredibly beneficial to wildlife. Deer, waterfowl, turkey, and squirrels depend heavily on hard mast such as acorns and pecans (McShea & Schwede, 1993; Norman & Steffen, 2003); soft mast trees such as hackberry (*Celtis occidentalis* L.) provide fruit for wintering species such as cedar waxwing (*Bombycilla cedrorum*); even trees and shrubs that do not produce fruit provide habitat structure, which is considered a crucial determinant of bird use (Hamel, 2003; James, 1971; Twedt & Best, 2004). Particularly for songbirds, floodplains are valuable habitats for obligate riparian species and for some declining species (Inman, Prince, & Hayes, 2002). Bottomland locations may provide habitat for a variety of songbirds of conservation interest throughout successional stages, from Henslow's sparrow, which may use early reforesting grasslands, to cerulean warbler (*Dendroica cerulea*), which uses mature bottomland forests, both of which have been petitioned for federal threatened or endangered species listing (Burhans, 2001; Burhans, Dearborn, Thompson, & Faaborg, 2002). In addition, other birds using floodplains are showing long-term population declines or are considered as Partners-in-Flight Priority Species (Rich et al., 2004).

While the advantages to wildlife of mast-producing trees such as oak and pecan may seem self-evident, recent research indicates that wildlife may benefit also from tree species that do not produce fruit commonly consumed by wildlife. More substantively, researchers in this system (Hamel, 2003; Twedt & Best, 2004; Twedt & Portwood, 1997; Twedt, Wilson, Henne-Kerr, & Grosshuesch, 2002) have stressed the structural benefits of faster growing tree stands to forest songbirds. Twedt and Portwood (1997) showed that the early onset of structure in such plantings allowed more songbird species to breed in young eastern cottonwood (*Populus deltoides* W. Bartram ex Marshall) plantings compared with similarly aged oak plantings, noting 36 species holding territories in cottonwoods compared with only nine species on oak plantings. Similarly, Hamel (2003) noted that for wintering songbirds, twice as many species were present in stands of fast-growing cottonwood plots than others; he attributed this to the addition of canopy-dwelling species in the latter stands. In another study, Twedt et al. (2002) noted that oak-dominated stands were not used by forest songbirds until the 10th yr after planting. Twedt et al. (2002) recommended planting oaks in combination with rapid-growing trees that promote quick stand development for early use by forest songbirds. Using fast-growing tree species, agroforestry practices can create early successional forest structure to function as buffer zones for bottomland forests (Dosskey et al., 2012). This approach can turn smaller, less-viable patches into effective interior habitat.

In early stages, however, bottomland restorations may provide habitat for songbirds and wildlife other than those using mature forest habitat. While Twedt et al. (2002) found lowered use of young oak plantings by songbirds overall, they noted that songbirds in the young oak plantings tended to be grassland species, which are considered some of the most important species for conservation (Robinson, 1997). Young oak-dominated plantings had less total "conservation value" for songbirds (as determined by an index using Partners-in-Flight prioritization scores), but birds using young oak stands, especially dickcissel, tended to be high-priority conservation species. Although Twedt et al. (2002) speculated that the use of young oak stands by species like dickcissel would similarly occur if the stands were simply left unplanted to old field succession, researchers in lower Missouri River floodplains (Burhans & Root, 2003, unpublished data) found that grassland songbird use of unmanaged bottomland old fields was limited to 1–2 yr after soil preparation. In contrast, fields planted in oaks using a cover crop of redtop grass were inhabited by grassland birds more than 5 yr post-planting. Additionally, the former study showed that oaks planted with the cover crop grew best (Dey, Lovelace, Kabrick, & Gold, 2004). Due to eventual succession and tree growth, oak-planted grasslands would obviously not remain ideal habitat for grassland birds indefinitely. Managers who wish to promote diversity for forest songbirds might wish to consider the recommendations of Twedt et al. (2002) and Hamel (2003); those who wish to provide habitat for earlier successional species such as grassland or shrubland songbirds could consider adding cover crops or shrubs that suit those wildlife species in the short to intermediate term.

One combination approach for establishing an early closed-canopy forest while retaining hard mast would be to plant oaks in combination with fast-growing species such as cottonwood. Stanturf and Gardiner (2000) and Stanturf, Schoenholtz, Schweitzer, and Shepard (2001) suggested the example of interplanting oaks between rows of cottonwood 1 or 2 yr after planting and then harvesting the cottonwood at age 10 yr to release the oaks. Twedt (2006) found that clusters of eastern cottonwood and American sycamore (*Platanus occidentalis* L.) planted within oak plantations led to increased stem density and greater maximum tree heights around clusters; however, this particular study (Twedt, 2006) did not test wildlife response to the cluster plantings.

It should be noted that stands of taller trees may be linked to higher incidence of cowbird brood parasitism. Twedt et al. (2002) found higher abundance of brown-headed cowbird and greater parasitism rates in older reforesting cottonwood stands than in oak stands and younger cottonwood stands. Findings by Clotfelter (1998), Hauber and Russo (2000), and Saunders, Arcese, and O'Connor (2003) indicated that cowbird parasitism declines with distance from potential perches such as snags or tall trees, which may explain why cowbird parasitism generally appears to be higher in forests in certain landscapes (Burhans, 1997; Hahn & Hatfield, 1995). Burhans and Root (2003, unpublished data, University of Missouri) also found less cowbird parasitism with increased distance to trees in lower Missouri River bottomland restorations. Presumably, because cowbirds appear to favor plantation forests having taller trees over other habitats, increases in cowbird abundance and parasitism such as these are not specific to the tree species planted. However, factors of scale also need to be considered in relation to cowbird parasitism and nest predation; studies have shown that regional and landscape effects, such as the amount of regional forest cover, may constrain parasitism at lower scales (Donovan, Thompson, & Faaborg, 2000; Thompson, Donovan, DeGraaf, Faaborg, & Robinson, 2002). Similarly, connecting small bottomland forest patches with tree corridors across croplands can help alleviate issues with fragmentation, facilitate the movement of forest wildlife between patches, and effectively increase the habitat viability (Dosskey et al., 2012).

Species of conservation concern including various bat species and swamp rabbits may benefit from agroforestry sites in bottomland forests. These forests are important roosting habitat for Rafinesque's big-eared bats (*Corynorhinus rafinesquii*) and southeastern myotis (*Myotis austroriparius*), species of concern whose current population statuses are unknown (Fleming, Jones, Belant, & Richardson, 2013). In southeastern Missouri, more bats were captured per hour on agroforestry sites (0.310 h^{-1}) than "natural" forest (0.204) and strip cover sites (0.217) (Warwick, 2003). Mean species richness for bats was also greater within agroforestry sites (4.61) than "natural" forest (4.39) and strip cover sites (3.84) (Warwick, 2003). Based on latrine site locations, Warwick suggested that swamp rabbits use agroforestry sites for food and natural forest remnants for other activities (e.g., loafing and resting). He concluded that, at his study site, agroforestry plantings were an important habitat component for swamp rabbits.

Greentree Reservoirs

Greentree reservoirs are bottomland hardwood forests that are artificially flooded during late fall and winter. Managing oaks for mast production in these systems can provide feeding areas for waterfowl (Fredrickson, 1980; Fredrickson & Heitmeyer, 1988). Greentree reservoirs are used extensively as a management tool by national wildlife refuges throughout the Mississippi Alluvial Valley and southeastern United States and now exist in more than 20 states. In a review of greentree reservoir management, Wigley and Filer (1989) reported that most greentree reservoirs were <100 ha in size, dominated by oaks, and provided <100 d of hunting. In these situations, bottomland hardwood forest stands are typically flooded every year through the waterfowl hunting season. At the end of the hunting season, the areas are drained. The flooding attracts species such as mallards and wood ducks (*Aix sponsa*) because of the availability of fallen acorns and other seeds during the hunting seasons. Additionally, invertebrate densities may increase in greentree reservoirs compared with naturally flooding bottomlands (Wehrle, Kaminski, Leopold, & Smith, 1995). Often, water control structures are installed to aid water management. Greentree reservoirs are typically managed so the area is flooded when the trees are dormant; however, in some areas, some trees are not dormant when flooding starts (Guttery & Ezell, 2006), which can result in the loss of mature trees. In southeastern Missouri, greentree reservoir managers are trying to mimic the natural hydrologic cycle in an attempt to keep trees drier during the growing season (Krekeler, Kabrick, Dey, & Wallendorf, 2006).

The majority of studies investigating wildlife use of greentree reservoirs involve waterfowl. In Arkansas, Christman (1984) reported that birds that forage in the understory were absent or in reduced density in greentree reservoirs compared with control plots. Birds that typically use the overstory canopy to forage were found in equal or higher densities than in control plots (Christman, 1984). These results led Christman (1984) to conclude that greentree reservoirs will typically have lower densities of nongame birds than unmanipulated habitats. It is generally agreed that greentree reservoirs are beneficial to wintering waterfowl, particularly during years when flooding does not occur (Rudolph & Hunter, 1964). Wood ducks and mallards in particular use greentree reservoirs extensively during the winter (Fredrickson, 1978; Fredrickson & Heitmeyer, 1988; Kaminski, Alexander, & Leopold, 1993). There are also associated waterfowl harvest benefits to greentree reservoirs (Rudolph & Hunter, 1964). Thus, there is a potential for income generated through lease hunt opportunities in bottomland situations. However, one must also consider potential impacts to timber production.

There are many adverse effects of greentree reservoir management to timber production, such as increased tree mortality, decreased mast production, reduced regeneration, and altered species composition (Deller, 1997; Wigley & Filer, 1989). For example, in Mississippi (Allen, 1980) and Arkansas (Guttery & Ezell, 2006) managers noted an increase in the density of overcup oak (*Quercus lyrata* Walter), a tree that is not commercially valuable; it also produces acorns that are not preferred waterfowl foods (Allen, 1980). Such issues have caused some (Young, Karr, Leopold, & Hodges, 1995) to suggest a 2-yr flooding cycle or even longer depending on tree species objectives. Thus, when considering the integration of agroforestry with greentree reservoir management, there are additional challenges associated with timber management and the potential to kill trees before they mature and become harvestable.

Income Associated with Wildlife in Agroforesty Systems

By considering and managing for wildlife habitat in agroforestry programs, landowners can diversify and increase income as well as reduce costs when implementing the practices. For example, agroforestry activities often overlap with the conservation practices of incentives programs that reward landowners for implementing them. In addition, increasing the abundance of wildlife on agroforestry lands enables landowners to lease hunting rights. Both approaches can provide revenue prior to and alongside income from timber management. In fact, profit potential from agroforestry practices has, in some cases, been competitive with agricultural crops and production forestry on marginal agricultural lands (Dosskey et al., 2012).

Financial Assistance for Conservation on Private Lands

The potential for landowners to offset costs when implementing agroforestry practices and potentially generate revenue for the wildlife habitat provided by those practices has expanded greatly with the recent emphasis on programs incentivizing conservation on private lands. With more than two-thirds of the United States held in private ownership, it is well understood by federal, state, and nongovernmental organizations that private lands are critically important to the conservation of the nation's soil, water, plant, fish, and wildlife resources. It is this realization that provides the basis for the array of conservation programs within and centered around what is known as the Farm Bill.

Overview of the Farm Bill

The Farm Bill is a series of acts periodically reauthorized by Congress to enhance agricultural productivity, rural economies, food security, and conservation on private lands. The first Farm Bill (the Agricultural Adjustment Act) was drafted in 1933, in the wake of the Great Depression and the Dust Bowl, to address the needs of America's farmers at that time. Provisions for habitat conservation were included for the first time in the 1985 version (Food Security Act). It created financial incentives for agricultural producers through the Conservation Reserve Program (CRP), which paid farmers to put sensitive and highly erodible land aside via 10–15-yr contracts. To date, the Farm Bill has seen many changes in the number and variety of conservation programs it offers. However, the voluntary, incentive-based program model has remained constant (North American Bird Conservation Initiative, 2015).

The additions and changes to conservation programs since 1985 reflect an evolution toward two overarching approaches to conserving private lands: environmental protection of lands and implementation of conservation on working lands. The first approach involves environmental protections through purchase of easements or long-term "rental payments" to landowners who remove land out of production and maintain and restore wetlands, grasslands, or other land conservation. Major programs under this approach have included CRP and the Wetland Reserve Program, which was initiated nationwide in the 1996 Farm Bill and helps landowners voluntarily restore, protect, and enhance wetlands and wildlife habitat. The second approach includes financial and technical assistance (often called *cost share*) to landowners for establishing and enhancing conservation management practices on working lands. The Environmental Quality Incentives Program (EQIP) is the most notable and widespread cost-share program (Figure 13–4). It started in 1996 with an emphasis on working lands programs, seeking to address environmental problems stemming from active agricultural production by offering incentives, both financial and technical, to farmers in exchange for adoption of conservation practices (Reimer & Prokopy, 2014). As of 2013, Farm Bill conservation programs together have resulted in the application of conservation practices to tens of millions of hectares of agricultural working lands, retirement of up to 14 million ha of marginal cropland, and restoration of >2.2 million ha of wetlands (Ciuzio et al., 2013).

Although agroforestry by definition is probably not listed under any conservation program, specific agroforestry practices that benefit wildlife do overlap with many of the incentivized practices across programs, and most of these practices can be utilized in more than one conservation program. The USDA is responsible for implementing the different Farm Bill programs through the Natural Resources Conservation Service (NRCS) and the Farm Service Agency (FSA). Below are listed details for some of the major conservation programs and how they may apply to agroforestry. Landowners interested in obtaining specific information on cost-share programs within their state should contact their local NRCS office.

Conservation Reserve Program. The CRP is a land conservation program administered by the FSA. In exchange for a yearly rental payment, farmers enrolled in the program agree to remove environmentally sensitive land from agricultural production and plant species that will improve environmental health and quality. Contracts for land enrolled in the CRP are 10–15 yr in length. The long-term goal of the program is to reestablish valuable land cover to help improve water quality, prevent soil erosion, and reduce loss of wildlife habitat (USDA, 2019). The CRP is the longest running conservation program, impacting millions of hectares of habitat in the last 25 yr.

In 2008, the CRP began targeting conservation practices related to agroforestry and wildlife habitat including tree plantings, shelter belts, riparian buffers, wildlife food plots, and upland bird habitat buffers (Figure 13–5). Within the CRP, the FSA also oversees initiatives related to selected

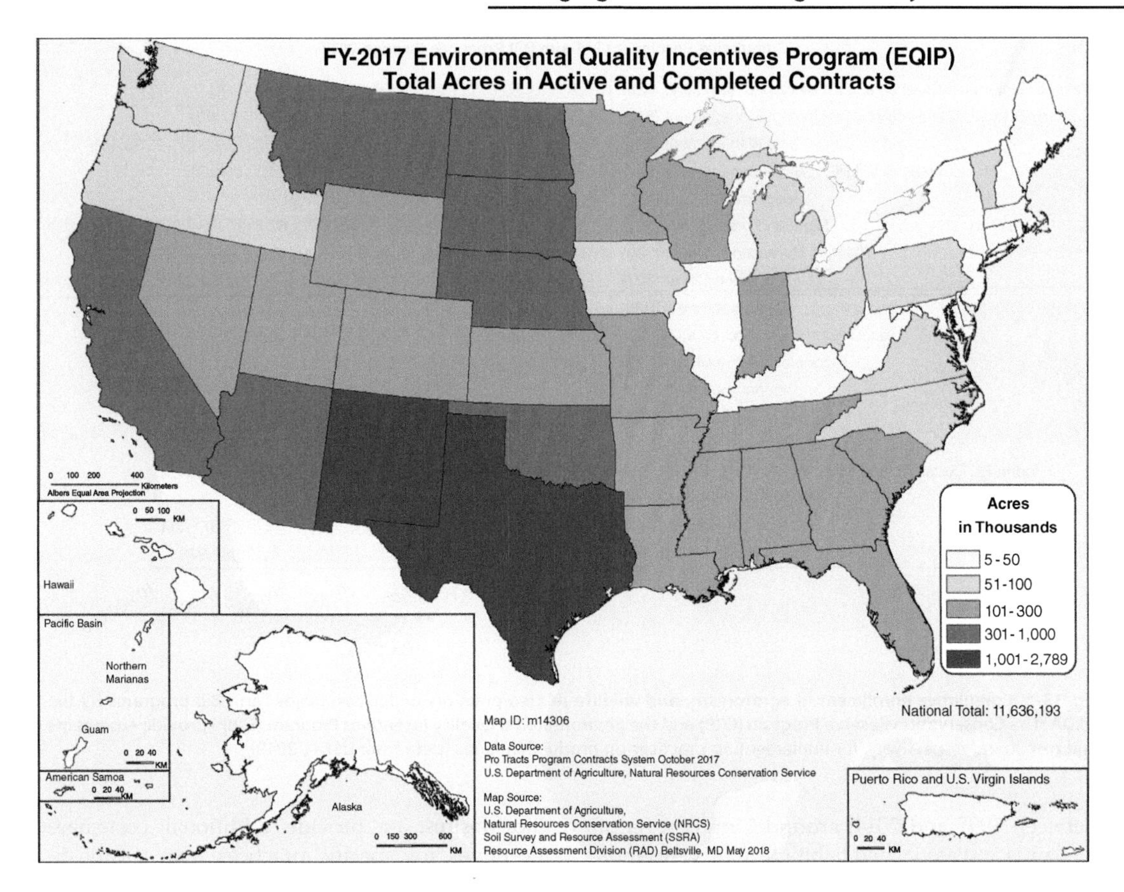

Fig. 13–4. Total acreage of land enrolled by state in the Environmental Quality Incentives Program (EQIP) in 2017. EQIP provides financial and technical assistance to agricultural producers through contracts up to a maximum term of 10 yr in length. These contracts provide assistance to help plan and implement conservation practices that address natural resource concerns to improve and conserve soil, water, plant, animal, air, and related resources on agricultural land and nonindustrial private forestland (USDA, 2017).

practices such as the Longleaf Pine Initiative or the Bottomland Hardwood Tree Initiative, which seek to improve the environment by restoring hardwood forests within floodplains. With this initiative, the financial benefits to restoring bottomland hardwoods include 10–15 yr of annual rental payments with an additional 20% incentive, payments covering 90% of the eligible costs of establishing the bottomland forest restoration practice, and a sign-up incentive up to $370/hectare (USDA, 2019).

Environmental Quality Incentives Program. This program provides financial and technical assistance to agricultural producers through contracts up to a maximum term of 10 yr in length. These contracts provide assistance to help plan and implement conservation practices that address natural resource concerns to improve and conserve soil, water, plant, animal, air, and related resources on agricultural land and nonindustrial private forestland. The EQIP has grown into the second largest conservation program in the United States, with more than $10 billion spent through 2017 and >80 million ha affected cumulatively since 1996 (USDA, 2019).

The EQIP focuses on many of the same conservation practices as the CRP, just on working lands, creating a potential opportunity for agroforestry systems to receive cost-share for implementing them. For example, since 2008 similar hectarages have been enrolled in wildlife and forest practices under EQIP and CRP (Figure 13–5).

Wetland Reserve Program. The Wetland Reserve Program (WRP) offers landowners the opportunity to protect, restore, and enhance wetlands, including bottomland hardwoods and riparian forests, on their property. The Agricultural Act of 2014 moved authority for the WRP easements to the Agricultural Conservation Easement Program (USDA, 2019). The WRP offered landowners an opportunity to establish long-term conservation and wildlife practices and protection, including riparian forest buffers and early successional habitat development.

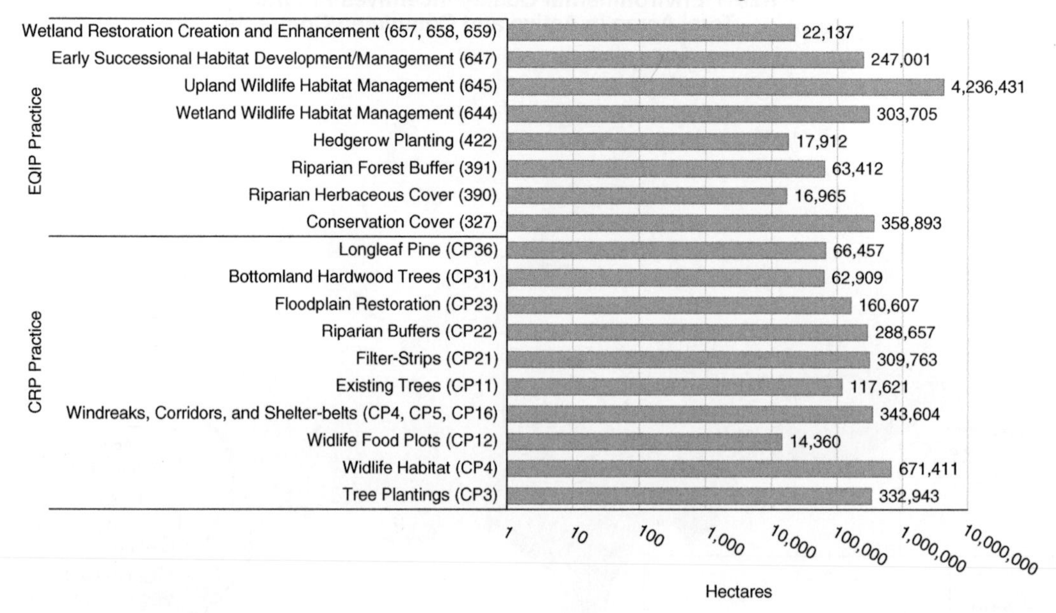

Fig. 13–5. Cumulative enrollment in agroforestry and wildlife related practices under two major Farm Bill programs by the USDA. The Conservation Reserve Program (CRP) and the Environmental Quality Incentives Program (EQIP) provide easements and cost share, respectively, for implementing practices on production lands (data from USDA, 2019).

Between 1995 and 2014, around 1 million ha of wetlands and associated habitats on private lands were restored through the WRP.

Wildlife Habitat Incentive Program. The Wildlife Habitat Incentive Program (WHIP) is a voluntary program for conservation-minded landowners who want to develop and improve wildlife habitat on agricultural land, nonindustrial private forest land, and tribal land. The NRCS administers WHIP to provide both technical assistance and up to 75% cost-share assistance to establish and improve fish and wildlife habitat. The Agricultural Act moved authority for WHIP to the Environmental Quality Incentives Program (EQIP).

Private Programs and Support

Over the years, several nationally recognized private groups have been organized to provide support for wildlife habitat management activities. A few of the prominent national groups include: National Wild Turkey Federation, Quail Unlimited, Ducks Unlimited, and Pheasants Forever. The main focus of these groups is habitat development, management, and maintenance for the wildlife species of interest. Often support is provided by placing interested landowners in touch with professionals, such as private land biologists working for the state, who will assist in designing habitat appropriate to those landowners' needs and desires, or provide additional cost-share incentives for specific practices (e.g., establishment of native warm-season grasses). In addition, these private groups often have seed mixes available at reduced costs and provide cost support for the development of local wildlife habitats.

Impacts of Conservation Incentives on Forests and Wildlife

Although a comprehensive review of the impacts of agroforestry practices is provided elsewhere in this chapter, there are examples where practices, implemented as part of conservation programs, benefited wildlife (see Johnson, 2005; Jones-Farrand, Johnson, Burger, & Ryan, 2007; Reynolds, 2005). For example, the loss of bottomland hardwoods diminished the Louisiana black bear's population to roughly 100 bears by the 1950s. Listed federally as a threatened species in 1992, the WRP helped reverse the bear's decline, with the first documented litter found on a WRP easement in 2004. With continued utilization of WRP habitat, the bear's population in Louisiana, Arkansas, Mississippi, and Georgia is now estimated at 500 bears and growing steadily (USDA, 2019). Beatty et al. (2014) found that midcontinent mallards selected wetlands near WRP easements and wetlands with high conservation easement area in the surrounding area, demonstrating that conservation easements have the potential to provide habitat for migratory birds throughout the nonbreeding period in the midcontinent region. Blank (2013) reported that bobwhite abundance was strongly positively associated with proportion of cover on CRP land in the landscape. Finally, the Conservation Effects Assessment Project is a multiagency effort to quantify the effects of USDA conservation practices and programs on fish and wildlife in landscapes in the United States (USDA, 2019). Much of the literature covered in this chapter stems from this effort. A full list of the evaluations and results can be found on the USDA website.

Income through Wildlife Hunting

Landowners willing to invest in wildlife habitat management and expend the energy in establishing pay-to-hunt arrangements can often generate an immediate revenue stream while waiting for timber harvest. Depending on location, hunting lease income may often pay annual property taxes with some money left over (Kays, 2000). However, there are several challenges for the landowner to consider before implementing a pay-to-hunt operation or lease hunting. Most importantly, a landowner must have reasonable and flexible expectations. The type and quantity of game animals depend on the existing local conditions (e.g., surrounding landscape conditions), and the available type and quantity of game species will evolve as the agroforestry trees move through different successional phases and change the surrounding habitat (Figure 13–2). Also, the most profitable pay-to-hunt operations are often the most labor and management intensive for the landowner. Our objective, therefore, is to provide a brief description of the range of available lease hunting options and considerations for establishing hunting leases.

Lease Hunting, Hunting Leases, and Pay-to-Hunt

Lease hunting is a scenario in which hunters purchase or exchange commodities to obtain trespass rights from a landowner. Depending on locale or tradition, it can be regarded as a pay-to-hunt operation, game farm, hunt club, or game lease. Although lease hunting has been popular in southern states and states containing minimal public hunting lands, it is becoming more popular across the entire United States as landowners look for greater income diversity (Masters, Bidwell, Anderson, & Porter, 1996) and hunters desire unique, predictable, or more high-quality hunting opportunities (Miller & Vaske, 2003; Schulz et al., 2003).

It is important for a landowner to consider how lease hunting opportunities stemming from their agroforestry programs could provide a valuable service to hunters, given recent patterns. Most hunters live in or near urban areas, and these urban residents have multiple demands on their available leisure time (Cordell & Betz, 2000). Survey data show that the remaining upland bird hunters travel farther and farther from home and make numerous trips out of state to find suitable hunting opportunities (Brown, Decker, Siemer, & Enck, 2000; Duda, Bissell, & Young, 1998). Public hunting land is also increasingly overcrowded, especially in states with little public land (Schulz et al., 2003). In addition, as smaller traditional working farms are consolidated into larger corporate farms or subdivided in smaller rural estates outside of a major metropolis, opportunities to hunt free on private lands are rapidly diminishing. Most hunters, private landowners, and state fish and game staff are just beginning to appreciate the importance of fee hunting to the future of hunting overall. Fee hunting could become a more important tool to help slow or reverse the long-term decline in hunter numbers (Brown et al., 2000) by providing additional hunting-related outdoor opportunities (Cordell & Betz, 2000; Cordell & Super, 2000), and providing opportunities near expanding urban and suburban sectors of the population would be advantageous (Schulz et al., 2003).

The second most important factor determining the success of the lease hunting endeavor is the "quality" of the experience provided by the landowner. Although difficult to define specifically, a quality lease hunting experience may consider one or more of the following items: cost of the lease, proximity to where hunters live or travel time to leased land, abundance and/or variety of game animals, real or perceived competition with other hunters, available trophy class animals, hunter safety, opportunities for companionship or fellowship, facilities to clean game animals, amount of hunter restrictions, camping or lodging facilities, and numerous others.

Third, a landowner should plan in advance for changing leasing opportunities throughout the timber rotation component of the agroforestry operation (Figure 13–2). Early during tree establishment the landowner will have greater opportunities to focus on small game species (e.g., mourning doves) that can be legally concentrated using managed lure crops like sunflowers. As the trees grow and agricultural land becomes more shaded by the timber, hunting opportunities will begin to shift away from early successional game animals to species like wild turkey or white-tailed deer along with a decrease in the intensity in management of wildlife habitat.

Types of Hunting Leases

A hunting lease is usually defined as an agreement between a landowner and a hunter (or group of hunters) where the right to trespass and hunt is granted for a particular time and fee (Masters et al., 1996). It is a simple business agreement between the person who owns and controls the land and another person who wants to use the land, or it can be a complex legal document requiring professional assistance and development. It can take numerous forms ranging from an informal oral agreement over a handshake to a clear and explicit legal document signed before a notary public. General categories include non-fee access, an exchange of services, a daily hunting lease, short-term or seasonal hunting lease, annual or multiyear lease, and broker/outfitter lease (Masters et al., 1996; Rempe & Simons, 1999; Stribling, 1994).

Non-fee access arrangements provide access to land for hunting with informal verbal or written permission; it is the easiest and often the first option for many landowners. For the landowner, this option may help manage nuisance animal populations (e.g., white-tailed deer) and help foster goodwill in the local community. Also, the reduction in deer populations may help provide a corresponding reduction in damage to agroforestry plantings, particularly those early in development. These ad hoc arrangements provide several non-monetary landowner benefits such as reduction in vandalism because of increased presence on rural land, potential help with seasonal farm work, and increased social networking. These arrangements are often more common in rural or rural–small town areas where hunters are more aware of farm-related issues and concerns.

Exchange of services arrangements (i.e., quid pro quo) are situations where something is given in return for something else. As the name implies, this type of arrangement allows hunting in exchange for monitoring the land for trespassers, helping with farming operations (e.g., baling hay, fixing fences, or posting signs). Without money changing hands, both parties benefit by exchanging or bartering services with each other. The major difference between exchange of services and non-fee access is the explicit expectation of rendering a specific service by the hunter for the privilege to hunt. These arrangements can be informal or formal agreements.

Masters et al. (1996) and Mozumder, Starbuck, Barrens, and Alexander (2007) described four general groups or categories of fee or lease hunting:

Daily lease. Daily leases work best for hunting situations with a relatively short season that can accommodate numerous hunters on a relatively small patch of ground (e.g., mourning dove hunting or pen-reared bird hunting). Daily leases are much more labor intensive for the landowner (Kays, 2000) but have greater opportunity for increased income.

Short-term or season lease. Short-term or seasonal leases work best for deer or turkey hunting and involve considerably less direct or immediate land or hunter management by the landowner. Landowners can have separate leases for different seasons and game species and for different groups of hunters.

Annual or multiyear lease. Annual or multiyear leases usually involve an agreement between a single landowner and a hunt club or group of friends willing to share the cost of having a long-term hunting spot available.

Broker or outfitter lease. Broker or outfitter leases involve a middle-man or broker who rents all the hunting rights from a landowner (or series of landowners) and subleases to individual hunters by species or season. For many landowners it is easier to deal with one individual who then manages (or subleases) all the hunting-related details of the lease hunting opportunities to individual hunters (e.g., LeaseHunting.com, Hunting Lease Magazine). Several nongovernmental organizations provide this service (e.g., Quail Unlimited) along with numerous private organizations or networks of landowners (Lease Network).

Although popular among many hunters and landowners, an informal oral hunting agreement often leads to misunderstandings and potential legal difficulties later. A signed and professionally prepared legal document stating all the payments, terms, expiration dates, and mutual agreements is the best way to ensure that the rights and privileges of the hunter and landowner are understood by all parties (Stribling, 1994). In other words, pay hunting is a business and a hunting lease is a business arrangement outlining the terms of the agreement.

Liability and Insurance

Although less of an objective of many U.S. citizens today, private land ownership is an important component of our American heritage. Outdoor recreation is still a major part of our culture, and access to private land plays a critical role now and into the future. However, landowners may have numerous concerns about allowing access to their lands, including past problems or negative perceptions about recreational users, proximity of residence to the location of recreational activities, illegal hunting activities, personal safety, and inconsistencies with long-term land management objectives (e.g., agroforestry, crop farming, ranching). Landowners may also be concerned with certain recreational activities (e.g., hunting, all-terrain vehicle trail riding), and potential liability issues (Jagnow et al., 2006; Wright & Kaiser, 1986; Wright, Kaiser, & Nicholls, 2002; Zhang, Hussain, & Armstrong, 2006). Despite the long list of landowner concerns about granting access to their lands, the primary and overriding concern is the fear of being sued or being held liable for injuries sustained while on the land (Jagnow et al., 2006; Wright et al., 2002). The most often used justification by landowners for restricting access to their private land is the potential threat of liability (Brown, Decker, & Kelley, 1984; Mozumder et al., 2007; Wright et al., 2002).

Common-law tort and property law regulate a landowner's obligations to recreational users of private land. Most states, however, currently have laws limiting a landowner's liability when access to private property is granted without a fee (Mozumder et al., 2007; Wright et al., 2002). In most cases, however, the liability protection is limited when a fee is charged. Under most state laws, the degree of landowner liability is dependent on the status of the visitor or user (i.e., trespasser, licensee, or invitee; Copeland, 1998; Wright et al., 2002); among the three groups, invitees have the greatest legal protection, licensee moderate protection, and trespasser little to no protection. Depending on the state, the category designation of the user is important because the specific category establishes the legal obligations of the landowner.

Although the myth and perception of landowner liability appears to be greater than the actual risk (Wright et al., 2002), the issue should not be handled in a cavalier fashion. Given the variety of legal differences among states (Gentle, Bergstrum, Cordell, & Teasley, 1999; Wright et al., 2002), a qualified lawyer and insurance agent should be consulted before entering into any hunting lease agreement or purchasing liability insurance. Local resource management professionals and university extension specialists, while offering free advice about basic aspects, may not be aware of the complexities and intricacies of landowner liability statutes.

Voluntary Nontoxic Shot Requirement

Another liability landowners need to consider when leasing hunting on agroforestry lands is the impacts of spent lead shot on the environment and wildlife and human health. Environmentally, deposition of lead shot on some heavily hunted agroforestry soils can achieve densities similar to public shooting ranges. During 2006, 11 public hunting areas in Missouri accommodated >8,000 hunters who hunted >25,000 h, fired >180,000 rounds of ammunition, and harvested >33,000 doves (Bonnot et al., 2011; Schulz, Fleming, & Gao, 2011). Dove hunting on these areas occurs near fields managed to attract feeding doves (Baskett 1993; Millspaugh et al., 2009), resulting in high concentrations of spent lead shot pellets. During 2005–2011, the average number of no. 8 lead pellets deposited per year on five popular public hunting areas in Missouri ranged from 35,624 to 128,632 lead pellets ha^{-1} or 1.0–3.6 kg ha^{-1} annually (Schulz et al., 2011). Other investigators have documented varying amounts of spent lead shot deposited on managed fields resulting from dove hunting (Best, Garrison, & Schmitt, 1992; Castrale, 1989; Gerstenberger & Divine, 2001; Kendall et al., 1996; Lewis & Legler, 1968; Locke and Bagley, 1967; Schulz et al., 2002).

Research has documented the effects of spent lead hunting ammunition on numerous wildlife species (Scheuhammer & Norris, 1995; Tranel & Kimmel, 2009), identifying exposure through ingestion of lead shotgun pellets, bullet fragments in gut piles, unretrieved animal carcasses, or ingestion of lost fishing tackle (Finkelstein et al., 2012; Grade, Pokras, Laflamme, & Vogel, 2018; Haig et al., 2014; Schulz, Gao, Millspaugh, & Bermudez, 2007; Schulz, Millspaugh, et al., 2006). Progress has been made in reducing lead poisoning in waterfowl (Anderson, Havera, & Zercher, 2000; Schulz, Padding, & Millspaugh, 2006) but the problem continues for California condors (*Gymnogyps californianus*), bald eagles (*Haliaeetus leucocephalus*), mourning doves, common loons (*Gavia immer*), and swans (*Cygnus* spp.). Despite these data, limited action has occurred.

It is important to recognize that lead poisoning is not only a wildlife health issue. Like the wildlife exposure pathway, human health is impacted by lead exposure through ingestion of lead bullet fragments in processed ground venison (Hunt et al., 2009; Iqbal et al., 2009). Individuals ingesting game meat shot with lead ammunition are susceptible to chronic and acute effects of lead exposure (Buenz & Parry, 2018; Lidsky & Schneider, 2006; Jones et al., 2009; Knott, Gilbert, Hoccom, & Green, 2010; Mateo et al., 2014; Rosen, 1995). Evidence shows that blood lead levels in humans <5 μg dl^{-1} is a biomarker for impaired cognition, suggesting no safe level of human lead exposure (Bellinger et al.,1991; Council on Environmental Health, 2016; Earl, Burns, Nettelbeck, & Baghurst, 2016). Not only are low blood lead levels a biomarker for impaired cognitive abilities, recent findings demonstrate that low-level exposure is a largely overlooked risk factor for cardiovascular disease mortality in the United States, corresponding to approximately 256,000 annual deaths due to cardiovascular disease and 185,000 deaths per year from coronary artery heart disease (Lanphear, Rauch, Auinger, Allen, & Hornung, 2018). Therefore, landowners concerned about the effects of lead exposure to wildlife and hunters using their lands may consider requiring the use of nontoxic shot in lease agreements.

Recommendations for Providing Wildlife Benefits

Based on the general considerations, we offer the following advice to those interested in promoting general wildlife benefits within agroforestry settings. If a landowner has refined, species-specific objectives (e.g., management specifically for white-tailed deer), the management of agroforestry plantings should be tailored for that species. In such cases, we recommend that landowners consult with local extension offices, which can provide information for that species (e.g., Walter & Pierce, 2007). The recommendations below are general in nature, and are designed to increase overall wildlife benefits and diversity.

Establish Realistic Goals

When considering wildlife benefits in an agroforestry context, a landowner must remember that not all wildlife will benefit from their practices. For example, alley cropping and windbreaks might create edge that will benefit white-tailed deer, but other species such as neotropical migrants might not be benefited. Such trade-offs should be considered when there is a goal of improving wildlife benefits. Also, because wildlife benefits are often secondary to timber production, there are limitations to the wildlife benefits that can be attained. In other words, changes to the tree species selected and planting options are somewhat limited given logistical and economic constraints.

Additionally, the scale of the agroforestry operation has an important impact on the realized wildlife benefits. Whereas a 5-ha agroforestry field might result in the harvest of a few hundred mourning doves if planted to sunflower, we cannot reasonably expect the same number of quail to be harvested within that area. Wildlife benefits are attainable with minor alterations to the timing of planting, cover crop management, shrub selection, tree spacing and configuration, and other factors, but setting goals will help a landowner better determine proper management methods and reach their objectives. Another consideration is that some wildlife may cause tree damage (e.g., rabbits and deer) and may be counterproductive to the long-term timber management objective.

Promote Structural Diversity and Manage Habitat Edges

In general, wildlife will be benefited by creating "soft" boundaries, promoting plant diversity (i.e., tree selection, cover crop, shrub selection), increasing the overall width of plantings within and between rows, and offering diversity in the ages of plantings (Figure 13–6). Slight alterations such as changing the spacing of tree plantings within a row could help improve cover and reduce predator success due to increased visual obstruction. Widening plantings also improves understory production by allowing more light to reach the ground; such increases could improve habitat quality for ground-nesting birds and improve cover for species such as bobwhite quail. Mixed species composition would greatly enhance within-stand diversity, but it does require careful planning to ensure that shade-tolerant plants are mixed appropriately with shade-intolerant species (Allen et al., 1996). Creating "soft" boundaries may provide several advantages over "hard" boundaries by increasing the variety of vegetation, the diversity of horizontal structure, and the overall amount of habitat available. An example of a change in a traditional alley cropping planting could include the incorporation of a native grass strip followed by shrubs, then a mix of trees, shrubs, and ground cover. The native grass strip would provide additional habitat while creating a buffer zone between other habitats and the tree plantings. The incorporation of a shrub component (both within and alone) again adds diversity, which would provide more habitat; when selected appropriately, shrubs would also provide food resources for wildlife.

Since agroforestry represents a suite of practices that are planned and designed by the landowner, there is an opportunity to create habitat edges that are more transitional rather than abrupt. By intentionally combining shrubs and taller warm-season grasses on the outer edges of practices that have traditionally focused on tree species alone, these practices can become structurally diverse components on agricultural

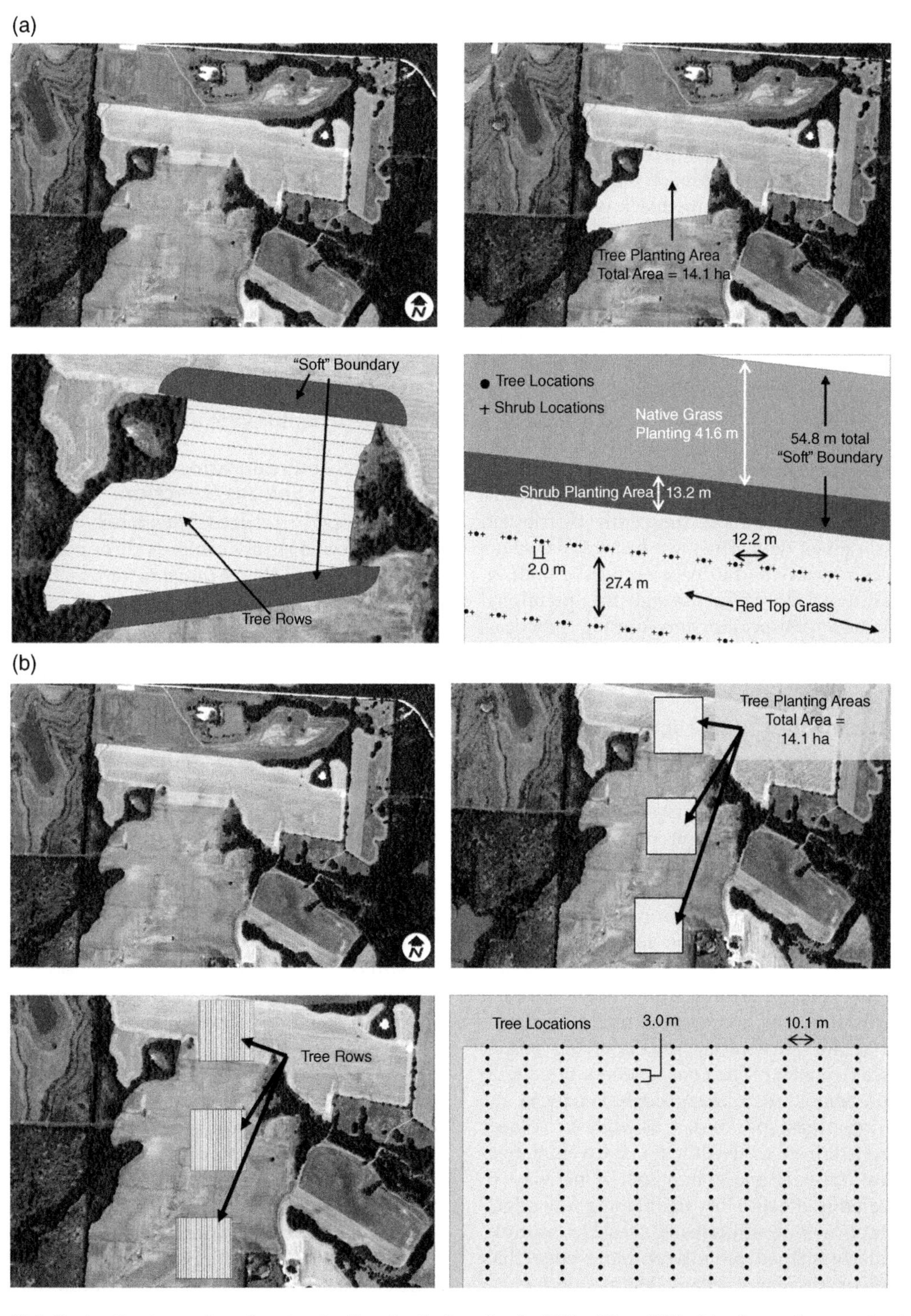

Fig. 13–6. Contrasting types of agroforestry plantings for the benefit of wildlife: (A) a wildlife-friendly agroforestry planting on the landscape because of the following attributes: an appropriate landscape context that connects previously fragmented environments; inclusion of a "soft" boundary between the tree planting area and the existing agricultural sites; and wide spacing between trees and tree rows, inclusion of shrub plantings on either side of each tree, offset tree rows to increase visual obstruction of predators, redtop grass planted within the tree planting area to limit rabbit herbivory, and wide "soft" boundary that includes a shrub planting area and native grass planting; and (B) a poor example of wildlife-friendly agroforestry plantings because of the following attributes: poor site configuration, which creates islands of fragmented agroforestry plots; hard boundaries, with no transition among habitat types; and tight spacing between trees and tree rows, which may lead to little or no ground cover and no visual obstruction.

landscapes. For example, in an alley cropping configuration, eastern black walnut may be planted to grow a future high-value walnut log. However, when open grown, black walnut tends to retain its lower branches, resulting in boles that must be pruned to maximize the tree's value potential. A different approach might be to plant shrubs on either side of the black walnut. These shrubs could include species such as wild plum (*Prunus americana* Marshall), roughleaf dogwood (*Cornus drummondii* C.A. Mey.) or false indigo (*Amorpha* L.). At the same time that a higher value walnut log is being produced, these shrubs would serve to both shade the bole to encourage natural pruning and provide a softer edge, or transitional zone, where use by a greater diversity of birds might exist. Management could be further enhanced by planting a desirable grass along the outer edge. Such a design is similar to the windbreak practice but more frequently distributed across a given field. Thus, we believe that structural complexity and dimension can be built on agricultural landscapes through the intentional integration of specific agroforestry practices. Design, however, will be specific to the wildlife species of interest.

Consider Landscape Context

The studies we cite above offer strong evidence for the importance of landscape context and issues of edge and fragmentation (Figure 13–6). When considering location of an agroforestry site, those sites adjacent to agriculture can be expected to promote generalist species (e.g., deer, cottontail rabbits) and will probably not promote area-sensitive species like forest neotropical migrant birds or habitat specialists like swamp rabbits. Expected small game wildlife hunting opportunities (e.g., more quail) might be diminished if the agroforestry site lacks the proper landscape context. The consequences of edge to wildlife should also be considered early in the planning stages; species such as white-tailed deer will benefit from edge while forest-dwelling neotropical migrant birds may not. One way to reduce edge effects is by maintaining low edge/area ratios within agroforestry sites. For example, irregularly shaped plots have more edge than square or circular habitats. Maintaining fewer large areas that are well connected to other similar sites would typically be more beneficial to wildlife than smaller and more scattered sites. Predator traps are possibilities where habitats are isolated, fragmented, and contain a high edge/interior ratio. In fragmented landscapes, landowners should view agroforestry as an opportunity to connect fragmented habitats.

Limiting Factors in the Environment

When attempting to improve wildlife benefits, often the greatest impact will be made by meeting needs that current land management practices are not addressing. One way to assess habitat is to gauge what element is most limiting to making a given land area more useable for a specific wildlife species. For example, if hard mast is a limiting factor, a landowner might consider planting oak trees to benefit deer and turkey. This recommendation inherently implies that the landowner must establish goals. Consideration of area-specific limiting factors not only influences site location and arrangement but will also drive the selection of plants, shrubs, and cover crops conducive to the site.

Site Preparation and Tree Species Selection

When properly selected, integrated, and managed, trees and shrubs can provide opportunities for attracting wildlife at the same time as they are developing their own potential products, such as wood value (i.e., timber grown for lumber). Selection of plant materials should reflect their potential to enhance land use by a desired wildlife species. Secondarily, this list of desirable plant materials must be suitable for the site to be planted. Landowners should also be creative in their approach to tree, shrub, and cover crop selection. For example, many shrubs such as wild plum, blackberry (*Rubus* sp.), and wild indigo offer good wildlife food and cover and do not compete significantly with adjacent plantings (Walter & Pierce, 2007). Planting shrubs next to bareroot tree stock will also stimulate the growth of the tree due to the competition. Thus, improvements may be made to tree growth while promoting wildlife benefits of the area.

Location, Location, Location

Dissimilarity in adjacent habitat types limits opportunities for wildlife to traverse the areas (Allen et al., 1996). Boundaries of different land uses differ in their permeability to wildlife (Wiens, Crawford, & Gosz, 1985). Thus, it is advisable to consider transitions among habitat types and to the agroforestry site. Such habitat barriers can create problems for wildlife movement to and from the site. Providing transitions and similarity in habitat types will enhance wildlife use of the area.

Interspersion of agroforestry plots within older, more established and connected forests has the potential to improve wildlife benefits (Figure 13–6). During establishment while trees are young, trees provide early successional habitat and a higher quality and quantity of herbaceous material that may be suitable for a diversity of wildlife. As the stand matures, a different wildlife component will become evident. However, these benefits are predicated on the necessary landscape context and general similarity in terms of habitat types and not necessarily age of development.

Grassland Habitats and Agroforestry Considerations

Agroforestry sites within grasslands will probably have mixed wildlife benefits. Despite the utility of shelter belts and tree cover to some wildlife species in grasslands (e.g., white-tailed deer and insects), grassland birds may not benefit from developing agroforestry fields in unconnected, open grasslands.

Control for Wildlife Damage

In areas with healthy deer and rabbit populations, steps should be taken to reduce damage. Tree guards, physical exclusions, repellents, and habitat management can be effective depending on the situation. If feral hogs are an issue, landowners can call their local wildlife damage biologists.

Share the Cost

National conservation programs are focusing more on incentivizing conservation practices on working forested lands. This trend increases the opportunity for landowners to offset some of the costs of establishing an agroforestry system. Some situations may even qualify for rental payments for maintaining forested lands. Therefore, landowners should reach out to their local NRCS office to inquire about opportunities.

Lease Hunting Opportunities

Given the potential loss of income during the development of agroforestry plots, a landowner can generate funds through lease hunting. With time, hunting opportunities will change (Figure 13–2). Greentree reservoirs represent another economically feasible option for leasing waterfowl hunting privileges, provided that trees are not damaged during the course of annual flooding.

When managing a lease hunting opportunity, planning should consider the numerous options for lease hunting and several important factors such as costs, liability, and cover crop management. Additionally, there are no guarantees with wildlife habitat management because by default the landowner is dealing with an unpredictable wild animal species. It is expected that all landowners will have good years and less successful seasons. We encourage landowners interested in pursuing lease hunting opportunities to speak with other landowners and state biologists to help determine demand, costs they incur, tips on planting, and other issues that are site-specific.

More Research is Needed

More research is needed to understand the relationship between agroforestry and wildlife benefits. Although attention is growing, there are still relatively few studies that investigate wildlife response to many of the common agroforestry practices. Consequently, we used review materials and recommendations from other, related habitat restoration techniques. However, agroforestry environments are different from traditional agriculture and forest environments and unique enough that further work is warranted. For example, unlike traditional tree plantations, tree spacing is wider and can be staggered for wildlife benefits. Also, more sunlight reaches the ground level, which changes the understory dynamics. There is also a great diversity in planting opportunities in agroforestry areas when compared with either natural forests or traditional tree plantations. All of these unique aspects of agroforestry point to the need to better understand wildlife dynamics in these environments. Although we may draw general principles from the ecological and wildlife management literature, technique-specific information is warranted. We encourage researchers to partner with landowners to obtain information about wildlife benefits from agroforestry situations. For example, data from lease hunting ventures would add some replication to better understand the integration of agroforestry and lease hunting.

References

Abernethy, Y., & Turner, R. E. (1987). US forested wetlands: 1940-1980: Field-data surveys document changes and can guide national resource management. *BioScience, 37,* 721–727. doi:10.2307/1310469

Adams, H., L., Burger, L. W., Jr., & Riffell, S. (2015). Edge effects on avian diversity and density of native grass conservation buffers. *The Open Ornithology Journal, 8,* 1–9. doi:10.2174/1874453201508010001

Allen, A. W., Bernal, Y. K., & Moulton, R. J. (1996). Pine plantations and wildlife in the southeastern United States: An assessment of impacts and opportunities (Inf. & Technol. Rep. 3). Washington, DC: U.S. Fish and Wildlife Service.

Allen, C. E. (1980). Feeding habits of ducks in a green-tree reservoir in eastern Texas. *Journal of Wildlife Management, 44,* 232–236. doi:10.2307/3808376

Allen, J. A. (1990). Establishment of bottomland oak plantations on the Yazoo National Wildlife Refuge complex. *Southern Journal of Applied Forestry, 14,* 206–210. doi:10.1093/sjaf/14.4.206

Anderson, D. R. (1997). *Corridor use, feeding ecology, and habitat relationships of black bears in a fragmented landscape in Louisiana* (Master's thesis). University of Tennessee.

Anderson, W. L., Havera, S. P., & Zercher, B. W. (2000). Ingestion of lead and nontoxic shotgun pellets by ducks in the Mississippi Flyway. *Journal of Wildlife Management, 64,* 848–857. doi:10.2307/3802755

Baskett, R. K. (1993). Shooting field management. In T. S. Baskett, M. W. Sayre, R. E. Tomlinson, & R. E. Mirarchi (Eds.), Ecology and management of the mourning dove (pp. 495–506). Harrisburg, PA: Stackpole.

Beatty, W. S., Kesler, D. C., Webb, E. B., Raedeke, A. H., Naylor, L. W., & Humburg, D. D. (2014). The role of protected area wetlands in waterfowl habitat conservation: Implications for protected area network design. *Biological Conservation, 176,* 144–152. doi:10.1016/j.biocon.2014.05.018

Bellinger, D., Leviton, A., Sloman, J., Rabinowitz, M., Needleman, H., & Waternaux, C. (1991). Low-level lead exposure and children's cognitive function in the preschool years. *Pediatrics, 87,* 219–227 [erratum: 93(2), A28].

Bentrup, G. (2014). A win–win on agricultural lands: Creating wildlife habitat through agroforestry. *The Wildlife Professional, 8,* 26–40.

Berges, S. A., Schulte Moore, L. A., Isenhart, T. M., & Schultz, R. C. (2010). Bird species diversity in riparian buffers, row crop fields, and grazed pastures within agriculturally dominated watersheds. *Agroforestry Systems, 79,* 97–110. doi:10.1007/s10457-009-9270-6

Best, T. L., Garrison, T. E., & Schmitt, C. G. (1992). Availability and ingestion of lead shot by mourning doves (*Zenaida macroura*) in southeastern New Mexico. *The Southwestern Naturalist, 37,* 287–292. doi:10.2307/3671871

Black, H. C. (1992). *Silvicultural approaches to animal damage in Pacific Northwest forests* (Gen. Tech. Rep. PNW-GTR-287). Portland, OR: U.S. Forest Service, Pacific Northwest Research Station. doi:10.2737/PNW-GTR-287

Blank, P. J. (2013). Northern bobwhite response to Conservation Reserve Program habitat and landscape attributes. *Journal of Wildlife Management, 77,* 68–74. doi:10.1002/jwmg.457

Bobryk, C. W., Rega-Brodsky, C. C., Bardhan, S., Farina, A., He, H. S., & Jose, S. (2016). A rapid soundscape analysis to quantify conservation benefits of temperate agroforestry systems using low-cost technology. *Agroforestry Systems, 90,* 997–1008. doi:10.1007/s10457-015-9879-6

Bollinger, E. K., Bollinger, P. B., & Gavin, T. A. (1990). Effects of hay-cropping on eastern populations of the bobolink. *Wildlife Society Bulletin, 18,* 142–150.

Bonnot, T. W., Schulz, J. H., & Millspaugh. J. J. (2011). Factors affecting mourning dove harvest in Missouri. *Wildlife Society Bulletin, 35,* 76–84. doi:10.1002/wsb.22

Bowler, D. E., Mant, R., Orr, H., Hannah, D. M., & Pullin, A. S. (2012). What are the effects of wooded riparian zones on stream temperature? *Environmental Evidence, 1,* 3. doi:10.1186/2047-2382-1-3

Brandle, J., Hodges, L., & Zhou, X. (2004). Windbreaks in sustainable agriculture. *Agroforestry Systems, 61,* 65–78.

Bromley, P. T., Wellendorf, S. D., Palmer, W. E., & Marcus, J. F. (2002). Effects of field borders and mesomammal reduction on northern bobwhite and songbird abundance on three farms in North Carolina. In S. J. DeMaso, W. P. Kuvlevsky, Jr., F. Hernández, & M. E. Berger (Eds.), Quail V: Proceedings of the Fifth National Quail Symposium, Corpus Christi, TX (p. 71). Austin, TX: Texas Parks and Wildlife Department.

Brown, T. L., Decker, D. J., & Kelley, J. W. (1984). Access to private lands for hunting in New York: 1963--1980. *Wildlife Society Bulletin, 12,* 344–349.

Brown, T. L., Decker, D. J., Siemer, W. F., & Enck, J. W. (2000). Trends in hunting participation and implications for management of game species. In W. C. Gartner & D. W. Lime (Eds.), *Trends in outdoor recreation, leisure and tourism* (pp. 145–154). Wallingford, UK: CAB International. doi:10.1079/9780851994031.0145

Buenz, E. J., & Parry, G. J. (2018). Chronic lead intoxication from eating wild-harvested game. *The American Journal of Medicine, 131,* e181–e184. doi:10.1016/j.amjmed.2017.11.031

Bugg, R. L., Sarrantonio, M., Dutcher, J. D., & Phatak, S. C. (1991). Understory cover crops in pecan orchards: Possible management systems. *American Journal of Alternative Agriculture, 6,* 50–62. doi:10.1017/S0889189300003854

Burger, L. W., Jr., Kurzejeski, E. W., Dailey, T. V., & Ryan, M. R. (1993). Relative invertebrate abundance and biomass in Conservation Reserve Program plantings in northern Missouri. In K. E. Church & T. V. Dailey (Eds.), Quail III: National Quail Symposium, Kansas City, MO (pp. 102–108). Pratt, KS: Kansas Department of Wildlife and Parks.

Burhans, D. E. (1997). Habitat and microhabitat features associated with cowbird parasitism in two forest edge cowbird hosts. *Condor, 99,* 866–872. doi:10.2307/1370136

Burhans, D. E. (2001). *Conservation assessment: Henslow's sparrow* Ammodramus henslowii (Gen. Tech. Rep. NC-226). St. Paul, MN: U.S. Forest Service, North Central Research Station.

Burhans, D. E., Dearborn, D., Thompson, F. R., III, & Faaborg, J. (2002). Factors affecting predation at songbird nests in old fields. *Journal of Wildlife Management, 66,* 240–249. doi:10.2307/3802890

Cable, T. T. (1991). Windbreaks, wildlife and hunters. In J. E. Rodiek & E. G. Bolen (Eds.), *Wildlife and habitats in managed landscapes* (pp. 35–55). Washington, DC: Island Press.

Campbell, T. A., & Long, D. B. (2009). Feral swine damage and damage management in forested ecosystems. *Forest Ecology and Management, 257,* 2319–2326. doi:10.1016/j.foreco.2009.03.036

Capel, S. W. (1988). Design of windbreaks for wildlife in the Great Plains of North America. *Agriculture, Ecosystems & Environment, 22–23,* 337–347. doi:10.1016/0167-8809(88)90029-1

Cassel, J. F., & Wiehe, J. W. (1980). Uses of shelterbelts by birds. In *Workshop proceedings: Management of western forests and grasslands for nongame birds, Salt Lake City, UT* (Gen. Tech. Rep. INT-86, p. 78–87). Ogden, UT: U.S. Forest Service, Intermountain Forest and Range Experiment Station.

Castrale, J. S. (1989). Availability of spent lead shot in fields managed for mourning dove hunting. *Wildlife Society Bulletin,* 17:184–189.

Chalfoun, A. D., Thompson, F. R., III, & Ratnaswamy, M. J. (2002). Nest predators and fragmentation: A review and meta-analysis. *Conservation Biology, 16,* 306–318. doi:10.1046/j.1523-1739.2002.00308.x

Christman, S. P. (1984). Breeding bird response to greentree reservoir management. *Journal of Wildlife Management, 48,* 1164–1172. doi:10.2307/3801777

Ciuzio, E., Hohman, W. L., Martin, B., Smith, M. D., Stephens, S., Strong, A. M., & Vercauteren, T. (2013). Opportunities and challenges to implementing bird conservation on private lands. *Wildlife Society Bulletin, 37,* 267–277. doi:10.1002/wsb.266

Clark, W. R., & Reeder, K. F. (2005). Continuous enrollment Conservation Reserve Program: Factors influencing the value of agricultural buffers to wildlife conservation. In J. B. Haufler (Ed.), *Fish and wildlife benefits of Farm Bill conservation programs: 2000--2005 update* (Tech. Rev. 05-2, pp. 93–114). Bethesda, MD: The Wildlife Society.

Clark, W. R., & Reeder, K. F. (2007). Agricultural buffers and wildlife conservation: A summary about linear practices. In J. Haufler (Ed.), *Fish and wildlife response to Farm Bill conservation practices* (Tech. Rev. 071, pp. 45–56). Bethesda, MD: The Wildlife Society.

Clotfelter, E. D. (1998). What cues do brown-headed cowbirds use to locate red-winged blackbird nests? *Animal Behaviour, 55,* 1181–1189. doi:10.1006/anbe.1997.0638

Conover, M. (2002). *Resolving human–wildlife conflicts: The science of wildlife damage management*. Boca Raton, FL: CRC Press.

Copeland, J. D. (1998). *Recreational access to private lands: Liability, problems and solutions*. Fayetteville, AR: National Center for Agricultural Law Research and Information, University of Arkansas.

Cordell, H. K., & Betz, C. J. (2000). Trends in outdoor recreation supply on public and private lands in the US. In W. C. Gartner & D. W. Lime (Eds.), *Trends in outdoor recreation, leisure and tourism* (pp. 75–89). Wallingford, UK: CABI Publishing. doi:10.1079/9780851994031.0075

Cordell, H. K., & Super, G. R. (2000). Trends in American's outdoor recreation. In W. C. Gartner & D. W. Lime (Eds.), *Trends in outdoor recreation, leisure and tourism* (pp. 133–144). Wallingford, UK: CABI Publishing. doi:10.1079/9780851994031.0133

Council on Environmental Health. 2016. Prevention of childhood lead toxicity. *Pediatrics 138*(1), e20161493 [errata: 140(2), e20171490; 145 (6) e20201014]. doi:10.1542/peds.2016-1493

Crawford, J. A., & Semlitsch, R. D. (2007). Estimation of core terrestrial habitat for stream-breeding salamanders and delineation of riparian buffers for protection of biodiversity. *Conservation Biology, 21,* 152–158. doi:10.1111/j.1523-1739.2006.00556.x

Davros, N. M., Debinski, D. M., Reeder, K. F., & Hohman, W. L. (2006). Butterflies and continuous Conservation Reserve Program filter strips: Landscape considerations. *Wildlife Society Bulletin, 34,* 936–943. doi:10.2193/0091-7648(2006)34[936:BACCRP]2.0.CO;2

Deller, A. S. (1997). *Effect of greentree reservoir management on the vegetative and breeding bird communities on the Montezuma National Wildlife Refuge* (*New York*) (Master's thesis). State University of New York. Masters Abstracts International 35–04.

Dey, D. C., Lovelace, W., Kabrick, J. M., & Gold, M. A. (2004). Production and early field performance of RPM seedlings in Missouri floodplains. In Proceedings of the 6th Walnut Council research symposium (Gen. Tech. Rep. NC-243, pp. 59--65). St. Paul, MN: U.S. Forest Service, North Central Research Station.

Dijak, W. D., & Thompson, F. R., III. (2000). Landscape and edge effects on the distribution of mammalian predators in Missouri. *Journal of Wildlife Management, 64,* 209–216. doi:10.2307/3802992

Donovan, T. M., Beardmore, C. J., Bonter, D. N., Brawn, J. D., Cooper, R. J., Fitzgerald, J. A., . . . Wigley, T. B. (2002). Priority research needs for the conservation of neotropical migrant landbirds. *Journal of Field Ornithology, 73,* 329–339. doi:10.1648/0273-8570-73.4.329

Donovan, T. M., Thompson, F. R., III, & Faaborg, J. R. (2000). Cowbird distribution at different scales of fragmentation: Trade-offs between breeding and feeding opportunities. In J. N. M. Smith, T. L. Cook, S. I. Rothstein, S. K. Robinson, & S. G. Sealy (Eds.), *Ecology and management of cowbirds* (pp. 255–264). Austin, TX: University of Texas Press.

Dosskey, M. G., Bentrup, G., & Schoeneberger, M. (2012). A role for agroforestry in forest restoration in the Lower Mississippi alluvial valley. *Journal of Forestry, 110,* 48–55.

Duda, M. D., Bissell, S. J., & Young, K. C. (1998). *Wildlife and the American mind: Public opinion on and attitudes toward fish and wildlife management*. Harrisonburg, VA: Responsive Management.

Dugger, S., Dey, D. C., & Millspaugh, J. J. (2003, 24–28 Feb.). Vegetation cover affects mammal herbivory on planted oaks and success of reforesting Missouri river bottomland fields. In K. F. Connor (Ed.) Proceedings of the 12th Biennial Southern Silvicultural Research Conference, Biloxi, MS (Gen. Tech. Rep. SRS-71, pp. 3–6). Asheville, NC: U.S. Forest Service, Southern Research Station.

Earl, R., Burns, N., Nettelbeck, T., & Baghurst, P. (2016). Low-level environmental lead exposure still negatively associated with children's cognitive abilities. *Australian Journal of Psychology, 68,* 98–106. doi:10.1111/ajpy.12096

Elliott, A. G., & Root, B. G. (2006). Small mammal responses to silvicultural and precipitation-related disturbance in northeastern Missouri riparian forests. *Wildlife Society Bulletin, 34,* 485–501. doi:10.2193/0091-7648(2006)34[485:SMRTSA]2.0.CO;2

Evans, K. O., Burger, L. W., Jr., Riffell, S. K., Smith, M. D., Twedt, D. J., Wilson, R. R., . . . Heyden, K. (2014). Avian response to conservation buffers in agricultural landscapes during winter. *Wildlife Society Bulletin, 38,* 257–264. doi:10.1002/wsb.405

Faulkner, S., Barrow, W., Jr., Keeland, B., Walls, S., & Telesco, D. (2011). Effects of conservation practices on wetland ecosystem services in the Mississippi alluvial valley. *Ecological Applications, 21,* S31–S48. doi:10.1890/10-0592.1

Finkelstein, M. E., Doak, D. F., George, D., Burnett, J., Brandt, J., Church, M., . . . Smith, D.R. (2012). Lead poisoning and the deceptive recovery of the critically endangered California condor. *Proceedings of the National Academy of Sciences, 109,* 11449–11454. doi:10.1073/pnas.1203141109

Fitzgerald, J. A., Herkert, J. R., & Brawn, J. D. (2000). Partners in Flight bird conservation plan for the Prairie Peninsula (Physiographic Area 31). American Bird Conservancy: The Plains, VA.

Fitzgerald, J. A., & Pashley, D. N. (2000). Partners in Flight bird conservation plan for the Ozark/Ouachitas (Physiographic Area 19). American Bird Conservancy: The Plains, VA.

Fleming, H. L., Jones, J. C., Belant, J. L., & Richardson, D. M. (2013). Multi-scale roost site selection by Rafinesque's big-eared bat (*Corynorhinus rafinesquii*) and southeastern myotis (*Myotis austroriparius*) in Mississippi. *The American Midland Naturalist, 169,* 43–55. doi:10.1674/0003-0031-169.1.43

Fredrickson, L. H. (1978). Lowland hardwood wetlands: Current status and values to wildlife. In P. E. Greeson, J. R. Clark, & J. E. Clark (Eds.), *Wetland functions and values: The state of our understanding* (pp. 296–306). Minneapolis, MN: American Water Resource Association.

Fredrickson, L. H. (1980). Management of lowland hardwood wetlands for wildlife: Problems and potential. *Transactions of the North American Wildlife and Natural Resources Conference, 45,* 376–386.

Fredrickson, L. H., & Heitmeyer, M. E. (1988). Waterfowl use of forested wetlands of southern United States: An overview. In M. W. Weller (Ed.), *Waterfowl in winter* (pp. 307–323). Minneapolis, MN: University of Minnesota Press.

Fretwell, S. D., & Shipley, F. S. (1981). Statistical significance and density-dependent nest predation. *Wilson Bulletin, 93,* 541–542.

Gardiner, E. S., & Oliver, J. M. (2005). Restoration of bottomland hardwood forests in Lower Mississippi alluvial valley, USA. In J. Stanturf & P. Madsen (Eds.), *Restoration of boreal and temperate forests* (pp. 235–251). Boca Raton, FL: CRC Press.

Geis, A. D. (1954). Rabbit damage to oak reproduction at the Kellogg Bird Sanctuary. *Journal of Wildlife Management, 18,* 423–424. doi:10.2307/3797042

Gentle, P., Bergstrum, J., Cordell, K., & Teasley, J. (1999). Private landowner attitudes concerning public access for outdoor recreation: Cultural and political factors in the United States. *Journal of Hospitality & Leisure Marketing, 6,* 47–65.

Gerstenberger, S., & Divine, D. D. (2001). Lead shot deposition and distribution in southern Nevada. *Nevada Journal of Public Health, 3,* 8–13.

Giuliano, W. M., & Daves, S. E. (2002). Avian response to warm-season grass use in pasture and hayfield management. *Biological Conservation, 106,* 1–9. doi:10.1016/S0006-3207(01)00126-4

Gordon, H. G. (2015). *Evaluating productivity of southern agroforestry for fiber, biofuels, and wildlife habitat* (Master's thesis). Mississippi State University. https://ir.library.msstate.edu/handle/11668/20288

Grade, T. J., Pokras, M. A., Laflamme, E. M., & Vogel, H. S. (2018). Population-level effects of lead fishing tackle on common loons. *Journal of Wildlife Management, 82*, 155–164. doi:10.1002/jwmg.21348</jrn>

Grossman, B. C., Gold, M. A., & Dey, D. C. (2003). Restoration of hard mast species for wildlife in Missouri using precocious flowering oak in the Missouri River floodplain, USA. *Agroforestry Systems, 59*, 3–10. doi:10.1023/A:1026124717097

Guttery, M. R., & Ezell, A. W. (2006). Characteristics of a bottomland hardwood forest under greentree reservoir management in east central Arkansas. In K. F. Connor (Ed.), Proceedings of the 19th Biennial Southern Silvicultural Research Conference (Gen. Tech. Rep. SRS-92, pp. 409–411). Asheville, NC: U.S. Forest Service, Southern Research Station.

Hagy, H. M., Linz, G. M., & Bleier, W. J. (2010). Wildlife conservation sunflower plots and croplands as fall habitat for migratory birds. *The American Midland Naturalist, 164*, 119–135. doi:10.1674/0003-0031-164.1.119</jrn>

Hahn, D. C., & Hatfield, J. S. (1995). Parasitism at the landscape scale: Cowbirds prefer forests. *Conservation Biology, 9*, 1415–1424. doi:10.1046/j.1523-1739.1995.09061415.x

Haig, S. M., D'Elia, J., Eagles-Smith, C., Fair, J. M., Gervais, J., Herring, G., . . . Schulz, J. H. (2014). The persistent problem of lead poisoning in birds from ammunition and fishing tackle. *Condor, 116*, 408–428. doi:10.1650/CONDOR-14-36.1

Hamel, P. B. (2003). Winter bird community differences among methods of bottomland hardwood forest restoration: Results after seven growing seasons. *Forestry, 76*, 189–197. doi:10.1093/forestry/76.2.189

Hauber, M. E., & Russo, S. A. (2000). Perch proximity correlates with higher rates of cowbird parasitism of ground nesting song sparrows. *Wilson Bulletin, 112*, 150–153. doi:10.1676/0043-5643(2000)112[0150:PPCWHR]2.0.CO;2

Haugen, A. O. (1942). Life history studies of the cottontail rabbit in southwestern Michigan. *The American Midland Naturalist, 28*, 204–244. doi:10.2307/2420701

Hawkes, V. C., & Gregory, P. T. (2012). Temporal changes in the relative abundance of amphibians relative to riparian buffer width in western Washington, USA. *Forest Ecology and Management, 274*, 67–80. doi:10.1016/j.foreco.2012.02.015

Haynes, R. J. (2004). The development of bottomland forest restoration in the Lower Mississippi River alluvial valley. *Ecological Restoration, 22*, 170–182. doi:10.3368/er.22.3.170

Hoover, J. P., Tear, T. H., & Baltz, M. E. (2006). Edge effects reduce the nesting success of Acadian flycatchers in a moderately fragmented forest. *Journal of Field Ornithology, 77*, 425–436. doi:10.1111/j.1557-9263.2006.00074.x

Hunt, W. G., Watson, R. T., Oaks, J. L., Parish, C. N., Burnham, K. K., Tucker, R. L., . . . Hart, G. (2009). Lead bullet fragments in venison from rifle-killed deer: Potential for human dietary exposure. *PLOS ONE, 4*, e5330. doi:10.1371/journal.pone.0005330

Husak, A. L. (2001). *Wildlife and economic benefits from agroforestry systems* (Master's thesis). Mississippi State University.

Inman, R. L., Prince, H. H., & Hayes, D. B. (2002). Avian communities in forested riparian wetlands of southern Michigan, USA. *Wetlands, 22*, 647–660. doi:10.1672/0277-5212(2002)022[0647:ACIFRW]2.0.CO;2

Iqbal, S., Blumenthal, W., Kennedy, C., Yip, F. Y., Pickard, S., Flanders, W. D., . . . Brown, M. J. (2009). Hunting with lead: Association between blood lead levels and wild game consumption. *Environmental Research, 109*, 952–959. doi:10.1016/j.envres.2009.08.007

Jagnow, C. P., Stedman, R. C., Luloff, A. E., San Julian, G. J., Finley, J. C., & Steele, J. (2006). Why landowners in Pennsylvania post their property against hunting. *Human Dimensions of Wildlife, 11*, 15–26. doi:10.1080/10871200500470944

James, F. C. (1971). Ordinations of habitat relationships among breeding birds. *Wilson Bulletin, 83*, 215–236.

Janke, A. K. (2016). *Windbreaks for wildlife* (Publ. 221). Ames, IA: Iowa State University Extension. http://lib.dr.iastate.edu/extension_pubs/221

Janke, A. K., & Gates, R. J. (2013). Home range and habitat selection of northern bobwhite coveys in an agricultural landscape. *Journal of Wildlife Management, 77*, 405–413. doi:10.1002/jwmg.461

Johnson, C. W., & Buffler, S. (2008). *Riparian buffer design guidelines for water quality and wildlife habitat functions on agricultural landscapes in the Intermountain West* (Gen. Tech. Rep. RMRS-GTR-203. Fort Collins, CO: U.S. Forest Service, Rocky Mountain Research Station. doi:10.2737/RMRS-GTR-203

Johnson, D. H. (2005). Grassland bird use of Conservation Reserve Program fields in the Great Plains. In J. B. Haufler (Ed.), *Fish and wildlife benefits of Farm Bill conservation programs: 2000--2005 update* (Tech. Rev. 05-2, pp. 17–32). Bethesda, MD: The Wildlife Society.

Johnson, R. J., & Beck, M. M. (1988). Influences of shelterbelts on wildlife management and biology. *Agriculture, Ecosystems & Environment, 22–23*, 301–335. doi:10.1016/0167-8809(88)90028-X

Johnson, R. J., & Timm, R. M. (1987). Wildlife damage to agriculture in Nebraska: A preliminary cost assessment. In N. R. Holler (Ed.), *Proceedings of the Third Eastern Wildlife Damage Control Conference, Gulf Shores, AL* (pp. 57–65). https://core.ac.uk/download/pdf/17222753.pdf

Jones, R. L., Homa, D. M., Meyer, P. A., Brody, D. J., Caldwell, K. L., Pirkle, J. L., & Brown, M. J. (2009). Trends in blood lead levels and blood lead testing among U.S. children aged 1 to 5 years, 1988–2004. *Pediatrics, 123*, e376–e385. doi:10.1542/peds.2007-3608

Jones-Farrand, D. T., Johnson, D. H., Burger, L. W., Jr., & Ryan, M. R. (2007). Grassland establishment for wildlife conservation. In J. Haufler (Ed.), *Fish and wildlife response to Farm Bill conservation practices* (Tech. Rev. 07-1, pp. 25–44). Bethesda, MD: The Wildlife Society.

Jose, S. (2009). Agroforestry for ecosystem services and environmental benefits: An overview. *Agroforestry Systems*, 76, 1–10. doi:10.1007/s10457-009-9229-7

Kaminski, R. M., Alexander, R. W., & Leopold, B. D. (1993). Wood duck and mallard winter microhabitats in Mississippi hardwood bottomlands. *Journal of Wildlife Management, 57*, 562–570. doi:10.2307/3809283

Kays, J. S. (2000). *Hunting lease enterprise*. Rural Enterprise Series. Keedysville, MD: Maryland Cooperative Extension and Agricultural Experiment Station.

Kelly, T. D., Dyer, A. R., & Chesemore, D. L. (1990). Impact of agroforestry plantations grown with agricultural drainwater on avian abundance in the San Joaquin Valley, California. *Transactions of the Western Section of The Wildlife Society, 2*, 97–103.

Kendall, R. J., Lacher, T. E., Jr., Bunck, C., Daniel, B., Driver, C., Grue, C. E., . . . Whitworth, M. (1996). An ecological risk assessment of lead shot exposure in non-waterfowl avian species: Upland game birds and raptors. *Environmental Toxicology and Chemistry, 15*, 4–20. doi:10.1002/etc.5620150103

King, S. L., Twedt, D. J., & Wilson, R. R. (2006). The role of the Wetland Reserve Program in conservation efforts in the Mississippi alluvial valley. *Wildlife Society Bulletin, 34*, 914–920. doi:10.2193/0091-7648(2006)34[914:TROTWR]2.0.CO;2

Knott, J., Gilbert, J., Hoccom, D. G., & Green, R. E. (2010). Implications for wildlife and humans of dietary exposure to lead from fragments of lead rifle bullets in deer shot in the UK. *Science of the Total Environment, 409*, 95–99. doi:10.1016/j.scitotenv.2010.08.053

Krekeler, N., Kabrick, J. M., Dey, D. C., & Wallendorf, M. (2006). Comparing natural and artificial methods for establishing pin oak advance regeneration in bottomland forests managed as greentree reservoirs. In K. F. Connor (Ed.), *Proceedings of the 19th Biennial Southern Silvicultural Research Conference* (Gen. Tech. Rep. SRS-92, pp. 224–228). Asheville, NC: U.S. Forest Service Southern Research Station.

Lanphear, B. P., Rauch, S., Auinger, P., Allen, R. W., & Hornung, R. W. (2018). Low-level lead exposure and mortality in US adults: A population-based cohort study. *The Lancet Public Health, 3*, e177–e184. doi:10.1016/S2468-2667(18)30025-2

Lewis, J. C., & Legler, E. (1968). Lead shot ingestion by mourning doves and incidence in soil. *Journal of Wildlife Management*, 32, 476–482. doi:10.2307/3798925

Lidsky, T. I., & Schneider, J. S. (2006). Adverse effects of childhood lead poisoning: The clinical neuropsychological perspective. *Environmental Research, 100*, 284–293. doi:10.1016/j.envres.2005.03.002

Locke, L. N., & Bagley, G. E. (1967). Lead poisoning in a sample of Maryland mourning doves. *The Journal of Wildlife Management, 31*, 515–518. https://doi.org/10.2307/3798133

Manning, D. H. (2005). *Upland game species use of no-till sites harvested by steers in a pasture and agroforestry setting in east central Mississippi* (Master's thesis). Mississippi State University.

Marczak, L. B., Sakamaki, T., Turvey, S. L., Deguise, I., Wood, S. L. R., & Richardson, J. S. (2010). Are forested buffers an effective conservation strategy for riparian fauna? An assessment using meta-analysis. *Ecological Applications, 20*, 126–134. doi:10.1890/08-2064.1

Marzluff, J. M., Gehlbach, F. R., & Manuwal, D. A. (1998). Urban environments: Influences on avifauna and challenges for the avian conservationist. In J. M. Marzluff & R. Sallabanks (Eds.), *Avian conservation: Research and management* (pp. 283–299). Washington, DC: Island Press.

Masters, R., Bidwell, T., Anderson, S., & Porter, M.D. (1996). *Lease hunting opportunities for Oklahoma landowners* (Fact Sheet F-5032). Stillwater, OK: Oklahoma State University Cooperative Extension.

Mateo, R., Vallverdú-Coll, N., López-Antia, A., Taggart, M. A., Martínez-Haro, M., Guitart, R., & Ortiz-Santaliestra, M. E. (2014). Reducing Pb poisoning in birds and Pb exposure in game meat consumers: The dual benefit of effective Pb shot regulation. *Environment International, 63*, 163–168. doi:10.1016/j.envint.2013.11.006

McShea, W. J., & Schwede, G. (1993). Variable acorn crops: Responses of white-tailed deer and other mast consumers. *Journal of Mammalogy, 74*, 999–1006. doi:10.2307/1382439

Meiners, S. J., & Martinkovic, M. J. (2002). Survival of and herbivore damage to a cohort of *Quercus rubra* planted across a forest–old field edge. *The American Midland Naturalist, 147*, 247–255. doi:10.1674/0003-0031(2002)147[0247:SOAHDT]2.0.CO;2

Miller, C., & Vaske, J. (2003). Individual and situational influences on declining hunter effort in Illinois. *Human Dimensions of Wildlife, 8*, 263–276. doi:10.1080/716100421

Millspaugh, J. J., Schulz, J. H., Mong, T. W., Burhans, D., Walter, W. D., Bredesen, R., . . . Dey, D. D. (2009). Agroforestry wildlife benefits. In H. E. Garrett (Ed.), *North American agroforestry: An integrated science and practice* (pp. 257–285). Madison, WI: ASA.

Moore, M. (2012). *Silvopasture establishment and economics: Modeling the cost of wildlife browse damage to stand establishment and cattle introduction on Redstone Arsenal* (Master's thesis). Auburn University. http://hdl.handle.net/10415/3347

Mozumder, P., Starbuck, C. M., Barrens, R. P., & Alexander, S. (2007). Lease and fee hunting on private lands in the U.S.: A review of the economic and legal issues. *Human Dimensions of Wildlife*, 12, 1–14. doi:10.1080/10871200601107817

Norman, G. W., & Steffen, D. E. (2003). Effects of recruitment, oak mast, and fall-season format on Wild Turkey harvest rates in Virginia. *Wildlife Society Bulletin*, 31, 553–559.

North American Bird Conservation Initiative. 2015. *2014 Farm Bill field guide to fish and wildlife conservation.* http://amjv.org/wp-content/uploads/2019/02/2014_Farm_Bill_Guide-to-Fish-and-Wildlife-Conservation.pdf

Nupp, T. E., & Swihart, R. K. (2000). Landscape-level correlates of small-mammal assemblages in forest fragments of farmland. *Journal of Mammalogy, 81*, 512–526. doi:10.1644/1545-1542(2000)081<0512:LLCOSM>2.0.CO;2

Olson, D. H., Anderson, P. D., Frissell, C. A., Welsh, H. H., Jr., & Bradford, D. F. (2007). Biodiversity management approaches for stream–riparian areas: Perspectives for Pacific Northwest headwater forests, microclimates, and amphibians. *Forest Ecology and Management, 246*, 81–107. doi:10.1016/j.foreco.2007.03.053

Peak, R. G., & Thompson, F. R., III. (2006). Factors affecting avian species richness and density in riparian areas. *Journal of Wildlife Management, 70*, 173–179. doi:10.2193/0022-541X(2006)70[173:FAASRA]2.0.CO;2

Peak, R. G., Thompson, III, F. R., & Shaffer, T. L. (2004). Factors affecting songbird nest survival in riparian forests in a Midwestern agricultural landscape. *The Auk, 121*, 726–737. https://doi.org/10.1093/auk/121.3.726

Perkins, D. W., & Hunter, M. L., Jr. (2006). Effects of riparian timber management on amphibians in Maine. *Journal of Wildlife Management, 70*, 657–670. doi:10.2193/0022-541X(2006)70[657:EORTMO]2.0.CO;2

Peterman, W. E., & Semlitsch, R. D. (2009). Efficacy of riparian buffers in mitigating local population declines and the effects of even-aged timber harvest on larval salamanders. *Forest Ecology and Management, 257*, 8–14. doi:10.1016/j.foreco.2008.08.011

Pierce, R. A., III, Farrand, D. T., & Kurtz, W. B. (2001). Projecting the bird community response resulting from the adoption of shelterbelt agroforestry practices in eastern Nebraska. *Agroforestry Systems, 53*, 333–350. doi:10.1023/A:1013371325769

Potter, M. J. (1988). Treeshelters improve survival and increase early growth rates. *Journal of Forestry, 86*, 39–41. https://doi.org/10.1093/jof/86.8.39

Pulliam, H. R. (1988). Sources, sinks, and population regulation. *The American Naturalist, 132*, 652–661. doi:10.1086/284880

Reimer, A. P., & Prokopy, L. S. (2014). Farmer participation in U.S. Farm Bill conservation programs. *Environmental Management, 53*, 318–332. doi:10.1007/s00267-013-0184-8

Rempe, D., & Simons, C. (1999). *Leasing your land to hunters: A profit center?* Presented at the Agricultural Economics Risks and Profit Conference, Manhattan, KS.

Reynolds, R. E. (2005). The Conservation Reserve Program and duck production in the U.S. Prairie Pothole Region. In J. B. Haufler (Ed.), *Fish and wildlife benefits of Farm Bill conservation programs: 2000--2005 update* (Tech. Rev. 05-2, pp. 33–40), Bethesda, MD: The Wildlife Society

Ribic, C. A., & Sample, D. W. (2001). Associations of grassland birds with landscape factors in southern Wisconsin. *The American Midland Naturalist, 146*, 105–121. doi:10.1674/0003-0031(2001)146[0105:AOGBWL]2.0.CO;2

Rich, T. D., Beardmore, C. J., Berlanga, H., Blancher, P. J., Bradstreet, M. S. W., Butcher, G. S., . . . Will, T. C. (2004). *Partners in Flight North American landbird conservation plan.* Ithaca, NY: Cornell Lab of Ornithology.

Robinson, S. K. (1997). The case of the missing songbirds. *Consequences, 3,* 3–15. http://www.gcrio.org/CONSEQUENCES/vol3no1/songbirds.html

Robinson, S. K., Thompson, F. R., III, Donovan, T. M., Whitehead, D. R., & Faaborg, J. (1995). Regional forest fragmentation and the nesting success of migratory birds. *Science, 267,* 1987–1990. doi:10.1126/science.267.5206.1987

Roseberry, J. L., & Klimstra, W. D. (1984). *Population ecology of the bobwhite.* Carbondale, IL: Southern Illinois University Press.

Rosen, J. F. (1995). Adverse health effects of lead at low exposure levels: Trends in the management of childhood lead poisoning. *Toxicology, 97,* 11–17. doi:10.1016/0300-483X(94)02963-U

Rudolph, R. R., & Hunter, C. G. (1964). Green trees and greenheads. In J. F. Linduska (ed.), *Waterfowl tomorrow* (pp. 611–619). Washington, DC: U.S. Fish and Wildlife Service.

Ryan, M. R., Pierce, R. A., Suedkamp-Wells, K. M., & Kerns, C. K. (2002). Assessing bird population responses to grazing. In W. Hohman (Ed.), *Migratory bird responses to grazing* (pp. 16–34). Washington, DC: Natural Resources Conservation Service.

Saunders, C. A., Arcese, P., & O'Connor, K. D. (2003). Nest site characteristics in the song sparrow and parasitism by brown-headed cowbirds. *Wilson Journal of Ornithology, 115,* 24–28. doi:10.1676/02-057

Scheuhammer, A. M., & Norris, S. L. (1995). *A review of the environmental impacts of lead shotshell ammunition and lead fishing weights in Canada* (Occasional Pap. 88). Ottawa, ON, Canada: Canadian Wildlife Service.

Schlossberg, S., & King, D. I. (2008). Are shrubland birds edge specialists. *Ecological Applications, 18,* 1325–1330. doi:10.1890/08-0020.1

Schulz, J. H., Fleming, J., & Gao, X. (2011). 2011 Mourning dove harvest monitoring program annual report. Jefferson City, MO: Missouri Department of Conservation, Resource Science Division.

Schulz, J. H., Gao, X., Millspaugh, J. J., & Bermudez, A. J. (2007). Experimental lead pellet ingestion in mourning doves (*Zenaida macroura*). *The American Midland Naturalist, 158,* 177–190. doi:10.1674/0003-0031(2007)158[177:ELPIIM]2.0.CO;2

Schulz, J. H., Millspaugh, J. J., Bermudez, A. J., Gao, X., Bonnot, T. W., Britt, L. G., & Paine, M. (2006). Acute lead toxicosis in mourning doves. *Journal of Wildlife Management, 70,* 413–421. doi:10.2193/0022-541X(2006)70[413:ALTIMD]2.0.CO;2

Schulz, J. H., Millspaugh, J. J., Washburn, B. E., Wester, G. R., Lanigan, J. T., III, & Franson, J. C. (2002). Spent-shot availability and ingestion on areas managed for mourning doves. *Wildlife Society Bulletin, 30,* 112–120.

Schulz, J. H., Millspaugh, J. J., Zekor, D. T., & Washburn, B. E. (2003). Enhancing sport-hunting opportunities for urbanites. *Wildlife Society Bulletin, 31,* 565–573.

Schulz, J. H., Padding, P. I., & Millspaugh, J. J. (2006). Will mourning dove crippling rates increase with nontoxic-shot regulations? *Wildlife Society Bulletin, 34,* 861–865. doi:10.2193/0091-7648(2006)34[861:WMDCRI]2.0.CO;2

Schroeder, R. L., Cable, T. T., & Haire, S. L. (1992). Wildlife species richness in shelterbelts: Test of a habitat model. *Wildlife Society Bulletin, 20,* 264–273.

Schwartz, C. W., & Schwartz, E. R. (1995). *The wild mammals of Missouri.* Columbia, MO: University of Missouri Press.

Semlitsch, R. D. (2000). Principles for management of aquatic-breeding amphibians. *Journal of Wildlife Management, 64,* 615–631. doi:10.2307/3802732

Sharrow, S. H. (2001). Effects of shelter tubes on hardwood tree establishment in western Oregon silvopastures. *Agroforestry Systems, 53,* 283–290. doi:10.1023/A:1013319308930

Shaw, G. W., Dey, D. C., Kabrick, J., Grabner, J., & Muzika, R. M. (2003). Comparison of site preparation methods and stock types for artificial regeneration of oaks in bottomlands. In W. Van Sambeek, J. O. Dawson, F. Ponder, Jr., E. F. Lowenstein, and J. S. Fralish (Eds.), *Proceedings of the 13th Central Hardwood Forest Conference, Urbana, IL* (Gen. Tech. Rep. NC-234, pp. 186–199). St. Paul, MN: U.S. Forest Service, North Central Research Station.

Soule, J. D., Carre, D., & Jackson, W. (1990). Ecological impact of modern agriculture. In C. R. Vandermeer and P. Rosset (Eds.), *Agroecology* (pp. 165–188). New York: McGraw-Hill.

Stainback, G. A., & Alavalapati, J. R. R. (2004). Restoring longleaf pine through silvopasture practices: An economic analysis. *Forest Policy and Economics, 6,* 371–378. doi:10.1016/j.forpol.2004.03.012

Stanturf, J. A., & Gardiner, E. S. (2000). Restoration of bottomland hardwoods in the Lower Mississippi Alluvial Valley. Paper presented at *Sustaining forests: The science of forest assessment, Durham, NC.* Asheville, NC: Southern Valley Forest Resource Assessment. https://www.srs.fs.fed.us/sustain/conf/ppr/stanturf-ssf2000.pdf

Stanturf, J. A., Schoenholtz, S. H., Schweitzer, C. J., & Shepard, J. P. (2001). Achieving restoration success Myths in bottomland hardwood forest. *Restoration Ecology, 9,* 189–200. doi:10.1046/j.1526-100x.2001.009002189.x

Stribling, H. L. (1994). Hunting leases and permits (ANR-0541). Auburn, AL: Alabama Cooperative Extension System.

Swihart, R. K., & Yahner, R. H. (1983). Browse preferences of jackrabbits and cottontails for species used in shelterbelt plantings. *Journal of Forestry, 81,* 92–94.

Templeton, A. R., Brazeal, H., & Neuwald, J. L. (2011). The transition from isolated patches to a metapopulation in the eastern collared lizard in response to prescribed fires. *Ecology, 92,* 1736–1747. doi:10.1890/10-1994.1

Tewksbury, J. J., Garner, L., Garner, S., Lloyd, J. D., Saab, V., & Martin, T. (2006). Tests of landscape influence: Nest predation and brood parasitism in fragmented ecosystems. *Ecology, 87,* 759–768. doi:10.1890/04-1790

Thompson, F. R., III, Donovan, T. M., DeGraaf, R. M., Faaborg, J., & Robinson, S. K. (2002). A multi-scale perspective of the effects of forest fragmentation on birds in eastern forests. In T. L. George & D. S. Dobkin (Eds.), *Effects of habitat fragmentation on birds in western landscapes: Contrasts with paradigms from the eastern United States* (Stud. Avian Biol. 25, pp. 8–19). Camarillo, CA: Cooper Ornithological Society.

Tranel, M. A., & Kimmel, R. O. (2009). Impacts of lead ammunition on wildlife, the environment, and human health—A literature review and implications for Minnesota. In R. T. Watson, M. Fuller, M. A. Pokras, & G. Hunt (Eds.), *Ingestion of lead from spent ammunition: Implications for wildlife and humans* (pp. 318–337). Boise, ID: The Peregrine Fund. doi:10.4080/ilsa.2009.0307

Twedt, D. J. (2006). Small clusters of fast-growing trees enhance forest structure on restored bottomland sites. *Restoration Ecology, 14,* 316–320. doi:10.1111/j.1526-100X.2006.00134.x

Twedt, D. J., & Best, C. (2004). Restoration of floodplain forests for the conservation of migratory landbirds. *Ecological Restoration, 22,*194–203. doi:10.3368/er.22.3.194

Twedt, D. J., & Cooper, R. (2005). *Bird use of reforestation sites: Influence of location and vertical structure* (Tech. Note 190-34). Washington, DC: USDA–NRCS.

Twedt, D. J., & Loesch, C. R. (1999). Forest area and distribution in the Mississippi alluvial valley: Implications for breeding bird conservation. *Journal of Biogeography, 26,* 1215–1224. doi:10.1046/j.1365-2699.1999.00348.x

Twedt, D. J., & Portwood, J. (1997). Bottomland hardwood reforestation for neotropical migratory birds: Are we missing the forest for the trees? *Wildlife Society Bulletin, 25,* 647–652.

Twedt, D.J., Wilson, R. R., Henne-Kerr, J. L., & Grosshuesch, D. A. (2002). Avian response to bottomland hardwood reforestation: The first 10 years. *Restoration Ecology, 10,* 645–655. doi:10.1046/j.1526-100X.2002.01045.x

USDA. (2019). *Conservation*. Washington, DC: USDA. https://www.usda.gov/topics/conservation.

U.S. Fish and Wildlife Service. (2010). *Recovery plan for the ivory-billed woodpecker* (Campephilus principalis). Atlanta, GA: U.S Fish and Wildlife Service, Southeast Region. www.fws.gov/ivorybill/pdf/IBWRecoveryPlan2010.

Vander Haegen, W. M., & DeGraaf, R. M. (1996). Predation on artificial nests in forested riparian buffer strips. *Journal of Wildlife Management*, *60*, 542–550. doi:10.2307/3802071

Van Sambeek, J. W., & Garrett, H. E. (2004). Ground cover management in walnut and other hardwood plantings. In C. H. Michler, P. M. Pijut, J. W. Van Sambeek, M. V. Coggeshall, J. Seifert, K. Woeste, F. Ponder, Jr. (Eds.), *Black walnut in a new century: Proceedings of the 6th Walnut Council Symposium, Lafayette, IN* (Gen. Tech. Rep. NC-243, pp. 85–100). St. Paul, MN: U.S. Forest Service, North Central Research Station.

Vesely, D. G., & McComb, W. C. (2002). Salamander abundance and amphibian species richness in riparian buffer strips in the Oregon Coast Range. *Forest Science*, *48*, 291–297.

Walter, D., & Pierce, R. (2007). Agroforestry and wildlife. In Training manual for applied agroforestry practices (pp. 131–152). Columbia, MO: University of Missouri, Center for Agroforestry.

Warner, R. E., & Etter, S. L. (1985). Farm conservation measures to benefit wildlife, especially pheasant populations. *Transactions of the North American Wildlife and Natural Resources Conference*, *50*, 135–141.

Warrington, B. M., Aust, W. M., Barrett, S. M., Ford, W. M., Dolloff, C. A., Schilling, E. B., . . . Bolding, M. C. (2017). Forestry best management practices relationships with aquatic and riparian fauna: A review. *Forests*, *8*, 331. doi:10.3390/f8090331

Warwick, J. A. (2003). Distribution and abundance of swamp rabbits and bats in fragmented wetland forests of southeast Missouri (Master's thesis). University of Missouri. http://merlin.mobius.umsystem.edu:80/record=b5115999~S8

Washburn, B. E., Barnes, T. G., & Sole, J. D. (2000). Improving northern bobwhite habitat by converting tall fescue fields to native warm-season grasses. *Wildlife Society Bulletin*, 28, 97–104.

Wehrle, B. W., Kaminski, R. M., Leopold, B. D., & Smith, W. P. (1995). Aquatic invertebrate resources in Mississippi forested wetlands during winter. *Wildlife Society Bulletin*, 23, 774–783.

West, A. S., Keyser, P. D., Lituma, C. M., Buehler, D. A., Applegate, R. D., & Morgan, J. (2016). Grasslands bird occupancy of native warm-season grass. *Journal of Wildlife Management*, *80*, 1081–1090. doi:10.1002/jwmg.21103

Wiens, J. A., Crawford, C. S., & Gosz, J. R. (1985). Boundary dynamics: A conceptual framework for studying landscape ecosystems. *Oikos*, *45*, 421–427. doi:10.2307/3565577

Wigley, T. B., Jr., & Filer, T. H., Jr. (1989). Characteristics of greentree reservoirs: A survey of managers. *Wildlife Society Bulletin*, *17*, 136–142.

Wilcove, D. S. (1985). Nest predation in forest tracts and the decline of migratory songbirds. *Ecology*, *66*, 1211–1214. doi:10.2307/1939174

Williams, S. C., & Short, M. R. (2014). Evaluation of eight repellents in deterring eastern cottontail herbivory in Connecticut. *Human–Wildlife Interactions*, *8*(1), 12.

Winter, M., Johnson, D. H., & Faaborg, J. (2000). Evidence for edge effects on multiple levels in tallgrass prairie. *The Condor*, *102*, 256–266. doi:10.1093/condor/102.2.256

Wright, B. A., & Kaiser, R. A. (1986). Wildlife administrators' perceptions of hunter access problems: A national overview. *Wildlife Society Bulletin*, *14*, 30–35.

Wright, B. A., Kaiser, R. A., & Nicholls, S. (2002). Rural landowner liability for recreational injuries: Myths, perceptions, and realities. *Journal of Soil and Water Conservation*, *57*, 183–191.

Young, G. L., Karr, B. L., Leopold, B. D., & Hodges, J. D. (1995). Effect of greentree reservoir management on Mississippi bottomland hardwoods. *Wildlife Society Bulletin*, *23*, 525–531.

Zhang, D., Hussain, A., & Armstrong, J. B. (2006). Supply of hunting leases from non-industrial private forest lands in Alabama. *Human Dimensions of Wildlife*, *11*, 1–14. doi:10.1080/10871200500470910

Study Questions

1. What is the tradeoff, from a wildlife standpoint, for placing a shelterbelt in a grassland setting?

2. List the factors you should consider when deciding on the location and placement of an agroforestry site to maximize wildlife benefits.

3. Describe two ways the surrounding landscape can influence the benefits of agroforestry for wildlife.

4. Sometimes agroforestry plantings attract wildlife that we do not want. Discuss the methods that could be used to minimize potential damage to crop or tree plantings.

5. What are some of the ways managers can diversify habitat?

6. Before initiating a lease hunting program, what are the factors you should consider?

7. Describe the two overarching approaches to habitat incentives programs issued by the federal government.

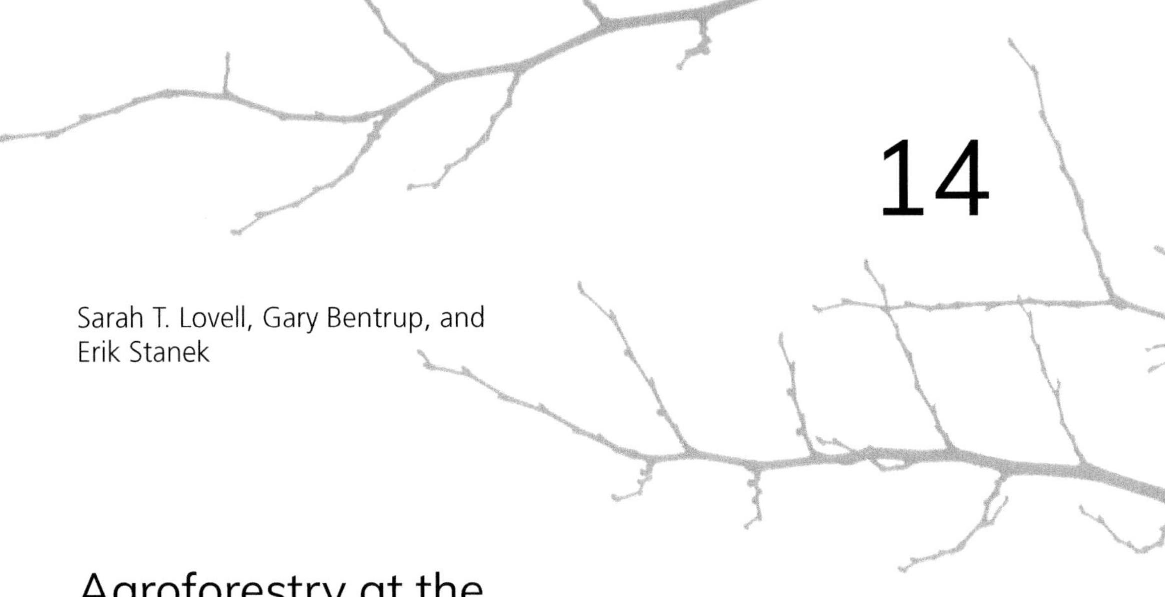

14

Sarah T. Lovell, Gary Bentrup, and
Erik Stanek

Agroforestry at the Landscape Level

Much of the work on agroforestry in North America has focused on six distinct practices—windbreaks, silvopasture, alley cropping, riparian and upland buffers, forest farming, and urban food forests—considering their functions, applications, and designs. These practices are typically planned and implemented for sites on individual farms and ranches to meet producer objectives, and therefore a site-scale perspective is the dominant lens through which these interventions are viewed. Incorporating a landscape perspective, however, can add significant value when designing agroforestry practices for enhancing multifunctionality. Understanding how individual sites function in the larger landscape can help identify where and how to design agroforestry practices to more effectively produce ecosystem services. A landscape framework involves looking beyond field boundaries and property lines to determine how a site is influenced by off-site conditions and how the site affects the surrounding landscape. At this broader scale, connections with other land uses become relevant, and more comprehensive land-based solutions can be developed.

This chapter focuses on the application of a landscape-level perspective to enhance delivery of ecosystem goods and services from agroforestry in order to create more resilient agricultural landscapes. The first section provides an overview of landscape ecology, the patch–corridor–matrix model, recent approaches for improving matrix quality, and the multifunctional landscape framework. The second section explores opportunities for agroforestry to contribute to a culture-based food supply that could improve human health and build on local knowledge. In the next section, an introduction to methods for assessing landscapes offers guidance for strategically placing agroforestry practices based on landscape and site conditions. Merging theory and application, a case study of the Upper Sangamon River Watershed demonstrates the value of research methods applied at the landscape scale. The final section describes how expanding landscape-scale agroforestry research could help to bring greater benefits and broader implementation.

Background

Landscape Ecology Framework

The discipline of landscape ecology—the study of biological, physical, and human interactions of a geographical area—offers a basis for integrating agroforestry into the landscape (Forman & Godron, 1986). Landscape ecology considers the spatial patterns of the fixed landscape elements that create landscape structure, which influences the movement of

North American Agroforestry, Third Edition. Edited by Harold E. "Gene" Garrett, Shibu Jose, and Michael A. Gold.

things such as animals, water, and nutrients. Landscape functions (the flow of things between landscape elements) and changes with time are characterized using various landscape ecology approaches.

While broader concepts guiding the discipline are more than a century old, the term *landscape ecology* was first used in 1939 by Carl Troll, a German geographer (Turner, 1989, 2005). In the 1980s, landscape ecology found a foothold in North America, and the patch–corridor–matrix model emerged to provide a framework, general principles, and a language for describing landscape structure (Forman, 1995a). The model describes the landscape as a "mosaic" consisting of three different types of spatial elements (patches, corridors, or matrix), each of which can be either natural or human in origin (Forman, 1995a, 1995b). The variation in plant communities across the landscape creates the patchwork heterogeneity that can be viewed in aerial images.

The *matrix* consists of the dominant landscape type or the background in the mosaic. The *patches* are areas of vegetation distinct from the surrounding landscape. Patches often consist of seminatural habitats that may be large enough to protect interior species and water networks, but they can also be smaller "stepping stones" that may support dispersal of species between large patches (Forman, 1995b). The *corridors* are linear features, often narrow strips of vegetation, that can be isolated elements but are more commonly found connecting patches of similar vegetation.

The patch–corridor–matrix concept has been widely used for sustainable landscape planning, in both urban and rural contexts (Lovell & Johnston, 2009b). For applications in temperate agroforestry, the matrix would typically consist of cropland. Treed habitats such as those involving forest farming could serve as patches, while riparian buffers and windbreaks might serve as corridors. Small patches of trees can serve as "stepping stones" that also increase connectivity (Forman, 1995b). Agroforestry elements typically perform six different landscape functions (Fig. 14–1).

This model, however, is less effective in describing agroforestry practices that are highly integrated into the landscape, such as alley cropping and silvopasture. In those cases, the resolution at which the treed features become distinct from the crops is so fine that an alternative approach is needed to fully capture the landscape heterogeneity and related ecological functions.

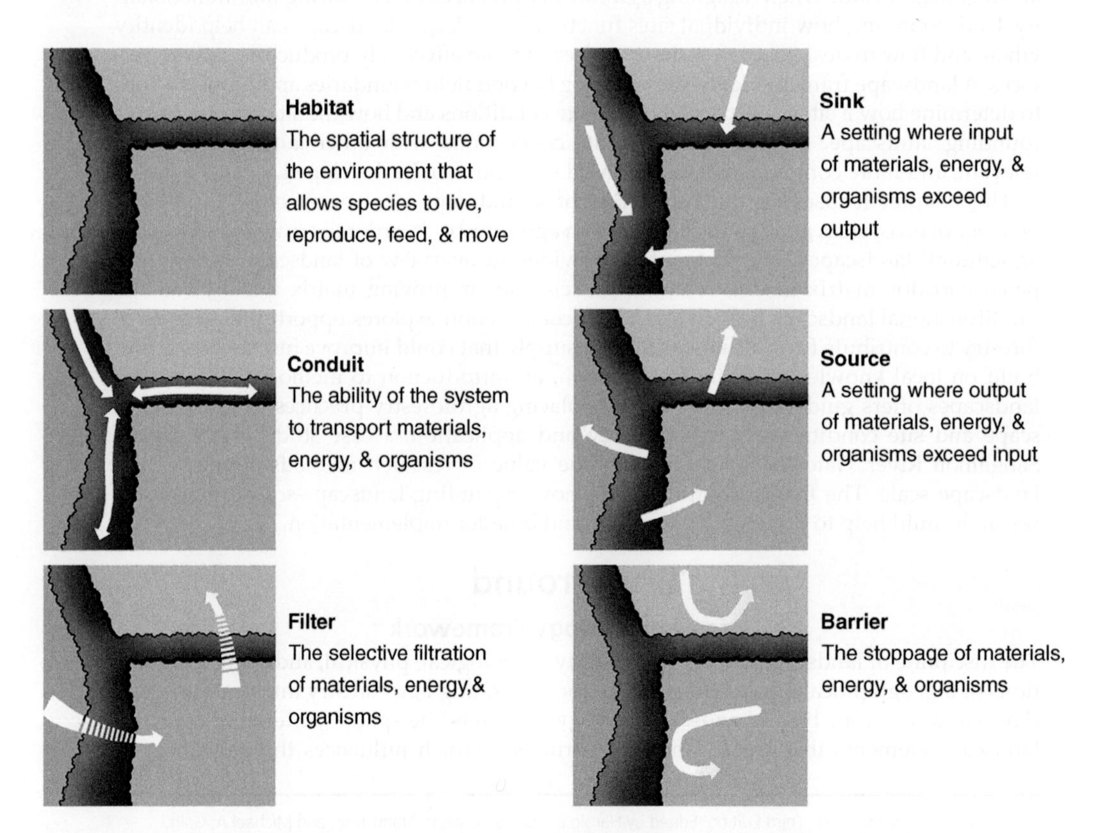

Habitat
The spatial structure of the environment that allows species to live, reproduce, feed, & move
Sink
A setting where input of materials, energy, & organisms exceed output
Conduit
The ability of the system to transport materials, energy, & organisms
Source
A setting where output of materials, energy, & organisms exceed input
Filter
The selective filtration of materials, energy,& organisms
Barrier
The stoppage of materials, energy, & organisms

Fig. 14–1. Six landscape-level functions of agroforestry provided within the patch–corridor–matrix model.

Matrix Quality

Consistent with the patch–corridor–matrix model, conservation efforts have historically emphasized the protection of large, high-quality patches and the development of corridors to connect them. An alternative approach is to improve the quality of the entire matrix by incorporating smaller features at a finer scale or transitioning to an alternative land use altogether (Perfecto & Vandermeer, 2002; Vandermeer & Perfecto, 2007). Increasing the complexity of landscape structure through the addition of perennial habitats has been proposed as the best alternative for reversing the negative impacts of landscape simplification and the resulting loss of ecosystem services. These perennial habitats would support and conserve the biodiversity that serves as the foundation of ecosystem services (Landis, 2017). Additionally, improving the complexity of the landscape can enhance cultural functions including visual quality and recreation (Angileri & Toccolini, 1993; de la Fuente de Val, Atauri, & de Lucio, 2006; Dramstad et al., 2001).

The integration of seminatural habitats such as agroforestry features will result in greater fine-scale heterogeneity, which will increase the connectivity of the entire landscape (Bailey, 2007; Lovell & Johnston, 2009a). Large forest patches are key to supporting the conservation of interior species that are often at risk (Robles, Flather, Stein, Nelson, & Cutko, 2008). By creating the necessary tree structure to function like a forested buffer, agroforestry can enlarge the area of effective interior forest habitat and turn smaller, less viable patches into more effective interior habitat (Dosskey, Bentrup, & Schoeneberger, 2012). Small isolated patches and scattered trees can also play an important role by serving as seed sources for the surrounding landscape (Benayas, Bullock, & Newton, 2008). Agroforestry habitats cover only a small proportion of most landscapes, yet the contributions to ecosystem functioning can be disproportionately large (Boutin, Jobin, & Belanger, 2003; León & Harvey, 2006). By focusing on the quality of the matrix, we move beyond the simple valuation of habitats as either "suitable" or "uninhabitable" (Baudry et al., 2003) to capture those agroforestry practices that truly combine trees with crops or livestock, such as alley cropping and silvopastoral systems.

Multifunctional Landscape Framework

With a focus on matrix quality as an appropriate goal for agroforestry implementation, a multifunctional landscape approach could offer a framework for developing more comprehensive land-based solutions. Multifunctional landscapes demonstrate that agriculture has the potential to provide not only production functions (goods) but also ecological functions such as biodiversity, nutrient cycling, and carbon sequestration and cultural functions including recreation, cultural heritage, and scenic beauty. The idea of landscape "functions" is consistent with the "ecosystem services" framework, where the provisioning, regulating, and cultural services are provided by different ecosystems or landscapes (Madureira, Rambonilaza, & Karpinski, 2007). The multifunctional landscape approach encourages a focus on the overall performance of agroecosystems, in which multiple functions can be combined or stacked, instead of considering only production (Dosskey, Wells, Bentrup, & Wallace, 2012; Jordan & Warner, 2010; Lovell et al., 2010; Lovell & Johnston, 2009a) (Fig. 14–2). Similarly, approaches that seek to bundle ecosystem

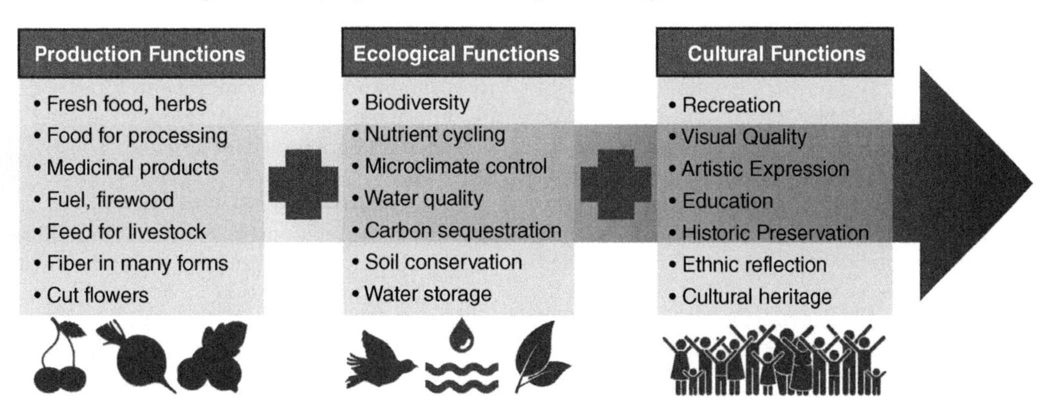

Increasing landscape performance by stacking functions
Production Functions
• Fresh food, herbs
• Food for processing
• Medicinal products
• Fuel, firewood
• Feed for livestock
• Fiber in many forms
• Cut flowers
Ecological Functions
• Biodiversity
• Nutrient cycling
• Microclimate control
• Water quality
• Carbon sequestration
• Soil conservation
• Water storage
Cultural Functions
• Recreation
• Visual Quality
• Artistic Expression
• Education
• Historic Preservation
• Ethnic reflection
• Cultural heritage

Fig. 14–2. Multifunctional landscape framework for increasing landscape performance.

services could result in comparable outcomes (Huang et al., 2015).

In many regions throughout the world, agricultural policy has promoted landscape multifunctionality (also referred to as "multifunctionality of agriculture"). In Europe, for example, agri-environmental schemes provide public funds that pay farmers for the benefits they provide to society including biodiversity conservation, water quality, carbon sequestration, and rural tourism (Sutherland, 2004). Examples of objectives from these programs that could cover agroforestry include "preservation of landscape and historical features such as hedgerows, ditches and woods" and "conservation of high-value habitats and their associated biodiversity" (European Commission, 2017). Similar policies exist in the United States, such as the Conservation Reserve Program administered by the USDA. These policies, however, are often written in a manner that intentionally separates production from conservation by discouraging food production or specifically restricting harvest (Raymond, Reed, Bieling, Robinson, & Plieninger, 2016). Furthermore, little consideration is given to cultural functions, and funding is often not provided to support these efforts (Raymond et al., 2016). Despite the challenges, agricultural policies are starting to offer a vision for redesigning the landscape to support the "transition to sustainable food systems" (Gliessman, 2010), while at the same time expanding the suite of goods and services provided (Jordan & Warner, 2010).

Even with a goal of improving landscape performance, tradeoffs between functions will inevitably exist (Lovell et al., 2010), and these will be most intense on highly productive portions of a farm. For instance, soil fertility is a crucial determinant of the location and extent of treed habitats, with areas of high fertility more likely to be deforested for cultivated crops (Seabrook, McAlpine, & Fensham, 2007). One alternative is using agroforestry practices on sensitive or marginal areas instead of targeting fields with highly productive soils that consistently produce the best crop yields (Jordan & Warner, 2010). These marginal or sensitive areas are often located in floodplains or wet spots, where they are hotspots for the loss of nutrients such as nitrogen and phosphorus. With such a strategy, the establishment of perennial vegetation in agroforestry would have the greatest benefits in ecological functions while minimizing losses in productivity of conventional crops (Lovell et al., 2018).

Several multifunctional landscape approaches are available to help guide agroforestry planning and navigate the inevitable trade-offs. These approaches often emphasize whole-farm planning and design (Huang et al., 2015), working at the scale at which management and land use decisions are rendered (Rigby, Woodhouse, Young, & Burton, 2001; Rotz et al., 2005). Strategies to integrate or conserve trees on farms are more successful when they take into account the preferences and values of farmers and rural residents (Seabrook, McAlpine, & Fensham, 2008). Farm design methods that allow landowners to compare different alternatives for the landscape can be quite valuable for assessing overall performance in the multifunctional landscape framework (Stanek & Lovell, 2020; Stanek, Lovell, & Reisner, 2019). The Multifunctional Landscape Assessment Tool proposed by Lovell et al. (2010) is an example of an approach that allows the comparison of alternative scenarios for the design of an agroecosystem. Overall, the multifunctional agriculture approach provides a mechanism for agroforestry to do more than only produce food. The framework can also support the broad range of ecosystem services that improve the environmental and social health of communities, even offering a delivery mechanism for training growers in new skills, educating the public about food systems, and retaining cultural identity for communities (Leakey, 2014).

Contributing to a Culture-Based Food Supply

Incorporating a landscape perspective in agroforestry offers an opportunity to build on the multifunctional agricultural framework and create a culture-based food supply that supports human health and cultural values in addition to enhancing food security (Fig. 14–3). For instance, nuts and berries grown in agroforestry practices can play an important role in providing key vitamins and nutrients for human health in addition to supporting cultural connections to food. To have meaningful impact on these functions and objectives, landscapes incorporating agroforestry need be informed by cultural knowledge and be designed and implemented at scale.

Improving Human Health

Regarding the relationship between our food system and human health, we have historically focused on undernutrition in developing countries, often neglecting the issues of overconsumption and poor nutritional quality of food found in North America and other temperate zones. As the quality of the human diet deteriorates, new health risks emerge in our

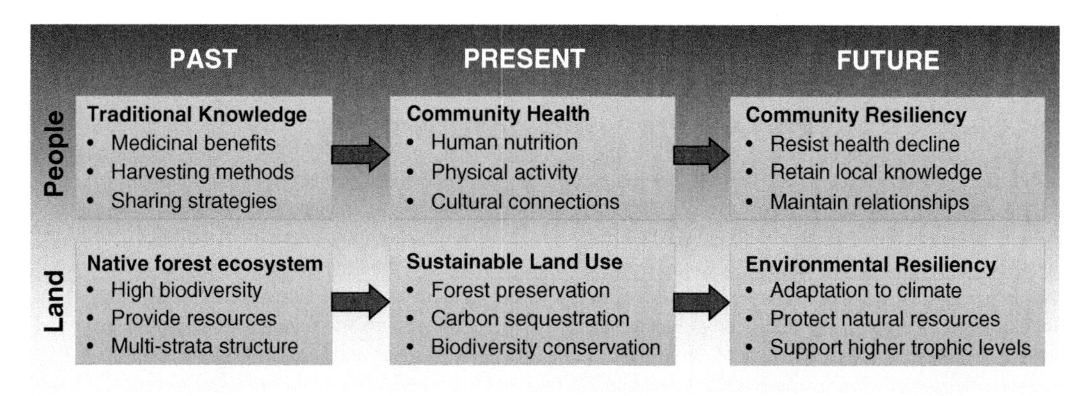

Fig. 14–3. For a culture-based food supply, contributions from land and people of the past can lead to a favorable present and future.

communities, including food allergies and intolerances, resistance to antibiotics, and conditions such as metabolic syndrome (Gordon, Negri, & Snyder, 2017). To reduce these risks, landscapes incorporating agroforestry can be purposefully designed to provide nutritional value for improving human health (Gordon et al., 2017). Agroforestry has the potential to reduce the prevalence of metabolic syndrome and improve community health by supplying natural foods such as berries (Park et al., 2015) and by encouraging physical activity in the care of tree crops (Baceviciene et al., 2013). Healthy diets, particularly those high in plant-based foods and lower in animal-based products, not only improve the longevity of individuals in a community but new evidence shows benefits for the biosphere. These consumption patterns result in lower greenhouse gas emissions, lower water use, and less energy consumption (Aleksandrowicz, Green, Joy, Smith, & Haines, 2016).

Building on Traditional Ecological Knowledge

Although indigenous communities may call agroforestry by a different name (e.g., traditional management, forestry, food security), their traditional ecological knowledge offers important lessons for how to create and manage agroforestry systems to support culture-based food systems (Rossier & Lake, 2014). In addition, agroforestry landscapes can offer a valuable connection to local traditional knowledge and cultural values. In the past, the study of indigenous agroforestry might have been considered less "scientific" than experimental research (Olofson, 1983), but we now recognize that these agroforestry systems and their associated landscape management practices can serve as a model for modern, locally adapted food systems (Schoeneberger, Bentrup, & Patel-Weynand, 2017). Traditional ecological knowledge facilitates the implementation of practices such as controlled burning, pruning, sowing, tillage, water management, and sustainable plant and animal harvesting (Box 14–1).

These agroforestry systems produced a range of traditional subsistence foods such as wild berries, nuts, and herbs that provided important health benefits with their high nutrient levels and medicinal values (Lila et al., 2014; Rossier & Lake, 2014). Plants also provided materials for basketry, dyes, and other culturally significant products. The cultivation and gathering of such items is recognized as an important physical and social activity for native communities (Baceviciene et al., 2013; Flint et al., 2011). The rules for sharing traditional ecological knowledge and resources and the arrangements for common property use are important components of sustaining productive landscapes, even today. The indigenous peoples of the Gwich'in region in the Northwest Territories of Canada offer insight into the deep level of knowledge and importance of retaining it. Elders in the community contribute to an important knowledge base on the abundance and distribution of native "dark" berries that offer nutritional and medicinal value. The landscape supports a wide range of species including cranberry (*Vaccinium vitis-idaea* L.), blueberry/bilberry (*Vaccinium uliginosum* L.), cloudberry (*Rubus chamaemorus* L.), red currant (*Ribes triste* Pall.), and black currant (*Ribes hudsonianum* Richardson). For the system to remain successful and sustainable, harvesters must be knowledgeable of the spatial distribution and maturity timing, at the same time respecting the property rights and unwritten rules of sharing (Teetl'it Gwich'in Renewable Resources Council, Parlee, & Berkes, 2006).

Box 14–1. The Karuk tribe, traditional ecological knowledge, and agroforestry

In Northern California, the Karuk tribe traditionally managed entire watersheds and ecosystems on their ancestral lands to meet their dietary, cultural, and spiritual needs. Their diverse, multistory agroforestry systems produced forest-grown foods, such as tanoak acorn, pine nut, huckleberry, hazelnut, camas bulb, gooseberry, raspberry, deer, elk, and mushroom, as well as salmon and eel from the Klamath River and its tributaries. In addition, these systems produced medicinal herbs like wormwood, plus willow, bear grass, and hazel for basket making. The tribe used prescribed burns at different intensities and frequencies, as well as a variety of pruning, gathering, and hunting techniques to manage their tanoak and Douglas-fir dominated forests. Fire cleared oak groves of encroaching conifers and killed the weevils that ruin acorns. The controlled burning removed dense underbrush, making it easier to harvest these resources while renewing the meadow grasses for grazing deer and elk. Fire also allowed willow and hazel trees to produce the straight shoots needed for baskets. For salmon dependent on clean, cold water, the burns helped maintain water quality and the amount of water returned to streams where salmon spawn. While these resources are still present, they are scarce compared with historic levels, and efforts are underway to use traditional ecological knowledge to help manage these forests. The Six Rivers National Forest is partnering with the Karuk tribe to combine their holistic vision and traditional ecological knowledge with lidar and GIS to help restore these resources and improve forest health (U.S. Forest Service, 2018).

Sources: (adapted from Lake, 2013; Lake, Tripp, & Reed, 2010)

Ron Reed of the Karuk Food Crew collecting gooseberries in a landscape managed by traditional ecological knowledge (photo credit: Colleen Rossier)

For indigenous communities, agroforestry can offer unique opportunities but also challenges related to climate change. Some evidence indicates that climate change will negatively impact the productivity of important wild berries in areas such as Alaska, where berry abundance is declining or becoming more variable (Hupp, Brubaker, Wilkinson, & Williamson, 2015). On the positive side, traditional practices offer insight for adaptation and mitigation in the face of variable climate conditions (Schoeneberger et al., 2017). The diversity of nutrient-rich species can serve as an adaptation strategy to improve the health of communities (Lila et al., 2014) and to spread the risk of inconsistent harvests (Altieri & Nicholls, 2017). The mitigation potential of traditional agroforestry can be found in the increase in carbon storage from above- and belowground biomass, as well as lower greenhouse gas emissions resulting from fewer inputs from pesticides, fertilizers, and energy (Altieri & Nicholls, 2017).

Assessing Landscapes for Agroforestry

While agroforestry practices offer an important application of landscape multifunctionality including a culture-based food supply, the potential to integrate trees into productive landscapes can be complicated in North America, particularly in those areas where annual crops are well established and highly productive. At the landscape scale, a goal might be to strategically place these treed systems where they could most effectively optimize the targeted benefits. This strategy also makes sense in terms of using public funds to incentivize agroforestry practices because the ecological and cultural functions can serve the broader community while any reduction in production functions (yield) primarily impacts the landowner. Using landscape analysis techniques, trained planners can play important roles in purposefully designating space for different uses, including the placement of permanent vegetation and habitat types such as agroforestry.

One tool to aid in strategically planning and designing agroforestry practices is landscape assessments. Landscape assessments spatially describe resource conditions within a larger planning area and identify opportunities to produce environmental and production benefits with targeted management activities, including agroforestry practices (McHarg, 1995). Landscape assessments provide a way to understand the relationships between landscape structure and functions, environmental problems, and agroforestry opportunities (Box 14–2). Assessments can be used to identify key conditions and processes in a landscape:

- problems and where they occur
- opportunities and where they occur
- sources and causes of the problems
- where and how resources flow across the landscape
- structure of the landscape and how it controls sources and movement

Assessing landscapes is greatly enhanced with the use of GIS—databases for managing, processing, and analyzing spatial information in a visual manner that aids decision-making. With the increasing availability of spatial data and enhanced computing power, the use of GIS has become commonplace in land use planning and natural resource management. For example, remotely sensed LiDAR (light detection and ranging) data can yield a variety of high-resolution data products such as 1-m digital elevation models that could be used for evaluating flow paths in riparian forest buffers (Wallace et al., 2018) or vegetation canopy maps that can guide tree thinning operations in silvopasture systems (Hung et al., 2005). The Soil Survey Geographic (SSURGO) database provides soils information mapped at scales ranging from 1:12,000 to 1:63,360, suitable for planning at the farm to watershed scales (USDA–NRCS, 2016). Most states and federal agencies have GIS data clearinghouses that provide free or low-cost access to a broad selection of spatial data that can be useful for planning agroforestry practices.

Despite the availability of publicly available data sets and the benefits of using GIS, the use of these analysis and planning tools in temperate agroforestry applications has been limited (Bentrup & Leininger, 2002; Carver, Danskin, Zaczek, Mangun, & Williard, 2004; Ellis, Bentrup, & Schoeneberger, 2004; Fagerholm et al., 2016; Mattia, Lovell, & Fraterrigo, 2018). One way to use GIS in agroforestry is to rank and combine data layers based on suitability for a particular function (Fig. 14–4). A primary advantage of using GIS is the ability to merge the assessments to identify where multiple functions or objectives can be simultaneously achieved with agroforestry practices (Fig. 14–5).

Landscape Scale Applications

The following is a cross-section of examples where a landscape assessment approach can be used to augment the benefits derived from agroforestry. This list is not inclusive but rather illustrates the range of opportunities a landscape strategy can provide, and the real optimization comes from combining approaches to derive multifunctional systems.

Riparian Zones

Of the six agroforestry practices, riparian buffers typically receive the most attention in terms of placement within the landscape due to their prominent role in capturing and filtering runoff as well as their value for biodiversity functions. For water quality functions, the watershed or sub-watershed is a common boundary of analysis, and several different strategies have been used to determine the best placement and design of these practices. Fixed width buffers are the most straightforward, designating the protection of a given distance from the stream edge or other water body (e.g., a 50-m buffer zone on each side). Using GIS and remote sensing, buffer

Box 14–2. Landscape Assessment for the Prairie States Forestry Project

Severe dust storms of the 1930s in the U.S. Great Plains region spurred the creation of the Prairie States Forestry Project (PSFP), a federal program designed to combat soil erosion by planting windbreaks. Through the PSFP, more than 29,900 km (18,600 miles) of windbreaks were planted on 30,000 farms from North Dakota into Texas (Williams, 2005).

(a)

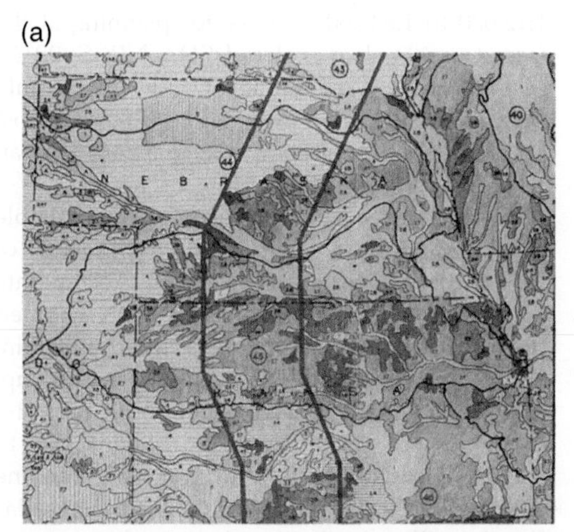

(b)

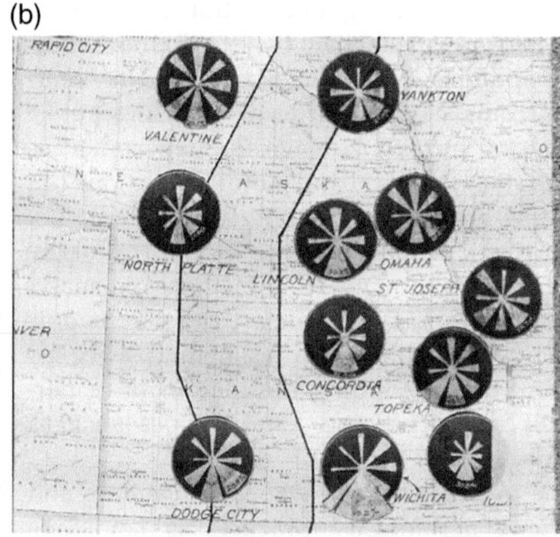

(A) *The erosion reconnaissance survey map depicts erosion types and severity;* **(B)** *summer wind direction is illustrated using wind roses. The boundary line outlines the 100-mile-wide and 1150-mile-long project area that roughly follows the 99th meridian.*

One of the innovative aspects of the PSFP was the use of biophysical and economic spatial data to determine where to locate the windbreaks. Planners concentrated efforts within a 161-km (100-mile) wide zone running from North Dakota to the Brazos River in Texas based on a minimal annual precipitation of 51 cm (20 inches) needed to establish the plantings. Within this zone, 14 resource maps were used to identify high-priority target areas for windbreak plantings. The resource maps included soils and natural vegetation, seasonal wind direction, erosion reconnaissance, regional farming systems, and land values. This suitability assessment process represents one of the earliest examples of analyzing landscapes for the placement of an agroforestry practice.

(a)

(b)

(A) *A dust storm rolls across eastern Colorado during the 1930s* (photo from the USDA–NRCS), *and* **(B)** *landowners tending to their windbreak planted with the Prairie States Forestry Project* (photo from the U.S. Forest Service).

Step 1: Determine the Question

Question X

Step 2: Identify Data Variables

Question X can be answered by slope and soil drainage

Step 3: Map Data Variables

A B C

A C B

Slope Map
A 0 to 3%
B 4 to 9%
C 10 to 15%

Soil Drainage Map
A Excessively Drained
B Well Drained
C Poorly Drained

Step 4: Develop Ranking Criteria

Slope			Soil Drainage		
A	B	C	A	B	C
3	2	1	2	3	1

3 - High Suitability
2 - Moderate Suitability
1 - Low Suitability

Step 5: Map Rankings for Each Variable

3 2 1

2 1 3

Slope Map

Soil Drainage Map

Step 6: Overlay the Maps to Obtain Composite Rankings

5 4 3 4 3 2 6 5 4

Highest numbers indicate best suitability to answer Question X

Fig. 14–4. The basic suitability assessment process begins with a question that can be answered by spatial data.

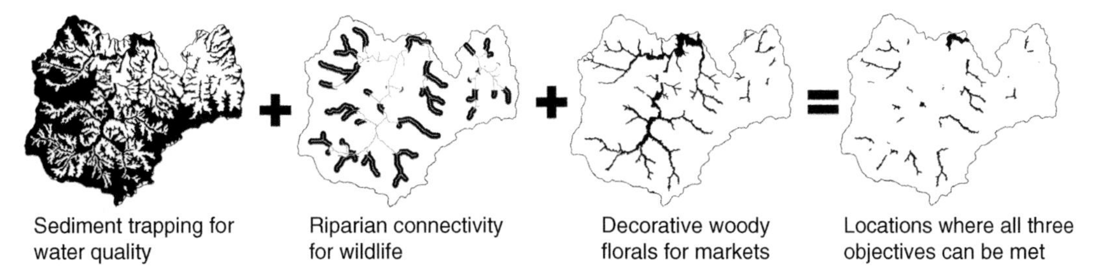

Fig. 14–5. An example of individual landscape assessments that identify suitable locations where riparian forest buffers can trap sediment, enhance connectivity for wildlife, and produce woody florals for markets. Combining the three assessments can pinpoint locations where all three objectives can be achieved simultaneously with a riparian forest buffer or similar agroforestry practice (source: Bentrup, Dosskey, Wells, & Schoeneberger, 2012).

effectiveness can be significantly enhanced by varying the width based on site conditions such as upslope land use (Basnyat, Teeter, Flynn, & Lockaby, 1999; Xiang, 1993). More complex approaches such as the soil survey technique and terrain analysis seek to place buffers in zones where runoff could best be intercepted (Qiu & Dosskey, 2012; Tomer et al., 2009; Wallace et al., 2018). These complex techniques tend to be more cost effective than simple riparian buffer approaches because they achieve higher water quality performance and often require less land to be converted (Qiu & Dosskey, 2012; Tiwari et al., 2016).

For biodiversity functions, riparian assessments can be used to identify locations where riparian buffers can enhance wildlife movement (Bentrup & Kellerman, 2004) and determine what plant species may be most optimal (Carver et al., 2004). Assessments can also be conducted to determine where riparian buffers can most effectively shade streams and manage water temperatures for cold-water habitat, a function that is becoming increasingly important due to climate change (Bentrup & Dosskey, 2017).

Marginal Land

For the placement of many agroforestry practices, a common strategy is to target lands that are considered sensitive or "marginal" for conventional crop production (Lovell et al., 2018; Tsonkova, Bohm, Quinkenstein, & Freese, 2012). Lands with highly erodible soils, for example, are more sensitive to the loss of topsoil, a problem that could be reduced substantially by converting the land into the perennial continuous cover of agroforestry (Garrett, McGraw, & Walter, 2009). Trees and shrubs would reduce environmental impacts and economic risks, as these woody systems can stabilize the soil and retain nutrients (Molnar, Kahn, Ford, Funk, & Funk, 2013). The use of marginal lands also makes sense from an economic perspective because marginal lands are often lower yielding and higher risk for conventional crop production (Rhoads, Lewis, & Andresen, 2016), so the landowner might be more open to alternative land use systems (Lovell et al., 2018).

A number of approaches and data are available for identifying marginal lands for the placement of agroforestry practices (Gopalakrishnan, Negri, & Snyder, 2011; Holzmueller & Jose, 2012; Mattia, Lovell, & Fraterrigo, 2018). With the high resolution and accuracy of modern maps, these spatial analyses could also identify problem "wet spots" within a field that can be sources of water quality problems as well as be low performing (Agnew et al., 2006; Brandes et al., 2016). Maps of crop productivity are also available for most regions, as with the cropland data layer provided by the USDA. In some cases, these analyses may yield small marginal areas in the middle of fields or other locations that can make management difficult and that will require an iterative design process with producers to develop a workable solution (Mattia, Lovell, & Fraterrigo, 2018).

Woody Crop Optimization

For agroforestry systems that contain a productive tree or shrub crop (e.g., timber, nuts or berries, woody florals, bioenergy, bio-oils), an entirely different approach to placement might be used. Geospatial tools and economic analyses can be combined to determine the best areas for optimizing the products derived from woody plants used in agroforestry practices (Bentrup & Leininger, 2002). Several woody crop optimization assessments are available to serve as examples (Ahmad, Goparaju, & Qayum, 2019; Reisner, de Filippi, Herzog, & Palma, 2007; Wallace & Young, 2008), and this landscape strategy can be particularly valuable because woody crop suitability will probably shift under climate change (de Sousa et al., 2017). These types of spatial assessments could help guide the creation of nutrient-rich landscapes that focus less on crop volume and calorie production and more on the nutritional content of a diverse set of plants for improved human health (Wood, 2018).

This approach could particularly make sense for an alley cropping system in which the tree productivity would be added onto the yield of the alley crop. Wolz and DeLucia (2019) tested this approach for the U.S. Midwest, modeling black walnut (*Juglans nigra* L.) systems grown for timber as a plantation (trees alone) or as rows in an alley cropping system with existing crops compared with the current corn (*Zea mays* L.)–soybean [*Glycine max* (L.) Merr.] rotation. Alley cropping systems had the highest economic rates of return on 23% of the land, some of which would not be considered marginal (Wolz & DeLucia, 2019). The approach is currently limited by the availability of high-resolution data on all of the necessary variables—soil suitability, timber prices, crop productivity, cash rents, and land cover.

Alley Crop Compatibility Zones

In a similar vein, agroforestry practices (particularly alley cropping) might be placed in regions with existing crops that are most compatible with trees, the distribution of which could be analyzed and mapped. For example, cool-season crops such as wheat (*Triticum aestivum* L.) have been successfully integrated with tree rows due to less competition for light during the active growing season between trees and the alley crops. Long-term studies in France have demonstrated the success of growing winter wheat in combination with hybrid walnut trees (Lovell et al., 2018). Forage hay crops also have good potential for the alley plantings, although less information is available on how the trees impact the productivity of the forage (Garrett et al., 2009). Also worth noting is the incompatibility of certain crops for use in an alley cropping system due to competition for resources or management constraints.

Understanding the spatial distribution of genetically modified, herbicide-resistant crop cultivars could be useful because these crops allow applications of various broad-spectrum herbicides late in the season, which could severely damage trees and shrubs grown in close proximity (Garrett et al., 2009).

Sensitive and Drought-Prone Areas

The placement of agroforestry practices can also be considered on a broader regional scale. For example, practices such as windbreaks can be particularly valuable in regions that are prone to drought or other severe conditions because the woody vegetation provides more favorable microclimate conditions. Windbreaks can offer an economic advantage for sensitive crops, as demonstrated in the arid cold temperate region of southern Patagonia, Argentina. In this very windy region, dense windbreaks reduce the wind speed by 85% at a ground distance equivalent to the height of the trees. Protection of sensitive crops such as strawberry (*Fragaria* sp.) and cherry (*Prunus avium* L.) can increase production substantially (Peri & Bloomberg, 2002). A shelterbelt system has also been developed for the Three-North Region of northern China. Initiated in 1978 and intended to go through 2050, the program covers about 40% of China's territory, providing windbreaks to improve microclimate conditions for sensitive crops, particularly to reduce drought stress in zones that are most limited by moisture (Zheng, Zhu, & Xing, 2016). The assessment process in the Prairie States Forestry Project (Box 2) offers a conceptual framework of what a regional-scale analysis could look like for addressing drought-prone areas.

Livestock Considerations

Just as sensitive crops can benefit from the protection of trees in harsh climates, the performance and comfort of livestock can also be improved. Silvopastoral systems are particularly valuable in areas where heat stress reduces livestock productivity. In the U.S. Southeast, for example, the addition of trees has been shown to provide a milder microclimate than open pasture (Karki & Goodman, 2015), and grazing is more consistent across the land (Karki & Goodman, 2010). The shading component in silvopasture systems has also been shown to improve forage quality, for example in cool-season grasses by increasing protein content while reducing fiber (Kallenbach, Kerley, & Bishop-Hurley, 2006). Windbreaks can also reduce stress on sensitive animals by providing more comfortable conditions (Dronen, 1988). Rows of trees can be placed so they shelter pastures, feedlots, and other livestock holding areas from cold winter winds, hot summer winds, or both. Windbreaks provide an additional benefit of mitigating odors and reducing the movement of particulate matter when placed downwind of livestock areas (Tyndall & Colletti, 2007; Willis et al., 2017). The prevailing winds, topographic setting, canopy cover, and other biophysical and social data can be used to locate and manage agroforestry practices to address livestock considerations (Adhikari et al., 2018; Hung et al., 2005; Mor-Mussery, Leu, & Budovsky, 2013).

Case Study: Application of Agroforestry in the Upper Sangamon River Watershed

A case study of the Upper Sangamon River Watershed (USRW) in Illinois demonstrates the potential application of agroforestry at the landscape scale. This research was conducted by a team at the University of Illinois from 2013–2018. The USRW study site is located in central Illinois, where intensively managed field corn and soybean occupy the majority of the landscape. The multistep study included a survey of stakeholders to examine adoption potential, mapping of marginal lands that would be targets for a transition to agroforestry, and development of design scenarios for individual landowners (Fig. 14–6). From this work, various solutions emerged for supplying food and promoting healthy land through agroforestry applications that included trees and shrubs that were themselves productive.

The landowner survey was designed to investigate motivators and barriers for implementing new multifunctional perennial cropping systems, including agroforestry, on marginal land. While the survey was sent to the majority of landowners in the counties of the USRW, the response rate was relatively low (9.2%), probably resulting in a sample somewhat biased toward an interest in these systems. Of those who responded, approximately one-third indicated they would be willing to convert their marginal land to multifunctional perennial cropping systems. The most popular perennial cropping options were bioenergy crops, hay, and fruit and nut trees. To follow up the results of the survey, a focus group revealed important barriers including the need for markets and infrastructure, as well as a labor force to manage the systems (Mattia, Lovell, & Davis, 2018).

Data from the survey were used to classify different types of landowners for predicting the potential to adopt multifunctional perennial cropping systems. Multivariate analysis revealed six categories of landowners from the pool of

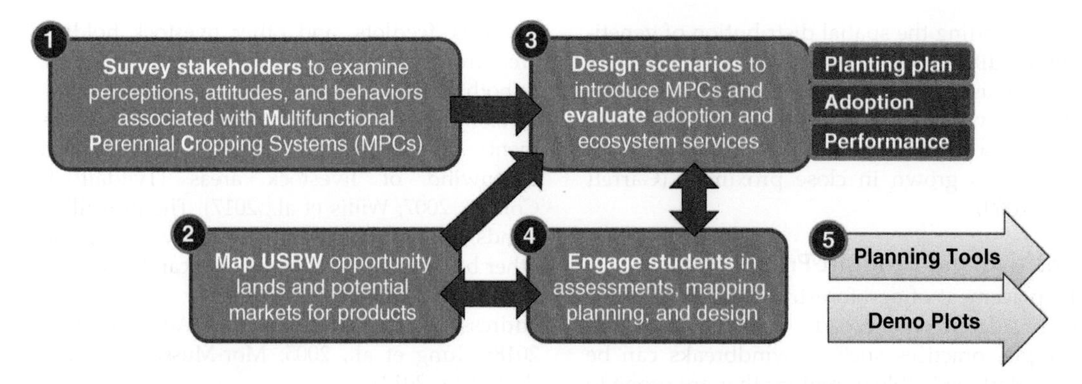

Fig. 14–6. Process for evaluating the Upper Sangamon River Watershed (USRW) in central Illinois for potential multifunctional perennial cropping systems.

survey respondents. Educated Networkers were high probability adopters, and they were characterized by high education and having farming as a secondary occupation. The Young Innovators group, which had the youngest average age but high decision-making involvement, also showed high adoption potential. Groups with medium adoption potential included Small Conventional and Large Conventional landowners, both of which identified most with conventional farming but differed in the size of operation. The lowest adoption potential was related with the Money Motivated (mostly cash rent lease arrangements) and Hands-Off (preference for low labor and time requirements) groups. From the landscape-scale perspective, focusing efforts on the needs of the high potential adopters could help with getting systems on the ground in the short term (Mattia, Lovell, & Davis, 2018).

To better understand the landscape of the USRW, the marginal land was identified and mapped using a suitability assessment model. Soil erosion potential, crop productivity, and other land traits were spatially analyzed to determine suitability. Target areas were often distributed in areas with low crop productivity and high potential for erosion. The total land classified as marginal was 7% of the agricultural land in the USRW or 18,685 ha (Fig. 14–7). Running the model with a scenario in which the identified marginal land would be converted to multifunctional perennial crops showed that simulated soil erosion could be reduced by 56% across the watershed. The approach demonstrates the potential for the conversion of a relatively small portion of land to have a large impact on the environmental health of the landscape (Mattia, Lovell, & Fraterrigo, 2018).

With the marginal land maps and survey results as a guide, 15 rural landowners were engaged in a collaborative design process to develop potential solutions involving multifunctional perennial crops, such as agroforestry, specific to their properties. Interviews were conducted with participants at the beginning of the design process to determine their preferred locations for these systems and to explore their perceptions of different system and species types. In the next step of the design process, three scenarios were developed for each landowner, distinguished by their focus on production, conservation, or cultural functions. Each of the scenarios was developed as a scaled planting plan and visualized in color (Fig. 14–8). In the final phase of the process, landowners assessed the three scenarios, providing rankings and comments on preferences for designs, as well as ideas for motivators and barriers in implementing each alternative.

The results of the collaborative design process revealed several important findings that apply broadly to our understanding of integrating agroforestry at the landscape scale. In the pre-design interviews, landowners expressed strong interest in edible food production, even above timber production, which is a more common use of trees in agroforestry. In fact, systems that were strictly timber-based were ranked lowest by this set of landowners. In ranking the design scenarios, production scenarios were most preferred because of the profit potential, but aspects of each of the scenarios were appreciated by most landowners. In working toward a final design, landowners often included some aspects of cultural and conservation focuses, in addition to production functions. These results, therefore, provide strong support for the concept of truly "multifunctional" designs, as discussed above (Stanek et al., 2019).

The combined efforts of the work in the USRW help to distill the types of agroforestry practices

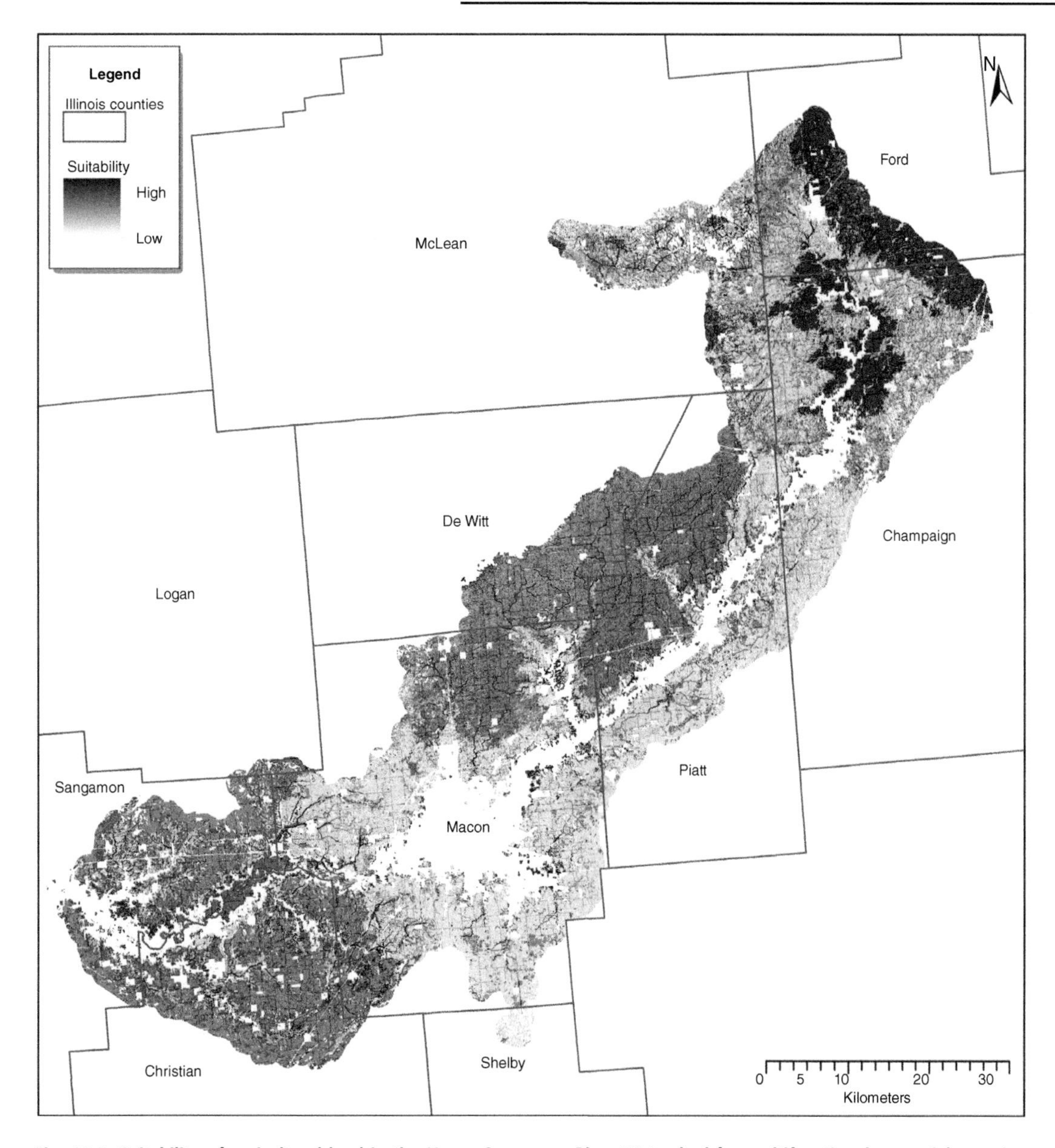

Fig. 14–7. Suitability of agricultural land in the Upper Sangamon River Watershed for multifunctional perennial cropping systems based on marginal land characteristics emphasizing areas of high soil erosion and low crop productivity.

that have the greatest potential for landscapes in which conventional annual crops dominate. Because of the strong drive for land to be productive, agroforestry solutions that offer marketable products such as fruits and nuts could be popular, particularly if they are combined with conservation goals on marginal land. Much of the land in this region could benefit from the addition of windbreaks strategically placed to protect sensitive crops, livestock, or human establishments. Typically, the windbreaks (including those established through the Conservation Reserve Program) are designed with several rows of trees and shrubs. By selecting species that would eventually yield nuts or fruits, landowners might be more likely to retain the plantings as they become more productive over the years. Even narrowing a species list down to those native to the region still allows the production of high-value, locally specific goods. Thinking at the landscape scale allows a productive agroforestry approach that can supply healthy food and promote land stewardship, contributing to culture-based food supply. Ongoing work to establish demonstration plots and develop planning tools could support agroforestry adoption.

Fig. 14–8. Example scenario developed for a landowner depicting a multifunctional agroforestry border around a farmstead.

Conclusions

Despite the growing awareness of the importance of landscape-scale analysis and planning for agroforestry adoption and implementation, further research is needed. Of particular importance is the need to gain a better understanding of the opportunities, synergies, and trade-offs of designing for multifunctional landscapes using agroforestry. Robust but easy-to-use methods are needed to conduct these evaluations to aid in designing high-performing agroforestry systems. This is an evolving and dynamic process as new goals and objectives continually get added into the mix, such as creating culture-based food systems that support human health. In addition, emerging insights from traditional ecological knowledge should be better integrated.

Another key research area is to investigate and expand the plant material available for agroforestry using a landscape approach. Agroforestry plant species could be collected, tested, and selected across a wide range of landscape conditions including marginal lands that are typically considered less productive (Schoeneberger et al., 2017). At the same time, crop species that have not traditionally been integrated into agroforestry might be evaluated for utility and compatibility (e.g., tolerance to interspecific competition, potential synergistic relationships). New and underutilized specialty crops could be particularly appropriate for this application because they can offer an economically viable alternative (Mori, Gold, & Jose, 2017). Climate envelope models and other scenario-based strategies may be necessary to understand the suitability and viability of plant material under climate change (Schoeneberger et al., 2017). Future research will also need to include a breeding effort to improve these crops for unique situations and changing climatic conditions. Leakey (2014) recommended collecting and selecting varieties with preferred traits from wild plants found locally, even exploring the potential to domesticate new food crops from native ecosystems.

An opportunity to broaden the implementation of agroforestry at the landscape level is to consider applications in nontraditional settings, particularly in urban areas. Agroforestry could be included as a key component of urban green infrastructure and greenway planning initiatives that are gaining attention across the globe. Opportunities exist to integrate working trees into the city landscape through applications such as urban food forests (see Chapter 10), riparian buffers along urban streams, forest farming in green spaces, and productive windbreaks to protect urban farms. Agroforestry may also be used to reduce the often common zone of conflict between urban and rural land uses by providing a buffer (Schoeneberger, Bentrup, & Francis, 2001). More research is needed, however, to better understand the potential for agroforestry to provide a wide range of ecosystem services in these nontraditional settings. In particular, social scientists should be engaged to better understand the cultural functions due to the tight connection to residents in urban areas (Lovell, 2010; Lovell & Taylor, 2013).

The potential exists to significantly expand the use of agroforestry throughout North America by strengthening and expanding landscape-scale research efforts. Particularly important is the study of agroforestry at this broader scale where connections with other land uses become relevant and comprehensive land-based solutions can be developed. Working effectively at the landscape scale will require collaboration between physical and social scientists, landowners, residents, as well as landscape designers and planners. Experts in these planning fields are trained across disciplines to respect and reflect the value of nature and also of culture in the proposed solutions (Dramstad, Olson, & Forman, 1996). They also play an important role in developing, articulating, and communicating the range of possible solutions. Engaging this broad group of stakeholders throughout the research, planning, and design process will help identify the best solutions for agroforestry integration and ultimately improve the utility of multifunctional landscapes across the world.

References

Adhikari, K., Owens, P. R., Ashworth, A. J., Sauer, T. J., Libohova, Z., Richter, J. L., & Miller, D. M. (2018). Topographic controls on soil nutrient variations in a silvopasture system. *Agrosystems, Geosciences & Environment, 1*, 1–15. https://doi.org/10.2134/age2018.04.0008

Agnew, L. J., Lyon, S. W., Gerard-Marchant, P., Collins, V. B., Lembo, A. J., Steenhuis, T. S., & Walter, M. T. (2006). Identifying hydrologically sensitive areas: Bridging the gap between science and application. *Journal of Environmental Management, 78*, 63–76. https://doi.org/10.1016/j.jenvman.2005.04.021

Ahmad, F., Goparaju, L., & Qayum, A. (2019). FAO guidelines and geospatial application for agroforestry suitability mapping: Case study of Ranchi, Jharkhand state of India. *Agroforestry Systems, 93*, 531–544. https://doi.org/10.1007/s10457-017-0145-y

Aleksandrowicz, L., Green, R., Joy, E. J. M., Smith, P., & Haines, A. (2016). The impacts of dietary change on greenhouse gas emissions, land use, water use, and health: A systematic review. *PLOS ONE, 11*, 16.

Altieri, M. A., & Nicholls, C. I. (2017). The adaptation and mitigation potential of traditional agriculture in a changing climate. *Climate Change, 140*, 33–45. https://doi.org/10.1007/s10584-013-0909-y

Angileri, V., & Toccolini, A. (1993). The assessment of visual quality as a tool for the conservation of rural landscape diversity. *Landscape and Urban Planning, 24*, 105–112. https://doi.org/10.1016/0169-2046(93)90089-V

Baceviciene, M., Luksiene, D. I., Cesnaitiene, V. J., Raubaite, S., Peasey, A., & Tamosiunas, A. (2013). Dose– response association between physical activity and metabolic syndrome. *Central European Journal of Medicine, 8*, 273–282. https://doi.org/10.2478/s11536-012-0123-8

Bailey, S. (2007). Increasing connectivity in fragmented landscapes: An investigation of evidence for biodiversity gain in woodlands. *Forest Ecology and Management, 238*, 7–23. https://doi.org/10.1016/j.foreco.2006.09.049

Basnyat, P., Teeter, L. D., Flynn, K. M., & Lockaby, B. G. (1999). Relationships between landscape characteristics and nonpoint source pollution inputs to coastal estuaries. *Environmental Management, 23*, 539–549. https://doi.org/10.1007/s002679900208

Baudry, J., Burel, F., Aviron, S., Martin, M., Ouin, A., Pain, G., & Thenail, C. (2003). Temporal variability of connectivity in agricultural landscapes: Do farming activities help? *Landscape Ecology, 18*, 303–314. https://doi.org/10.1023/A:1024465200284

Benayas, J. M. R., Bullock, J. M., & Newton, A. C. (2008). Creating woodland islets to reconcile ecological restoration, conservation, and agricultural land use. Frontiers in Ecology and the Environment, 6, 329–336. https://doi.org/10.1890/070057

Bentrup, G., & Dosskey, M. (2017). Appendix B: Evaluation of agroforestry for protecting coldwater fish habitat. In: M. M. Schoeneberger, G. Bentrup, & T. Patel-Weynand (Eds.), Agroforestry: Enhancing resiliency in U.S. agricultural landscapes under changing conditions (pp. 207–209). General Technical Report WO-96. Washington, DC: U.S. Forest Service.

Bentrup, G., Dosskey, M., Wells, G., & Schoeneberger, M. M. (2012). Connecting landscape fragments through riparian zones. In J. Stanturf, D. Lamb, & P. Madsen (Eds.), Forest landscape restoration: Integrating natural and social sciences (pp. 93–109). Dordrecht, the Netherlands: Springer. https://doi.org/10.1007/978-94-007-5326-6_5

Bentrup, G., & Kellerman, T. (2004). Where should buffers go? Modeling riparian habitat connectivity in northeast Kansas. *Journal of Soil and Water Conservation, 59*, 209–215.

Bentrup, G., & Leininger, T. (2002). Agroforestry: Mapping the way with GIS. *Journal of Soil and Water Conservation, 57*, 148A–153A.</jrn>

Boutin, C., Jobin, B., & Belanger, L. (2003). Importance of riparian habitats to flora conservation in farming landscapes of southern Quebec, Canada. *Agriculture, Ecosystems & Environment, 94*, 73–87. https://doi.org/10.1016/S0167-8809(02)00014-2

Brandes, E., McNunn, G. S., Schulte, L. A., Bonner, I. J., Muth, D. J., Babcock, B. A., . . . Heaton, E. A. (2016). Subfield profitability analysis reveals an economic case for cropland diversification. *Environmental Research Letters, 11*, 014009. https://doi.org/10.1088/1748-9326/11/1/014009

Carver, A. D., Danskin, S. D., Zaczek, J. J., Mangun, J. C., & Williard, K. W. J. (2004). A GIS methodology for generating riparian tree planting recommendations. *Northern Journal of Applied Forestry, 21*, 100–106. https://doi.org/10.1093/njaf/21.2.100

de la Fuente de Val, G., Atauri, J. A., & de Lucio, J. V. (2006). Relationship between landscape visual attributes and spatial pattern indices: A test study in Mediterranean-climate landscapes. *Landscape and Urban Planning, 77*, 393–407. https://doi.org/10.1016/j.landurbplan.2005.05.003

de Sousa, K., van Zonneveld, M., Imbach, P., Casanoves, F., Kindt, R., & Ordonez, J. C. (2017). Suitability of key Central American agroforestry species under future climates: An atlas. Turrialba, Costa Rica: ICRAF.

Dosskey, M., Wells, G., Bentrup, G., & Wallace, D. (2012). Enhancing ecosystem services: Designing for multifunctionality. *Journal of Soil and Water Conservation, 67*, 37A–41A. https://doi.org/10.2489/jswc.67.2.37A

Dosskey, M. G., Bentrup, G., & Schoeneberger, M. (2012). A role for agroforestry in forest restoration in the Lower Mississippi alluvial valley. *Journal of Forestry, 110*, 48–55.

Dramstad, W. E., Fry, G., Fjellstad, W. J., Skar, B., Helliksen, W., Sollund, M. L. B., . . . Framstad, E. (2001). Integrating landscape-based values: Norwegian monitoring of agricultural landscapes. *Landscape and Urban Planning, 57*, 257–268. https://doi.org/10.1016/S0169-2046(01)00208-0

Dramstad, W. E., Olson, J. D., & Forman, R. T. T. (1996). Landscape ecology principles in landscape architecture and land-use planning. Washington, DC: Island Press.

Dronen, S. I. (1988). Layout and design criteria for livestock windbreaks. *Agriculture, Ecosystems & Environment, 22–23*, 231–240. https://doi.org/10.1016/0167-8809(88)90022-9

Ellis, E. A., Bentrup, G., & Schoeneberger, M. M. (2004). Computer-based tools for decision support in agroforestry: Current state and future needs. *Agroforestry Systems, 61–62*, 401–421.

European Commission. (2017). *Agri-environment measures.* Agriculture and Environment.

Fagerholm, N., Oteros-Rozas, E., Raymond, C. M., Torralba, M., Moreno, G., & Plieninger. T. (2016). Assessing linkages between ecosystem services, land-use and well-being in an agroforestry landscape using public participation GIS. Applied Geography, 74, 30–46. https://doi.org/10.1016/j.apgeog.2016.06.007

Flint, C. G., Robinson, E. S., Kellogg, J., Ferguson, G., BouFajreldin, L., Dolan, . . . Lila, M. A. (2011). Promoting wellness in Alaskan villages: Integrating traditional knowledge and science of wild berries. *EcoHealth*, *8*, 199–209. https://doi.org/10.1007/s10393-011-0707-9

Forman, R. T. T. (1995a). Land mosaics: The ecology of landscapes and regions. Cambridge, UK: Cambridge University Press. https://doi.org/10.1017/9781107050327

Forman, R. T. T. (1995b). Some general principles of landscape and regional ecology. *Landscape Ecology*, *10*, 133–142. https://doi.org/10.1007/BF00133027

Forman, R.T. T., & Godron, M. (1986). Landscape ecology. New York: John Wiley & Sons.

Garrett, H. E., McGraw, R. L., & Walter, W. D. (2009). Alley cropping practices. In H. E. Garrett (Ed.), *North American agroforestry: An integrated science and practice* (2nd ed., pp. 133–162). Madison, WI: ASA. https://doi.org/10.2134/2009.northamericanagroforestry.2ed

Gliessman, S. (2010). Landscape multifunctionality and agriculture. *Journal of Sustainable Agriculture*, *34*, 465. https://doi.org/10.1080/10440046.2010.484660

Gopalakrishnan, G., Negri, M. C., & Snyder, S. W. (2011). A novel framework to classify marginal land for sustainable biomass feedstock production. *Journal of Environmental Quality*, *40*, 1593–1600. https://doi.org/10.2134/jeq2010.0539

Gordon, L. J., Bignet, V., Crona, B., Henriksson, P. J. G., Van Holt, T., Jonell, M., . . . Queiroz, C. (2017). Rewiring food systems to enhance human health and biosphere stewardship. *Environmental Research Letters*, *12*, 100201. https://doi.org/10.1088/1748-9326/aa81dc

Holzmueller, E. J., & Jose, S. (2012). Biomass production for biofuels using agroforestry: Potential for the North Central Region of the United States. *Agroforestry Systems*, *85*, 305–314. https://doi.org/10.1007/s10457-012-9502-z

Huang, J., Tichit, M., Poulot, M., Darly, S., Li, S. C., Petit, C., & Aubry, C. (2015). Comparative review of multifunctionality and ecosystem services in sustainable agriculture. *Journal of Environmental Management*, *149*, 138–147. https://doi.org/10.1016/j.jenvman.2014.10.020

Hung, I. K., McNally, B. C., Farrish, K. W., & Oswald, B. P. (2005). *Using GIS for selecting trees for thinning*. Nacogdoches, TX: Stephen F. Austin State University.

Hupp, J., Brubaker, M., Wilkinson, K., & Williamson, J. (2015). How are your berries? Perspectives of Alaska's environmental managers on trends in wild berry abundance. *International Journal of Circumpolar Health*, *74*, 28704. https://doi.org/10.3402/ijch.v74.28704

Jordan, N., & Warner, K. D. (2010). Enhancing the multifunctionality of US agriculture. *BioScience*, *60*, 60–66.

Kallenbach, R. L., Kerley, M. S., & Bishop-Hurley, G. J. (2006). Cumulative forage production, forage quality and livestock performance from an annual ryegrass and cereal rye mixture in a pine walnut silvopasture. *Agroforestry Systems*, *66*, 43–53. https://doi.org/10.1007/s10457-005-6640-6

Karki, U., & Goodman, M. S. (2010). Cattle distribution and behavior in southern-pine silvopasture versus open-pasture. *Agroforestry Systems*, *78*, 159–168. https://doi.org/10.1007/s10457-009-9250-x

Karki, U., & Goodman, M. S. (2015). Microclimatic differences between mature loblolly-pine silvopasture and open-pasture. *Agroforestry Systems*, *89*, 319–325. https://doi.org/10.1007/s10457-014-9768-4

Lake, F. K. (2013). Historical and cultural fires, tribal management and research issues in Northern California: Trails, fires and tribulations. Occasion: *Interdisciplinary Studies in the Humanities*, 5.

Lake, F. K., Tripp, W., & Reed, R. (2010). The Karuk tribe, planetary stewardship, and world renewal on the middle Klamath River, California. *Bulletin of the Ecological Society of America*, *91*, 147–149.

Landis, D. A. (2017). Designing agricultural landscapes for biodiversity-based ecosystem services. *Basic and Applied Ecology*, *18*, 1–12. https://doi.org/10.1016/j.baae.2016.07.005

Leakey, R. B. (2014). Twelve principles for better food and more food from mature perennial agroecosystems. In C. Batello, L. Wade, S. Cox, N. Pogna, A. Bozzini, & J. Choptiany (Eds.), *Perennial crops for food security: Proceedings of the FAO Expert Workshop* (pp. 282–306). Rome: FAO.

León, M. C., & Harvey, C. A. (2006). Live fences and landscape connectivity in a neotropical agricultural landscape. *Agroforestry Systems*, *68*, 15–26. https://doi.org/10.1007/s10457-005-5831-5

Lila, M. A., Kellogg, J., Grace, M. H., Yousef, G. G., Kraft, T. B., & Rogers, R. B. (2014). Stressed for success: How the berry's wild origins result in multifaceted health protections. In O. VanKooten & F. Brouns (Eds.), X International Symposium on Vaccinium and Other Superfruits (pp. 23–43). Leuven, Belgium: ISHS.

Lovell, S. T. (2010). Multifunctional urban agriculture for sustainable land use planning. *Sustainability*, *2*, 2499–2522. https://doi.org/10.3390/su2082499

Lovell, S.T., DeSantis, S., Nathan, C. A., Olson, M. B., Mendez, V. E., Kominami, H. C., . . . Morris, W. B. (2010). Integrating agroecology and landscape multifunctionality in Vermont: An evolving framework to evaluate the design of agroecosystems. *Agricultural Systems*, *103*, 327–341. https://doi.org/10.1016/j.agsy.2010.03.003

Lovell, S. T., Dupraz, C., Gold, M., Jose, S., Revord, R., Stanek, E., & Wolz, K. J. (2018). Temperate agroforestry research: Considering multifunctional woody polycultures and the design of long-term trials. *Agroforestry Systems*, *92*, 1397–1415. https://doi.org/10.1007/s10457-017-0087-4

Lovell, S. T., & Johnston, D. M. (2009a). Creating multifunctional landscapes: How can the field of ecology inform the design of the landscape? *Frontiers in Ecology and the Environment*, *7*, 212–220. https://doi.org/10.1890/070178

Lovell, S. T., & Johnston, D. M. (2009b). Designing landscapes for performance based on emerging principles in landscape ecology. *Ecology and Society*, *14*(1), 44. https://doi.org/10.5751/ES-02912-140144

Lovell, S. T., & Taylor, J. R. (2013). Supplying urban ecosystem services through multifunctional green infrastructure. *Ecology and Society*, *28*, 1447–1463.

Madureira, L., Rambonilaza, T., & Karpinski, I. (2007). Review of methods and evidence for economic valuation of agricultural non-commodity outputs and suggestions to facilitate its application to broader decisional contexts. *Agriculture, Ecosystems & Environment*, *120*, 5–20. https://doi.org/10.1016/j.agee.2006.04.015

Mattia, C. M., Lovell, S. T., & Davis, A. (2018). Identifying barriers and motivators for adoption of multifunctional perennial cropping systems by landowners in the Upper Sangamon River Watershed, Illinois. *Agroforestry Systems*, *92*, 1155–1169. https://doi.org/10.1007/s10457-016-0053-6

Mattia, C. M., Lovell, S. T., & Fraterrigo, J. W. (2018). Identifying marginal land for multifunctional perennial cropping systems in the Upper Sangamon River Watershed, Illinois. *Journal of Soil and Water Conservation*, *73*, 669–681. https://doi.org/10.2489/jswc.73.6.669

McHarg, I. (1995). *Design with nature*. Hoboken, NJ: John Wiley & Sons.

Molnar, T. J., Kahn, P. C., Ford, T. M., Funk, C. J., & Funk, C. R. (2013). Tree crops, a permanent agriculture: Concepts from the past for a sustainable future. *Resources, 2*, 457–488. https://doi.org/10.3390/resources2040457

Mori, G. O., Gold, M., & Jose, S. (2017). Specialty crops in temperate agroforestry systems: Sustainable management, marketing and promotion for the Midwest region of the U.S.A. In F. Montagnini (Ed.), *Integrating landscapes: Agroforestry for biodiversity conservation and food sovereignty* (pp. 331–366). Cham, Switzerland: Springer. https://doi.org/10.1007/978-3-319-69371-2_14

Mor-Mussery, A., Leu, S., & Budovsky, A. (2013). Modeling the optimal grazing regime of *Acacia victoriae* silvopasture in the Northern Negev, Israel. *Journal of Arid Environments, 94*, 27–36. https://doi.org/10.1016/j.jaridenv.2013.02.001

Olofson, H. (1983). Indigenous agroforestry systems. *Philippine Quarterly of Culture and Society, 11*, 149–174.

Park, J. H., Kho, M. C., Kim, H. Y., Ahn, Y. M., Lee, Y. J., Kang, D. G., & Lee, H. S. (2015). Blackcurrant suppresses metabolic syndrome induced by high-fructose diet in rats. *Evidence-Based Complementary and Alternative Medicine, 2015*, 385976. https://doi.org/10.1155/2015/385976

Perfecto, I., & Vandermeer, J. (2002). Quality of agroecological matrix in a tropical montane landscape: Ants in coffee plantations in southern Mexico. *Conservation Biology, 16*, 174–182. https://doi.org/10.1046/j.1523-1739.2002.99536.x

Peri, P. L., & Bloomberg, M. (2002). Windbreaks in southern Patagonia, Argentina: A review of research on growth models, windspeed reduction, and effects on crops. *Agroforestry Systems, 56*, 129–144. https://doi.org/10.1023/A:1021314927209

Qiu, Z., & Dosskey, M. G. (2012). Multiple function benefit: Cost comparison of conservation buffer placement strategies. *Landscape and Urban Planning, 107*, 89–99. https://doi.org/10.1016/j.landurbplan.2012.05.001

Raymond, C. M., Reed, M., Bieling, C., Robinson, G. M., & Plieninger, T. (2016). Integrating different understandings of landscape stewardship into the design of agri-environmental schemes. *Environmental Conservation, 43*, 350–358. https://doi.org/10.1017/S037689291600031X

Reisner, Y., de Filippi, R., Herzog, F., & Palma, J. (2007). Target regions for silvoarable agroforestry in Europe. *Ecological Engineering, 29*, 401–418. https://doi.org/10.1016/j.ecoleng.2006.09.020

Rhoads, B. L., Lewis, Q. W., & Andresen, W. (2016). Historical changes in channel network extent and channel planform in an intensively managed landscape: Natural versus human-induced effects. *Geomorphology, 252*, 17–31. https://doi.org/10.1016/j.geomorph.2015.04.021

Rigby, D., Woodhouse, P., Young, T., & Burton, M. (2001). Constructing a farm level indicator of sustainable agricultural practice. *Ecological Economics, 39*, 463–478. https://doi.org/10.1016/S0921-8009(01)00245-2

Robles, M. D., Flather, C. H., Stein, S. M., Nelson, M. D., & Cutko, A. (2008). The geography of private forests that support at-risk species in the conterminous United States. *Frontiers in Ecology and the Environment, 6*, 301–307. https://doi.org/10.1890/070106

Rossier, C., & Lake, F. (2014). *Indigenous traditional ecological knowledge in agroforestry* (Agroforestry Notes 44). Lincoln, NE: USDA National Agroforestry Center.

Rotz, C. A., Taube, F., Russelle, M. P., Oenema, J., Sanderson, M. A., & Wachendorf, M. (2005). Whole-farm perspectives of nutrient flows in grassland agriculture. *Crop Science, 45*, 2139–2159. https://doi.org/10.2135/cropsci2004.0523

Schoeneberger, M., Bentrup, G., & Francis, C. A. (2001). Ecobelts: Reconnecting agriculture and communities. In C. Flora (Ed.), *Interactions between agroecosystems and rural communities* (pp. 239–260). Boca Raton, FL: CRC Press.

Schoeneberger, M. M., Bentrup, G., & Patel-Weynand, T. (2017). *Agroforestry: Enhancing resiliency in U.S. agricultural landscapes under changing conditions*. Washington, DC: U.S. Forest Service.

Seabrook, L., McAlpine, C., & Fensham, R. (2007). Spatial and temporal analysis of vegetation change in agricultural landscapes: A case study of two brigalow (*Acacia harpophylla*) landscapes in Queensland, Australia. *Agriculture, Ecosystems & Environment, 120*, 211–228. https://doi.org/10.1016/j.agee.2006.09.005

Seabrook, L., McAlpine, C., & Fensham, R. (2008). What influences farmers to keep trees? A case study from the Brigalow Belt, Queensland, Australia. *Landscape and Urban Planning, 84*, 266–281. https://doi.org/10.1016/j.landurbplan.2007.08.006

Stanek, E., & Lovell, S. T. (2020). Building multifunctionality into agricultural conservation programs: Lessons learned from designing agroforestry systems with central Illinois landowners. *Renewable Agriculture and Food Systems, 35*, 313–321.

Stanek, E., Lovell, S. T., & Reisner, A. (2019). Designing multifunctional woody polycultures according to landowner preferences. *Agroforestry Systems, 93*, 2293–2311. https://doi.org/10.1007/s10457-019-00350-2

Sutherland, W. J. (2004). A blueprint for the countryside. *Ibis, 146*, 230–238. https://doi.org/10.1111/j.1474-919X.2004.00369.x

Teetl'it Gwich'in Renewable Resources Council, Parlee, B., & Berkes, F. (2006). Indigenous knowledge of ecological variability and commons management: A case study on berry harvesting from Northern Canada. *Human Ecology, 34*, 515–528 (erratum: *36*, 143). https://doi.org/10.1007/s10745-006-9038-9

Tiwari, T., Lundstrom, J., Kuglerova, L., Laudon, H., Ohman, K., & Agren, A. M. (2016). Cost of riparian buffer zones: A comparison of hydrologically adapted site-specific riparian buffers with traditional fixed widths. *Water Resources Research, 52*, 1056–1069. https://doi.org/10.1002/2015WR018014

Tomer, M. D., Dosskey, M. G., Burkart, M. R., James, D. E., Helmers, M. J., & Eisenhauer, D. E. (2009). Methods to prioritize placement of riparian buffers for improved water quality. *Agroforestry Systems*, 75, 17–25. https://doi.org/10.1007/s10457-008-9134-5

Tsonkova, P., Bohm, C., Quinkenstein, A., & Freese, D. (2012). Ecological benefits provided by alley cropping systems for production of woody biomass in the temperate region: A review. *Agroforestry Systems*, 85, 133–152.

Turner, M. G. (1989). Landscape ecology: The effect of pattern on process. *Annual Review of Ecology and Systematics, 20*, 171–197. https://doi.org/10.1146/annurev.es.20.110189.001131

Turner, M. G. (2005). Landscape ecology in North America: Past, present, and future. *Ecology, 86*, 1967–1974. https://doi.org/10.1890/04-0890

Tyndall, J., & Colletti, J. (2007). Mitigating swine odor with strategically designed shelterbelt systems: A review. *Agroforestry Systems, 69*, 45–65. https://doi.org/10.1007/s10457-006-9017-6

USDA–NRCS. (2016). *Gridded Soil Survey Geographic (gSSURGO): Database for the conterminous United States*. Lincoln, NE: National Soil Survey Center.

U.S. Forest Service. (2018). *Somes Bar Integrated Fire Management Project: Final environmental assessment* (R5-MB-312). Eureka, CA: U.S. Forest Service, Pacific Southwest Region, Six Rivers National Forest.</bok>

Vandermeer, J., & Perfecto, I. (2007). The agricultural matrix and a future paradigm for conservation. *Conservation Biology, 21*, 274–277. https://doi.org/10.1111/j.1523-1739.2006.00582.x

Wallace, C. W., McCarty, G., Lee, S., Brooks, R. P., Veith, T. L., Kleinman, P. J. A., & Sadeghi, A. M. (2018). Evaluating concentrated flowpaths in riparian forest buffer contributing areas using LiDAR imagery and topographic metrics. *Remote Sensing*, *10*, 614. https://doi.org/10.3390/rs10040614

Wallace, D. C., & Young, F. J. (2008). Black walnut suitability index: A Natural Resources Conservation Service national soil information system based interpretive model. In D. F. Jacobs & C. H. Michler (Eds.), Proceedings of 16th Central Hardwood Forest Conference (Gen. Tech. Rep. NRS-P24, pp. 589–595). Newtown Square, PA: U.S. Forest Service, Northern Research Station.

Williams, G. W. (2005). *The USDA Forest Service: The first century*. Washington, DC: U.S. Forest Service.

Willis, W. B., Eichinger, W. E., Prueger, J. H., Hapeman, C. J., Li, H., Buser, M. D., . . . Yao, Q. (2017). Particulate capture efficiency of a vegetative environmental buffer surrounding an animal feeding operation. *Agriculture, Ecosystems & Environment*, *240*, 101–108. https://doi.org/10.1016/j.agee.2017.02.006

Wolz, K. J., & DeLucia, E. H. (2019). Black walnut alley cropping is economically competitive with row crops in the Midwest USA. *Ecological Applications*, *29*, e01829. https://doi.org/10.1002/eap.1829

Wood, S. A. (2018). Nutritional functional trait diversity of crops in south-eastern Senegal. *Journal of Applied Ecology*, *55*, 81–91. https://doi.org/10.1111/1365-2664.13026

Xiang, W. N. (1993). A GIS method for riparian water-quality buffer generation. *International Journal of Geographical Information Systems*, *7*, 57–70. https://doi.org/10.1080/02693799308901939

Zheng, X., Zhu, J. J., & Xing, Z. F. (2016). Assessment of the effects of shelterbelts on crop yields at the regional scale in Northeast China. *Agricultural Systems*, *143*, 49–60. https://doi.org/10.1016/j.agsy.2015.12.008

Study Questions

1. Briefly describe the differences between a landscape-level perspective and a site-specific perspective when designing agroforestry systems. How can the use of a landscape-level perspective enhance the overall delivery of ecosystem goods and services?

2. The development of the patch–corridor–matrix model in the 1980s allowed ecologists to efficiently describe landscape structure, especially within heterogeneous lands. Within the context of this model, describe the patch, corridor, and matrix elements of an example agroforestry system of your choosing and briefly list the landscape functions it provides.

3. Describe the difference between "high-quality" and "low-quality" patches in a landscape? Which agroforestry practices are most likely to serve as high-quality patches?

4. In what ways does the multifunctional landscape framework differ from traditional, widely utilized approaches to cropping systems?

5. Profitability is often the underlying measure of success for an agricultural system in modern times. When profit is the sole focus of a system, what landscape benefits may be excluded and how can a multifunctional approach avoid this?

6. What is traditional ecological knowledge? How can traditional ecological knowledge be used to improve the cultural and environmental benefits provided by agroforestry systems at a landscape scale?

7. Provide three examples of how a landscape-scale assessment can be leveraged to improve the design or function of agroforestry systems.

8. In some circumstances, a site-scale perspective is more appropriate than a landscape-scale approach when designing or assessing the use of agroforestry. Provide an example of when this may be the case and describe why a site-scale perspective is more appropriate.

9. How can the use of modern geospatial technology aid in the development and effectiveness of landscape-level agroforestry design?

10. Landscape-level agroforestry design has been acknowledged to provide a wide range of benefits, yet it is still relatively underutilized. What are some of the major challenges and opportunities for future work in this area?

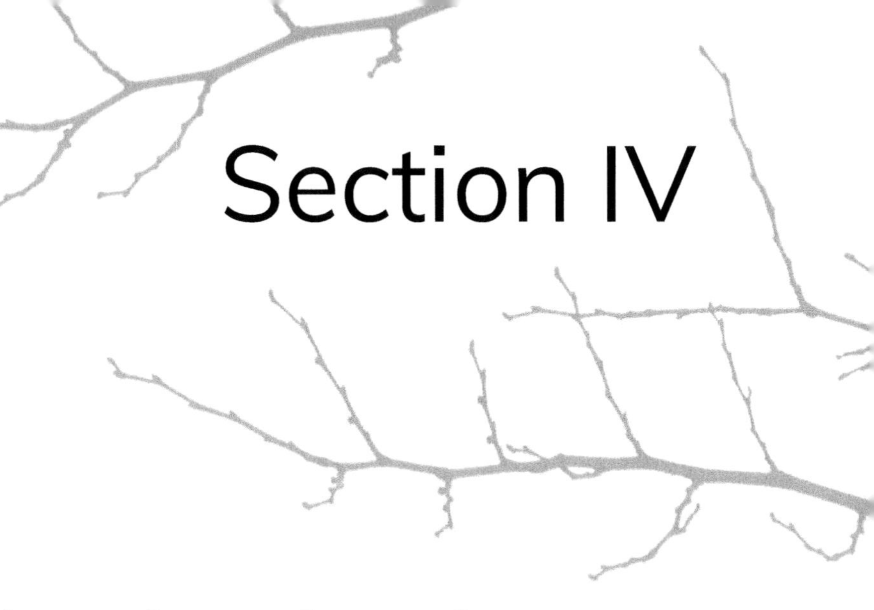

Section IV

Human Dimensions of Agroforestry

North American Agroforestry, Third Edition. Edited by Harold E. "Gene" Garrett, Shibu Jose, and Michael A. Gold.

15

Zhen Cai, Michael A. Gold,
Michaela M. "Ina" Cernusca, and
Larry D. Godsey

Agroforestry Product Markets and Marketing

A decade has passed since the second edition of *North American Agroforestry: An Integrated Science and Practice* (Garrett, 2009) was published. During these years, the science of agroforestry has started to move from research to practical implementation. The number of socioeconomic publications on agroforestry between 2001 and 2016 are 4.6 times those published between 1981 and 2000 (Himshikha, 2016). In the early 2000s, agroforestry research was mostly focused on products and production systems, as only 15 of 750 presentations at the first World Congress of Agroforestry dealt directly with trees, markets, and marketing (Nair, Allen, & Bannister, 2005). This chapter provides updates by adding the latest scientific research findings, relevant market statistics, and new examples on how to apply marketing strategies to better market agroforestry products.

Why discuss markets and marketing in an agroforestry textbook? For many, the assumption has been made that the value of the production systems will be enough to convince people to adopt agroforestry. However, Chamberlain and Hammett (1999) linked this approach to an unsuccessful strategy designed to "push" products to potential adopters, an "if we grow it they will buy it" strategy. Instead, they suggested an alternative market-oriented strategy that will "pull" temperate agroforestry along by encouraging product market opportunities. The need for a market-oriented "pull" strategy thus helps to explain why it is useful to discuss markets and marketing in an agroforestry textbook.

It is now widely recognized in both temperate and tropical areas that widespread adoption and active use of agroforestry systems and practices may never fulfill their potential without the "pull" of market forces. Widespread adoption of agroforestry in North America is lagging. This is due, in part, to risk-averse producers' understandable reluctance to establish agroforestry practices in the absence of readily available market information. The comprehensive multidisciplinary scientific knowledge underpinning agroforestry practices is essential but will never find widespread application unless that knowledge can be applied to products that are sought in the marketplace (Gold, Godsey, & Josiah, 2004; Russell & Franzel, 2004). Thus, understanding markets and marketing is a key ingredient in the success of profitable agroforestry enterprises that produce commercially valuable products.

North American Agroforestry, Third Edition. Edited by Harold E. "Gene" Garrett, Shibu Jose, and Michael A. Gold.

Promising Market Trends

Within the past 20 years, a number of changes, both disturbing and promising, have taken place worldwide and within the United States. The horror of the attack on the World Trade Center in September of 2001 and volatile oil prices have resulted in global insecurity over dependence on foreign oil for long-term energy supplies and increased national and international efforts to develop renewable energy alternatives. There has been enormous growth in the U.S. market for locally grown and organic food products in both fresh and value-added forms (Agriculture Marketing Service, 2014; Organic Trade Association, 2018). Environmental issues, including nonpoint-source pollution control, restoration of wildlife habitat, climate change, and carbon sequestration, have become part of daily conversations. Finally, the use of computers, the internet, cell phones, and tablets has evolved to the point where these devices now connect most Americans and billions of people throughout the world, leading to direct market buy–sell opportunities among individuals, irrespective of location. From the agroforestry perspective, all of these changes are emerging as promising trends that provide strong market opportunities for agroforestry products, whether they be commodities or niche crops.

Trends that Encourage Production and Use of Agroforestry Products

In the past 15 years there have been massive investments in both corn (*Zea mays* L.) ethanol and soy [*Glycine max* (L.) Merr.] biodiesel facilities throughout the U.S. Midwest. In the past decade, the surge in demand for biofuels is one of the main causes of the increase in commodity prices for corn and soybean (Kim & Moschini, 2018). The 2018 Agriculture Improvement Act (Farm Bill) provides funding for several bioenergy initiatives, such as the Bioenergy Program for Advanced Biofuels, the Biodiesel Fuel Education Program, the Biobased Markets Program, the Biomass Crop Assistance Program, and the Feedstock Flexibility Program for Bioenergy Producers, among others. Private sector venture capital is pouring into renewable energy options, including cellulosic ethanol. Cellulosic feedstock conversions are handled in three ways—thermochemically, biochemically, and chemically—using solid (e.g., pellets), liquid (e.g., ethanol and methanol), and gaseous forms of energy. All conversion processes can utilize perennial warm-season grasses (e.g., switchgrass [*Panicum virgatum* L.] and giant miscanthus [*Miscanthus giganteus* J.M. Greef & Deuter ex Hodk. & Renvoize]) and woody biomass (Eaglesham, 2007; Perlack et al., 2005). Windbreak, alley cropping, and riparian forest buffer agroforestry practices are well suited to accommodate large-scale production of these types of cellulosic feedstocks. One focus of the 2007 Energy Independence and Security Act was to increase biofuel production using cellulosic feedstocks (Energy Independence and Security Act of 2007). In the future, more ethanol will be made from renewable biomass sources than from corn starch, paving the way for a strong, domestic farm-based bioenergy industry in the United States. (Eaglesham, 2007; Schubert, 2006).

Both within the United States and globally, concerns about industrial agriculture practices, food quality, and links to human health have fostered interest in new, alternative, local, and more sustainable agricultural practices. Organic and locally grown foods are perceived by consumers as healthier and safer for both people and the environment. According to the Organic Trade Association (2021), the U.S. organic industry grew from US$3.4 billion to $61.9 billion between 1997 and 2020. Organic foods are one of the fastest growing market segments within the food industry, with sales growing at an annual rate of 6.4% in 2017. The pace of conversion of cropland from conventional to organic has failed to keep pace with the growth in sales. From 2013 to 2016, the United States imported $1,476 million in organic products annually (Demko, Dinterman, Marez, & Jaenicke, 2017). This trend provides an ongoing opportunity for U.S. farmers to enter this market and is reflected in a major increase in the number of certified organic operations and land devoted to organic production in recent years (USDA Agricultural Marketing Service, 2017).

Parallel to growth trends in organic sales, consumers are also strongly interested in consuming products that are locally grown (Brown, 2003; Cai, Gold, & Brannan, 2018; Kirby, Jackson, & Perrett, 2007; Loureiro & Hine, 2002; Mohebalian, Aguilar, & Cernusca, 2013). A nationwide survey conducted by the Leopold Center (Pirog & Larson, 2007) indicated that American consumers are skeptical about the safety of the global food system, and many believe that local foods are safer and better for their health than foods from abroad. Respondents placed high importance on food safety, freshness, and pesticide use, with 85% stating that local foods were somewhat safe or safe compared with 53% who perceived foods grown elsewhere in the world as somewhat safe or safe.

The latest USDA agriculture census, conducted in 2017, showed the number of full-time farmers in Oregon to have almost doubled from 13,884 in 1974 to 27,726 in 2017 (USDA National

Agricultural Statistics Service, 2019a). Part-time farming, with many growers specializing in farmers markets and other buy-local niches, is increasing. Growers and industry analysts attribute the increase in Oregon farmers to a growing number of small- and medium-sized operations designed to meet the increasing demand for local food. This stands in sharp contrast to farmers' plight nationwide, where about 300,000 farms have disappeared since 1980. Missouri, for example, lost approximately 15,000 farms between 1997 and 2017 (USDA National Agricultural Statistics Service, 2019c).

Nationwide, farmers markets have increased from 1,755 in 1994 to 8,771 in 2019, growing more than 396% since 1994 (USDA Agricultural Marketing Service, 2019). Numerous surveys have reported that consumers shop at farmers markets primarily because of product quality and the fact that the food is locally grown (USDA, 2017). All six recognized temperate agroforestry practices (i.e., alley cropping, riparian and upland forest buffers, windbreaks, forest farming, silvopasture, urban food forests), intensively managed to incorporate a diverse number of crops, can be designed to produce locally grown and/or organic crops in both fresh and value-added form for these growing markets. Sparse populations and the difficulty of linking rural places to national and international markets present obstacles for new rural businesses. The percentage of farms with internet access has risen from approximately 15% in 1997 to approximately 71% in 2017, but much of rural America still lacks high-speed internet access, with 29% of operators using digital subscriber lines in 2017, and only 17% of farms have mobile internet service (USDA National Agricultural Statistics Service, 2019a). Levine, Locke, Searls, and Weinberger (2001) described the positive impact of the internet on markets as companies with high-speed internet spur more rural entrepreneurs who can communicate directly with their markets. The internet is enabling conversations among people that were simply not possible in the era of traditional mass media. As a result, markets are becoming more informed and better organized. Grewal, Iyerb, and Levya (2004) indicated that online retailing (e-tailing) fulfills several consumer needs more effectively and efficiently than conventional store-based retailing. Internet retailing sites enable consumers to conveniently view the entire assortment carried by the retailer with minimal effort, inconvenience, and time investment. In addition, consumers can efficiently obtain critical educational knowledge about firms, products, and brands and thereby increase their competency in making sound purchase decisions. Finally, consumers can compare product features, availability, prices, and other factors across a range of e-tailers more efficiently.

Widespread concerns about environmental issues, including nonpoint-source pollution, loss of habitat, and climate change, have resulted in an array of mitigation efforts. Riparian forest buffers and windbreaks are agroforestry practices widely known for their positive environmental impacts; however, all six recognized agroforestry practices, when properly implemented, directly address each of these major environmental issues. Godsey, Mercer, Grala, Grado, and Alavalapati (2009) and Alavalapati and Mercer (2004) described the values of non-market goods and services that can be realized through increased use of agroforestry practices. For instance, Bauer and Johnston (2017) estimated the economic values provided by riparian buffers in the Great Bay watershed to be $34 million. Alam et al. (2014) estimated the total value of the ecosystem services provided by tree-based intercropping in southern Quebec, Canada, to be $2,645 ha^{-1} yr^{-1}. The U.S. Farm Bill incentive programs will continue to provide cost sharing for landowners to establish agroforestry practices on their land. The Natural Resources Conservation Service in Missouri has an Environmental Quality Incentives Program that provides financial assistance to Missouri farmers with agroforestry practices. Stavins and Richards (2005) mentioned the adoption of agroforestry practices as a means to increase carbon sequestration on forestland. Kulshreshtha and Kort (2009) estimated the non-market value provided by riparian buffers in the Canadian Prairie provinces, Canada, and found that carbon sequestrated by the buffer had an economic value of CDN$46 million. As carbon credit trading matures, an increasing number of landowners will tap these markets through the use of agroforestry practices.

Awareness and understanding of the major trends described above provide landowners with substantial market-based opportunities to grow and market a diversity of products across the value chain, from producer to consumer.

Products Produced through the Application of Agroforestry Practices

Unlike other types of conservation practices, where land is taken out of production, agroforestry is "productive conservation." Agroforestry practices enable landowners to generate income from the production of a wide range of conventional and specialty products while simultaneously protecting and conserving soil, water, and other natural resources. Many observers have examined the potential of these dual-purpose, market-driven conservation systems in North America, including Campbell (1991); Chamberlain and Hammett (1999); Garrett et al. (1994); Garrett and Harper (1999); Gold, Godsey, and Josiah (2004); Kays (1999); Gordon, Newman, and Coleman (2018); Kurtz, Garrett, Slusher, and Osburn (1996); Kurtz, Thurman, Monson, and Garrett (1991); Josiah, Brott, and Brandle (2004); Josiah, Pierre, Brott, and Brandle (2004); Kuyah, Öborn, and Jonsson (2017); and Mbow et al. (2014). Agroforestry practices must shift from a "product and selling" orientation to a "customer and marketing" orientation. This means that determining the needs and wants of buyers and final consumers and satisfying these needs more effectively and efficiently than competitors are critical to success (Kotler, 1994).

Products produced through agroforestry practices, including specialty or non-timber forest products, are produced from trees, within forests, or in myriad combinations with trees or shrubs, crops, and/or animals (Chamberlain et al., 2009; Gold & Garrett, 2009; Gold, Godsey, & Josiah, 2004). Many of these products have proven economic value but have been ignored by, or are unknown to, agricultural and forest landowners. In North America such products include edibles (e.g., mushrooms, nuts, berries), herbal medicinals (e.g., ginseng [*Panax quinquefolius* L.], goldenseal [*Hydrastis canadensis* L.], and witch hazel [*Hamamelis virginiana* L.]), specialty wood products (e.g., diamond willow [*Salix* spp.] for canes, eastern redcedar [*Juniperus virginiana* L.] for closet liners, and walnut [*Juglans nigra* L.] for gunstock blanks], floral and greenery products (e.g., curly [*Salix matsudana* 'Tortuosa'] and pussy willow [*Salix discolor* Muhl.]), ferns, salal (*Gaultheria shallon* Pursh.), fiber and mulch (e.g., cedar pet bedding and pine straw), and recreation (e.g., agritourism and fee hunting) (Chamberlain et al., 2009).

In the tropics, there is an enormous range of indigenous and exotic species producing commercially valuable products that are traditional to or have been introduced into agroforestry systems. Shade-grown coffee (*Coffea arabica* L.) is produced in the highland tropics worldwide for the international marketplace in two- and three-tier agroforestry systems, with additional markets for managed overstory timber trees (Hull, 1999). In the community of Tomé-Açu, Pará, Brazil, more than 55 different crops are grown in intensively managed, highly diverse agroforestry systems. They are commercially marketed, with the primary profits derived from long-lived woody perennial crops, such as black pepper (*Piper nigrum* L.), cacao (*Theobroma cacao* L.), coffee, passion fruit (*Passiflora edulis* Sim.), papaya (*Carica papaya* L.), rubber (*Hevea brasiliensis* Müll. Arg.), and timber (Yamada & Gholz, 2002).

Marketing diverse products from agroforestry practices presents a unique set of challenges for the producer. Diversification is a well-accepted strategy for reducing risks from price and output uncertainty (Godsey et al., 2009). It can increase profitability by making more efficient use of labor and other resources. Yet, diversification increases the complexity of production and marketing decisions. Agroforestry enterprises that produce multiple niche products and commodity goods in complex integrated production systems require meticulous planning and coordination to grow, harvest, and profitably market those goods. Producers are often faced with the dual challenge of identifying markets and successfully marketing agricultural commodities and niche-market products.

There is a growing recognition in both temperate and tropical agroforestry that as a product moves through the value chain, that is, from the producer in the field through various levels of processing, manufacturing, and distribution toward the wholesaler, retailer, and consumers, the product increases in value. Either an individual producer must choose to be vertically integrated and perform all value-added steps on their own, or they must develop deliberate linkages across the value chain to increase the competitive advantage of all market participants (Phi, Duong, Quang, & Vang, 2004; Porter, 1980; te Velde et al., 2006). In many cases, this requires the formation of grower co-ops (e.g., the Midwest Elderberry Cooperative) or "clusters" (i.e., highly specialized economies in a geographic area) comprising groups of companies that provide similar goods and services, resulting in better access to suppliers, services, and labor markets; information and innovation (Drabenstott & Sheaff, 2001); and active collaboration with brokers to reach additional markets.

Because there is limited experience using market research methodologies to examine agroforestry product markets, market forces, and marketing, and because detailed market information is lacking for many products produced in agroforestry practices, the remainder of this chapter focuses on analyzing marketing opportunities, researching and selecting target markets, and developing and implementing marketing strategies and plans. The chapter closes with examples that demonstrate the application of competitive market analysis and strategy development to agroforestry product markets.

Marketing Agroforestry Products

What is Marketing?

The term *marketing* has changed and evolved with time. According to the American Marketing Association, marketing is a set of processes for creating, communicating, and delivering value to customers and for managing customer relationships in ways that benefit the organization and its stakeholders (American Marketing Association, 2013). Marketing starts before a product exists. To create value to customers, their wants and needs must be identified. This involves doing market research on customers, analyzing their needs, and then making strategic decisions about product design, pricing, promotion, and distribution. Marketing continues throughout the product's life to find new customers and retain existing ones by improving product appeal and performance, learning from product sales results, and managing repeat performance (Kotler & Armstrong, 2001). Customers are becoming more active participants in the exchange process, and maintaining long-term relationships with them is an important part of the marketing process. The satisfaction of other stakeholders should also be a consideration in any marketing exchange. *Stakeholders* are individuals and organizations, including employees, suppliers, buyers, and the community, that influence or are influenced by what the organization does (Hult, Mena, Farrell, & Ferrell, 2011).

Marketing involves the process of planning and implementing a strategy for exchanging goods. Marketing is more than just letting a buyer know that you have a product to sell; it is a process that requires planning, research, and creativity. Similarly, with planning, research, and creativity, agroforestry provides landowners with the potential to earn income from a wide range of alternative products. The key to earning income from agroforestry practices is understanding what products to sell and how to successfully sell those products—in other words, understanding the "ins and outs" of markets and marketing.

The "Black Box"

Agroforestry enterprises often produce niche products for markets about which little is known (Gold, Godsey, & Josiah, 2004). All that may be known about a product's market is that it is produced and eventually purchased and consumed (Thomas & Schumann, 1993). What happens to the product along the value chain between producer and consumer and why the consumer is buying the product is unknown, and it is commonly referred to as a "black box." From a producer's perspective, the list of unanswered questions is long. How many times does the product change hands before it reaches the final consumer? How do I get into the market? What are my costs and potential returns? Who are my customers? Who are my competitors? What strategy should I use to be successful in this market? These and many other questions complicate the decision to produce and market niche products (Bogash, 1998; Sullivan & Greer, 2000). Planning is important for farms and all small businesses striving to increase profitability and competitiveness and become economically successful. Planning encourages forward, proactive thinking, focused on where the farm business or firm is going and how to get there. Unlike commodity markets with readily available market information, the challenge for farms and small businesses that engage in agroforestry enterprises is the lack of information about alternative product markets. To successfully exploit niche product markets, agroforestry entrepreneurs must perform organized market research to open the "black box" and overcome the information asymmetry inherent in these niche markets. If this challenge is met through organized market research, agroforestry entrepreneurs are provided with an opportunity and competitive advantage (i.e., gaining an edge over competitors) to successfully market their product(s).

The Marketing Planning Process

Marketing planning is essential, especially for new farm enterprises. The process of gathering information for developing a marketing plan is complex but, in the end, worth the effort. According to Kotler (1994), the marketing planning process involves analyzing marketing opportunities (the industry and economic conditions, the customer and other marketing system participants, and the competition in the market), researching and selecting target markets, product and service positioning (grouping consumers on the basis of individual characteristics, selecting the most profitable market segments), developing marketing strategies (formulation of competitive marketing strategies and marketing mix to market the product[s] or services), developing the marketing plan (tying it all together), and implementing and evaluating this plan (dynamic markets require the continual adaptation of plans to new changes in the environment) (Figure 15–1).

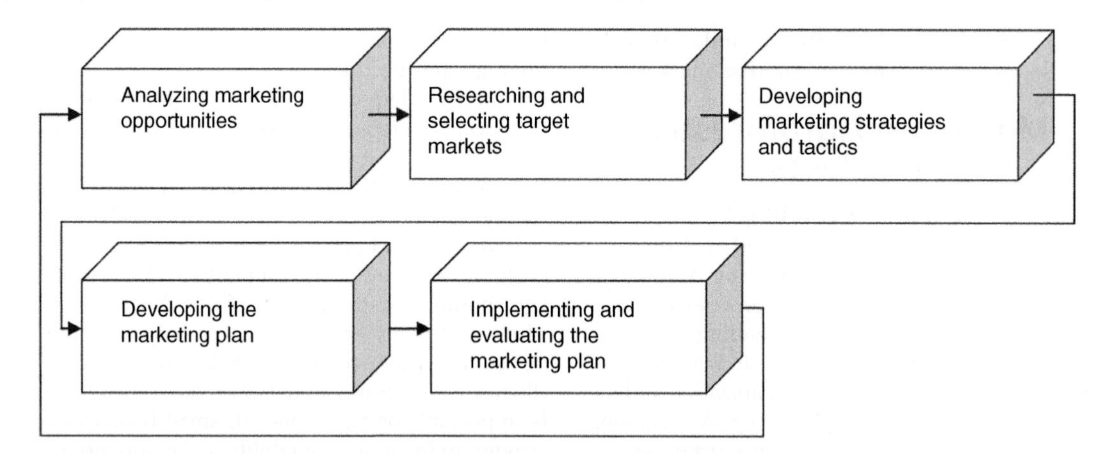

Fig. 15–1. The marketing planning process (adapted from Kotler, 1994).

Analyzing Marketing Opportunities—Market Research and Analysis

The planning process must have a market-oriented focus directed toward the goal of achieving customer satisfaction within a constantly changing environment. Market orientation means customer focus (i.e., understanding customer needs and delivering customer satisfaction) and competitor orientation (i.e., recognition of competitors' moves, creation of competitive advantages).

To analyze marketing opportunities, a farm business must understand the environment in which it operates to continually identify new opportunities and threats, understand and monitor its competition, and understand consumers and their buying behavior. To gain this understanding, there is a need to collect and manage information. Market research provides a tool to help gather and analyze the needed information. Market research is the starting point in understanding the market, providing a broad yet detailed overview of the industry being examined. The American Marketing Association defines market research as the systematic gathering, recording, and analyzing of data with respect to a particular market (American Marketing Association, 2013). As markets are constantly changing, market research is an ongoing process that influences all aspects of a marketing strategy and continues throughout the life of a business.

The first step in any market research process is data collection. It starts by investigating publicly available data relevant to the specific market being evaluated. This is known as *secondary data* and includes published reports and market studies, trade magazines and journals, newspapers, books, literature from competitors, and business directories. Internet search, libraries, government agencies, professional associations, offices of the local chamber of commerce, and universities are good sources for obtaining this kind of information. Secondary data are important but insufficient for developing detailed market research. While secondary data provide market insight, this information must be accompanied by information gathered specifically for the market(s) of interest. This requires primary data, which can be derived from direct consultations with existing or prospective customers, observations of and conversations with competitors, discussions with other businesses outside the potential market area, observations of how and where potential product(s) are currently being marketed, and discussions with brokers, retailers, and distributors. Methods that can be used to gather primary data include surveys, observations, focus groups, and interviews (University of Missouri Center for Agroforestry, 2007).

Understanding the Business Environment—SWOT Analysis

There are many questions in the mind of a producer trying to launch a new enterprise: How difficult is it to enter the market and what resources are needed? Who can provide the needed supply? Who has more control in the supply chain? Who is going to buy the product(s)? Would it be better to deliver directly to consumers or to use intermediaries? Who are the intermediaries? How much control does the buyer exert? Who are the final customers and what are their needs? Who are the competitors and what are their competitive advantages?

Several methods are commonly used to analyze marketing opportunities and create successful marketing strategies and marketing plans. Strengths, Weaknesses, Opportunities, Threats (SWOT) analysis is recommended for formulating marketing strategies (Kotler, 1994; Shrestha, Alavalapati, & Kalmbacher, 2004). A SWOT analysis includes an examination of both internal factors (strengths and weaknesses) and external factors (opportunities and threats). Analyzing internal factors involves looking inside the farm business and identifying its strengths and weaknesses. *Strengths* refers to the farm's competencies that are the basis for its competitive advantage (e.g., the ability to provide quality products, market knowledge, consistent year-round supply, excellent customer service, competitive price, production skills). *Weaknesses* are areas where the farm can be vulnerable to competitors. Weaknesses that are identified (e.g., lack of consistent supply because of the seasonality of the product) can be transformed into strengths (e.g., building indoor facilities to extend production time). In general, a farm must find ways to minimize the impact of its weaknesses on its business operations. An example of SWOT analysis for a farmer who would like to start growing American elderberry (*Sambucus canadensis* L.) is provided in Figure 15–2 (Cernusca et al., 2018).

External factors belong to the marketing environment. As with any organization, a farm's marketing environment consists of all factors that affect its ability to develop and maintain successful transactions and relationships with its customers (Kotler, 1994). From a strategic perspective, the marketing environment has both a micro and a macro component.

The microenvironment includes suppliers, intermediaries, customers, competitors, and the public at large, all factors that help or affect the firm's ability to produce and sell products (Kotler, 1994). For a specialty-crop mushroom producer, the microenvironment includes spawn and media suppliers, distributors that supply retail stores, retail stores, consumers, other firms that produce mushrooms, and the public-at-large. Microenvironment analysis provides information that will help a farm business or firm react by creating competitive advantages and/or formalizing arrangements with channel partners (suppliers or buyers) to achieve better market coordination.

The macroenvironment consists of a collection of forces and conditions facing a farm business that are not under the direct control of the producer, including demographic, economic, natural, technological, political, and cultural forces. Factors in the macroenvironment include:

- federal and state regulations and consumer legislation (political and legal factors)
- shifts in consumer tastes and preferences impacting consumption of fresh, locally produced, or organic produce (socio/cultural factors)
- geographic distribution and population density, including regions undergoing population growth or decline (demographic factors)

Strengths	Weaknesses
Previous farming experience Available size of land Number of children returning to the farm Location (e.g. close to a big city)	Moderate finances Seasonal and perishable product Lack of marketing skills
Opportunities	**Threats**
New cultivars made available Cooperative (e.g. Mid-west Elderberry Co-op) Increased interest in locally produced products	Competition from imports Increased federal and state regulations Extreme climate events

Fig. 15–2. A Strengths, Weaknesses, Opportunities, Threats (SWOT) analysis of starting an elderberry orchard (modified from Gold, Cernusca, & Godsey, 2015).

- innovative approaches to preserve fresh produce or create value-added products and the increasing role of the internet (technological factors)
- evolution of imports, inflation, interest rates, income changes, and consumer spending patterns (economic factors)
- shortage in raw materials, energy cost changes, extreme climate events, and levels of pollution (natural factors).

By analyzing the macroenvironment, important trends can be discovered and used to a firm's advantage. Based on SWOT analysis, strategies can be created to transform a farm's strengths into competitive opportunities and react to market threats in a way that will lead to the accomplishment of the farm business' goals and objectives.

Understanding the Competition— Porter's Five Forces Model

An approach proven successful in shedding light on the "black box" is the model developed by Porter (1980) that describes the forces driving industry competition, known as the Porter's Five Forces Model (PFFM). The framework of the PFFM provides information about microenvironment market dynamics by focusing both on the actual situation (bargaining power of buyers and suppliers and rivalry among existing firms) and on foreseeable developments (barriers to entry and threat of substitute products) (Figure 15–3). Porter's strategic forces help evaluate the ease of market entry and exit, buyer and seller power, power of substitute products, and competitive rivalry and provide a general view of the industry. With respect to agroforestry, the method is especially useful for farm businesses that plan to enter new markets.

Potential Entrants (Barriers to Entry)

For producers considering entry into new agroforestry product markets, risks and uncertainties may deter them from looking at the market possibilities. Porter (1980) identified six major entry barriers, including economies of scale, capital requirements, switching costs, access to distribution channels, product differentiation, and cost advantages independent of scale. Not all of these barriers will be evident in all markets. However, every market will have some entry barrier, either real or perceived, that must be overcome. Market risks and uncertainties can be overcome by understanding what critical resources are needed to successfully operate in the market and what entry barriers exist.

Economies of scale (i.e., as the volume of production increases, the cost of producing each unit decreases) usually create advantages for existing firms. There are industries where a minimum size is required for an operation to be profitable.

Capital requirements can be an important barrier to entering the industry if financing cannot be obtained or economies of scale prevent entry through a process of starting small and scaling

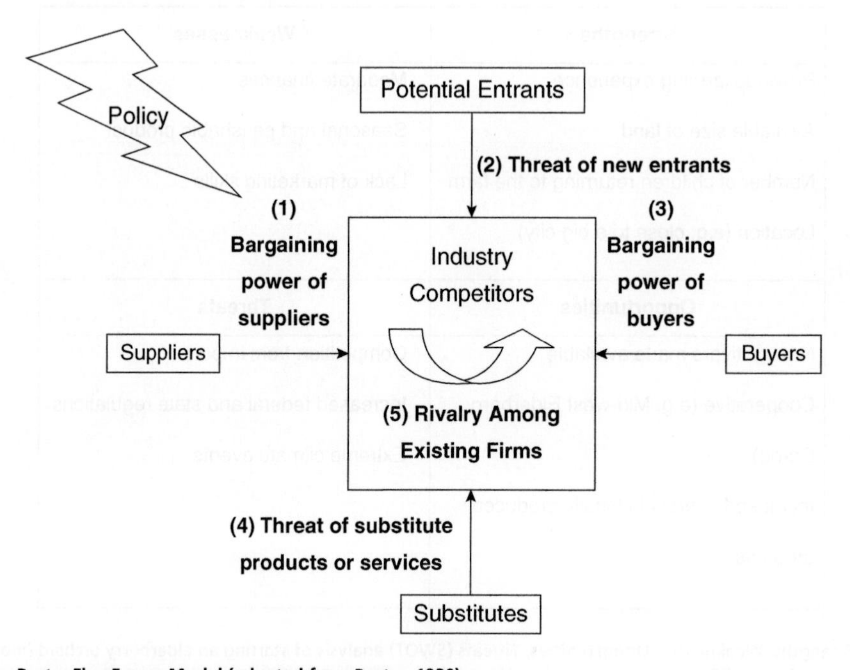

Fig. 15–3. The Porter Five Forces Model (adapted from Porter, 1980).

up. Usually, agroforestry enterprises, such as log-grown shiitake mushroom [*Lentinus edodes* (Berk.) Sing.] production, pine straw production, or woody floral production, can be started with personal finances. Companies already in the market often recommend starting small and growing the business after becoming familiar with the production process and creating a loyal customer base.

In some markets, consumers face substantial *switching costs* when switching between suppliers, brands, or products. The costs can be monetary, effort and time based, or psychological. Incumbent firms create strong relationships and brand loyalty from advertising, customer service, product differences, or just by being first in the market. Time and money are required for a new firm to create its own relationships, establish a recognized brand, and gain loyal customers.

Restricted access to distribution channels and difficulty in reaching final consumers are other challenges posed to new entrants. In the mushroom industry, many large "white button" mushroom [*Agaricus bisporus* (J.E. Lange) Pilát] firms diversify their production into different kinds of specialty mushrooms, including shiitake. These firms already have production facilities and distribution channels in place that make the production and marketing of specialty mushrooms easier, thus creating an advantage over small farms producing log-grown shiitake in agroforestry settings.

Proven methods exist to overcome *product differentiation* entry barriers. Reaching customers is commonly achieved through advertising, which can be costly. Publicity is more accessible to producers because it uses free media coverage of the farm and its products. Producers of agroforestry products can then take advantage of "face" and "place" as publicity tools to help reach their customers. These businesses are marketing products that are unique. If the producers are enthusiastic about what they do, they must get the word out by attending trade shows, giving presentations and bringing products for demonstration and sale, holding field days, and contacting journalists. Face and place stories are often compelling and attract new readers to the magazine publishing the story and new customers to the business telling the story. As a result, many magazines are willing to write articles about producers and their farm businesses (Gugino, 2007).

To secure their position in the market, existing producers may be reluctant to share their sources for products, methods of production, or potential markets. This knowledge about production and marketing acquired with time through practical experience (i.e., the learning curve) creates advantages to existing firms. In agroforestry, the range of such knowledge available to new entrants varies from market to market. For example, prospective shiitake growers have access to literature on production practices and marketing strategies, but they still face a learning curve due to the amount of labor, level of care, and attention mushroom cultivation demands. In contrast, a grower establishing a chestnut (*Castanea* spp.) orchard faces more uncertainty. Little research has been done on specific regional cultivars, orchard management practices, and value-added opportunities. Producers must learn as they go and continually experiment.

Suppliers, Buyers, and Bargaining Power

The goal of any entrepreneur is to have bargaining power (i.e., control) in the market. Bargaining power fluctuates through time as market participants enter and exit the market and consumer demands change. However, all market participants are both buyers and sellers in a market at different points in the business process. Therefore, it is important to understand how to minimize pressure on profits that can be exerted through bargaining power. Suppliers with bargaining power can extract greater profit by increasing input costs. Likewise, buyers with significant bargaining power can extract excess profit by putting downward pressure on output prices. Wal-Mart is an example of overwhelming bargaining power in the hands of the buyer, dictating price to suppliers and switching suppliers whenever a better price can be extracted (Munson, Rosenblatt, & Rosenblatt, 1999). Agricultural commodity producers are squeezed by both supplier and buyer bargaining power. The U.S. farmer has a limited number of large firms supplying farm inputs and a limited number of large firms buying the commodities. As a result, the farmer's profits are continually being eroded by higher input costs and lower market prices. To avoid this situation, agroforestry niche market producers must focus on products that retain value to buyers through uniqueness, relative scarcity, and value-added approaches. New products and new markets must be developed continually. By taking these steps, niche producers can retain bargaining power at the supply level and retain more profit through value-added production, thereby controlling more of the value chain and retaining more of the profit.

Substitutes

Substitute products can replace other products with limited impact to the consumer. If the market perceives Product B (e.g., margarine) to be a close substitute to Product A (e.g., butter), then small price increases in Product A will increase demand for Product B. As a result, close substitutes can have an effect on market prices by providing an option to consumers when prices fluctuate.

Substitute products can play an important role in new markets where demand is increasing rapidly or there are few direct competitors. When supply cannot keep up with demand, substitutes can be used to "fill the gap." Yet, in the process of filling the gap, these substitutes limit the profit potential of the original product. This describes the current situation in the organic marketplace where domestic supply has not kept pace with demand, drawing production from overseas and impacting U.S. producer prices. However, in periods of excess capacity, in highly competitive markets, in markets where product supply is limited (e.g., pine nut markets) (Sharashkin & Gold, 2006), or in markets lacking substitutes, substitutes have little or no effect on market prices.

Gold, Godsey, and Cernusca (2005) examined the question of potential substitutes for eastern redcedar as part of a comprehensive market analysis. Western redcedar (*Thuja plicata* Donn ex D. Don) is often sold as "redcedar," but it does not have the same suite of marketable properties as aromatic eastern redcedar. Different uses of eastern redcedar have different substitutes. For example, western redcedar or chromated copper arsenate (CCA) treated lumber are used as more common substitutes for eastern redcedar in the area of home construction or dimensional lumber. For novelties, other cypress species are often substituted. Many pine species are used for pet bedding and mulch in place of eastern redcedar. However, eastern redcedar is often considered to be cheaper, more available, and unique in quality when compared with those substitutes. The bottom line is that eastern redcedar's unique combination of desirable qualities—color, rot resistance, and pleasant "cedar" aroma—place it in a niche by itself. Information regarding potential substitutes in the redcedar market can help identify new marketing niches.

Industry Competitors

Rivalry within an industry is a function of the number of competitors, the size range of competitors, homogeneity of products, level of fixed investments, and volatility of market demand. In markets where rivalry is intense, business strategies are often implemented that focus more on competitive posturing and less on profit generation. For example, a producer may cut prices for their goods to undercut the prices of a competitor (Oster, 1999). In other markets, hazelnuts (*Corylus avelana* L.) or chestnuts, for example, rivalry may not be a factor, and market participants actually work together to further the demand for their products and increase profitability for all market participants.

Government Policy

Although governmental policies are not explicitly listed as one of the five competitive forces in the PFFM, they do have a dramatic impact on the market and often act as an implicit market force along with the five forces proposed by Porter's model (Gold, Godsey, & Josiah, 2004). Governmental policies that impact markets include:

- Intellectual property rights—patents, trademarks, and copyrights can be entry barriers or methods of product differentiation
- Regulations—specific laws dictating how a market must operate can reduce rivalry among market participants since all firms must follow the same guidelines
- Quotas and tariffs—price-fixing strategies such as quotas and tariffs can increase the number of market participants by allowing inefficient producers enough income incentive to stay in the market

Governmental policies affect all aspects of the market (Drabenstott & Sheaff, 2001), from rivalry to entry barriers. Niche products will typically have fewer governmental policies impacting the market than agricultural commodities because of the scale of the market. However, niche product markets are often impacted by policies that were not designed for that market but have an indirect application. For example, CCA-treated lumber, widely used due to its rot resistance, is being phased out by order of the U.S. Environmental Protection Agency because of public health concerns about arsenic. This phase-out has the potential to stimulate market demand for eastern redcedar lumber due to its natural rot resistance. Federal and state grants, low-cost loans, and tax breaks offered to farms in specific industries can be helpful in launching new agroforestry products and in easing market entry and development.

Understanding the Customer—Consumer Analysis

The ultimate purpose of marketing is to understand customers' needs, desires, and motivations. Customers are divided into two groups: consumers and business organizations. Agroforestry products are often sold to the final consumer, retail store, or restaurant chef, but they can also be distributed wholesale. It is very important to understand the specific requirements of each buyer in terms of quality, quantity, delivery frequency, packaging, and sizing.

Whatever distribution strategies are used, the product will ultimately reach the consumer. Market research techniques will help understand consumers' needs and ways to achieve consumer satisfaction. Techniques ranging from simple observation and interviews to more complex focus groups and surveys will provide an insight into consumers' preferences and the way consumers make their purchase decisions.

There are many factors that influence consumers' purchase behavior, including cultural (e.g., subculture and social class), social (e.g., reference groups, family), personal (age, occupation, personality, lifestyle), and psychological (motivation, perception, beliefs, attitudes) (Kotler, 1994). The product attributes also influence consumers' willingness to buy. A study performed in Michigan evaluated the effect on consumers' preferences of color combination and species of containerized edible flowers (Kelley, Behe, & Biernbaum, 2002), and another in Alabama researched fruit quality characteristics that affect consumers' preferences for Satsuma mandarins (*Citrus reticulata* L.) (Campbell et al., 2004). Another example is an analysis of consumers' preferences for value-added seafood products derived from crayfish (Harrison, Stinger, & Prinyawiwatkul, 2002). Consumers' familiarity with and preferences for specialty crops (e.g., chestnut, elderberry, and pawpaw [*Asimina triloba* (L.) Dunal] have also been examined by Aguilar, Cernusca, and Gold (2009); Aguilar, Cernusca, Gold, and Barbieri (2010); Cernusca, Aguilar, and Gold (2012); Mohebalian, Cernusca, and Aguilar (2012); Mohebalian et al. (2013); and Cai et al. (2018). For unfamiliar products such as chestnuts, longitudinal studies should be performed to allow consumers to become more familiar with the product and compare how their preferences change with time (Gold, Cernusca, & Godsey, 2005). Cernusca, Gold, and Godsey (2008) examined the change in consumers' familiarity with and knowledge about chestnuts from 2003 to 2006 using longitudinal data.

Target Markets—Finding a Niche

As stated above, marketing involves the process of planning and implementing a strategy for exchanging goods. This includes idea development, pricing, promotion, and distribution right through to the exchange of a product for cash. More than just selling products, marketing is focused on solving problems for customers whose needs are at the center of the business' activities. Agroforestry practices often combine production of both commodity and niche products. Producers must understand the differences between commodity and niche product markets and take advantage of how both market categories meet the needs of their customers.

Commodity Markets

The nature of modern farming and forestry has created an environment for marketing commodities requiring a minimum amount of effort from the producer. For example, marketing no. 2 yellow corn (*Zea mays* L.) can be done by selling to the local grain elevator. In commodity markets, the buyer seeks the lowest price. Success in commodity markets requires high-volume production and relies on advantages of size and scale. Examples of commodity markets include major grain and fruit crops, beef and dairy, and lumber and veneer.

In commodity markets, an infrastructure has been established that reduces the risk inherent in marketing by simplifying the processes of information gathering, bargaining and decision making, and supervision and enforcement. For agricultural commodities, established markets such as livestock auctions, grain elevators, and the Chicago Board of Trade minimize the time and risk involved in the information gathering process. The Chicago Board of Trade, along with the USDA, provides objective market information that minimizes the risks associated with the bargaining and decision-making processes. Further, accepted standards for agricultural commodities have been established that are easily enforced by marketing institutions. Timber markets are similar to agricultural markets. Grading standards are established, market prices are publicly available, and market participants are often brought together by an established bidding process.

Niche Product Markets

Markets for specialty products often differ from markets for more common agricultural commodities. They are usually characterized as "niche" or "product" markets—small, volatile, specialized, and with relatively few buyers. The scale of product markets is normally much smaller than that of commodity markets. Product markets differ from commodity markets in that they rely on the value-added concept, represent unique niche products "with a face and a place," and seek to increase benefits to the customer. Customers purchasing niche products are attracted not only by quality and consistency but by the individual(s) or company behind the product(s) and their story along with the location (origin) of the product. In other words, niche product markets are based more on trust and authenticity. Trust and authenticity are competitive advantages that favor smaller firms (Sustainable Agriculture Network, 2003). Further, the unique, personalized, customer-oriented approach presents an inherent competitive advantage for those who are in niche product markets compared with mass markets.

In contrast to commodity markets, specialty products often lack established marketing institutions, accessible market information, and established grade or quality standards. The resultant lack of market information creates understandable disincentives to individuals considering niche product-based enterprises. Thus, producers often continue to produce commodities with known markets at a loss instead of pursuing other novel but potentially lucrative opportunities where markets are not well known or established. Further, because knowledge of the value chain is so valuable in niche markets, producers may be reluctant to share their sources of information, methods of production, or current and potential markets. These barriers to entry can make it difficult for newcomers to succeed in the niche marketplace (Vollmers & Vollmers, 1999).

Market Segmentation and Target Market Selection

The process of identifying customers' preferences and dividing the larger market into niche markets is known as *market segmentation*. The most attractive market segment(s) should be chosen as the target market(s). Marketers use consumer demographic (e.g., age, gender, family size, income, occupation, education), psychographic (e.g., lifestyle, personality), behavioral (e.g., usage patterns, brand loyalty, readiness stage, attitudes toward quality or environment), or geographic characteristics to establish market segments. An example is a study that identified market segments for elderberry products and attributes that influence purchasing decisions for each segment. According to this study, current elderberry consumers are younger, more educated, and less price sensitive than consumers who have not yet purchased elderberry products (Mohebalian et al., 2012). The most common method for segmenting markets is geography. Targeting local buyers is especially favorable for producers of agroforestry products. Local markets facilitate timely delivery, especially of perishable products, and provide more opportunities to create and maintain strong relationships based on communication and trust. Besides geographic segmentation, niche producers have used consumers' concerns about health and the environment as a basis for market segmentation and identifying a target market (Phillips & Peterson, 2001). Health- and environment-conscious consumers are a good target market for many products that are grown in agroforestry settings. Consumers' beliefs and value systems affect consumption patterns; strongly held beliefs influence patterns of consumption. This presents unique opportunities for niche markets (Heiman, Just, McWilliams, & Zilberman, 2004).

Once market segments have been identified, the producer must evaluate these opportunities and decide how many and which one(s) to target. Kotler (1994) suggested evaluating the segments by looking at the segment size and growth, segment structural attractiveness, and producer objective and resources. Individual producers can select smaller segments that show growth potential. To determine the long-run profit attractiveness of a segment, the PFFM can be applied. A segment is attractive if (Kotler, 1994):

- it doesn't contain numerous, strong, or aggressive competitors (i.e., segment rivalry)
- the product the farm business is offering is differentiated from the ones offered by competitors (i.e., threat of substitute products)
- suppliers don't have the power to raise prices and reduce quantity (i.e., bargaining power of suppliers)
- buyers don't possess the power to negotiate or switch suppliers (i.e., bargaining power of buyers)
- the segment has high entry barriers to provide high, stable returns

The final factor to consider is to determine whether the producer has the needed skills and resources to succeed in that segment. Once the target market is selected, the producer must find the best way to satisfy this market. Specific marketing goals and objectives should be set and strategies to attain them should be developed.

Developing Marketing Strategies

A marketing strategy is a plan to achieve a market goal (Kohls & Uhl, 1998). A marketing strategy uses the information provided by primary and secondary research and provides a framework to identify the products to sell as well as the methods to sell them in the most profitable way and satisfy the needs of the target market. Marketing strategies take into account the producer's and competitors' capabilities, market threats and opportunities, target market needs, goals, and financial resources.

There are two major components to a marketing strategy—how the firm addresses the competitive marketplace (competitive strategies) and how the firm will actually develop and provide the products to its customers (product, distribution, pricing, and promotion strategies).

Competitive Strategies

According to Porter (1980), to cope successfully with the five competitive forces and become profitable, a firm should develop competitive strategies. At the broadest level, he identified three generic competitive strategies: cost leadership, differentiation, and focus, which can be used by themselves or in combination (Figure 15–4).

Cost leadership strategies focus on producing products or services for a wide consumer group at the lowest possible cost. This strategy has been the traditional strategy used in the production of agricultural commodities, with the intention of lowering costs and increasing income by maximizing volume.

Differentiation strategies involve creation of a product or service that is perceived industry-wide as being unique (Porter, 1980). In agroforestry, approaches to differentiating can take many forms: design or brand image (e.g., Ozark Forest Mushrooms, a brand recognized for quality in the Saint Louis, MO, area), technology (e.g., Chestnut Growers Inc., a for-profit cooperative in Michigan that purchased a commercial peeler unique in the U.S. industry, enabling production of value-added products such as peeled, frozen chestnuts and chestnut chips, differentiating its product from the fresh in-shell chestnut market), and customer service. Differentiation is well suited to agroforestry-derived products and can be a viable strategy for earning above-average returns because it creates a defensible position for coping with the five competitive forces (i.e., because of customer brand loyalty, the differentiation strategy provides entry barriers, improves positioning vis-à-vis substitutes, and protects against competitive rivalry). Differentiation yields higher margins that make it easier to deal with supplier and buyer power, since the buyer, lacking comparable alternatives, is less price sensitive (Porter, 1980).

The *focus strategy* concentrates on a particular buyer group, product line segment, or geographic market. Different from either the cost leadership or differentiation strategies, which are aimed at achieving their objectives industry-wide, the focus strategy strives to serve a particular target market very well. As a result of serving a limited number of customers, a farm can strive to better meet the needs of the target market or lower the cost in serving this target (Dalton, Holland, Hubbs, & Wolfe, 2007; Porter, 1980). Most farmers use a combination of marketing methods, finding that in marketing as well as in production, diversity helps provide stability and sustainability.

Product Strategy

The most common marketing strategy for farmers producing agroforestry products is product differentiation to appeal to a focused group of consumers (Phillips & Peterson, 2001). Compared with the cost leadership strategy employed in commodity markets, the focus-differentiation (niche) strategies can be very innovative. Farmers have the opportunity to implement many creative marketing ideas to differentiate their products and services in response to the needs of their customers.

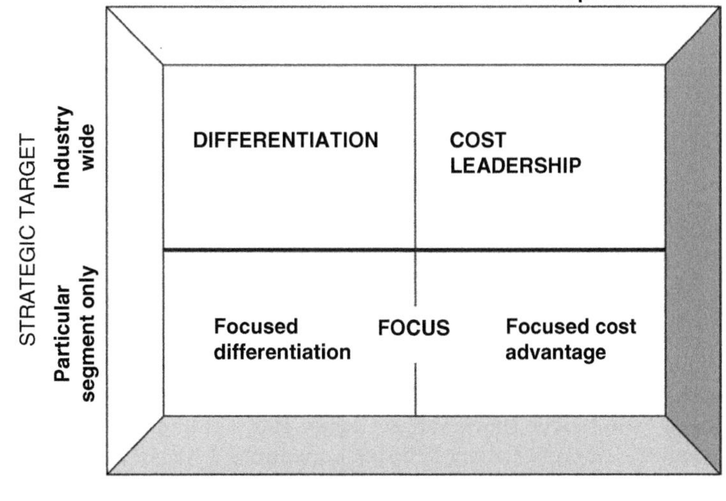

Fig. 15–4. Generic competitive strategies (adapted from Porter, 1980).

Differentiation creates competitive distinction via product features—some visually identifiable and some claimed by reference to real or suggested hidden attributes that promise results or values different from those of competitor products (Levitt, 1980). Differentiation equates to adding value to the products. Customer value is derived from perceived benefits as a function of price. Value can be added by determining the customers purchasing criteria and increasing the bundle of benefits to the consumer. Eight different approaches to adding value by product differentiation and addressing diverse niche market needs include the following:

1. Processing fruits and vegetables allows the farmer to extend the marketing season of fresh products and diversify their marketing outlets. There are many value-added opportunities for food products (e.g., dried fruits or vegetables, wines, juices, spreads, sauces, syrups, preserves, salsa). Value-added craft products (e.g., dried and evergreen wreaths, baskets, dried floral materials, twig furniture) can provide good returns. Herbal products offer various opportunities for adding value (e.g., dried herbs, vinegars, spices, herbal teas).
2. Offering high-quality, locally produced products is a way to differentiate in markets dominated by imports. Locally produced chestnuts in the Columbia, MO, area are highly valued by consumers and preferred over imports (Aguilar et al., 2009, 2010). Missouri Northern Pecan Growers, LLC, is a cooperative in Missouri that differentiated the Missouri northern pecan [*Carya illinoensis* (Wangenh.) K. Koch.] from the more commonly sold southern pecan. Cooperative members claim that the Missouri northern pecan has a unique taste (due to the cooler climate and shorter growing season in the northern region). The unique local characteristics are combined with organic certification and a brand name, "Sweeter by Nature—Missouri Northern Pecans," to further differentiate the product.
3. Introduction of new products (nontraditional commodities or traditional commodities with special characteristics desirable to customers) is important in a market economy, especially for the dynamics of the market structure. Innovative firms that introduce new products will be those who grow in the medium and long term (Roder, Herrmann, & Connor, 2000). In niche markets, variety and small quantities are more appealing to many consumers than bulk bargains (Andreatta, 2000).
4. Season extension offers time convenience to customers and also brings benefits to the producer. The first produce at the farmers market in the season will command a price premium. Early crops can also create opportunities for innovative value-added products.
5. Customer relationships offer other ways to differentiate because consumer loyalty and trust are increasingly important in small markets (Boehlje, Hofing, & Schroeder, 1999). When it comes to food, consumers want to know more about what they buy, where it comes from, what production methods were used, and if the food is nutritious and environmentally friendly. The disclosure of product nutrition information can increase consumer preferences (Mohebalian et al., 2013).
6. Offering extra convenience differentiates a product from its competitors. Place convenience can be created by transporting products to the location where they are demanded by customers (e.g., farmers markets, chefs). Offering recipes along with the product creates convenience and is a way to increase sales. Since cooking skills and knowledge restrict food choices, there is potential gain from providing recipes and cooking demonstrations. Busy people value time convenience (e.g., offering ready-to use foods, offering fresh products out of season).
7. The product image offers additional opportunities for adding value and diversification. A product produced in a manner perceived by consumers to be healthy or environmentally friendly can differentiate it from competitors. Growers that sell heirloom seed types promote flavor and history. Organic practices and high quality will provide additional differentiation and bring higher prices (Andreatta, 2000; Aguilar et al., 2009; Cai et al., 2018; Mohebalian et al, 2013). Audubon certified beef (beef from farms and ranches with good grassland stewardship) may obtain higher consumer preferences than beef that is not certified (National Audubon Society, 2018). Agroforestry product producers must strive to preserve their product's identity and their unique story as the product moves along the value chain from producer to consumer. Producers must take full advantage of "face" and "place" by marketing the farm. Higher prices are paid for identity-preserved products that come with a story (Hartman, 2003).
8. Differentiating a product by using the label of origin is more common in Europe than in the

United States. Protected Designation of Origin labels are assigned to products that are traditionally produced, processed, and prepared in a specific geographic area. One example is Prosciuto di Parma ham, a dry-cured ham produced in Parma, Italy. Examples of the protected designation of U.S. origin labels are Vidalia onions (*Allium cepa* L.), Sunkist oranges [*Citrus sinensis* (L.) Osbeck], and California almonds [*Prunus dulcis* (Mill.) D.A. Webb] and prunes (*Prunus domestica* L. var. *domestica*) (Hayes, Lence, & Stoppa, 2004). Empirical studies have found that the region of origin label can significantly affect consumer purchasing preferences (Cai et al., 2018; Mohebalian et al., 2013). The Mississippi River Hills area is serving as the pilot region for the Missouri Regional Cuisines Project, which seeks to market Missouri wine and food products using distinct labels of origin based on ecological regions of the state (see https://www.mississippiriverhillswinetrail.com/). Ideally, a firm differentiates itself in multiple ways.

Distribution, Pricing, and Promotion Strategies

Distribution

Distribution strategies describe where, how, and when products enter the market. With specific customers and products in mind, it is essential to determine how to distribute products from the farm to the customer(s). Distribution strategies commonly describe scope, movement, packaging, scheduling, and handling. The farmer has two major alternatives to consider, direct marketing distribution or intermediary distribution (DiGiacomo, King, & Norquist, 2003).

Direct marketing to consumer is the process of selling a product or service directly to the end user, and it is the alternative most suited to agroforestry product producers. The number of direct-sales farms increased by 68% between 2007 and 2010 (Ahearn & Sterns, 2013). Selling direct provides the grower the opportunity to capture a larger share of consumers' spending and the opportunity to educate the consumer about the farm and its production methods (e.g., about the advantages to buy organic or locally grown products). Direct marketing is growing in popularity because consumers now demand safer, high-quality products. Buyers place a value on coming face to face with the producer and their production location (farm, farmers market, on-farm retail store) and obtaining more information about the products produced.

Outlets for direct marketing niche products to consumers include farms, farmers markets, pick your own (PYO), farm to family, community-supported agriculture (CSA), home delivery service, festivals, garden and trade shows, county and state fairs, and online or catalog mail-order sales. These outlets provide a place to sell products directly to the consumer and also to promote the product(s) and the business.

Farmers markets are growing in number and diversity (USDA Agricultural Marketing Service, 2018). Growth in farmers markets is driven by consumer demand and the potential for increased farm income (Brown et al., 2007). They offer an outlet where vendors can sell a large variety of products. Customers develop loyalty to farms they encounter regularly at a farmers market. In milder climates, covered structures can provide a profitable market outlet all year long.

Community-supported agriculture is a method in which members of a community invest in a local farm by paying in advance for a share of the harvest. Participating in CSA enables consumers to share in the responsibility along with the producer. This creates a social responsibility toward the people involved in the food system and an ecological responsibility for land stewardship (Andreatta, 2000).

Farm outlets that provide PYO and farm stands bring the consumer to the production place. Some PYO operations have diversified and constructed gourmet specialty stores and offer agritourism activities.

Agritourism provides a way to diversify the farm operation. Entertainment and tourism-based enterprises can take on many forms (e.g., bed and breakfast, food and craft shows, harvest festivals, petting zoos, horseback riding, guided nature walks, boating and canoeing, camping and picnicking, weddings, corporate picnics, family reunions, historical tours or hayrides, hunting, bird watching, and school and educational tours). Location is very important in deciding which agritourism enterprise(s) to choose. Close-to-town locations are more suited to festivals, children's activities, and educational tours, but remote locations can be better suited to a bed-and-breakfast or picnicking and camping (Minnesota Institute for Sustainable Agriculture, 2007).

Direct marketing of farm products online is another method of direct-to-consumer marketing. Computer use and high-speed internet access have become much more common in the farm sector since 2005. A total of 72% of small farms (sales between \$10,000 and \$100,000) had access to a computer in 2017, and 68% of U.S. small farms had internet access (USDA National

Agricultural Statistics Service, 2019d). Farmers are willing to sell their farm products and purchase supplies online or shop for prices online before purchasing conventionally (Batte & Ernst, 2007). This trend encourages use of the internet as a sales outlet. Khanal, Mishra, and Koirala (2015) found that small farmers can benefit from internet access as it increases income from the farm business. Shipping fresh produce can be challenging, especially for perishable products. An alternative is to produce processed products. The internet provides a good market for jams, honey, dried herbs, and baked goods. Various options exist to access the internet, including the development of one's own web page or using a site created by others.

Farm to Family is another method to help farmers to sell their products and help customers to purchase fresh and locally produced food. Farm to Family businesses source high-quality agricultural products from local farms and sell them online. Customers order products online and pick them up at a nearby delivery location on the delivery date. This marketing method provides transparency in the food chain so that customers know the source and quality of the products.

MarketMaker (https://foodmarketmaker.com/) is a growing national network of websites hosted by Global Food and Ag Network, LLC, to provide access to databases of growers, wineries, food processors, wholesalers, retailers, farmers markets, and restaurants across the United States as well as demographic data from the U.S. Census. It has an interactive mapping system that locates business and markets across the state, providing an important link between producers and buyers.

Direct marketing to institutions includes sales to restaurants, schools, hospitals, and senior-care facilities. Selling high-quality products to chefs at high-end restaurants and food co-ops commands a premium price. Consistent quality and quantity help build long-term relationships between chefs and growers.

Sales to institutions other than restaurants provide additional market opportunities. Farm-to-School programs connect schools with local farms with the objectives of serving healthy meals in school cafeterias, improving student nutrition, providing health and nutrition education opportunities, and supporting local farmers. There are an estimated 42,000 schools involved in these programs in the United States, and the number is increasing (www.farmtoschool.org). However, selling to schools can be challenging because of budget limitations, rigid regulations, and the involvement of multiple decision-makers. In addition, institutions require continuous, large quantities of goods, and producers must become knowledgeable about local, state, and federal regulations that impact purchase orders and may have to compete with preexisting suppliers and their distribution channels (Chase, 2007; Severson, 2007).

Intermediate distributors provide other market outlets and include wholesalers, brokers, cooperatives, and retailers. One of the challenges in the wholesale marketplace is the need to provide a constant, dependable supply of large quantities of quality goods. Wholesalers may only accept a few weeks' worth of products at any one time, forcing the producer to incur storage and multiple delivery costs. Another challenge is to maintain premium product integrity along the value chain (i.e., maintain freshness or maintain organic integrity at each stage of the product's journey to the market) (Dimitri & Greene, 2002).

New value chain configurations can be created through the establishment of new partnerships among farmers, processors, and retailers. Farmers can work together to accomplish marketing goals by forming horizontal alliances or cooperatives. Cooperatives offer members the opportunity to access otherwise unreachable markets (i.e., help market products by offering consistent supply at higher volume, which can be challenging for individual producers), share risk, and negotiate favorable prices through increased bargaining power (DiGiacomo et al., 2003). A market survey conducted by the Center for Agroforestry at the University of Missouri found that 48% of the surveyed chestnut growers marketed their products through cooperatives in 2016 (Cai & Gold, 2017).

By vertical integration, the farm gains better control over levels of inputs or distribution of outputs, decreasing the bargaining power of suppliers or buyers. Vertical integration occurs when a firm is acquiring a supplier or buyer or expanding its operation (i.e., performs activities traditionally undertaken by suppliers or distributors).

Pricing

Developing pricing strategies can be challenging because pricing for niche products produced through agroforestry practices is not regulated as in commodity markets. Producers who sell commodities are normally price takers, sellers that have no market control and must "take" or accept the going market price. Traditional pricing strategies are based on production costs and competitors' prices. Price is set based on the income needed to cover fixed and variable costs, including production, marketing and promotion, and a return for time and investment. In addition, prices must be competitive with the price charged by other sellers (Hall, 2002).

For differentiated (value-added) niche products, one or more pricing strategies can be considered, depending on the target market and product strategy. Current market prices—what customers are willing to pay and what direct competitors charge—can be a base for developing a pricing strategy. Niche product prices may be difficult to locate, as markets are poorly organized and little information is available. The internet, including the websites of individual companies, is a good starting point to search for niche product pricing. Once prevailing market prices and an individual's production costs are known, a pricing strategy can be developed. A variety of pricing strategies exist, and the choices depend on the producer's ability to set prices. DiGiacomo et al. (2003) described the strengths and weaknesses of eight differentiated pricing strategies. *Cost-oriented pricing* is the most straightforward strategy and is based on production costs and the producer's decision to price their product at a given percentage (i.e., 10, 50, or 100%) above production costs. Quality products with attractive packaging and labels that give brand identity or third-party certification are usually perceived as having more value. This pricing method also takes into consideration intangible attributes that are valued by consumers (e.g., practice of good environmental stewardship, preference for locally produced products). This strategy requires marketing research to identify the specific product attributes for which the consumer is willing to pay a premium price (Aguilar et al., 2009; Brown, 2003; Cai et al., 2018; Darby, Batte, Ernst, & Roe, 2006; Gold, Cernusca, & Godsey, 2004, 2005; Mohebalian et al., 2013).

Different prices are provided by different market outlets. Intermediaries often have the power to set the price. Marketing direct to the consumer or to chefs at high-end restaurants will return higher prices than using intermediaries (Andreatta, 2000; Darby et al., 2006), creating the opportunity for farms to capture a greater share of consumers' budgets. Bargaining can be used as a price strategy in farmers markets. Bargaining usually increases the volume sold and provides satisfaction to both producer and consumer by allowing them to agree on a mutually acceptable price (Brown et al., 2007).

Promotion

Promotion strategies answer the question: How and what will be communicated to buyers and consumers? Promotion is essential to gain product recognition among customers. A broad strategic approach can concentrate on promoting either the image of the business, the product, or both. Promotion strategies should include a promotional method, specific promotion tools, and ideas for actual promotion (DiGiacomo et al., 2003). Marketing communication is a tool used by producers to connect with their customers and their distribution channel partners (i.e., the organized network of institutions that link producers with end customers to accomplish the marketing task) and to create and strengthen relationships. Information can be transmitted to actual and potential customers through advertising, publicity, and direct marketing.

Local producers can advertise in newspapers, magazines, flyers and catalogs; on radio, TV, and billboards; within health food stores; and online. To increase product awareness, publicity is more commonly used than advertising. Publicity is more convenient than advertising because it uses unpaid media coverage of the firm and its products. Methods used to generate publicity include free sample offerings, participation in festivals and fairs, collaboration with charities, sponsoring of community events, and news releases. News releases to the media are a low-cost method to get promotion. Using a product in a fundraiser for a local charity will capture the interest of the media, who will write about the event and provide free promotion. Offering free samples is a commonly used practice that helps establish local markets (Nickels & Wood, 1996). Organizing workshops, giving talks and farm tours, attending farmers markets, collaborating with local CSA groups, and word-of-mouth are ways to raise awareness, inform and educate consumers, and build trust and understanding. Word-of-mouth communication is very useful for stimulating positive attitudes toward a product, leading to trial and adoption (Nickels & Wood, 1996). Consumers are often unfamiliar with niche agroforestry products. Therefore, the more information provided about the product's benefits, the more likely people are to try the product. Communicating "freshness," "local," or "small-scale production methods" is an important promotion strategy (Darby et al., 2006). If paid advertising media are used, partnering with other farms can reduce individual advertising costs.

Brands are created to identify a business' product and distinguish it from the competition. While expensive, many niche producers concentrate promotional efforts on image advertising, such as promoting the concepts of heart-healthy, locally grown, or "green" products (DiGiacomo et al., 2003).

Marketing strategies must continually adapt. Differentiating attributes become commoditized with time (e.g., once considered a niche product, organic food has become more available and affordable for consumers in conventional grocery stores), and the successful farmer must constantly examine new opportunities for diversification (Boehlje et al., 1999; Stevens-Garmon, Huang, & Lin, 2007).

Applying Marketing Strategies: Examples

Example 1. Ozark Forest Mushrooms

Ozark Forest Mushrooms (Timber, MO; http://www.ozarkforest.com/) specializes in outdoor production of organic, log-grown shiitake mushrooms. In addition to sales of fresh mushrooms to local chefs and upscale grocers and creating a well-known regional brand, Ozark Forest Mushrooms has created many value-added products including organic gourmet seasoned mushroom meal mixes. After years of seasonal outdoor production, Ozark Forest Mushrooms built a greenhouse with a wood furnace that allowed them to extend production and introduce new products, such as oyster mushrooms [*Pleurotus ostreatus* (Jacq.) Quélet]. The greenhouse is fueled with recycled "spent" shiitake logs previously used in mushroom production. Extending the production season strengthened the relationship with the company's customers, helped spread out the seasonal workload, and provided year-round employment to their workforce. In addition to mushroom production, the company offers ecological holidays in the Ozark region, providing tours, camping areas, and vacation home rentals.

Example 2. Ozark Mountain Ginseng

Ozark Mountain Ginseng (www.ozarkmountainginseng.com) specializes in growing ginseng (*Panax quinquefolius* L.). Ginseng is a wonderful herb with medicinal properties and economic value. Ginseng values vary by production method and quality. One pound of dried wood-grown or wild simulated ginseng could sell for prices between \$181 and \$340 kg^{-1} (Wild Ozark, 2019), depending on the wood quality. About 70% of the ginseng goes to China, where the demand for American ginseng is huge (U.S. Fish and Wildlife Service, 2016).

Ozark Mountain Ginseng grows ginseng with many other medicinal herbs, growing just like they do in a natural setting in the forest, including goldenseal (*Hydrastis canadensis* L.), bloodroot (*Sanguinaria canadensis* L.), mayapple (*Podophyllum peltatum* L.), ladyslipper orchid (*Cypripedium*), wildginger (*Asarum* L.), black cohosh (*Actaea racemose* L.), and some wood lily or white trillium [*Trillium grandiflorum* (Michx.) Salisb.]. They also grow shiitake mushrooms right next to medicinal herbs. Ozark Mountain Ginseng sells everything from ginseng seed to woods-grown transplanting rootlets. There are also root buyers who purchase fresh (green) ginseng and dried ginseng roots. They advertise their ginseng by welcoming potential buyers to come and visit the farm to see medicinal herbs growing in a natural forest setting. Ozark Mountain Ginseng also sends their product information by mail upon request and sells ginseng seeds and transplanting rootlets online.

Example 3. Walter's Pumpkin Patch

Walter's Pumpkin Patch (Burns, KS; https://www.thewaltersfarm.com/) demonstrates innovation by creating a whole array of differentiating features. Starting as a pumpkin farm, it has become a regional attraction. The farm produces many varieties of pumpkin (*Cucurbita pepo* L.), squash (*Cucurbita maxima* Duchesne), zucchini (*Cucurbita pepo* L.), and gourds (*Cucurbita* spp.), has a gift shop with a large variety of pumpkin-related products, and a Pumpkin Pantry (i.e., a certified kitchen where they produce pumpkin pie, pumpkin bread, and pumpkin salsa). By selecting an early maturing type of pumpkin, the owners were able to time the ripening of pumpkins with the tomato (*Lycopersicon esculentum* Mill.) and pepper (*Capsicum annuum* L., *Capsicum frutescens* L.) season and create their signature salsa. Innovation did not stop there. They package the salsa in Jack-O-Lantern jars that can be reused as a candle holder after the salsa is consumed. Pick-your-own pumpkins, hayrides, corn maze, play areas, barrel train, and a huge underground slide for children and adults helped to further differentiate the farm and make it a family agritourism destination (Sustainable Agriculture Network, 2003).

Example 4. Stouffer's Cedar Hill Farms

The Cedar Hill Farm (https://www.cedarhillfarms.com/) in Saline County, Missouri, was established by the Adkisson family in 1858. The family initially claimed the farm as a producing farm, growing everything from livestock to hemp and row crops of corn and wheat. Bill and Sue Ellen Stouffer moved to the farm in 1989 and, in 2009, they retired from large-scale commodity farming to lead a simpler life, growing specialty crops of chestnut and elderberry on this 53-ha (130-acre) farm. They planted chestnut trees on the farm in the spring of 2009. Additional trees have been subsequently added, bringing the total count to

1,500 chestnut trees in 2018. The Chinese chestnut trees (*Castanea mollissima* Blume) they planted, while native to China, thrive in central Missouri and can tolerate the high summer heat and humidity typical of this part of the country. These trees are resistant to the chestnut blight that devastated the native American chestnut species and produce a large and sweet-tasting chestnut. Specific cultivars grown in the orchard include 'Qing', 'Peach', 'Gideon', 'Kohr', and 'Sleeping Giant'.

To market their chestnuts, the Stouffers have chosen to focus their attention on selling their crop direct to consumers online to maximize their revenue per pound. They invested in an ecommerce website to capture orders, accept payment, and facilitate sales. In terms of their orchard production in the next 20 years, their strategy is to methodically build a loyal following of customers that will enable them to drive down their costs as a percentage of revenue over time.

The Marketing Plan

A marketing plan can be compared with a road map because it provides direction on how to get from a given situation to a destination. A comprehensive marketing plan provides valuable information pertaining to a product's potential in the marketplace. Constructing the marketing plan is analogous to putting together a jigsaw puzzle. Many individual components provide specific pieces of information, but none of the individual pieces provide sufficient information to clearly visualize the big picture. The objective is to combine the individual pieces to clearly see the big picture, resulting in a plan to market the products that a business will be producing (Dalton et al., 2007). The marketing plan also provides a framework for implementation and for reviewing the progress. Many planning guides are available that provide detailed guidance for the development of marketing plans (Dalton et al., 2007; DiGiacomo et al., 2003).

Structure of a Marketing Plan

Marketing plans can be broken into six major sections (Figure 15–5) (Dalton et al., 2007). The *introduction* includes a detailed description of the business and the products or services involved. The description reveals how the product(s) or service(s) fit into the current business environment and where the farm would like to be in the future. *Market research and analysis* summarizes the results, analysis, and implications of market research and is an essential component of the marketing plan. Specifically, information in this section provides details on consumers, competitors, industry trends, the target market selected, and other pertinent information collected in the market research and analysis. The *marketing objectives* section defines detailed and specific sales goals and market objectives for a value-added product. Marketing goals and objectives should be quantified and easily measurable. Goals are helpful in keeping on task and focused. For example, specific goals, such as "to sell 4,000 kg (~10,000 lb) of fresh chestnuts and 450 kg (~1,000 lb) of chestnut flour by the end of the fiscal year," should be defined. The *marketing strategies* section defines and describes the broad competitive strategy along with specific operational strategies (product, diversification, price, and promotion) that will be used to reach the specific marketing and financial objectives and goals outlined in the previous section. The *marketing tactics* or *marketing mix* section provides a means of describing the specific tools that the farm will use to pursue its objective(s) in the target market. The marketing mix is normally segmented into four categories that are generally thought of as the four P's of marketing—Product, Price, Place,

Structure of a Marketing Plan

1. Introduction
2. Market research and analysis
    - Environmental analysis – opportunities and threats Posed by the external environment as well as internal strengths and weaknesses
    - Product/service analysis
    - Target market identification and description
    - Comperitor analysis
3. Marketing and financial objectives and goals
4. Marketing strategies
5. Marketing tactics/marketing mix: product, price, place, promotion
    - Product – Packaging, product characteristics and image
    - Price – price for different markets, different outlets, different quantities and different seasons
    - Distribution plans – mix of marketing channels
    - Promotion mix – activities and tactics that will be used to promote the product
6. Marketing budget – estimated costs of the planned marketing tactics, the cost associated with market analysis and an explanation of the source of funds (e.g., borrowed, savings, percent of revenue).

Fig. 15–5. Structure of a marketing plan (adapted from Dalton et al., 2007).

and Promotion—and describe specific ways of implementing the operational strategies (Dalton et al., 2007). This section will describe product characteristics, packaging and image, price tactics for different market segments, distribution plans, and planned activities that will be used to promote the product. Finally, the *marketing budget* estimates the costs of the planned marketing tactics, the cost associated with market analysis, and an explanation of the source of funds (e.g., borrowed, savings, percentage of revenue).

Implementing and Evaluating the Plan

The success of the whole marketing strategy and marketing plan depends on timely and efficient implementation. Dynamic markets require constant adaptation of plans to new changes in the environment. The marketing activities in the marketing plan should be continually monitored and evaluated. Without evaluation, a farm business will have no idea if it has reached its goals or is headed in the right direction.

Monitoring involves tracking the progress of a farm business and taking action when necessary to make sure it remains on course to meet identified goals. The marketing plan identifies a series of concrete tactics and activities to accomplish along with specific dates of implementation. Timely evaluation will determine how effective the marketing strategy and plan are at any given point in time (Dalton et al., 2007; DiGiacomo et al., 2003). Questions that are answered include: How close did the farm business come to meeting its measurable objectives? Is the business headed in the right direction to meet its economic goals? Are the goals and measurable objectives realistic? Were the customers the ones who were expected? Were their needs met? Were the chosen market outlets suitable? What was done to attract customers? Was adequate attention paid to promotion, quality, and public relations? Did the customers receive the product they wanted in the form they wanted it? Were products priced appropriately?

Checklists are developed that identify each tactic and activity involved in the marketing plan and the method(s) to monitor and evaluate success. For example, if one of the promotion tactics was to develop news releases for the local media at a specific point in time, monitoring the use of articles and overall sales within a week of the date of the news releases provides a means to measure the effectiveness of this tactic (Dalton et al., 2007).

Finally, marketing plans are revised and refined based on the evaluation results. Doubtless, certain objectives will not have been reached. The evaluation identifies areas that need to be changed.

Putting Porter's Five Forces Model and Marketing Strategy Development to Work

Two examples are presented to show the broad application and effectiveness of the PFFM as a market analysis tool, and a third case exemplifies the application of the marketing planning process. The first example, eastern redcedar, applies the PFFM and reveals the model's effectiveness in helping understand the competitive forces underlying redcedar markets. The second example, shiitake mushrooms, demonstrates a successful approach to specialty product niche market development. Finally, a third, real-world example of market analysis, marketing strategy development and implementation for Midwest Elderberry Cooperative is highlighted.

Example 1. Applying the PFFM to the Eastern Redcedar Industry

Still considered by many to be a "trash" or "nuisance" tree, eastern redcedar is a prime example of a "black box" market. Due to diverse priorities among landowners and land management agencies (e.g., wild life conservation, prairie restoration, reduction of fire hazard, watershed rehabilitation, grazing), along with a lack of knowledge and understanding of the markets for redcedar, the species has been subject to management practices that call for its eradication. Responding to the above-named priorities, the Oklahoma Natural Resource Conservation Service, Oklahoma Department of Wildlife Conservation, Oklahoma Chapter of the Nature Conservancy, and other nonprofit organizations in Oklahoma include redcedar eradication among their top priorities and have adopted policies that assist with eradication of eastern redcedar from state and private lands by means of burning, chaining, or herbicide application (Newlin, 2016).

The policy of providing cost-share dollars to destroy a resource that has a market value reveals genuinely conflicting goals and objectives among stakeholders. This inconsistency indicates, in part, that traditional approaches to understanding markets sometimes fail to describe important aspects of small or niche markets. Several aspects of the redcedar market were unanswered by traditional market-analysis approaches. First, there was no aggregated information about who the market participants were or what the scale of the industry was. It appeared that the redcedar market had developed as a disjointed group of producers who operated within the boundaries of very narrow local market areas, each oblivious to the others' activities. Second, information about the factors that impact profitability and the competitive environment, such as output prices and bargaining power of suppliers or buyers, was unknown. Finally, because eastern redcedar was not considered a major commercial timber species, even though it has great potential for certain lumber applications, information regarding its future growth potential and current availability was unknown.

Porter's Five Forces Model of competitive market forces was used to develop and analyze these and other market characteristics of eastern redcedar (Gold, Godsey, & Cernusca, 2005). Primary and secondary data regarding market participants were collected through meetings with producer groups, internet searches, retail market visits, and personal contacts. This information identified market participants and different products that were bought and sold along the eastern redcedar value chain. From the list of market participants collected, a mail survey (n = 187) was developed and sent to businesses and individuals across the United States who were actively participating in the eastern redcedar market. The survey was designed to create a basic understanding of the size, location, position, and production of the eastern redcedar market and to obtain a basic understanding of supply and demand perceptions and trends.

One practical output from this market documentation was a directory that included all those who were surveyed and wished to be listed. From this eastern redcedar market directory, networks of buyers and sellers were created. For example, a producer of eastern redcedar cants in southwest Missouri was connected with a fencing company in St. Louis, and a working relationship was developed. From these mail survey respondents, a subsample of respondents (n = 25) was selected to participate in a follow-up phone survey. The phone survey consisted of a directed series of questions based on competitive forces within the eastern redcedar market. Results from both surveys were analyzed using the PFFM. Data from both surveys identified the level of production throughout the country and provided valuable information for developing market strategies. From these surveys, it was determined that the eastern redcedar industry generated nearly $60 million in gross revenue annually. More importantly, the surveys identified the potential for future growth in the market.

By focusing survey questions on areas that were perceived to be barriers to entry, critical resources for successful entry into the redcedar market at all levels of the value chain were identified. These critical resources are: access to labor, market knowledge, access to the raw material resource, financial resources, equipment, and the cultivation of personal relationships among players in the market value chain.

A closer look at the bargaining power of suppliers and buyers led to the need for further analysis drawn from U.S. Forest Service Forest Inventory and Analysis (FIA) data. The FIA data revealed that four states—Arkansas, Tennessee, Kentucky, and Missouri—possessed more than half of the nation's current and future growing stock for eastern redcedar. The inventory data coincide precisely with the primary locations of the eastern redcedar industry. This information is critical to any current or potential industry participant concerned about the location, cost, and availability of current and future supplies of eastern redcedar.

Other important observations resulting from the application of the PFFM were the ban on CCA-treated lumber and the use of federal cost-share money to eradicate redcedar, as framed from an industry perspective. The impacts of these policies were shown to affect the demand for redcedar products or the supply of raw redcedar resources. By looking at the redcedar market through the lens of the PFFM, key issues that impact the market were identified. The opening of the "black box" of the eastern redcedar market permits industry participants to pursue pragmatic strategies that are based on solid information. Weaknesses in the market, such as a lack of market networking, were identified, and strategies, including the development of a market directory, were used to overcome these problems. Results from the PFFM redcedar market analysis have been shared with landowners and industry representatives from the Aromatic Redcedar Association and are stimulating a closer look at the status and potential of the redcedar market.

According to Kaur, Joshi, Will, and Murray (2020), the introduction of new redcedar-based bioproduct industries including particleboard, mulch, and oil would contribute $96 million per year to the economy of Oklahoma, generating 319 new jobs.

Example 2. Analysis of competition forces in the shiitake market

To get information about the dynamics of the U.S. shiitake mushroom market, the PPFM was used as a framework to organize a national survey of producers to analyze the forces that influence the market (Figure 15–6; Gold, Cernusca, & Godsey, 2008). Barriers to entry included start-up and

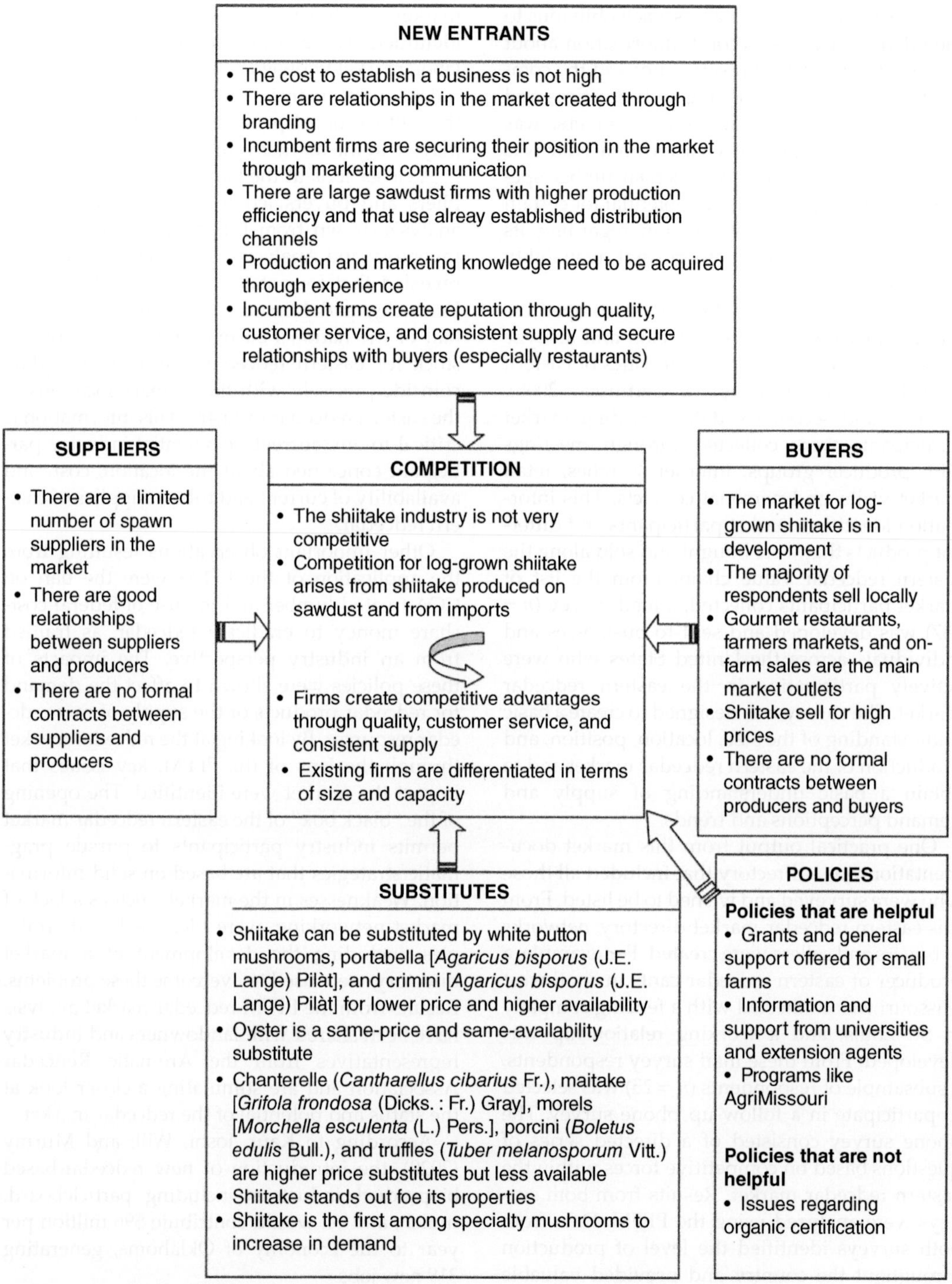

NEW ENTRANTS

- The cost to establish a business is not high
- There are relationships in the market created through branding
- Incumbent firms are securing their position in the market through marketing communication
- There are large sawdust firms with higher production efficiency and that use alreay established distribution channels
- Production and marketing knowledge need to be acquired through experience
- Incumbent firms create reputation through quality, customer service, and consistent supply and secure relationships with buyers (especially restaurants)

SUPPLIERS

- There are a limited number of spawn suppliers in the market
- There are good relationships between suppliers and producers
- There are no formal contracts between suppliers and producers

COMPETITION

- The shiitake industry is not very competitive
- Competition for log-grown shiitake arises from shiitake produced on sawdust and from imports
- Firms create competitive advantages through quality, customer service, and consistent supply
- Existing firms are differentiated in terms of size and capacity

BUYERS

- The market for log-grown shiitake is in development
- The majority of respondents sell locally
- Gourmet restaurants, farmers markets, and on-farm sales are the main market outlets
- Shiitake sell for high prices
- There are no formal contracts between producers and buyers

POLICIES

Policies that are helpful

- Grants and general support offered for small farms
- Information and support from universities and extension agents
- Programs like AgriMissouri

Policies that are not helpful

- Issues regarding organic certification

SUBSTITUTES

- Shiitake can be substituted by white button mushrooms, portabella [*Agaricus bisporus* (J.E. Lange) Pilàt], and crimini [*Agaricus bisporus* (J.E. Lange) Pilàt] for lower price and higher availability
- Oyster is a same-price and same-availability substitute
- Chanterelle (*Cantharellus cibarius* Fr.), maitake [*Grifola frondosa* (Dicks. : Fr.) Gray], morels [*Morchella esculenta* (L.) Pers.], porcini (*Boletus edulis* Bull.), and truffles (*Tuber melanosporum* Vitt.) are higher price substitutes but less available
- Shiitake stands out foe its properties
- Shiitake is the first among specialty mushrooms to increase in demand

Fig. 15–6. The influence of the Porter's Five Forces Model on the U.S. shiitake market (Gold et al., 2008).

growing costs for a 4-yr shiitake mushroom production cycle, including the cost of logs, estimated at $3,000 (Szymanski, Hill, & Woods, 2003). Following establishment, the biggest barrier was deemed to be the availability of affordable and dependable labor. According to survey respondents, the costs to establish a shiitake business are modest and can be personally financed (72% of respondents financed their businesses from personal funds). However, a lack of funds can make it difficult for a potential shiitake producer to enter this market. Strong relationships and brand identity create product loyalty for existing farm businesses, while time and money are required for a new farm business to establish its own relationships and brand. More than one-half of survey participants sell their shiitake products under a brand name. Survey results indicated that the need to establish a brand name increased with the development of value-added products and with the market outlet; that is, there is more use of branding when selling value-added products compared with fresh shiitake mushrooms and when selling to upscale stores or online compared with restaurants or farmers markets (Gold et al., 2008).

Another way incumbent firms secure their position in the market is by improving marketing communication. Fifty-seven percent of respondents use advertising to increase awareness about their products and services, and 33% plan to advertise in the future. Eighty-three percent of respondents use some form of publicity to increase awareness, to inform and educate consumers, and to stimulate positive attitudes toward the products they sell.

Large firms that grow shiitake on sawdust produce mushrooms with higher production efficiency than small firms focused on log-grown shiitake. Synthetic logs may produce three to four times as many mushrooms as natural logs in one-tenth of the time (Royse, 2001). Large-volume commodity production can create barriers to entry for log-grown producers if they both attempt to compete in the same target markets. Many large white button mushroom firms diversified their production to include different kinds of specialty mushrooms, including shiitake. These firms already have production facilities and distribution channels in place, making the production and marketing process much easier, thus creating an advantage over small firms. However, log-grown niche producers successfully compete in a variety of direct market outlets, including sales to farmers markets and restaurant chefs (UVM Center for Sustainable Agriculture, 2013).

Another advantage identified for existing producers is the knowledge about production and marketing acquired through practical experience (i.e., the learning curve) because log-grown shiitake mushroom cultivation demands a higher degree of attention to detail than commodity mushroom production (Chen, 2000; UVM Center for Sustainable Agriculture, 2013).

There were no major issues revealed in relation to the bargaining power of suppliers. Survey results indicated that a limited number of major spawn suppliers in the market have developed good relationships with producers. Most of the respondents stated that the spawn available through supply channels is of good quality and readily available. They also appreciate the information and support they get from suppliers.

The market for log-grown shiitake mushrooms is still developing. The majority of respondents sell their shiitake mushrooms locally. Gourmet restaurants, farmers markets, and on-farm sales are the primary shiitake market outlets. Shiitake mushrooms are a high-value crop that sells for a high price. Shiitake's selling points are aromatic flavor and nutritional value—rich protein content, high in vitamins and minerals, low in calories and fat, and other positive health benefits (Mattila, Suonpaa, & Piironen, 2000; UVM Center for Sustainable Agriculture, 2013). Market prices for fresh shiitake vary by season and market outlet. According to the USDA National Agricultural Statistics Service (2019b), the price per kilogram received by shiitake growers between 2016 and 2019 (average price producers receive at point of first sale) ranged from \$1.46–2.01 kg^{-1} (\$3.21–4.44 lb^{-1}). The USDA Agricultural Marketing Service provides daily reports about wholesale prices at different terminal markets. According to the information provided by the USDA Agricultural Marketing Service (2020), average wholesale prices reported at the main terminal market in Pennsylvania in 2020 ranged from \$1.81–2.12 kg^{-1} (\$4.00–4.67 lb^{-1}). For survey respondents, wholesale prices varied from \$2.50–3.18 kg^{-1} (\$5–7 lb^{-1}), with the highest prices obtained through sales to restaurants, online, or at the farm gate.

Compared with sawdust-grown shiitake, the literature indicates that prices should be higher for log-grown shiitake because shiitake grown on logs are meatier, tastier, have a longer shelf life, and have better medicinal and nutritional properties (Sabota, 2007). In contrast to Japanese consumers, who are more familiar with shiitake, U.S. retail consumers do not appreciate the difference and generally will pay the same price for both categories.

To strengthen their position with buyers, producers can create and sustain relationships by consistently providing good quality mushrooms, ensuring that quality is maintained through the supply channel, and constantly communicating with buyers. To maintain a steady supply and to help market the mushrooms, producers can create partnerships with other producers (i.e., become a broker for other growers or create marketing cooperatives). A threat of substitutes exists if there are alternative products with lower prices, better quality, better nutritional benefits, and better availability that can be used for the same purposes. There are substitutes for shiitake, but this mushroom stands out for its flavor, nutritional properties, and availability and is experiencing the most rapid increase in demand among specialty mushrooms (Augostini, 2002). Better promotion of the additional properties of log-grown shiitake would encourage consumers to seek out log-grown shiitake.

Nationally, the log-grown shiitake industry remains non-competitive. The number of log-grown shiitake mushroom producers increased slightly during the past 15 years, from 176 growers in 2006 to 185 in 2018 (Gold et al., 2008; USDA National Agricultural Statistics Service, 2019b). However, compared with outdoor production on logs in 2005–2006, production in 2017–2018 (measured in number of logs) decreased dramatically from 383,000 to 152,000 (Gold et al., 2008; USDA National Agricultural Statistics Service, 2019b). Competitiveness is influenced by the size and production capacity of the producer and the number of producers in the area. In wholesale markets, competition for log-grown shiitake arises from shiitake produced on sawdust (Royse, 2001; USDA National Agricultural Statistics Service, 2019b) or from imports.

Recent trends in specialty mushroom sales have focused on retail markets. This trend is driven by increased interest in specialty mushrooms and by the convenience packaged products offer to the consumer. In some retail markets, only 10% of the customers buy 90% of the specialty types (Agricultural Marketing Resource Center, 2018).

Marketing of specialty mushrooms in the United States is still a relatively new enterprise. For most individual producers, growing shiitake mushrooms will be easier than selling them. As shiitake becomes more popular throughout the United States, growers must be excellent at both marketing and producing to succeed financially (Agricultural Marketing Resource Center, 2018). To compete, a small-scale start-up producer may wish to develop a niche market for high-quality fresh products or produce a value-added product. In addition, mushroom-marketing cooperatives may be an important asset for growing new businesses by purchasing wild-harvested mushrooms and those grown by hobbyists and small commercial producers (Agricultural Marketing Resource Center, 2018). High quality standards, excellent service, and consistent supply will help differentiate competitors and create strong relationships with local customers.

Federal and state government grants and general support offered for small farms and direct marketing were mentioned among the policies that help producers establish a shiitake business. Information and support offered by university specialists, extension agents, and state programs (e.g., AgriMissouri, a state government program designed to promote Missouri products) were also helpful.

Some respondents mentioned a recent controversy regarding the use of cheese wax and organic certification that is affecting their ability to remain organic. However, a petition to add cheese wax to the National List of Allowed Substances was received and reviewed by the National Organic Program of the USDA. The National Organic Program ruled that cheese wax was not permitted, and this has impacted both growers' production methods and increased production costs.

Example 3. Application of the Marketing Planning Process to Midwest Elderberry Cooperative

The Midwest Elderberry Cooperative (MEC) (originally named Minnesota Elderberry Cooperative; http://www.midwest-elderberry.coop/) was formed in 2012 by Christopher J. Patton, who acts as its president. American native elderberry has long been a hobby crop in North America. Its commercial potential was recognized by University of Missouri horticultural and marketing researchers and by longtime organic farmer Terry Durham. While successful in selecting new cultivars and establishing best production practices for commercial elderberry production, growers found that something was missing to successfully adopt elderberry as a commercial crop.

Chris Patton met with Terry Durham at a conference in La Crosse, WI, in February 2011. After talking with Terry, Chris correctly perceived the missing key component to successful agroforestry implementation: marketing. Chris established the MEC to assist small organic or natural farmers in elderberry production and processing in the Midwest. The MEC offers environmentally and humanly sustainable—and often certified organic and local—elderberry flower and elderberries.

Step 1: Analyzing marketing opportunities. Elderberry has multiple uses. Elderberry flowers have diuretic, laxative, and anti-inflammatory properties, and elderberries are anti-inflammatory, antiviral, anti-oxidant, and antibacterial (Mohebalian et al., 2012). Elderberry flowers are used in medicine, cosmetics, and food. Elderberries are used to produce wine, concentrate, jam, jelly, and nutraceutical products.

The elderberry industry has high growth prospects. According to a national elderberry producer survey in the United States (Cernusca, Gold, & Godsey, 2012), the elderberry industry is composed of mostly small-scale participants, and the demand trends for elderberry products are favorable. Sales of elderberry supplements were ranked the fifth best-selling supplements in "natural" retail markets in 2017, an increase of 21% over 2016 (Smith, Kawa, Eckl, Morton, & Stredney, 2018).

However, more than 95% of the elderberry consumed in the United States is imported European elderberry (*Sambucus nigra* L.) and the domestic supply of elderberry is low. Most of the imports take the form of concentrate ingredients, extracts, powders, dried berries or flowers, and supplement products. This provides an opportunity for American elderberry to substitute for the imports.

Barriers to increased U.S. production exist. For instance, elderberry is primarily prized for its potential health benefits and use as a food colorant. Both federal and state regulations require expensive proprietary research in order for any health claims for food and beverage products to be made. This encourages elderberry producers and nutraceutical companies to use European elderberry rather than American elderberry because European species are better supported by medical research (Cernusca et al., 2012).

Step 2: Researching and selecting target markets. The MEC targets health- and environment-conscious consumers as stated in its mission and vision:

Mission: Growing to make a positive impact on the health of people by supplying the highest quality elderberry flower and elderberry ingredients using environmentally and socially sustainable practices.

Vision: Through profitable, environmentally and humanly sustainable—and often Certified Organic—elderberry flower and elderberry production, MEC will contribute to the economic welfare of its growers and ingredient customers—from local to global food and beverage producers. Ingredient nutritional and taste quality, consumer satisfaction and a naturally healthy environment take precedence over excessive return on invested capital, which must nonetheless be robustly positive for our operations to be sustainable and provide acceptable return to our investor partners.

Step 3: Developing marketing strategies. The MEC, with the help from Terry Durham, developed a focus-differentiation (niche) strategy. To differentiate from imported elderberry products, MEC adds value to elderberry by introducing new elderberry products and offers high-quality, locally grown, organic, and/or sustainably grown elderberry products. Locally and/or organically grown elderberry products can attract consumer preferences and obtain price premiums (Mohebalian et al., 2012).

Step 4: Developing a marketing plan. By understanding the current market for elderberry, MEC developed a marketing plan to create and address a niche market for elderberry flower and elderberry products. The plan emphasizes their expertise and passion for elderberry, and helps MEC achieve its mission.

Step 5: Implementing the plan. The biggest challenge in marketing elderberry was to inform the public about elderberry products and their health potential without running afoul of the regulations. This challenge was mostly met by in-store and event sampling of the products, product promotions (sales), along with some advertising regionally through Chris Patton's River Hills Harvest Marketers, LLC. Both Chris and Terry attended agricultural trade shows to recruit growers and spread product knowledge. For instance, River Hills Harvest Marketers rented booth space at national trade shows sponsored by national distributor KeHE, the Specialty Food Association, and New Hope Network.

The initial elderberry market has also been built through direct sales to supplement manufacturers. The MEC has also established connections with a few regional and national fruit brokers. In terms of marketing value-added elderberry products to differentiate with imported elderberry products, MEC sells frozen raw elderberry juice in bulk and retails elderberry juice, cordial, jelly, jam, and syrup through River Hills Harvest Marketers.

Member growers of the MEC are all environmentally sustainable and are trending toward certified organic practices. Small-scale MEC member growers sell their products to local markets, such as farmers markets and local grocery stores, among others, which offers convenience to their customers. Larger scale MEC member growers are encouraged to develop the infrastructure for commercial-scale production and supply of high-quality native elderberry flower and elderberry ingredients. They can also share in the cooperative profits and losses, hedging their financial risks.

Step 6: Re-evaluating the plan. The MEC has successfully increased U.S. farmers' knowledge of growing and marketing elderberry and encouraged farmers to grow elderberries. The MEC now has about 100 member growers, with the majority of its members being small-scale farms. At the end of 2018, MEC had about 18 producing members located in the Upper Midwest from North Dakota to Iowa, including Wisconsin. In January 2019, it launched expansion plans to include the central part of the continental United States from Indiana to New Mexico, with scattered affiliated partners from California to Connecticut.

MEC has successfully supported U.S. elderberry growers in marketing their products and competing against the European imports. The MEC sells only what they or their networked American farmers grow. In 2008, MEC had to turn down potential orders for more than 137,000 kg (300,000 lb) of certified organic elderberries due to high demand and lack of supply. In the same year, they sold out all their frozen and dried elderberries in October. Focusing on environmentally and humanly sustainable elderberry flower and elderberry, along with value-added products, MEC has recognized the niche market and developed successful strategies.

Summary

Widespread adoption and active use of agroforestry systems and practices will never achieve their potential without the "pull" of market forces. The scientific knowledge underpinning agroforestry practices is essential but will never achieve widespread application unless that knowledge is applied to products sought after in the marketplace. Thus, understanding markets and marketing is an essential ingredient in the success of profitable agroforestry enterprises that produce commercially valuable products. The key to earning income from agroforestry practices is to move from a "product" and "selling" orientation to a "customer" and "marketing" orientation. This means determining the needs and wants of buyers and final consumers and satisfying these needs more effectively and efficiently than competitors.

After presenting promising trends that encourage the production and use of products grown in agroforestry practices, this chapter presented the major steps in the marketing planning process. The planning process must have a market-oriented focus directed toward the goal of achieving customer satisfaction within a constantly changing environment. The chapter focused on analyzing marketing opportunities, researching and selecting target markets, and developing and implementing marketing strategies and plans. It also presented examples that demonstrate the application of competitive market analysis and strategy development to agroforestry product markets.

References

Agricultural Marketing Resource Center. (2018). *Mushrooms profile*. Ames, IA: AgMRC, Iowa State University. Retrieved from https://www.agmrc.org/commodities-products/specialty-crops/mushrooms-profile/

Agriculture Improvement Act of 2018, Pub. L. no. 115-334, 132 STAT. 4490 (2018). https://www.agriculture.senate.gov/imo/media/doc/Section-by-sections%20(Committee%20Print).pdf

Aguilar, F. X., Cernusca, M. M., & Gold, M. A. (2009). A preliminary assessment of consumer preferences for chestnuts (*Castanea* sp.) using conjoint analysis. *HortTechnology, 19*, 216–223.

Aguilar, F. X., Cernusca, M. M., Gold, M. A., & Barbieri, C. E. (2010). Frequency of consumption, familiarity and preferences for chestnuts in Missouri. *Agroforestry Systems, 79*, 19–29.

Ahearn, M., & Sterns, J. (2013). Direct-to-consumer sales of farm products: Producers and supply chains in the Southeast. *Journal of Agricultural and Applied Economics, 45*(3), 497–508.

Alavalapati, J. R. R., & Mercer, D. E. (Ed.) (2004). *Valuing agroforestry systems: Methods and applications*. Dordrecht, the Netherlands: Kluwer Academic.

Alam, M., Olivier, A., Paquette, A., Dupras, J., Revéret, J., & Messier, C. (2014). A general framework for the quantification and valuation of ecosystem services of tree-based intercropping systems. *Agroforestry Systems, 88*, 679–691.

American Marketing Association. (2013). *Definitions of marketing*. Chicago: AMA.Retrieved from https://www.ama.org/AboutAMA/Pages/Definition-of-Marketing.aspx.

Andreatta, S. L. (2000). Marketing strategies and challenges of small-scale organic producers in central North Carolina. *Culture & Agriculture, 22*, 40–50.

Augostini, N. (2002). Specialty mushrooms survey. Raleigh, NC: North Carolina State Extension Service, New Crops and Organics. Retrieved from https://newcropsorganics.ces.ncsu.edu/wp-content/uploads/2016/12/MushroomSurvey.pdf?fwd=no.

Batte, M. T., & Ernst, S. (2007). Net gains from 'net purchases? Farmers' preferences for online and local input purchases. *Journal of Agricultural and Resource Economics Review, 36*, 84–94.

Bauer, D. M., & Johnston, R. J. (2017). *Buffer options for the bay: Economic valuation of water quality ecosystem services in New Hampshire's Great Bay Watershed*. Retrieved from https://www.bufferoptionsnh.org/wp-content/uploads/2017/12/BOB_Economic_Assessment.docx.pdf.

Boehlje, M. D., Hofing, S. L., & Schroeder, R. C. (1999). *Financing and supplying inputs to the 21st century producer* (Staff Paper 99-11). West Lafayette, IN: Dep. of Agricultural Economics, Purdue University. Retrieved from https://www.researchgate.net/publication/228356478_Financing_and_Supplying_Inputs_to_the_21st_Century_Producer;.

Bogash, S. M. (1998). Marketing of agricultural and natural resource income enterprises: Learning from and sharing with entrepreneurs. In J. S. Kays, G. R. Goff, P. J. Smallidge, et al. (Eds.), *Proceedings and invited papers: Natural Resources Income Opportunities on Private Lands Conference*, Hagerstown, MD (pp. 21–25). College Park, MD: Maryland Cooperative Extension Service.

Brown, C. (2003). Consumers' preferences for locally produced food: A study in southeast Missouri. *American Journal of Alternative Agriculture, 18*, 213–223.

Brown, C., Miller, S. M., Boone, D. A., Boone, H. N., Gartin, S. A., & McConnell, T. R. (2007). The importance of farmers markets for West Virginia farm marketers. *Renewable Agriculture and Food Systems, 22*, 20–29.

Cai, Z., & Gold, M. (2017). Annual chestnut market survey continues to reveal steady growth in the chestnut industry. *The Chestnut Grower, 18*(3).

Cai, Z., Gold, M., & Brannan, R. (2018). An exploratory analysis of US consumer preferences for North American pawpaw. *Agroforestry Systems, 93*, 1673–1685. https://doi.org/10.1007/s10457-018-0296-5

Campbell, B. L., Nelson, R. G., Ebel, R. C., Dozier, W. A., Adrian, J. L., & Hockema, B. R. (2004). Fruit quality characteristics that affect consumer preferences for Satsuma mandarins. *HortScience, 39*, 1664–1669.

Campbell, R. D. (1991). High value windbreaks for cash crop lands. In S. Finch & C. S. Baldwin (Eds.) *Proceedings of the Third International Windbreak and Agroforestry Symposium* (pp. 177–178). Ridgetown, ON, Canada: Ridgetown College.

Cernusca, M. M., Aguilar, F. X. & Gold, M. A. (2012). Post-purchase evaluation of U.S. consumers' preferences for chestnuts. *Agroforestry Systems*, 86(3), 355–364.

Cernusca, M. M., Gold, M. A., & Godsey, L. D. (2008). Influencing consumer awareness through the Missouri Chestnut Roast. *Journal of Extension 46*(6), 6RIB7.

Cernusca, M. M., Gold, M. A., & Godsey, L. D. (2012). Using the Porter model to analyze the U.S. elderberry industry. *Agroforestry Systems, 86*, 365–377.

Cernusca et al., (2018). https://centerforagroforestry.org/wp-content/uploads/2021/07/Gold-Marketing-Agroforestry-1.pdf.

Chamberlain, J. L., & Hammett, A. L. (1999). Marketing agroforestry: An alternative approach for the 21st century. In L. E. Buck & J. P. Lassoie (Eds.) *Exploring the opportunities for agroforestry in changing rural landscapes: Proceedings of the 5th Biennial Conference on Agroforestry in North America* (pp. 208–214). Ithaca, NY: Cornell University.

Chamberlain, J. L., Mitchell, D., Brigham, T., Hobby, T., Zabek, L., & Davis, J. (2009). Forest farming practices. In H. E. Garrett (Ed.), *North American agroforestry: An integrated science and practice* (2nd ed.). Madison, WI: ASA.

Chase, C. (2007). *Northeast Iowa local food survey summary report*. Ames, IA: Leopold Center for Sustainable Agriculture, Iowa State University. Retrieved from https://lib.dr.iastate.edu/leopold_grantreports/324/.

Chen, M. (2000). *A potential new profit from oak woodlands*. Integrated Hardwood Range Management Program, Oak Fact Sheet no. 34. Retrieved from https://oaks.cnr.berkeley.edu/a-potential-new-profit-from-oak-woodlands/.

Dalton, A., Holland, R., Hubbs, S., & Wolfe, K. (2007). *Marketing for the value-added agricultural enterprise* (UT Extension Service PB 1699). Spring Hill, TN: University of Tennessee Center for Profitable Agriculture. Retrieved from https://extension.tennessee.edu/publications/Documents/PB1699.pdf.

Darby, K., Batte, M. T., Ernst, S., & Roe, B. (2006). *Willingness to pay for locally produced foods: A customer intercept study of direct market and grocery store shoppers* [Paper presentation]. American Agricultural Economics Association Annual Meeting, Long Beach, CA. Retrieved from https://ageconsearch.umn.edu/bitstream/21336/1/sp06da03.pdf.

Demko, I., Dinterman, R. Marez, M., and Jaenicke, E. (2017). *US organic trade data: 2011 to 2016*. Retrieved from https://ota.com/sites/default/files/indexed_files/OTATradeReport_10-30-2017.pdf.

DiGiacomo, G., King, R., & Norquist, D. (2003). *Building a sustainable business: A guide to developing a business plan for farms and rural businesses* (Handbook Series Book 6). College Park, MD: Sustainable Agriculture Research and Education. Retrieved from https://www.sare.org/Learning-Center/Books/Building-a-Sustainable-Business.

Dimitri, C., & Greene, C. (2002). *Recent growth patterns in the U.S. organic food market* (AIB-777). Washington, DC: USDA Economic Research Service. Retreived from https://www.ers.usda.gov/webdocs/publications/42455/12915_aib777_1_.pdf?v=7169.6.

Drabenstott, M., & Sheaff, K. H. (2001). *Exploring policy options for a new rural America: A conference summary*. Kansas City, MO: Center for the Study of Rural America. Retrieved from www.kansascityfed.org/PUBLICAT/ECONREV/Pdf/3q01drab.pdf.

Eaglesham, A. (Ed.). (2007). *Summary Proceedings: Fourth Annual World Congress on Industrial Biotechnology and Bioprocessing: Linking biotechnology, chemistry and agriculture to create new value chains, Orlando, FL*. Ithaca, NY: National Agricultural Biotechnology Council. Retrieved from http://nabc.cals.cornell.edu/Publications/WC_Proc/WCIBB2007_proc.pdf.

Energy Independence and Security Act of 2007, Pub. L. no. 110-140, 121 Stat. 1492 (2007). https://www.gpo.gov/fdsys/pkg/BILLS-110hr6enr/pdf/BILLS-110hr6enr.pdf.

Garrett, H. E. (Ed.). (2009). *North American agroforestry: An integrated science and practice* (2nd ed.) Madison, WI: ASA. https://doi.org/10.2134/2009.northamericanagroforestry.2ed

Garrett, H. E., Buck, L. E., Gold, M. A., Hardesty, L. H., Kurtz, W. B., Lassoie, J. P., . . . Slusher, J. P. (1994). *Agroforestry: An integrated land-use management system for production and farmland conservation: A comprehensive assessment of U.S. agroforestry*. Washington, DC: USDA Soil Conservation Service.

Garrett, H. E., & Harper, L. S. (1999). The science and practice of black walnut agroforestry in Missouri, U.S.A.; A temperate zone assessment. In L. E. Buck, J. P. Lassoie, & E.C. M. Fernandes (Eds.) *Agroforestry in sustainable agricultural systems* (pp. 97–110). Boca Raton, FL: CRC Press.

Godsey, L. D., Mercer, D. E., Grala, R. K., Grado, S. C., & Alavalapati, J. R. R. (2009). Agroforestry economics and policy. In H. E. Garrett (Ed.), *North American agroforestry: An integrated science and practice* (2nd ed., pp. 315–338). Madison, WI: ASA.

Gold, M. A., Cernusca, M. M., & Godsey, L. D. (2004). Comparing consumer preferences for chestnuts with eastern black walnuts and pecans. *HortTechnology, 14*, 583–589.

Gold, M. A., Cernusca, M. M., & Godsey, L. D. (2005). Update on consumers' preferences for chestnuts. *HortTechnology, 15*, 904–906.

Gold, M. A., Cernusca, M. M., & Godsey, L. D. (2008). A competitive market analysis of the U.S. shiitake mushroom marketplace. *HortTechnology, 18*, 489–499.

Gold, M. A., & Garrett, H. E. (2009). Agroforestry nomenclature, concepts and practices. In H. E. Garrett (Ed.), *North American agroforestry: An integrated science and practice* (2nd ed., pp. 45–56). Madison, WI: ASA.

Gold, M. A., Godsey, L. D., & Cernusca, M. M. (2005). Competitive market analysis of eastern redcedar. *Forest Products Journal, 55*, 58–65.

Gold, M. A., Godsey, L. D., & Josiah, S. J. (2004). Markets and marketing strategies for agroforestry specialty products in North America. *Agroforestry Systems, 61*, 371–382.

Gordon, A. M., Newman, S. M., & Coleman, B. (Eds.). (2018). *Temperate agroforestry systems*. Wallingford, UK: CABI Press.

Grewal, D., Iyer, G. R., & Levy, M. (2004). Internet retailing: Enablers, limiters and market consequences. *Journal of Business Research, 57*, 703–713. https://doi.org/10.1016/S0148-2963(02)00348-X

Gugino, S. (2007, 30 November). Cooking with chestnuts. *Wine Spectator*, pp. 19–20.

Hall, C. R. (2002). *Direct marketing guide for producers of fruits, vegetables and other specialty products* (PB 1711). Knoxville, TN: Universityu of Tennessee Agricultural Extension Service. Retrieved from https://extension.tennessee.edu/publications/Documents/PB1711.pdf.

Harrison, R. W., Stinger, T., & Prinyawiwatkul, W. (2002). An analysis of consumer preferences for value-added seafood products derived from crawfish. *Journal of Agricultural and Resource Economics, 31*, 157–170.

Hartman, H. (2003). *Reflections on a cultural brand: Connecting with lifestyles*. Bellevue, WA: The Hartman Group.

Hayes, D. J., Lence, S. H., & Stoppa, A. (2004). Farmer-owned brands? *Agribusiness, 20*, 269–285.

Heiman, A., Just, D., McWilliams, B., & Zilberman, D. (2004). Religion, religiosity, lifestyles and food consumption. Giannini Foundation of Agricultural Economics, University of California. *Agricultural and Resource Economics Update, 8*, 9–11. Retrieved from https://s.giannini.ucop.edu/uploads/giannini_public/aa/b1/aab113bb-0eda-4d6f-82f1-9e4e40b9a586/v8n2_4.pdf

Himshikha, D. (2016). Socio-economic research in agroforestry: A review of ninety articles. *International Journal of Agricultural Science and Research, 6*(6), 283–300.

Hull, J. B. (1999). Can coffee drinkers save the rain forest? *Atlantic Monthly, 284*, 19–21.

Hult, G. T. M., Mena, J. A., Ferrell, O. C., & Ferrell, L. (2011). Stakeholder marketing: A definition and conceptual framework. *AMS Review, 1*, 44–65. https://doi.org/10.1007/s13162-011-0002-5

Josiah, S. J., Brott, H., & Brandle, J. (2004). Producing woody floral products in an alleycropping system in Nebraska. *HortTechnology, 14*, 203–207.

Josiah, S. J., St. Pierre, R., Brott, H., & Brandle, J. (2004). Productive conservation: Diversifying farm enterprises by producing specialty woody products in agroforestry systems. *Journal of Sustainable Agriculture, 23*, 93–108.

Kaur, R., Joshi, O., Will, R. E., & Murray, B. D. (2020). Sustainable management of unused eastern redcedar: An integrated spatial and economic analysis approach. *Resources, Conservation & Recycling, 158*, 104806. 10.1016/j.resconrec.2020.104806

Kays, J. S. (1999). Improving the success of natural-resource based enterprises. In S. J. Josiah (Ed.), *Proceedings of the North American Conference on Enterprise Development through Agroforestry: Farming the forest for specialty products, Minneapolis, MN* (pp. 171–174). St. Paul, MN: Center for Integrated Natural Resources and Agricultural Management, University of Minnesota.

Kelley, K. M., Behe, B. K., & Biernbaum, J. A. (2002). Combination of colors and species of containerized edible flowers: Effect on consumer preferences. *HortScience, 37*, 218–221.

Khanal, A. R., Mishra, A. K., & Koirala, K. H. (2015). Access to the Internet and financial performance of small business households. *Electronic Commerce Research, 15*(2), 159–175.

Kim, H., & Moschini, G. (2018). *The dynamics of supply: US corn and soybeans in the biofuel era* (18-wp579). Ames, IA: Center for Agricultural and Rural Development, Iowa State University.

Kirby, L. D., Jackson, C., & Perrett, A. (2007). *Growing local: Expanding the western North Carolina food and farm economy*. Asheville, NC: Appalachian Sustainable Agriculture Project. Retrieved from https://asapconnections.org/downloads/growing-local-expanding-the-western-north-carolina-food-and-farm-economy-full-report.pdf/.

Kohls, R. L., & Uhl, J. N. (1998). *Marketing agricultural products* (8th ed.). Upper Saddle River, NJ: Prentice Hall.

Kotler, P. (1994). *Marketing management: Analysis, planning, implementation, and control* (8th ed.). Englewood Cliffs, NJ: Prentice Hall.

Kotler, P., & Armstrong, G. (2001). *Principles of marketing* (9th ed.). Englewood Cliffs, NJ: Prentice Hall.

Kulshreshtha, S., & Kort, J. (2009). External economic benefits and social goods from prairie shelterbelts. *Agroforestry Systems, 75*, 39–47.

Kurtz, W. B., Garrett, H. E., Slusher, J. P., & Osburn, D. B. (1996). *Economics of agroforestry* (MU Guidesheet G5021). Columbia, MO: University of Missouri. Retrieved from https://mospace.umsystem.edu/xmlui/bitstream/handle/10355/50927/g5021-1996.pdf?sequence=1&isAllowed=y.

Kurtz, W. B., Thurman, S. E., Monson, J. J., & Garrett, H. E. (1991). The use of agroforestry to control erosion: Financial aspects. *The Forestry Chronicle, 67*, 254–257.

Kuyah, S., Öborn, I., & Jonsson, M. (2017). Regulating ecosystem services delivered in agroforestry systems. In *Agroforestry* (pp. 797–815). Singapore, Springer.

Levine, R., Locke, C., Searls, D., & Weinberger, D. (2001). *The cluetrain manifesto*. Cambridge, MA: Perseus Publishing.

Levitt, T. (1980). Marketing success through differentiation—of anything. *Harvard Business Review, 58*, 83–91

Loureiro, M. L., & Hine, S. (2002). Discovering niche markets: A comparison of consumer willingness to pay for a local (Colorado-grown), organic, and GMO-free product. *Journal of Agricultural and Applied Economics, 34*, 477–487.

Mattila, P., Suonpaa, K., & Piironen, V. (2000). Functional properties of edible mushrooms. *Nutrition, 16*, 694–696.

Mbow, C., Noordwijk, M. V., Luedeling, E., Neufeldt, H., Minang, P. A., & Kowero, G. (2014). Agroforestry solutions to address food security and climate change challenges in Africa. *Current Opinion in Environmental Sustainability, 6*, 61–67.

Minnesota Institute for Sustainable Agriculture. (2007). *Marketing local food*. St. Paul, MN: Minnesota Institute for Sustainable Agriculture. Retrieved from https://www.misa.umn.edu/publications/localregionalfoodsystems/marketinglocalfood

Mohebalian, P. M., Aguilar, F. X., & Cernusca, M. M. (2013). Conjoint analysis of US consumers' preference for elderberry jelly and juice products. *HortScience, 48*, 338–346.

Mohebalian, P. M., Cernusca, M. M., & Aguilar, F. X. (2012). Discovering niche markets for elderberry juice in the United States. *HortTecnology, 22*, 556–566.

Munson, C. L., Rosenblatt, M. J., & Rosenblatt, Z. (1999). The use and abuse of power in supply chains. *Business Horizons, 42*, 55–65.

Nair, P. K. R., Allen, S. C., & Bannister, M. E. (2005). Agroforestry today: An analysis of the 750 presentations to the 1st World Congress of Agroforestry, 2004. *Journal of Forestry, 103*, 417–421.

National Audubon Society, (2018). *Conservation ranching*. Retrieved from https://www.audubon.org/conservation/ranching.

Newlin, L. (2016, 15 Aug.). Waging war on cedar trees. *High Plains Journal*. Retrieved from https://www.hpj.com/ag_news/waging-war-on-cedar-trees/article_7b8c7ae8-204d-56db-8d39-24bfd9a5fcf4.html.

Nickels, W. G., & Wood, M. B. (1996). Marketing: Relationships, quality, value. Irving Place, NY: Worth Publishers.

Organic Trade Association. (2021, 25 May). *U.S. organic sales soar to new high of nearly $62 billion in 2020*. Retrieved from https://ota.com/news/press-releases/21755.

Oster, S. M. (1999). Modern competitive analysis (3rd ed.) New York: Oxford University Press.

Perlack, R. D., Wright, L. L., Turhollow, A. F., Graham, R. L., Stokes, B. J., & Erbach, D. C. (2005). *Biomass as feedstock for a bioenergy and bioproducts industry: The technical feasibility of a billion-ton annual supply*. Oak Ridge, TN: Oak Ridge National Laboratory. Retrieved from https://www1.eere.energy.gov/bioenergy/pdfs/final_billionton_vision_report2.pdf.

Phi, L. T., Duong, N. V., Quang, N. N., & Vang, P. L. (2004). *Making the most of market chains: Challenges for small-scale farmers and traders in upland Vietnam*. Small and Medium

Forest Enterprises Ser. 2. London: International Institute for Environment and Development.

Phillips, J. C., & Peterson, C. (2001). *Segmentation and differentiation of agri-food niche markets: Examples from the literature* (Staff paper 2001–2005). East Lansing, MI: Michigan State University, Department of Agricultural Economics. Retrieved from http://ageconsearch.umn.edu/record/11481/files/sp01-05.pdf

Pirog, R., & Larson, A. (2007). *Consumer perceptions of the safety, health, and environmental impact of various scales and geographic origin of food supply chains.* Ames, IA: Leopold Center for Sustainable Agriculture. Retrieved from https://lib.dr.iastate.edu/cgi/viewcontent.cgi?article=1171&context=leopold_pubspapers.

Porter, M. E. (1980). *Competitive strategy: Techniques for analyzing industries and competitors.* New York: The Free Press.

Roder, C., Herrmann, R., & Connor, M. J. (2000). Determinants of new product introductions in the U.S. food industry: A panel model approach. *Applied Economics Letters, 7,* 743–748.

Royse, D. J. (2001). *Cultivation of shiitake on natural and synthetic logs.* University Park, PA: Pennsylvania State University. Retrieved from https://www.americanmushroom.org/clientuploads/Consumers/cutlivation_of_shiitake.pdf.

Russell, D., & Franzel, S. (2004). Trees of prosperity: Agroforestry, markets and the African smallholder. *Agroforestry Systems, 61,* 345–355.

Sabota, C. (2007). *Shiitake mushroom production on logs* (UNP 25). Auburn, AL: Alabama Cooperative Extension System. Retrieved from https://mushroomcompany.com/resources/shiitake/ala-logs.pdf.

Schubert, C. (2006). Can biofuels finally take center stage? *Nature Biotechnology, 24,* 777–784.

Severson, K. (2007, 17 Oct.). Local carrots with a side of red tape. *New York Times.* Retrieved from http://www.nytimes.com/2007/10/17/dining/17carr.html.

Sharashkin, L., & Gold, M. (2006). Pine nuts (*pignolia*): Species, products, markets, and potential for U.S. production (pp. 53–64). In D. Fulbright (ed.) *Northern Nut Growers Association 95th annual report: Proceeding for the 95th annual meeting, Columbia, Missouri, August 16-19, 2004.* Retrieved from https://pinenut.com/growing-pine-nuts/ringing-cedar-nuts-oil.shtml.

Shrestha, R. K., Alavalapati, J. R. R., & Kalmbacher, R. S. (2004). Exploring the potential for silvopasture adoption in south-central Florida: An application of SWOT-AHP method. *Agricultural Systems, 81,* 185–199.

Smith, T., Kawa, K., Eckl, V., Morton, C., & Stredney, R. (2018). Herbal supplement sales in US increased 8.5% in 2017, topping $8 billion. *HerbalGram, 119,* 62–71.

Stavins, R. N., & Richards, K. R. (2005). *The cost of U.S. forest-based carbon sequestration.* Arlington, VA: Pew Center on Global Climate Change. Retrieved from https://www.c2es.org/document/the-cost-of-u-s-forest-based-carbon-sequestration/.

Stevens-Garmon, J., Huang, C. L., & Lin, B. H. (2007). Organic demand: A profile of consumers in the fresh produce market. *Choices, 22,* 109–115. Retrieved from http://www.choicesmagazine.org/2007-2/grabbag/2007-2-05.htm.

Sullivan, P., & Greer, L. (2000). Evaluating a rural enterprise: Marketing and business guide. Fayetteville, AR: ATTRA. Retrieved from https://s3.wp.wsu.edu/uploads/sites/2073/2014/09/Resources-Evaluating-a-Rural-Enterprise.pdf.

Sustainable Agriculture Network. (2003). *Marketing strategies for farmers and ranchers.* College Park, MD: USDA-SARE Sustainable Agriculture Network. Retrieved from https://www.sare.org/Learning-Center/Bulletins/Marketing-Strategies-for-Farmers-and-Ranchers.

Szymanski, M., Hill, D., & Woods, T. (2003). *Kentucky shiitake production workbook: Potential profits from a small-scale shiitake enterprise* (FOR-88). Lexington, KY: University of Kentucky College of Agriculture, Cooperative Extension Service. Retrieved from http://www.ca.uky.edu/agc/pubs/for/for88/for88.pdf.

te Velde, D. W., Rushton, J., Schreckenberg, K., Marshall, E., Edouard, F., Newton, A., & Arancibia, E. (2006). Entrepreneurship in value chains of non-timber forest products. *Forest Policy and Economics, 8,* 725–741.

Thomas, M. G., & Schumann, D. R. (1993). *Income opportunities in special forest products: Self-help suggestions for rural entrepreneurs* (Agriculture Information Bull. 666). Washington, DC: U.S. Forest Service.

University of Missouri Center for Agroforestry. (2007). *Redcedar market research 2007.* Columbia, MO: UMCA. Retrieved from https://centerforagroforestry.org/wp-content/uploads/2021/05/RedCedarMailSurvey.pdf.

USDA. (2017). *Top reasons to shop at a farmers market.* https://www.usda.gov/media/blog/2012/07/02/top-reasons-shop-farmers-market.

USDA Agricultural Marketing Service. (2017). *2016 Count of certified organic operations shows continued growth in U.S. market.* Retrieved from https://www.ams.usda.gov/press-release/2016-count-certified-organic-operations-shows-continued-growth-us-market.

USDA Agricultural Marketing Service. (2019). *National count of farmers market directory listings.* Available at https://www.ams.usda.gov/sites/default/files/media/NationalCountofFarmersMarketDirectoryListings082019.pdf.

USDA Agricultural Marketing Service. (2020). Philadelphia terminal prices as of 22-APR-2020. *Specialty Crops Market News.* https://www.ams.usda.gov/mnreports/na_fv020.txt.

USDA National Agriculture Statistics Service. (2019a). *Census of agriculture, 2017: Oregon, state and county data. Vol. 1, Part 37.* Retrieved from https://www.nass.usda.gov/Publications/AgCensus/2017/Full_Report/Volume_1,_Chapter_1_State_Level/Oregon/orv1.pdf.

USDA National Agriculture Statistics Service. (2019b). *Mushrooms.* Retrieved from https://downloads.usda.library.cornell.edu/usda-esmis/files/r781wg03d/6682xg610/cv43p8639/mush0819.pdf.

USDA National Agriculture Statistics Service. (2019c). *Census of agriculture, 2017: Missouri: Historical Highlights. Table 1.* Retrieved from https://www.nass.usda.gov/Publications/AgCensus/2017/Full_Report/Volume_1,_Chapter_1_State_Level/Missouri/st29_1_0001_0001.pdf

USDA National Agriculture Statistics Service. (2019d). *Farm computer usage and ownership.* Retrieved from https://www.nass.usda.gov/Publications/Todays_Reports/reports/fmpc0819.pdf

U.S. Fish and Wildlife Service. (2016). *U.S. exports of american ginseng 1998–2015.* Retrieved from https://www.fws.gov/international/pdf/report-us-exports-of-american-ginseng-1998-2015.pdf.

UVM Center for Sustainable Agriculture. (2013). *Best management practices for log-based shiitake cultivation in the northeastern United States.* Burlington, VT: University of Vermont Extension. Retrieved from https://www.uvm.edu/sites/default/files/media/ShiitakeGuide.pdf.

Vollmers, C., & Vollmers, S. (1999). Designing marketing plans for specialty forest products. In S. J. Josiah (Ed.), North American Conference on Enterprise Development through Agroforestry: Farming the Forest for Specialty Products (pp. 175–182). St. Paul, MN. University of Minnesota, Center for Integrated Natural Resources and Agricultural Management.

Wild Ozark. (2019). 2019 Ginseng prices. Kingston, AR: Wild Ozark Ginsent Nursery. Retrieved from https://www.wildozark.com/2019-ginseng-prices/.

Yamada, M., & Gholz, H. L. (2002). An evaluation of agroforestry systems as a rural development option for the Brazilian Amazon. *Agroforestry Systems, 55,* 81–87.

Study Questions

1. Why discuss markets and marketing in the context of agroforestry?

2. Describe three global trends that encourage the production and use of products grown in agroforestry practices.

3. What is meant by the "black box" of agroforestry enterprises? How does this impact potential agroforestry market opportunities?

4. What is the Porter Five Forces Model? What can be learned through its application?

5. What are marketing strategies? Elaborate on the components that are contained within marketing strategies.

6. Contrast commodity markets and niche product markets. Why is this differentiation important to landowners practicing agroforestry?

7. What are marketing plans? Why are they important?

8. Describe three approaches that are used to add value to a niche product through product differentiation.

9. Describe the steps used in the Midwest Elderberry Cooperative example and explain how they exemplify the successful use of the marketing planning process.

16

Zhen Cai, Larry D. Godsey,
D. Evan Mercer, Robert K. Grala,
Stephen C. Grado, and
Janaki R. R. Alavalapati

Agroforestry Economics and Policy

Essentially, every living organism on Earth has applied the basic concepts of economics. That is, every living entity has to use a limited set of resources to meet a minimum set of needs or wants. Although the study of economics is often confused with that of marketing or finance, economics is simply a social science that studies the choices humans make. As a social science, economics is the study of human motivations, or, as Landsburg (1993) put it, "most of economics can be summarized in four words: 'People respond to incentives'." Those incentives reflect the value of the trade-offs made between a limited set of resources and an unlimited set of wants, needs, and desires.

Agroforestry, an intensive land management system, combines trees and/or shrubs with crops and/or livestock and seeks to increase economic and environmental benefits (University of Missouri, Center for Agroforestry, 2012a). Agroforestry systems not only provide ecosystem goods for human consumption (market benefits) but also provide other functions to society (nonmarket benefits) including: maintaining ecosystem health, providing wildlife habitat, and providing recreational and aesthetic opportunities (McAdam, Burgess, Graves, Rigueiro-Rodríguez, & Mosquera-Losada, 2009; Noel, Qenani-Petrela, & Mastin, 2009). All these services have economic values. Prices are often used as proxy values for market goods and services (e.g., timber, crops); however, it is difficult to value nonmarket goods and services (e.g., recreation values, biodiversity values) (Sagoff, 2004).

Agroforestry economics uses models to help farmers, educators, researchers, and policy makers, among others, make their decisions (Buongiorno & Gilles, 2003). Capital budgeting, linear programming, production frontier analysis, and risk analysis can be used to help landowners determine whether agroforestry systems are suitable for them (Mercer, Frey, & Cubbage, 2014). Atangana, Khasa, Chang, and Degrande (2014, pp. 291–322) discussed seven different economic analysis methods used in agroforestry, which include: benefit–cost analysis, environmental economics, farm budgeting, risk assessment, econometrics, policy analysis matric models, and regional economic models. Alavalapati and Mercer (2004) also discussed diverse economic models and their applications to agroforestry (Table 16–1). Some were designed to assess simple costs and benefits of outputs and inputs for which markets are fairly well established, while others are amenable to a variety of environmental services and damages for which there are no established markets. Furthermore, some methodologies were more appropriate for assessing issues at a farm or household level, and others were applicable at regional and national scales.

North American Agroforestry, Third Edition. Edited by Harold E. "Gene" Garrett, Shibu Jose, and Michael A. Gold.

Table 16–1. Economic methodologies commonly applied to assess agroforestry systems.

Economic approach or model	Nature, scope, and scale of the issue for investigation
Enterprise and farm budget models	estimate the profitability of a farm or enterprise by calculating indicators such as net present value, benefit/cost ratio, and internal rate of return
Risk assessment models	incorporate probabilities of events occurring and estimate the expected profitability of agroforestry
Policy analysis matrix models	assess the profitability at a farm or regional level from both the individual and society perspective (similar to farm budget models but also include market failures)
Faustmann and Hartmann models	optimization models to estimate land expectation values assuming that the land will be used for agroforestry (the best possible productive use) in perpetuity
Linear and nonlinear programming models	estimate optimum resource allocation subject to various constraints faced by the decision maker
Econometric models	estimate the relationships among variables under investigation for forecasting, policy analysis, and decision making
Nonmarket valuation models	stated and revealed preference models to estimate values for environmental goods and services such as reducing soil erosion, improving water quality, and carbon sequestration (examples include hedonic pricing and contingent valuation models)
Regional economic models	estimate changes in income, employment, and price levels at regional or national levels in response to a policy or program change and explicitly incorporate intersectoral linkages

Note. Adapted from Alavalapati and Mercer (2004).

The goal of this chapter is to present the concepts of economics and explain how they apply to natural resource management decisions and subsequently those related to agroforestry. This chapter discusses many tools used by economists to measure and determine how choices are made at the farm level. Financial concepts are the basis for many tools that economists use to measure market- and nonmarket-based values in monetary terms at the individual level. These tools include benefit–cost, discounted cash flow (DCF), and willingness-to-pay (WTP) analysis. To give a flavor of how the tools of economics can be used to analyze agroforestry, in the next sections we describe and provide examples of using two of the approaches in Table 16-1: enterprise or farm budget models and nonmarket valuation models. Relevant publications using these methods to address agroforestry issues were also reviewed and provided. These are followed by an overview of policies and incentives to encourage landowners to adopt agroforestry systems.

Budgeting in an Agroforestry Context

Implementation of agroforestry as well as other farm practices is often constrained by the available resources such as land, financial capital, equipment, production technology, and labor (Kurtz, 2000). Farmers and landowners allocate these limited resources in a way that allows them to attain their objectives in the most efficient manner. When the objective is to maximize financial returns on farm production, this implies that you are comparing costs and returns, and selecting, from a financial perspective, the most promising practices (Alavalapati & Mercer, 2004; Kurtz, 2000).

In this section, we describe methods and tools commonly used to assess the financial viability of agroforestry practices as well as the entire farm production.

Farm Budgeting

The process of financially determining the most effective farm operations is not an easy task. Increasing production costs and changing demands for farm products require continuous reevaluation of farm management objectives and adjustments of farm production to make it profitable. Possible adjustments might include lowering farm production costs, improving production technology, using current software programs to facilitate management, and introducing new farm operations such as agroforestry practices. These changes can have a significant impact on the financial viability of the entire farm. Thus, such enterprises have to be well planned and examined from a financial perspective to ensure that only alternatives improving overall profitability are implemented. This requires a consistent examination of short- and long-term financial effects of proposed changes in farm management. Farm budgeting is a method used to evaluate the attainment of farm financial goals by comparing revenues and costs associated with farm production. There are several types of farm budgeting, and we briefly discuss three: whole-farm, enterprise, and partial budgeting.

Whole-Farm Budgeting

A *whole-farm budget* is a snapshot describing the entire production on the farm. It identifies individual farm components, called *enterprises*, and shows how they contribute to the overall profit generated by farm production (Doye, 2007). A farm enterprise consists of any type of farm production such as corn (*Zea mays* L.), soybean [*Glycine max* (L.) Merr.], wheat (*Triticum aestivum* L.), tomato (*Solanum lycopersicum* L.), and cattle (*Bos taurus*) production, as well as agroforestry systems (Chase, 2006). Therefore, a typical farm will include several enterprises.

Whole-farm budgeting serves as a guideline to accomplish the owner's objectives, given limited resources, and to monitor progress in the attainment of these objectives (Doye, 2007). A whole-farm budget can be fairly extensive and complicated, depending on farm size and the number of farm enterprises involved. Typically, it lists all of the farm's physical and financial assets and describes how they are allocated to whole-farm production as well as particular farm enterprises (Doye, 2007). The pivotal part of the budget is the summary of costs associated with conducting outlined farm operations, expected revenues from selling farm products, and estimated net income generated by the entire farm (Smathers, 1992). A whole-farm budget can be used to determine the net value of farm production and how it will be affected by changes in costs, product prices, and expected crop yields. By comparing several alternative management plans, landowners can use this budget to determine farm potential and negotiate financing from lending institutions (Smathers, 1992).

Enterprise Budgeting

An *enterprise budget* describes costs and revenues associated with a specific farm enterprise and explains how farm resources are allocated in the production of farm products (Chase, 2006). As with whole-farm budgets, enterprise budgets also vary in format and the amount of information provided. Most often they include information on revenues generated from the enterprise and costs such as planting, fertilizing, weed control, labor, machinery, land and building costs, and overhead (Doye, 2007; Smathers, 1992). In addition, these budgets often include break-even prices per unit of production and sensitivity analyses (Smathers, 1992). Enterprise budgets can be used in several ways to aid decision making. Most commonly, they are used to identify the most profitable farm enterprises and determine if current crop or livestock operations can be replaced with more profitable alternatives, such as agroforestry (Chase, 2006). For example, Tiamiyu, Adagba, and Shaahu (2014) used the enterprise budgeting method to analyze the economic profitability of shea (*Vitellaria paradoxa* C.F. Gaertn.)-based enterprises in Nigeria. They found that a shea collection and processing business brings positive economic returns to processors.

Partial Budgeting

A *partial budget* is used in situations where a change in the farm operation affects only a part of farm production (Lessley, Johnson, & Hanson, 1991). Thus, instead of developing an extensive whole-farm or enterprise budget, it is possible to examine only the costs and revenues affected by the change. For instance, partial budgeting has been used by Barungi, Edriss, and Mugisha (2013) to examine the economic returns of applying soil erosion control technologies during farm management in the highlands of eastern Uganda. Wanjiku and Kimenye (2007) used partial budgeting to compare the profitability of using different soil fertility technologies (e.g., manure and chemical fertilizers vs. manure combined with biomass transfer using tithonia [*Tithonia diversifolia* (Hemsl.) A. Gray]) in kale (*Brassica* spp.) and tomato production.

To determine the net outcome of the proposed change, it is necessary to identify associated positive and negative effects as outlined by the partial budget methodology. Increased revenues and reduced costs resulting from the change are considered positive effects, whereas lost or decreased revenues and increased costs are negative effects. If positive effects exceed negative ones, overall farm income increases. In contrast, if negative effects are greater than positive ones, farm income will decrease (Doye, 2007).

Consider Chase, Smith, and Delate's (2006) example of a situation where a farmer considers switching from organic soybean production to organic corn production on 16 ha of farmland. Current revenue associated with organic soybean production on this parcel of land is US$20,160, whereas the cost is $3,150. If the land is shifted to organic corn production, it is expected that it will generate revenue of $27,000 at a cost of $6,530. The positive effects in this case include increased revenue of $27,000 generated from corn production and reduced costs of $3,150, amounting to a total of $30,150. The reduced cost of $3,150 is considered a positive effect because it is associated with soybean production. Since soybean will be replaced with corn, this cost won't be incurred again and thus represents additional savings. The negative effects include lost income of $20,160 and increased costs of $6,530 totaling $26,690.

Again, since soybean won't be cultivated, lost soybean revenue has to be accounted for as a negative effect. Similarly, the cost associated with corn production is considered as a new cost and consequently also a negative effect. The net outcome of the proposed change from organic soybean to organic corn production equals $3,460 (positive effects – negative effects = $30,150 – $26,690). This value represents an amount by which overall farm income will increase if organic corn production is implemented versus organic soybean production.

Interest Rates, Compounding, Discounting, and Discounted Cash Flow Method

A common feature of agroforestry alternatives is that they involve long investment periods (often more than a decade), which requires special approaches in financial evaluation. Comparing costs and revenues simply as they appear is misleading and will lead to an incorrect decision on the financial viability of an agroforestry alternative.

Proper evaluation of agroforestry investments requires that all cash flows (i.e., costs and revenues) are brought to the same point in time. The reason is that costs and revenues typically occur at different times (Godsey, 2000). As a result, the value of such cash flows cannot be compared directly because each cash flow has a different "time value."

Most often the financial analysis of an investment is expressed in terms of the starting year of the project [net present value (NPV)] through the process of discounting. The analysis can be conducted also in terms of the project's end year (net future value) through the process of compounding. An analysis in terms of any intermediate year in the duration of an agroforestry project is also possible; however, this would require both compounding and discounting.

Interest Rates

To conduct discounting or compounding, interest rates are needed. You have probably heard the term *interest rate* many times. When you step into a bank to open a new savings account, purchase a certificate of deposit, or take out a loan, you are informed about current interest rates. So, what does this mean? When you open a savings account or purchase a certificate of deposit at your local bank, you are the lender and the bank is the borrower. The bank borrows money from you for a specified length of time and promises to pay you back more than it borrowed at the end of an agreed-upon period. The interest rate tells you how much the bank will pay you for using your money. If you are taking out a loan, the term interest rate is used to indicate the price paid by borrowers to lenders for borrowing their money and is expressed as a percentage (expressed as a decimal for financial analyses) (Gunter & Haney, 1984). This is the price you will pay to use their money, again to be paid back under the terms of an agreed-upon time frame.

Various terms are used to indicate the interest rate depending on the context; sometimes they are used interchangeably. Here we use three of them: interest rate, discount rate, and minimum acceptable rate of return (MARR). The interest rate is used to calculate the future value of cash flows as well as return on financial instruments, such as savings accounts and certificates of deposit. The discount rate is used to calculate the present value of future cash flows, whereas MARR indicates the minimum rate of return required on a specific investment such as those in agroforestry (Klemperer, 2003).

An interest rate has three unique components: time preference, risk, and inflation (Gunter & Haney, 1984). Time preference refers to an individual's preference for current rather than future consumption (Price, 1993). Gunter and Haney (1984) described this as "increased future gratification" that will make current and future consumption equivalent. So, how does it work? Suppose you deposit money into a bank account for 1 yr. During that period you cannot use your money to buy items you might need. Essentially, you have agreed to postpone your current consumption in exchange for a certain amount of money (i.e., interest) that will be paid to you by your bank at a future date. With the extra money (after taxes) you will be able to buy more of the items you desire, buy items you couldn't previously afford, or reinvest this money. Individuals or institutions have different time preferences. To some, a 5% interest rate might be acceptable, but to others such an interest rate might be too low to induce them to save their money. For these individuals or institutions, the extra 5% is not worth waiting 1 yr, and they would prefer to use their money now. They value current consumption more than those accepting a 5% interest rate and thus require larger compensation (i.e., a higher interest rate) to postpone current consumption. A higher time preference indicates a higher preference for current consumption (Gunter & Haney, 1984) and results in a higher interest rate required to induce further savings.

Most investments are associated with some risk (Peterson & Fabozzi, 2002). If you were wondering how to invest some money, you would find that there are numerous options available, each offering a different rate of return. It seems that selecting an option with the highest return would be the most reasonable action. However, when you look at how others invest their money, you will see that they don't always select the opportunity with the highest possible return. This is because each investment bears a different level of risk. Some investments, such as a savings account or a certificate of deposit, are relatively safe investments (Gunter & Haney, 1984). People can be certain that when they come to the bank the next day, their money will be there and they will be able to collect it. Other investments are considered more risky, yet people are still willing to engage in these investments if a rate of return is high enough to compensate them for the risk involved. To account for risk, the discount rate is adjusted by adding a risk premium and is called a risk-adjusted discount rate (Peterson & Fabozzi, 2002). The amount of risk premium indicates an additional return required due to risk and depends on the riskiness of the investment, investor aversion toward risk, and duration of the investment (Klemperer, 2003).

Compounding and Discounting

Most of us have been exposed directly or indirectly to the process of compounding. For example, when you deposit money into a savings account or purchase a certificate of deposit, this is the process that will be used to determine how much money will accumulate in that account after a specified time period. If you deposit \$2,000 into a savings account that pays an annual interest rate of 4%, after 5 yr you will be able to withdraw \$2,000 × (1 + 0.04)^5 = \$2,433.31 (using Equation 16–1, below). You gain an additional \$433.31 by investing now and cashing in after 5 yr instead of doing so today. Figure 16–1 shows how much money will accumulate in the above account if you decided to make a deposit for a period longer than 5 yr. You can see that the initial amount will double after 18 yr, whereas after 50 yr you will accumulate about \$14,213.37 (a sevenfold increase). In other words, you are lending your money to the bank, which will compensate you for the fact that you cannot use it during this period. The longer you leave your deposit in this account, the more you will be able to withdraw at a later date.

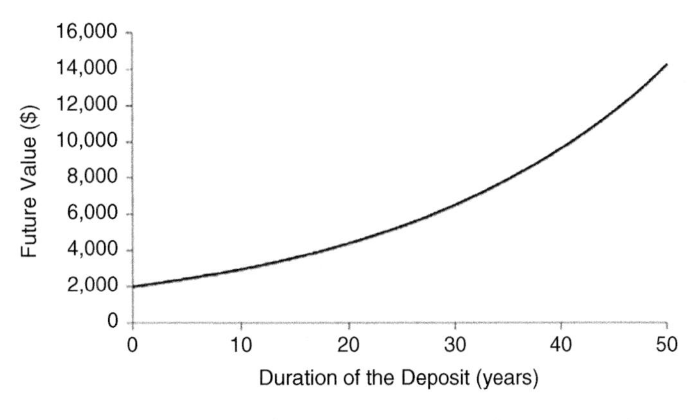

Fig. 16–1 Future value of a US\$2,000 deposit accumulated at a 4% annual interest rate with time.

Future value can be calculated as (Klemperer, 2003)

$$V_n = V_0 \times (1+i)^n \qquad 16\text{–}1$$

where Vn is value in the nth year (future value), V_0 is the initial value (present value), i is the interest rate, and n is the number of years of compounding.

Discounting is the process of calculating the present value of project cash flows. By using a discounting factor (Price, 1993), a present-value equivalent is calculated for cash flows expected to occur in the future. As the term suggests, this process decreases the value of future cash flows and, as a result, the present value is always smaller than a future value. The further in the future a cash flow is expected to occur, the smaller its present value will be (assuming other factors will stay the same). An increase in the interest rate will also result in a smaller present value. Present value can be calculated as (Klemperer, 2003)

$$V_0 = V_n / (1+i)^n \qquad 16\text{–}2$$

where V_0 is the initial value (present value), Vn is the value in the nth year (future value), i is the interest rate, and n is the number of years of discounting.

Discounting has important implications for financial evaluations of agroforestry alternatives. Imagine a simplified agroforestry investment in which there was only one expense of \$1,500 now and expected revenue of \$2,000 in 10 yr. At first you might think that it is a good investment because it will generate a profit of \$500 (\$2,000 – \$1,500). However, when you account for opportunity forgone (i.e., a savings account earning a 4% interest rate), the \$2,000 is now worth only \$1,351.13 (calculated using Equation 16–2), thus indicating that the aforementioned agroforestry investment would generate a loss of \$148.87 (\$1,351.13 – \$1,500.00 = –\$148.87) in present-value terms.

Discounted Cash Flow Method

The financial viability of agroforestry investments can be determined by their NPVs. Typically, agroforestry projects are long-term investments with cash flows occurring in different years. Consequently, a project's net value cannot be determined by comparing nominal values of costs and revenues because they have different time values. Instead, the value of each cash flow has to be recalculated in terms of the same point in time (the same base year). The DCF method considers all of a project's cash flows expressed in terms of the starting year, Year 0 (i.e., the present time) (Bright, 2001). A present value of each cash flow is determined through discounting by using Equation 16–2 to calculate present values. When all cash flows for a particular agroforestry project are discounted to Year 0, they are comparable, and the project's NPV can be established.

When calculating NPV, it is useful to prepare a DCF table that provides a summary of all project activities and corresponding present values. Table 16–2 illustrates DCFs of an elderberry [*Sambucus nigra* L. ssp. *canadensis* (L.) R. Bolli] orchard based on the Elderberry Financial Decision Support Tool developed by Dr. Godsey (University of Missouri, Center for Agroforestry, 2012b). The University of Missouri Center for Agroforestry has generated financial decision support tools for elderberry, black walnut (*Juglans nigra* L.), and chestnut [*Castanea dentata* (Marshall) Borkh.]. All these tools were developed to help growers better understand the economic profitability of growing these specialty crops under different establishment, management, harvesting, and marketing scenarios. Present costs and revenues in the elderberry example are expressed in terms of 2012 U.S. dollars. Projected activities, corresponding years of occurrence, and cash flow values are listed. Costs are assigned negative values, whereas revenues positive ones. The last two columns show the formulas used to calculate the present value of each cash flow, and the corresponding present value calculated at a 5% MARR. The last column is crucial because it permits us to determine the project's NPV. When cash flow present values are summed, NPV amounts to \$13,357.39 ha^{-1}. The positive value indicates that investment costs are more than offset by revenues and, therefore, the investment should be profitable. The use of NPV and other financial indicators to evaluate agroforestry alternatives is explained below.

Financial Indicators

So far we have discussed how to calculate present and future values of single cash flows and determine the NPV of agroforestry investments. Now, we use financial tools to determine if the investment is profitable or not. There are various financial criteria available for examining the profitability of agroforestry alternatives. We will focus on the five most commonly used criteria: NPV, annual equivalent value (AEV), benefit/cost ratio (BCR), internal rate of return (IRR), and land expectation value (LEV), discussing the basics of each criterion as well as guidelines for accepting and rejecting agroforestry alternatives according to each criterion.

Net Present Value

Net present value is often used to determine the financial viability of an investment. It is calculated by subtracting the present value of an investment's total costs from the present value of the investment's total revenues (Bullard & Straka, 1998; Gunter & Haney, 1984; Klemperer, 2003). The general formula for calculating NPV is

$$\text{NPV} = \text{Present value of all revenues} - \text{Present value of all costs} \qquad 16\text{–}3$$

As this formula indicates, an investment's NPV is determined by discounting all revenues (R) and costs (C) to the present (i.e., Year 0 in the life of the project) with interest rate i. Net present value is calculated by summing the present values of costs and revenues. Calculated NPV can be a positive or negative dollar value or zero.

A positive dollar value (NPV > 0) indicates that discounted revenues exceed discounted costs and a profit is generated. In such a case, an agroforestry alternative should be accepted because it is profitable. A negative dollar value (NPV < 0), on the other hand, indicates that discounted costs exceed discounted revenues. Such an agroforestry alternative should be rejected because it doesn't generate enough revenue to offset costs and will result in a monetary loss. A zero dollar value (NPV = 0) indicates that

Table 16–2. Projected discounted cash flows for a 10-yr elderberry orchard calculated at a 5% minimum acceptable rate of return (MARR) using 2012 as the analysis year.

Activity	Year	Cash flow	Formula	Present value (V_0) at 5%
		$/ha ($/acre)		$/ha ($/acre)
		Costs		
Establishment	0	−7,902.15 (−3,199.25)	$V_0 = \frac{-3199.25}{(1+0.05)^0}$	−7,902.15 (−3,199.25)
Management				
Composting	1–10	−691.60 (−280.00)	$V_0^a = \sum_{n=1}^{10} \frac{(-280.00)}{(1+0.05)^n}$	−5,340.36 (−2,162.09)
Pruning	1–10	−247.00 (−100.00)	$V_0 = \sum_{n=1}^{10} \frac{(-100.00)}{(1+0.05)^n}$	−1,907.26 (−772.17)
Mowing	1–10	−123.5 (−50.00)	$V_0 = \sum_{n=1}^{10} \frac{(-50.00)}{(1+0.05)^n}$	−953.64 (−386.09)
Deer control (fencing)	1–10	−863.12 (−349.44)	$V_0 = \sum_{n=1}^{10} \frac{(-349.44)}{(1+0.05)^n}$	−6,664.75 (−2,698.28)
Harvest	1	0	$V_0 = \frac{0}{(1+0.05)^1}$	0
	2	−451.57 (−182.82)	$V_0 = \frac{(-182.82)}{(1+0.05)^2}$	−409.58 (−165.82)
	3	−1,136.99 (−460.32)	$V_0 = \frac{(-460.32)}{(1+0.05)^3}$	−982.17 (−397.64)
	4	−1,686.42 (−682.76)	$V_0 = \frac{(-682.76)}{(1+0.05)^4}$	−1,387.42 (−561.71)
	5	−2,118.40 (−857.65)	$V_0 = \frac{(-857.65)}{(1+0.05)^5}$	−1,659.82 (−671.99)
	6	−2,450.12 (−991.95)	$V_0 = \frac{(-991.95)}{(1+0.05)^6}$	−1,828.32 (−740.21)
	7	−2,697.22 (−1,091.99)	$V_0 = \frac{(-1091.99)}{(1+0.05)^7}$	−1,916.87 (−776.06)
	8	−2,873.87 (−1,163.51)	$V_0 = \frac{(-1163.51)}{(1+0.05)^8}$	−1,945.15 (−787.51)
	9	−2,992.78 (−1,211.65)	$V_0 = \frac{(-1211.65)}{(1+0.05)^9}$	−1,929.17 (−781.04)
	10	−3,065.12 (−1,240.94)	$V_0 = \frac{(-1240.94)}{(1+0.05)^{10}}$	−1,881.72 (−761.83)
		Revenues		
Elderberry yields	1	0	$VO = \frac{0}{(1+0.05)^1}$	0
	2	2,257.80 (914.09)	$VO = \frac{914.09}{(1+0.05)^2}$	2,047.90 (829.11)

Table continued.

Activity	Year	Cash flow	Formula	Present value (V_0) at 5%
	3	5,685.00 (2,301.62)	$VO=\frac{2301.62}{(1+0.05)^3}$	4,910.90 (1,988.22)
	4	8,432.04 (3,413.78)	$VO=\frac{3413.78}{(1+0.05)^4}$	6,937.04 (2,808.52)
	5	10,592.00 (4,288.26)	$VO=\frac{4288.26}{(1+0.05)^5}$	8,299.13 (3,359.97)
	6	12,250.60 (4,959.76)	$VO=\frac{4959.76}{(1+0.05)^6}$	9,141.59 (3,701.05)
	7	13,486.13 (5,459.97)	$VO=\frac{5459.97}{(1+0.05)^7}$	9,584.34 (3,880.30)
	8	14,369.40 (5,817.57)	$VO=\frac{5817.57}{(1+0.05)^8}$	9,725.77 (3,937.56)
	9	14,963.88 (6,058.25)	$VO=\frac{6058.25}{(1+0.05)^9}$	9,645.84 (3,905.20)
	10	15,325.61 (6,204.70)	$VO=\frac{6204.70}{(1+0.05)^{10}}$	9,408.60 (3,809.15)

Note. The financial decision support tool has different management practice options. In this analysis, we used the tool's default management practices.

[a] $V_0=\sum_{n=1}^{10}\frac{(-280.00)}{(1+0.05)^n}=\frac{(-280.00)}{(1+0.05)^1}+\frac{(-280.00)}{(1+0.05)^2}+\frac{(-280.00)}{(1+0.05)^3}+\frac{(-280.00)}{(1+0.05)^4}+\frac{(-280.00)}{(1+0.05)^5}+\frac{(-280.00)}{(1+0.05)^6}+\frac{(-280.00)}{(1+0.05)^7}+\frac{(-280.00)}{(1+0.05)^8}$

$+\frac{(-280.00)}{(1+0.05)^9}+\frac{(-280.00)}{(1+0.05)^{10}}=-\$2,162.09.$

discounted revenues equal discounted costs. The agroforestry alternative should be accepted because it still generates enough revenues to offset costs—this is referred to as the financial break-even point.

Annual Equivalent Value

Annual equivalent value is an indicator that expresses NPV in annual equivalents distributed equally over each year of an investment's lifespan. Since AEV is calculated based on NPV, it is positive when NPV is positive and negative when NPV is negative. Annual equivalent value is useful in an agroforestry context because it allows comparison of alternatives on an annual basis, which is particularly helpful when comparing long-term tree investments with annual agricultural crop production (Bullard & Straka, 1998). The formula for calculating AEV is

$$\text{AEV}=\text{NPV}\left[\frac{i(1+i)^n}{(1+i)^n-1}\right] \qquad 16\text{–}4$$

where i is the interest rate and n is the project life in years. A 10-yr agroforestry investment with a NPV of $910 ha^{-1} calculated at an 8% MARR will have an AEV of $118.89 ha^{-1}:

$$\text{AEV}=\$910\ \text{ha}^{-1}\left[\frac{0.08(1+0.08)^{10}}{(1+0.08)^{10}-1}\right]=\$118.89\ \text{ha}^{-1}$$

Benefit/Cost Ratio

Benefit/cost ratio is calculated by dividing the sum of investment discounted revenues by the sum of discounted costs. It is also referred to as the profitability index because it indicates a return generated for each dollar invested in the project (Bullard & Straka, 1998; Gunter & Haney, 1984; Klemperer, 2003). The formula for calculating BCR is

$$\text{BCR}=\frac{\text{Present value of all investment revenues}}{\text{Present value of all investment costs}} \qquad 16\text{–}5$$

A BCR value greater than one (BCR > 1) indicates that each dollar invested in the agroforestry alternative generates more than $1 in return in present-value terms. Therefore, the alternative should be accepted. However, if the BCR value is less than 1 (BCR < 1), the alternative should be rejected because each dollar invested generates less than $1 in return, indicating a loss on each dollar invested. A BCR equal to one (BCR = 1) indicates that each dollar invested generates one dollar in return. This means that an agroforestry alternative has broken even and also should be accepted.

Internal Rate of Return

Internal rate of return is a discount rate at which an investment's NPV equals zero (Bullard & Straka, 1998; Gunter & Haney, 1984; Klemperer, 2003). This is the maximum discount rate at which an agroforestry alternative can break even. The IRR is determined by an iterative process, in which an investment's NPV is calculated at various discount rates. Two interest rates at which the NPV changes sign from positive to negative (one of which the NPV is positive and the other at which it is negative) need to be selected to calculate the IRR. The reason for using this iterative process is that there is an inverse relationship between NPV and the discount rate used to calculate NPV. More specifically, NPV decreases as the discount rate used to calculate it increases. When the discount rate is increased sufficiently high, NPV will become negative (the opposite will hold if the discount rate is decreased). This means that between the two discount rates (resulting in positive and negative NPVs, respectively) there is one that will result in an NPV equal to zero. This is the IRR, and it can be approximated by (Bright, 2001)

$$\text{IRR} = \text{Discount rate resulting in negative NPV} + \left(\text{Difference between discount rates} \times \frac{\text{Positive NPV}}{\text{Incremental NPV}}\right) \quad \text{16–6}$$

As an example, we calculate an IRR for an alley cropping system that generates a NPV of $650 ha^{-1} at a 6% discount rate. To determine the IRR for this investment, we need to find the rate of return at which the NPV will become negative. Since NPV was positive at 6%, this means that we need to increase the discount rate. If the discount rate is increased to 8%, NPV is still positive at $320 ha^{-1}. Consequently, the discount rate needs to be increased even further. At 10%, it generates a NPV of $130 ha^{-1}, but at 12% NPV drops to −$28 ha^{-1}. To determine the IRR, we need to select the highest discount rate at which the NPV is still positive and another where the NPV becomes negative. This is 10 and 12%, respectively. Now, we know that the IRR is >10% but <12%. By inserting this information into the above formula, we can approximate that the IRR for this agroforestry alternative is 11.65%:

$$\text{IRR} = 10\% + \left(2\% \times \frac{\$130\ \text{ha}^{-1}}{\$130\ \text{ha}^{-1} + \$28\ \text{ha}^{-1}}\right) = 11.65\%$$

How can the IRR be used to determine if an agroforestry alternative is financially acceptable? If the farmer's acceptable rate of return is less than or equal to the IRR, the alternative should be accepted. However, if the MARR is greater than the IRR, the alternative should be rejected. This is because a MARR that is higher than the IRR will result in a negative NPV and a financial loss for the agroforestry alternative.

If the MARR for the alley cropping system mentioned above is 6%, then it would be a good investment because the actual IRR is well above that acceptable rate. In fact, this alley cropping system would break even at a MARR as high as 11.65%.

Table 16–3 gives a summary of how the indicators of NPV, IRR, and BCR can be used to assist in the decision-making process. The landowner should accept investments with NPVs ≥0, a BCR that is ≥1, and an IRR that is greater than or equal to the MARR.

Land Expectation Value

Land expectation value is a financial tool used to estimate land value based on all expected future costs and revenues continually generated from the use of this land. The LEV (known also as the Faustmann formula) has been used primarily to calculate land parcel values for which timber production was determined to be the best land use. Its major assumption is that timber production will be continued on a particular parcel of land in perpetuity under the same management regime (Bullard & Straka, 1998; Klemperer, 2003). However, the LEV can also be used to establish the value of a specific land parcel based on costs and revenues associated with both tree and agricultural production. In this case, the LEV is interpreted as the maximum amount a landowner can pay for the land and still earn the MARR on an agroforestry investment. The LEV

Table 16–3. Guidelines for accepting or rejecting agroforestry alternatives according to net present value (NPV), benefit/cost ratio (BCR), and internal rate of return (IRR).

	Decision rule		
Accept the investment	NPV ≥ 0	BCR ≥ 1	MARR[a] ≤ IRR
Reject the investment	NPV < 0	BCR < 1	MARR > IRR

[a] Minimum acceptable rate of return.

can be computed in several ways. We calculate the LEV for a silvopastoral system based on its NPV of \$5,395.04 ha^{-1} by using

$$\text{LEV} = \frac{\text{NPV}(1+i)^t}{(1+i)^t - 1} = \frac{\$5,395.04\ \text{ha}^{-1} \times (1.05)^{30}}{(1.05)^{30} - 1}$$

$$= \$7,019.10\ \text{ha}^{-1}$$

The value of \$7,019.10 ha^{-1} represents the maximum amount a landowner can pay for this land and still earn a 5% MARR on this silvopastoral system, cycle after cycle in perpetuity.

Economic Profitability Estimates

The literature has estimated economic returns from agroforestry systems by mostly focusing on silvopasture and alley cropping practices. Most, but not all, of these studies have indicated a positive NPV from silvopasture. For instance, Broughton, Bukenya, and Nyakatawa (2012) evaluated the economic profitability of a silvopastoral practice in Alabama and found that this practice was profitable at a 6% discount rate. Broughton and Bukenya (2010) estimated economic returns from a silvopasture production system in the Alabama Black Belt region. They found that it brings positive returns with a BCR of 1.108, a NPV of \$6,182 ha^{-1}, an AEV of \$264 ha^{-1}, and an IRR of 5.99%. However, there is another study that investigated the economic profitability in a silvopastoral pecan [*Carya illinoinensis* (Wangenh.) K. Koch] and cow–calf production practice in southeastern Kansas (Ares, Reid, & Brauer, 2006). Results indicated that under the nut and beef market prices in the study year, this silvopastoral system was not profitable.

Other studies have compared economic returns between a silvopasture system and traditional ranching. Stainback and Alavalapati (2004) reported that silvopasture was more profitable than either traditional ranching or traditional forestry using the LEV. Orefice, Smith, Carroll, Asbjornsen, and Howard (2019) estimated the economic profitability in a silvopastoral system in the state of New York. Their results indicated that silvopasture can generate higher economic returns than open pasture (IRR: 6.4 vs. 2.6%; NPV: \$1,277 vs. −\$77 ha^{-1}). Chizmar (2018) compared the economic returns between silvopasture and a conventional cattle system (cattle and pasture) in North Carolina and found that, with 25% of forest cover within the fields, the NPV from a silvopasture system was higher than that from a conventional cattle system.

In terms of the economic returns from an alley cropping system, Cubbage et al. (2012) had conducted an alley cropping trail in Goldsboro, NC, and estimated its economic returns. Their results indicated that longleaf pine (*Pinus palustris* Mill.) with pine straw harvests had an IRR of 5.5% compared with 3.5% without pine straw harvests. However, crop yields at this site were very poor due to droughts and floods, leading to net economic losses for the demonstration site. Susaeta, Lal, Alavalapati, Mercer, and Carter (2012) used the LEV to compare the profitability of alley cropping of loblolly pine (*Pinus taeda* L.) and switchgrass (*Panicum virgatum* L.) in the southern United States. They found that when switchgrass prices were greater than \$30 Mg^{-1}, landowners would benefit more by adopting intercropping than a loblolly pine monoculture. Stamps, McGraw, Godsey, and Woods (2009) found that alley cropping alfalfa (*Medicago sativa* L.) with black walnut can bring economic profitability to landowners by using NPV and AEV. Their estimated NPV ranged from \$94 to \$107 ha^{-1}.

Limitations of Financial Indicators

So far we have discussed five economic criteria used to evaluate agroforestry alternatives. However, we haven't discussed how to choose among these criteria. When deciding if a particular investment is acceptable or not, the choice is easier than it may appear. Any of the three commonly referred to criteria—NPV, BCR, and IRR—can be used to evaluate an alternative because they will provide the same recommendation (Bullard & Straka, 1998; Gunter & Haney, 1984). If the agroforestry alternative is acceptable according to one criterion, it will also be acceptable according to the remaining two because all three criteria involve comparing the present value of revenues with the present value of costs. For example, a NPV of \$750 ha^{-1} indicates that the investment is acceptable according to the NPV because it is >0. A positive NPV means that discounted revenues are greater than discounted costs. This indicates that the BCR is >1 and the investment is acceptable. Furthermore, a positive NPV indicates that the MARR used to calculate the NPV for this agroforestry alternative is smaller than the IRR. As a result, the alternative is also acceptable according to the IRR criterion (Bullard & Straka, 1998).

Landowners often not only need to decide if an agroforestry project is acceptable but also must select the best agroforestry alternative among several financially acceptable alternatives. One reason is that capital, labor, or land available to landowners is often limited (Klemperer, 2003), allowing for implementation of only one or a limited number of viable investment alternatives. Second, even if all considered alternatives can be financed, some of them might be mutually exclusive, thereby allowing the implementation of only one alternative. For example, if a landowner decides to use part of their farmland for the production of organic soybean, this part of the farm cannot be used for the cultivation of other crops during the same season on the same land parcel.

When the goal of a financial analysis is to rank the available alternatives to determine the best land use, the process of selecting an appropriate financial indicator is more challenging. This is because NPV, BCR, and IRR might provide conflicting recommendations with regard to the ranking of a particular agroforestry alternative. Furthermore, they do not specify the investment scale or the amount of capital required to implement it. Consequently, it is not difficult to imagine a situation in which even acceptable alternatives might not be undertaken due to, for example, limited funding.

So, which indicator should be used to rank agroforestry alternatives? Generally, it is recommended that NPV should be used to rank agroforestry alternatives to select, from a financial viewpoint, those that are most viable and feasible (Bullard & Straka, 1998; Gunter & Haney, 1984). However, it is also recommended that a two-step ranking process be used to ensure that the best financial decision is made. In the first step of the ranking process, NPV should be used to determine if the considered alternatives are acceptable. Unacceptable alternatives should be removed from further consideration. In the second step, a list of acceptable alternatives should be created and mutually exclusive alternatives identified. If mutually exclusive alternatives are present, the best alternative from each set of mutually exclusive alternatives should be selected (e.g., by computing the AEV). The most poorly performing alternatives are removed from further analysis because their implementation would be inefficient. The final ranking list consists of only acceptable and mutually nonexclusive alternatives ranked in order of decreasing NPV. The ranking also should include information on the cost required to implement each alternative. This two-step analysis process provides the information needed to determine which agroforestry alternatives can be implemented given an existing budget and a landowner's financial objectives (Bullard & Straka, 1998; Gunter & Haney, 1984).

Valuation of Nonmarket Benefits from Agroforestry

One of the key factors in determining agroforestry adoption is the profitability of the practice in comparison with other land use practices. However, profitability from a landowner's perspective, generally termed *private profitability*, can be different from that of a social perspective, often referred to as *social profitability*. The exclusion or inclusion of social benefits and costs and nonmarket goods and services (e.g., biodiversity and carbon sequestration), also known as *externalities*, largely differentiates the private and social profitability.

Marginal Costs and Marginal Benefits

Research suggests that ecosystem goods and services associated with agroforestry are significant and that failure to incorporate them will result in gross undervaluation of agroforestry (Alavalapati, Shrestha, Stainback, & Matta 2004; Obeng, Aguilar, & McCann, 2018). Following Alavalapati et al. (2004), we present a graphical analysis to illustrate the effect of internalizing environmental services associated with silvopasture adoption (Figure 16–2). The horizontal axis measures the extent of trees and buffer strips on ranchlands, a proxy for silvopasture, while the vertical axis

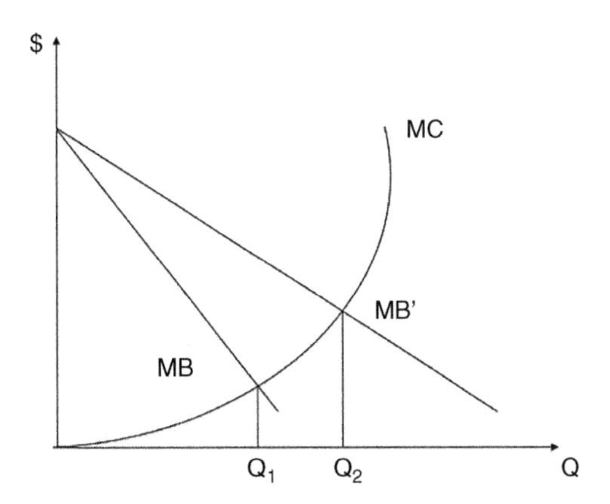

Fig. 16–2. Effect of internalizing environmental services associated with silvopasture adoption; MC, marginal cost; MB, private marginal benefit; MB′, marginal benefit to society (adapted from Alavalapati et al., 2004).

measures the costs and benefits of maintaining tree cover and buffer strips. Trees and buffer strips on ranchlands are considered beneficial to ranchers by providing shade to cattle and additional revenue from timber. These benefits to the rancher are reflected in Figure 16–2 as private marginal benefits (MB). However, these activities also provide environmental benefits to the public by improving water and air quality (Holmes, Bergstrom, Huszar, Kask, & Orr, 2004; Zinkhan & Mercer, 1997). Consequently, maintaining trees and buffer strips on ranchlands is a cost paid by individual ranchers, yet some of the benefits are enjoyed by everybody. Furthermore, the cost of fencing, tree seedlings, site preparation, and pruning is expected to increase with an increase in tree density. These costs are also reflected in Figure 16–2 as marginal costs (MC). In the absence of benefits from environmental services of silvopasture, ranchers will equate the private marginal cost of maintaining trees on ranchlands (MC) and private marginal benefit of trees (MB) and therefore maintain only *Q1* amount of tree cover on their ranches. However, if markets exist for ecosystem services and if ranchers can capitalize on water quality improvement and carbon sequestration, the marginal benefit of trees would be higher. The marginal benefit to society is reflected in Figure 16–2 as MB′. This increased marginal benefit motivates ranchers to maintain Q_2 amount of tree cover on their ranchlands.

In the absence of incentives for the provision of positive externalities and penalties for causing negative externalities, a rational landowner is less likely to produce services at a societal optimum. Institutional economics suggests that under exclusive, transferable, and enforceable property rights, individuals and institutions can be rewarded for positive externalities and penalized for negative ones. Furthermore, Coase (1960, 1992) stated that under zero transaction costs, markets allocate resources such that externalities are produced at optimum levels regardless of the initial assignment of property rights to either buyer or seller.

Consider, for example, a watershed with a group of ranchers upstream and a settlement of households downstream. Ranchers pollute a nearby stream through the application of chemical fertilizers and pesticides on pasture land. Households, who depend on the stream for clean water and other economic activities such as fishing, are affected negatively from water pollution. Ranchers, however, can reduce pollution by maintaining tree cover and buffer strips, but it would cost them. On the other hand, households would benefit if ranchers reduce pollution. In Figure 16–3, the horizontal axis represents the quantity of tree cover and buffer strips, and the vertical axis represents the costs or benefits of reducing pollution. The MC curve represents the marginal cost to ranchers, and the MB curve represents the marginal benefits to households from reducing pollution.

First, let's assume that ranchers are not required to maintain tree cover and buffer strips—ranchers have a right to pollute. However, most ranchers would be willing to maintain tree cover and buffer strips if the public would cover the costs. In this scenario, the public would have to pay ranchers an amount equal to area OBD, the entire area under the MC curve, to maintain maximum tree cover for reducing water pollution. Since the amount that the public must pay ranchers to plant more trees than C is higher than the benefits they derive, the public would not want to pay for tree cover and buffer strips beyond C. Alternatively, if the public has a right to clean water, then ranchers would have to pay an amount equal to area ABO, the entire area under the MB curve, for the public to put up with water pollution. Given that the amount ranchers must pay households is initially much higher than the cost of withholding pollution, ranchers would want to maintain tree cover and buffer strips up to C. At this point, the amount that ranchers must pay equals the marginal cost of withholding pollution. The Coase (1960, 1992) theorem suggests that if property rights are defined, the parties will negotiate according to their benefits and costs and reach an optimal solution. Regardless of which one has the property rights, with zero transaction costs, the solution is reached at C.

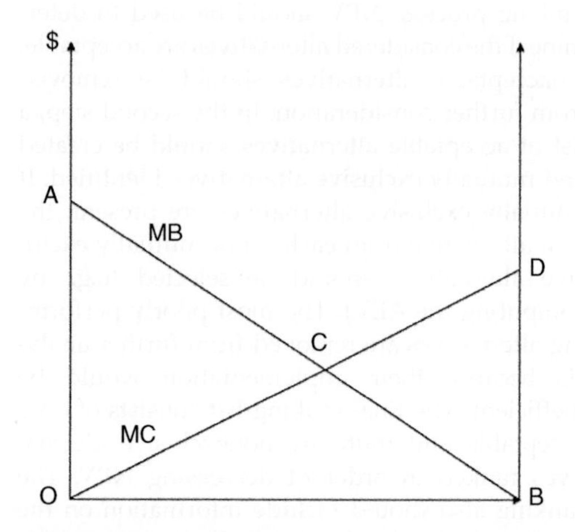

Fig. 16–3. Optimal silvopasture adoption with zero transaction costs; MB, marginal benefit; MC, marginal cost (adapted from Alavalapati et al., 2004).

To realize optimum policy solutions to the above problem, a variety of information is needed, including the environmental benefits and costs of silvopasture from both households' and ranchers' perspectives. In the recent past, advancements in environmental economics have produced a range of economic tools to generate information for policy making.

Economic Values of Ecosystem Services

Economic values of ecosystem services from agroforestry have been recently estimated in the literature using different environment valuation methods. Most of these studies focused on the estimation of ecosystem service values from silvopasture, riparian buffers, and windbreaks. Shrestha and Alavalapati (2004a) assessed the value of environmental services associated with silvopasture in Florida using a choice experiment. Assuming that silvopasture has the potential to reduce phosphorus runoff, sequester additional carbon dioxide, and improve wildlife habitat in the Lake Okeechobee watershed, Shrestha and Alavalapati tested whether the public have a willingness to pay (WTP) to realize those benefits. They found that the average household would have a WTP \$137.97 yr^{-1} for 5 yr for a moderate level of improvement in all three environmental attributes. With 1.34 million households in the watershed, the total WTP for environmental services would be \$924.40 million. This value reflects the total demand for environmental services associated with silvopasture in the Lake Okeechobee watershed.

As indicated above, the environmental services associated with silvopasture are external to cattle ranchers. As such, they may have little or no motivation to adopt silvopasture unless they are compensated for those environmental services. Shrestha and Alavalapati (2004b) assessed ranchers' willingness to accept (WTA) to produce the above environmental services by adopting silvopasture practices. In particular, they assessed the effect of a premium on beef prices and a direct payment on adoption of silvopasture using a contingent valuation approach. It was found that, on average, a price premium of \$0.07 kg^{-1} of beef or a direct payment of \$23.03 ha^{-1} was required for ranchers to adopt silvopasture practices. With approximately 2.4 million ha of pasture and ranchlands, and at the rate of \$23.03 ha^{-1}, the direct payment policy would cost about \$66 million annually.

Ranchers in Florida are increasingly managing their lands for recreational hunting. This is a supplemental economic activity for many ranchers, wherein they sign a lease agreement with hunting clubs or other interested parties allowing them to hunt on their lands. Land attributes, such as the distance to urban centers, scenic view, and nature of the habitat, influence hunter preferences and thus the lease price. Therefore, ranchers may be interested to know the effect of their land attributes on lease price. Shrestha and Alavalapati (2004c) estimated the effect of ranchland attributes on recreational hunting in Florida using a hedonic price analysis (a revealed preference approach). They found that trees and vegetative cover on ranchlands have a positive impact on hunting revenues, indicating opportunities for silvopasture. In particular, ranchers in Florida who maintain about 22% trees and other vegetative cover on ranchlands could charge \$39.91 ha^{-1} yr^{-1} for hunting.

The economic values of environmental services provided by riparian buffers and windbreaks have also been estimated in the literature. Qiu, Prato, and Boehrn (2006) conducted a survey in Dardenne Creek watershed in St. Louis, Missouri, to examine the economic values of riparian buffers. Their results showed that, on average, local residents had a WTP of \$1,625 more for a house that was close to an openly accessible tree buffer compared with other houses. This WTP price premium may reflect the economic value of services provided by riparian buffers. Bauer and Johnston (2017) estimated the total economic values provided by riparian buffers to be \$34 million in the Great Bay watershed of New Hampshire. Hunting opportunities provided by riparian buffers had also been estimated at \$30 per visit per party of four hunters in north-central Iowa (Grala, Colletti, & Mize, 2009). In terms of windbreaks, local residents had a mean WTP of \$4.77–8.50 to support the establishment of windbreaks in Iowa (Grala, Tyndall, & Mize, 2012).

Policy Implications

Federal Cost-Share Programs

The success of federal cost-share programs promoting tree planting and forest management by nonindustrial private forest landowners is evidence of the potential of federal land-use policy in promoting agroforestry. For example, the Conservation Reserve Program (CRP), aiming to help farmers offset the costs of practices that improve water and soil quality and protect wildlife habitat, had enrolled more than 323,750 ha in 2015 (USDA Farm Service Agency, 2016). A wide array of federal, state, and private programs provide financial incentives to landowners for adopting agroforestry. Federal funding for agroforestry is administered by the U.S. Forest Service, USDA Farm Service Agency (FSA), USDA Natural

Resource Conservation Service (NRCS), and USDA Fish and Wildlife Service. The USDA Agroforestry Strategic Framework was developed in 2012 to increase public awareness and support for agroforestry in the United States.

Most federal programs providing incentives to landowners to manage forests and trees to produce environmental benefits have been authorized under the Farm Bill, the primary federal tool for developing U.S. policies and programs affecting agriculture, rural lands, and food consumers. The first Farm Bill, developed in the 1920s, focused primarily on agricultural commodity programs such as price supports, agricultural exports, farm credit, and agricultural research. The Farm Bill is reviewed and amended every 6 yr by the U.S. Congress. Reacting to concerns regarding the environmental impacts of rural land use, Congress first introduced resource conservation policies and programs in the 1985 Farm Bill. A forestry title (Title XII, The Forest Stewardship Assistance Act) was first included in the 1990 Farm Bill and authorized the Forest Legacy Program, Forest Stewardship Program, Forestry Incentives Program, and the Stewardship Incentives Program. Several tree-planting initiatives were also included in the conservation title (Title XIV) of the 1990 Farm Bill. Since then, forestry stakeholders have used the Farm Bill as the primary avenue for renewing or promoting new forestry incentive programs.

USDA Farm Service Agency

The FSA provides incentives for adopting agroforestry practices on private lands through the Biomass Crop Assistance Program, the CRP, the Continuous Conservation Reserve Program, and the Conservation Reserve Enhancement Program. These programs provide soil rental payments, cost sharing, and other financial incentives to landowners who agree to retire or convert agricultural lands to alternative uses including riparian buffers, windbreaks, and tree planting.

USDA Natural Resource Conservation Service

In addition to providing technical assistance to landowners interested in agroforestry and other conservation practices, the NRCS provides funding for tree planting (including agroforestry) through the Agricultural Conservation Easement Program, the CRP (the NRCS provides technical assistance for land eligibility, planning, and practice implementation), Environmental Quality Incentives Program (EQIP), and Conservation Stewardship Program. The EQIP provides incentive payments for alley cropping, riparian buffers, and windbreaks as well as cost-share for tree planting. In 2017, the Missouri NRCS started to offer a funding pool for agroforestry and woody crops under EQIP. The Agricultural Conservation Easement Program is a new program initiated under the 2014 Farm Bill. It consolidates three programs, including the Wetland Reserve Program, Grassland Reserve Program, and Farmland Protection Program.

Program Effectiveness and Barriers

A number of studies have examined the social and economic efficiency of public financial incentive programs for private forest investments such as agroforestry. One hypothesis has been that these programs substitute government payments for private capital investments. Several studies have shown that cost-share assistance programs are effective in improving forest land productivity (Mills, 1976; Royer & Moulton, 1987). By 2006, the CRP had restored 1.01 million ha of buffers, and 1.09 million ha of trees had been planted (Cowan, 2010). Besides the environmental impacts, these cost-share programs have been determined to generate positive economic impacts. The USDA NRCS (2017) estimated the economic impacts of conservation programs (costs around $135 million) in Minnesota and found that these programs had generated a total value of $160,488,824 for the local economy and 1,837 jobs in 2016. In 2008, the Oklahoma Cooperative Extension Service (2017) found that a dollar spent on certain farm management practices (e.g., pasture and hay planting, prescribed grazing, and range planting) through the conservation programs created an additional of $0.91 to $1.18 of economic activity in Oklahoma.

An important aspect of cost-share and management assistance programs is the interaction between landowners and land managers. Generally, landowners are required to develop management plans before receiving cost-share or lease payments. Plans are generally developed by public or private professionals, often with the participation of the landowner. Direct contact with professional land managers has been identified as a leading factor in landowner decisions to adopt conservation practices such as agroforestry. Several studies have found that programs that put landowners in direct contact with a forester or other natural resource professional are most influential in encouraging landowners to adopt sustainable forestry practices (Greene, Daniels, Jacobson, Kilgore, & Straka, 2005; Kilgore & Blinn, 2004; Kilgore, Greene, Jacobson, Straka,

& Daniels, 2007; Ma, Butler, Kittredge, & Catanzaro, 2012). Garbach, Lubell, and DeClerck (2012) found that technical assistance from a professional has a positive influence on the adoption of silvopastoral conservation practices. Esseks and Moulton (2000) found that two-thirds of Forest Stewardship Program (FSP) participants had never had contact with a professional forester before developing the required management plan. A similar number began managing their land for multiple purposes and using new practices due to the FSP. In addition, participation in the FSP prompted owners to spend an average of $2,767 of their own funds for forest management activities. However, without their involvement in the FSP and receiving cost-share assistance, nearly two-thirds of participating owners said they would not have made these expenditures.

There are some barriers for promoting participation in conservation programs. Lack of regulatory assurance is one of the stated reasons for not participating (Bennett et al., 2014). Landowners may perceive that participation in cost-share programs may lead to legal and regulatory pressure. The complicated enrollment process also leads to a lower participation rate (Bennett et al., 2014). Funding is also a crucial barrier in promoting sustainable land use practices such as agroforestry. Lack of enough funding increases competition during application for these programs.

Summary

Agroforestry is a way for landowners to manage scarce natural resources that balances environmental stewardship, financial feasibility, and social responsibility. Because it is a balance of these three objectives, it requires the landowner to make complex decisions. Economic analysis uses a set of tools that can identify trade-offs that are made in the decision process. This chapter gave a broad overview of the decision tools that economists use, including budgeting methods, financial methods, and nonmarket valuation methods. Real-world applications of these methods illustrate the applicability and importance to the decision process.

Governmental policies have an impact on land management decisions. Management of privately owned natural resources can have an impact on society in both positive and negative ways. Therefore, land management policies have been developed that provide monetary incentives for land use practices, such as agroforestry, that protect, conserve, and improve the natural resource base.

References

Alavalapati, J. R. R., & Mercer, D.E. (Eds.). (2004). *Valuing agroforestry systems: Methods and applications*. Dordrecht, the Netherlands: Kluwer Academic Publishers.

Alavalapati, J. R. R., Shrestha, R. K., Stainback, A., & Matta, J. R. (2004). Agroforestry development: An environmental economic perspective. *Agroforestry Systems, 61*, 299–310.

Ares, A., Reid, W., & Brauer, D. (2006). Production and economics of native pecan silvopastures in central United States. *Agroforestry Systems, 66*, 205–215.

Atangana A., Khasa, D., Chang, S., & Degrande, A. (2014). Tropical agroforestry). Dordrecht, the Netherlands: Springer.

Barungi, M., Edriss, A., & Mugisha, J. (2013). Profitability of soil erosion control technologies in eastern Uganda Highlands. *African Crop Science Journal, 21*(1), 637–646.

Bauer, D. M., & Johnston, R. J. (2017). *Buffer Options for the Bay: Economic valuation of water quality ecosystem services in New Hampshire's Great Bay watershed*. Retrieved from http://www.bufferoptionsnh.org/wp-content/uploads/2017/12/BOB_Economic_Assessment.docx.pdf.

Bennett, D., Nielsen-Pincus, M., Ellison, A., Pomeroy, A., Burright, H., Gosnell, H., . . . Gwin, L. (2014). *Barriers and opportunities for increasing landowner participation in conservation programs in the interior Northwest*. Eugene, OR: Institute for a Sustainable Environment, University of Oregon. Retrieved from https://scholarsbank.uoregon.edu/xmlui/bitstream/handle/1794/19385/WP_49.pdf;jsessionid=B5B8D0F2832B426A3410CED0781B6020?sequence=1.

Bright, G. (2001). *Forestry budgets and accounts*. Wallingford, U.K.: CAB International.

Broughton, B., & Bukenya, J. O. (2010). Financial feasibility of simultaneous production of pine sawlogs, forage, and meat goats on small farms in Alabama: A preliminary analysis. Paper presented at the *Southern Agricultural Economics Association 42nd Annual Meeting, Orlando, FL* (no. 1370-2016-108782).

Broughton, B., Bukenya, J. O., & Nyakatawa, E. (2012). Economic feasibility of simultaneous production of pine sawlogs and meat goats on small-sized farms in Alabama. *Journal of Life Sciences, 6*(1), 80–90.

Bullard, S. H., & Straka, T. J. (1998). *Basic concepts in forest valuation and investment analysis* (2nd ed.). Wetumtka, AL: Preceda.

Buongiorno, J., & Gilless, K. (2003). *Decision methods for forest resource management*. San Diego, CA: Academic Press.

Center for Agroforestry, University of Missouri. (2012a). *What is agroforestry?* Retrieved from http://www.centerforagroforestry.org/practices.

Center for Agroforestry, University of Missouri. (2012b). *Profit in agroforestry*. Retrieved from http://www.centerforagroforestry.org/profit.

Chase, C. (2006). *Using enterprise budgets to make decisions* (Ag Decision Maker, File A1-19, Publ. FM 1875). Ames, IA: Iowa State University Extension.

Chase, C., Smith, M., & Delate, K. (2006). Organic crop production enterprise budgets (Ag Decision Maker, File A1-18, Publ. FM 1876). Ames, IA: Iowa State University Extension.

Chizmar, S. J. (2018). *A comparative economic assessment of silvopasture systems in the Amazonas region of Peru and in North Carolina, USA* (Master's thesis). North Carolina State University. Retrieved from http://www.lib.ncsu.edu/resolver/1840.20/35059

Coase, R. H. (1960). The problem of social cost. *Journal of Law and Economics, 3*, 1–14.

Coase, R. H. (1992). The institutional structure of production. *The American Economic Review, 78*, 54–64.

Cowan, T. (2010). *Conservation Reserve Program: Status and current issues.* Washington, DC: Congressional Research Service. Retrieved from http://www.nationalaglawcenter.org/wp-content/uploads/assets/crs/RS21613.pdf.

Cubbage, F., Glenn, V., Mueller, J. P., Robison, D., Myers, R., Luginbuhl, J. M., & Myers, R. (2012). Early tree growth, crop yields and estimated returns for an agroforestry trial in Goldsboro, North Carolina. *Agroforestry Systems, 86*, 323–334.

Doye, D. (2007). Budgets: Their use in farm management (Fact Sheet F-139). Norman, OK: Oklahoma State University Cooperative Extension Service.

Esseks, J. D., & Moulton, R. J. (2000). *Evaluating the forest stewardship program through a national survey of participating forest land owners.* De Kalb, IL: Social Science Research Institute, Northern Illinois University.

Garbach, K., Lubell, M., & DeClerck, F. A. (2012). Payment for ecosystem services: The roles of positive incentives and information sharing in stimulating adoption of silvopastoral conservation practices. *Agriculture, Ecosystems & Environment, 156*, 27–36.

Godsey, L. D. (2000). Economic budgeting for agroforestry practices (Agroforestry in Action, 3-2000). Columbia, MO: University of Missouri Center for Agroforestry.

Grala, R. K., Colletti, J. P., & Mize, C. W. (2009). Willingness of Iowa agricultural landowners to allow fee hunting associated with in-field shelterbelts. *Agroforestry Systems, 76*, 207–218

Grala, R. K., Tyndall, J. C., & Mize, C. W. (2012). Willingness to pay for aesthetics associated with field windbreaks in Iowa, United States. *Landscape and Urban Planning, 108*, 71–78.

Greene, J., Daniels, S., Jacobson, M., Kilgore, M., & Straka, T. (2005). *Existing and potential incentives for practicing sustainable forestry on non-industrial private forest lands* (Final Report, Research Project C2). Washington, DC: National Commission on Science for Sustainable Forestry.

Gunter, J. E., & Haney, H. L. (1984). Essentials of forestry investment analysis. Corvallis, OR: Oregon State University Bookstores.

Holmes, T. P., Bergstrom, J. C., Huszar, E., Kask, S.,B & Orr, F., III. (2004). Contingent valuation, net marginal benefits, and the scale of riparian ecosystem restoration. *Ecological Economics, 49*, 19–30.

Kilgore, M. A., & Blinn, C. R. (2004). Policy tools to encourage the application of sustainable timber harvesting practices in the United States and Canada. *Forest Policy and Economics, 6*, 111–127.

Kilgore, M. A., Greene, J. L., Jacobson, M. G., Straka,T. J., and Daniels, S. E. (2007). The influence of financial incentive programs in promoting sustainable forestry on the nation's family forests. *Journal of Forestry, 105*, 184–191.

Klemperer, W. D. (2003). Forest resource economics and finance. Blacksburg, VA: W. David Klemperer.

Kurtz, W. (2000). Economics and policy of agroforestry. In H. E. Garrett, W. J. Rietveld, and R. F. Fisher (Eds.), *North American agroforestry: An integrated science and practice.* Madison, Wisconsin: ASA.

Landsburg, S. E. (1993). *The armchair economist: Economics and everyday life.* New York: The Free Press.

Lessley, B. V., Johnson, D. M., & Hanson, J. C. (1991). *Using the partial budget to analyze farm change* (Fact Sheet 547). College Park, MD: University of Maryland Cooperative Extension.

Ma, Z., Butler, B. J., Kittredge, D. B., & Catanzaro, P. (2012). Factors associated with landowner involvement in forest conservation programs in the US: Implications for policy design and outreach. *Land Use Policy, 29*, 53–61.

McAdam, J. H., Burgess, P. J., Graves, A. R., Rigueiro-Rodríguez, A., & Mosquera-Losada, M. R. (2009). Classifications and functions of agroforestry systems in Europe. In A. Rigueiro-Rodróguez, J. McAdam, & M. R. Mosquera-Losada (Eds.), *Agroforestry in Europe* (Advances in Agroforestry 6, pp. 21–41). Dordrecht, the Netherlands, Springer.

Mercer, D. E., Frey, G. E., & Cubbage, F. W. (2014). Economics of agroforestry. In S. Kant & J. R. R. Alavalapati (Eds.), Handbook of forest resource economics (pp. 188–209). New York: Routledge.

Mills, T. J. (1976). Cost effectiveness of the 1974 forestry incentives program (Research Paper RM-175). Fort Collins, CO: U.S. Forest Service, Rocky Mountain Forest and Range Experiment Station.

Noel, J. E., Qenani-Petrela, E., & Mastin, T. (2009). A benefit transfer estimation of agro-ecosystems services. *Western Economics Forum, 8*(1), 18–28.

Obeng, E. A., Aguilar, F. X., & McCann, L. M. (2018). Payments for forest ecosystem services: A look at neglected existence values, the free-rider problem and beneficiaries' willingness to pay. *International Forestry Review, 20*, 206–219.

Oklahoma Cooperative Extension Service. (2017). Economic impact of conservation dollars in Oklahoma (CR-1016). Retrieved from http://factsheets.okstate.edu/documents/cr-1016-economic-impact-of-conservation-dollars-in-oklahoma.

Orefice, J., Smith, R. G., Carroll, J., Asbjornsen, H., & Howard, T. (2019). Forage productivity and profitability in newly-established open pasture, silvopasture, and thinned forest production systems. *Agroforestry Systems, 93*, 51–65. https://doi.org/10.1007/s10457-016-0052-7

Peterson, P. P., & Fabozzi, F. J. (2002). *Capital budgeting: Theory and practice.* New York: John Wiley & Sons.

Price, C. (1993). *Time, discounting and value.* Cambridge, MA: Basil Blackwell.

Qiu, Z., Prato, T., & Boehrn, G. (2006). Economic valuation of riparian buffer and open space in a suburban watershed. *Journal of the American Water Resources Association, 42*, 1583–1596.

Royer, J. P., & Moulton, R. J. (1987). Reforestation incentives: Tax incentives and cost sharing in the South. *Journal of Forestry, 85*, 45–47.

Sagoff, M. (2004). *Price, principle and the environment.* New York, Cambridge Univ Press.

Shrestha, R. K., & Alavalapati, J. R. R. (2004a). Valuing environmental benefits of silvopasture practice: A case study of the Lake Okeechobee watershed in Florida. *Ecological Economics, 49*, 349–359.

Shrestha, R. K., & Alavalapati, J. R. R. (2004b). Estimating ranchers' cost of agroforestry adoption: A contingent valuation approach. In J. R. R Alavalapati and E. Mercer (Eds.), *Valuing agroforestry systems: Methods and applications* (pp. 183–199). Dordrecht, the Netherlands: Kluwer Academic Publishers.

Shrestha, R. K., & Alavalapati, J. R. R. (2004c). Effect of ranchland attributes on recreational hunting in Florida: A hedonic price analysis. *Journal of Agricultural and Applied Economics, 36*, 763–772.

Smathers, R. L. (1992). Understanding budgets and the budgeting process (Publ. CIS 945). Boise, ID: University of Idaho Cooperative Extension System.

Stainback, G. A., & Alavalapati, J. R. R. (2004). Restoring longleaf pine through silvopasture practices: An economic analysis. *Forest Policy and Economics, 6*, 371–378.

Stamps, W. T., McGraw, R. L., Godsey, L., & Woods, T. L. (2009). The ecology and economics of insect pest management in nut tree alley cropping systems in the midwestern United States. *Agriculture, Ecosystems & Environment, 131*,:4–8.

Susaeta, A., Lal, P., Alavalapati, J., Mercer, E., & Carter, D. (2012). Economics of intercropping loblolly pine and switchgrass for bioenergy markets in the southeastern United States. *Agroforestry Systems, 86*, 287–298.

Tiamiyu, S. A., Adagba, M. A., & Shaahu, A. (2014). Profitability analysis of shea nuts supply chain in selected states in Nigeria. *Journal of Agricultural and Crop Research, 2*, 222–227.

Tyndall, J. C., & Grala, R. K. (2009). Financial feasibility of using shelterbelts for swine odor mitigation. *Agroforestry Systems, 76*, 237–250.

USDA Farm Service Agency. (2016). *USDA announces Conservation Reserve Program results.* Retrieved from https://www.usda.gov/media/press-releases/2016/05/05/usda-announces-conservation-reserve-program-results.

USDA Natural Resources Conservation Service. 2017. Economic impact of conservation practices in Minnesota. Retrieved from https://www.nrcs.usda.gov/Internet/FSE_DOCUMENTS/nrcseprd1329946.pdf.

Wanjiku, J., & Kimenye, L. N. (2007). Profitability of kale and tomato production under different soil fertility replenishment technologies in western Kenya. *Journal of Sustainable Agriculture, 29*(3), 135–148.

Zinkhan, F. C., & Mercer, E.D. (1997). An assessment of agroforestry systems in the southern U.S.A. *Agroforestry Systems, 35*, 303–321.

Appendix 16-1—Inflation

Inflation exists in almost every economy and has a significant effect on the economic viability of investments (Price, 1993), including agroforestry alternatives. If inflation is not accounted for in financial analyses when evaluating agroforestry investments, this would likely lead to an incorrect decision about accepting or rejecting a particular investment opportunity. For example, it might result in acceptance of an alternative that seems financially viable in nominal terms; however, when inflation is accounted for, it might generate a smaller monetary return or even a loss.

Inflation is measured using a specially constructed index that expresses prices in relation to a base year for which the value of 100 is assigned (Gunter & Haney, 1984; Klemperer, 2003). There are two types of indices: the Consumer Price Index (CPI) indicates an increase in prices of a basket of goods and services consumed by consumers, whereas the Producer Price Index (PPI) measures an increase in wholesale or producer input costs. Both indices can be used to calculate an inflation rate, depending on the analysis goal. In both cases, you need to know the start and end value of the index to calculate inflation for a specific period. For example, to calculate inflation for 2010 through 2016 using the CPI, you would need to obtain the CPI values for years 2010 and 2016. They can be obtained from numerous sources available online, such as the Bureau of Labor Statistics, U.S. Department of Labor. Assuming that the CPI is 172.2 and 201.6 for 2010 and 2016, respectively, the annual inflation rate can be calculated as

$$f = \sqrt[n]{\frac{\text{CPI}_n}{\text{CPI}_0}} - 1$$

where f is the annual inflation rate, CPI_0 and $\text{CPI}n$ are the CPI values for the start and end years, respectively, and n is the number of years in the period for which inflation is calculated. By inserting the CPI values and number of years into the formula, we can calculate that the annual inflation rate during 2010–2016 averaged 2.66%:

$$f = \sqrt[n]{\frac{\text{CPI}_{2016}}{\text{CPI}_{2010}}} - 1 = \sqrt[6]{\frac{201.6}{172.2}} - 1 = 0.0266$$

The calculated inflation rate indicates only an average annual price increase for goods and services included in the market basket used to measure inflation. For some goods and services, the rate of price increases will be high, while other goods and services might have a decrease in prices (Gunter & Haney, 1984). Similarly, inflation also will differ each year depending on economic conditions.

It is important to understand the concepts of current and real dollar values when inflation is discussed. Current dollar values include values of prices and costs reported as they occur in a given year, and hence they include inflation (Klemperer, 2003). For example, prices that you pay in the store are current values. Likewise, your salary is also a current value. On the other hand, a real dollar value is a value that doesn't include inflation (Klemperer, 2003). The process of removing the effect of inflation from current value is called deflating and works exactly as when you are calculating present value except that current value is discounted at the inflation rate instead of discount rate (Klemperer, 2003).

A simple example using an agricultural land purchase and resale will help illustrate this process. The property was purchased in 2006 for $200,000 and sold in 2016 for $300,000. Suppose the inflation rate during that period averaged 2.53%. When we compare the nominal values of the purchase and sale, it seems that there is a $100,000 profit. However, we need to account for the inflation that consumed part of that profit. To compare these two values, we need to express them in terms of the same base year. Consequently, the 2016 sale value of $300,000 needs to be deflated to 2006 and will be comparable with the purchase price of $200,000. The formula for calculating deflated value (Vn) is (Klemperer, 2003)

$$V_n = \frac{I_n}{(1+f)^n}$$

where In is the inflated (current) value (in this case, $300,000), f is the annual inflation rate, and n is the number of years in the period being analyzed.

By inserting information referring to the agricultural land sale into the formula, we obtain the following result:

$$V_n = \frac{\$300{,}000}{(1+0.0253)^{10}} = \$233{,}675$$

The nominal sale price deflated to 2006 is $233,675 and can now be compared with the purchase price of $200,000. The real gain on this investment is $33,675. We can see that of the $100,000 that constituted a nominal gain, $66,325 was due to inflation. This means that even though the sales price increased by $100,000, the purchasing power of the sales price increased by only $33,675. Of note, although inflation eroded profit, we need to realize that one benefit of this investment was that it more than kept pace with the inflation rate.

Study Questions

1. A farmer considers planting a shelterbelt around the farmstead to decrease heating and cooling costs. A projected total cost of planting the shelterbelt is $10,000, whereas the expected average savings are estimated to be $1,800 yr^{-1} during a 30-yr shelterbelt lifespan. Is this investment financially justifiable if the farmer's minimum acceptable rate of return is 10%?

2. Historical data show that an average return on an alley cropping system was 12% during a 10-yr period, whereas inflation during that time averaged 3%. What was the real rate of return on the alley cropping system? Should a farmer engage in this investment if it requires a real minimum acceptable rate of return of 10%?

3. A landowner considers introducing mushroom production into their forest farming operations and would like to know what rate of return can be expected on this investment. The landowner was able to collect accurate information on costs associated with mushroom production but was unsure about the revenues because it was expected that the price of mushrooms will vary significantly. How can the landowner decide if mushroom production is a worthwhile investment?

4. A financial analysis of a proposed agroforestry alternative generated a BCR of 0.7. Should this alternative, given its BCR, be accepted?

5. A landowner considers purchasing a land parcel to install a silvopastoral system. The LEV calculated for this investment based on projected cash flows and a MARR of 10% is $2,350 ha^{-1}. What is the maximum dollar amount the landowner can pay for this land and earn the required 10% rate of return on this investment? How would the rate of return be affected if the landowner paid $2,800 ha^{-1}? How would the rate of return be affected if the landowner paid less than $2,350 ha^{-1}?

6. A report indicated that the price of mushrooms increased 4% above an annual inflation rate during the last 5 yr. Calculate the inflated price at the end of this period knowing that the starting price was $8 kg^{-1} and inflation was 3% yr^{-1}.

7. A financial analysis shows that an alley cropping system generates an NPV of $750 ha^{-1} at a 12% MARR. Explain the meaning of the NPV criterion for this investment.

8. Explain why landowners whose MARR is 9% shouldn't accept an agroforestry alternative with an IRR of 6%.

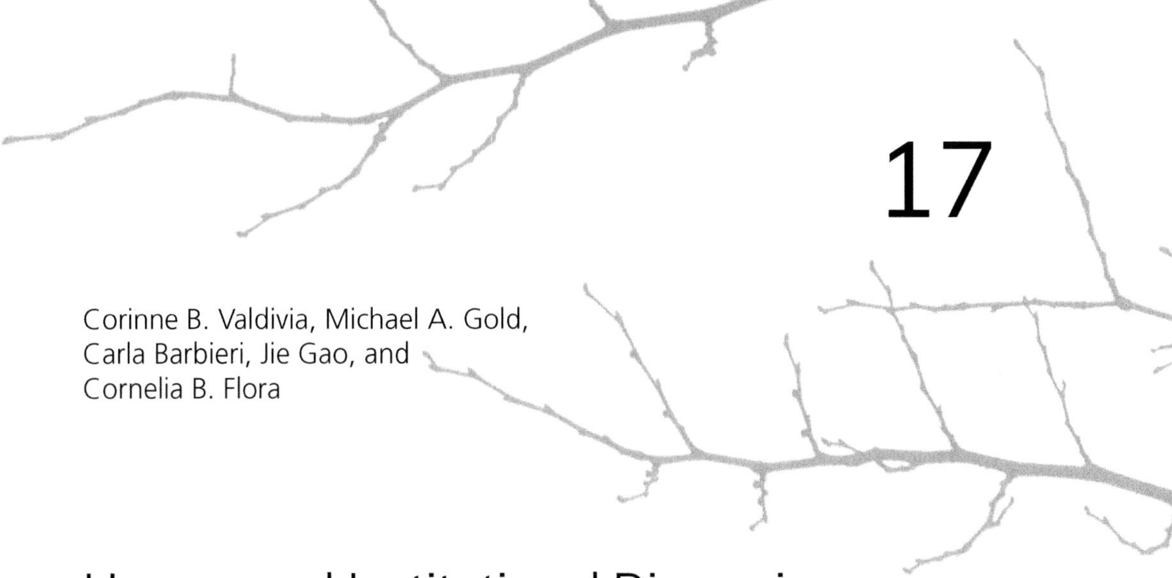

17

Corinne B. Valdivia, Michael A. Gold,
Carla Barbieri, Jie Gao, and
Cornelia B. Flora

Human and Institutional Dimensions of Agroforestry

In Pinantan Lake, BC, Canada, Jennifer and Chris Cunningham manage Jay Springs Lamb Co., an evolving story of a four-family, three-generation farm business that mixes agriculture, forestry, agroforestry, and direct marketing (Valdivia, Gold, Zabek, Arbuckle, & Flora, 2009). They produce free-range lambs from a silvopasture operation in the Southern Interior of British Columbia. Their journey began with Jennifer's parents, who purchased Jay Springs Ranch in 1964, evolving as livestock management progressed from a combination of pasture and relatively passive forest grazing systems to a mix of extensive and intensive silvopasture practices. Jennifer's parent, Geoff Boden, was one of the founding members of a local naturalist's group and an avid birder. Ecological and social values were an intrinsic part of their family life; thus, the concept of agroforestry found fertile ground with Jennifer and Chris as they became more involved with the operation. The original settler ranched with sheep because these were more economic and easier to handle for Jennifer's mom, Shirley, and the children could help. Jennifer and her two siblings, Jocelyn and Dean, grew up herding sheep on horseback in a very scenic part of the province, which ultimately contributed to their decisions to return to the farm and facilitated their current-day intimate knowledge of every corner of the ranch. In 1985, a woodlot license was awarded to Geoff, adding a forestry dimension to their agriculture business. The ranch consists of a mixture of private and Crown lands that require different management according to the land tenure. Jennifer and Chris, along with various members of the family at times, supplemented ranch income via off-farm jobs. A legal partnership created in 1993 formally brought Jennifer, Jocelyn, and Dean (and their families) into the management and operation of the ranch and woodlot. As happens frequently in many areas across North America, the children left the farm to pursue other careers; however, all three came back, driven by family, natural resources, aesthetics, and a tie to the land that would be hard to match anywhere else. Their education, experiences, and world perspectives contributed to shaping the family business in new ways while keeping the feel of "the way it was." This story was at the center of the human dimensions chapter of the 2nd edition of this book, developed by L. Zabeck (Valdivia et al., 2009).

Another story of a family farm and agroforestry is that of Early Boots Farm in Minnesota (Robinson, 2013a). Beginning farmer Tyler Carlson and wife Kate Droske started in 2011 on land purchased by family in the 1960s near Sauk Centre, MN. They are using silvopasture to raise gourmet, grass-fed and finished beef, as it was a relatively low-cost way to get started. The goals of the farm include: producing food in ways that are sustainable and regenerate

North American Agroforestry, Third Edition. Edited by Harold E. "Gene" Garrett, Shibu Jose, and Michael A. Gold.

the land; managing risks by placing diversity back on the landscape; and using perennial cropping systems to mitigate drought, heat, windstorms, and other extreme events. Carlson is using an ultra high stocking density grazing system (aka "mob" grazing), where cattle are moved en masse multiple times a day, while working toward organic certification. He also planted thousands of red oak (*Quercus rubra* L.) and red and white pine (*Pinus resinosa* Aiton and *P. strobus* L., respectively) saplings in 8 ha of old crop fields. The trees will shade pasture to improve feed value and shelter a growing herd of Australian Lowline Angus beef cattle. He received cost-share assistance from USDA's Environmental Quality Incentives Program (EQIP) to establish fencing for rotational grazing and help pay for trees; planting assistance information and mentorship came from the University of Minnesota Extension and the Land Stewardship Project. He is also involved in research on the benefits of agroforestry for producing food, managing risks, and increasing landscape resiliency. The latter is pursued by integrating trees, fields, and water with an eye to connecting habitat corridors for pollinators, other insects, and wildlife. Carlson is planning his landscape design to help reconnect a fragmented agricultural landscape, intentionally planting trees to link up areas of cover. His planning considerations also include managing the edge effects to create viable refugia, especially for birds.

Creating a new elderberry [*Sambucus nigra* L. ssp. *canadensis* (L.) R. Bolli] enterprise on a historic farm provides another example of how agroforestry fits the goals of a farm family seeking to identify new sources of income (Robinson, 2013b). East Grove Farms near Salem, IA, specializes in elderberry and other heritage crops and has further diversified by establishing a winery in 2014 and further specialized by focusing exclusively on mead production in 2017. This case highlights management practices, lessons learned, and the marketing and economic concerns of the operation.

East Grove Farms is a family business managed by Kurt Garretson on Iowa's oldest continuously settled farm, established in 1837 by his ancestors. The goals are to create a viable and enduring business on historic family land and build community with family, neighbors, and other elderberry growers. Garretson started farming in 2009, working with other family members to create a farm business that initially specialized in elderberry products and other heritage crops. The family started out small to learn what works and to properly prepare the soil for planting. In 2010, Kurt started with about 1,200 elderberry plants on about half a hectare, had 2 ha in production in 2013, and added more hectarage in subsequent years. Elderberry was also planted along property boundaries and in two new shelter belts, along with hazelnut (*Corylus americana* Walter) and wild plum (*Prunus americana* Marshall). The family worked with River Hills Harvest, a Missouri-based elderberry marketing and processing cooperative that also provides training and networking. They also worked on a license for wine making and organic certification. They worked off the farm when needed, providing farm-related services to others, cared for land and soil, and valued family and community. The Garretsons want to show people that farming is not all corn and soybean. Kurt chose elderberry as a native plant that grows well in his area, even in places where row crops won't grow, liked that it's a perennial that doesn't require the ground to be plowed up every year, and is also a product with growing demand in the markets due to its health benefits (high in antioxidants, vitamins, and minerals). East Grove Farms also includes cultivars of white peach, native persimmon (*Diospyros virginiana* L.), and 'Green Gage' plum (*Prunus domestica* L.). Elderberry is in growing demand due to research that has shown the berries to have anti-viral, immunity-boosting properties. The family mead business sells in numerous market outlets across Iowa.

Initially, East Grove Farms' products included elderberry juice, cordial, and jelly and elderberry and elderflower wines. By 2017, the focus shifted exclusively to selling mead from a variety of different fruit sources. As mentioned, their marketing outlets are diverse and include the Iowa grocery chain, Hy-Vee, and River Hills Harvest Co-op, as well as the winery, with a focus on farm-related ecotourism. They restored the 114-yr-old Victorian house on the land that serves as the farm's headquarters and the winery.

Connections to government programs include the USDA Natural Resources Conservation Service (NRCS), EQIP, and the Conservation Reserve Program (CRP). Garretson says, "the philosophy is the more elderberries, the better. Elderberries are such a new thing that everybody's learning. The more growers we have nearby, the more we can help each other, share equipment costs, and save by shipping together" (Robinson, 2013b).

The life stories of these family farms show how diverse the decision makers are in North America and share the common thread of how these landowners have been able to make the shift to agroforestry. They exhibit the four As: (a) an Asset-based approach; (b) Adaptability by constant adjustments; (c) Awareness of alternative value webs and rules, regulations, and legislation; and (d) Associations—local, regional, national, and international (Cooperrider, Sorenson, Whitney, & Yager, 2000; Pirog & Bregendahl, 2012), and the process is dynamic, with the intermingling of the household and family and the business, taking into account family goals and relationships and how these shape the nature of the business (Glover, 2010).

These stories show family farms that look to diversify activities and increase income, expecting to be part of their community for many years; landowners who are not farming and may have never farmed, but are seeking to improve the landscape, address flooding concerns, and enjoy wildlife and the many non-economic services the land provides; those who are entering farming full time but focusing on alternative products for high-end markets, including tourism; and those seeking alternative markets and relying on family, networks, and new markets. These are recent experiences of agroforestry practices in North America, while there have been other experiences of indigenous communities that managed diversified landscapes and livelihoods, such as Native Americans in the United States and Canada, for millennia (see Chapter 1; Rossier & Lake, 2014). Indigenous communities in California managed landscapes for centuries as diversified, complex, and integrated systems of animals, plants, trees, and fungi (see Chapter 2; MacFarland et al., 2017).

How do we understand the decision maker(s) and what motivates them? How does agroforestry fit into their livelihood strategies? Are decisions made in a vacuum, or are they shaped by what we know and value and by what we possess? In which ways do markets and the knowledge of institutions, policies, and social relations influence decisions? This chapter addresses the human and institutional dimensions of agroforestry, framed by sustainable livelihoods, the assets that decision makers possess or can access, and the values and relationships that influence the decisions to adopt agroforestry practices that fit with the goals of farmers and family businesses. It also addresses how the field of social relations, institutions, and organizations emerges in support of the practice of agroforestry.

Agroforestry

With the changing structure of agriculture, concerns for the environment, and the need for new income opportunities, farmers in North America must continually search for new enterprises that contribute to their livelihood goals and are consistent with their resources, assets, and values (Valdivia et al., 2009). Linking farmers and consumers in the local community, diversifying production away from commodity crops, engaging in organic farming, developing new niche products, and connecting with consumers seeking to enjoy the landscape are among the many alternatives available. Agroforestry provides commercial opportunities, conservation benefits, and aesthetic and environmental services in tandem with the concept of multifunctionality (Barbieri & Valdivia, 2010a, 2010b; Dobbs & Pretty, 2004; Godsey, Mercer, Grala, Grado, & Alavalapati, 2009; Holderieath, Valdivia, Godsey, & Barbieri, 2012; Lovell et al., 2017; Plieninger, Muñoz-Rojas, Buck, & Scherr, 2020). It also provides landscape and open space amenities linked to land use practices (Abler, 2004; Gao, Barbieri, & Valdivia, 2014b), as well as benefits to the environment, such as clean water and air, reduced erosion, and wildlife benefits. Agroforestry, consistent with these multiple functions and benefits, "seeks to help bridge the gap between production agriculture and natural resource management" (see Chapter 2). However, on-the-ground adoption of agroforestry in the United States remains limited according to available statistics from the USDA National Agricultural Statistics Service, which show that only 1.5% of farmers engage in agroforestry (Romanova, 2020).

In North America, agroforestry is defined as "... land use management that optimizes the benefits (physical, biological, ecological, economic, social) from biophysical interactions created when trees and/or shrubs are deliberately combined with crops and/or livestock" (see Chapter 2). While commonly described as intentional, integrated, interactive, and intensive (see Chapter 2), the British Columbia and U.S. family farms showcased in this chapter provide scope for both extensive and intensive examples, expressions of the diversity of contexts shaped by landscapes, markets, and policies, and the goals of the family farm.

Understanding if an agroforestry practice will interest rural producers requires knowledge of the context in which farmers, operators, and landowners make decisions regarding the activities they pursue, as well as knowledge of the characteristics, benefits, and complexity of practices in order to determine how practices align with the landowner's (or land manager's) objectives. *Compatible management* (Haynes, Monserud, & Johnson, 2003; Titus et al., 2004) reflects a continuum of activities and intensities, from wild harvest through extensive and/or intensive forest farming, to riparian buffers (see Chapter 9) and silvopasture (see Chapter 6). The more active phases of compatible management are consistent with extensive or intensive agroforestry practices, as both the timber and non-timber resources are explicitly co-managed (see Chapter 9). "Agroforestry is a way for landowners to manage scarce natural resources that balances environmental stewardship, financial feasibility, and social responsibility" (see Chapter 16).

Agroforestry can be adopted for multiple purposes that include production, protection, diversification to reduce risk or increase income, provision of ecosystem goods and services (improvement of air and water quality, for example), recreational uses such as hunting, bird watching, increased wildlife habitat, and enjoyment by tourists (Barbieri & Valdivia, 2010b). Ecosystem services accrue not only to the landowner but to society as a whole; but if no markets exist to value these services, tangible economic benefits can't be accrued (Holderieath et al., 2012; Udawatta, Gantzer, & Jose, 2017).

Why are Agroforestry Practices a Potential Benefit?

Benefits from agroforestry can be realized at multiple levels. At the field, farm, watershed, regional, national, and global levels, agroforestry practices can help address public concerns about the impacts of nonpoint-source pollution in the waterways. This is why governments concerned with these problems develop programs to protect resources. In the United States, the USDA CRP and EQIP have been implemented to help protect against floods that threaten to erode prime cropland. Small family farms received 75% of CRP payments in 2018 (Whitt, MacDonald, & Todd, 2019).

Agroforestry has the potential to be a source of environmental and economic benefits in the United States, where more than 90% of farms are small family farms (USDA National Agricultural Statistics Service, 2020). The other 10% are farmers engaged in traditional monoculture cropping, such as corn (*Zea mays* L.) and soybean [*Glycine max* (L.) Merr.], requiring roughly 600 ha (1,500 acres) to be economically viable and even larger hectarages to become more profitable (Westhoff, Meyer, & Zimmel, personal communication, 2020), often renting a large proportion of the land operated. Small family farms, on the other hand, are very diverse and include family farms with off-farm occupation, as well as low-sales farms, and "second career" retirement farms. To remain viable, these farms must find ways to incorporate new income-generating activities into their farming activities. Agroforestry provides such an opportunity.

At the farm level, agroforestry practices can enhance crop yields (e.g., windbreaks), reduce costs of production (e.g., silvopasture), or increase profits by reaching growing niche markets, such as log-grown shiitake mushroom [*Lentinus edodes* (Berk.) Singer] (e.g., forest farming) or chestnuts (*Castanea mollissima* Blume) (e.g., alley cropping), with potentially lucrative profits per hectare (see Chapter 15). Many professionals see agroforestry not only as an approach for sustainable land management but also as a way to increase or stabilize farm income through diversification (Garrett et al., 1994; Gold & Hanover, 1987; MacFarland et al., 2017; Prokopy et al., 2020; Valdivia, Barbieri, & Gold, 2012; Valdivia & Konduru, 2003). While extension professionals may consider agroforestry practices to be sustainable land use strategies that are alternative or complementary to traditional cropping systems, it is only during the past 20 yr that attention has been paid to how landowners and operators view agroforestry (Keeley et al., 2019; Matthews, Pease, Gordon, & Williams, 1993; Raedeke, Green, Hodge, & Valdivia, 2003), and more recently, how potential consumers view it as recreation (Valdivia & Barbieri, 2014; Gao et al., 2014b).

The Decision Makers

In the two decades since the first chapter on social dimensions was written for the first edition of this book (Rule, Flora, & Hodge, 2000), many studies on the interest and adoption of agroforestry have been conducted (Jose, Gold, & Garrett, 2018; Rhodes, Aguilar, Jose, & Gold, 2018; Romanova, 2020; Trozzo, Munsell, & Chamberlain, 2014a; Trozzo, Munsell, Chamberlain, & Aust, 2014b). Several frameworks have been used to study landowners' and farm operators' attitudes about the concept of agroforestry and regarding specific agroforestry practices. Common to all are the social, cultural, environmental, organizational, and demographic factors influencing decisions (Romanova, 2020; Rule et al., 2000; Valdivia et al., 2009, 2012; Valdivia & Barbieri, 2014). Here we first discuss a *sustainable livelihoods framework* that identifies the assets and capital that decision makers manage and invest to develop strategies in pursuit of their multiple goals. *Capital* is assets invested long term to create other assets. Short-term use of assets to create new assets is speculation (Flora, Flora, & Gasteyer, 2016). Through the stories of family farmers, we learn about the factors influencing decisions in diverse settings, from privatized to leased and licensed lands, introducing the context in which decisions are made (structures and institutions), adaptive capabilities of family businesses and their symbolic capital, and the social relations contained in the practice of agroforestry. For definitions and clarification of the terms used to describe the sustainable livelihoods framework see Table 17–1.

Sustainable Livelihoods

A seminal study by Chambers and Conway (1992) developed the concept of sustainable livelihoods as one of the "practical concepts for the 21st Century" to assess how individuals choose strategies to achieve well-being that include the environment as a resource in production and as an outcome through the services it provides. The decision maker—an individual, household, or group—, the assets or capital accessed or controlled, and the context within which decisions are made (e.g., markets, policies, and social organizations) are elements of this framework (Figure 17–1) (Valdivia et al., 2009). A livelihood encompasses people and their capabilities, means of making a living, income, and tangible and intangible assets. The tangible assets include resources and stores of wealth, and intangible assets consist of the relationships or claims to access to resources. A livelihood is sustainable when it is able to maintain or enhance the assets on which it depends, and *sustainability* is defined in terms of "the ability to maintain and improve livelihoods while maintaining or enhancing the local and global assets and capabilities on which livelihoods depend" (Chambers & Conway, 1992), providing opportunities for the next generation.

Table 17–1. A primer on concepts in human dimensions terminology.

Term	Definition	References
Agency/human agency	the ability to act or negotiate with markets, organizations, institutions, and policies—the "hinge" between a livelihood and the structures	de Haan (2000), Bebbington (1999), Valdivia & Gilles (2001)
Capability	the ability of an individual, household, or group to use capital (tangible and intangible assets) in ways that produce a livelihood; the result of agency, negotiation with markets, organizations, institutions, and policies	Chambers & Conway (1992), Bebbington (1999)
Capital	tangible and intangible resources, owned or controlled, that can be accessed and used to create a stream of benefits	Bebbington (1999)
Diversification	*Farm*: the reallocation and recombination of farm resources (i.e., land, labor, or capital) into new unconventional crops, animals, practices (e.g., agroforestry) or into non-agricultural enterprises developed on the farm *Household*: along with on-farm diversification, it also considers off-farm income-earning activities	Barbieri et al. (2008), Valdivia et al. (1996), Ellis (1998)
Field	the dominant social relations in which the individual engages in pursuing a livelihood; the network of objective relations consists of social relations representing the external dimensions or the relations that make the "game" possible; describes the rules of the game	Raedeke et al. (2003), Bourdieu (1990)
Habitus	the set of values upheld in a field, the norms that influence human behavior, the "taken for granted" social processes, values, beliefs, and behaviors shared	Raedeke et al. (2003)
Household Portfolio	*Of activities and enterprises*: includes the set of productive or income-generation activities *Of assets*: consists of the various forms of capital and assets that the household commands	Valdivia (2004), Chen & Dunn (1996)
Institution	"regularized practices" or patterns of behavior in society, accepted norms (both formal and informal) that are part of the process of social negotiation, embedding power relations; the rules of the game; embeds social relations and power structures at any given time	Scoones (1998)
Livelihood	encompasses income-generating activities that the household members pursue, along with the mechanisms or means to accessing resources (assets and capital) through the life cycle (stages in the life of the family)	Ellis (1998), Valdivia (2004)
Multifunctionality	Recognizes that agriculture provides different services to society in addition to the production of food and fiber; a continuum ranging from weak to strong productivist action and thought; the strong end of the continuum promotes environmental sustainability, enhanced food quality and shortens distribution channels, fostering local or regional embeddedness, in spite of its reduced productivity	Barbieri & Valdivia (2010a); Marsden & Sonnino (2008); Ploeg et al. (2000); Wilson (2008)
Structure	includes three dimensions: (a) social, which consists of the rules that govern common norms, (b) economic, which is defined by the "rules of the game" that govern markets, and (c) political, which expresses existing power relations; often determines the outcome of the livelihood strategies pursued, although agency can change this	de Haan (2000), Scoones (1998)

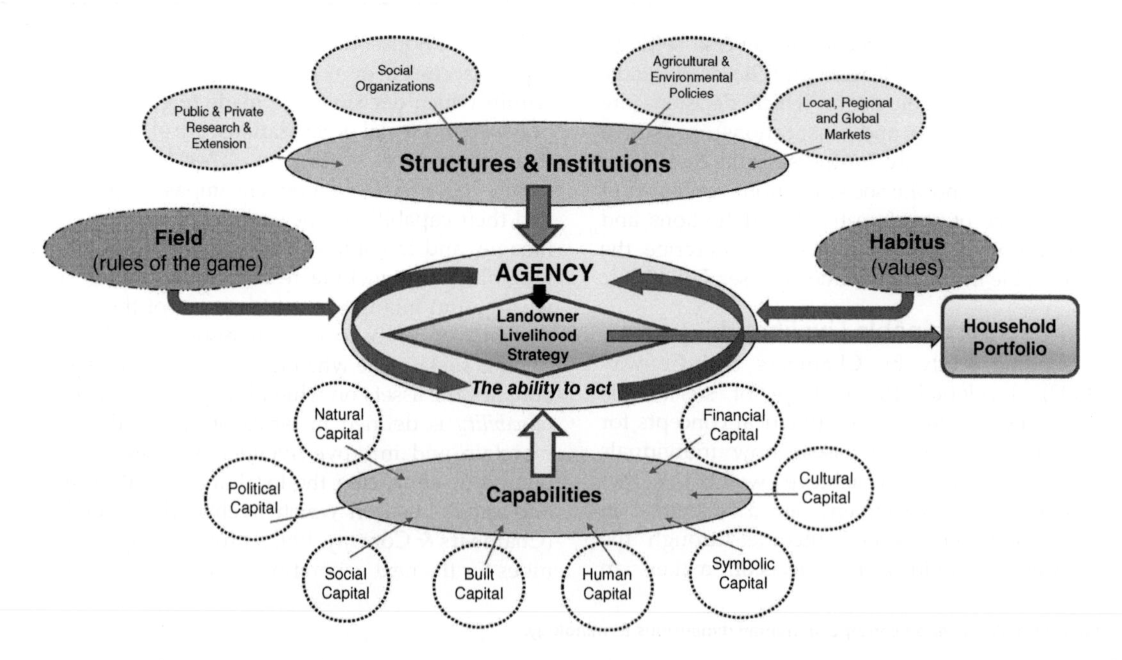

Fig.17–1. Landowner capital, livelihood strategies, and the practice of agroforestry: Field, habitus, and agency.

The decisions people make do not happen in a vacuum. People use their capital and capabilities to implement their livelihood strategies (Bebbington, 1999; Valdivia, 2004) and negotiate within networks of social relationships as well as economic and political structures in a given "field" (de Haan, 2000), shaping the outcomes of the landowner strategy (Valdivia et al., 2009, 2012). The landowner's capital and capabilities in the context of societal structures result in particular livelihood strategies, such as the cases of Jay Spring Lamb, Early Boots Farm (Box 17–1), and East Grove Farms (Box 17–2). Each family farm has unique characteristics and histories that have led to the development of diversified family businesses in which agroforestry plays a unique role, in line with their values and objectives.

Capital and capabilities (Figure 17–1, bottom) include human (the skills and abilities of people), social (social relations and networks involving trust and reciprocity), cultural (shared values and norms), symbolic, which "contains the prestige and renown attached to a family and a name" (Bourdieu, 1990), natural (soil, plant, and environmental diversity, resource stocks and services), financial capital (economic resources that can be invested) for investment in successful agroforestry practices, and political capital (Flora et al., 2016; Glover, 2010; Rule et al., 2000; Valdivia et al., 2009). The latter captures the influence that can be mobilized by producers to set the norms and values that lead to setting standards and rules and their enforcement in both the public and private sectors. For example, the rules regarding the leasing structure of Crown land at Jay Springs Lamb (Valdivia et al., 2009) are a manifestation of public-sector political capital. The green label is an example of private-sector political capital in action. Landowners draw on these types of capital and their capabilities when they make decisions regarding investments and how they will develop their family farm business, including those who pursue agroforestry. Thus, *capital* refers to the resources potentially available to the decision maker.

Decisions are also influenced, and in many cases constrained, by structures and institutions (Figure 17–1, top). The public and private research and extension agencies, social organizations, markets (local and global) for products and inputs, and the agricultural and environmental policies established by governments constitute the *field*. The field of farming, especially commodity farming in the United States, is composed of mainstream organizations, policies, and markets that share similar values (Raedeke et al., 2003). Similarly, the field for forestry is also well defined, while agroforestry has been emerging (Valdivia et al., 2012). The field and the values define the practice and shape landowners' decisions and, in turn, their livelihood strategies. In the case of Early Boots Farm (Box 17–1), young farmer Tyler Carson was able to connect with The Land Stewardship Project where he participated in the Farm Beginnings training course, was

Box 17–1. Silvopasture at Early Boots Farm

Early Boots Farm (https://www.earlyboots.com) is managed by Tyler Carlson in Minneapolis, with support from his wife Kate Droske. Tyler adopted silvopasture management—an agroforestry practice of incorporating trees into pasture settlings—on the farm and raises gourmet grass-fed and furnished beef. This young farmer, who began farming in 2011, benefited from a Farm Beginnings training course sponsored by the Land Stewardship Project, which resulted in the development of a business plan. He has also gained valuable information and mentorship from the University of Minnesota Extension that aided the design of his fairly complex farming system while also participating in their research on the benefits of agroforestry. In addition, Early Boots Farm has been a recipient of funding assistance from the Environmental Quality Incentives Program (EQIP) to undertake planting of trees and fencing.

Early Boots Farm has set out to accomplish the dual goals of producing food in a sustainable manner and regenerating land. Tyler Carlson has adopted a rotational grazing system known as ultra high stocking density or "mob grazing," where he has high concentrations of cattle per hectare per day and targets about one to two moves per day. In that way an opportunity is provided for the restoration of grass health. In addition, he has planted thousands of red oak and white pine saplings on his field. However, prior to the trees growing big enough to hold off rubbing from the livestock, electric fences are being used on the farm to keep the cattle away. After maturing, the trees will provide shade for the livestock in the summer, as well as a windbreak and shelter in the winter. Furthermore, the shades will help the pastures stay cooler and extend the seasons for their prime growth, thus providing higher feed value for his livestock. In general, this perennial cropping strategy helps in mitigating the impacts of various environmental conditions.

Early Boots Farm is working toward receiving organic certification. Initially, their sales went to family and friends, but they have expanded their marketing channels to include local area farmers' markets and online sales. Early Boots Farm travels to the Twin Cities for farmers markets two to three times a month. In season, Early Boots Farm also sells honeyberry (*Lonicera caerulea* L.) and sour cherry (*Prunus cerasus* L.) fruit at farmers markets. Down the line, Early Boots Farm plans to trade some of its thinnings to producers of paper pulp and fence posts and eventually sell mature trees as timber.

The Early Boots herd of lowline angus has been born on pasture, rotationally grazed on a diverse sward of forage plants, and fed free-choice organic minerals and grass and legume hay in the winter months. The lambs come to their farm each spring to graze and grow until autumn. Animal welfare is very important to Early Boots Farm, and they believe in letting cows be cows and sheep be sheep, with their instincts and autonomy intact. They practice low-stress handling techniques, and their animals are never given subclinical antibiotics or kept in confinement conditions.

Lessons learned: During his farming years, Tyler has realized the importance of having larger trees because these have a better ability to withstand rubbing and browsing. He also notes the benefits of preparing pasture and soils prior to planting trees as well as providing sufficient spacing between trees for the smooth operation of farm machinery. Additionally, he sees the need for a landscape design that connects habitats of flora and fauna such as insects, birds, and wildlife—hence his strategic integration of trees, field, and water.

Source: Robinson (2013a).

mentored by the University of Minnesota Extension in the design of the farming system, and received funding assistance from the EQIP (Robinson, 2013a). In a similar way, the Garretsons' East Grove family farm (Box 17–2) benefited from USDA–NRCS programs such as EQIP and CRP (Robinson, 2013b).

Capabilities refers to the ability of an individual, household or group to act, using capital in ways that produce a livelihood. Human agency captures this ability to act, to negotiate in markets and with organizations, institutions, and existing policies. The latter are components of the field (Figure 17–1, top). This negotiation or ability to act, described as the "hinge" by de Haan (2000), is expressed or captured in the livelihood strategies the decision maker is able to pursue by exercising "agency" (Figure 17–1, center), through the adaptive (Glover, 2010) and dynamic capabilities (Alonso, Kok, & O'Shea, 2018) of the family farm business. Dynamic capabilities include sensing, sizing, and transforming (Alonso et al., 2018). The *values* of the decision maker (cultural and symbolic capital) and their relationships with other

Box 17–2. Elderberry business at East Grove Farms

At East Grove Farms, Iowa (http://eastgrovemead.com/), the Garretson family specializes in elderberry along with other fruits and serves as a grower for a processing and marketing cooperative, River Hills Harvest. The family farm was established in 1837, and Kurt Garretson started managing the farm in 2010. Kurt has benefited from participation in the EQIP and the CRP. He has also been able to secure funding from the USDA–NRCS.

The growing demand for elderberry due to its associated health benefits as well as its perennial nature gives the plant greater appeal to Kurt. Additionally, the plant grows well in his area. As a result, selected hectarage has been devoted solely to elderberry cultivation, but the plant has also been sown along the property boundary together with hazelnut and wild plum. Moreover, other crops are found on the farm such as cultivars of white peach and native persimmon. The farm has also experimented with Aronia berry [*Aronia melanocarpa* (Michx.) Elliott] and chestnut (*Castanea mollissima* Blume). The Garretsons have been able to identify local types of elderberry that meet their planting needs. The family often restores and maintains old farm machinery but has also invested in newer technologies like a mulch layer, a wheel planter, and a greenhouse.

Besides the profit objective of East Grove Farms, Kurt is passionate about preserving and creating topsoil. Furthermore, the business strives to build community with family and neighbors. In previous years, elderberry marketing was run through the River Hills Harvest cooperative, and elderberry products were sold through the Hy-Vee grocery chain in the state. However, East Grove Farms have evolved and now focus exclusively on mead production and sales.

The Garretsons' plans to establish a winery on their land in order to promote farm-related ecotourism were realized in 2014 when the East Grove Farms' winery opened for business. East Grove Wine specializes in a variety of honey wines (aka mead) made from various fruits including elderberry, persimmon, peach, apple, and grape. East Grove Farms has also expanded to include a tasting room and created a stage to host on-farm folk music and mead festivals.

Lessons learned: For Kurt Garretson, cultivating perennial crops has taught him the virtue of patience and the value of starting small and learning through the process. He also notes the importance of the cooperative model in which he is involved, highlighting that more elderberry growers within a cluster leads to greater potential knowledge spillovers, as well as more opportunities for sharing production and marketing costs. As the Garretsons moved into more diverse fruit production, they were faced with the question of how to maximize their return on a crop not normally associated with the corn and soybean fields of Iowa. They came up with mead in order to be able to take their fruit and turn it into a product that would offer the most return. In addition, there was no local competition for mead production.

Source: Robinson (2013b).

individuals and groups through shared norms of reciprocity (social capital) inform and shape the strategies expressed in the household portfolio of economic activities (Valdivia et al., 2009).

What are the determinants of a livelihood? How does a person, household, or family operation decide what to do for a living? For a farmer born on a family farm, their livelihood may be defined by what their parents and grandparents did (cultural capital), the history of place (symbolic capital), and the ability of those parents to provide the means to follow that occupation through education (human capital) and access to land (financial capital). Perhaps they view farming as a "way of life," their livelihood. For Jennifer Cunningham, it was her parents' farm business (financial capital), her experience growing up on a ranch, and her belief that there was "no place like home" (cultural capital) that influenced her decision to go back to the family farm and build a business for her and her family's livelihood. The choices people make about how they want to make a living depend largely on the types of capital (assets or resources) they possess, their ability to use them and negotiate, and on the economic and social relations and environment in which they live.

Figure 17–1 describes the individual characteristics and context that shape (incentivize and constrain) the strategy a decision maker may choose. The various kinds of capital or assets that the landowner or household possesses or can access are indicated and coupled with the landowner capabilities, defining the livelihood strategies (Figure 17–1, bottom). The livelihood strategies are also shaped by the context, which includes the social, economic, and political structures within which decisions are made (top of Figure 17–1). In the case of British Columbia, for example, varying land tenures impact how the land can be used and managed. From the social side, the Jay Springs Lamb Co. case study (Valdivia et al., 2009) illustrates development of a viable livelihood consistent with family values and the diversity of families in the business. It is built on both social capital to develop the market networks to sustain a label and political capital to enforce the norms and values that the label represents in terms of ecofriendly production of lamb and timber. The producers decided to create business opportunities that would make the ranch and the woodlot work for them, focusing on financial viability and diversity. They targeted a niche market where sheep would not be a commodity but instead a branded niche product with a "face and place" (see Chapter 15; Gold, Godsey, & Josiah, 2004). But they did not accomplish that entirely by themselves. Without the links to other producers and consumers who looked for specific characteristics, they could not have implemented their livelihood strategy.

Similarly, East Grove Farms (Box 17–2) pursued an elderberry business in a historic family farm (symbolic capital), focusing on elderberries. Two key benefits, market demand for associated health benefits and the perennial nature of the plant, appealed to Kurt Garretson (Robinson, 2013b). Symbolic capital, sensing and seizing opportunities, and the relationships with the cooperative and community are examples of capital, agency, and institutions.

Decisions are made within a structure that includes a *social aspect*, the rules that govern common norms (social capital and cultural capital); an *economic aspect*, the rules of the game in markets, of supply and demand; and a *political aspect*, the power relations in government and markets, expressed in public policies and prices.

Four elements of structures and institutions impact landowner strategies (Figure 17–1) that incorporate an agroforestry practice as an activity. The elements are: (a) public–private research and extension, creating the information and knowledge for the practice to succeed; (b) social organizations such as informal or formal landowner groups (e.g., Chestnut Growers of America, River Hills Harvest marketing cooperative, farmers markets); (c) agricultural and environmental policies such as USDA–NRCS program incentives for silvopasture or riparian buffers in the United States and the leasing rules and regulations for public lands in Canada; and (d) the local, regional, and global markets affecting agroforestry products, such as niche markets, and the markets for other products that compete for the use of assets such as land or labor.

The institutions and organizations that conform to the structure provide signals and often determine how successful are the livelihood strategies pursued. The decision maker's capabilities contribute the landowner's negotiation and ability to cut through red tape to get things done (i.e., their human agency) (de Haan, 2000). When discussing the human dimensions of agroforestry, the term *institutions* has a unique meaning. Institutions are "regularized practices" or patterns of behavior, the "rules of the game," a manifestation of political capital (North, 1990, 1993). These are the accepted norms, both formal and informal, and part of the process of social negotiation, that contain power relations (Scoones, 1998). Institutions are important because these may serve as barriers or may provide opportunities for livelihood strategies to develop.

Picture human agency as the "hinge" (Figure 17–1, center) between the individual capabilities (or the households or the community) and structure (e.g., local market or government conservation programs). *Agency* (Table 17–1) defines the landowner strategy, as it enables individual capabilities or the ability to negotiate with structures and institutions (Figure 17–1). In the case of Jennifer Cunningham's family in British Columbia, their agency resulted in a specialty product that capitalized on a number of factors: (a) lamb raised and managed in a green-tag certified forest with sustainable management practices and no pesticides; (b) the political capital that produced the green-tag certification; (c) the ability to act with other like-minded producers on the knowledge developed, creating a valued product (the lamb); and (d) the ability to access information from sector organizations (structure). Thus, they created their own market linkages (agency) through direct marketing (structure), moving away from the notion of commodity production. For Jay Springs Lamb, it was important to develop a differentiated product from their sheep that embeds both the value of their natural capital and their own values regarding their way of life.

The ability to negotiate with others in the market and to capture profits through investment of human, social, cultural, financial, and political capital in economically viable enterprises throughout life is shaped and mediated by culture, society, and policies, as well as local and global markets. The Jay Springs Lamb Co. evolved such that their products captured the concept of lifestyle, their *cultural capital*, which is shared by some other producers and, most importantly, their customers. In the process, they were able to capture the value of their labor by developing a niche and local concept for their product (agency), turning the concept of lifestyle into a business through a set of market relations (social and financial capital) and resulting in a viable livelihood. It evolved in terms of family involvement, developing cooperative approaches with other like-minded family farms selling under the Jay Springs Lamb label.

Livelihood Strategies

A livelihood encompasses the income-generating and expense-reducing activities pursued by a household and its individuals and the mechanisms or means to access resources through the various stages in life, defined as the *life cycle* (Ellis, 1998; Valdivia, 2004). Pathways are developed with time, conditioned on changes throughout the life cycle from young individuals and households to older households with children, those who do not have children, and the elderly. The activities are decided within the household or, in the larger context, the community, market, and society. This is apparent in the three examples, where there is a process of intergenerational households, with the land being passed on to the children, who are developing new pathways, identifying and sensing opportunities that distinguish them from other producers (Robinson, 2013a, 2013b), or multiple families working together in different aspects of the business in British Columbia.

In defining their rural livelihood strategies (Bebbington, 1999; Chambers & Conway, 1992; Ellis, 1998; Scherr, 1995; Shucksmith, 1993; Valdivia, Raedeke, Hodge, Green, & Godsey, 2000), households seek to maximize net returns or minimize costs, along with other noneconomic objectives (Scherr, 1995; Valdivia & Gilles, 2001). Important criteria shaping the activities, outputs, and outcomes are stages in the life cycle of the family, as well as specific objectives related to their life in rural areas, such as the importance of the environment, the quality of resources, wildlife, their children, and lifestyle.

In temperate regions such as North America and Europe, contingency markets for risky events exist, such as crop insurance, disaster insurance, and government-provided safety nets to address price and yield risks. Natural resources, such as soils, trees, and streams, plus the household's or individual's access to assets, shape the types of agricultural and nonagricultural activities that may be pursued on or off the farm. While a motive for diversifying may be to reduce risk, another may be to make a living consistent with the values and capital of the decision maker.

Capital and Capabilities

When individuals, households, or groups access and use resources, these resources become capital that creates a stream of benefits (Valdivia & Gilles, 2001). Access gives individuals the capability (Bebbington, 1999) to build a livelihood. For Jay Springs Lamb, their human capital (education, knowledge, and search for new information), cultural capital (value of the way of life), and natural capital (the land and forest they manage), plus their social capital (e.g., industry associations) have been the means to a goal—"where a sustainable life style motivates our business." This capital and the producers' capability (ability to act) lead to the accumulation of assets (economic and noneconomic) that contribute to their target outcomes in life.

The literature has grown to include many forms of capital, including economic, human, social, cultural, and natural (see, e.g., Bebbington, 1999; Flora, 2001; Winters, Davis, & Corral, 2002). *Economic* or *financial capital* includes cash, credit and debt, savings, and other economic assets like basic infrastructure. *Human capital* includes skills, knowledge, ability, good health, and physical capabilities (Scoones, 1998). *Social capital* is described as "features of social organization, such as networks, norms, and trust that facilitate coordination and cooperation for mutual benefit; social capital enhances the benefits of investment in physical and human capital" (Putnam, cited by Flora, 2001). *Culture*, especially indigenous knowledge, has also been recognized as a capital valuable in the development of niche markets linked to biodiversity. This concept is revisited below in the presentation of *habitus* and *field*.

A critical capital, in terms of the practice of agroforestry, is the natural capital. *Natural capital* is defined by stocks and services of the environment (i.e., the natural resources in and around the farm), which include plants (including trees), soil, water, and air (Scoones, 1998). Natural capital heavily influences what types of practices are feasible.

The Household Portfolio and Farm Diversification

Households or individuals make decisions by allocating labor and other resources to farm enterprises and off-farm income-generating activities to make a living, to profit from a business, and for well-being motives. A portfolio of activities is pursued based on access and control of resources and assets, capabilities of the decision maker, and the enabling environment. This portfolio is useful to evaluate how a new practice or activity can be integrated into a livelihood strategy (Shucksmith, 1993; Valdivia, Dunn, & Jetté, 1996). The approach considers the household's multiple objectives in pursuing their livelihoods—some households engage in farming solely for commercial gain, while others seek to maximize net returns along with other objectives consistent with their life-cycle stage. These may include valuing the environment, concern for the quality of resources, wildlife, and lifestyle choices, such as enjoying the countryside or valuing rural life. Economics, along with other motives, inform the allocation of labor and other resources.

Diversity is closely linked with resilience in the context of human and biological systems, with the idea that having options available makes these systems the less vulnerable to the natural disasters and market fluctuations that often impact on farming and the rural landscape. To achieve resilience through agroforestry, it is necessary to understand how an agroforestry practice "fits" with a livelihood strategy, how it diversifies the household portfolio, and what are the multiple benefits it provides to humans and the environment.

Diversification

How agroforestry practices may contribute to the rural livelihoods of farm operators and landowners is based on their ongoing activities and use of resources. Diversification can be measured in several ways, such as within the farm pluriactivity (Loughrey, Donnellan, Hennessy, & Hanrahan, 2013) or farm diversification. Another form has to do with on-farm and off-farm diversification, defined as household diversification, also known as farm-gate pluriactivity (Loughrey et al., 2013).

Diversification can be studied through the household's portfolio of activities. There are many reasons why households make decisions to diversify. For instance, households may seek to reduce risk (Hoppe, 2001; Knutson, Smith, Anderson, & Richardson, 1998; Skees, Harwood, Somwaru, & Perry, 1998). Incorporating new activities, different from existing ones, can reduce the risk of the portfolio. Off-farm employment also reduces the risk of a portfolio that includes agriculture because the risks faced by each is not the same. Markets and climate have very different impacts on agriculture than on employment—a reason why non-farm income reduces the variability in total household income (Mishra & Sandretto, 2003).

Diversification may also be a mechanism to maximize the use of all available resources (Ellis, 1998; Scherr, 1995). The different types of household labor used on and off the farm defines a livelihood strategy (Shucksmith, 1993). In areas where farming is highly profitable, the farm operator, and maybe family members, are employed on the farm. On the other hand, in areas where agriculture is not as profitable, the operator and spouse may be employed part-time or full-time off the farm. In the case of Jay Springs Lamb, four families and their differing labor skills, availability, and interests came together to address the varying farm needs. As it is a dynamic process, these responsibilities have and will continue to shift with the changing life-cycle stages of the various family members.

A study of the nature and extent of farm and ranch diversification in North America (Barbieri, Mahoney, & Butler, 2008) identified a lack of scientific information regarding the types of enterprises being developed. Policies in Canada and the United States have been enacted to support diversification, assisting small farms in the United States and promoting environmentally sustainable production in Canada through a renewal strategy that focuses on profitability, "value-added," and networks of support (Barbieri et al., 2008). Diversification includes a multifaceted income or value-generating activity incorporating concepts of conservation, quality, locality, efficiency, and new marketing approaches.

Barbieri et al. (2008) found eight enterprises were being used by farmers and ranchers across North America to diversify the operation within farming and agriculture. These included:

- the introduction of nontraditional crops or livestock (reported by 71% of respondents)
- value added in processing and packaging (65%)
- new marketing and distribution strategies (53%)
- integration of recreation tourism and hospitality (50%)

- historic preservation and adaptive reuse (48%), placing a value on culture
- expertise, consulting, and education (30%), consisting of educational tours, workshops, and seminars relevant to new practices like agroforestry
- leases, easements, and time-shares of farm and resources (11%)
- contracting and services in holistic health counseling, plant breeding, boarding horses, training, composting, maintenance of orchards, and others (8%)

The study found that the more diversified respondents were the younger farmers (often second- and third-generation farmers), who hoped to keep the operation in the family and who listed farming or ranching as the primary occupation. Women farmers represented almost one-third of the diversified farms.

Understanding whether the differentiated livelihood strategies of households have an effect on the nature of the agroforestry practice of interest and how the enabling environment (social, economic, and political structures, institutions, and policies) shapes household decisions (de Haan, 2000; Valdivia & Gilles, 2001; Winters et al., 2002) can inform on pathways for adoption. These may include a new economic opportunity that may reduce risk in the enterprise portfolio, a practice to improve the environment (natural capital), and/or the generation of new income sources that complement ongoing rural and urban household enterprises. The motives are several, and the concepts of *habitus* and *typologies* (Glover, 2010; Raedeke et al., 2003; Shucksmith, 1993) provide a framework to identifying these pathways.

Pathways

Disposition to Act—Use of *Habitus* in Understanding Strategies

While commodity crops dominate market relations in farming, the Jay Springs Lamb Co. case study illustrates an alternative marketing strategy that allowed the producers to command a higher price for their product. They capitalized on the characteristics of the land they managed and the production methodology, creating a unique product consistent with both family values and customer lifestyle choices. Early Boots Farm and East Grove Farms (Boxes 17–1 and 17–2, respectively) also exemplify how values shaped the way these businesses incorporated agroforestry and the way these evolved. Present in these cases are sensing, seizing, and transforming, key elements in the processes of change and adaptation of family businesses with time (Alonso et al., 2018).

An early study proposed that motives may trigger different rural household strategies of farmers in Scotland (Shucksmith, 1993) in response to a new agricultural policy in the 1990s. The behavior model developed in this study consisted of internal resources, the capital and capabilities, as well as the external context of the structures and institutions (markets and policies), and cultural and social values (Figure 17–1). *Behavior* is understood to be an outcome of the interplay between the disposition of the members to act, their material resources (cultural capital, life cycle, skills, land, and economic capital), and the external context (cultural and social norms, markets, and government policies) (Shucksmith, 1993).

The *habitus* concept of Bourdieu (1990) informed the "disposition to act," a product of socialization and interaction. The individual internalizes the dominant mode of thought and experience to which he or she is exposed, providing the basis for practice or behavior (Shucksmith, 1993). Individuals making decisions have goals and interests and pursue livelihood strategies, but the freedom to act is constrained by habitus. Following this argument, Shucksmith (1993) stated that "many options potentially open to farmers (including many unusual forms of diversification) may never be seriously considered because they are literally 'unthinkable'." In other words, the ability to "think outside the box" may be constrained by habitus.

Three types of behaviors leading to potential paths of farm households were identified: (a) the *accumulator*, characterized as a risk taker, willing to engage and willing to depart from set ways (thinks outside the box); (b) the *conservative*, characterized by a traditional outlook, conservative in farming techniques, and resistant to change (stays firmly within the box); and (c) the *disengager*, characterized by a decreasing commitment to agriculture, such that this activity plays the residual role. Further research by Shucksmith and Herrmann (2002), using cluster analysis with the main components of habitus that included the initial circumstances of the farm and household, the farmer's socialization, and the farmer's attitudes, found more pathways to adapt to the changes brought about by policies. The greater diversity of strategies included hobby farmers, pluriactive successors, struggling and contented monoactives, potential diversifiers, and agribusinessmen (Shucksmith & Herrmann, 2002).

This framework was used to examine the potential for agroforestry practices in U.S. farming (Raedeke et al., 2003; Valdivia, 2007), arguing that considering new activities like agroforestry practices requires that farmers be interested in identifying new markets and learning new methods. Farmers who view farming as a way of life may express different behaviors, looking for opportunities to pursue their rural lives by increasing the number of activities on the farm. This may include alternative farming activities, such as agritourism, providing hunting leases, carbon sequestration, and managing new crops or animals, or it may include off-farm activities, a new network of social relations, a new *field*, with new institutions and organizations, rental and familial relations, and their corresponding values. Off-farm employment is an example of an external source of income that allows a family to remain on the farm or a means to gradually move out of farming. This type of diversification is possible as the interaction with other *fields* increases.

Agroforestry practices can incorporate new economic opportunities that entail benefits, markets, and production risks different from commodity farming (Barbieri & Valdivia, 2010a, 2010b; Gao, Barbieri, & Valdivia, 2014a; Gold et al., 2004). These practices may also be a combination of environmental improvement (natural capital) and commercial venture or the appreciation by consumers of the environment through agritourism (Gao et al., 2014a, 2014b). Alley cropping, forest farming, urban food forests, and silvopasture are viewed as commercial, profit-seeking practices, while windbreaks and riparian buffers have been traditionally thought of as environmentally focused practices. While this is true, depending on the policy environment, the latter practices can also be designed to yield short-, medium-, and long-term economic benefits. For agroforestry practices to realize multiple benefits, these must be part of the possibilities contemplated by a decision maker, and for this to happen, existing networks and relations play a critical role.

Practice and *Field* in Agroforestry

The practices of farming, forestry, and/or agroforestry in Bourdieu's framework embody the give-and-take (dialectical) relations of *habitus* and *field* (Raedeke et al., 2003; Valdivia et al., 2012; Valdivia, Hodge, & Raedeke, 2002). Again, *habitus* is the taken-for-granted social processes, values, beliefs, and behaviors shared. *Field* is the network of social relations representing the rules of the game (institutions).

These concepts were applied to assess how agroforestry can be integrated with farming or forestry or as a new practice in Missouri (Arbuckle, Valdivia, Raedeke, Green, & Rikoon, 2008; Dorr, 2006; Flower, Valdivia, & Dorr, 2005; Raedeke et al., 2003; Valdivia et al., 2012; Valdivia & Poulos, 2009) in studies that included operators and landowners contemplating practicing agroforestry, as well as more recently a study of farmers engaged in agroforestry through the USDA Sustainable Agriculture Research and Education (SARE) program in the United States (Romanova, 2020). When landowners progressively decrease their reliance on an income and livelihood from farming (because the existing market does not secure enough income to make a living from farming), decision makers are more open to other alternatives and thinking outside the box (Robinson, 2013a, 2013b; Shucksmith & Herrmann, 2002).

The field of farming is defined by the economic, family, and rental relations, as well as the buyers, equipment dealers and input suppliers, financial institutions, past and future generations, and current and future landlords (Raedeke et al., 2003). These relationships also include different positions of power; for example, bankers and landlords have positions of influence and impact farmers' attitudes and behaviors when they are seeking loans or securing land for farming. Raedeke et al. (2003) found that landowners and bankers had a large influence on farmers' decisions in Missouri.

The field of forestry is defined by the government agencies, timber buyers, forest product companies, loggers, and forest professionals. Farmers have expressed concerns about getting fair prices in this field (forestry) from the entire industry. Commercial buyers and timber loggers were listed by farmers as their least trusted source of information about trees, while they trusted university extension, with its history of serving the field of farming (Raedeke et al., 2003). The habitus in farming differed from that of forestry. To the farmer, farming was active and was about farming the land, producing crops. In contrast, forestry was seen as passive, leaving the land idle, with trees in the landscape. "Planting trees in farming space conflicts with the habitus of farming" (Raedeke et al., 2003). The underlying logic relates to maximizing profits within the practice of farming, and decisions about the use of farming space are consistent with, and constrained by, the values in this field. However, a forester would view forest management as both active and economic.

The field of agroforestry on the other hand, has been developing through relationships for more than 20 years, such as the University of Missouri Center for Agroforestry, the Savanna Institute, the Leopold Center for Sustainable Agriculture, the regional agroforestry working groups, and the SARE program funded by the USDA National Institute for Food and Agriculture. These centers, in collaboration with programs such as the NRCS EQIP, have provided technical and financial assistance for agroforestry and the establishment of woody crops (Box 17–3). In 2010, to increase support for agroforestry, the USDA–NRCS, U.S. Forest Service, and many

Box 17–3. The evolution of an NRCS EQIP special funding pool for agroforestry and woody crop establishment: A Missouri case study

The USDA–NRCS is responsible for assisting landowners in addressing resource concerns on private land to improve our soil, water, air, plants, and animals (both domestic and wild) and improve energy efficiency. The vision of NRCS is to ensure productive lands in harmony with a healthy environment.

In 2010, the USDA–NRCS and U.S. Forest Service, along with other partners and stakeholders, developed the USDA Agroforestry Strategic Framework to increase awareness and support for agroforestry across the country. In Missouri, a multiyear set of intersecting events lead to the establishment, in 2017, of an NRCS "dedicated funding pool for Agroforestry and Woody Crop Establishment within the Environmental Quality Incentives Program (EQIP)". It is instructive to document the key set of individuals and events, stretching back more than 20 yr, to understand how this change in policy occurred in the state of Missouri.

In the late 1990s, a group of University of Missouri, and University of Missouri Center for Agroforestry, and Missouri State University researchers and extension staff initiated an elderberry improvement program to develop better local cultivars and explore the market opportunities to support future growth of a regional elderberry specialty crop industry. Terry Durham, a long-time local organic vegetable grower, recognized the opportunity, began collaborating with the research and extension team, and eventually became the leading force promoting the growth of this specialty crop to local growers. By about 2010, many area farmers began to establish elderberry, research continued, and local training events were underway to support the growth of this "new" niche crop. From this time onward, farmers across Missouri began to contact their local NRCS offices seeking cost-share funding for the establishment of elderberry. These calls were often redirected to Lauren Cartwright in the NRCS state office. Responding to these calls, two extension publications were released by the Center for Agroforestry, *Using NRCS Technical and Financial Assistance to Establish Elderberries* (Cartwright, Goodrich, Cai, & Gold, 2017) and *Growing and Marketing Elderberries in Missouri* (Byers et al., 2014).

Three key events took place in 2013. First, University of Missouri researchers convened an international research conference and farmer workshop to pull together the most up-to-date body of scientific and practical knowledge on elderberry (Thomas et al. 2013). Second, the Center for Agroforestry established a weeklong "train the trainer" crash course known as the Agroforestry Academy. Third, a multiday elderberry workshop was hosted at Lincoln University, led by Durham, Thomas, Byers, Godsey, and Gold. Lauren Cartwright, the NRCS state economist in charge of EQIP funding program, attended all of these events and was inspired by the potential to assist Missouri farmers interested in both new crops (e.g., elderberry) and new production approaches (i.e., agroforestry practices).

By 2013, a small number of successful EQIP contracts supported elderberry growers, but these growers competed directly with commodity producers (the core NRCS constituency) for EQIP funding. Calls to the state office from elderberry growers increased in frequency. In 2016, Cartwright approached the Missouri NRCS state technical committee with the idea of creating a special funding pool, using 1% of the general EQIP allocation, directed specifically to support niche crop producers. Because agroforestry and woody crop establishment is consistent with NRCS goals and vision, the special funding pool was approved starting in 2017 and an updated Center for Agroforestry extension guide was created (Cartwright et al. 2017) to support these landowners. The core conservation practices associated with the EQIP Agroforestry and Woody Crop Establishment funding pool are: alley cropping; multi-story cropping; riparian forest buffers;

silvopasture; tree/shrub establishment; and windbreak/shelterbelt establishment. These conservation practices may be utilized individually or in combination in different parts of the landscape depending on the landowner's goals.

Applications to and financial commitments from the agroforestry and woody crop establishment funding pool have increased each year since 2017, growing from $42,000 in 2017 to $95,000 in 2019. Landowners interested in agroforestry and woody crop establishment on their property have many resources at their disposal through the NRCS Conservation Technical Assistance and EQIP financial assistance opportunities.

Lessons learned: A combination of producer interest, knowledge acquisition and subsequent motivation and responsiveness by key NRCS staff, and support from the state techinical committee and partners led to the successful establishment of a dedicated EQIP special funding pool for agroforestry and woody crop establishment.

partners and stakeholders developed the USDA Agroforestry Strategic Framework 2011–2016 (updated for 2019–2024). The University of Missouri Center for Agroforestry (UMCA) has been collaborating with many institutions since its establishment in 1998. In 2013, with USDA SARE support, UMCA and a network of collaborators supporting the practice of agroforestry established the Agroforestry Academy train-the-trainer program. Concurrently, UMCA collaborated closely with the Missouri NRCS to develop funding programs through EQIP inclusive of agroforestry practices. In 2017 these agents of change (Cartwright, Goodrich, Cai, & Gold, 2017) formalized EQIP funding for multistory cropping, riparian buffers, silvopasture, tree and shrub establishment, and windbreak and shelter belt programs. These significant changes have contributed to existing organizations and institutions in support of agroforestry practices, finding a path to address the challenge agroforestry faced of being neither solely forestry nor solely farming (Valdivia et al., 2012).

Along with the creation of an institutional setting in support of the practices, there is also a need for understanding the multifunctionality of agroforestry and its economic benefits, such as agritourism. This was identified by East Grove Farms in Iowa as important, recognizing the value of scenic beauty, the symbolic capital of the historic farm, and the interests of consumers who look for tourism experiences in the countryside (Box 17–4).

Additionally, producer attitudes and views are regionally influenced, depending on individual exposure to both farm and timber management opportunities. In British Columbia, the woodlot license program originated in 1948 (Federation of British Columbia Woodlot Associations, 2007), designed to allow farmers to acquire farm woodlots of limited land area, and involved specific management requirements and data collection of the licensees (Valdivia et al., 2009). The program evolved such that there were 826 woodlot licenses throughout British Columbia in 2007, forming a small (~1.5% of the province's total annual allowable cut) but integral component of the British Columbia forest sector. Because being a farmer is no longer a requirement, many of the licensees are involved in other activities such as ranching and other businesses (Federation of British Columbia Woodlot Associations, 2007), thereby influencing the awareness and knowledge of both agricultural and forest producers about each other's sectors. In addition to the deliberate retention or introduction of trees and/or shrubs in agricultural production systems, an equivalent opportunity for agroforestry in BC arises from the introduction or enhancement of other crops or livestock to forest production systems (Valdivia et al., 2009). Currently, agroforestry has expanded all across Canada, and there is interest in the contributions this practice can make to carbon sequestration (Baah-Acheamfour, Chang, Bork, & Carlyle, 2017).

During the 20 years since the first agroforestry interviews with landowners and farm operators were conducted in Missouri, the number of farms has decreased. The Agricultural Census of 2017 listed a total of 95,320 farms, with 33% of operators selling less than $10,000 annually and 32.6% selling between $10,000 and $100,000. In terms of hectarage, 64.6% of operators manage less than 73 ha. Several landowner perspectives and strategies have developed as a result of the changes in agriculture, including social relations and values that are making it possible to explore new alternatives, such as practices involving planting trees on land. According to the 2017 Census of Agriculture, the number of operators in Missouri practicing silvopasture and alley

Box 17–4. Maximizing landowner's gains from agroforestry: The consumer's perspective

Ample evidence indicates that agroforestry brings many benefits to landowners (e.g., income diversification) and the entire society (e.g., landscape beautification). It is critical that landowners capitalize on these benefits when placing their products in the market, especially when interacting directly with consumers (e.g., farm visitors). Strategic marketing requires understanding consumers' product awareness and preferences. Thus, residents from Missouri, Pennsylvania, and Texas (250 per state) were surveyed to gauge their awareness of agroforestry. These states were selected because of their distinctive landscapes and different levels of on-farm direct sales (lowest in Missouri and highest in Texas) while holding similar agricultural (e.g., farm size distribution) and demographic (e.g., household income) attributes. Two primary takeaways stemmed from survey results that landowners can use in their marketing and managerial decisions.

Landowners should deliver the message of agroforestry benefits in their marketing strategies. Survey results showed that consumers have little awareness of the environmental (e.g., wildlife conservation) and socioeconomic (e.g., maximization of agricultural land use) benefits of agroforestry. Yet, the increasing number of consumers willing to support sustainable agricultural systems, such as agroforestry, through purchasing behaviors indicates a competitive advantage for landowners practicing agroforestry in selling their products. To account for this growing market, landowners should incorporate messages related to agroforestry benefits in their product promotion efforts (e.g., brochures, ads). Results also showed that women and young consumers are the most aware of the benefits that agroforestry farms deliver to society. Thus, landowners implementing agroforestry should strategically promote their products to these market segments. This is particularly important because women tend to be the decision makers of household purchases, and millennials are more socially and environmental conscious in their purchasing behaviors.

Landowners should capitalize on the visual appeal of agroforestry when receiving farm visitors. The number of individuals visiting farms for education, recreation, or direct purchase, commonly referred to as *agritourism*, has increased over the past few years. Likewise, the number of landowners offering agritourism has increased due to the many benefits this activity produces (e.g., increased profits, branding of farm products). Enhancing the farm visual appeal is key to strengthening agritourism offerings and enticing visits. Study results indicated that farm animals and planted trees or shrubs are consumers' favorite agricultural features to see when visiting a farm, and wildlife is their most favored natural feature to encounter. In addition, women like very much seeing native plants, flowers, or grasses in the farm landscape. These results indicate that key agroforestry elements (e.g., incorporating animals and trees or shrubs) and its ecological benefits (e.g., conservation of wildlife and native flora) can enhance the farmland's visual appeal. Thus, landowners should (re) design their managerial practices (e.g., attracting wildlife with feeders, propagating native plants or grasses) to capitalize on and augment these aesthetic benefits to better serve current visitors and attract more potential visitors.

In conclusion, adopting agroforestry practices adds more values to the farmland and creates more assets for landowners who sell directly to final consumers. To capitalize on these values and assets, landowners should strategically stimulate the purchase of their products by communicating agroforestry benefits to consumers, making a great effort to reach female and young consumers. Furthermore, landowners seeking to diversify through agritourism should also emphasize the aesthetic gains of agroforestry.

cropping grew from 141 in 2012 to 1,311 in 2017 (USDA National Agricultural Statistics Service, 2020). Missouri and British Columbia highlight the unique nature of each practice in terms of the values, institutions, and rules. In a similar way today, Early Boots Farm with silvopasture in Minnesota and East Grove Farms with elderberry in Iowa are examples of new pathways for family farms that involve agroforestry (Robinson, 2013a, 2013b).

The Changing Practice of Farming—Agroforestry

In the agency-structure dimension, non-productivist action and thought is a key element in understanding change (Barbieri & Valdivia, 2010a; Burton & Wilson, 2006; Raedeke et al., 2003; Shucksmith, 1993; Wilson, 2008). The practice of farming, which revolves around the institutions and organizations in farming, rental and familial relations, and the corresponding values of farming, may be changing because at present farming is the main or sole source of income for a few large and very large farmers, while 86.24% of the privately owned lands in Missouri are in the hands of "small" farmers, operators with sales of less than $100,000 a year (USDA National Agricultural Statistics Service, 2020). As individuals increase their participation in more than one field, they are more likely to accept new fields such as agroforestry.

In predicting change toward inclusion of agroforestry, Raedeke et al. (2003) proposed three possible paths. First was incorporation of agroforestry into farming practices that complement the field by protecting natural resources, land, water, and production through windbreaks, riparian buffers, and silvopastoral systems. Farmers' attitudes toward conservation correlate positively with the adoption of conservation practices (Dorr, 2006; Ervin & Ervin, 1982; Valdivia & Poulos, 2009). In a second path, other practices remain in the field but with different approaches, and the third path consists of a new field, where those engaging would define a new activity with new technologies and sets of relations, a spectrum of possibilities that can lead to multifunctional agriculture (Raedeke et al., 2003; Romanova, 2020; Wilson, 2008).

The use of environmentally friendly practices doubled between 1991 and 2001 in Canada, as did the value placed on ecological goods and services by producers and consumers (Barbieri et al., 2008). This suggested a possible pathway for agroforestry, as Barbieri et al. (2008) found that "diversification was associated with the propensity to be engaged in environmental and cultural resource stewardship," services that agroforestry provides.

The Multiple Benefits and Functions of Agroforestry

Product diversification, environmental impact mitigation such as carbon sequestration and water quality improvement, land rehabilitation, land use conversion, food production, protection of marginal lands through sustainable use or set-asides for enhanced wildlife habitats, and scenic beauty are among the multiple services provided by agroforestry (Buck, 1995; Jose et al., 2018; Kremen & Merenlender, 2018; Williams, Gordon, Garrett, & Buck, 1997). Some of the economic benefits may be direct, while others are indirect. For example, forested riparian buffers can be designed to provide harvestable products for economic returns along with the environmental benefits they provide (see Chapter 8). Indirectly, agroforestry can enhance the farm appeal to draw visitors to the farm (Gao et al., 2014a). Household decision making is at the center of the success of adopting new technologies in agroforestry, as the examples of Early Boots and East Grove farms show (Robinson, 2013a, 2013b). Learning what motivates the decision makers is essential to design effective policies (Cartwright et al., 2017; Mercer & Miller, 1998; Romanova, 2020) (see Box 17–3). Despite the many benefits of agroforestry, Gao et al. (2014b) suggested that landowners put more effort into increasing the public awareness of these benefits to entice direct purchase of farm products (see Box 17–4).

Interest and Adoption in Agroforestry

Understanding the perceptions and knowledge about agroforestry practices contributes to defining possible pathways for the adoption of agroforestry (Dorr, 2006; Fleming et al., 2019; Flower, 2004; Fregene, 2007; Mattia, Lovell, & Davis, 2018; Raedeke et al., 2003; Romanova, 2020; Valdivia et al., 2012), which differs from its individual components of agriculture and forestry (Raedeke et al., 2003). The discussion above on habitus and field revealed that it is not simply a combination of both individual fields, as each has distinct values as well as social relations in the economic and social spheres. A full-time commodity farmer with a high degree of investment in the field of agriculture, in assets and relationships with farming organizations, will be less likely to be interested in agroforestry. Households with more experience or interest in conservation practices, on the other hand, would be more likely to be interested in resource conservation or wildlife and therefore likely to be interested in practices that enhance ecological goods and services. Decision makers who are less able to make a living solely from commodity farming and prefer to remain in a rural community may also be more likely to pursue other activities and include agroforestry in their portfolio of household activities. Educated children returning to the farm bring with them new perspectives and skills as well as values about thinking differently about the family farm. It means, as the examples have captured, a process of learning different values and ideals and a field for the practice that brings together organizations and policies in support of the innovators (Romanova, 2020).

Knowledge Gained during Two Decades

Several studies of the adoption and interest in U.S. agroforestry have been completed in the past 20 yr (Garrett, 2009; this volume). Here we review the lessons learned, particularly related to temperate agroforestry. Figure 17–1 also depicts the factors that have been hypothesized or found to influence interest in and adoption of agroforestry practices. The concepts of sustainable livelihoods, such as capital, structures, field, and habitus, are significant factors in the ability to act. Human capital, including life-cycle factors, are considered along with social, economic, and natural capital. Individual motives, values (*habitus*), and attitudes of the decision maker are internal factors, while structures, institutions, and the field are external to the individual and constitute elements of the *field* (Raedeke et al., 2003; Valdivia et al., 2009).

A study of the adoption of agroforestry (Pattanayak, Mercer, Sills, & Yang, 2003) identified five critical factors: (a) farmer preferences, (b) resource endowments, (c) market incentives, (d) biophysical factors, and (e) risk and uncertainty. Adoption and diffusion of agroforestry practices depend on a complex set of factors, such as the physical characteristics of the natural resources (natural capital), social and economic characteristics of the decision maker, culture, and policy incentives and institutions (Pattanayak et al., 2003). Awareness of innovation, perceptions of feasibility and worthiness, and consistency with the decision maker's objectives are conditions required for adoption. The new practice must be more profitable, and any uncertainties that exist must be addressed (Pannell, 1999). A number of studies have shown that profitability continues to be important (Romanova, 2020; Borremans et al., 2016; Faulkner, Owooh, & Idassi, 2014; Rois-Díaz et al., 2018; Tsonkova, Mirck, Böhm, & Fütz, 2018). Monetary and non-monetary factors have been shown to be significant for landowners whose livelihoods depend on agriculture (Barbieri & Valdivia, 2010a, 2010b; Koontz, 2001; Romanova, 2020). Environmental motivations were also significant in multiple studies (Arbuckle et al., 2008; Dorr, 2006; Romanova, 2020; Trozzo et al., 2014b; Valdivia & Poulos, 2009).

Offsets of up-front costs and cost sharing were more effective in encouraging adoption than increasing rental rates in the case of riparian buffers in Chesapeake Bay (Lynch & Brown, 2000). In Pennsylvania, existing production practices, reasons for owning land, and issues in land management were used to identify farmers' adoption potential (Strong & Jacobson, 2005). Several of the motives identified related to the importance of the knowledge of institutions as well as the costs involved in changing to a new economic activity (Valdivia et al., 2012).

In the tension between agriculture and forestry, transaction costs and profitability are two important factors. Transaction costs are the costs of adjustment to a new practice, of access to information, and of negotiating, monitoring, and enforcing contracts (Valdivia et al., 2012). These can represent a barrier to adopting a new agroforestry practice. Profitability is also an important motivation, as seen in previous studies. Using these two concepts, the importance of transaction costs and profitability factors, an analysis of data from farm operators in Missouri resulted in resulted in dividing those operators into three groups: environmentalists, agriculturalists, and the disengaged (Valdivia et al., 2012).

The results of other variables that capture stage in the life cycle, such as type of farm operation, age, gender, and farm size, have been mixed. Studies have found gender (female) and age (younger) to be positive significant factors in diversification and interest in new practices (Barbieri et al., 2008; Valdivia & Poulos, 2009) and in farmer potential for adoption (Strong & Jacobson, 2005).

Attitudes, such as the importance of land stewardship to the landowner, had the strongest effect on a willingness to adopt agroforestry practices in Ontario (Matthews et al., 1993). Noneconomic motives (Fleming et al., 2019; Koontz, 2001) and non-farming opportunities (Lynch & Brown, 2000) were also criteria in land-use decisions. In British Columbia, the factors deemed most important in decisions to plant trees and shrubs for harvest or to adopt agroforestry practices were primarily socially and environmentally oriented (Thevathasan et al., 2012). However, factors preventing these activities were of economic nature.

The awareness of agroforestry in Pennsylvania was low (Strong & Jacobson, 2005). Farmers had difficulty accessing specific information and, therefore, assessing if agroforestry practices would fit and be profitable. The reasons for owning land was a useful question to identify farmers with adoption potential. Windbreaks and riparian buffers, aligned with conservation, have been integrated successfully into the field of farming. In British Columbia, there was generally low to moderate knowledge of agroforestry practices but a moderate to high level of interest in implementing them (Thevathasan et al., 2012). Producers were most familiar with riparian management systems and silvopasture. These systems also garnered the greatest degree of interest in adoption, followed by windbreaks and forest farming systems.

Values and motives also significantly impact adoption. Differences in the values and motives of landowners, compared with the institutions that work with or for them (e.g., extension), can lead to ineffective programs. Research on the perceptions of landowners and extension workers about agroforestry in the southeastern United States found differences between the objectives and concerns of these two groups. Extension professionals were more concerned with water quality, while the landowners' interest was in aesthetics (Workman, Bannister, & Nair, 2003). In contrast to this experience, recent work of the USDA with indigenous communities (MacFarland et al., 2017) found multiple reasons for their interest in agroforestry, such as sustaining culturally significant food and fiber systems, diversification of operations and income sources, increased food security, and economic development. The study also highlighted lessons learned from the adaptation strategies of native communities in the United States.

Factors that limit agroforestry adoption have included farm size, age of the decision maker, education, lack of information and technical assistance about the practice, lack of networks and institutions of support, and lack of knowledge about markets. For example, the British Columbia survey indicated that market-oriented support services were high priorities for stakeholders, and barriers to adoption included both a lack of market-related resources and the degree of labor intensity required for production (Thevathasan et al., 2012). These adjustment costs are some of the transaction costs in shifting to agroforestry. Pattanayak et al. (2003) highlighted factors that needed more attention, such as biophysical characteristics, resources, and risk and uncertainty, while preferences and resource endowments had been often included in adoption studies. While Pattanayak et al. (2003) included only two studies in North America, many of the findings and areas of research recommendations were consistent with studies about the interest and adoption of agroforestry in Missouri (Arbuckle et al., 2008; Dorr, 2006; Flower, 2004; Flower et al., 2005; Fregene, 2007; Valdivia & Poulos, 2009).

Building on these studies, a profile of "typical" potential adopters seeking to diversify would include the following characteristics: younger, female, educated, and informed through trusted sources and networks. Economic issues and the existing degree of investments and profits in ongoing economic activities are also factors that weigh in decisions about new enterprises.

Considering habitus, the field of farming and the field of agroforestry may increase our understanding of which households are interested in agroforestry as a new diversification strategy. Landowners who are not traditional commodity farmers may have a greater interest in exploring this new field. Lifestyle farmers (Valdivia et al., 2000) are a group interested in new enterprises that include trees. These landowners seek livelihoods in rural areas that combine farming with off-farm full- or part-time employment. They are often conservation oriented, interested in new commercial opportunities such as niche markets, and have nonfarming capital to invest. The Early Boots and East Grove farms are examples of innovative young farmers with new ideas, and farms that have been in the family for many years, interested in the environment and building relationships to grow new products.

Studies about interest in and adoption of agroforestry, framed by habitus and field (Shucksmith, 1993), explore these values and attitudes, along with other factors. Three major areas that contribute to understanding interest in and adoption of the practice of agroforestry in North America include: (a) attitudes and values reflecting habitus along with individual characteristics captured in the human and social capital; (b) the social relations, networks, and information in the fields of farming, agroforestry, forestry, and conservation; and (c) the physical properties of the natural and symbolic capital.

Landowners and Adoption Research

Landowner and farm operator surveys in Missouri elicited information about land use practices, household characteristics, attitudes toward trees, and perceptions and values of farming. The surveys also asked specific questions about interest in, knowledge of, and adoption of the initial five agroforestry practices. Both operator and nonoperator landowners were surveyed in several regions of the state at two different times. In 1999, the survey included a high proportion of farm operators. In 2006, the survey included a larger proportion of landowners who did not farm, reflecting structural changes taking place in the United States.

Attitudes and Values: Interest and Disposition to Change

Shucksmith (1993) and Shucksmith and Herrmann (2002) argued that the actions of farm households may be understood not only in terms of the structure (markets and policies) but also as an expression of the values and motivations that underlie behavior (habitus). Figure 17–1 captures the habitus effect on strategy and depicts the individual characteristics that have an effect on interest and adoption. The five studies conducted in Missouri, quantitative analysis models of interest and adoption of agroforestry practices, sought to determine which motives are related to interest in the concept of agroforestry and in specific agroforestry practices (Arbuckle et al., 2008; Dorr, 2006; Flower et al., 2005; Fregene, 2007; Valdivia & Poulos, 2009). These studies operationalized the concepts of habitus or attitudes informed by values and social networks that pertain to the various fields (Raedeke et al., 2003; Valdivia et al., 2012), i.e., farming, forestry, and conservation. Three general groups of variables were identified in a study by Flower et al. (2005): attitudinal (habitus), internal, and external variables. Internal variables include the social, psychological, economic, and personal characteristics and the assets of the decision maker (cultural, economic, human, and social capital in Figure 17–1). External variables include institutional and market incentives (structures related to the concept of field), and ecological and biophysical factors (the natural capital) in Figure 17–1.

Valdivia and Poulos (2009) and Arbuckle et al. (2008), studying operators and non-operators (landowners), respectively, included the five types of factors affecting interest, defined in the framework of the capital in sustainable livelihoods (Figure 17–1), informed by field and habitus. Valdivia and Poulos (2009) studied the relationship between diversification and how attitudes influence interest in two specific agroforestry practices: riparian buffers and forest farming. A logit model to study interest in riparian buffers and forest farming among farm operators included variables that measured economic and non-economic motives. The former were part-time farming and diversification measures, and the latter included concerns with scenic beauty and the environment for future generations. Those with environmental concerns and with more education were more interested in riparian buffers and forest farming, consistent with the case studies from British Columbia. Older farmers were less interested, while being a part-time farmer and the level of diversification had no effect on interest.

The study of landowners (Arbuckle et al., 2008) who were not farming the land included attitude/habitus variables to capture values in the field of farming and how these influenced their interest in agroforestry using ordinary linear regression analysis. Nonfarming landowners who maintained relationships in the field of farming and those who had economic reasons for owning land were not interested in agroforestry. On the other hand, landowners who had recreation and environmental reasons for owning land were interested in agroforestry. These finding show the relationship between values and belonging to a field.

The research of Flower (2004) focused on interest in riparian buffers and forest farming among farm operators using logit analysis and attitudinal variables in the vein of Shucksmith (1993): (a) landowners invested in the field of farming ("conservatives"), (b) landowners who enjoy the environmental and aesthetic values of the land ("lifestyle"), and (c) those who are willing to develop new ventures ("accumulators") in their livelihood strategies. The three attitudinal variables were developed using survey responses about preferences related to trees. Flower found that "lifestyle" and "accumulator" were significant attitudinal variables that explained interest. "Conservative," on the other hand, was not significant. Other studies of attitudes in the context of habitus and field include Dorr's (2006) study of agroforestry practices, with a sample of landowners who were operating or not operating a farm. This study found that the percentage of assets invested in farming had a negative effect on the likelihood of being interested in riparian buffers, forest farming, silvopasture, and alley cropping. In other words, the more invested in traditional farming, the less likely a landowner would be to have an interest in agroforestry. In contrast, landowners who valued trees for scenic beauty and future generations were more likely to be interested in these four agroforestry practices (Dorr, 2006). Fregene (2007) also found that landowners who perceive economic benefits from agroforestry practices are adopting this practice. On the other hand, landowners concerned with the costs of establishing and managing trees are less likely to adopt this practice, highlighting the fact that transaction costs are a barrier.

On-the-ground adoption of agroforestry remains limited (Romanova, 2020). Findings from research on the perspectives of those who are early adopters of agroforestry and participate in SARE programs, using the diffusion of innovation theory, consider several stages in the adoption process from early adopters. The characteristics of early adopters in the SARE sample analyzed by Romanova (2020) include partial engagement in farming, with off-farm income, and land ownership for some time (29 years); tenure was important in the establishment of trees. She found that both economic and noneconomic factors are equally important. Internal personal factors are a driving force and include belief in sustainable agricultural practices, conservation, and visual aesthetics.

Social Relations: Knowledge Networks and Information

Pattanayak et al. (2003) pointed to knowledge and information as a mechanism to reduce risk and uncertainty. At the same time, human capital (i.e., knowledge of agroforestry practices) was also a very powerful variable in all the studies in Missouri for both farming and nonfarming landowners. The effect of prior knowledge of agroforestry practices on increasing the likelihood of interest and adoption was positive, large, and always statistically significant. College education has also been found to be significant but not with the same consistency (Dorr, 2006; Flower et al., 2005). Knowing someone who is implementing agroforestry has a profound effect on agroforestry interest and adoption (Flower, 2004; Fregene, 2007). Other forms of building human capital are seeking professional advice (Arbuckle et al., 2008) and advice from other farmers, which also implies that networks exist among farmers and are important. Social networks or capital is significant in accessing information for riparian buffers (Flower, 2004), for alley cropping (Dorr, 2006), and for adoption of agroforestry (Fregene, 2007). Social capital within a field is defined as the relationships with other farmers and connections to institutions that provide information, such as university extension services, and it has a positive effect on both interest and adoption. Finally, in the field of human capital, landowners who subscribe to magazines about conservation and alternative farming have also been more likely to be interested in riparian buffers (Flower et al., 2005) and adoption of agroforestry (Fregene, 2007).

As an illustration, Jay Springs Lamb Co. was born from the ideas, knowledge, and education of the family members (human capital), who assessed the capital they possessed and their values. They sought further information through workshops, joined industry-related organizations, and learned about government programs and the policies and regulations relevant to their business (social capital). Thus, they used knowledge and information as a mechanism to assess and manage risk and uncertainty and continue to do so in an iterative process. Similar relationships have been established in other cases, such as the examples of the elderberry farm and silvopasture.

It is through the cultivation of relationships to develop the field of agroforestry that a new field for the practice of agroforestry is developing. Romanova (2020) found that on-the-ground adoption is still limited. Farmers engaged in agroforestry practices in 2017 (according to the USDA–National Agricultural Statistics Service, 2020) compose 1.5% of U.S. farmers (Romanova, 2020). Findings from survey research on the actual perspectives of those who are early adopters of agroforestry and participate in USDA SARE grant programs have shown that personal views are more influential and drive adoption, while economic benefits may be more important for early and late majority groups (Romanova, 2020).

Given the current early stage of U.S. agroforestry adoption, more practitioners need to be involved to set the stage for wider adoption by majority groups. To promote further adoption, relevant agencies need to promote agroforestry messages to targeted groups of people who share similar traits with agroforestry early adopters. The recommended strategy for agroforestry adoption agents would be to provide agroforestry information and involve groups of landowners and part-time farmers who value productive conservation and enjoy landscapes with trees. Targeted messaging should include information not only on the economic and noneconomic benefits of agroforestry adoption but also its relation to personal values. Highlighting personal gratification from working with trees and teaching others sustainable ways of living can help motivate adoption decisions (Romanova, 2020).

Physical Properties: Biophysical Characteristics and Problems

Landscape characteristics and specific problems also play a role in creating interest in agroforestry. Dorr (2006) found that the presence of managed timber on the landowner's land increased the likelihood of interest in forest farming for landowners in general. Valdivia and Poulos (2009) found the same for farm operators. Dorr (2006) also found that unmanaged timber increased interest in silvopasture and that landowners who perceived erosion as a problem were more likely to be interested in windbreaks. Surprisingly, in the case of riparian buffers, Dorr (2006) did not find soil erosion to be a motivating factor. This contrasts with Flower's (2004) findings for farm operators and Valdivia and Poulos' (2009) findings for farming landowners concerned with stream bank erosion.

Field

Recall that *field* refers to the social relationships established among different actors in the practice. For example, farmers may be members of co-ops for the purposes of marketing their products and purchasing inputs. Various studies have identified the effect of social relations among farmers, between farmers/landowners and professionals in conservation at the local and federal levels, and with professionals in the field of forestry. Relations in a field are not necessarily equal (horizontal). They may be unequal (vertical) where there are differences in power relations. For instance, if a farmer is seeking to participate in a federal program, he or she will have to comply with the requirements, as happens with farmers seeking loans or cost sharing or renting land from other landowners (Raedeke et al., 2003). Incentive programs reflect knowledge and contact with institutions in the field of farming and conservation. However, farmers who participated in these programs did not show more interest in or adopt agroforestry more readily than farmers who did not participate in cost-share programs. This is consistent with the results of a study by Valdivia and Konduru (2003), who found no differences among part-time and full-time farmers in relation to interest in agroforestry practices. Batie (2001) found that participation by farmers in government programs was constrained by inadequate funding, targeting problems, complexity of the programs, and reduced capacity of agencies. Arbuckle et al. (2008) explored the influence of the field of farming on interest in agroforestry, especially the influence of the landowners, bankers, and other farmers' opinions. They found that the only potential influence was from bankers, which had a positive effect on interest in agroforestry. Valdivia et al. (2012) identified transaction costs and profitability concerns as the main barriers for adoption.

While adoption of agroforestry remains low (Romanova, 2020; Tsonkova et al., 2018) adoption rates are higher in regions that have organizations actively engaged in agroforestry education and extension, as is the case of Missouri. While recent studies only found adoption rates of 1.5%, in Missouri adoption rates were higher. This is also where UMCA and the University of Missouri research experiment stations were conducting research about agroforestry practices. Related to the field, factors affecting adoption also include the role of social support and individual resources to adopt a new technology (Francesconi et al., 2014; Hayden, 2014; Martini, Roshetko, & Paramita, 2017; Romanova, 2020; Stutzman, Barlow, Morse, Monks, & Teeter, 2019).

Diverse Agroforestry Practices and Decision Makers

Here we elaborate on agroforestry practices from the perspective of the decision maker. Can the technology adjust to the needs and resources of the landowners? Does it blend with the portfolio of activities and characteristics of the household or family farm? Does it match the landowner's goals? Will the technology require a certain amount of land and labor to be viable? Will it yield the environmental benefits or aesthetic values the landowner seeks? Does it require a certain level of investment to be operational? And finally, are the institutions and information available to support the practices?

Characteristics of Agroforestry Practices

Is there a minimum amount of land and investment required? Are there risks to establishment and mechanisms to protect against these risks? Does it require intensive management? Can it generate income streams throughout the life of the practice? Are there institutions and policies to support implementation? Are there markets to sell the products? These are some of the questions about the nature of the technology, and the answers are critical in adoption decisions.

Environmental benefits (services to society—e.g., climate, carbon management, biodiversity, and water quality), scenic and aesthetic beauty, and personal or individual benefits (e.g., from economic returns to a sense of place, values, and well-being for future generations) are among the multiple potential benefits of agroforestry. How does the decision maker benefit, what are the costs, and how does it "fit" in his or her livelihood strategy? This knowledge, accessed by the decision maker, also has an effect on interest and adoption.

Possible Barriers

Four broad categories of barriers to the widespread incorporation of agroforestry practices can be described:

Economic. The availability of specific financial information for multi-enterprise agroforestry practices is often lacking, including known returns for specialty crops and the impacts of the complementary and competitive interactions on the yield of both the tree and non-tree components (Van Vooren et al., 2016). This is further complicated if the enterprise includes specialty crops lacking in readily accessible market information (see Chapter 15).

Landowners will often face barriers in accessing credit to launch these enterprises.

Cultural. Because agroforestry practices run against the monoculture tradition of North American farmer beliefs, values, attitudes, perceptions, and knowledge, ingrained resistance (and healthy skepticism) is found among "traditional" farmers who have farmed for generations (Mattia et al., 2018).

Policy. Lack of incentives and, where present, poorly designed and/or promoted incentives remain major barriers to wider incorporation of agroforestry practices. Until recently, incentive programs have often been too rigid to actually promote the practices that are permitted under their purview; however, this is evolving and incentive programs are becoming more flexible (USDA National Agricultural Statistics Service, 2020; Box 17–3). For example, to successfully establish trees and shrubs, areas of surrounding vegetation must be controlled for at least the first 3 yr. Some federal incentive programs often restrict follow-up management, and agroforestry practices suffer as a consequence as the plants do not thrive and fail to achieve their intended purposes. In addition, incentive programs may also dictate specific numbers of trees per unit area, which often requires planting too many trees for agroforestry practices. Furthermore, incentive programs can prohibit landowners from earning income within the areas set aside for incentive payments. However, agroforestry practices almost always require active management throughout, conflicting with the incentive guidelines and damaging the productivity of the agroforestry practices. Additionally, national incentive programs are administered at the state level (Box 17–3), and if the individuals in charge do not place a high priority on agroforestry, the resulting incentives can be too small to attract the interest of landowners. Finally, professional field-level staff are often unaware of or unfamiliar with agroforestry practices and thus fail to recommend their use to landowners when evaluating options for the application of incentives on the farm. While changes are underway, this may help to explain how these institutional program payments have failed to promote interest in agroforestry until very recently.

Technical Assistance. Knowledgeable social networks must be created to break through the above barriers. Farmers, university extension agents, natural resource professionals, seed and implement dealers, Farm Bureau personnel, loan officers, land appraisers, state conservationists, and crop, animal, and forest commodity groups must all become much better educated with regard to the potential of agroforestry practices. How? Education takes many forms and needs to include a stronger research base, better economic data, better market information, and more on-the-ground demonstrations on landowner property tied together with substantial targeted outreach events (see Chapter 19).

The decision-making process of agroforestry adoption requires examining all costs—along with benefits—together. According to Valdivia et al. (2012), adoption barriers can be due to transaction costs (related to information access and perceived establishment costs) and profitability concerns (associated with perceptions of the effects of agroforestry on farm profitability and agricultural production). Although transaction costs appear to be a greater barrier to implementation across all landowners, environment-concerned landowners are more likely to adopt agroforestry.

Multifunctionality

Valdivia et al. (2012) was based on the lessons learned from more than a decade of research focused on understanding decision makers and the adoption of agroforestry, as well as the culture and values of landowners in rural communities where farming is still an important way of life. The article acknowledged the clearly defined habitus and field institutions of both agriculture and forestry. The question focused on agroforestry—was the practice well defined? At the time the answer was "not yet." The what, how, and why is agroforestry becoming a pathway in farming were answered and multifunctionality and transaction costs were key concepts in the answer. Data for the research came from 360 landowners in Missouri with concerns about the costs of establishing and the time to manage trees. Lack of experience in terms of implementing agroforestry on the farm was also an important concern. Using principal component factor analysis about the barriers perceived by landowners related to transaction costs, the barriers were associated with the economic expenses derived from planting trees. In terms of profitability concerns, the barriers associated with low profitability were connected to multiple factors about producing trees. Transaction costs included: lack of tree management experience, the costs of establishing and managing trees, lack of access to technical information, the time required to manage trees, and the amount of effort needed to clear the land. The profitability concerns included obstacles in the use of farm equipment, long time lags to obtain returns to the investment, the negative effects that trees have on crops, and inadequate market prices for timber. The concerns with transaction costs were greater than those of profitability. The notion of *field* relates closely to transaction costs because it captures the difficulties in access to information, missing institutions (rules or norms), and organizations that can support the practice of agroforestry. Transaction costs capture these institutional barriers in the implementation of agroforestry. Profitability concerns relate to *habitus*. Using cluster analysis with these two concepts, the study identified three types of landowners: environmentalists, agriculturalists, and the disengaged. Environmentalists had moderate concerns about transaction costs and profitability in their decision to plant trees, although transaction costs are more important. Environmentalists were more concerned with scenic beauty and wildlife conservation, while carbon sequestration was more important for agriculturalists, who had greater transaction costs and profitability concerns than the other groups. Those that were called the disengaged did not have either of these concerns and perceived lower benefits of planting trees than the other two groups. In terms of incentives to support planting trees, environmentalists preferred cost sharing and exhibited the lowest preference for rental and incentive programs, which relates to conservation practice programs like EQIP and the Wildlife Habitat Incentives Program (WHIP).

These three landowner types exhibited different perceptions about barriers to agroforestry adoption. These differences were related to the attributes of their farmlands, multifunctionality indicators, and their advantageous resources for the adoption of agroforestry. Environmentalists appear to be the group most likely to adopt agroforestry, and as the case studies have shown, this concern informed how families would pursue their businesses, such as Early Boots Farm, incorporating silvopasture management because of the benefits of trees for the pasture in terms of shade during the summer heat and shelter from cold temperatures and wind in winter. Early Boots Farm is also marketing at local farmers' markets and online, selling a product that is unique.

In terms of established networks and access to information, the institutions and businesses to which these three groups relate, and economic or business factors facilitating information access to reduce transaction costs and/or capitalize on the values that are important, the disengaged group had significantly less contact with conservation agencies. Landowners in the disengaged group also had fewer established connections with federal agricultural agencies, e.g., the NRCS and Soil and Water Conservation Districts, compared with the other two groups. The disengaged group also networked less with other farmers in contrast to the environmentalists and agriculturalists. A smaller proportion in this group (7.6%) participated in agribusiness field days, while 18% of environmentalists and 23% of agriculturalists did. Environmentalist landowners preferred university extension and state agencies for information about planting and managing trees compared with the other groups. Agriculturalists preferred federal agencies and other farmers as sources of information. The disengaged were less interested in contacting agencies to learn about planting or managing trees as well as the lowest interest in obtaining that information from other farmers.

Valdivia et al. (2012) used a multifunctional framework that took into account the multiple motives and goals a landowner places on their land, recognizing a diversity of interests and knowledge about the functions of agriculture ranging from weak to strong multifunctionality (Wilson, 2008). For example, part-time farmers had a lower time commitment in agriculture and were therefore only moderately multifunctional. Lifestyle farms were often in the strong end of the multifunctionality spectrum, as their livelihood does not depend on the land as a production asset. The diversity of relationships to the land is captured in the role of farming in their income generation and in the different types of barriers to the adoption of practices, as Barbieri and Valdivia (2010a) found, with different types of landowners having varying levels of understanding and willingness to adopt agroforestry practices.

Agri-Environmental Policies and Agroforestry

Studies on the adoption of agroforestry practices have increased in the last decade, in tandem with changes in government policies that have included a diversity of agroforestry practices in their programs. The 2017 U.S. Census of Agriculture reported that about 1.5% (30,853) farms self-identified as using one or more agroforestry practices (USDA National Agricultural Statistics Service, 2020). The accuracy of the results, however, should be viewed with reservations due to possible misidentification, given that only brief descriptions of agroforestry practices were provided in the Census of Agriculture forms.

In the study of barriers and incentives to adopt agroforestry practices in Missouri (Valdivia et al., 2012), environmentalists and agriculturalists shared values about the market that reduce transaction costs (i.e., trustworthy buyer, established relationship, reputation, and highest price) that were not shared by the disengaged. Environmentalists and agriculturalists also shared similar values, such as their support of the local community, and both have more established relationships with public conservation and education agencies and social networking. Overall, landowners in these clusters were more connected and had higher social capital bonding and bridging than those in the disengaged cluster, which is consistent with their different levels of involvement in the practice of farming. It is important to highlight that the environmentalists trusted not only horizontal (farmer-to-farmer) but also vertical (organizations) information sources to learn about innovations and technologies related to trees. Horizontal and vertical relationships in social networks, rather than information, have been suggested as contributing to the decision to adopt innovations and management practices (Prokopy, Floress, Klotthor-Weinkauf, & Baumgart-Getz, 2008), which points to the importance of investing in programs that foster connections with others.

In sum, the socio-economic characteristics of environmentalists and agriculturalists in terms of farm size and acres farmed, portfolio diversity, overall awareness of environmental problems, and their horizontal and/or vertical social networks suggest that policies fostering agroforestry adoption should target both groups, as these characteristics appear to be determinants of incorporating trees on the landscape among other conservation management practices (Prokopy et al., 2008). However, it is also necessary to recognize differences between different types of landowners in diffusion efforts, as has been suggested (Barbieri & Valdivia, 2010a). For example, although environmentalists and agriculturalist landowners are well networked and share values related to the benefits of trees and their concerns about conservation, it is imperative to recognize that more landowners in the agriculturalist cluster are actively farming compared with their environmentalist counterparts.

Current and Future Farm Bill Policies: Implications for the Practice of Agroforestry in the Landscape

Incentive structures are powerful motivators for changing practices. According to the findings of Romanova (2020) and Valdivia et al. (2012), access to information and cost-share programs would reduce the barriers to increased adoption of practices that incorporate trees on the landscape. For those more engaged in farming, incentives must consider reduction of costs and increased profitability. Widespread adoption and use of agroforestry will require multiple, integrated, deliberate, and opportunistic approaches including: market-driven and targeted funding, top-down (government) and bottom-up (landowners) efforts, high-tech (research breakthroughs) and high-touch (one-on-one technology transfer) efforts, and active partnerships (Jose et al., 2018). There are a number of promising trends in support of the growth of agroforestry including consumer-driven demand for healthier food and a healthier environment, national efforts to reduce nonpoint-source pollution and increase wildlife habitat, the reality of climate change, and the cost and insecurity of a dependence on fossil fuels (Jose et al., 2018).

Nationwide efforts to reduce nonpoint-source pollution and increase wildlife habitat are supported by national policy through Farm Bill programs (e.g., CRP, EQIP, and WHIP) that support the use of agroforestry practices like riparian forest buffers, windbreaks, and alley cropping. Climate change is causing a shift in interest toward long-term carbon storage, increased use of perennials, and agroecosystem resilience. Despite huge fluctuations in fossil fuel prices, the U.S. military maintains a strong interest in the long-term potential of lignocellulosic biomass and bioenergy as alternative energy sources for reasons of national security (Lovell et al., 2017).

Continued policy changes are needed in future U.S. Farm Bill programs to further stimulate landowner adoption of agroforestry through market-based incentives. A "new generation" of incentive programs is needed in which landowners are expected to pursue alternative market opportunities made available through the establishment of agroforestry practices (Chenyang et al., 2021). Policies should be developed that support perennial crop establishment (Box 17–3), allowing and encouraging landowners to generate income from the trees, shrubs, or alternative crops, as incentive payments are reduced accordingly. Programs that reduce up-front establishment costs and provide income while alternative crops come into production will increase their appeal to landowners who are seeking to earn on-farm income. In turn, such programs will reduce long-term costs to the federal government as landowners are weaned off cost-share programs and become financially self-sufficient.

The Way Forward: Using Agroforestry to Transform the U.S. Landscape

Agroforestry will play a part in a re-envisioned midwestern landscape. It can balance sustainability goals, build the adaptive capacities of rural communities, and strengthen resiliency against "system shocks," e.g., COVID-19 (Prokopy et al., 2020). Wider application of agroforestry will also help to address climate variability and changing consumer demands. This wider application is made possible on the one hand by the changing relationships that have an effect on the rules of the game, as expressed in the way policies change to facilitate the integration of agroforestry practices (Box 17–3). It is also made possible by the values that are in line with sustainability, including work with indigenous communities to support their traditional practices, valuing culture, along with the new generation of family farm businesses such as those profiled in this chapter.

In times of transformational changes, new pathways are identified by the agency of individuals and the development of relationships and values that create the *field* of agroforestry as seen in the changes in public policies in support of agroforestry practices. To transform our landscape, the diversity of agricultural systems at farm, landscape, and market or supply chain levels must be increased to become more resilient for farmers, rural communities, and the environment. Multifunctional working landscapes should include strategic deployment of an array of agroforestry and other continuous living cover practices on large swaths of flood-prone and highly erodible landscapes, while at the same time being a source of income, such as tourism and carbon sequestration. To be successful, landscape diversification must be paired with the development of local and regional processing infrastructure and direct farm-to-consumer markets that enable farmers and food businesses to be financially viable. Reliance on lengthy commodity supply chains and consolidated markets must be curtailed. Finally, governmental and nongovernmental policy tools that incentivize diversification and de-prioritize the current culture of monocropping will be essential (Prokopy et al., 2020).

Summary

Agroforestry practices are very diverse and must be applied strategically across an equally diverse biophysical landscape. Of equal importance, such diversity is a strength in a context where family farms and households in agriculture are also diverse. When individuals, households, or groups access and use resources, these resources become capital that creates a stream of benefits. Access gives individuals the capability to build a livelihood. For Jay Springs Lamb, Early Boots Farm, and East Grove Farms, their human capital (education, knowledge, and search for new information), cultural capital (value of their way of life, and how the seen is connected with the unseen), natural capital (the land and forest they manage), and social capital (e.g., industry associations) have been the means to similar goals. For Jay Springs Lamb, silvopasture was consistent with the goals of the family . . . "a sustainable lifestyle motivates our business." For Early Boots Farm, silvopasture was consistent with their dual goals of "producing food in a sustainable manner and regenerating lands." For East Grove Farms, perennial fruit crops were consistent with their goals of "profit, soil preservation, and building community." These forms of capital and the producers' capability (ability to act) lead to the accumulation of assets (economic and noneconomic), where agroforestry contributed to their target outcomes in life. The field of agroforestry is developing in ways that are supportive of a diversity of farmers who are seeking new ways of making a living or keeping their sustainable traditions, as is the case for indigenous communities in North America. The changing policies emergent from experiences of working together facilitate learning, adoption, and adaptation. *Agency* is at the center of this process.

Acknowledgments

We thank L. Zabek for permission to use the British Columbia story as an example published in the 2nd edition and Ann Robinson from the Leopold Center, and Kelly Paul and Barituka Bekee, graduate research assistants in Agricultural and Applied Economics for their work. This work was supported by Hatch funds, Accession no. 1006044 from the USDA National Institute of Food and Agriculture.

References

Abler, D. (2004). Multifunctionality, agricultural policy, and environmental policy. *Agricultural and Resource Economics Review*, *33*, 8–17. https://doi.org/10.1017/S1068280500005591

Alonso, A. D., Kok, S., & O'Shea, M. (2018). Family businesses and adaptation: A dynamic capabilities approach. *Journal of Family and Economic Issues*, *39*, 683–698.

Arbuckle, J.G., Jr., Valdivia, C., Raedeke, A., Green, J., & Rikoon, J. S. (2008). Non-operator landowner interest in agroforestry practices in two Missouri watersheds. *Agroforestry Systems*, *75*, 73–82.

Baah-Acheamfour, M., Chang, S. X., Bork, E. W., & Carlyle, C. N. (2017). The potential of agroforestry to reduce atmospheric greenhouse gases in Canada: Insight from pairwise comparisons with traditional agriculture, data gaps and future research. *The Forestry Chronicle*, 93(2), 180–189, 10.5558/tfc2017-024

Barbieri, C., Mahoney, E., & Butler, R. (2008). Understanding the nature and extent of farm and ranch diversification in North America. *Rural Sociology*, *73*, 205–229.

Barbieri, C., & Valdivia, C. (2010a). Recreational multifunctionality and its implications for agroforestry diffusion. *Agroforestry Systems*, *79*, 5–18.

Barbieri, C., & Valdivia, C. (2010b). Recreation and agroforestry: Examining new dimensions of multifunctionality in family farms. *Journal of Rural Studies*, *26*, 465–473.

Batie, S. S. (2001). *Public programs and conservation on private lands*. Paper presented at Private lands, public benefits: A policy summit on working lands conservation. Washington, DC: National Governor's Association.

Bebbington, A. (1999). Capitals and capabilities: A framework for analyzing peasant viability, rural livelihoods and poverty. *World Development*, *27*, 2021–2044.

Borremans, L., Reubens, B., Van Gils, B., Baeyens, D., Vandevelde, C., & Wauters, E. (2016). A sociopsychological analysis of agroforestry adoption in Flanders: Understanding the discrepancy between conceptual opportunities and actual implementation. *Agroecology and Sustainable Food Systems*, *40*, 1008–1036. https://doi.org/10.1080/21683565.2016.1204643

Bourdieu, P. (1990). *The logic of practice* (R. Nice, Trans.). Stanford, CA: Stanford University Press.

Buck, L. E. (1995). Agroforestry policy issues and research directions in the U.S. and less developed countries: Insights and challenges from recent experience. *Agroforestry Systems*, *30*, 57–73.

Burton, R. J. F., & Wilson, G. A. (2006). Injecting social psychology theory into conceptualisations of agricultural agency: Towards a post-productivist farmer self-identity? *Journal of Rural Studies*, *22*, 95–115. https://doi.org/10.1016/j.jrurstud.2005.07.004

Cartwright, L., Goodrich, N., Cai, Z., & Gold, M. (2017). *Using NRCS technical and financial assistance for agroforestry and woody crop establishment through the Environmental Quality Incentives Program (EQIP)*. Columbia, MO: Center for Agroforestry, University of Missouri. Retrieved from https://centerforagroforestry.org/wp-content/uploads/2021/05/NRCS_AgroforestryandWoodyCrop.pdf

Chambers, R., & Conway, G. R. (1992). Sustainable rural livelihoods: Practical concepts for the 21st century (IDS Discussion Paper 296). Brighton, UK: Institute of Development Studies.

Chen, M. A., & Dunn, E. (1996). Household economic portfolios: Assessing the impact of microenterprise services (AIMS). Washington, DC: Management Systems International. http://www.uncdf.org/mfdl/readings/HHecoPF.pdf

Chenyang, L., Currie, A., Darrin, H., & Rosenberg, N. (2021). Farming with Trees: Reforming U.S. Farm Policy to Expand Agroforestry and Mitigate Climate Change. *Ecology Law Quarterly*, *48*(1). https://dx.doi.org/10.2139/ssrn.3717877

Cooperrider, D., Sorenson, P., Jr., Whitney, D., & Yager, T. (2000). *Appreciative inquiry: Rethinking human organization toward a positive theory of change*. Champaign, IL: Stripes Publishing.

de Haan, L. J. (2000). Globalization, localization and sustainable livelihood. *Sociologia Ruralis*, *40*, 339–365.

Dobbs, T., & Pretty, J. N. (2004). Agri-environmental stewardship schemes and "multifunctionality". *Review of Agricultural Economics*, *26*, 220–237.

Dorr, H. (2006). Non-operator and farm operator landowner interest in agroforestry in Missouri (Master's thesis). Dep. of Agricultural Economics, University of Missouri, Columbia. Retrieved from http://link.umsl.edu/resource/MAFOW7RUgHw/

Ellis, F. (1998). Household strategies and rural livelihood diversification. *Journal of Development Studies*, *35*, 1–38.

Ervin, C. A., & Ervin, D.E. (1982). Factors affecting the use of soil conservation practices: Hypothesis, evidence, and policy implications. *Land Economics*, *58*, 277–292.

Faulkner, P. E., Owooh, B. & Idassi, J. (2014). Assessment of the adoption of agroforestry technologies by limited-resource farmers in North Carolina. *Journal of Extension*, *52*(5), 5RIB7.

Federation of British Columbia Woodlot Associations. (2007). *Report on British Columbia's woodlot license program*. Retrieved from https://woodlot.bc.ca/wp-content/uploads/2014/01/FBCWA_Report_BCsWoodlotLicenseProgram.pdf.

Fleming, A., O'Grady, A. P., Mendham, D., England, J., Mitchell, P., Moroni, M., & Lyons, A. (2019). Understanding the values behind farmer perceptions of trees on farms to increase adoption of agroforestry in Australia. *Agronomy for Sustainable Development*, *30*, 9. 10.1007/s13593-019-0555-5

Flora, C. B. (2001). Access and control of resources: Lessons from the SANREM CRSP. *Agriculture and Human Values*, *18*(1), 41–48.

Flora, C. B., Flora, J. L., & Gasteyer, S. P. (2016). Rural communities: Legacy and change (5th ed.). Boulder, CO: Westview Press.

Flower, T. (2004). *Characteristics of farm operator attitudes and interest in agroforestry in Missouri* (Publication no. 1426055) [Master's thesis, University of Missouri–Columbia]. Proquest Dissertations Publishing.

Flower, T., Valdivia, C., & Dorr, H. (2005, 24–27 July). *Habitus and interest in agroforestry practices in Missouri*. Paper presented at the American Agricultural Economics Association Annual Meeting, Providence, RI.

Francesconi, W., Nair, P. R., Stein, T. V., Levey, D. J., Daniels, J. C., & Cullen, L. J. (2014). Agroforestry dissemination and the social learning theory in Pontal do Paranapanema, São Paulo, Brazil. *International Journal of Environmental Sustainability*, *9*. https://doi.org/10.18848/2325-1077/CGP/v09i04/55101

Fregene, E. (2007). *Policy and program incentives and the adoption of agroforestry in Missouri* (Master's thesis). Dep. of Agricultural Economics, University of Missouri, Columbia. Retreived from http://link.umsl.edu/resource/VwXyE-p8Jgc/

Gao, J., Barbieri, C., & Valdivia, C. (2014a). A socio-demographic examination of the perceived benefits of agroforestry. *Agroforestry Systems*, *8*, 301–309.

Gao, J., Barbieri, C., & Valdivia, C. (2014b. Agricultural landscape preferences: Implications for agritourism development. *Journal of Travel Research*, *53*, 366–379.

Garrett, H. E. (Ed.). (2009) *North American agroforestry: An integrated science and practice* (2nd ed.). Madison, WI: ASA.

Garrett, H. E., Kurtz, W. B., Buck, L. E., Hardesty, L. H., Gold, M. A., Pearson, H. A., . . . Slusher, J. P. (1994). *Agroforestry: An integrated land-use management system for production and farmland conservation.* Washington, DC: USDA Soil Conservation Service.

Glover, J. L. (2010). Capital usage in adverse situations: Applying Bourdieu's theory of capital to family farm businesses. *Journal of Family Economic Issues, 31,* 485–497.

Godsey, L. D., Mercer, D. E., Grala, R. K., Grado, S. C., & Alavalapati, J. R. R. (2009). Agroforestry economics and policy. In H. E. Garrett (Ed.), *North American agroforestry: An integrated science and practice* (2nd ed., pp. 315–338). Madison, WI: ASA.

Gold, M. A., Godsey, L. D., & Josiah, S. J. (2004). Markets and marketing strategies for agroforestry specialty products in North America. *Agroforestry Systems, 61,* 371–382.

Gold, M. A., & Hanover, J. W. (1987). Agroforestry systems for the temperate zone. *Agroforestry Systems, 5,* 109–121.

Hayden, T.P. (2014) Using the medical loss ratio to incentivize the adoption of innovative medical technology. *Vanderbilt Journal of Entertainment and Technology Law, 17*(1), 239+.

Haynes, R. W., Monserud, R. A., & Johnson, A. C. (2003). Compatible forest management: Background and context. Chapter 1. *In* R.A. Monserud, R.W. Haynes, and A.C. Johnson (eds.) *Compatible forest management* (pp. 3–32). Dordrect, the Netherlands: Kluwer Academic Publishers. Retrieved from https://www.fs.fed.us/pnw/pubs/journals/pnw_2003_haynes001.pdf

Holderieath, J., Valdivia, C., Godsey, L., & Barbieri, C. (2012). The potential for carbon offset trading to provide added incentive to adopt silvopasture and alley cropping in Missouri. *Agroforestry Systems, 86,* 345–353. 10.1007/s10457-012-9543-3

Hoppe, R. A. (Ed.). (2001). *Structural and financial characteristics of US Farms: 2001 family farm report* (Information Bull. 768). Washington, DC: USDA Economic Research Service. Retrieved from https://www.ers.usda.gov/publications/pub-details/?pubid=42391.

Jose, S., Gold, M. A., & Garrett, H. E. (2018). Temperate agroforestry in the United States: Current trends and future directions. In A. M. Gordon, S. M. Newman, & B. Coleman (Eds.), *Temperate agroforestry systems* (2nd ed., pp. 50–71). Wallingford, UK: CAB International.

Keeley, K. O., Wolz, K. J., Adams, K. I., Richards, J. H., Hannum, E., von Tscharner Fleming, S., & Ventura, S. J. (2019). Multi-party agroforestry: Emergent approaches to trees and tenure on farms in the Midwest USA. *Sustainability, 11,* 2449. 10.3390/su11082449

Knutson, R. D., Smith, E. G., Anderson, D. P., & Richardson, J. W. (1998). Southern farmers' exposure to income risk under the 1996 farm bill. *Journal of Agricultural and Applied Economics, 30,* 35–46.

Koontz, T. M. (2001). Money talks—but to whom? Financial versus nonmonetary motivations in land use decisions. *Society and Natural Resources, 14,* 51–65.

Kremen, C., & Merenlender, A. M. (2018). Landscapes that work for biodiversity and people. *Science, 362,* eaau6020. 10.1126/science.aau6020

Loughrey, J., Donnellan, T., Hennessy, T., & Hanrahan, K. (2013). *The role of pluriactivity in farm exit and labour supply decisions.* Factor Markets Working Paper no. 67. Retrieved from http://aei.pitt.edu/58611/1/Factor_Markets_67.pdf.

Lovell, S. T., Dupraz, C., Gold, M., Jose, S., Revord, R., Stanek, E., & Wolz, K. (2017). Temperate agroforestry research: Considering multifunctional woody polycultures and the design of long-term field trials. *Agroforestry Systems, 92,* 1397–1415. 10.1007/s10457-017-0087-4

Lynch, L., & Brown, C. (2000). Landowner decision making about riparian buffers. *Journal of Agricultural and Applied Economics, 32,* 585–596.

MacFarland, K., Elevitch, C., Friday, J. B., Friday, K., Lake, F. L., & Zamora, D. (2017). Human dimensions of agroforestry systems. In M. M. Schoeneberger, G. Bentrup, & T. Patel-Weynand (Eds.), *Agroforestry: Enhancing resiliency in U.S. agricultural landscapes under changing conditions* (Gen. Tech. Rep. WO-96, pp. 73–90). Washington, DC: U.S. Forest Service. 10.2737/WO-GTR-96

Martini, E., Roshetko, J. M., & Paramita, E. (2017). Can farmer-to-farmer communication boost the dissemination of agroforestry innovations? A case study from Sulawesi, Indonesia. *Agroforestry Systems, 91,* 811–824. https://doi.org/10.1007/s10457-016-0011-3

Matthews, S., Pease, S. M., Gordon, A. M., & Williams, P. A. (1993). Landowner perceptions and the adoption of agroforestry practices in southern Ontario, Canada. *Agroforestry Systems, 21,* 59–68.

Mattia, C. M., Lovell, S. T., & Davis, A. (2018). Identifying barriers and motivators for adoption of multifunctional perennial cropping systems by landowners in the Upper Sangamon River Watershed, Illinois. *Agroforestry Systems, 92,* 1155–1169.

Mercer, D. E., & Miller, R. P. (1998). Socioeconomic research in agroforestry: Progress, prospects, priorities. *Agroforestry Systems, 38,* 177–193.

Mishra, A. K., & Sandretto, C. L. (2003). Stability of farm income and the role of nonfarm income in U.S. agriculture. *Review of Agricultural Economics, 24,* 208–221.

North, D. C. (1990). *Institutions, institutional change and economic performance.* Cambridge, UK: Cambridge University Press.

North, D. C. (1993). *The new institutional economics and development.* Retrieved from https://econwpa.ub.uni-muenchen.de/econ-wp/eh/papers/9309/9309002.pdf.

Pannell, D. J. (1999). Social and economic challenges in the development of complex farming systems. *Agroforestry Systems, 45,* 393–409.

Pattanayak, S. K., Mercer, D. E., Sills, E., & Yang, J. (2003). Taking stock of agroforestry adoption studies. *Agroforestry Systems, 57,* 173–186.

Pirog, R., & Bregendahl, C. (2012). *Creating change in the food system: The role of regional food networks in Iowa.* East Lansing, MI: MSU Center for Regional Food Systems. Retrieved from https://www.canr.msu.edu/resources/creating-change.

Plieninger, T., Muñoz-Rojas, J., Buck, L. E., & Scherr, S. J. (2020). Agroforestry for sustainable landscape management. *Sustainability Science, 15,* 1255–1266. 10.1007/s11625-020-00836-4

Prokopy, L. S., Floress, K., Klotthor-Weinkauf, D., Baumgart-Getz, A. (2008). Determinants of agricultural best management practice adoption: Evidence from the literature. *Journal of Soil and Water Conservation, 63,* 300–311.

Prokopy, L. S., Gramig, B. M., Bower, A., Church, S. P., Ellison, B., Gassman, P. W., . . ., Ulrich-Schad, J. D. (2020). The urgency of transforming the midwestern U.S. landscape into more than corn and soybean. *Agriculture and Human Values, 37,* 537–539. 10.1007/s10460-020-10077-x

Raedeke, A. H., Green, J. J., Hodge, S. S., & Valdivia, C. (2003). Farmers, the practice of farming and the future of agroforestry: An application of Bourdieu's concepts of field and habitus. *Rural Sociology, 68,* 64–86.

Rhodes, T. K., Aguilar, F. X., Jose, S., & Gold, M. A. (2018). Factors influencing the adoption of riparian forest buffers in the Tuttle Creek Reservoir watershed of Kansas, USA. *Agroforestry Systems, 92,* 739–757. 10.1007/s10457-016-0045-6

Robinson, A. Y. (2013a). *Agroforestry case studies: Silvopasture at Early Boots Farm* (Publications and Papers 51). Ames, IA: Leopold Center, Iowa State University. Retrieved from http://lib.dr.iastate.edu/leopold_pubspapers/51

Robinson, A. Y. (2013b). *Agroforestry case studies: Elderberry business on historic farm* (Publications and Papers 55). Ames, IA: Leopold Center, Iowa State University. Retrieved from http://lib.dr.iastate.edu/leopold_pubspapers/55

Rois-Díaz, M., Lovric, N., Lovric, M., Ferreiro-Domínguez, N., Mosquera-Losada, M. R., den Herder, M.,. Burgess, P. (2018). Farmers' reasoning behind the uptake of agroforestry practices: Evidence from multiple case-studies across Europe. *Agroforestry Systems, 92*, 811–828. https://doi.org/10.1007/s10457-017-0139-9

Romanova, O. (2020). *Factors influencing practitioner adoption of agroforestry: A USDA SARE case study* [Unpublished master's thesis]. University of Missouri.

Rossier, C., & Lake, F. (2014). *Indigenous traditional ecological knowledge in agroforestry* (Agroforestry Note 44). Lincoln, NE: USDA National Agroforestry Center. https://www.fs.usda.gov/nac/assets/documents/agroforestrynotes/an44g14.pdf

Rule, L. C., Flora, C. B., & Hodge, S.S. (2000). Social dimensions of agroforestry. In H. E. Garrett, W. J. Rietveld, & R. F. Fisher (Eds.) North American agroforestry: An integrated science and practice (p. 361–386). Madison, WI: ASA.

Scherr, S. J. (1995). Economic factors in farmer adoption of agroforestry: Patterns observed in western Kenya. *World Development, 23*, 787–804.

Scoones, I. (1998). *Sustainable rural livelihoods: A framework for analysis* (IDS Working paper 72). Brighton, UK: Institute of Development Studies.

Shucksmith, M. (1993). Farm household behaviour and the transition to post productivism. *Journal of Agricultural Economics, 44*, 466–476.

Shucksmith, M., &Herrmann, V. (2002). Future changes in British agriculture: Projecting divergent farm household behaviour. *Journal of Agricultural Economics, 53*, 37–50.

Skees, J. R., Harwood, J., Somwaru, A., & Perry, J. (1998). The potential for revenue insurance policies. *Southern Journal of Agricultural Applied Economics, 30*, 47–61.

Strong, N., & Jacobson, M. (2005). Assessing agroforestry adoption potential utilizing market segmentation: A case study in Pennsylvania. *Small-Scale Forest Economics, Management and Policy, 4*, 215–231.

Stutzman, E., Barlow, R. J., Morse, W., Monks, D., & Teeter, L. (2019). Targeting educational needs based on natural resource professionals' familiarity, learning, and perceptions of silvopasture in the southeastern U.S. *Agroforestry Systems, 93*, 345–353. https://doi.org/10.1007/s10457-018-0260-4

Thevathasan, N. V., Gordon, A. M., Bradley, R., Cogliastro, A., Folkard, P., Grant, R., . . . Zabek, L. (2012). Agroforestry research and development in Canada: The way forward. In P. K. R. Nair & D. Garrity (Eds.), Agroforestry: The future of global land use (Advances in Agroforestry 9, pp. 247–285). Heidelberg, Germany: Springer.

Titus, B. D., Kerns, B. K., Cocksedge, W., Winder, R. S., Pilz, D., Kauffman, G., . . . and Ballard, H. L. (2004, 2–6 Oct.). Compatible (or co-) management of forests for timber and non-timber values. In Proceedings of the Canadian Institute of Forestry and Society of American Foresters Joint 2004 Annual General Meeting and Convention, Edmonton, AB [CD-ROM]. Bethesda, MD: Society of American Foresters. Retrieved from https://d1ied5g1xfgpx8.cloudfront.net/pdfs/25245.pdf

Trozzo, K. E., Munsell, J. F., & Chamberlain, J. L. (2014a). Landowner interest in multifunctional agroforestry riparian buffers. *Agroforestry Systems, 88*, 619–629.

Trozzo, K. E., Munsell, J. F., Chamberlain, J. L., & Aust, W. M. (2014b). Potential adoption of agroforestry riparian buffers based on landowner and streamside characteristics. *Journal of Soil and Water Conservation, 69*, 140–150.

Tsonkova, P., Mirck, J., Böhm, C., & Fütz, B. (2018). Addressing farmer-perceptions and legal constraints to promote agroforestry in Germany. *Agroforestry Systems, 92*, 1091–1103. https://doi.org/10.1007/s10457-018-0228-4

Udawatta, R. P., Gantzer, C. J., & Jose, S. (2017). Agroforestry practices and soil ecosystem services. In M.M. Al-Kaisi & B. Lowery (Eds), Soil health and intensification of agroecosystems (pp. 305–333). London: Academic Press. 10.1016/B978-0-12-805317-1.00014-2

USDA National Agricultural Statistics Service. (2020). 2017 census of agriculture. Washington, DC: USDA–NASS. Retrieved from https://www.nass.usda.gov/Quick_Stats/CDQT/chapter/2/table/43/year/2017

Valdivia, C. (2004). Andean livelihoods and the livestock portfolio. *Culture & Agriculture, 26*, 69–79.

Valdivia, C. (2007, 10–13 June). The effect of land fragmentation on habitus, field, and agroforestry in the Midwest. In A. Olivier and S. Campeau (Eds.) When trees and crops get together: Economic opportunities and environmental benefits from agroforestry: Proceedings of the 10th North American Agroforestry Conference, Québec, Canada (pp. 621–633). Quebec, Canada: Université Laval. https://giraf.fsaa.ulaval.ca/proceedings_congres_qc2007.pdf

Valdivia, C., & Barbieri, C. (2014). Experiential agritourism: A sustainable strategy for adapting to climate change in the Andean Altiplano. *Tourism Management Perspectives, 11*, 18–25.

Valdivia, C., Barbieri, C., & Gold, M. (2012). Between forestry and farming: Policy and environmental implications of the barriers to agroforestry adoption. *Canadian Journal of Agricultural Economics, 60*(2), 155–175.

Valdivia, C., Dunn, E., & Jetté, C. (1996). Diversification as a risk management strategy in an Andean agropastoral community. *American Journal of Agricultural Economics, 78*, 1329–1334.

Valdivia, C., & Gilles, J. L. (2001). Gender and resource management: Households and groups, strategies and transitions. *Agriculture and Human Values, 18*, 5–9.

Valdivia, C., Gold, M., Zabek, L., Arbuckle, J., & Flora, C. (2009). Human and institutional dimensions of agroforestry. In H. E. Garrett (Ed.), *North American agroforestry: An integrated science and practice* (2nd ed., pp. 339–367). Madison, WI: ASA. https://acsess.onlinelibrary.wiley.com/doi/abs/10.2134/2009.northamericanagroforestry.2ed.c13

Valdivia, C., Hodge, S., & Raedeke, A. (2002, 17–20 Nov.). Rural livelihoods and agroforestry practices in the Missouri flood plains. In *Small Farms in an Ever Changing World: Meeting the challenges of sustainable livelihoods and food security in diverse rural communities: Proceedings of the 17th Symposium of the International Farming Systems Association, Lake Buena Vista, FL* [CD-ROM]. Gainesville, FL: University of Florida, Institute of Food and Agricultural Sciences.

Valdivia, C., & Konduru, S. (2003, 22–25 June). Interest in agroforestry practices, farmer diversification strategies and government transfers in Missouri's northeast and southeast. In S. Sharrow (Ed.), *Proceedings of the 8th Biennial Conference on Agroforestry in North America, Corvallis, OR* (pp. 281–296). Columbia, MO: Association for Temperate Agroforestry.

Valdivia, C., & Poulos, C. (2009). Factors affecting farm operators' interest in incorporating riparian buffers and forest farming practices in northeast and southeast Missouri. *Agroforestry Systems, 75*, 61–67.

Valdivia, C., Raedeke, A., Hodge, S., Green, J., & Godsey, L. (2000). *The economic and social value of flood plain agroforestry to rural development: A baseline economic and social profile of producers in northeast and southeast of Missouri, the Fox–Wyaconda watershed and Scott County.* Agricultural Economics Working Paper AEWP 2000-4. Columbia, MO: Dep. of Agricultural Economics, University of Missouri.

Van Vooren, L., Reubens, B., Broekx, S., Pardon, P., Reheul, D., van Winsen, F., . . . Lauwers, L. (2016). Greening and producing: An economic assessment framework for integrating trees in cropping systems. *Agricultural Systems, 148*, 44–57. 10.1016/j.agsy.2016.06.007

Whitt, C., MacDonald, J., & Todd, J. E. (2019). *America's diverse family farms: 2019 edition* (Economic Information Bulletin 214). Washington, DC: USDA Economic Research Service. Retrieved from https://www.ers.usda.gov/publications/pub-details/?pubid=95546.

Williams, P. A., Gordon, A. M., Garrett, H. E., & Buck, L. (1997). Agroforestry in North America and its role in farming systems. In A. M. Gordon and S. M. Newman (Eds.), *Temperate agroforestry systems* (pp. 9–84). New York: CAB International.

Wilson, G. A. (2008). From 'weak' to 'strong' multifunctionality: Conceptualising farm-level multifunctional transitional pathways. *Journal of Rural Studies, 24*, 367–383.

Winters, P., Davis, B., & Corral, L. (2002). Assets, activities and income generation in rural Mexico: Factoring in social and public capital. *Agricultural Economics, 27*, 139–156.

Workman, S. W., Bannister, M. E., & Nair, P. K. R. (2003). Agroforestry potential in the southeastern United States: Perceptions of landowners and extension professionals. *Agroforestry Systems, 59*, 73–83.

Study Questions

1. What did the owners of Jay Springs Lamb Co. do to help realize their personal and business goals?

2. What did the owners of Jay Springs Lamb Co. do to strengthen their direct marketing approach?

3. What factor(s) led Jay Springs Lamb Co. to begin practicing silvopasture and shift to direct marketing?

4. Identify three factors that shape or influence human livelihood decisions.

5. Describe what is meant by capital and capabilities in the context of landowner decision-making. What other factors influence decision-making?

6. Describe the four elements of structures and institutions that impact landowner strategies with regard to agroforestry practices.

7. Explain the major differences that exist between the structures that influence decisions in rural livelihood strategies in North America compared with developing countries.

8. What are the roles of diversification relative to livelihood strategies and agroforestry?

9. What is meant by the term *field* relative to human dimensions and agroforestry?

10. Identify barriers that can influence more widespread use of agroforestry practices.

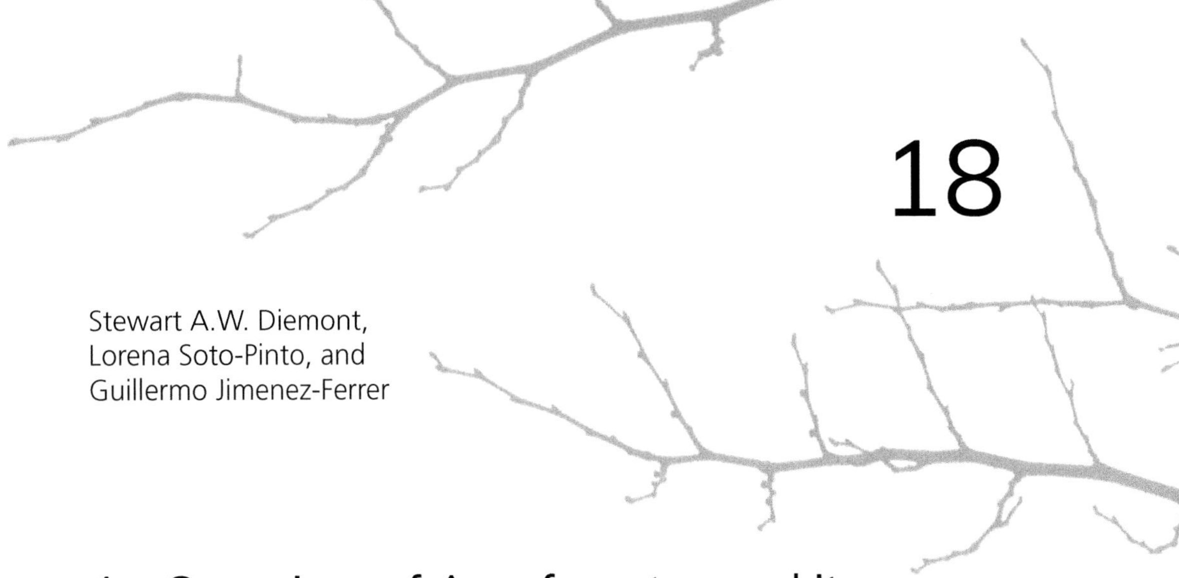

18

Stewart A.W. Diemont,
Lorena Soto-Pinto, and
Guillermo Jimenez-Ferrer

An Overview of Agroforestry and its Relevance in the Mexican Context

Mexico offers a unique set of landscapes, politics, cultures, and history, which makes it a critical part of North American agroforestry. Studies of systems like the *milpa*, home gardens, and silvopastoral systems have historically led to framing Mexican agroforestry as a traditional, integrated agroecosystem (Hernández-Xolocotzi, 1977). One of the original centers of agriculture, beginning ~10,000 yr ago with rotational *milpa* agroforestry, Mexico has long been skillful in the use of plant, fungus, and animal resources and natural classifications of soils and the living world (Schmitter, Mariaca-Méndez, & Soto-Pinto, 2016). Since the mid-20th century, much has been published in Mexico about the origins of agriculture, ethnobiology, domestication, and archeology describing the composition, structure, and functions of ancient and traditional agroecosystems (Coe, 1964; Cutler, 1968). In the 1970s, this work expanded to the agroecology of traditional agroforestry (Hernández-Xolocotzi, 1977; Gliessman, Garcia, & Amador, 1981). The concept of agroforestry systems for Mexico was soon after formalized through the International Centre for Research in Agroforestry (ICRAF) and by Centro Agronómico Tropical de Investigación y Enseñanza (CATIE) in Mexico, Central America, and the Caribbean (Budowsky, 1979a, 1979b). Currently in Mexico, development agencies, research centers, and producer organizations are re-envisioning agroforestry in a modern context while embracing its history. Agroforestry is a tool that makes it possible for individuals, families, and communities to support themselves and conserve their surrounding environment (Noponen et al., 2017; Soto-Pinto & Anzueto-Martinez, 2016).

Agroforestry as a Human–Nature System

Agroforestry in Mexico has historically moved beyond the technical stage to a sacred place of worship (Gomez-Pompa, Salvador, & Fernandez, 1990) and community building (Klooster & Masera, 2000). It is based on an intimate relationship between humans and their environment (Martin, Roy, Diemont, & Ferguson, 2010). Humans are not separated from the environment; instead, livelihoods, rituals, and day-to-day existence are intimately melded with the environment. Ceremony and ritual related to the forest are still part of peoples' lives (Nigh, 1976; Nigh & Diemont, 2013). In some areas, the agroforest is believed to still be alive with spirits that guide how the forest is managed (Terán & Rasmussen, 2009). This relationship is widespread and deeply felt (Tiedje, 2018). It affects how people manage the forest and their decisions (Solorio et al., 2017).

North American Agroforestry, Third Edition. Edited by Harold E. "Gene" Garrett, Shibu Jose, and Michael A. Gold.

This spiritual relationship forms the basis for an interaction with forests that creates a human–nature system, where humans become inseparable from the nature around them (Atran et al., 2002). In understanding the significance of agroforestry in Mexico, an understanding of this human–nature connectivity is critical. And, in designing change utilizing techniques that are indigenous to the area and in determining how agroforestry can move forward, it is critical to embrace this understanding. In designing the agroforest, there is a need to be cognizant of the lives of the people who interact with it (Martin et al., 2010).

Overview of Agroforestry Past in Mexico

Because agroforestry in Mexico is derived directly from a rich agricultural past (Ford & Nigh, 2016), those cultures and their history are briefly described. An exhaustive description of the cultures, economies, and landscapes that are located in present-day Mexico would require volumes, and much has been written (Longhena, 1998; Rojas & Sanders, 1985; Staller & Carrasco, 2010). An important source of current agroforestry knowledge comes directly from the indigenous peoples, notably from Mayan groups in southern Mexico.

The Maya civilization dates back to the first century, and the Classic period of the Maya lasted until after the ninth century, when Mayan civilizations in southern Mexico, Guatemala, Honduras, and Belize rapidly declined. At roughly this time, city states in other parts of the Yucatan Peninsula in Mexico increased in population and survived until Spanish *conquistadors* arrived in the 1500s (Sharer & Traxler, 2005). The Maya are still a part of today's cultures. More than 15 Mayan languages are spoken in southern Mexico (Instituto Nacional de Estadística y Geografía, 2005), and Mayan land management today derives from the ancient Maya (Diemont & Martin, 2009; Ford & Nigh, 2016; Nigh & Diemont, 2013).

Ancient Maya utilized a combination of landscapes to provide food for city states (Fedick, 1996). The *milpa* with the associated forest was a spiritually important form of production (Barrera-Bassols & Toledo, 2005). The *milpa* was a successional agroforestry system with corn (*Zea mays* L.) as the primary crop in the field stage. The home garden agroforest was a critical part of people's day-to-day production. Silvopastoral systems, through the use of foraged leaves and fruits of *Brosimum alicastrum*, have been used for many centuries by the Maya and their descendants to feed a diverse array of animals, such as deer. These systems were linked into one overall system (Fedick, 1996; Fedick, Allen, & Gomez-Pompa, 2003; Ford & Nigh, 2016).

Colonization by Europeans led to a mix of technologies with Mayan traditions. Plants, tools, and knowledge were added, all the while maintaining traditions. For instance, the use of animals for traction, the use of the hoe, new crops such as wheat (*Triticum aestivum* L.), cruciferous species for leaves, citrus fruits, deciduous fruit trees, as well as sheep, horses, pigs, cattle, and chickens were introduced. These new elements enriched the systems, the availability and diversity of food, and other use values. Mexico has a large number of domesticated plants derived from agroforestry systems. Avocado (*Persea americana* Mill.), corn, bean (*Phaseolus vulgaris* L., *P. lunatus* L.), chili (*Capsicum annuum* L.), agave (*Agave* spp.), prickly pear (*Opuntia* spp.), cocoa (*Theobroma cacao* L.), and tomato (*Solanum lycopersicum* L.), for example, were domesticated. With the appropriation of communal lands and the ownership of land in the hands of the individuals, the traditional systems withdrew to the marginal areas, but they subsisted.

After the Mexican war for independence from Spain in the 19th century, the lands were mainly in the hands of the Spanish colonizers, militaries, and church, forming the *haciendas* and plantations; little was held by indigenous groups. The indigenous people supplied the labor force for the Spanish agricultural companies. Likewise, the massive use of indigenous workers (*peones*) affected the capacity and autonomy of food production of the original communities.

After the Mexican Revolution of the early 20th century, private properties were divided and the lands returned to the campesinos and indigenous people, predominantly in the form of *ejidos* land tenure system—a communal resource-holding institution where a community was granted land by the federal government. The *ejido* land is regulated by an assembly and individually managed by each farmer (often through family decision making) (López-Cruz, Soto-Pinto, Salgado-Mora, & Huerta-Palacios, 2020). *Haciendas* and plantations, mainly devoted to commercial production, were returned to *milpa* agroforestry.

Agroforestry as a Cultural Driver

As the historical center of the landscape, traditional agroforestry is a central part of Mexico's culture (De Frece & Poole, 2008). This nature-centered self-awareness drives how nature is tended, approached, left, and passed to the next generation (Salmón, 2000). Humans are made of corn in traditional Mayan belief (Bassie-Sweet, 2008; Goetz & Morley, 1950). Ancient Maya had corn deities (Bassie, 2002; Taube, 1985), which were a part of the very recent past. In Naha, Chiapas, in southern Mexico, chapter author Stewart Diemont recalls participating in the annual corn ceremony with other men in the community, a ceremony in which corn and corn mush were eaten to worship and pray to the god of corn. This experience was fewer than two decades ago.

This connection between people and nature, and indeed between people and their agroforestry, cannot be overstated. When the people are managing and caring for the agroforestry system, when they plant corn that makes up more than half of the field crops in agroforestry (Diemont & Martin, 2009), when they harvest and prepare tortillas, they are managing and caring for planting, harvesting, and preparing themselves for a better life (Bassie-Sweet, 2008). In mistreating, harming, degrading, taking without giving back, denuding, or doing anything that reduces productivity, they are hurting themselves.

Family as the Center of Agroforestry

The forest is part of the Mexican heritage and therefore it is at the center of what represents the family. Family is also at the center of agroforestry. The family in Mexico revolves around agroforestry in much of the country, and agroforestry revolves around the family. Every member of the family participates. Children learn about the environment at the same time they are learning about themselves (Falkowski, Martinez-Bautista, & Diemont, 2015). A child will go with a parent to plant, weed, tend, and harvest. Children will participate with parents in fire ceremonies. They will learn early how to safely use a machete. It is not unusual to see a 10-yr-old boy *limpiando*, or cleaning, a *milpa* with a machete half his size. Parents require participation in agroforestry and meeting the needs of the *milpa* (swidden), *cafetal* (coffee [*Coffea arabica* L.] farm), traditional forest–livestock (López-Carmona, Jiménez-Ferrer, De Jong, Ochoa-Gaona, & Nahed-Toral, 2001), and *huerto* (home garden) systems.

Agroforestry and the Economy

A large portion of food and raw materials is produced locally from agroforestry systems. Corn is central to the local economy, and concern over this food staple was at the heart of the 1994 Zapatista uprising (Nations, 1994). However, corn is not the only product that comes from agroforestry. Sugar, coffee, milk, cacao, fruits, and meat are just a few of the additional foods that come directly to families from agroforestry. It could be argued that the local economy disincentivizes participation in a global economy. Agroforestry that is local and centered in the family creates and encourages an economy based in local community (Altieri & Toledo, 2011).

Traditionally, people needed little from outside the community, as most all family needs were provided through their relationship with nature (Nations & Nigh, 1980). Foods, raw materials for construction and clothes, medicines, spiritual enlightenment, and entertainment were all provided by areas within a half-day's walk (Diemont, Martin, & Levy-Tacher, 2006; Falkowski et al., 2015). For many in Mexico today, agroforestry still meets most of their economic needs.

Agroforestry and the Environment

As a natural system, it may seem obvious that agroforestry in Mexico is a driver of environmental change. Because it is such a large part of the land area of Mexico, agroforestry's effects on the environment are equally extensive. These systems both enhance and degrade ecosystems and their associated services. The effect depends on the way that agroforestry practitioners manage their land.

The overall diversity of Mexico accounts for the greatest richness of birds (Peterson & Navarro-Sigüenza, 2016), insects, mammals, plants, and human languages (CONABIO, 2014; Instituto Nacional de Lenguas Indígenas, 2008) in North America. Each of these diversities is directly attributable to agroforestry. Human languages may seem out of place in this list, but human language diversity correlates with biodiversity globally (Gorenflo, Romaine, Mittermeier, & Walker-Painemilla, 2012). Agroforestry in Mexico provides the basis for communication about the natural world. This communication leads to conservation of birds, insects, and mammals. While it is conservation through use, as many animals are used as food, it creates a positive interaction between the people and nature. But people care for their larder. The natural world provides them with sustenance for themselves and their families. Therefore, the incentive is enormously high to care for their forests.

The mosaic of agroforestry in Mexico is a critical source of diversity. Each form of agroforestry adds complementarily to Mexico's diversity. Each agroforestry type from *milpa* to coffee to cocoa to home garden to silvopasture offers unique landscapes and niches. Some of these areas are rich in grains, others in fruit, and others present dense overstories rich in timber, fuelwood, fiber, and other materials for domestic purposes and enhanced opportunities for granivore, frugivore, and insectivore birds, mammals, and insects (Diemont, 2006; Falkowski, 2018). Within successional stages of agroforestry, niches are equally rich due to the high diversity of the plant community in traditional management.

Unfortunately, the story of Mexico's agroforestry is replete with examples that degrade the environment (Soto-Pinto, Castillo-Santiago, & Jiménez-Ferrer, 2012). *Milpa* preparation in particular has a public face that can be negative. During late spring, smoky skies are a reminder of agroforestry on the other side of the mountain range. Yet, the fault does not rest with the traditional *milpa*. It is more a result of land distribution patterns that can lead to limited land areas for individuals and therefore overly aggressive management.

Types of Agroforestry in Mexico

Mexico has a long history of tree uses in family agriculture, and therefore numerous forms of agroforestry have developed. Because many cultures make up the history of Mexico, agroforestry has the nuance of that influence. Agroforestry differs from landscape to landscape in terms of plant community and wildlife, planting, tending, and harvest. A desert of the central plains would not have the same plant community as the humid lowlands of southern Mexico, but some locations in the desert zones are very rich in plants, knowledge, and culture, such as the Tehuacan–Cuicatlan area (Moreno-Calles, Casas, Toledo, & Vallejo, 2016).

The *milpa*, home gardens, coffee farms, cocoa farms, and silvopastoral or agrosilvopastoral systems are all dominant forms of agroforestry in Mexico. The elements of these agroforestry types are described, such as structure and management, plants and wildlife that these systems support, to give a sense of the current complexity of Mexico's agroforestry. Some types are presented in greater detail than others, not due to the relative importance of expanded types but to represent the complexity of all forms of Mexico's agroforestry.

Home Garden Systems

The home garden is a common part of the Mexican landscape that provides easily harvested and readily accessible fruits and vegetables close to the home. The home garden is more than a kitchen garden; it has high diversity, pervasiveness, and a widespread inclusion of trees. Home garden agroforestry includes many types of plant communities, orientations in relation to the house, size, shape, and type of tending (Figure 18–1).

The plant community of the typical home garden of Mexico is rich in herbaceous species, vegetables, and fruit trees. Farmers use home gardens to provide medicine and food for daily use. The home garden supplements or complements the *milpa* and associated forests.

Fig. 18–1. A home garden in the highlands of Chiapas, Mexico.

By planting a rich plant community in the home garden, farmers ensure food security in times of scarcity and complement or substitute for medicine from local clinics, pharmacies, and hospitals. The home garden is a spatially diverse landscape. The fruit trees provide an overstory and are not normally dense. Other parts of the garden are mixed and integrated. Many types of fruit trees are in the home gardens of Mexico. In subtropical and tropical areas, citruses such as orange [*Citrus×sinensis* (L.) Osbeck], lemon [*Citrus limon* (L.) Burm. f.], lime [*Citrus aurantiifolia* (Christm.) Swingle], tangerines (*Citrus reticulata* Blanco), and grapefruit (*Citrus paradisi* Macfad) are found mixed with avocado, peach [*Prunus persica* (L.) Batsch], and guava (*Psidium guajava* L.) (Figure 18–2). In higher elevation landscapes, apple (*Malus* Mill.) is common.

In Mexico, traditional medicine is part of healing. The home garden will typically contain many types of remedies from root to shoot to leaf to flower. These medicinal plants can include trees. Components of magnolias (*Magnolia* spp.), for example, are a traditional remedy for stomach pains. Traditional pharmacies can be found in some cities where medicines have been lightly processed to a dried form or a tincture. For the most part, these medicines are collected from the medicinal garden of the home garden, the *milpa* field, or the *milpa* forest.

In the understory of fruit trees, herbs are a part of most home gardens. Typical herbs found might be epazote (*Dysphania ambrosioides* L.), chamomile (*Matricaria chamomilla* L.), cilantro (*Coriandrum sativum* L.), and rosemary (*Rosmarinus officinalis* L.). The distinction between cultivar and wild may be less defined than in other parts of North America. *Quelites*, or wild herbs, are an important part of the Mexican diet.

Fruits and vegetables are a part of most home gardens. Unlike the herbs, the vegetables and fruits of the home garden are typically cultivars. Squash (*Cucurbita* spp.) are grown alongside melons (*Cucumis* spp. and *Citrullus lanatus* (Thunb.) Matsum. & Nakai]. Whereas in the *milpa* field these may be extensive, in the home garden they are for convenience. For example, runners of cucumber (*Cucumis sativus* L.) sit alongside pepper (*Capsicum annuum* L.), which can be easily harvested in preparing a meal.

A Mexican agroforestry researcher was under the misapprehension that every part of the home garden was for harvest use. One day passing the house of the mother of a friend, he asked what the flower was for, this beautiful yellow and orange flower. The friend replied simply, "My mother thinks they are pretty." Flowers are everywhere in the home garden. Some are herbs, some are medicines, some are fruits and vegetables, but some are just pretty.

Family dogs and cats spend time in the home garden, a cool spot outdoors, where a cat might find mice or insects to eat and a dog will eat food scraps, as will chickens, turkeys and ducks,

Fig. 18–2. A home garden showing extensive tree species.

which are a common sight running and pecking through a home garden. Occasionally a pig, sheep, or goat will be found in the home garden, but these are more controlled, as they could easily damage the plants. Goats and sheep are often pastured separately, and pigs will generally receive feed and scraps. Rodents are part of any home garden, as are birds. The mosaic that home gardens provide is an important nesting and feeding site for both tropical and migratory birds. Homes are often located near water, so the home garden provides shelter in the fruit trees, food in the garden, and access to water.

The home garden is a shared space for the family. It is a place to teach, learn, play, and do other domestic tasks, such as process seeds, fertilizer, and harvests, make food, remedies, and ornamentals, and grow seedlings. Each day kids play in the home garden and observe their parents weeding, harvesting, and shooing a dog or a chicken. It is full of daily life. It is a place where family life in all its forms, from love to ceremony, take place. It is a place where children take their first steps and where those same children learn about traditional healing, what to eat, and how to care for nature. It is also the same place where they sit with their lovers and where they pass the evening with their spouses. A chair or bench may rest against the house in the home garden, under a fruit tree. Some chairs may be grouped together, and one might ask, "Is this the porch, the patio, or the garden"? It is all three.

The economics of the home garden are local and familial. Occasionally a family will grow a product for sale, but this would be an exception. The home garden is a place where families grow supplements to what they grow or gather from the *milpa* and the little that they purchase.

Because local, native, and adapted species comprise the home garden, it is found throughout Mexico. The species vary from north to south, from the coasts to inland, from lowlands to highlands, but the general form does not change. In nearly all areas, it is an agroforest with understory and overstory. Home gardens range in size from a half hectare down to a few square meters. Together, linked across the landscape, they comprise a considerable part of the Mexican landscape.

In two communities in Central Valley Chiapas, 194 useful species were found in the home gardens and 72 in two communities in Highland Chiapas. In the areas around the houses were home gardens with a large number of trees mimicking forest in structure and function. Domesticated species such as *Spondias*, *Annona*, and *Persea* have been managed for centuries in home gardens. By contrast, *milpas* contained 26 species in the Central Valley and 121 in the Highlands, while the pastures under silvopastoral management contained 45 and 71 species in the Central Valley and Highlands, respectively. Fallows (fallow stages of the *milpas*) had 144 and 121 species, respectively.

Milpa Systems

The *milpa* (Figure 18–3), as mentioned, is symbolic of Mexico. It is a complex agroforest with a temporal cycle that varies in complexity. The primary components of *milpa* are an ecological initiating event, typically fire (Nigh & Diemont, 2013; Soto-Pinto & Anzueto-Martinez, 2016) but occasionally now slashing only (Figure 18–4), a period of field production that is often very diverse and can include trees, and a fallow period that is commonly called *descanso*, or rest. This rest period is anything but rest in many cases and can involve a very rich forest full of activity by humans and a richness of harvested plants and animals.

The plant community of the *milpa* agroforest is dynamic and rich. This dynamism can span decades, and the richness of the system should be understood at this time scale. Patches of field and forest at different stages of growth can be positioned next to each other in the *milpa*. This spatial arrangement creates enormous opportunities for movement and exchange among all biological taxa. Birds can feed and nest in spaces next to each other, feeding in the field and nesting in the forest. Forested land near a field can provide animal-carried or wind-blown seed to jump-start or accelerate succession.

Fire is in most cases the event that initiates succession in the *milpa*. Farmers generally understand the great responsibility they have in preventing the escape of fire from their own fields and to limit air pollution. Fire is used through cultural practices and in some cases ceremonies. Groups of people are organized to monitor the fire and to keep it from spreading. Teamwork is critical, and only with complex social organization can it be achieved. Fire controls pests and weeds, germinates seeds, and prevents fuel storage in a landscape that could lead to large forest fires. The fire in the *milpa* also provides potash and other nutrients to the *milpa* field for growing corn and other crops (Nigh & Diemont, 2013).

Succession in *milpa* agroforestry follows ecologically based stages of change. All ecosystems go through succession in their natural forms. Conventional agriculture often works to stop succession. Perhaps understanding and accepting succession are the best lessons that can

Fig. 18–3. *Milpa* field stage in Motozintla, Chiapas, Mexico.

Fig. 18–4. *Milpa* with trees. The farmers no longer utilize fire in this *milpa*.

be learned from the *milpa* agroforest. Succession is a very powerful part of nature and can be used productively. The field stage, following the burn, is the successional stage akin to a grass stage in a parcel of land that is unmanaged. Following the grasses, shrubs will invade and begin to produce shade, which will be increased by a temporal series of trees with increasingly higher canopies. In *milpa* agroforests, each of these stages is useful for food, medicine, and raw material production. In each of these stages, the agroforest provides ecological services beyond provisioning services and ecological function to the landscape as a whole.

The *milpa* agroforest provides numerous provisioning ecosystem services to humans at all stages of the successional trajectory. In the *milpa* stage, corn is the most common plant grown. From a distance, the *milpa* or field stage looks to be only corn. But this view belies an often rich agroecosystem that includes squash, bean, pepper, and watermelon [*Citrullus lanatus* (Thumb.) var. *lanatus*] in the understory and papaya (*Carica papaya* L.) and banana (*Musa* ×*paradisiaca* L.) above the corn. Tens of species have been found in the field of the *milpa* (Diemont & Martin, 2009). As succession continues, these crops change to pineapple [*Ananas comosus* (L.) Merr.] and sugarcane (*Saccharum* spp.), then to avocado and mango (*Mangifera indica* L.) (Martinez-Bautista, 2017). The advanced forest is no less productive than the field (Falkowski, 2018). From field to forest, products are harvested. Trees for lumber and firewood are felled. Vines and tree bark for rope are gathered. In some cases, even clothes are made from bark (Nations & Nigh, 1980). Hammocks, canoes, houses, a pharmacy of medicines, and food for the family both seasonally in the field and throughout the year in the forest and in the field are all provided by the *milpa* agroforest (Falkowski, 2018). In a complete *milpa* agroforestry system, all stages exist at the same time as in a forest mosaic (Diemont, Bohn, Rayome, Kelsen, & Cheng, 2011). Therefore, nutrient and calorie richness is always present for human consumption (Falkowski, 2018).

The *milpa* agroforest is an important part of healing for many communities in Mexico. The *milpa* field will contain medicinal plants, such as chamomile flower, used for stomach aches and tension, and wild medicinals, which are left in the field for future treatment of ailments. The forest and shrub stages, due to the size of the land area relative to the home garden and many niches provided by a forest, are an important source of medicine (Diemont & Martin, 2009).

Raw materials provided by the *milpa* tend to come from the forest stages. Daily harvests of firewood are common. Communities build their homes using wood grown in the forest. Baskets and hammocks that they weave and macramé from bark are a part of traditional craftwork. Roofs may even come from the leaves of palms growing in the agroforest. The *milpa* agroforest can and does provide everything that a community needs for survival.

The *milpa* has been shown to be important in maintaining wildlife health in Mexico. Bird populations are in some cases identical in biodiversity to conservation areas and preserves (Bohn, Diemont, Gibbs, Stehman, & Vega, 2014; Falkowski, 2018). Harvest of wildlife is critical to meeting the dietary needs of communities. Harvested wildlife includes many taxa of animals (Nations & Nigh, 1980).

The *milpa* agroforestry system is central to the cultural identity of many of Mexico's citizens. The *milpa* represents a rite of passage, where a man is welcomed into manhood when he plants his own *milpa*. Also, traditionally, processing corn and making tortillas, which are a staple part of the Mexican diet, is a part of womanhood. It would be far too simple to distill the *milpa* down to only corn and the other food it provides. The *milpa* cycle is a means by which people follow the changing seasons, the rain, and the desiccation of the landscape. People are tied to nature through the *milpa*. Rituals spring from this relationship with nature—such as the corn ceremony at harvest and the water ceremonies in the spring, calling the rains in particular. These religious ceremonies speak to the way in which *milpa* is both a means for production and a cultural identity.

The *milpa* localizes the economy. Farmers may trade outside the community but agroforestry, by providing for a family's needs directly, makes trade less necessary than specialized work. Commonly purchased items include clothes, tools, and equipment (Diemont, Martin, & Levy-Tacher, 2006; Guillen Trujillo, 1998). The more regularly consumed items, such as food and energy for cooking, come from the agroforest.

The *milpa* system is more common in the southern part of Mexico than it is in northern Mexico. In current Mayan and Zapotec areas in particular, the *milpa* dominates the landscape. Much of the states of Oaxaca, Chiapas, Tabasco, Quintana Roo, and the Yucatan have landscapes with extensive *milpa*. This trend continues south into Guatemala.

Case Study: Lacandon Maya *Milpa*

Although now practiced in its complete form by only a handful of people, elements of Lacandon Maya agroforestry are widely practiced in the villages of Naha, Metzobok, and Lacanja Chansayab in the state of Chiapas, Mexico. These villages are all located in the eastern lowlands of Chiapas where the native forest is moist (Diemont & Martin, 2009; Falkowski, 2018; Nations & Nigh, 1980). Lacandon Maya, with a population in all villages of fewer than 1000 people, are probably descendants of Maya who escaped the Spanish conquest.

The agroforestry of the Lacandon Maya is a complex, multistage, successional system that at its most complex includes seven successional stages and lasts for more than 60 yr. These stages include: the *kor*, which is a field stage dominated by corn and has 60 or more other crops in a polyculture; the *robir*, a grassy stage lasting for 2 yr with crops such as sugarcane; the *jurup che*, an early woody secondary forest with trees just beginning to straighten and undergrowth beginning to clear, lasting for 2 yr; *pak che kor*, literally meaning planted tree *milpa* and containing many fruit trees, lasting 5–10 yr; the *mehen che* "small tree" in Lacandon lasting 5–10 yr; and many years of *nu kux che*, (Figure 18–5) or "large tree" (Diemont & Martin, 2009; Falkowski, 2018). At the successional stage chosen by the farmer, the land is prepared for *kor* by using fire. *Nu kux che* can be prepared for *kor*, but *pak che kor* or one of the other stages can be burned as well. No more than the amount of land needed to provide for the family is burned. A family will typically have only 1–2 ha in *kor* at a given time. Sometimes products are sold within the community to a family that does not have enough land for *milpa* or is working outside of the farming economy, most commonly in tourism. The product from the *milpa* is rarely sold outside of the community. While the successional cycle of Lacandon agroforestry takes place, the Lacandon also preserve primary forest in an area called *tam che*.

Each stage of the *milpa* agroforest is productive for the family. The *kor* will have staples, the most important of which is corn. Farmers will also grow beans, squash, and numerous other crops (Diemont & Martin, 2009). They will harvest crops that follow the changing seasons of sunlight and rain from the *kor* throughout the year. In each successional stage, families will harvest food, medicine, and raw materials. In each stage, as many as 100 plants and animals are part of the family diet and used in the household (Diemont & Martin, 2009; Falkowski, 2018). Houses are constructed of lumber from the trees of the advanced forest. Medicine is provided by the entire system.

Lacandon will plant trees early in the system. A *kor* will have mahogany [*Swietenia mahagoni* (L.) Jacq.] and cedar (*Cedrela odorata* L.) that will be harvested in the *nu kux che*. It will also contain several trees like balsa [*Ochroma pyramidale* (Cav. ex Lam.) Urb.] that the Lacandon farmers will plant to accelerate the restoration of the forest. They thus leapfrog time needed for regrowth through selective planting (Diemont, Martin, Levy-Tacher, Nigh, et al., 2006).

Wildlife are common in Lacandon Maya agroforestry. Local farmers have noted up to five different species of large cats. Howler monkey calls reverberate throughout the forest. Spider monkeys climb through the higher branches. Rare birds are not as common as in the forest preserves, but the biodiversity of birds in an advanced agroforestry forest is as high as in forest preserves (Falkowski, 2018).

Fig. 18–5. ***Nu kux che*, or advanced secondary forest stage of Lacandon Maya *milpa*, Lacanja Chansayab, Chiapas, Mexico.**

Soil recovers rapidly in the Lacandon agroforestry successional stages (Diemont & Martin, 2009; Falkowski, 2018; Falkowski, Diemont, Chankin, & Douterlungne, 2016). Nutrients and nematodes have been shown to differ little in middle forests from forest preserves and not at all between advanced forests in the Lacandon agroforestry system and the forest preserves.

Coffee Agroforestry Systems

Coffee is grown in Mexico through a gradient of management scenarios: rustic coffee, traditional polyculture, commercial polyculture, shaded monoculture, and unshaded coffee—systems with different structures and functions depending on the production techniques of the individual farm (peasant families to businessmen).

Originally, coffee was grown within Mexico's forests. Subsequently, its cultivation was expanded to plantations, which underwent a process of modernization due to the growing demand in the market. Currently, the majority of Mexican coffee grows in diverse, complex agroforestry systems on family farms. Coffee farms combine between 2,000 and 3,000 coffee plants per hectare with different varieties, mainly of the species *Coffea arabica* L. Today, these varieties are being slowly changed to those derived from the Timor hybrid (*C. arabica* × *C. canephora*). Coffee is mixed with shade trees traditionally ranging in density from 200 to 400 trees per hectare, including palms and shrubs. Legumes and other botanical families, as well as plants with different uses (food, timber, fuel, fodder, fiber, handicrafts, medicine, and shade) are part of each design. The composition of the shade vegetation of the coffee farm is a result of secondary succession. Producers will introduce species chosen based on their locally known characteristics, ecosystem type, and cultural functions.

Farmers manage shade trees through natural succession. They select some species known as good species for shade depending on several criteria: deciduousness, foliage density, impact on coffee yields, amount of litter and decomposition rate, leaf size, crown shape and branch extension, height, growth rate, impact on microclimate, and impact on pest and disease incidence. Farmers also select for other characteristics such as the role of tree species in weed control, wind resistance, branch hardness, root strength, moisture maintenance, and additional goods and services offered by trees (Soto-Pinto et al., 2007).

Coffee farms have high complexity, diversity, and multifunctionality, especially traditional polyculture systems. Coffee systems in Chiapas, Oaxaca, Veracruz, and Puebla, for example, were found to provide multiple benefits due to dense coverage, high tree density, quality of coffee grown, high species diversity, and number of vegetative strata (Figure 18–6). They accumulate organic matter and provide nutrient cycling, pest and disease control, and microclimate stability.

Fig. 18–6. Edible understory species intercropped in coffee agroforest.

They produce soil organic matter, habitat for flora and fauna, high diversity and abundance of pollinators, and also better coffee quality than simplified systems. The tree species of the coffee agroforests often have multiple uses. The trees provide food, fuelwood, and construction timber and offer multiple cultural, recreational, and aesthetic services. In a recent study in the Sierra Madre de Chiapas, a total of 112 plant species were found in the coffee plots, of which 57 are for shade production: 37 trees, 16 shrubs, and 4 herbs taller than 1 m high. Of the 112 plant species, 34.8% were trees, 15.2% were shrubs, and 50% were herbs. These species belonged to 53 botanical families and had various uses but mainly were used to produce food and timber (Escobar, 2017).

Cocoa Agroforestry Systems

Cocoa agroforests are multilayered agrosilvicultural systems where cocoa and shade trees are combined with timber, legumes, fruit trees, and both native and introduced tree species (Figure 18–7). For example, most families (74.1%) in the municipality of Acacoyagua, Chiapas, intersperse coffee plants with cocoa of the *forastero* and *trinitario* types. Within the agroforestry systems, 48 useful species were recorded; one-fifth serve multiple uses (e.g., food, shade, timber, firewood, remedies, fertilizer, and so on). Among the fruit trees, mamey [*Pouteria sapota* (Jacq.) H.E. Moore & Stearn] stands out for its food and commercial value. Native timber trees in cocoa agroforests are crafted into furniture, used for construction, or are considered a source of funds for medical or domestic emergencies. The highest valued timber species are the *primavera* [*Roseodendron donnell-smithii* (Rose) F. Miranda] and cedar.

Most of the products grown in cocoa agroforestry systems, such as avocado, citrus fruits, banana, chayote (*Sechium edule* Jacq.), palm flowers, guanabana (*Annona muricata* Macfad.), and chili pepper, are used by the family, given to friends, and, to a lesser extent, sold at local markets run primarily by women. In a study in southern Mexico, 16.7% of tree species being used were recognized by the producers as creating "good shade." These were mainly leguminous species such as *Inga* sp. and *Lonchocarpus* sp. Almost 23% of the tree species were identified as creating "bad shade or shade that adversely affects the crop." Examples are the palo de chiche (*Aspidosperma* sp.), cedar, and mango (López-Cruz et al., 2020). However, these species are left in the system for timber or to produce fruits for sale, even though they are not considered suitable to produce cocoa shade.

Cocoa has been used and has been a part of the culture in Mexico since pre-Columbian times. Cocoa has been grown in the Soconusco region in Chiapas since 1900 BCE, as has been demonstrated by a ceramic vessel found at Paso de la Amada in what is now Mazatán, Chiapas (Gasco, 2021;

Fig. 18–7. Chol cocoa system in southern Mexico.

Powis, Cyphers, Gaikwad, Grivetti, & Cheong, 2011). Grown under the canopy of tropical species in forest gardens, it is used for rituals, food, tribute, and currency and has been pervasive throughout Mexican history. At the beginning of the 18th century, in 13 towns within 297 tributaries of Soconusco, most inhabitants (83%) had cocoa orchards (Gasco, 1996).

Today cocoa is a crop with high economic, environmental, and social importance. However, fungal diseases such as those caused by *Moniliophtora roreri* (Ciferri & Parodi) Evans, et al. and *Phytophthora palmivora* (E. J. Butler) E. J. Butler have reduced its prominence.

However, like coffee, cocoa agroforests have intensified recently, thus losing structure and function. Three types of systems can currently be found: one characterized by its complex structure and high diversity of shade trees, shrubs, and palms; a second one dominated by the shade of legumes, specially *Inga* spp. and *Lonchocarpus* sp.; and a third system, with scattered shade trees and cocoa trees, mainly of the varieties *forastero* and *trinitario*, planted in rows with interspersed shade trees randomly distributed.

Silvopastoral Systems and Practices

From the 16th century to the end of the 19th century, ranching in Mexico, primarily with pigs, cows, and goats, was managed in extensive, adapted systems where natural landscapes and agricultural landscapes interacted (Hernández, 2001). Cattle in particular were managed with high biodiversity in semiarid zones, forests, and jungles.

Today in Mexico, a high diversity of silvopastoral systems and practices exists. From traditional management systems utilized by local and indigenous communities that are oriented toward the production of food and products for the family (López-Carmona et al., 2001), to silvopastoral systems focused on holistic management for the production of meat and milk for regional markets or export (Ferguson et al., 2013; Marinidou, Jiménez-Ferrer, Soto-Pinto, Ferguson, & Saldivar, 2017), many fodder trees and agrosilvopastoral practices are found distributed throughout Mexico (Figure 18–8).

Living Fences

Living fences are an ancient traditional agroforestry practice that provide a range of services (shade, pasture boundaries, agricultural area) and products (lumber, firewood, forage, fruits) to the farmer (Esponda, 1888). Living fences contribute to biological conservation and play an important role in the "connectivity" of ranching and farming landscapes. The combination of species with different objectives, such as lumber, forage, or fruits, can transform living fences into small biological corridors that contribute to multiple ecosystem services and conservation. Some species that are important in tropical zones in

Fig. 18–8. A silvopastoral system in southern Mexico.

southern Mexico are: mata ratón, cocoite, or shan'te [*Gliricidia sepium* (Jacq.) Kunth ex Walp.]; ñanguipo (*Cordia dentata* Poir.); pito, mote, or uku'm (*Erythrina* sp.); guaje or guash [*Leucaena leucocephala* (Lam.) de Wit]; ramón or osh (*Brosimum alicastrum* Sw.); guacimo (*Guazuma ulmifolia* Lam.); jobo (*Spondias mombin* L.); and cuajilote (*Parmentiera* sp.).

Trees Scattered in Pastures

This practice consists of letting scattered trees or shrubs grow in pastured areas, which provide diverse services and products such as wood, fruit trees, shade for animals and shelter for wildlife. Some species have a high capacity for fire survival and are an important source of wood for families. For example, in the Marqués de Comillas subregion of the Lacandon jungle (Chiapas), it is common to manage different species in the grazing areas. Different types of palms [*Sabal yapa* C. Wright ex Becc., *Acrocomia aculeata* (Jacq.) Lodd. ex Mart.] are established, as are timber species, such as *popiste* (*Blepharidium mexicanum* Standl.), to colonize grazing areas.

Grazing in *Milpa* Fallow Areas

In many regions of central, southern, and northern Mexico, farmers make use of the *milpa* fallow, taking advantage of the natural regeneration of trees and shrubs and grazing cattle in forest pastures. Both the fallow and the forest are strategic to the survival of the herd in the dry season. In some mountain areas, such as in northern Chiapas or in the border area with Guatemala, livestock producers use the fallow as a transition area between the rainy season and the dry season for their livestock to take advantage of the agricultural crop residues, grasses, vegetation, and fruits that are available. This practice also helps control fire through the use of foraging in the forest. Among the arboreal species managed in these systems are the *guacimo* or *a'kit* (*Guazuma ulmifolia*), *timbre* [*Acacia angustissima* (Mill.) Britton & Rose], hawthorn [*Acacia pennatula* (Schltdl. & Cham.) Benth.], *huizache* (*Prosopis* sp.), *no'k* [*Cordia alliodora* (Ruiz & Pav.) Oken], and *guash* or *guashin* (*Leucaena leucocephala*), which also provide other services such as timber, firewood, and charcoal production.

Grazing in Plantations and Orchards

The grazing of large and small ruminants in forest plantations and fruit orchards is practiced extensively in many states of central and southern Mexico for a variety of purposes including the production of soil fodder and agricultural residues and for weed control. In the tropical lowlands of Mexico, grazing of bovine and ovine Pelibuey is common. However, cattle are most frequently observed in plantations of African palm (*Elaeis guineensis* Jacq.), coconut palm (*Cocos nucifera* L.), and rubber [*Hevea brasiliensis* (Willd. ex A. Juss.) Müll. Arg.] and fruit orchards of mango, orange, and banana. In transition zones (1000–1500 m asl), small and medium producers in coffee growing areas raise Pelibuey sheep under their shade coffee plantations to control weeds, and in high areas (>2000 m asl), as in the Tzotzil Maya zone of Chiapas, indigenous producers manage flocks of Creole sheep, taking advantage of agricultural residues of crops (corn and pumpkin [*Cucurbita pepo* L.]) in fruit orchards of peach, plum (*Prunus domestica* L.), pear (*Pyrus communis* L.), and apple.

Pastures in Alleys of Trees

In hillside and mountain areas of central and southern Mexico, planting of trees in parallel strips with forage in the alleyways is common to improve soil fertility, prevent erosion, and reduce the trampling of trees by animals. Species most commonly used in alleys are N_2–fixing trees such as *Gliricidia sepium* or *Leucaena* sp.; others include fruit trees such as mulberry (*Morus alba* L.).

Windbreaks

Windbreaks are simple or dense strips of trees (double row) designed to minimize wind effects on pastures, agricultural crops (corn, bean), and animals. In many areas of Oaxaca and Chiapas, many plants are used for this purpose. For example, in the border area of Chiapas, bell curtains (*Acacia angustissima*), *quebracho* (*Acacia farnesiana* (L.) Willd.), hawthorn (*Acacia pennatula*), and reed (*Arundo donax* L.) are used to protect agricultural areas of corn, fruit trees, and cattle. In the highlands of Chiapas, winds and frosts occur in the winter months, so pine (*Pinus* spp.), poplar (*Populus* spp.) and noc'k (*Alnus acuminnata* Kunth) are planted.

Banks of Protein and/or Energy Storage

Bush fodder plants are cultivated at high densities in medium to small areas. The main objective is to provide high quality forage (high protein content and good digestibility) and plenty of dry matter. Protein banks are a strategy to better accommodate meat and milk production and release land for other agroforestry uses. This practice has been very successful in medium livestock systems, mainly for milk producers. These areas can be planted with *cocoite*, *caulote* (*Guazuma ulmifolia* Lam.), *guash* (*Leucaena* sp.), mulberry (*Morus alba* L.), or daisy [*Thitonia diversifolia* (Hemsl.) A. Gray]. Energy banks are designed as a type of cutting and hauling system with the production of high sugar, starch, or oil useful in animal feed. Sugarcane or cassava (*Manihot esculenta* Crantz) are species of choice. In livestock regions of the humid tropics (Chiapas), it is customary to plant rows of *cuajilote* trees [*Parmentiera edulis* (Kunth) Seem.] and take advantage of the fruits as energy supplements for cattle.

Case Study: Sheep in the Tzotzil Maya Region of Chiapas

The region of the Highlands of Chiapas is an area with a large indigenous population, mainly Tzotzil and Tzeltal Maya. This area is a region of about 5,000 km^2 with a population of more than 600,000 people. It is mountainous (1,200–2,898 m asl) and cold, has high annual precipitation (1,000–1,500 mm), concentrated between May and September, a dry season, and frosts in the months from November to February. San Juan Chamula is the municipal seat and is a center of economic, political, and cultural importance in this municipality. The main agricultural activity is the cultivation of corn, in combination with bean (*Phaseolus coccineus* L., *P. vulgaris* L.), potato (*Solanum tuberosum* L.), broad bean (*Vicia faba* L.), or *chilacayote* (*Cucurbita ficifolia* Bouché), and the production of fruits and flowers and other vegetables (Alemán, López, Martínez, & Hernández, 2002). Livestock agriculture is focused on sheep farming, an activity of vital importance for the families of the municipality of Chamula (Figure 18–9).

Despite the changes in indigenous agriculture in recent decades due to strong dependence on the use of external inputs (fertilizers, herbicides, etc.) for the production of basic crops, vegetables, and flowers, the region still has an old and solid tradition of polycultures, soil conservation practices, domestication of plants and animals, and a broad ethnobotanical culture. A wide variety of trees and shrubs are used in their systems (Soto-Pinto, Jimenez, & de Jong, 1997). The central activity of the system is directed by women, who breed sheep that have a cultural, religious, and economic importance. Nahed (2001) and Alemán et al. (2002) mentioned that within each Tzotzil family, sheep breeding has different functions and is a priority.

When moving through the crop rotation (corn–fallow–grazing), the foliage of trees and shrubs, weeds from the cultivation of plots, and agricultural residues are fed to the sheep. In this spatial rotation, manure is used as a fertilizer for a variety of crops and vegetables. It is common to move the sheep around in small mobile pens. After moving the animals to a new location, vegetables are planted. The sheep in this system come from the Spanish breeds Merino, Lacha, and Churra (Gómez, 1978), which were adapted to form a group called Borrego Chiapas, with characteristics of rusticity and adaptation to the conditions of the Highlands of Chiapas. Perezgrovas and Pedraza (1989) stated that although wool productivity of the sheep of

Fig. 18–9. A silvopastoral system in San Juan Chamula, Chiapas, Mexico.

Chiapas is low (0.8–1.6 kg of wool annually), any other more productive breed of sheep would not be able to survive under such difficult environmental and management conditions. In this agrosilvopastoral system, numerous trees and fodder shrubs are important because of their nutritional value for sheep and their alternative uses for the Tzotzil family. Some of the species used are: *Acacia pennatula*, *Alnus acuminata*, *Buddleia parviflora* H.B.K., *B. skutchii* C.V. Morton, *Erythrina chiapasana* Krukoff, *Holodiscus argenteus* (L.f.) Maxim., and *Leucaena brachycarpa* Urb., among others (Nahed-Toral, López-Tirado, Alemán-Santillán, Aluja-Schunemann, & Parra-Vázquez, 1997; Soto-Pinto, 1997).

For families in Chamula, sheep raising is integral to the family's well being. The central objective is the production of wool and the sale of animals. Wool production is important for traditional clothing and handicrafts. The sale of these products provides income for survival and the purchase of agricultural inputs.

Challenges for Mexican Agroforestry

Cultural, economic, political, and environmental drivers are all affecting agroforestry in Mexico. The Mexican population is increasingly one of movement. Many communities in Mexico are suffering from the loss of young people to emigration, mostly to the United States, but also movement to larger cities in Mexico (Valentine, Barham, Gitter, & Nobles, 2017) and areas where seasonal jobs are available. The generational change and its complexities can be seen in livestock management in tropical areas. It is common for livestock activity to be directed by the elderly and young people who return from the United States with economic resources. However, these young people are not interested in holistic agroforestry management. They favor very intensive practices, using agrochemicals. In some areas, such as on the border with Guatemala or part of eastern Mexico, serious problems of illegal movement of animals and drug trafficking exist (Soberanes, 2018).

Cultural shift also comes from changes in priorities even within communities. People's lives have changed in recent years, especially in rural areas. Neoliberal economic policies have reduced the value of rural products, leaving people without options for profitable agriculture, which increases poverty and changes the life of rural communities (Castillo Fernández & Arzate Salgado, 2016). In previous times, and still in some indigenous communities, people lived in collective spaces with a strong relationship among different generations. Family life and the teaching of practices, among them agriculture, took place. Children learned in the field and forest. Their grandparents and parents taught them the ways of agriculture, children observed, following the practices of adults and also innovating on their own (Cervantes, Estrada, & Bello, 2017). Because education is now typically taking place in the schools instead of the *milpas*, young people are learning how to read and write, which is good, but are not learning how to manage the farm. This trend will create a future for agroforestry in Mexico that is less accommodating than currently.

Visual media, in all of its forms, has been a game changer for tradition. The internet took the rural communities of Mexico that were wobbly after television diverted their attention to *lucha libre* and *telenovelas* instead of the day-to-day life of the farm and hit a "knockout punch" for many young people as they now spend an inordinate amount of time on Facebook and with their cell phones. This change is keeping Mexican youth out of the fields, where they once learned all the names of the plants and their uses; now instead they learn from a new community, a new culture that is not local.

Rural areas, a complex socio-environmental space, are now often in a state of abandonment and permeated by poverty, migration, deterioration of the social fabric, drug trafficking, and external food dependence. For example, in Chiapas more than three-fourths (4.1 million people) of its population live in poverty, and of these, almost 1.5 million live in extreme poverty, according to the National Council for the Evaluation of the Policy of Social Development (CONEVAL, https://www.coneval.org.mx/coordinacion/entidades/Chiapas/Paginas/Pobreza-2016.aspx).

Policy changes are affecting communities from global to local levels. Global environmental policy has a direct effect on Mexico's rural communities. The United Nations policy REDD+ is incentivizing the conservation of forests. Conserving forests for carbon storage and to mitigate climate change seems reasonable. This policy, however, often targets traditional agroforestry as a means of storing carbon. The policy in effect gives money to communities to reduce burning the forest and eliminate traditional agroforestry.

At the federal level, reforestation policy leads to incentivizing the growing of forests to reap the benefit of global incentives and satisfying policy. Agencies are increasingly being assigned the responsibility in the southern areas of Mexico to restore forests. The effect of this restoration is inconsistent and is a source of waste resulting in limited actual reforestation. Reforestation project leaders tend to work little with communities in planning and setup and do not rely on their agroforestry traditional ecological knowledge. It is a top down approach that centers around expert knowledge from outside the communities. As these projects are brought to communities, members of the community will accept monthly cash payments that lead to: (a) planting trees that have been grown in greenhouses, (b) abandoning traditional *milpa* agroforestry, or perhaps (c) shortening the time that the forest is allowed to develop. Management of the systems is short term, and projects are often abandoned after the payments from the federal and state agencies dry up. Sometimes, the payments never arrive, and the farmer is left with trees that will never be planted and indecision about how to proceed.

Currently, global development agencies are interested in promoting agroecological strategies, including agroforestry systems. This approach has been promoted by the Food and Agriculture Organization of the United Nations (FAO) and is an important policy to improve food production in the context of the Sustainable Development Goals established by the United Nations. However, the FAO has allowed transnational companies (e.g., Monsanto, Bayer, Nestle) to appropriate the "green discourse" and are promoting a food and agriculture policy of transgenics and the use of high inputs with agroecological and agroforestry practices (Giraldo & Rosset, 2018).

Contrast between the development strategies among rural development government institutions exacerbates the problem. Governmental agencies, such as the National Commission for the Use and Knowledge of Biodiversity (CONABIO), promote silvopastoral systems in the Lacandon rainforest (Chiapas) through conservationist approaches, while in the same region other governmental agencies, such as Mexico's Secretariat of Agriculture and Rural Development (SAGARPA), promote extensive livestock systems (deforestation) and oil palm (*Elaeis* Jacq.) production, putting traditional systems at risk (Soto-Pinto et al., 2012).

Mexico's economy since implementation of the North American Free Trade Agreement (NAFTA) in 1994 has become increasingly global. Multinational corporations once uncommon in cities are increasingly a part of the landscape. These businesses push out local businesses and change the local community. The global economic landscape also provides a beacon for young people looking to better their lives, leading to a further increase in migration (Otero, 2011). The stress of this migration on local economies is especially felt in the field of agroforestry. How to enter this economy without disrupting the long chain of knowledge transfer that makes Mexico a wealth of traditional ecological knowledge, cultural identity, and local community recognition is an enormous economic challenge. One possibility, and a possibility that is occurring in some other parts of the world (Pimbert, 2015), is re-localizing Mexico's rural economy centered again on agroforestry.

Mexico is highly biodiverse and thus is a place where we see conservation efforts in large force (CONABIO, 2014). Southern Mexico is home to four United Nations biosphere reserves. The biosphere reserve network is a critical part of biodiversity conservation globally. Locally, however, biosphere reserves can become a place of conflict. Communities are forcibly removed from biosphere reserves, leaving them in some cases with little or no place to go (Trench & Kohler, 2012). Studies (e.g., Diemont et al., 2011; Falkowski, 2018) have indicated that traditional agroforestry may provide similar biodiversity conservation benefits as preserve management. Agroforestry conserves biodiversity while providing food and raw materials to local people.

Energy is a critical driver of change in the rural villages of Mexico. One form of alternative energy, purported to be a sustainable form of agroforestry, that is leading to widespread degradation of the agroforestry lands of southern Mexico, is African palm oil. African oil palm (*Elaeis guineensis* Jacq.) plantations were initially sold as a sustainable form of agroforestry that would positively contribute to the ecosystem services of the area. In reality, these plantations are high-water-demanding systems that leave degraded soils for more than 5 yr after the site is abandoned. Farmers have reported that *milpa* cannot be planted within 100 m of these plantations because the plantation sucks the water from the ground, drying out the entire area (Adolfo Chankin, Lacanja Chansayab, Chiapas, Mexico, personal communication, 2018). In Chol Maya communities, cultivating oil palm has been noted by farmers to decrease soil quality, leading people to abandon *milpa* completely (Rodríguez family, Arroyo Palenque, Chiapas, México, personal communication, 2014).

Current Opportunities for Mexico's Agroforestry

Mexico can move forward with agroforestry in a way that embraces past innovations spanning millennia to address current local and global needs. Environmental improvement and food production are important contributions that agroforestry can make in Mexico. Change can occur due to grass-root initiatives and policy recognition of the potential benefits of agroforestry.

As has already been described, agroforestry has the potential to contribute to improving the environment. It can do so while contributing to the local economy and the well-being of the people of Mexico. Agroforestry in Mexico can also serve as a model or inspiration for how humans and the environment can be part of the same system without diminishing the health of either. Two particular areas of influence in the environment for which agroforestry can provide a remedial element are climate change and biodiversity conservation.

Agroforestry provides a means by which carbon is absorbed into nature. Some evidence (Laurenceau & Soto-Pinto, 2015; Nigh & Diemont, 2013; Orihuela-Belmonte et al., 2013; Soto-Pinto & Aguirre-Davila, 2015; Soto-Pinto, Anzueto-Martinez, Mendoza, Jimenez-Ferrer, & de Jong, 2010) suggests that through carbon sequestration in soil and active charring during low-intensity burns, this movement could have a much longer term storage in soil, immobilizing the carbon as charred or black carbon. Furthermore, agroforestry foliage of fodder trees and shrubs has been shown to be useful in the mitigation of CH_4 emissions caused by livestock. Results found in recent research conducted in southeastern Mexico demonstrates that the inclusion of foliage [*Leucaena leucocephala*, *Bursera simaruba* (L.) Sarg.] and pods [*Enterolobium cyclocarpum* (Jacq.) Griseb., *Samanea saman* (Jacq.) Merr.) of trees and shrubs from silvopastoral systems can reduce CH_4 enteric production by 23–40% (Ku-Vera et al., 2018; Valencia-Salazar et al., 2018; Piñeiro-Vazquez et al., 2017).

Traditional ecological knowledge and local knowledge of farmers could be valuable as we adapt to a changing climate. Farmers mark seasonal and weather changes with biological activity changes and physical changes around them. Insects, birds, mammals, and even halos around the sun are all part of a consortia of change that indicates a necessary response in how they are managing their ecosystem (Zeiger, 2019).

Several forms of agroforestry have been used to show how traditional and novel forms of agroforestry in Mexico are conserving biodiversity (e.g., Falkowski, 2018). Understanding these systems and learning from the land managers and farmers involved is critical to conserving the biodiverse environment of Mexico. In many cases, the plants and animals are adapted specifically to the land management that is part of the landscape of Mexico. By removing these forms of agroforestry, the very organisms that are striving to be protected are put at risk. The landscape designed over thousands of years of practice could be the very key to restoring degraded areas. Agroforestry in its many forms could drive conservation and biodiversity restoration along multiple pathways—a truly collaborative undertaking.

Agroforestry can help improve the lives of people throughout Mexico. By combining food and forest, services of many types come to the agroforestry manager. A local form of food production, seemingly simple, in many cases can be extremely well adapted as a distributed form of food production. It provides food directly to the people. *Milpa* agroforests and home gardens have provided food sovereignty and security to Mexico for thousands of years (Ford & Nigh, 2016). The systems provide balanced nutrients and minerals and are productive throughout the year (Falkowski, 2018).

Both grassroots initiatives and policy changes are necessary to preserve and enhance agroforestry application in Mexico. Mexico has a rich social history centered around rural organizations that fight for production and agrarian ways of life. In Chiapas, Mexico, thousands of indigenous producers are still leaders in the production of organic coffee, an agroforestry system that allows management with multiple social and environmental benefits. The *Scolel Te'* project (tree that grows, in the Tzeltal language) is the oldest project of its kind worldwide (https://www.planvivo.org/scolelte). For two decades they have promoted agroforestry systems such as the taungya system, coffee with diversified shade, silvopastoral systems, and improvement of fallow with agroforestry, all of them with native species and propagation strategies (nurseries) that provide employment in the communities. At the national level, research and development is being conducted on agroforestry at various institutions. Moreover, there is a widespread membership that has been established in the Mexican Agroforestry Network, which is a collection of civil society and government institutions to influence research, teaching, and governance of agroforestry systems in Mexico.

These efforts show that local people can take charge of their own production. Movements for food sovereignty that began in Latin America have spread around the world and speak to the rights of local people to have access to good, healthy food that they grow themselves. Traditional agroforestry is very much a part of these movements. Agroforestry systems have a high diversity associated with both basic crops for self-sufficiency as well as crops destined for the market. Nonetheless, conflict still exists between sustaining agrobiodiversity, maintaining the quality of local foods, and the changes being recommended. These changes are brought about by society and market trends that force the simplification of diverse systems in favor of production systems that sacrifice food quality and sovereignty.

Policy changes need to recognize agroforestry for the complex system it is, a system that contributes tangibly to food security, food sovereignty, biodiversity, economic well-being, and cultural identity. Agroforestry in Mexico is critical to the future of agriculture as we embrace grassroot changes and write new policies that recognize the many benefits that agroforestry provides.

References

Alemán, T., López, J., Martínez, A., & Hernández, L. (2002). Retos de un sistema productivo indígena: Altos de Chiapas. *LEISA, 6*(4), 17–20.

Altieri, M. A., & Toledo, V. M. (2011). The agroecological revolution in Latin America: Rescuing nature, ensuring food sovereignty and empowering peasants. *Journal of Peasant Studies, 38*, 587–612.

Atran, S., D. Medin, N. Ross, E. Lynch, V. Vapnarsky, E. Ucan Ek', . . . M. Baran. 2002. Folkecology, cultural epidemiology, and the spirit of the commons: A garden experiment in the Maya Lowlands, 1991–2001. *Current Anthropology, 43*, 421–450. https://doi.org/10.1086/339528

Barrera-Bassols, N., & Toledo, V. M. (2005). Ethnoecology of the Yucatec Maya: Symbolism, knowledge and management of natural resources. *Journal of Latin American Geography, 4*, 9–41.

Bassie, K. 2002. Corn deities and the complementary male/female principle. In L. S. Gustafson & A. N. Trevelyan (Eds.), *Ancient Maya gender identity and relations* (pp. 169–190). Westport, CT: Bergin & Garvey.

Bassie-Sweet, K. 2008. *Maya sacred geography and the creator deities*. Norman, OK: University of Oklahoma Press.

Bohn, J. L., Diemont, S. A. W., Gibbs, J. P., Stehman, S. V., & Vega, J. M. (2014). Implications of Mayan agroforestry for biodiversity conservation in the Calakmul Biosphere Reserve, Mexico. *Agroforestry Systems, 88*, 269–285.

Budowsky, G. 1979a. National, bilateral and multilateral agroforestry projects in Central and South America. In T. Chandler & D. Spurgeon (Eds.), *International Cooperation in Agroforestry: Proceedings of an International Conference*. Nairobi, Kenya: ICRAF.

Budowsky, G. 1979b. *Sistemas Agroforestales en América Tropical* (Documento Técnico). Turrialba, Costa Rica: Programa de Recursos Naturales, CATIE.

Castillo Fernández, D., & Arzate Salgado, J. (2016). Economic crisis, poverty and social policy in Mexico. *Critical Sociology, 42*, 87–104. https://doi.org/10.1177/0896920513501352

Cervantes, E., Estrada, E. I. J., & Bello, E. (2017). Kinship practices and the configuration of spaces for collective living in the Tseltal coffee area, Tenejapa, Chiapas. (In Spanish with English abstract.) *Relaciones. Estudios de Historia y Sociedad, 150*, 281–315.

Coe, M. (1964). The chinampas of Mexico. *Scientific American, 211*(1), 90–99.

CONABIO. (2014, 30 Oct). *Biodiversidad mexicana*. www.biodiversidad.gob.mx

Cutler, H. (1968). Origins of agriculture in the Americas. *Latin American Research Review, 3*(4), 3–21.

De Frece, A., & Poole, N. (2008). Constructing livelihoods in rural Mexico: Milpa in Mayan culture. *Journal of Peasant Studies, 35*, 335–352.

Diemont, S. A. W. (2006). *Ecosystem management and restoration as practiced by the indigenous Lacandon Maya of Chiapas, Mexico* (Doctoral dissertation, Ohio State University). Retrieved from https://etd.ohiolink.edu/!etd.send_file?accession=osu1154582623&disposition=inline

Diemont, S. A. W., Bohn, J. L., Rayome, D. D., Kelsen, S. J., & Cheng, K. (2011). Comparisons of Mayan forest management, restoration, and conservation. *Forest Ecology and Management, 261*, 1696–1705.

Diemont, S. A. W., & Martin, J. F. (2009). Lacandon Maya ecosystem management: Sustainable design for subsistence and environmental restoration. *Ecological Applications, 19*, 254–266. https://doi.org/10.1890/08-0176.1

Diemont, S. A. W., Martin, J. F., & Levy-Tacher, S. I. (2006). Emergy evaluation of Lacandon Maya indigenous swidden agroforestry in Chiapas, Mexico. *Agroforestry Systems, 66*, 23–42. https://doi.org/10.1007/s10457-005-6073-2

Diemont, S. A. W., Martin, J. F., Levy-Tacher, S. I., Nigh, R. B., Lopez, P. R., & Golicher, J. D. (2006). Lacandon Maya forest management: Restoration of soil fertility using native tree species. *Ecological Engineering, 28*, 205–212.

Escobar, C. S. (2017). *Las plantas comestibles en el agroecosistema de café: Uso, conocimiento y diversidad en el ejido La Rinconada, Bella Vista, Chiapas* (Master's thesis, El Colegio de la Frontera Sur, San Cristobal de las Casas, Chiapas, Mexico).

Esponda, J. M. (1888). *Manual Práctico del Nuevo Ganadero Mexicano*. México: Oficina Tip. de la Secretaría de Fomento.

Falkowski, T. B. (2018). *Assessing the socioecological restoration potential of successional Lacandon Maya agroforestry in the Lacandon rainforest of Chiapas, Mexico* (Doctoral dissertation, State University of New York, College of Environmental Science and Forestry). Retrieved from https://digitalcommons.esf.edu/etds/20/

Falkowski, T. B., Diemont, S. A. W., Chankin, A., & Douterlungne, D. (2016). Lacandon Maya traditional ecological knowledge and rainforest restoration: Soil fertility beneath six agroforestry system trees. *Ecological Engineering, 92*, 210–217.

Falkowski, T. B., Martinez-Bautista, I., & Diemont, S. A. W. (2015). How valuable could traditional ecological knowledge education be for a resource-limited future?: An emergy evaluation in two Mexican villages. *Ecological Modelling, 300*, 40–49.

Fedick, S. L. (1996). *The managed mosaic: Ancient Maya agriculture and resource use*. Salt Lake City, UT: University of Utah Press.

Fedick, S. L., Allen, M., & Gomez-Pompa, A. (Eds). (2003). *The lowland Maya area: Three millennia at the human–wildland interface*. Boca Raton, FL: CRC Press.

Ferguson, B., Diemont, S. A. W., Alfaro-Argüelles, R., Martin, J., Nahed-Toral, J., Álvarez-Solís, D., & Pinto-Ruiz, R. (2013). Sustainability of holistic and conventional cattle ranching in the seasonally dry tropics of Chiapas, Mexico. *Agricultural Systems, 120*, 38–48. https://doi.org/10.1016/j.agsy.2013.05.005

Ford, A., & Nigh, R. (2016). *The Maya forest garden: Eight millennia of sustainable cultivation of the tropical woodlands*. Walnut Creek, CA: West Coast Press.

Gasco, J. (2021). Etnoagroforestería y cacao en la región del Soconusco en el pasado y el presente. In: A. I. Moreno-Calles (Ed.), *Los Sistemas Agroforestales de México: Avances, experiencias, acciones y temas emergentes en México* (pp. 186–216). Mexico City: Universidad Nacional Autónoma de Mexico (in press).

Gasco, J. (2006). Etnoecología histórica en la region del Soconusco, Chiapas. In *Anuario del Instituto de Estudios Indígenas X* (pp. 25–47). Tuxtla Gutiérrez, Chiapas, Mexico: Universidad Autónoma de Chiapas.

Giraldo, O., & Rosset, P. (2018). Agroecology as a territory in dispute: Between institutionality and social movements. *Journal of Peasant Studies, 45*, 545–564.

Gliessman, S., Garcia, R. E., & Amador, M. A. (1981). The ecological basis for the application of traditional technology in the management of tropical agroecosystems. *Agro-Ecosystems, 7*, 173–185. https://doi.org/10.1016/0304-3746(81)90001-9

Goetz, D., & Morley, S. G. (1950). *Popol Vuh: The sacred book of the ancient Quiché Maya* (from the Spanish translation by A. Recinos). Norman, OK: University of Oklahoma Press.

Gómez, J. (1978). *Perspectivas del desarrollo ovino en el estado de Chiapas* (Bachelor's thesis, Facultad de Medicina Veterinaria y Zootecnia, Universidad Nacional Autónoma de Mexico, Mexico City.

Gomez-Pompa, A., Salvador, F. J., & Fernandez, M. A. (1990). The sacred cacao groves of the Maya. *Latin American Antiquity, 1*, 247–257.

Gorenflo, L. J., Romaine, S., Mittermeier, R. A., & Walker-Painemilla, K. (2012). Co-occurrence of linguistic and biological diversity in biodiversity hotspots and high biodiversity wilderness areas. *Proceedings of the National Academy of Science, 109*, 8032–8037.

Guillen Trujillo, H. A. (1998). *Sustainability of ecotourism and traditional agricultural practices in Chiapas, Mexico* (Doctoral dissertation, University of Florida). Retrieved from https://ufdc.ufl.edu/AA00062180/00001

Hernández, L. (2001). *Historia ambiental de la ganadería en México*. Xalapa, México: Instituto de Ecología.

Hernández-Xolocotzi, E. (Ed.). (1977). *Agroecosistemas de México: Contribuciones a la enseñanza, investigación y divulgación agrícola*. Chapingo, México: Colegio de Posgraduados.

Instituto Nacional de Lenguas Indígenas. (2008). Catálogo de las lenguas indígenas nacionales: Variantes lingüísticas de México con sus autodenominaciones y referencias geoestadísticas. *Diario Oficial de la Federación, 652*(9), 22–78.

Instituto Nacional de Estadística y Geografía. 2005. II Conteo de Población y Vivienda. Aguascalientes, México: INEGI.

Klooster, D., & Masera, O. (2000). Community forest management in Mexico: Carbon mitigation and biodiversity conservation through rural development. *Global Environmental Change, 10*, 259–272.

Ku-Vera, J., Valencia-Salazar, S., Piñeiro-Vázquez, T., Molina-Botero, I. C., Arroyave-Jaramillo, J., Montoya-Flores, M. D. Solorio-Sánchez, F. J. (2018). Determination of methane yield in cattle fed tropical grasses as measured in open-circuit respiration chambers. *Agricultural and Forest Meteorology, 258*, 3–7. https://doi.org/10.1016/j.agrformet.2018.01.008

Laurenceau, M., & Soto-Pinto, L. (2015). Sistemas agroforestales para la adaptación al cambio climático en el área protegida La Frailescana, Chiapas, México. *Sociedades Rurales Producción y Medio Ambiente, 15*(30), 19–49.

Longhena, M. (1998). *Ancient Mexico: The history and culture of the Maya, Aztecs and other pre-Columbian peoples*. New York: Stewart, Tabori, & Chang.

López-Carmona, M., Jiménez-Ferrer, G., de Jong, B., Ochoa-Gaona, S., & Nahed-Toral, J. (2001). El sistema ganadero de montaña en la región norte-tzotzil de Chiapas, México. *Veterinaria México, 32*(2), 93–102.

López-Cruz, A., Soto-Pinto, L., Salgado-Mora, M. & Huerta-Palacios, G. (2020). Simplification of the structure and diversity of cocoa agroforests does not increase yield nor influence frosty pod rot in El Soconusco, Chiapas, Mexico *Agroforestry Systems*. https://doi.org/10.1007/s10457-020-00574-7

Martin, J., Roy, E., Diemont, S. A. W., & Ferguson, B. (2010). Traditional ecological knowledge (TEK): Ideas, inspiration, and designs for ecological engineering. *Ecological Engineering, 36*, 839–849.

Martinez-Bautista, I. (2017). *Soil fertility, emergy evaluation, and improvements to milpa in indigenous Zapotec agroforestry systems* (Doctoral dissertation, State University of New York, College of Environmental Science and Forestry). Retrieved from https://digitalcommons.esf.edu/etds/10/

Marinidou, E., Jiménez-Ferrer, G., Soto-Pinto, L., Ferguson, B., & Saldivar, A. (2017). Agroecosystems services assessment silvopastoral experiences in Chiapas, Mexico: Towards a methodological proposal. *Experimental Agriculture, 55*, 1–17. https://doi.org/10.1017/S0014479717000539

Moreno-Calles, A. I., Casas, A., Toledo, V., & Vallejo, R. M. (2016). *Etnoagroforesteria en Mexico*. Morelia, Mexico: Universidad Nacional Autonoma de Mexico.

Nahed, J., Villafuerte, L., Grande, D., Pérez-Gil, E., Alemán, T., & J. Carmona. 1997. Fodder shrub and tree species in the Highlands of southern Mexico. *Animal Feed Science and Technology, 68*, 213–223. https://doi.org/10.1016/S0377-8401(97)00052-7

Nahed-Toral, J., López-Tirado, Q., Alemán-Santillán, T., Aluja-Schunemann, A. & Parra-Vázquez, M. 2001. Los ovinos en la agricultura integral de los tzotziles. *LEISA, 16*(3), 23–25.

Nations, J. D. (1994). The ecology of the Zapatista revolt. *Cultural Survival Quarterly, 18*(1), 31–33.

Nations, J. D., & Nigh, R. B. (1980). The evolutionary potential of Lacandon Maya sustained-yield tropical forest agriculture. *Journal of Anthropological Research, 36*, 1–30.

Nigh, R. B. (1976). Evolutionary ecology of Maya agriculture in highland Chiapas, Mexico (Doctoral dissertation, Stanford University).

Nigh, R., & Diemont, S. A. W. (2013). The Maya milpa: Fire and the legacy of living soil. *Frontiers in Ecology and the Environment, 11*, e45–e54. https://doi.org/10.1890/120344

Noponen, M. R., Góngora, C., Benavides, P., Gaitán, A., Hayward, J., Marsh, C., . . . Wille, C. (2017). Environmental sustainability: Farming in the Anthropocene. In: B. Folmer (Ed.), *The craft and science of coffee* (pp. 81–107). San Diego, CA: Academic Press.

Orihuela-Belmonte, D. E., de Jong, B. H. J., Mendoza-Vega, J., Vander Wal, J., Paz-Pellat, F., Soto-Pinto, L., & Flamenco-Sandoval, A. (2013). Carbon stocks and accumulation rates in tropical secondary forest at the scale of community, landscape and forest type. *Agricultural, Ecosystems & Environment, 171*, 72–84. https://doi.org/10.1016/j.agee.2013.03.012

Otero, G. (2011). Neoliberal globalization, NAFTA, and migration: Mexico's loss of food and labor sovereignty. *Journal of Poverty, 15*, 384–402.

Perezgrovas, R., & Pedraza, P. (1989). *Ovinocultura indigena. Desarrollo corporal del borrego "Chiapas"*. Chiapas. Mexico: Universidad Autónoma de Chiapas.

Peterson, A. T., & Navarro-Sigüenza, A. G. (2016). Bird conservation and biodiversity research in Mexico: Status and priorities. *Journal of Field Ornithology, 87*(2), 121–132.

Piñeiro-Vazquez, A. T., Jiménez-Ferrer, J.G.O., Alayón, J. A., Chay-Canul, A. J., Ayala-Burgos, A. J., & Aguilar-Pérez, C. F. (2017). Effects of quebracho tannin extract on intake, digestibility, rumen fermentation, and methane production in crossbred heifers fed low-quality tropical grass. *Tropical Animal Health and Production, 50*, 29–36. https://doi.org/10.1007/s11250-017-1396-3

Pimbert, M. (2015). Food sovereignty and autonomous local systems. In M. Pimbert, R. Shindelar, & H. Schösler (Eds.), Think global, eat local: Exploring foodways (pp. 37–43). RCC Perspectives 2015, no. 1. Munich, Germany: Rachel Carson Center for Environment and Society.

Powis, T. G., Cyphers, A., Gaikwad, N. W., Grivetti, L., & Cheong, K. (2011). Cacao use and the San Lorenzo Olmec. *Proceedings of the National Academy of Sciences, 108*, 8595–8600. https://doi.org/10.1073/pnas.1100620108

Rojas, T., & Sanders, W. (Eds.). (1985). *Historia de la agricultura: Época prehispánica, siglo XVI*. Tomo I y II. Mexico City: Instituto Nacional de Antropología e Historia.

Salmón, E. (2000). Kincentric ecology: Indigenous perceptions of the human–nature relationship. *Ecological Applications, 10*, 1327–1332.

Schmitter, S. J. J., Mariaca-Méndez, R., & Soto-Pinto, L. (2016). Una breve historia del conocimiento y uso de la biodiversidad en la Frontera Sur de México. *Sociedad y Ambiente*, 4(11), 160–173.

Sharer, R. J., & Traxler, L. P. (2005). The ancient Maya (6th ed.). Palo Alto, CA: Stanford University Press.

Soberanes, R. (2018). Illegal cattle ranching deforests Mexico's massive Lacandon jungle. Retrieved from https://news.mongabay.com/2018/03/illegal-cattle-ranching-deforests-mexicos-massive-lacandon-jungle/.

Solorio, F., Ramírez, L., Basu, S., Trenchard, L., Sarabia, L., Wright, J., . . . Ku, J. (2017). Native trees and shrubs for ecosystems services and the redesign of resilient livestock production systems in the Mexican neotropics. In M.R. Ahuja & S.M. Jain (Eds.), Biodiversity and conservation of woody plants (pp. 489–511). Sustainable Development and Biodiversity Book Series 17. Cham, Switzerland: Springer. https://doi.org/10.1007/978-3-319-66426-2_16

Soto-Pinto, L. (1997). Plantas útiles no convencionales para el desarrollo de los sistemas productivos. In M. Parra & B.M. Diaz (Eds.), Los Altos de Chiapas: Agricultura y crisis rural (pp. 119–147). San Cristóbal de Las Casas, Chiapas: ECOSUR.

Soto-Pinto, L., & Aguirre-Davila, C. M. (2015). Carbon stocks in organic coffee systems in Chiapas, Mexico. *Journal of Agricultural Science, 7*, 117–128.

Soto-Pinto, L., & Anzueto Martinez, M. (2016). Los acahuales mejorados: Una práctica agroforestal innovadora de los maya tseltales. In A. I. Moreno Calles, A. Casas, V. M. Toledo, & M. Vallejo Ramos (Eds.), Etnoagroforestería en México (pp. 221–236). Mexico City, Mexico: Universidad Nacional Autónoma de México.

Soto-Pinto, L., M. Anzueto-Martinez, J.V. Mendoza, G. Jimenez-Ferrer, & B. de Jong. 2010. Carbon sequestration through agroforestry in indigenous communities of Chiapas, Mexico. Agrofor. Syst. 78(1):39–51.

Soto-Pinto, L., Castillo-Santiago, M. A., & Jiménez-Ferrer, G. (2012). Agroforestry systems and local institutional development for preventing deforestation in Chiapas, Mexico. In P. Moutinho (Ed.), Deforestation around the world (pp. 333–350). London: InTech Open.

Soto-Pinto, L., Jimenez, G., & de Jong, B. (1997). La agroforesteria en Chiapas: El caso de los altos de Chiapas. In M. Parra & B.M. Diaz (Eds.), Los Altos de Chiapas: Agricultura y crisis rural (pp. 167–186). San Cristóbal de Las Casas, Chiapas: ECOSUR.

Soto-Pinto, L., Villalvazo, V., Jimenez-Ferrer, G., Ramírez-Marcial, N., Montoya, G., & Sinclair, F. (2007). The role of local knowledge in determining shade composition of multistrata coffee systems in Chiapas, Mexico. *Biodiversity and Conservation, 16*, 419–436.

Staller, J. E., & Carrasco, M. D. (2010). Pre-Columbian foodways in Mesoamerica. In J. E. Staller & M. D. Carrasco (Eds.), Pre-Columbian foodways: Interdisciplinary approaches to food, culture, and markets in Mesoamerica. New York: Springer.

Taube, K. A. (1985.) The Classic Maya maize god: a reappraisal. In V. M. Fields (Ed.), *Proceedings of the Fifth Palenque Round Table Conference, June 12–18, 1983, Palenque, Chiapas, Mexico*. San Francisco, CA: Pre-Columbian Art Research Institute.

Terán, S., & Rasmussen, C. H. (2009). *La milpa de los Mayas* (2nd ed). Mexico City, Mexico: Universidad Nacional Autónoma de México.

Tiedje, K. (2018). The indigenous pastoral in the Huasteca, Mexico. In E. Berry and R. Abro (Eds.), *Church, cosmovision and the environment: Religion and social conflict in contemporary Latin America*. Abingdon, UK: Routledge.

Trench, T., (Producer) & Köhler, A. (Director) (2012). *¿No existe Nuevo Villaflores?* [Documentary film].

Valencia-Salazar, S., Piñeiro-Vázquez, A., Molina-Botero, I., Lazo, F., Segura-Campos, M., Ramírez-Avilés, L.,.Ku-Vera, J. (2018). Potential *of Samanea saman* pod meal for enteric methane mitigation in crossbred heifers fed low-quality tropical grass. *Agricultural and Forest Meteorology, 258*, 108–116.

Valentine, J.L., Barham, B., Gitter, S., & Nobles, J. (2017). Migration and the pursuit of education in southern Mexico. *Comparative Education Review, 61*(1), 141–175.

Zeiger, J. (2019). Traditional indicators of rainfall in the Selva Lacandona, Chiapas, Mexico (Master's thesis, State University of New York, College of Environmental Science and Forestry, Syracuse). Retrieved from https://digitalcommons.esf.edu/cgi/viewcontent.cgi?article=1078&context=etds.

Study Questions

1. What are the similarities and differences among agroforestry systems in Mexico?
2. How does local and traditional knowledge help with the design of agroforestry in Mexico?
3. How do traditional agroforestry farmers use an understanding of ecology in designing their systems?
4. How could agroforestry practices of indigenous communities help to improve food security?
5. How could agroforestry assist with adaptation to and mitigation of climate change?
6. What are agroforestry contributions to human health in rural communities?
7. How does agroforestry contribute to cultural ecosystem services?
8. What is the relationship between families and agroforestry in Mexico?
9. What can Western science scholars learn from traditional agroforestry management?
10. What do you think are the challenges that rural communities in Mexico will face in continuing their agroforestry?

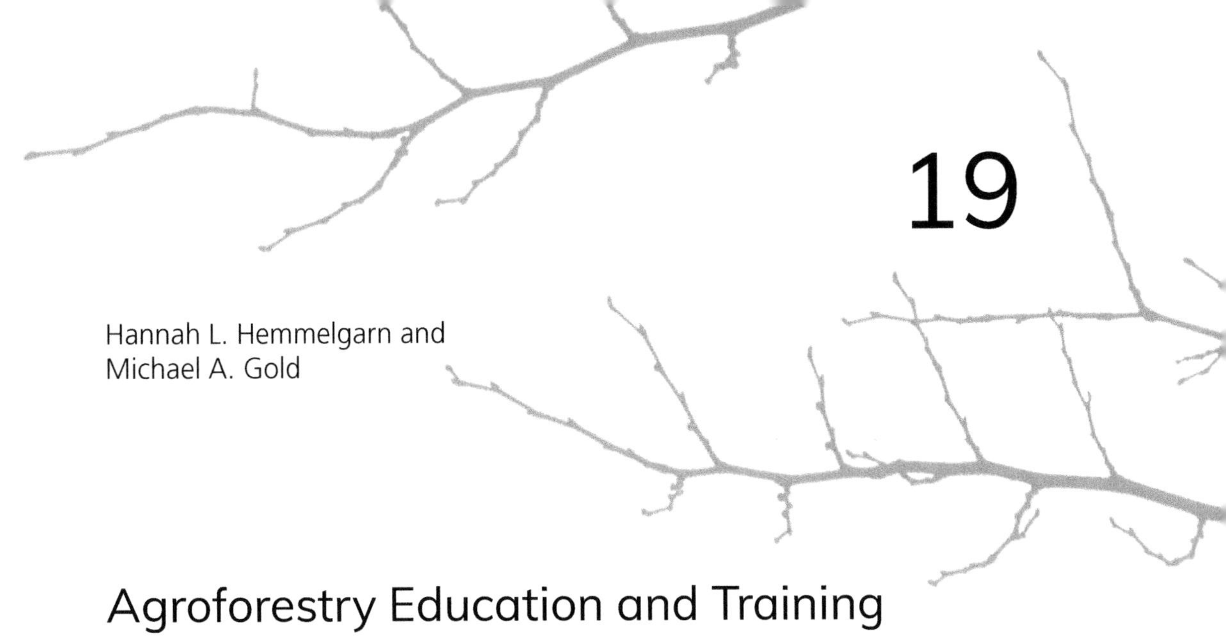

19

Hannah L. Hemmelgarn and
Michael A. Gold

Agroforestry Education and Training

Agroforestry education and training programs require a complex, dynamic, and context-specific process that requires attention to the particularities of *who*, *why*, *how*, and *what* is learned and the relationship between such learning and behavioral change. Agroforestry practices are knowledge intensive, and to achieve widespread adoption of agroforestry in the future will require more thinking workers and more working thinkers (Ikerd, 2016). Public and private universities and colleges, and land grant universities in particular, are largely responsible for educating future agroforestry professionals, while on-the-ground application requires outreach, education, and engagement at multiple levels. The teaching of agroforestry courses offers the opportunity to help meet the interests of students for an interdisciplinary, problem-solving education, which is challenging to provide due to the demands for scientific rigor within discipline-based curricula (Gold & Jose, 2012; Lassoie, Huxley, Buck, 1994; "Towards a comprehensive education and training program in agroforestry", 1990). Agroforestry also provides a model for teaching holistic approaches to land use management and may attract students from a wide variety of disciplines within the agricultural and natural resource sciences.

Despite acceptance of the environmental, economic, and social benefits of agroforestry within the research community, adoption of agroforestry remains low in North America and other temperate regions. The agroforestry literature indicates the importance of education as a tool for empowering agroforestry adoption (Jose, Gold, & Garrett, 2018; Strong & Jacobson, 2005; Valdivia, Barbieri, & Gold, 2012). Multiple forms of education, including formal, non-formal, and informal platforms, may be central to a comprehensive approach for a culture, awareness, and knowledge network that can support implementation of agroforestry practices on the ground (Romanova, 2020; Hemmelgarn, Ball, Gold, & Stelzer, 2018; Kueper, Sagor, & Becker, 2013; "Towards a comprehensive education and training program in agroforestry", 1990).

Farmers in agricultural and rural communities throughout the United States and Canada face ongoing financial challenges. The long-term economic and environmental sustainability of family farms is a serious concern. Adoption of sustainable agricultural practices, including agroforestry, and promotion of locally produced specialty crops through market-based conservation approaches will provide new opportunities for farms to be both profitable and ecologically resilient. Agroforestry practices, by retaining some of the structural and functional characteristics of natural ecosystems, offer solutions to many of the environmental issues associated with modern agriculture.

North American Agroforestry, Third Edition. Edited by Harold E. "Gene" Garrett, Shibu Jose, and Michael A. Gold.

Across North America, the majority of farmers, land managers, and natural resource professionals lack even a basic familiarity with agroforestry and related approaches for sustainable land use management, lack the specific technical knowledge to successfully implement, or are not aware of federal, state, and non-governmental organization (NGO) programs that support adoption of these practices. This situation is the result of a lack of educational infrastructure for farmers, land managers, and natural resource professionals specifically regarding agroforestry practices. To address this, expanded education, training, and outreach efforts are needed to raise awareness of and technical proficiency in agroforestry and related sustainable land use practices, in conjunction with economic opportunities through specialty crops and farm bill programs that support adoption of these practices.

Since the publication of the second edition of North American Agroforestry: An Integrated Science and Practice (Garrett, 2009), formally accredited online graduate certificate and master's degree programs have been established. Further, many universities have created agroforestry outreach and extension programs to train educators and landowners (Gold, Hemmelgarn, & Mendelson, 2019). In addition, state and federal agencies (e.g., USDA National Agroforestry Center, 2015), NGOs (e.g., Savanna Institute), and the private sector (e.g., Iroquois Valley Farmland REIT), and multiple specialty crop and livestock cooperatives (e.g., elderberry, chestnut, hazelnut) actively engage with landowners as part of a knowledge exchange network. This chapter reviews the history of agroforestry education and training in North America, showcases current effective programs, and discusses future agroforestry education needs.

Background: Agroforestry Education and Training in the United States

The introduction of the term *agroforestry*, as described by Smith (1929) and defined by Bene, Beall, and Cote (1977) and Lundgren (1982), did not mark the inception of its practice and science. For thousands of years, Indigenous peoples managed land—and continue to manage land—in the ways we now describe as agroforestry, sharing knowledge and insights from observation and experimentation within and beyond their respective communities (Rossier & Lake, 2014). Indeed, traditional ecological knowledge continues to inform and support Western agroforestry science and practice. These knowledge networks are vital models of the kind of education systems needed for enduring landscape stewardship; however, conventional agriculture and natural resources structures require novel approaches to continue to advance the implementation of agroforestry suitable to today's farms and farmers. Beginning in the mid-1980s, temperate zone agroforestry, as a unique discipline recognized by Western scientists, started a course of steady progress with biophysical and socioeconomic science, academic literature, professional associations, degree programs, and training programs. The knowledge infrastructure that might support the development of agroforestry has blossomed in that time.

Both education and training programs have been developed and the status in 2022 is one in which there are robust offerings yielding a strong foundation for on-the-ground practice. This chapter is focused on *temperate agroforestry* education and outreach, a specific sector of land-use disciplines dealing explicitly with the set of practices that defines temperate agroforestry. While conventionally described as alley cropping, silvopasture, forest farming, windbreaks, riparian and upland buffers, urban food forests, and other special applications such as biomass production, the hard boundaries of this definition are becoming more embracing of a wide range of creative approaches. In each unique context, the intentional integration of trees and shrubs in a whole farm system can manifest as any number of innovative combinations or modifications of this basic set of practices. In temperate North America and elsewhere globally, regenerative agriculture, permaculture, agroecology, and other related frameworks have emerged as major parallel movements that often incorporate agroforestry in their "toolbox" for on-the-ground implementation. These movements can and do work in tandem while occupying unique niches in the realm of education.

In any of these contexts, education can be formal, non-formal, or informal in nature, often serving as interrelated mechanisms. Formal education is characterized by an organized curriculum and structured learning environments wherein a student is intentional about their undertaking of a program toward an earned credential (Werquin, 2007). Formal education offers the opportunity for a receipt of completion, such as a recognized degree or certification. Academic settings generally adhere to this classification and are unique in their intent and outcomes. Non-formal education, while also highly organized and typically led by an experienced instructor, results in enrichment of skills and knowledge without a formal receipt of completion and is often driven more by the learner's interests rather than by curriculum standards (Werquin, 2007).

Examples of formal education include high school, college, or graduate school courses, while non-formal learning takes place in extracurricular groups and organizations, via extension outreach programs, and other organized workshops. Informal learning, on the other hand, can take place anytime, without an organized educational program. For instance, informal learning happens when a peer shares information on social media or when a neighbor demonstrates a unique practice. Informal learning often goes unnoticed but can be an equally powerful means of education, often occurring during the "coffee break" moments of formal and non-formal programs (Kueper et al., 2013). Agroforestry education opportunities now fully occupy each of these arenas for learning.

Formal Agroforestry Education

Undergraduate agroforestry coursework and degree programs are well positioned to enhance professional readiness for agroforestry implementation (Gold, 2015). While sustainable agriculture and agroecology degree programs may sometimes integrate agroforestry-related content, the presence of agroforestry-specific courses and programs has been more limited. By the late 1980s and early 1990s, agroforestry educators were active at a number of North American universities including Berkeley, Cornell, Florida, Guelph, Hawaii, Idaho, Kentucky, Laval University (Canada), Michigan State, Minnesota, Missouri, Purdue, Virginia Tech, Washington State, and Yale. At that time, both research and education efforts were heavily weighted toward tropical agroforestry and international development (Lassoie et al., 1994; Wright, 2017).

A survey of 39 Society of American Foresters (SAF) forestry accredited institutions in 1988 revealed that at least 14 schools offered an agroforestry course (Warren & Bentley, 1990). In 2017, a more extensive survey was administered to 127 land-grant and SAF accredited institutions in the United States, indicating overall growth to 27 institutions offering agroforestry courses and three offering degree programs (Wright, 2017). Despite the general increase in course offerings, eight of the institutions that had offered agroforestry courses in 1988 were no longer offering these courses in 2017. Those institutions that no longer offer agroforestry courses cited lack of resources, lack of student enrollment, change in curriculum, or the departure of faculty. Where tropical agroforestry content dominated course content in 1988, the agroforestry courses currently offered have shifted to greater representation of temperate agroforestry (Warren & Bentley, 1990; Wright, 2017).

Agroforestry courses typically attract highly qualified students, often coming with extensive international agroforestry experience, including the Peace Corps (Gold & Jose, 2012) or students familiar with permaculture, agroecology, and sustainable or regenerative agriculture. In the past, such interest was limited to graduate students seeking careers in international development. More recently, however, both undergraduate and graduate students have been attracted to agroforestry courses, probably reflecting growing interests in courses and careers dealing with issues of sustainability. Employment opportunities where agroforestry credentials are valued are increasing. Many federal agencies (e.g., the NRCS), global, national, and regional conservation organizations (e.g., Heifer International, The Nature Conservancy, National Wild Turkey Federation, Trees Forever), along with NGOs specifically dedicated to agroforestry (e.g., Savanna Institute) are hiring individuals with agroforestry backgrounds.

Short courses, workshops, and conferences are essential forms of agroforestry outreach and education. However, working professionals and landowners across the United States, Canada, and overseas also seek more in-depth and comprehensive formal education programs in the form of undergraduate courses, graduate certificates, and/or graduate degrees in agroforestry. Prior to 2010, only one university, Université Laval in Canada, had established a graduate degree program in agroforestry (Khasa, Olivier, Atangana, & Bonneville, 2017). In 1996, Université Laval implemented a French language agroforestry graduate program offering both a Master of Science (M.Sc.) with thesis and non-thesis M.S. The thesis M.Sc. includes a series of 21 course-based credits and a compulsory internship where students learn the biophysical, socioeconomic, and cultural dimensions of agroforestry, plus an additional 24 research-based credits. The non-thesis (essay-based) M.S. consists of 33 course-based credits including the same required internship, plus 12 essay-based credits. Between 1996 and 2014, 35 M.Sc. theses and 19 M.S. essays were completed. Fourteen essays and 18 theses investigated biophysical aspects of agroforestry and five essays and 17 theses focused on socioeconomic aspects. Of the 54 theses and essays that were completed, 61% focused on tropical countries and 39% on temperate countries. The program also published more than 55 peer-reviewed articles, two book chapters, and one textbook. The program is expanding in tropical French-speaking countries through international projects on training in the management of natural resources and with the use of digital learning tools to reach more students online (Khasa et al., 2017).

Distance education programs, increasingly available, provide working professionals with a chance at a graduate education, reach those disadvantaged by limited time or distance, and update the knowledge base of workers at their places of employment. For universities, these programs serve to increase enrollment and make more efficient use of internal resources and existing facilities. Online education offers decentralized access to courses throughout the state, region, nation, and internationally. The quality of distance learning has greatly improved in recent years, as both students and educators have become more comfortable with the technology (Gold & Jose, 2012).

In 2011, the University of Missouri Center for Agroforestry (UMCA) created a 100% asynchronous online graduate certificate and master's program in agroforestry to help address current and future needs of the agroforestry profession. With initial funding support from the University of Missouri System (a four-campus system), the Center for Agroforestry developed an initial suite of eight online courses to create the interdisciplinary online graduate program in agroforestry. Participating faculty were drawn from the University of Missouri's College of Agriculture, Food and Natural Resources (CAFNR) and School of Natural Resources (SNR), which includes all Center for Agroforestry faculty. As part of the UM System funding, all participating faculty attended a 2-day eLearning workshop, which consisted of presentations, illustrations, and demonstrations of the fundamental principles of online course design and development (Gold & Jose, 2012). The online agroforestry M.S. courses became fully operational in 2013. Since its inception, additional faculty have come online to offer more breadth and depth to the course options. Currently, the University of Missouri maintains three graduate degree offerings: (a) campus M.S., (b) online M.S., and (c) online graduate certificate. A Ph.D. program is currently under development. During the 5-yr period from 2015–2019, a total of 48 students completed agroforestry graduate programs through the University of Missouri. Of these 48 graduate students, 31 M.S. graduate programs were 100% online and 17 were on campus in the traditional thesis/dissertation format.

The University of Missouri online agroforestry graduate program currently serves as a model for similar programs elsewhere in the world. One important outcome of the online M.S. program is that it is catalyzing additional interest at other U.S. colleges and universities (e.g., an online agroforestry M.S. graduate from the University of Missouri created and taught two agroforestry courses at Sterling College in Vermont). As of late 2021, no other fully online English language agroforestry M.S. programs have been established.

Toward Continuity of Formal Education: Curricula and Teacher Training

While opportunities for agroforestry education have expanded in post-secondary academic settings throughout the past 20 yr (Warren & Bentley, 1990; Wright, 2017), student pursuit of this content may be limited to those previously exposed to agroforestry in personal life experiences or academic endeavors. Agriculture education at the high school level serves as a primary context for equipping young agricultural professionals with the skills and knowledge that will guide their careers. Until very recently, agroforestry content was not explicitly included in any known high school agriculture curricula in the United States (Hemmelgarn et al., 2018).

In Missouri, where the first statewide high school agroforestry curriculum was developed in 2016, approximately 6,000 students graduate annually from high school agriculture programs, and more than 60% of these graduates pursue careers in agriculture (Missouri Department of Elementary and Secondary Education, 2016). Nationally, the number of college-educated farmers has increased substantially in the last 10 yr, but for the 70% of U.S. farm operators who have not completed a college degree (USDA Economic Research Service, 2012), the high school agriculture program and extracurricular organizations (e.g., FFA and 4-H) may also be among the most essential spheres of early professional training.

Teachers, as stewards of knowledge exchange in formal education settings, hold the key to training a new generation of farmers and foresters with the skills to integrate agroforestry in their careers. Where unfamiliar content is introduced, as was the case in Missouri with an agroforestry curriculum for high schools, professional development for agriculture and science teachers served as an initiation of momentum for ongoing content integration.

By the 1990s, sustainable agriculture education had become a prominent theme in the Journal of Agricultural Education. The "sustainable agriculture" detailed in this literature is exemplified by practices such as soil testing, soil erosion control, crop rotation, reduced use of chemicals, integrated pest management, and the use of cover crops (Agbaje, Martin, & Williams, 2001; Alonge & Martin, 1995), although it rarely includes reference to agroforestry practices. By integrating agroforestry education at all levels and in various platforms, an established culture and awareness is likely to magnify the impact of individual programs.

Non-Formal and Informal Agroforestry Training

Agroforestry-related organizations and working farms have presented the public with creative and innovative non-formal agroforestry outreach and education programs that serve farmer-landowners, natural resource professionals, and many others. Each of these programs can result in ongoing informal networks whose reach may be difficult to measure but that have potential for lasting impact. Examples include conferences, intensive training programs, seasonal farmer-mentored apprenticeships, in-field workshops and farm visits, and a plethora of diverse media and events that bring people together in a learning environment.

In North America, the history of agroforestry conferences and workshops closely allied with agroforestry concepts dates back to the late 1970s and early 1980s. In 1988, the first major conference that specifically focused on agroforestry education, and included a focus on the temperate zone, was convened in Gainesville, FL (Nair, Gholz, & Duryea, 1990). Specific agroforestry training and education programs commenced in the late 1980s and increased in quantity and scope throughout the 1990s and 2000s.

The earliest U.S. agroforestry focused conferences included the 1976 International Hill Land Symposium at West Virginia University, Morgantown, WV; the 1980 Tree Crops for Energy Co-Production on Farms workshop at the Solar Energy Research Institute in Estes Park, CO; and the 1983 Foothills for Food and Forests conference at Oregon State University, Corvallis, OR. From 1985–2000, notable agroforestry conferences and workshops that were specifically agroforestry focused included the 1986 First International Windbreak Symposium at the University of Nebraska, Lincoln; the 1986 International Agroforestry Short Course at Colorado State University, Ft. Collins; the 1990 Mid-South Conference on Agroforestry Practices and Policies in West Memphis, AR; and the 1994 Agroforestry and Sustainable Systems conference at Colorado State University, Ft. Collins. By the late 1990s, the scope of agroforestry trainings broadened to include specialty crop markets, notably in the 1998 North American Conference on Enterprise Development through Agroforestry: Farming the Forest for Specialty Products at the University of Minnesota, Minneapolis (Gold, 2015).

The first truly comprehensive North American agroforestry conference to include all dimensions of temperate agroforestry, both biophysical and socioeconomic, took place in 1989. This first North American Agroforestry Conference (NAAC) was organized at Guelph University, Guelph, ON, Canada (Williams, 1991). This conference subsequently became the preeminent biennial agroforestry research conference for North America. Seventeen biennial NAACs (1989–2021) have been held across the United States and Canada, and since 1993 all NAACs have been sponsored by the Association for Temperate Agroforestry (AFTA), which was established in 1993 (https://www.aftaweb.org/).

In 2011, the USDA released its Agroforestry Strategic Framework as a roadmap for advancing the science, practice, and application of agroforestry (USDA, 2011). The framework's first goal was to increase agroforestry adoption by landowners and communities by expanding learning partnerships with stakeholders and educating professionals. A major concern was that professionals (e.g., NRCS and state agency conservation staff) were not equipped to provide the technical, financial, and marketing assistance needed to plan and apply agroforestry.

In response to the needs identified in the USDA Strategic Framework and a recognition that lack of familiarity with agroforestry among educators (e.g., NRCS, university extension) limited landowner adoption, a week-long pilot Agroforestry Academy was created by UMCA and the Mid-American Agroforestry Working Group (MAAWG, organized in 2009) to train natural resource professionals, extension agents, and other agricultural educators who work with landowners. Initial funding was supplied by a grant from the North Central Region Sustainable Agriculture Research and Extension Professional Development Program (Gold, Cernusca, & Jose, 2013; Gold et al., 2019).

Since 2013, the *Agroforestry Training Manual* (Gold, Hemmelgarn, Mori, & Todd, 2018), the primary text developed for the Agroforestry Academy, has been updated annually and a second core publication, a *Handbook for Agroforestry Planning and Design* (Gold, Cernusca, & Hall, 2013), was created to provide the framework for the weeklong training program. In 2018, a third publication, *Perennial Pathways—Planting Tree Crops: Designing and Installing Farm-Scale Edible Agroforestry* (Wilson, Lovell, & Carter, 2018) was included as core reading material. The Agroforestry Academy is an intensive short course that provides advanced training on the six recognized temperate zone agroforestry practices with in-class seminars, field visits, and an applied planning and design exercise.

A comprehensive evaluation in 2019 of the impact of the Agroforestry Academy on trainees' (n = 175) knowledge and practice revealed a strongly positive experience, with the majority of farmer attendees indicating agroforestry adoption in their operations (64%) and the majority of professional trainees applying their agroforestry skills in their careers (90%) and in their personal contexts (65%). The most common write-in reason for non-adoption was land access, followed by resource constraints (e.g., money, time, and labor capacity) (Mendelson, 2020; Mendelson, Gold, Lovell, & Hendrickson, 2021).

The 2013–2019 program evaluation also addressed needs for extended training beyond the week-long intensive academy, including the need for additional practice-specific workshops and targeted resources. Silvopasture was the highest ranked topic for additional training experiences; desired resource materials included a regulatory toolkit and an agroforestry resource guide for accessing the growing cadre of updated agroforestry information sources. Continued hands-on practice, professional support to ensure a complete understanding of financial and land management options, and opportunities to connect with farmer-landowners who have been successful were cited as critical gaps between knowledge and practice. Forty-one percent of farmers and 54% of non-farmers also called for access to a community of practice, with a focus on in-person group opportunities for exchange (Mendelson, 2020; Mendelson, Gold, Lovell, & Hendrickson, 2021).

Informal Learning Outcomes of Non-Formal Training

Non-formal educational contexts are common approaches for land-based disciplines, including workshops, field days, extension programs, and conferences. Examples of non-formal agroforestry education are shared here to demonstrate the likely outgrowth of these events into informal learning opportunities. Informal learning occurs from experience and interaction, outside of organized programs, although this does not diminish its impact on the learner. In fact, informal learning can be a powerful tool for in-depth understanding (Jacobson & Kar, 2013; Kueper et al., 2013).

The identification of highly respected "farmer champions" (National Wildlife Federation, 2012, 2013; Shelton, Wilke, Franti, & Josiah, 2009) plays a critical role in the introduction, transfer, and implementation of new agroforestry techniques. In addition, the promotion of community involvement facilitates the transfer of information, not simply to introduce information, but also to promote social exchange and interaction, strengthen preexisting informal source networks, and increase social proximity among farmers. Specialized information introduced into the network by farmer champions may also reinforce the social capital of these individuals. However, original and accessible information is also produced by on-farm experimentation. Informal advice networks, which may be reinforced during public events, are instrumental in the successful transfer of this available information throughout the farming community and provide a foundation for community-based adaptive management (Isaac, Erickson, Quashie-Sam, & Timmer, 2007).

The annual Agroforestry Symposium, first held in 2010 at the University of Missouri, is one such public event that is designed to open doors for ongoing exchange (https://centerforagroforestry.org/annual-symposium/). This themed mini-conference engages diverse regional stakeholders with internationally recognized presenters to share relevant information for farmer-landowner, academic, and professional audiences. Annual themes have included climate change resilience, soil health, pollinators, medicinal plants, and other topics that bring together people from within and outside agroforestry circles. Each year, the event is livestreamed for widespread access, but the local connections that occur in person have resulted in the event's more long-lasting outcomes, including recruited agroforestry graduate students, new market exchanges between producers and buyers, and research relationships between departments and campuses.

University extension programs have also served to advance non-formal agroforestry education into informal networks. Outreach leaders with agroforestry knowledge are key in this process. The Cornell Small Farms Program is a prime example of a university extension effort with strong agroforestry ties. At Cornell, Steve Gabriel and others have initiated learning networks for agroforestry such as the specialty mushroom cultivation webinar series and resource catalog. The Cornell Small Farms Program also hosts 3-day Agroforestry in Practice trainings for service providers and veterans, supported, like others at the University of Missouri, by the USDA. At North Carolina State University Extension Service, Jeanine Davis and others have developed a rich source of forest farming knowledge exchange focused on sustainable cultivation of medicinal plants and other non-timber forest products. Together with extension providers at these and other universities including Virginia Tech, University of Minnesota, Iowa State, Penn State, and University of Missouri, the forest farming extension community (online) offers centralized knowledge and technology transfer.

Association for Temperate Agroforestry

In 1989, the first North American Agroforestry Conference (NAAC) convened at the University of Guelph, Guelph, ON, Canada, principally organized by Dr. Andrew Gordon and Peter Williams. Participants recognized the uniqueness of the event as the first meeting exclusively focused on all topics related to temperate zone agroforestry. There was a further realization that no professional organization existed to serve as a "home" for all scientists and practitioners with an interest in all aspects of temperate agroforestry.

In 1991, the second NAAC was organized by Gene Garrett and the University of Missouri in southwestern Missouri. The key outcome of the second NAAC was that participants agreed to establish what became known as AFTA as a nonprofit association focused on temperate zone agroforestry. Its purpose was to organize, catalyze, and network all individuals with a common interest in temperate agroforestry, with the primary geographic focus to be temperate North America. In 1993, AFTA developed its goal and mission statements, identified a list of association objectives, developed the association bylaws, created a logo, and launched its newsletter, *The Temperate Agroforester*.

In 2004, AFTA launched its website (https://www.aftaweb.org/). *The Temperate Agroforester* newsletter shifted from print to an online journal. In addition, the AFTA website serves as the repository for past issues of the newsletter, general articles on a wide array of agroforestry subject matter, links to the regional agroforestry working groups, and serves as the contact point for information about agroforestry events including the biennial NAAC series.

Regional Working Groups

In addition to the National Agroforestry Center and the Association for Temperate Agroforestry, during the past decade (2010–2020) a number of regional agroforestry working groups have been established to bring agroforestry practitioners together. These informal networks serve as venues for the exchange of knowledge and experiences among practitioners, cooperatives, researchers, outreach professionals, and NGOs. These regional networks share a number of common goals including:

- Identification of key barriers, gaps, and opportunities for enhancing the adoption of tree- and shrub-based perennial agricultural systems.
- Collaboration on regional trainings, technical resource development, and research efforts for resolving specific core issues.
- Establishment of region-wide networks of technical service providers, producers, and landowners to collaborate, share, and enhance the communication on agroforestry developments and resource opportunities,
- Inspiration of innovation in perennial tree- and shrub-based agricultural systems through regional networking and resource sharing
- Enhancement of the economic diversity and potential for land-based economies through tree- and shrub-based perennial agriculture.

These groups work across social, economic, technological, and environmental disciplines for increasing awareness and application of agroforestry as a multifunctional approach toward sustainability in agriculture and natural resource management. The growing list of key regional agroforestry working groups includes:

- Northeast and MidAtlantic Working Group (NEMA). The NEMA Working Group is a network of researchers, technical service providers, agency staff, farmers, and producers focused on educating, promoting, and implementing agroforestry systems in the Northeast and Mid-Atlantic regions of North America. NEMA's purpose is to collaborate across political and geographic borders to enhance the science, adoption, and efficacy of agroforestry systems through education, research, and outreach to natural resource professionals, technical assistance providers, producers, and landowners.

- Mid-American Agroforestry Working Group (MAAWG). MAAWG was formed in 2009. Its goals include identifying core issues for advancing agroforestry in the region, initiating and coordinating actions to address and resolve these core issues, and communicating effectively with key audiences about the group's purpose, core issues, and findings. Initial members included the Agricultural Marketing Resource Center (Iowa State University), the Center for Agroforestry (University of Missouri), the Center for Integrated Natural Resources and Agricultural Management (University of Minnesota), Green Lands Blue Waters (co-located at the University of Minnesota), the Leopold Center for Sustainable Agriculture (Iowa State), NRCS, Trees Forever (Iowa and Illinois), University of Wisconsin extension, U.S. Forest Service, and the USDA National Agroforestry Center.
- Pacific Northwest Agroforestry Working Group (PNAWG). The PNAWG was formed in May 2013 to assess the status of Pacific Northwest agroforestry education, research, and outreach activities. PNAWG's mission is to work as a collaborative group of academics, extension agents, and land stewards to promote agroforestry awareness and adoption, as well as facilitate research and demonstration in the Pacific Northwest. PNAWG holds monthly meetings from September through June at Oregon State University, including allowance for remote call-in. Monthly meetings involve updates on agroforestry efforts, presentations by one or more of the members to facilitate feedback from the group, occasional field trips, and agroforestry opportunities including research, grant writing, and publications.
- Southwest Agroforestry Action Network (SWAAN). In 2019, SWAAN members were forming a chartered organization for agroforestry in the southwestern United States. Once formalized, a new website will be built to accommodate the mission, vision, action plans, and related research. SWAAN seeks to share information, connect potential collaborators and partners, and generate ideas, research, and initiatives that can advance adoption of agroforestry in the U.S. Southwest by agricultural producers, forest landowners, and communities.
- The 1890 Agroforestry Consortium. The 1890 Agroforestry Consortium is composed of a group of 1890 land-grant institutions and USDA government agency partners. Its mission is to develop and advance agroforestry research, teaching, and extension among the 1890 land-grant institutions and Tuskegee University. The consortium consists of multidisciplinary teams of faculty and staff working in partnership with government agencies and other entities.

Nonprofit Organizations

In response to regional interests and needs, nonprofit organizations have also formed to collect, curate, and convene for agroforestry implementation, resource exchange, and knowledge development. The following are exemplary organizations who have succeeded in coalescing around specific agroforestry practices or unmet regional needs, bolstered by strong partnerships and member support.

- Appalachian Beginning Forest Farmers Coalition (ABFFC). The ABFFC is a network of forestland owners, universities, and governmental and non-governmental organizations that share a common goal of improving agroforestry production opportunities and farming capabilities among forest farmers. Their collective aim is to increase awareness of forest-grown medicinal plants through education and relationship building and to support conservation efforts through stewardship of existing plant populations and forest farming of these native botanicals. The coalition focuses on education, training, and support for beginning forest farmers and the improvement of a forest farm inventory and medicinal plant habitat management services for beginning forest farmers. Using a participatory approach to coalition building, with input from multiple sectors and via multiple communication pathways, the ABFFC is advised by forest farmer members and technical service provider members.
- Savanna Institute. The Savanna Institute is a 501(c)(3) nonprofit organization making efforts to lay the groundwork for widespread agroforestry in the U.S. Midwest. The organization works in collaboration with farmers and scientists to develop perennial food and fodder crops within multifunctional polyculture systems grounded in ecology and inspired by the savanna biome. The Savanna Institute strategically enacts this mission with research, education, and outreach. One of their primary research activities is a cooperative and participatory case study program that explores the potential of agroforestry practices as ecologically sound, agriculturally productive, and economically viable alternatives to annual row crop agriculture.

Farmer-led Cooperatives

Many agroforestry practitioners seek to affiliate with professional organizations, associations, and/or cooperatives focused on both long-established and newer, emergent, specialty crops. A number of organizations have been established that support a wide variety of specialty crops that are suitable for use in agroforestry practices (Table 19–1).

Private Sector

Iroquois Valley Farmland Real Estate Investment Trust (REIT) is an organic farmland finance company providing farmer-friendly leases and mortgages to the next generation of organic farmers according to triple bottom-line principles: social responsibility, environmental soundness, and economic viability. The company is committed to scaling organic agriculture in the United States by getting more organic farmers on the land, funded by socially responsible investors. Iroquois Valley has directed US$50 million in investments in more than 50 farms in 14 states comprising more than 4,000 ha (nearly 10,000 acres). Since its establishment in 2007, Iroquois Valley's goal is to make organic agriculture the norm, not the exception, in America to benefit the health of the soil and of future generations (https://iroquoisvalley.com/about/).

RSF Social Finance is a financial services organization that has created a growing community of motivated, values-driven investors, donors, and entrepreneurs committed to transforming an extractive economy into one that brings healing and regeneration. RSF Social Finance focuses on relationships over transactions, integration over fragmentation, and mutual support over competition. RSF Social Finance maintains a Food System Transformation Fund that provides loan capital to enterprises that provide critical infrastructure services to support the development of

Table 19–1. Specialty crops for which there are existing associations and co-ops, among others not listed.

Specialty crop	Association, Co-op	Association, Co-op
Aronia	Midwest Aronia Association (https://midwestaronia.org/)	Midwest Aronia Association (https://midwestaronia.org/)
Chestnut	Chestnut Growers of America (http://www.chestnutgrowers.org/) Route 9 Cooperative (https://route9cooperative.com/) Chestnut Growers, Inc. (https://www.chestnutgrowersinc.com) Prairie Grove Chestnut Growers (http://prairiegrovechestnutgrowers.com/)	Chestnut Growers of America (http://www.chestnutgrowers.org/) Route 9 Cooperative (https://route9cooperative.com/) Chestnut Growers, Inc. (https://www.chestnutgrowersinc.com) Prairie Grove Chestnut Growers (http://prairiegrovechestnutgrowers.com/)
Eastern black walnut	Heartland Nuts 'N More (https://heartlandnutsnmore.com/ Hammons Black Walnuts (https://black-walnuts.com/)	Heartland Nuts 'N More (https://heartlandnutsnmore.com/ Hammons Black Walnuts (https://black-walnuts.com/)
Elderberry	Midwest Elderberry Coop (https://www.midwest-elderberry.coop/)	Midwest Elderberry Coop (https://www.midwest-elderberry.coop/)
Hazelnuts	NOTE: The majority of hazelnut production in the United States occurs in Oregon Oregon Organic Hazelnut Cooperative http://oregonorganichazelnuts.org/ Hazelnut Growers of Oregon https://www.hazelnut.com/ Upper Midwest Hazelnuts https://www.midwesthazelnuts.org/)	NOTE: The majority of hazelnut production in the United States occurs in Oregon Oregon Organic Hazelnut Cooperative http://oregonorganichazelnuts.org/ Hazelnut Growers of Oregon https://www.hazelnut.com/ Upper Midwest Hazelnuts https://www.midwesthazelnuts.org/)
Maple syrup	Many states have their own maple syrup associations Vermont Maple Sugar Makers' Association https://vermontmaple.org/ Ohio Maple Producers Association https://www.ohiomaple.org/ NYS Maple Producers Association https://nysmaple.com/ Wisconsin Maple Syrup Producer's Association https://wismaple.org/ Massachusetts Maple Producers Association https://www.massmaple.org/ Minnesota Maple Syrup Producers' Association https://www.mnmaple.org/	Many states have their own maple syrup associations Vermont Maple Sugar Makers' Association https://vermontmaple.org/ Ohio Maple Producers Association https://www.ohiomaple.org/ NYS Maple Producers Association https://nysmaple.com/ Wisconsin Maple Syrup Producer's Association https://wismaple.org/ Massachusetts Maple Producers Association https://www.massmaple.org/ Minnesota Maple Syrup Producers' Association https://www.mnmaple.org/
Medicinals and herbs	Appalachian Beginning Forest Farmers Coalition https://www.appalachianforestfarmers.org/	Appalachian Beginning Forest Farmers Coalition https://www.appalachianforestfarmers.org/
Mushrooms	North American Mycological Association https://namyco.org/mushroom_cultivation_resources.php	North American Mycological Association https://namyco.org/mushroom_cultivation_resources.php
Pecan	Missouri Northern Pecan Growers https://mopecans.com/	Missouri Northern Pecan Growers https://mopecans.com/

resilient, regional food systems in the United States. They support organizations that provide aggregation, processing, distribution, and/or market access for farming enterprises and also work with organizations that provide farmers with access to farmland and training services (https://rsfsocialfinance.org/).

Future Needs and Gaps in Agroforestry Education and Training

To increase agroforestry adoption in the United States, a critical mass of professionals must have agroforestry expertise. Therefore, the future of agroforestry education and training also includes the need for the development of a Professional Agroforester credential (Mason, 2016; USDA, 2019). Since AFTA's establishment in 1993 there have been ongoing discussions about the need to create a formal agroforestry credential for practicing agroforesters similar to that offered to certified crop advisors, certified horticulturalists, and professional foresters. A national agroforestry credentialing program with broad participation by agriculture and forestry professionals would advance the agroforestry profession and increase the adoption of agroforestry (Mason, 2016).

A 2012 commentary in the *Journal of Forestry* "Advancing Agroforestry through Certification of Agroforesters: Should SAF have a Role?" (Mason et al., 2012) addressed this issue. In 2015, the Society of American Foresters Certification Review Board (CRB), which oversees the Certified Forester credential, contacted the American Society of Agronomy (ASA) to propose the formation of a joint SAF–CRB–ASA task force to explore the costs and benefits of establishing a national joint agroforestry credential, certificate, or certified agroforester program. Certification can draw on existing agroforestry academies and graduate certificate programs designed to meet the needs of working professionals. When brought to fruition, the establishment of a Certified Agroforester program will not only create new job opportunities but, more importantly, will also expand the growth of agroforestry practices "on the ground" with professional guidance.

Simultaneously, the advancement of innovative farmers and peer networks may increase the capacity to disseminate agroforestry resources from trained professionals. This activity, while likely to occur as a result of informal learning and exchange, can be facilitated by the ongoing organized education and outreach programs that connect natural resource professionals with practitioners.

Conclusion: The Future of Agroforestry Education and Training

The incredible growth of agroforestry education and training programs in the last 40 years continues to unfold, with new connections and support filling remaining gaps in the agroforestry knowledge network infrastructure. Just as agroforestry is defined in terms of the four I's (*intentional integration* of trees and shrubs into agricultural systems managed *intensively* for beneficial *interactions*), these core tenets can also be applied to the education and training foundation necessary to bring agroforestry practices to life.

An *intentional* focus on education and outreach, with special attention to needs assessment and program evaluation, will guide the continued development of *intensive* training programs for innovators in the field. *Interactions* among practitioners, researchers, and diverse agroforestry supporters will yield an *integrated* knowledge network that addresses the challenges and circumstances of unique biophysical and socioeconomic contexts. Given the (now) relatively abundant resources and communities prepared to respond according to the four I's, collaboration and partnerships are ripe for amplifying our efforts.

References

Agbaje, K. A. A., Martin, R. A., & Williams, D. L. (2001). Impact of sustainable agriculture on secondary school agriculture education teachers and programs in the North Central Region. *Journal of Agricultural Education*, *42*, 38–45. https://doi.org/10.5032/jae.2001.02038

Alonge, A. J., & Martin, R. A. (1995). Assessment of the adoption of sustainable agriculture practices: Implications for agricultural education. *Journal of Agricultural Education*, *36*, 34–42. https://doi.org/10.5032/jae.1995.03034

Bene, J. G., Beall, H. W., & Cote, A. (1977). Trees, food and people. Ottawa, ON, Canada: International Development Research Centre.

Garrett, H. E. "Gene" (Ed.). (2009). North American agroforestry: An integrated science and practice (2nd ed). Madison, WI: ASA. https://doi.org/10.2134/2009.northamericanagroforestry.2ed

Gold, M. A. (2015). Evolution of U.S. agroforestry research and formalization of agroforestry education. *Inside Agroforestry*, 23(3), 10–11. Retrieved from https://www.fs.usda.gov/nac/assets/documents/insideagroforestry/IA_Vol23Issue3.pdf

Gold, M. A., Cernusca, M. M., & Hall, M. M. (Eds.). (2013). *Handbook for agroforestry planning & design*. Columbia, MO: University of Missouri Center for Agroforestry. Retrieved from http://www.centerforagroforestry.org/pubs/training/handbookp&d13.pdf

Gold, M. A., Cernusca, M. M., & Jose, S. (2013). Creating the knowledge infrastructure to enhance landowner adoption of agroforestry through an agroforestry academy. In L. Poppy, J. Kort, B. Schroeder, T. Pollock, & R. Soolanayakanahally (Eds.), *Agroforestry—Innovations in agriculture: Proceedings of the 13th North American Agroforestry Conference*, Charlottetown, PEI, *Canada* (pp. 92–96).

Gold, M. A., Hemmelgarn, H. L., & Mendelson, S. E. (2019). Academy offers professional development to boost agroforestry. *Forestry Source, 24*(3), 12–13.

Gold, M. A., Hemmelgarn, H. L., Mori, G. O., & Todd, C. (Eds.). (2018). *Training manual for applied agroforestry practices*. Columbia, MO: University of Missouri Center for Agroforestry. Retrieved from http://www.centerforagroforestry.org/pubs/training/index.php.

Gold, M. A. & Jose, S. (2012). An interdisciplinary online certificate and master's program in agroforestry. *Agroforestry Systems, 86*, 379–385. https://doi.org/10.1007/s10457-012-9522-8

Hemmelgarn, H. L., Ball, A., Gold, M. A., & Stelzer, H. (2019). Agroforestry education for high school agriculture science: An evaluation of novel content adoption following educator professional development programs. *Agroforestry Systems, 93*, 1659–1671. https://doi.org/10.1007/s10457-018-0278-7

Ikerd, J. (2016). *Family farms of North America*. Working Paper no. 152. Brasilia: International Policy Center for Inclusive Growth. Retrieved from http://www.fao.org/3/a-i6354e.pdf.

Isaac, M. E., Erickson, B. H., Quashie-Sam, S. J., & Timmer, V. R. (2007). Transfer of knowledge on agroforestry management practices: The structure of farmer advice networks. *Ecology and Society, 12*(2), 32. https://doi.org/10.5751/ES-02196-120232

Jacobson, M., & Kar, S. (2013). Extent of agroforestry extension programs in the United States. *Journal of Extension, 51*(4), rb4. Retrieved from https://joe.org/joe/2013august/rb4.php.

Jose, S., Gold, M. A., & Garrett, H. E. (2018). Temperate agroforestry in the United States: Current trends and future directions. In A. M. Gordon, S. M. Newman, & B. Coleman (Eds.), Temperate agroforestry systems (2nd ed.). Wallingford, UK: CAB International.

Khasa, D. P., Olivier, A., Atangana, A.R., & Bonneville, J. (2017). Two decades of agroforestry training, education and research at Universite Laval, Quebec, Canada. *Agroforestry Systems, 91*, 825–833. https://doi.org/10.1007/s10457-015-9871-1

Kueper, A. M., Sagor, E. S., & Becker, D. R. (2013). Learning from landowners: Examining the role of peer exchange in private landowner outreach through landowner networks. *Society & Natural Resources, 26*, 912–930. https://doi.org/10.1080/08941920.2012.722748

Lassoie, J. P., Huxley, P., & Buck, L. E. (1994). Agroforestry education and training: A contemporary view. *Agroforestry Systems, 28*, 5–19. https://doi.org/10.1007/BF00711984

Lundgren, B. O. (1982). What is agroforestry? *Agroforestry Systems, 1*, 7–12.

Mason, A. (2016). Professionalizing agroforestry in the U.S.: What will it take? *Temperate Agroforester, 22*(2). Retrieved from https://www.aftaweb.org/latest-newsletter/newsletter-test/listid-1/mailid-42-temperate-agroforester-volume-22-number-2.html.

Mason, A., Blanche, C., Crowe, T., Gold, M., Jacobson, M., Jose, S., . . . Wight, B. (2012). Advancing agroforestry through certification of agroforesters: Should the Society of American Foresters have a role? *Journal of Forestry, 110*(8):466–467. https://doi.org/10.5849/jof.12-069

Mendelson, S. E. (2020). Educating for adoption: Understanding long-term outcomes of the Agroforestry Academy (Master's thesis). University of Missouri.

Mendelson, S., Gold, M., Lovell, S., & Hendrickson, M. (2021). The agroforestry academy: Assessing long-term outcomes and impacts of a model training program. *Agroforestry Systems, 95*:601–614. https://doi.org/10.1007/s10457-021-00604-y

Missouri Department of Elementary and Secondary Education. (2016). *Agricultural education in Missouri*. Retrieved from https://dese.mo.gov/sites/default/files/Ag-Ed-in-Missouri-2016.pdf.

Nair, P. K. R., Gholz, H. L., & Duryea, M. L. (Eds.). (1990). Agroforestry education and training: Present and future. *Agroforestry Systems, 12*(1).

National Wildlife Federation. (2012). *Roadmap to increased cover crop adoption*. Reston, VA: National Wildlife Federation. Retrieved from https://www.nwf.org/~/media/PDFs/Global-Warming/Policy-Solutions/Cover_Crops_Roadmap%20Report_12-12-12.ashx.

National Wildlife Federation. (2013). *Cover crop champions*. Reston, VA: National Wildlife Federation. Retrieved from https://www.nwf.org/~/media/PDFs/Wildlife/NWF_CoverCropChampions_2013.pdf.

Romanova, O. (2020). *Factors influencing practitioner adoption of agroforestry: A USDA SARE case study* [Unpublished master's thesis]. University of Missouri.

Rossier, C. & Lake, F. (2014). Indigenous traditional ecological knowledge in agroforestry (Agroforestry Notes 44). Lincoln, NE: USDA National Agroforestry Center. Retrieved from https://www.fs.usda.gov/treesearch/pubs/47452

Shelton, D.P., Wilke, R. A., Franti, T. G., & Josiah, S.J. (2009). "Farmlink": Promoting conservation buffers farmer-to farmer. *Agroforestry Systems, 75*, 83–89. https://doi.org/10.1007/s10457-008-9130-9

Smith, J.R. (1929). *Tree crops: A permanent agriculture*. New York: Harcourt Brace and Company.

Strong, N. A., & Jacobson, M. G. (2005). Assessing agroforestry adoption potential utilizing market segmentation: A case study in Pennsylvania. *Small-scale Forest Economics, Management and Policy, 4*, 215–228. https://doi.org/10.1007/s11842-005-0014-9

Towards a comprehensive education and training program in agroforestry. (1990). *Agroforestry Systems, 12*, 121–131. https://doi.org/10.1007/BF00055583

USDA Economic Research Service. (2012). *Share of principal farm operators with college degrees has increased*. Washington, DC: USDA–ERS. Retrieved from https://www.ers.usda.gov/data-products/chart-gallery/gallery/chart-detail/?chartId=76128.

USDA. (2011). USDA agroforestry strategic framework, fiscal year 2011–2016. Washington, DC: USDA. Retrieved from https://www.usda.gov/sites/default/files/documents/AFStratFrame_FINAL-lr_6-3-11.pdf.

USDA National Agroforestry Center. (2015). 25th anniversary issue. *Inside Agroforestry Newsletter, 23*(3). Lincoln, NE: National Agroforestry Center. Retrieved from https://www.fs.usda.gov/nac/assets/documents/insideagroforestry/IA_Vol23Issue3.pdf.

USDA. (2019). Agroforestry strategic framework, fiscal year 2019–2024. Washington, DC: USDA. Retrieved from https://www.usda.gov/sites/default/files/documents/usda-agroforestry-strategic-framework.pdf.

Valdivia, C., Barbieri, C., & Gold, M. A. (2012). Between forestry and farming: Policy and environmental implications of the barriers to agroforestry adoption. *Canadian Journal of Agricultural Economics, 60*(2), 155–175. https://doi.org/10.1111/j.1744-7976.2012.01248.x

Warren, S. T., & Bentley, W. R. (1990). Expanding opportunities for agroforestry education in the U.S. and Canadian Universities. *Agroforestry Systems, 12*, 115–120. https://doi.org/10.1007/BF00055582

Werquin, P. (2007). Terms, concepts and models for analysing the value of recognition programmes EDU/EDPC(2007)24. Paris: Organisation for Economic Co-operation and Development. Retrieved from http://www.oecd.org/education/skills-beyond-school/41834711.pdf.

Williams, P. (Ed.). (1991). *Agroforestry in North America: Proceedings of the First Conference on Agroforestry in North America, 13–16 August 1989*. Guelph, ON, Canada: Ontario Ministry of Agriculture and Food.

Wilson, M. H., Lovell, S. T., & Carter, T. (2018). *Perennial pathways—Planting tree crops: Designing and installing farm-scale edible agroforestry*. Madison, WI: Savanna Institute. Retrieved from http://www.savannainstitute.org/tree-crops.html

Wright, M. (2017). Agroforestry education: The status and progress of agroforestry courses in the US (Master's thesis, Virginia Polytechnic Institute and State University). Retrieved from http://hdl.handle.net/10919/77521

Study Questions

1. What aspects of agroforestry make the development of agroforestry education and training programs unique?

2. Why is agroforestry education known to be one of the keys to increased adoption?

3. Why are there only a limited number of comprehensive agroforestry education and training programs in the United States?

4. What are the differences between formal, non-formal, and informal approaches to education?

5. What are the reasons for including agroforestry education at the high school level?

6. What is the importance of the North American Agroforestry Conference series?

7. What is the role of the Association for Temperate Agroforestry in relation to agroforestry education and training?

8. Why are regional working groups useful in the promotion of agroforestry?

9. For what purpose was the Agroforestry Academy created?

10. List some of the future needs of agroforestry education and training.

Index

North American Agroforestry, Third Edition. Edited by Harold E. "Gene" Garrett, Shibu Jose, and Michael A. Gold.

Printed and bound by CPI Group (UK) Ltd, Croydon, CR0 4YY

16/04/2025

14658588-0003